D1415234

SOURCE BOOK ON MATERIALS SELECTION

Volume I

A discriminative selection of
outstanding articles from the
periodical and reference literature

Compiled by
Consulting Editor:

Russell B. Gunia
Consulting Metallurgist

AMERICAN SOCIETY FOR METALS
Metals Park, Ohio 44073

First printing, March 1977

Library of Congress Cataloging in Publication Data
Source book on materials selection

Includes index.
1. Materials.
I. Gunia, Russell B.
TA 403.S62 620.1'1'08 77-1347

Library of Congress Catalog Card Number: 77-1347

PRINTED IN THE UNITED STATES OF AMERICA

Preface

SOURCE BOOK ON MATERIALS SELECTION is especially timely because of the increased emphasis currently placed on judicious selection of all materials used in manufacturing. Engineers and technical managers concerned with product design, specification and the use of materials need dependable information on available materials to assist them in manufacturing products that are both economical and reliable.

This Source Book is being issued as a two-volume set, an expedient required to accommodate a representative cross-section of the available periodical and reference literature on a broad subject without sacrificing coverage or comprehensive treatment. The two volumes contain approximately 1000 pages of useful data and commentary pertaining to virtually all metals and alloys of industrial importance and some widely used nonmetallics and mixtures, including plastics, ceramics and cermets. Coverage is also afforded composites, electrodeposited metals, and metallic and nonmetallic coatings.

The acknowledged classic on the subject of selection of metals, Vol. 1 of *Metals Handbook* (8th Edition), was published by ASM in 1961. A number of its outstanding articles have been included in the present collection which, additionally, has adhered to one of the Handbook's principal aims. That aim, which originally was restricted to the selection of metals and now has been broadened to include nonmetallics and other materials of special interest, is to provide practical information that will help readers to select the most suitable materials for specific purposes.

Volume I of SOURCE BOOK ON MATERIALS SELECTION has been arranged in eight principal sections, all but two of which are devoted exclusively to specific types of steels and irons. The exceptions deal with materials systems (Section I) and tool and die materials, including tool steels (Section VI). A brief summary of the contents of this volume by sections follows.

Materials Systems. This section contains eight articles that outline the development of materials systems from three common vantage points — the over-all assembly and its operation, a class of related assemblies and their performance requirements, and materials selection for reliable performance requirements in extreme environments. The systems examined include aircraft, aerospace, automotive and nuclear applications, and structures intended for cryogenic service.

Carbon and Alloy Steels. This 238-page section, the largest in the volume, contains a total of 50 articles and data sheets on the carbon and alloy steels produced in the highest tonnages and serving the major sectors of our economy. Included are low-carbon and hardenable carbon steels, carburizing steels, free machining steels, high-strength low-alloy steels, boron steels, silicon steels, and a variety of steels for special service. Emphasis is placed on key selection factors — properties, processing characteristics, cost, and suitability for general and special service applications.

Cast Irons and Steels. This section comprises a total of ten articles and data sheets covering gray, white, malleable and ductile irons and cast carbon and low-alloy steels. Specifications, structure and properties, processing characteristics and applications are emphasized. The substitution of castings for forgings and weldments is also considered.

Wrought Stainless Steels. Because of the comprehensive coverage afforded stainless steels in the *ASM Source Book on Stainless Steels*, no attempt was made to duplicate that coverage in this section. Rather, it was assumed that the earlier volume, published in 1976, would be retained as a basic reference and guide to selection. Consequently, this section serves as a supplement, providing additional information on ferritic and austenitic, nitrogen-bearing, and high-strength stainless steels for special applications.

Cast Stainless Steels. This section presents basic information on the ACI cast alloys, widely used in pumps, valves and other components for chemical and petroleum processing plants and other applications for which resistance to corrosion is a primary requirement.

Tool and Die Materials. This section opens with the comprehensive *Metals Handbook* article on tool steels, a basic reference for metallurgists, materials specialists, designers and manufacturing engineers. There follows one of the most varied and useful collections of data and commentary on tool and die materials ever assembled in a single volume. Among the materials covered are superalloys, maraging steels, stainless alloys, plastics, tungsten and titanium carbides, graphitic tool steels and ceramics. Applications run the gamut of manufacturing processes requiring tool and die materials.

Ultrahigh Strength Steels. A relatively new class of ultrahigh strength alloys, including the maraging steels, are the basis for this section, which serves to augment and update property and selection data published in *Metals Handbook* (Vol. 1) and other standard references. All of the articles and data sheets derive from the periodical literature.

Selected Data Sheets. A limited assortment of data sheets related to selection that serve to supplement the section on carbon and alloy steels comprise the concluding section of this volume.

Mr. Russell B. Gunia, a distinguished metallurgical and materials engineer, served as consulting editor of this, and its companion, volume, and to him the American Society for Metals extends its grateful appreciation. Materials selection is a specialized discipline to which Mr. Gunia devoted much of his time and attention throughout his career in industry. His knowledge and experience were invaluable in guiding the selection of articles for these Source Books. Mrs. Mary Ann Podboy, Coordinator of ASM's Publications Development activity, contributed to, and supervised, all aspects of the production of both volumes. Most grateful acknowledgment is also extended to the many authors whose articles appear in these volumes; our debt to each of them is shared by all who benefit from the use of their works.

Paul M. Unterweiser
Staff Editor
Manager, Publications Development
American Society for Metals

Allen G. Gray
Technical Director
American Society for Metals

Contributors to This Source Book*

S. L. AMES
Allegheny Ludlum Indus. Inc.

E. C. BAIN
United States Steel Corp.

ALAN M. BAYER
Teledyne Vasco

C. M. BERGER
Taussig Associates

JOHN E. BEVAN
Climax Molybdenum Co. of Mich.
Subs. AMAX Inc.

PAUL R. BORNEMAN
Allegheny Ludlum Indus. Inc.

E. F. BRADLEY
Pratt & Whitney Aircraft
Div. United Aircraft Corp.

EDWARD V. BRAVENEC
Armco Steel Corp.

DALE H. BREEN
International Harvester Co.

J. P. BRESSANELLI
Crucible Stainless Steel
Colt Industries

KENNETH G. BRICKNER
Applied Research Laboratory
United States Steel Corp.

ROBERT L. BROOKS
Lukens Steel Co.

ROBERT A. CARY
Vasco, a Teledyne Co.

VIJAY K. CHANDHOK
Crucible Steel Corp.

HARRY E. CHANDLER
American Society for Metals

W. P. CHERNOCK
Combustion Engineering Inc.

SEYMOUR K. COBURN
United States Steel Corp.

HARRY H. CORNELL
Continental Copper & Steel Indus. Inc.

M. J. DONACHIE
Pratt & Whitney Aircraft
Div. United Aircraft Corp.

JOSEPH A. DOUTHETT
Armco Steel Corp.

EDWARD J. DULIS
Crucible Steel Corp.

R. N. DUNCAN
Combustion Engineering Inc.

J. F. ENRIETTO
Westinghouse Electric Corp.

GEORGE T. ELDIS
Climax Molybdenum Co. of Mich.
Subs. AMAX Inc.

FERDINAND L. EWALD
Budd Co.

LEROY E. FINCH
ESCO Corp.

STEPHEN FLOREEN
International Nickel Co. Inc.

WILLIAM H. GORMAN JR.
Lukens Steel Co.

SHERMAN GREENBERG
Argonne National Laboratory

JOHN J. GRENAWALT
Youngstown Sheet and Tube Co.

JOHN H. GROSS
United States Steel Corp.

JOHN D. GROZIER
Graham Research Laboratory
Jones & Laughlin Steel Corp.

A. M. HALL
Battelle Memorial Institute

ALLEN G. HAYNES
International Nickel Ltd.

ROBERT L. HICKEY
Armco Steel Corp.

RICHARD J. HENRY
Vasco, a Teledyne Co.

J. M. HODGE
Applied Research Laboratory
United States Steel Corp.

ALBERT J. HOERSCH JR.
Lukens Steel Co.

DENNIS D. HUFFMAN
Latrobe Steel Co.

RONALD M. JAMIESON
Stelco Research Center
Steel Company of Canada

ALEXANDER H. JOLLY JR.
Wisconsin Steel Div.
International Harvester Co.

HERBERT S. KALISH
Adamas Carbide Corp.

CARL J. KEITH JR.
International Harvester Co.

W. L. KENNICOTT
Kennametal Inc.

ROY F. KERN
Kern Engineering Co.

MICHAEL KORCHYNSKY
Jones & Laughlin Steel Corp.

VICTOR A. KORTESOJA
Ford Motor Co.

JAMES KOSMALA
Dexter Corp.

JACK E. LaBELLE
General Motors Corp.

PETER B. LAKE
Youngstown Sheet and Tube Co.

E. J. LANE
Carpenter Technology Corp.

BERNARD S. LEVY
Inland Steel Co.

REMUS A. LULA
Allegheny Ludlum Indus. Inc.

E. E. LULL
A. Finkl & Sons Co.
Subs. Republic Steel Corp.

*List does not include names of contributors to Metals Handbook

JACOB L. LYTELL JR.
International Harvester Co.

JOHN C. MAJETICH
Applied Research Laboratory
United States Steel Corp.

F. A. MALAGARI
Allegheny Ludlum Indus. Inc.

E. GARY MARSHALL
Lukens Steel Co.

L. J. McEOWEN
Republic Steel Corp.

JAMES J. McCARTHY
Bethlehem Steel Corp.

WILLIAM H. McFARLAND
Inland Steel Co.

GEORGE F. MELLOY
Bethlehem Steel Corp.

JOHN L. MIHELICH
Climax Molybdenum Co.

J. W. MORROW
Climax Molybdenum Co. of Mich.
Subs. AMAX Inc.

F. J. MILLER
Allegheny Ludlum Indus. Inc.

ARTHUR W. MORSE
Sandvik Steel Inc.

A. MOSKOWITZ
Crucible Stainless Steel Div.
Colt Industries

WILLIAM J. MURPHY
United States Steel Corp.

ALEXANDER NAGY
Universal-Cyclops Specialty Steel Div.
Cyclops Corp.

PHILIP A. NAGY
Timken Roller Bearing Co.

ROBERT J. NEVILLE
Dominion Foundries & Steel Ltd.

ARNE OMSEN
Uddeholm AB

SUMANT PATEL
Standard Pressed Steel Co.

GERHARD PERSSON
Sandvik Steel Works

RICHARD B. PETERSON
Ren Plastics Inc.

E. H. PHELPS
United States Steel Corp.

K. E. PINNOW
Crucible Stainless Steel Div.
Colt Industries

B. G. REISDORF
United States Steel Corp.

WILLIAM H. RICE
ESCO Corp.

JOHN A. REINBOLT
Republic Steel Corp.

CLIFFORD E. ROWDEN
Republic Steel Corp.

R. J. SCHMITT
United States Steel Corp.

JOHN SCHUYTEN
Vulcan Foundry Co.

FRED R. SCHWARTZBERG
Martin Marietta Corp.

ERIC C. SHARPLESS
Ametek Inc.

EMIL G. SIGNES
Bethlehem Steel Corp.

BENGT E. SKOOG
Uddeholm AB

PAUL R. SLIMMON
Bethlehem Steel Corp.

JAMES H. SMITH
Youngstown Sheet & Tube Co.

J. H. SMITH
Interprovincial Steel & Pipe Corp.

YANCEY E. SMITH
Climax Molybdenum Co. of Mich.
Subs. AMAX Inc.

CARL E. SPAEDER JR.
Applied Research Laboratory
United States Steel Corp.

CHARLES H. SPERRY
International Harvester Co.

JOHN T. SPONZILLI
International Harvester Co.

ROBERT H. STERNE JR.
Lukens Steel Co.

J. R. STEWART
Combustion Engineering Inc.

GEORGE L. SWALES
International Nickel Ltd.

STUART E. TARKAN
Chromalloy American Corp.

EDWARD TAYLOR
Standard Pressed Steel Co.

HAROLD L. TAYLOR
Inland Steel Co.

SETH R. THOMAS
Teledyne Rodney Metals

G. TITHER
Climax Molybdenum Co. of Mich.
Subs. AMAX Inc.

JOHN J. VETT
Youngstown Sheet & Tube Co.

GORDON H. WALTER
International Harvester Co.

DONALD WEBSTER
Lockheed Missiles & Space Co.

R. D. WELTZIN
IBM General Systems Div.

CARL R. WEYMUELLER
American Society for Metals

RALPH B. G. YEO
Union Carbide Corp.

WILLIAM YOUNG
Crucible Specialty Metals Div.
Colt Industries

DEWEY G. YOUNGER
Ford Motor Co.

Table of Contents

Special Applications

Section III: Cast Irons and Steels

Section IV: Wrought Stainless Steels

Section V: Cast Stainless Steels

Section VI: Tool and Die Materials

Section VII: Ultrahigh Strength Steels

Section VIII: Selected Data Sheets

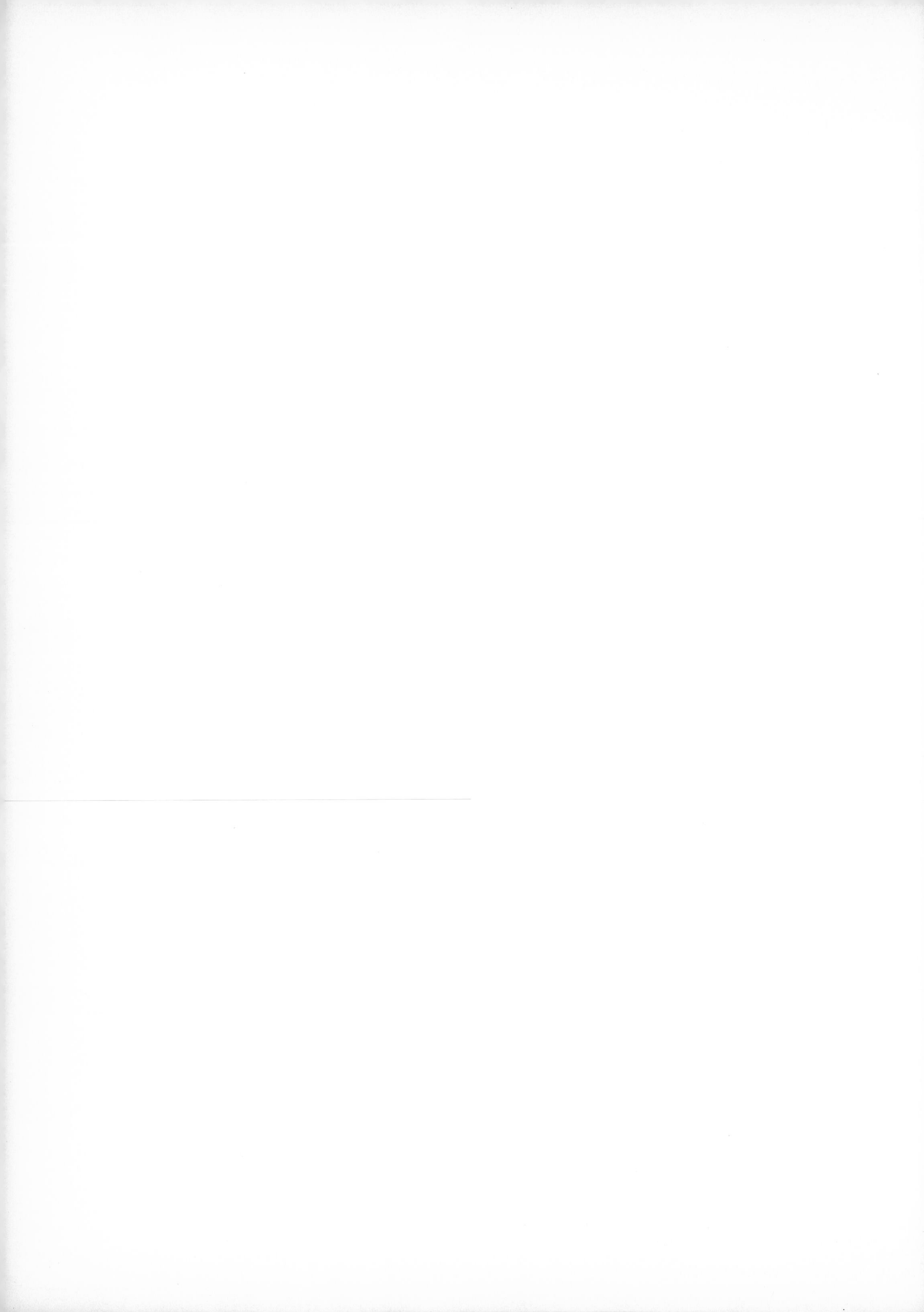

Section I: Materials Systems

Changes and Evolution of Aircraft Engine Materials

E. F. BRADLEY AND M. J. DONACHIE

AIRCRAFT engine materials and design are subject to continuous review because of the constant demand for improved performance; however, changes are not made easily. Because it is necessary to thoroughly evaluate each revision, time to accomplish a change can be agonizingly long.

New designs and material changes have resulted in spectacular gains, as shown by a comparison of engines in service today with those engines which initiated the jet age. Specific fuel consumption has been more than cut in half, thrust-to-weight ratios have been tripled, physical size has increased in some cases, and most dramatically the time between overhauls has been increased from under 100 h to over 12,000 h. Engines now operate at up to 1645 K (2500°F) with life for many engine components up to 30,000 h.

Materials developments which have been largely evolutionary, not revolutionary, have played a significant role in these advances. During this period, substantial changes in gas turbine materials have occurred; over 80 pct of the components in the JT9D engine are of a different material than the corresponding part in the JT3 engine of twenty years ago. The need for material changes is not likely to diminish as there is continual pressure to further improve engine performance and reduce costs. This paper will review some of the important material changes of the past two decades, considering both the cold and hot sections of the engine.

Several factors are involved in making a material change. First the new material must be fully characterized—we must know as much about it as the old material to preclude service problems. Of course, the material must do the job technically, but it also must be cost effective, available, and reliable. Social/environmental interactions are becoming more important in the current era of environmental control since the atmosphere must be protected.

A fundamental rule in the aricraft engine business is not to change a material that is working. This rule is generally followed, but sometimes other pressures exist that cause change even when the current material is performing well. The following two recent examples—one based on economy, the other on availability—relate to this situation. First, copper-manganese-cobalt alloy has replaced some gold-nickel and silver-palladium braze alloys used in compressor stators on the basis of cost only—the rapidly escalating prices of precious metals made this change imperative. Second, it has been necessary to change the material of several brackets from SAE 4130 and 8630 low alloy steel to Type 410 stainless steel solely because of availability. The low alloy steels are not warehouse stocked in suitable sizes for these parts, but the Type 410 steel is available.

E. F. BRADLEY and M. J. DONACHIE are associated with Pratt & Whitney Aircraft, Div. of United Aircraft Corp., East Hartford, Conn.

The modern JT9D engine which powers the Boeing 747 wide body transport and the older JT3D powerplant of the earlier Boeing 707 and DC8 transports are pictured in Fig. 1, which shows the obvious external size differences. Fig. 2 indicates the steady increase in turbine temperatures over the years; this trend has demanded new alloys with improved high temperature strength and oxidation resistance.

Fig. 3 shows a cutaway view of the F100 engine, which is used in the military F15 air superiority aircraft. This figure shows the three major engine areas (Fan-compressor, turbine, and burner) which will be considered relative to material changes. Turbine disks will be included in the cold section discussion along with fan and compressor parts; the hot section comprises the burner and turbine components.

Fig. 4 compares strength/density ratios of high strength steels and titanium alloys, showing why titanium replaced steel in the lower temperature regions of the engine. Originally, AMS 5616 (12 pct chromium-3 pct tungsten-2 pct nickel corrosion resistant steel) was used for compressor blades, AMS 5613 (12 pct chromium corrosion resistant steel) was used for compressor vanes, and plated low alloy steels (AMS 6415 with cadmium plate, and the higher temperature grade AMS 6304 with diffused nickel-cadmium plate) were used for compressor disks. The development of titanium in the 1950's with its superb corrosion resistance and attractive strength/weight characteristics made it a natural replacement for many of these parts.

Because titanium is more expensive than steel, it was first used in military engines. As time went on, however, significant engineering advantages of titanium forced replacement of steel in the compressors of all jet engines, both military and commercial. The use of steel for brazed compressor vane stator assemblies has persisted in many applications because titanium has poor brazing properties. Steel has also remained in use in some of the latter compressor stages for both disks and blades because of temperature considerations.

Elevated temperatures have been a disappointing limitation for titanium ever since its introduction into turbine engines. Two reasons for this limitation persist—namely, the affinity of titanium for interstitial elements, and its inadequate creep strength at even the moderate temperature level of about 811 K (1000°F).

Titanium was not developed without problems and change. Three early serious technical problems were encountered with titanium, and these problems completely altered melting practice. The problems were 1) hydrogen in amounts above 150 parts per million caused titanium alloys to be brittle, 2) segregation of heavy alloying elements such as molybdenum resulted in defective parts, and 3) undissolved tungsten from fixed melting electrodes often got into finished parts causing rejection. The solution to these problems was achieved via one sweeping change—titanium pro-

Fig. 1—Jet engines have increased greatly in size in the last decade. The JT9D is used for the Boeing 747, while the JT3D is the power plant for the Boeing 707.

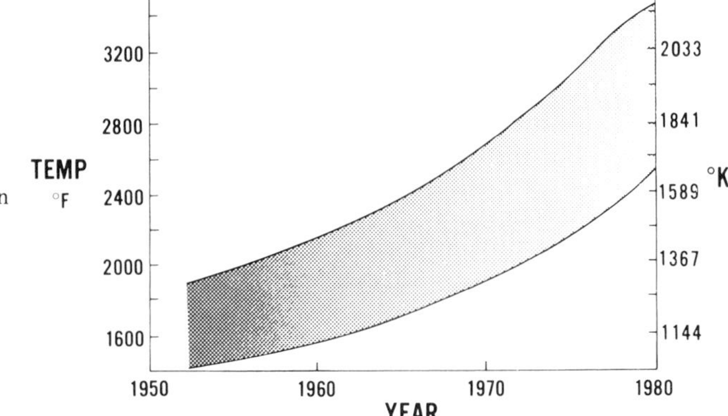

Fig. 2—Through the years turbine inlet temperatures have risen appreciably.

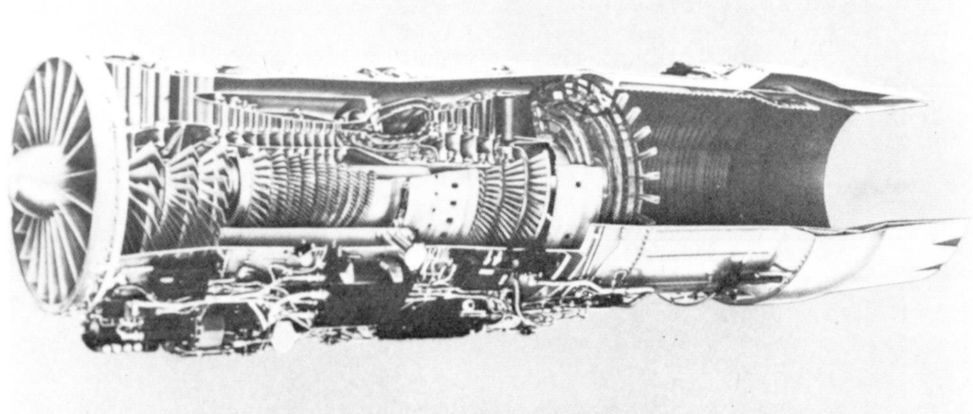

Fig. 3—Cutaway of F100 turbofan engine displays three major engine areas–fan-compressor, turbine, and burner.

ducers were directed to double melt in vacuum using the consumable electrode process. More recently, the evolutionary change to triple melting was implemented to improve quality for highly stressed fan disks.

The use of a fan engine was perhaps the single most important design step taken in the past twenty years. Fan engines would not have been possible without the availability of titanium alloys. These low density, high strength alloys are uniquely suited to fan blades, and they also permit lower weight in almost all stages of the compressor.

Fig. 5 shows the evolutionary changes that have occurred in turbine disk materials as temperature demands have increased through the years. At the beginning of the jet age, AMS 5616 corrosion resistant steel was used for disks exposed to relatively low temperatures. Increases in operating temperature, however, soon forced a material change to the nickel base superalloys—first Incoloy 901, then Waspaloy, Astroloy, and currently IN-100 made via powder metallurgy.

High temperature strength improvement of one alloy over another comes mainly from increased amounts of the strengthening elements, aluminum and titanium, which form the principal gamma prime hardening constituent $Ni_3(Al, Ti)$. This entire alloy development was made possible by the advent of vacuum melting, which allowed these reactive elements to be completely effective as strengtheners. IN-100 is an alloy that is generally used only in the cast condition because of its high strength at elevated temperatures. Use of powder metallurgy techniques, however, allow production of IN-100 forged disks, which are beginning to see service in advanced military engines.

The hot section, particularly first-stage turbine blades, historically has paced aircraft turbine engine development. Fig. 6, which gives an overview of hot section materials, shows why the gamma-prime strengthened nickel alloys are used for highly stressed turbine blades. These remarkable alloys show very high strength from 922 K to 1255 K (1200°F to 1800°F). The carbide-strengthened cobalt alloys, of lower strength than the gamma prime nickel alloys, are used for turbine vanes (stationary and lower stressed than rotating blades) because of lower cost and good thermal shock characteristics. The low-strength solid solution strengthened nickel and cobalt alloys are used as sheet metal components of the combustor because of their excellent fabricability (forming and welding), oxidation resistance, and hot corrosion resistance.

The single line of Fig. 6 represents the strength of the oxide dispersion strengthened nickel alloys. These materials are of relatively low strength at low temperatures, but retain their strength effectively as

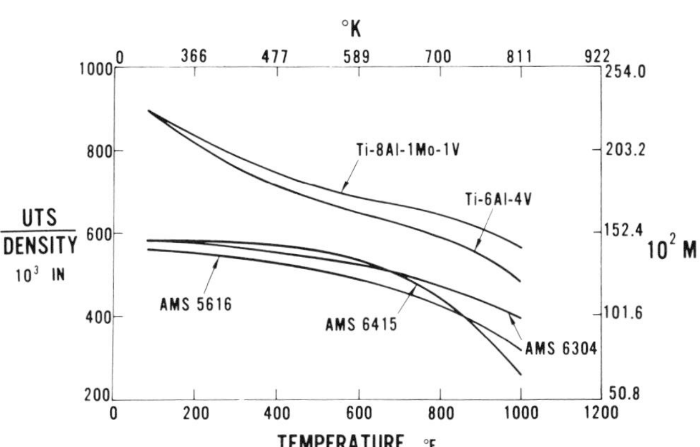

Fig. 4—Titanium alloys demonstrate higher strength/density ratios than steels.

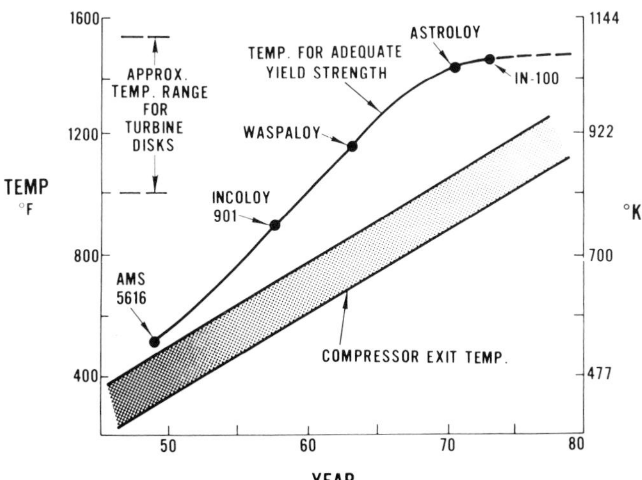

Fig. 5—Continuing increases in compressor exit temperatures have called for better superalloys for turbine disks.

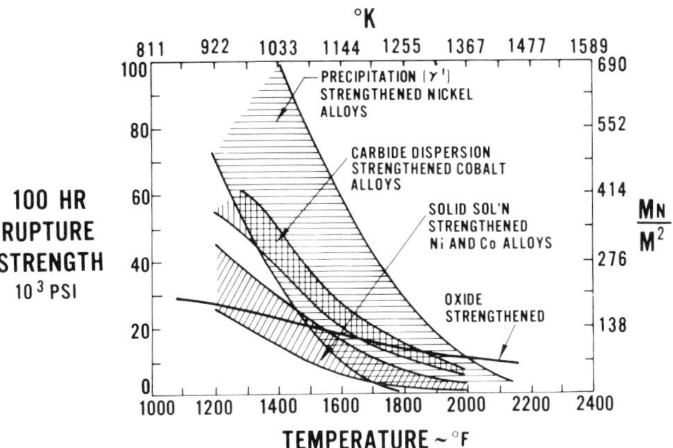

Fig. 6—Their high rupture strengths make the gamma-prime strengthened nickel-base alloys favorites for turbine blades of jet engines.

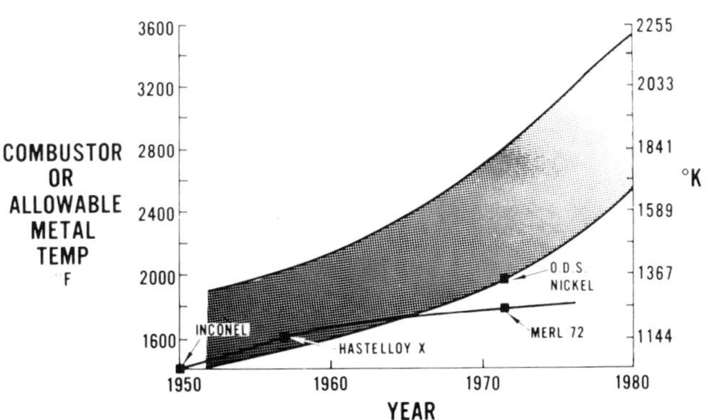

Fig. 7—Work on burner materials is concentrated on raising temperature resistance because combustor exit temperatures continue to rise, as shown by the band.

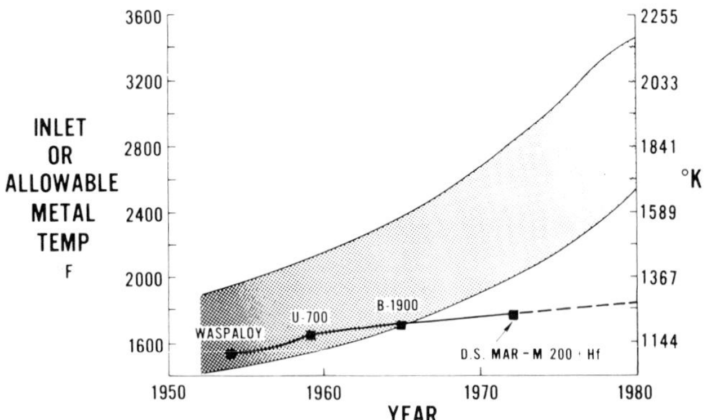

Fig. 8—As turbine inlet temperatures continued to rise (band), cast alloys—B-1900, for example—have become favored for turbine blades.

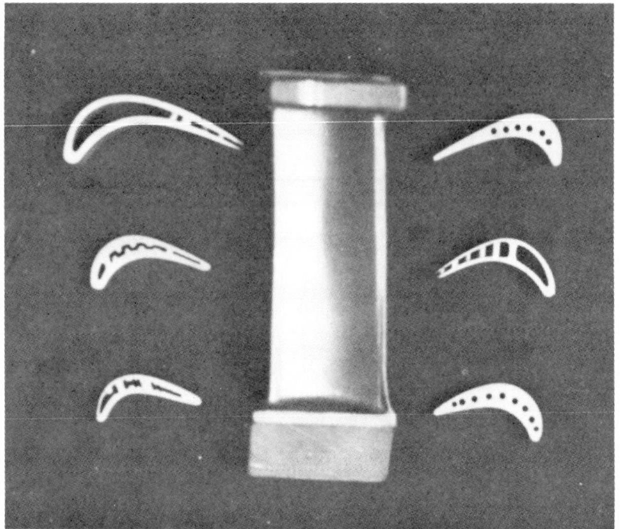
Fig. 9—Cast turbine blades include air passages for cooling.

temperatures increase. It appears that this class of alloys will be useful at very high temperatures above 1370 K (2000°F). The reason for this improved retention of strength is that the dispersed strengthening phase remains in the structure until the alloy is melted. The gamma prime strengthener, on the other hand, tends to dissolve in the solid state at elevated temperatures so that the strength curve of these alloys falls off sharply at very high temperatures.

Fig. 7 shows the progression of burner materials development. Hastelloy X replaced Inconel as a burner material in the second half of the 1950's, and continues to be used today. As combustor temperatures increased, it was necessary to adequately cool the Hastelloy X components to ensure satisfactory life. An experimental cobalt alloy containing chromium, aluminum, and yttrium (identified as MERL 72), and dispersion strengthened nickel alloys are currently being developed for advanced engines.

Fig. 8 shows the turbine blade alloys that have been used throughout the years. All alloys shown are gamma prime strengthened nickel superalloys made by vacuum melting. Forged Waspaloy was replaced in the late 1950's by the higher strength forged U-700, but in time it became apparent that the engine operating temperatures were becoming too hot for forgings. As temperatures increased, the next step was to replace forged U-700 with the higher strength cast B-1900 alloy. This was indeed a significant change because it was necessary to painstakingly develop satisfactory casting quality to provide hardware which was as reliable as forgings. An evolutionary revision that occurred in this time period was the addition of small amounts of hafnium (1 to 2 pct) to the nickel base superalloys to improve intermediate temperature ductility and thermal fatigue characteristics.

The next significant change was to utilize air cooling designs to allow continued use of B-1900 castings

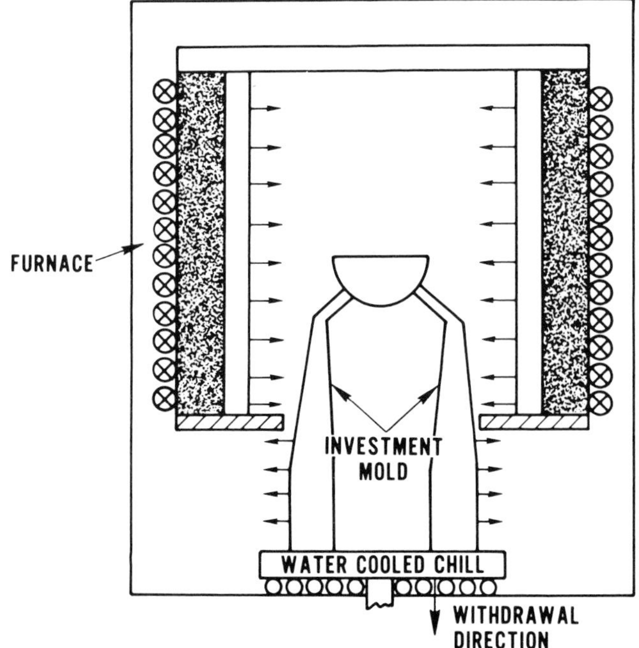

Fig. 10—Directional solidification improves turbine blades. The method is diagrammed on the left, and conventional (left) and directionally cast blades are at the right.

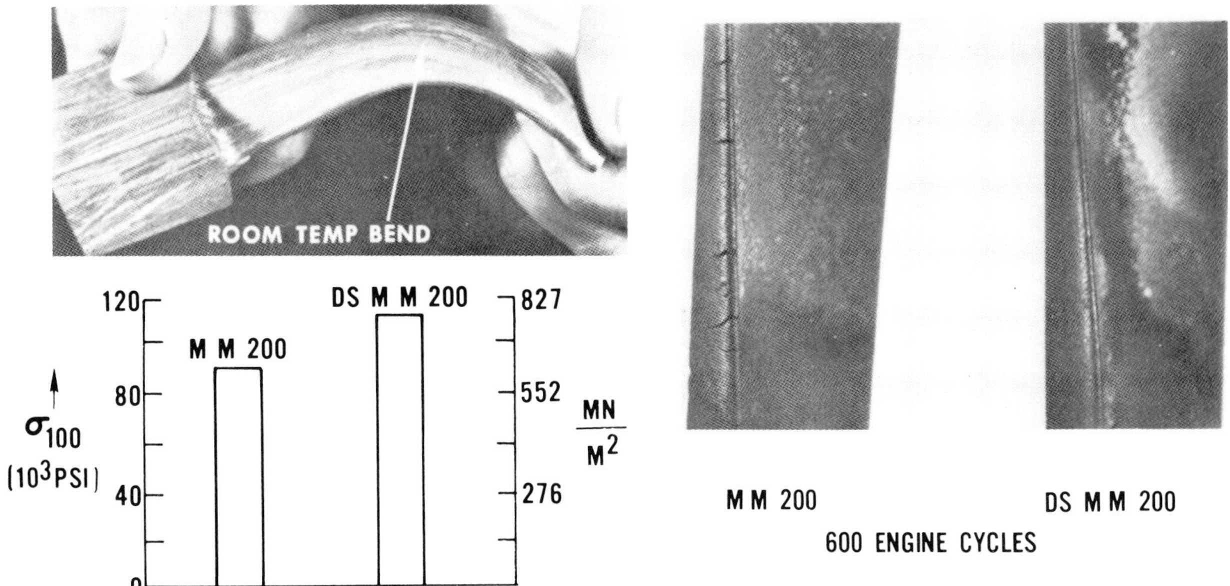

Fig. 11—Directional solidification raises the strength and ductility of blades cast of Mar-M-200.

at even higher operating temperatures. Fig. 9 shows some of the air cooled turbine blade designs required for high temperature operation. An additional step was the introduction of directional solidification. The difference in structure between conventionally cast equiaxed material and the directionally solidified material with grains running only in the longitudinal direction is shown in Fig. 10, which also illustrates schematically how directionally solidified blades are made utilizing a water cooled chill at one end of the mold and a mold withdrawal procedure.

The most advanced turbine blade material now available is directionally solidified Mar-M-200 alloy containing hafnium. The advantages of directional solidification (including improved ductility, higher strength, and improved thermal fatigue resistance) are summarized in Fig. 11. The excellent thermal fatigue characteristics result in part from the elimination of transverse grain boundaries, which usually serve as sites for the start of thermal fatigue cracking, as well as producing a preferred texture orientation (100) which is accompanied by low modulus of elasticity.

Another feature which has developed in the past dozen years is the application of coatings to turbine airfoils. A pack aluminum coating has been successfully used for years to protect turbine hardware. A recent development in coatings is the deposition of overlay coatings of the NiCoCrAlY type from an electron beam cloud. Such coatings afford considerable flexibility in the tailoring of coating compositions for gas turbine applications.

Gas turbine engine materials development has led to many evolutionary and occasionally revolutionary changes in the past two decades. Changes continue to be made in all areas. The extent to which future advances will occur is highly dependent on the economic climate and the aggressive development efforts of materials engineers.

Materials for Aerospace Forgings

The selection of material for a forged part usually requires some compromise between opposing factors — for instance, strength vs toughness, stress-corrosion resistance vs weight, manufacturing cost vs useful load-carrying ability, production cost vs maintenance cost, and so on.

Material selection also involves consideration of melting practices, forming methods, machining operations, heat treating procedures, and deterioration of properties with time in service, as well as the conventional mechanical and chemical properties of the alloy to be forged.

Structural material for forgings is first appraised for selection on the basis of strength at room temperature — either yield strength or tensile strength. For airborne vehicles, a more meaningful comparison is strength per unit weight, commonly designated as strength-density ratio. Figure 1 shows strength-density ratios of aerospace forging alloys tested at room temperature; note the increase in ratio for aluminum alloys, reflecting progressive alloy development over a 38-year period.

Material Requirements

An efficient forging design obtains maximum performance from the minimum amount of material consistent with the loads to be applied, producibility, and desired life expectancy. With weight penalties on performance of aerospace vehicles, efficient design requires materials to operate at the high stress levels.

To match a material to its design component, the material is first appraised for strength and toughness, and then qualified for stability to temperature and environment. Optimum materials are then analyzed for producibility and finally for economy.

Requirements of a forged component include:

1 **Pattern of Applied Loads**
  - (*a*) **Uniaxial Loads.** Tensile or compressive, or reversible with changes in operating conditions.
  - (*b*) **Multiaxial or Combined Loads.** Tensile, compressive, shear, bending, torsion, and bearing. These may be either parallel to a central axis or at an angle. Stress concentration should be minimized in design by specifying smooth, contoured fillets at changes of configuration. Where stress concentration cannot be avoided, notch toughness of the material is usually important in material selection.
  - (*c*) **Cyclic Loads.** These may be either high-cycle or low-cycle loads.
  - (*d*) **Sustained Loads.** If these loads are tensile, they may accelerate stress corrosion. Interference fits and residual stresses may give rise to sustained loading.
  - (*e*) **Thermal Loads.** These are caused by variations in temperature.

2 **Load Magnitudes and Conditions of Loading**
  - (*a*) **Magnitudes**
  - (*b*) **Rate of Load Application.** Gradual or impact.
  - (*c*) **Temperature.** The major time accumulations should be estimated for minimum, normal and maximum temperatures.
  - (*d*) **Environment.** Cyclic periods of atmospheric condensation, chemical composition of environment, circumstances of corrosion, abrasion, erosion or other wear.

3 **Special** mechanical, physical or chemical requirements, if any.

4 **Life Expectancy or Reliability.** Service life, including downtime, should be estimated. Repairability should be considered.

Property requirements can be matched to candidate materials by consulting references on material properties available for established materials in MIL-HDBK-5A, Department of Defense, and other governmental publications, ASM Metals Handbook, SAE Aerospace Material Specifications, and the publications of technical and engineering societies.

Failure analyses are a useful data source for matching material properties to requirements. Failure of a component can occur during operation within the design stress range. One cause of premature failure is lack of proper orientation of a critical design stress with the preferred grain flow of a forging (see Chapter 3).

Unpredicted failure also may occur because of deterioration of material properties with time and service. For instance, stress-corrosion cracking, which results from sustained tensile stress, may occur even in a typical ambient atmosphere. Under these conditions, failure is most likely to occur at locations in the forging that coincide with exposed end grain (see Chapter 3, and Example 98 in Chapter 12).

Failure analyses may uncover other causes of premature failure, such as excessive grain growth, inclusions of nonmetallic impurities, grain flow folding from improper forging practice, lack of a wrought metallurgical structure, and from the inadvertent production of stress raisers by machining to an overly sharp fillet or by poor fit in assembly.

High-Strength Steel

High-strength steels equal or exceed the strength-density ratios of high-strength alloys of aluminum and titanium. They are in competition for selection for aerospace structural forgings such as landing-gear components, rocket cases, and airframe fittings. The dividing line above which a steel is designated "high-strength" is commonly regarded as 180-ksi yield strength. High-strength steels that approximate or surpass this strength differ in kind

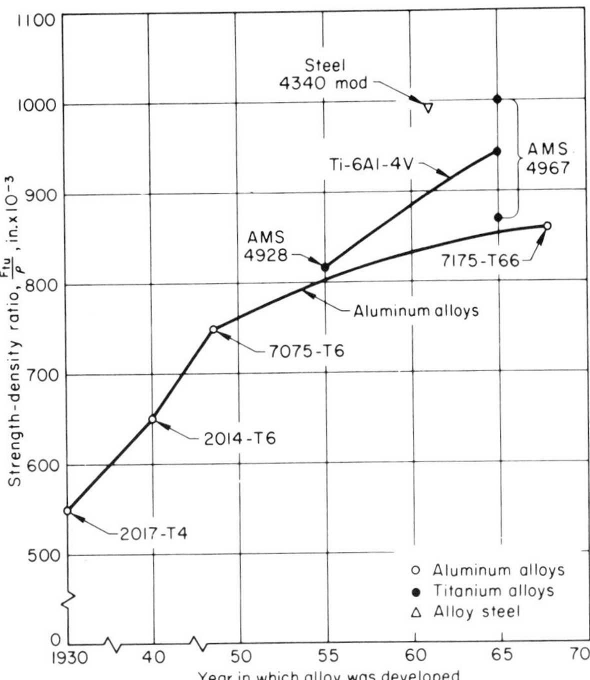

Fig. 1. Room-temperature strength-density ratios of aerospace materials developed over a period of about 40 years.

and quantity of alloying elements. Alloy content ranges from a few per cent up to about one-third the weight of the steel.

With the exception of the high-alloy maraging steels, all the commercially significant alloy steels are strengthened by conventional quench hardening. Alloying elements are added to prevent or retard the formation of nonmartensitic microconstituents during quenching. The maximum attainable strength level is determined by the carbon content.

In order to improve the ductility and toughness of hardened steel, it is reheated (tempered) for a relatively short time at a moderate temperature. Quenched-and-tempered steels are available in a wide range of strength levels. Thus, quenched 4340 steel is customarily given an appropriate tempering treatment to attain tensile strength levels in one of five different ranges: 140/160, 160/180, 180/200, 200/220 or 260/280 ksi. Many high-strength steels are variations of 4340. Small or light forgings are also produced from 4330, but the strength required in large airframe forgings is often supplied by 4340 or modifications; it is commonly modified by silicon (1.45 to 1.80%) and by vanadium (0.05% minimum) to enhance strength and toughness (MIL-S-8844, class C).

D-6ac, a deeply hardenable steel (AMS 6431) containing carbon, chromium, molybdenum and vanadium, is ordinarily hardened by conventional quenching and tempering. When D-6ac is also thermal-mechanical treated (ausformed), tests have shown high combinations of strength and toughness (Ault, McDowell and Hendricks, Technical Report AFML-TR-66-276, September 1966).

Steels containing alloying elements in amounts significantly greater than in 4340 can be transformation hardened by air cooling instead of liquid quenching. The slower cooling rate permits improved control of distortion and residual stresses, compared with conventional oil or water quenching. An example is 5Cr-Mo-V, a high-strength forging-quality version of H11 chromium hot work die steel.

A steel containing 9% Ni and 4% Co was developed specifically for improved toughness at high strength. Heat treatment of this steel depends on carbon content. Steels with 0.20 to 0.30% carbon are heat treated by oil quenching followed by subzero cooling and double tempering. This treatment produces a tempered martensite microstructure. Steels with 0.45% carbon are recommended for isothermal transformation (austempering) to produce a bainitic microstructure, which is normally tougher than martensite at a given strength. With any of these carbon contents, 9Ni-4Co steel has better toughness at high strength than do the lower-alloy steels. (Bulloch, Eichenberger and Guthrie, Technical Report AFML-TR-68-57, March 1968.)

Maraging steels are the only high-strength steels that do not depend primarily on carbon content for strength. Heat treatment of these steels consists of cooling from the austenite range to form iron-nickel martensite containing only about 0.02% carbon. The martensite is age hardened to precipitate intermetallic phases. As formed, the low-carbon martensite is soft and easily machined. The aging treatment is not accompanied by any appreciable dimensional changes, so forgings can be finish machined in the soft condition. The maraging steels, of which the 18% Ni grade 300 develops the highest strength, have toughness superior to that of low-alloy steels.

Stainless steels are used in high-strength applications where corrosion resistance is a controlling factor. The martensitic, age-hardenable martensitic, and semi-austenitic grades are used for small forgings. Austenitic stainless steels that depend on cold working for hardness are not useful for high-strength forgings. The discussion that follows is confined to the alloy steels of highest strength and strength-weight ratio and does not include stainless steels.

Melting. High-strength steel is especially sensitive to nonmetallic inclusions because these inclusions become stress raisers that reduce fracture toughness and related measures of ductility. To minimize the content of nonmetallic inclusions, high-strength steels are typically remelted under vacuum. Air-melted electric-furnace steel, deoxidized with aluminum and silicon, is poured into ingots that later serve as consumable electrodes. After cooling and conditioning, these as-cast ingots are remelted as consumable electrodes in a vacuum-arc process. Second-cast ingots prepared in this manner are then rolled to typical mill products, including bars or billets that are later forged and heat treated to high strength.

A measure of the advantage of vacuum-arc remelting is shown in Fig. 2. The transverse ductility of large billets of H11 steel heat treated to a high tensile strength in the range of 280 to 310 ksi shows significant improvement for consumable-electrode vacuum-remelted steel over air-melted electric-furnace steel. Reduction in area is raised from an original 7 to 12% in the air-melted steel, to 22 to 35%. Microscopic examination confirms a decrease in nonmetallic inclusions in the vacuum-remelted steel. Gas analysis has shown a corresponding reduction in hydrogen from 3 to 6 parts per million (ppm) down to 0.6 to 1 ppm, in oxygen from 20 to 75 ppm down to 9 to 25 ppm, and in nitrogen from 50 to 300 ppm down to 15 to 40 ppm. The increased ductility in tension tests is attributed to the lower gas content and fewer inclusions of the vacuum-remelted steel. There is added assurance of effective homogenization when steel is vacuum-arc remelted. Not only are nonmetallics and gases decreased, but more uniform distribution of metallic alloying elements (less segregation) is obtained.

Vacuum remelting of steel also permits improved deoxidation. In a vacuum furnace, carbon is an excellent deoxidizer, and the deoxidation product, carbon monoxide gas, is continuously removed under vacuum. Conventional deoxidation with aluminum and silicon, as is typical in electric-furnace air melting, yields solid oxides that often become undesirable inclusions. Steel intended for carbon deoxidation during vacuum remelting is therefore not deoxidized during air melting in the electric furnace. Instead it is poured in the so-called "open" or semikilled condition, with low silicon and manganese contents. When the cast electrodes of air-melted steel are remelted under vacuum and deoxidized with carbon, the product has significantly improved purity.

Consumable-electrode vacuum-arc remelted steel with carbon deoxidation has shown improved toughness when tested as wrought products. Figure 3 illustrates this effect for a steel with differing deoxidation practice and a range in yield strength from 200 to 250 ksi; toughness is expressed as nominal notch strength. Experimental heats of 9Ni-4Co-0.45C high-strength steel were made using three melting practices:

1 Air-melted in an electric furnace using aircraft quality practice and deoxidized with aluminum and silicon.
2 Air-melted in an electric furnace using aircraft quality practice, and deoxidized with aluminum and silicon, and then remelted in a consumable-electrode vacuum-arc furnace.
3 Air-melted in an electric furnace using aircraft quality practice except that the steel was not deoxidized, and then remelted in a consumable-electrode vacuum-arc furnace and deoxidized with carbon in the vacuum furnace.

These trial heats did not differentiate between air-melting and vacuum remelting processes when the air-melted steel was deoxidized with aluminum and silicon, but showed superior notch strength for the vacuum-arc remelted steel that was deoxidized with carbon. The notch strength was measured longitudinal to the grain direction, with flat specimens 0.180 in. thick, notched by fatigue cracking at a center-drilled aperture. At a yield strength of 225 ksi, the notched strengths for the Al-Si-deoxidized heats showed only 80 ksi, compared with 160 ksi for the carbon-deoxidized heats.

Table 1 shows static tensile properties averaged for a large number of heats of air-melted 4340 and vacuum-remelted and carbon-deoxidized 9Ni-4Co-0.45C. The higher elongation of the latter steel at the same strength level is considered statistically significant.

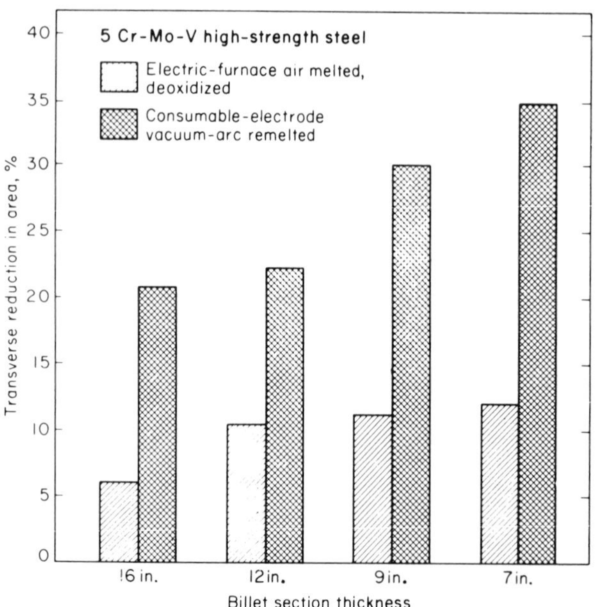

Fig. 2. Effect of melting method on transverse reduction in area for specimens cut from billets of H11 (5Cr-Mo-V) steel with a tensile strength of 280 to 310 ksi.

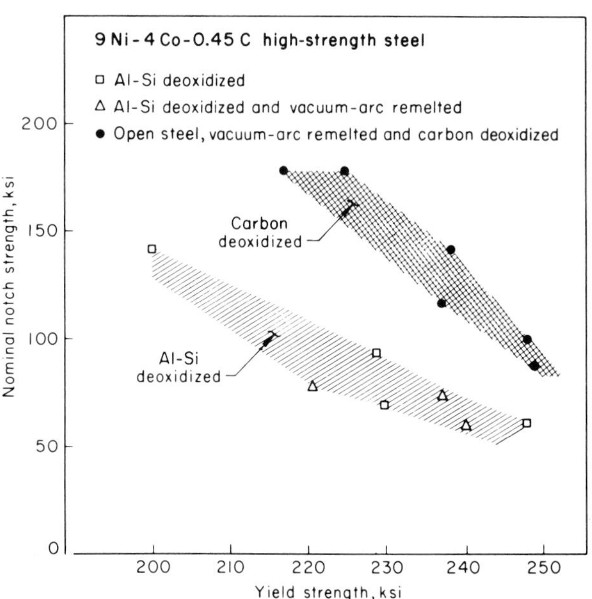

Fig. 3. Effect of deoxidation practice on relation of yield strength to notch strength for 9Ni-4Co-0.45C steel. (Matas, "Influence of Impurities and Related Effects on Strength and Toughness of High-Strength Steels", ASM Metals Engineering Quarterly, p 48 to 57, May 1964. The paper gives 58 references.)

Fatigue strength of several high-strength steels is summarized in Fig. 4. All of the steels were vacuum-arc remelted. Specimens of the flat type shown in Fig. 4(a) were cut from billets 3 by 9 in. in section, and all tests represent the long-transverse grain direction. Fatigue testing was axial, with stress ratio R of 0.06 and notch concentration

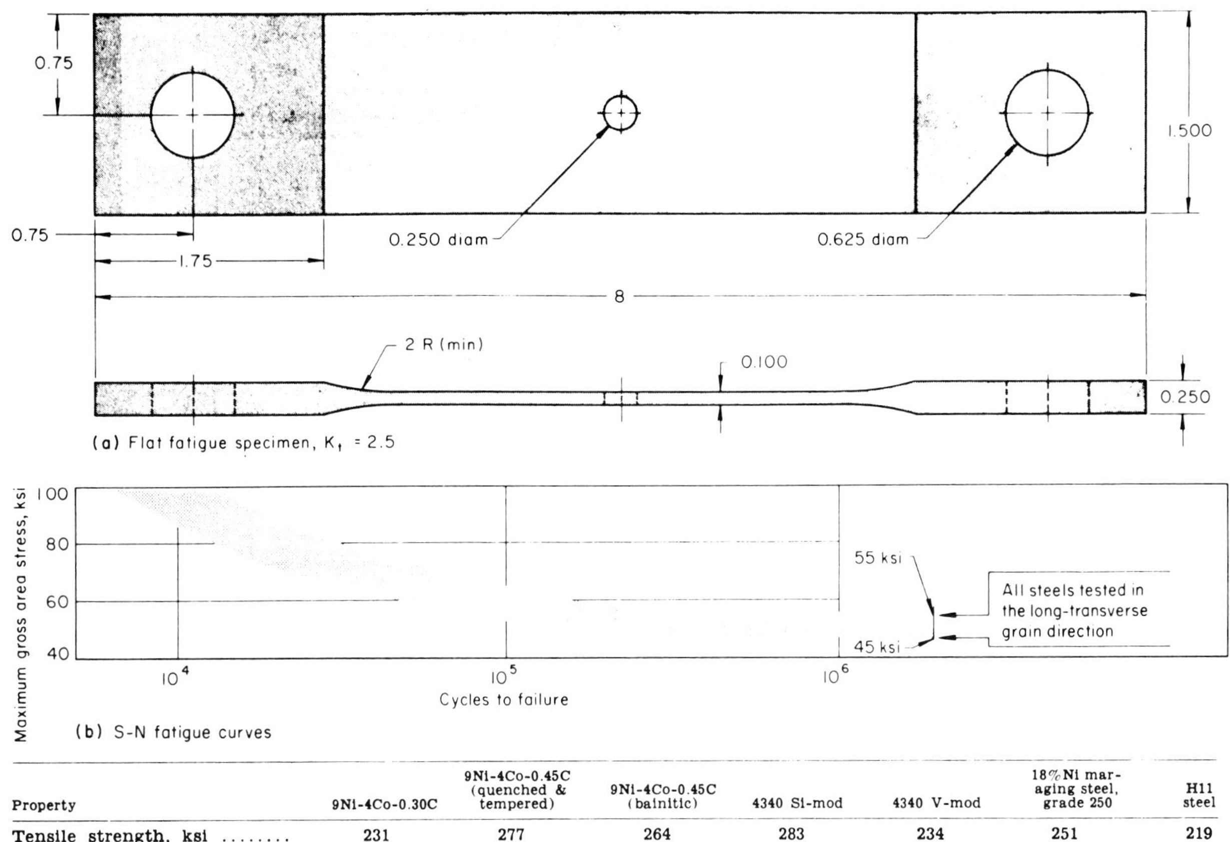

Property	9Ni-4Co-0.30C	9Ni-4Co-0.45C (quenched & tempered)	9Ni-4Co-0.45C (bainitic)	4340 Si-mod	4340 V-mod	18%Ni maraging steel, grade 250	H11 steel
Tensile strength, ksi	231	277	264	283	234	251	219
Yield strength, ksi	201	237	220	238	191	240	188

Fig. 4. Fatigue specimen (a) *and results of fatigue tests on several vacuum-arc remelted steels* (b). *(Bulloch, Eichenberger and Guthrie, Technical Report AFML-TR-68-57, March 1968)*

factor K_t of 2.5. All the steels fall within a narrow band as is shown in Fig. 4(b); at 10^6 cycles, the variation for all the steels extends approximately from 45 to 55 ksi. This corresponds to a variation of 219 to 283-ksi tensile strength and 188 to 240-ksi yield strength for specimens cut from the same billets.

Table 1. Mechanical Properties of Air-Melted 4340 Steel and Vacuum-Arc Remelted 9Ni-4Co-0.45C Steel(a)

Symbol(b)	4340 steel(c) L	4340 steel(c) T	9Ni-4Co-0.45C steel(d) L	9Ni-4Co-0.45C steel(d) LT	9Ni-4Co-0.45C steel(d) ST	9Ni-4Co-0.45C steel(d) L-LT-ST
Number of Tests						
N	301	463	14	15	14	43
Tensile Strength, Ksi						
A	261	259	....	....	....	260.4
$\overline{X}$	270.8	273.1	272.7	271.1	275.1	272.9
S	3.81	5.90	4.37	4.56	2.98	4.28
Yield Strength, Ksi						
A	214	206	....	....	....	204.7
$\overline{X}$	223.2	228.0	226.2	225.2	229.7	227.0
S	3.75	8.79	6.21	10.35	4.82	7.66
Elongation, %						
A	....	....	....	....	....	7.5
$\overline{X}$	9.6	7.3	....	....	....	11.3
S	2.16	1.97	....	....	....	....

(a) Steels were heat treated to a minimum tensile strength of 260 ksi. (b) N = number of tests; A = lower limit of 99% population band; $\overline{X}$ = arithmetic average; S = standard deviation. L = longitudinal; T = transverse; LT = long transverse; ST = short transverse; L-LT-ST = a combination of these test results.
(c) Specimens were cut from bar and billet stock from 1-in.-diam round to 8-by-8-in. round-corner square, and then rough machined, heat treated, finished and tested. (d) Specimens were cut from forging stock and from subblocks from forgings, and then rough machined and heat treated to a bainitic condition, finished and tested. Maximum section of billet and forgings was about 3 in. (Bulloch, Eichenberger and Guthrie, Technical Report AFML-TR-68-57, March 1968)

Toughness. In high-strength steel high strength is attained at the expense of reduced ductility. The most common mechanical tests to assure that forging stock and forgings of high-strength steel will meet ductility and toughness requirements are: *(a)* the reduction in area and percent elongation in a transverse tension-test specimen, and *(b)* Charpy impact tests on both standard and precracked specimens taken from the forged material.

These tests, even though they show relative degrees of ductility and toughness, may not give the designer an accurate indication of the resistance of the material to plastic deformation under a complex loading system. As the state of stress approaches that of hydrostatic tension, even the most ductile alloy may fracture in a brittle manner. Flaws or cracks within a material often create these complex stress states. Depending on the material and its relative strength, a "critical crack length" or "critical flaw size" is required at a particular stress level for crack propagation. This critical crack length has decreased with the development of the high-strength steels. High-strength steel that contains a microscopic flaw or crack or other local stress concentration may fail in a brittle or catastrophic manner even when the base metal has considerable ductility.

Fracture toughness testing attempts to predict the behavior of a material under conditions in which a flaw or notch is present. The ultimate goal is to determine the critical flaw size below which brittle fracture will not occur, for a structure of specified design, using material with a specified strength level, and with a specified design load. The resistance of a material to brittle fracture may be expressed in terms of its plane-strain fracture toughness (K_{Ic}), which is measured in terms of the stress intensity factor (K_I).

K_I defines the elastic stress field intensification in the vicinity of a crack tip, where the applied load is perpendicular to the crack plane. K_{Ic} represents a critical value of K_I and characterizes the susceptibility of the material to unstable crack propagation under conditions of high elastic constraint. Since K_{Ic} is sensitive to many processing and testing variables, and is subject to various methods of computation, care is necessary in comparing test results from different investigations.

Values of fracture toughness derived from tests of a series of high-strength steels of various compositions are presented in Fig. 5. The steels were tested under identical conditions. The values represent tests from three billets (3 by 9 by 24 in.) of each alloy. The specimens were flat, typically about 1½ by ½ in. in section and 7½ in. long. A transverse notch was machined across one of the long ½-in. sides. The root of the notch was precracked by cycling in fatigue. The coupon was then bent to failure in four-point loading in a testing machine so that the cracked tip was placed in tension of measured intensity.

In Fig. 5, critical stress intensity (K_{Ic}) is plotted against tensile strength at room temperature. The K_{Ic} values were calculated from parameters depending on the size of specimen, depth of notch and crack, and the loads and deflections recorded by the testing machine. Failure is recorded at the load and deflection at which the specimen separates at the precracked notch by brittle fracture. Figure 5 shows K_{Ic} values for tests at room temperature and at −65 F. K_{Ic} is always smaller at −65 F that at room temperature. The tests shown in Fig. 5 were taken from specimens cut from the long-transverse grain direction of the billets.

Ideally, material selection on the basis of fracture toughness requires K_{Ic} values in all three principal grain directions — longitudinal, long-transverse, and short-transverse. Further insight is gained by securing K_{Ic} values at subzero and elevated temperatures when these temperatures will be encountered in service. Values are most useful for material selection when the tensile strengths of the materials being compared are the same.

Stability in Environments; Corrosion. High-strength steels generally are suitably protected from exfoliation corrosion by the application of organic finishes or metallic plating. However, these steels are sometimes subject to stress-corrosion or hydrogen-stress cracking; then other preventives are needed, in addition to coatings. Both types of cracking occur in a plane normal to the direction of stress or load (but below the yield strength), thereby reducing the design section, and both types

High-strength steel	No. heats	No. suppliers
H11	3	1
4330 V-mod	3	2
9Ni-4Co-0.30C	3	1
9Ni-4Co-0.45C	3	1
18% Ni maraging steel, gr 250	3	3
4340 Si-mod	2	2

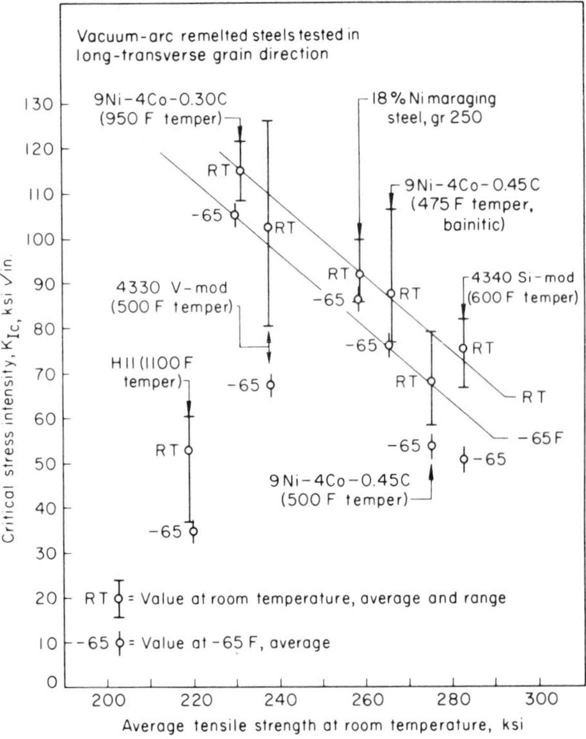

Fig. 5. Relation between level of tensile strength at room temperature and fracture toughness expressed as critical stress intensity, K_{Ic}. **(Source: Bulloch, Eichenberger and Guthrie, Technical Report AFML-TR-68-57, March 1968)**

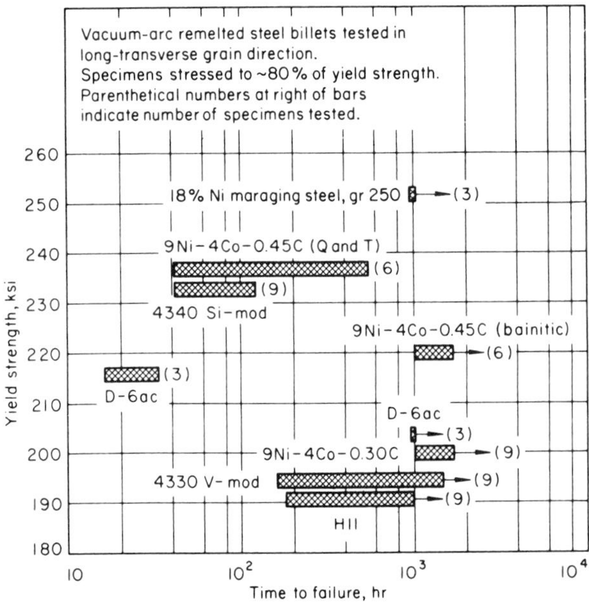

Fig. 6. Resistance to stress corrosion of various steels. Smooth bend specimens were tested after alternate immersion in a 3.5% NaCl solution at room temperature. **(Source: Same as for Fig. 5)**

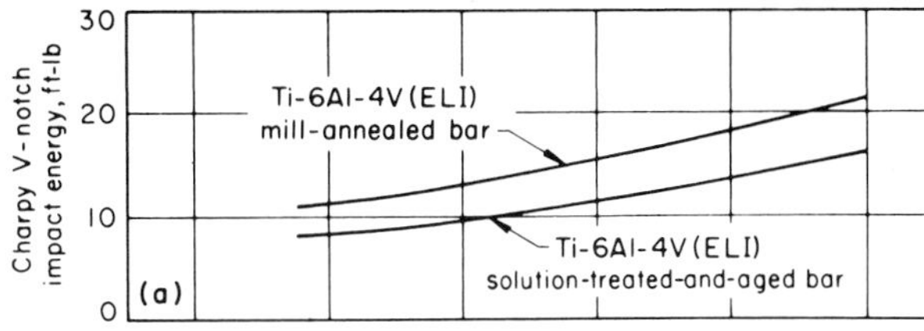

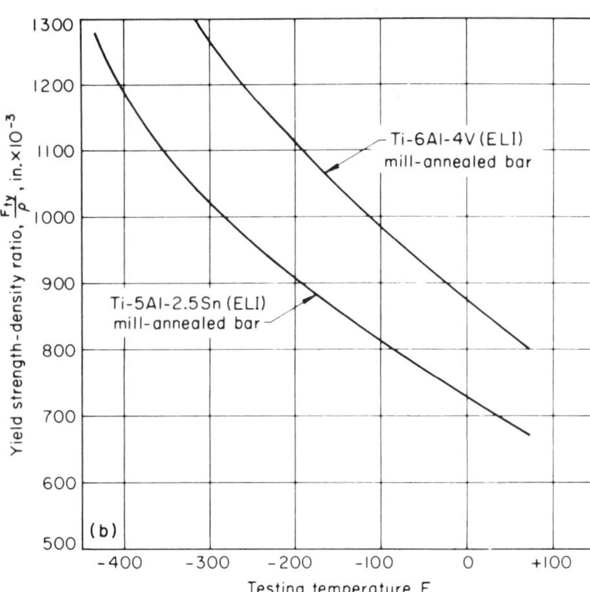

Fig. 7. Notch toughness and ratio of yield strength to density for Ti-6Al-4V ELI and Ti-5Al-2.5Sn ELI at cryogenic temperatures.

Table 2. Maximum Amount of Iron and Interstitial Elements in Ti-5Al-2.5Sn and Ti-6Al-4V

Element, %	Ti-5Al-2.5Sn Standard	Ti-5Al-2.5Sn ELI(a)	Ti-6Al-4V Standard	Ti-6Al-4V ELI(a)
Iron	0.50	0.15(b)	0.25	0.15(b)
Carbon	0.08	0.08	0.08	0.08
Nitrogen	0.05	0.05	0.05	0.05
Oxygen	0.20(c)	0.12(b)	0.20(c)	0.13(b)
Hydrogen:				
Bar	0.0200	0.0125(b)	0.0125	0.0125
Billet	0.0175	0.0125(b)	0.0100	0.0100

(a) Extra-low interstitial. (b) Maximum allowable percentages are reduced in comparison to standard grade. (c) Or by agreement of producer and purchaser. Typically not specified.

proceed with the enlargement of flaws to accelerate further cracking and eventual brittle fracture. Material selection for high-strength steel therefore ideally discriminates first by fracture toughness tests in dry laboratory air, and then by resistance to simulated stress-corrosion and hydrogen-stress cracking in service environments. Factors that affect the stress-corrosion susceptibility of high-strength steels are their composition, structure, strength level, applied stress, residual stress, environment, and time. (Fletcher, Berry, Elsea, DMIC Report 232, Battelle, 29 July 1966; and Bulloch, Eichenberger, Guthrie, Technical Report AFML-TR-68-57, March 1968.)

Figure 6 presents test results showing degree of resistance to stress-corrosion for various steels and heat treatments, representing yield strengths from 190 to 250 ksi. Tests were made under controlled and similar conditions. Each sample was stressed to approximately 80% of its yield strength (applied stress), each was alternately dipped in and removed from a solution of 3.5% NaCl at room temperature, after a uniform surface preparation (environment), and each was tested for 1000 hr or until failure occurred in a shorter time.

Titanium Alloys

Titanium alloys generally contain two or more alloying elements. Many contain four or more of the following elements: aluminum, vanadium, tin, zirconium, chromium, molybdenum, iron, manganese, silicon, tungsten and palladium. All titanium alloys contain minute residual quantities of the interstitials, carbon, oxygen, nitrogen and hydrogen. Most of the alloys are hardenable by solution treating and aging. Titanium alloys are used from cryogenic temperatures of −423 F to elevated temperatures close to 1000 F. No single alloy, however, is suitable for the entire range.

Titanium alloys are usually produced by multiple vacuum-arc melting — that is, consumable electrodes are melted under vacuum at least twice in series. This melting practice provides advantages of degassing and homogenization under vacuum.

The first-stage consumable electrode is prepared from sections or segments called "compacts", which are pressed forms of the sponge pellets of titanium through which pellets of other materials are uniformly mixed for alloying. Compacts are welded together to form an electrode. Most electrode joining by welding is done by a plasma arc or other high-temperature heat source in an inert atmosphere. If oxidation of titanium occurs during welding, the oxides cannot be reduced during subsequent vacuum-arc remelting and appear in the forging as low-density inclusions. The first electrode is melted and solidified to form a second-stage electrode, which, in turn, is remelted and solidified to a production ingot. Ingot molds are water-cooled copper. The cast electrodes and final ingots are progressively chilled to minimize dendritic segregation. Homogenization occurs during the period of mixing of the molten pool below the vacuum arc; chilling inhibits segregation during final solidification.

Titanium alloy ingots have been produced from 30 to 36 in. in diameter and weighing 10,000 to 15,000 lb. Ingots 37 to 48 in. in diameter weighing 16,000 to 30,000 lb are anticipated. Forgings of Ti-6Al-4V alloy weighing more than 5000 lb have been produced.

Cryogenic Temperatures. For pressure vessels used at temperatures to −320 F and lower, two titanium alloys are available. These are the extra-low-interstitial (ELI) modifications of Ti-5Al-2.5Sn and Ti-6Al-4V. Ti-5Al-2.5Sn ELI is used at −423 F, the temperature of liquid hydrogen. However, titanium alloys are *not* recommended for containment or other use with either liquid or gaseous oxygen in cryogenic service. This restriction avoids the possibility of explosion from reaction on any freshly fractured surface of titanium. Titanium in contact with liquid oxygen results in a reaction at impact energy levels as low as 10 ft-lb. With gaseous oxygen, a partial pressure of about 50 psi is sufficient to ignite a fresh titanium surface over the temperature range from −250 F to room temperature or higher (MIL-HDBK-5A).

Interstitial elements are maintained at low maximums, even in standard alloys. Extra-low-interstitial alloys are further reduced in iron, as well as in the interstitials oxygen and hydrogen. Table 2 compares maximum iron and interstitials in standard and ELI alloys.

Reduction of iron and oxygen in the ELI alloys creates an increment of toughness but at some sacrifice in yield and tensile strengths. Toughness is shown in Fig. 7(a), in which Charpy V-notch impact values are plotted against temperature from the cryogenic range to room temperature. Figure 7(b) shows the ratio of yield strength to density over the same temperature range. The values in Fig. 7(a) and (b) are typical for bar stock of ½ to 4-in. diameter, tested longitudinally.

Airframe Structural Forgings. Material selection of titanium alloys for use at atmospheric temperatures up to about 500 F is dominated by Ti-6Al-4V. This range of operating temperature encompasses all subsonic airframe forgings and those for supersonic aircraft up to Mach 2.75. From −50 to 500 F, Ti-6Al-4V and its alternates maintain adequate stability.

Beginning at about 300 F and continuing to between 500 and 700 F, standard grades of titanium alloys exhibit strain aging that inhibits creep beyond the initial yield strength. However, below 300 F and down to room temperature these alloys may creep significantly at a stress greater than 75% of the yield strength (MIL-HDBK-5A).

Figure 8 shows typical ranges of mechanical properties of Ti-6Al-4V, from room temperature to 600 F, both annealed and solution treated and aged. The wide bands display the distribution of properties for bars ranging from ½ to 4 in. in diameter, in both the longitudinal and transverse directions.

In the temperature range of −50 to 500 F, Ti-6Al-4V is the alloy most often used; Ti-6Al-6V-2Sn is the principal alternate. Minimum strength values at room temperature for annealed forgings of Ti-6Al-4V are: tensile strength, 130 ksi, and yield strength, 120 ksi (AMS 4928). Corresponding values for Ti-6Al-6V-2Sn (less than 2 in. thick) are tensile strength, 150 ksi, and yield strength, 140 ksi. Heat treated bars 1 to 2 in. in diameter, of Ti-6Al-4V, also have minimum values of tensile strength, 150 ksi, and yield strength, 140 ksi (AMS 4967). Corresponding minimum values for Ti-6Al-6V-2Sn are: tensile strength, 170 ksi, and yield strength, 160 ksi (AMS 4971). However, the minimum tensile strength of 170 ksi for alloy Ti-6Al-6V-2Sn is difficult to maintain in a section thickness greater than 2 in., which accounts in part for the unsuitability of this alloy for some applications.

When design permits the relatively reduced strength of annealed Ti-6Al-4V, these values can be guaranteed to sections of 5 in. and sometimes in greater sizes. The helicopter rotor forging shown in Fig. 9 demonstrates this. Figure 9(b) indicates section sizes up to 8½ in. Test results from specimens taken in various directions from a prototype

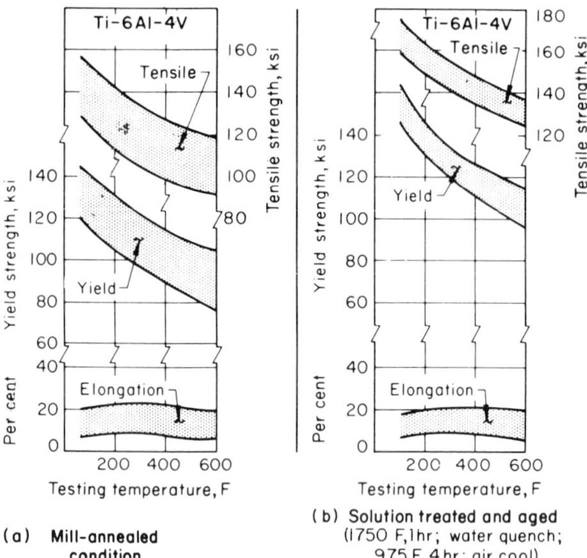

(a) Mill-annealed condition

(b) Solution treated and aged (1750 F, 1 hr; water quench; 975 F, 4 hr; air cool)

Fig. 8. Typical ranges of mechanical properties of alloy Ti-6Al-4V from room temperature to 600 F.

forging have shown the tensile properties given in the table accompanying Fig. 9.

For improved hardenability (depth of hardness) and at a high strength level, Ti-13V-11Cr-3Al, a beta-stabilized alloy at room temperature, is available. A landing gear cylinder forging of this alloy, with a plan area of 350 sq in., weighing 250 lb, and of section thickness to 9 in. as forged, is shown in Fig. 10(a). The cylinder was forged solid and then bored (approx 3½-in. diam) before heat treatment. Boring was from the shaft end, and left the trunnion end blind. The heat treatment was 1350 F for 2 hr and air cool, then 1450 F for ½ hr and water quench, and age at 900 F for 20 hr, followed by air cooling. Yield strength values from specimens taken at the locations shown in Fig. 10(b) were in excess of 174 ksi, and tensile elongation at all but three locations was 4% or greater. Production cylinders machined from these forgings have had satisfactory service records.

The hardenability effect is shown for several alloys in Fig. 11.

Directionality of properties in bars or billets of titanium alloys is generally not apparent in tensile and yield strength values. There is, however, directionality in ductility as measured by elongation and reduction in area of tension-test specimens. The usual guaranteed minimums for annealed Ti-6Al-6V-2Sn are taken at an elongation of 10% for longitudinal specimens and 8% for transverse specimens. Corresponding reduction in area is 20% longitudinal and 15% transverse. Heat treated wrought products of titanium alloys also commonly reflect a decrease in elongation of 2% when testing is in the transverse direction. This effect may be accentuated for forgings when testing is in the short-transverse direction.

Fatigue strength of titanium alloys has been thoroughly explored in the laboratory. Alloy Ti-6Al-4V has a smooth-bar rotating-beam endurance limit, in both the annealed and the solution-treated-and-aged conditions, that is 50% or more of the tensile strength, and the notched endurance limit at K_t = 3.5 is approximately 25% of the tensile strength. Meaningful data on fatigue strength are best secured by testing actual components.

Fracture Toughness. Increased use of titanium alloys for large airframe components, and at higher strength levels, has heightened interest in toughness. Fracture toughness testing is being used to supplement the more common measures of ductility that are appraised for material selection. Fracture toughness testing of titanium alloys is also being correlated with forging practice, particularly with respect to "beta forging".

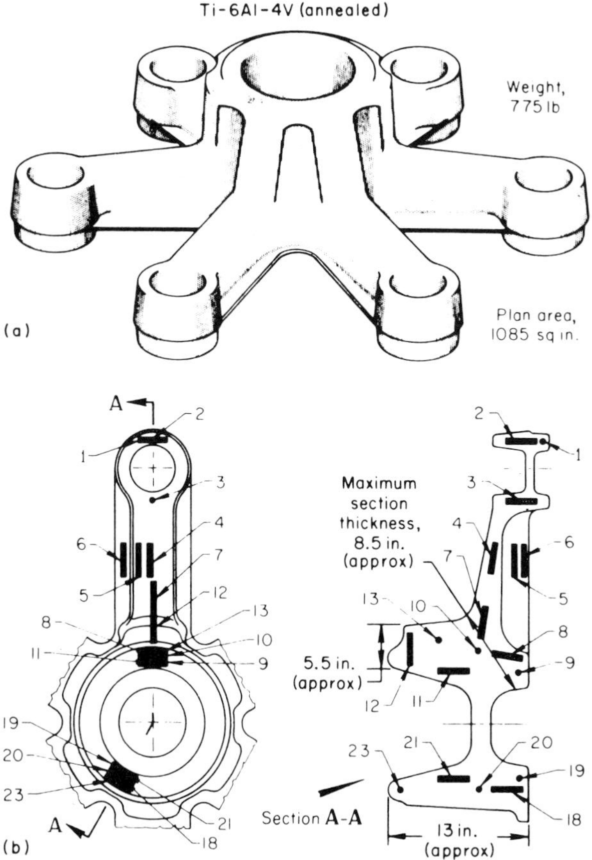

Specimen location	0.2% F_{ty}, ksi	F_{tu}, ksi	Elongation, % in 4D	Reduction in area, %
Mechanical Properties				
1	141.4	152.8	14.0	41.9
2	134.8	147.0	12.0	32.7
3	133.0	145.6	11.0	32.4
4	135.6	146.8	14.0	33.0
5	139.8	148.6	14.0	37.6
6	141.2	149.6	16.0	39.1
7	129.8	141.4	15.0	27.5
8	128.8	142.2	12.0	27.2
9	137.6	149.8	15.0	35.8
10	131.8	141.8	12.0	33.7
11	134.6	144.0	11.0	28.6
12	135.8	147.0	13.0	32.8
13	129.0	140.8	10.0	20.9
18	132.4	143.2	12.0	33.8
19	136.0	146.6	14.0	40.4
20	127.0	140.0	12.0	36.7
21	132.4	140.6	12.0	33.7
23	125.2	138.6	10.0	25.3
Summary				
Number ...	18	18	18	18
Average ...	134	145	12.7	33
Range	125 to 141	139 to 153	10 to 16	21 to 42

Fig. 9. Helicopter main rotor forging of Ti-6Al-4V, and views showing location of test bars. Mechanical properties of the test bars are given in the table. (Erbin, Los Angeles meeting of AIME, Feb 1967)

Beta forging refers to the forging of alpha-beta alloys at higher temperatures where the alloy consists entirely of the beta phase, because of increased forgeability at the higher temperatures or to develop notch toughness. The use of forging temperatures in the beta range can result in lower power requirements, permitting the forging of more complex shapes to closer dimensional tolerances, provided certain basic requirements are observed. One requirement is that a minimum reduction of 50% must be obtained in the forging operations performed *below* the beta transus temperature, following a similar reduction during initial forging *above* the beta transus temperature.

Thus, in one test program,* the alpha-beta alloys were forged in a two-step sequence on a 1500-ton press, using 8½-in.-diam slugs. The first reduction, performed in the beta range, was from 3.5 to 1.75 in. This produced a pancake about 11 in. in diameter. The pancake was then cut in half, and each half was forged in the alpha-beta range to a thickness of 0.625 in. this corresponds to reductions in height of 50 and 65%, respectively.

Tests on these alloys showed that, given the same heat treatment, the beta-forged materials had lower yield strength and ductility than conventionally forged alloys. However, beta-forged alloys had superior notch tensile strength and usually superior fracture toughness. The beta transus temperature for alloy Ti-6Al-4V is 1810 F; for Ti-7Al-4Mo, 1870 F; for Ti-6Al-6V-2Sn, 1710 F; and for Ti-6Al-2Sn-4Zr-2Mo, 1810 F. These alloys were beta forged at 2100 F.

Although beta forging offers some advantages in producibility, it creates processing problems; the final stages of deformation, in the alpha-beta range, are extremely important in developing a satisfactory degree of structural uniformity with acceptable mechanical properties. Because microstructure depends on the amount of deformation, sections of forgings that receive relatively little deformation during final forging retain beta-annealed structures on cooling to room temperature. Accordingly beta forging is most successful with forging designs that permit a rather uniform and appreciable amount of hot working of cross sections.

Based on alpha-beta forging standards, a well-wrought microstructure consists of a controlled amount (15 to 30%) of equiaxed alpha phase in a matrix of transformed beta phase. An excessive amount of equiaxed alpha leads to reduced notch toughness, and insufficient equiaxed alpha is usually related to lower tensile ductility. The notch toughness of forgings containing excessive alpha usually can be restored by heat treatment at temperatures 30 to 60 F below the beta transus, followed by normal heat treating practice. Low duc-

*G. H. Heitman, J. E. Coyne, and R. P. Galipeau, The Effect of Alpha + Beta and Beta Forging on the Fracture Toughness of Several High-Strength Titanium Alloys, *Metals Engineering Quarterly*, Vol 8, Aug 1968, p 15-18.

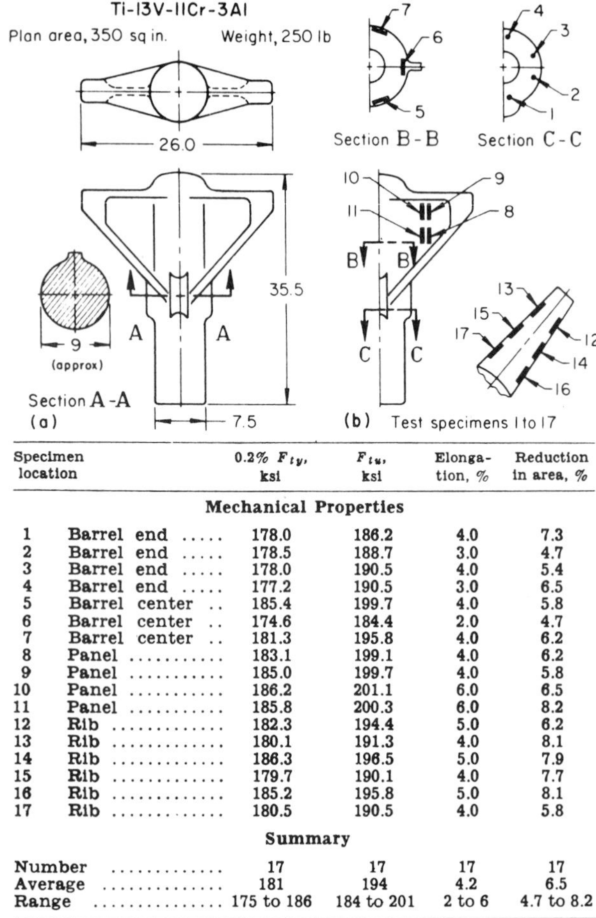

Specimen location		0.2% F_{ty}, ksi	F_{tu}, ksi	Elongation, %	Reduction in area, %
Mechanical Properties					
1	Barrel end	178.0	186.2	4.0	7.3
2	Barrel end	178.5	188.7	3.0	4.7
3	Barrel end	178.0	190.5	4.0	5.4
4	Barrel end	177.2	190.5	3.0	6.5
5	Barrel center	185.4	199.7	4.0	5.8
6	Barrel center	174.6	184.4	2.0	4.7
7	Barrel center	181.3	195.8	4.0	6.2
8	Panel	183.1	199.1	4.0	6.2
9	Panel	185.0	199.7	4.0	5.8
10	Panel	186.2	201.1	6.0	6.5
11	Panel	185.8	200.3	6.0	8.2
12	Rib	182.3	194.4	5.0	6.2
13	Rib	180.1	191.3	4.0	8.1
14	Rib	186.3	196.5	5.0	7.9
15	Rib	179.7	190.1	4.0	7.7
16	Rib	185.2	195.8	5.0	8.1
17	Rib	180.5	190.5	4.0	5.8
Summary					
Number		17	17	17	17
Average		181	194	4.2	6.5
Range		175 to 186	184 to 201	2 to 6	4.7 to 8.2

Fig. 10. Landing gear cylinder forging of Ti-13V-11Cr-3Al, and views showing location of test bars. Mechanical properties of the test bars are given in the table. (Source: Same as Fig. 9)

tility resulting from the absence of primary alpha can be restored only by further hot deformation.

Jet Engine Forgings. Material selection of titanium alloys for application to jet engines involves service at temperatures up to about 950 F. This is seen graphically in Fig. 12. Design stress is that for 0.2% yield strength or 0.2% creep in 10,000 hr. Creep is the controlling strength criterion over the range of temperature from 600 to 1000 F.

Material selection for this temperature range passes from Ti-6Al-4V to the higher-strength Ti-6Al-2Sn-4Zr-2Mo, as shown in Fig. 12. Alloys intermediate to these (not shown) are Ti-7Al-4Mo and Ti-8Al-1Mo-1V. Figure 13 shows a Larson-Miller parameter plot for typical 0.1% creep deformation for five creep-resistant titanium alloys.

Stability and Corrosion. Titanium alloys have generally good corrosion resistance. This is most apparent in resistance to sea water at atmospheric temperature. There are two important exceptions to the inert action of titanium alloys, particularly on any freshly machined surfaces: (*a*) the danger of explosion from a chemical reaction with liquid or gaseous oxygen, as discussed in the paragraph on Cryogenic Temperatures on page 6 in this chapter, and (*b*) the danger of explosion from a chemical reaction with anhydrous red fuming nitric acid.

Selection of titanium alloys for resistance to stress-corrosion cracking proceeds by first ranking the materials by laboratory tests and then confirming by service testing. Testing for resistance to stress-corrosion cracking, like fracture toughness testing, is not standardized, but Fig. 14 places both these parameters on a single graph. Fracture toughness testing involves short-time tests in air. Testing for resistance to stress-corrosion cracking

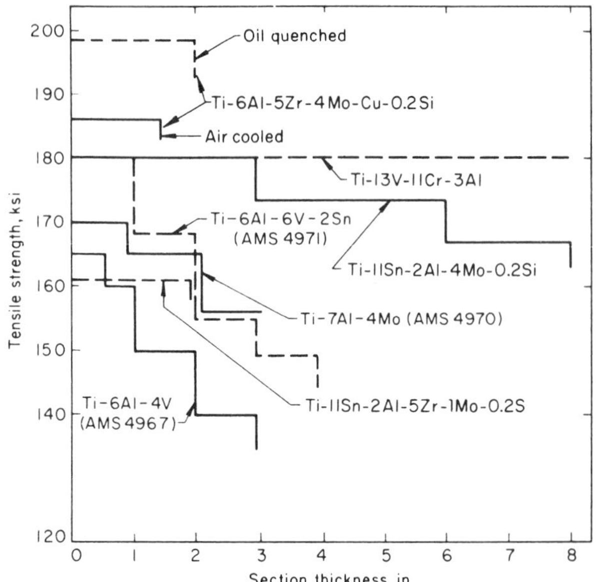

Fig. 11. Effect of section thickness on room-temperature tensile strength (hardenability effect) for several titanium alloys.

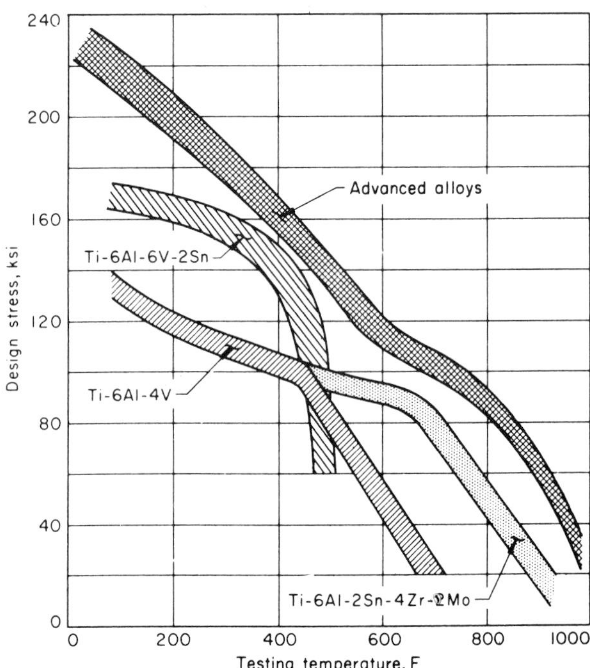

Fig. 12. Typical design stress in relation to temperature, for high-strength titanium alloys for jet engine forgings. Design stress is for 0.2% yield strength or 0.2% creep in 10,000 hr. (L. P. Jahnke, Metal Progress, *Sept 1968)*

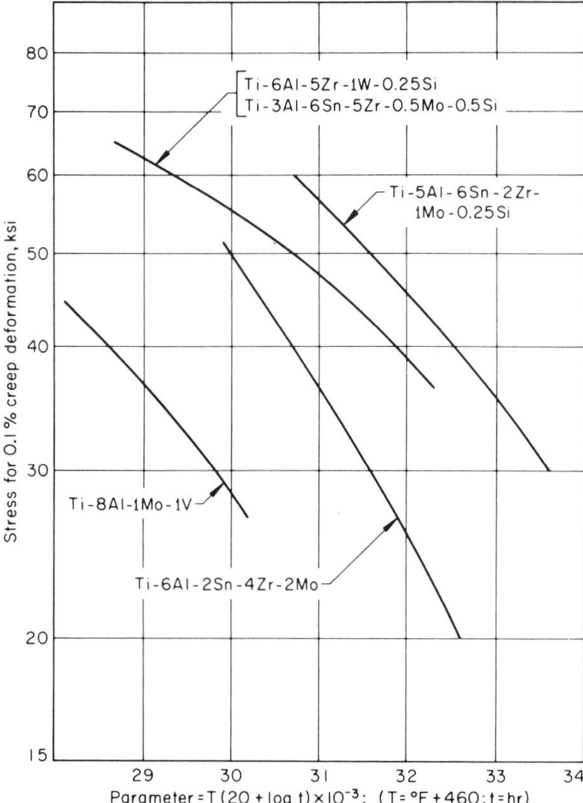

Fig. 13. Larson-Miller parameter plot showing typical elevated-temperature creep strength of five titanium alloys.

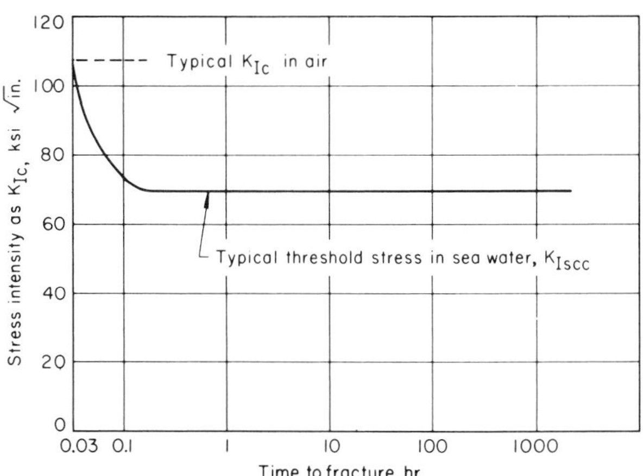

Fig. 14. Typical stress-corrosion curve for precracked and cantilever-beam-loaded mill-annealed samples of Ti-6Al-4V from 1-in.-thick plate.

adds the variables of time and corrosive environment. Under laboratory control, a threshold stress is established for resistance to sea water at room temperature to compare with K_{Ic} for air, using annealed 1-in. plate of Ti-6Al-4V (Fig. 14). Curves of this kind can be useful in material selection.

Resistance to stress-corrosion cracking in hot salt is also measured and recorded under laboratory control. This refers to dry sodium chloride at temperatures above 550 F. This type of cracking has been observed only in the laboratory.

Table 3. Minimum Mechanical Properties of Aluminum Alloys With High Strength-Density Ratios

Aluminum alloy and temper	Tensile strength (F_{tu}), ksi	Yield strength (F_{ty}), ksi	Elongation, %	Strength-density ratio, in. × 10^{-3}
7075-T6(a):				
Longitudinal	75	65	7	750
Transverse	71	62	3	710
7175-T66(b):				
Longitudinal	86	76	7	860
Transverse	77	66	4	770
7075-T73(a):				
Longitudinal	66	56	7	660
Transverse	62	53	3	620
7175-T736(b):				
Longitudinal	76	66	7	760
Transverse	71	62	4	710
7079-T6(c):				
Longitudinal	72	62	8	720
Transverse	70	60	3	700

(a) Adapted from specification MIL-A-22771, and for section thickness to 3 in. (b) Table values and text adapted from private communication, Air Force Materials Laboratory, WPAFB, 1968; and Alcoa Technical Information, May 1968. (c) Adapted from specification MIL-A-22771, and for section thickness to 6 in.

Aluminum Alloys

More than any other single class of materials, the age-hardenable high-strength aluminum alloys have been depended on to provide the majority of forged structural components and fittings for aircraft. All of the high-strength aluminum alloys offer favorable strength-density ratios. Some, such as 7075-T6 and 7079-T6, are well represented among the numerous airframe forgings for which aspects of design are discussed in this Handbook.

In recent years, stronger aluminum alloys with even higher strength-density ratios have been introduced. Alloy 7175-T66, for example, with a longitudinal yield strength of 76 ksi, exceeds the longitudinal yield strength of 7075-T6 by almost 17%, with no sacrifice in ductility. With a strength-density ratio of 860, alloy 7175-T66 leads the field of high-strength aluminum alloys, as shown in the comparison of minimum mechanical properties in Table 3. However, alloy 7175 in both the T6 and the T66 tempers is intended for applications that involve relatively low sustained tensile stresses in the short-transverse direction, and is not preferred for applications where enhanced resistance to stress-corrosion cracking is a controlling factor.

With a longitudinal yield strength of 56 ksi in the T73 temper, alloy 7075-T73 is frequently selected for good resistance to stress corrosion, particularly in the transverse direction, although the T73 temper incurs a 14% decrease in longitudinal yield strength when compared to the T6 temper (see Table 3). However, the newer alloy, 7175, in the T736 temper offers stress-corrosion resistance equivalent to that of 7075-T73 at a strength level equivalent to that of 7075-T6 (Table 3). Service tests of alloy 7175-T736 substantiate that, in the short-transverse direction, the alloy has a stress-corrosion cracking threshold of 35 ksi or more; in addition, these tests show that the elongation and notch toughness of alloy 7175 in the T736 temper are equivalent to those of alloy 7075-T73, and in the T66 temper to those of 7075-T6.

Heat-Resisting Alloys

Heat-resisting alloys useful at temperatures above 1200 F are based on iron, on nickel and on cobalt and contain elements that form precipitates that harden the matrix after solution treating and aging. Structural stability and resistance to oxidation and corrosion at elevated temperatures are required of these alloys.

Iron-base (actually, iron-chromium-nickel-base) alloys are the least costly and are applied in the lower temperature range, 1200 to 1500 F. Nickel-base and cobalt-base alloys are both applicable within the range of 1500 to 2000 F, and at temperatures below 1500 F as well. The hardening phase in nickel-base alloys is a nickel-aluminum-titanium phase called gamma prime. The hardening phase in cobalt-base alloys is a complex carbide.

Vacuum melting has gained acceptance in the production of heat-resisting alloys, as it permits close control of cleanness and homogeneity. These qualities become more critical as ingot size increases. Size estimates for the near future include billets 20 to 24 in. in diameter made from ingots 30 to 36 in. in diameter. As alloying elements are increased to provide high-temperature strength in these alloys, so is segregation during solidification, especially on slow cooling.

Vacuum melting permits accurate adjustment of composition and deoxidation with carbon, thus permitting oxygen removal in gaseous combination with carbon and inhibiting the formation of solid oxides in the bath. Under vacuum, gaseous hydrogen and nitrogen are removed to trace residuals. Vacuum melting also removes volatile metals, such as lead and zinc. Final additions of reactive metals are facilitated by the absence of any reaction of the bath with either air or slag.

During a run of 100 heats for more than 1,200,000 lb of ingots of Inconel 718, aluminum was held within the range of 0.5 to 0.6%, for a control of ±0.05%. In 97 of the heats, titanium varied only 0.90 to 1.10%; the aim for columbium content was 5.25%, and 99% of the values fell between 5.05 and 5.40%. In 95 of the heats, carbon ranged from 0.04 to 0.05%. (Aufderhaar, *Metal Progress,* June 1968.)

For the most complex alloy systems, powder metallurgy is employed to prevent gross segregation. The alloy is melted in a conventional way and atomized while still in the liquid state, to form spheres, which are ground to fine powders of homogeneous chemical composition. The powders are compacted into preforms, sintered and then forged in the conventional way to produce segregation-free forgings.

A great many cast and wrought heat-resisting alloys are available. Figures 15 and 16, on the next two pages, show rupture strengths for about 40 different compositions as a function of temperature.

Iron-base heat-resisting alloys are only slightly more alloyed than stainless steels. They maintain useful strength within the lower range of temperatures, up to 1200 F; some are used at up to 1500 F. Greek Ascoloy, with the curve shown at the extreme left in Fig. 15 over a temperature range of 900 to 1200 F, is a martensitic chromium stainless steel containing about 80% iron. It represents a norm against which to appraise other iron-base alloys.

Table 4. Chemical Composition and Heat Treatments for Inconel 718 Nickel-Base Alloy Forgings
(See Fig. 17, on page 14, for mechanical-property data)

Composition		Heat treatment
Carbon	0.05%	**Condition A**
Manganese .	0.10	
Silicon	0.10	Solution treat at 1950 F, air cool,
Chromium .	18.8	age at 1400 F for 10 hr, furnace
Nickel	52.3	cool to 1200 F, hold for 20 hr
Molybdenum	3.0	
Columbium .	5.07	
Aluminum .	0.65	**Condition B**
Titanium ...	1.03	
Iron	18.8	Solution treat at 1800 F, air cool,
Copper	0.10	age at 1325 F for 8 hr, furnace
Boron	0.00004	cool to 1150 F, hold for 18 hr

Source (and for Fig. 17): Malin and Schmidt, "Tensile and Fatigue Properties of Alloy 718 at Cryogenic Temperature", National Metal Congress, ASM, Oct 1967.

Curves for the iron-base alloys in Fig. 15 and 16 end at 1800 F. Those for nickel-base and cobalt-base alloys extend to 2000 F.

The rupture strength shown by the curves of Fig. 15 and 16 is not a design criterion, because deformation at rupture far exceeds that allowable in design. Also, the curves shown are typical and reflect neither statistical distribution nor specified minimums. Variations in composition, melting, forging and heat treatment are not reflected by these smoothed, typical curves. The curves therefore provide only a first approximation for material selection. Creep characteristics, microstructural stability, and resistance to corrosion by sulfur-containing gas at high temperature must also be taken into consideration.

Cryogenic Properties. Although developed originally for use at high temperature, some heat-resisting alloys have also been used at cryogenic temperatures, as forged components for handling liquid oxygen and liquid hydrogen.

Mechanical-test results for Inconel 718 at room and cryogenic temperatures are shown in Fig. 17 (page 14) for specimens cut from forged components of over-all dimensions 4 by 9 by 15 in. The forgings were produced from 6-in.-diam billets broken down from an 18-in.-diam ingot.

As shown in Table 4, above, the composition of Inconel 718 is about half nickel, and one-fifth each of chromium and iron, with significant amounts of molybdenum, columbium, aluminum, and titanium. The two solution and aging treatments defined in Table 4 were used for the specimens reported in Fig. 17.

Test results shown in Fig. 17 include tensile, notch-tensile and Charpy impact values. Each plotted point is an average of four tests. Testing was at room temperature, at −110 F in gaseous nitrogen, at −320 F in liquid nitrogen, and at −423 F in liquid hydrogen. The test values that concern ductility (elongation, notch-tensile/smooth-tensile ratio, and impact toughness) are shown for both longitudinal and transverse directions. Longitudinal bars were machined parallel to the 15-in. dimension of the forging; long-transverse direction bars were machined parallel to the 9-in. dimension; and the short-transverse direction bars, parallel to the 4-in. dimension.

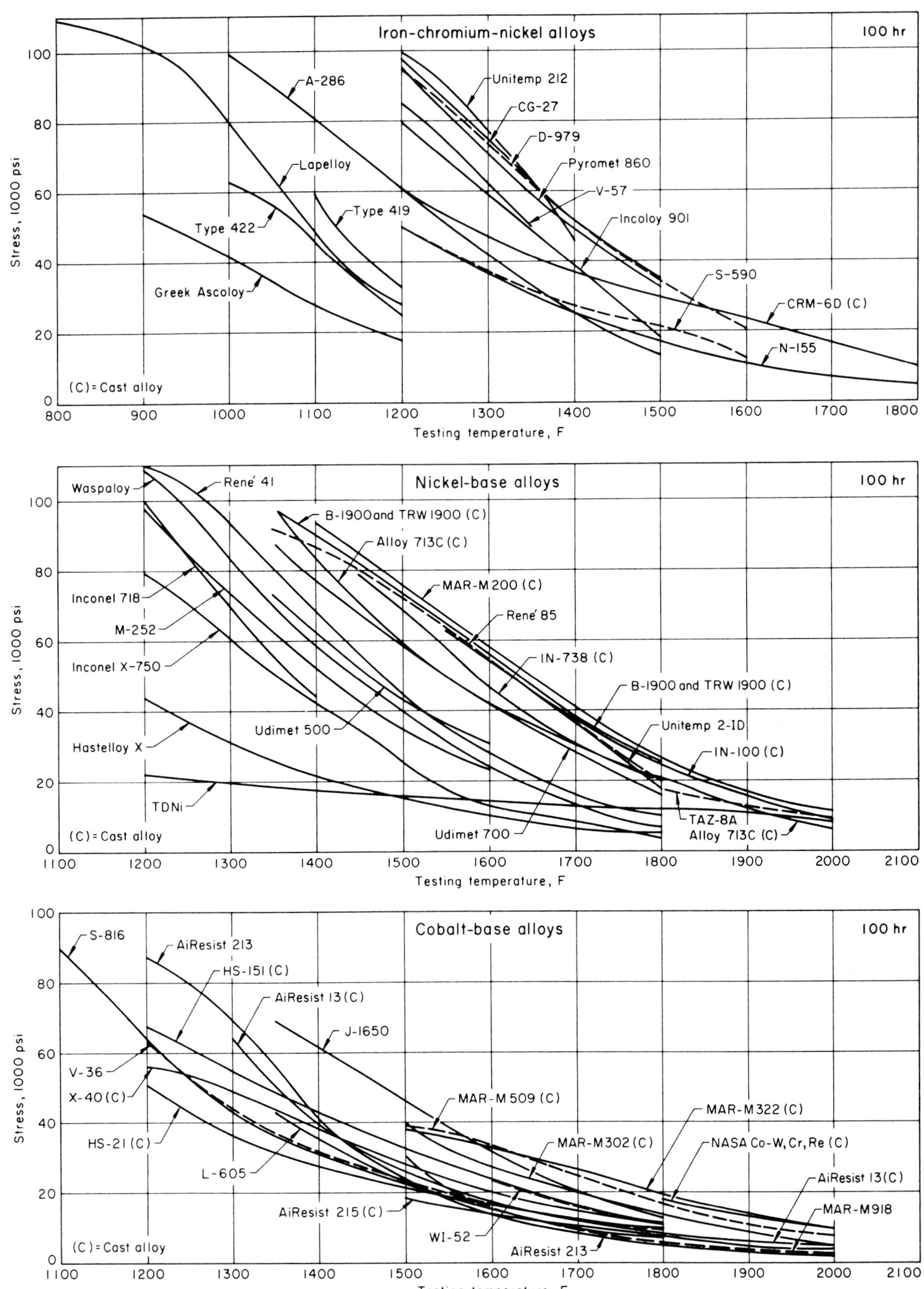

Fig. 15. Effect of temperature on 100-hr rupture stress for iron-chromium-nickel, nickel-base and cobalt-base heat-resisting alloys. (Simmons and Wagner, Metal Progress, *June 1967; Simmons, DMIC Memo 236, May 1, 1968)*

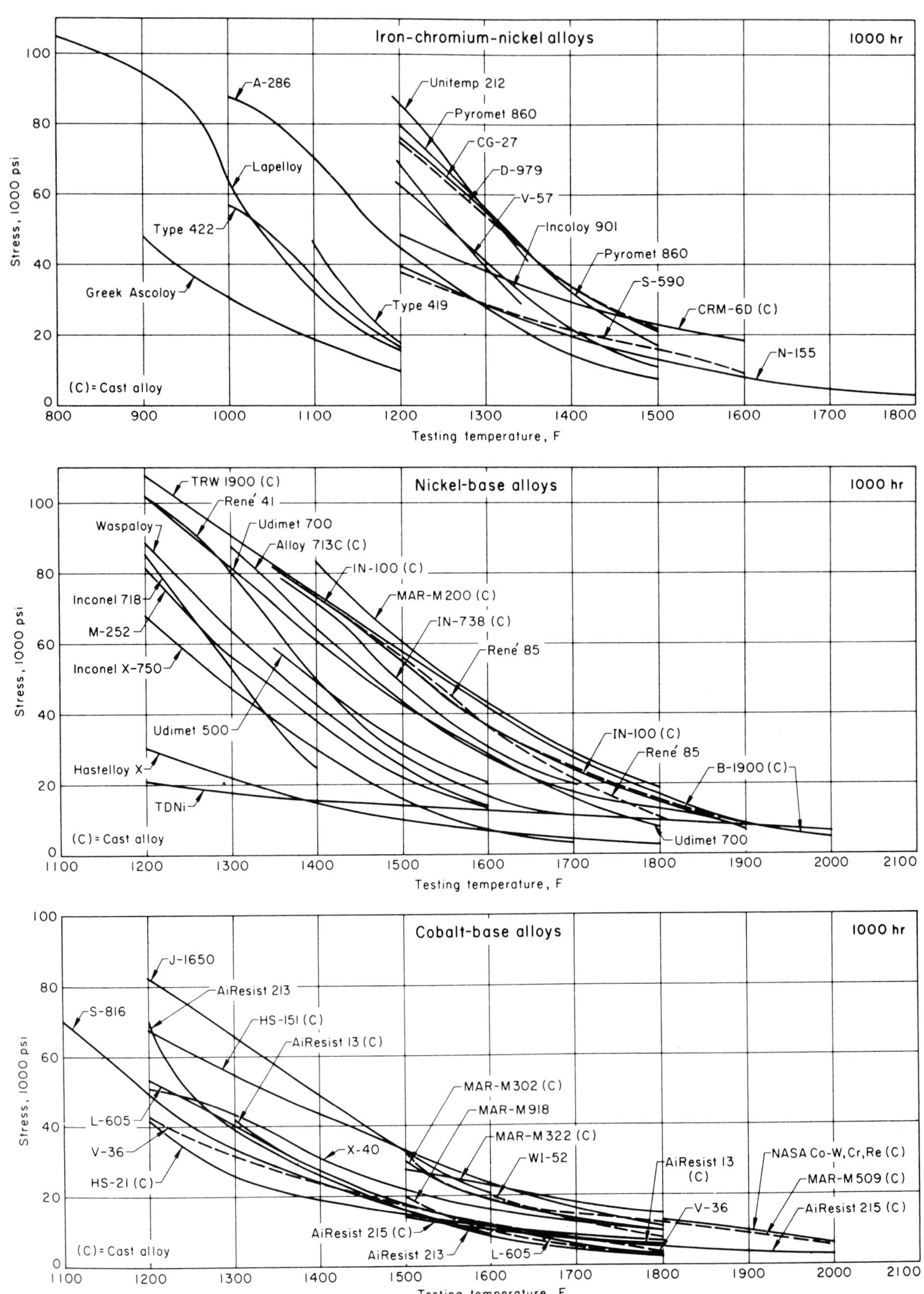

Fig. 16. Effect of temperature on 1000-hr rupture stress for iron-chromium-nickel, nickel-base and cobalt-base heat-resisting alloys. (Simmons and Wagner, Metal Progress, *June 1967; Simmons, DMIC Memo 236, May 1, 1968)*

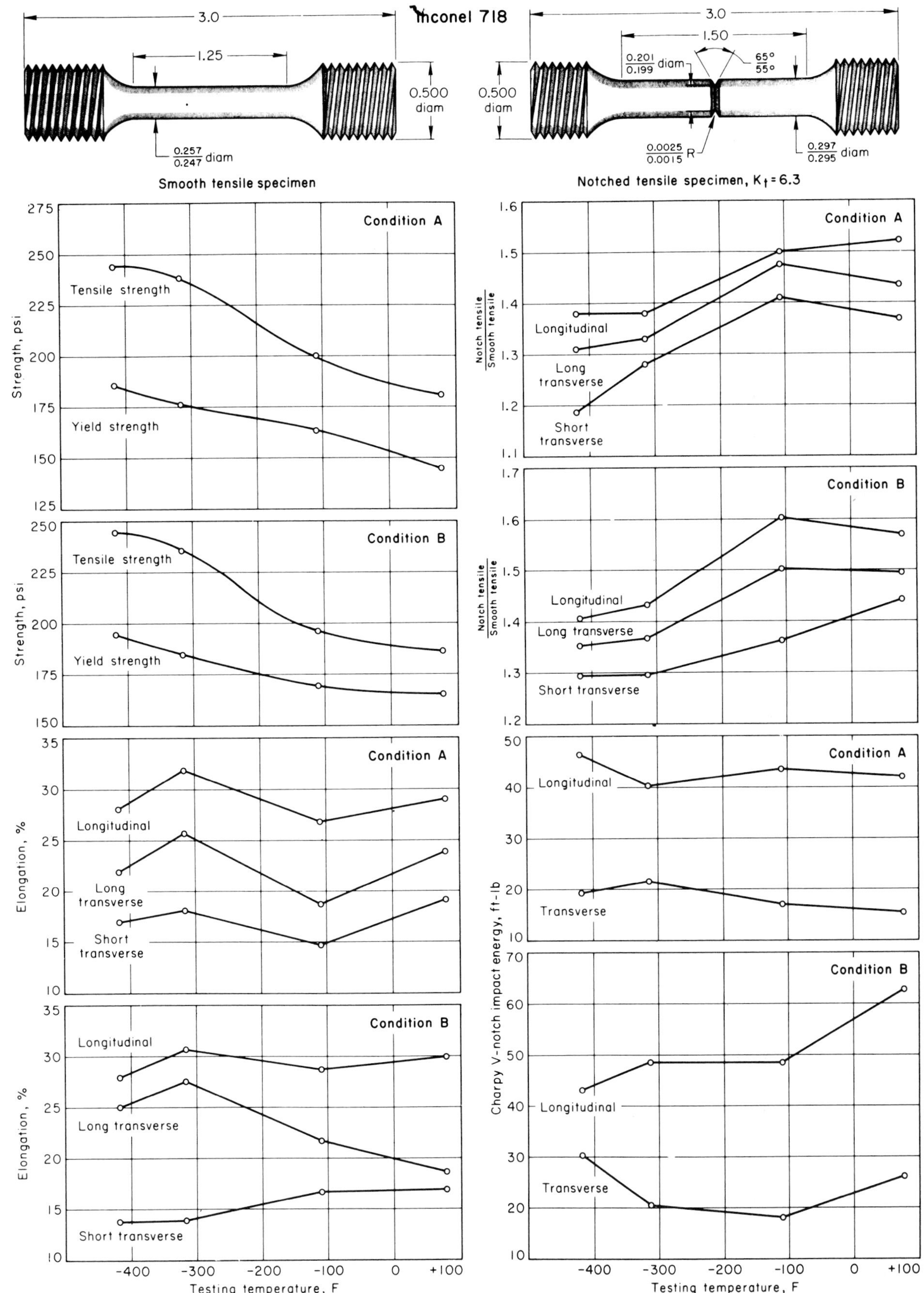

Fig. 17. Effect of heat treatment and temperature on mechanical properties of Inconel 718. Composition of Inconel 718, times and temperatures for heat treatment to conditions A and B, and source of data are given in Table 4, page 11.

The Diesel Truck Engine

By JACK E. LaBELLE

"Three hundred thousand miles of troublefree operation between major overhauls." That's what diesel engine users demand. Primary material selection criterion for the diesel engine metallurgist is consequently durability.

THE DIESEL ENGINE is an exceptionally efficient method for converting petroleum fuel into mechanical power. It has a large edge over its nearest competitor, the gasoline engine. For this reason it is the first choice for power plants of ships, locomotives, heavy-duty trucks, buses, and a number of other applications where fuel costs are a major expense.

High compression ratios — from 16:1 up to 20:1 — require rugged construction which makes the diesel more costly than the gasoline engine. The complex, precision fuel injection system also adds to the diesel's cost premium.

However, users have requirements in addition to fuel economy. Most important are durability and reliability, which make for low maintenance. Naturally, initial engine cost is important, but a lower purchase price means little without the other attributes. This is where the diesel shines.

In heavy-duty truck applications, for example, owners demand 300 000 or more miles (483 000 km) of troublefree engine operation between major overhauls. At this point they expect to be able to replace wearing parts such as bearings, pistons, and cylinder sleeves and then operate another 300 000 or more miles to the next overhaul. This user demand for durability makes material selection for diesel components a challenge.

Two-Cycle Engine — The 6V-71 and 8V-71 are two of eight Series 71 engines of 71 in^3. (1165 cm^3) displacement per cylinder made by Detroit Diesel Allison Div., General Motors Corp. They range from an in-line two (2-71) to a sixteen cylinder V (16V-71).

Detroit Diesel engines are distinctive among diesels in that all models are two-cycle engines as opposed to four-cycle. In other words, every down stroke of the piston is a power stroke. This is accomplished by exhausting combustion gases and refilling cylinders with fresh air at the bottom of each stroke. A blower, located between the cylinder banks, substitutes for a four-cycle engine's exhaust and intake strokes. Two-cycle operation provides substantially more power, size-for-size, than four-cycle operation in the 200+ horsepower category.

Although the system discussed here is for the 6V and 8V-71 truck engines, many of the material characteristics apply to other 71 series engines as well.

Total Cost — The 6V and 8V-71 material system is based on the premise that primary selection criteria for each component are dura-

---

Mr. LaBelle is chief metallurgist, Detroit Diesel Allison Div., General Motors Corp., Detroit.

All Series 71 pistons are two-part pearlitic malleable iron assemblies comprised of the dome casting (left) and skirt casting (center). The skirt is plated with 0.001 in. (0.025 mm) of tin for scuff resistance.

bility and reliability. Lowest possibile total cost is a secondary, but continuously sought after, objective.

One definition of durability is "resistance to failure." It can be more clearly defined by considering the kinds of failure that must be resisted by each component. These include fatigue, wear, corrosion, and deterioration from heat.

There are, of course, combinations such as corrosive-wear and thermal-fatigue; and there are a few parts, notably hardened highly stressed parts, where other phenomena such as hydrogen embrittlement must be dealt with.

Failures can also originate from deficient or defective materials. These problems come under the category of reliability and are controlled by the adequacy of the manufacturer's quality assurance system.

Critical Zone — The small space between piston and cylinder head at the time of firing is a critical diesel engine zone. Primary components are the head fire deck, valves, valve seats, fuel injector tip, piston dome, piston rings, and the top of the cylinder walls.

These parts must withstand high firing forces, high temperatures, and potentially oxidizing or corrosive exhaust gases. Valves, valve seats, cylinders, and piston rings must also resist wear. Each of the four primary types of failure must be counteracted in these few critical components.

Piston — Pearlitic malleable iron is specified for all Series 71 pistons. Selection was based on the ability of the cast material to handle the heat in the dome without deterioration; the wear of the piston rings as they seal against the ring grooves; and the stresses imposed by the heat and firing loads.

Piston skirts are plated with 0.001 in. (0.03 mm) of tin to provide a bearing surface and significantly improved scuff resistance.

In a four-cycle engine, forces on the connecting rod are tensile on the intake stroke and compressive on the firing stroke. This kind of stress reversal means that fatigue resistance must be considered in designing a four-cycle engine connecting rod. Piston weight is an important factor. Consequently, a lightweight piston is desired in a four-cycle engine and aluminum is commonly specified.

In the two-cycle engine, there is no intake stroke and forces on the rod are always compressive. No significant stress reversal takes place. An aluminum piston is not needed and the more rugged and wear-resistant pearlitic malleable iron can be specified.

Valves — Each 6V or 8V-71 cylinder has four exhaust valves. The press-forged, Inconel 751 head is friction welded to a medium-carbon semistainless steel stem which, in turn, is hardened and tempered for sliding wear resistance. The stem tip is also induction hardened. The extra wear resistance is needed to handle the force of the valve bridge.

The Inconel 751 head is solution treated and age hardened. Good hot hardness, to combat seat pitting, and a high level of thermal fatigue resistance for long life and temperature resistance are provided. It should be noted that deterioration from oxidation and microstructural degradation are functions of both temperature and time. Inconel 751's long-term temperature stability insures a long, troublefree life.

Inserts — A cast, heat resisting steel — 1.35 C, 4 Cr, 6.5 Mo, 5.5 W — is specified for valve seat inserts. The material is hardened and tempered to produce an Rc 42 to 52 martensitic structure.

The insert is used for two reasons: 1. It provides a means of replacing the valve seat — a limited life area — without having to replace the more expensive cylinder head. 2. The relatively hard, temperature - resistant material lasts many times longer than if cylinder head material were used as the valve seat.

Evolution — The cylinder liner has undergone more material changes than any other component except the crankshaft. The liner's now made of gray cast iron, hardened for wear resistance. A dry wall design — no contact with cooling water — avoids cavitation erosion which is sometimes experienced with wet liners.

The 11 in. (208 mm) long, $4\frac{1}{4}$ in. (110 mm) ID part is martempered because its center ring of port holes makes distortion control and crack avoidance difficult using conventional oil quenching.

Parts are heated to 1600 F (870 C) in an endothermic atmosphere with a 20 to 45 F (−7 to +7 C) dew point, quenched in molten nitrite-nitrate salt at 475 to 500 F

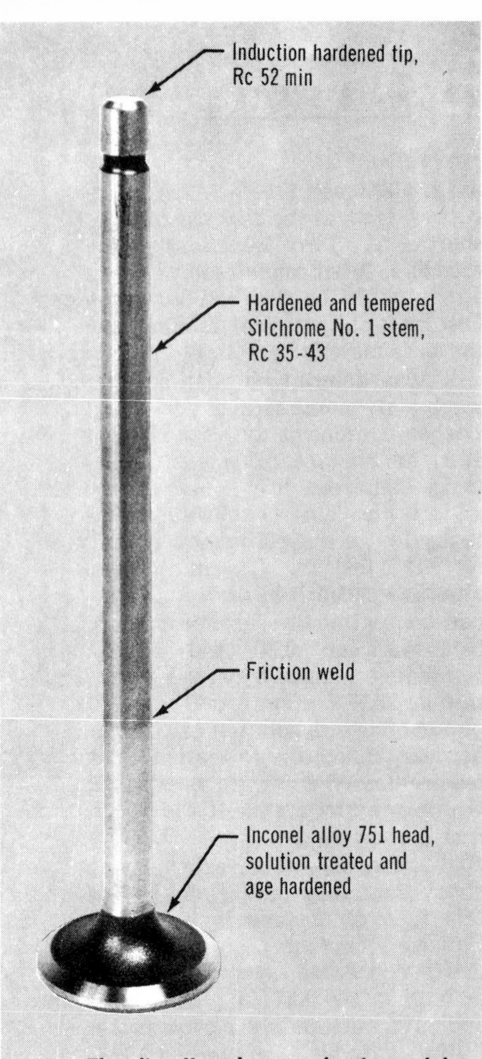

The diesel's exhaust valve is a mini-material system. The stem is hardened and tempered for sliding wear resistance, and the tip is induction hardened to handle the force of the valve bridge. Press forged Inconel 751 is selected for the head to provide longterm pitting and thermal fatigue resistance.

Gray cast iron Model 71 cylinder liners are shown at the charging end of Detroit Diesel's automatic martempering furnace. The cycle: atmosphere heat to 1600 F (870 C); salt quench, 1 min at 475 F (245 C); air cool, 1 h; wash; and temper, 1 h at 700 F (370 C).

(245 to 260 C) for 1 min, and then air cooled to room .temperature. After washing the salt off, liners are tempered at 700 F (370 C) for 1 h.

The liner flange is induction annealed after hardening to provide machinability. The flange is then finish machined, OD and ID are ground, and the ID is honed. Included in this sequence is a fillet rolling operation that increases flange strength, eliminating breakage during assembly and in service.

Graphite — A long cylinder liner wear life depends on a proper combination of matrix hardness and graphite type, size, and amount. Typically, graphite constitutes about 2.65% of the weight of the iron. It must consist of close to 100% type A Graphite in the ASTM 4 to 6 size range on the wearing surface for optimum long term wear and score resistance.

The matrix, after hardening, is typically martensitic and contains about 0.7% C with a hardness of Rc 63 or more (converted from Tukon Microhardness determinations). Composite (normal) hardness usually runs from Rc 40 to 45.

Some variations from these values are permitted; however, the allowable deviation is quite small to prevent any sacrifice in cylinder life, piston ring life, score resistance, or strength.

Piston Rings — Hardened and tempered SAE 9254 spring steel, chromium plated on the OD, is specified for compression piston rings.

The top, fire ring also has to seal against high gas pressures just below the piston rim. Because forces on this ring are high, it is also cold worked on the steel OD before chromium plating to provide maximum fatigue resistance.

Atomizer — The fuel injector, a reciprocating pump powered by the camshaft, develops 20 000 psi (140 MPa) pressure to atomize fuel as it passes through tiny, 0.006 in. (0.2 mm) in diameter holes in the injector spray tip. Materials selected for this precise mechanism are conventional except in the high pressure zone.

The plunger (piston) and bushing (cylinder) are the injector's pumping parts. When pressure on the fuel builds up to a critical level, a needle valve lifts from its seat and allows the compressed fuel charge to pass through the spray tip and out into the cylinder. The plunger moves up and down at every revolution of the engine, and must fit in the bushing with a clearance of only 0.00005 to 0.00006 in. (0.0013 to 0.0015 mm). The precise fit is necessary so that pressure can be developed without

Compression piston rings are fabricated from coiled SAE 9254 spring steel, hardened and tempered to Rc 42-48, and then chromium plated on the OD. The cross section shows the relative depth of the hard chromium plate.

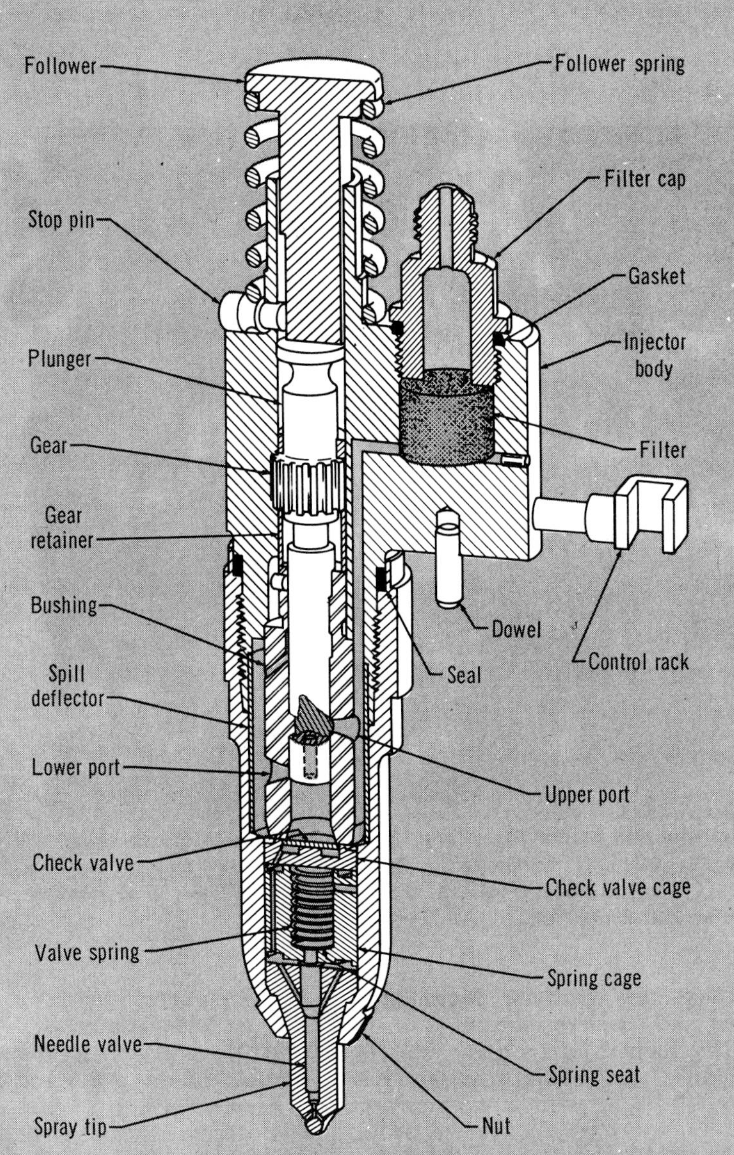

This cross section amply demonstrates the complexity of the Model 71 fuel injector. Plunger, bushing, needle valve, and spray tip are made of nitrided type 501 stainless steel to combat wear, fatigue, and erosion.

leakage. The plunger must last for about 1 billion strokes without showing significant wear, and using only diesel fuel as a lubricant.

Type 501 — Selection of nitrided type 501 semistainless steel for both plunger and bushing was made after hundreds of tests on a variety of materials. The heat-resisting grade has a nominal composition of 0.10 C min, 4.00-6.00 Cr, and 0.40-0.65 Mo.

The needle valve functions similarly, continuously opening and closing on the valve seat in the spray tip. Seat abrasion and erosion can result. Again, the best material for both spray tip and seat is nitrided type 501. The high pressure pulsing action in the spray tip can cause most other materials to split by fatigue.

Type 501's high hardness (Rc 70) and its semi - corrosion - resistant properties also prevent erosion by fuel in the tiny spray tip holes.

It is a rare occurrence when one material and one heat treatment combine to overcome such a variety of potential problems in such a small area. In the fuel injector, nitrided type 501 resists wear, erosion, scoring, and fatigue.

Camshaft — SAE 8620 steel, carburized to a depth of 0.060 in. (1.5 mm) and hardened to Rc 60, was formerly specified for Detroit Diesel camshafts.

Subsequent modifications for improved engine performance and fuel economy caused the Hertz stresses on the camshaft to increase. At the extreme point of experimentation, stresses in excess of 275 000 psi (1895 MPa) were encountered, at the expense of camshaft life. Two questions were raised: 1. What magnitude of Hertz stress could be handled with the best material practice? 2. What was the best material practice?

It was determined that specific steel alloy content had no effect. Carbon content at the case surface and the amount of retained austenite appeared to be of primary importance. This conclusion eventually led to a specification calling for an 0.060 in. (1.5 mm) deep effective carburized case with carbon content at the finished surface held between 0.90 and 1.10%. Hardening temperature was specified at 1550 F min (845 C min) to obtain complete solution of the carbon and induce the formation of as much retained austenite as possible within the limitations of the specified carbon range.

High Stresses — The exact Hertz stress durability for infinite life of this type of surface is unknown. However, current designs operate under calculated average stresses as high as 240 000 psi (1655 MPa) and give outstanding performance.

The material used now is SAE 1513 steel. Low-carbon provides sufficient core ductility to permit straightening the shaft after hardening. Carburizing is done in conventional, batch furnaces under a natural gas-enriched endothermic atmosphere. Carbon potential controls are employed. Both carburizing and diffusion cycles are used to produce a flat carbon content curve in the carburized case (near the surface) to accommodate variable grinding depths.

The shaft is slow cooled from the carburizing temperature and, after some finish machining operations to remove case in selective areas, is then either furnace or low-frequency induction hardened. It's quenched in a well-agitated caustic solution and tempered at 350 F (175 C).

Crankshafts — Steel forgings, of a composition similar to SAE 5046, were specified for crankshafts for the original 71 series engines (produced in 1938). The material was hardened and tempered to Bhn 229-269 as a rough forging. Bearing journals were induction hardened to Rc 50-55 for wear resistance.

By 1948, the incidence of fatigue failures in heavy duty applications indicated a need for improved durability. Without changing geometry, steel, or heat treatment, it was determined that a fillet rolling operation would increase shaft endurance limit by about 30%. A carbon steel (SAE 1049) was also substituted for the alloy steel with no effect on performance.

This process was used until 1968 when the need developed to further upgrade shaft fatigue resistance. A process first tested in 1950 — induction hardening the shaft fillet — was selected. The process extended the former induction hardening pattern, used primarily for journal bearing wear resistance, through the fillets.

At the time induction hardened fillets were introduced it was believed that if further improvement was called for, nitriding would get the nod. A different steel would also be needed.

Since 1968, however, development of better process methods and control procedures has enhanced the process to the point that shafts with induction hardened fillets are now equal to nitrided shafts in fatigue. They exceed nitrided shafts in serviceability since they can be reground up to 0.060 in. (1.5 mm) undersize without affecting either wear or fatigue resistance. Nitrided shafts are limited to regrinds of 0.015 in. (0.38 mm) undersize.

A detailed discussion of these ". . . better process methods and control procedures . . ." appeared in *Metal Progress*, October 1974, p. 105.

Aluminum — Another interesting material aspect of the 6V and 8V-71 engines is the use of aluminum castings. The two blower rotors are examples. The primary advantage here is weight reduction to reduce inertia. Aluminum is also used for blower housings, end plates, and end covers.

Aluminum may also be specified to increase payload via engine weight reduction. Examples include flywheel housings, front covers, and many optional components on bus, truck, and boat engines. However, these same parts are specified in gray cast iron for off-the-road vehicle engines. Weight is needed in the construction industry to provide traction.

Costs — Aluminum sand castings cost roughly three times as much as their cast iron counterparts. By going to permanent mold casting, this differential can be reduced significantly; and, by specifying die castings where the design lends itself to the process, the cost differential can frequently be overcome or swung in favor of aluminum, particularly if machining operations are eliminated.

In manufacturing aluminum and cast iron parts of essentially the same geometries, it's often advantageous to tailor equivalent machinability in both metals so that they can be worked interchangeably on the same machines with the same tools.

One method devised for doing this involves a lead-bismuth addition to aluminum which has the same general effect as lead in steel. By using a premium cutting fluid, these modified alloys exhibit essentially the same machining characteristics as cast iron in all types of machining operations on parts as complex as cylinder blocks.

These few examples only scratch the surface of the 6V and 8V-71 material system. Similar case studies could be written for gears, bearings, springs, gaskets, cylinder blocks, cylinder heads, and a variety of other parts.

Each has its engineering requirements which have undergone a series of revisions. These, in turn, have resulted in material revisions to enhance the performance of the part.

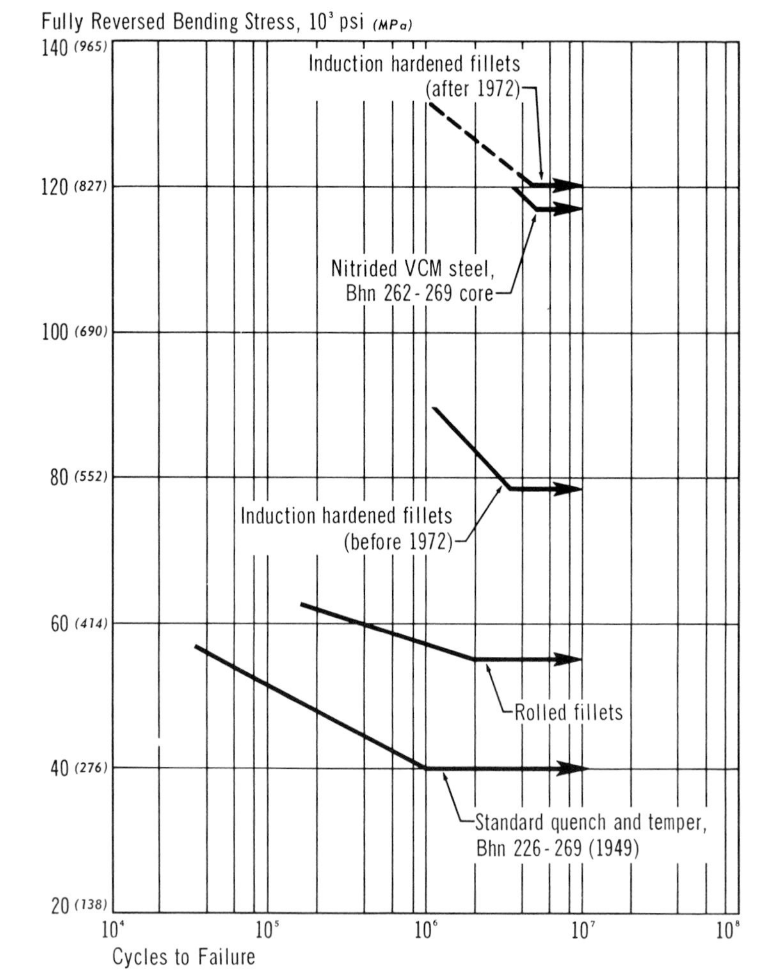

Data compare fatigue resistance of various materials and surface treatments used for diesel engine crankshafts. Horizontal lines with arrows represent endurance limit or allowable stress with infinite life. Metallurgical advances alone have accounted for a three-fold increase in fatigue strength; from 1949's quenched and tempered shaft to today's shaft boasting induction hardened fillets.

This 29 ft sphere of 5083 aluminum, being lowered into the lower half of a steel outer shell, will hold 90,000 gal of liquid hydrogen at —423 F. Welded seams give the vessel the soccer-ball look.

Selecting Structural Materials for Cryogenic Service

By FRED R. SCHWARTZBERG

Aluminum, titanium, stainless steels, nickel-base and cobalt-base superalloys, and a relative newcomer, 9% Ni steel, offer designers a wide choice. Composites also look promising.

Applications of cryogenic technology have increased at a phenomenal rate during the past decade. They range from exotic, such as systems in spacecraft and launch vehicles, to the commonplace, such as the transport of liquid methane. The key motive for utilizing this technology is that at cryogenic temperatures liquid gases occupy much less volume than their pressurized counterparts. Containment vessels for liquid gases may be smaller, thinner (because of lower pressure), and less costly. We will review the behavior, fabricability, and applications of various structural materials in cryogenic service.

Crystal Structure Determines Behavior

At temperatures below ambient, a metal's behavior is characterized somewhat by crystalline structure. The yield and tensile strengths of metals that crystallize in the body-centered cubic (bcc) form—iron, molybdenum, columbium, vanadium, and chromium—depend greatly on temperature. And the metals generally display a marked loss of ductility in a narrow temperature region below room temperature. This is called the ductile-brittle transition temperature.

The tensile strength of metals with face-centered cubic (fcc) structures—aluminum, copper, nickel, and austenitic stainless steel—is more temperature-dependent than their yield strength; and the metals often increase in ductility as temperatures decrease. The tensile and yield strengths of hexagonal close-packed (hcp) metals—titanium, magnesium, and

Mr. Schwartzberg is head, Mechanics of Materials Unit, Denver Div., Martin Marietta Corp., Denver.

zinc—display a marked temperature dependency; and they usually suffer a severe loss of ductility at subzero temperatures. Some alpha-titanium alloys, however, have good ductility down to cryogenic temperatures.

Transformation occurring in compositions that are normally stable at room temperature, but metastable at cryogenic temperatures, can greatly alter behavior. For example, the combination of gross plastic deformation and cryogenic temperatures can cause a normally ductile and tough stainless steel (such as types 301, 302, 304, and 321) to partially transform to a bcc structure, resulting in an impairment of ductility and toughness. A fully stable stainless steel (type 310) cannot be transformed at cryogenic temperatures.

Minor variations in composition can affect ductility in certain materials. Thus an increase in oxygen from 0.10 to 0.20% in a ductile titanium alloy can lower ductility at —423 F from 15 to almost 0%.

Summary of Mechanical Properties

The Data Sheet on p. 71 lists typical unnotched tension properties and typical applications for low-temperature and cryogenic materials at temperatures between +70 and —423 F.

Aluminum Alloys—The unalloyed grades and the manganese grades, such as 1100 and 3003, have relatively low strength in the annealed condition and must be cold worked to obtain adequate strength. These alloys find application because they come in a variety of forms. The heat-treatable Al-Cu alloys (2000 series) are widely used for aerospace parts because they have the highest strength of the weldable alloys. Alloy 2014 has been used extensively where high strength is essential. Alloy 2219, which is easier to weld than 2014, has slightly lower strength and has been used in cryogenic aerospace service. A new candidate, 2021, appears to exhibit higher strength properties than the former grades (see *Metal Progress*, May 1969, p. 68). Type 2024 has good strength, but it cannot be welded. Its cryogenic applications are limited.

The 5000 series alloys exhibit a combination of properties which make them popular for most applications. Their moderate strength (obtained through cold working), good toughness, and good weldability have resulted in their selection for service in ocean-going tankers, tank trailers, stationary storage containers, processing equipment, and propellent tanks. Typical alloys are 5052, 5083, 5086, and 5456. The 6000 alloys, characterized by 6061, offer the advantages listed for 5000 series alloys except that in the as-welded condition their strength is low.

The high-strength 7000 series—7075, 7079, and 7178—display the highest strength of all aluminum alloys. But they lose toughness below liquid nitrogen temperature (—320 F). They are generally nonweldable and see only limited application. Two newer alloys—7039 and X7007—show promise for cryogenic service because they are readily welded and retain adequate toughness at all temperatures. Strengths are below those of the nonweldable types but are similar to those of the 2000 series.

Stainless Steels—The 300 series steels offer a fine combination of toughness and weldability for service to the lowest temperatures. In the annealed condition, their strength properties are adequate for ground-based equipment but inadequate for flightweight structures. For aerospace applications, fabricators can take advantage of the alloy's strain-hardening characteristics and use them in a highly cold-worked condition. (Atlas and Centaur space vehicles are type 301 worked to its extra-full-hard condition.) The principal shortcomings of cold-worked materials are: low weld-joint efficiencies caused by annealing during welding and the transformation to martensite that occurs during cryogenic exposure. Selection of a fully stable grade, type 310, overcomes the transformation problem.

Two new alloys offer higher strengths. The 21Cr-6Ni-9Mn alloy 21-6-9 in its annealed condition exhibits outstanding toughness and weldability. Precipitation-hardening A286 stainless has even higher strength when cold worked before aging.

Alloy Steels—The only alloy steel recommended for cryogenic service is 9% Ni steel. It is satisfactory for service down to —320 F and is used extensively for transport and storage of cryogens because of its low cost and ease of fabrication. Other alloy steels are suitable for service in the low-temperature range. The A212 and A201 carbon steels and A517 ("T-1") alloy steel can suffice to —50 F; 2¼% Ni steel to —75 F; and 3½% Ni steel to —150 F.

Titanium Alloys—The three cryogenic materials are Ti-6Al-4V (ELI), Ti-5Al-2.5Sn (ELI), and unalloyed titanium. Ti-6Al-4V is an alpha-beta composition (alpha plus a minor amount of beta phase). It combines strength and toughness down to —320 F. Below —320 F, the small beta content causes a

Cryogenic or Low Temperature?

In physics, these two terms are often used interchangeably. However, a materials engineer should make a distinction. In materials work, the "low-temperature" region is associated with the "warmer" liquefied gases: ammonia (—28 F), propane (—44 F), carbon dioxide (—110 F), and ethane (—129 F). The low-temperature range extends down to —150 F; depending on the specific temperature, carbon steel, 2¼% Ni steel, or 3½% Ni steel is applicable.

The "cryogenic" range is normally associated with colder liquefied gases (cryogens): methane (—259 F), oxygen (—296 F), argon (—302 F), fluorine (—304 F), nitrogen (—320 F), hydrogen (—423 F), and helium (—454 F). Typical containment materials are aluminum alloys, stainless steels, and 9% Ni steel.

significant loss of toughness. For service down to —423 F, the lower-strength all-alpha Ti-5Al-2.5Sn composition is preferred because of its better retention of toughness. Where material forms of either of these alloys are difficult to obtain, the low-strength unalloyed material is normally used.

Superalloys and Other Materials — Nickel and cobalt-base superalloys perform well in cryogenic service because they are weldable and strong. For certain critical applications where the strength of stainless steels is insufficient, these alloys are used. Inconel alloy 718 has good response to cold working before aging and can have room temperature strengths exceeding 200,000 psi. Other face-centered cubic materials—such as copper, brass, and nickel—do not provide high strength, but they exhibit high toughness and good fabricability. They are often used for cryogenic hardware.

Designers Consider Fracture Behavior

Most cryogenic materials exhibit moderate strength and high toughness at room temperature. As temperatures decrease, the characteristic rise in strength is normally accompanied by a loss of toughness. And some moderate-strength materials exhibit cryogenic behavior similar to that displayed by high-strength materials at ambient temperature. As a result, designers must assess and characterize a material's toughness to intelligently evaluate its performance in cryogenic service.

For many years, the transition-temperature approach, as measured in a Charpy impact test (more recently in a dropweight test), has been used to determine the behavior of alloy steels at low temperatures. This approach has limited value for cryogenic service because designers do not use these materials below their transition temperature. The reasons: the transition temperature approach does not provide data useful in design or stress analysis in cryogenic service (this also applies to low-temperature service). Also, the materials most frequently used, stainless and aluminum, don't display transition behavior. An alternate approach to characterize resistance to frangible behavior is needed.

Designers have turned to fracture mechanics to predict structural performance. Two sources of information are available to the designer: *Fracture Toughness Testing and Its Applications*, ASTM STP 381, and *ASME Metals Engineering Design Handbook*, 2nd Edition.

Fracture mechanics is a stress-analysis approach to the problem of low-strength fracture in the presence of sharp defects. It involves the amount of strain energy available and released in initiating and maintaining fast fracture.

Fracture strength follows an inverse square root law; σ is a function of $\frac{1}{\sqrt{a}}$ where σ is stress and a is crack length. The theory of linear elastic-fracture mechanics is an extension of Griffith's concept of crack propagation in brittle solids (a defect will propagate rapidly if the elastically stored energy of the system equals or exceeds the energy required to form the additional crack surface). The extension of this concept (derived by Orowan and Irwin) to account for the energy of plastic deformation in the formation of two new surfaces is the basis for current theories of fracture mechanics.

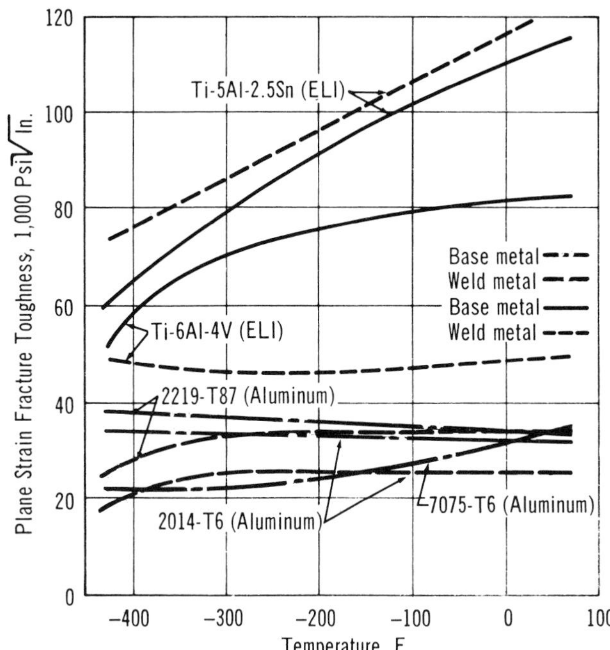

Fig. 1 — Designers should consider the fracture toughness of both the base metal and weld in designing cryogenic hardware.

The ASTM Special Committee on Fracture Testing of High-Strength Materials, which was formed in 1959, has evaluated techniques for determining the strength of metals in the presence of sharp defects. It has made recommendations for specimen designs, boundary conditions, and testing techniques. Complete agreement on specimen design and test methods does not exist, but good engineering work can be done within the framework of our current understanding of fracture mechanics.

The problem of valid testing is particularly acute for thin cryogenic materials — the medium-strength, tough materials with a face-centered cubic structure. This combination of high toughness and thinness makes testing particularly difficult.

Figure 1 summarizes the limited fracture-toughness data of several aluminum and titanium alloys. Type 2219-T87 aluminum has the edge on 2014. The marked decrease in toughness for 7075 characterizes the 7000 series alloys. The toughness of the alpha alloy, Ti-5Al-2.5Sn (ELI), exceeds that of the alpha-beta grade, Ti-6Al-4V (ELI).

Thermal Behavior Is Important

Designers must consider thermal expansion when specifying a material for cryogenic service. Changes in dimension between ambient and service tempera-

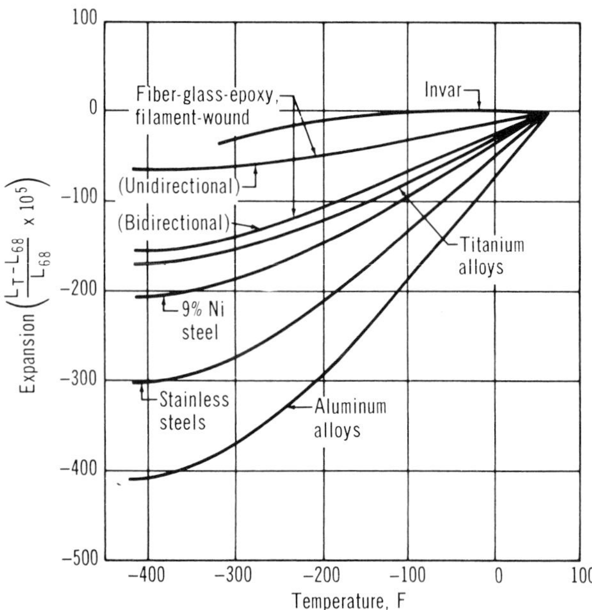

Fig. 2 — Thermal expansion of materials plays an important role in cryogenic design.

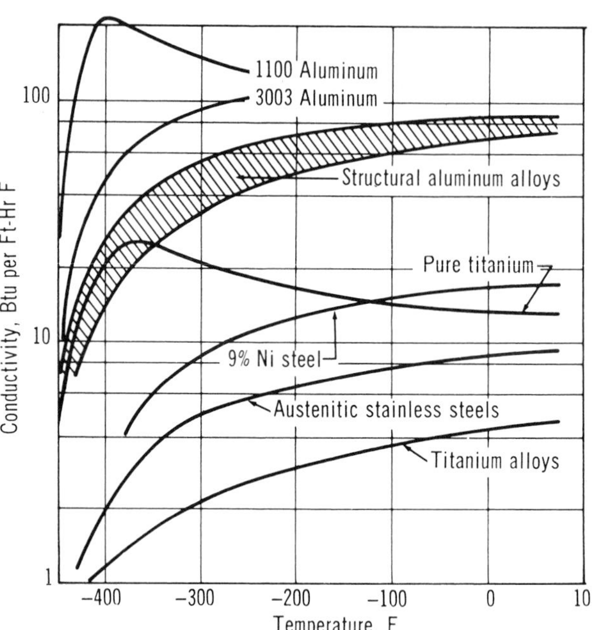

Fig. 3 — Selection of materials with low thermal conductivity can improve the operation efficiency of cryogenic equipment.

ture can lead to problems unless this behavior is accounted for in design. Figure 2 compares the thermal expansion behavior of the common cryogenic materials. Titanium alloys exhibit the least change in dimension; aluminum, the most. Invar is frequently used for applications governed by thermal expansion. Unidirectional filament-wound material exhibits little contraction when cooled. The bidirectional form of filament material — the type usually found in tankage — has a thermal expansion that closely matches that of titanium, a popular candidate for liners (see box on p. 56). There are problems associated with the use of dissimilar metals in cryogenic service. However, differential expansion can act favorably to provide sealing, fastener tensioning, and similar desirable mechanical situations.

Cryogens have relatively low specific heats, and designers must frequently design cryogenic hardware to minimize thermal conduction. Materials with low conductivity are desirable for plumbing and supports. Stainless steel is often specified because its conductivity is lower than that of aluminum (Fig. 3). And recently materials reinforced with high-strength glass filaments have been used because of their low conductivities.

Fabrication of Cryogenic Structures

Most cryogenic materials are relatively formable and weldable if adequate care and precautions are taken.

A material is commonly formed in its condition of application. Many times, annealing or stress relieving is not required. Higher-strength materials, however, are often formed in their annealed or solution-treated condition, then heat treated to their final level before welding.

Most fabricators weld cryogenic vessels. Brazing finds limited applications, primarily where joints are long and design stresses low, such as in heat exchangers.

The desire for thinner sections to reduce heat transfer and weight has prompted the development of joining technology. The Atlas and Centaur boosters represent an interesting example. Constructed from 0.010 to 0.030 in. type 301 (XFH) stainless steel, the parts are joined without filler metal. Also, the joint strength had to be raised from the 100,000 psi value to an amount closer to the 200,000 psi strength of the cold-rolled sheet. The solution: reinforce the joint with resistance-welded doubler strips that increase the load-carrying area across the fusion weld.

In fabricating efficient aerospace structures of aluminum, a similar technique is used. Designers call for a thickened zone at the weld so that the stress in the weld area is decreased in proportion to the increased joint area. The remainder of the part is chemically milled to cut thickness and weight.

For most ground-based applications, high design efficiency is not required, and conventional fabrication techniques are employed.

Welding must frequently meet stringent requirements. Joints must be free of defects, inclusions, and porosity. And they must be vacuum tight to assure that the 10^{-6} torr vacuum, which separates the inner and outer containers in a double-walled vessel, is maintained for long periods.

Fabricators must develop joining techniques for

Are Composites in the Cryogenic Race?

"Definitely," point out **Arthur Feldman** (right) and **Donald A. Stang,** Martin Marietta Corp., Denver. "One key difficulty remains — the development of a thin metal liner."

Glass, boron, and carbon-epoxy composites (see table) have high strength-to-density properties at room temperature, and the development of pressure vessels using these composites is an attractive technological goal. At —320 F, these ratios are even more enticing for boron and glass, though there is some dropoff for carbon.

While a sphere is the most efficient configuration for metals, a cylinder with dome ends that are somewhat flatter than hemispheres is the best composite design. Also, metal will generally develop equal strength in a vessel and in a flat coupon test, but several investigators have found that composites lose some strength in the pressure vessel configuration. Performance factor at burst, defined as pressure times volume divided by weight, is therefore the meaningful comparison and can be related theoretically to density and material burst strength. Future values in the table are based on anticipated improvements in materials and vessel-manufacturing techniques.

The development of a thin metal liner is needed to get acceptance of composite tankage for cryogenic storage. The permeable materials require a liner for both gas and liquid containment. And elastomeric and polymeric materials will not maintain their impermeability at cryogenic temperatures at the strains needed to effectively develop the strength of the composites. It means that metal liners must be as thin as manufacturing techniques will permit.

To develop the full capability of glass and other filaments, the metal liners must first be strained plastically by pressure loading. But if this is done, depressurization forces the metal into compression, causing either buckling or compressive yield, both undesirable.

Several solutions are under study. One is a cryogenic adhesive capable of preventing a paper-thin metal liner from buckling away from the composite tank wall. An adhesive of this type is especially required with glass-epoxy vessels. Another approach is to reduce the strain requirements on the metal liner by using higher modulus overwrapping materials such as boron or carbon filaments. This approach enables the liner to function within its elastic limits, eliminating the need for an adhesive.

In an attempt to press glass-epoxy composites into early use for cryogenic tankage without compounding the problem of a thin metal liner, NASA has also taken a less efficient but still attractive approach: the development of a metal liner that is thick enough not to buckle while bearing a large part of the load and operating elastically from high compression to high tension. The glass-fiber overwrapped metal tankage has employed Inconel X-750 liners, and Ti-5Al-2.5Sn types are being tested. In these vessels, the composite reinforces the metal tank and extends the elastic range of the metal. The concept is an important stepping-stone to full use of composites in cryogenic tankage.

Predicted Pressure Vessel Burst Performance at Cryogenic Temperatures

Material	Fiber, Vol. %	Specific Strength* α/ρ 10^6 in.		Configuration	Material Efficiency in a Pressure Vessel Configuration, %	Predicted Cryogenic Performance Factor at Burst =PV/W	
		at 70 F	at −320 F			Formula*	10^6 in.
S-glass-epoxy	70	3.5	4.2	Cylinder with domes	90	$\alpha/3\rho$	1.3
Boron-epoxy	50	2.7	3.1	Cylinder with domes	68 (90)†	$\alpha/3\rho$	0.7 (0.9)
Carbon-epoxy	50	2.0 (4.0)	2.0 (4.0)	Cylinder with domes	60 (90)†	$\alpha/3\rho$	0.4 (1.2)
Ti-5Al-2.5Sn	—	0.75	1.12	Sphere	100	$2\alpha/3\rho$	0.75

Note: Liner weight not considered.
* α is unidirectional tensile strength of material; ρ is density.
† Parenthetical values are based on anticipated improvements in materials and vessel-manufacturing techniques.

the many cryogenic structures which contain dissimilar metals. For example, many large storage dewars have aluminum inner shells and carbon steel outer shells. To support the inner vessel and provide the venting for liquid handling and pressurization, piping must protrude through both walls. The requirement for dissimilar metal joints has many solutions, such as mechanical fastening, adhesive bonding, and metallurgical bonds known as transition joints. One successful transition joint is made by coextruding dissimilar metal tubes.

Most cryogenic structures are operated in the as-welded condition. The philosophy is that the cryogenic materials (as opposed to low-temperature steels) are relatively tough at the service temperature and are not frequently subject to failures attributable to residual stresses resulting from welding.

What's to Come

One word — growth — summarizes the future for the structural application of materials in cryogenic service. Cryogens in aerospace, nuclear, petrochemical, steelmaking, food processing, and the utility industries will gain wider usage requiring the fabrication of more and more containment vessels. For example, supertankers for liquid methane are under construction. The probable replacement of liquid nitrogen spray freezing for the refrigerated freezing of some food products represents another new application.

Liquid methane fuel for the SST could result in a 30% cost saving over jet fuel. We might even see the ultimate fuel, liquid hydrogen, used in aircraft.

A most interesting future application may be in the electrical power industry. Cryogenically cooled motors and power transformers could take advantage of the lower resistivity of conductors in the cryogenic range. Or we might see great size reductions in equipment because of the superconductivity of materials at liquid helium temperature.

A big payoff could come in cryogenic power-transmission lines. They could carry the ever-rising power needs of industry at higher voltages. For example, the resistivity of conductors (copper or aluminum) decreases tenfold at liquid nitrogen temperature (—320 F) and several hundredfold at liquid hydrogen temperature (—423 F). A typical cable could consist of thin tubular conductors in which liquid hydrogen circulates. The tubes would be surrounded by another tube and a vacuum drawn on the space between them would serve as insulation in the same manner as dewars are constructed.

Another area where cryogenics could grow is in metal processing operations. Cryogenic stretch forming (see p. 64) is used to manufacture lightweight, high-strength pressure vessels of stainless steel. The process takes advantage of the transformation of metastable austenite to martensite that occurs under the influence of gross plastic straining at cryogenic temperatures. The resultant vessel exhibits strengths approaching 300,000 psi compared with the untransformed strength of the annealed material of less than 100,000 psi. An undersize preform vessel is submerged in liquid nitrogen, and the desired stretch is obtained by pressurizing it. Advantages include the elimination of heat treatment and the attendant distortions, high weld-joint efficiency, and relatively low cost.

Cryogenic quenching (see p. 62) has proved successful in reducing distortion in the heat treatment of aluminum, particularly sheet gages. The process uses liquid nitrogen as the quenching media. It is successful because the cooling progresses at a uniform rate rather than by stages as in conventional quenching. Ultrasonic agitation reportedly increases the cooling rate.

Another processing development is in the stress relieving of aluminum parts before aging. A solution-treated part is cooled to —320 F and immediately subjected to a high-velocity steam blast. This treatment has reduced residual stresses by more than 80%.

A Steel With Cryogenic Potential

At —320 F, an austenitic stainless steel containing 18 Cr, 15 Mn, 5.5 Ni, and 0.38 N has twice its room temperature strength. It also displays good toughness down to —423 F.

By CARL E. SPAEDER JR., JOHN C. MAJETICH, and KENNETH G. BRICKNER

DESIGNERS OF CRYOGENIC ASSEMBLIES base their stress calculations on the room-temperature properties of the material. The reason is that it is the highest temperature the material will encounter. And it stands to reason that if a higher-strength material that stands up to supercold conditions were available, designers might specify it.

We have recently developed and proved the capability of an austenitic stainless steel to satisfy high-strength needs at cryogenic temperatures. At 80 F, the steel has tensile and yield strengths that are 25,000 psi greater than the corresponding strengths for type 304 stainless. At —320 F, its

Mr. Spaeder is senior research metallurgist, Mr. Majetich is research metallurgist, and Mr. Brickner is section supervisor, Applied Research Laboratory, U.S. Steel Corp., Monroeville, Pa. This article summarizes the authors' presentation at Westec '69 in Los Angeles. The complete article will appear in the August 1969 issue of ASM's *Metals Engineering Quarterly*.

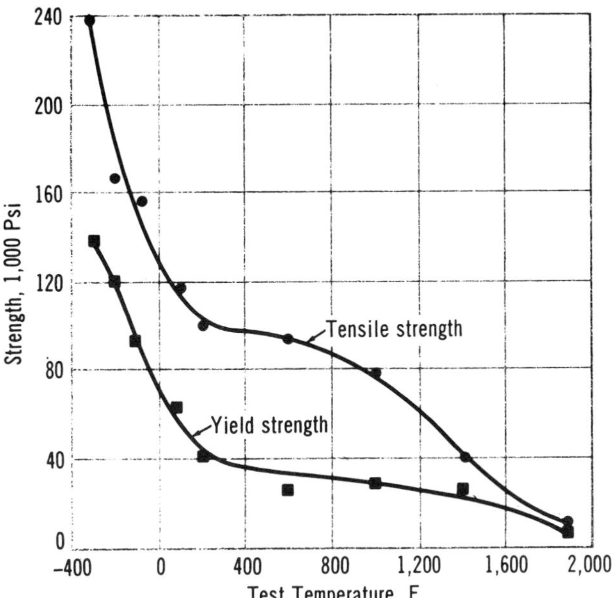

Fig. 1 — The tensile strength of 1/2 in. Cr-Mn-Ni-N plate at —320 F is 80,000 psi higher than that of type 304 stainless at the same temperature. Its yield strength is 40,000 psi greater.

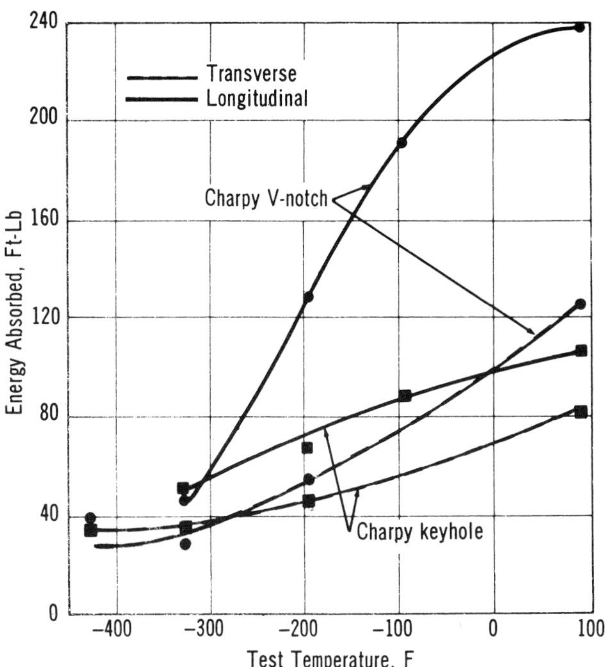

Fig. 2 — At —320 F, the longitudinal and transverse impact properties of the alloy are about equal.

tensile and yield strengths exceed those of type 304 by 80,000 and 40,000 psi respectively. In the annealed condition, its minimum transverse toughness at —320 F, as measured by Charpy V-notch and keyhole impact tests, is 20 ft-lb. And it is weldable.

Composition — To determine the composition that gives the best properties, we first investigated five heats of steel which were essentially the same except that their nickel contents ranged from 0.10 to 8.05%. A heat containing 4% Ni exhibited the best combination of room-temperature strength and cryogenic toughness.

Further studies on a 22-ton electric-furnace heat of steel established limits for commercial products that give the best combination of properties. Typical compositions of two thicknesses of material are:

	½ In. Plates	0.080 In. Sheets
C	0.072%	0.080%
Mn	16.0	15.8
P	0.020	0.020
S	0.008	0.008
Si	0.41	0.38
Ni	5.85	5.55
Cr	17.8	18.0
N	0.36	0.40
Fe	Bal	Bal

Physical Properties — The Cr-Mn-Ni-N alloy is slightly less dense at 0.281 lb per cu in. than type 304. Its mean coefficient of expansion between —310 and +70 F is 5.42×10^{-6} in. per in. per F. The value for type 304 is 7.3×10^{-6} in. per in. per F.

Mechanical Properties — Figure 1 gives the tensile and yield strength curves for specimens removed from ½ in. thick commercial plates. As 0.080 in. sheets, the steel has a slightly higher yield — 165,000 psi yield — at —320 F.

Figure 2 shows the longitudinal and transverse impact strengths of the steel as determined on Charpy V-notch and keyhole specimens.

Weldability — In tests in accordance with procedures of the ASME Boiler and Pressure Vessel Code, Section IX, Welding Qualifications, the alloy has displayed satisfactory weldability.

Butt welds in ½ in. plates were made by the shielded metal-arc method, using AWS E310-15 stainless steel covered electrodes and by the gas metal-arc method, using AWS ER310 bare electrodes.

Transverse specimens from the manually made weldment broke in the weld at a tensile strength of 111,800 psi. The Charpy V-notch impact strength was 54 ft-lb at +80 F and 26 ft-lb at —320 F. Side-bend specimens also withstood the 180° bend over a 1½ in. diameter mandrel.

The gas metal-arc weld broke at a tensile strength of 111,500 psi. Its impact strength was 93 ft-lb at +80 F and 53 ft-lb at —320 F. And it passed the 180° bend test.

Applications — The material's combination of high strength, good toughness, and weldability should prompt designers to specify it for welded pressure vessels for the storage of cryogens. ASTM includes the steel in specification A412.

MATERIAL SELECTION FOR PWR NUCLEAR POWER PLANTS

J. F. Enrietto
Manager, Materials and Process Engineering

Westinghouse Electric Corporation

The Nuclear Power Industry has achieved a safety record that has never been matched in the history of American Industry. Further, this record was attained while developing a new method for energy conversion and making this new technology competitive with more traditional ways of generating electrical power. A major contributor to the success of nuclear power has been the selection of the materials that are used to construct the nuclear steam supply system. This paper will describe the materials used to fabricate some of the structural components of the NSSS, why a specific material was chosen for a particular job, and how some materials have evolved from those used in earlier plants to those used today. The approach is based on Westinghouse experience, but the reasoning is similar to that used through the industry.

Overall Requirements

Most of the materials used in the construction of the NSSS must operate as part of a pressure retaining boundary. This fact immediately imposes certain technical restrictions on material selection. For example, the material must conform to the ASME Boiler and Pressure Vessel Code. This means that the material must be code approved. It also means that the material must possess certain specified levels of mechanical properties. However, when all the technical jargon is removed, all we are saying is that the materials used must posses mechanical properties dictated by design and Code rules.

Materials used in the construction of the NSSS must have other conventional attributes such as the capability of being fabricated, i.e. rolled, spun, cast, forged, drawn, welded, etc. And, since the apparatus has to be commercially competitive, the materials should be available at a reasonable cost.

To the extent described above, the criteria for the selection of materials for nuclear power plant applications do not differ greatly from those utilized for other, more conventional uses. The factor that makes nuclear applications unique is the necessity to contain a highly radioactive environment. This, of course, adds a new dimension to the technical requirements for materials. Over the life of a plant, some of the structural portions of a reactor vessel will be exposed to neutron fluences of the order of 10^{19} nvt and components that support the core will be subjected a fluence of the order of 10^{21} nvt. The materials used to construct these components cannot significantly degrade in a metallurgical or mechanical sense under this irradiation.

In addition to its direct effect on material properties, the presence of a radioactive environment plays an indirect, but crucial, role in material selection. Special limitations are imposed on the incidence of service failures since repair or replacement of radioactive components is at best difficult and expensive. Furthermore, stringent safety requirements demand that the probability of gross material failures be reduced to such a low value that, for all practical purposes, these failures may be considered incredible. Thus, a heavy emphasis is placed on material reliability. It is not enough that materials be strong, tough, ductible, etc. They must also be uniform, virtually defect free, inspectable, and above all, pedigreed. By the last term, I mean that records and histories of the material--its composition, heat treatments, manufacturing process, tests, past performance, etc.--are known and maintained. We strive to avoid surprises in the nuclear industry.

Specific Applications

A schematic representation of a typical 2-loop PWR primary coolant system is shown in Figure 1.

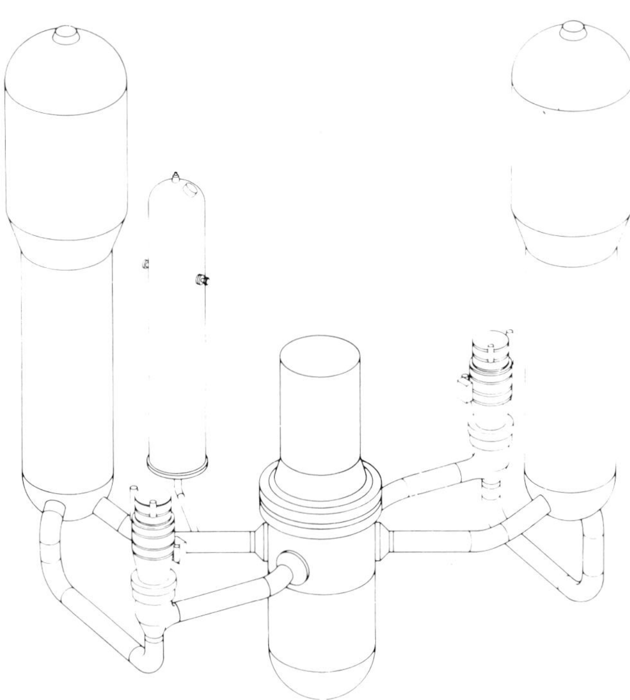

Figure 1. Westinghouse PWR Two-Loop NSSS

The principal components in this system are the Reactor Vessel, the Reactor Coolant Pump, the Pressurizer, the Reactor Coolant Piping, and the Steam Generator. Design requirements are that the system operate at an internal pressure of 2250 psi and a temperature of 600°F. The loop contains high purity water with special additives such as boric acid and lithium hydroxide. Although the environment is not particularly corrosive to meet structural materials, only corrosion resistant materials like stainless steel are permitted to come into contact with the primary water. The reason for this is not a concern over material failures due to corrosion. Rather, it is to prevent corrosion products from entering the system, plating on the core and thus interfering with the thermal-hydraulic operation of the plant.

Reactor Vessel

A sketch of a reactor vessel is shown in Figure 2. The sheer size of the vessel, 14 feet diameter by 40 feet high, coupled with the pressure to be contained dictates that a steel must be used. Economics precludes the use of any high strength non-ferrous material. But we must now answer the question, "Which of the myriad grades of steel shall we choose and why?"

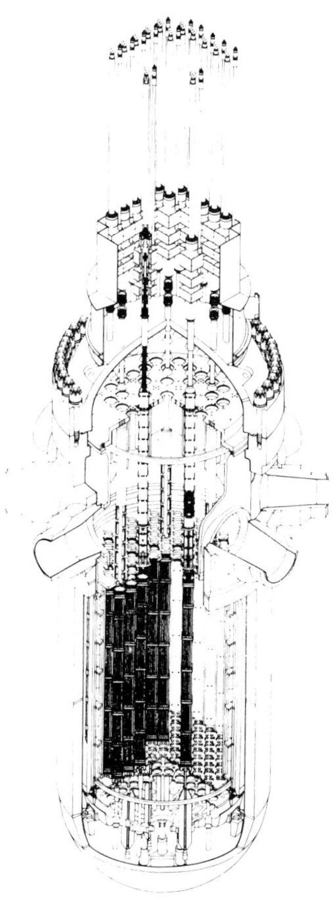

Figure 2. Reactor Vessel

Fortunately, the choice is not as unrestricted as it appeared at first glance. Those steels having yield strengths below about 50,000 psi can be eliminated from consideration because of the enormous section sizes required. Even with a 50,000 psi minimum yield strength steel, we have 11 5/8 inch thickness requirements for the plate used in some parts of the reactor vessel. In a similar manner, although for different reasons, we can discard steels possessing a yield strength in excess of about 100,000 psi. One reason for this is that because of ASME Code restrictions, it is difficult to take advantage of a high yield strength. The Code allowable stress is 1/2 the yield strength or 1/3 the tensile strength, whichever is smaller. Unfortunately, as illustrated in Figure 3, an increase in the yield strength does not result in a proportional increase in tensile strength. Another reason for avoiding very high strength steels is their susceptibility to brittle fracture and stress corrosion. Finally, there has been little or no experience with the use of high strength steels in the construction of large pressure vessels.

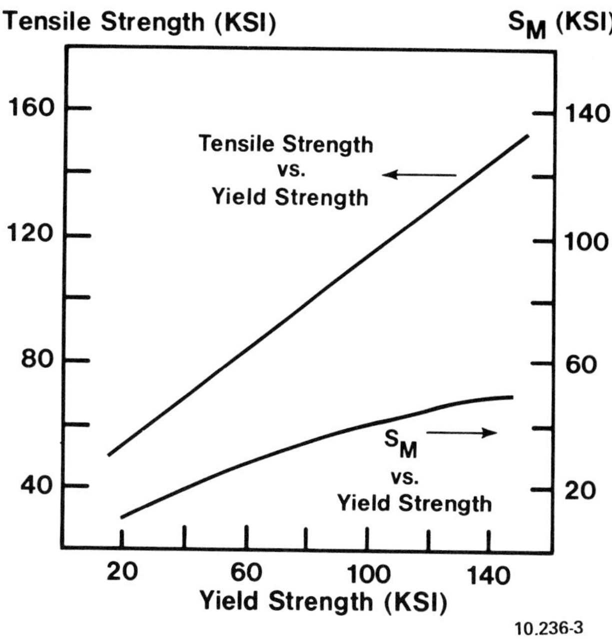

Figure 3. Tensile Strength (T.S.) and Code Allowable Stress (S_m) as a Function of Yield Strength

In practice, then, the original choice for a reactor vessel steel was limited to those steels having a yield strength between 50,000 and 100,000 psi. To attain this strength level in section sizes over 6 inches, it is necessary to quench and temper a low alloy steel. Fortunately, a grade of steel, which had long been used in the construction of pressure vessels, was available. This steel is the well-known A302 Grade B, whose properties and composition are shown in Tables I and II respectively.

Table I

Mechanical Properties of A302 B

Yield Strength	Tensile Strength	Elongation (2")
50,000 PSI Min.	80-100,000 PSI	18%

Table II

Chemistry of A302 B and A533 B

	A302 B	A533 B
C	0.25 Max.	0.25 Max.
Mn	0.95/1.30	1.15/1.50
P	0.035 Max.	0.035 Max.
S	0.040 Max.	0.040 Max.
Si	0.15/0.30	0.15/0.30
Mo	0.45/0.60	0.45/0.60
Ni	--	0.40/0.70

A302B not only met the strength requirements for a reactor vessel, but, because of its previous applications, its fabricating characteristics were also well-known. However, the sine qua non of reactor vessel design was, and still is, that the vessel cannot undergo a brittle fracture. Hence, modifications in the chemistry of A302 B were made, particularly the addition of Ni, to improve its

properties and characteristics of this material has been developed. As a result, I do not foresee any major change in the reactor vessel material unless a major advantage, economic or technical, can be proven with no compromise of integrity.

Reactor Coolant Piping

We have seen that the material for the reactor vessel was selected primarily on the basis of design and metallurgical factors. Reactor coolant piping presents a slightly different situation. In the Westinghouse design of the NSSS, the piping has an internal diameter of about 27 inches. Under the conditions of temperature, pressure, and environment previously described, the material choices are limited to three:

1. Wrought austenitic stainless steel.
2. Cast austenitic stainless steel.
3. Stainless clad low alloy steel.

In this case, the selection, cast austenitic stainless steel, is based primarily on economic considerations since there are technical advantages and disadvantages associated with all three.

The principal difficulty associated with centrifugally cast CF8A (the casting grade of type 304) is that of inspectability. The preferred technique for conducting the in-service inspection of nuclear component weldments is the pulse-echo ultrasonic method. In all austenitic stainless steels, the shear waves commonly used in angle beam testing are rapidly attenuated by virtue of the crystallographic nature of the material. This problem is accentuated by a grain morphology found in cast products and the relatively thick section size, about 3 inches, of the pipe.

The development of a refracted longitudinal wave inspection technique along with control of grain morphology in the casting enabled this problem to be solved.

Steam Generator Tubing

The final example of material selection to be discussed is that for the steam generator tubing. A cut-away sketch of a steam generator is depicted in Figure 6. The tube bundle shown in Figure 6 consists of several thousand U-bend tubes about 3/4 inches in diameter. Primary water flows through the tubes, giving up part of its heat to the secondary side in the process. On the external surface of the tubes, secondary water is converted to steam, which is piped to the turbine. The secondary side operates at a temperature of about 550°F and a pressure of about 1000 psi. The environment is high purity, deoxygenated water at a pH of 9. However, unlike the primary system, the secondary side is not a closed circuit. Secondary water can come into contact with the outside environment; this fact places severe limitations on material selection as we shall see a little later.

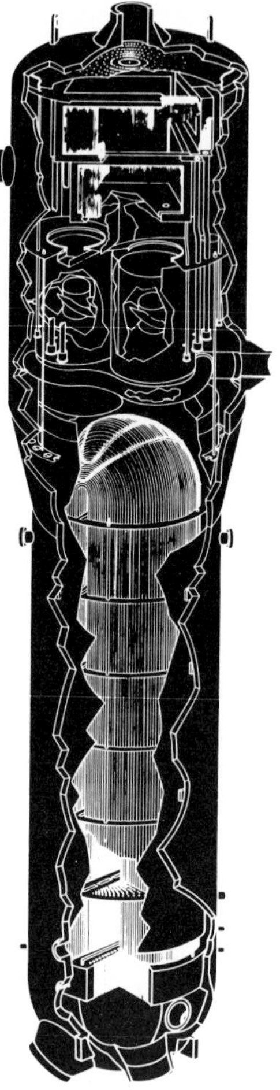

Figure 6. Steam Generator

The nature and function of steam generator tubes imposes requirements in addition to those described for other nuclear components. Since the tubing comprises approximately 96% of the total area wetted by the primary coolant, the release of corrosion products that can become radioactive must be held to a minimum. The reason for this is that any cobalt released into the system becomes activited to Co_{60}, an isotope with a long half life that contaminates the primary side and renders it highly radioactive. In addition, the thermal conductivity of the tube material must be high because of its heat tranfer function. Also, for heat transfer purposes it is desirable to have a thin walled tube, and this means that the material should possess as high a yield strength as compatible with other requirements. For fabrication purposes, the material must be capable of being drawn into long, of the order of 70 feet, tubes and seal welded to a tube sheet.

In addition to the foregoing, the steam generator tube has to be capable of operating in a potentially corrosive environment--this requirement is unique for PWR materials. Since by necessity heat transfer and boiling takes place at the secondary side tube surfaces, it is possible for contaminants in the secondary water to become concentrated in localized areas. Through careful hydraulic design and control of secondary water chemistry the possibility of corrodents concentrating to a dangerous level is reduced to a minimum. Nevertheless, the metallurgist must select a tube material that can withstand brief excursions when corrodents are in-

toughness. The result was A533 B, whose composition is also shown in Table II. Today, A533 B is the plate material used throughout the world for reactor vessel construction.

Over the past decade, further modifications were made to A533 B. The effect of copper and phosphorous on radiation induced embrittlement was discovered. Figure 4 shows how a change in the copper content from 0.30% to 0.10% can dramatically reduce the rise in transition temperature caused by radiation. Accordingly, copper and phosphorous are limited in reactor vessel steels to 0.10 and 0.012% respectively for ladle analyses. This chemistry restriction enables the steel to retain its toughness under neutron bombardment.

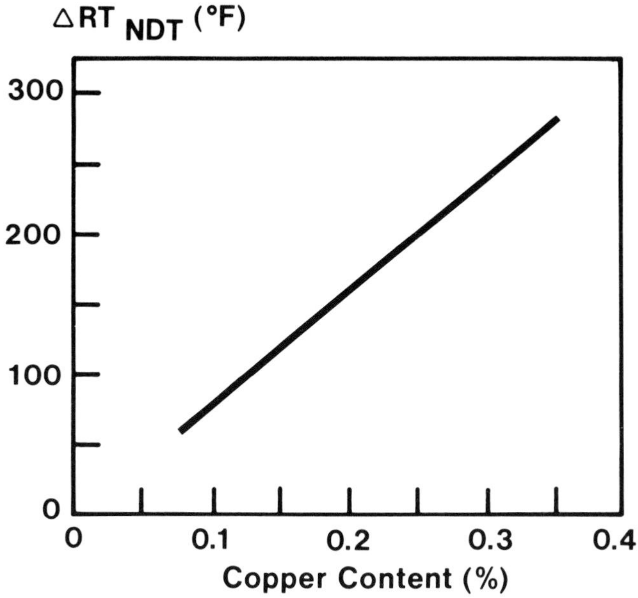

Figure 4. Effect of Copper on the Shift of RT_{NDT}. RT_{NDT} is a measure of a steel's ductile to brittle transition temperature.

While A533 B is the material used for plates for reactor vessels, portions of the vessel, e.g. nozzles and flanges, and in the case of some suppliers the entire vessel, are constructed of forgings. The steel used is A508, Class 2, which is very similar in composition to A533 B (See Table III).

Table III

Chemistry of A508, Class 2 and A533 B

	A508 Class 2	A533 B
C	0.27 Max.	0.25 Max.
Mn	0.50/0.90	1.15/1.50
P	0.025 Max.	0.035 Max.
S	0.025 Max.	0.040 Max.
Si	0.15/0.30	0.15/0.35
Ni	0.50/0.90	0.40/0.070
Cr	0.25/0.45	--
Mo	0.55/0.70	0.45/0.60
V	0.05 Max.	--

However, a problem was encountered with this material, as described in the following paragraphs.

Earlier, I alluded how the nuclear industry strives to avoid surprises. Here is an illustration on how even minor, seemingly innocuous, changes can have unexpected results. The reactor vessel is clad on its inner surfaces with stainless steel weld metal, e.g. type 308. This cladding is commonly done by a submerged arc two or three wire welding process. About 1970, one of the vessel suppliers started using a 2 1/2 inch strip for the weld metal feed. After several vessels had been constructed, it was discovered that small cracks occurred in the low alloy steel under the weld metal cladding deposited by this process. A typical crack is shown in Figure 5.

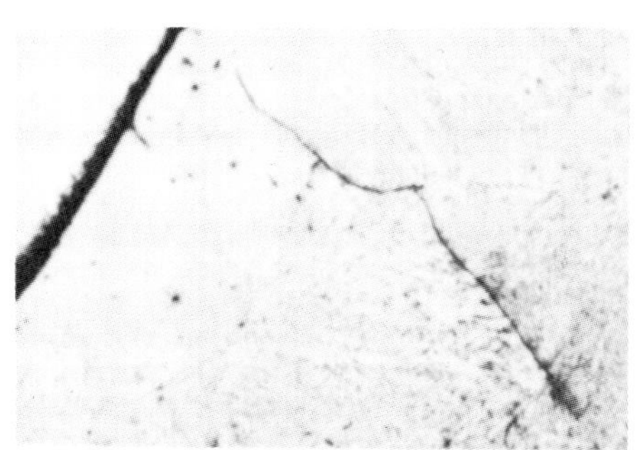

Figure 5. Typical Underbead Crack in A508 Cl 2

These intergranular cracks were very tiny, the maximum depth was 0.10 inches, and were located in the heat affected zone. Subsequent investigations demonstrated that this underbead cracking was really a form of reheat cracking and occurred only when cladding was deposited on A508 Class 2 with a high heat input process, e.g. strip cladding or 6-wire cladding.

Extensive tests and fracture mechanics analyses conclusively showed that the underbead cracks were innocuous and in no way compromised the integrity of the vessel. Nevertheless, immediate steps were taken to prevent their occurrence. The cladding process was modified and, in some cases, a less susceptible material, A508 Class 3, was utilized.

Two points were brought out by the incident just described. The first is the extraordinary measures the nuclear industry is willing to take to ensure the integrity of its components. Even though all available data and analyses showed that vessels containing small underbead cracks were perfectly safe, millions of dollars were spent to prevent their future occurrence. The second point is that change, any change, can bring about unexpected ramifications. For the reactor vessel, an acceptable material has evolved. Over the years, a vast data bank on the

advertently introduced into the secondary side of the steam generator.

A wide variety of candidate materials are commercially available and could be considered for use as steam generator tubes. A partial list would include the ferritic steels, the austenitic stainless steels, the monels, the inconels, the incoloys, the hastalloys, etc. However, none of these are absolutely perfect. They all suffer from one disadvantage or another. The ferritic stainless steels can embrittle in service (885° embrittlement), the martensitic stainless steels are difficult to weld, the austenitic stainless steels are susceptible to chloride stress corrosion, the hastalloys contain cobalt, etc. The best compromise material was judged to be Inconel 600.

Inconel 600 has a high yield strength, about 40,000 psi, and reasonably good thermal conductivity. It is easily drawn into tubes and welded. Its general corrosion rate is low as is its release rate of elements, such as cobalt, that can become radioactive. Inconel 600 is immune to chloride stress corrosion cracking and highly resistant to caustic cracking. Extensive corrosion testing has demonstrated that Inconel 600 has the best corrosion resistance to those environments expected to be found on the secondary side of a steam generator. Hence, it is the preferred choice today.

However, Inconel 600 is not perfect. It is attacked by hot phosphoric acids and it is susceptible to stress corrosion by lead compounds and concentrated caustic. As a result, research has been and still is underway to improve the corrosion resistance of Inconel 600 and to develop new candidate materials. Here is a case where change is going on now, but in a careful and deliberate manner, supported by extensive laboratory and operating plant experience.

Concluding Remarks

This paper has demonstrated that within an overall framework the basis for the choice of a particular material for a nuclear component varies for the application. In the case of the reactor vessel, the reasons were metallurgical. Economics dictated the final decision for the main reactor coolant piping. And functional requirements, heat transfer and corrosion resistance, led to the selection of a steam generator tube material. The choice of materials for other nuclear components was based on equally diverse factors.

The basic philosophy has been to learn more about the materials currently in use rather than to develop new materials. To accomplish this, a tremendous amount of effort has been, and continues to be, expended. Figure 7 shows fracture toughness data accumulated for A533 Grade B Class 1, the reactor vessel steel plate. Information of this type is used by the metallurgists and design engineers to establish and maintain the integrity of the vessel. However, it is very expensive and time consuming to run these experiments. The gathering of these data alone cost about one million dollars. This is an illustration of a small portion of the large body of information that has been obtained on the materials in use today. To accumulate a comparative amount of knowledge on a new material would be a vast undertaking indeed. The incentives, whether technical or economic, would have to be extremely powerful to justify a substantial material change.

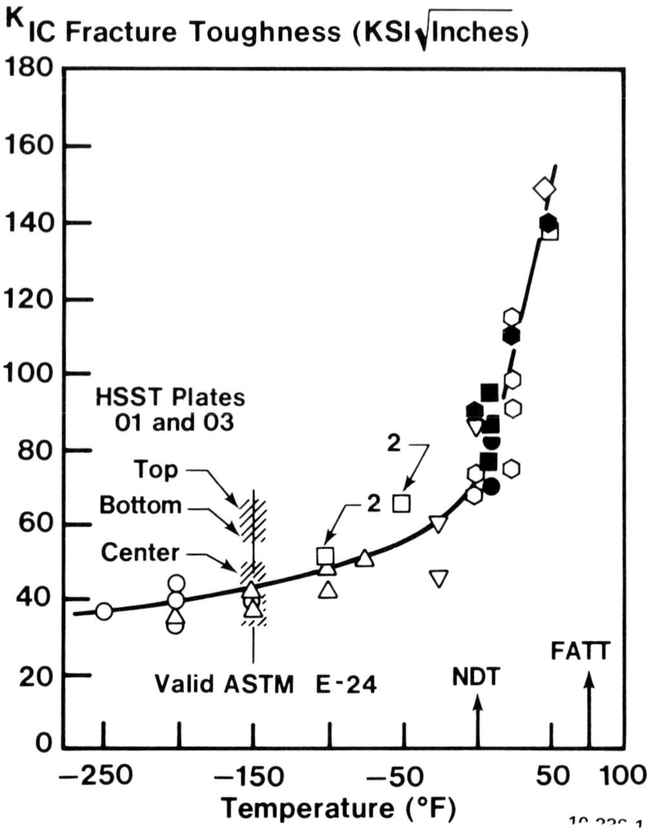

Figure 7. Fracture Toughness of A533 B plate as a function of temperature.

The one common thread that runs through the material selection process for nuclear power plants is reliability. We have seen that wherever possible tried and true materials are used, i.e. materials that have performed under similar conditions. Once selected, the characteristics of the material are continually evaluated and up-dated. Although improvements are constantly implemented, radical changes are made with great caution, and then only if a clear and substantial incentive exists.

RECENT DEVELOPMENTS IN MATERIALS FOR LIGHT WATER REACTOR APPLICATION

W. P. Chernock
Vice President, Development

R. N. Duncan
Director, Fuels Development

J. R. Stewart
Director, Materials and Chemistry Development

Combustion Engineering, Inc.
Nuclear Power Systems

INTRODUCTION

This paper will address itself to materials utilized for fuel rods, fuel assemblies, pressure vessels and core internals. For the most part, the discussion will center upon the impact of neutrons on the properties of materials exposed to such neutron environments.

FUEL ASSEMBLY

The fuel which is in universal use in the United States in light water-cooled and -moderated power reactors is uranium dioxide of low enrichment in the form of pressed and sintered high density ceramic pellets which are placed in zirconium alloy cladding tubes. The products of U-235 fission are highly energetic fission fragments and fast neutrons. Primary containment of the fission products is within the fuel itself and leads to considerations of volume changes, swelling and fission gas retention. Secondary containment is provided by means of the fuel rod cladding which must be capable of accommodating volume changes within the fuel and retaining fission products while at the same time the cladding is subjected to fast neutron irradiation which alters its properties.

Cladding and Components

The main functions of fuel rod cladding are to prevent the escape of fission products or fissionable material from the fuel rod and to contribute toward structural support for the fuel assembly (fuel rods typically in a 16 x 16 array).

Zirconium-base alloys, called Zircaloy, have been specifically developed for fuel rod cladding because of their low neutron absorption characteristics (more neutrons are made available for continuing the fission process), excellent corrosion resistance in the high temperature water utilized to cool and moderate light water reactor cores and retention of mechanical integrity and ductility in a neutron environment. These alloys contain about 1.5% Sn, 0.1% to 0.2% Fe and 0.1% Cr and less than 0.5% Ni. Zircaloy-4, the low nickel variant of Zircaloy-2, contains less than 70 ppm Ni and is used in pressurized water reactors to decrease the pickup of hydrogen generated during corrosion in the high temperature primary coolant.

Zircaloy-2 and -4 (HCP crystal structure) tubing fabrication processes are selected to enhance a preferred orientation which combines enhanced hoop ductility with minimum potential for radial orientation of zirconium hydride precipitates. These hydride precipitates can, if improperly oriented, reduce burst ductility.

Low parasitic absorption of neutrons is the theme employed in selection of structural materials in fuel assemblies. As an example, the Combustion Engineering (C-E) fuel assembly design employs Zr-4 for guide tubes (within which control rods move) which, combined with the grids at various elevations, form the main structure of a fuel assembly. Grids are used to prevent fuel rods from vibrating during operation. C-E utilizes Zr-4 grids, whereas other light water reactor designs utilize various Inconel alloys for the grids with attendant higher neutron absorption.

Production of "reactor grade" zirconium by the Kroll process involves the reaction of zirconium tetrachloride (formed during a chemical separation process to remove the high neutron absorption cross section impurity, hafnium) with magnesium to form zirconium (called sponge zirconium) and magnesium chloride. The sponge zirconium is multiple vacuum arc melted using consumable electrodes of compacted zirconium sponge and alloying elements. The multiple vacuum arc melting steps remove the volatile impurities, particularly magnesium and chlorine, and produce homogeneous Zircaloy ingots. After hot working these ingots, seamless tubing is produced by extruding billets into tube shells and then tube reducing these shells with the appropriate reduction schedules to produce high tolerance seamless Zircaloy tubing with the desired texture. Vacuum annealing, either for stress relief or recrystallization, must be used at this stage since the thin sections of Zircaloy are very reactive. After final cleaning in a mixture of nitric and hydrofluoric acid, the finished high tolerance seamless tubing is 100% quality control checked for dimensions and flaws.

Prior to assembly into fuel rods, the Zircaloy tubing surfaces are sometimes treated to produce uniform, thin initial oxide films. The oxide films are produced by exposure to water, steam or air at high temperature or by anodizing. In other cases, the tubing is used "as pickled" or with the inside diameter grit blasted to remove any remnants of fluorides which may be left during the pickling operation. The specific surface treatments used vary between fuel manufacturers.

Sheet or bar stock is produced from the hot worked billets by conventional rolling techniques. As the section size decreases, processes which may cause contamination or reaction must be minimized and vacuum annealing is used.

Welding of Zircaloy is quite readily accomplished using tungsten inert gas (TIG) or magnetic force welding (MFW) techniques. Again, during the welding operation, the atmosphere must be controlled, either by vacuum or inert gas, to prevent contamination of the material and subsequent degradation of its corrosion properties in high temperature water.

Zircaloy exhibits enhanced creep in reactor due to an increase in the mobility of line defects. The intense radiation within a reactor core produces a supersaturation of point defects within the material, allowing dislocations to climb and glide more easily over obstacles in the presence of the stress field (1). Radiation-enhanced creep of Zircaloy has been shown to be dependent on stress, flux, temperature and time (2). The stress and flux dependencies are often expressed as power functions with exponents of 1 to 2 for stress and about 0.85 for flux. In-reactor creep rates exhibit a relatively weak temperature dependence. One form for an irradiation-enhanced creep expression is shown below (3).

$$\dot{\epsilon} = A\,\sigma^{n}\,\phi^{m}\,e^{-Q/RT}\,(1 - B^{e^{-kt}})$$

where:

σ = Stress

T = Temperature

ϕ = Fast flux

t = Time

and:

A, B, k, n, m, R = Constants

The mechanical design of fuel rods and the Zircaloy assembly structures acknowledge enhanced in-reactor creep of Zircaloy.

Zircaloy also exhibits dimensional changes in reactor (2) due to preferential alignment of radiation-produced dislocation loops on particular crystallographic planes and the preferred alignment of active dislocations during fabrication. As indicated above, Zircaloy is an anisotropic material and forms a strong basal texture during fabrication of tubular mill products. This orientation results in axial growth of Zircaloy fuel rod cladding and Zircaloy structurals. A large number of measurements have been made on irradiated fuel rods, Zircaloy structural components and empty cladding tubes such that the magnitude of the irradiation-induced axial growth has been quantified (4). Figure 1 shows the magnitude of the axial growth of fuel rods which are cold worked and stress relieved as well as for structural components which have been recrystallization annealed. The scatter in the fuel rod data is attributed to a variable component of pellet-to-clad mechanical interaction in the axial direction. Figure 1 also shows test reactor results obtained on Zircaloy with various degrees of cold work (5, 6). As the degree of cold work increases, the magnitude of the axial growth for a given fast neutron exposure increases. A statistically derived best fit expression for use in design for the axial growth of fuel rods with cold worked and stress relieved Zircaloy cladding is given below (4).

$$\%\frac{\Delta L}{L} = 0.1\,(\phi t/10^{21})^{0.7}$$

where: ϕ = Fast flux (E > 0.82 Mev)

t = Time

The dimensional changes of Zircaloy in reactor have been well characterized and are fully accommodated in the design of various assembly types.

Fuel and Fuel Rods

The fissile isotope (U-235) content of natural uranium is about 0.7%. Fuels used in current light water reactors utilize U-235 contents in the range of 1% to 4% which is achieved via diffusion enrichment of UF_6. UF_6 is converted to ceramic grade UO_2 powder by various chemical conversion methods and the UO_2 powder is cold pressed into right circular cylinders. These cold pressed compacts are sintered in a reducing atmosphere such as hydrogen or cracked ammonia to a density in the 93% to 97% of theoretical range. The sintering process removes residual fluorides and produces a near stoichiometric UO_2 ceramic body. The cylindrical pellets are ground to final size and inspected prior to loading into the Zircaloy cladding tubing.

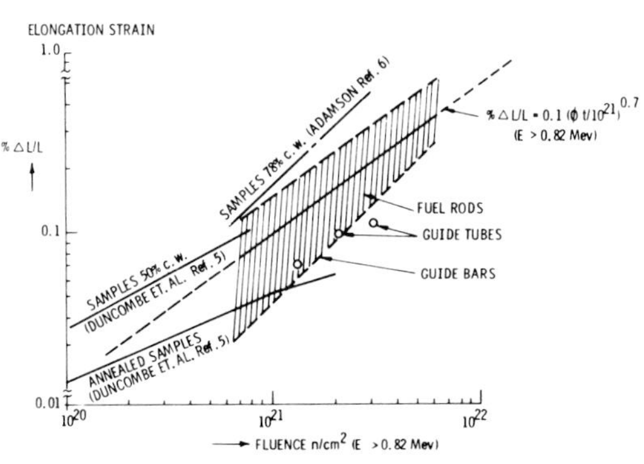

ELONGATION OF LWR FUEL RODS AND ASSEMBLY STRUCTURAL MEMBERS
Figure 1

In-reactor densification of the UO_2 pellet can occur at temperatures significantly less than the normal sintering temperature for the fuel (7). This low temperature in-reactor densification is due to the interaction of fission fragments with fine porosity. Fission product spikes which overlap fine pores result in the annihilation of fine porosity. In short periods of time (less than 1,000 hours), significant densification of the fuel can occur (8). In a fuel rod with pellets with a high degree of fine porosity, the fuel column height can decrease due to the shrinkage of the fuel, leaving gaps along the length of the rods. The cladding may collapse into these gaps. In addition to this effect, the UO_2 pellet shrinkage results in an increase in the diametral gap between pellet and clad, thereby increasing the average temperature of the UO_2 fuel.

In-reactor densification can be precluded by the control of the UO_2 pellet microstructure during fabrication. Experiments have shown that a coarse porosity distribution and large UO_2 grain size prevent the in-reactor densification from occurring (7). Minimizing the small pore size fractions, less than 4 microns, results in a delay in annihilation of porosity and, on a macro basis, the bulk UO_2 pellet density does not change appreciably in reactor. Pore former additions can be used to obtain a non-densifying structure. UO_2 densification in reactor is balanced by the matrix swelling induced by the generation of solid fission products in the UO_2 lattice. Figure 2 is a schematic representation of the net effect of UO_2 pellet volume change as a function of burnup. For densifying fuels, i.e., fuels with a high fraction of the porosity in the small size class, densification will occur rapidly and at low burnups; the net effect is a major reduction in fuel volume with attendant axial column and diametral dimension changes. On the other hand, for non-densifying fuel, i.e., containing coarse porosity, the magnitude of the densification decreased as does the rate. The net effect is either no fuel volume decrease or a very minor decrease. These two cases are schematically shown on Figure 2.

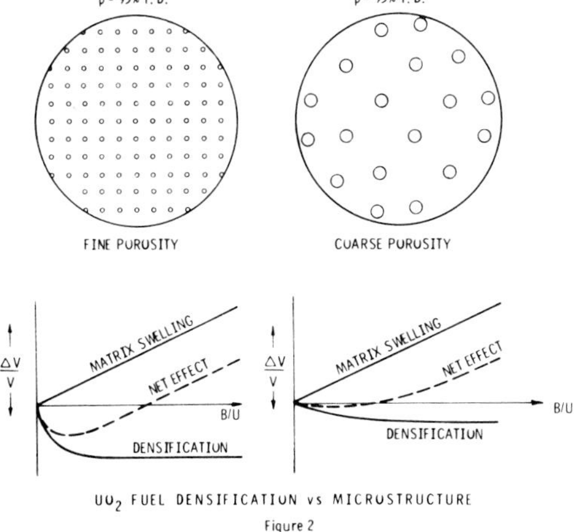

UO_2 FUEL DENSIFICATION vs MICROSTRUCTURE

Figure 2

Densification of UO_2 is no longer of concern. Current production controls provide fuels with essentially no propensity for in-reactor densification. The microstructure of the UO_2 pellets is controlled such that pore size and grain size distributions will result in minor in-reactor densification.

One effect that occurs in modern light water reactor fuel is interaction between the hard UO_2 ceramic pellet and the Zircaloy cladding. During operation, the UO_2 fuel pellet is at a significantly higher temperature than the cladding. As a result of the thermal expansion of the UO_2, as well as matrix swelling (Figure 2), pellet stresses are produced in the Zircaloy cladding. This effect is termed Pellet-Clad Interaction (PCI). In addition to a source of general and localized cladding stresses due to PCI, Zircaloy is susceptible to the nucleation and propagation of cracks in some environments when stressed. In the specific case of a fuel rod, these cracks originate from the ID as the result of chemical reactions when stressed in the presence of volatile fission products (iodine or cesium) (9). This response is analogous to stress corrosion cracking (SCC) in other materials when stressed in certain specific environments. Although PCI and SCC are not major problems in modern light water reactor performance, this is an area of current intensive investigation. Numerous tests are being performed to relate the complex synergistic effects of the fuel rod duty cycle to the resultant clad stresses produced by PCI as well as establishing requisite inventories of active fission product chemical species to nucleate and propagate cracks by an SCC mechanism (10, 11).

Overall, light water reactor fuel has currently achieved a very high level of reliability, including effects of PCI. As an example, United States pressurized water reactor reliabilities are currently in the 99.98% to 100% range (12, 13, 14). This means that typically, in a modern large pressurized water reactor which contains 30,000 to 40,000 individual fuel rods, less than eight fuel rods have perforations due to all causes. The reliability of light water reactor fuel is attributed to the conservative design and the stringent material specifications utilized for the production of nuclear fuel. The very high levels of quality control and rigorous process control have resulted in a reliable high quality energy source.

PRESSURE VESSEL

Materials for light water reactor pressure vessels are primarily ferritic Mn-Mo steels (ASTM A533 Grade B and ASTM 508 Class 2 or Class 3) which exhibit both high strength and ductility. Room temperature Charpy-V notch (C_V) impact energy absorption values are in the vicinity of 80 ft-lb. Since these alloys are ferritic, they exhibit a marked change in their impact energy absorption characteristics in the temperature range of about -40°F to +40°F. The plates forming the beltline region (in the vicinity of the nuclear core) are exposed to neutrons which may alter the temperature at which this transition in impact energy absorption occurs. The changes in properties associated with neutron exposure are dependent upon the neutron fluence, residual element content in the steel and the operating temperature of the vessel.

Measurement of neutron-induced property changes in pressure vessel plates, weldments and heat affected zones is achieved via inclusion of surveillance samples which are placed in holders affixed to the stainless steel cladding on the inside surface of the reactor vessel. These samples are removed from the reactor vessel at specific times and are tested to determine fracture toughness.

The strong influence of composition on the transition temperature shift in A302B (an early pressure vessel steel of the Mn-Mo class) was dramatically illustrated in low residual element plate material (0.03 w/o Cu, 0.001 w/o P, 0.004 w/o S) which, after being irradiated at $550^{\circ}F$ to a fluence of 3×10^{19}, exhibited no transition temperature change (15). Additional data (16,17) confirmed the importance of Cu and P control and investigations by the Naval Research Laboratory (18) involved the preparation of a 30 ton heat of A533B which was divided into two ingots containing 0.13 and 0.03 w/o Cu with P and S at 0.008 and 0.007 w/o in both ingots. At a neutron fluence of 2.8×10^{19} at $550^{\circ}F$, the transition shift in the higher copper material was on the order of $125^{\circ}F$ to $140^{\circ}F$, whereas that for the lower copper samples was on the order of $40^{\circ}F$ to $65^{\circ}F$. Earlier data on material with even higher copper contents indicated transition temperature increases on the order of $240^{\circ}F$ at this fluence and temperature. Thus, significant reductions in transition temperature shift are achieved even with modest control of copper content (0.13%). Other data on residual element effects have been obtained (19,20) via controlled irradiation tests as well as detailed analyses of surveillance data.

More recently, a detailed analysis of the effect of residual elements on neutron-induced property changes in pressure vessel steels formed the basis of a joint program sponsored by Combustion Engineering and the Atomic Energy Commission (now the Nuclear Regulatory Commission) and the Naval Research Laboratory (21) and included commercially produced ASTM A533 Grade B steel plate, weld deposits and heat affected zones with the following three levels of copper content: (1) greater than 0.15%, which is typical of commercially produced A533 Grade B plate prior to 1971; (2) 0.10% maximum, representative of improved copper control production and (3) 0.06% copper maximum which is considered the practical lower limit for commercially produced heavy steel plates and forgings. A summary of results from this program is shown on Figure 3 in which the neutron-induced changes in the 30 ft-lb C_V value for A533B pressure vessel steel are plotted as a function of neutron fluence as well as residual copper content. These data are also compared with earlier data obtained for A302B steels (21,22) as well as laboratory heats of A533B containing very low Cu contents. Two distinct scatter bands associated with "low" and "high" Cu are apparent on Figure 3.

An apparent anomaly is noted on Figure 3 in that a weldment exhibited greater irradiation sensitivity than its copper content would suggest. Examination

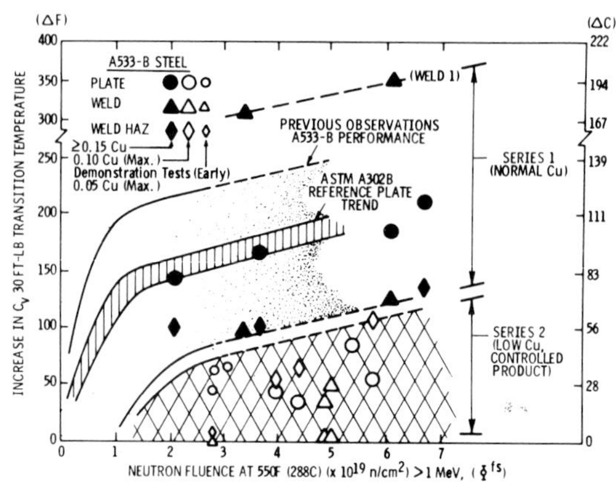

INCREASE IN CHARPY V TRANSITION TEMPERATURE (30 FT-LB) FOR A533B STEEL AS A FUNCTION OF FLUENCE AND COPPER CONTENT (REFERENCE 21)

Figure 3

of previously irradiated weld metal in the HSST (Heavy Section Steel Technology) program also revealed abnormally high transition temperature shifts. Results of a detailed analysis (23) of welding processes, weld composition and irradiation history suggest that Ni and Si increase the tendency for carbon atoms to form carbon-vacancy complexes which can result in ductility reduction (24). Mn, Mo and Cr counteract the influence of Ni and Si and the ratio of Ni plus Si to that of Mn plus Mo plus Cr in a deposited weld metal is an important consideration in addition to Cu and P content and post-weld heat treatment time in controlling neutron-induced transition temperature shifts in welds of A533B pressure vessel steel (23). Additional details of the influence of metallurgical structure in the weld and of post-weld heat treatment on transition temperature are discussed in Reference 23. A summary of the influence of composition is provided in Figure 4 (23) which shows the influence of Cu as well as the Ni + Si/Mn + Mo + Cr ratio on 30 ft-lb C_V transition temperature shift.

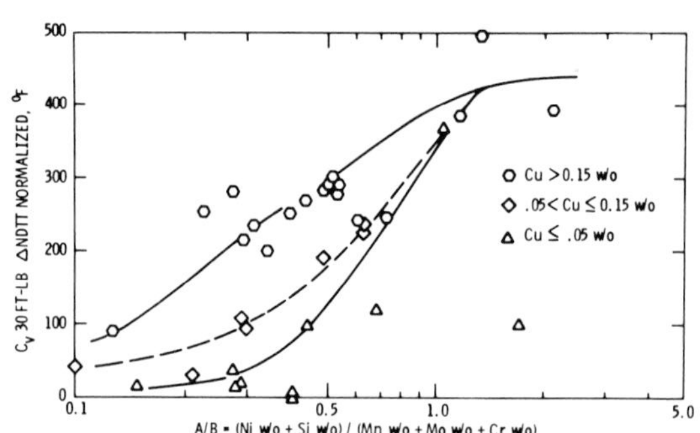

INCREASE IN CHARPY V TRANSITION TEMPERATURE (30 FT-LB) AS A FUNCTION OF A/B RATIO (A = w/o Ni + Si AND B = w/o Mn + Mo + Cr) AND COPPER CONTENT (REFERENCE 23)

Figure 4

Steel suppliers have been responsive to the requirements of the industry in reducing Cu and P in materials designated for the beltline region of the pressure vessel. Vessel manufacturers have developed new welding electrodes which reduce the amount of Cu, Ni and Si in the deposited weldment. Thus, developing technology has succeeded in achieving substantial reductions in neutron-induced transition temperature shifts in pressure vessel plates, weldments and heat affected zones.

CORE INTERNALS

Austenitic stainless steel is the material most generally utilized for core internals and as cladding on the pressure vessel, piping and other components in contact with the primary coolant in pressurized water reactors.

Type 304 stainless steel is the most commonly used material for core internals. It has good corrosion resistance and is not susceptible to radiation-induced damage which would adversely affect its design properties (25). It does not exhibit a ductile-to-brittle transition as a result of irradiation.

The experience with austenitic steels in pressurized water reactors has been excellent. Although use of sensitized stainless steels in pressurized water reactor environments presents no problem, precautions aimed at reducing or eliminating sensitization of austenitic stainless steels have been generally employed. Recent studies (26,27) have shown that precautions are required during fabrication of stainless components to prevent fluoride contamination. Fluorides are present in welding rod fluxes and can be deposited on stainless surfaces as a result of volatilization during welding.

The fluoride attack is intergranular in nature. Attack can occur at room temperature in aerated water. It is strongly accelerated by applied stress in excess of yield; however, it has been observed on samples with only residual stresses due to welding. Rather specific metallurgical conditions are necessary in order to initiate this type of attack. The material must be sensitized and have a coarse grain size. In addition, oxygen (greater than 30 ppb) and fluoride ions must be present in the aqueous solution. Prevention of fluoride attack is readily achieved via control of welding procedures since elimination of any one of the above prerequisites will negate the potential for fluoride attack. With such controls, continued successful use of austenitic stainless steels for internal reactor structures is assured.

SUMMARY

Application of materials within light water reactors requires consideration of the influence of neutron-induced changes in properties. Requirements for materials within the core include low neutron absorption for structural components, a fuel material which is dimensionally stable and chemically compatible with the cladding, and cladding and structural materials which exhibit good corrosion resistance in high temperature water. The understanding and quantification of irradiation effects on materials properties have been utilized in the selection of materials for core components. The successful operation of light water reactors provides ample evidence that materials engineers and reactor designers have adequately utilized quantified irradiation effects data to create a safe and reliable energy source.

REFERENCES

1 Dollins, C. C. and Nichols, F. A., "Mechanisms of Irradiation Creep in Zirconium-Base Alloys," Zirconium in Nuclear Applications, ASTM STP 551, 1974, 11. 229-248.

2 Fidleris, V., "Summary of Experimental Results on In-Reactor Creep and Irradiation Growth of Zirconium Alloys," Atomic Energy Review, Vol. 13, March 1975, p. 51.

3 Franklin, D. G. and Fisher, H. D., "Requirements for In-Reactor Zircaloy Creep Measurements for Application in the Design of PWR Fuel," to be published in "Proceedings - International Colloquium on Measurement of Irradiation Enhanced Creep in Nuclear Materials," Petten, The Netherlands, May 5-6, 1976.

4 Duncan, R. N., Fuhrman, N., LaVake, J. C., Knaab, H. and Manzel, R., "Dimensional Stability of Water Reactor Fuel," in Proceedings of Joint Topical Meeting on Commercial Nuclear Fuel Technology Today, April 28-30, 1975, Toronto, Canada, 75-CNA/ANS-100.

5 Duncombe, E., Meyer, J. E. and Coffman, W. A., "Comparisons with Experiment of Calculated Dimensional Changes and Failure Analysis of Irradiated Bulk Oxide Fuel Test Rods Using the CYGRO-1 Computer Program," WAPD-TM-583 (1966) p. 35.

6 Adamson, R. B., "Irradiation Growth of Zircaloy," to be published in ASTM Symposium on Zirconium in the Nuclear Industry, Quebec, August 10-12, 1976.

7 Harkness, S. D., Pati, S. R., Andrews, M. G. and Chernock, W. P., "In-Pile Densification of UO_2," Proceedings of European Nuclear Conference, Paris (April 1975); to be published by Pergamon Press;for abstract, see Trans. Am. Nucl. Soc. 20, 215 (1975).

8 Stehle, H. and Assmann, H., "The Dependence of In-Reactor UO_2 Densification on Temperature and Microstructure," Journal of Nuclear Materials, 52, 303 (1974).

9 Cox, B. and Wood, J. C., "Iodine Induced Cracking of Zircaloy Fuel Cladding -- A Review" in Corrosion Problems in Energy Conversion and Generation, C. S. Tedmon, Ed. The Electrochemical Society, Inc., Princeton, New Jersey (1974).

10 Fuhrman, N. and Pasupathi, V., "Joint CE/EPRI Fuel Performance Evaluation Program, Task C -- Evaluation of Fuel Rod Performance in Maine Yankee Core I," Report on Work Performed up to July 1975, CENPD-221, December 1975.

11 Fuhrman, N., Hollowell, T. E., Pasupathi, V. and Scott, D. B., "Effect of UO_2 Densification on the Performance of Non-Pressurized PWR Fuel Rods," presented at 78th Annual Meeting of the American Ceramic Society, Cincinnati, Ohio, May 5, 1976.

12 Smerd, P. G. and Stehle, H., "CE/KWU Operating Experience with LWR Fuel," Trans. Am. Nucl. Soc., 23, 256 (1976).

13 Mayer, J. T. and Montgomery, M. H., "Performance of B&W PWR Fuel," Trans. Am. Nucl. Soc., 23, 255 (1976).

14 Kramer, F. W., "PWR Fuel Experience -- The Westinghouse View," in Proceedings of Joint Topical Meeting on Commercial Nuclear Fuel Technology Today, April 28-30, 1975, Toronto, Canada, 75-CNA/ANS-100.

15 Hawthorne, J. R., et al., "Irradiation Effects on Reactor Structural Materials," NRL Memorandum Report 1753, February 15, 1967.

16 Potapovs, U. and Russell, J. R., "The Effect of Residual Elements on 550°F Irradiation Response of Selected Pressure Vessel Steels and Weldments," NRL Report 6803, November 22, 1968.

17 Steele, L. E., et al., "Irradiation Effects on Reactor Structural Materials," NRL Memorandum Report 1937, November 15, 1968.

18 Hawthorne, J. R., "Demonstration of Improved Radiation Embrittlement Resistance of A533B Steel Through Control of Selected Residual Elements," NRL Report 7121, May 29, 1970.

19 Potapovs, U. and Hawthorne, J. R., "Nuclear Applications," Vol. 1, 1969, pp. 27-46.

20 Smidt, F. A., Jr. and Sprague, J. A., "Effects of Radiation on Substructure and Mechanical Properties of Metals and Alloys," ASTM STP 529, 1973, pp. 78-91.

21 Hawthorne, J. R., Koziol, J. J. and Groeschel, R. C., "Evaluation of Commercial Production A533-B Plates and Weld Deposits Tailored for Improved Radiation Embrittlement Resistance," ASTM STP 570, 1976, pp. 83-102.

22 Hawthorne, J. R., "Irradiation Effects on Structural Alloys for Nuclear Reactor Applications," ASTM STP 484, 1971, pp. 96-127.

23 Byrne, S. T. and Biemiller, E. C., "An Evaluation of the Effect of Chemical Composition on the Irradiation Sensitivity of Reactor Vessel Weld Metal," to be published in ASTM STP 611, 1976.

24 Beeler, J. R., Jr. and Beeler, M. F., "Attrition and Stabilization of Void Nuclei: Critical Nucleus Size," Effects of Radiation on Substructure and Mechanical Properties of Metals and Alloys, ASTM STP 529, 1973, pp. 289-302.

25 Shober, F. R., "The Effect of Nuclear Radiation on Structural Metals," DMIC Report 166, Battelle Memorial Institute, 1961.

26 Ward, C. T., Mathis, D. L. and Staehle, R. W., "Intergranular Attack of Sensitized Austenitic Steel by Water Containing Fluoride Ions," Corrosion, Vol. 25, No. 9, September 1969.

27 Habicht, P. R. and Bryant, P. E. C., "Fluoride Induced Intergranular Corrosion of Sensitized Austenitic and Austenoferritic Stainless Steels," IAEA Work Shop on Stress Corrosion Cracking, Palo Alto, California, Spring 1976. To be published by IAEA Work Shop on Stress Corrosion Cracking, Palo Alto, California, 1976.

How Structural Alloys Perform in EBR-II

After 671 days of exposure in this experimental breeder reactor operated by Argonne National Laboratory, only beryllium copper and tantalum appeared to be unsuitable for service in contact with sodium coolant.

EBR-II, THE ONLY high-power-density fast flux breeder reactor now in operation in this country, marked its tenth anniversary last August. At that time, the Idaho facility, which is operated by Argonne National Laboratory, had generated the equivalent of two years of electricity consumption by a city of 38,000. More importantly, experience indicates: 1. This type reactor performs reliably on a year-in, year-out basis. 2. Sodium coolant can be reliably handled with existing technology. Since 1965, emphasis in the EBR-II program has been switched to irradiation studies of oxide fuels and structural materials for liquid metal fast breeder reactors.

One study, called SURV-I, involved the exposure of 15 structural alloys and shield graphite canned in type 304 stainless steel for 671 days at an estimated average fluence of 3×10^{19} n/cm^2 and energy levels exceeding 0.82 MeV. The materials ranged from aluminum bronze to tool steel. Service temperature was 370 C (700 F). Post-radiation examinations included weight change, metallography, hardness change, strength and ductility, impact strength, bend test of welded type 304, spring measurement, and tube-burst tests.

General findings: only beryllium copper and tantalum experienced appreciable weight loss and appeared to be unsuitable for use in contact with sodium under conditions of the test. Types 304 and 347 stainless steel and the graphite assemblies experienced no significant changes in mechanical or physical properties.

The other materials experienced appreciable changes in mechanical properties and/or metallurgical structure; but, Argonne comments, "in the absence of thermal-effects control samples, the relative importance of the effects of neutron dosage and long exposures at reasonably elevated temperature cannot be established conclusively for these materials. Ample metallurgical evidence exists, however, to indicate that these materials could experience long-term thermal effects such as annealing, aging, and overaging."

Procedure — Materials placed in the reactor were contained in three-section subassemblies. The upper section contained graphite canned in type 304; the center section, duplicate samples of 15 alloys used in the EBR-II primary systems; the bottom section, a labyrinth and Stellite sleeve similar to those used on EBR-II control-rod drive mechanisms where they pass through the reactor cover.

Cans — Graphite was canned (Fig. 1) under a helium atmosphere at a pressure of 150 to 200 torr. One held plain graphite, one borated (3% B_4C) graphite, and one half plain and half borated. A can wall thickness of 1.27 mm (0.050 in.)

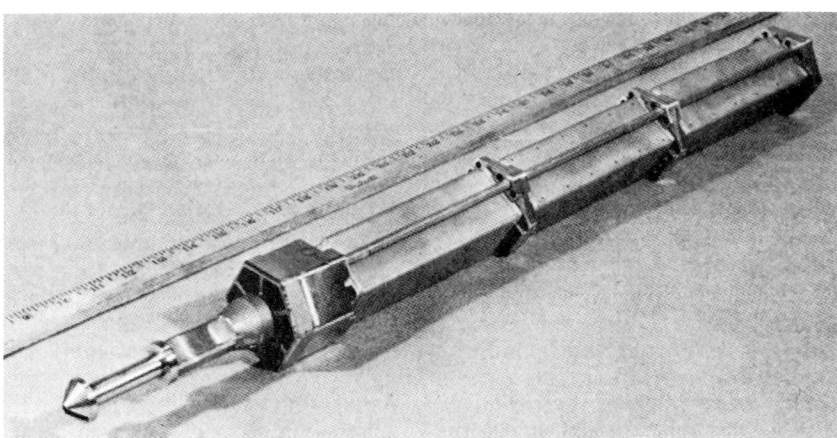

Fig. 1 — Individual graphite cans and the partially assembled section.

Based on an Argonne National Laboratory report, "The EBR-II Materials Surveillance Program: I. Program and Results of SURV-1," compiled by Sherman Greenberg.

was chosen to give stresses approximately equal to those in the walls of full-sized shield cans. Bottom end plates of the section were made of 12.7 mm (0.5 in.) type 304 stainless.

Metallurgical Samples — These materials included: Ampco Grade 18 aluminum bronze, Stellite 6B, Inconel X-750, type 420 stainless steel, T1 tool steel, type 347 stainless steel, type 416, Berylco 25 beryllium copper, type 304 with boron, type 17-4 PH stainless steel, type 304, type 304 welded with type 308, tantalum, and a type 304 cover plate.

Compositions and mechanical properties of the alloys are set forth in Table I. The metallurgical condition of samples is summarized in Table II. Samples included tensile test, hardness and corrosion, bend test, Izod impact, coil spring, and Belleville spring specimens. Many of the materials were exposed directly to sodium in the EBR-II primary tank.

Before being loaded, samples were cleaned ultrasonically in alkaline detergent solution; rinsed successively in hot tap water, alcohol, and acetone; and air dried. Subsequent handling was with gloves or tweezers.

About 20% of the welded type 304 bend samples had known weld defects, as indicated by radiography. Many of the multinotch Izod-impact samples of Inconel X-750 and type 17-4 PH had out-of-specification notches because original defects had to be removed. Test load ranges for coil spring samples were based on service conditions of EBR-II springs. Three nominal preloads were specified: 1.09, 2.41, and 3.72 kg (2.4, 5.3, and 8.2 lb). Each Belleville-spring sample was preloaded to a compression of 0.15 mm (0.006 in.).

Labyrinth Section — This assembly includes an aluminum-bronze labyrinth and a Stellite sleeve with a stainless steel outer cylinder (Fig. 2). Precise diametrical measurements were taken at three points on each land of the labyrinths and along 11 areas of the Stellite sleeve (opposite the lands). In this test, resistance to corrosion was of particular interest.

Results — Following dismantling, sodium trapped within the subassembly was removed by reaction with isoamyl alcohol and then with ethanol to preclude high-temperature chemical attack of the metallic surfaces by sodium hydroxide and undesired high-temperature metallurgical changes.

Table I — Composition of Test Samples

Material	C	Cr	Cu	Fe	Mn	Mo	Ni	P
					Composition, %*			
Ampco Grade 18 aluminum bronze	—	—	Bal	3.6	—	—	—	—
Stellite 6B	1.06	29.98	—	0.80	1.24	0.19	1.79	—
Inconel X-750	0.04	14.82	0.11	6.39	0.57	—	Bal	—
	0.03	14.70	0.09	6.52	0.55	—	Bal	—
	0.05	15.08	0.10	6.51	0.50	—	Bal	—
	0.03	15.56	0.05	6.42	0.58	—	Bal	—
Type 420	0.36	13.47	—	Bal	0.39	0.02	0.22	0.01
	0.33	13.47	0.10	Bal	0.40	0.03	0.25	0.01
T-1	0.73	3.90	—	Bal	0.22	—	—	—
Type 347	0.07	17.93	0.15	Bal	1.48	0.13	9.70	0.01
Type 416	0.10	12.91	0.08	Bal	0.39	0.04	0.19	0.01
Berylco 25 beryllium copper	—	—	Bal	—	—	—	—	—
Type 304 with boron	0.01	18.33	0.09	Bal	1.49	0.05	9.36	0.01
17-4 PH	0.031	15.99	3.40	Bal	0.23	—	4.34	0.01
	0.046	16.04	3.48	Bal	0.31	—	4.16	0.01
	0.041	16.28	3.45	Bal	0.32	—	4.08	0.01
Type 304	0.08	18.38	0.18	Bal	0.89	0.21	10.00	0.01
Type 304 welded	0.02	18.90	0.04	Bal	0.68	0.37	9.01	0.02
with type 308	0.04	20.50	—	Bal	1.32	—	9.80	—
Tantalum†	—	—	—	—	—	<20	<20	—
	—	<5	<1	7	—	21	< 5	—
EBR-II Cover Plate (type 304‡)	0.056	18.57	—	Bal	1.17	0.16	9.46	0.01

Note: Specification called for use of material still stocked from the manufacture of the original reac
parts or purchased from the same vendor with the same specification as for the original reactor p
*From certified analysis received from vendor. Mechanical properties are not necessarily those

Observation during dismantling operations revealed that deposits of corrosion product and/or precipitated metal had accumulated to varying degrees on the stainless steel used in various parts of the subassembly.

Example: the upper extension shaft used for removing the subassembly was covered with a dark-brown, fairly adherent coating. Several large blobs of white material (probably sodium-alcohol reaction products) also were attached to the gripper extension.

Example: the outer stainless steel cylinder of the Stellite sleeve was covered with a very dark-brown coating.

In general, the subassembly had a normal appearance when compared with standard EBR-II fueled subassemblies.

Weight Change — Hardness samples exposed to sodium were also used for weight-change measurements. Except for beryllium copper and tantalum, weight losses were small. Results are summarized in Table III.

Metallographic Examination — Hardness samples exposed to sodium and unexposed control samples were examined by optical metallography. Only beryllium copper showed surface attack. Results are summarized in Table IV.

Hardness Change — All hardness tests, including those on unirradiated samples, were made with the same Tukon microhardness tester. Reproducibility (to ±3%) was verified by hardness measurements on a calibrated standard test block before and after testing each sample.

Samples were exposed to primary sodium during irradiation, but most exhibited no surface hardening or softening. The most striking result was the reduction in hardness of beryllium copper samples. This is a precipitation-har-

	Si	Other	Hardness	Mechanical Properties* Tensile Strength, psi (MPa)	Yield Strength, psi (MPa)	Elongation, %
-	—	10.7 Al	—	95 000 (655)	47 000 (324)	14
-	0.69	4.51 W, bal Co	Rc 36	152 200 (1048)	93 950 (648)	10
07	0.32	0.70 Al, 2.47 Ti, 0.84 Cb+Ta	Rc 39	185 000 (1276)	135 000 (931)	28
07	0.27	0.78 Al, 2.54 Ti, 0.94 Cb+Ta	Rc 39	187 000 (1289)	135 500 (938)	27
07	0.26	0.68 Al, 2.27 Ti, 0.79 Cb+Ta	—	—	—	—
07	0.36	0.61 Al, 2.17 Ti, 0.87 Cb+Ta	—	—	—	—
10	0.42	—	Rc 52	—	—	—
07	0.38	—	—	—	—	—
-	0.24	18.20 W, 1.11 V	—	—	—	—
09	0.69	0.90 Cb+Ta	Bhn 248	—	—	—
47	0.69	—	Rc 38	—	—	—
-	—	1.9 Be, 0.25 Co	—	110 000 (758)	90 000 (620)	15
05	0.64	2.10 B	—	—	—	—
14	0.58	0.27 Cb, 0.02 Ta	Rc 43	193 000 (1331)	192 000 (1324)	20.2
19	0.69	0.25 Cb, 0.01 Ta	Rc 44	203 000 (1400)	200 000 (1379)	14.0
18	0.62	0.32 Cb, 0.02 Ta	Rc 43	204 000 (1407)	203 000 (1400)	14.2
20	0.68	—	Rb 93	—	—	—
28	0.69	—	—	—	—	—
-	0.44	—	—	—	—	—
-	—	16 C; 50 O; 29 N; 3 H; 185 W; 340 Cb; <20 Zr, Ti, V, Co	Rb 35.5-38.5	—	—	—
-	60	15 C; 14 O; 13 N; 60 Cb; <25 Al; <5 Ti; 170 W	—	—	—	—
09	0.57	0.013 Sn, 0.003 Pb	Bhn 149	81 000 (559)	35 000 (241)	62.0

samples because some material was heat treated after preliminary or final machining. †Impurities ppm. ‡When the original batch of type 304 wrought specimens was made, no original ERB-II stock s believed available. Subsequently, ERB-II stock was located, and specimens were prepared from it.

dening alloy, and it was previously reported to increase in hardness during irradiation. Possibly the sodium environment is responsible for the softening; surface corrosion was observed on all four samples. Microstructures also changed; and there was severe overaging, as evidenced by agglomerization.

Samples of T-1 tool steel became softer. This was probably due to decarburization in sodium at the 370 C (700 F) operating temperature.

Inconel X-750, borated type 304, and regular type 304 did not harden appreciably.

Aluminum bronze, Stellite, type 420, type 347, type 416, type 17-4 PH, and tantalum hardened appreciably. Generally, their hardness increased with neutron exposure. However, correlations aren't possible because of uncertainties in hardness data and neutron exposure values.

Strength-Ductility — Stress-strain curves to approximately 0.5% elongation were obtained at room temperatures.

The strength and ductility of annealed and moderately work-hardened type 304 were not affected by irradiation. The ductility of type 308 weld metal was reduced slightly.

Hardened Inconel X-750 and type 17-4 PH overaged, with accompanying losses in ductility; Inconel X-750 was weakened, but type 17-4 PH was strengthened. The tensile strength of tantalum was almost doubled, but its ductility was reduced considerably.

Aluminum bronze apparently overaged similarly to Inconel X-750. Both tantalum and aluminum bronze exhibited an approximate 50% increase in modulus of elasticity.

Analyses for carbon, oxygen, and nitrogen in tantalum exposed to sodium (made in an attempt to explain the change in properties in terms of increases in concentration of those elements) showed no appreciable increase in bulk concentration as a result of reactor exposure.

The softening of hardened type 420 is believed to be an effect of long-time annealing.

There was no conclusive evidence that the sodium or helium environments had any effect on strength or ductility. In addition, there was no consistent relationship between hardness and tensile strength. Example: the hardening of type 17-4 PH reflected its strengthening, but aluminum bronze hardened and lost strength.

Impact Strength — A Baldwin impact tester with a maximum impact-energy capacity of 22 J (16 ft-lb) delivered at 3444 mm/s (11.3 ft/s) was used — it was modified for remote operation.

Results obtained with Inconel X-750 and type 17-4 PH stainless are compared in Fig. 3. Although sample fluence ranged from 0.6×10^{19} to 1.1×10^{19} (more than 0.82 MeV), scatter of the data masked any effect of neutron dosage as well as any difference between the effects of exposure in sodium or helium.

The significant reduction in impact strength of Inconel X-750 is probably due to radiation hardening, but may be a result of aging at 370 C (700 F). Resolution of the contribution of the different factors in the environment was impossible in the absence of thermal-effects control samples. There was no discernible effect of sodium on impact strength.

The change in impact strength of hardened type 17-4 PH (H900 condition) was more pronounced. The exposed material was hardened considerably from dph 298 to 363. The result is believed to be due to aging during long-time exposure at 370 C (700 F). Again, there was no discernible effect of sodium on impact strength.

Bend Test — Beam bend-test specimens were supported on two round pins 64 mm (2½ in.) apart as the bending force was applied perpendicularly at the midpoint to the 95 by 11 mm (3.75 by 0.425 in.) face. Deflections were measured by compressometer with a maximum travel of 25.4 mm (1 in.). Maximum load was reached at about 12.7 mm (½ in.) deflection, after which the load tapered off with additional deflection.

The maximum load sustained by

Table II — Metallurgical Condition of Test Samples *

Ampco Grade 18 aluminum bronze	No heat treatment required.
Stellite 6B	Wrought condition.
Inconel X-750	Tensile, impact, and hardness samples: anneal starting at 800 F (427 C) max; heat to 2100 F (1149 C) at 300 F/h(165 C/h); hold at 2100 F (1149 C) for 4 h; air cool and rough machine. Age harden by heating from 800 F (427 C) max to 1625 F (885 C) max for 24 h; air cool. Reheat from 800 F (427 C) max to 1300 F (704 C) at 300 F/h (165 C/h); hold at 1300 F (704 C) for 20 h; air cool and grind to finish dimensions. Final hardness should be Rc 37 to 40.
	Spring samples: heat treat at 1350 F ±25 F (732 C ±14 C) for 16 h; cool in air. Heat press at 800 F (427 C) for 1 h with 100 lb (45 kg) load.
Type 420	Hardness of Rc 40 to 45.
T1	Hardness of Rc 55 to 60.
Type 347	Before machining, give full solution anneal at 1850 to 2050 F (1010 to 1121 C) and water quench. After machining, stress relieve at 875 to 925 F (468 to 496 C) for 2 h.
Type 416	Harden. Temper at 900 F (482 C) for hardness of Rc 30 to 34.
Berylco 25 beryllium copper	Harden and temper to Rc 41 to 45. †
Type 304 with boron	None indicated.
17-4 PH	Harden to Rc 36 to 41. Stress relieve at 875 to 925 F (468 to 496 C) for 2 h.
Type 304	Stress relieve after machining by holding at 875 to 925 F (468 to 496 C) for 2 h.
Type 304 welded with type 308	After machining, stress relieve at 875 to 925 F (468 to 496 C) for 2 h.
Tantalum	As received.
Type 304 cover plate	Annealed. Stress relieve at 875 to 925 F (468 to 496 C) for 2 h.

* Specifications as originally provided for EBR-II parts. † More properly stated as solution treated and aged.

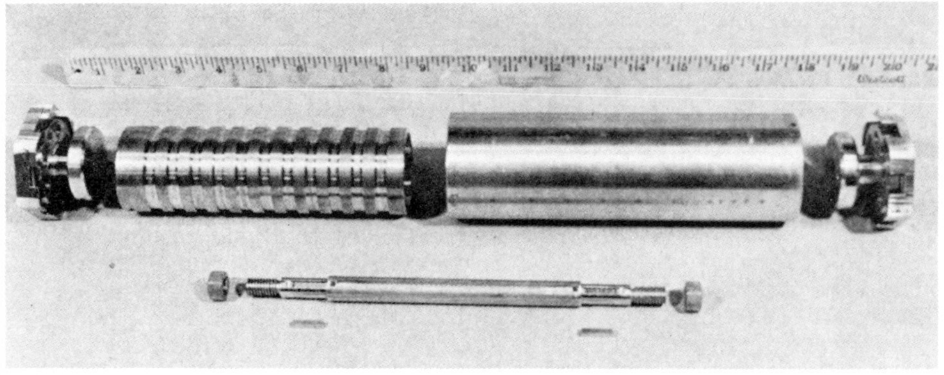

Fig. 2 — Labyrinth (left) and guide tube.

Table III — Weight Changes of Hardness Samples

Material	Average Initial Weight, g	Weight Loss, mg* Average	Range	Comments
Ampco Grade 18 aluminum bronze	16.06	1.7	0.5-2.6	—
Stellite 6B	18.09	1.4	1.1-1.7	—
Inconel X-750	17.74	1.6	0.2-3.3	—
Type 420	16.50	1.8	1.2-2.6	Hardened
T-1	18.68	4.2	0.6-8.9	—
Type 347	16.94	1.1	0.0-1.7	Three samples
Type 416	16.29	2.4	1.6-2.9	Hardened
Berylco 25 beryllium copper	17.97	151.2	126.2-177.0	Three samples†
Type 304 with boron	16.78	3.1	2.0-4.2	—
17-4 PH	16.69	1.1	0.7-1.5	Hardened
Type 304	16.85	1.3	0.0-2.1	Three samples
Tantalum	36.20	13.1	3.8-22.4	Two samples

Note: Results are based on four samples of each, unless otherwise indicated.
*Exposed surface area was 7.8 cm². †A fourth sample lost only 3.4 mg and is not included in the average.

exposed samples varied from 120 to 131 kg (265 to 288 lb). Load at 25.4 mm (1 in.) deflection varied from 105 to 114 kg (232 to 252 lb). Corresponding loads for unexposed samples were 119 to 127 kg (264 to 282 lb) and 91 to 101 kg (200 to 222 lb). There was no consistent relationship between relative loads at 12.7 mm (½ in.) and 25.4 mm (1 in.), probably because of friction effects of the three load points.

Visual examination revealed no large cracks or other abnormalities. The slight graininess of surface metal in bend areas was equivalent to that noted in unexposed type 304 welded control samples that had been bent under similar conditions. There was no difference in bending characteristics between samples exposed in sodium or helium. Known weld effects in two of the samples had no effect.

Springs — The load and spring rate of exposed and unexposed Inconel X-750 springs was measured. The average reduction in preload of unexposed springs (stored at room temperature) was 16% compared with 26% for those that were exposed. Spring rate increased an average of 3.7% for exposed springs with a light (1 kg or 2.4 lb) and medium (2 kg or 5.3 lb) preload. The increase for unexposed springs was only 0.3%. There was too much scatter in data at the preload level of 4 kg (8.2 lb) to make meaningful comparisons. There was no discernible difference in the behavior of springs exposed in sodium or those sealed in helium. Belleville springs were not examined.

Tube Tests — Demineralized water was used. Tests were conducted with 102 mm (4 in.) lengths of type 304 blanket-rod tubes used as containers for metallurgical samples. Tubes were from two different environments and in two different stress rates: 1. Unstressed tubes with flowing sodium outside and essentially static sodium inside. 2. Nominally stressed tubes with flowing sodium outside and helium inside.

Ultimate tensile strengths and ductilities of exposed tubing at room temperature were compared with those of unexposed blanket-rod tubes from two sources — those used as containers to ship SURV-1 specimens from Idaho to Illinois, and tubing taken directly from ship stock.

Effects of exposure on ultimate strength were clouded by scatter in data for unexposed specimens, but two conclusions were apparent.

Neither strength nor ductility was significantly affected by the 1.2 by 10^{19} n/cm^2 dosage; the length of the longitudinal split and the bulge diameter at splitting (both $\approx$1.5 times the diameter) were essentially the same for all sections tested (29 unexposed and 37 exposed). In addition, there were no observable effects of differences in environment, radiation dose, or thermal gradient along the tube axis.

Labyrinth Section — The lower hard-chromium-plated adapter was sectioned to determine the effects of sodium in the crevices of two grid-plate bearing areas. Plating was still present to a depth of 0.065 mm, and the outer surface was smoother on a microscale. Imperfections in the plating were similar to those on an unexposed specimen. Measurements of the diameters of the labyrinth and guide tube showed increases of less than 0.050 mm in all cases. All changes were in the direction of increased material dimensions; but both parts were easily disassembled

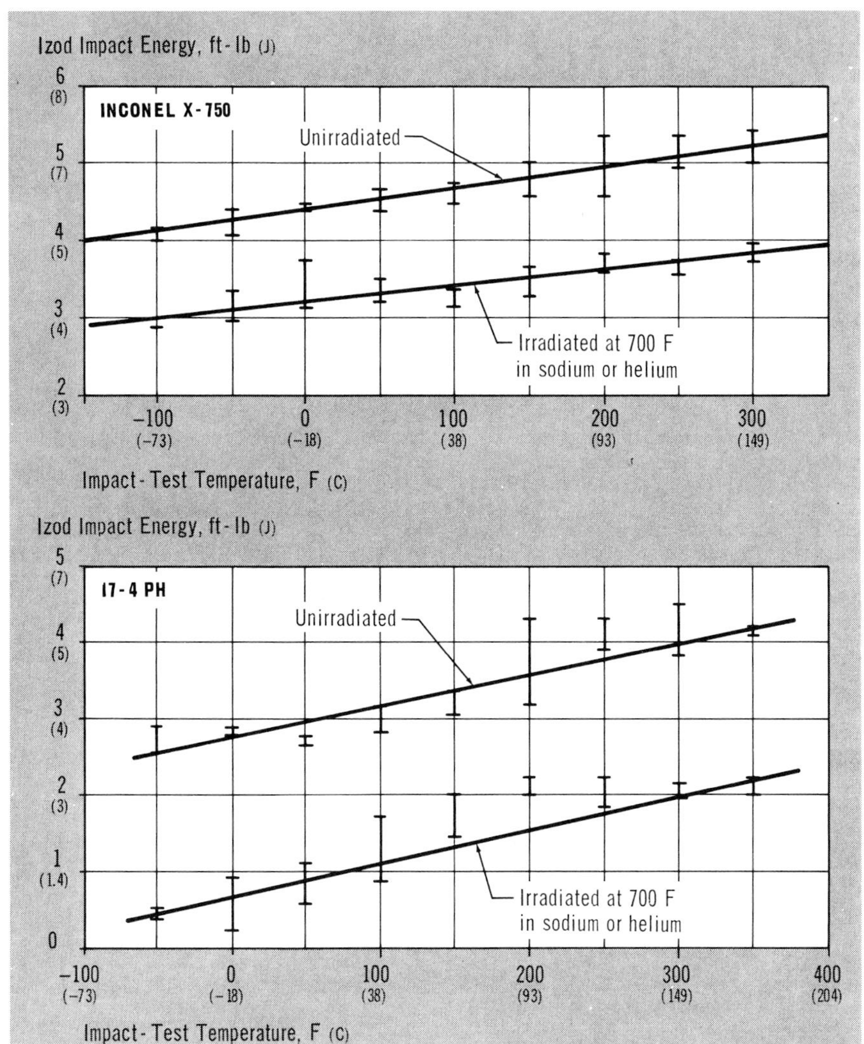

Fig. 3 — Izod impact strength of Inconel X-750 and 17-4 PH stainless steel; data for all irradiated samples, whether exposed to sodium or helium, are plotted on a single line for each material.

Table IV — Effect of EBR-II Exposure on Microstructure

Material	Effect
Ampco Grade 18 aluminum bronze	Dispersion of γ_2-phase during exposure; probably due to peritectoid reaction at exposure temperature.
Stellite 6B	Martensitic transformation.
Inconel X-750	No change. Exposed sample was less sensitive to etchant.
Type 420	Grain-boundary precipitation and precipitate agglomerization.
T-1	No change.
Type 347	No change. Exposed sample was more sensitive to etchant.
Type 416	No change.
Berylco 25 beryllium copper	Irregular surface dissolution. Considerable fine precipitation apparently due to overaging, since time and temperature far exceeded standard aging conditions.
Type 304 with boron	No change.
17-4 PH	No change.
Type 304	No change except for possible transformation at surface.
Tantalum	No change.

Section II: Carbon and Alloy Steels

Low-Carbon Steel Sheet

*By the ASM Committee on Low-Carbon Steel Sheet**

HOT ROLLED AND COLD ROLLED low-carbon steel sheets and strip are used primarily in consumer goods. These applications require materials that are serviceable under a wide variety of conditions, and also adaptable to low-cost techniques of mass production and to consideration of sales appeal in the finished article. These flat rolled products must therefore incorporate, in various degrees and combinations, ease of fabrication, predetermined strength after fabrication, ductility, and attractive appearance before and after fabrication, as well as compatibility with other materials and various coatings and processes. To attain these characteristics under the most economical conditions for production, the bulk of the flat rolled steel is of low carbon content (0.15% max).

In producing sheets, rimmed steel is ordinarily used. Typical ladle analyses will be approximately 0.05 to 0.10% C, 0.25 to 0.50% Mn, 0.04% P max and 0.05% S max. Where only mild forming is involved, a certain latitude is permissible, especially for carbon, phosphorus and sulfur. In sheets for deep drawing, the phosphorus and sulfur in particular are held as low as possible. Where nonaging characteristics are desired, aluminum, vanadium, titanium or boron may be added. Further variations in mechanical properties and finish necessary to fit the requirements are usually secured by processing treatments such as normalizing, open annealing, box annealing, cold rolling, pickling or combinations of these.

Production of Hot Rolled Sheets

Before the development of rolling by the continuous method on wide strip mills, all hot rolled sheets were produced from sheet bars on hand sheet mills. As the wide strip mills were developed, greater tonnages of hot rolled sheets were produced on these new mills, until at the present time only a very small percentage of the hot rolled sheets, called hot rolled annealed sheets, are produced on hand mills.

Hot rolling includes rolling operations in which the metal is heated before it is passed between the rolls. Most continuous hot strip mills are capable of producing hot rolled sheets in thicknesses down to and including 16 gage (0.0598 in.).

Qualities. Hot rolled sheets are suitable where normal surface oxide and minor imperfections are acceptable, and are generally used for parts where finish is of secondary importance. Minimum ductility requires that commercial quality sheet withstand bending flat on itself in any direction at room temperature in a standard bend test. Improved ductility is obtained by specifying drawing quality, which is usually negotiated prior to purchase, with the supplier guaranteeing the material to form an identified part within an established breakage allowance. Additional quality in hot rolled sheets is available at extra cost in the form of special killed, special soundness or special surface. Hot rolled sheets may be normalized or box annealed, and pickled.

Production of Cold Rolled Sheets

Cold rolled sheets are characterized by improved surface finish or special temper and properties. The cold reduction process is used also to make gages lighter than the hot strip mill can produce economically and to produce more uniform thickness and closer width.

Sheets and strip can be either hot rolled directly to near the desired thickness, or hot rolled to an intermediate thickness, pickled, and then cold reduced to the desired final thickness. These are all called cold rolled sheets.

Generally, hot rolled coils are pickled on a continuous pickler and cold reduced through three or four mills in tandem.

The amount of cold reduction is governed largely by the individual application. For example, cold rolled sheets and strip intended for drawing applications are cold reduced from 30 to 70% in thickness. Increasing amounts of cold reduction will make the resulting sheets and strip finer-grained and stiffer, and slightly harder after annealing.

Qualities. Cold rolled sheets are manufactured to as heavy as 11 gage (0.1196 in.), but are most widely available in thicknesses less than 16 gage (0.0598 in.) to about 28 gage (0.0149 in.).

Cold rolled sheets have closer size tolerances and a better finish than a pickled surface on hot rolled steel. The normal cold rolled finish is a suitable base for paints, enamels, lacquers and electroplated coatings. Extra smoothness and freedom from surface imperfections are essential. An improved cold rolled luster finish is available at extra cost for parts that are to be formed and electroplated to a high luster finish such as decorative chromium.

Commercial quality cold rolled sheets are suitable for moderate draws and, if temper rolled, are suitable for exposed parts requiring good surface finish. The hardness is ordinarily less than Rockwell B 60 at the time of shipment. The sheets may be specified to be free from surface disturbances such as fluting or stretcher straining during fabrication if they are roller leveled before fabrication. Cold rolled commercial quality sheet should withstand bending flat on itself in any direction at room temperature by the standard bend test.

Drawing Quality. When greater ductility or more uniform properties are necessary, drawing quality or drawing quality special killed is specified. When drawing quality in cold rolled sheets is specified, the supplier guarantees the steel to fabricate an identified part within an established breakage allowance. The identification of the part is included in the purchase order. Special killed steels are specified if freedom from significant changes in mechanical properties or freedom from stretcher strains without roller leveling prior to use is expected.

Physical quality is applicable when specific mechanical properties are required. Minimum values of tensile strength ranging to 80,000 psi, or intermediate temper ranges with maximum and minimum Rockwell B hardness, are available.

Fully annealed (dead soft) cold rolled steel is the product of the conventional box anneal without temper rolling and represents the softest available grade. It has a greater tendency to show stretcher strains and fluting and is used for unexposed parts where these conditions are not objectionable — for instance, floor pans for automobiles. A temper pass would decrease ductility, particularly if the sheets are not used immediately.

The harder grades of steel, such as stainless, alloy, electrical and higher-carbon compositions, are carefully cold reduced to finished

*Emil C. Blocks, *Chairman,* Research Metallurgist, Ekco Products Co.; Robert B. Chamillard, Chief Tool Engineer, Larson Tool & Stamping Co.; T. P. Hanford, Chief Metallurgist, The Stanley Works; R. H. Heyer, Supervising Research Metallurgist, Armco Steel Corp.; M. F. Kroesch, Chief Product Engineer, Pressed Steel Tank Co.; Charles J. Machala, Engineering Supervisor, Acklin Stamping Div., Tecumseh Products Co.; C. C. Mahlie, Superintendent, Service and Quality Control, Inland Steel Co.; H. Earle Ross, Plant Metallurgist, Geuder, Paeschke & Frey Co.; Fred C. Valentin, Manufacturing Engineer, Worcester Pressed Steel Co.; J. K. Waterman, Republic Steel Corp.

gage in two stages, with a softening anneal between the first and second stages of cold reduction.

Deoxidation Practice

Except for special compositions or special requirements, rimmed steel is used for the majority of sheet and strip applications. Rimmed steel is cheaper because of the higher yield of finished sheets.

In a rimmed steel the dissolved oxygen is allowed to evolve in the form of carbon oxide gases (CO and CO_2) as the liquid metal freezes. The rimming action lowers the average carbon content of the ingot and causes segregation of the remaining carbon and of the sulfur and phosphorus toward the center and top of the ingot during solidification. This action progresses until the ingot is capped or until it freezes over the top naturally.

When properly carried out, rimming produces an ingot with a heavy "rim" that is markedly lower in carbon, sulfur and phosphorus, and a segregate that is much higher in these metalloids. This segregation of metalloids accounts for variations in composition and mechanical properties from sheet to sheet and in different parts of the same sheet. Hence, rimmed steel sheets are not subject to rigorous check analyses, nor are they noted for uniformity in properties.

Killed steels are made by adding a sufficient quantity of a strong deoxidizer, usually aluminum, to the ladle or ingot molds. The deoxidizing materials have a greater affinity for oxygen than carbon and form stable oxides. Little gas is evolved, either by reaction or through coming out of solution; hence, the top of the ingot freezes rapidly and the ingot solidifies as a comparatively homogeneous mass with only slight segregation. The composition of a killed steel ingot is more uniform than a rimmed steel ingot.

The terms "killed" and "special killed" are synonymous in the low-carbon sheet industry, and refer to aluminum-killed steels. The term "special killed" has been adopted for steels of drawing quality to distinguish them from constructional steels, which are usually silicon killed. Aluminum-killed steels are most widely used for cold rolled sheets that will be subjected to severe forming or drawing and also for sheets that will be stored for long periods before use.

A principal advantage of special killed steels is minimum strain aging. Special killed steels are usually cold rolled, although such material is occasionally ordered as hot rolled for a particular application, usually to obtain a fine-grained sheet or to produce uniform properties or to improve drawability.

Another application of killed hot rolled sheet is to achieve superior mechanical properties by retaining a finer grain in welding. Killed hot rolled steel is sometimes preferred for drawing operations that are too severe for rimmed steel.

The characteristic differences in the two major types of steel have definite effects on the sheets and strip produced from them. The rimmed steel, with its thicker skin, results in a sheet comparatively free from surface disturbances. A killed steel generally has a more "sensitive" skin that is more likely to be ruptured in the mills during hot reduction from the ingot to the bar or slab. If a rupture or crack occurs, the "wound" becomes oxidized, and it may not heal as readily as with rimmed steels. This results in surface defects called seams. For this reason, when surface is important, rimmed steels are generally used, although there are exceptions, for thousands of tons of killed steels (where required for other reasons) are successfully used where surface conditions are important.

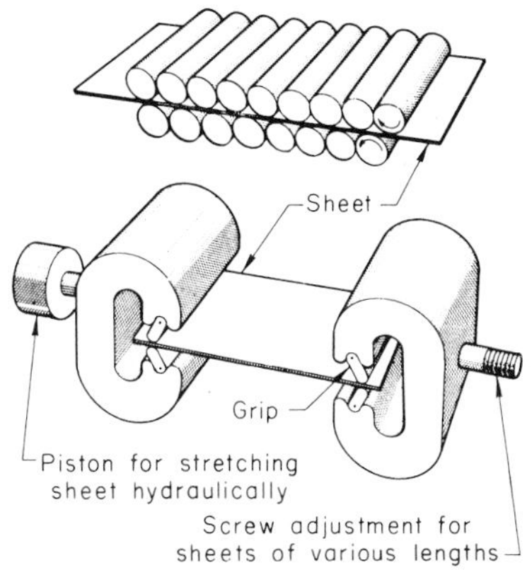

Fig. 1. Diagrammatic sketches of a roller leveler (top) and a stretcher leveler (bottom)

Two modifications of rimmed steel that are finding increasing application are the so-called "chemically capped" and mechanically capped (or "bottle top") steels.

Chemically capped steel is poured into conventional molds, allowed to "rim" for 1 to 3 min and the action halted by the addition of sizable quantities of shot aluminum or ferrosilicon to the top. This reduces surface action and promotes rapid solidification of the top.

Bottle top steel is poured into a specially designed mold, with a relatively small opening in the top. The metal is deoxidized in such a manner that it becomes a "rising" steel. A heavy cast iron cap is used to close the top opening as soon as teeming is completed. The metal then rises until it strikes the cap when gas evolution is stopped by the pressure buildup.

Both chemically and mechanically capped types have less segregation than normal rimmed steel but have the surface benefits of the low-metalloid skin.

Specifications sometimes require special mechanical properties or check analyses. Silicon-killed steels are often specified or required because the composition of the sheets will be more uniform and will correspond closely to the ladle analysis.

Although, with present methods of control, greater uniformity from heat to heat is now possible, in rimmed steels there still remains the variation within the ingot (top to bottom as well as rim to core) and the variation from ingot to ingot within a heat. (These variations are summarized numerically in the section headed Mechanical Properties in this article.) Differences may be lessened somewhat, if desired, by diverting the top portion or the bottom portion of the ingot to other work. However, such diversion is not economical, and it is more practical to use a capped steel to minimize segregation.

Perhaps the most objectionable characteristic of rimmed steel in cold reduced sheets is the aging that occurs after the final temper rolling. This is manifested by the return of the tendency to stretcher strain, by an appreciable increase in hardness and yield strength, and by a decrease in ductility. Aging begins after the final cold or temper rolling but it is usually not serious until after two or three weeks. The extent to which it affects performance depends largely on the properties of the steel at the time of temper rolling and the severity of the application: for example, a 20-gage sheet with a hardness of Rockwell B 45 and an Olsen cup value of 0.400 in. may age harden to Rockwell B 48 and an Olsen value of 0.375 in.

These matters are dealt with in two other articles in this Handbook — "Selection of Sheet Steel for Formability" and "Selection of Sheet Steel for Deep Drawing".

Annealing and Normalizing

Cold reduced sheets and strip must be annealed unless a hard temper is desired. This may not be necessary on some hot rolled sheets and strip finished at 1500 to 1600 F and coiled while hot enough to obtain some annealing in the coil. If the sheets are reduced to gage below the recrystallization temperature, the grain structure is fragmented and strain hardening results.

There are a number of softening operations used in the production of sheet and strip; the most important are: (*a*) cold strain, low-temperature recrystallization annealing, (*b*) normalizing and (*c*) low-temperature stress-relief annealing.

The cold strain, low-temperature recrystallization anneal may be used on material cold reduced more than 30% and on hot rolled material finished at a temperature low enough to give strain equivalent to that of cold rolled material. This type of annealing, also known as "process annealing" or "box annealing", is carried out by placing coils, or stacks of sheets, on a bottom plate and then enclosing them with a cover within which a protective gas atmosphere is maintained. An outer cover consisting of a bell-type heating furnace is then placed over the atmosphere container.

The charge may vary from 5 to 200 tons or more, depending on the size of the sheets and the height of the stack, or the number of coil bases under one cover. After heating to the predetermined temperature of approximately 1100 to 1400 F, the charge is allowed to soak until the temperature is uniform throughout, after which the heating furnace is removed and the charge is allowed to cool in the protective atmosphere before being uncovered.

At the present time (1960) more cold rolled steel is batch annealed in coils under a protective atmosphere than by any other method. However, since about 1949, an increasing tonnage of sheet steel has been annealed or normalized by the continuous method, which provides a fully recrystallized steel that is high in ductility and, if normalized, is nearly free from directionality.

Normalizing. The normalizing operation consists of heating the sheet or strip to a temperature above the Ac_3 point (about 1650 F for a steel that contains 0.15% C) and cooling to room temperature. This treatment recrystallizes and refines the grain structure by phase transformation. The normalizing operation is utilized on hot reduced material which, if box annealed, might show critical grain growth resulting from a low residual strain in the hot reduced product. Normalizing may also be utilized for both hot and cold rolled sheets that will be porcelain enameled. For a given chemical composition, normalizing does not soften the material as much as annealing because of the faster cooling rate. However, it does produce steel with nonscalloping properties and a uniform grain size.

Low-Temperature Stress Relief. If softness equivalent to batch annealing is required, it is necessary to anneal at low temperature in coils or stacks, usually at 1100 to 1200 F in a controlled atmosphere.

Annealing and other processing operations are not well standardized throughout the industry and availability of various grades depends on mill facilities.

Strain Aging

Strain aging of rimmed and capped steels after temper rolling is an important factor influencing the drawing quality of sheet and strip. Sheets in the box-annealed or dead soft condition maintain their as-annealed properties indefinitely. Cold working after annealing, which is done principally by temper rolling or skin rolling, changes the yielding characteristics of the sheet.

The sharp yield point, with its yield point elongation and the tendency for the sheet to stretcher strain and flute, disappears with small amounts of cold work performed on a temper mill. This small amount of cold work also initiates a phenomenon known as strain aging. That is, with time, after temper rolling, the yield point elongation and the tendency to stretcher strain and flute return; and as aging progresses, the hardness and yield strength increase and ductility decreases.

Depending on the degree of cold work, after complete aging the yield point increases, the ductility as measured by elongation decreases, and the Rockwell hardness slightly increases. Yield point elongation, because of its importance in measuring the tendency to stretcher strain and flute, has been used as a principal criterion for strain aging.

Proper roller leveling after aging will eliminate the tendency of the sheets to stretcher strain and to flute if used immediately after roller leveling but such leveling does not restore softness and ductility. In fact, roller leveling after aging further work hardens the sheet and hence further reduces ductility.

Temper rolling (also called "temper passing" or "skin passing") is the slight amount (usually ½ to 1½%) of cold rolling applied to sheets or strip at the mill, to forestall stretcher strains.

In rimmed steel, stretcher strains are almost certain to return to some degree in a week or less. Steel sheets for exposed parts are temper rolled to an extent dictated by the severity of the draw and the requirements for a smooth surface free from strain markings. When the latter requirement is unimportant, the steel may be specified dead soft.

Flattening. Sheets are ordinarily sold to two standards of flatness: (*a*) commercial flatness, in which the sheets are to be severely drawn, so that the original flatness does not have much effect on the shape of the drawn article, and (*b*) the stretcher-level standard of flatness, in which the sheets are used for panel work, where little forming is done on the sheet and the finished product is desired flat and free from waves and "oil can".

Simply because sheets are flat before moderate forming does not necessarily mean that they will be flat after fabrication, unless the fabricator takes care that no buckling occurs in the sheets during forming. Commercial flatness can ordinarily be produced by roller leveling or by temper rolling and roller leveling, but where very flat sheets are required, producers may resort to stretcher leveling or other processes to produce this standard of flatness.

The permissible variations for flatness of hot and cold rolled sheets have been established by the Technical Committee of the American Iron and Steel Institute, and are stated in the AISI Steel Products Manual.

Roller Leveling. A roller-leveling machine may consist of two sets of several horizontal rolls of small diameter, held in a housing and arranged so that the top and bottom rows are offset, permitting the rollers to mesh (Fig. 1). Consequently the sheet passing through the leveler is flexed alternately up and down. Near the exit end the rollers are not meshed as deeply as at the entry end, so the action is chiefly that of straightening.

Stretcher strains can be eliminated from temper-passed rimmed steel and from insufficiently temper-passed killed steel by roller leveling in a machine capable of flexing the sheet in bending enough to remove the sharp yield point and the yield point elongation.

The performance of some annealed steel used for very difficult unexposed parts may be improved by a single pass through the roller leveler.

Dead soft annealed sheets are not roller leveled for an exposed part because the rolls kink the sheet severely. The deformed areas or kinks will not deform further on stretching and will appear as raised welts after forming.

Stretcher Leveling. Stretcher leveling is a more positive means of producing flatness by stretching the sheet lengthwise between jaws (Fig. 1). The stretching may vary from about 1 to 3% elongation, which exceeds the elastic limit of the steel and hence results in some permanent elongation. The sheet must be of a killed or a capped steel having nearly uniform properties, so that it will spring back uniformly across its

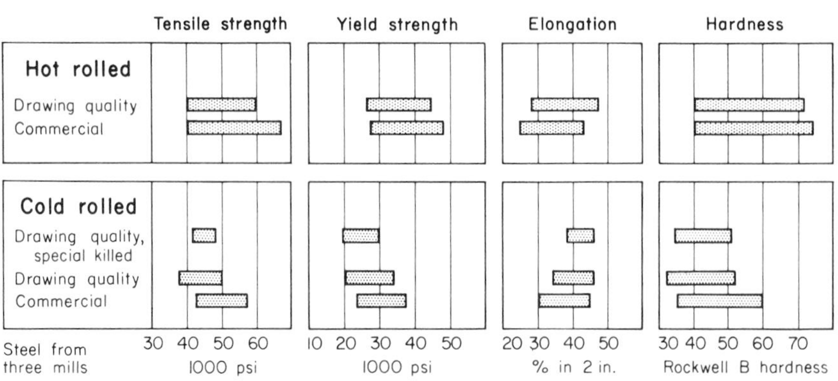

Fig. 2. Typical range of mechanical properties of low-carbon steel furnished by three mills. Hot rolled sheet thicknesses from 0.0598 to 0.1345 in. (16 to 10 gage, inclusive); cold rolled from 0.029 to 0.0598 in. (22 to 16 gage, inclusive). All cold rolled grades include a temper pass. All grades were rolled from rimmed steel except the one labeled special killed.

full width and remain flat. It is also necessary to use killed or capped steel having nearly uniform properties so that, after stretching, strain markings do not develop.

Other types of equipment suitable for producing the stretcher-level standard of flatness are also being used by some manufacturers.

Surface Characteristics

The surface texture of sheets and strip can be varied between rather wide limits. For chromium plating and similar finishes, a smooth, bright surface is necessary, but for porcelain enameling and many drawing operations, a certain dullness or roughness of surface is more suitable. In porcelain enameling, roughness tends to improve the uniformity of coating, and in certain drawing operations where heavy pressures are developed, the duller type of surface retains more lubricant, helping the sheet to draw better. This type of surface also seems to aid in distribution of the draw.

Minor surface imperfections and slight strains are less noticeable on a dull surface than on a bright one. However, the surface should not be so dull that subsequent paint finishes will not cover. A smooth or bright sheet surface may be obtained by grinding and polishing the rolls, and the dull surface may be obtained by either etching or grit blasting the rolls.

Defects of some types do occur on strip and sheet made by the continuous process. The following is a partial list of the imperfections that may be found during inspection. Some of these defects warrant rejection of the steel, while others cause the material to be allocated into a lower class, depending on the severity of the defect and the end

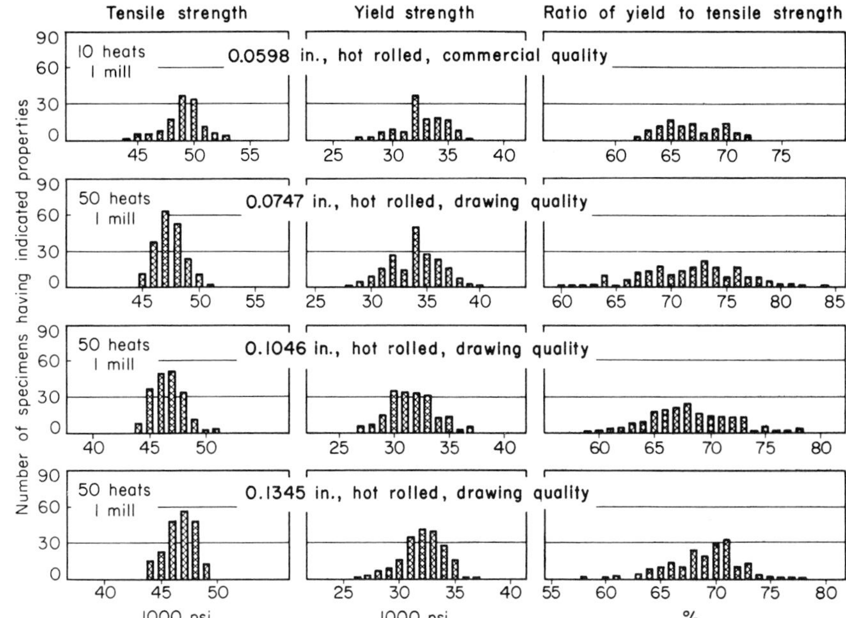

Fig. 3. Distribution of mechanical properties of hot rolled low-carbon steel sheet.

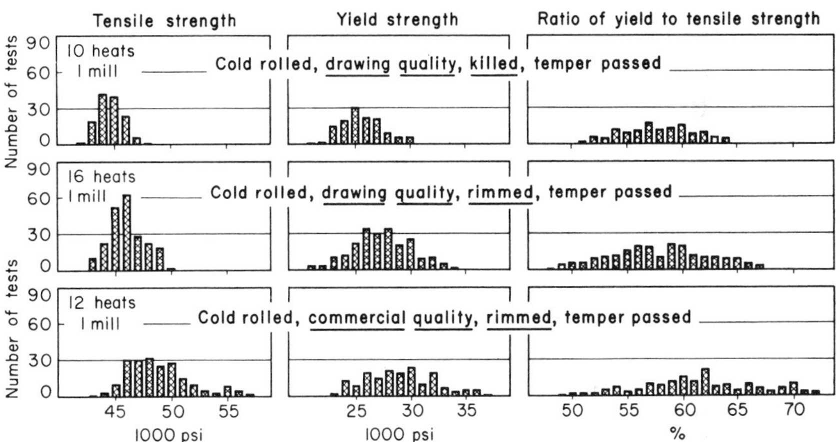

Fig. 4. Distribution of mechanical properties for three classes

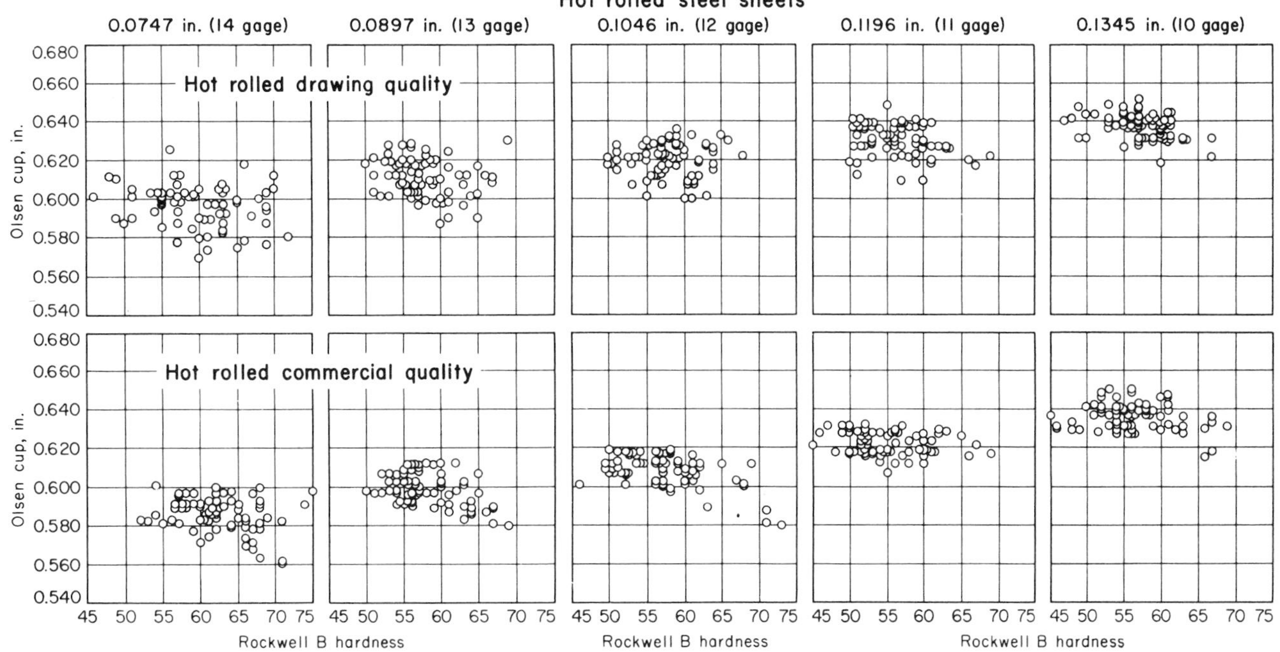

Fig. 5. Range of scatter in Olsen cup values for two qualities of hot rolled low-carbon steel sheets

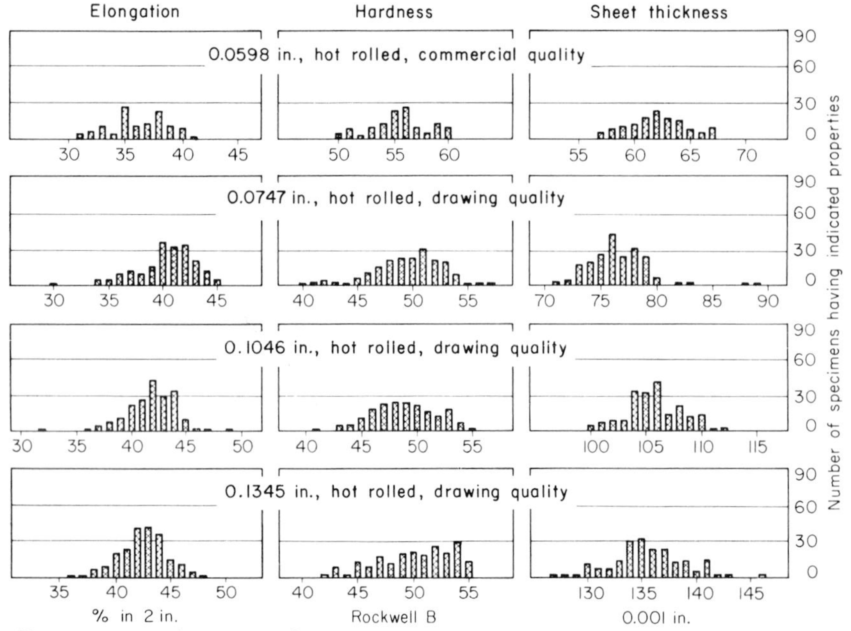

Top row, range from one mill; lower rows, ranges from another mill.

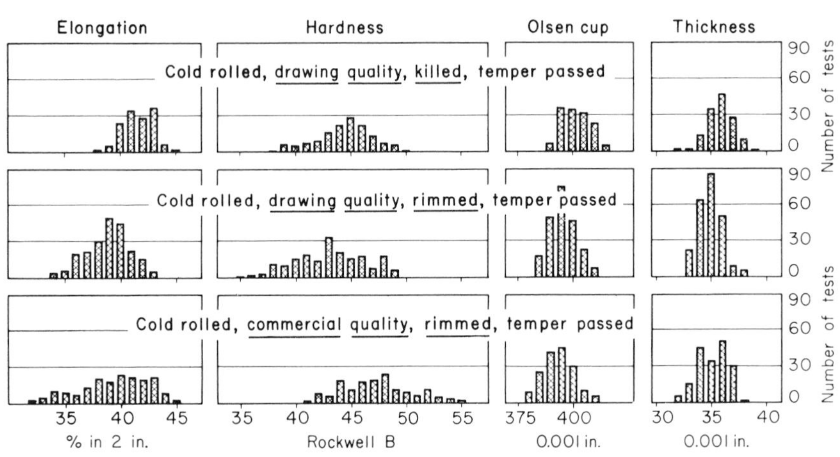

of 20-gage (0.0359 in.) cold rolled sheet furnished by one mill.

use. This list is from "Classification of Major Visible Imperfections in Sheet Steel", Spec Publ 145, SAE.

Buckles	Pipes
Coil breaks	Ragged edge
Coil weld	Rolled-in-dirt
Floppers	Rolled-in-metal
Friction gouges or scratches	Rolled-in-scale
Healed-over scratch	Rust
Large pits and holes	Skin laminations
Orange peel	Slitter-damaged edge
Oxidized surface	Slivers
Pickle stain	Small pits
Pinchers	Sticker breaks
	Stretcher strains
	Torn edge

Mechanical Properties

The mechanical properties of low-carbon steel sheets are not readily related to their performance in fabrication, and are not ordinarily used in specifications unless special strength properties are required in the fabricated product. As a matter of general interest, however, the ranges of mechanical properties typical of sheets produced by three mills in five qualities are given in Fig. 2. The bands would be wider if the product of the entire industry were represented.

It will be noted that the ranges are broader and the sheets harder for the hot rolled than for the cold rolled grades, and that cold rolled, drawing quality, special killed sheets are produced to a narrower range of mechanical properties than cold rolled drawing quality, a rimmed steel grade. There is a great deal of overlapping in properties between commercial quality and drawing quality sheets.

As examples of the distribution of properties within single grades, data for hot rolled commercial quality and hot rolled drawing quality sheets from one mill are given in Fig. 3. For each gage both longitudi-

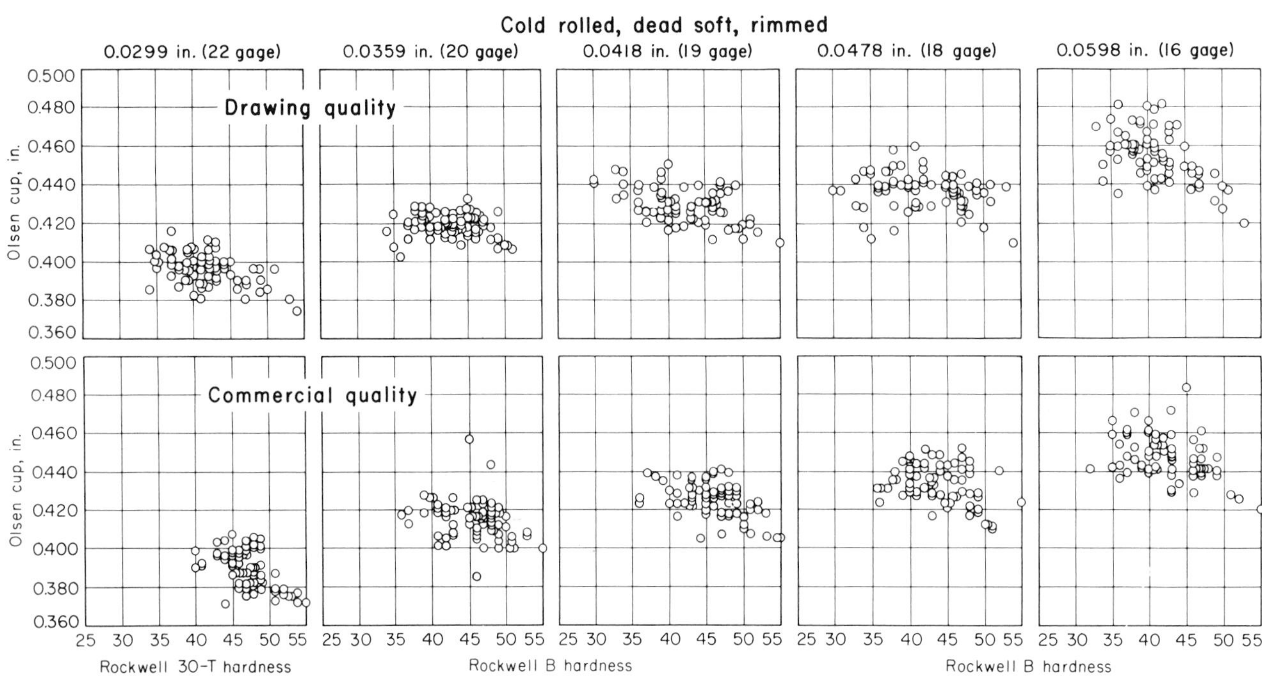

Fig. 6. Range of scatter in Olsen cup values for dead soft cold rolled low-carbon steel sheets

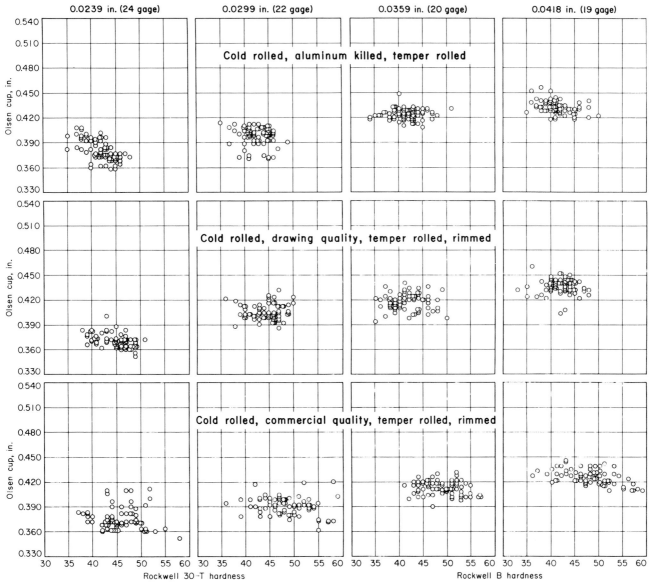

Fig. 7. Range of scatter in Olsen cup values for

nal and transverse tension test results are included.

Figure 4 shows similar distributions of mechanical properties and sheet thickness variation for three classes of 20-gage cold rolled low-carbon steel. These data compare two rimmed and one special killed grade. Among the rimmed steels, drawing quality has slightly lower strength and hardness and increased Olsen cup height, compared with commercial quality. The special killed, drawing quality material shows further improvement over the commercial and drawing quality grades, with lower yield and tensile strengths. Formability increases in the following order: commercial rimmed, drawing quality rimmed, and drawing quality special killed.

In comparing 20-gage cold rolled *commercial* quality, rimmed and temper passed, with 20-gage cold

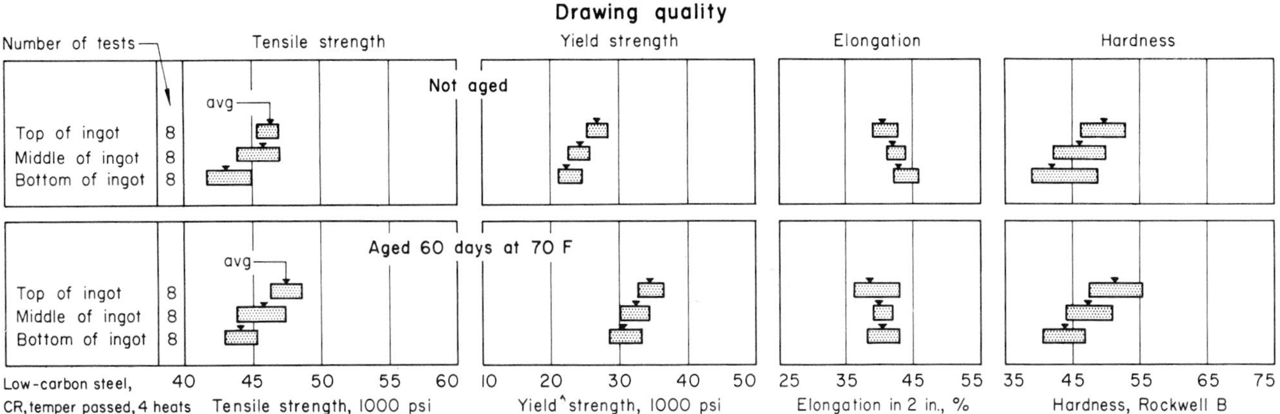

Fig. 8. Effect of aging on two classes of cold rolled low-carbon rimmed steel. The samples not aged were tested immediately after the temper pass. The eight tests that make up one horizontal bar on a chart, each showing the range of a mechanical

0.0478 in. (18 gage) | 0.0598 in. (16 gage) | 0.0673 in. (15 gage) | 0.0747 in. (14 gage)

Cold rolled, aluminum killed, temper rolled

Cold rolled, drawing quality, temper rolled, rimmed

Cold rolled, commercial quality, temper rolled, rimmed

Olsen cup, in.: 0.330, 0.360, 0.390, 0.420, 0.450, 0.480, 0.510, 0.540

Rockwell B hardness: 30, 35, 40, 45, 50, 55, 60

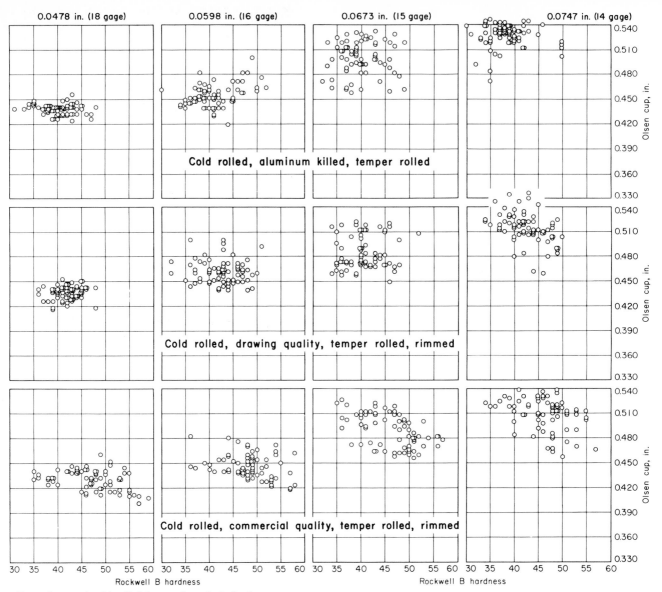

three classes of cold rolled low-carbon steel sheets

rolled *drawing* quality, rimmed and temper passed, a considerable overlapping of mechanical properties is evident. However, the lower hardness, yield strength and tensile strength, together with the greater elongation and Olsen ductility, indicate superior drawing properties in the drawing quality material. Also, greater uniformity of properties can be expected in the drawing quality steel.

These features are of particular importance in the selection of steel for drawn parts. Users frequently secure one or several small lots of commercial quality and have reasonably good results on the press, but eventually the nonuniformity of the commercial quality steel is likely to cause poorer die performance, particularly on the more severe drawing jobs.

In comparing 20-gage cold rolled

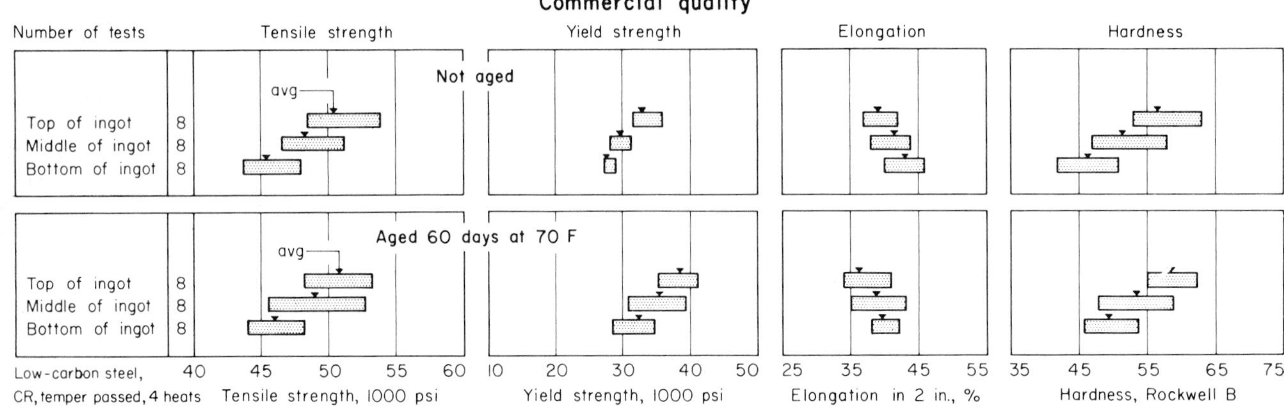

property, were made on duplicate specimens from four heats. The sheets were temper passed and cooled prior to sampling. The test coils were identified in relation to the top, middle or bottom of the ingot. Extreme edges of the sheets were not sampled.

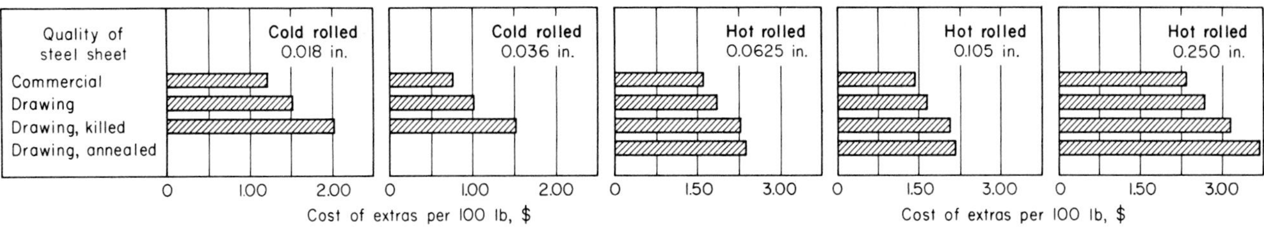

Fig. 9. Relative magnitudes of typical cost extras for mill lots

drawing quality, *rimmed* and temper passed, with 20-gage cold rolled drawing quality, *killed* and temper passed, the lower yield and tensile strengths, together with the higher percentage elongation and greater Olsen ductility, indicate still further improvement in drawing properties of the aluminum-killed steel over rimmed steel.

Olsen Cup Value. Figures 5, 6 and 7 show the relations among hardness, Olsen cup value and sheet thickness. Formability increases as sheet thickness increases. Olsen tests for materials thicker than 0.062 in. require special balls and rings or dies. Test results apply only to the specific materials tested because they are special. However, they serve as an indication of the formability of the specific material.

Formability relationships of sheets thicker than 0.062 in. are shown in the distribution charts, Fig. 5 and 7. It will be noted that both the drawing quality and the commercial quality hot rolled steels show slightly better formability properties than cold rolled steels of similar thickness (0.0747 in.). Assuming that the proper Olsen ball and die ring were used for testing sheet thicker than 0.062 in., it is indicated that the slightly harder hot rolled sheets possessed better formability properties than the cold rolled sheets. Some fabricators compare Olsen cup values and reduction of area values for sheets thicker than 0.062 in. Hot rolled sheet has a higher reduction of area than cold rolled.

In drawing thicker material (over 0.062 in.) some plants have used one-half the reduction of area as a guide for initial reduction in drawing the parts. In most instances cold rolled material is better for deep drawing than hot rolled. However, there are notable exceptions; for instance, in severe draws the lubricant often does not adhere as well to the cold rolled surfaces as to the hot rolled material, and the hot rolled material may be superior for this reason.

Effect of aging on the formability of cold rolled rimmed steels is variable, and may be unpredictable on the basis of tests. One rimmed steel may not age appreciably; another may make the most difficult draws as received and yet not make minimum draws after 30 days aging.

Figure 8 illustrates the effects on mechanical properties from aging commercial and drawing qualities of rimmed steel. Significant aspects are the trend toward increased yield strength and hardness in the aged material and the narrower ranges of the drawing quality grade. After blanking, forming or finishing operations, strain aging is more pronounced than for unworked steel. It is therefore advisable to complete the sequence of operations without

Table 1. Range of Typical Costs for Low-Carbon Steel Sheet

Cost item	Cold rolled 26 gage (0.018 in.)	Cold rolled 20 gage (0.036 in.)	Cold rolled 16 gage (0.0625 in.)	Hot rolled 16 gage (0.0625 in.)	Hot rolled 12 gage (0.105 in.)	Hot rolled Plate (0.250 in.)
Base price, $ per 100 lb	6.275	6.275	6.275	5.10	5.10	5.30
Extras for Quality Class						
Drawing quality	0.25	0.25	0.25	0.25	0.25	0.35
Drawing quality, killed	0.75	0.75	0.75	0.65	0.65	0.80
Drawing quality, annealed	....	....	....	0.75	0.75	1.35
Extras, Same for All Quality Classes						
Gage and width(a)	0.85	0.45	0.45	0.80	0.80	1.30
Length(a)	0.40	0.30	0.20	0.45	0.35	0.05
Pickling and oiling	....	....	....	0.55	0.45	0.90
Packaging	0.025	0.025	0.025	0.025	0.025	0.10

NOTE: Quantity extras are not included and do not apply to mill shipments greater than 40,000 lb. Cost of freight, shearing to specified size, and other extras are not included. Costs are typical as of September 1958.

(a) Mandatory extras, all other optional; width extras for sheet 36 to 48 in. wide; length extras for cold rolled sheet 60 to 180 in. long and hot rolled sheet 60 to 240 in. long.

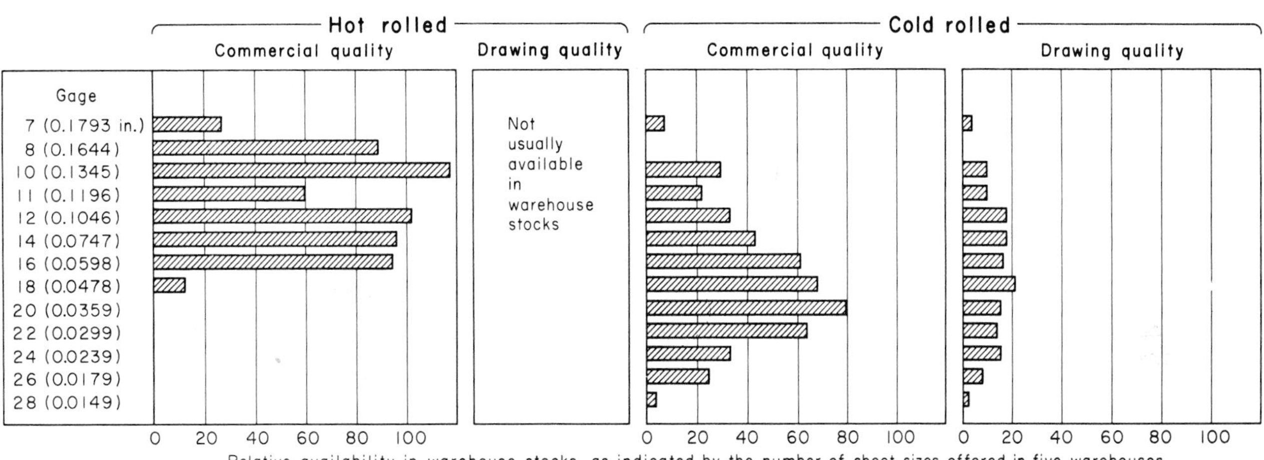

Fig. 10. Typical availability of grades and sizes of low-carbon steel sheet from warehouse stocks. Composite survey of five midwestern cities. A few odd-numbered gages have been added to the nearest even-numbered gage.

intervening storage unless artificial aging tests indicate the absence of aging for the particular steel.

Efficient production requires killed steels when the material must be stored for any length of time, when production is intermittent rather than continuous, or when heat is applied before drawing, as when baked finishes are applied to the flat steel. This is especially important in a transfer press where severe initial draws induce high frictional temperatures, or where the work is allowed to cool to room temperature before further reduction. The higher temperature would cause immediate aging in a rimmed steel. Killed steel is preferred also for progressive die operations with a particularly severe initial draw.

Cost and Availability

Hot rolled steel is largely limited to thicknesses greater than 16 gage (0.0598 in.). It is difficult to control the dimensions of thinner gages during hot rolling; consequently, the thickness cost extra increases as sheet thickness decreases.

Typical mill costs for various qualities of hot and cold rolled sheets are summarized in Fig. 9. Details of cost extras are given in Table 1.

Relative availability from warehouses is summarized in Fig. 10, which gives an approximate indication of relative usage of the several grades and sizes by a large variety of users who purchase sheet in less than mill lots.

Hardenable Carbon Steels

*By the ASM Committee on Carbon Steel Bars and Forgings**

CARBON STEELS are produced in greater tonnage and have wider use than any other metal because of their versatility and low cost.

For about a decade before 1940 there was a trend toward the greater use of alloy steels to replace carbon steels. The scarcity of alloying elements that began about 1940 caused a reappraisal of carbon steels, and in many instances users reverted to the carbon grades. There were several reasons why carbon steels proved satisfactory on reappraisal: (*a*) their hardenability, though less than that of alloy steels, was adequate for many parts, and for some parts shallower hardening was actually an advantage because of minimized quench cracking; (*b*) refinements in heat treating methods, such as induction hardening, flame hardening, and "shell quenching", made it possible to obtain higher properties from carbon steels than previously; and (*c*) new compositions were added to the carbon steel group, permitting more discriminating selection.

There are now almost 50 grades available in the nonresulfurized series 1000 carbon steels and nearly 30 grades in the resulfurized series 1100 and 1200. In 1935 there were less than one third as many grades of carbon steels. The versatility of the carbon steel group has also been extended by availability of the various grades with lead additions.

Carbon steels can be divided into three arbitrary classifications based on carbon content.

Steels with 0.10 to 0.25% C

Three principal types of heat treatment are used for this group of steels: (*a*) conditioning treatments, such as process annealing, that prepare the steel for certain fabricating operations, (*b*) case hardening treatments, and (*c*) quenching and tempering to improve mechanical properties. The improvement in mechanical properties that can be gained by straight quenching and tempering of the low-carbon steels is usually not worth the cost.

An example of process annealing is in the treatment of low-carbon cold headed bolts made from cold drawn wire. Sometimes the strains introduced by cold working weaken the heads so much that they break through the most severely worked portion under slight additional strain. Process annealing overcomes this condition. Since the temperatures used are close to the lower transformation temperature, this treatment results in considerable reduction of the normal mechanical properties of the shank of the cold headed bolt.

A more suitable treatment is stress relieving at about 1000 F, in order to retain much of the strength acquired in cold working and to provide ample toughness. A common practice is to combine a stress-relieving treatment with a quench from the upper transformation

*RAYMOND H. HAYS, *Chairman,* Metallurgist, Caterpillar Tractor Co.; RAYMOND BLOM, Chief Metallurgist, Ladish Co.; P. W. CARBAUGH, Chief Metallurgist, Oliver Corp.; CHARLES R. FUNK, Chief Metallurgist, Eastern Div., Colorado Fuel & Iron Corp.; H. L. HOPKINS, Chief Engineer and Assistant Works Manager, National Screw & Mfg. Co.; WENDELL K. HUNT, Chief Metallurgist, John Deere Harvester Works; L. R. SCHMIDT (deceased); W. SIMON, Materials Engineering Dept., Westinghouse Electric Corp.; J. W. SUTLIFF, Manager, Chemistry and Metallurgy Laboratory, Clarkwood Road Plant, Thompson Ramo Wooldridge, Inc.; H. F. WALKER, Metallurgist, Tonawanda Forge and Foundry, Chevrolet Motor Div., General Motors Corp.

temperature, or slightly above, producing mechanical properties that approach those of cold drawn stock. A common quenching medium is a water solution of soluble oil, the use of which produces two desirable results: (*a*) the surface of the parts acquires a pleasing black color accepted as a commercial finish, and (*b*) the speed of the quench is slowed to the point where fully quenched hardness is not produced, so it is not necessary to temper the parts.

Heat treatments are frequently employed to improve machinability. The generally poor machinability of the low-carbon steels, except those containing sulfur or other special alloying elements, results principally from the fact that the proportion of free ferrite to carbide is high. This situation cannot be changed fundamentally, but the machinability can be improved by putting the carbide in its most voluminous form, pearlite, and dispersing this pearlite evenly throughout the ferrite mass. Normalizing is commonly used with success, but best results are obtained by quenching the steel in oil from 1500 to 1600 F. With the exception of steels 1024 and 1025, no martensite is formed, and the parts do not require tempering.

Steels with 0.25 to 0.55% C

Because of their higher carbon content, these steels are usually used in the hardened and tempered condition. By selection of quenching medium and tempering temperature a wide range of mechanical properties can be produced. They are the most versatile of the three groups of carbon steels and are most commonly used for crankshafts, couplings, tie rods and many other machinery parts where the required hardness values are within the range from 229 to 447 Bhn.

In this group of steels there is a continuous change from water-hardening to oil-hardening types; the hardenability is very sensitive to changes in chemical composition, particularly to the content of manganese, silicon and residual elements, and to grain size; the steels are sensitive to section changes.

The rate of heating parts for quenching has a marked effect on hardenability under certain conditions. If the structure is nonuniform, as a result of severe banding or lack of proper normalizing or annealing, extremely rapid heating, such as may be obtained in liquid baths, will not allow sufficient time for diffusion of carbon and other elements in the austenite. As a result, nonuniform or low hardness will be produced unless the duration of heating is extended. In heating the steels that contain free carbide (for example, spheroidized material), sufficient time must be allowed for the solution of the carbides; otherwise the austenite at the time of quenching will have a lower carbon content than is represented by the chemical composition of the steel, and disappointing results may be obtained. This condition may be produced deliberately (see discussion of higher-carbon steels, below, for an example).

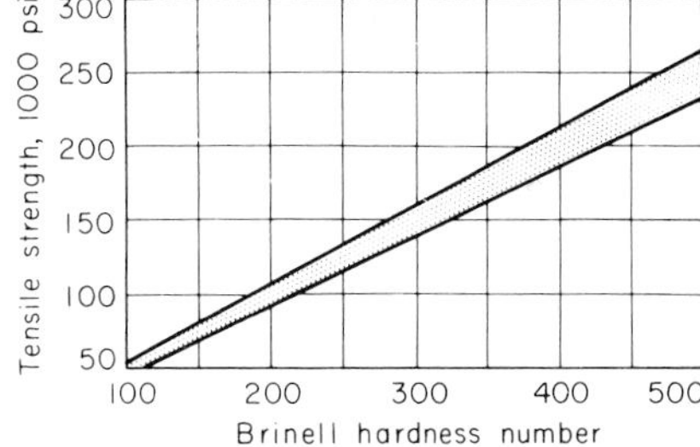

Fig. 1. Relation of tensile strength and hardness for hardened and tempered, as-rolled, annealed, and normalized steels (SAE Handbook)

These medium-carbon steels should usually be either normalized or annealed before hardening, in order to obtain the best mechanical properties after hardening and tempering. Parts made from bar stock are frequently given no treatment prior to hardening, but it is common practice to normalize or anneal forgings.

Most bar stock, both hot finished and cold finished, is machined as received, except the higher-carbon grades and small sizes, which require annealing to reduce the as-received hardness. Forgings are usually normalized, since this treatment avoids the extreme softening and consequent reduction of machinability that result from annealing.

In some instances a "cycle treatment" is used. In this practice the parts are heated as for normalizing, and are then cooled rapidly in the furnace to a temperature somewhat above the nose of the S-curve — that is, within the transformation range that produces pearlite. Then the parts are held at temperature or cooled slowly until the desired amount of transformation has taken place; thereafter they are cooled in any convenient manner. Specially arranged furnaces are usually required. Details of the treatments vary widely and are frequently determined by the furnace equipment available.

Cold headed products are commonly made from these steels, especially from the ones containing less than 0.40% C. Process treating before cold working is usually necessary because the higher carbon decreases the workability. For certain uses, these steels are normalized or annealed above the upper transformation temperature, but more frequently a spheroidizing treatment is used. The degree of spheroidization required depends on the application. After shaping operations are finished, the parts are heat treated by quenching and tempering.

These medium-carbon steels are widely used for machinery parts for moderate duty. When such parts are to be machined after heat treatment, the maximum hardness is usually held to 321 Bhn, and is frequently much lower.

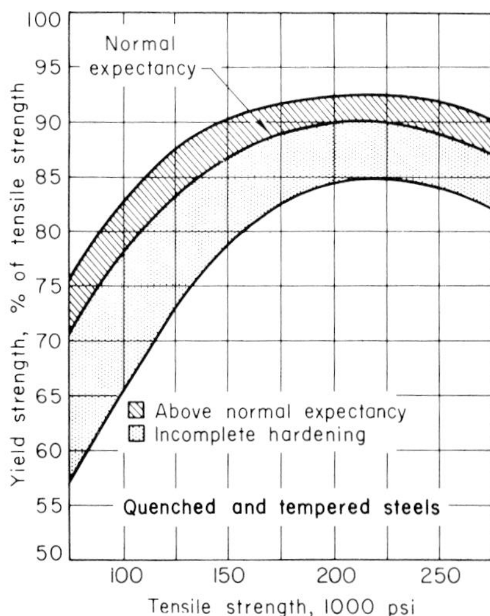

Fig. 2. Relation of tensile strength and yield strength for quenched and tempered steels (SAE Handbook)

Water is the quenching medium most commonly used because it is the cheapest and easiest to install. Caustic soda solution (5 to 10% NaOH by weight) is used in many instances with improved results. Compared with water, it is a faster and, therefore, a more thorough and more uniform quench, producing better mechanical properties in all but light sections. Because of its rapid action, more scale is removed from the parts.

When used hot (130 to 160 F), caustic soda solution frequently makes possible drastic quenching of parts that could not be water quenched without breakage. The disadvantages of caustic soda are that it can be used only in a closed system in which provisions are made for cooling; operators must be protected against contact with it; the solution must be checked constantly and the proper concentration maintained; and woolen clothing disintegrates rapidly after contact with it. Cotton clothing is not affected.

Salt solutions (brine) are often successfully used. Up to 100 F they produce almost as good results as caustic, but are much less effective when hot. Like caustic, they require a closed system. Salt solutions are not dangerous to operators but their corrosive action on iron or steel parts of equipment is very serious.

When the section is light or the properties required after heat treatment are not high, oil quenching is often used. This nearly always eliminates the breakage problem and is very effective in reducing distortion.

A wide range in austenitizing temperatures is made necessary in order to meet required conditions. Lower temperatures should be used for the higher-manganese steels, light sections, coarse-grained material and water quenching; higher temperatures are required for lower manganese, heavy sections, fine grain and oil quenching.

From these steels are made many

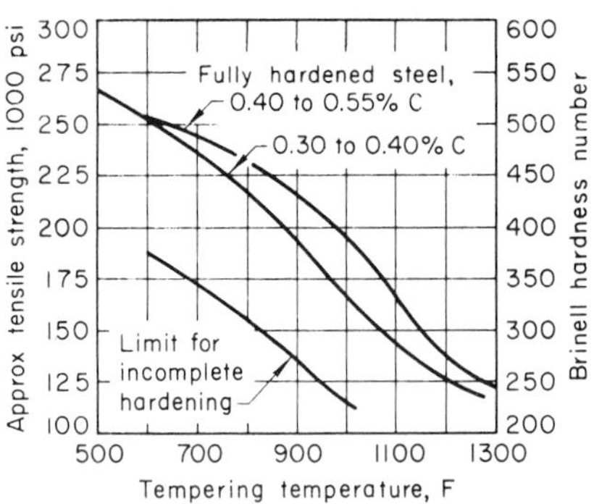

Fig. 3. Effect of tempering temperature on the tensile strength and hardness of carbon and alloy steels with 0.30 to 0.55% C (SAE Handbook)

common hand tools, such as pliers, open-end wrenches, screw drivers, and a few edged tools — for example, tin snips and brush knives. The cutting tools are necessarily quenched locally on the cutting edges, in water, brine or caustic, and are subsequently given suitable tempering treatments. In some instances the edge is time quenched; then the remainder of the tool is oil quenched for partial strengthening. When made of these grades of steel, pliers, wrenches and screw drivers are usually quenched in water, either locally or completely, and are then suitably tempered.

Steels with 0.55 to 1.00% C

Carbon steels with these higher carbon contents are more restricted in application than the 0.25 to 0.55% C steels since they are more costly to fabricate, because of decreased machinability, poor formability and poor weldability. They are also more brittle in the heat treated condition. Higher-carbon steels such as 1070 to 1095 are especially suitable for springs where resistance to fatigue and permanent set are required (see the article on Selection of Steel for Springs). They are also used in the nearly fully hardened condition (Rockwell C 55 and higher) for applications where abrasion resistance is the primary requirement, as for agricultural tillage tools such as plowshares, and knives for cutting hay or grain.

Forged parts should be annealed because refinement of the forging structure is important in producing a high-quality hardened product, and because the parts come from the hammer too hard for cold trimming of the flash or for economical machining. Ordinary annealing practice, followed by furnace cooling to 1100 F, is satisfactory for most parts.

Most of the parts made from steels in this group are hardened by conventional quenching. However, special technique is necessary at times. Both oil and water quenching are used — water, for heavy sections of the lower-carbon steels and for cutting edges; oil, for general use. Austempering and martempering are often successfully applied; the principal advantages from such treatments are considerably reduced distortion, elimination of breakage, in many instances, and greater toughness at high hardness.

For heavy machinery parts, such as shafts, collars and the like, steels 1055 and 1061 may be used, either normalized and tempered for low strength, or quenched and tempered for moderate strength. Other steels in the list may be used, but the combination of carbon and manganese in the two mentioned makes them particularly well adapted for such applications.

It must be remembered that even with all hardenability factors favorable, including the use of a drastic quench, these steels are essentially shallow hardening, as compared with alloy steels, because carbon alone, or in combination with manganese in the amounts involved here, does not promote deep hardening to any significant extent. Therefore, the sections for which such steels are suited will be definitely limited. In spite of this limitation the danger of breakage is real and must be carefully guarded against when such parts are being treated, especially whenever changes in section are involved.

Hand tools made from steels in this group include open-end wrenches, Stillson wrenches, hammers, mauls, pliers and screw drivers; and cutting tools, such as hatchets, axes, mower knives and band knives. The combination of carbon and manganese in the steels used may vary widely for the same type of tool, depending partly on the equipment available for manufacture and partly on personal experience with, or preference for, certain combinations. A manganese content lower than standard will be used in some tools. This is justified when it makes a particular carbon range easier to handle, but it should be understood that for many applications, a combination of lower carbon and higher manganese would serve just as well. Hand cutting tools, discussed later, are an exception.

For hardening wrenches (except the Stillson type), screw drivers, pliers and similar tools, oil quenching is generally used, followed by tempering to the required hardness range. Even when no reduction of as-quenched hardness is desired, stress relieving at 300 to 375 F is desirable, to help prevent sudden service failures. In Stillson-type wrenches, the jaw teeth are really cutting edges and are nearly always quenched in water or brine to produce a hardness of Rockwell C 50 to 60. Either the jaws may be locally heated and quenched or the parts may be heated all over and the jaws locally time-quenched in water or brine. The entire part is then quenched in oil for partial hardening of the remainder. In this way considerable structural strength is obtained.

Hammers must possess high hardness on the striking face and somewhat lower hardness on claws. They are usually locally hardened and tempered on each end, depending on their type. The striking face is always quenched in water or brine. Satisfactory service depends on getting the proper depth of fully hardened (martensitic) surface on this face, and then stress relieving at about 350 F. Final hardness on the striking face is usually Rockwell C 50 to 58; on claws, 40 to 47.

Hand cutting tools, particularly axes and hatchets, must possess high hardness and high relative toughness in their cutting edge, as well as the ability to hold a keen edge. Since nothing is so effective as carbon in imparting the latter property, the carbon content is always higher than if hardness and toughness alone were to be considered. Many such tools are given an ordinary furnace anneal after forging, but high-quality tools are prepared for hardening by spheroidization, which may be performed as a separate operation after regular annealing. Most frequently, however, the refining and

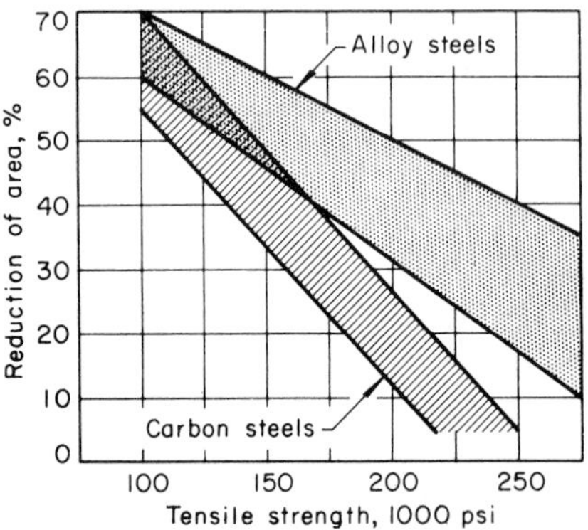

Fig. 4. Relation of tensile strength and reduction of area for quenched and tempered steels (SAE Handbook)

spheroidizing treatments are accomplished by quenching in oil from 1600 F, and then tempering at 1250 to 1375 F. The quench keeps the carbide in a finely divided state and from this condition spheroidization takes place rapidly at the temperatures specified.

For hardening, the cutting edges of such tools are usually heated in liquid baths to the lowest temperature at which the piece can be hardened, and are then quenched in brine. The quick heating of the liquid bath, and the low temperature, fail to put all the spheroidal carbon into solution. As a result the cutting edge of the tool consists of martensite with less carbon than indicated by the chemical composition of the steel and containing many embedded particles of cementite. In this condition the tool is at its maximum toughness, relative to its hardness, and the embedded carbides promote long life of the cutting edge. Final hardness is Rockwell C 55 to 60.

Manufacturers of agricultural implements make much use of the steels in this group. Braces, control rods, shafts and similar parts are often made of high-carbon steels, untreated, in order to obtain increased strength at low cost. The principal heat treated parts are plowshares, moldboards, coulters, cultivator shovels, disks for harrows and plows, mower and binder knives, ledger plates and band knives. Those parts used for cutting or turning soil must be moderately tough and must have the ability to resist abrasion. They are made with various combinations of carbon and manganese that will permit full hardening.

Plowshares and cultivator shovels are usually quenched in water or brine to obtain 500 to 600 Bhn. Plowshares, moldboards and cultivator shovels are sometimes made from "soft center steel", which consists of a layer of dead soft low-carbon steel between two equally thick layers of high-carbon steel. This material is rolled from specially cast composite ingots. Such parts are heat treated to produce high hardness in the outer layer, leaving the soft interior layer to provide toughness. The final product is similar in mechanical properties to a case hardened part; in fact, these parts are sometimes case hardened.

The grass-cutting and grain-cutting tools are usually made of 1090 or 1095, because of the effect of carbon in providing the desired edge that will last long in service. These parts are made from strip stock by blanking, and no annealing is done, except probably some process annealing performed in the rolling mills, at or below the transformation range. Spheroidization is, therefore, only occasionally obtained. Local hardening is done either by induction heating or in continuous furnaces provided with fixtures that permit the pieces to be heated on the cutting edges only. Upon discharge, the parts are quenched all over in oil and are tempered at a low temperature. Final hardness on the cutting edges is in the neighborhood of Rockwell C 55 to 60.

Mechanical Properties

When quenched to martensite and tempered to the same hardness, carbon and alloy steels have similar tensile properties in that portion of the cross section that reacts to the quench. If carbon steel has the hardenability required by the critical section of the part and the quench used, the resulting tensile strength, yield strength and elongation in the fully hardened zone will be in the same range as in a similar zone in an alloy steel quenched and tempered to the same hardness. The similarity in properties of the hardened zone holds, regardless of the depth of hardening, but the strength of the piece will be governed by the thickness of the hardened zone (depth of hardening).

Figure 1 shows the relation between hardness and tensile strength for constructional steels, both carbon and alloy. Because of the effect of cold working, this relationship is unreliable for cold drawn steels. Figure 2 shows the relation between tensile strength and yield strength; the effect of tempering temperature on tensile strength and hardness is shown in Fig. 3.

An important exception to this similarity of properties is the relationship between tensile strength and reduction of area. For any tensile strength the reduction of area is less for carbon steels than for alloy steels (Fig. 4). In applications where this property is important the difference must be considered in steel selection but it should be noted that (*a*) values do not drop to questionable levels until very high strength is reached and (*b*) in many applications (springs, for example) low values of ductility do not impair service life. In fact, within the limits of section-hardenability relationship, more springs are made from carbon steel than from alloy steel. Thus it can be concluded that, in spite of some difference in resulting properties, the principal difference between carbon and alloy steels is hardenability.

One other and sometimes important difference between carbon and alloy steels is that, for the same hardness levels, fully quenched alloy steels require higher tempering temperatures than carbon steels. This higher tempering temperature is presumed to reduce the stress level in the finished parts without impairing mechanical properties.

Property relations given in Fig. 1 to 4 illustrate general correlations among mechanical properties of steels. No details were reported regarding the size, number or preparation of the specimens used in obtaining these data. However, such data are usually obtained from specimens that are closely controlled in chemical composition, grain size, section size, and condition of heat treatment. Normal variations in composition and grain size from heat to heat and even within one heat produce a considerable scatter of results in sections of the same size.

Mechanical properties of carbon

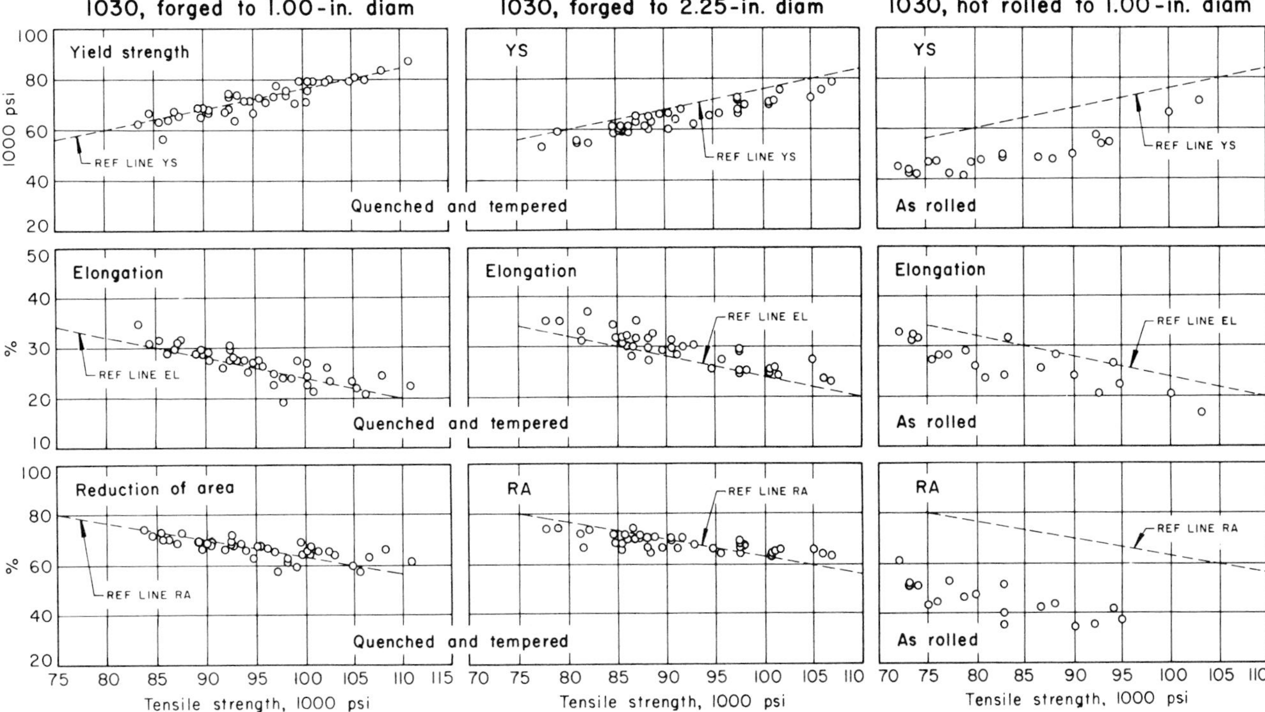

Fig. 5. Correlation of mechanical properties for specimens of 1030 steel forged to 1 and 2.25-in. diam quenched and tempered and for as-rolled 1030 of 1-in. diam. Heat treated specimens were water quenched from 1600 F and tempered at 1000 to 1200 F. Specimens were taken from the center of 1-in. bars and at half-radius from the 2.25-in. bars.

steels (particularly when quenched and tempered) are influenced more by changes of section size than alloy steels because of the lower hardenability of carbon steels. In addition to the effect of section size on specific properties, the relation of one property to another is affected by size of the heat treated section. This is implied by the tensile and yield strength relationship given in Fig. 2. As the section size increases, incomplete hardening will lower the ratio of yield strength to tensile strength.

Figures 5 to 8 give detailed tensile property relations for eight different carbon steels in the range from 1030 to 1050. Most of those data are for steels in the quenched and tempered condition although tensile relations are given for as-rolled 1030 and 1040 (Fig. 5 and 7). For any given composition and section size, the scatter of results, especially for reduction of area, is usually greater in the hot rolled condition than in the quenched and tempered condition.

Increasing carbon content brings about a consistent increase in tensile and yield strength and decrease in elongation and reduction of area, regardless of whether the steel is in the hot rolled or quenched and tempered condition (provided the ranges of tempering temperature are the same). For example, compare the data on hot rolled 1030 and 1040 steels in Fig. 5 and 7, and those on the 1-in. diam bars of quenched and tempered 1030 (Fig. 5), with 1-in. diam quenched and tempered 1035 (Fig. 6).

Data given in Fig. 5 to 8 present the effect of section size on mechanical properties and property correlations for several carbon steels, including tensile-hardness relations for 1050 (Fig. 8). The tensile strength decreases as the section size increases, for a given composition and heat treatment, and there is some lowering of the ratio of yield to tensile strength. For example, in Fig. 5, compare the yield strengths

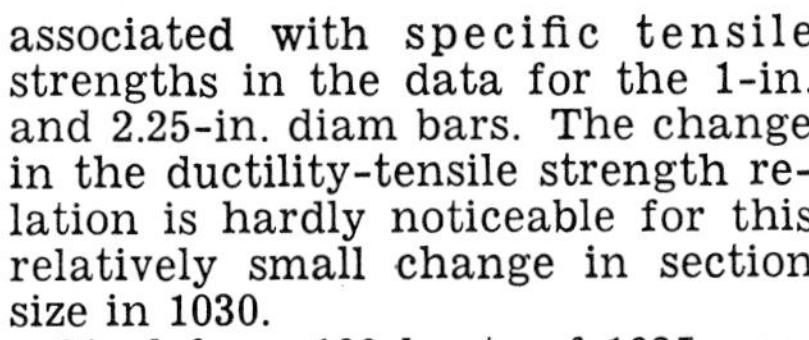

associated with specific tensile strengths in the data for the 1-in. and 2.25-in. diam bars. The change in the ductility-tensile strength relation is hardly noticeable for this relatively small change in section size in 1030.

Steel from 100 heats of 1035 produced to a range of sizes was tested and the results plotted statistically in Fig. 6, upper graph. The specimens were quenched and tempered in sizes ranging from 0.5 to 1.75-in. diameter. The scatter in tensile strength was from 117,000 to 138,000 psi and the yield strength varied from 100,000 to 124,000 psi. Hardness range was Rockwell C 23 to 30.

Variations in chemical composition within a specific grade contribute to the scatter of mechanical properties. This is illustrated by the test data in Fig. 8 (lower graphs), where the properties for two heats of quenched and tempered 1050 steel are compared for a tempering range of 600 to 1200 F.

Tempering. When carbon steels are quenched to obtain an almost completely martensitic structure, the hardness will decrease as the tempering temperature is increased, in a straight-line relation, for tempering temperatures to 1200 F or higher (Fig. 9 and 10). The rate of hardness decrease with tempering temperature is greater for carbon steels than for alloy steels.

Cold Drawn Bars

Cold drawing markedly increases the strength of carbon steels; the magnitude of the increase depends on the amount of draft and whether or not cold drawing is followed by stress relieving. Cold drawing also increases the ratio of yield strength to tensile strength but lowers ductility and notch toughness. Therefore, property relations are different for cold drawn steels than for similar compositions quenched and tempered to the same tensile strength. Section size also influences the properties of cold drawn steels. The properties of hot rolled and cold drawn 1137 steel for several different size ranges are compared in Fig. 11.

Cold drawn and stress-relieved steels are often used in place of quenched and tempered steels. Several grades such as 1137, 1141 and 1144 respond readily to this process, with resulting high strength values. Because their structures are still pearlitic they machine more easily than their quenched and tempered counterparts. Therefore, even though these grades cost more than nonresulfurized grades, significant savings can often be realized in manufacturing costs, chiefly through elimination of heat treating. However, even though the strength of cold drawn and stress relieved steel may be equal to that of quenched and tempered steel, it is not reliable to translate other properties from one condition to the other. For example, specimens from 1-in. diam cold drawn and stress-

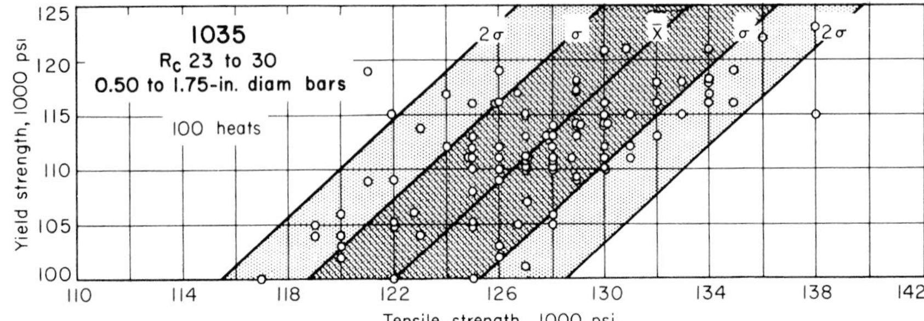

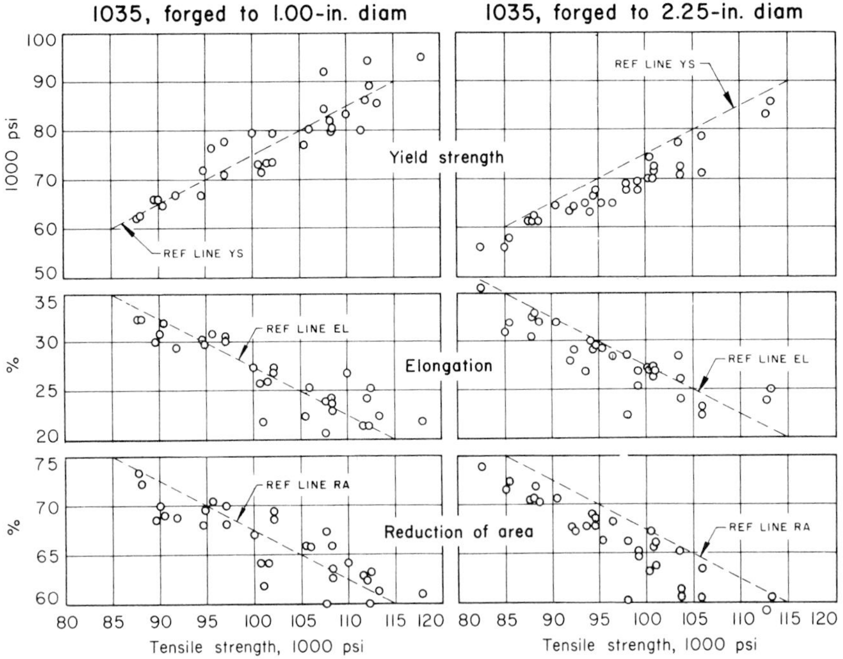

Fig. 6. (Upper graph) *Correlation of tensile and yield strength for heat treated 1035 steel. All specimens were heated in a protective atmosphere at 1600 F, water quenched and tempered. The tempering temperature used for the 100 heats ranged between 960 and 1000 F; the exact tempering temperature depended upon the expected response to softening for each heat as it was processed in the laboratory. All heats were fine grained. Standard tensile specimens were taken at half radius where bar size permitted. Yield strength was determined by the 0.2% offset method. The line of best fit ($\overline{X}$), and also sigma and two-sigma standard deviations, were found by correlation analysis.*
(Lower graphs) *Relations among mechanical properties for 1035 steel water quenched from 1600 F and tempered at 1000 to 1200 F. Standard specimens were taken from the center of 1-in. diam bars and at half-radius in the 2.25-in. diam bars.*

relieved 1144 steel were compared with specimens from the same heat that had been quenched and tempered to the same strength level (150,000 psi).

Although strength and hardness were the same, reduction of area was less than half as much for the cold drawn steel as for the same steel in the quenched and tempered condition. Notch toughness of the cold drawn steel, evaluated by the Charpy V-notch method, was decreased to one-third the value of the quenched and tempered material. Also, the difference between longitudinal and transverse properties is significantly greater for cold drawn and stress-relieved steel than for a similar steel in the quenched and tempered condition.

The scatter in mechanical properties for 41 heats of 1144 steel, cold drawn and stress relieved at 1050 F, is shown in Fig. 12. A part of this scatter is attributed to the variation in composition (0.41 to 0.52% C, 1.33 to 1.68% Mn, and 0.220 to 0.336% S), which is slightly greater than the specified ranges for 1144 steel.

Because of the size of the equipment required for heavy drafts, the practical size limit of steels cold drawn to high strength values is usually about 2½ in. in diameter, although larger sizes are sometimes available. The highest strengths can be obtained only in sizes under 1 in. in diameter. (Additional information is given in the article on Cold Finished Steel, in this Handbook.)

Orientation Effects in Forgings

Section thickness and orientation act individually or together to affect mechanical properties of quenched and tempered carbon steel forgings. The effect of these variables on the properties of a billet of modified 1050 steel, square forged with round corners and quenched and tempered, is shown in Fig. 13. Neither direction nor distance from the surface has a significant effect on yield strength, tensile strength or hardness. However, there is a marked effect on elongation and reduction of area, caused by direction and distance from the surface. The greatest difference in these properties is between the transverse center and surface specimens (specimens denoted as TC and TS in Fig. 13).

Effects of orientation and section thickness are shown also for tensile specimens taken from a section of a 150-lb wheel forging (Fig. 14).

Hardenability of Carbon Steels

The most important factor influencing a choice between a carbon and an alloy steel is the required hardenability. As of June 1, 1960, it is not possible to purchase carbon steels to a specified hardenability band, as it is alloy steels. This situation causes a more definite distinction between carbon and alloy steels than is caused by the actual hardenability values.

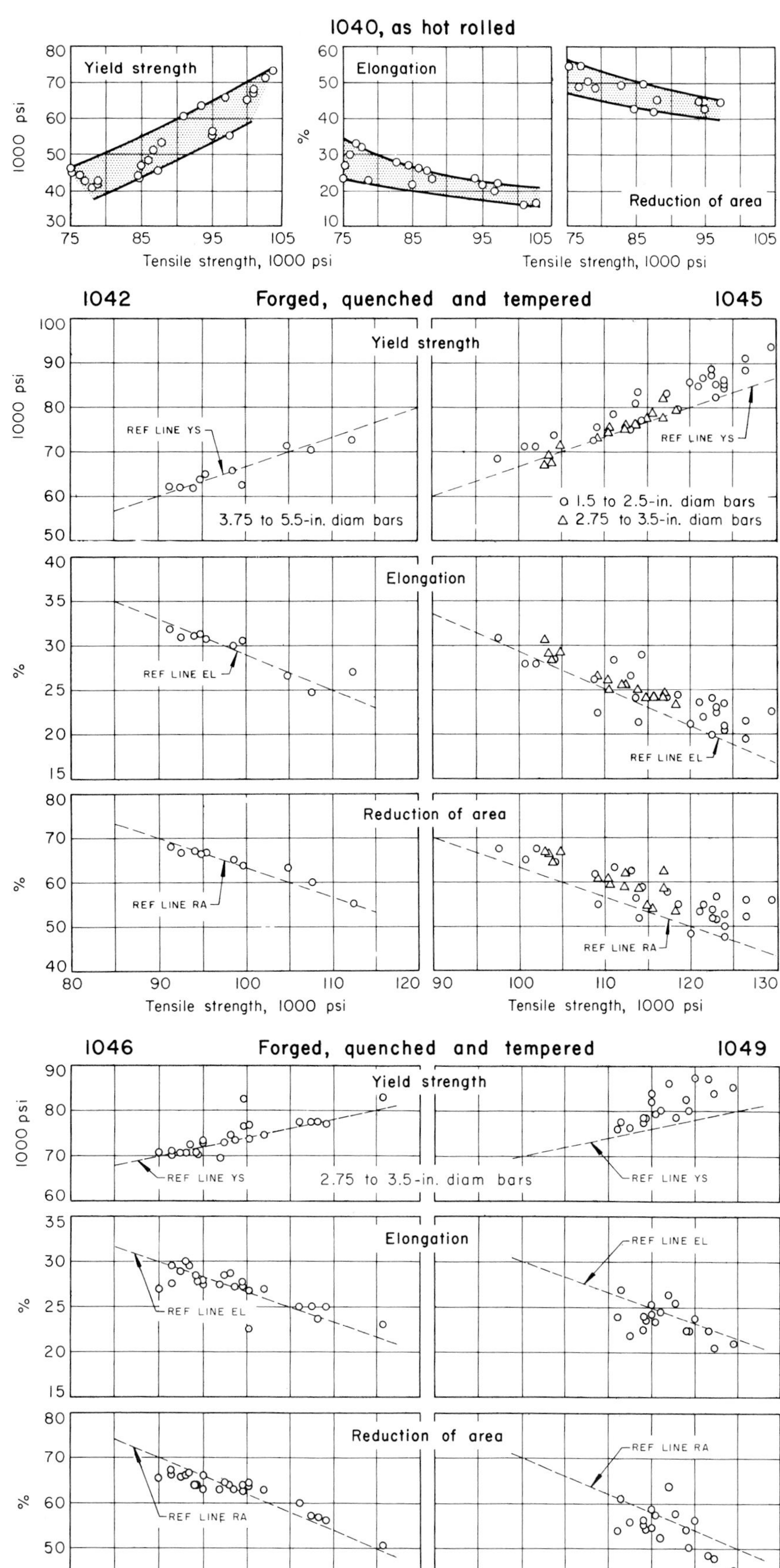

Fig. 7. Relations among mechanical properties of medium-carbon steels. (Upper row of graphs) *1-in. diam hot rolled 1040.* (Center graphs) *1042 and 1045 forged to indicated diameters, water quenched from 1550 F and tempered at 1000 to 1200 F.* (Lower graphs) *1046 and 1049 forged to indicated diameters, water quenched from 1550 F and tempered at 1000 to 1200 F. Specimens were taken at half-radius from bars larger than 1-in. diam.*

Carbon steels with low manganese (some grades as low as 0.30%) and a virtual absence of residual nickel, chromium and molybdenum, are the lowest in hardenability of all steels. This holds true for nearly all carbon levels, because in the absence of higher manganese or other alloying elements, carbon functions almost entirely to control maximum hardness and has only a minor effect on hardenability.

The majority of the series 1000 carbon steels in the hardenable class have 0.60 to 0.90% Mn although there are numerous exceptions, several grades having higher and some lower manganese content. A notable example is 1041 with 1.35 to 1.65% Mn. (The effect of increasing manganese content on the tensile strength of steels in the hot rolled condition is shown in Fig. 15.)

Manganese has a marked effect on hardenability. Even a difference of 0.25% makes a significant difference on the end-quench hardenability of 0.50% C steel (Fig. 16).

Considering the range of manganese that is available in carbon steels, it follows that a wide range of hardenability can exist. For example, 1041 steel frequently shows end-quench hardenability values higher than the minimum of the hardenability band for 1340 steel. Thus, there is a gradual transition in hardenability from the carbon to the alloy grades.

Steels are not necessarily "better" because they are higher in hardenability. They are better only when the hardenability is required. There are many applications where minimum, rather than maximum, hardenability is needed, which accounts for the many low-manganese grades melted. For example, it is often desirable to produce thin layers of maximum hardness on shaft bearings or cam contours. This is usually accomplished by induction or flame hardening, but if the hardened zone is too deep, an unfavorable stress pattern will be established, with resultant cracking in quenching or premature failure in service. In one instance cams were made from standard 1050 steel (0.60 to 0.90% Mn) and induction hardened to Rockwell C 60 to a depth of about 1/16 in. If the hardened zone

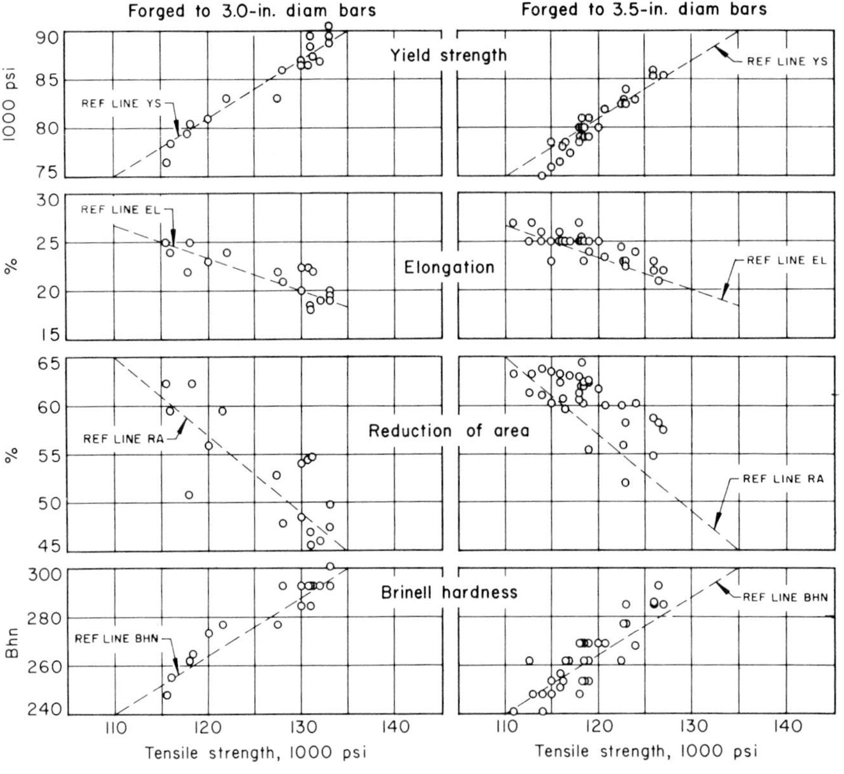

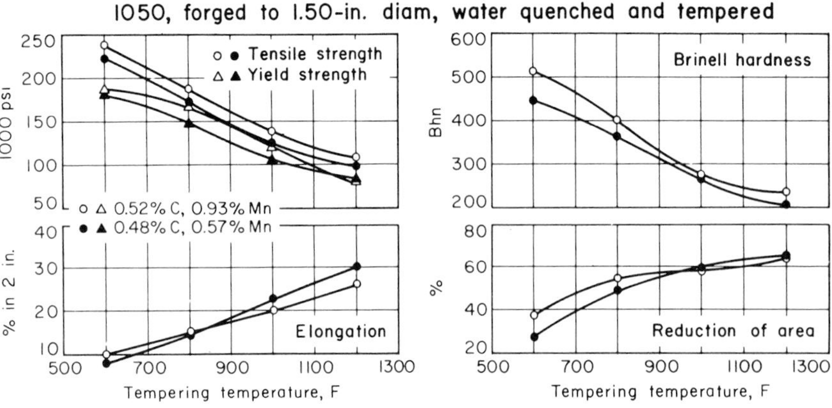

Fig. 8. (Upper graphs) *Relations among mechanical properties for modified 1050 steel forged to two diameters. All specimens were quenched in agitated water from 1600 F. The 3-in. diam specimens were tempered at 980 to 1130 F and the 3.5-in. specimens at 1020 to 1110 F. Tensile specimens were taken ½ in. below surface, and hardness readings were taken 1/16 in. below forged surface. Carbon ranged from 0.48 to 0.55% and manganese from 0.80 to 1.07%.*
(Lower graphs) *Effect of carbon, manganese, and tempering temperature on mechanical properties of two heats of water-quenched 1050 steel.*

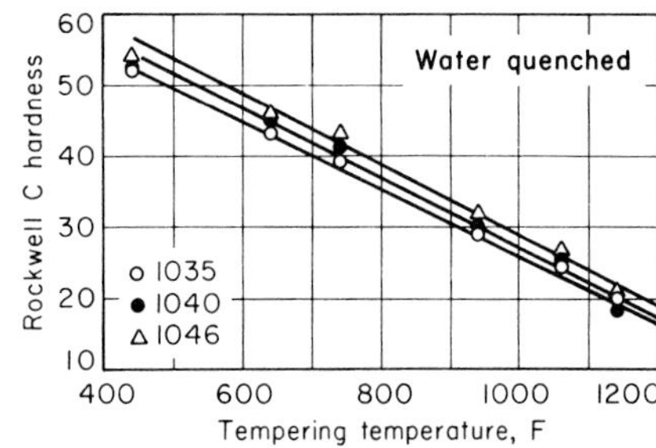

Fig. 9. Effect of tempering temperature on the hardness of three medium-carbon steels. The 1035 specimens were water quenched from 1550 F and showed an as-quenched hardness of Rockwell C 58. The 1040 specimens were water quenched from 1600 F and showed an as-quenched hardness of Rockwell C 57. The 1046 specimens were water quenched from 1560 F and showed an as-quenched hardness of Rockwell C 59. All specimens showed more than 90% martensite after quenching. Specimens were tempered for 2.25 hr at the indicated temperatures. Hardness readings were taken 0.015 in. below the surface. Data represent one heat for each steel and were taken from 0.875, 0.75 and 1-in. diam bars for 1035, 1040 and 1046, respectively.

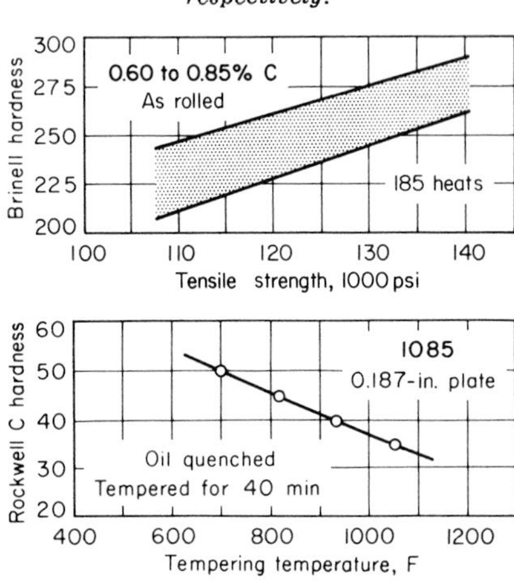

Fig. 10. The relation between Brinell hardness and tensile strength in the as-rolled condition for 185 heats of steel containing from 0.60 to 0.85% C is shown in the upper graph. In the lower graph, the effect of tempering temperature on the hardness of 1085 steel plate, oil quenched from 1600 F. This heat had 0.87% C, 0.80% Mn and 0.15% residual nickel.

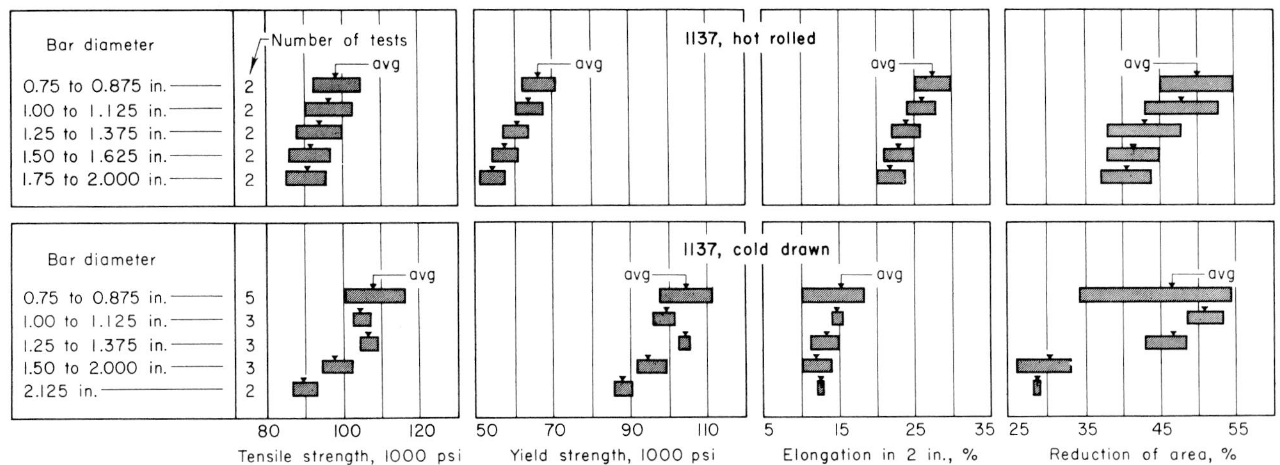

Fig. 11. Mechanical property ranges for several sizes of hot rolled and cold drawn 1137

became as deep as ⅛ in., a significant number of parts cracked. Cracking was eliminated by using a modified grade of 1050 steel (0.30 to 0.60% Mn), which resulted in a shallower hardened zone after induction hardening.

It is more economical to use carbon steels whenever possible. The higher-manganese grades cost more than lower-manganese grades (maximum difference is about $0.60 per 100 lb), but less than the lowest-cost alloy grades. For example, 1340H costs $1.00 per 100 lb more than 1041.

Control of Hardenability. Despite the fact that carbon steels are not made to prescribed hardenability bands, users employ various means of controlling hardenability within their own plants. One method is to probe test specimens that have been made to simulate the size and shape of actual parts.

The end-quench test is used for testing hardenability of carbon steels. This test involves heating a test bar 1 in. in diameter to the proper austenitizing temperature, placing the bar in a special hardenability fixture, and quenching only the end surface with water. The hardness is then measured below the surface, at intervals of 1/16 in. from the water-quenched end to determine how far from this end the hardness extends. Typical end-quench hardenability data for several carbon steels are given in the article on Selection of Steel for Hardenability, in this Handbook.

SAC (Rockwell-Inch) Test. Some industries have found the SAC (surface-area-center) test to be more discriminating than the end-quench test for determining hardenability of shallow-hardening steels, because of the sharp gradient on the end-quench curve.

Method. The SAC test surveys hardnesses on a heated and quenched cross section. The specimen is 5½ in. long by 1 in. in diameter (Fig. 17). After normalizing at the specified temperature for 1 hr and cooling in air, it is austenitized at 1550 F ± 10 F for ½ hr and quenched in water at 75 F ± 10 F, where it is allowed to remain until the temperature is the same throughout the specimen.

After quenching, a cylinder 1 in. long is cut from the middle of the hardened specimen (Fig. 17). The cut faces of the cylinder are carefully ground parallel to remove any burning or tempering that might result from cutting and to assure that a flat face will be presented to the anvil or fixture of the hardness testing machine. Etching with 10% nital is recommended to check for any evidence of burn or tempering before the hardness tests are made. Hardness is measured on the cylindrical surface of the specimen at a minimum of two points 90° to each other. The average of these two readings then becomes the surface reading.

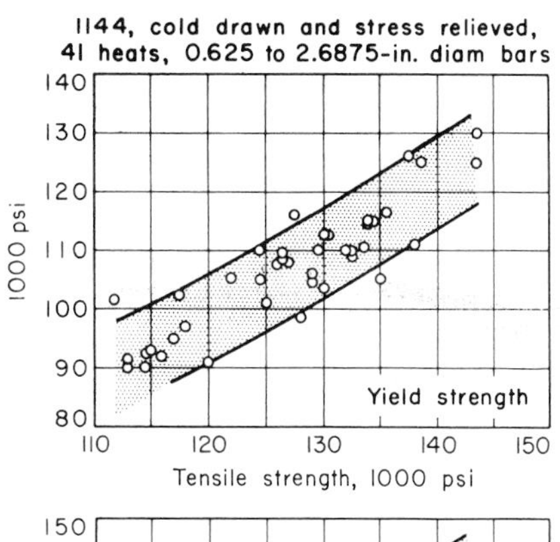

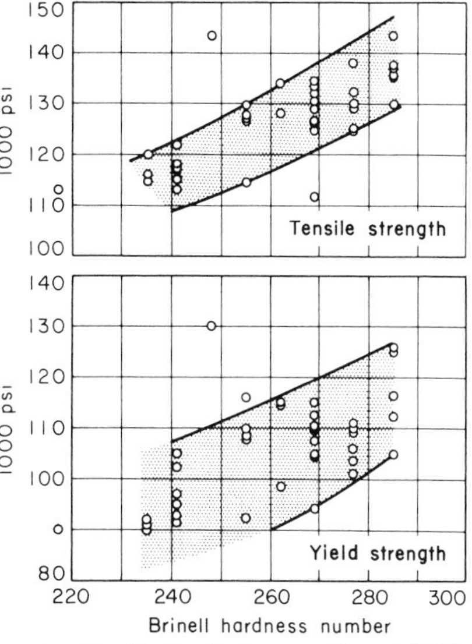

Fig. 12. Mechanical properties of 1144 steel cold drawn and stress relieved at 1050 F. Range of composition for 41 heats was 0.41 to 0.52% C, 1.33 to 1.68% Mn and 0.220 to 0.336% S.

The specimen is positioned (etched face up) in a holding fixture designed to permit lateral adjustment while maintaining a centerline relationship of the specimen to the diamond indenter. Then a series of Rockwell C readings is taken on the cross section in steps of 1/16 in. from the surface to center. From these readings a quantitative value can be computed and designated by a code known as the SAC number. This code consists of a set of three two-digit numbers: first, the surface hardness; second, the Rockwell-inch area; and last, the center hardness. While all portions of the code are valuable, the Rockwell-inch area is the most significant and is normally used for relating hardenability and response to heat treatment. The formula for calculating Rockwell-inch hardenability is as follows:

$$\text{Rockwell-inches} = \tfrac{1}{8}(S/2 + h_1 + h_2 + h_3 + h_4 + h_5 + h_6 + h_7 + C/2) \text{ where}$$

S = hardness at the surface as measured on the Rockwell C scale

h_1 to h_7 = average Rockwell C readings at 1/16-in. intervals from surface to center

C = the average of Rockwell C readings at the center and at four locations 1/16 in. from the center

For example, the quenched specimen in Fig. 17 shows a surface reading of Rockwell C 59, then 61.5, 60, 56.75, 50, 41.75, 36.5 and 34.5 (for h_1 through h_7), and finally Rockwell C 36 at the center. According to the formula this becomes ⅛ (29.5 + 61.5 + 60 + 56.75 + 50 + 41.75 + 36.25 + 34.75 + 17.50, or 48.5 Rockwell-inches.

In steels with shallow hardenability, the SAC traverse can be interpreted more accurately than an end-quench curve. The sharp inflection of the end-quench curve poses difficulties that are not overcome by changing the method of plotting from Cartesian to logarithmic coordinates.

The sharp descent of the end-quench curve in plotting the hardenability of 1045 steel is indicated in Fig. 18 along with the gentler curve obtained from an SAC test on a specimen from the same bar of steel. A correlation between points on the SAC and end-quench curves is indicated by straight connecting lines. Generally, such correlation is complicated by cooling rate differentials and the effects of chemical segregation, which are more pronounced in shallow-hardening steels than in higher-alloy, deep-hardening steels. As demonstrated by the curves, the cooling rate differential between the surface and any particular test distance is markedly different and does not vary equivalently. The effects of segregation are more accurately reflected in the SAC test, since hardness results are taken across a wrought bar, thus traversing the original ingot segregation. Often there is a slight increase in hardness at the center of the bar due to center segregation of the ingot.

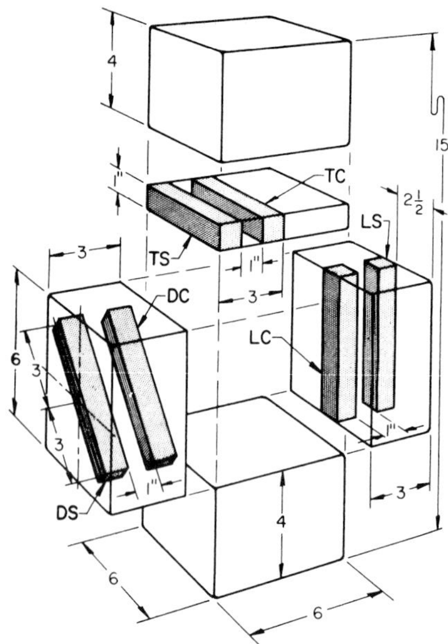

Location	Tensile strength, psi	Yield strength, psi	Elongation, %	RA, %	Bhn
TC ...	123,000	74,500	8.5	18.0	229
TS ...	118,000	74,000	19.0	36.7	235
LC ...	127,000	76,250	23.5	55.8	248
LS ...	119,000	74,500	23.0	52.6	229
DC ...	122,500	73,500	15.5	38.9	235
DS ...	120,000	75,000	22.5	50.7	229

Fig. 13. Effect of test specimen location on mechanical properties of a 6 by 6 by 15-in. round-cornered square billet of modified 1050. After forging, the billet was normalized at 1600 F, reheated to 1600 F, water quenched and tempered at 1050 F. Check analysis ½ in. below the surface showed 0.44% C and 1.14% Mn; ½ in. from center, 0.55% C and 1.19% Mn, compared with the ladle analysis of 0.48% C and 1.17% Mn. Mechanical properties are related to test bar location in the above tabulation. Neither direction nor distance from the surface has a significant effect on tensile strength, yield strength or hardness; however, elongation and reduction of area are affected by these variables. The effect of location on the latter values is sometimes far more pronounced than the effect of hardness.

A comparison of SAC and end-quench results obtained on 94 heats of carbon steel containing from 0.36 to 0.48% C and 0.60 to 0.94% Mn appears in Fig. 19.

Figure 20 presents additional SAC data based on a total of 24 heats of 1045. The results obtained on 12 heats containing low amounts of residual elements are compared with those from a similar number of heats with slightly higher residuals. These data indicate the sensitivity of the SAC test to even minor differences in the content of residual alloying elements in a shallow-hardening carbon steel.

Hardenability Examples

Example 1 (Fig. 21). Forged hollow cylinders with an OD of 8$\frac{5}{32}$ in. and an ID of 6$\frac{3}{8}$ in. made from 26 heats of 1042 steel were heat treated to correlate hardenability (previously determined) compared with depth of zone hardened to Rockwell C 50, Fig. 21(a). The cylinders were furnace hardened using an austenitizing temperature of 1530 F for 70 min and quenched in water at 95 F for 82 sec. Manifold-type fixture quenching was used so that the quench was drastic on the outside, while the inside was protected with compressed air. None of the cylinders was tempered. Hardenability and hardened depth were plotted for the first and last ingots of each of the 26 heats. Several pieces were hardened from each heat; hence, a range in hardened depth is plotted for each heat.

A second series of forged hollow cylinders from 30 heats of 1046 steel of known hardenability was heat treated in the same manner. The cylinders had a 10$\frac{5}{8}$-in. OD and a 7$\frac{5}{16}$-in. ID. Hardenability by SAC versus depth of hardening is plotted in Fig. 21(b). In this instance the average hardenability of the first and last ingots for each mill heat was used.

While greater depths of hardening were obtained with heats having higher Rockwell-inch values, a limiting hardenability was approached, above which quench cracking became severe. Therefore, it was necessary to establish Rockwell-inch specifications that would provide the highest possible yield of good parts with a minimum sacrifice in service performance.

The foregoing illustrates the value of the Rockwell-inch test in predicting the response to hardening.

A distribution of SAC hardenability data for 250 heats of the same grade of steel received from four different mills is shown in Fig. 22.

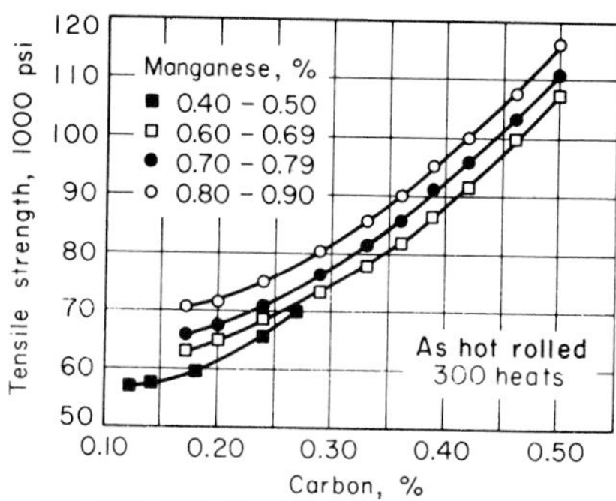

Fig. 15. Effect of manganese and carbon contents on the tensile strength of carbon steels in the hot rolled condition

Distortion in Heat Treatment*

Carbon steels usually distort more in heat treatment than alloy steels because carbon steels require a water or brine quench to develop full hardness (at least in sections thicker than about ⅜ in.). This often eliminates selection of carbon steels for critical parts.

Several factors contribute to the total distortion that occurs during heat treatment. These include residual stresses that may be present as a result of machining or other cold working operations, the method of placing in the furnace, the rate of heating, and the natural volumetric changes that take place with phase transformations. However, the most important single factor is the uneven cooling rate during quenching, caused mainly by the shape of the parts. Symmetrical parts with little or no variation in section may have almost no distortion, whereas complex parts with wide variations in section may distort so much that they cannot be used or at least require excessive finishing operations.

Other factors being equal, the distortion in carbon steels will increase

*See also the article "Distortion in Tool Steels", page 654 in this Handbook, for discussion and data on the factors influencing both size and shape distortion.

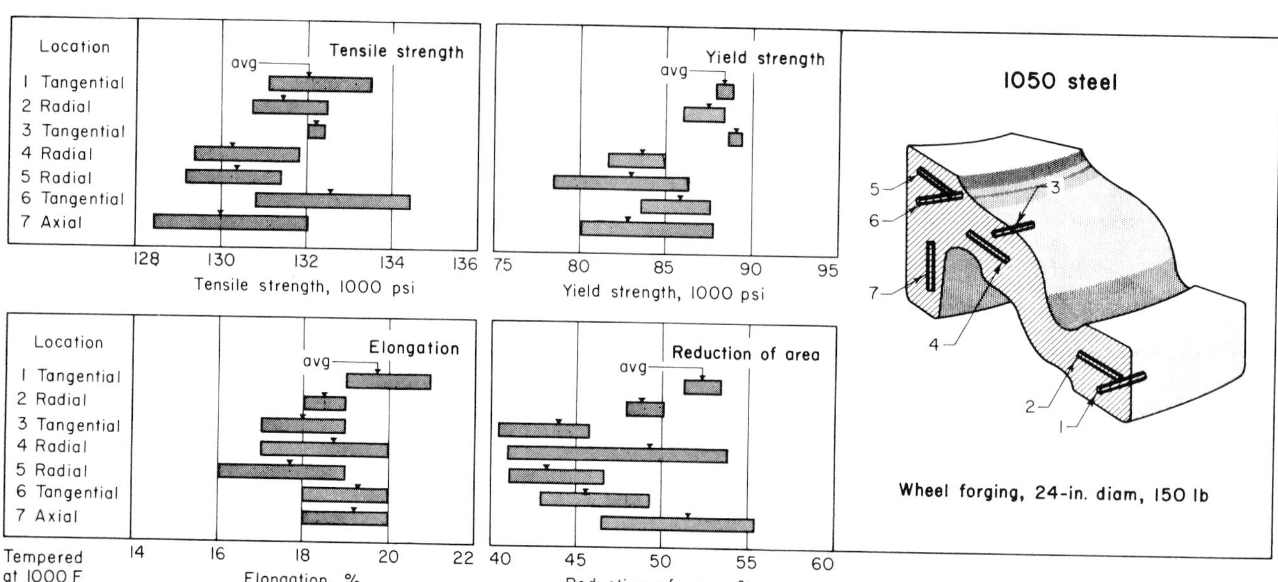

Fig. 14. Effect of test specimen location on mechanical properties of a 150-lb wheel forging of 1050 steel, water quenched from 1550 F and tempered at 1000 F. Each bar on the chart represents maximum, minimum and average values for three tests.

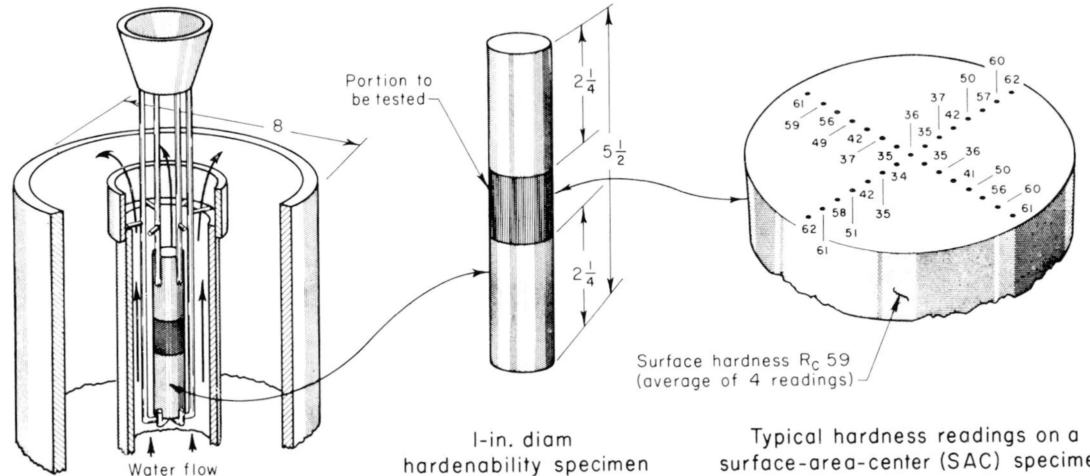

Fig. 17. The SAC (Rockwell-inch) hardenability specimen, the apparatus used to quench it, and a section 1 in. long cut from the hardened specimen indicating typical hardness readings on which the calculation of hardenability is based

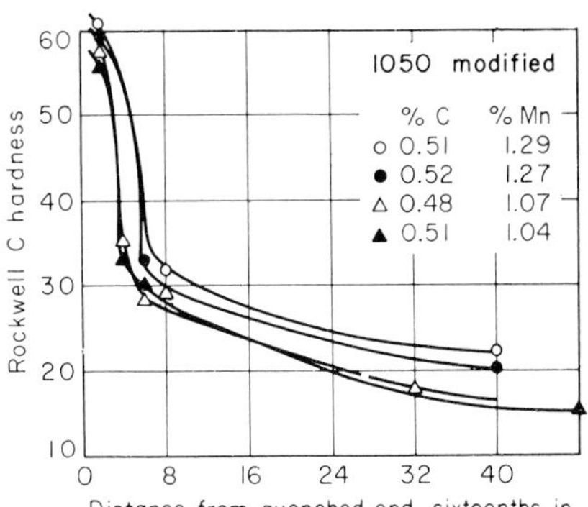

Fig. 16. Effect of variations in carbon and manganese contents on end-quench hardenability of modified 1050 steel. The steels with 1.29 and 1.27% Mn contained 0.06% residual chromium. Steels with 1.07 and 1.04% Mn contained 0.06 and 0.08% residual chromium, respectively. No other residual elements were reported.

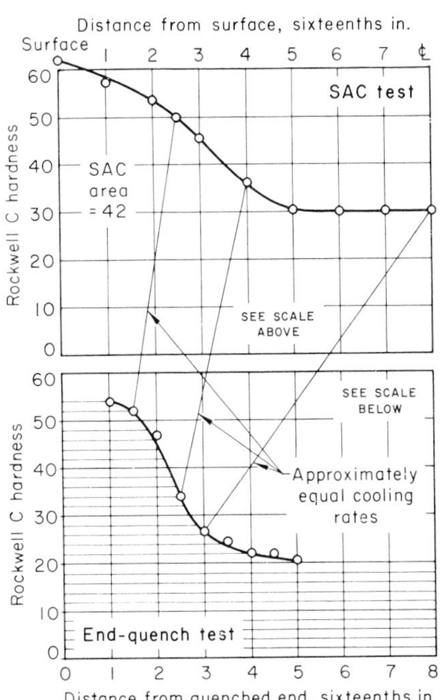

Fig. 18. Correlation between SAC (Rockwell-inch) and end-quench hardenability test results on specimens cut from the same bar of 1045 steel

as the carbon content increases, because of the gradual lowering of M_s temperature with increasing carbon.

There is also a significant variation in the magnitude of distortion and direction of dimensional change among different heats of the same grade of steel, even though other variables are kept to a minimum. This happens because of several factors, including minor variations in composition and grain size, but mainly because of the history of the steel with regard to hot working, cold working and heat treatment.

Because of the several variables that contribute to the total distortion in a specific instance, prediction of distortion in actual parts, based on the behavior of small test pieces, is seldom reliable.

The most practical approach toward providing information on distortion is to make studies on pilot lots of actual pieces that have been heat treated under production conditions. This procedure eliminates the shape variable so that direction and magnitude of distortion can be plotted as ranges that incorporate most of the other variables. After a quantity of such data has been secured by this procedure a series of "guideposts" is established, and it becomes possible to predict distortion behavior for similar parts with reasonable accuracy. However, it must be emphasized that any such study is accurate only when a quantity of parts made from different heats supplied by several mills is included, even though the grade specification remains the same.

Example 2 (Fig. 23) involves data obtained by measuring 25 sections of 1018 and 1024 carburized steel tubing. The ID, OD and length of each tube were measured before heat treatment, marked for later identification, and remeasured after each of two stages of heat treatment. In this instance a carburizing operation further complicated the distortion problem. Direction (plus or minus) and total ranges of dimensional variation are given in the bar charts of Fig. 23. Tubes were carburized at 1700 F, cooled, reheated to 1460 F, and water quenched.

Example 3 (Fig. 24) is a similar study made on 1046 steel shafts, with runout in six locations as the main consideration.

Induction and Flame Hardening

The relatively low hardenability of carbon steels is often a reason for their selection where localized hardening is done by induction or flame. One example that demonstrated the advantage of a low manganese content for a specific application was discussed here under hardenability.

One of the oldest rules for selecting steels for heat treating is to

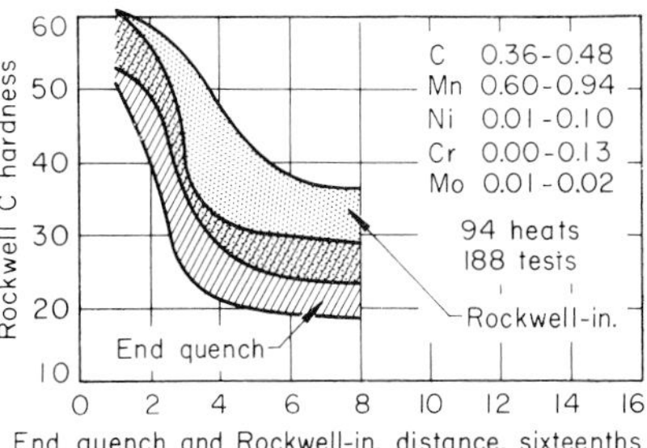

Fig. 19. Correlation between SAC and end-quench test results based on 94 heats of carbon steel with chemical analyses within the ranges shown

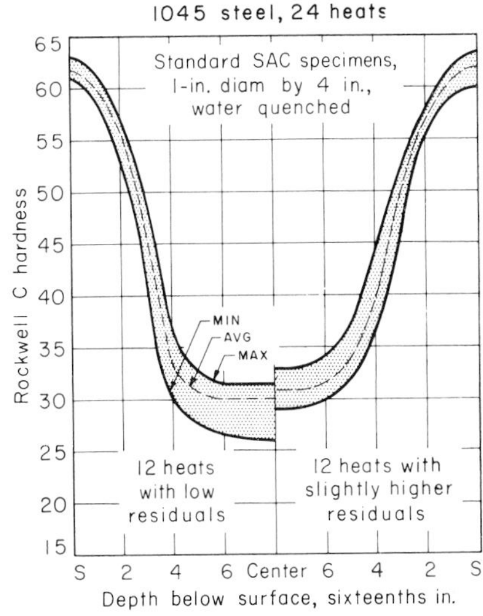

Fig. 20. Effect of low and higher residuals on the SAC hardenability test results of 24 heats of 1045 steel

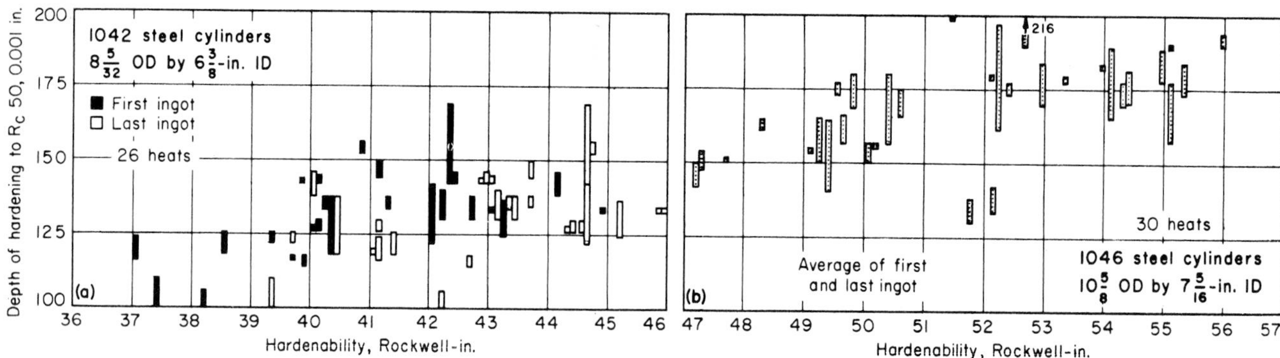

Fig. 21. (a) *Correlation of hardenability in Rockwell-inches with depth of hardening to Rockwell C 50 for 26 heats of 1042 steel. After determining hardenability, the steel was made into cylinders that were heated to 1530 F and water quenched in fixtures so that the outside received a drastic quench while the inside was protected by compressed air.* (b) *Similar data (same heat treatment) for 30 heats of 1046 steel except that average hardenability for the first and last ingots of each heat is plotted. Data from several tests are represented by each vertical bar plotted.* (Example 1) *The Rockwell-inch test is used to predict response to hardening under actual production conditions.*

choose grades that are no higher in carbon content than is essential to develop required properties. This rule remains valid in selecting steels to be heat treated by induction or flame processes.

When the outer peripheries of steel parts are heated rapidly and quenched, the tendency toward cracking depends mainly on a combination of three factors: (*a*) final surface hardness, (*b*) temperature to which the surface was heated and (*c*) depth of hardened zone.

The optimum heat pattern for either induction or flame heating depends on the type of steel in addition to the mass and shape of the part. The ideal heat pattern for any specific part will provide a hardened shell to a depth that will strengthen the part, by means of establishing a favorable stress pattern. However, if the hardened zone is too deep for the specific section thickness, high tension stresses are established in the surface layers, and these may cause cracking.

Excessive depth of the hardened zone can be caused by technique (as overheating) or by choice of a steel with too much hardenability. However, excessive carbon can aggravate other contributing factors and become the basic cause for cracking. The M_s temperature decreases as the carbon content increases. It is lowered further by higher austenitizing temperature and, as the M_s temperature is lowered, probability of surface cracking increases.

Further discussion of these variables, and examples involving both carbon and alloy steels, are included in the article on Selection of Steel for Minimum Quench Cracking, in this Handbook.

Weldability

Low-hardenability steels are less likely to crack because of martensite formed during welding than high-hardenability steels. Therefore, carbon steels are preferred for weldments unless other properties such as hardenability or corrosion resistance demand the use of an alloy steel. However, the production of sound, crack-free welded joints becomes more complex as the carbon content of the steel increases.

Low-carbon steels with less than 0.30% C are easily welded by all of the commercial welding processes. High-quality joints can usually be obtained without the necessity for preheating or postheating. For steels of this low-carbon range, selection of electrode material for arc welding is not critical. Standard all-purpose electrodes used in the manufacturer's recommended heat ranges will be low in crack sensitivity in welding steels with less than 0.30% C.

Medium-carbon steels with 0.30 to 0.50% C can also be satisfactorily welded by all of the common methods, but because of the formation of hard constituents in the welded zone, preheating and postheating may be necessary, especially for the steels with 0.40 to 0.50% C. There is usually less need for preheating and postheating with gas welding methods than arc welding methods because in gas welding a larger area is heated and the cooling rate is slower.

For arc or other welding methods where the welded zones are rapidly cooled, a preheating temperature of 300 to 500 F has been found effective in eliminating hard, brittle areas. By heating to 1100 to 1200 F after welding it is possible to restore ductility in the weld-affected areas. Modifications in welding technique, for example a larger V, will decrease the cooling rate and subsequent tendency to weld cracking.

Selection of welding electrodes for arc welding becomes more critical as the carbon content of the steel increases. Steels with higher carbon content are more susceptible to toe and root cracking from electrodes that contain hydrogen. The E7018, low-hydrogen electrodes are frequently used to minimize cracking caused by hydrogen, especially where close heat control is not practical as in operations performed on an incentive basis. As the carbon content of the steel being welded increases to nearly 0.50%, the use of low-hydrogen high-tensile electrodes such as E7018 or E10076 becomes mandatory.

High-carbon steels with more than 0.50% C are more difficult to weld and require closer control, chiefly because of the increased susceptibility to weld cracking and the susceptibility to deleterious effects such as grain growth from overheating.

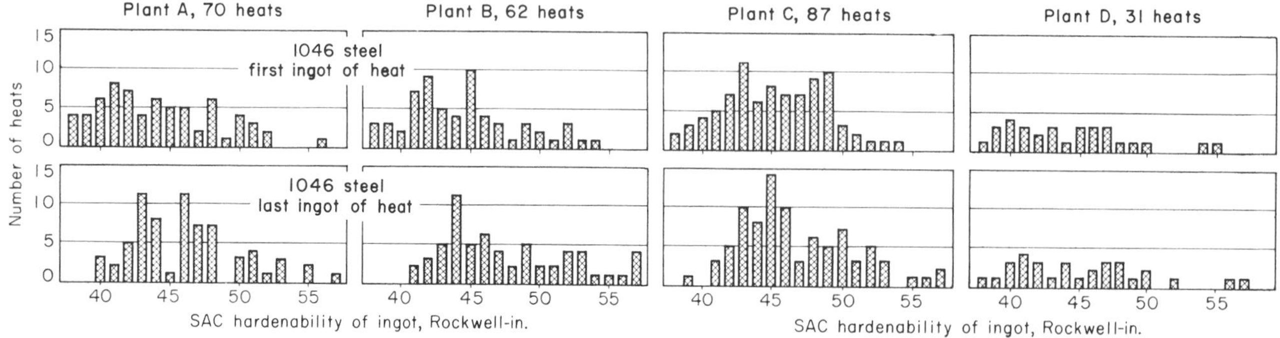

Fig. 22. Distribution of SAC hardenability results for each of 250 heats of 1046 steel produced by four different mills

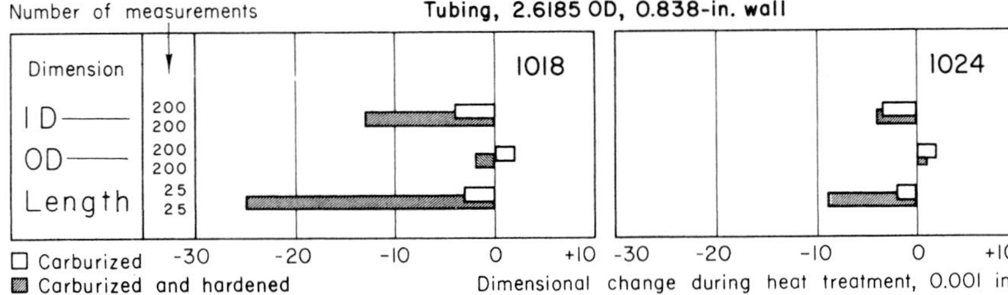

Fig. 23. Direction and magnitude of dimensional changes for heat treated tube sections made from 1018 and 1024 carbon steel. Dimensions before heat treating were in the range from 8.230 to 8.240-in. length, 2.617 to 2.620-in. OD and 1.778 to 1.783-in. ID for 25 tubes from each grade. Each tube was measured before heat treatment, and marked for later identification. Tubes were carburized at 1700 F, cooled to room temperature, reheated to 1460 F and water quenched. They were not tempered. (Example 2)

In gas welding the high-carbon grades, it is advantageous to use an excess of acetylene gas to provide a carburizing flame. High-carbon filler rods are often used, although satisfactory results and an increase in ductility can be obtained with medium-carbon filler rods. However, some strength will be sacrificed. Preheating to at least 500 F is recommended. Postheating to about 1200 F is also recommended for improving the strength of the welds.

The arc welding of steels containing more than 0.50% C is more critical than gas welding. Excessive hardness and brittleness are probable in the fusion zone. For best results the same electrodes mentioned above under medium-carbon steels are recommended for high-carbon steels. Preheating to at least 500 F and postheating at 1200 to 1450 F should be employed.

Mild steel electrodes of the shielded-arc type can be used for welding high-carbon steels, but the deposited metal will absorb some carbon from the base metal and thus lose considerable ductility. Austenitic steel electrodes are often used for joining high-carbon steels. The resulting welds will have good mechanical properties, but the fusion zone may still be hard and brittle, and a post-treatment should be used.

Weld cracking correlated with carbon content is illustrated graphically in Fig. 25. This graph summarizes a study made in one large manufacturing plant. Since there are many causes for weld cracking other than carbon content, it was desirable to analyze the causes for cracks that appeared in a variety of welded components. Cracks occurring in steels of lower carbon content were attributed chiefly to design restraint and unbalanced thermal stresses. However, as the carbon content increased, it was evident that hardening became the major cause of weld cracking.

Underbead cracking can occur in steels of all carbon contents. Design restraint, unbalanced thermal stresses in unequal sections, lack of weld throat, improper welding sequence in restraint designs, and lack of bead penetration to the root of the joint are common causes of underbead cracking.

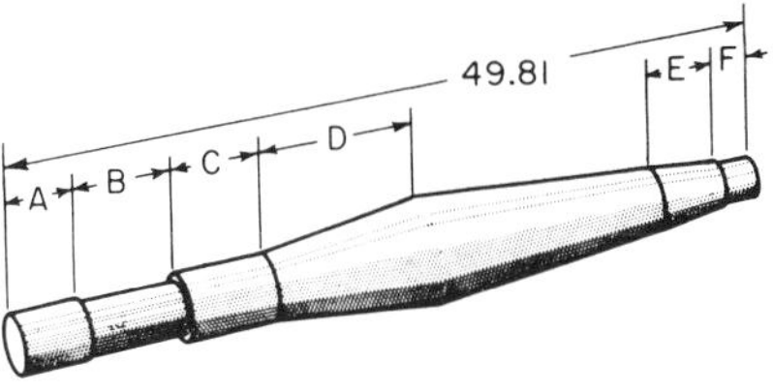

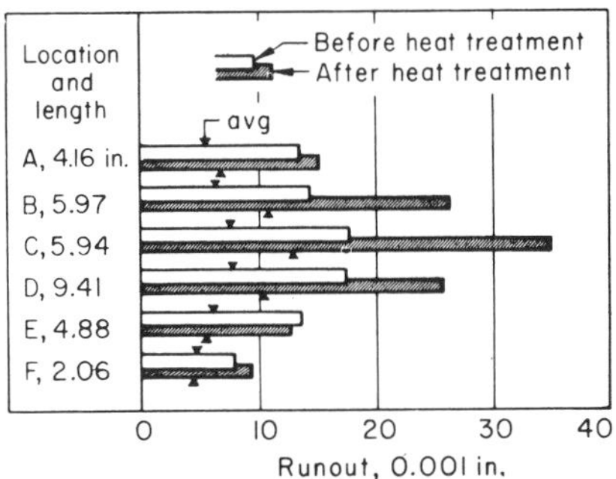

Fig. 24. Comparison of runout at six locations for 50 carbon steel shafts before and after heat treatment. Shafts were furnace heated at 1550 F, water quenched and tempered at 1000 F. Runout was measured on shafts mounted between centers. (Example 3)

Forgeability

Where carbon steels can meet property requirements, they are preferred for forging because they have lower hot strength than alloy steels, and the cooling rate from the forging hammer or press is less critical than for alloy steels.

Forgeability of the various grades is influenced somewhat by the carbon content. The lower-carbon grades can be more easily forged than the higher-carbon grades because higher forging temperatures can be used and they are therefore more plastic. The recommended maximum forging temperatures for carbon steels are as follows:

Carbon, %	Maximum forging temperature, F
0.10	2400
0.20	2375
0.30	2350
0.40	2300
0.50	2300
0.60	2250
0.70	2225
0.90	2150
1.10	2075

The finishing temperature should be enough above the transformation range to prevent rupture of the forging and excessive die wear, but low enough to prevent grain growth. A finishing temperature of about 2000 F is satisfactory for all steels except those with more than 0.80% C, for which a finishing temperature of about 1900 F is recommended.

The cost of forging even a simple shape of mild severity, such as Part 1 in Fig. 26, increases as the carbon content increases because of decreased rate of production and die life. The same forging is seldom made from steels that vary appreciably in carbon content; hence, direct comparisons are not available. However, one large forging plant uses the following figures in estimating the cost of forging steels of various carbon contents. With 1030 steel rated at 100%, the percentages for higher-carbon steels show how efficiently they may be produced in comparison with 1030.

Steel	Production, %	Die life, %
1030	100	100
1040	98	96
1045	96	92
1050	94	88
1060	90	81
1070	86	74
1080	82	67
1090	78	61
1095	76	58

All steels with less than 0.30% C are rated the same as the 0.30% C grade. The column headed "production" shows that the rate of production decreases as the carbon content increases; that is, forgings can be produced from 1095 steel at only 76% of the rate for 1030 steel. The expected die life (third column) must also be considered when estimating forging costs. For example, 58% die life is anticipated when forging 1095, against 100% life for 1030.

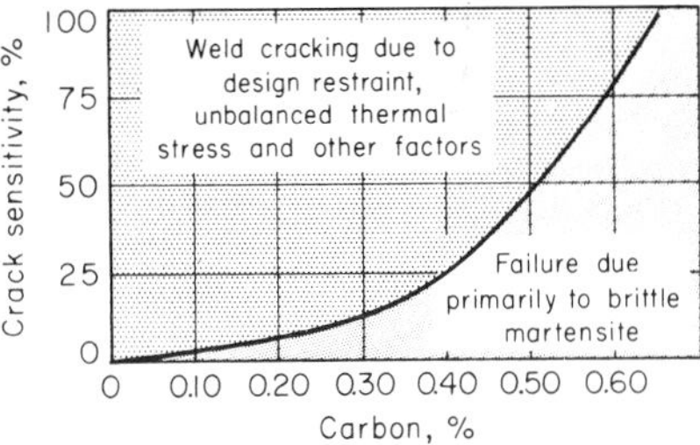

Fig. 25. Summary of investigation in a large fabricating plant to determine causes of cracking in carbon steel weldments. Brittle martensite was a minor cause of weld cracking in steels of lower carbon content but the major contributing factor for steels in the higher carbon range.

Shape of the forging affects both rate of production and die life. All carbon steels can be forged to shapes equaling the degree of severity of Part 2 in Fig. 26, but both rate of production and die life will be drastically reduced, regardless of the carbon content of the steel being forged. Forgings similar in severity to Part 1 can be made in about three hammer blows, but at least 10 blows would be required for forging Part 2, and it would probably require one reheating to assure a high enough finishing temperature. Assuming the same forging conditions and carbon level, it is estimated that the die life would be only about 20% as high for producing forgings like Part 2 as for forgings like Part 1.

Machinability

Carbon steels nearly always have better machinability than alloy steels when carbon content and strength are the same.

Carbon content has a strong effect on the machinability of carbon steels, chiefly because it governs strength and ductility. Low-carbon steels with less than about 0.15% C are low in tensile strength in the annealed condition and machine poorly because they are soft and gummy and adhere to the cutting tools. The machinability of these grades can be improved by quenching and tempering to raise the strength level and lower the ductility of chips. Steels in the carbon range from 0.15 to 0.30% usually machine satisfactorily in the annealed or normalized condition (pearlitic structure).

The medium-carbon grades, to about 0.55% C, machine best if the prior annealing treatment has been such that a mixture of lamellar pearlite and spheroidite is produced, because the strength will be too high for best machinability if the structure is not partially spheroidized. For steels with carbon higher than about 0.55% a completely spheroidized structure is preferred.

Where considerable machining is required, grades with additions of sulfur or lead should be considered. These grades are discussed in the sections that follow. For a detailed discussion of machinability and the free-maching grades, see the article on Selection of Steel for Economy in Machining, in this Handbook.

Resulfurized Grades

There are about 24 steels in the 1100 series of free-cutting carbon steels; eleven of these are within the range of 0.27 to 0.55% C and belong strictly to the hardenable group.

A few of the grades in the 1100 series may contain as much as 0.330% S, although most of them contain from 0.080 to 0.130% S.

Sulfur is added to steel for the sole purpose of decreasing machining costs, either by increasing productivity through greater machining speeds and improved tool life, or by eliminating secondary operations through an improvement in finish.

Response to heat treatment is not changed significantly by the addition of sulfur; the same techniques are used for the series 1100 steels as for the their series 1000 counterparts. However, there may be no exact counterpart between the two series because of manganese content, which must be considered in heat treating, since hardenability increases with increasing manganese content.

Cold Forming. The resulfurized steels should not be chosen for parts that require cold forming operations such as heading, swaging or staking, because the sulfide inclusions cause splitting and crumbling.

Joining. The resulfurized grades can be welded, although they are not preferred for welding. Gases from the sulfur are likely to produce a more porous and less sound weld than in the series 1000 steels.

Tests of furnace-brazed assemblies involving both resulfurized and nonresulfurized steels showed no detectable difference between the two steels in joint characteristics for either copper brazing at 2050 F or silver brazing at 1450 F. Furthermore, there is apparently no significant difference between the free-cutting steels and their nonresulfurized counterparts in joinability with the soft tin-lead solders.

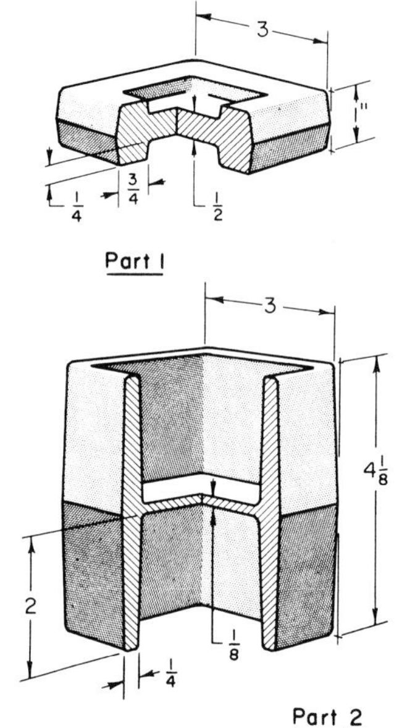

Fig. 26. Extremes of severity in closed-die forging. See text for discussion of forgeability of steels to these shapes.

Leaded Grades

The addition of lead to carbon steels is another means of increasing machinability and improving finish. Nearly all carbon steels of both the 1000 and 1100 series are now available with 0.15 to 0.35% Pb.

Mechanical Properties. Several studies have verified that there is no significant or discernible difference in static strength, ductility or notch sensitivity between fine-grained leaded carbon steels and their nonleaded counterparts. Several investigations report that 0.25% Pb slightly refines the austenitic grain size. Because of the grain refinement, lead may cause the hardenability to decrease slightly, but not to any significant extent. Some reports have shown that lead lowers the impact transition temperature of carbon steel; this also is attributed to the grain-refining effect.

Fabrication. All reports have indicated that the leaded grades can be forged, cold formed, swaged, welded, brazed and soldered as well as similar nonleaded grades. Leaded grades are seldom shaped by closed-die forging or cold forming because there is rarely any justification for the higher cost of the steel for processes that minimize machining. However, some plants have found it economical to use leaded grades for large open-die forgings on which a great deal of machining will be done.

Welding the leaded grades has been controversial because of the possibility of releasing toxic fumes from the lead. Reports disagree on the extent of danger to operators. However, any welding operation with leaded steels should be conducted with good ventilation so that operators will not be exposed to fumes.

The machinability of leaded steel is discussed in detail, with numerous examples, in the article on Selection of Steel for Economy in Machining.

Selection of Bar Finish

The difference in price between hot rolled and cold finished bars ($2.00 to $3.00 per 100 lb) is of major importance to most manufacturing plants, and a careful cost study for each different part is warranted.

The principal reasons for choosing cold finished bars instead of hot finished are: (*a*) closer dimensional tolerances, (*b*) scale-free surfaces, (*c*) straighter bars, (*d*) increased strength and (*e*) better machinability. All of these factors are important in the lower-carbon grades and one or more of them may justify the higher cost of cold finished bars. However, as carbon increases above about 0.25% the last two factors have little if any significance.

Careful analysis of the use of cold drawn medium-carbon steels for machining shows that their major appeal arises from the uniformity of section thickness, with cleaner surfaces and straightness being secondary factors. Surface finish is especially important when it permits a reduction in the number of machining operations. If finish is not critical and tooling can adjust to the greater size differences encountered, then hot rolled steels will be more economical.

While the cold drawn steels have better machining characteristics, this factor alone is only rarely a criterion for selection. The extra strength obtained with cold drawn steel is more important from a cost standpoint, since it is often high enough to eliminate the need for heat treatment.

By specifying hot rolled, pickled and machine-straightened bars, the user of steel can overcome most of the objections to plain hot rolled steel. There still remains the dimensional variation, but the major steel producers seldom use more than half the allowable tolerance. Of course, one producer may use the upper half and another the lower half, but at least the consumer is in a position to know the approximate dimensional range and to correlate shipments of steel with specific machine tools, selecting special collets or chucks in advance as needed. The extra cost for pickled, oiled and machine-straightened bars is small ($1.10 per 100 lb) compared with the extra of $2.00 to $3.00 per 100 lb for cold drawn bars.

Example 4. In one large factory 97 parts were surveyed to determine the extent to which a change from cold drawn to hot rolled, pickled and machine-straightened bar would be profitable. As a result of this study 35 of the 97 parts are now made from hot rolled material. Some of the parts and the amounts saved are:

Part	Steel	Net savings per 100 pcs
Splined drive shaft	1045	$16.64
Tightener shaft	1045	8.13
Splined shaft	1045	19.56
Spindle	1045	10.48
Roll stud	1045	14.01
Jack shaft	1025	8.37
Spindle	1025	5.66

The savings were entirely from cost of material since no operations were changed. The cost difference, however, does not reflect all of the difference in steel cost because in some instances it was necessary to use special collets and these extra tool costs have been deducted from the total savings to give the net savings shown.

Example 5. In another plant, pieces were manufactured from carbon-restored cold finished 1049 steel bars 1.312 to 1.316 in. in diameter. A change was made to bars of machine-straightened 1049 steel 1 13/32 in. in diameter (1.406 in.). The hot finished pieces were turned and polished. Despite the necessity for removing about 0.093 in. from the diameter (reflected in labor cost), a significant saving was realized. Details are tabulated below:

Item	Cold finished, per piece	Hot rolled, per piece
Steel cost	$0.504	$0.385
Handling	0.032	0.026
Labor	....	0.053
Burden	....	0.032
Total	$0.536	$0.496

Example 6 (Table 1) gives an additional cost summary showing hot rolled steel selected in preference to cold finished steel. For these lawn-mower tie rods, hot rolled and pickled bars had a satisfactory finish and machine-straightened bars could be processed in screw machines. The higher strength of cold drawn steel was of no value in this instance, and the hot rolled grade proved more economical.

Selection of Quality Level

Killed versus Semikilled Steel. Killed steels should usually be specified in ordering carbon steels with more than 0.25% C. For ranges below 0.25% C, semikilled steels are often used, chiefly because they are about 4% cheaper. The evolution of gases during the solidification of semikilled steel has an adverse effect on quality, manifested in decreased internal soundness and a greater variation in composition throughout the ingot, compared with killed steels.

For making parts from low-carbon steels, especially if they are not heat treated, it may be economical to take advantage of the lower cost of the semikilled product. Since cost of the two qualities is usually identical for steels higher than 0.25% C, there is rarely any economic reason for specifying or permitting the use of semikilled medium-carbon grades. However, there have been occasions when, because availability was a factor, it became necessary to use semikilled steel.

Example 7. An instance of this sort caused a comparative study of the two grades in one large manufacturing plant. A specific section had been rolled from ingots of fully killed steel but because of a shortage of hot topped steel, semikilled steel was purchased. After one year of using semikilled steel in production the following facts became evident: *(a)* cold shearing losses increased to 7%, compared with insignificant losses when fully killed steel was used, *(b)* carbon varied as much as 0.17% within some heats of 1045 steel (0.36 to 0.53%), *(c)* these carbon variations together with similar variations in manganese resulted in erratic hardenability and often failure to meet the specified as-quenched hardness of Rockwell C 27 to 34, oil quenched, *(d)* parts made from semikilled steel cracked more frequently than those made from the fully killed steel. Because of these disadvantages the cost of the finished parts was higher using semikilled steel and, as soon as killed steel was available, use of the semikilled grade was discontinued.

Other Quality Restrictions. Bars and billets can be obtained in different quality levels including restricted chemical composition, internal soundness and surface limitations. Requirements of the end product influence the amount of extra payment that can be justified for quality restrictions.

Example 8. In one instance, 1010 steel of merchant-bar quality was used for the hot forging of nut blanks. Surface seams caused cracks in approximately 12% of the forged blanks, making sorting necessary. A change to special-bar quality reduced the number of cracked blanks almost to zero, so that sorting was no longer required. Comparative costs were:

Item	Special-bar quality	Merchant-bar quality
Cost per 100 lb ..	$6.715	$6.365
Cost of steel per 100 good blanks	8.058	8.682
Cost of sorting ..	....	0.220
	8.058	8.902
Scrap recovery ..	....	0.246
Net cost per 100 blanks	$8.058	$8.656

It is seldom economical to select merchant-bar quality for parts to be cold headed or cold extruded.

Cold heading quality is produced by controlled mill practices and is subject to mill inspection for internal soundness and surface seams. Grades containing 0.30% C or more are specified in the annealed or spheroidize-annealed condition, which is necessary for cold heading and cold forging.

Example 9. In one plant, under emergency conditions, it was necessary to use 1038 steel of merchant-bar quality instead of cold heading quality. Parts were produced on a 1/2-in. boltmaker where the operator was able to sort out defective pieces amounting to approximately 4%. This caused a decrease in machine productivity, with an increase of 10% in machine-hour rate. The following tabulation shows that parts made from merchant-bar quality cost about 34% more than identical parts produced from cold heading quality steel. Costs other than those shown were the same.

Item	Cold heading	Merchant bar
Steel cost per 100 lb ...	$5.975	$5.175
Steel cost per 100 lb of blanks	5.975	5.434
Added inspection	....	0.25
Increased machine rate per hr	....	2.392
Net cost per 100 lb of blanks	$5.975	$8.076

Cold extrusion quality is available in various carbon levels. In order to obtain maximum workability and to minimize age hardening this quality level is produced only from fully killed fine-grained steels that have been inspected for internal soundness and surface seams. Steels conforming to cold extrusion quality are usually spheroidize annealed to attain maximum softness.

It is impractical to use merchant-bar quality for manufacturing parts by cold extrusion. In one plant where merchant-bar quality was tried, even after selecting only those bars with the best surface, virtually every part showed hairline cracks extending in from the extruded section, even when only those bars with the best surface were selected for processing. Furthermore, it was impossible to make any parts from merchant bar without leaving a flash that had to be removed in a secondary operation. Thus, it was not only impractical, but impossible, to produce satisfactory parts from steel of merchant-bar quality.

Table 1. Cost Comparison of Cold Drawn and Hot Rolled 1212 Steel, Used for Tie Rods (Example 6)

	1212 CD	1212 HR
Unit weight, lb	1.673	1.673
Work standard, hr per 100	2.5	2.5
Material cost, $ per 100 lb	$ 10.20	$ 7.50
Direct material, lb ...	41,825	41,825
Material cost	$4266.25	$3136.87
Direct labor, hr	625	625
Direct labor cost	$ 905.00	$ 905.00
Total saving using hot rolled		$1129.38
Unit saving per piece using hot rolled		$ 0.045

For making lawn-mower tie rods from 5/8-in. diam 1212 steel, based on annual production of 25,000 pieces

Surface Defects and Segregation

Under this same heading in the article on Hot Finished Carbon Steel, seams, decarburization, depth of surface defects and segregation are discussed. Decarburization is greater in the larger bar sizes. Machining allowances for bars of various diameters are tabulated and discussed in the article on Selection of Steel for Economy of Manufacture.

Hardenable Alloy Steels

By J. M. Hodge and E. C. Bain*

BY FAR the largest tonnage of alloy steels is of the types containing generally 0.25 to 0.55% C, or less if for carburized parts — widely used in automotive and other machinery and almost always quenched and tempered for high strength and toughness. Manganese, silicon, nickel, chromium, molybdenum, vanadium, aluminum and boron are commonly present in these steels to enhance the properties obtainable after quenching and tempering.

These alloy steels are ordinarily quench-hardened and tempered to the level of strength desired for the application. Even though the strength level at which the steel is used may be as low as, or lower than, that which could be achieved by the microstructure (fine pearlite or upper bainite) developed by a simple cooling from forging or normalizing temperature, the steels are quench-hardened and tempered, indirectly reflecting the engineering and economic basis of the demand for this type of steel.

The microstructure (tempered martensite or bainite) produced by quenching and tempering these alloy steels is characterized by a greater toughness or capacity to deform without rupture at any strength level. Similarly, under the adverse state of stress below a notch in bending, the tempered martensite may flow considerably at a testing temperature far below that at which a pearlitic steel of equal strength would break in a brittle manner; the Charpy or Izod values are thus improved. The basic phenomenon of developing this favorable microstructure by heat treatment is manifest in plain carbon steels, but only in small sections; thus the most important effect of the alloying elements in these steels is to permit the attainment of this microstructure, and the accompanying superior toughness, in larger sections.

Alloying Elements Dissolved in Austenite

The general effect of elements dissolved in austenite is to decrease the rates of transformation of the austenite at subcritical temperatures. The only one among the common alloying elements to behave exceptionally in this regard is cobalt. Since the desirable products of transformation in these steels are martensite and lower bainite, formed at low temperatures, this decreased transformation rate is essential; it means that pieces can be cooled more slowly, or larger pieces can be quenched in a given medium, without transformation of austenite to the undesirable high-temperature products, pearlite or upper bainite. This function of decreasing the rates of transformation, and thereby facilitating hardening to martensite or lower bainite, is known as hardenability and is the most important effect of the alloying elements in these steels. Thus, by increasing hardenability, the alloying elements greatly extend the scope of enhanced properties in hardened and tempered steel to the larger sections involved in many applications.

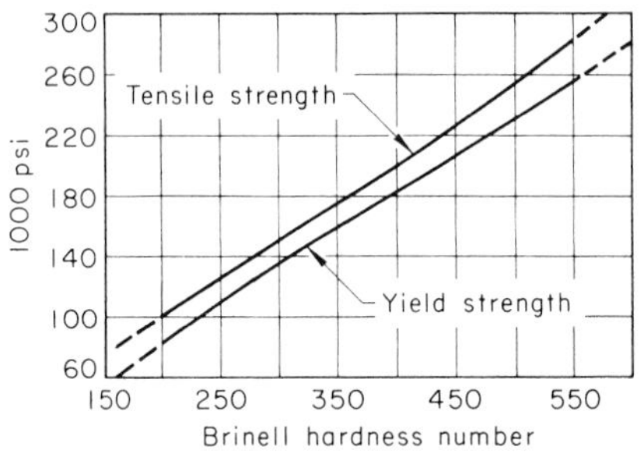

Fig. 1. Statistical mean of the relationships among Brinell hardness, tensile strength and yield strength in quenched and tempered alloy steels

The several elements commonly dissolved in austenite prior to quenching, increase hardenability in approximately the following ascending order: nickel, silicon, manganese, chromium, molybdenum, vanadium and boron. The effect of aluminum on hardenability has not been accurately evaluated, but at 1% Al, as used in "nitralloy" steels, the effect on hardenability seems to be relatively small. Further, it has been found that the addition of several alloying elements in small amounts is more effective in increasing hardenability than the addition of much larger amounts of one or two.

In order to increase hardenability effectively, it is essential that the alloying elements be in solution in the austenite, and the steels containing the carbide-forming elements — chromium, molybdenum and vanadium — require special consideration in this respect. These elements are present predominantly in the carbide phase of annealed steels, and such carbides dissolve only at higher temperatures and more slowly than iron carbide. Hence, for effectiveness, heating schedules should be set up in such a way as to dissolve an adequate proportion of these elements. It is uneconomical to heat at such low temperatures that the carbide-forming elements are largely undissolved.

Since the basic function of the alloying elements in these steels is to increase hardenability, the selection of a steel and the choice of suitable austenitizing conditions should be based primarily on the assurance of adequate hardenability. More than adequate hardenability is rarely a disadvantage, except in cost.

Alloying Elements in Quenching

Since the sections treated are often relatively large, and since the alloying elements have the general effect of lowering the temperature range at which martensite is formed, the thermal and transformational stresses set up during quenching tend to be greater in these alloy steel parts than those involved in quenching the necessarily smaller sections of plain carbon steels. In general, this means greater distortion and risk of cracking.

The alloying elements, however, have two functions that tend to offset these disadvantages. The first and probably the most important of these functions is that of permitting the use of a lower carbon content for a given application. The decrease in hardenability accompanying the decrease in carbon content may be offset very readily by the hardenability effect of the added alloying elements, and the lower-carbon steel will exhibit a much lower susceptibility to quench cracking. This lower susceptibility results from the greater plasticity of the low-carbon martensite and from the generally higher temperature range at which martensite is formed in the lower-carbon materials. Quench cracking is seldom encountered in steels containing 0.25% C or less, and the susceptibility to cracking increases progressively with increasing carbon content.

The second function of the alloying elements in quenching is to permit slower rates of cooling for a given section, because of increased hardenability, and thereby generally to decrease the thermal gradient and, in turn, the cooling stress. It should be noted, however, that this is not altogether advantageous, since the direction, as well as the magnitude, of the stress existing after the quench, is important in relation to cracking.

In order to prevent cracking, the

*J. M. Hodge is Research Consultant, Steel Product Development, Applied Research Laboratory, United States Steel Corp.; E. C. Bain, Past President, American Society for Metals, is retired Vice President, United States Steel Corp.

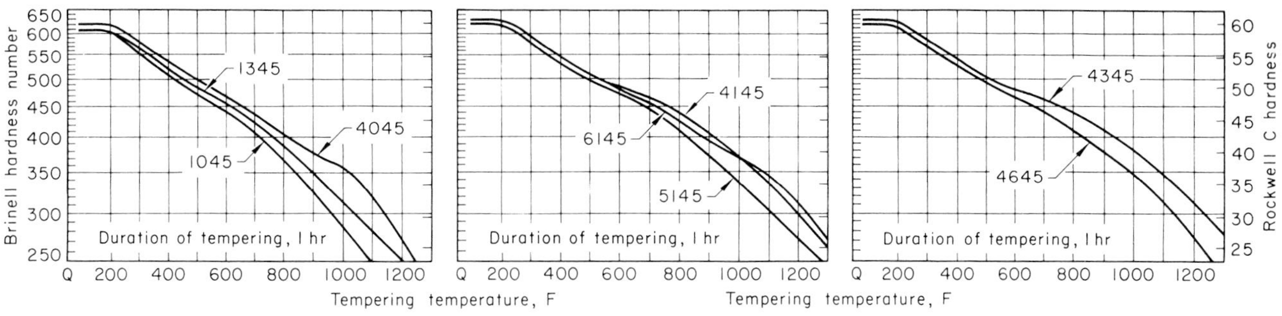

Fig. 2. Tempering characteristics of nine 0.45% C alloy steels at various temperatures

surface stresses after quenching should be either compressive or at a relatively low tensile level and, under certain circumstances, lowering the cooling rate may lead to increased tensile stresses at the surface, thus increasing the tendency to crack. In general, though, unless a study of the particular piece being quenched indicates that it falls within this category of increased susceptibility with decreased quenching rates, the use of a less drastic quench suited to the hardenability of the steel will result in lower distortion and greater freedom from cracking.

Furthermore, the increased hardenability of these alloy steels may permit heat treatment by "austempering" or "martempering", and thereby the level of adverse residual stress before tempering may be held to a minimum. In "austempering", the workpiece is cooled rapidly to a temperature in the lower bainite region and is allowed thereafter to transform completely at some chosen temperature. Since this transformation occurs at a relatively high temperature and proceeds rather slowly, the stress level after transformation is quite low and distortion is held to a minimum.

In "martempering", the piece (*a*) is cooled rapidly at the surface to a temperature that permits very little martensite to form, if any; (*b*) is equalized at this temperature; and (*c*) is then cooled slowly so that transformation throughout the whole section occurs more or less simultaneously, thereby holding transformational stresses at a very low level and minimizing distortion and danger of cracking.

Alloying Elements in Tempering

Hardened steels are softened by reheating, but this effect is not the one actually sought in tempering. The real need is for increasing the capacity of the piece to flow moderately without fracture, and this is inevitably accompanied by a loss of strength. Since the tensile strength is very closely related to hardness in this class of steels, as heat treated, it is satisfactory to follow the effects of tempering by measuring the Brinell or Rockwell hardness. The statistical mean of the relationships among Brinell hardness, tensile strength and yield strength, is shown in Fig. 1, drawn from many data.

Figure 2 shows the softening pattern of nine 0.45% C steels with increase in the tempering temperature, for one hour. Somewhat shorter or longer intervals at temperature would show little difference in hardness values. As a first approximation, the softening pattern of steels similar but differing in carbon content through the range from 0.25 to 0.55% C may be estimated from Fig. 3. This illustrates the general effect of carbon content on this softening pattern in terms of Rockwell C hardness units to be added to or subtracted from the 0.45% C value at different levels of tempering temperature, but accurate data that would permit a strictly quantitative relationship of this type are not available.

It will be noted that the effect of carbon content on the hardness of the tempered steels is much greater for the lower tempering temperatures than for 1200 F and higher, and that the effect likewise decreases when there is more than 0.50% C. Figures 2 and 3, used together, make it possible to estimate the strength from a given tempering treatment applied to a given type of steel.

It will be seen that the general effect of the alloying elements is to retard the softening rate, so these alloy steels will require a higher tempering temperature to obtain a given hardness than carbon steel of the same carbon content. However, the individual elements show significant differences in the magnitude of their retarding effect. Nickel, silicon, aluminum and, to a large extent, manganese, which have little or no tendency to occur in the carbide phase, and merely remain dissolved in ferrite, have only a minor effect on the hardness of the tempered steel — an effect that would be expected from the general pattern of solid-solution hardening.

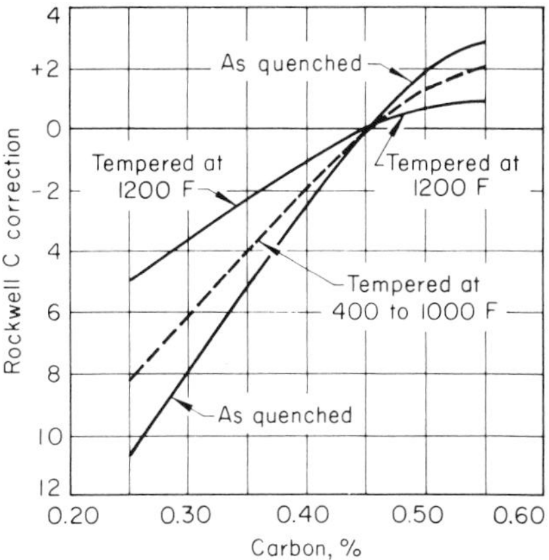

Fig. 3. Effect of carbon content on the hardness of tempered steel. Rockwell C hardness units to be added to or subtracted from the value for 0.45% C at different tempering temperatures

Chromium, molybdenum and vanadium, on the other hand, which migrate to the carbide phase when diffusion is possible, bring about a retardation of softening, particularly at the higher tempering temperatures. These elements do not merely raise the tempering temperature; when they are present in higher percentages, the rate of softening is no longer a continuous function of the tempering temperature. That is, the softening curves for these steels will show a range of tempering temperature in which the softening is retarded or, with relatively high alloy content, in which the hardness may actually increase with increase in tempering temperature. This characteristic behavior of the alloy steels containing the carbide-forming elements is known as "secondary hardening" and results presumably from a delayed precipitation of fine alloy carbides.

As mentioned previously, the primary purpose of tempering is to impart a degree of plasticity or toughness to the steel in contrast with the brittleness of the as-quenched martensite, and the loss in strength is only incidental to this very important increase in toughness. The increase in toughness after tempering reflects two effects of tempering: (*a*) the relief of residual stress set up by the quenching operation, and (*b*) a precipitation, coalescence and spheroidization of iron and alloy carbide resulting in a microstructure of greater plasticity.

In addition to their effects on the microstructure, as discussed above in connection with the softening behavior, the alloying elements have a secondary function. The higher tempering temperatures for a given hardness, which has been seen to be a characteristic of these alloy steels, particularly of the ones containing the carbide-forming elements, will presumably permit greater relaxation of the stress and thereby improve the resulting properties. Furthermore, as discussed in the section on quenching, the hardenability of these steels may permit the use of less drastic quenching practices,

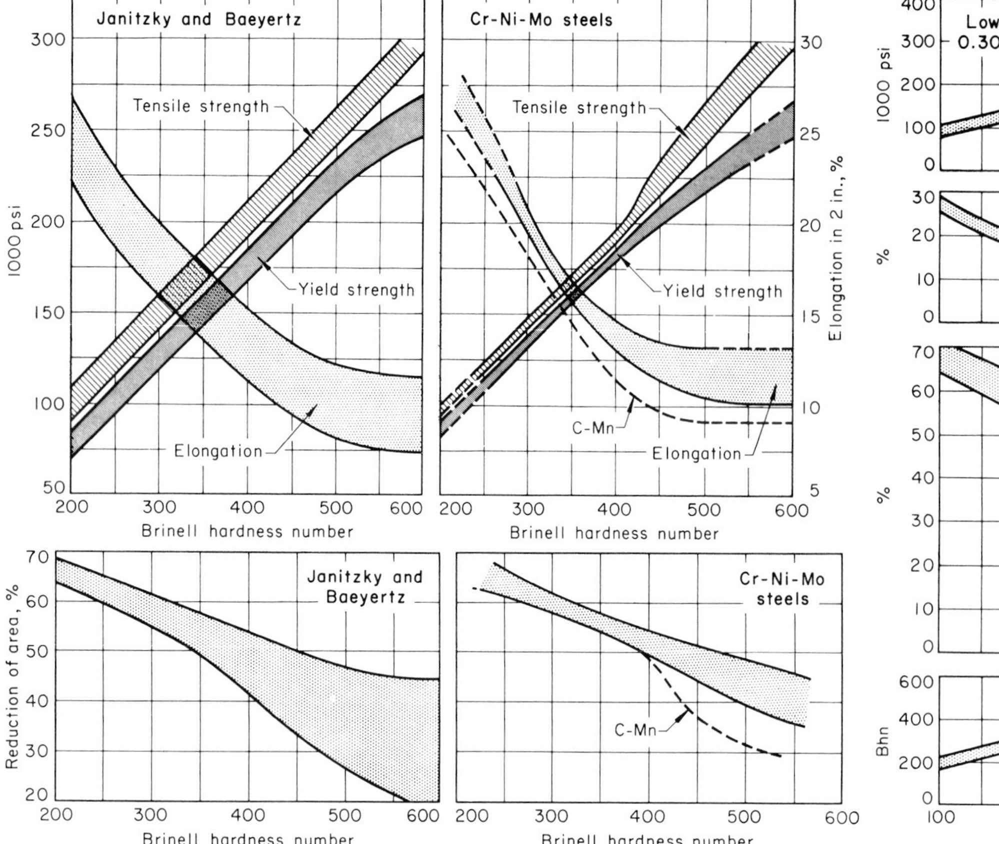

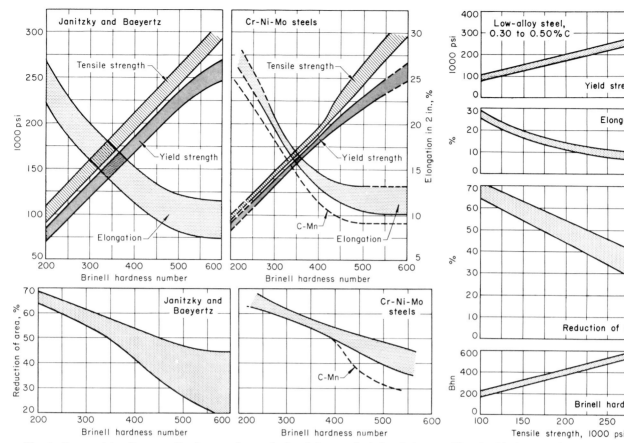

Fig. 4. Range of tensile properties in several quenched and tempered steels. Data of Janitzky and Baeyertz and more recent data for Cr-Ni-Mo steels.

Fig. 5. Most probable properties of tempered martensite (Patton)

so that the stress level before tempering will be lower, thereby permitting the use of these steels at a higher level of hardness, since the higher temperatures are not then required for relief of quenching stresses. It should be noted, however, that this latter is only a secondary function of the alloying elements in tempering; the effect primarily reflects the hardenability function of the alloying elements.

Still another secondary function of the alloying elements in tempering is to permit the use of steels with lower carbon content for a given level of hardness, since adequate tempering may be assured by taking advantage of the retardation of softening brought about by the alloying elements. This results in greater freedom from cracking and a generally improved plasticity at the given hardness. Here again, however, the function of the alloying elements in tempering is a secondary function; their primary function is to increase the hardenability sufficiently to offset the effect of the decreased carbon content.

The increase in plasticity on tempering is discontinuous in these alloy steels and their behavior under the stress condition of a notch shows a characteristic irregularity at approximately 500 to 600 F. The quenched martensitic steel gains toughness, as reflected in the notched-bar impact test, by tempering at temperatures as high as 400 F, but after tempering at higher temperatures this type of steel loses toughness until it may be less tough than the untempered martensite. Still higher temperatures for tempering restore greater toughness.

The mechanism of this behavior is not fully understood, but it seems to be associated with the first precipitation of carbide particles, and is presumably a grain-boundary phenomenon since fractures of steels tempered in this region tend to be intercrystalline. Thus, there is a range of tempering temperatures (about 400 to 700 F) never used for these steels; the tempering temperature is either below 400 F or above 700 F. Although this phenomenon is common to all these alloy steels, the alloying elements have a secondary function in this connection; a combination of carbon and alloy contents of suitable hardenability may be chosen that would permit tempering to the desired strength at temperatures outside this undesirable range.

Temper brittleness is another example of a discontinuous increase in plasticity after tempering steels of these types. This phenomenon is manifested as a loss of toughness observed after slow cooling from tempering temperatures of 1100 F or higher, or after tempering in the temperature range between approximately 850 and 1100 F. Thus, a steel that is susceptible to this type of embrittlement may lose much of its plasticity, as indicated by a notched-bar impact test, during slow cooling from a tempering temperature of 1150 F, although the same steel will be very tough if it is quenched from the same tempering temperature.

This expedient of quenching from the tempering temperature is a common practice to insure freedom from loss of toughness. However, in steels that are susceptible to temper brittleness, embrittlement will also be observed after tempering at 850 to 1100 F, particularly if the tempering times are protracted, and under such circumstances quenching from the tempering temperature will never restore the toughness completely.

High manganese, phosphorus and chromium appear to accentuate this behavior; and molybdenum has a definite retarding effect. Here again, the carbon and alloying elements may be chosen so that the susceptibility to temper embrittlement is minimized, or the desired strength level may be obtained by tempering either below 850 F or above 1100 F, and then quenching.

Marked Similarities Among the Steels

This article has emphasized hardenability as the primary function of the alloying elements in alloy steels and it has been found that, if the hardenability and heat treatment are such as to obtain a tempered martensitic microstructure, the mechanical properties of steels are markedly similar regardless of the alloy combination used to obtain the requisite hardenabilities.

This similarity among the properties of heat treated steels was noted early by Janitzky and Baeyertz and has since been verified by many investigators. Figure 4 shows these property bands as determined by Janitzky and Baeyertz and also as more recently determined on a wide range of Mn-Ni-Cr-Mo steels. For comparison, the more recent plots include the values for a simple carbon-manganese steel (treated to tempered martensite).

This should not be interpreted as implying that these tempered martensitic steels are alike in every respect, regardless of composition, because the composition is responsible for differences in preservation of strength at elevated temperatures, for differences in abrasion resistance, in resistance to corrosion and even, to a certain extent, in toughness. However, the similarities are sufficiently marked to permit reasonably accurate predictions of the mechanical properties of tempered martensitic steels, regardless of composition, and thereby to justify the emphasis on hardenability as the most important function of the alloying elements.

One of the most comprehensive studies of this nature has been made by Patton. The average values of the mechanical properties of tempered martensite as found in this study are shown in Fig. 5. Although representing a compilation of data concerning a very wide variety of alloy steels from a wide range of sources, with inevitable differences in treating and testing, the results still fell within quite narrow bands. Therefore these curves can be used to predict the properties of any of the common alloy steels within plus or minus 10%, if the hardenability and heat treatment are such as to obtain a microstructure consisting essentially of tempered martensite.

Hardness Number Conversion Chart

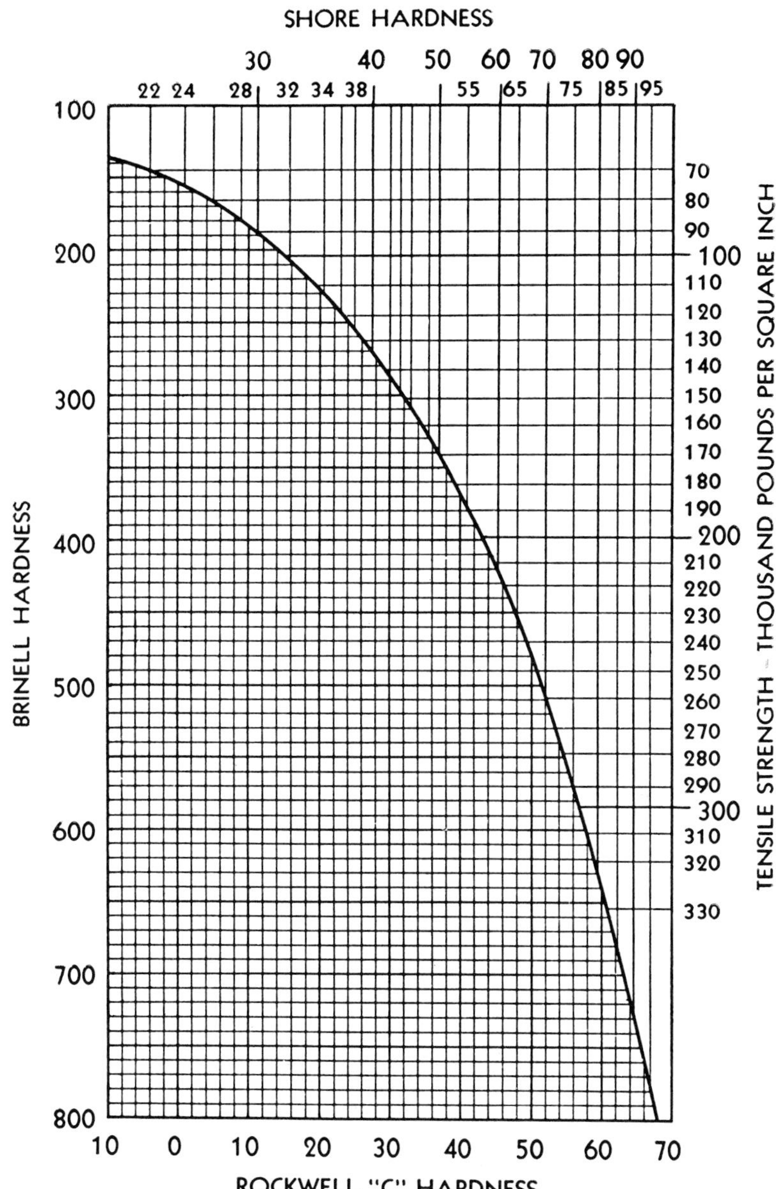

(Reproduced by courtesy of International Nickel Co.)

A System for Classifying

Heat-Treatable Steels

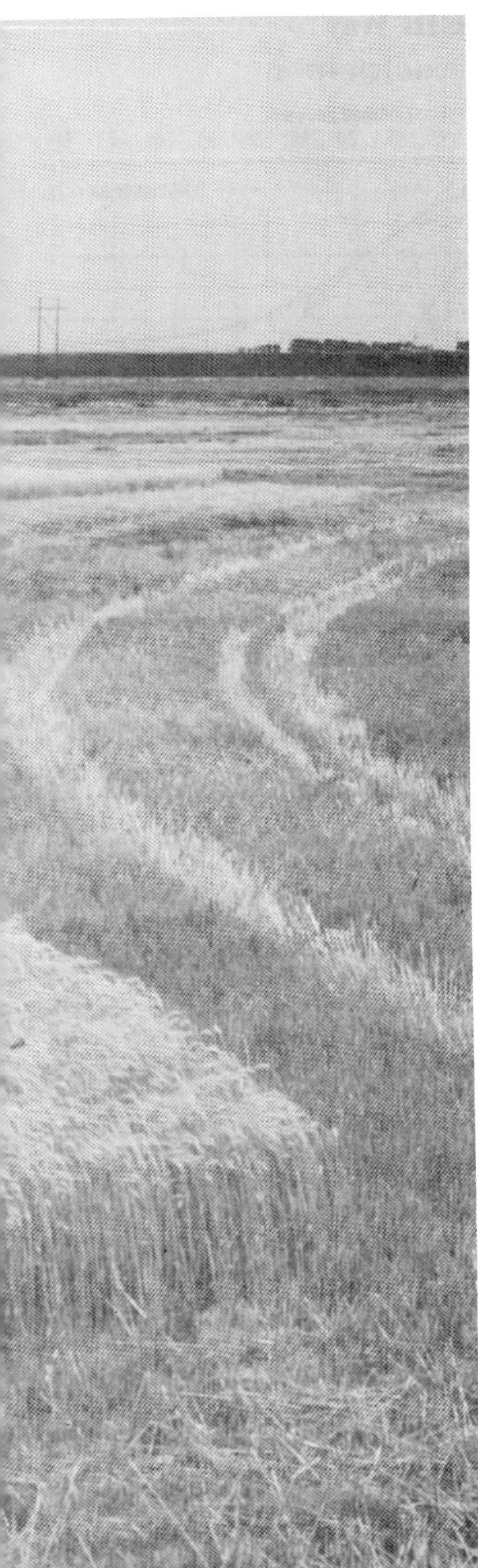

By John T. Sponzilli, Charles H. Sperry, and Jacob L. Lytell Jr.

SINCE MID-1973 International Harvester has been using a simplified system to specify steels for heat-treated components going into agricultural, industrial, construction, and transportation equipment. Because this system allows use of multiple grades with comparable hardenabilities for given heat-treated parts, its usage has grown as IH engineers and steel suppliers have become familiar with it.

Basis of the System

At IH it is common to specify more than one standard SAE/AISI grade of steel on a component drawing. Benefits include flexibility of manufacture, cost savings, and inventory control.

This method of steel classification permits our manufacturing facilities to select a steel from one of several which can be heat treated to the same hardness gradients. (The code is not applicable to parts such as gears, shafts, and bearings which are not usually applicable for multiple grades.)

Specific heats of steel of similar grades that meet the code

Many steel components in the combine (left) and hydraulic excavator (below) are heat treated for strength and durability.

How to Code Hardenable Steels the IH Way

8640H IH Code: C37-44E7-15

Distance From Quenched End, mm

5 10 15 20 25 30 35 40 45 50

C37-44*

Rc 60 max at $J = \frac{1}{16}$ in. (1.6 mm)

Rc 53 min at $J = \frac{1}{16}$ in. (1.6 mm)

Hardness, Rc

65 60 55 50 45 40 35 30 25 20

50% martensite

1 2 4 6 8 10 12 14 16 18 20 22 24 26 28 30 32

Distance From Quenched End, $\frac{1}{16}$ in.

Rc 41 min at $J = \frac{7}{16}$ in. (11 mm) for 50% martensite (0.37% C)

Rc 44 max at $J = \frac{15}{16}$ in. (24 mm) for 50% martensite (0.44% C)

E7-15

*Coded from 0.37-44% C, the range for 8640H and 4140H.

Diesel engines include a number of heat-treated parts.

requirements may also be employed for an application, and overseas-made grades can be applied where comparable North American steels are not available.

The system also permits outside suppliers to have maximum freedom in steel selection consistent with maintaining required properties, allows for development and use of more economical steels, and facilitates steel procurement during times of shortage.

One of the most important requirements for steels of heat-treated components is that they develop the minimum microstructures and strength properties required. To do this, the steels must have the proper carbon content and hardenability. In general terms, steels with equivalent carbon content and hardenability can be expected to produce components with equal microstructures and hardnesses. Thus, by specifying carbon content and hardenability requirements on a component drawing, we can permit the use of different grades of steel somewhat irrespective of alloy content.

In the IH system, the hardenability criterion is the Jominy distance (J distance) to a 50% martensitic microstructure. This point is thought to provide the best measure of

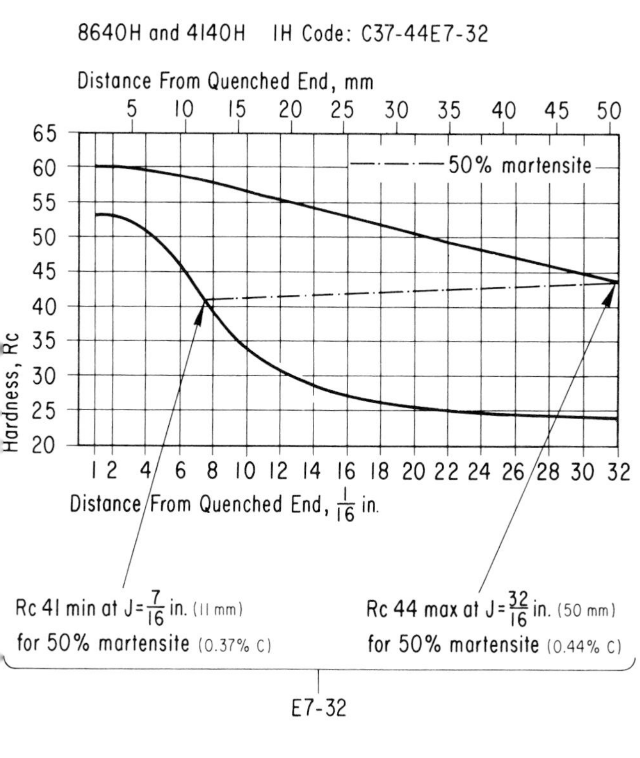

comparative hardenability; it also approximates the inflection point on a Jominy hardenability curve.

In practice, it is more convenient to work with hardnesses on Jominy hardenability bands rather than with microstructures. Fortunately, hardnesses are directly related to microstructures, the hardness of a 50% martensitic microstructure being governed primarily by carbon content though the quantity and type of the remaining microconstituents affect hardness to some extent. Similarly, the maximum attainable hardness on a Jominy bar at the quenched end (1/16 in. [1.6 mm] position = J1 position) is governed primarily by carbon content; it equals the hardness of a fully martensitic microstructure.

Relationships among carbon content, average hardness, and microstructure are listed in Table I. Note: Although the maximum J1 hardness depends primarily on carbon content, it also varies according to whether carbon is intended to be the minimum or maximum of a specified range. This dependency reflects the many standard SAE/AISI steel hardenability bands whose J1 hardness varies from the theoretical value, depending upon whether the carbon level is a minimum or a maximum of a specified range.

Coding Is Simple

To show how the system works, we will use Table I and the illustrated curves to establish a code for steels with hardenability limits equal to those of 8640H (specified carbon range, 0.37-0.44%). The hardenability curve of 8640H (lefthand curve) shows a J1 hardness range of Rc 53 to 60. Converted to the corresponding hardness in Table I, this range is coded C37-44. Also from Table I, a 50% martensitic structure with a carbon content of 0.37% has a hardness of Rc 41, and 0.44% C indicates Rc 44 for the 50% martensite hardness.

Reference to the 8640H curve shows that the Jominy distances corresponding to these hardnesses are approximately 7/16 and 15/16 in. (11 and 24 mm); this is denoted E7-15. (E indicates that Jominy distances are given in English units, sixteenths; M is used instead if Jominy distances are in millimetres.) Thus, the coded hardenability of 8640H is called out by C37-44E7-15. Although this code describes the hardenability of 8640H, it also permits application of all steels whose hardenability curve falls within the hardenability band described by the code.

Similarly a code could be developed for the 4140H steel. Using the procedures outlined above, the code would be C37-44E11-32.

A code can be expanded to accommodate additional grades. For example, 8640H and 4140H – both contain 0.37 to 0.44% C – could be described by the code C37-44E7-32. As the righthand curve shows, the Jominy distance is expanded to include both.

Codes are expanded because materials engineering practice dictates use of the widest range of materials which will give the required engineering properties in the end product. A low-side heat of 8645H or 4145H that meets code requirements could also be used interchangeably with 8640H and 4140H. Since most heats of 4135H have Jominy hardenabilities that fall within this code, they could also be used for a component whose steel requirements are specified in this manner. This is one of the most important and useful features of the code – it does not restrict manufacturing to the use of any particular steel grade.

The code can also be used to specify a restricted hardenability or carbon content. For example, a restricted 8640H hardenability band could be described by C37-44E9-15. Restricted hardenability bands, however, generally involve price penalties and should be avoided if at all possible.

Codes that describe the carbon content and hardenability levels of some North American and overseas steels are given in Table II. Whenever the steel's hardenability level is so high that 50% martensite is not reached until beyond the

Table I — Carbon Content/Hardness/Microstructure Data for Steels

Specified Carbon Content, %	Hardness, Rc J Distance, 1/16 in. or 1.6 mm		
	Carbon Specified as Min	Carbon Specified as Max	50% Martensite for Specified Carbon Content
0.07	36	—	25
0.08	36	—	25
0.09	36	—	26
0.10	37	—	26
0.11	37	—	27
0.12	38	42	27
0.13	38	43	28
0.14	39	43	28
0.15	40	44	29
0.16	40	44	30
0.17	41	45	30
0.18	42	45	31
0.19	43	46	31
0.20	44	46	32
0.21	44	47	32
0.22	45	48	33
0.23	46	48	34
0.24	47	49	34
0.25	48	50	35
0.26	49	51	35
0.27	49	51	36
0.28	50	52	36
0.29	50	53	37
0.30	50	54	37
0.31	51	55	38
0.32	51	55	38
0.33	51	56	39
0.34	52	56	40
0.35	53	57	40
0.36	53	57	41
0.37	53	58	41
0.38	54	58	42

Specified Carbon Content, %	Hardness, Rc J Distance, 1/16 in. or 1.6 mm		
	Carbon Specified as Min	Carbon Specified as Max	50% Martensite for Specified Carbon Content
0.39	55	58	42
0.40	55	58	43
0.41	55	59	43
0.42	56	59	43
0.43	56	60	44
0.44	57	60	44
0.45	57	61	45
0.46	58	62	45
0.47	59	62	45
0.48	59	63	46
0.49	59	63	46
0.50	60	63	47
0.51	60	64	47
0.52	60	64	48
0.53	60	64	48
0.54	60	65	48
0.55	60	65	49
0.56	—	65	49
0.57	—	65	50
0.58	—	65	50
0.59	—	65	51
0.60	—	65	51
0.61	—	65	51
0.62	—	65	51
0.63	—	65	52
0.64	—	65	52
0.65	—	65	52
0.66	—	—	52
0.67	—	—	52
0.68	—	—	53
0.69	—	—	53
0.70	—	—	53

J32 position on the Jominy bar, a hardness at the J32 position on the Jominy bar is specified. For example, the code for 4142H is C39-46E13-32 (50); (50) indicates a hardness of Rc 50 at J32.

Special Considerations

In some applications, the 50% martensite hardness criteria may not adequately describe the requirements of a steel suitable for a particular component. Then 50% martensite hardnesses are replaced by minimum and maximum hardnesses, which indicate that hardnesses for structures other than 50% martensite are being used to define the steel's hardenability.

These hardnesses are placed in parentheses after Jominy positions. Thus, for a particular application it may be decided that a 0.42-0.49C steel will be used, the hardness at J = 14/16 in. (22 mm) should be Rc 32 minimum, and at J = 22/16 in. (34 mm) the hardness should be Rc 42 maximum. The code would then be written C42-49E14(32)-22(42).

The code system can be applied to carburizing grades of

able II — Typical Codes of North American and Overseas Steels

Mean arbon ntent, %	Code*	Hardness Range, Rc for J Distance J=1/16 in or 1.6 mm	J_{min}/J_{max}	J_{min}/J_{max} Distance† 1/16 in.	mm	J_{min}/J_{max} Distance, 1/16 in. (mm)
S.	C34-43E1.5-3.5	51-58	40-44	1.5-3.5	2-5.5	1038H
	C35-45E3.5-7	53-60	40-45	3.5-7	5.5-11	1541H
	C37-44E5.5-11	53-60	41-44	5.5-11	9-17	5140H
	C37-44E7.5-15	53-60	41-44	7.5-15	12-24	50B40H, 8640H, EX 36
40	C35-45E9-17	53-61	40-45	9-17	14-27	15B41H
	C37-44E11-32	53-60	41-44	11-32	17-50	4140H
	C39-46E13-32(50)	55-62	42-50	13-32†	21-50†	4142H →
	C37-44E30-32(56)	53-60	41-56	30-32†	48-50†	4340H →
	C42-51E2-3.5	55-62	43-47	2-3.5	3-5.5	1045H
	C42-51E4.5-7.5	56-63	43-47	4.5-7.5	7-12	15B46H
45	C42-49E8-17	56-63	43-46	8-17	13-27	50B44H
	C42-49E9-19	56-63	43-46	9-19	14-30	8645H, EX 38
	C42-49E15-32(54)	56-63	43-54	15-32†	24-50†	4145H →
reign	C37-44M9.5-22	53-60	41-44	6-14	9.5-22	530H40
	C38-44M10-25	54-61	42-44	6.5-16	10-25	41Cr4
	C38-45M11-21	54-61	42-45	7-13	11-21	42MnV7
	C37-44M11-23	53-60	41-44	7-15	11-23	40NCD2H
40	C38-45M16-50(46)	54-61	42-46	10-32†	16-50†	42CrMo4 →
	C37-44M17-50	53-60	41-44	11-32	17-50	SCM4H
	C38-45M20-50(50)	54-61	42-50	13-32†	20-50†	42CD4 →
	C39-46M20-50(50)	55-62	42-50	13-32†	20-50†	708H42 →
45	C41-48M11-22	55-63	43-46	7-14	11-22	45C4

Distance scale (1/16 in.): 2 4 6 8 10 12 14 16 18 20 22 24 26 28 30 32; (mm): (6.3) (13) (19) (25) (32) (38) (44) (50)

Numbers in parentheses indicate maximum hardness at J = 32/16 in. or 50 mm.

† J_{min} = J distance for 50% martensite on min of hardenability band. J_{max} = J distance for 50% martensite on max of hardenability band. Exception: where J_{max} is more than 32/16 in. or 50 mm, max hardness is that for J = 32/16 in. or 50 mm.

steel to control core hardenability. Though this may be satisfactory, many applications make it necessary to specify a minimum case hardenability in addition to the code. Calculated ideal critical diameters for case compositions, denoted D_{Ic} values, can then be employed. (For a method to calculate D_{Ic} from chemical composition, see "Determining Hardenability From Composition," by C. F. Jatczak, *Metal Progress,* September 1971.) For example, a code constructed around an 8620H steel would read C17-23E3.5-6, minimum D_{Ic} = 4.45 in. (113 mm).

Users of a steel classification system such as this one must consider some important factors in addition to hardenability requirements. Probably one of the most important of these is tempering response of alloys. Others include grain size requirements and chemical analysis limits.

Mr. Sponzilli is materials research stress engineer, Mr. Sperry is chief engineer, materials engineering, and Mr. Lytell is materials engineer, Materials Research and Engineering, Engineering Research, International Harvester Co., 7 South 600 County Line Rd., Hinsdale, Ill.

New Developments in Carburizing Steels

YANCEY E. SMITH AND GEORGE T. ELDIS

The effects of the major alloying elements on carburizing steel properties are discussed. Potential difficulties which may arise in designing new or alternate steel grades strictly on the basis of core hardenability criteria are noted. These difficulties include inadequate case hardenability, poor annealability, formation of carbide networks and bainite in the case, and excessive surface oxidation during carburizing. The properties of some standard SAE carburizing steels are presented, and compared with those of lower-cost EX alternate grades. Finally, development of a new carburizing steel is outlined—it features high hardenability, lower cost, and more toughness than possessed by available standard grades.

IN recent years, considerable effort has been directed toward development of alternate carburizing steels to replace standard SAE carburizing grades, resulting in the current list of EX ("exchange") steels. Most of the compositions on this list were designed on the basis of empirical considerations of hardenability. That is, knowing the effects of the various alloy elements on core and case hardenability, engineers could propose compositions which would attain the same levels of as-quenched hardness as a particular standard carburizing steel, while simultaneously taking better advantage of the current cost and/or availability of the various alloy elements.

Designing a carburizing steel strictly on the basis of hardenability, however, has some shortcomings. It ignores other properties of the steel such as annealability, tendency toward formation of alloy carbides or small quantities of bainite in the carburized case, and toughness of the heat-treated material, which can also be affected by alloy content. These properties can be just as important as hardenability in deciding

YANCEY E. SMITH is Research Supervisor and GEORGE T. ELDIS is Senior Research Associate, Research Laboratory Climax Molybdenum Company of Michigan, a subsidiary of AMAX Inc., Ann Arbor, Michigan.

whether or not a given steel can be used for a specific application or under a particular set of processing conditions.

In this paper, the effects of alloy elements on various carburizing steel properties will be briefly discussed. Also, the characteristics of some standard SAE grades, some EX grades, and some experimental steels will be presented to illustrate these effects. A new grade of carburizing steel will be introduced, and its development briefly outlined. This steel features higher hardenability and more toughness than available in standard grades, which are costlier as well.

INFLUENCE OF ALLOYING ELEMENTS

The primary elements used to develop hardenability in carburizing steels are manganese, chromium, nickel, and molybdenum. Relative effects of these elements on core hardenability have been quantitatively described by deRetana and Doane.[1] On a weight percent basis, molybdenum and manganese influence core hardenability most strongly, with the chromium effect being moderate. The individual influence of nickel is relatively weak until 1.5 pct is approached. However, a molybdenum-nickel interaction has a strong positive effect on hardenability, which becomes significant around 0.75 pct Ni or somewhat higher, depending on the specific steel composition.

Jatczak has made an extensive evaluation of elemental effects on case hardenability.[2] Though the relative effects of the alloy elements are similar to those just discussed for core hardenability, these effects depend upon quenching temperature. Quenching from 1700°F (925°C) causes the effect of molybdenum to be notably stronger than that of the other elements. Quenching from 1525°F (830°C) causes the molybdenum, chromium, and manganese effects, to be more similar than at 1700°F (925°C), with nickel remaining the least effective.

Although manganese produces a large increase in core hardenability and is certainly the most cost-effective element in this respect, it also has several adverse effects on the steel properties. For example, a modest increase in manganese content can significantly lengthen the time required for isothermal annealing. This can be a problem, especially in steels of medium to high hardenability which require cycle annealing prior to machining.

Fig. 1 and Table I illustrate this point. The curves in Fig. 1 represent partial TTT diagrams for Steels *A*, *B*, and *C* (Table I), showing plots of time required for 100 pct transformation of austenite to ferrite + pearlite at various transformation temperatures. Such curves are a good measure of annealability, and can be readily determined with a minimum of effort. First, several thin specimens of the steel are

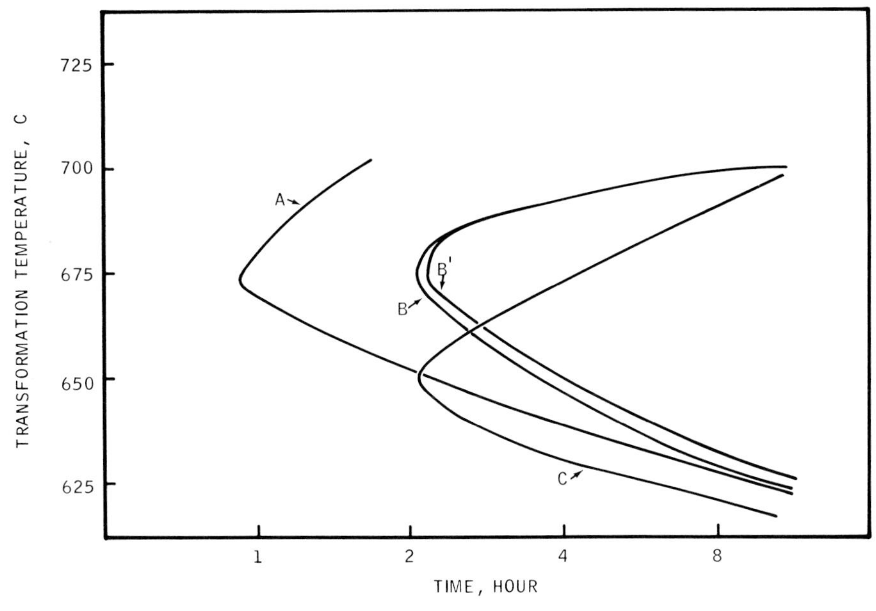

Fig. 1—Approximate isothermal transformation curves (100 pct transformation) for Steels *A*, *B*, and *C* (Table I).

Table I. Steels for Comparison of Annealability

Steel	Composition, Wt Pct						ASTM Grain Size Number	Core D_I*	Transformation Characteristics†		
	C	Mn	Si	Ni	Cr	Mo			Min Time, h	Optimum Temperature °F	(°C)
A	0.19	0.78	0.28	0.66	0.75	0.55	8	3.40	1/2 to 1	1250	(675)
B‡	0.22	1.09	0.22	0.60	0.69	0.58	7-8	4.50	2 to 4	1250	(675)
C	0.20	0.68	0.32	1.59	0.51	0.45	8	3.60	2 to 4	1200	(650)

*Ideal critical diameter, in inches, calculated from known core hardenability factors; assuming a grain size of 7.
†Complete transformation to ferrite plus pearlite.
‡B' has 0.26 pct C.

austenitized and quenched into an isothermal bath (usually salt or lead). At various times, individual specimens are removed, and quenched into water. After repeating this process for a few different bath temperatures, the investigator has a group of specimens isothermally transformed at different temperatures for different times. These are then metallographically examined to determine the time for complete transformation to ferrite plus pearlite.

Steels presented in Table I are some experimental compositions on which comparative hardenability and annealability studies have been performed in our laboratory. Steels *A* and *C* have equivalent core hardenabilities (D_I values). Substitution of Steel *A* for Steel *C* would result in a substantial savings in material costs. Processing costs would also be reduced because, as Table I and Fig. 1 show, the minimum annealing time for Steel *A* is less than half that of Steel *C*.

A comparison of Steels *A* and *B* illustrates what may happen if one attempts to attain markedly increased hardenability in the most economical way, namely by a large increase in manganese. Although the hardenability is about 32 pct higher in Steel *B*, the minimum annealing time is also more than twice as high. (It should also be noted here that the three-point increase in carbon on going from Steel *A* to *B* also contributes substantially to the total D_I increase, accounting for more than half of the 32 pct. But, as a comparison of curves *B* and *B′* in Fig. 1 shows, an increase in carbon content of a few points at these overall alloy contents does not notably add to the minimum annealing time: *B′* has the same composition as *B*, plus four more points of carbon, 0.026 pct instead of 0.22 pct.

Fig. 1 also illustrates another important point regarding annealability–namely, the temperature range for relatively rapid transformation to ferrite plus pearlite is quite narrow. Because of the sharpness of the "nose" of the transformation curve, an error of only 45°F (25°C) in annealing temperature–either above or below–can easily double the required annealing time.

Certainly, manganese is not the only element affecting annealability. As the data in Table I and Fig. 1 show, alloy composition can markedly affect both the vertical and horizontal positions of the nose of the transformation curve. These alloy effects must be considered if processing problems are to be avoided in new medium-to-high hardenability steels. Unfortunately, there is currently no means to quantitatively predict the isothermal transformation characteristics as a function of alloy content, although research toward this end is in progress.[3] For the present, one can at least be aware of these effects when proposing new compositions, and make use of certain qualitatively observed trends to avoid pitfalls. As illustrated, one of these pitfalls can arise if increased hardenability is obtained merely by adding manganese.

Another potential difficulty with high manganese contents relates to manganese control problems in steelmaking. Of the four major alloying elements, manganese is the hardest to control, which is why this element has the broadest specification ranges. The higher the specified manganese level in a given steel, the more difficult it is to control the concentration range; hence, the range of hardenability is broader. Steels made to a low specified manganese content can be held to within a narrower range of hardenability, which can be important with regard to controlling the maximum variation in core hardness of the final carburized and heat-treated part.

Another difficulty with manganese arises from its tendency toward oxidation during the carburizing process, which can lead to detrimental case microstructures, depending on the intended use of the carburized component. This familiar phenomenon, called surface oxidation, was reviewed recently by Chatterjee-Fischer.[4] Because all endothermic carburizing atmospheres contain some oxygen, alloying elements less noble than iron–and these include silicon and chromium as well as manganese–are oxidized at the steel surface, forming oxide particles on the surface of the material, at austenite grain boundaries, and within the austenite grains. As oxidation progresses, these alloying elements become depleted from solid solution in subsurface regions of the part. Our laboratory observations show that a commercially carburized high-manganese, high-chromium steel can develop an alloy depletion zone of as much as 0.002 in. (0.05 mm); Fig. 2. The resulting decrease in case hardenability in the alloy depletion zone may result in formation of high temperature transformation products in the outermost case of the specimen during quenching, especially if there is some delay between removing the carburized part from the furnace and immersing it into the quenching medium. Fig. 3 shows the type of case microstructure that might be obtained. Formation of these high temperature transformation products can be avoided by ensuring a sufficient concentration of nonoxidizable alloying elements to maintain adequate case hardenability in the steel. Formation of intergranular alloy oxides can be minimized by reducing

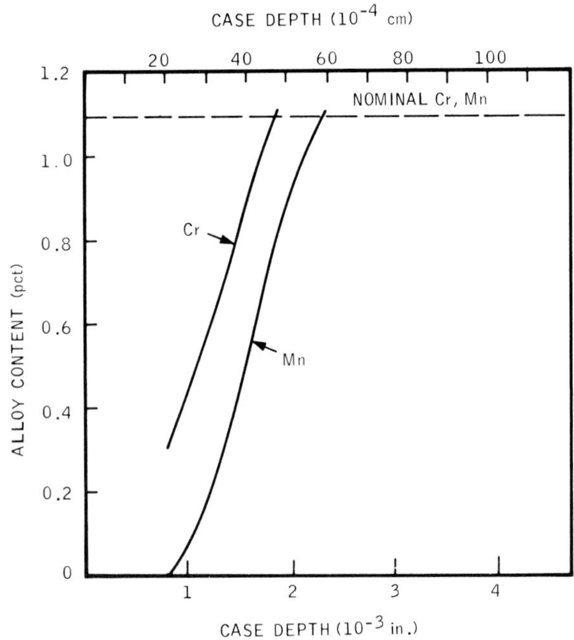

Fig. 2—Cr and Mn concentration profiles near the specimen surface developed during gas carburizing of a 1.1 pct Mn-1.1 pct Cr steel.

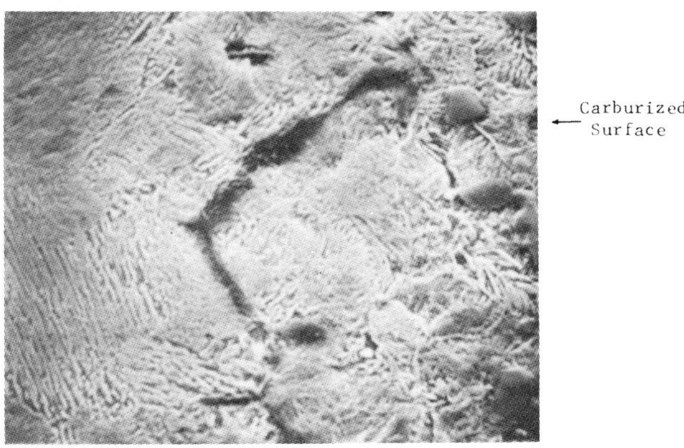

Fig. 3—Microstructure in outer regions of carburized case of 1.1 pct Mn–1.1 pct Cr carburizing steel. Scanning electron micrograph, 5000 times.

Table II. Some Standard SAE Grades and EX Alternates

	Nominal Composition, Pct					
Steel	C	Mn	Si	Ni	Cr	Mo
8260	0.20	0.80	0.28	0.55	0.50	0.20
EX24	0.20	0.87	0.28	–	0.55	0.25
EX10	0.22	1.10	0.28	0.30	0.32	0.07
EX15	0.20	1.05	0.28	–	0.50	0.17
4320	0.20	0.55	0.28	1.85	0.50	0.25
EX29	0.20	0.87	0.28	0.55	0.55	0.35
4817	0.17	0.50	0.28	3.50	–	0.25
EX31	0.17	0.80	0.28	0.85	0.55	0.52

the concentration of oxidizable alloying elements.

As just noted, chromium is also susceptible to surface oxidation, which limits the amount of chromium that can be used in carburizing steels. Another limitation on chromium content arises from its tendency to form large carbide particles near the specimen surface, where the carbon content is highest. This excess carbide may take the form of an intergranular carbide, which can be detrimental to mechanical properties. The problem can usually be avoided by maintaining chromium at less than 0.7 pct.

During the past few years, certain parts producers have been obliged to learn a new criterion of case hardenability, quite different from the typical Jominy hardness measurement. One heavy equipment manufacturer has written specifications which forbid bainite in the outer region of the carburized case; the new hardenability criterion is "distance to first bainite," or DFB. Carburized end quench bars can be used for measuring DFB. A bar is carburized with a companion bar of equal diameter that is used to determine the carbon gradient. Longitudinal flats are then surface gound on the end-quenched bar to a depth corresponding to a preselected carbon level. The flats are polished, etched, and scanned in the optical microscope to determine the distance from the quenched end of the bar at which bainite first appears (DFB). An extensive study of the DFB hardenability criterion over a wide range of alloy compositions has recently been completed.[5] Results indicate that both chromium and molybdenum markedly increase DFB, while manganese and nickel have a negligible effect as single alloy elements. Some examples of typical DFB values in commercial steels will be given below.

Core toughness, also an important property of carburizing steels, is the most difficult property to predict by theory. It is a function of the specific microconstituents present in the material and their relative volume fractions as well as their chemical composition. Thus, it is difficult to make general statements about the effect of individual elements on toughness, and designing a steel with a given level of toughness in mind remains an exercise in empiricism. An example of such an effort is presented below.

RECENTLY DEVELOPED CARBURIZING STEELS

Having briefly discussed some of the effects of different alloying elements on carburizing steel properties and indicated some of the many points that should be considered in steel design, we now compare some recently developed carburizing steels with some of the standard SAE grades. Ideally, comparisons should be chosen to illustrate the application of some of the points just discussed. Unfortunately, this is not possible because, as already noted, virtually all of the new carburizing grades have been designed purely from the standpoint of hardenability, effectively leaving the other steel properties to chance. However, a comparison is still instructive as it illustrates some of the pitfalls that may be encountered. Such comparisons also show how a qualitative knowledge of past performance can be effectively used to design steels for a particular set of properties, even though specific quantitative expressions for these properties may not presently exist.

Table II and Figs. 4 through 6 present data for standard grades SAE 8620, 4320, and 4817, and their recently developed lower-cost alternates, EX24, EX29, and EX31. In all three figures, hardenability bands are those published by the SAE, and the remaining data are from laboratory measurements. Hardenability was determined in accordance with SAE Standard J406a. Impact energy data were obtained from Charpy V-Notch specimens pre-machined to 0.015 in. (0.38 mm) oversize, blank carburized for 8 h at 1700°F (925°C), cooled to 1550°F (845°C), stabilized, oil quenched, and tempered 1 h at 375°F (190°C). (This treatment was selected because of its popularity in the automotive industry.) The specimens were then machined to standard impact specimens and tested in accordance with ASTM Specification E23-66.

The core hardenability curve for a nominal composition heat of EX24 (Fig. 4) falls slightly above the curve for a nominal heat of 8620 and in the middle of the published SAE hardenability band. Case hardenabilities of both steels are also virtually identical, with the EX24 having a slight advantage. Thus, hardenability requirements for 8620 are clearly met by EX24, and at a significant cost advantage; the 0.55 pct nominal nickel content of the 8620 has been replaced by an increase of about 0.05 pct in each ot the other elements. Fig. 4 also shows that the alloy substitution has not resulted in a sacrifice of impact properties. As seen from the CVN impact energy transition curves, the EX24 has a marginal advantage over the 8620 above 25°F (–3°C), while at lower temperatures the EX24 impact energies are only marginally lower than those of 8620.

Core and Case Hardenability

Impact Transition Curve

Fig. 4—Comparison of hardenability and Charpy V-Notch impact energy of 8620 and EX24. Normalizing temperature 1700°F (925°C), quenching temperature temperature 1700°F (925°C).

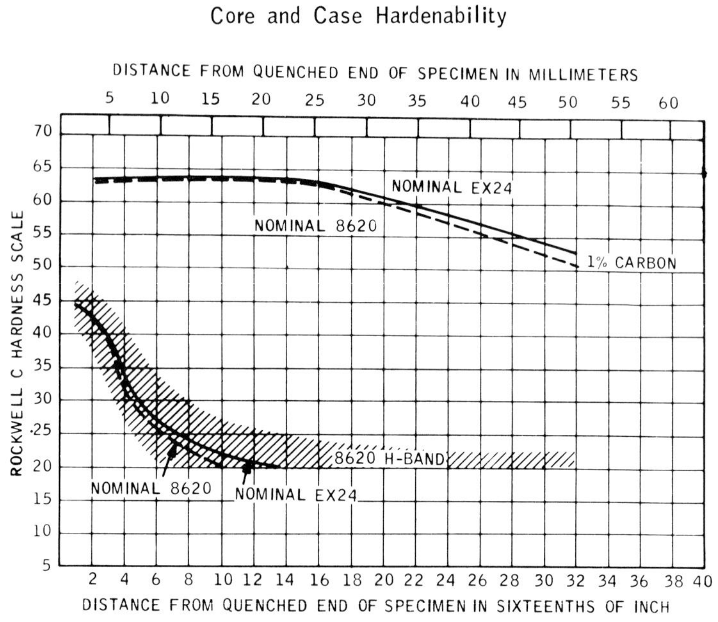

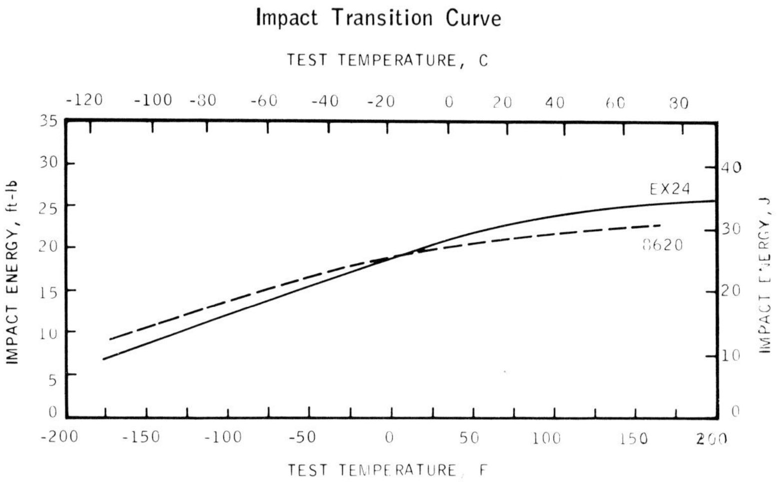

Table III. Properties of 8620 and Alternate Grades

Steel	Core D_I*	Case D_I† 0.9 Pct Carbon	DFB‡ In 1/16 in. (mm) 0.9 Pct Carbon	
8620	1.85	6.2	5	(8)
EX10	1.65	4.4	2	(3)
EX15	1.85	5.85	3	(4.5)
EX24	1.90	6.45	5	(8)

*Calculated values, based on composition and an assumed grain size of 7 (Ref. 1).

†Calculated values, based on composition (Ref. 2).

‡Laboratory data on end-quench bars direct quenched from 1700°F (925°C).

Other proposed alternates to 8620 are EX10 and EX15, compositions of which are also shown in Table II. Table III contrasts hardenabilities and DFB with those of 8620 and EX24. Note that designing from the standpoint of core hardenability alone (EX10) resulted in an acceptable level for this parameter, but case hardenability was unacceptable. Consideration of both core and case hardenability in design (EX15) resulted in a third parameter, DFB, being significantly lower than in the 8620.

It should also be noted that EX24 has given a few years of satisfactory service as a bearing and gear steel, production in 1973 being over 30,000 tons. Since it is less expensive than 8620, tonnage is expected to continue to increase.

The data of Fig. 5 for 4320 and EX29 show the same basic trends as those for EX24 and 8620. Core and case hardenabilities are equal for the two grades, and both have the same toughness characteristics below −50°F (−45°C). Above −50°F (−45°C), EX29 shows slightly superior toughness. Again, EX29 offers a significant cost advantage over 4320 because 0.33 pct Mn, 0.05 pct Cr and 0.10 pct Mo replace approximately 1.3 pct Ni.

As a replacement for 4817, EX31 provides significant cost saving and somewhat better properties (Fig. 6). Both core hardenability and impact properties above about −100°F (−73°C) are better in EX31, while 4817 has superior toughness below −100°F (−73°C). Both steels have equally good case hardenability. The lower case hardness of the 4817 results from the

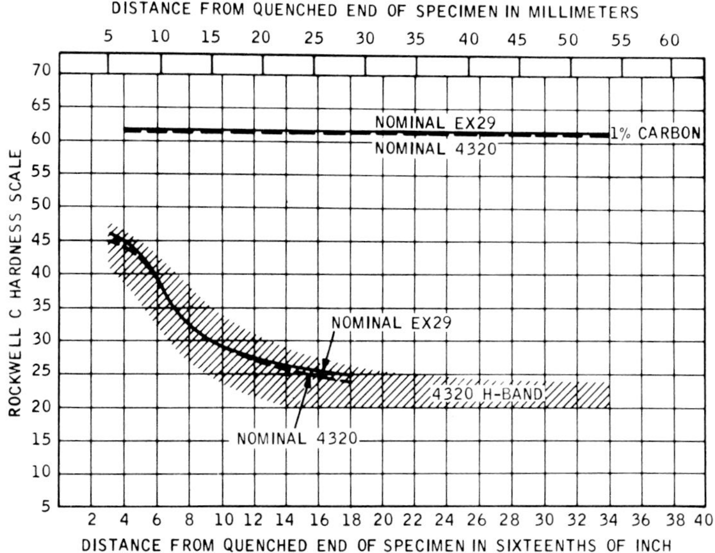

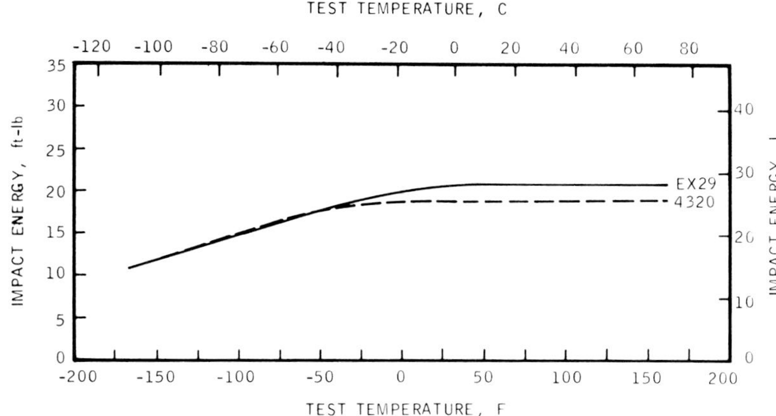

Fig. 5—Comparison of hardenability and Charpy V-Notch impact energy of 4320 and EX29.

large amount of retained austenite in the case (1 pct C). Normally, it would be used at a lower case carbon level so that retained austenite would be lower and case hardness higher.

Recently, a carburizing steel design problem was encountered that involved more than just "equivalent hardenability at lower cost." One manufacturer requested a new steel of very high hardenability plus better core toughness than that which he was currently obtaining with 4817. As an additional requirement, the new steel could not be more expensive than 4817.

Since hardenability can be readily calculated from composition, it is not difficult to attain a desired hardenability level under certain economic restrictions. As noted above, however, so many variables affect toughness in an unpredictable manner that trial-and-error testing is still important in improving this property. A clue as to how to proceed with this problem was provided by the toughness data just presented. The higher toughness that occurred in EX29 and EX31 steels when molybdenum, manganese, and chromium were raised and nickel reduced suggested a potential area for investigation. This prospect offered the additional advantage of more economical hardenability because of the larger unit contribution of the other elements (relative to that of nickel) to both core and case hardenability. Another advantage was that, since relatively modest increases in three alloying elements would be involved rather than a very large increase in any one element, certain problems such as surface oxidation or excess carbide formation would be minimized.

With these points in mind, a trial heat of the composition shown in Table IV (Heat 5115) was made in the laboratory. Fig. 7 shows that this steel exhibits higher core hardenability than does SAE 4817 or 9310. A second trial heat, heat 5124 (Table IV), was prepared to the same nominal composition as heat 5115, but with 0.12 pct C. As Fig. 7 shows, this trial heat meets hardenability requirements of SAE 9310.

On the basis of these encouraging results, a composition range was selected for this new steel, and EX numbers 55 and 56 were asigned to both the high and low carbon variations, respectively (Table IV).

Table IV. Test Compositions for a High-Hardenability Carburizing Steel

Material	Composition, Wt Pct					
	C	Mn	Si	Ni	Cr	Mo
Heat 5115	0.18	0.86	0.30	1.89	0.43	0.73
Heat 5124	0.12	0.87	0.30	1.94	0.48	0.76
EX55 (Range)	0.15/0.20	0.70/1.00	0.20/0.35	1.65/2.00	0.45/0.65	0.65/0.80
EX56 (Range)	0.08/0.13	0.70/1.00	0.20/0.35	1.65/2.00	0.45/0.65	0.65/0.80
EX55 (Commercial Heat)	0.17	0.86	0.33	1.81	0.50	0.73
4817 (Commercial Heat)	0.18	0.60	0.27	3.15	–	0.27

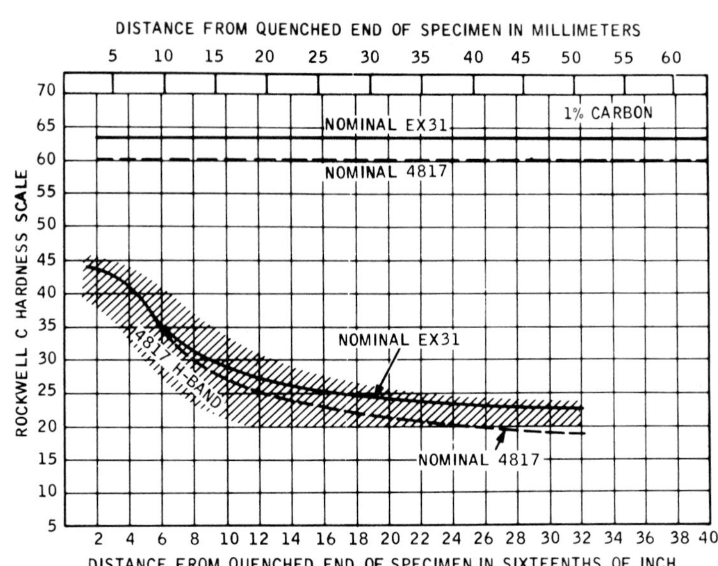

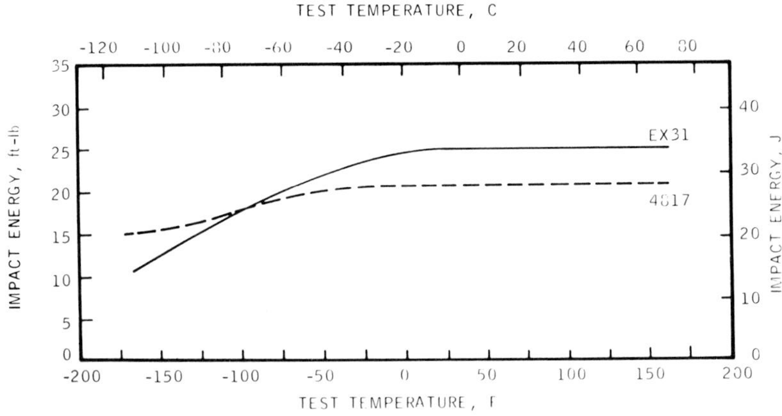

Fig. 6—Comparison of hardenability and Charpy V-Notch impact energy of 4817 and EX31.

For further evaluation, a small commercial heat of EX55 was ordered. Its composition is also shown in Table IV, along with that of a commercial heat of 4817 used for comparison purposes. Metallographic specimens of both of these steels passed ASTM Specification E45 for the highest level of cleanliness, the SAE 4817 being slightly cleaner than the EX55.

Charpy impact specimen blanks of both steels were prepared, blank carburized for 8 h at 1700°F (925°C), and cooled in still air. The specimens were then reheated to their respective austenitizing temperatures, 1470°F (800°C) for the 4817* and 1600°F (870°C) for the EX55.† After oil quenching and tempering for 1 h, as indicated in Table V, the specimens were machined to final size and tested at room temperature. As the data of Table V show, toughness levels of the EX55 were 50 to 65 pct higher, depending on

*This austenitizing temperature was the one specified by the manufacturer for whom the new high hardenability steel was developed.

†Preliminary laboratory studies showed this to be the minimum austenitizing temperature for optimum core properties with this composition.

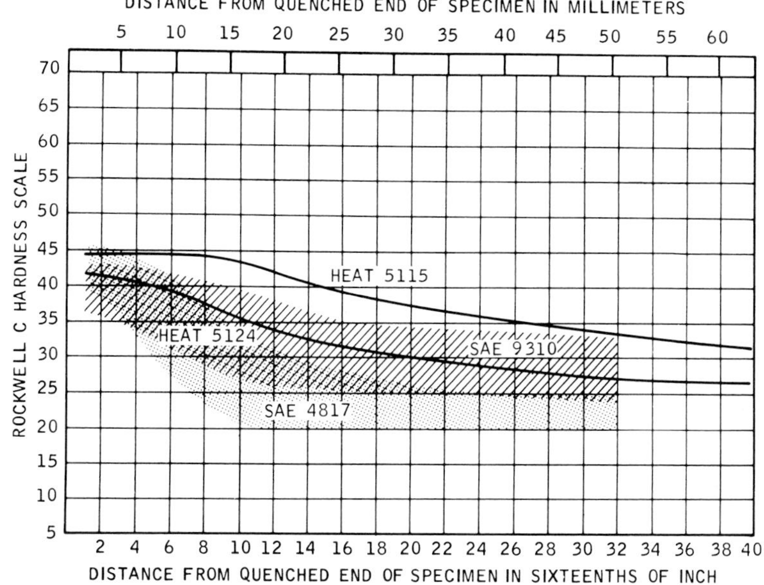

Fig. 7—Comparison of core hardenability of two laboratory steels (Table IV) with SAE grades 4817 and 9310; ASTM grain size number, 7.5.

Table V. Toughness Comparison of EX55 with SAE 4817

Steel Designation	Tempering Temperature °F	(°C)	Hardness, HRC	Impact Energy ft-lb	(J)
EX55	300	(149)	39.0	45.0	(61.0)
	320	(160)	40.5	47.5	(64.5)
	360	(182)	39.0	54.0	(73.0)
	380	(193)	39.5	51.5	(70.0)
SAE 4817	320	(160)	40.0	33.0	(44.5)
	340	(171)	42.0	33.0	(44.5)
	360	(182)	39.5	33.5	(45.5)

tempering temperature, than those of 4817. The goal of a very high hardenability carburizing steel with better toughness than 4817, and at moderate cost, was attained.

DISCUSSION AND SUMMARY

In the foregoing paragraphs, the authors emphasize that consideration of core hardenability alone in the design of carburizing steels can lead to severe difficulties with overall steel performance. Alloying elements used to achieve desired core hardenability levels can also markedly affect other properties of the steel, and in general, relative magnitudes of these elemental effects may be quite different for different properties. Thus, as has been illustrated, it is possible to have two or more steels of nearly identical core hardenability which exhibit markedly different case hardenability, annealability, and susceptibility to bainite formation in the case.

As reviewed here, the effects of manganese, nickel, chromium, and molybdenum which should be considered in designing a carburizing steel can be summarized as follows:

1) On an alloy element weight percent basis, core hardenability of carburizing steels is most strongly increased by molybdenum and manganese. Chromium has a moderate effect on core hardenability, and nickel a weak effect.

2) Effects of alloying elements on case hardenability are sensitive to quenching temperature. When quenching is from 1700°F (925°C), molybdenum has a notably stronger hardenability influence than the other alloy elements. Quenching from 1525°F (830°C) reduces the effectiveness of molybdenum to a level that is more similar to the manganese and chromium effects, with the nickel effect remaining weakest.

3) Susceptibility to bainite formation in the carburized case is most strongly reduced by molybdenum and chromium.

4) Although manganese is the most cost-effective element where core hardenability is concerned, excessive use of this element can create problems in annealability and control of hardenability bandwidth. Since manganese is also oxidizable, it tends to be depleted from solution at the specimen surface, resulting in lower hardenability, and possibly formation of nonmartensitic transformation products at the surface.

5) Excessive use of chromium can also involve difficulties with surface oxidation and formation of excess carbides in the case.

These alloy effects were taken into consideration, and balanced in the design of a high hardenability, high toughness carburizing steel that is lower in cost than standard steels of notably lower hardenability.

REFERENCES

1. A. F. deRetana and D. V. Doane: *Metal Progr.*, 1971, vol. 100, p. 65.
2. C. F. Jatczak: *Met. Trans.*, 1972, vol. 4, p. 2267.
3. W. Peter and H. Finkler: *Arch. Eisenhuettenw.*, 1974, vol. 45, p. 533.
4. R. Chatterjee-Fischer: *Härt.-Techn. MITT.*, 1973, vol. 28, p. 259.
5. G. T. Eldis and Y. E. Smith: *The Effect of Composition on Distance to First Bainite in Carburizing Steels*, to be published.

The EX Steels and Equivalent Standard Grades

Originated by the Iron & Steel Technical Committee (ISTC) of the Society of Automotive Engineers in 1963, the EX numbering system constitutes a uniform means for designating new grades of wrought steels on a temporary basis. The EX number, which indicates only the chronological order of its acceptance by the ISTC, is assigned to a new grade upon written request by the developer and subsequent approval by the SAE staff and chairman of the appropriate ISTC division. An EX grade will be removed from this temporary listing if: (1) its sponsor requests, in writing, that the grade be discontinued; (2) the SAE division responsible for the grade takes formal action to drop it because interest is nil; or (3) the grade becomes popular enough to be listed as a standard SAE steel. Once assigned, an EX number is exclusive to that particular grade; it is never reassigned.

EX No.	Composition, %					Equivalent SAE Grade
	C	Mn	Cr	Mo	Other	
1*	0.15-0.21	0.35-0.60	—	0.20-0.30	4.80-5.30 Ni	9310
9*	0.19-0.24	0.95-1.25	0.25-0.40	0.05-0.10	0.20-0.40 Ni	8620
10	0.19-0.24	0.95-1.25	0.25-0.40	0.05-0.10	0.20-0.40 Ni	8620
11*	0.38-0.43	0.75-1.00	0.25-0.40	0.05-0.10	0.20-0.40 Ni, 0.0005 B min	8640
12	0.38-0.43	0.75-1.00	0.25-0.40	0.05-0.10	0.20-0.40 Ni, 0.0005 B min	8640
13*	0.66-0.75	0.80-1.05	0.25-0.40	0.05-0.1C	0.20-0.40 Ni	—
14*	0.66-0.75	0.80-1.05	0.25-0.40	0.05-0.10	0.20-0.40 Ni	—
15	0.18-0.23	0.90-1.20	0.40-0.60	0.13-0.20	—	8620
16	0.20-0.25	0.90-1.20	0.40-0.60	0.13-0.20	—	8622
17	0.23-0.28	0.90-1.20	0.40-0.60	0.13-0.20	—	8625
18	0.25-0.30	0.90-1.20	0.40-0.60	0.13-0.20	—	8627
19	0.18-0.23	0.90-1.20	0.40-0.60	0.08-0.15	0.0005 B min	94B17
20	0.13-0.18	0.90-1.20	0.40-0.60	0.13-0.20	—	8615
21	0.15-0.20	0.90-1.20	0.40-0.60	0.13-0.20	—	8617
24	0.18-0.23	0.75-1.00	0.45-0.65	0.20-0.30	—	8620
27	0.25-0.30	0.75-1.00	0.45-0.65	0.20-0.30	—	8627
29	0.18-0.23	0.75-1.00	0.45-0.65	0.30-0.40	0.40-0.70 Ni	4320
30	0.13-0.18	0.70-0.90	0.45-0.65	0.45-0.60	0.70-1.00 Ni	4815
31	0.15-0.20	0.70-0.90	0.45-0.65	0.45-0.60	0.70-1.00 Ni	4817
32	0.18-0.23	0.70-0.90	0.45-0.65	0.45-0.60	0.70-1.00 Ni	4820
33	0.17-0.24	0.85-1.25	0.20 min	0.05 min	0.20 Ni min	4027
34	0.28-0.33	0.90-1.20	0.40-0.60	0.13-0.20	—	8630
35	0.35-0.40	0.90-1.20	0.45-0.65	0.13-0.20	—	8637
36	0.38-0.43	0.90-1.20	0.45-0.65	0.13-0.20	—	8640
37	0.40-0.45	0.90-1.20	0.45-0.65	0.13-0.20	—	8642
38	0.43-0.48	0.90-1.20	0.45-0.65	0.13-0.20	—	8645
39	0.48-0.53	0.90-1.20	0.45-0.65	0.13-0.20	—	8650
40	0.51-0.59	0.90-1.20	0.45-0.65	0.13-0.20	—	8655
41	0.56-0.64	0.90-1.20	0.45-0.65	0.13-0.20	—	8660
42	0.13-0.18	0.95-1.25	0.25-0.40	0.05-0.10	0.20-0.40 Ni	8615
43	0.13-0.18	0.95-1.25	0.25-0.40	0.05-0.10	0.20-0.40 Ni, 0.0005 B min	—
44	0.15-0.20	0.95-1.25	0.25-0.40	0.05-0.10	0.20-0.40 Ni	8617
45	0.15-0.20	0.95-1.25	0.25-0.40	0.05-0.10	0.20-0.40 Ni, 0.0005 B min	—
46	0.20-0.25	0.95-1.25	0.25-0.40	0.05-0.10	0.20-0.40 Ni	8622
47	0.23-0.28	0.95-1.25	0.25-0.40	0.05-0.10	0.20-0.40 Ni	8625
48	0.25-0.30	0.95-1.25	0.25-0.40	0.05-0.10	0.20-0.40 Ni	8627
49	0.28-0.33	0.95-1.25	0.25-0.40	0.05-0.10	0.20-0.40 Ni	8630
50	0.33-0.38	0.95-1.25	0.25-0.40	0.05-0.10	0.20-0.40 Ni	8635
51	0.35-0.40	0.95-1.25	0.25-0.40	0.05-0.10	0.20-0.40 Ni	8637
52	0.38-0.43	0.95-1.25	0.25-0.40	0.05-0.10	0.20-0.40 Ni	8640
53	0.40-0.45	0.95-1.25	0.25-0.40	0.05-0.10	0.20-0.40 Ni	8642
54*	0.19-0.25	0.70-1.05	0.40-0.70	0.05 min	—	4118

*All steels contain (1) 0.035 P max except EX 1 (0.040 P max), and EX 13 and EX 14 (0.025 P max); (2) all contain 0.040 S max except EX 13 and EX 14 (0.025 S max); and (3) all contain 0.20 to 0.35 Si except EX 9, EX 11, and EX 13 (0.050 Si max), and EX 54 (0.35 Si max).

Selecting Steels for Carburized Gears

By ROY F. KERN

The engineer must take into account the imposed loads, compressive and bending stresses, effects of case crushing, and operating temperatures with their effect on lubrication.

Gears made of alloy steel, carburized, have great load-carrying capacity. To obtain minimum cost and maximum performance for such gears, the engineer must thoroughly understand the factors involved in selecting steel and heat treating procedures.

Briefly, he needs a grade that can be efficiently machined and heat treated with predictable distortion from heat to heat. The steel must be capable of providing, after carburizing, a strong surface to support contact and bending loads. Where occasional severe overloading is known to take place, a high-nickel steel will probably be needed. The final proof, however, of proper selection is the number of hours that the gear set operates in its environment.

Consider the Essentials

In gear engineering, it is important to concentrate on the essentials such as the quality of the case microstructure, the case depth to Rc 50, and the steel's alloy content, machinability, and cost. The engineer must also be sure that operating demands on a gear tooth are recognized and satisfied.

A gear tooth is, he should realize, a structural member of a mechanical device. Therefore, it requires adequate static strength to support imposed loads without deflecting excessively. Active surfaces of a tooth are, furthermore, subject to high contact loads in rolling and sliding, and the direction of slide abruptly changes at the pitch line.

Loads imposed on a gear tooth make it a cantilever beam in that contact stresses cause a bending stress, which is concentrated at the root fillet; this effect results in bending fatigue. Relative velocity at the interface of two mating gears can be very high, generating heat, which causes the lubricant film to break down, leading to scoring.

High contact stresses and the tendency to score, plus the presence of abrasive materials, require that mating surfaces be very hard to resist wear. Because a gear set will usually undergo at least one severe overload in its lifetime, toughness (an outstanding property of a carburized gear) is essential too.

Selection of steel includes consideration of availability, formability (required when cold forging is employed), machinability, and consistency of dimensional change in heat treatment from heat to heat. These engineering needs relate primarily to properties of the surface and immediate subsurface zones of the gear tooth. Mechanical properties at the radial centerline of the tooth are of little concern because stresses are low, resembling those at the neutral axis of a beam.

Complete gear design should include determination of the following factors:

1. The diametral pitch to provide the necessary static strength and stiffness.
2. The compressive stress at the lowest point of single tooth contact (often referred to as Hertz stress) on the driving gear.
3. The bending stress in the root fillet, particularly that of the pinion.
4. The case crushing load at the lowest point of tooth contact in the pinion.
5. The interface temperature of the gear set at several anticipated speeds.
6. The level and rate of application of the overloads. (This factor can be determined with strain gages and high-speed strain measuring equipment.)

Methods of calculating these factors (except for load application rates) are in the literature, an excellent source being the Gear Handbook, by D. W. Dudley, McGraw-Hill, 1962.

If the designer is not limited by space, weight, or cost, he can overdesign the gear to bypass the need for good metallurgical quality. Most of the time, however, either the cost will be excessive or one or more of the operating requirements will be too close to the following typically recommended maximums (calculated per AGMA methods where applicable):

Compressive stress	220,000 psi
Bending stress	90,000 psi
Case crushing (at any point below active surface)	55% max of shear yield strength
Scoring temperature	500 F with SAE 50 bulk oil at 200 F

These values apply to gears of alloy steel of basic open-hearth or basic oxygen quality, containing chromium, nickel, or molybdenum, present singly or together. Plain carbon, carbon-manganese, or carbon-boron steels can carry approximately 80% of the stresses shown.

The Selection Process

When choosing a gear steel for maximum load-carrying ability, the designer should first exclude all grades known to be troublesome in machining or heat treatment with the facilities available, and those of known erratic performance. Examples are:

1. Steels known to contain numerous, large nonmetallic inclusions such as those typical of semikilled

Mr. Kern is president, Kern Engineering Co., Peoria, Ill. This article is adapted from Mr. Kern's presentation at the Metal Progress Workshop "Selecting and Using the Economical Alternative Materials," 22-23 June 1971, Cleveland.

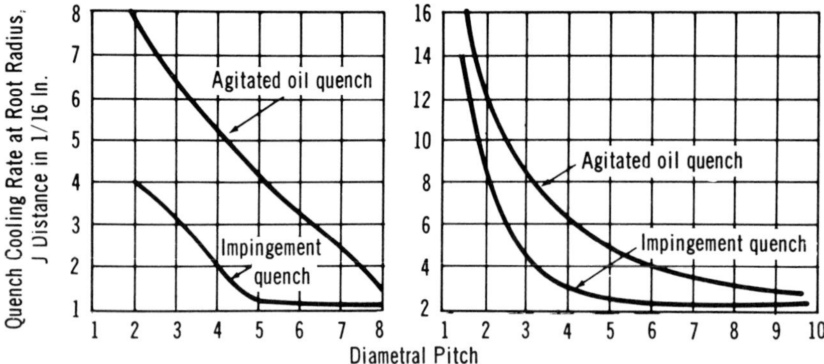

Fig. 1 — Surface cooling rates (actually determined 0.060 in. below the surfaces) of web type gears (left) and pinion gears vary with the gear pitch and the quenching method. These figures can be used for selecting steels and processing methods for gears.

melting or those which are resulfurized (over 0.04%S).

2. Grades having poor machinability such as AISI 1524 and other similar high-manganese grades.

3. Coarse-grain or aluminum-killed steels (a silicon-killed fine-grain steel is preferred; aluminum is used in amounts only to produce fine grain). Coarse-grain steels lack toughness, while steels killed only with aluminum tend to be abnormal in carburizing (that is, areas of ferrite and carbide in the microstructure resist solution, resulting in soft spots).

4. Grades such as 5120 and 6120 that contain poor balances between carbide formers and matrix strengtheners, particularly with reheat hardening. Grades such as 4118 can be used, providing that adequate atmosphere control is available to keep the carbon content below 1.10% in the case.

5. Grades showing erratic dimensional change in heat treating.

The next step is to determine the quench cooling rate as closely as possible to the surface in the root fillet. To estimate this cooling rate, use Fig. 1 (web type gears, left; and solid pinions, right). In these graphs, a distance of 0.060 in. beneath the surface is considered to represent the surface because Rockwell hardnesses cannot be determined accurately at closer distances on gear cross sections.

Note the large difference between results for agitated oil quenching and impingement quenching. Cooling rates during quenching can be determined by the method described on p. 25 of my book, Selecting Steels and Designing Parts for Heat Treatment (ASM, 1969), or on p. 70 of my article, "Selecting Steels for Heat-Treated Parts – Part I," Metal Progress, November 1968.

At this point, a decision must be made regarding the level of metallurgical quality required. Design limits shown previously are those recommended where the carburized case is fully hardened to be free of upper bainite and carbides in continuous networks. Small percentages of these constituents significantly reduce static strength as well as both rolling contact and bending fatigue. Carbide networks also contribute to chipping on tooth ends of gears subject to clash.

If it is decided that maximum load-carrying ability is not necessary, case hardness becomes the major criteria. Then the steel can be chosen

Table I — Gear Steel Selection According to Hardness

Cooling Rate, J Distance (1/16 in.)*	Direct-Quenched Gears			Reheated Gears		
	Case Carbon					
	1.10%	0.90%	0.80%	1.10%	0.90%	0.80%
1	1018	1018	1018	1018	1018	1018
2	1524	1524	1524	1524	1524	1524
3	4026	4026	4026	8620	8620	4118
4	4118	4118	4118	8620	8620	8620
5	4118	4118	4118	8620	8620	8620
6	8620	8620	8620	8720	8620	8620
7	8620	8620	8620	8720	8720	8720
8	8720	8720	8720	8822†	8822	8822
9	8822†	8822	8822	8822	8822	8822
10	8822	8822	8822	8822	8882	8822
11	8822	8822	8822	4320	8822	8822
12	8822	8822	8822	4320	4320	4320
13	8822	8822	8822	4320	4320	4320
14	8822	8822	8822	4320	4320	4320
15	8822	8822	8822	4820	4820	4820
16	8822	8822	8822	4820	4820	4820
18	4320	4320	4320	4820	4820	4820
20	4320	4320	4320	4820	4820	4820
Over 20	4320	4820	4820	9310	9310	9310

*Cooling rate required for Rc 60 min, as quenched.

†8822 may need a lower carbon content if used for gears requiring machining after hardening.

Fig. 2 — Determinations of shear yield strengths (top curve) and net stresses at various depths below the surface of a carburized gear indicate where the gear will fail by case crushing. In this instance, the case will fail in the zone below 0.050 in. from the surface because, as Pederson and Rice found, the stress-to-strength ratio within that zone is greater than 0.55.

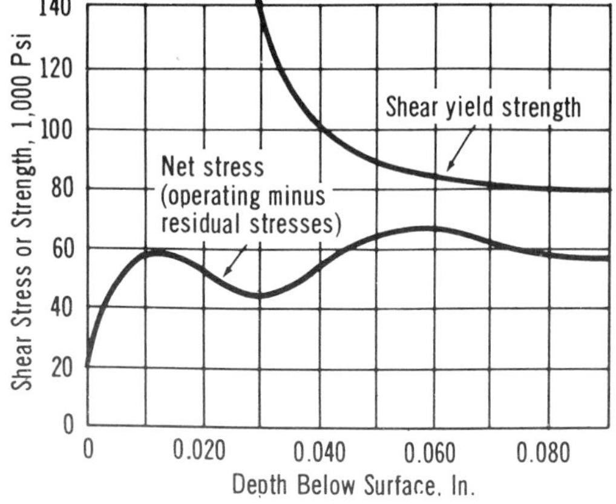

1,550 F. (The lower quench temperature is mainly used to control retained austenite.)

The load-carrying ability and bending fatigue strength of gears properly made by either hardening method are equal. Direct quenching usually offers two benefits: (1) steels require less alloy to obtain a bainite-free case microstructure, and (2) there is less heat treating distortion. However, there is greater susceptibility to microcracking, particularly in steels containing substantial percentages of carbide-forming alloys, and in structures containing much austenite and long needles of martensite. When microcracks occur close to the surface, fatigue properties are substantially reduced.

Microcracking can usually be controlled by limiting the case carbon content and hardenability level. However, nickelfree compositions in which hardenability is developed with manganese, chromium, or molybdenum are quite susceptible to this problem.

Cases of direct quenched carburized gears usually contain 15 to 30% retained austenite. As long as the final hardness is at least Rc 57, however, austenite seldom causes performance problems. As a matter of fact, some manufacturers consider its presence to this amount to be desirable. Several investigators have found that gear teeth treated to contain 25% or more retained austenite and heavily shotpeened have maximum resistance to long-cycle bending fatigue.

When carburized gears are reheated for hardening, the engineer applies either steel of higher alloy content or more vigorous quenching to obtain satisfactory microstructures. With large reheating furnaces, however, heat treating cost may be equal to, and in some instances even lower than, that of direct quenching – due to limitations on tray loading in direct quenching. While direct quenching may produce microcracks, reheating sometimes poses problems with carbide network or decarburization. Though most carburized gears are direct quenched, choice of practice depends on local conditions, obviously affecting steel selection.

Surface of gears carburized in atmospheres containing CO and small amounts of CO_2 exhibit surface and intergranular oxidation to a depth of approximately 0.002 in. max. This effect is due mainly to oxidation of the steel's silicon and (to a lesser degree) manganese. Though not considered to be significantly detrimental to most gears, this condition can be minimized through the use of low-silicon steel which is vacuum degassed and carbon deoxidized. Most production carburizing done today utilizes carrier gas rich in CO.

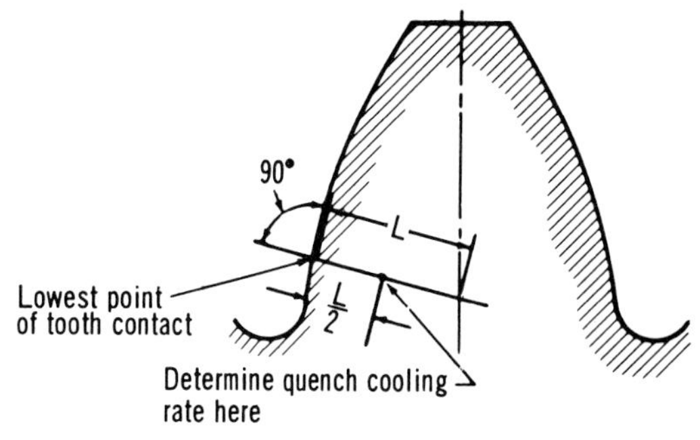

Fig. 3 — The case depth needed to preclude case crushing is determined by first finding the cooling rate at the indicated point. The steel, carbon level, and required depth to Rc 50 are then ascertained as described in the text.

Core Hardness and Structure

Core hardness, particularly at the centerline of the tooth, is relatively unimportant, assuming of course, that case crushing has been forestalled. Where the gear needs great bending fatigue strength, the designer should make sure that surface compressive stresses are as high as is practicable. To assure this effect, a base steel with the lowest possible carbon and a case as shallow as possible are recommended. Core microstructure should reflect complete austenitization to provide maximum toughness, and blocky ferrite should be avoided.

Interface Temperature

Although interface temperature in the operation of a gear set is not strictly a matter of steel selection, the materials engineer is often confronted with scoring problems that designers sometimes assign to metallurgical factors. When a gear tooth surface is both indentation and file hard, and also free of partial decarburization and upper bainite, little can be done with either material or heat treatment to eliminate scoring. Scoring is related to the bulk oil temperature, the relative velocities of the tooth faces and their finish, the coefficient of friction of the two materials, and the load per inch of face. One widely used formula for determining interface temperatures is as follows:

$$T_t = \frac{T_B + 1.297f\left(\frac{W}{\cos\Theta}\right)^{3/4}(V_1 - V_2)}{\left(1 - \frac{F}{50}\right)\left(\frac{P_P \quad P_G}{P_P + P_G}\right)^{1/4}}$$

Here, T_t is interface temperature (F), T_B is bulk oil temperature (F), f is coefficient of friction, W is load (lb per in. of face), V_1 and V_2 are surface velocities at the contact point on the gear teeth (in. per sec), Θ is pressure angle (degrees), F is surface finish after run-in (RMS micro-in.), P_P is pinion radius of profile (in.), and P_G is gear radius of profile (in.). The T_t should be calculated for several anticipated operating speeds to insure that the aforementioned 500 F limit is not exceeded for SAE 50 oil. Another method for doing this calculation can be found in the Gear Handbook previously cited.

Case Hardness

Surface hardness of a carburized gear should normally be Rc 62 to 63 and file hard. Crude as it may seem, a file test will usually detect partial decarburization or upper bainite, whichever is present in the case.

When maximum scoring resistance and wear life are required, specify a minimum Rc 60, file hard. This aim usually requires tempering at 300 F or less, instead of the normal 350 F. (Some automotive hypoids are used untempered.) Gears made in this manner require care to prevent grinding cracks. Some large gear manufacturers have, on the other hand, found it desirable to temper at 375 to 400 F to Rc 55. Gears made this way will "wear in" slightly spreading the loading more evenly across tooth faces and reducing operating noise.

Toughness and Short Cycle Fatigue

When a gear is expected to operate for only a finite life with bending stresses substantially exceeding 100,000 psi at the root fillet, a con-

from Table I, a chart that utilizes typical data from carburized end quench specimens. For example, consider a five-pitch, web-type gear quenched in agitated oil. Figure 1, left, shows that the cooling rate is 4.2 (in 1/16 in. increments) at the surface. Turning to Table I, we see that the choice is 4118 with direct quenching or 8620 with reheat hardening to assure a minimum of Rc 60 at the root. For gears of the highest quality (no bainite is permitted in the case), steels with more case hardenability are required.

Case Crushing Must Be Prevented

After the steel has been chosen to provide the gear surface with either proper hardness or both hardness and microstructure, the case has to be made thick enough to resist case crushing (also known as spalling, core yielding, and core deformation). This phenomenon is considered to be a subsurface fatigue failure in shear. Therefore, carburizing provides resistance to case crushing because it increases subsurface strength. Increasing the vigor of the quench has a similar effect.

To determine the strength required, the reader is referred to the work of R. Pederson and S. L. Rice, "Case Crushing of Carburized and Hardened Gears," presented at the SAE National Farm, Constructional, and Industrial Meeting, 12 to 15 September 1960. In this method, subsurface stress and shear yield strengths are plotted as shown in Fig. 2. Messrs. Pederson and Rice found that case crushing could be avoided if the ratio of the subsurface shear yield stress to the shear yield strength at any point below the surface did not exceed 0.55.

An alternate method to approximate the necessary minimum depth to Rc 50 to prevent case crushing is as follows:

1. Determine the quench cooling rate at a point midway between the radial centerline of the tooth along a line drawn normal to the surface at the lowest point of single tooth contact midway between the sides of the gear (Fig. 3, p. 62).
2. Choose a steel series that has adequate hardenability to provide the required microstructure at the tooth surface, employing methods previously described.
3. Select a carbon level within the series that will provide a core hardness of Rc 30 to 45. (Restrict the carbon to 0.30% maximum if the gear requires the maximum endurance limit in bending fatigue strength.)
4. Calculate the required depth to Rc 50 as follows:

$$CD = \frac{12 \times 10^{-6}\, Wt}{F \cos \Theta}$$

Here, CD is case depth to Rc 50 (in.), Wt is tangential tooth load (lb), F is face width (in.), and Θ is pressure angle.

The maximum subsurface hardness that can be expected primarily depends on the carbon content either in the steel or produced by carburizing. Subsurface hardness also increases with higher alloy content (with the same carbon content) and the vigor of the quench employed (which determines the quench cooling rate). To obtain a certain hardness at a given depth below the surface, we could use, for example, a deep as-carburized case, a medium alloy, and a vigorous impingement quench.

A second choice might be a shallower as-carburized case, a steel of higher hardenability, and a vigorous agitated quench. When the gears are very large and only a few are made per day, a high-alloy steel with a mild quench is perhaps the most practical route.

Depending on his experience with gears and the flexibility of his furnace equipment, the engineer can vary the as-carburized case depths to achieve adequate subsurface strength. For medium-alloy steels (such as 4320 and 4820), increases of up to 3 Rc points can be expected for each 0.10% additional carbon at a point approximately 0.06 in. beneath the surface, providing that the quench cooling rate corresponds to J3 to J9 inclusive. To assure a properly diffused case, this requires approximately a 0.020 in. increase in as-carburized case depth. With less diffusion, such as is necessary on very fine pitch gears, it approaches an additional 0.010 in.

For a given carbon content within the carburized layer (with these steels), the subsurface hardness increase per unit of actual Grossmann D_I is approximately 3 + 2/3 of the cooling rate (given as J distance in 1/16 in.) For example, at a point where the subsurface carbon content is 0.40%, an actual D_I increase of 1 in. with a cooling rate of J9 would result in a hardness increase of 3 + (2/3 × 9), or approximately 9 Rc points. This applies to steels with nominal base carbon contents of 0.17 to 0.25%. As the hardenability approaches a D_I of 6.0 in., however, the hardness increase per unit D_I increase is much less. Subsurface hardness can also be increased through more severe quenching.

As can be seen, it is difficult to decide the course to follow in providing adequate subsurface hardness. Carburizing to deeper cases increases direct costs and uses up valuable furnace time. Use of a higher-alloy steel not only increases material costs but also can reduce machinability.

As an alternative, core hardenability can be increased by using a grade with more carbon. (For example, 8622 might be used instead of 8620.) Increasing the severity of the quench must be carefully studied to determine if fixtures or more violent agitation is practical and economical. (Nonuniform agitation results in increased distortion, even cracking.) Local conditions, such as available equipment, also affect this decision.

Selecting the Carbon Content

Proceeding further, we know by experience that most gears need a minimum of 0.70% C at the surface – preferably, it should be 0.85 to 1.00%. The amount depends on the extent of slide taking place between the teeth in operation; the greater the amount of slide, the higher the carbon content should be to prevent scoring. (There is evidence that the case should have 15 to 25% austenite to improve resistance to sliding wear.)

To minimize the case-core metallurgical notch, a smooth gradient in carbon content should be maintained from case to core. A slope of at least 0.10% C decrease per 0.010 in. depth is recommended, especially for heavily loaded gears with deep cases (over 0.050 in. carburized depth). For fine pitch gears, this procedure is not practical, but should be considered as a goal.

Microstructure After Treatment

After the gear is hardened, the case should be free of continuous network carbide. For nickelfree alloy steels, reheated to 1,550 F for hardening, this usually requires that the carbon potential of the carburizing atmosphere be restricted to approximately 0.90% maximum. Network carbide is practically never a problem, however, with direct quenching.

Requirements for case microstructure are related to hardening practice – the gear may be reheated for hardening, or direct quenched either from the carburizing temperature or from a lower temperature, such as

dition of short-cycle fatigue exists. With shift shock in transmissions, for example, it is not uncommon for tooth loading to exceed eight times the steady state torque. Fatigue life with this level of stress is usually under 200,000 cycles. The metallurgical requirements for this situation require both strength and toughness. This situation is complicated because gears also need hardness to resist wear and scoring.

Though these requirements would appear to be diametrically opposed, one low-cost approach is a compromise involving tempering the gear, for example, down to Rc 55 to 60. It must be remembered that this step reduces contact and bending fatigue strength as well as wear and scoring resistance. Lower hardnesses, however, are often employed for parts that operate intermittently. Nickel-bearing steels can also be applied when good toughness is needed.

Pitting

Carburized gear teeth can break at the roots, but this problem can often be solved by metallurgical means. If the failure occurs up to a few hundred thousand cycles, tempering at a higher temperature, use of high-nickel steels, or shot-peening may be the answer. If failure occurs after millions of cycles, changing to a lower-carbon, higher-alloy steel coupled with vigorous quenching will result in higher residual compressive stress in the carburized case, thereby raising the endurance limit. If endurance limit fatigue failures are associated with large inclusions (usually of the alumina type), extra clean steel should be applied.

When a gear fails by pitting, however, it must be redesigned to reduce contact stresses. (This solution assumes, of course, that heat treatment and gear alignment are satisfactory.) Excessive contact stress can be due to an error in Hertz stress calculation, but it is often caused by misalignment with the mating gear, causing end loading of a tooth.

The mechanism of pitting appears to be a fatigue failure which starts at the point of the maximum tension component of the Hertz stress. It is also enhanced by the stress due to friction and heat at the lowest point of contact. Pitting also can start due to a nonmetallic stringer. However, there is substantial evidence that the destruction will often "heal" in operation, unless this inclusion is very large.

Guide to Gear Steels and Gearmaking Processes

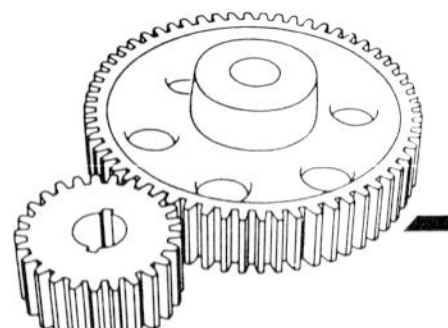

Spur

	Grades (a)	Sizes, In. Pitch Diameter (b)	Sizes, In. Face Width	Forming Processes	Heat Treatment	Applications
A	EX 33, 8627 4027 4118	3.25-6.25 5.6 2.727-3.090	1.0-1.8 2.4 1.902-1.925	Hot forge Hot forge Hot forge	Normalize (1,650 F), carburize (1,650-1,700 F), 300 F oil quench. Case depth: 0.035-0.050 in.	Automotive transmission gears Truck two-speed differential gears
B	8720H 8822H	2-20 (0.2-0.6) (0.4-1.0)	1-6	Machine from upset forging	Carburize (1,700 F), reheat (1,560 F), oil quench, and temper (300 F). Case depth: 0.9-1.2 mm to 2.0-2.4 mm	Final drive for tractors
	1039	15-30 (0.5-0.7)	3-6	Machine from upset forging	Heat (1,550 F), water quench, temper (300 F), and induction harden surface to 0.090-0.100 in., temper (300 F)	Final drive for tractors
	4118 mod 8620 8720H 8822H	2-12 (0.1-0.4) (0.1-0.4) (0.3-0.5) (0.5-0.7)	1-4	Machine from upset forging or bar stock	Carburize (1,700 F), reheat (1,560 F), oil quench, and temper (300 F); or Carburize (1,700 F), direct quench, and temper (300 F); or Carburize (1,700 F) and direct quench; tempered by residual heat (150-200 F). Case depth: 0.9-1.2 mm to 2.0-2.4 mm	Transmission for tractors
	1018 mod 1019	1-4 (0.2-0.4)	0-1	Machine from bar stock	Carbonitride (1,600 F), drop to 1,450 F, oil quench, and temper (360 F). Case depth: 0.3-0.8 mm	Accessory drive for tractors
	ASTM B426	1-2 (0.1-0.4)	1-2	Powder metal	None	Accessory drive for tractors
C	1524 8620	0.5-48.0	0.385-3.00	Hob and shave, or hob	Carburize (1,700 F), oil quench, and temper (360 F). Effective case depth: 0.050-0.060 in.	Speed change gears, planet pinions for farm equipment
	1524 1022	0.5-48.0	0.385-3.00	Hob and shave, or hob	Carburize (1,700 F), air cool, induction heat, water quench, temper (360 F)	Speed change gears, planet pinions for farm equipment
D	EX 16	Various	Various	Forge and machine	Carburize (1,750 F), oil quench, and temper (380-400 F). Effective case depth: 0.030-0.080 in.; larger gears have heavier cases	Truck transmissions
E	4027H	3.0 (6.5)	0.87	Hob and shave	Carburize 0.042-0.055 in. effective case depth, and temper at 350 F	Four-speed manual truck transmission
	EX 33	6.1 (6.5)	1.1	Hob and shave	Carburize 0.042-0.055 in. effective case depth, and temper at 350 F	Four-speed manual truck transmission
	4027H	2.5 (6.5)	1.1	Shear speed and shave	Carburize 0.042-0.055 in. effective case depth, and temper at 350 F	Four-speed manual truck transmission
F (e)	9310 1118 4820 5130 5140 5150 5155 8620	3.0-18.0 2.5-8.0 6.0-14.0 1.0-16.0 1.0-16.0 4.0-14.0 4.0-14.0 0.5-16.0	0.12-3.00 0.40-1.00 1.5-6.0 0.50-4.00 0.50-4.00 0.50-2.50 0.50-2.50 0.25-4.00	Hob, grind, hone Hob, shave Hob, shave, grind Hob, shave Hob, shave Hob, shave Hob, shave Hob, shave, grind, hone	Carburize (1,500-1,550 F) Carbonitride (1,500-1,700 F) Carburize (1,500-1,700 F) Carburize (1,500-1,700 F) Carburize (1,500-1,700 F) Induction harden (1,475-1,550 F) Induction harden (1,475-1,550 F) Carburize (1,525-1,575 F)	Aircraft gas turbine engines Transmissions for heavy-duty transport — off-highway equipment, buses, trucks, ordnance, etc.
G	4023 4023, 4027, 4028 1045			Machine from hot-rolled tubing or bar Machine from hot-rolled bar Machine from butt welded, hot-rolled bar	Carburize, oil quench, and temper to Rc 58 min. Case depth: 0.015-0.025 in. Carburize, oil quench, and temper to Rc 58 min. Case depth: 0.030-0.045 in. Induction harden teeth, temper to Rc 46-52 min. Case depth: 0.100-0.200 in.	Truck transmission gears (4023, 4027, 4028) Automotive starter drive gears (1045)
H	1020 Low-carbon: 8620, 4320, 4620, 4820, 9310, 2320, etc. Medium-carbon: 1045, 4140, 4150, 4340, etc.	¾ in. diam—16 DP (c) to 134 in. diam.—0.75 DP (c) Max. face width, 54 in.		Bar stock, closed die forgings, open frame forgings, rolled rings with welded webs and hubs. Steel and iron castings. Hob teeth and shave when required	1020: Carburize (1,750 F) water quench from 1,550 F, and temper to desired hardness. Medium-carbon: Heat (1,550-1,575 F), oil quench, and temper (600-1,200 F), depending on desired hardness — some gears are flame or induction hardened. Low-carbon, shallow and deep cases: see footnote (f) for heat treatments	Steel mill equipment, paper machinery, mining machinery, earthmoving equipment, printing presses.

Guide to Gear Steels and Gearmaking Processes (Continued)

	Grades (a)	Sizes, In. Pitch Diameter (b)	Face Width	Forming Processes	Heat Treatments	Applications
I	Pinions: 4140 4340 4350	20-35 0.375-1 DP (c)	10-21	Forge and machine teeth	Heat (1,600 F), water quench, and temper to Bhn 277-331	Mining machinery
	Gears: Chromium-molybdenum cast steel; ASTM A148	50-600 0.375-1 DP (c)	10-21	Cast and machine teeth	Heat (1,600 F), water quench, and temper to Bhn 229-277	Mining machinery

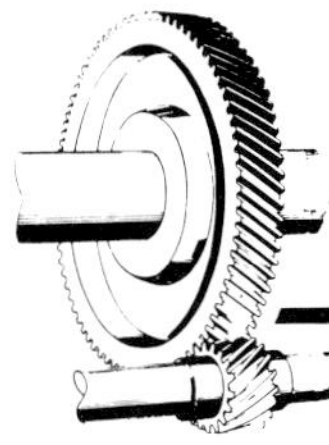

Single Helical

A	EX 33	1½-6	⅝-1	Forge	Anneal, carburize (1,650 F, 0.025-0.040 in. case depth), and shot-peen for fatigue strength	Gears for manual transmissions
	EX 33 and 8620 4027	2.1-6.7 3.5-6.4	1.0 approx. 1.1-1.6	Forge Forge	Normalize (1,650 F), carburize (1,650-1,700 F), oil quench and temper (275-425 F, depending on desired hardness). Case depth: 0.035-0.050 in.	Automotive transmission gears
	5130	0.962	0.746	Cold form and machine from coil	Cycle anneal blank, carbonitride 0.010-0.016 in., hot oil quench, and temper	Automotive transmission gears
	5130	2.044	3.700	Machine from seamless tubing	Carbonitride 0.012-0.016 in., hot oil quench, and temper	Automotive transmission gears
B	Nitralloy N	5.8 (0.3-0.5)	5-7	Machine from forging	Solution treat (1,650 F), oil quench, temper (1,250 F), and gas nitride (980 F).	Final drive for marine transmissions
	4118 mod 8620	4-10 (0.1-0.5)	1-3	Machine from upset forging	Carburize (1,750 F), reheat (1,560 F), oil quench, and temper (300 F); or Carburize (1,700 F), direct quench, and temper (300 F); or Carburize (1,700 F) and direct quench; tempered by residual heat (150-200 F). Case depth: 0.9-1.2 mm to 2.0-2.4 mm.	Accessory drive for tractors
C	1022, 1524	0.5 to 18	3	Hob and shave	Carburize (1,700 F), air cool, induction harden, and temper (360 F)	Transmissions for farm equipment
	1524, 4023	0.5 to 18	3	Hob and shave	Carburize (1,700 F), oil quench, and temper (360 F). Effective case depth: 0.050-0.060 in.	Low range pinion for farm equipment
D	EX 16	Various	Various	Forge and machine	Carburize (1,750 F), oil quench, and temper (380-400 F). Effective case depth: 0.030-0.080 in.; larger gears have heavier cases	Truck transmissions
E	4027H	4.5 (8.0)	0.9	Shaper cut and shave	Carburize (effective case depth 0.042-0.055 in.), and temper at 350 F	Four-speed manual truck transmissions
	8620H 4027H 4620H	2.0 (9.3) 3.2 (9.3) 3.2 (8.3)	0.9 0.8 0.75	Hob and shave	Carburize (effective case depth 0.035-0.045 in.), and temper at 350 F	Three-speed manual transmissions
	9310	2.1 (9.2)	0.8	Shaper cut and shave	Carburize (effective case depth, 0.035-0.045 in.), deep freeze −120 F temper at 350 F	Four-speed manual dragster transmission
	8620H	1.8 (9.4) 2.1 (9.4) 2.5 (9.4) 2.8 (10.6)	1.4 0.8 0.8 0.7	Hob and shave Shaper cut and shave Shaper cut and shave Hob and shave	Carburize (effective case depth, 0.035-0.045 in.), temper at 350 F.	Four-speed manual transmission
		2.3 (15.5)	0.8	Hob and shave	Carbonitride (case depth, 0.018-0.023 in.), and temper at 350 F	Sun gear in automatic transmission
F (e)	9310 4140 4145 5140 8620	3.0-18.0 1.0-11.0 1.0-11.0 6.0-14.0 6.0-14.0	0.12-3.0 2.5 max 2.5 max 1.0-4.0 1.0-4.0	Hob and grind Hob, shave, and hone Hob, shave, and hone Hob and shave Hob and shave	Carburize (1,500-1,550 F) Nitride (1,550-1,600 F) Nitride (1,550-1,600 F) Carburize (1,500-1,550 F) Carburize (1,525-1,575 F)	Aircraft gas turbine engines Truck engines Truck engines Transmissions for heavy-duty transport — buses, trucks,
G	4023			Machine from hot-rolled tubing or bar	Carburize, oil quench, and temper to Rc 58 min. Case depth: 0.015-0.025 in.	Automotive manual and automatic transmissions (4023, 4027)
	4023, 4027			Hot forged	Carburize, oil quench, and temper to Rc 58 min. Case depth: 0.030-0.045 in.	

Guide to Gear Steels and Gearmaking Processes (Continued)

	Grades (a)	Sizes, In. Pitch Diameter (b)	Sizes, In. Face Width	Forming Processes	Heat Treatments	Applications
H	1020 Low-carbon: 8620, 4320, 4620, 4820, 9310, 2320, etc. Medium-carbon: 1045, 4140, 4150, 4340, etc.	¾ in. diam—16 DP (c) to 134 in. diam—1 DP (c) Max. face width, 54 in.		Bar stock, closed die forgings, open frame forgings, rolled rings with welded webs and hubs. Steel and iron castings. Hob teeth and shave when required	1020: Carburize (1,750 F) water quench from 1,550 F, and temper to desired hardness. Medium-carbon: Heat (1,550-1,575 F), oil quench, and temper (600-1,200 F), depending on desired hardness — some gears are flame or induction hardened. Low-carbon, shallow and deep cases: see footnote (f) for heat treatments	Steel mill equipment, paper machinery, mining machinery, earthmoving equipment, printing presses
I	Pinions: 4140 4340 4350	7-15 1-3 DP (c)	5-11	Forge and machine teeth	Heat (1,600 F), water quench, and temper to Bhn 277-331	Mining machinery
	Gears: Chromium-molybdenum cast steel; ASTM A148	32-170 1-3 DP (c)	5-11	Cast and machine teeth	Heat (1,600 F), water quench, and temper to Bhn 229-277	Mining machinery
J	4142H	2-12	1-4	Hob and shave	Heat (1,550 F), oil quench, temper, and gas nitride (975 F)	Industrial drives

Herringbone

	Grades (a)	Pitch Diameter (b)	Face Width	Forming Processes	Heat Treatments	Applications
H	1020 Low-carbon: 8620, 4320, 4620, 4820, 9310, etc. Medium-carbon: 1045, 4140, 4150, 4340, etc.	¾ in. diam—16 DP (c) to 134 in. diam.—0.75 DP (c) Max. face width, 54 in.		Bar stock, closed die forgings, open frame forgings, rolled rings with welded webs and hubs. Steel and iron castings. Hob teeth and shave when required	1020: Carburize (1,750 F), water quench from 1,550 F, and temper to desired hardness. Medium-carbon: Heat (1,550-1,575 F), oil quench, and temper (600-1,200 F), depending on desired hardness — some gears are flame or induction hardened. Low-carbon, shallow and deep cases: see footnote (f) for heat treatments	Steel mill equipment, paper machinery, mining machinery, earthmoving equipment, printing presses

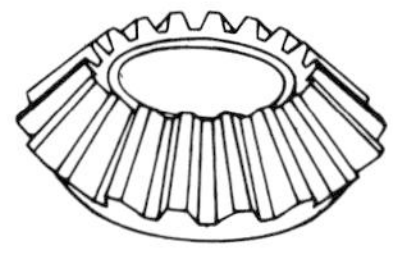

Mitre

	Grades (a)	Pitch Diameter (b)	Face Width	Forming Processes	Heat Treatments	Applications
H	1020 Low-carbon: 8620, 4320, 4620, 4820, 9310, etc. Medium-carbon: 1045, 4140, 4150, 4340, etc.	¾ in. diam—16 DP (c) to 134 in. diam.—0.75 DP (c) Max. face width, 54 in.		Bar stock, closed die forgings, open frame forgings, rolled rings with welded webs and hubs. Steel and iron castings. Hob teeth and shave when required	1020: Carburize (1,750 F), water quench from 1,550 F, and temper to desired hardness. Medium-carbon: Heat (1,550-1,575 F), oil quench, and temper (600-1,200 F), depending on desired hardness — some gears are flame or induction hardened. Low-carbon, shallow and deep cases: see footnote (f) for heat treatments	Steel mill equipment, paper machinery, mining machinery, earthmoving equipment, printing presses

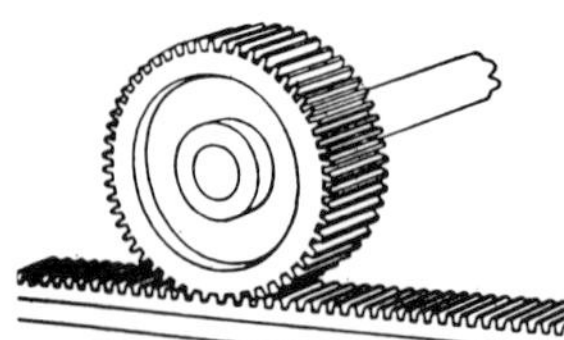

Spur Rack and Pinion

	Grades (a)	Pitch Diameter (b)	Face Width	Forming Processes	Heat Treatments	Applications
H	1020 Low-carbon: 8620, 4320, 4620, 4820, 9310, etc. Medium-carbon: 1045, 4140, 4150, 4340, etc.	¾ in. diam—16 DP (c) to 134 in. diam.—0.75 DP (c) Max. face width, 54 in.		Bar stock, closed die forgings, open frame forgings, rolled rings with welded webs and hubs. Steel and iron castings. Hob teeth and shave when required	1020: Carburize (1,750 F), water quench from 1,550 F, and temper to desired hardness. Medium-carbon: Heat (1,550-1,575 F), oil quench, and temper (600-1,200 F), depending on desired hardness — some gears are flame or induction hardened. Low-carbon, shallow and deep cases: see footnote (f) for heat treatments	Steel mill equipment, paper machinery, mining machinery, earthmoving equipment, printing presses.

	Grades (a)	Sizes, In. Pitch Diameter (b)	Sizes, In. Face Width	Forming Processes	Heat Treatments	Applications

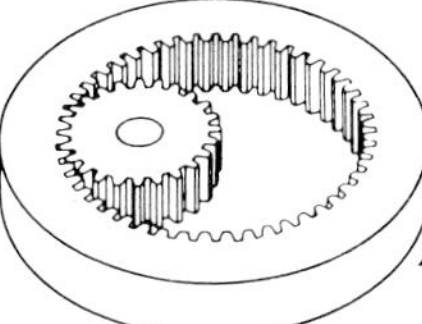

Internal

	Grades (a)	Pitch Diameter (b)	Face Width	Forming Processes	Heat Treatments	Applications
A	EX 33	2.75-4.5	1.0-1.25	Hot-rolled annealed tubing	Carburize (1,650-1,700 F), oil quench, and temper (275-425 F, depending on desired hardness). Case depth: 0.035-0.050 in.	Automotive transmission gears
	5130	3.968	1.090	Machine and broach from seamless tubing	Carbonitride 0.012-0.016 in., hot oil quench, and temper	Automotive transmission gears
	8627	2.75-4.5	1.0-1.25	Hot forge	Normalize (1,650 F), carburize (1,650-1,700 F), oil quench, and temper (275-425 F, depending on desired hardness). Case depth: 0.035-0.050 in.	Automotive transmission gears
	Pearlitic Malleable Iron	0.666	1.000	Machine and broach from casting	Oil quench (Bhn 197-241), induction harden, and temper to Rc 50-56	Automotive transmission gears
B	4118 mod 8620 8822H	5-30 (0-0.3) (0.3-0.7)	1-4	Machine from upset forging, or ring roll forging	Carburize (1,700 F), reheat (1,560 F), oil quench, and temper (300 F); or Carburize (1,700 F), direct quench, and temper (300 F); or Carburize (1,700 F) and direct quench; tempered by residual heat (150-200 F). Case depth: 0.9-1.2 mm to 2.0-2.4 mm.	Planetary final drive and transmission gears for tractors
	4140H 41L40H	5-20 (0-0.2)	1-4	Machine from upset forging, or ring roll forging	Heat (1,560 F), oil quench, temper (1,150 F), and gas nitride (980 F)	Transmissions for tractors
C	1053 ring 1566 ring 1524	3-18		Shape or broach	Anneal, stress relieve, induction harden teeth, temper (360 F) Carburize (1,700 F), oil quench, and temper (360 F). Effective case depth: 0.050-0.060 in.; or carburize (1,700 F), air cool, induction harden, and temper (360 F)	Final drive ring gears, farm equipment Planet pinions for farm equipment
E	4028H	4.8 (15.5)	1.2	Broach	Carbonitride (case depth: 0.018-0.023 in.)	Planetary ring gear in automatic transmission
	1137	3.5 (9.4)	0.7	Broach	None	Pump gear in automatic transmission
F (e)	9310	17.0 max	2.1 max	Shape and grind	Carburize (1,500-1,550 F)	Aircraft gas turbine engines
	4140	6.0-18.0	0.50-3.0	Shape and hone	Nitride (1,550-1,600 F)	Transmissions, engines for heavy-duty equipment.
	4145	6.0-18.0	0.50-3.0	Shape and hone	Nitride (1,500-1,600 F)	Truck engines
	5140	5.0-12.0	0.50-2.0	Shape and shave	Carburize (1,500-1,600 F)	Transmissions for heavy-duty transport — buses, trucks, etc.
H	Low-carbon: 8620, 4620, etc. Medium-carbon: 1050, 4140, etc.	156 1½ DP (c)	9½	Rolled rings or forgings. Teeth are shaped on internal gears, hobbed on planet gears	Medium-carbon: Heat (1,550-1,575 F), oil quench, and temper (600-1,200 F), depending on desired hardness — some gears are flame hardened. Low-carbon, shallow and deep cases: see footnote (f) for heat treatments	Steel mill equipment, paper machinery, mining machinery, earthmoving equipment, printing presses

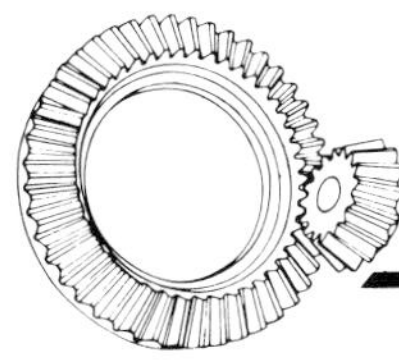

Straight Bevel

	Grades (a)	Pitch Diameter (b)	Face Width	Forming Processes	Heat Treatments	Applications
A	EX 33 8627H	2.124-3.398	0.800	Machine from bar stock, hot forge	Normalize (1,650 F), carbonitride (1,550 F), 300 F oil quench. Case depth: 0.012-0.015 in.	Passenger automotive differential gears
	EX 33	2.214-4.64	0.875-0.950	Machine from bar stock, hot forge	Normalize (1,650 F), carbonitride (1,550 F), 300 F oil quench	Truck differential gears
B	4118 mod 8620 8822H	3-25 (0.1-0.2) (0.3-0.5)	0-3	Machine from upset forging or cold finished bar stock	Carburize (1,700 F), reheat (1,560 F), oil quench, and temper (300 F). Case depth: 0.9-1.2 mm to 2.0-2.4 mm	Final drive for tractors

	Grades (a)	Sizes, In.		Forming Processes	Heat Treatments	Applications
		Pitch Diameter (b)	Face Width			
C	1524 8615 1118	3			Carburize (1,700 F), oil quench, and temper (360 F). Effective case depth: 0.050-0.060 in.	Differential bevel gears and pinions for farm equipment
D	94B17			Machine from bar stock	Carburize (1,750 F), oil quench, and temper (380-400 F). Effective case depth: 0.030-0.080 in.; larger gears have heavier cases.	Truck rear axles, differential pinions
F (e)	9310 8620	6.0 max 9.0 max	1.0 max 2.0 max	Cut and grind Cut and grind	Carburize (1,500-1,550 F) Carburize (1,525-1,575 F)	Aircraft gas turbine engines Transmissions, engines for heavy-duty equipment
G	4023, 4027, 1518, 4615 1518, 1526	1.5-5.0	0.5-1.0	Cold formed Cold formed (1518), or hot forged (1526)	Carburize, quench and temper to Rc 58 min. Case depth: 0.030-0.045 in.	Automotive rear axle differential pinions (4023, 4027, 1518, 4615) and side gears (1518, 1526)
H	**Low-carbon: 8620, 4620, 4820, 9310, 2320, etc. Medium-carbon: 1045, 4140, 4150, 4340, etc.**	94 in. OD 5½ CP (d) Carburized limited to 28 in. OD	13 max	**Forgings, steel castings; generate teeth**	**Medium-carbon: Heat (1,550-1,575 F), oil quench, and temper (600-1,200 F), depending on desired hardness — some gears are flame or induction hardened. Low-carbon, shallow and deep cases: see footnote (f) for heat treatments**	Steel mill equipment, paper machinery, mining machinery, earthmoving equipment, printing presses

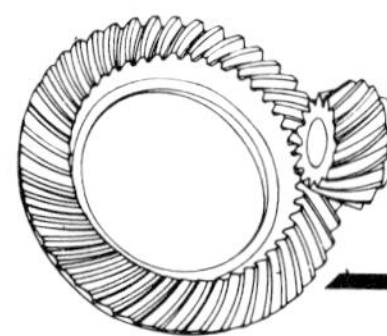

Spiral Bevel

	Grades (a)	Pitch Diameter (b)	Face Width	Forming Processes	Heat Treatments	Applications
B	8720 mod 4118 mod	10-20 (0.1-0.4)	1-3	Machine from upset forging	Carburize (1,700 F), reheat (1,560 F), oil quench, and temper (300 F). Case depth: 0.9-1.2 mm to 2.0-2.4 mm	Final drive for tractors
C	1524, 8620	16 in. OD		Machine	Carburize (1,700 F), oil quench, and temper (360 F). Effective case depth: 0.050-0.060 in.	Pinions and shafts for farm equipment
F (e)	9310 4820 8620	7.0 max 15.0 max 10.0 max	1.5 max 3.5 max 3.0 max	Cut and grind Cut Cut	Carburize (1,500-1,550 F) Carburize (1,550-1,700 F) Carburize (1,525-1,575 F)	Aircraft gas turbine engines Transmissions for heavy-duty transport—trucks, buses, etc.
H	**Low-carbon: 8620, 4620, 4820, 9310, etc. Medium-carbon: 1045,4140, 4150, 4340, etc.**	94 in. OD 5½ CP (d) Carburized limited to 28 in. OD	13 max	**Forgings, steel castings; generate teeth**	**Medium-carbon: Heat (1,550-1,575 F), oil quench, and temper (600-1,200 F), depending on desired hardness — some gears are flame hardened. Low-carbon, shallow and deep cases: see footnote (f) for heat treatments**	Steel mill equipment, paper machinery, mining machinery, earthmoving equipment, printing presses

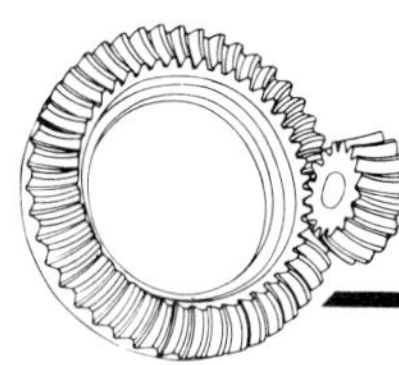

Zerol Bevel

	Grades (a)	Pitch Diameter (b)	Face Width	Forming Processes	Heat Treatments	Applications
B	8720 mod 4118 mod	10-20 (0.1-0.4)	1-3	Machine from upset forging	Carburize (1,700 F), reheat (1,560 F), oil quench, and temper (300 F). Case depth: 0.9-1.2 mm to 2.0-2.4 mm	Final drive for tractors
F (e)	9310	6.0 max	1.0 max	Cut and grind	Carburize (1,500-1,550 F)	Aircraft gas turbine engines
H	**Low-carbon: 8620, 4620, 4820, 9310, etc. Medium-carbon: 1045, 4140, 4150, 4340, etc.**	94 in. OD 5½ CP (d) Carburized limited to 28 in. OD	13 max	**Forgings, steel castings; generate teeth**	**Medium-carbon: Heat (1,550-1,575 F), oil quench, and temper (600-1,200 F), depending on desired hardness — some gears are flame hardened. Low-carbon, shallow and deep cases: see footnote (f) for heat treatments**	Steel mill equipment, paper machinery, mining machinery, earthmoving equipment, printing presses

	Grades (a)	Sizes, In. Pitch Diameter (b)	Sizes, In. Face Width	Forming Processes	Heat Treatment	Applications
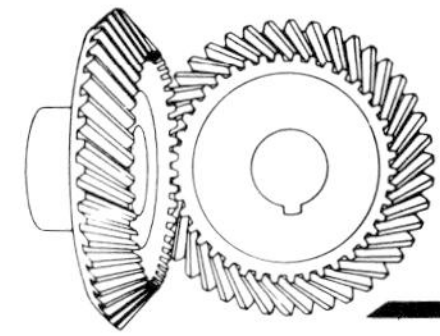	**Skew Bevel**					
E	8620H	3.1 (6)	2.0	Machine	Carburize (effective case depth: 0.042-0.055 in.), temper at 350 F	Marine transmission
	8620H	5.5 (6)	2.0	Machine	Carburize (effective case depth: 0.042-0.055 in.), temper at 350 F	Marine transmission
H	Low-carbon: 8620, 4620, 4820, 9310, etc. Medium-carbon: 1045, 4140, 4150, 4340, etc.	94 in. OD 5½ CP (d) Carburized limited to 28 in. OD	13 max	Forgings, steel castings; generate teeth	Medium-carbon: Heat (1,550-1,575 F), oil quench, and temper (600-1,200 F), depending on desired hardness — some gears are flame hardened. Low-carbon, shallow and deep cases: see footnote (f) for heat treatments	Steel mill equipment, paper machinery, mining machinery, earthmoving equipment, printing presses
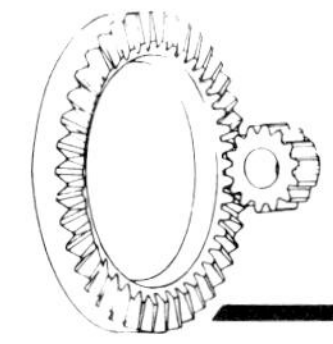	**Face**					
B	8720 mod 8822H	1-4 (0.1-0.3)	1-3	Machine from upset forging	Carburize (1,700 F), reheat (1,560 F), oil quench, and temper (300 F). Case depth: 0.9-1.2 mm to 2.0-2.4 mm	Differential drive gear for tractors
F (e)	9310	4.6	0.3 max	Hob and grind	Carburize (1,500-1,550 F)	Aircraft gas turbine engines
	**Hypoid**					
A	EX 33 1526 4626	6-8½ 9⅜		Forge and machine	Carburize (case depth: 0.035-0.050 or 0.045-0.060 in.), and induction anneal threaded end	Rear axle ring and pinion gears
	1527 EX 33	8.50 10.5-15.0	1.240-1.375 1.552-2.0	Hot forge	Carburize (1,650-1,700 F), 300 F oil quench. Case depth: 0.035-0.050 in.	Passenger differential gears Truck differential gears
D	Pinions: 4817 Gears: EX 17	Large		Forged and machine	Carburize (1,750 F), oil quench, and temper (380-400 F). Effective case depth: 0.030-0.080 in.; larger gears have heavier cases	Truck rear axles
G	Ring: 4023, 8615, 4620	7-9	1-1½	Hot forging	Carburize, press quench, and temper to Rc 58 min. Case depth: 0.040-0.055 in.	Automotive rear axle drive pinions (4023, 8615, 4422, 4620) and gears (4023, 8615, 4620).
	Pinions: 4023, 8615, 4422, 4620	3-4	1-1½		Carburize, mass quench, and temper to Rc 58 min. Case depth: 0.040-0.055 in.	
H	Low-carbon: 8620, 4620, 4820, 9310, etc. Medium-carbon: 1045, 4140, 4150, 4340, etc.	94 in. OD 5½ CP (d) Carburized limited to 28 in. OD	13 max	Forgings, steel castings; generate teeth	Medium-carbon: Heat (1,550-1,575 F), oil quench, and temper (600-1,200 F), depending on desired hardness — some gears are flame hardened. Low-carbon, shallow and deep cases: see footnote (f) for heat treatments	Steel mill equipment, paper machinery, mining machinery, earthmoving equipment, printing presses

Supplier	Grades (a)	Sizes, In.: Pitch Diameter (b)	Sizes, In.: Face Width	Forming Processes	Heat Treatments	Applications

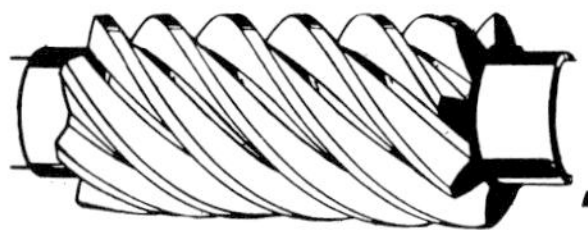

Cylindrical Worm

Supplier	Grades (a)	Pitch Diameter (b)	Face Width	Forming Processes	Heat Treatments	Applications
F (e)	1117	1.0 max	—	Grind	Carburize (1,500-1,700 F)	Truck engines
H	1020 8620 4140	12 in. max lead 20½ OD 1 DP (c)		Bar stock or forgings, cut teeth	1020: Carburize (1,550 F), water quench, and temper to desired hardness. Medium-carbon: Heat (1,550-1,575 F), oil quench, and temper (600-1,200 F), depending on desired hardness. Low-carbon alloy, shallow and deep cases: See footnote (f).	Steel mill equipment, paper machinery, mining machinery, earthmoving equipment, printing presses

Single Enveloping Worm

Supplier	Grades (a)	Pitch Diameter (b)	Face Width	Forming Processes	Heat Treatments	Applications
B	Gear: 10L45	8-10 (0.2-0.4)	1.2	Machine from upset forging	None	Steering gears for tractors
	Worm: 1040	1-3 (0.2-0.4)		Machine from forging	Heat (1,600 F), water quench, and temper to Bhn 241-285.	
	1215	2-4 (0.3-0.5)		Machine from bar stock	None	
G	Drive gear: 8620			Machine from hot-rolled bar	Carburize, oil quench, and temper to Rc 58 min. Case depth: 0.020-0.030 in.	Automotive gas turbine regenerator drive gears
H	1020 8620 4140	12 in. max lead 20½ OD 1 DP (c)		Bar stock or forgings, cut teeth	1020: Carburize (1,550 F), water quench, and temper to desired hardness. Medium-carbon: Heat (1,550-1,575 F), oil quench, and temper (600-1,200 F), depending on desired hardness. Low-carbon alloy, shallow and deep cases: See footnote (f).	Steel mill equipment, paper machinery, mining machinery, earthmoving equipment, printing presses

Crossed Axis Helical

Supplier	Grades (a)	Pitch Diameter (b)	Face Width	Forming Processes	Heat Treatments	Applications
B	1040 1144	1-3 (0-0.2)	0-2	Machine from forging, or cold finished steel	Heat (1,550 F), water quench, temper (300 F), and induction harden surface to 0.090-0.100 in., temper (300 F); or surface induction harden and furnace temper	Accessory drive for tractors
	4118 mod	1-3 (0-0.2)	0-2	Machine from forging	Carburize (1,700 F), reheat (1,560 F), oil quench, and temper (300 F). Case depth: 0.9-1.2 mm to 2.0-2.4 mm	
	4140H	1-3 (0-0.2)	0-2	Machine from forging	Heat (1,550 F), oil quench, and temper (300 F)	
F (e)	1117	1.4 max	0.75 max	Hob	Carburize (1,500-1,700 F)	Truck engines

(a) AISI or SAE steels except where specified. (b) Dimension in parentheses for supplier B is for circular thickness (chordal thickness for bevel gears); dimension in parentheses for supplier E is for normal pitch.
(c) Diametrical pitch—ratio of number of teeth to number of inches of pitch diameter.
(d) Circular pitch—length of arc of the pitch circle between corresponding points on adjacent teeth.
(e) Case depths of carburized gears vary with diametral pitch (number of teeth per inch of circumference) as follows:

Diametral Pitch	Before Finish	After Finish
4.5-8	0.040-0.050 in.	0.035-0.050 in.
8-10	0.030-0.040	0.025-0.040
10-13	0.020-0.030	0.016-0.030
13-16	0.015-0.025	0.012-0.025
16-20	0.013-0.020	0.010-0.020

Carburized gears requiring further finishing must meet "After Finish" sizes; others must meet "Before Finish" sizes. For all other case-hardening treatments, case depths vary with time, as empirically determined by knowledge of the gear's propensity for distortion during treatment.
(f) Low-carbon, shallow cases (0.030-0.035 in.): Carburize (1,700 F), lower to 1,525 F, oil quench, and temper to desired hardness. Low-carbon, deep cases (0.200 in.): Carburize (1,800 F for 90 hr), cool, and equalize at 1,500 F, cool to room temperature, reheat to 1,525 F, oil or water quench according to grade, and temper at 350 F. Gears of high-nickel steels are cooled to −120 F to reduce retained austenite.

Source: Warner Gear Div., Borg-Warner Corp.; Bucyrus-Erie Co.; Caterpillar Tractor Co.; General Mfg. Div., Chrysler Corp.; Deere & Co.; U. S. Electrical Motors Div., Emerson Electric Co.; Chevrolet Motor Div., and Detroit Diesel Allison Div., General Motors Corp.; Truck Div., International Harvester Co.; Tool Steel Gear & Pinion Co.; Illinois Gear Div., Wallace-Murray Corp.

A Group of Boron-Treated Steels

By JOHN A. RINEBOLT

Produced by either conventional steelmaking practice with standard silicon or by vacuum carbon deoxidation practice with 0.05 Si max, these grades combine excellent cold formability with good toughness and strength after heat treatment.

Republic Steel recently announced a new series of boron-treated steels which combine cold formability with strength and toughness in heat-treated parts. Compared to grades which they may replace (such as 1038, 1040, or 1045), these steels have lower carbon plus slightly higher manganese and boron (Table I). The combination of boron and higher manganese provides increased hardenability which permits reduction in carbon levels without lowering strength. Less carbon results in lower as-rolled hardness, and permits extensive cold forming in the as-rolled condition. Because these low-carbon boron steels are self-tempering, quench cracking and distortion are minimized during heat treatment. These steels are produced with both normal and reduced silicon levels. Lowering the silicon enhances cold formability and helps toughness after heat treatment.

Hot-rolled bars also have lower hardnesses when silicon is reduced. The combined effect of lower silicon and carbon provides maximum opportunity to eliminate annealing

Mr. Rinebolt is a metallurgical product engineer, Hot Rolled Bar & Semi-finished Products Div., Republic Steel Corp., Cleveland.

Table I—Composition of Standard and Special Steels

Grade*	C	Mn	Si	B (Min.)
AISI/SAE 1018	0.15-0.20	0.60-0.90	0.15-0.30	—
RSB-18-SS	0.15-0.20	0.80-1.10	0.15-0.30	0.0005
RSB-18-LS	0.15-0.20	0.80-1.10	0.15 max	0.0005
RSB-18-LSV	0.15-0.20	0.80-1.10	0.05 max	0.0005
AISI/SAE 1021	0.18-0.23	0.60-0.90	0.15-0.30	—
RSB-21-SS	0.18-0.23	0.80-1.10	0.15-0.30	0.0005
RSB-21-LS	0.18-0.23	0.80-1.10	0.15 max	0.0005
RSB-21-LSV	0.18-0.23	0.80-1.10	0.05 max	0.0005
AISI/SAE 1022	0.18-0.23	0.70-1.00	0.15-0.30	—
RSB-22-SS	0.17-0.23	1.00-1.30	0.15-0.30	0.0005
RSB-22-LS	0.17-0.23	1.00-1.30	0.15 max	0.0005
RSB-22-LSV	0.17-0.23	1.00-1.30	0.05 max	0.0005
AISI/SAE 1023	0.20-0.25	0.30-0.60	0.15-0.30	—
RSB-23-SS	0.17-0.23	1.10-1.40	0.15-0.30	0.0005
RSB-23-LS	0.17-0.23	1.10-1.40	0.15 max	0.0005
RSB-23-LSV	0.17-0.23	1.10-1.40	0.05 max	0.0005
AISI/SAE 8620	0.18-0.23	0.70-0.90	0.20-0.35	—
RS-8620-LSV	0.18-0.23	0.70-0.90	0.05 max	—
RS-86B20-LSV	0.18-0.23	0.70-0.90	0.05 max	0.0005
AISI/SAE 8640	0.38-0.43	0.75-1.00	0.20-0.35	—
RS-8640-LSV	0.38-0.43	0.75-1.00	0.05 max	—
RS-41B21-SS	0.18-0.23	0.75-1.00	0.20-0.35	0.0005
RS-41B21-LSV	0.18-0.23	0.75-1.00	0.05 max	0.0005

* SS indicates standard silicon; LS, low silicon; LSV, low silicon, vacuum deoxidized. All 8620 types contain 0.40-0.70 Ni, 0.40-0.60 Cr and 0.15-0.25 Mo; RS-41B21-SS contains 0.25-0.40 Cr and 0.15-0.25 Mo; and RS-41B21-LSV contains 0.25-0.40 Cr. and 0.15-0.40 Mo.

prior to cold forming. In addition, the reduced silicon results in lower hardness after cold working. Therefore, the low-silicon steels require less force for cold deformation resulting in lower tool and die costs.

One fabricator, for example, made 7/16 in. diameter axle bolts for bicycles of a low-silicon grade, RSB-21-LS. Engineers found that the die life extended far beyond that of dies used to make the bolts of standard AISI 1038.

These low-carbon boron steels develop excellent strength-toughness relationships compared to higher carbon steels. The low silicon levels provide additional toughness at all strength levels. Silicon, a ferrite hardener, is detrimental to cold forming even in relatively small amounts. It tends to increase the load required for cold deformation, raises both anneal hardness and as-rolled hardness, contributes little to hardenability, and usually penalizes toughness after heat treatment.

Once a problem, boron steels can be made with reproducible hardenability by using proprietary addition agents and advanced steelmaking techniques. Steels with very low silicon levels are achieved with good internal quality by vacuum carbon deoxidation in our induction-stirred ladle degasser.

Tendencies toward distortion and quench cracking during heat treatment are minimized by the self-tempering characteristics of these steels. Because of the lower carbon content, the M_S transformation during quenching is raised to a point where tempering occurs just after transformation. This characteristic, known as self-tempering, results in relief of transformation stresses which normally contribute to distortion and quench cracking.

Properties Reveal Merits

Figure 1 compares relative deformation loads required to cold upset hot rolled (or annealed) RSB-21 containing 0.15 to 0.30, 0.15 max, and 0.05 max Si. Note that lowering the silicon reduced loads required for deformation.

On the basis that 100 be assigned as the force required to yield 30% deformation of hot-rolled spheroidize annealed SAE 1038, the RSB-21-SS steel, unannealed, required a force of 112; the RSB-21-LS (0.15 Si max) a force of 93; and the RSB-21-LSV (0.05 Si), a force of less than 85. Lower deformation loads also act to improve tool life and increase equipment capability.

Figures 2 and 3 illustrate the effect of tempering on strength, ductility, and toughness of the boron-treated steels. Note that tempering the RSB-21-LSV at 400 F develops a minimum tensile strength of 200,000 psi. When comparing the ductility of the RSB-21-LSV steel tempered at 400 F to that of AISI 1038 tempered at 700 F (to develop an equivalent tensile strength), engineers found that the former had approximately 63% reduction in area while that of AISI 1038 was only 55%.

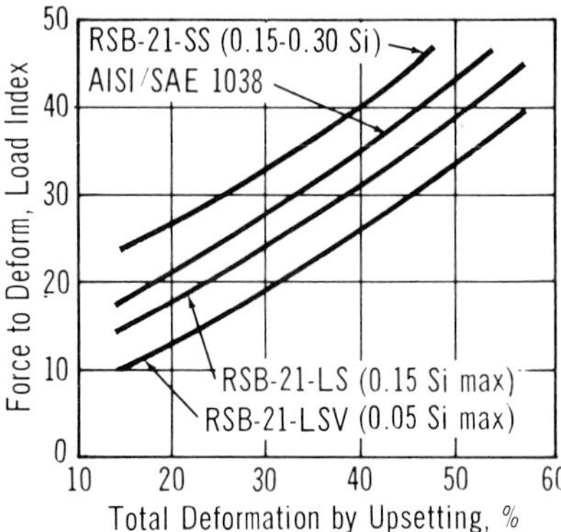

Fig. 1 — Given the same amount of deformation (30% upsetting for these tests), it takes less force to form low-silicon steels than high-silicon steels. Except for containing different silicon contents, these three types have the same composition, and were tested in the hot-rolled condition.

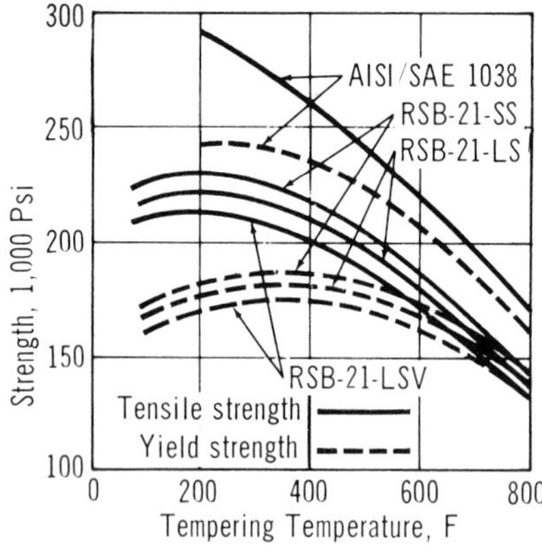

Fig. 2 — Strengths of low-carbon, boron-treated steels with different silicon contents vary with the tempering temperature. All specimens were austenitized at 1,600 F, quenched in water, and tempered at the indicated temperatures for 2 hr.

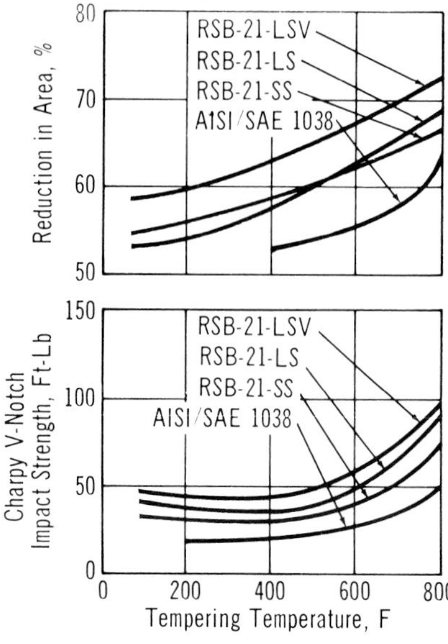

Fig. 3 — Ductility and impact strength of boron-treated grades rise with tempering temperature. Austenitizing temperature, 1,600 F; quenchant, water; tempering time, 2 hr.

Note in Fig. 4 that toughness of the RSB-21 steels as measured by impact strength is higher at all strengths and hardness levels compared to that of the AISI 1038. The lower silicon content further improves this relationship.

Fig. 4 — At equivalent strengths, the low-carbon, boron-treated steels have better impact properties than AISI 1038. Specimens were austenitized at 1,600 F, water-quenched, and tempered at 400, 600, and 800 F.

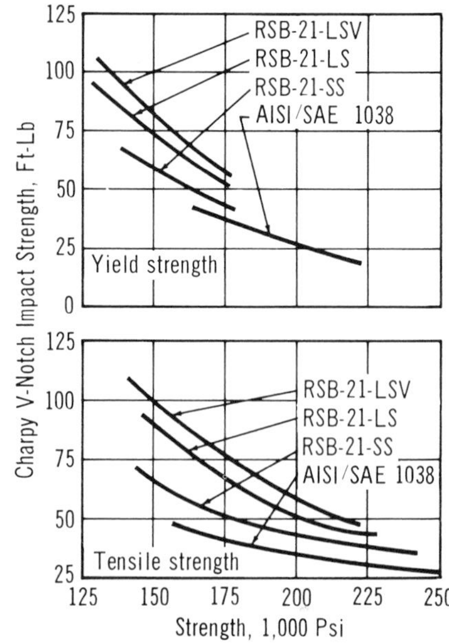

Leaded Grades: Answer to Free-Machining Needs

Greater machining feeds at higher speeds can improve productivity by up to 30% while dimensional tolerances and surface smoothness are maintained. Properties generally equal those of nonleaded counterparts.

Fig. 1 — Valve control shaft of 41L40 is being machined. Switch from 1045 improved machinability by 25% and doubled lathe tool life for the manufacturer, Limitorque Corp. Load-carrying capacity of the finished part was also raised 25%.

Table I — Machinability Ratings of Standard Steels

Grade	Rating, %*
11L17	100
11L37	85
11L41†	85
11L44†	93
11L53†	75
12L13	175
12L14	170
12L15	170
40L47†	80
41L30	85
41L40†	91
41L47†	80
41L50†	75
61L50†	72
52L100†	52
86L20	80
86L30	85
86L40	80
86L50†	75

*Machinability of B1112 = 100%; †Annealed

If you are looking for a free-machining steel to help increase production, why not try a leaded grade?

Here's what engineers at Copperweld Steel Co. report on user experiences:

- A pump maker switched from 4150 to 41L50, and boosted production of injection cylinders from 43 to 63 parts per h. Spindle speed went from 249 to 369 rpm without sacrificing dimensional precision; finish also improved.
- The lead in 52L100 helped a bushing manufacturer to raise speeds from 65 to 90 surface ft/min (20 to 27 surface m/min) and feeds from 0.004 to 0.006 in./min (0.10 to 0.15 mm/min), raising bushing production from 98 to 130 pieces per h. Machining another bushing of the same grade lifted production from 87 to 124 parts per h.
- A maker of piston racks turned to 11L26 when he ran into problems with counterboring — long, continuous chips were wrapping around tools, cutting their life down to 25 to 50 pieces. The move helped to raise tool life to 800 pieces per tool at the same speeds and feeds. Figure 2 illustrates such a part.
- Switching to 86L20 allowed an appliance manufacturer to improve machining quality, cut tool changes by 50%, and reduce in-plant rejects of wringer drive sleeves. Operations include boring, counterboring, cutting, milling, and drilling.

These and other examples of success with leaded steels demonstrate that these grades can enable partsmakers to raise production of quality components with efficiency and economy (Fig. 1). Tools last longer because lead provides built-in lubrication. Lead particles also act as chipbreakers, helping to reduce downtime otherwise needed to remove stringy chips and clean out clogged lubricating systems.

The upshot: fewer tool regrinds; tighter part tolerances; smoother surface. In fact, additional machining operations, such as deburring and grinding, may be eliminated through switching to leaded steels.

A Word on Segregation — At one time, potential users shied away from leaded grades because of lead segregation within ingots. Though producers have since solved these problems, some old timers have long memories. Results of some Copperweld tests should help dispel their fears. Data show uniformities of lead content throughout seven typical heats (Fig. 3). Lead contents were determined from tops to bottoms of the first, middle, and last ingots of each heat, labeled by letters A through G on the diagram. Note that values generally fell between 0.19 and 0.22% Pb; the entire range was

Fig. 2 — Parts of leaded steel in a typical steering gear mechanism include pump rotor (left) of 51L20, piston rack (right) of 11L26, and small hose coupling of 12L14.

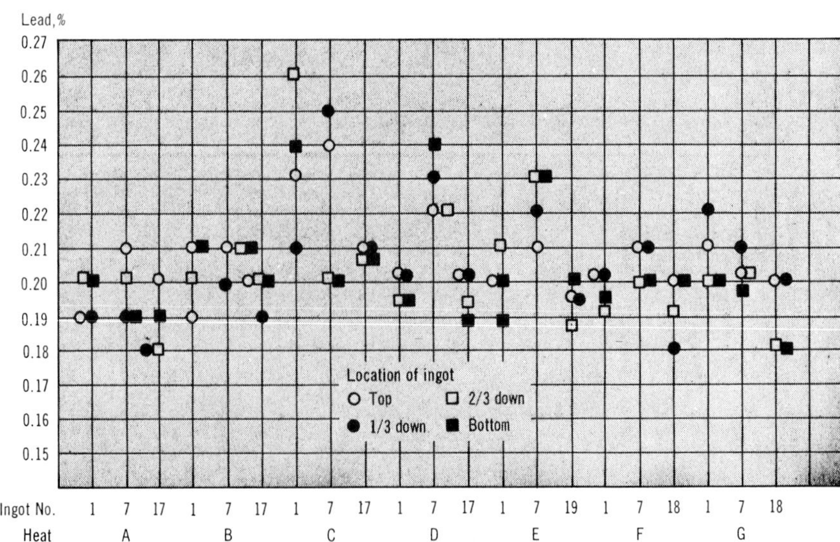

Fig. 3 — Analyses of the first, middle, and last ingots of seven Copperweld heats indicate that lead contents in leaded steels are generally uniform from tops to bottoms of ingots and throughout heats.

0.18 to 0.26% Pb.

Copperweld also runs lead sweat tests to ascertain the extent of lead segregation within ingots. Billet sections, 1 in. (25.4 mm) thick, are heated to 1290 F (700 C) for 30 min. Lead "sweats" to the surface, forming beads. Billets displaying excessive segregation are scrapped or cropped back to sound material.

These tests demonstrate that leaded steels have improved appreciably over the years. During one past period, 1221 sweat tests of 86 heats were rated 43% "Excellent". With improvements in processing, the "Excellent" rating rose to 87%, as determined by 802 tests on 137 heats.

Results from the field tend to confirm Copperweld's findings. Recently, informants say, an automaker inspected 4.6 million lb (2.1 million kg) of parts machined of the company's leaded steel. Rejection rate for lead stringers and seams was a mere 0.04%.

Properties are good — Table I shows that lead provides both carbon and alloy steels with good machinability, as rated in terms of the machinability of B1112, a standard resulfurized grade. It should be noted that the medium and high-carbon grades are annealed for best machinability. Also, this list contains only a few representative grades; virtually any steel, standard or special, can be leaded.

Checks of mechanical properties also reveal that lead has no effect on them, generally indicating that components made of leaded steels can be freely substituted for those of nonleaded steels without danger. Table II lists comparative longitudinal and transverse properties of typical heats of 4140 and 41L40. All test specimens were processed as 1 in. (25.4 mm) diameter rounds, being held 1 h at the austenitizing temperature and 2 h at the tempering temperature. Results are averages.

Leaded steels will generally satisfy all application requirements that their nonleaded counterparts do. The only exceptions are high-temperature (over 500 F [or 260 C]) applications and parts which undergo contact fatigue, such as high speed bearings.

Leaded steels are unsuitable for high temperature uses because lead melts at about 500 F (260 C). At higher temperatures, molten lead exists in the structure, weakening the steel and shortening fatigue life.

Table II — Properties of Leaded and Nonleaded 4140 Steel

Tempering Temperature, F(C)*	Yield Strength, psi (MPa)	Tensile Strength, psi (MPa)	Elongation, %	Reduction in Area, %	Hardness, Bhn	Impact Energy, ft-lbf(J)†
41L40 Longitudinal Properties						
1000 (540)	144 000 (990)	157 500 (1285)	16.3	56.7	321	48 (65)
1100 (590)	118 750 (820)	136 000 (938)	18.8	61.6	277	65 (88)
1200 (650)	98 250 (676)	120 000 (827)	21.5	64.7	248	73 (99)
AISI 4140 Longitudinal Properties						
1000 (540)	140 750 (972)	154 500 (1065)	16	56	321	52 (38)
1100 (590)	121 250 (835)	137 750 (950)	19	64.1	277	65 (88)
1200 (650)	101 750 (703)	121 750 (840)	20.5	67	248	74 (100)
41L40 Transverse Properties						
900 (480)	176 750 (218)	183 250 (1264)	4	8.6	388	9 (12)
1000 (540)	151 500 (1045)	158 250 (1090)	4.5	9.6	341	12 (16)
1100 (590)	128 750 (888)	143 250 (988)	9.5	17	302	17 (23)
1200 (650)	115 500 (796)	132 500 (914)	10.3	22.3	277	21 (28)
AISI 4140 Transverse Properties						
900 (480)	177 750 (1225)	182 250 (1257)	3	4.4	388	6 (8)
1000 (540)	156 000 (1076)	165 750 (1142)	4	7.7	341	7 (9)
1100 (590)	131 750 (909)	145 000 (1000)	6.5	10.5	302	9 (12)
1200 (650)	116 000 (800)	131 000 (903)	7	13.4	277	11 (15)

*Austenitized at 1600 F (870 C); †Izod; average results of separate breaks

Properties and Applications of High Strength Cold-Rolled Steels

PETER B. LAKE AND JOHN J. GRENAWALT

Four families of low-carbon, high-strength cold-rolled (HSCR) sheet exhibiting minimum yield strengths of 40 to 140 ksi are discussed. Relative merits of the various grades are analyzed in terms of important fabricating and selection criteria: strengthening capability, planar anisotropy, formability, impact loading response, weldability, and cost reduction. Paralleling the increase in strengthening capability in HSCR was an increased planar anisotropy and variation in formability within the various families. Impact loading of all cold-rolled grades showed increased yield strengths and energy absorption. Weldability of the different grades varied widely, depending upon strengthening mechanism, strength level, and composition. Cost reductions possible with HSCR stock decreased as the degree of formability required in the finished part increased. Generally, gauge reductions of 6 to 34 pct were necessary to offset cost extras for different grades of HSCR sheet. Continued work towards development of a full range of HSCR sheet with varying degrees of formability and cost appears necessary in light of present and future needs of steel buyers and designers.

RECENT developments in low carbon steelmaking have included development of the technology necessary to produce high-strength low-alloy (HSLA) hot-rolled sheet with yield strengths of 40 ksi (276 MPa) to 100 ksi (689 MPa).[1-6] Many excellent discussions and examples of the applications of hot-rolled, high-strength sheet have been advanced in recent publications[7-9] for saving vehicle weight, increasing strength, reducing costs, and increasing steel utilization. A natural consequence of the processing and marketing of various precipitation and solid-solution strengthened hot-rolled sheets is the analysis and characterization by the steel producer of the various high-strength, cold-rolled sheet products that can be processed directly from the same steels.

Cold rolling and annealing provide the steelmaker with the capability of producing from the hot band a wider range of even higher yield strength products exhibiting varying degrees of strength, formability, weldability, and cost reduction. These new families of lighter gage, high-strength cold-rolled (HSCR) sheets should be of increasing interest to sheet fabricators for cost and weight savings similar to those obtained with HSLA hot-rolled grades.

Three basic strengthening mechanisms are now advanced in defining four different families of HSCR sheet produced at the Youngstown Sheet and Tube Company. These families of steels are then subsequently characterized with respect to important sheet fabricating and selection criteria such as: mechanical properties, anisotropy of mechanical properties, dynamic deformation properties, formability parameters, spot weldability, and cost reduction.

PETER B. LAKE is a Research Supervisor and JOHN J. GRENAWALT is a Project Engineer, Research and Development Dept., Youngstown Sheet and Tube Co., Youngstown, Ohio.

STRENGTH CONTROLLING FACTORS

To obtain and maintain the yield strength level desired in cold-rolled sheet, four basic processing factors can be applied. They can be varied independently to obtain a broad range of feasible commercial products with yield strengths ranging from 40 ksi (276 MPa) to 140 ksi (965 MPa). They are: hot band yield strength, percentage of cold reduction, annealing cycle, and temper reduction.

Hot band yield strength can vary in magnitude from 30 ksi (207 MPa) to 100 ksi (689 MPa). Recent development of plain carbon, HSLA titanium hot-rolled steels[1-3] has provided the bases for new, higher strength cold-rolled and annealed steels. The three specific hot-band strengthening mechanisms discussed here include:

- •Precipitation strengthening and grain refinement, via titanium additions to an aluminum-killed steel, to obtain hot band yield strengths up to 100 ksi (689 MPa)
- •Precipitation strengthening and grain refinement in a semi-killed steel through the addition of columbium to obtain yield strengths up to 60 ksi (413 MPa)
- •Solid solution strengthening in a rimmed steel by additions of phosphorus to obtain hot band yield strengths of 40 ksi (276 MPa) to 50 ksi (345 MPa).

A plain carbon rimmed or aluminum-killed steel will normally exhibit yield strengths of 30 ksi (207 MPa) to 35 ksi (241 MPa).

Cold reduction is normally maintained between 50 and 75 pct for fully recrystallized plain carbon sheet steels. This range has been shown to be most feasible for optimizing formability through grain size and texture control. However, the percent reduction used can also be applied as a strengthening variable; the increase in yield strength as a function of percent cold reduction is shown in Fig. 1. The applicability of the relationship is maintained over the range of 10 to 70 pct reduction despite significant differences in hot band work-hardening rates. Use of this relationship enables one to produce a wide range of strengths, as demonstrated by two unique products. One, a cold reduction of 60 pct applied to a 100 ksi (689 MPa) HSLA hot band product will provide a 160 ksi (110 MPa) as-rolled yield strength product. Two, a low (32 pct) cold reduction of a 35 ksi (241 MPa) mild rimmed steel will result in a 75 ksi (517 MPa) as-rolled yield strength. Hence, at this point, using hot band yield strength and percent reduction, the capability exists for producing as-rolled yield strengths from as low as 75 ksi (517 MPa) to over 160 ksi (1103 MPa).

The *annealing cycle* now becomes the deciding influence on the final yield strength, once a reproducible as-cold-rolled sheet is obtained. The generalized response to the annealing process can now be reviewed; all of the materials exhibit a recovery and recrystallization behavior such as depicted schematically in Fig. 2. The steels may be processed practically via normal tight-wound facilities for batch annealing at temperatures in the 800°F (427°C) to 1330°F (722°C) range for soak periods usually in excess of 20 h. Alternately, they can be continuously annealed at temperatures of 1000°F (538°C) to 1650°F (899°C) for times measured in seconds. Whichever process suits the product, cold-rolled sheet can be annealed with practical, reasonable control of properties in recovery-annealed (BC) or fully-recrystallized (DE) regions, as shown in Fig. 2.

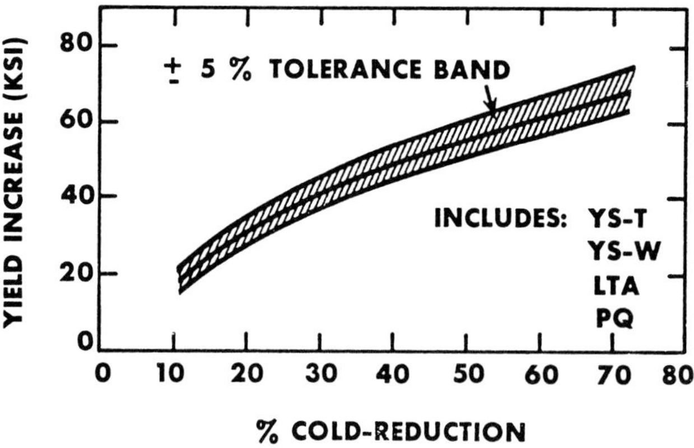

Fig. 1—Effect of cold reduction on yield strength of as-rolled product.

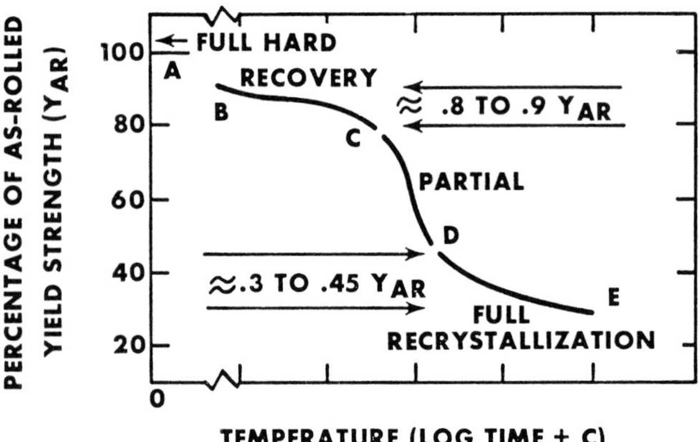

Fig. 2—Recovery and recrystallization of as-cold-reduced steel, plotted as a function of a generalized time/temperature parameter.

The recovery-annealing region (BC) can be conveniently defined as one in which:

- The annealing cycle reduces the as-rolled yield strength to between 80 and 90 pct of the yield as-rolled strength (YAR).
- There is no appreciable recrystallization (<5 pct).
- A significant increase (one to three times) in the ductility (expressed as percent total elongation).

It is the unique application of a low temperature anneal to obtain a stress-relieved condition which allows one to take fullest advantage of the variation in yield strength as a function of the percent cold reduction. The first example in the previous section, a 160 ksi (1103 MPa) yield strength as-rolled product, can be recovery-annealed back to produce a 140 ksi (965 MPa) minimum yield strength sheet. Similarly in the second example, the application of the same type of low temperature anneal will reduce the yield strength to about 65 ksi (448 MPa) with a coincident increase to about 15 pct in elongation. It is for yield strengths below the 65 ksi (448 MPa) range that full recrystallization annealing is required.

The kinetics of partial recrystallization (region CD in Fig. 2) preclude practical annealing in this strength region except in very sluggishly recrystallizing steels, such as the titanium-bearing aluminum-killed grades discussed later. The "fully-recrystallized" steel (region DE) can be characterized as a structure with more than 80 pct stress-free crystals, and as a steel with yield strengths that are normally 30 to 45 pct of YAR. The recrystallized cold-rolled steels usually exhibit yield strengths at or slightly below the original hot band strength, as evidenced with most high-strength, cold-rolled steels.[10,11]

Temper-rolling for over 5 pct in a second reduction has not been applied as an additional mode of strengthening for the steels in this discussion. Any final operation involving significant cold-working to obtain additional strength levels has been shown historically to give reduced transverse ductility in stretching or bending.[12,13] The final temper pass has been maintained within extensions of 0.3 to 1.0 pct for usual control of final shape, surface, and elimination of discontinuous yielding where required.

In summary, cold-rolled and annealed yield strength can be varied from mold steel values up to the 150 ksi (1034 MPa) range by varying the hot band strength, percent reduction, and annealing cycle.

FOUR FAMILIES OF HSCR SHEET

The four families of HSCR sheet are developed from four distinct types of hot-rolled products. The first two families use, respectively, titanium and columbium as strengtheners; and the third uses phosphorus. In the fourth, strength is obtained by applying combinations of thermal and mechanical processes to low-carbon mild steels.

Titanium Grades-YS-T

These steels are aluminum-killed, and are generally produced in the 0.07/0.10 pct C and 0.10/0.25 pct Ti range. The titanium acts in two distinctly beneficial ways. First, the interphase precipitation of titanium carbonitrides during controlled hot rolling provides strengthening through precipitation and grain refinement to yield strength levels of 50 ksi (345 MPa) to 100 ksi (689 MPa) in the hot band. Second, manganese sulfide stringers are displaced by harder, nonelongated titanium carbo-sulfide inclusions which provide a significant degree of sulfide inclusion shape control for improved transverse bendability in both hot and cold-rolled sheet products. As-cold-rolled yield strengths of over 160 ksi (1103 MPa) can be reduced to the 140 ksi (965 MPa) range by recovery annealing. Further annealing at higher temperatures produces recrystallized sheet in the 55 ksi (379 MPa) to 70 ksi (482 MPa) yield strength range. Over the entire range of strengths available with YS-T steels, ductility (expressed by total elongation, E_t) decreases as yield strength increases, which is expected. This relationship is shown in Fig. 3 for the YS-T series. Typical compositions for the YS-T steels are listed in Table I.

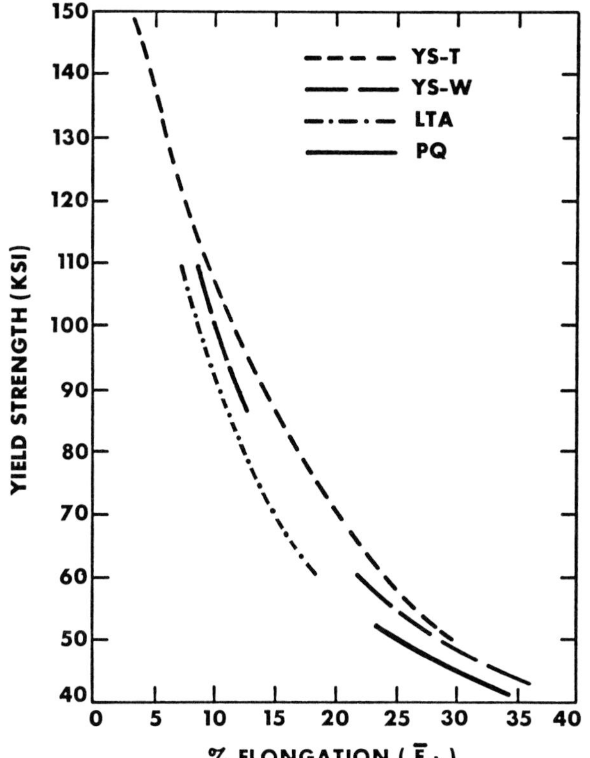

Fig. 3—Relationships between average yield strength and average total elongation for four families of high-strength sheet.

Table I. Typical Compositions of HSCR and CQ/DQSK Steels

Grade	Carbon, Pct	Manganese, Pct	Phosphorus, Pct	Sulfur, Pct
YS-T*	0.05/0.11	0.40/0.60	0.010 max	0.025 max
YS-W*	0.09/0 18	0.55/0.90	0.020 max	0.030 max
LTA	0.05/0.10	0.25/0.50	0.020 max	0.025 max
PQ	0.05/0.12	0.40/0.60	0.05/0.12 max	0.030 max
DQSK*	0.08 max	0.23/0.33	0.010 max	0.022 max
CQ	0.08 max	0.25/0.35	0.010 max	0.025 max

*YS-T also has 0.08/0.28 Ti and 0.025/0.075 Al; YS-W, 0.012/0.035 Cb; and DQSK, 0.025/0.075 Al.

Columbium Grades-YS-W

This family consists of semikilled steels generally containing 0.012/0.035 pct Cb. Hot band strengths normally range from 40 ksi (276 MPa) to 60 ksi (412 MPa) via precipitation strengthening and grain refinement from the columbium carbonitride. Columbium gives no inclusion shape control; zirconium or rare earth additions would provide inclusion shape control, but would require that the columbium steel be fully killed. Typical compositions are shown in Table I. The stress-relief-annealed product provides yield strengths in the 95 ksi (655 MPa) to 115 ksi (793 MPa) range; yield strengths of fully-recrystallized sheet lie in the 50 ksi (345 MPa) to 60 ksi (413 MPa) range. The relationship between yield strength and elongation for the YS-W products can be compared with that for the YS-T series in Fig. 3. Note that, at comparable strength levels, there is little difference in tensile elongation between the two precipitation-strengthened steels.

LTA Grades

These steels are rimmed or killed grades of 1008 or 1010 plain-carbon steel which have been given a combination of hot-rolling, cold-rolling, and low temperature annealing such as to produce sheets with minimum yield strengths ranging from 60 ksi (413 MPa) to 90 ksi (620 MPa) without the purposeful addition of precipitation-strengthening agents. Heretofore, the above range of yield strengths were usually obtainable only by means of precipitation or solid-solution strengthening through costly alloying additions to fully or semikilled steels. A 60 ksi (413 MPa) minimum yield strength grade of this unique product is produced by cold-reducing a low carbon rimmed steel less than 40 pct, and then applying the low temperature anneal. By its nature, the 60 ksi (413 MPa) grade is limited to thicknesses greater than 0.045 in. (0.114 cm), but is nevertheless a useful condidate for less severe cold-rolled applications presently requiring cold-worked structural quality steels, renitrogenized-prestrained sheet, or low carbon columbium or titanium-alloyed steels.

The 75 ksi (517 MPa) or 90 ksi (620 MPa) minimum yield strength LTA grades are produced by applying the same low temperature anneal after using either or both greater percent cold reductions and higher strength hot bands. Recovery annealing can be provided by batch annealing in the 800°F (427°C) to 1000°F (537°C) range for cold-rolled products, or in the 1000°F (537°C) to 1300°F (704°C) range for continuously-annealed tin mill or galvanized products.[12,14,15,16] These LTA steels are characterized by improved bending properties associated with their striated, stress-relieved structure,[12,14] particularly when compared with the bending properties of full-hard or double-reduced (prestrained) grades. The yield-elongation relationship is shown in Fig. 3; ductility is significantly less than that for precipitation-strengthened steels at comparable yield strength levels. Typical compositions are also listed in Table I.

PQ Grades

The "PQ" steels are rimmed, plain-carbon grades strengthened by the addition of 0.05 to 0.12 pct P, one of the most effective solid-solution strengtheners.[17]

Hot band and cold-rolled and fully-recrystallized sheets vary in yield strength from 40 ksi (276 MPa) to 50 ksi (345 MPa), and yield strength-elongation relationship in Fig. 3 indicates good stretchability. Another favorable characteristic of these steels is the observed increase in drawability ($\bar{r}$ values > 1.1) obtained through the addition of phosphorus.[17]

IMPORTANT FABRICATING AND SELECTION CRITERIA

The four families of steels are now compared with one another and with CQ and DQSK mild steels, with respect to the following fabricating and selection criteria:

Mechanical Properties

The application of these strength controlling factors leads to the arbitrary definition of various YS-T, YS-W, LTA, and PQ strength grades, based entirely on minimum yield strength.

All the grades and their respective mechanical properties are shown in Table II. They have the following characteristics:

- All material tested was obtained from full-scale production heats; the steel was mill processed on typical production processing schedules.
- The status of the grades shown range from experimental to commercial.
- The quasi-static tensile tests were performed at a strain rate of 5×10^{-2} min.
- Values of uniform elongation (E_u) were determined as the value of elongation at maximum load. Yield strengths were obtained using a 0.2 pct offset method, avoiding any upper yield point that may have existed.

The following observations are made concerning the HSCR grades in Table II:

- Steel in the minimum yield range of 40 ksi (276 MPa) to 60 ksi (413 MPa) can be processed via three methods. Deciding criteria for a particular application would be the required formability and cost. Both of these subjects are treated in subsequent paragraphs.
- Steel in the 60 ksi (413 MPa) to 90 ksi (620 MPa) range of minimum yield strength can be processed by two methods. At comparable strength levels, the YS-T grades exhibit far superior ductility to the LTA series. Again the critical differences are primarily cost and formability.
- The only grades produced significantly above the 100 ksi (689 MPa) level are YS-T 120 and YS-T 140. These levels have only been obtained with the titanium-bearing steels, which, in many instances, can exist as alternatives to competitive materials such as the "martensitic" sheet grades.

Anisotropy of Mechanical Properties

When designing for a particular application, planar anisotropy of mechanical properties should also be considered. The following trends have been noted for a particular grade of mild or HSCR steel when comparing the mechanical properties in the longitudinal (rolling), transverse, and diagonal (45 deg) directions.

The lowest yield strengths, the highest elongations, and the best r and n values usually fall in the same sheet direction.

Depending upon the direction of measurement,

Table II. Mechanical Properties of HSCR and CQ/DQSK CR Steels

Grade		Mechanical Properties*						
					Elongation (Pct in 2 in.)			
		Yield Strength (ksi)	Tensile Strength (ksi)	Hardness, Rb (Rc)	Eu Uniform	Et Total	Min. Bend Radius	Remarks
YS-T	50	53	68	70	18	27	Flat	Fully aluminum-killed, inclusion controlled, titanium-bearing sheet: excellent bending characteristics.
	50	65	78	77	15	24	Flat	
	70	74	82	84	13	21	Flat	
	80	82	88	91	12	19	Flat	
	100	110	119	98	9	14	Flat	
	120	130	135	(32)	7	11	1T	
	140	145	147	(34)	2	4	1T	
YS-W	45	48	60	66	20	30	1T	Semi-killed, no inclusion control, columbium-bearing medium strength steel.
	50	53	65	70	18	28	1T	
	55	58	70	73	16	24	1T	
	60	64	75	78	11	17	1T	
	90	109	112	97	6	10	2T	
LTA	60	71	78	86	7	13	1T	Rimmed plain-carbon steel, good bending properties, most economical.
	75	82	87	90	5	10	2T	
	90	109	112	97	4	8	2T	
PQ	40	44	60	69	20	31	1T	Rimmed steel with phosphorus added, good stretchability and drawability.
	45	47	64	73	18	29	1T	
DQSK		27	44	42	27	43	Flat	AL–killed, drawing quality rimmed,
CQ		28	43	40	26	42	Flat	commercial quality.

*Typical averages: $\bar{X} = \frac{X_L + X_T + 2X_{45}}{4}$. All values obtained from production cold-rolled sheet.

yield strength and percent elongation can vary over 20 pct from the average.

•The diagonal (45 deg) direction: in CQ and DQSK steels, exhibits the highest yields, the lowest values of E_t, n, and r; in YS-T and YS-W steels, exhibits the lowest yields, the highest values of E_t, n, and r; in LTA steels, exhibits the intermediate values of yield, E_t, n, and r.

•The degree of anisotropy of mechanical properties increases with strength level.

Deviations of yield strength and elongation values from their respective averages have been plotted as percentages on polar coordinates, the method used being defined in Fig. 4. Examples of the property variation in four different steels are shown in Fig. 5; they include DQSK, YS-W 50, YS-T 80, and LTA-90. A 45 deg shift in planar anisotropy in the precipitation-strengthened steel grades exists at all strength levels, and is apparently related to orientation of the inter-phase precipitation that occurs during controlled hot rolling. Cold rolling and annealing textures that develop are similar to those observed in the hot band, although the magnitude of the differences changes as a function of percent recrystallization and the thermal history. This type of 45 deg effect has also been observed in other vanadium and columbium steels.[10]

The advantages to the steelmaker and the user in determining qualitative magnitudes of the mechanical properties in various directions are:

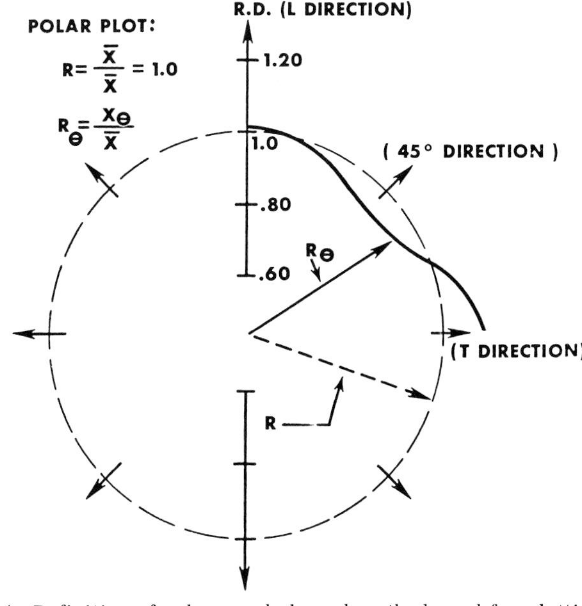

Fig. 4—Definition of polar symbols and method used for plotting anisotropy of yield strength and percent total elongation.

(a)

(b)

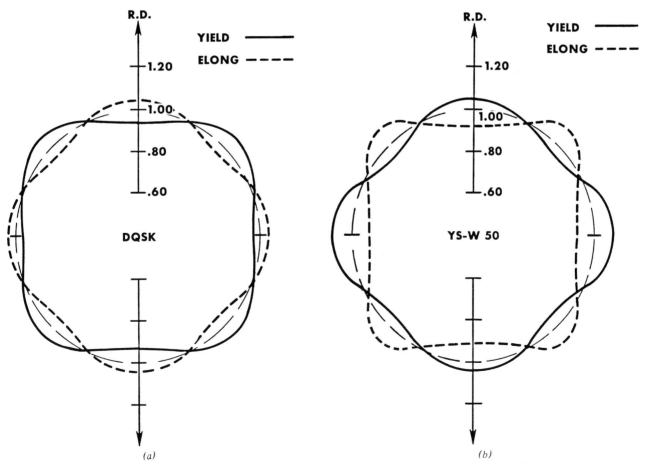

Fig. 5—Yield strength and total elongation anisotropy: (a) DQSK, (b) YS-W 50, (c) YS-T 80, (d) LTA-90.

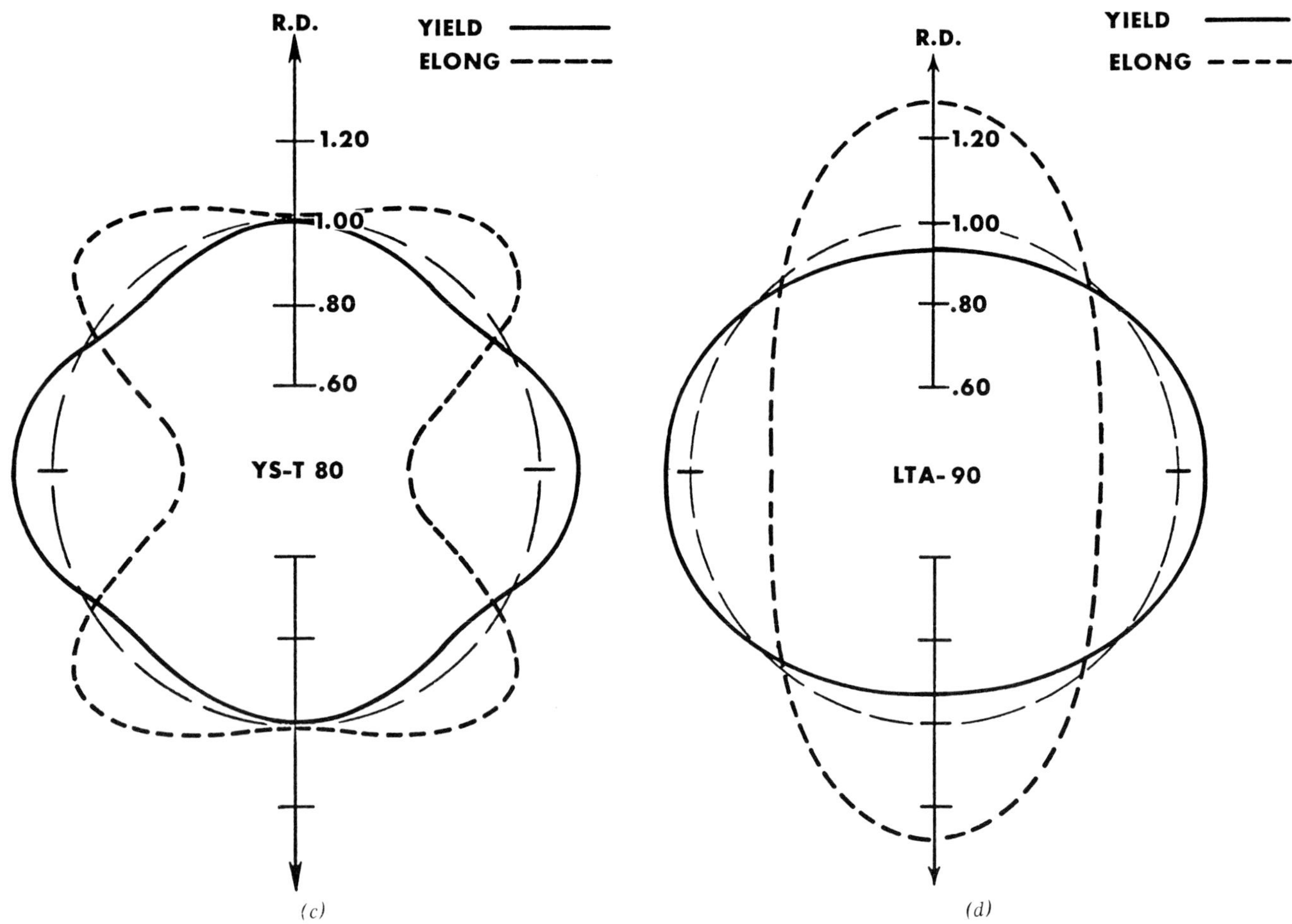

Fig. 5—Continued

•To achieve better coordination and understanding of the relevance of the release or acceptance testing criteria.

•To provide more design information concerning the strongest or most ductile sheet for setting up manufacture of various parts.

•To make use of optimum forming orientations when laying out blanking operations for parts with severe deformation in a particular direction or localized area.

•To enable one to be cognizant of these differences when comparing the various sheet materials with one another in forming limit curves (FLC), mechanical tests, or any other testing procedure where sample orientation can be arbitrarily chosen.

As the strength levels of HSCR steels increase, definitions of typical or minimum properties become more important. It may be noted that, for the HSCR grades in Table II, typical properties and minimum yield strengths are all given as average values.

Dynamic Deformation Properties

The additional hardware required to meet Federal safety and damageability standards has contributed to much of the increase in vehicular weight. As a result, many of the applications under consideration for weight reduction through gage reduction and increased material yield strengths are safety-related. It then becomes incumbent upon the designer to know and appreciate the relationship between dynamic and quasi-static properties of the HSCR steels available for his use.

Tensile impact properties of HSLA hot-rolled steels have been investigated and discussed by others.[18,19,20] A similar investigation of the high strain rate properties of mild steels and of four families of HSCR: YS-T, YS-W, LTA, and PQ was performed[21] to determine significant changes in key material parameters. To determine the effect of impact loading on the yielding, ductility, and energy absorption in the cold-rolled steels, a variety of subsized steel tensile specimens were deformed in tension at strain rates ranging from 0.05 to 31,000 in./in./min. (The latter rate corresponds to a 30 mph deformation over a 1 in. gage length.)

The trends noted in the test materials (YS-T 50/65/80/120; YS-W 50; LTA 60/90; PQ 45; CQ rimmed; and DQSK—the last two had 28 ksi, 193 MPa, yield strengths—were as follows:

•As the strain rate was increased, yield strength, tensile strength, and amount of energy absorbed increased.

•As the strain rate was increased, total elongation remained essentially constant.

•Sensitivity of yield strength to strain rate decreased with increasing quasi-static yield strength. Increase in yield strength as a function of strain rate is shown for YS-T, LTA, and mild steel in Fig. 6. Rate sensitivity of the yield strength is determined by a ratio of the values at the highest/lowest strain rates; results are shown in Fig. 7. It should be noted that the LTA series steels show a linear relationship plus reduced sensitivity with yield strength. The YS-W 50 and PQ-45

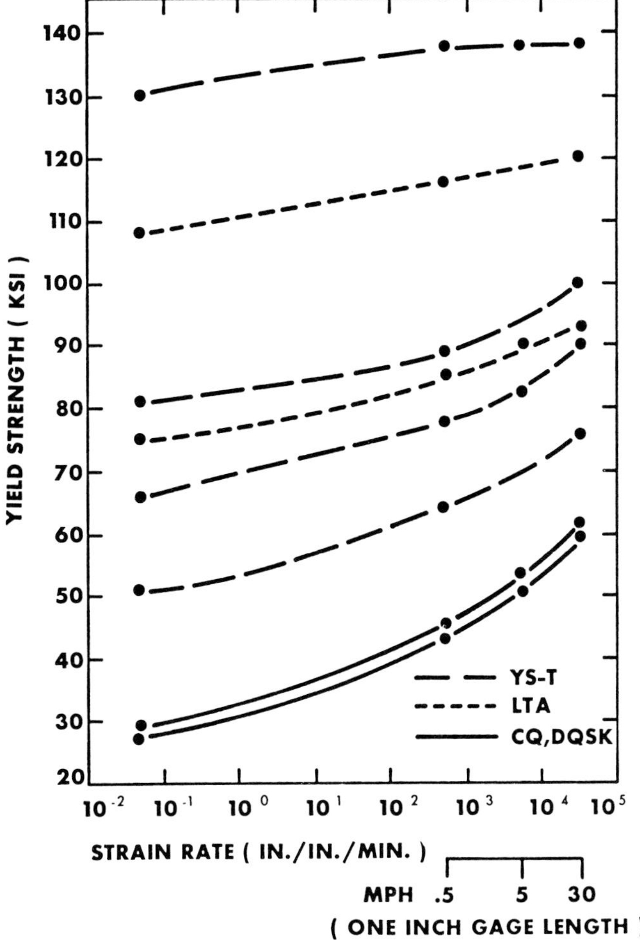

Fig. 6—Effect of strain rate upon yield strength of YS-T, LTA, CQ and DQSK steels.

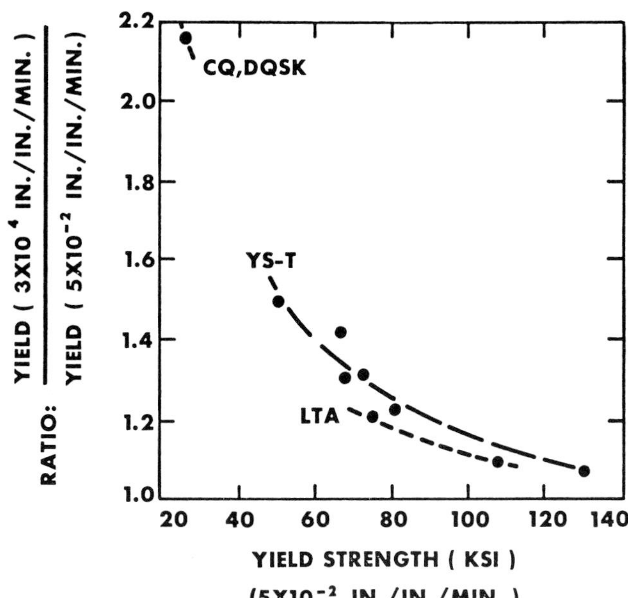

Fig. 7—Dynamic sensitivity of yield strength for various steels.

steels reacted similarly to the YS-T 50, and are not shown. In comparison, cold-rolled and hot-rolled grades from 30 ksi (207 MPa) to 80 ksi (552 MPa) showed similar yield sensitivity values.[18]

•In the partially and fully-recrystallized steels, the uniform elongation and the tensile/yield ratio decreased as strain rate increased. These effects were most exaggerated in the DQSK mild steel, which is shown in Fig. 8. Uniform elongation decreased from 65 pct of the total to 25 pct in the mild steels, and from 65 pct to about 45 pct in the 50 ksi (345 MPa) and 65 ksi (448 MPa) steels. Again, sensitivity of the uniform elongation to strain rate decreased as the yield strength (and ASTM Grain Size Number) increased.

•In the stress-relief-annealed YS-T or LTA steel, the uniform elongation (≈60 pct of total elongation) and tensile-to-yield ratio were unaffected by strain rate. In these steels, the increase in energy absorption (as measured by the area under the load-elongation curve) closely paralleled the increase in yield strength with strain rate.

•Sheet anisotropy of mechanical properties appears to be maintained from quasi-static through dynamic strain rates. YS-T 65 specimens were dynamically strained in the L,T, and 45° directions, with the resulting values of yield, elongation, tensile strength, etc., exhibiting the same relative magnitudes as were observed in quasi-static tests.

These results indicate that high-strength, cold-rolled plain-carbon sheet steels are sensitive to strain

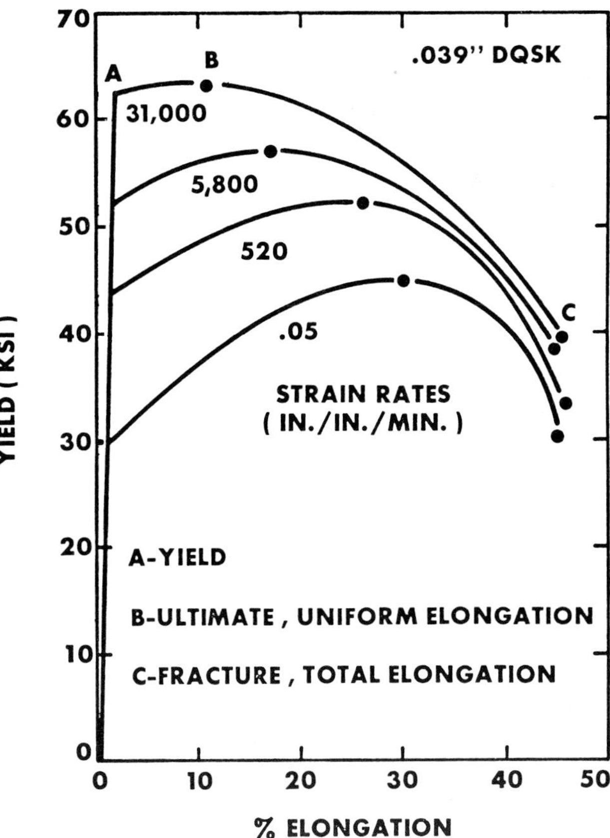

Fig. 8—Change in yield, uniform elongation, and tensile strength for DQSK steel. Typical of observed trends in recrystallized mild and high-strength grades.

rate in such a manner as to make them attractive for use in safety-related structures and for improved resistance to damageability.[22] The industry-wide investigation of high-strength cold-rolled sheets for use in tension-loaded door intrusion beams and like applications appears to be well directed.

Formability

In choosing a HSCR steel for a particular application, the type and severity of the forming operation will often dictate the final selection of material. At a given yield strength level, one or more of the materials covered by this discussion may be considered. The following material properties (expressed as $\bar{X}$, or $\frac{1}{4}(2X_{45} + X_L + X_T)$ correlate well with performance in actual forming operations:[23,24]

$\bar{E}_t$–total elongation from a tensile test
$\bar{r}$–coefficient of anisotropy of Lankford value (measure of drawability)
$\bar{n}$–strain hardening exponent (measure of stretch formability).

Total elongation ($\bar{E}_t$) as a function of yield strength for the four HSCR families has already been presented as Fig. 3. Coefficient of anisotropy ($\bar{r}$) and strain hardening exponent ($\bar{n}$) are similarly presented as functions of yield strength in Figs. 9 and 10, respectively. Again, these formability parameters decrease in value as yield strength increases. At yield strengths greater than 50 ksi (345 MPa), the YS-T series exhibits superior values in all categories of formability.

The variation in r and n values with direction are tabulated in Table III for several of the HSCR steels, and for a DQSK mild steel. The directional variations in r and n change in a manner consistent with increasing ductility and decreasing strength values.

The ASTM grain size numbers and grain shape eccentricity ratios are also shown. As strength increases, $\bar{r}$, $\bar{n}$, and $\bar{E}_t$ generally decrease with decreasing grain size, as expected. The effect of phosphorus in the PQ grade on drawing texture as well as strength is evident from the $\bar{r}$ of 1.26 and the grain size of 9.0. The 45 deg rotation in planar anisotropy is evidenced by the Δr values of +0.64 and +0.25 in DQSK and PQ grades, as compared to −0.32 and −0.34 in YS-W and YS-T steels, where $\Delta r = 1/2(r_L + r_T - 2r_{45})$.

The most significant difference in formability at yield strengths over 60 ksi (413 MPa) is in transverse bendability, the YS-T inclusion-controlled grades being superior. An example of an actual production part in which YS-T cold-rolled steel exhibited formability superior to that of LTA or YS-W of lower strengths is illustrated in Fig. 11. Sulfide control of the YS-T 70 and resultant excellent elongation and bendability in the transverse direction enabled this grade to be successfully roll formed and bent flat on itself in the channel section, while both YS-W 50 and LTA 65 were rejected due to transverse cracking in the bends.

An interesting observation is the good bendability noticed in the stress-relief-annealed LTA series of steels relative to the strength levels achieved. Commercial grades ranging in yield strength from 70 ksi (483 MPa) to 110 ksi (758 MPa) have been bent flat on themselves in very light gages, and will usually take a 1t bend in the heavier gages. This has been related in the past[12,14,25] to the nature of the stress-relieved, striated grain morphology, in terms of a more diffuse

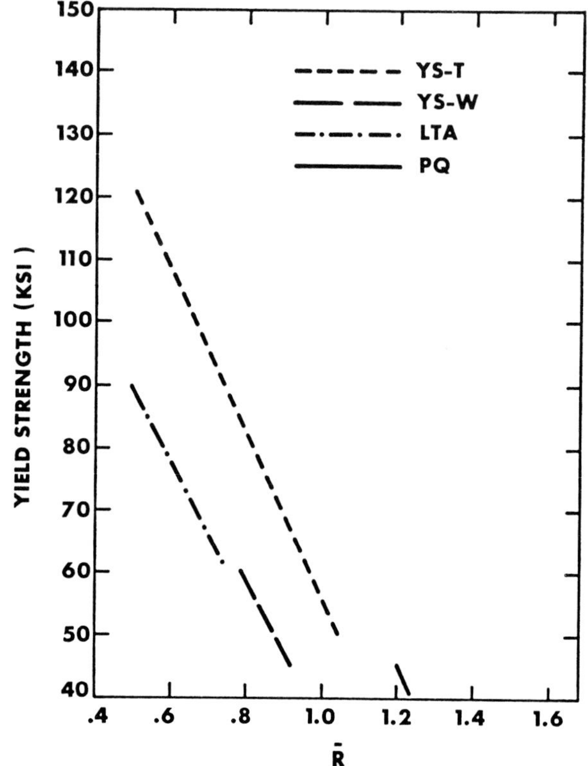

Fig. 9—Average yield strength *vs* the average Lankford value for four families of HSCR.

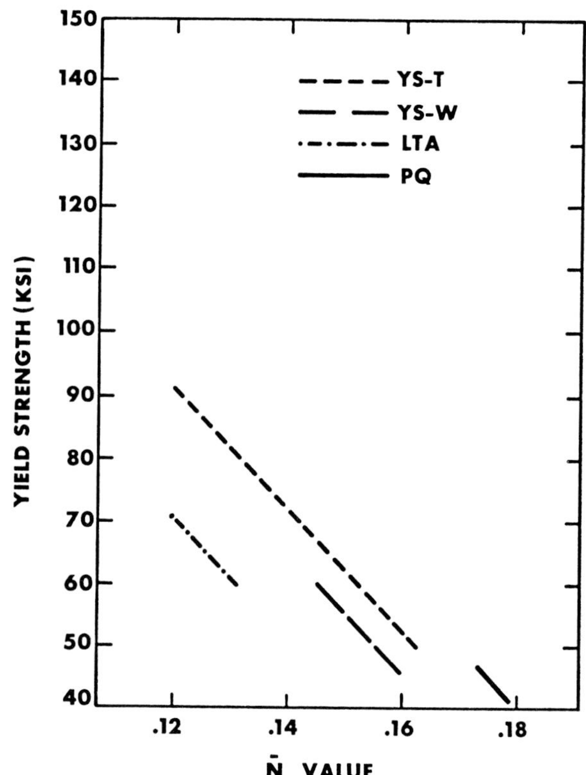

Fig. 10—Average yield strength *vs* average strain-hardening exponent for four familes of HSCR.

crack propagation path and lower concentrations of residual stress.

Another useful method of expressing formability of sheet steels is by use of the forming limit curve (FLC).[26] Fig. 12 shows a group of FLC's for CQ rim and five grades of HSCR steels. These FLC's were generated by strain analysis of electroetched circular grids. On the stretch side, lubrication of square blanks was varied; blank width was varied on the draw side. In Fig. 12, performance in stretching and drawing operations are indicated, respectively, by portions of the FLC line which lie to the right and left of the vertical axis.

Although the high-strength steels have performance levels obviously lower than the CQ rim material, their values of formability are considered quite good, especially on the stretch side, for yield strengths up to 80 ksi (552 MPa).

As would be expected, HSCR steels have greater springback, and thus require greater press loads than mild steel grades. Adjustment in such factors as die geometry, liberalized draw radii, and modification of draw bead locations, etc., by automotive die makers has led to successful stampings of such items as door inners/outers, deck lids, and roofs from 50 ksi (345 MPa) YS-T and PQ grade sheets. Many other less complex stampings and roll-formed shapes have been formed on experimental and production bases from HSCR steels in the four families.

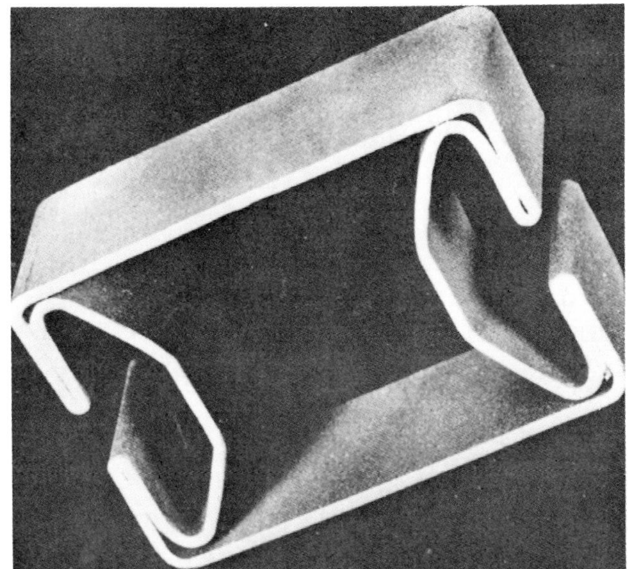

(a)

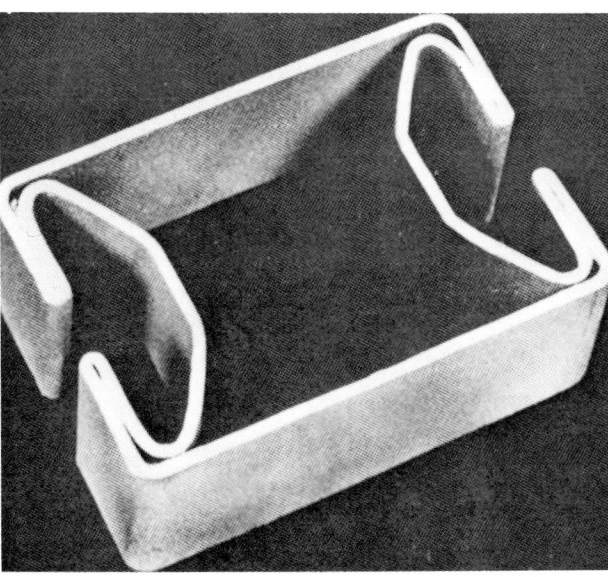

(b)

Fig. 11—A 0.066 in. roll-formed support channel processed with a flat transverse bend. This part was successfully formed in production with YS-T 70 steel, but failed at the bend with similar gage YS-W 50 and LTA-65 sheet.

Spot Weldability

The four HSCR steel families all have low carbon contents, most being below 0.10 pct C. Such steels have low carbon equivalents, and are readily joined by various fusion welding operations, such as the MIG and TIG processes. A large area of demand for such steels, however, has been from the automotive industry for applications that require steels which are readily joined by resistance welding techniques, of which spot welding is the most prevalent method. Each steel in this discussion has been subjected to spot welding tests to characterize and, where possible, optimize spot weldability.

One of the more explicit tests detailing acceptance criteria of light-gage sheet steels for spot weldability is outlined for various gage ranges in Table IV. The test, in which hold or quench time is kept constant at 30 cycles, establishes a range of spot welding current for the second weld produced on a series of small test coupons. After two successive spot welds are made between two coupons, one weld is peeled apart, and the button diameter is measured. The useful current range at a given weld time is bounded by the lowest current which will produce the minimum "accep-

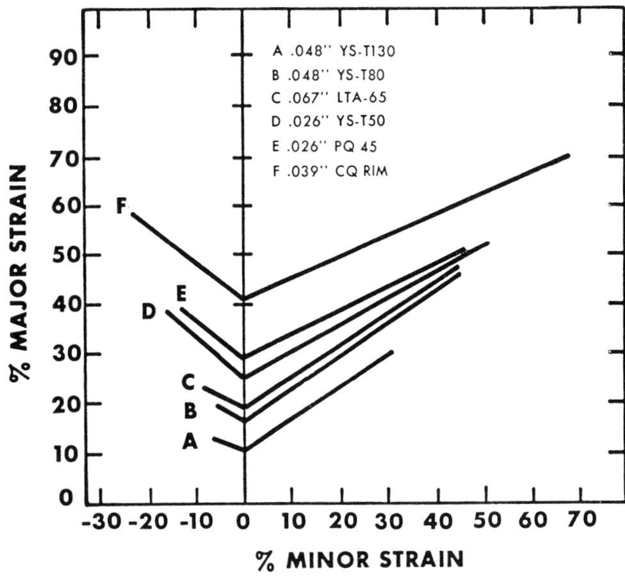

Fig. 12—Forming limit curves for six selected cold-rolled steels.

table'' button diameter and by the highest current which does not cause flashing—that is, expulsion of molten metal.

When useful current range is plotted for a range of weld times, characteristic curves, or lobes (Figs. 13 through 16) result.[27] Fig. 13, for three 0.066 in. (0.168 cm) steels, shows that the useful spot welding ranges of LTA-65 and CQ rimmed steels nearly overlap, while a YS-T 65 material has a narrower current range and is offset to lower current values. The relative sizes and positions of these useful welding ranges can largely be explained by composition. Despite their yield strength differences, the LTA and CQ products are essentially low-carbon, low-manganese steels with no purposeful alloy addition. The YS-T 65 steel, however, contains considerably more manganese plus the titanium addition, both of which increase electrical resistivity. Requiring less current for a given heat input than a steel with lower resistivity,[28] the YS-T curve is shifted to the left.

Fig. 14 compares materials of 0.028 in. (0.071 cm) thickness. The YS-T steel is shifted to lower current values than the CQ, but the shift is less than that of the previous figure due to the lower titanium in the YS-T

Table III. Formability Parameters* for HSCR Grades

Grade	Dir	Strength: Yield Y (ksi)[1]	Strength: Tensile U (ksi)	Elongation (Pct in 2 in.): Uniform[2] Eu	Elongation (Pct in 2 in.): Total Et	r	Δr	n[3]	Δn	Average Grain Size: ASTM No.	Average Grain Size: Eccentricity Ratio (Long.)
DQSK	L	27.0	43.5	29	45	1.68		0.22			
0.039 in.	T	28.5	43.0	30	47	2.00		0.21			
	45	28.7	44.6	25	39	1.20		0.20			
	$\bar{X}$	28.2	43.9	27	43	1.51	+ 0.64	0.21	+0.01	8	1.4
PQ 45	L	45.8	63.1	18	29	1.16		0.18			
0.030 in.	T	47.3	64.1	18	29	1.62		0.18			
	45	48.7	65.3	18	29	1.14		0.18			
	$\bar{X}$	47.6	64.5	18	29	1.26	+0.25	0.18	0	9	2.1
YS-W 50	L	52.6	68.1	17	26	0.55		0.15			
0.047 in.	T	57.7	73.4	17	26	0.79		0.13			
	45	53.2	68.1	18	28	0.98		0.15			
	$\bar{X}$	54.2	69.4	18	27	0.82	−0.32	0.15	−0.01	9.5	2.8
YS-T 50	L	58.6	76.9	16	25	0.74		0.16			
0.036 in.	T	67.9	80.2	15	23	1.10		0.14			
	45	60.4	74.2	17	26	1.26		0.16			
	$\bar{X}$	61.8	76.4	16	25	1.09	−0.34	0.15	−0.02	11	1.6
YS-T 80	L	81.2	90.7	15	18	0.38		0.13			
0.048 in.	T	86.5	91.8	14	17	0.34		0.12			
	45	78.9	83.3	16	25	0.82		0.13			
	$\bar{X}$	81.4	87.3	15	21	0.60	−0.48	0.13	−0.01	12	2.7
YS-T 120	L	128.8	131.2	7	11						
0.048 in.	T	141.8	144.1	4	6						
	45	126.4	131.4	8	13						
	$\bar{X}$	130.9	134.6	7	11					13.5	"Striated"

$^*\bar{X} = \frac{X_L + X_T + 2X_{45}}{4}$; $\Delta X = \frac{X_L + X_T - 2X_{45}}{2}$

[1]Yield strength, Y, calculated using 0.2 pct offset.

[2]Uniform elongation, Eu, taken as elongation at maximum load.

[3]Strain hardening exponent, n, determined using the Nelson-Winlock method.

Table IV. Typical Spot Selding Schedule Standards for Low Carbon Steel*

Minimum Thickness of Thinnest Sheet (in.)	Electrode Tip Size (in.)	Minimum Values: Electrode Force (lbs)	Minimum Values: Weld Time Cycles (60 per s)	Average Values: Secondary Welding Current (amps)	Button Diam (in.): Minimum	Button Diam (in.): Minimum Set-Up
0.026	0.18	450	8	9,500	0.12	0.18
0.031	0.25	550	9	10,500	0.15	0.25
0.036	0.25	700	10	12,500	0.18	0.25
0.043	0.25	800	12	13,000	0.18	0.25
0.054	0.25	980	15	14,250	0.18	0.25
0.068	0.31	1220	18	15,000	0.22	0.31
0.082	0.31	1450	24	16,000	0.22	0.31

*Direct welding applications: holdting time, 30 cycles.

50. The rephosphorized PQ-45 lies to the right of both YS-T and CQ, and was often found to have erratic spot welding characteristics.

Figs. 15 and 16 show the current ranges of YS-T and CQ steels welded to themselves and to each other. Note, first, that the current ranges for YS-T steels in Figs. 13 through 16 contract as the yield strength increases from 50 ksi (345 MPa) to 100 ksi (689 MPa). Secondly, in the last two figures, magnitudes and positions of the current ranges for welds of YS-T to CQ lie between the magnitudes and position of pure CQ to CQ or YS-T to YS-T ranges. The phenomenon wherein welding a HSLA steel to mild steel greatly expands the useful welding range has been previously documented.[27-29] Preliminary work with spot welding of cold-rolled YS-T steels shows that useful current ranges can also be expended by using longer weld times and/or shorter hold times, as already noted for the hot-rolled YS-T steels.[29] In both of the precipitation-strengthened steels, YS-T and YS-W, lowering the carbon contents has been shown to improve spot weldability.

The YS-T, LTA, and PQ steels have generally met automotive company welding requirements, whether welded to themselves or to CQ steels. The YS-W steels with carbon contents of 0.12 to 0.16 pct are presently less desirable for critical spot welding.

Cost Reduction

A most important consideration for the steel user is the cost of the high-strength, cold-rolled product, which is usually measured in terms of cost per piece or per unit area. When a weight savings is contemplated via gage reduction and yield strength increase, most users require at least a "washout" if not a reduction in piece cost. The four families of HSCR defined here not only differ in strengthening capability and formability, but, just as importantly, in their unit costs. As a rule of thumb, at a given strength level, the cost will increase as formability increases.

Table V shows the range of useful gage reductions

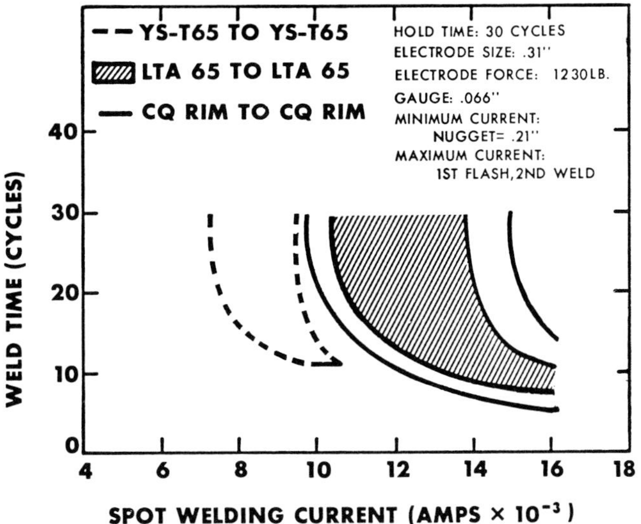

Fig. 13—Heavy gage YS-T 65, LTA-65, and CQ steels welded to themselves.

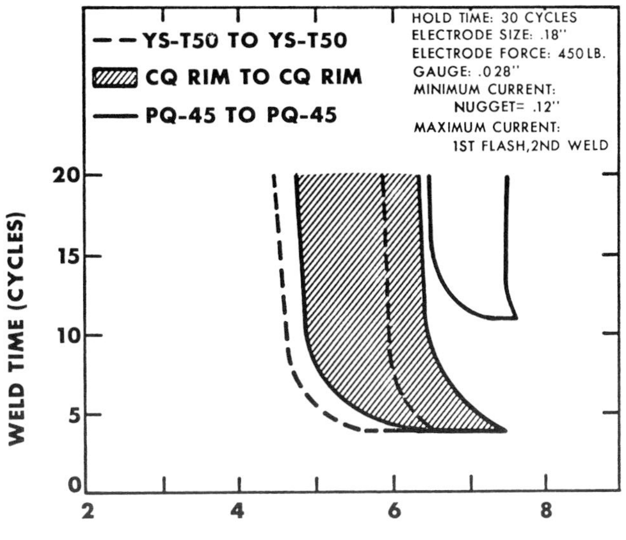

Fig. 14—Light gage YS-T 50, PQ-45, and CQ steels welded to themselves.

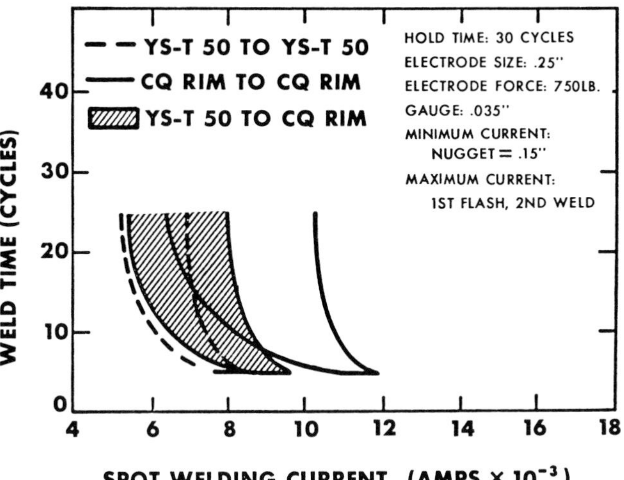

Fig. 15—YS-T 50 and CQ steels welded to themselves and one another.

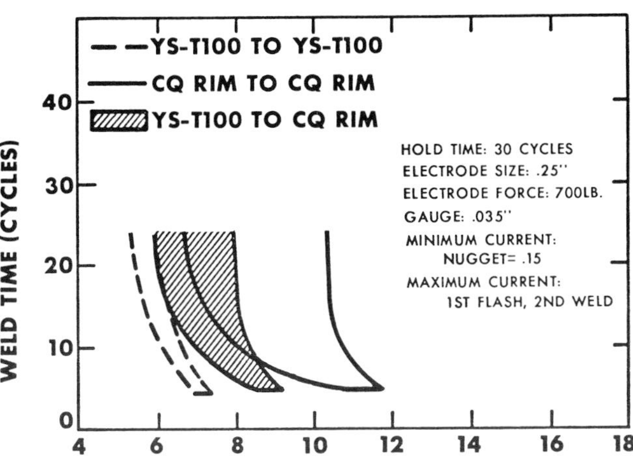

Fig. 16—YS-T 100 and CQ steels welded to themselves and one another.

Table V. HSCR—Range of Gage Reductions (Based on CQ Class 1 CR)

Grade¶		Minimum Pct Reduction*	Optimum Pct Reduction†	Maximum Pct Reduction‡
YS-T	50	22	23	40
	60	25	29	50
	70	28	35	57
	80	30	39	62
	100	32	45	70
	120	33	50	75
	140	34	54	78
YS-W	45	18	18	33
	50	19	23	40
	55	21	26	45
	60	23	29	50
	90	30	42	67
LTA	60	9	29	50
	75	10	37	60
	90	11	42	67
PQ	40	5	13	25
	45	6	18	33

*Based on cost¶ to washout: $\frac{\text{Gage (CQ)}}{\text{Gage (HSCR)}} \cong \frac{\text{cost (HSCR)}}{\text{cost (CQ)}}$

†Based on stiffness after DeCello *et al.*[30]: $\frac{\text{Gage (CQ)}}{\text{Gage (HSCR)}} \cong \sqrt{\frac{\text{yield (HSCR)}}{\text{yield (CQ)}}}$

‡Based on load-bearing capability: $\frac{\text{Gage (CQ)}}{\text{Gage (HSCR)}} \cong \frac{\text{yield (HSCR)}}{\text{yield (CQ)}}$

¶Costs are based on commercial or estimated developmental prices from the Youngstown Sheet and Tube Company.

of the four families: YS-T, YS-W, LTA, and PQ. Using current base prices and grade extras for the commercial grades, and using estimated developmental grade extras for experimental and developmental grades, the table shows the range of feasible gage reductions in which the cost was also reduced. In all instances, the "base" is a 30 ksi (207 MPa) yield strength, cold-rolled, Class I, CQ sheet product. The *minimum* gage reduction (Column 1 in Table V), is that required to washout cost on a per-piece basis. The *optimum* gage reduction is arbitrarily chosen and calculated using a stiffness relationship (after DiCello *et al.*[30]) between yield strength and gage. The *maximum* percent reduction is calculated based on tensile load to yielding.

The most economical grades are the LTA and PQ steels; they exhibit minimum required gage reductions, in the range of 5-11 pct, and the largest spread between minimum and maximum (25-67 pct). The lower costs are in line with reduced formability and the lower cost of production by the steelmaker.

The YS-T steels are the most expensive, and thus require the largest gage reductions at any particular strength level for a cost washout. Minimum required gage reductions range from 22-34 pct. The high costs are justified by the fact that these are fully-killed, titanium-bearing, inclusion-controlled steels with the broadest strengthening capabilities and highest degrees of formability. However, even with the significant minimum reductions necessary in YS-T steels costwise, a wide range of useful weight and cost savings does exist at yield strength levels of 60 ksi (413 MPa) and above. Several of the most interesting and newest grades are YS-T 100/120/140, which can be used in tension loaded parts to obtain large gage reductions.

The YS-W grades provide intermediate levels of strength, formability, and cost between the rimmed grades and the fully-killed YS-T series.

This type of comparison between the four families of HSCR steels and mild steels shows quite emphatically that, for each particular application for a high-strenth, cold-rolled product, all the factors (gage reduction, stiffness, formability, and cost) must be reviewed to determine the optimum usefulness. In tension loaded parts where significant gage reductions are possible, all grades in the four families provide viable alternatives. As part loading tends toward bending and sheet formability requirements increase, the cost washout becomes a more critical factor. When reducing gages from the heavier hot-rolled gages into the cold-rolled category, additional cost increments will also increase the minimum reductions necessary for cost savings.

All cold-rolled, high-strength steels cannot be grouped loosely into one category normalized by yield strength, but must be considered separately for each application.

Applications

Proprietary considerations preclude discussion of specific case studies regarding gage reductions and piece savings, but generalizations can be made as to areas of application of the new families of HSCR.

In galvanized product, there exist many production items of the PQ, LTA, or YS-W families ranging in strength from 40 ksi (276 MPa) to 90 ksi (620 MPa) minimum yield, and in gage from 0.017 in. (0.043 cm) to 0.069 in. (0.118 cm). Most are roll-formed into simple light structural shapes.

For cold-rolled sheet, most of the stamping trials have involved lighter gages in the 40 ksi (276 MPa) to 60 ksi (413 MPa) yield strength range. YS-T, PQ, and YS-W grades have been successfully used in trial production of automotive deck lids, door inners and outers, hoods, and roofs. Other less complex stampings are also in production at this time.

The area receiving most user and supplier interest for HSCR steel is the 60 ksi (413 MPa) to 140 ksi (965 MPa) yield strength range of product for use in safety-related parts, such as the door impact-intrusion beams. These parts are found in every door, and can represent an additional 8 to 40 pounds per car, depending on vehicle weight and size. Beams are often tension-loaded during deformation, and as such are condidates for significant gage reduction. Spot welding is also involved, which had led to our emphasis of the development of HSCR steels with carbon levels below 0.10 pct C (YS-T, LTA, PQ). The sheet supplied is usually unexposed, in the heavier CR gages in the narrower widths, and will involve a minimum in deformation via roll-forming or stamping. Because more production experience is needed in processing and fabricating HSCR sheet, the door impact beam is proving to be a good starting application. While there are a number of other less critical nonautomotive applications for HSCR steel, the modern overweight vehicle provides the greatest source of impetus for the development of high-strength, cold-rolled sheets.

SUMMARY AND CONCLUSIONS

Four families of HSCR sheet have been developed by independently varying strength, percent cold reduction, and annealing cycle, with either a CQ, solid-solution strengthened, or precipitation-strengthened steel. Each family varies in strengthening capabilities, formability, weldability, and cost such as to provide the user with a wide variety of high-strength steels. All of the HSCR steels exhibit desirable dynamic loading characteristics, such as increased yield strength and energy absorption, for improved damageability resistance and vehicle safety. The growing need for reduced weight, improved steel utilization, and cost reduction continue to provide strong economic justification for development and evaluation of these steels by both supplier and user.

ACKNOWLEDGMENTS

The authors wish to thank the Youngstown Sheet and Tube Company for permission to publish this paper. Special thanks are extended to Dr. R. A. Bosch and Dr. E. J. Schneider of the Youngstown Sheet and Tube Company, for many helpful discussions, and also to Dr. C. L. Magee of the Ford Motor Company for his cooperative efforts in dynamic strain rate testing.

REFERENCES

1. R. A. Bosch and J. A. Straatman: U.S. Patent No. 3,625,780, December 7, 1971.
2. J. A. Straatman, R. A. Bosch, and W. H. Herrnstein: presented at 97th Annual AIME Meeting, 1968.
3. L. Meyer, F. Hasterkamp, and D. Lanterborn: *Processing and Properties of Low-Carbon Steel,* pp. 297-320, AIME, 1973.
4. J. D. Baird and R. R. Preston: *Processing and Properties of Low-Carbon Steels,* pp. 1-42, AIME, 1973.
5. J. M. Gray: *Processing and Properties of Low-Carbon Steel,* pp. 225-42, AIME, 1973.
6. M. Korchynsky and H. Stuart: *Symposium Low-Alloy, High-Strength Steels,* pp. 17-27, Nuremberg, BRD, May 1970.
7. J. A. Vaccari: *Mater. Eng.,* February 1974, pp. 24-28.
8. J. A. Vaccari: *Mater. Eng.,* May 1974, pp. 62-66.
9. D. G. Younger: *Proceedings of the 16th Mechanical Working and Steel Processing Conference,* pp. 63-90, AIME, January 1974.
10. C. R. Rarey and J. F. Butler: SAE Paper No. 740957, presented October 21, 1974.
11. W. Muschenborn, L. Meyer, and C. Strassburger: presented at Metals Engineering Congress, Chicago, October 1973.
12. P. B. Lake, R. P. Morgan, and E. J. Schneider: presented at Regional Meeting, AISI, Pittsburgh, October 1971.
13. R. G. Toth and H. N. Lander: *Yearbook of the AISI,* 1969, p. 271.
14. J. E. Hartmann, P. B. Lake, and E. K. Hutson: presented at Regional Meetings AISI, Chicago, October 1971, and San Francisco, November 1971.
15. J. E. Hartmann, A. LaCamera, and P. B. Lake: U.S. Patent No. 3,825,448, July 23, 1974.
16. R. H. Frazier and R. G. Toth: U.S. Patent No. 3,264,149, August 2, 1966.
17. F. B. Pickering and T. Gladman: BISRA, MG-Conference, pp. 10-20, 1963.
18. D. A. Chatfield and R. R. Rote: presented at SAE Automotive Engineering Congress, Detroit, February 1974.
19. R. G. Davies and C. L. Magee: Ford Motor Company Technical Report SR-73-114, October 8, 1973.
20. W. C. Leslie, R. J. Sober, S. G. Babcock, and S. J. Green: *Trans. ASM,* 1969, vol. 62, pp. 690-704.
21. C. L. Magee, P. B. Lake, and co-workers: unpublished work, August 1974.
22. T. E. Johnson, Jr., and W. O. Schaffnit: presented at the 15th Mechanical Working and Steel Processing Conference, AIME, Pittsburgh, January 24, 1973.
23. L. Lilet and W. Wybo: *Sheet Metal Ind.,* October 1974, p. 783.
24. D. J. Blickwede: *Trans. ASM,* 1968, vol. 61, p. 652.
25. P. B. Lake: unpublished work at Youngstown Sheet and Tube Company,
26. S. S. Hecker: *Metals Eng. Quart.,* August 1973, p. 42.
27. J. P. Maloney, R. A. Heimbuch, and L. J. Rose: *Automotive Engineering,* July 1974, p. 24.
28. *Welding Handbook*: 1969, Section Two, 6th ed., Chapt. 26.
29. R. L. Pastorek: private communication, the Youngstown Sheet and Tube Company, March 1974.
30. J. A. DiCello and R. A. George: SAE Paper No. 740081, presented February 1974.

Plate and Structural High-Strength Steels

By WILLIAM J. MURPHY and RALPH B. G. YEO

Alloys ranging from 36,000 to well over 100,000 psi in yield strength provide design engineers with a versatile material for constructions of all types and sizes. Most of the grades are covered by ASTM specifications, which list compositions, properties, and other important data.

OUR KNOWLEDGE OF the behavior of high-strength steels is quite extensive. In particular, the effects of composition and processing on properties are now on a quantitative basis. As we will see, this knowledge has resulted in a wide variety of steels, each with its own combination of properties and attributes.

Grades Are Classified

In this report on recent developments in plate and structural steels, we will range from hot-rolled carbon steels (the most economical and versatile construction material available) to quenched and tempered alloy steels with properties suitable for submarine-hull plates.

To begin, structural steels can be grouped into four general classifications according to minimum yield strength:

Hot-rolled carbon steels: 30,000 to 42,000 psi.

Hot-rolled, high-strength, low-alloy steels (HSLA): 42,000 to 80,000 psi.

Heat-treated carbon steels: 46,000 to 80,000 psi.

Heat-treated constructional alloy steels: 90,000 to 130,000 psi.

Carbon steels comprise most of the tonnage because they are suited to applications in which stresses are low, weight is not important, and rigidity is required. In general, however, the "yield-strength-to-price ratio" favors the stronger steels; their use often saves on both cost and weight.

Hot-rolled carbon steels meeting the ASTM A36 specification typically contain 0.20 C and 0.90 Mn. Developing a minimum yield point of 36,000 psi, grades in the A36 classification are the most widely used of structural steels. Thus, we can use A36 steel as the basis for subsequent comparisons.

The HSLA Types

A number of HSLA steels are produced to a minimum yield point of 50,000 psi. Many of them,

---

Dr. Murphy is division chief, Bar, Sheet, Railroad, & Wire Products, and Dr. Yeo is former section supervisor, Applied Research Laboratory, U. S. Steel Corp., Monroeville, Pa. Dr. Yeo is now manager, materials science, Mining & Metals Div., Union Carbide Corp., Niagara Falls, N. Y.

such as those covered by ASTM A440, A441, and A242, were developed some time ago.

The ASTM A440 steel contains higher carbon (0.22%) and manganese (1.40%) than A36 steel plus 0.3% Cu to double resistance to atmospheric corrosion. We should note that the steel is recommended primarily for bolted and riveted applications because the higher carbon and manganese contents raise the carbon equivalent and reduce weldability. The companion A441 steel, on the other hand, is weldable because strength is gained from adding 0.05% V, so that the carbon and manganese can be kept low enough to maintain a relatively low carbon equivalent.

Steels made to the A242 specification incorporate alloying elements (such as phosphorus, chromium, and silicon) that improve corrosion resistance up to eight times that of carbon steel. Such steels are being increasingly used unpainted for architectural applications.

In recent years, the growing use of uncoated corrosion-resistant steels for buildings and bridges has prompted development of weldable steels suitable for sections with greater thickness than that covered by A242. Specification A588, for example, covers seven HSLA steels which develop 50,000 psi minimum yield point in sections up to 4 in. thick (this strength level is maintained by A242 steels in thicknesses up to only ¾ in.).

All seven of these HSLA steels depend on precipitation hardening to achieve part of the strength; five contain vanadium, the sixth, columbium, and the seventh, titanium. Steels in this group achieve their properties by combinations of elements. Some elements (carbon, manganese, chromium) increase hardenability, and thus produce finer ferrite-pearlite microstructures. Others (vanadium, columbium, titanium) induce precipitation hardening, and still others (copper, chromium, silicon) impart corrosion resistance. For corrosion resistance, in fact, all seven contain copper and six contain chromium.

The A572 Grades

Another comparatively new group of HSLA steels is covered by the ASTM A572 specification. It includes those containing columbium or vanadium or both. With minimum yield strengths in the 42,000 to 65,000 psi range, these steels have attracted considerable attention because they are economical. They are also fascinating to the metallurgist, but a headache to producing mills.

Columbium is very useful for strengthening semikilled and killed steels. The remarkably powerful effect of about 0.03% Cb was first reported only about ten years ago. Concurrent studies had also shown that controlled hot rolling at low temperature improved the notch impact toughness of hot-rolled ship-plate steels. Subsequent application of controlled rolling to columbium steels markedly improved their toughness, which had been a drawback. Widespread adoption of this practice has led to a group of columbium steels (and vanadium steels) that provide the highest yield strength per dollar for constructional purposes.

According to AISI statistics, columbium and vanadium steels account for most of the grades classified as HSLA types. They are, to a great extent, responsible for the growth of domestic shipments of HSLA grades for structural and line-pipe applications. The total rose from about 500,000 tons in 1960 to over 2 million tons in 1967.

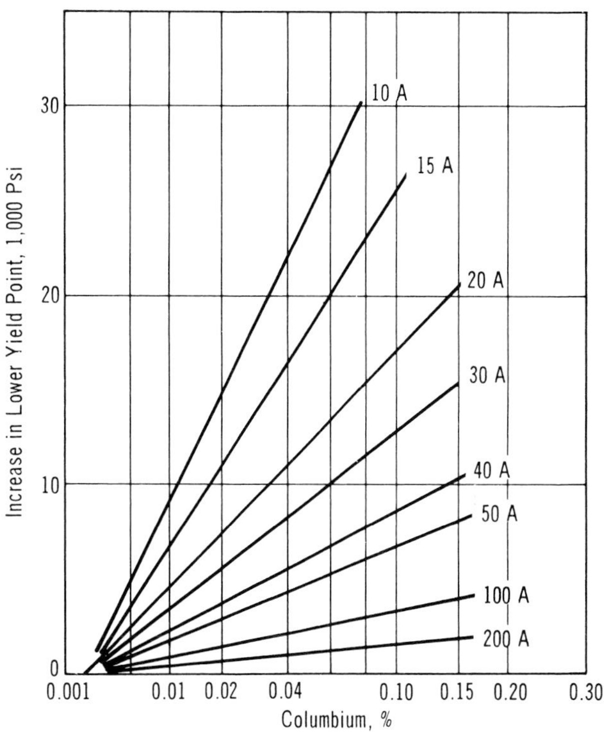

Fig. 1—For an equivalent amount of columbium, the smaller the particles of columbium carbides, the greater the strengthening effect.

Researchers explain the effects of composition and processing on the properties of columbium steels on the basis of precipitation, grain size, and solid-solution strengthening. The effect of a precipitated phase on strengthening depends mainly on the spacing between precipitate particles. This spacing is controlled by both the number and the size of the precipitated particles. Thus, for a given amount of precipitate, fine particles strengthen more effectively than large particles. As Fig. 1 reveals, 0.02% Cb will have almost no effect if precipitated as carbide particles 100 or 200 A in diameter, while the same amount precipitated as particles 10 A in diameter will raise the yield point by almost 15,000 psi.

Columbium precipitates will be in solution at normal soaking-pit temperatures (about 2,300 F) in a steel containing, for example, 0.2 C and 0.01 N. Though the solubility limit will be exceeded when

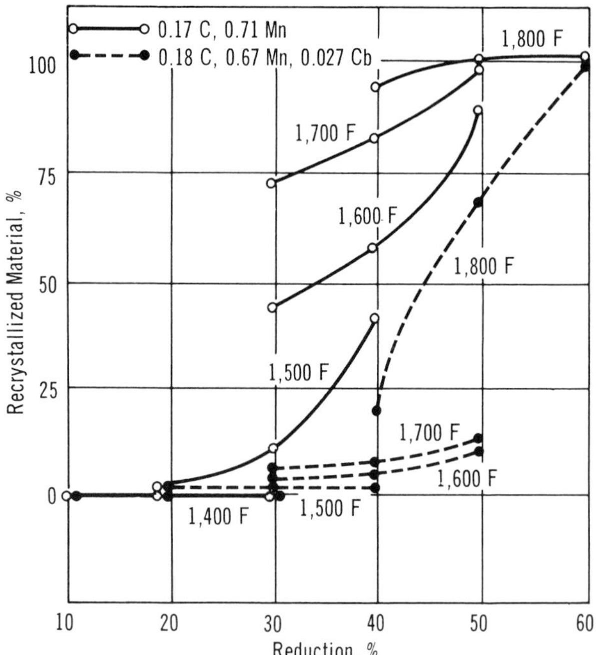

Fig. 2 — Addition of columbium retards the recrystallization of carbon steel appreciably. For these tests, steel plates were heated to 2,300 F, air cooled to the indicated temperatures, rolled to the indicated reductions in one pass, air cooled to 1,400 F, and quenched in water.

the temperature of the steel falls below about 2,100 F on hot rolling, the mode and extent of precipitation depend largely on rolling practice.

J. M. Gray and R. B. G. Yeo report (in "Columbium Carbonitride Precipitation in Low-Alloy Steels With Particular Emphasis on 'Precipitate Row' Formation," *ASM Transactions Quarterly*, Vol. 61, June 1968, p. 255) that columbium carbides or carbonitrides can precipitate: 1. In austenite. 2. During the transformation of austenite to ferrite. 3. In ferrite after transformation. Briefly, when columbium steels are rolled at finishing temperatures above about 1,800 F, few or no columbium carbonitrides precipitate in austenite. Most of the columbium, consequently, remains available to precipitation harden the subsequently formed ferrite.

Controlled rolling for improved toughness, however, not only reduces the grain size, but also accelerates the precipitation of columbium carbide in austenite (as discussed by J. M. Gray, R. B. G. Yeo, P. E. Repas, and A. G. Melville in "Strong, Tough Structural Steels," ISI Report No. 104, 1967, p. 249). If, for example, large precipitates (about 150 to 300 A) form in austenite, they would confer negligible strength, and furthermore, reduce the amount of columbium available to precipitation harden the ferrite. Therefore, the effect of reducing the hot-rolling finishing temperature is twofold. Rolling to lower temperatures refines the ferrite grains and reduces precipitation hardening in the ferrite. With regard to yield point, these effects counter each other so that the yield point remains relatively constant as the finishing temperature decreases. However, both effects improve toughness.

In addition to precipitation effects, columbium markedly retards recrystallization of austenite during rolling (R. B. G. Yeo, A. G. Melville, P. E. Repas, and J. M. Gray, "Properties and Control of Hot-Rolled Steels," *Journal of Metals*, Vol. 20, June 1968, p. 33). Figure 2 shows the effect of single-pass reductions at different temperatures (after reheating to 2,300 F) on the extent of austenite recrystallization for semikilled steels containing 0.17 to 0.18 C and 0.67 to 0.71 Mn, with and without 0.027 Cb. Note that the carbon steel recrystallized when deformed at temperatures down to somewhat above 1,400 F, while the columbium steel did not recrystallize when deformed below about 1,700 F. Furthermore, when the columbium steel was reduced 50% at 1,700 F, the unrecrystallized austenite transformed to a fine-grained ferrite-pearlite structure characteristic of good finishing practice at low temperatures.

As the steel cools after rolling, more columbium carbides precipitate during the austenite-ferrite transformation. According to recent research by Messrs. Gray and Yeo, precipitation occurs on the austenite-ferrite interface during its stationary periods (the interface moves intermittently during transformation). This mode of precipitation accounts for the precipitate layers characteristic of columbium (and also vanadium) steel. The precipitates are spaced closely enough together to harden the steel moderately. The degree of hardening depends mainly on the amount of columbium available for precipitation and on the transformation temperature.

By controlling transformation temperature, we can vary the properties. Lower transformation temperatures result in increasingly fine precipitates with correspondingly greater strengthening power and smaller ferrite grains which improve both strength and toughness. In plate steels, the transformation temperature of a given composition is controlled primarily by plate thickness. (Thicker plates cool more slowly, and transformation occurs at higher temperatures.) For a fixed thickness and columbium content, toughness and strength can be raised by adding alloying elements (such as manganese) which lower the transformation temperature.

Strip mills have an advantage in that cooling rates on the runout table can be fairly rapid if adequate water is applied. Also, columbium is utilized very efficiently because columbium carbonitride is not precipitated in austenite during the comparatively rapid passage of the steel through the mill. The combination of fine grain size achieved by accelerated cooling, plus the precipitation of columbium carbonitride that occurs as the coiled strip cools, produces steels with excellent combinations of strength and toughness. These steels are eminently suitable for applications such as electric-resistance-welded line pipe.

The A572 specification also includes steels made with vanadium, with or without columbium. Reports state that vanadium strengthens these hot-rolled steels, principally by the precipitation of its nitride (nitrogen is sometimes added to these steels). Since vanadium is a less effective strengthener than columbium, at least 0.05% V is added to steels made to the A572 specification. Vanadium carbide and nitride dissolve more readily than corresponding columbium compounds. Hence, there is less tendency for vanadium compounds to precipitate in austenite. During and after transformation, however, the precipitation of vanadium compounds appears to be qualitatively similar to that of columbium compounds.

The higher-strength HSLA steels are covered by A572 specifications, and have minimum yield strengths up to 65,000 psi. (Several proprietary grades have yield strengths up to 80,000 psi in light gages.)

An alloy containing about 0.05 C, 1 Cu, 1 Ni, and 0.03 Cb has recently been developed; it has a yield strength of about 65,000 psi in the hot-rolled condition and can be aged at about 1,050 F to a yield strength of about 85,000 psi. The steel strengthens first by precipitation of columbium carbide during cooling from the hot-rolling temperature, then by precipitation of particles of copper during the aging treatment. Presently, the strongest available steel that does not require quenching, this grade can be cold formed and welded in the hot-rolled condition. Then, aging simultaneously strengthens and stress-relieves it.

Heat-Treated Steels

When greater strengths or toughness or both are desired, heat treatment is necessary. Though normalizing — that is, air cooling from the austenitic range — produces fine ferrite-pearlite structures with good toughness, the highest minimum yield strength is about 60,000 psi. Quenching and tempering produce a better combination of properties than normalizing, even though the carbon steels ordinarily do not have sufficient hardenability to be quenched to a fully martensitic structure by spray-quenching facilities. As-quenched, these steels have ferrite-martensite microstructures, and tempering produces a uniform dispersion of carbides in ferrite.

ASTM specification A537 covers normalized as well as quenched and tempered carbon steels that are for applications such as pressure vessels, penstocks, and tanks for liquefied gases. Normalized steels up to 1¼ in. thick made to A537 Grade A have a minimum yield point of 50,000 psi, while

Table I — Effects of Heat Treatment on A537 Carbon Steel

Condition	Tensile Properties				Charpy V-Notch Impact Properties	
	Yield Strength (0.2% Offset), Psi	Tensile Strength, Psi	Elongation %	Reduction in Area %	15 Ft-Lb Transition Temperature, F	Energy Absorbed at −50 F, Ft-Lb
Hot-rolled	54,100	76,900	24.0	45	−5	6
Normalized (1,650 F)	53,100	75,200	28.0	66	−110	24
Quenched and tempered (1,200 F)	73,400	91,300	30.5	75	−150	84

quenched and tempered steels in the Grade B category have a yield strength of 60,000 psi. (Tempering temperature is about 1,200 F.) Users of either grade may specify a minimum average Charpy V-notch impact energy absorption of 15 ft-lb at —75 F transverse to the rolling direction.

These steels are made to fine-grain practice with aluminum, and contain up to 0.24 C and 0.65 to 1.4 Mn. Small amounts of copper, nickel, chromium, and molybdenum (up to 0.35 Cu, 0.25 Ni, 0.25 Cr, and 0.08 Mo) are also allowed. Table I shows how heat treating affects ½ in. thick plates of one of these steels (0.16 C, 1.25 Mn, 0.026 P, 0.029 S, 0.25 Si, 0.25 Cu, 0.25 Ni, 0.15 Cr, 0.058 Mo, and 0.03 Al).

Quenched and Tempered Steels

To achieve minimum yield strengths of 90,000 psi and above, structural steels must be quenched to essentially a fully martensitic structure before tempering. Alloy steels are generally used to obtain sufficient hardenability.

The first of the quenched and tempered constructional alloy steels was U. S. Steel's T-1, introduced about 15 years ago. This grade has a guaranteed minimum yield strength of 100,000 psi in thicknesses up to 2½ in. inclusive and 90,000 psi in thicknesses over 2½ to 6 in. A minimum transverse impact energy absorption (Charpy V-notch) of 15 ft-lb at —50 F is guaranteed in thicknesses up to 2½ in. inclusive. The composition is as follows: 0.10 to 0.20 C, 0.60 to 1.00 Mn, 0.035 P max, 0.040 S max, 0.15 to 0.35 Si, 0.15 to 0.50 Cu, 0.70 to 1.00 Ni, 0.40 to 0.65 Cr, 0.40 to 0.60 Mn, 0.03 to 0.08 V, and 0.002 to 0.006 B.

The availability of this steel has led to many changes in design and construction. Also many new welding electrodes have been developed to provide weld metal with adequate strength and toughness. Fabricators have also instituted more stringent welding procedures, including control of heat input, control of preheat and interpass temperature, and the routine use of low-hydrogen electrodes.

The growing use of thinner, lighter sections of these high-yield-strength steels, with attendant savings in costs, has led steelmakers to introduce other quenched and tempered grades with 100,000 psi yield strengths. Some have leaner compositions because hardenability as high as that exhibited by T-1 is not always necessary.

Specification A514 covers quenched and tempered plates for structural use, and A517 covers plates for pressure-vessel use. Of the ten grades included in these specifications, two are for plates up to 4 in. thick inclusive, three for plates up to 2 in. thick, and five for plates up to 1¼ in. thick. Though the steels differ somewhat in combinations of major alloying elements, seven of the ten contain chromium; and all contain molybdenum and boron (in addition to manganese and silicon) to provide hardenability. Two of the steels contain nickel to provide toughness in addition to hardenability. In all, carbon content is restricted to a maximum of 0.20 or 0.21% to insure good weldability.

Quenched and Tempered Carbon Steels

As a natural consequence of the trend toward leaner alloy steels, we are witnessing growing use of light-gage carbon grades which are quenched to essentially 100% martensite. In these, hardenability is imparted essentially by carbon, manganese and boron. These economical steels are usually tempered to minimum yield strengths of 90,000 to 100,000 psi, although their use at higher yield strengths is being considered — for example, as high as 160,000 psi. They are widely used — earth-moving equipment, for example.

An Extra-Strong Grade

A steel, designated HY130, for future submersibles is the strongest constructional grade we will discuss here. Its development involved more than meeting the basic goal: a minimum yield strength of 130,000 psi in sections up to 4 in. thick.

Also required were: high resistance to shear tearing (with a maximum fracture-transition-plastic temperature of 0 F); a maximum nil-ductility-transition temperature of —100 F; adequate ductility for forming; resistance to low-cycle, high-strain fatigue; resistance to stress-corrosion cracking and corrosion fatigue in sea water; adequate base-metal weldability; and development of suitable welding electrodes.

In development work, which was aimed at providing a satisfactory high-strength steel weldment, engineers considered the compatibility of the various requirements from the start.

Nominally, the steel contains 0:1 C, 0.75 Mn, 5.0 Ni, 0.55 Cr, 0.55 Mo, and 0.06 V. The major alloying elements were balanced to provide adequate strength, toughness, and weldability. Control of residual impurities is also essential because sulfur, oxygen, and nitrogen impair the impact strength. This grade has adequate fabricability, and engineers can be confident of weldments made of it. Experience gained during the development of HY130 will certainly be applied to create lower-cost steels with similar strengths.

Specifications for High-Strength Steels

By CARL R. WEYMUELLER

Compositions, mechanical properties, size ranges and general fabrication data are included in ASTM A36, A242, A440, A441, A529, A572 and A588, as well as SAE J410b and J368 (proposed for quenched and tempered grades).

No MATTER HOW GOOD a new constructional steel is, potential users are reluctant to employ it until the specification-issuing groups — especially the American Society for Testing & Materials (ASTM) and the Society of Automotive Engineers (SAE) — sanction it officially. In recent years, these organizations have issued new and revised specifications, including many of these grades. As you can see by the Data Sheet on p. 89, most of the listed steels bear specification numbers. though some categories are not yet covered. As more grades become included in similar specifications, the picture gets more complicated.

To explain how these steels are applied, we propose to review the ASTM and SAE specifications that concern the high-strength grades listed in the Data Sheet. Here, we will try to give information pertaining to composition, mechanical properties, and other data needed to select these important materials for structural applications.

The A36 Grades

Though the ASTM A36 steels are not considered high-strength steels — none is listed in the Data Sheet, you will note — we are starting with them because they constitute the basic structural grades. Made in shapes, plates and bars (general composition range: 0.25 to 0.28 C, 0.60 to 1.20 Mn, and, for plates only, 0.15 to 0.30 Si), ASTM A36 steels are required to have 36,000 psi min yield strength, 58,000 to 80,000 psi tensile strength, and minimum elongations of 20 to 23%.

These steels, which are used for bridges, buildings, and other structural applications, can be riveted, bolted, or welded. Specification A36 also notes: "When steel is to be used at temperatures where improved notch toughness is important, the material may be specified to be silicon-killed, fine-grain practice."

Mr. Weymueller is senior editor, *Metal Progress.*

The High-Strength, Low-Alloy Grades

Going up the strength ladder, we come to the HSLA steels covered by ASTM A242, A440, A441, A529, A572, and A588. These grades, with minimum yield strengths of 42,000 to 65,000 psi, constitute most of those listed on the Data Sheet.

Each of the specifications differs from the others, of course. For example, steels included in ASTM A242 (titled "High-Strength, Low-Alloy Structural Steel") have atmospheric corrosion resistances at least twice that of carbon structural steels with copper (equivalent to four times that of copperfree carbon structural steels).

Type I contains 0.15 C, 1.00 Mn max, and 0.20 Cu min, while Type II has 0.20 C max, 1.35 Mn max, and 0.20 Cu min. (If the silicon and chromium contents in the latter are each 0.50% or over, the copper requirement does not apply.) To provide the mechanical properties and corrosion resistance, the manufacturer can add other elements such as chromium, nickel, silicon, vanadium, titanium, and zirconium.

All plates and bars of A242 grades must meet the following minimum yield strength requirements: 50,000 psi for thicknesses ¾ in. and under; 46,000 psi for thicknesses over ¾ in. to 1½ in.; and 42,000 psi for thicknesses over 1½ to 4 in. Among A242 grades listed in the Data Sheet (Section V) are Ni-Cu-Ti and Kaisaloy 50CR.

Steels included under ASTM A440 are much like those of A242 except that corrosion resistance need only be twice that of carbon steels — half that of the minimum for A242 steels, in other words. These grades are intended primarily for riveted or bolted (not welded) bridges and buildings where weight savings are important. Carbon content is somewhat higher too (extra carbon usually lowers weldability). Thus A440 grades contain 0.28 C max, 1.10 to 1.60 Mn, and 0.30 Si max, plus 0.20 Cu min to get corrosion resistance. Mechanical property requirements are practically the same as those for A242 grades. Typical A440 steels are Man-Ten (A440), Jalten-3S, AW-440, Stenlite 440, and Kaisaloy 50 MM.

Next come the ASTM A441 steels which are, aside from being weldable, the same as those in the A440 group. We should also note that the specification covers material up to 8 in. thick (A440

Table I — Chemical Requirements for ASTM A514 and A588 Grades

Grade*	C	Mn	Si	Cr	Mo	Cu	B	Other
A514								
A (1¼ in.)†	0.15-0.21	0.80-1.10	0.40-0.80	0.50-0.80	0.18-0.28	—	0.0025 max	0.05-0.15 Zr
B (1¼ in.)	0.12-0.21	0.70-1.00	0.20-0.35	0.40-0.65	0.15-0.25	—	0.0005 to 0.005	0.03-0.08 V, 0.01-0.03 Ti
C (1¼ in.)	0.10-0.20	1.10-1.50	0.15-0.30	—	0.20-0.30	—	0.001 to 0.005	—
D (1¼ in.)	0.13-0.20	0.40-0.70	0.20-0.35	0.85-1.20	0.15-0.25	0.20-0.40	0.0015 to 0.005	0.04-0.10 Ti
E (4 in.)	0.12-0.20	0.40-0.70	0.20-0.35	1.40-2.00	0.40-0.60	0.20-0.40	0.0015 to 0.005	0.04-0.10 Ti
F (4 in.)	0.10-0.20	0.60-1.00	0.15-0.35	0.40-0.65	0.40-0.60	0.15-0.50	0.002 to 0.006	0.70-1.00 Ni, 0.03-0.08 V
G (2 in.)	0.15-0.21	0.80-1.10	0.50-0.90	0.50-0.90	0.40-0.60	—	0.0025 max	0.05-0.15 Zr
H (2 in.)	0.12-0.21	0.95-1.30	0.20-0.35	0.40-0.65	0.20-0.30	—	0.0005 to 0.005	0.30-0.70 Ni, 0.03-0.08 V
J (1¼ in.)	0.12-0.21	0.45-0.70	0.20-0.35	—	0.50-0.65	—	0.001-0.005	—
K (2 in.)	0.10-0.20	1.10-1.50	0.15-0.30	—	0.45-0.55	—	0.001-0.005	—
L (2 in.)	0.13-0.20	0.40-0.70	0.20-0.35	1.15-1.65	0.25-0.40	0.20-0.40	0.0015-0.005	0.04-0.10 Ti‡
A588								
A	0.10-0.19	0.90-1.25	0.15-0.30	0.40-0.65	—	0.25-0.40	—	0.02-0.10 V
B	0.10-0.20	0.75-1.25	0.15-0.30	0.40-0.70	—	0.20-0.40	—	0.25-0.50 Ni, 0.01-0.10 V
C	0.15 max	0.80-1.35	0.15-0.30	0.30-0.50	—	0.20-0.50	—	0.25-0.50 Ni, 0.01-0.10 V
D	0.10-0.20	0.75-1.25	0.50-0.90	0.50-0.75	—	0.30 max	—	0.05-0.15 Zr, 0.04 Cb
E	0.15 max	1.20 max	0.15-0.30	—	0.10-0.25	0.50-0.80	—	0.75-1.25 Ni, 0.05 V
F	0.10-0.20	0.50-1.00	0.30 max	0.30 max	0.10-0.20	0.30-1.00	—	0.40-1.10 Ni, 0.01-0.10 V
G	0.20 max	1.20 max	0.25-0.70	0.50-1.00	0.10 max	0.30-0.50	—	0.80 Ni, 0.07 Ti max

* All grades contain 0.035% P max and 0.04% S max.
† Numbers in parentheses are maximum thicknesses.
‡ Vanadium may be substituted for part or all of titanium content on a 1 for 1 basis.

applies only for materials up to 4 in. thick). Composition is 0.22 C max, 0.85 to 1.25 Mn, 0.30 Si max, 0.20 Cu min, and 0.02 V min. As before, mechanical property requirements are practically equivalent to those for A242 steels. Included in the A441 group are Jalten 1, Clay-Loy, Tri-Ten, and Stelco-Vanadium.

Specification ASTM A529 concerns steel with 42,000 psi min yield point made in plates and bars ½ in. and under in thickness. These steels are intended for use in buildings and similar riveted, bolted, or welded constructions. Composition is 0.27 C max, 1.20 Mn max, and (when called for) 0.20 Cu min. Mechanical properties are 60,000 to 85,000 psi tensile strength, 42,000 psi min yield strength, and 19% min elongation in 8 in. A typical grade: PX SK42.

The HSLA steels containing columbium and vanadium are included in ASTM A572. Six types are covered. Grades 42, 45, and 50 are for riveted, bolted, or welded structures; grades 55, 60, and 65 are for riveted or bolted bridges, as well as riveted, bolted, or welded structures in other applications.

All of these steels contain 1.35 Mn max and 0.30 Si max; maximum carbon varies from 0.21% (Grade 42) to 0.26% (Grade 65). Producers also employ columbium (0.005 to 0.05%), vanadium (0.01 to 0.10%), and nitrogen (0.015% max), singly and in combination, as strengthening agents. Minimum mechanical properties range from 42,000 psi yield strength, 60,000 psi tensile strength, and 20% elongation in 8 in. for Grade 42; to 65,000 psi yield strength, 80,000 psi tensile strength, and 15% elongation in 8 in. for Grade 65. Typical tradenames in this category are CB/V45, JLX-60, Kaisaloy 50-CV, V65, and INX-42.

Limited to material up to 8 in. thick, ASTM A588 covers HSLA steels which are intended primarily for use in welded bridges and buildings where savings in weight or extra durability are important. Compositions among these grades differ appreciably, as Table I reveals. Elements such as nickel, chromium, molybdenum, copper, vanadium, columbium, zirconium, and titanium are added to give the desired mechanical properties through heavy sections, plus atmospheric corrosion resistance at least four times that of carbon steel.

ASTM A588 steels have minimums of 50,000 psi yield strength and 70,000 psi tensile strength in thicknesses of 4 in. and under; 46,000 psi yield strength and 67,000 psi tensile strength in thicknesses between 4 and 5 in.; and 42,000 psi yield strength and 63,000 psi tensile strength in thicknesses from 5 to 8 in. Whatever the size of section, however, the minimum elongation in 2 in. is always 21%. Typical grades: Mayari R-50, Stelcoloy 50, Algo-Tuf 50, and Cor-Ten.

The Extra-Strong Steels

Titled "High-Yield-Strength, Quenched and Tempered Alloy Steel Plate, Suitable for Welding," ASTM A514 includes grades which are heat treated to a minimum yield strength of 90,000 to 100,000 psi, depending on the section thickness. These steels are designed mainly for use in welded bridges and other structures. As with the A588 grades, several compositions exist (Table I).

Note that the listing also indicates the maximum thicknesses in which the specified properties can be attained for each type. Mechanical properties: to ¾ in., 100,000 psi min yield strength, 115,000 to 135,000 psi tensile strength, 18% min elongation, and 40% min reduction in area; over ¾ to 2½ in., 100,000 min yield strength, 115,000 to 135,000 psi tensile strength, 18% min elongation, and 40% (or 50%) reduction in area; over 2½ to 4 in., 90,000 psi min yield strength, 105,000 to 135,000 psi tensile strength, 17% min elongation, and 50% min reduction in area. Typical grades: T-1 (Reg. Qual., A and B), and Jalloy S-100.

The SAE Specifications

The Society of Automotive Engineers has developed two specifications (called Recommended Practices) for high-strength steels. One is for HSLA grades, the other, for quenched and tempered steels. The first, SAE J410b, covers 14 grades, compositions of which are listed below:

Grade	C, %	Mn, %	Grade	C, %	Mn, %
942X	0.21	1.35	950D	0.15	1.00
945A	0.15	1.00	950X	0.22	1.35
945C	0.23	1.40	955X	0.25	1.50
945X	0.22	1.25	960X	0.26	1.65
950A	0.15	1.30	965X	0.26	1.65
950B	0.22	1.30	970X	0.26	1.65
950C	0.25	1.60	980X	0.26	1.65

Maximum contents are given; all grades contain 0.04% P max, except for 950D (0.15% max). It will be noted that 960X, 965X, 970X and 980X have the same composition. Their specified yield strengths, however, are 60,000, 65,000, 70,000 and 80,000 psi, respectively. (In line with this, we might note that the last two digits of the steels' specification numbers indicate the corresponding minimum yield strength in 1,000 psi increments.)

Taking the grades in turn, 942X, treated with columbium or vanadium, is similar to 945C, but is easier to form and weld. Grade 945A is readily welded, and has the best formability and notch toughness of the group. Grade 945C has satisfactory weldability, and may be used where more strength than 945A provides is required. Treated with columbium or vanadium, 945X is similar to 945C except that welding and forming properties are somewhat better. Stronger than the 945 steels, 950A has good weldability (both arc and resistance), notch toughness, and formability.

Grade 950B arc welds satisfactorily, and its notch toughness and formability are fairly good. The 950C grade is unsuitable for resistance welding, but it can be arc welded with special care. Though 950D is weldable and formable, the effect of high phosphorus (0.15%) on it must be considered when low-temperature properties are important. Grade 950X, treated with columbium or vanadium, is similar to 950C except that it has somewhat better welding and forming properties. Much like 945X and 950X, grades 955X, 960C, 965X, 970X, and 980X have higher strengths due to more carbon and manganese; nitrogen is also added in amounts up to 0.015% to increase strength. With these higher strengths, weldability and formability are lower, of course.

Quenched and Tempered Grades

The new recommended practice proposed by the SAE organization for quenched and tempered steels, numbered J368, covers carbon-manganese and carbon-manganese-boron grades. Compositions are:

Grade	C, %	Mn, %	Grade	C, %	Mn, %
Q980	0.20	1.35	Q990A	0.20	1.50
Q980B	0.20	1.50	Q9100A	0.21	1.50
Q980A	0.20	1.10	Q9110A	0.21	1.50
Q990B	0.20	1.50			

All grades contain maximums of 0.04 P and 0.05 S; Q980B and Q990B also have 0.0005%B min.

Numbers with A suffixes represent grades with one or more additional alloying elements employed to develop higher strengths in greater thicknesses. Table II lists mechanical properties specified for the grades.

These steels are characterized by strength, abrasion resistance, and (for certain compositions) good resistance to atmospheric corrosion. Designed for components of mobile equipment and other structures where lightness is desirable, these grades are typically applied for truck wheels and bodies, frames, scrapers, cranes, shovels, booms, chutes, and conveyors. High strength limits the formability of these steels, but moderate forming (preferably transverse to the rolling direction) is possible. Recommended minimum inside radii for cold bending in any direction are 2T for thicknesses to 1 in. inclusive and 3T for 1 to 2 in. thicknesses. Plate over 2 in. thick should not be bent cold. Readily weldable, these steels are commonly joined by low-hydrogen metal-arc processes. However, the SAE recommends that the steel's producer be consulted for the correct practice, because special procedures must sometimes be followed.

Table II — Mechanical Properties of SAE J368 Steels

Grade and Thickness	Yield Strength, Psi*	Tensile Strength, Psi	Min Elongation in 2 In., % †	Bend Test ‡
Q980 To ¾ in. incl.	80,000	95,000 min	18	2T
Q980B To 1½ in. incl.	80,000	95,000 min	18	2T
Q980A To 1¼ in. incl.	80,000	95,000 min	18	2T
Q990B To 1¼ in. incl.	90,000	100,000-120,000	18	3T
Q990A To 1 in. incl.	90,000	105,000-135,000	18	2T
Over 1 to 2 in. incl.	90,000	105,000-135,000	18	3T
Over 2 to 4 in. incl.	90,000	105,000-135,000	18	4T
Q9100A To 1 in. incl.	100,000	115,000-135,000	18	2T
Over 1 to 2½ in. incl.	100,000	115,000-135,000	18	3T
Q9110A To 1 in. incl.	110,000	115,000-135,000	18	2T
Over 1 to 2½ in. incl.	110,000	115,000-135,000	18	3T

* Yield strength minimum at 0.2% offset or 0.5% extension under load.
† Minimum percentage elongation for all material under ¼ in. thick, 15%.
‡ Ratios of bend diameter to thickness (T) of specimens.

Potentials for More HSLA in Autos

General Motors Corp. used about 115 000 tons (104 000 Mg) of microalloyed (high-strength, low-alloy) steels in the 1975 model year. The total in its 1976's could be three times higher. Such an increase is obviously impressive, but it should be interpreted with some caution for two reasons:

1. Application of HSLA in passenger cars is relatively new. GM's first major part, for example, was the rear-bumper reinforcement bar for the 1974 Oldsmobile Toronado. Automotive engineers are still easing ahead in low gear as far as new applications are concerned. By one estimate, Detroit is probably 20 to 30% of the way up the normal learning curve of understanding how to apply these steels. Formability and spot weldability have been considered drawbacks. Progress is reported in both areas.
2. Detroit's war on weight makes HSLA a prime candidate as a substitute for conventional hot and cold-rolled sheet steels, which account for 50 to 60% of a typical vehicle's weight. One study shows that a total saving of 880 lb (400 kg) is possible with HSLA in a standard car. However, two other materials – aluminum and plastics – are in the thick of this competition. At this time, it's not clear which material route or routes automakers will take after 1976.

Those observations are based on papers presented at Microalloying '75, Washington, sponsored by the Metals Div., Union Carbide Corp.; Foote Mineral Co.; Molycorp Inc.; and Shieldalloy Corp.

Portions of three papers follow:

"Present and Future Use of Microalloyed Steels in Automobiles," by Farno L. Green, executive engineer, Mfg. Dept., General Motors Technical Center.

A paper on formability, "Using Microalloyed Steels to Reduce Weight of Automotive Parts," by Subimal Dinda, materials development engineer, Chrysler Corp.; J. A. DiCello, now with National Steel Corp.; and A. S. Kasper, supervisor, Sheet Metal Forming Laboratory, Chrysler Corp.

"Resistance-Spot Welding of Microalloyed Steels for Automotive Applications," by J. W. Mitchell, manager, Welding Development Dept.; and Uck I. Chang, welding development engineer, Ford Motor Co.

Other papers relating to formability presented at Microalloying '75 included:

"Relationship Between Laboratory Material Characterization and Press Shop Formability," by Stuart P. Keeler, National Steel Corp., and William G. Brazier, Fisher Body Div., General Motors Corp.

"Stamping Potential of Hot-Rolled, Columbium-Bearing High-Strength Steels," by R. R. Hilsen and T. E. Fine, Inland Steel Research Laboratores; and G. J. Hansen, A. O. Smith Inc.

For copies of papers cited in this article, write: Michael Korchynsky, director-alloy development, Metals Div., Union Carbide Corp., 270 Park Ave., New York, N.Y. 10017. Proceedings of Microalloying '75 will be available in April 1976.

GM Assessment of Present and Future

The 1976 Chevette has chromium-plated front and rear bumpers made of HSLA with 55 000 psi (380 MPa) yield strength. Each weighs 21 lb (9.5 kg). The steel is 0.095 in. (2.4 mm) thick.

Two other approaches were evaluated:

An extruded, one-piece anodized aluminum bumper varying from 0.20 to 0.44 in. (5.1 to 11.1 mm) thick, with a yield strength of 43 000 psi (297 MPa).

An extruded aluminum reinforcing bar 0.12 to 0.24 in. (3.0 to 6.1 mm) thick, with a yield strength of 55 000 psi (380 MPa); it was combined with a chromium-plated HSLA steel face bar 0.060 in. (1.5 mm) thick. Yield strength is 55 000 psi (380 MPa).

Comparative weights, piece costs per bumper, and tooling and equipment costs are surprising. Using the anodized aluminum extrusion as the base, the weight of the HSLA bumper was −0.52 lb (0.24 kg) of base; piece cost per bumper was 68% of base; tooling and equipment costs were only 15% of base.

Such examples, GM's Mr. Green comments "... might lead you to believe that HSLA steel will be used in all bumpers. However, this is not necessarily so. Competitive materials are being carefully studied for various groups using renitrogenized and other steels, aluminum, and plastics. For example, a production door guard beam is now made of renitrogenized steel. After aging, it has a yield strength of 60 000 psi (414 MPa). If HSLA steel is not sufficiently competitive, choices of other materials will be made."

Considerations: HSLA steels have several pluses. For one thing, they provide design flexibility. Yield strength may be chosen in the range of 50 000 to 90 000 psi (345 to

HSLA in curved sections at ends of Chevette bumper had to be both stretched and compressed in forming. In addition, good surface appearance was necessary for plating.

620 MPa) compared with about 30 000 (207 MPa) for conventional steels.

Where yield strength is a principal design criterion in a structure, these steels may allow a weight reduction of 30% or more. In an experimental vehicle, 225 lb (100 kg) of HSLA provided a 40% weight saving over mild steel.

There is no advantage in conventional design configurations where stiffness of a single sheet of body steel (0.020 to 0.035 in [0.5 to 0.9 mm]) is the most important property. "If ribs or honeycomb structures are used to add stiffness," Mr. Green explains, "the additional material tends to offset weight savings and increase the cost of the thinner sheet."

The corrosion rate is similar to that of mild steels. This means that with thinner sections, protective coatings may be required where they are not needed for conventional mild steels.

Application Experience: At GM, HSLA bumpers have probably been studied and developed more extensively than any other part for these reasons:

First, requirements for impact and low damage are severe; and materials must have a higher strength-to-weight ratio than mild steel if the weight penalty is not to be excessive.

Second, the bumper is mounted on a lever arm extending from the chassis. This means the bumper should be as light as possible.

Third, because of the mounting arrangement for the bumper, it is isolated and can be analyzed more easily than other structural components like frames and door pillars, which are integral parts of the car structure.

New bumper applications on 1976 models in addition to the Chevette include Cadillac's Seville and the Delta 88 and Oldsmobile 98.

The Seville has a front bumper reinforcement made of 50 000 psi (345 MPa) HSLA steel. The part weighs 25 lb (11 kg).

The two Oldsmobile cars have an HSLA reinforcement with a very intricate configuration. The 80 000 psi (552 MPa) grade part is 0.10 in. (2.5 mm) thick and weighs 45 lb (20.5 kg). "When manufacturing studies were started at the GM Technical Center a number of a years ago," Mr. Green remarks, "many engineers would have said that a part with this geometry would never be formed successfully with 80 000 psi HSLA steel." Process engineers worked closely with the design group to develop dies and a formable design.

Other Applications: A number of small HSLA parts are going into GM cars for one or more reasons – to save weight, to save space, or to take a greater impact load.

For example, impact plates made of an 80 000 psi (552 MPa) grade have been resistance welded to more than 20 million Enersorbers used in bumper systems. Cadillac has released an engine-to-transmission coupling which specifies a 50 000 psi (345 MPa) grade. The same steel has been used in seat back head-restraint supports on some regular GM cars since 1971. The Chevette has four HSLA parts in seat back locking mechanisms.

Conclusion: Mr. Green states, "The future of microalloyed steel continues to show that there are many potential applications. The material has unique properties. We need to learn much more in applying this new material. Trends indicate that uses will increase particularly for structural applications in automobiles."

Chrysler Reports on Formability

"Although high-strength, low-alloy sheet steels have been available for years," comment the Chrysler authors, "they have seen little use in the automotive industry until recently. The limited formability of these materials was the major factor restricting their acceptance. Recently, the formability

Table I — Typical Mechanical Properties of Low-Carbon Mild Steel and HSLA

Material	Yield Strength, 10^3 psi (MPa)	Tensile Strength, 10^3 psi (MPa)	Elongation, %	Strain Hardening Exponent, n	True Stress at Unit Strain, K 10^3 psi (MPa)	Plastic-Strain Ratio, r
Low-carbon mild steel (hot-rolled, drawing quality)	30 (207)	45 (311)	38	0.22	80 (552)	1.00
High-strength, low-alloy steel (hot-rolled, killed)	50 (345)	65 (449)	30	0.18	110 (759)	0.90
High-strength, low-alloy steel (hot-rolled, killed)	80 (552)	90 (621)	20	0.13	145 (1001)	0.70
Ultrahigh-strength steel (cold-rolled, semikilled)	120-160 (827-1103)	130-170 (897-1173)	3	—	—	—

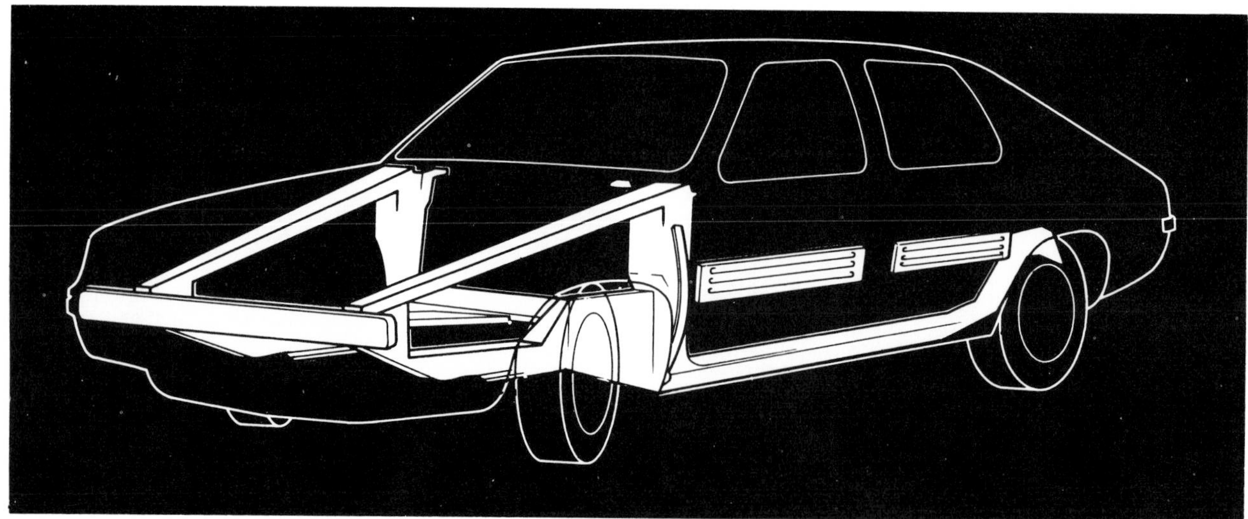

In this GM experimental vehicle, 225 lb (102 kg) of 80 000 psi (552 MPa) HSLA replaced 375 lb (170 kg) mild steel in blue areas. Gages ranged from 0.070 to 0.095 in. (1.8 to 2.4 mm). Assemblies were spot welded.

of new HSLA steels has improved through controlled processing and inclusion-shape control with additions of cerium, zirconium, or titanium. This improved formability has increased their potential usefulness and range of applications. The weight-reduction drive and critical steel shortages have also helped promote the use of HSLA steels."

Fundamentals: As Table I shows, HSLA has lower inherent formability than low-carbon mild steels. As yield strength increases, total elongation, n values, and $\bar{r}$ values decrease. But HSLA properties, especially in grades containing inclusion shape-control additives, are more uniform in all planar directions and sometimes form better than mild steels.

Table II — Typical Gage Reductions in Converting From Mild Steel to HSLA

	Gage		HSLA Yield Strength, 10^3 psi (MPa)
Chassis Parts	**Mild Steel, in. (mm)**	**HSLA, in. (mm)**	
Front side rail	0.105 (2.6)	0.089 (2.2)	50 (345)
Steering gear mount	0.139 (3.5)	0.110 (2.7)	50 (345)
Engine mount	0.152 (3.8)	0.122 (3.1)	50 (345)
Suspension crossmember	0.093 (2.3)	0.075 (1.9)	50 (345)
Bumper reinforcement	0.152 (3.8)	0.115 (2.9)	60 (414)
Torque converter plate	0.125 (3.1)	0.100 (2.5)	50 (345)
Diamond bracket	0.375 (9.5)	0.310 (7.8)	60 (414)
Brake strut plate	0.180 (4.5)	0.152 (3.8)	60 (414)
U-bracket (bumper)	0.230 (5.8)	0.189 (4.8)	60 (414)
Nerf strip	0.175 (4.4)	0.125 (3.1)	60-80 (414-552)
Body Parts			
Door (inner and outer)	0.033 (0.8)	0.028 (0.7)	50 (345)
Deck lid (inner and outer)	0.033 (0.8)	0.028 (0.7)	50 (345)
Hood (inner and outer)	0.033 (0.8)	0.028 (0.7)	50 (345)
Door hinge	0.199 (5.0)	0.157 (3.9)	50 (345)
Hood hinge	0.145 (3.6)	0.122 (3.1)	50 (345)
A-pillar	0.093 (2.3)	0.082 (2.0)	45 (380)
Door impact beam	0.069 (1.7)	0.035 (0.8)	150 (1034)

Designs must be modified to accommodate these steels. In fact, their formability characteristics are as important as their strength. Springback has been one of the major concerns. An 80 000 psi (552 MPa) grade, for instance, does not appear to be suitable for outer-body panel applications because of inadequate formability and excessive springback which cannot be reduced sufficiently with die modification. On the other hand, 40 000 to 50 000 psi (276 to 345 MPa) grades may be used.

Current Applications: The authors cite three areas where substitutions of HSLA for mild steel are possible in current applications:

1. Many highly stressed parts that do not require the ductility of low-carbon steels are prime candidates for conversion.
2. Flange or edge tearing encountered in forming mild steels can be eliminated in many parts by using HSLA steels processed to control inclusion shape. Only minor tool modifications are required.
3. Where safety or emission control requirements call for an increase in the thickness of mild steel parts, HSLA of the same thickness as the existing part can be substituted. Only minor tool modification is necessary.

Suitable candidates for chassis and body components are listed in Table II. Some of the parts are in production at Chrysler. In all instances, forming characteristics were determined in the lab. Selection of material and gage reduction were based on part geometry and other requirements. All parts were formed in production tools. Experience with two hot-rolled HSLA parts (front side rail and an alternator fan) and one cold-rolled HSLA part (door inner) is summarized below.

Front Side Rail: This part has been made from 0.089 in. (2.2 mm) low-carbon mild steel. Gage must be increased to 0.105 in. (2.6 mm) because of federal crush resistance

Table III — Selected Passed-and-Failed Ductility Tests of Spot Welded HSLA

Thickness							
in.	0.088	0.086	0.091	0.090	0.085	0.086	0.092
(mm)	(2.24)	(2.18)	(2.31)	(2.29)	(2.16)	(2.18)	(2.34)
Ductility Test							
5 cycles	Passed	Passed	Passed	Passed	Failed	Passed	Passed
30 cycles	Passed	Failed	Failed	Passed	Failed	Failed	Passed
Chemical Composition							
Carbon	0.07%	0.13%	0.126%	0.07%	0.13%	0.124%	0.09%
Manganese	0.81	0.44	0.37	0.87	0.70	1.50	0.86
Silicon	0.15	0.05	—	0.12	0.44	0.40	0.11
Titanium	0.18	0.22	—	—	—	—	—
Vanadium	0.005	0.01	—	—	—	—	0.03
Columbium	—	—	0.013	0.022	0.017	0.09	—
Chromium	0.01	0.01	0.01	0.01	0.04	0.04	0.02
Molybdenum	0.01	—	0.01	0.01	0.01	0.01	0.01
Aluminum	0.03	0.03	0.085	0.01	0.04	0.08	0.038

requirements and added car weight. The change requires new tooling. Rail weight is boosted about 13 lb (6 kg) per car.

As an alternative to increasing gage, prototype parts were made of 50 000 psi (345 MPa) steel in production tools without any adjustment. Higher springback was the only problem. It can be minimized by overbending the material in the die. Because the part has a wide-stretched flange, an HSLA grade with inclusion-shape control was used to avoid edge tearing; and a grade suitable for resistance welding was chosen because the part is joined to its top plate.

Strain levels at critical areas after forming were measured by grid analysis and compared with those for mild steel production parts. The HSLA part had better strain distribution than the mild steel version even though the former had a lower n value (0.17) than the latter (0.19). With the same blank-holding pressure, the HSLA blank had reduced stretch and was more drawn in forming the part shape.

The mild steel part was formed by stretching more than drawing because of its lower yield strength. As a result, stretch strains were lower in HSLA than in low-carbon steel. This finding is significant because laboratory-determined forming parameters and forming-limit diagrams do not always predict the actual press-performance of new materials. Press shop experience combined with shape analysis presents a more accurate view of formability.

Alternator Fan: A 50 000 psi (345 MPa) grade in 0.095 in. (2.4 mm) gage was specified to improve fatigue life and eliminate yielding problems in blades. In forming the part from mild steel, a fatigue crack originated at the sharp radius of blades, where the end is parallel to the rolling direction. The cause: stringer-type inclusions along the rolling direction.

Alternator fan: Chrysler demonstrated HSLA can solve a problem in forming of mild steel (fatigue cracking, starting at arrow). Grade: 50 000 psi (345 MPa).

Mild steel 0.115 in. (2.9 mm) thick could not be used because clearance limits and additional weight created problems for the shaft bearing. Fatigue life was substantially improved by using HSLA with inclusion shape control – properties were uniform in all directions.

Door Inner: This part was formed in production tools with 45 000 psi (310 MPa) cold-rolled HSLA 0.028 in. (0.7

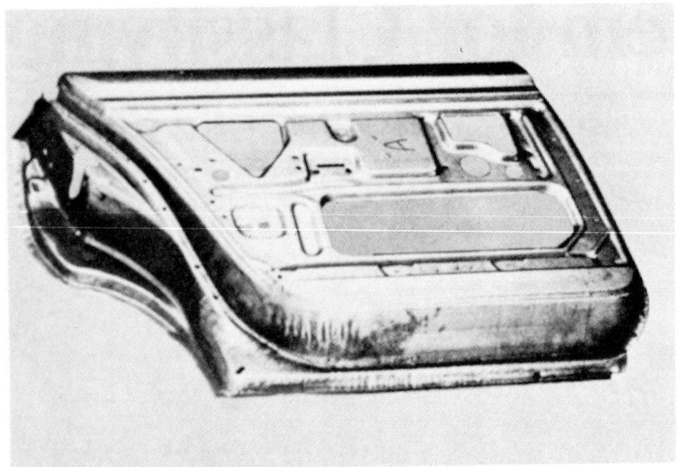

Door inner formed by Chrysler from 0.028 in. (0.7 mm) HSLA with 45 000 psi (310 MPa) yield strength.

mm) thick. The production material is 0.033 in. (0.8 mm) drawing quality mild steel. Blank size development is needed to form high-strength materials. In many areas where the material should be drawn-in rather than stretched, blank-holding pressures were reduced.

Strain levels in critical areas were measured and plotted in a forming-limit diagram. Because of lower n values, strains developed in the HSLA were higher than those in mild steel and were formed almost purely by stretch. Strain values for both materials were much below the critical level. A weight reduction of about 15% can be obtained, but HSLA is not directly cost competitive in this instance.

Ford's Views on Resistance Spot Welding

Ability to be spot welded is a key requirement of a new material in the auto industry because this is the chief method of fabricating the body and structural members of unitized cars. A typical auto contains 5000 spotwelds, about 50% of which are made by fully automatic multispot welders, usually of the portable gun type. HSLA steels in the 50 000 to 80 000 psi (345 to 552 MPa) range with proper shear-impact properties meet Ford's requirements.

The importance of impact properties stems from the fact that hardenability is the critical factor in the use of HSLA. Weld button or heat-affected-zone hardness with Rockwell C readings above 30 in thinner gages and 40 in heavier gages usually fail peel tests and result in low impact values.

The quality of a spot weld in automotive specifications and operations is based on the results of a peel test. Usually, welded pieces are forced apart with a tapered chisel, placing the edge of the spot weld in tension. If a weld button of the proper size is pulled from one of the two pieces of metal in this separation process, the weld is considered satisfactory.

Benefits of HSLA: Unlike other weight-saving materials, HSLA steels achieve their benefits through a reduction in thickness. In comparison with aluminum, there is a significant reduction in spot-weld energy requirements; and there is a slight reduction in power demand due to higher electric resistivity. These advantages offset higher welding pressures and slightly longer weld times.

A common substitution – 0.089 in (2.2 mm) HSLA for 0.12 in. (3 mm) mild steel – has a significant positive effect on tooling requirements and productivity. In this example, the weld diameter of mild steel is 0.35 in. (8.9 mm), while that of a 50 000 psi (345 MPa) HSLA grade is 0.31 in. (7.9 mm). Electrode force for mild steel is 1900 lb (865 kg); that for HSLA is 1650 lb (750 kg). Weld time for mild steel (a single pulse of five cycles preweld and nine cycles cool at 26 000 A) is 38 Hz; that for HSLA is 26 Hz. The mild steel requires a secondary current of 20 000 A; the value for HSLA is 13 000 A.

Surface condition, the Ford authors report, is also of prime importance in resistance welding. "To date," they state, "the HSLA steels that have been examined are often deficient in surface cleanliness, usually due to inadequate pickling procedures. While this is a problem that can be solved, it is extremely important that their surface condition be comparable to that found on hot-rolled, low-carbon steels which have been pickled and oiled."

Role of Carbon: In Ford studies, titanium-bearing steels were the most sensitive to carbon content in meeting the peel test, while vanadium steels were the least sensitive. The point: carbon plays a dominant role, but other elements also have an impact on the hardenability of a resistance spot weld. These influences vary considerably.

To date, Ford has tested 55 lots of HSLA from nine different producers. Thicknesses ranged from 0.028 to 0.135 in. (0.7 to 3.45 mm); and 29 lots were between 0.083 and 0.096 in. (2.1 to 2.44 mm). In the latter group, 16 lots – representing three different alloy systems – passed the ductility test. Results of all tests are summarized in Table III.

Analysis of the total lots, considering thickness, yield a tentative carbon-equivalent formula:

$$CE = C + \frac{Mn}{30} + \frac{Cr + Mo + Zr}{10} + \frac{Ti}{2} + \frac{Cb}{3} + \frac{V}{7} + \frac{TS\,(ksi)}{900} - \frac{T\,(in.)}{20}$$

"The formula," it's reported, "was an initial attempt to develop a procedure to reduce peel testing as a means of determining the ductility of the spot-weld. A carbon-equivalent value of 0.30 is considered the dividing line between pass and fail in the peel test. Applying the formula to the lots tested, a 92% reliability has been obtained compared with a reliability factor of 81% when only carbon was considered."

—Harry E. Chandler

How Ford Evaluates Conversions to HSLA

By DEWEY G. YOUNGER

It's predicted that in weight-saving applications the need for new designs and new tooling will delay adoption on a large scale. The 950X grade is expected to get the most attention.

HSLA STEELS have been used primarily as solutions to strength problems in automotive parts. Engineers now look to them for weight reductions. This goal presents a different engineering problem because a new set of rules comes into play.

A weight problem is not nearly so acute as a strength problem, so such considerations as manufacturing problems, cost increases, and other problems get a lot of attention when changes in grades are being studied. We have to prove that weight reductions are not only necessary but also can be accomplished economically.

Reducing Weight — Theoretically, any steel part which is designed to an allowable yield strength criterion can be reduced in weight by using a higher yield strength grade. The amount of permissible weight reduction, however, depends on how loads are transferred through the part.

Figure 1 illustrates theoretically determined potentials for weight savings with HSLA steels. Parts with the most efficient load paths fall along the left boundary.

Between these boundaries lies the region which covers most typical automotive structural parts. As shown by the lower curve, the weight saving potential of SAE 950X lies between 22.5 and 40%. On the average, the potential for weight saving in our cars would be about 30% for strength-designed parts using 950X steel.

If we use 980X in place of 1010 steel, the potential for weight reduction lies between 38.8 and 62.5%, if some new mode of failure, like excessive deflection or local cross-section instability, doesn't place a limit on these theoretical values.

Figure 2 relates levels of desired weight saving and numbers of sheet gage reductions that must be made to achieve given levels of weight savings. For example, a 20% weight saving in a given sheet metal part requires that the original gage be reduced by either two, three, or four gages, depending on the original gage of the low-carbon (SAE 1010) steel.

Likewise, a 40% weight saving requires a reduction in the original thickness by four to eight times. (Table I lists gage numbers and corresponding sheet thicknesses, as given in the AISI Steel Products Manual, "Carbon Sheet Steel," April 1974.)

Weight reduction involves two conflicting desires. On one hand, we desire to save weight in a given part; on the other, we desire to use our present tooling to minimize costs.

The conflict occurs because metal stamping and forming people inform us that, as a general rule, a two-gage reduction in the sheet used for a given part calls for new, or at least highly modified, tooling. For most formed parts, any attempt to retain tooling when there is more than one-gage reduction only leads to poor quality parts and excessive springback.

Thus, weight reductions of as high as 20% require new tooling. For some designs, in fact, a weight saving as low as 8 to 12% may require new tooling. In short, if we properly exploit the weight reduction potential of HSLA steels, new tooling will be required to assure quality parts.

Cost Calculations — Now, let's look at steel prices to determine the potential for cost reduction through the use of HSLA steels. In 1974, SAE 950 was about 24% more expensive than 1010, and SAE 960X, 970X, and 980X were, respectively, 29%, 36%, and 46% more expensive. Though prices have gone up, relative costs remain the same.

Using these relative prices, we developed the graph in Fig. 3. It

Mr. Younger is manager, System Research Development & Service Dept., Product Development Group, Ford Motor Co., Dearborn, Mich.

Table I — Gage Numbers and Equivalent Thicknesses

HOT AND COLD ROLLED SHEET

Manufacturers Standard Gage Number	Thickness Equivalent, in. (mm)	
3	0.2391	(6.073)
4	0.2242	(5.695)
5	0.2092	(5.314)
6	0.1943	(4.935)
7	0.1793	(4.554)
8	0.1644	(4.176)
9	0.1495	(3.797)
10	0.1345	(3.416)
11	0.1196	(3.038)
12	0.1046	(2.657)
13	0.0897	(2.278)
14	0.0747	(1.897)
15	0.0673	(1.709)
16	0.0598	(1.519)
17	0.0538	(1.367)
18	0.0478	(1.214)
19	0.0418	(1.062)
20	0.0359	(0.912)
21	0.0329	(0.836)
22	0.0299	(0.759)
23	0.0269	(0.683)
24	0.0239	(0.607)
25	0.0209	(0.531)
26	0.0179	(0.455)
27	0.0164	(0.417)
28	0.0149	(0.378)

GALVANIZED SHEET

Galvanized Sheet Gage Number	Thickness Equivalent, in. (mm)	
8	0.1681	(4.267)
9	0.1532	(3.891)
10	0.1382	(3.510)
11	0.1233	(3.132)
12	0.1084	(2.753)
13	0.0934	(2.372)
14	0.0785	(1.994)
15	0.0710	(1.803)
16	0.0635	(1.613)
17	0.0575	(1.460)
18	0.0516	(1.311)
19	0.0456	(1.158)
20	0.0396	(1.006)
21	0.0366	(0.930)
22	0.0336	(0.853)
23	0.0306	(0.777)
24	0.0276	(0.701)
25	0.0247	(0.627)
26	0.0217	(0.551)
27	0.0202	(0.513)
28	0.0187	(0.475)
29	0.0172	(0.437)
30	0.0157	(0.399)
31	0.0142	(0.361)
32	0.0134	(0.340)

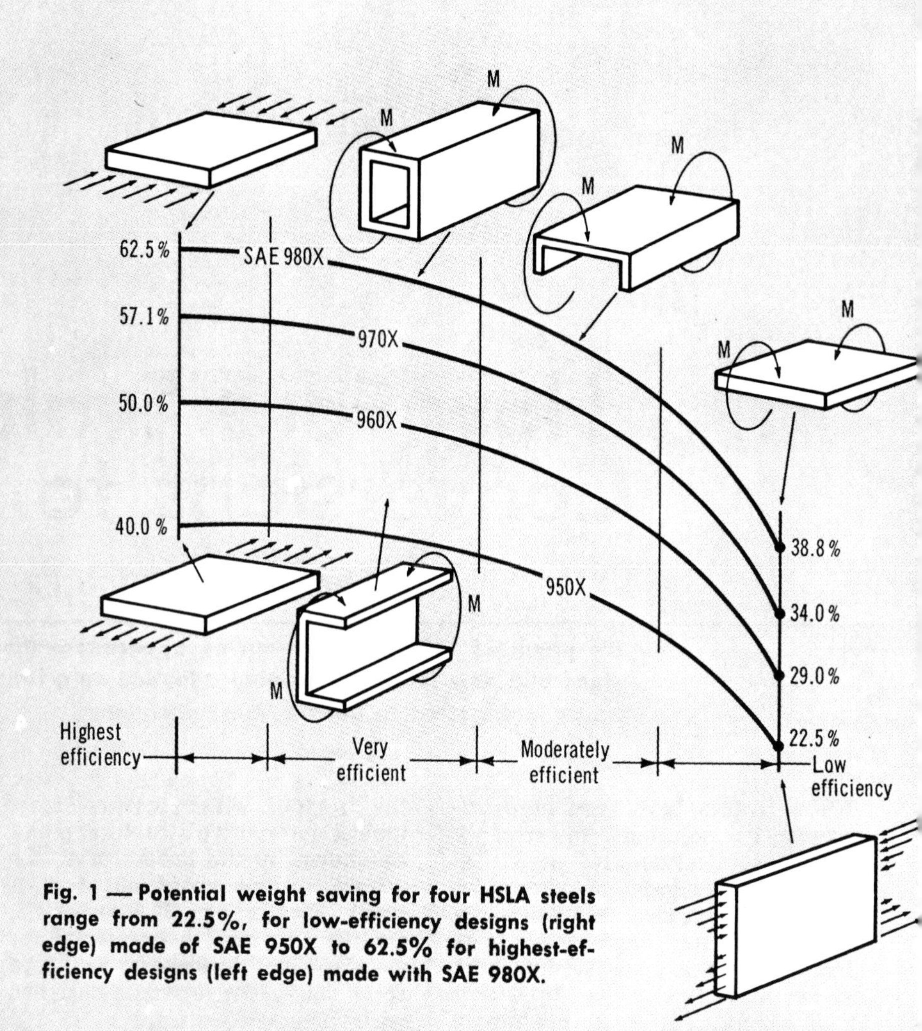

Fig. 1 — Potential weight saving for four HSLA steels range from 22.5%, for low-efficiency designs (right edge) made of SAE 950X to 62.5% for highest-efficiency designs (left edge) made with SAE 980X.

shows the weight saving required in a new HSLA steel part to break even with the material costs in the old part of 1010 steel.

Specifically, if we switch a given 1010 part to, say, SAE 950X, we need a weight saving of 19% in the new part to break even on material costs alone. Comparative values for SAE 960X, 970X, and 980X, are 23, 27, and 32%, respectively.

Processing Characteristics

Gage for gage, HSLA steels are tougher to form than 1010 steels because of their higher yield strengths.

But we don't look upon this as a major problem, especially with the newer, easier-to-form HSLA steels becoming available. It's more a matter of understanding the forming characteristics inherent in HSLA steel, and taking them into account in both component and tooling design.

In most instances, less press tonnage will be needed to form and shear HSLA steel parts than with the 1010 parts they replace. This is because tonnage requirements are based on the product of the material's thickness and tonnage factor. And even though HSLA steels have a higher tonnage factor, it is more than offset by the thinner gage permitted in the high-strength, low-alloy steel parts.

Weldability — Steel producers have adequately demonstrated that they can produce weldable HSLA steels that pass our ductile-spot-weld peel test. The titanium-bearing HSLA grades have had the toughest time passing this test. Small amounts of titanium make it mandatory that the carbon content be carefully controlled to low levels. When the

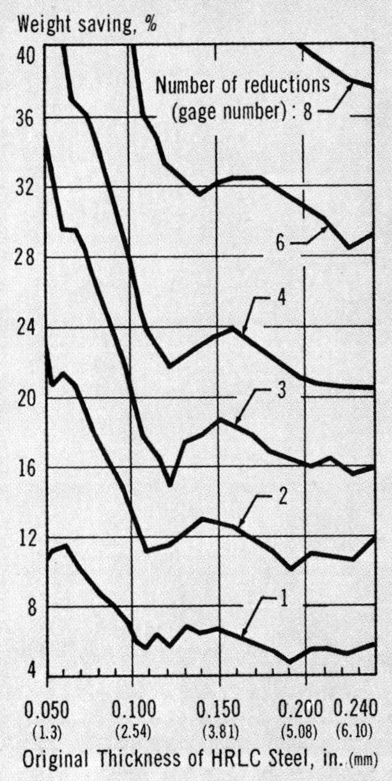

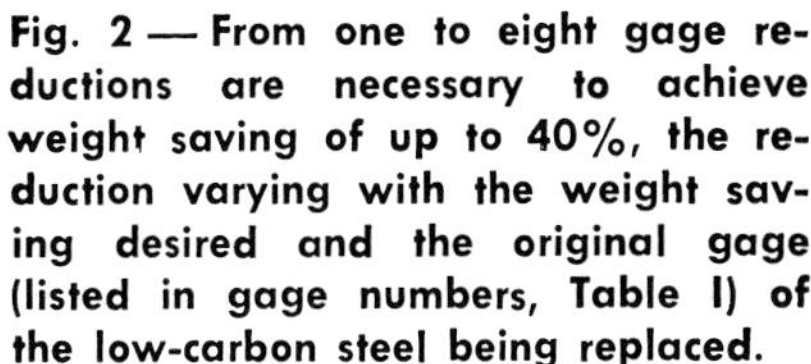
Fig. 2 — From one to eight gage reductions are necessary to achieve weight saving of up to 40%, the reduction varying with the weight saving desired and the original gage (listed in gage numbers, Table I) of the low-carbon steel being replaced.

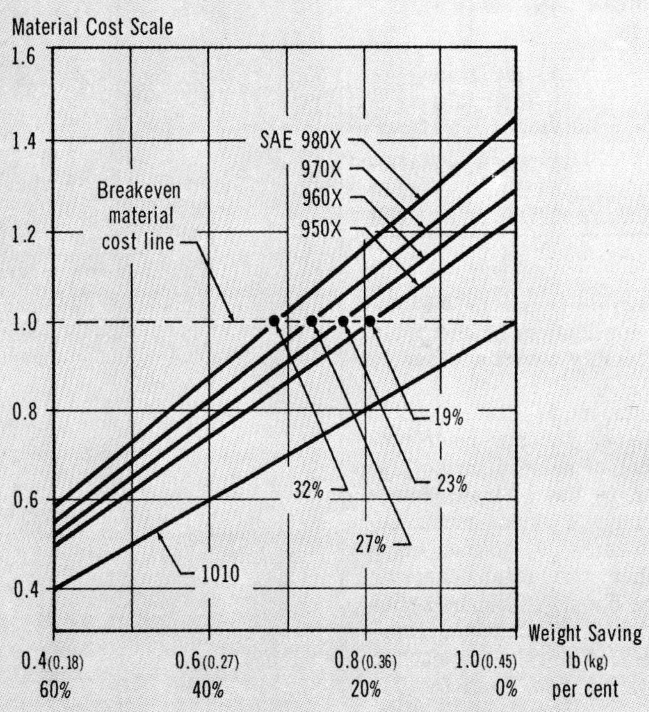

Fig. 3 — Because HSLA steels cost more, parts in which they replace SAE 1010 must weigh less. Percentages of weight saving needed to break even on material costs are shown by the four points.

HSLA steel meets our welding specification, the same resistance welding equipment and techniques used for 1010 steel will suffice. Major problems are not evident in fusion welding either.

Corrosion — It's known that HSLA steels are equal to, or better than, the 1010 steels in terms of corrosion resistance to typical automotive environments. On a gage-for-gage basis, the same preventative measures used for corrosion-critical areas in parts of 1010 steel can be used on HSLA.

Nevertheless, when thinner gages of HSLA are used to achieve weight reduction, we must expect to encounter corrosion problems that have not been experienced before — simply because the sheet is thinner.

In some areas, we may have to zinc-coat more underbody areas or use more tar-base emulsion coatings than before.

Fatigue Characteristics — The high strengths of HSLA grades are accompanied by sufficient ductility to give excellent fatigue strengths (Table II). We expect to find better fatigue resistance in HSLA parts, even after a 30% weight reduction, than we had before with 1010 steel.

What more could we ask for? HSLA can give us, simultaneously, a lighter part, a stronger part, and a more fatigue-resistant part.

Figure 4 compares strain-rate effects in 950X and 980X with those for 1010, 1019, and 1070. Strain rates shown along the abscissa cover 0.01 in. (0.25 mm) per min to 100 000 in. (2 540 000 mm) per min, which is equivalent to that experienced in a 100 mph (160 km per h) crash into a barrier. Lines for 950X and 980X slope less than those for the other three steels, indicating that high strain rates affect HSLA steels less than they do the others. The same general effect holds for tensile and compression yield strengths, as well as for response to inelastic bending and torsion.

Figure 5 shows minimum allowable stress-strain curves for 1010 and 950X steels that must be used in design. Note that 950X steel has a 67% higher yield strength and that its elastic range is about 33% greater. The latter is important because it enables weight to be taken out of many designs without introducing problems associated with permanent set. The 950X curve also displays a more generous knee, showing that its tangent modulus changes more slowly with increasing inelastic stresses.

Thus, the 950X steel will perform much better than 1010 steel in designs where it is important to delay the onset of inelastic buckling, as is the situation for door beams and bumper reinforcements.

Applications — Intrusion - resistant door beams and bumper reinforcement beams are good examples of strength-designed parts, although the designs are based more on critical buckling stress levels. Figure 6 shows a typical rear bumper assembly with a reinforcement to which we've applied 950X, achieving about a 15 lb (6.8 kg)

Table II — Fatigue Characteristics of HSLA Steels

Steel	Fatigue Strength, 10^3 psi (MPa) 10^3 Cycles	10^8 Cycles
1010	±36 (250)	±21 (145)
SAE 950X	±55 (380)	±35 (240)
SAE 980X	±83 (570)	±42 (290)

weight reduction in our 1974 Pinto. A similar application in the front bumper assembly saved another 15 lb (6.8 kg).

The failure mode that's critical to these heavy bumper reinforcements is that of section instability, or buckling, in the inelastic range of the material's stress-strain response. We must, of course, demonstrate that the reinforcements can pass the 5 mph (8 km/h) series of impact tests without buckling.

When HSLA and low-carbon steels are being compared for use in the same structural application, the questions are, "How much lighter can the HSLA beam be?", and "How different will the cross sections of the two beams be?"

It is assumed, of course, that the two beams are capable of supporting the same maximum bending moment. Design calculations demonstrate that, typically, the lightest 950X design will be 15 to 20% (as based on cross-section instability) lighter than the lightest 1010 steel design. Calculations also show that the two minimum-weight designs have distinctly different values of the "structural index," M/b^3 — b is a characteristic dimension of the beam's cross section, and M is the bending moment.

If these two beams are to carry the same maximum bending moment (M), their characteristic dimension (b) cannot be the same for both. Thus, if we want to save 15 to 20% in weight by using 950X instead of 1010, we will have to redesign the beam cross section so that the beam of thinner-gage 950X will not buckle when loaded with the same bending moment as that carried by the 1010 steel beam.

We can't, in short, get large weight reductions in a new design merely by substituting a thinner, stronger steel for a thicker, weaker one. To exploit the weight reduction potential of the HSLA steels properly, we must redesign the component and employ all-new

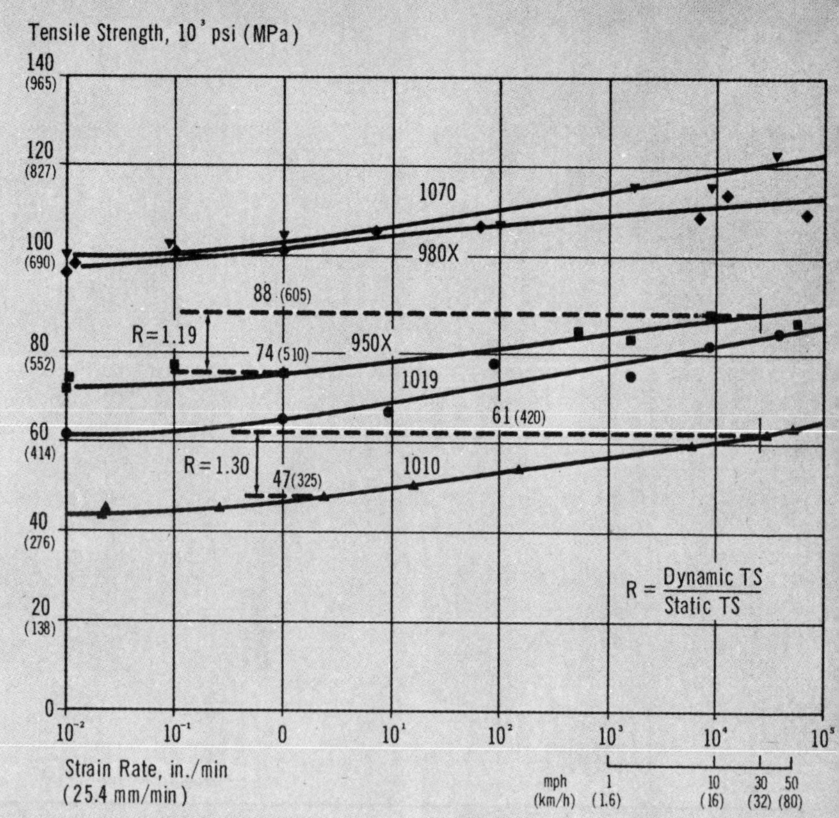

Fig. 4 — High strain rates have less effect on HSLA steels than on 1010, 1019, or 1070, indicating that high-speed crashes will damage high-strength, low-alloy steel parts to a lesser extent.

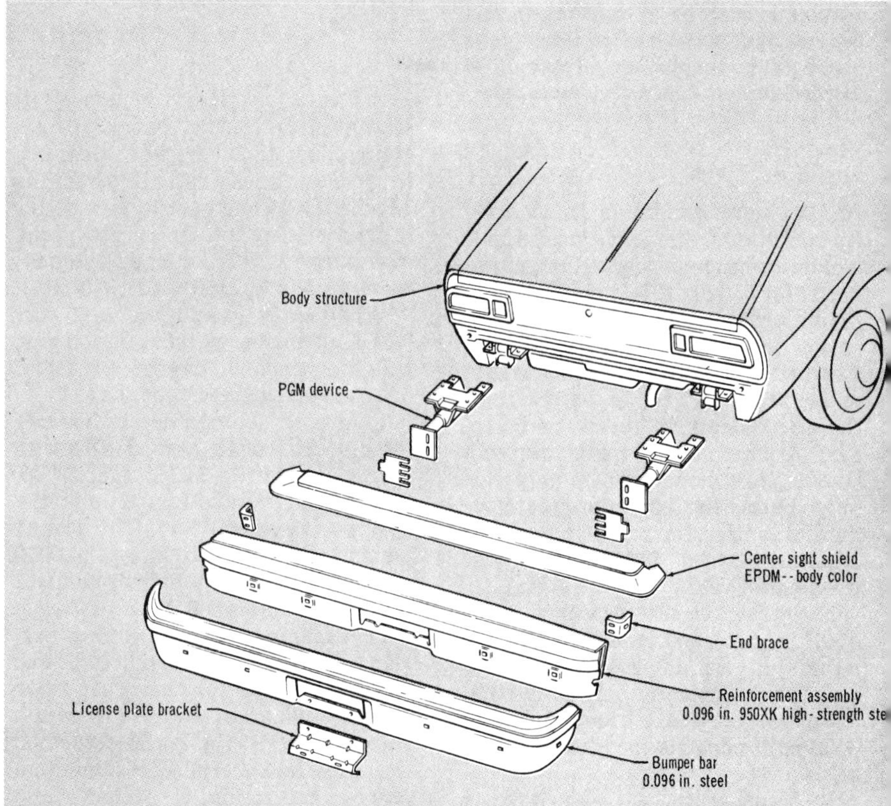

Fig. 6 — Reinforcement beams in bumpers of the 1974 Pinto are of HSLA steel, saving about 15 lb (6.8 kg) in weight apiece.

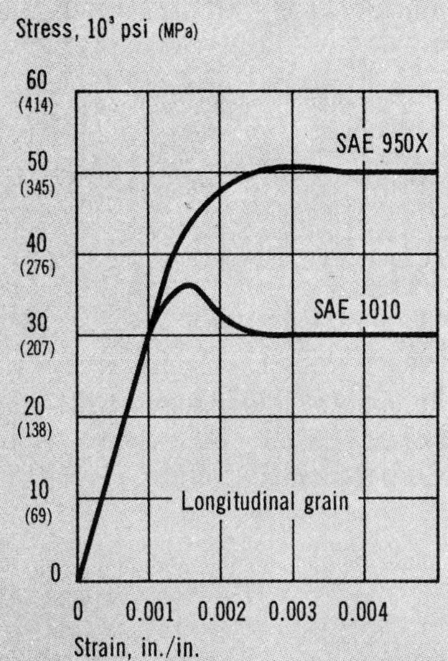

Fig. 5 — SAE 950X has a higher yield strength and a greater elastic range than 1010 steel.

(or highly modified) tooling in its manufacture. We understand this at Ford, so when the time is right for a tooling change you can expect to find more and more HSLA steel in our cars.

Limitations — In many structural areas of the automobile, the use of HSLA steels cannot be fully justified, or effectively applied, for weight reduction. To explain, many structural elements in automobiles are not designed primarily for strength or instability, but fall in another category — deflection-critical design.

A frame in a body-frame car is typical of a deflection-critical design. It is generally composed of at least 30 individual parts. The frame is a critical component in the body-frame structural systems. It has natural frequencies, mode shapes, and amplitudes, all of which are key considerations in the over-all vehicle design.

Thus, bending and torsional stiffnesses of the frame are very critical. Because center rails and crossmembers play an especially important role in the frame, we must maintain section properties.

For example, let's consider what happens to geometry when we take weight out of a frame center rail, which must retain the same moment of inertia. Assume that our frame rail is comprised of a closed-box cross section in which we want to reduce the wall thickness by 20%, thus making a 20% weight saving. However, the depth ratio must increase by 10% if the moment of inertia of the section is to remain constant. If the deflection-critical frame rail was originally 4 in. (100 mm) deep, our new lightweight design will have to be 10% deeper, or 4.4 in. (110 mm). Weight saving for this example has been computed to be 14.7%. Even though thickness reduction produced a 20% weight saving, we had to give back 5.3% of the saving when we increased the rail depth.

The 0.4 in. (10 mm) increase in the rail depth will, of course, force the designer to change some key dimensions of the over-all body/chassis package. Ground clearances, interior package, and over-all car height can all be significantly affected. Considerable design and engineering leadtime would be required to accommodate such major frame changes to achieve the weight reduction.

In this frame rail design exercise, the material strength requirement must also be considered. If the old-design and new-design frame rails carry the same bending moment, all that's needed is a steel that's 10% higher in yield strength. And this added strength is only required if we want to keep the same margin between allowable stress and design stress during the redesign. For this example, the yield strength requirement would go from 30 000 to 33 000 psi (210 to 230 MPa) — a 50 000 psi (345 MPa) steel, of course, is unnecessary.

Essentially, in short, deflection-critical components are not good candidates for applying HSLA steels. Though weight reductions can be achieved, they are not dependent on the use of HSLA steels.

Summation — Ford's position regarding use of HSLA steels to reduce weight can be stated as follows:

- They must be considered as definite contenders for replacing about 250 to 350 lb (110 to 160 kg) of the current 1010 steels. Sometime between 1978 and 1980, about 250 lb (100 kg) of HSLA steel could be incorporated in large Fords, and about 180 lb (80 kg) in compact cars. Associated weight savings will be about 100 lb (45 kg) for large cars and about 70 lb (32 kg) for small cars.
- The biggest obstacle to making large weight reductions earlier than 1978 is the need for new designs and new tooling. It behooves Ford to wait until new designs are dictated by other reasons, so that all of the new tooling costs don't have to be charged against the switch to HSLA steels. Nevertheless, some of the costs for new tooling will be charged against the new HSLA parts — the tooling will be somewhat more complex than that for the 1010 steel parts.
- If a new HSLA steel part has to stand alone from a cost standpoint, all elements of added cost have to be charged to the steel, including those for more-complex tooling. It will be necessary in this instance to shoot for weight reductions near 25% when using 950X and near 40% when using 980X.
- The future for 950X in Ford-built cars is excellent since cost effective weight reductions as high as 25% should be structurally manageable through redesign. As the material's yield strength and related price approaches that of the 980X steels, however, the required cost-effective weight reductions, 40%, are likely to lead to unmanageable structural problems, whether they be corrosion or section instabilities.

Thus, for every 100 lb (45 kg) of HSLA steel that eventually gets into Ford cars for weight reduction, 70 to 80 lb (32 to 36 kg) is likely to be 950X, with 980X accounting for no more than 5 or 10 lb (2.25 to 4.5 kg). The remainder would be distributed between 960X and 970X.

Selected HSLA Steels

Extra-Fine Grain Structure Makes Strong, Formable Steel

By JOHN D. GROZIER and MICHAEL KORCHYNSKY

Development of VAN-80 was sparked by growing needs of fabricators of trucktrailers, railroad equipment, and heavy construction machinery for strong steels with good fatigue properties that can be easily formed. The hot-rolled grade has minimums of 80,000 psi yield strength, 95,000 psi tensile strength, and 18% elongation in 2 in. Rolled in gages up to 0.375 in. on the 80 in. hot strip mill at J&L's Cleveland Works, the steel contains 0.18 C max, 1.50 Mn, max, 0.60 Si max, 0.05 V min, 0.02 Al min, and 0.005 N min. Plates have excellent formability and fatigue limits in excess of 40,000 psi in reverse bending.

Tensile properties of a ferrite-pearlite steel depend on its microstructure and the presence of elements that strengthen the ferrite. The most important factor is grain size; yield strengths approaching 50,000 psi can be obtained by decreasing grain size to ASTM 12. By adding ferrite-strengthening elements (such as silicon and manganese) and carbon (which forms pearlite), additional increases in strength can be attained. (Carbon and manganese also help to promote a fine grain size.) Elements such as vanadium and nitrogen which form precipitates can be added too; they increase strength if the steel is coiled at the temperature where precipitation occurs.

Combining these building blocks, engineers at this laboratory have devised VAN-80. Silicon, manganese, pearlite, vanadium nitride precipitation, and grain refinement act together. The extent of grain refinement (ASTM 11.7 to 12.5) results from the accelerated cooling rate (20 to 40 F per sec) of the strip as it travels under the water cooling system on the mill's runout table.

Dr. Grozier is staff metallurgist and Mr. Korchynsky is assistant director of research, Graham Research Laboratory, Jones & Laughlin Steel Corp., Pittsburgh.

(More details of the system are given in "Improving High-Strength, Low-Alloy Steels Through Controlled Cooling," *Metal Progress,* January 1966, p. 125.)

How Fine Grains Are Developed

Three main factors influence ferritic grain size. These are: 1. The austenitic grain size before transformation. 2. The cooling rate during transformation. 3. The coiling temperature. As this steel is processed on the 80 in. hot strip mill, the austenite grain size is refined by deformation (in each rolling stand) and subsequent recrystallization. For a given cooling rate, finer austenite grains will produce finer ferrite grains because there are more sites for grain-boundary nucleation.

The rate at which steel strip is cooled depends on its thickness and the efficiency of the cooling system. As an estimate, the rate at which the water system of the 80 in. mill cools the strip is about five times that of air in cooling the same thickness of steel. In general, a fivefold increase in cooling rate will reduce the ferritic grain size by about one ASTM number.

Controlled cooling makes it possible to cool the steel below the "pearlite start" temperature before coiling. To explain this advantage, if the steel is coiled above the pearlite start temperature, ferrite grains can grow at the expense of remaining austenite. In this steel, high coiling temperatures also lead to a loss in precipitation strengthening due to overaging of vanadium nitride precipitates.

Grain refinement also imparts good impact toughness. As Fig. 1 reveals, the impact energy (determined with half-sized longitudinal Charpy specimens) ranges from 5 ft-lb at −80 F to 50 ft-lb at +80 F.

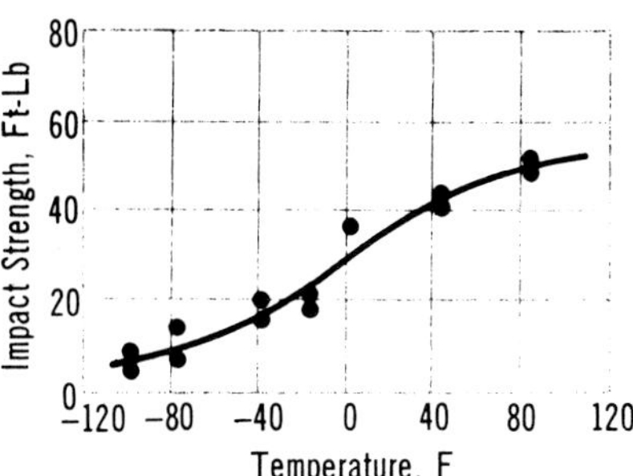

Fig. 1 — Because of its fine-grained structure, VAN-80 has excellent impact energy down to subzero temperatures.

The fatigue limit of a steel depends on its inherent strength and surface condition. Plates of this hot-rolled steel have little or no decarburization at the surface, and they have excellent fatigue strength — typically 44,000 psi at 10^7 cycles. As determined on specimens with polished surfaces, this steel has a reverse bending fatigue limit of 58,000 psi.

Because the steel is produced with a low carbon content (typically 0.15%), it exhibits excellent weldability. Joint efficiency of 100% is easily obtained when the electrode has strength equivalent to that of E100XX. Its low carbon equivalent (C.E. = %C + %Mn/6 + %Ni/20 + %Cr/10 − %V/10 = 0.31) makes the steel highly resistant to hot or cold cracking. Further, its low alloy content assures that hard, brittle microconstituents will not form in the heat-affected zone. Minimum hardness of a typical heat-affected zone in a multiple-pass weld is Rb 89, only 4 Rb points below that of the base metal (Rb 93).

Formability is a most important feature of a high-strength steel. Fabricators often make transverse bends with brake presses or roll-forming equipment. Since sulfur content of this steel is held to 0.025% max, it is possible to bend VAN-80 along or across its rolling direction with the bend radius as little as 2½T. Power requirements and elastic springback are equivalent to those obtained for a heat-treated steel of the same strength. Figure 2 illustrates a typical example, a truck frame crossmember.

Fig. 2 — This 33 in. crossmember for a truck frame demonstrates formability of the new grade.

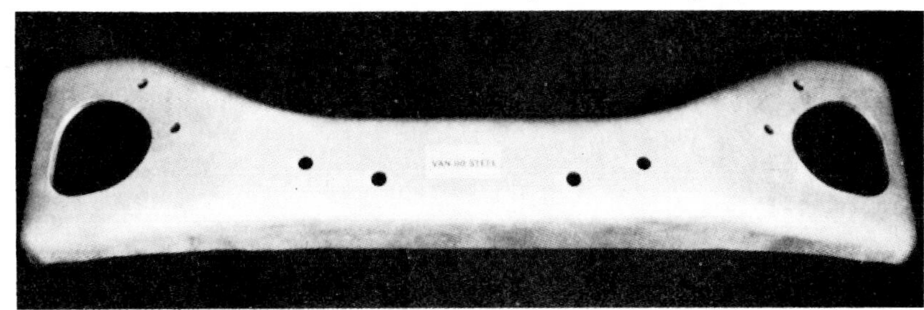

Plate Steels With Guaranteed Impact Properties

By RONALD M. JAMIESON

ADVANCES IN WELDING technology coupled with the availability of many HSLA steels have encouraged engineers to design larger structures. Since these structures often consist of welded plates which are required to resist dynamic loads at low temperatures, fabricators are demanding more notch toughness in such steels. Recognition of this need for constructional steels whose impact properties would be assured by the steel manufacturer encouraged engineers of this organization to explore and exploit controlled rolling.

Dr. Jamieson is senior research engineer, Stelco Research Center, Steel Co. of Canada, Hamilton, Ont.

As a result of this work, the company is prepared to guarantee the impact properties of certain grades of as-rolled plates. In fact, such alloys are now in service; Stelcoloy 50 (formerly called Stelcoloy S), in thicknesses of ½, ⅝ and 2¼ in., is used in built-up box sections in the tower legs of the Quebec City suspension bridge. This grade, incidentally, contains 0.15 C, 0.80 to 1.35 Mn, 0.15 to 0.30 Si, 0.20 to 0.50 Cu, 0.25 to 0.50 Ni, 0.30 to 0.50 Cr, and 0.01 to 0.10 V.

Applying Controlled Rolling

Controlled rolling is not new; European mills first used it some ten years ago. Recently, however, it has become economically justifiable to use the rolling mill to enhance the properties of steel in addition to its prime function, producing the required geometrical shape.

It is well established that ferrite grain refinement increases tensile properties (particularly yield strength) and lowers transition temperature, raising notch toughness. Thus, both strength and toughness are influenced in the desired manner. Any rolling technique which promotes grain refinement can either lower the impact transition temperature at a given strength level, or minimize the deterioration of notch toughness (increase in transition temperature) which occurs when strengths are raised by adding alloying elements.

During controlled rolling, grain refinement is achieved by performing predetermined amounts of reduction at or below certain specified temperatures. At the Stelco plant, initial trials were conducted on high-strength skelp for the manufacture of large-diameter pipe for gas transmission lines. This experience was subsequently extended to certain constructional plate grades.

Regarding guaranteed impact properties, plates of Stelcoloy 50 up to 2½ in. thick will have 15 ft-lb min average energy (Charpy V-notch, longitudinal) at +32, +20, 0, −10, and −20 F. Stelco-Vanadium offers the same properties in plates up to 1½ in. thick. Controlled schedules are used to process the steel. To meet these impact levels as the test temperature decreases from +32 F to −20 F, the finishing temperature is progressively lowered while the final reduction in the mill is increased.

ASTM A517-Grade J Maximizes Properties

I-Strength With Flatness

By GEORGE F. MELLOY

PLATES PRODUCED to ASTM A517 in thicknesses up to 1¼ in. must have a minimum yield strength of 100,000 psi and a range in tensile strength from 115,000 to 135,000 psi. Other than strength, factors of practical importance for these steels are notch toughness, weldability, formability, uniformity and consistency of properties, surface condition, and flatness.

Developed at Bethlehem's Homer Research Laboratories, RQ100A (ASTM A517-Grade J) was designed to optimize the maximum number of such properties. Containing types and amounts of alloying elements which contribute most to the desired properties, it is produced by a combination of advanced plate rolling and heat-treatment practices. Composition range: 0.12 to 0.21 C, 0.45 to 0.70 Mn, 0.20 to 0.35 Si, 0.50 to 0.65 Mo, and 0.001 to 0.005 B.

As a matter of interest, Grade J has less total alloy content (Fig. 1) than other A517 grades, a feature which minimizes problems associated with steelmaking. Further, it has a narrower range in composition, resulting in more uniform properties.

By modifying the heat treatment, we can use the same composition as an abrasion-resisting steel; it is designated as RQ321A (321 represents the minimum Brinell hardness). Because tensile strength is about 160,000 psi, forming and welding are more difficult than they are in material at the 100,000 psi yield strength level.

Making the Grade

Plates of this alloy are made by coordinating rolling and heat treating. The heated slab first passes through a two-high reversing mill, where fluted rolls, plus high-pressure steam, effectively break up and remove scale. From the roughing stand, the semifinished plate, free of scale, enters the four-high finishing stand. After being rolled to the desired thickness, the plate is leveled to obtain optimum flatness. (Removal of scale, uniformity of thickness, and over-all flatness are all prerequisites for optimum response to quenching.)

Optimum quenching efficiency is guaranteed by the use of a continuous roller quench unit. Described in "Quenching Steel Plate on the Fly," (*Metal Progress*, March 1967, p. 131), this system subjects the entire plate to maximum water quenching of uniform severity as it passes through restraining rollers. Balanced quenching severity and roller action maintain the original flatness; Fig. 2

Fig. 1 — Grade J (RQ100A) has a much lower alloy content than other A517 grades. Bars show ranges for all elements but carbon.

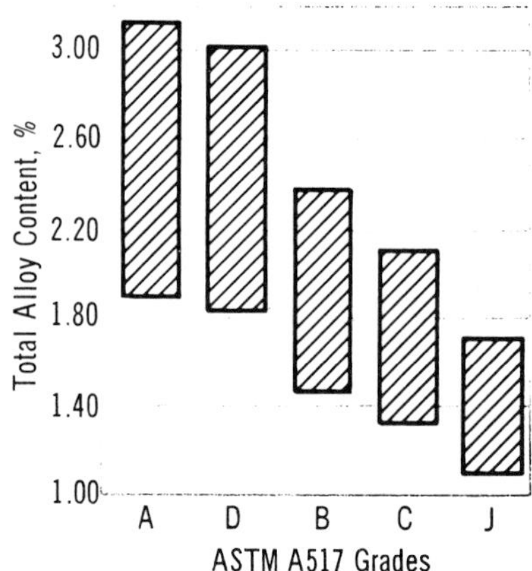

Mr. Melloy is section manager, Alloy Development Section, Homer Research Laboratories, Bethlehem Steel Corp., Bethlehem, Pa.

shows flat plate leaving the unit.

Because this plate has high mechanical properties of unusual uniformity, freedom from scale, and good flatness, it is suited to demanding applications. A 45-ton-capacity off-the-road dump truck represents a severe service use. Although individual rocks are usually less than a cubic yard in volume, the side and floor plates are subjected to severe impact when the truck is loaded. Plates ½ and ⅜ in. thick were used for side members where strength, notch toughness, and weldability are mandatory. Floor plates are of ¾ in. thick RQ321A for abrasion resistance.

Fig. 2 — After being rolled and leveled, plates of RQ100A (ASTM A517-Grade J) are quenched under pressure to assure optimum flatness. Here, flat plate emerges from the quenching unit.

Welding is simplified because the flat plates fit together easily, and the same composition is used for both parts of the body. Most of the welding was done at 250 amp by the gas metal-arc method, the shielding gas being argon and 5% O_2. Electrodes were 0.045 in. diameter Airco A632 type, with some manual welding with E10016 electrodes being done also. Where the high-strength RQ100A plate was joined to the supporting framework of lower-strength (about 50,000 psi yield strength) steel, gas metal-arc welding was also employed using 0.045 in. diameter AWS Class E60S-3 electrodes. When joining was complete, welds in critical areas were ground to smooth surfaces, providing optimum resistance to fatigue under the demanding service conditions.

ASTM A517-Grade J Maximizes Properties

II - Good Resistance to Fatigue

By EMIL G. SIGNES, PAUL R. SLIMMON, and GEORGE F. MELLOY

BECAUSE QUENCHED AND TEMPERED constructional steels are generally used where weight reduction is important, designers keep sizes of load-bearing members as low as practicable. The result is higher operating stresses and more flexible construction, making the steel's fatigue behavior of increased concern. As a consequence, inadeqate fatigue data or lack of knowledge about service conditions can lead to underdesign or overdesign.

Mr. Signes is engineer, Mr. Slimmon is supervisor, Welding Group, and Mr. Melloy is section manager, Alloy Development Section, Homer Research Laboratories, Bethlehem Steel Corp., Bethlehem, Pa.

Though extensive field tests are the most reliable way to establish material behavior, they are frequently impractical. Fortunately, laboratory tests serve adequately as a basis for design. While S-N curves are suitable for presenting basic fatigue data, the constant-life diagram is of more use to the designer. Such diagrams present information for all conditions of stress—from complete reversal to varying tensile loads. Further, constant-life diagrams provide data on specimens which represent the material as it will be used.

Six Grades Compared

ASTM A517 steels have a minimum yield strength of 100,000 psi. Because of their popularity, we

Table I — Composition Ranges of Several ASTM A517 Grades

Grade	C	Mn	Si	Cr	Mo	B	Other
B	0.15 to 0.21	0.70 to 1.00	0.20 to 0.35	0.40 to 0.65	0.15 to 0.25	0.0005 to 0.0050	0.01 to 0.03 Ti, 0.03 to 0.08 V
D	0.13 to 0.20	0.40 to 0.70	0.20 to 0.35	0.85 to 1.20	0.15 to 0.25	0.0015 to 0.0050	0.04 to 0.10 Ti, 0.20 to 0.40 Cu
E	0.12 to 0.20	0.40 to 0.70	0.20 to 0.35	1.40 to 2.00	0.40 to 0.60	0.0015 to 0.0050	0.04 to 0.10 Ti, 0.20 to 0.40 Cu
F	0.10 to 0.20	0.60 to 1.00	0.15 to 0.35	0.40 to 0.65	0.40 to 0.60	0.002 to 0.006	0.70 to 1.00 Ni, 0.03 to 0.08 V, 0.15 to 0.50 Cu
G	0.15 to 0.21	0.80 to 1.10	0.50 to 0.90	0.50 to 0.90	0.40 to 0.60	0.0025 max	0.05 to 0.15 Zr
J	0.12 to 0.21	0.45 to 0.70	0.20 to 0.35	—	0.50 to 0.65	0.001 to 0.005	—

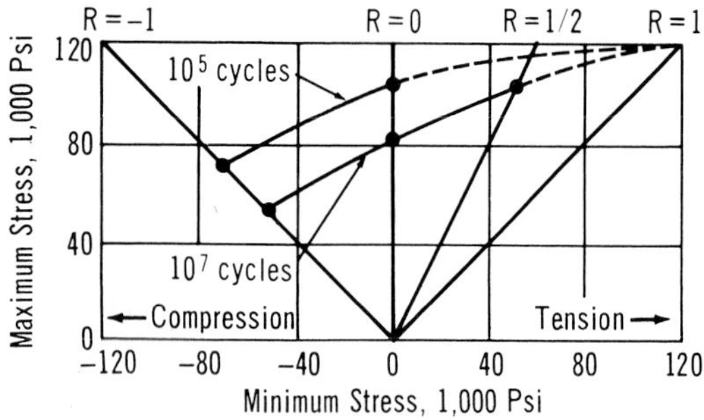

Fig. 1 — This constant-life diagram shows that round specimens of ASTM A517-Grade J (average tensile strength: 120,000 psi) lasted 10^7 cycles when cycled between 50,000 psi tension and 50,000 psi compression. Given the same number of cycles, the maximum stress for the zero-load-to-tension test (R=0) was 80,000 psi; and for the half-tension-to-full-tension test (R=½), 100,000 psi.

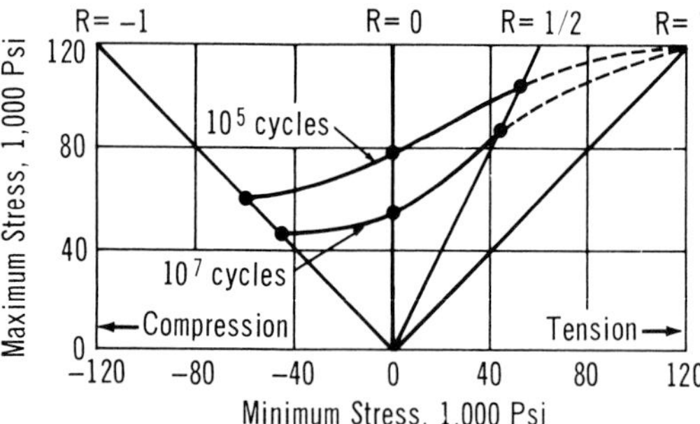

Fig. 2 — Compared with unwelded specimens (Fig. 1), welded ones of ASTM A517-Grade J had slightly lower fatigue resistance at the three stress ratios.

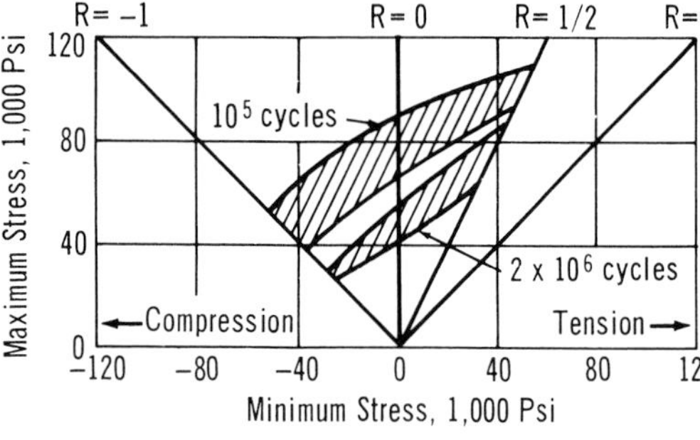

Fig. 3 — Flat specimens representing heat-treated plates of six ASTM A517 grades displayed lower fatigue strengths than round, smoothed specimens (Fig. 1). However, strengths remained adequate, especially for stress ratios of 0 and ½.

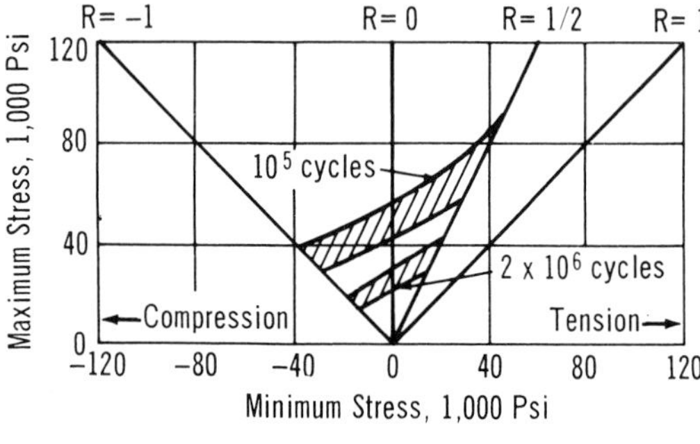

Fig. 4 — As-welded flat specimens of the six ASTM A517 grades had comparatively low fatigue strengths, largely because of the rough surfaces of the reinforcements. Compare with Fig. 3 and 5.

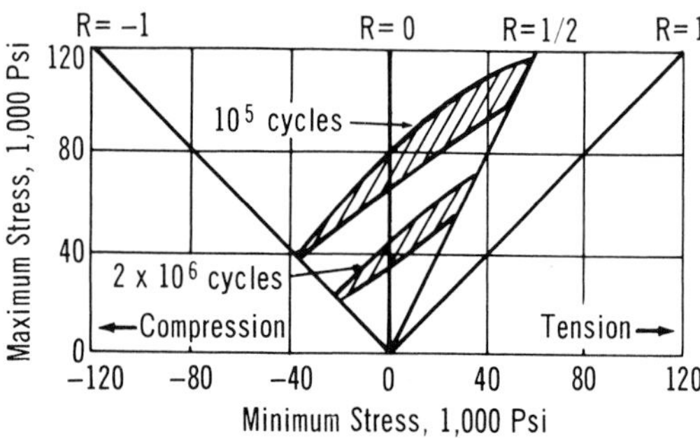

Fig. 5 — When reinforcements were removed (from flat welded specimens of the six grades), fatigue strengths rose appreciably. Compare with Fig. 4.

tested A517-Grade J (RQ100A) under several conditions and compared our results with those published for other grades. Table I lists composition ranges.

In all instances, we worked with specimens machined from 1 in. thick plates. Two types of specimens, 0.438 in. in diameter rounds and full-thickness flat plates, were tested.

Unwelded and welded round specimens were tested by the submerged-arc process with Linde 100 submerged-arc wire and a neutral flux, Linde 709-5. A heat input of 60,000 joules per inch at a preheat and interpass temperature of 200 F was chosen as representing the maximum heat input recommended for the grade. For unwelded and welded specimens, surface finish was about 15 rms.

We employed a 22,000 lb Amsler Vibrophore operating at 11,200 cycles per minute to develop constant-life diagrams for fatigue lives of 10^5 and 10^7 cycles. Stress ratios corresponded to loading conditions of full compression to full tension (R=—1); zero load to tension (R=zero); and half tension to full tension (R=½).

As Fig. 1 reveals, unwelded specimens had good fatigue strength even under the worst testing condition (R=—1). Figure 2 shows that welding slightly lowered the fatigue strength at all stress ratios.

Results With Flat Specimens

Because materials employed in commercial structures usually do not have smooth finishes, we also tested flat full-thickness specimens with: 1. As-heat-treated surfaces. 2. Welds that had intact reinforcements. 3. Welds with reinforcements removed. Welding was as described previously.

Machined sides of all specimens had a surface roughness of 25 to 30 rms; where the reinforcement was removed, surface roughness was 40 to 50 rms. Employing such specimens, we determined constant life diagrams (for fatigue lives of 10^5 and 2×10^6 cycles), with a 220,000 lb Amsler alternating-stress machine operating at 500 cycles per minute.

Figure 3 illustrates a constant-life diagram representing data for RQ100A and five other A517 grades tested in the unwelded condition with as-heat-treated surfaces. Comparison with Fig. 1 indicates that the relatively rough surface lowered fatigue strength. However, reasonably high fatigue strengths can still be realized, particularly at stress ratios (R) of 0 and ½.

In Fig. 4, the constant-life diagram for grades tested with the weld reinforcement intact, we see that fatigue strength is appreciably lowered, particularly when R = —1. However, as illustrated by Fig. 5, removing reinforcement restores fatigue strengths to nearly those of the unwelded material.

This information emphasizes that the designer must have knowledge of service conditions if he is to gain the maximum potential from these steels. Designers working with quenched and tempered steels of the ASTM A517 class should know that:

- Nominal compositional variations have a minor influence on fatigue behavior for the materials tested.
- In designs that call for dead loads which provide a substantial steady stress upon which smaller fluctuations of loads are imposed, higher-strength materials offer appreciable advantages.
- In structures which undergo complete reversals of stress, it is necessary to remove the weld reinforcement to improve fatigue strength.
- Surface imperfections and welding diminish the advantages of quenched and tempered steels, but desirable properties can be largely recovered by localized surface improvement. It is also useful to relocate the welded joint to an area of lower stress or apply residual compressive stresses at the surface.

A 75,000 Psi Heat-Treated Grade

By ALBERT HOERSCH JR.

Among the newer low-carbon, high-manganese structural steels is our LT-75HS, a heat-treated grade. Made to fine grain practice, it contains 0.22 C max, 1.10 to 1.60 Mn, 0.035 P max, 0.04 S max, 0.20 to 0.60 Si, and 0.20 Cu min (when specified). Plates over 1½ in. thick may be boron-treated too. This plate steel also has nickel, copper, chromium, and molybdenum in small amounts which are controlled so that heat treatment will develop the required strength. Heat treatment consists of water quenching from 1,650 F, followed by tempering from 1,050 to 1,175 F, depending upon individual melt composition and plate thickness. Properties are shown below:

	¼ to ¾ in. incl.	Over ¾ to 1½ in. incl.	Over 1½ to 2 in. incl.
Yield strength, psi	75,000 min	70,000 min	65,000 min
Tensile strength, psi	95-115,000	90-110,000	85-105,000
Elongation, %	19	19	19

Typical Uses

Large tonnages of this grade in most thicknesses have been welded by fabricators. Principally used for construction equipment (such as crane and shovel booms, and earthmoving machinery), the steel has proved easy to weld when basic rules are observed.

Mr. Hoersch is technical service engineer, Lukens Steel Co., Coatesville, Pa.

This steel is useful in a wide range of stationary structures and mobile equipment which has to perform in subzero temperatures, or under extreme stress concentrations, or both. Because its minimum yield strength, 75,000 psi, fills the gap between the 50,000 and 100,000 psi yield strength steels, this grade can be applied where full-alloy strengths don't afford enough rigidity, and the lower-strength steels call for increased thicknesses (and thus extra weight) in plates.

For example, a switch to LT-75HS enabled a power shovel manufacturer to make a definite cost saving in material for a dragline bucket. The digging part of a giant excavator, the welded bucket has LT-75HS plate in the sides, bottom and back in thicknesses ranging from ½ to 1½ in.

Engineers were able to maintain the same plate sections needed for a 100,000 psi yield strength steel they had been using. Before, they could utilize only about 70,000 psi of the available yield strength. To explain, full use of the alloy steel's strength would have reduced plate sections enough to result in high deflection under service loads. Since the modulus of elasticity is much the same for all steels, plates had to be thick enough to avoid deflection.

Low-temperature notch toughness also figured into the steel's selection because the power shovel components frequently had to withstand large stresses—such as those imposed by thousands of pounds of pull on the dipper—in freezing temperatures. With LT-75HS, the power shovel builder employed a minimum preheat of 75 F for torch-burning and 100 F (for plates up to 1½ in. thick) for welding. Electrodes were E9018 or semiautomatic McKay 105.

Essentially, the steel's microstructure consists of bainite, plus ferrite. In thinner sections, it is partially martensitic; greater thicknesses contain fine pearlite. Impact properties are good, the minimum being 15 ft-lb at —75 F for all thicknesses.

The material is readily cold formed in rolls or press brakes. Because its yield strength is 2 to 2½ times that of mild steel, power requirements and springback will be proportionately greater.

For cold forming, it is preferable that the bend axis be perpendicular to the major rolling direction. More severe bends can be made by interstage stress relieving 50 F below the tempering temperature (when tempering temperature is unknown, use 1,000 F). If practicable, forming between 700 and 1,000 F (or 50 F below tempering temperature) will reduce power required for forming, and increase the steel's ductility by 15 to 25%. Forming between 300 and 700 F is not advisable.

Results of Welding Tests

Because this steel derives its properties from heat treatment, it is logical to question the effect of welding. Tempering occurs in the heat-affected zone, but if welding energy inputs are reasonable, softening is minor and does not affect joint strength. Likewise, acceptable levels of toughness are maintained in the heat-affected zone by controlling the energy.

Weldability has been evaluated in terms of crack propensity, underbead crack tests, cruciform tests, and highly restrained vertical butt welds. In the underbead crack test, specimens welded with E9018M (low-hydrogen type) electrodes showed no evidence of any cracking, but those welded with E6010 (cellulosic type) electrode contained cracks averaging from 40 to 69% of the weld length. Obviously, low-hydrogen-type electrodes should be used.

Cruciform tests showed that heat-affected zones and weld metal are not apt to crack (even though restrained, and if transformation stresses and shrinkage stresses are high), provided that low-hydrogen-type electrodes are used.

Preheat temperatures suggested for use with coated low-hydrogen electrodes are 50 F to ½ in. thickness, 100 F in the ½ to 1 in. range, 150 F in the 1 to 1½ in. range, and 200 F in the 1½ to 2 in. range. Higher preheats may be necessary for highly restrained, complex structures; tee and corner joints; and inadequately dried filler materials and chamfers. Lower preheats may be tolerable for automatic welds, inert-gas metal-arc welds, and austenitic welds. Coated electrodes, submerged-arc fluxes, welding wires and chamfers should be kept clean and dry. Coated electrodes commonly used are AWS E9018M and 10018M.

This plate steel may also be welded by the carbon-dioxide-shielded arc process, using AWS E70T-5 electrodes for fillets and proprietary Mn-Ni-Mo types for full-strength butt joints.

Sheets Compete With Other Products

I-Low-Carbon Martensitic Grades

By WILLIAM H. McFARLAND
and HAROLD L. TAYLOR

LOW-CARBON (UNDER 0.20%) martensitic sheet steels, called MartINsite steels, develop 225,000 psi tensile strength. Typical values range from 130,000 (0.08% C) to 225,000 psi (0.15% C). The three grades also contain 0.45 Mn, 0.012 P max, 0.25 S max, and 0.01 Si max. Due to the high temperature at which martensite starts to form, these steels temper during the quench.

Because of its high strength-to-weight ratio and high modulus of elasticity, this sheet material can be considered for applications now served by metals such as aluminum, titanium, and stainless and maraging steels, or nonmetals such as wood, plastics and fiber glass, says the developer, Inland Steel Co. Also, it adds, fewer processing steps are required to produce a finished part, and finished parts have good uniformity of strength.

Potential uses include energy-absorbing crash bars, welded tubing, fasteners and small spring-type parts, and corrugated panels. In each of these categories, development work has been done to determine fabricating, manufacturing, and fastening techniques. And properties of the fabricated components have been evaluated.

Mr. McFarland is staff scientist and Mr. Taylor is associate manager, Research Dept., Inland Steel Co., Chicago.

This material is supplied in tensile strength grades of 130,000 minimum to 190,000 psi minimum (in 30,000 psi increments). It has a density of 0.283 lb per cu in., a modulus of elasticity of 30,000,000 psi, a minimum yield strength which is 95% of the tensile strength, a 1.5% minimum elongation in 2 in., a fatigue limit which is 35 to 50% of the tensile strength, and a tensile strength-to-density ratio of 520,000 to 800,000 in. Weldability is reported to be excellent, and formability is superior to that of full-hard cold rolled sheets, says Inland.

When applying this steel, the materials engineer must consider its forming characteristics because it exhibits considerable springback. However, it can be sheared with equipment used for ordinary cold-rolled steel. Naturally, since the material is supplied in the fully quenched condition, parts manufactured from it are already at full strength. Where extreme forming in localized areas is required or where several forming operations are to be performed, the sheet can be heated to improve formability. If the heated area is softened (as by a full anneal), the strength will revert to that of cold-rolled steel.

Some Appropriate Uses

Strip is being processed into continuous-seam welded tubing with radio-frequency welding equipment. The tubing is produced on conventional roll-forming equipment.

To determine relative strengths, tests have been conducted comparing this tubing to cold-rolled welded types of equal diameter and wall thickness. Researchers have made axial tension, compression, torsion, internal pressure, and circumferential force tests. In axial tension, compression, and torsion, the tubes have load-carrying capability which is directly proportional to the tensile strength of the parent sheet. Consequently, weight saving or increase in load-carrying capacity is possible.

To make fasteners and small spring-type parts, only a simple processing sequence is needed. The manufacturing sequence merely calls for blanking and forming. Finished parts have improved shape, increased uniformity of properties, and excellent fatigue resistance (due to the high fatigue limit of the martensite). Where additional corrosion protection is needed, the strip can be supplied as electrocoated with zinc so that final plating can be eliminated where cut-edge corrosion is not critical.

Corrugated panels have a strength-to-weight ratio which is attractive for such applications as air-cargo containers, truck panels, and railroad-car roofing and siding panels.

Sheets with tensile strengths of 200,000 psi in thicknesses of 0.005 to 0.125 in. can be used to achieve panel weights of 0.7 to 1.0 lb per sq ft of surface. For an 8 by 8 by 20 ft container, these densities would give a total skin weight of approximately 425 to 600 lb, or a weight-to-volume ratio of 0.22 to 0.47 lb per cu ft. Assuming an allowable container weight-to-volume ratio of 3.0 lb per cu ft, the panel provides an ample margin for the weight of the structural frame, floor, and miscellaneous hardware.

Sheets Compete With Other Products

II-Parts Strain Age During Paint Curing

By BERNARD S. LEVY

INLAND'S AA GRADE steel develops its final yield strength of 70,000 psi after fabrication. The feature is attractive because a large proportion of steel sheet is fabricated by cold working and painting. This steel is cold worked while it is formable. Strain aging takes place during paint curing (200 to 500 F for 7½ to 120 min) to strengthen the part. Thus, high yield strengths are reached by a combination of higher-than-usual initial strength, maximum work-hardening properties, and strain aging.

The grade is approved for the seat track rail in one 1970 model auto, and it will be used in a number of parts making up side reinforcing beam assemblies for several other 1970 models. Other potential applications (including brackets, wheel rims, reinforcement beams and columns, and guides) are being evaluated by automotive and appliance companies. The sheet is also being used for base plates of office swivel chairs.

Properties of the Steel

Essentially, the grade is AISI 1008 plus silicon and nitrogen. It can be cold-rolled in all gages, but properties of hot-rolled sheet in gages thicker than 0.150 in. have not yet been investigated. Typical as-delivered tensile properties are:

	Hot-Rolled	Cold-Rolled
Yield strength, psi	50,000	48,000
Tensile strength, psi	63,000	64,000
Elongation in 2 in., %	29.0	26.0

The steel has a formability in stretching operations that is somewhat less than that for the best commercial-quality sheet, judging by properties and experience on a variety of automotive stampings and roll-formed parts. Thus, it is suitable for average, commercial-quality applications where stretchability is the primary forming mode. The steel also has adequate drawability, as indicated by r values (approximately 0.9 for hot-rolled sheets, and between 0.9 and 1.2 for cold-rolled sheets) which class it with rimmed steel.

In a free bend, the sheet will bend flat on itself in any orientation to the rolling direction. (Exception: hot-rolled AA steel requires a 2 to 3t radius when the bending moment is transverse to the rolling direction.) Most commercial practices, of course, do not involve free bending due to die restraint. Because free bends probably represent the worst condition, some improvement in minimum radius might be expected if a substantial stretching component is experienced during bending. Finally, though AA steel has hole expansion characteristics about equal to those of commercial-quality rimmed steel, special changes must be made during its processing if difficulties are expected.

Whatever formability factors are involved, the forming operation can be considered as prestraining, and the part loading, as re-straining. Because commercial parts are not uniformly prestrained, they commonly have nonuniform strain distributions. The relation between re-strain strength and the amount of prestrain is also often complicated by differences in the direction of prestraining and re-straining, with the possibility of consequent Bauschinger-type effects. In most instances, the strength increase cannot be calculated; the actual component must be tested. When this is done, the trial part must be made much like the production part to produce similar strain distributions.

The strength increase that results during a paint curing treatment exhibits remarkably little variation. As long as the AA steel has been strained beyond the Lüders region, the added strength equals approximately 9,000 psi. When the 9,000 psi increase from aging is combined with the strength increase by a 10% prestrain (approximately 15,000 to 17,500 psi), a total increase of 24,000 to 26,500 psi can be expected after prestraining and aging.

It should be noted that the aged sheet has satisfactory ductility. Roughly, the reduction in area is the same as in a virgin tensile test, and total elongations (including prestrain) are 20 to 26%. Limited data also indicate that a fatigue-tensile strength ratio of about 0.55 is applicable. Thus, AA steel appears to have good long-life fatigue properties.

Laboratory work has shown that this steel can easily be joined by resistance spot welding, CO_2-shielded, bare wire, and consumable coated electrode methods. Welding techniques are similar to those for AISI 1008 steel, but higher-strength filler metal should be used for work with consumable electrodes. Resistance spot welds have been made under routine production assembly conditions. One such subassembly has been successfully tested on automotive proving grounds. In it, spot welds were as strong as the base steel. Laboratory studies indicate that welding can precede or follow aging. Similar studies are underway to evaluate the strength characteristics of CO_2-shielded, bare wire welds, the consumable coated electrode welds, and inert-gas tungsten-arc spot welds.

Mr. Levy is senior research engineer, Research Dept., Inland Steel Co., Chicago.

An HSLA Steel for Heavy Sections

By ROBERT L. HICKEY
and EDWARD V. BRAVENEC

OVERCOMING ONE PROBLEM of HSLA steels, limited thicknesses, Armco researchers have developed a new normalized steel. Designated Armco VNT, it contains vanadium and nitrogen for extra strength. Fully killed and made to fine-grain practice, the steel provides 60,000 psi minimum yield strength in thicknesses through 3¼ in. Further, there is ample indication that 60,000 psi minimum yield strength can be developed in thicknesses possibly as great as 6 in.

Armco VNT possesses a toughness approaching that of commercially produced quenched and tempered carbon steel with 60,000 psi yield strength. To date, the steel is the only grade of its type available with minimum Charpy V-notch toughness of 20 ft-lb (longitudinal) and 15 ft-lb (transverse) at —60 F through 3½ in. Thus, the grade is a natural for heavy-sectioned parts requiring notch toughness and high strength, but where distortion of parts from quenching and tempering cannot be tolerated. Its composition is 0.22 C max, 1.15 to 1.50 Mn, 0.04 P max, 0.05 S max, 0.30 Si max, 0.04 to 0.11 V, and 0.01 to 0.02 N.

Mr. Hickey is senior research metallurgist and Mr. Bravenec is senior metallurgical engineer, Armco Steel Corp., Middletown, Ohio.

Development of the grade was spurred by a need for a steel that would meet 60,000 psi yield strength in heavy sections and provide an impact strength of at least 20 ft-lb (Charpy V-notch) at —20 F for fittings and surge tanks to be used with large-diameter line pipe. Fabrication procedures involved torching plates to size, roll-forming shells at 300 to 400 F, welding seams, cold-drawing nozzles, and normalizing at 1,650 F. One important need, retention of strength during normalizing, was achieved by the precipitation of vanadium nitride particles in the structure.

Results Demonstrate Merit

Data representing eight production heats show that yield strength is over 65,000 psi in thicknesses up to 1¼ in. From 1¼ to 3¼ in. thicknesses, the yield strength drops to about 60,000 psi, and holds there for greater thicknesses. Tensile strengths are over 80,000 psi, and transition temperatures (determined with Charpy V-notch specimens) range from —75 to less than —100 F for longitudinal tests, and —60 F to less than —100 F for transverse tests.

Another feature of this steel is resistance to grain coarsening. For example, tests showed an ASTM grain size of 9.5 (as-normalized), and 7.0 after being heated at 2,200 F for 8 hr.

The alloy can be readily welded by any of the low-hydrogen processes. We recommend the E80XX electrodes, Classes C1 and C2 containing 2% and 3% Ni, for shielded metal-arc welding because their notch toughnesses match that of Armco VNT. Also available are a variety of electrode wires, fluxes and gases for submerged-arc and gas metal-arc welding.

High-Strength Grade Modified for Special Use

By CLIFFORD E. ROWDEN

PLATES OF X-70-W, a hot-rolled grade rated at 70,000 psi min yield strength, are currently being produced in Republic Steel's Southern District Plant at Gadsden, Ala.

General composition specifications for this grade (produced in gages up to ½ in. maximum) are: 0.26 C max, 1.65 Mn max, 0.040 P max, 0.050 S max, 0.30 Si max, and 0.01 Cb min and/or V. Yield point, as noted, is 70,000 psi min, tensile strength is 85,000 psi min, and elongation is 14% min.

Sometimes, the composition must be modified slightly to develop special properties. To demonstrate this, we will consider the fabrication of drive housings for trucks made by a manufacturer of heavy earth-moving equipment. Working as a team with the manufacturer, Republic metallurgists determined the steel attributes required for the housing.

Mr. Rowden is process metallurgist, Republic Steel Corp., Cleveland.

For example, the steel needed a minimum yield point of 65,000 psi in the as-rolled ½ in. plate, enough ductility to permit moderate forming, and an impact value (Charpy V-notch) of 15 ft-lb at —20 F because of expected service conditions. To cut fabrication costs, the development team also established the following specification:

1. Good weldability with a maximum carbon equivalent of 0.57%, $(\%C + \frac{\%Mn}{4})$.
2. No preheat or postheat for welding.
3. No special preparation for flame cutting.
4. Fine-grained structure in both the austenitic and as-rolled state.

All of the features required for drive housings could be satisfied by modifying the general composition to be consistent with the required carbon equivalent while assuring an adequate safety factor over the designed 65,000 psi minimum yield point requirement.

Features of the Mill

We should note that the Gadsden plant has electric and basic oxygen steelmaking facilities, and a recently installed 134 in. plate mill which is equipped with computer control. A great deal of the steel's success can be attributed to the development of a deoxidation practice which results in completely deoxidized ingots. These are slabbed on a 48 in. mill to the proper width, thickness, and length which is required to permit the predetermined reductions and directional rolling sequences on the plate mill.

At the plate mill, the computer control system is used to control those properties of X-70-W which are influenced by the degree and temperature at which mechanical deformation is performed. The rolling program includes steps to maintain the degree of directionality by crossrolling, reductions, the final pass reduction, and temperature at final reduction, as well as steel temperature before leveling. Because these program controls are vital in maintaining consistent mechanical properties and formability characteristics, they are a major contributor to the success of the product.

Titanium-Strengthened Grades

By JAMES H. SMITH
and JOHN J. VETT

CLASSICALLY, HIGH-STRENGTH, hot-rolled steels have utilized four principal mechanisms for strengthening the metallurgical structure. Solid-solution strengthening has been effected by alloying element additions such as phosphorus, chromium, and silicon. Precipitation-strengthened grades have been developed which utilize either a single alloying element precipitate such as copper or compound precipitates formed from the reaction of solid-solution alloying elements such as columbium or vanadium with carbon or nitrogen. Two additional strengthening mechanisms have become commercially significant with the installation of highly sophisticated process controls on modern hot rolling mills which have enabled the kinetics of deformation and transformation to be controlled more closely during production. The selection of the proper steel composition has permitted the production of grades which are strengthened by tailoring the microstructure to obtain the desired proportion of austenite decomposition products. In addition, these new process controls have led to the production of steels strengthened by a fourth mechanism, grain refinement. A finer transformed grain size permits the attainment of a given strength level with smaller alloy additions. Recently higher strength grades have been created by combining two or more of these classical strengthening mechanisms to produce hot-rolled steels with strengths approaching the 100,000 psi yield strengths of quench and tempered grades.

Strength is not the sole criterion for evaluation of high-strength, hot-rolled steels. Other characteristics such as toughness, weldability, formability, fatigue strength, and corrosion resistance can influence the material selection. For several years, our research staff has been concentrating on the formability aspect of high-strength steels to develop a product which could be formed with fabrication techniques similar to those used in mild steel. This work has culminated in the development of a new family of high-strength, hot-rolled steels with yield strengths presently covering the range of 50,000 to 85,000 psi.

Functions of Titanium

One of the most attractive means for strengthening hot-rolled steels is the addition of small amounts of titanium in combination with the proper hot rolling process control. The implications of this control can be seen from the work of Herrnstein, Straatmann, and Bosch – "Strengthening in Low-Carbon Titanium Steels," AIME Meeting, Pittsburgh, May 14, 1969 – as shown in Fig. 1 which is a continuous cooling transformation diagram for an 0.06 C, 0.02 Si, 0.30 Mn, 0.04 Al, 0.23 Ti hot-rolled steel. Hot-rolled steel processing sequences are superimposed on the diagram by the dashed lines. After the steel has been deformed at an elevated temperature such as 1,750 F it must be cooled rapidly at a rate on the order of 45 F per sec to preclude transformation at an elevated temperature. With the proper process control, the rapid cooling rate can be maintained until the "Start" of transformation to proeutectoid ferrite has commenced at a temperature slightly above 1,200 F. Subsequently, the cooling rate must be decreased to about 2 F per sec until the transformation to ferrite has been completed. This process sequence strengthens the steel by a combination of the titanium carbonitride precipitation during the phase transformation and a ferrite grain size refinement to produce a steel with a 105,000 psi yield strength. If precipitation is allowed to initiate in the austenite prior to the transformation to ferrite the strengthening benefit of titanium is diminished, as indicated by the upper dashed curve where the resul-

Fig. 1—In a 0.06% C steel containing 0.23% Ti, the continuous cooling process control following hot rolling can produce yield strengths of 105,000 psi.

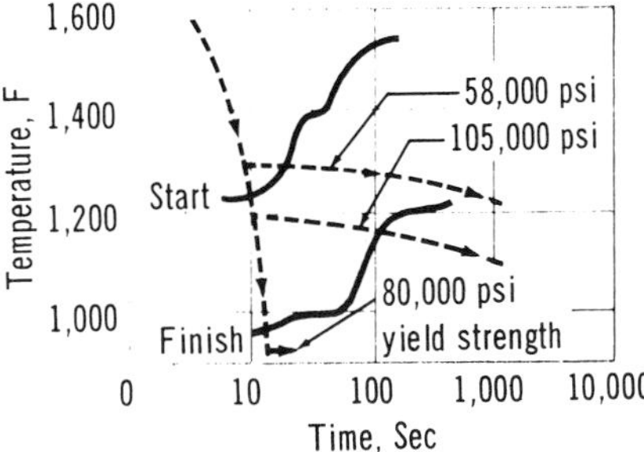

Dr. Smith is manager, Research Planning, and Mr. Vett is manager, Product Development, Youngstown Sheet & Tube Co., Youngstown, Ohio.

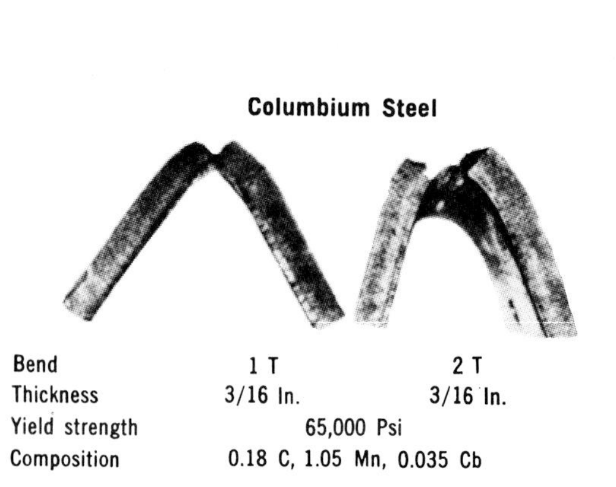

Fig. 2 — Comparison of transverse 180° cold bend test results for columbium and titanium high-strength steels.

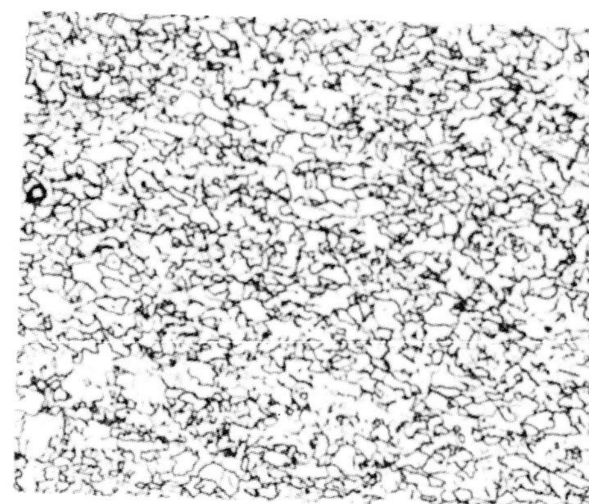
Fig. 3 — Fine-grained ferritic microstructure of high-strength steel. Etchant, Nital; 250×.

tant yield strength is 58,000 psi. Conversely, when a rapid cooling rate is maintained throughout the entire transformation temperature range, Fig. 1 shows that an intermediate strength level is developed (80,000 psi yield strength).

Thus, high strengths in a hot-rolled steel can be achieved with only small additions of titanium. Because these grades have low carbon and manganese concentrations similar to those in mild steels, they have low hardenabilities and no preheating is required for welding these steels. In fact these steels were subjected to the welding thermal cycle test described by Aronson – "The Weldability of Columbium-Bearing High-Strength, Low-Alloy Steel," *Welding Journal*, Welding Research Supplement, June 1966, p. 266 – and no loss in strength was observed. Preliminary tension-compression fatigue studies indicate that the fatigue behavior of these steels will be similar to that of other grades in the same strength range.

Steels Form Readily

However, a more important attribute of these high-strength steels is their excellent forming characteristics. For example, in the standard cold bend test a flat sample is bent around mandrels of various diameters until cracking is observed at the bend; frequently, the crack initiates along the sheared edge of the sample. A typical 50,000 psi yield strength conventional grade can be bent 180° around a mandrel diameter 1½ to 2 times the steel thickness (T), depending on the gage of the material. The new family of high-strength, hot-rolled sheet steels known as YST grades have been produced over a yield strength range of 45,000 to 85,000 psi and can be cold bent 180° flat without a mandrel using any hot-rolled sheet gage from 14 gage to ¼ in. This forming characteristic is shown in Fig. 2 where samples from a lower-strength columbium-strengthened steel are included for comparison. The ferrite grain size of these hot-rolled steels is characteristically fine as revealed in Fig. 3 which shows the grain size for the 85,000 psi yield strength grade.

Trial lots of these high-strength steels have been used successfully in truck frames and other difficult applications. Developments are now in the final stages for producing high-strength, low-alloy grades with these same superior forming characteristics which will have increased corrosion resistance similar to that now available in the present Yoloy family of steels.

Weathering Steel Favored by Architects

By SEYMOUR K. COBURN

During the past few years, weathering steels have been employed in such diverse applications as bridges, barges, railroad freight cars, pole line hardware, transmission towers, and trucks. As time goes on, more architects and engineers are specifying such grades as structural materials for building exteriors because they are relatively inexpensive and require little or no maintenance.

Two popular grades in this group are USS Cor-Ten A, a steel which maintains a 50,000 psi minimum yield point in thicknesses up to ½ in. inclusive; and USS Cor-Ten B steel, which meets the same requirement in thicknesses up to 4 in. inclusive.

Cor-Ten A contains 0.12 C max, 0.2 to 0.5 Mn, 0.07 to 0.15 P, 0.05 S max, 0.25 to 0.75 Si, 0.25 to 0.55 Cu, 0.65 Ni max, and 0.30 to 1.25 Cr. Cor-Ten B has 0.10 to 0.19 C, 0.90 to 1.25 Mn, 0.04 P max, 0.05 S max, 0.15 to 0.30 Si, 0.25 to 0.40 Cu, 0.40 to 0.65 Cr, and 0.02 to 0.10 V.

The protective film or oxide that forms when these steels are exposed to various atmospheres is relatively thin and adheres tightly to the base metal. As the steel weathers, its color changes. In a severe marine environment, for instance, the film becomes reddish brown; in an industrial environment, it will develop a gray-to-purple-tinted brown. Mois-

Mr. Coburn is associate research consultant, Applied Research Laboratory, U. S. Steel Corp., Monroeville, Pa.

ture from dew and rain stimulate formation of the protective film.

Corrosion resistance of steels in various atmospheres is determined by exposing specimens in different parts of the country. Each specimen is set at an angle of 30° facing south as prescribed by ASTM specifications, and inspected periodically. Figure 1 shows how three grades fared during 20 years of testing at Kearny, N.J. Thickness, it will be noted, is not a factor in determining corrosion resistance.

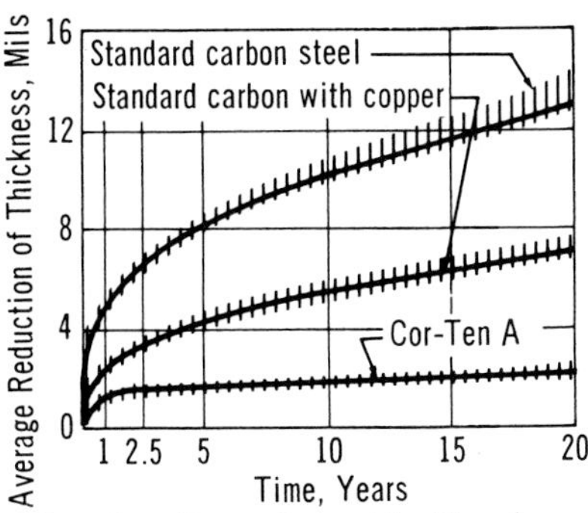

Fig. 1 — Once the oxide film develops, weathering steel resists further atmospheric corrosion.

More on High-Strength Steels

In the September 1969 issue of *Metal Progress*, William J. Murphy and Ralph B. G. Yeo will report on "Developments in Plate and Structural Steels" at U.S. Steel Corp.

They will cover such innovations as the use of columbium and other grain-refining additions, controlled rolling practices, and hardenability improvers that raise strength while maintaining weldability.

Four Typical Buildings

Eero Saarinen, the first proponent of Cor-Ten A steel in architectural applications, employed it for the Deere & Co. offices in Moline, Ill. Before the grade was selected, it was subjected to a two-year atmospheric-exposure test in a semirural wooded area adjacent to a corn field near the building site. Tests included structural sections that would involve welding, bolting, riveting, and painting. The building was opened for occupancy in April, 1964.

Sheathing for the Chicago Civic Center, a 31-story building nearly 680 ft tall, is also made of this grade. Steel exposed to the weather, which amounts to about 275,000 sq ft, includes column covers, spandrel panels, louvers, and window frames. Its companion grade, Cor-Ten B steel, was employed for three flag poles and a 50 ft sculpture based on a design created by Pablo Picasso.

The convention center in Fresno, Calif., is another interesting application. It consists of three large assembly-hall-type buildings. Two utilize 20-gage Cor-Ten A in a mansard-style roofing application, while the third has wall panels of the steel for vertical siding.

The Ford Foundation Building also employs Cor-Ten A steel for the exterior and interior faces. The faces consist of ⅛ in. thick spandrel pans that are 12 ft wide by 28 in. high by 7½ in. deep. Window frames employ the same steel. The interior court was allowed to remain open to the atmosphere for 20 months to weather the columns and spandrels. During this period, the film formed at a rapid rate, stimulated by the industrial atmosphere that characterizes New York City.

Steels for Farm and Construction Equipment

By ALEXANDER H. JOLLY, JR.

EMPLOYED IN MANY industrial applications, high-strength steels are especially useful for farm and construction equipment because of their weight-saving, shock-resisting, and load-bearing characteristics. For example, bar flats and plates of IH 50 are used for structural members of International Harvester's construction equipment. These include crawler tractors, off-highway trucks, end loaders, and scrapers where the alloy's low transition temperatures and good yield strength make it desirable. Our IH 65 steel is primarily used for pusher arms, main frames and C-frames of crawler tractors, we should also note. And high-strength steels in bar rounds and flats have been used in main sills and knife pitmans on threshers, and as hitch components on farm tractors.

Some Typical Properties

Research and development on our IH 50 and IH 65 steels have resulted in excellent welding capabilities, improved ductility, and good low temperature properties. (Based on test procedures developed by the U.S. Navy's Bureau of Ships, nil ductility transition temperatures are —60 to—100 F for IH 50 steels, and 0 to—60 F for IH 65 steel.) In addition, these grades have varying degrees of atmospheric corrosion resistance, and toughness superior to that of carbon steel.

They owe their properties to the presence of small percentages of one or more alloying elements. Though they cost more per pound, their higher yield points enable designers to reduce section sizes (compared to carbon steel), resulting in lighter, stronger components for less.

In addition to devising IH 50 and IH 65 steels, engineers of Wisconsin Steel Works have developed a series of high-strength steels designated as IH X-42 through IH X-70. Containing columbium or vanadium, these steels offer yield points of 42,000 to 70,000 psi minimum. Welding and forming characteristics are also good, and corrosion resistance can be raised by adding copper.

In all of these grades, properties vary with thickness and size. Also, mechanical properties, obtained in the as-rolled state, are influenced by various mill facilities as well as section thickness. Both IH and IH X steels can be welded by manual metal arc, submerged arc automatic, and CO_2 semiautomatic methods.

Mr. Jolly is associated with Wisconsin Steel Div., International Harvester Co., Chicago.

HSLA and Low Carbon Steel: Same Corrosion Resistance

By ROBERT J. NEVILLE

Generally, results were comparable in two years of under-car testing in southern Ontario where roads are sprayed with a salt-sand mixture (75 lb salt per ton of sand) about 60% of the time.

AUTOMAKERS are interested in substituting high-strength, low-alloy (HSLA) steels for low-carbon grades to make stronger, lighter, and potentially less expensive cars. But there is the question of durability. How do HSLA steels compare in under-car corrosion resistance? This is an important question because HSLA steels will generally be used in lighter gages.

Because under-car corrosion is particularly aggressive in the snowbelt areas of North America where de-icing salts are used, we elected to run tests in these areas. In general, these tests, begun in 1972, revealed:

- Under-car corrosion cannot be predicted by atmospheric corrosion tests.
- Most under-car corrosion occurs during the first winter, after which road deposits and oxides provide some protection.
- No known alloying element improves the under-car corrosion resistance of HSLA steels. Silicon is detrimental.
- Laboratory tests are not suitable for simulating under-car corrosion tests, which take years.

Though these results may appear to be negative, they clearly point out areas in which further work is needed.

Mr. Neville is research supervisor, Dofasco Research, Dominion Foundries & Steel Ltd., Hamilton, Ont.

First Road Test — Our first investigations were made to determine if under-car corrosion of HSLA steels was similar to their atmosphere corrosion performance, and to compare under-car corrosion of HSLA steels versus low-carbon steel.

The study involved corrosion of four commercially available HSLA steels and low-carbon steel under cars and in an industrial atmosphere. Both tests were carried out in southern Ontario (a snow-belt area) from 9 February to 29 May 1972 (110 days).

The steels (Table I) were selected either because of their known superior atmospheric corrosion resistance or potential usefulness for making car parts because of formability, strength, and cost advantages. Figure 1 shows how coupons were mounted under the car for this road test. Two privately owned cars were used, and each accumulated close to 5000 miles (8050 km) in the test period. Mileage was typically built up in driving to and from work with an occasional business or pleasure trip.

Roads in southern Ontario are sprayed with a salt-sand mixture (75 lb salt per ton [37.5 kg per Mg] of mixture) about 60% of the time. De-icing salt is used about 40% of the time in snow or freezing rain conditions.

Table II reveals that the rate of uniform corrosion of these steels is much higher under cars than in

Fig. 1 — Coupons of HSLA and mild steels were mounted under cars for road tests in the winter and spring of 1972. Corrosion rates were determined when the 110-day run was finished.

Table I — Compositions of Corrosion-Tested Steels (First Test Run)

Steel	Composition, %*											
	C	Mn	P	S	Si	Cu	Ni	Cr	Mo	Cb	Al	Other
HSLA (Si)	0.145	1.26	0.009	0.008	0.56	0.05	0.03	0.04	0.008	0.001	0.095	0.12 V
HSLA (Si, Cr)	0.145	0.76	0.007	0.013	0.65	0.02	0.02	0.43	0.03	0.001	0.032	0.02 Zr
HSLA (Cu)	0.14	0.83	0.006	0.021	0.047	0.24	0.035	0.025	0.02	0.024	0.003	—
HSLA (Ni, Cu, Cr, Si)	0.17	0.34	0.005	0.015	0.24	0.40	0.43	0.31	0.04	0.012	0.035	—
Low-carbon (mild) steel	0.135	0.34	0.007	0.019	0.005	0.06	0.015	0.015	0.01	—	0.009	—

*Underlined values indicate sufficient amounts of elements claimed to aid atmospheric corrosion resistance.

Table II — Corrosion of Test Steels

Steel	Average Weight Loss, mg/cm²	
	Industrial Atmosphere	Under Car
HSLA (Ni, Cu, Cr, Si)	11.2	36.4
HSLA (Cu)	11.3	44.2
Low-carbon (mild) steel	14.0	55.7
HSLA (Si)	17.3	49.4
HSLA (Si, Cr)	19.7	40.6

Table III — Pitting Corrosion of Test Steels

Steel	Average Pit Depth, mils (mm)	
	Industrial Atmosphere	Under Car
HSLA (Ni, Cu, Cr, Si)	3.38 (0.0859)	4.45 (0.113)
HSLA (Cu)	3.40 (0.0864)	4.65 (0.118)
Low-carbon (mild) steel	3.92 (0.0996)	5.45 (0.138)
HSLA (Si)	4.19 (0.106)	4.25 (0.108)
HSLA (Si, Cr)	4.26 (0.108)	4.73 (0.120)

Table IV — Compositions of Corrosion-Tested Steels (Second Test Run)

Steel	Composition, %*										
	C	Mn	P	S	Si	Cu	Ni	Cr	Mo	Al	V
HSLA (Si)	0.10	1.20	0.008	0.014	0.61	0.05	0.04	0.05	0.01	0.058	0.101
HSLA (Si, Cr)†	0.14	0.75	0.009	0.014	0.61	0.02	0.02	0.46	0.01	0.046	0.003
HSLA (Cu)	0.19	0.91	0.008	0.014	0.04	0.24	0.01	0.01	0.01	0.003	0.002
HSLA (Ni, Cu, Cr, Si)	0.13	0.75	0.007	0.015	0.25	0.40	0.45	0.29	0.04	0.049	0.003
HSLA (Ni, Cu, Cr, Si, P)	0.12	0.54	0.108	0.025	0.25	0.42	0.45	0.42	0.04	0.06	0.004
HSLA (Cu, Si)	0.12	0.98	0.007	0.012	0.19	0.25	0.01	0.02	0.01	0.05	0.036
Low-carbon (mild) steel	0.09	0.33	0.005	0.007	0.01	0.03	0.01	0.01	0.01	0.012	0.002

*Underlined values indicate sufficient amounts of elements claimed to aid atmospheric corrosion resistance. †Also contains 0.092 Zr.

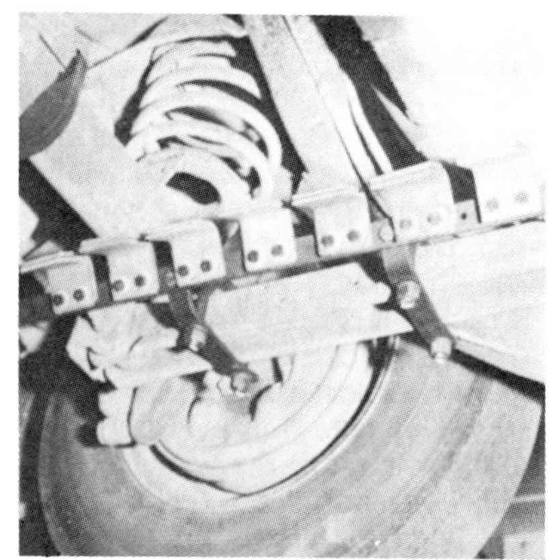

Fig. 2 — The second road test in southern Ontario, which lasted through the winter and spring of 1972-1973, used coupons bent to 90° angles.

Table V — Results of Corrosion Tests (Second Run)

Steel	Corrosion, mils (mm)		
	Uniform	Pitting	Crevice
HSLA (Cu)	1.4 (0.036)	7.30 (0.185)	8.75 (0.222)
HSLA (Cu, Si)	1.4 (0.036)	7.20 (0.183)	10.56 (0.268)
HSLA (Ni, Cu, Cr, Si)	1.4 (0.036)	7.61 (0.193)	11.47 (0.291)
Low-carbon (mild) steel	1.5 (0.038)	9.72 (0.247)	11.53 (0.293)
HSLA (Ni, Cu, Cr, Si, P)	1.4* (0.036)	7.15 (0.182)	11.71 (0.297)
HSLA (Si)	1.4 (0.036)	9.30 (0.236)	15.19 (0.386)
HSLA (Si, Cr)	1.4 (0.036)	9.55 (0.243)	17.39 (0.442)

*Result of retesting; first test showed 2.0 mils (0.051 mm), which was suspect.

the industrial atmosphere; also ranking of steels is different. Table III shows a similar result in pitting corrosion, although the rates are not as different.

From this investigation, we concluded that atmospheric corrosion properties cannot be used to select steels for use under cars. The presence of chloride ion and the lack of conditions for steel passivity in the road test are thought responsible for differing corrosion effects.

The road test also pointed out the existence of another type of corrosion, crevice corrosion. All steels were pitted on the back side of the coupons where the water, sand, and salt penetrated between the plastic liner in the rack and the coupons. Also HSLA (Si) and HSLA (Si, Cr) appeared more susceptible to crevice corrosion than the other steels.

We immediately scheduled a road test for the following winter designed to investigate crevice corrosion, as well as uniform, stress, and pitting corrosion.

Next Program — The second test, again in southern Ontario, included a full winter, lasting from 10 November 1972 to 22 June 1973 (224 days).

Coupons were nested and bent together to a 90° angle, as shown in Fig. 2. A given steel was represented by a pair of coupons. Five test cars, each with an identical rack, were used; exposure ranged from 5000 to 12 000 miles (8050 to 19 300 km).

The nested coupons represented

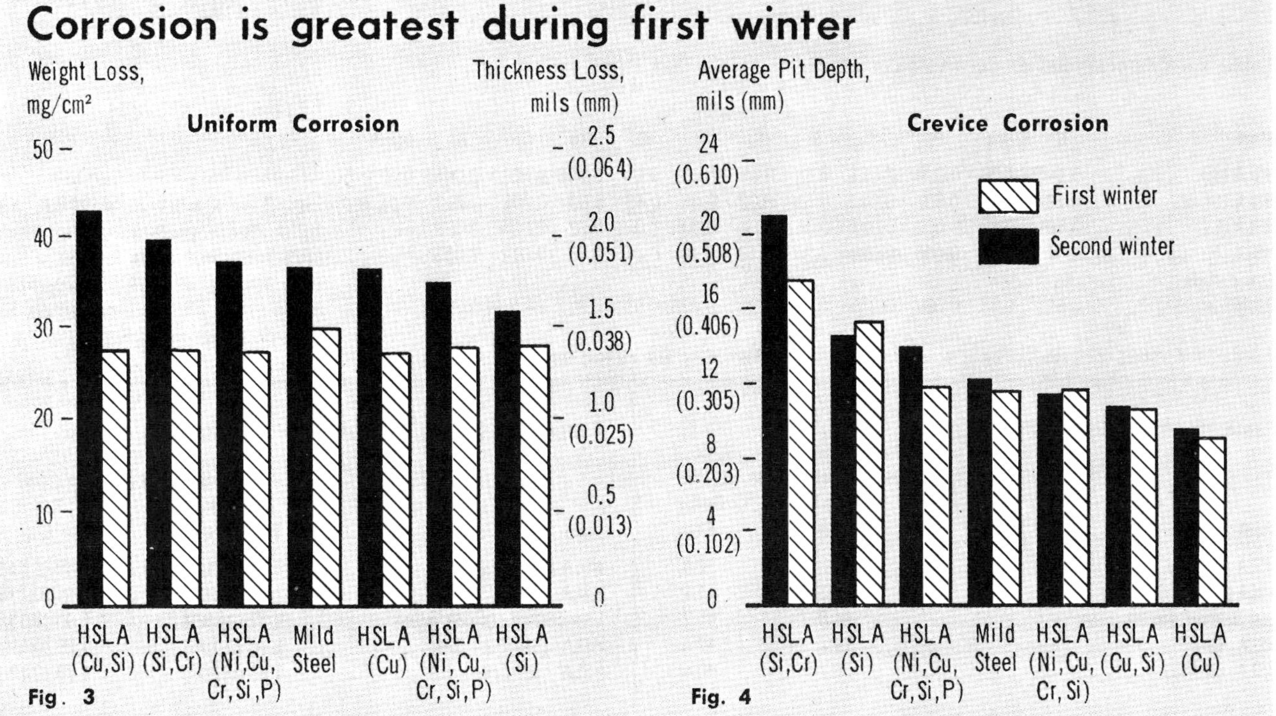

Fig. 3

Fig. 4

a lap joint to encourage crevice corrosion, and the 90° bend allowed for an evaluation of stress corrosion, if any.

Table IV lists composition of tested steels. This test included four of the HSLA steels studied in the original test plus two additional alloys and low-carbon steel.

Uniform corrosion was based on weight loss, and then thickness loss was calculated. Pitting corrosion was measured on outsides of coupon pairs, which were directly exposed. Crevice corrosion was measured as pit depth on insides of coupon pairs, where the crevice corrosion condition existed. Because pitting in or on bends was not more evident than that occurring on flat surfaces, we concluded stress corrosion was not a factor.

Table V lists results. We concluded that HSLA steels behaved similarly to low-carbon steel under cars with respect to uniform corrosion (weight loss), and that this type of corrosion was insignificant to automotive structure designers. As well, HSLA steels were equal to or better than low-carbon steel in pitting of exposed surfaces. The degree of pitting may be significant in designing a part typically 100 mils (2.54 mm) thick. Finally we observe, from Table V, that crevice corrosion is the most significant type of corrosion under cars, and that HSLA (Si) and HSLA (Si, Cr) were worse than mild steel in this respect.

Another road test the following winter gave surprisingly similar results in crevice corrosion, and two additional alloys [HSLA without beneficial alloys and HSLA (Mo, Si)] were worse than mild steel, along with HSLA (Si) and HSLA (Si, Cr).

Winter Corrosion Rates — Our one-winter results indicated disturbing levels of crevice corrosion, expressed as average pit depth, for some steels. Accordingly, a road test was run to determine if this corrosion rate was maintained through two successive winters of exposure. One-winter results were obtained using five cars, while two-winter results were based on only one car. Coupons, mounting method, and location, were identical for both.

As noted before, the one-winter test ran from 10 November 1972 to 22 June 1973 (224 days); mileage ranged from 5000 to 12 000 miles (8050 to 19 300 km) for the cars. Two-winter tests ran continuously from 7 November 1972 to 17 May 1974 (552 days), with the car accumulating 34 412 miles (55 381 km).

Figure 3 illustrates a dramatic decrease in the uniform corrosion rate, two-winter results being only slightly higher than one-winter results. Note that all steels only suffer 1.5 to 2.0 mils (0.038 to 0.051 mm) decrease in thickness. Clearly, uniform corrosion is not a concern to automotive structure designers.

The same trend appeared when we rated pitting corrosion of exposed surfaces. Crevice corrosion, the most damaging type found, also fell off dramatically the second year (Fig. 4).

It appears that buildup of dense road deposit and oxide eventually protects all of the steels. This is not to say corrosion can be discounted; significant crevice corrosion progressed before the slowdown.

Effect of Compositions — Analyzing the data from the various road tests, we pinpointed only two elements as possibly having an effect on the corrosion behavior of HSLA steels under cars.

Throughout the various road tests, copper-bearing steels always appeared to perform as well as or better than the reference, low-carbon steel. A systematic study was needed to confirm this result, and also to determine the optimum level of copper required. A single HSLA grade, uncomplicated by other additions, was chosen, and individual ingots within the heat were treated with varying levels of copper, via mold additions.

Coupons obtained after hot rolling and pickling were all mounted under one car for one winter. Results were disappointing. Varying the copper from 0.018 to 0.40% had no effect, and all the steels performed similarly to mild steel in all three corrosion criteria.

As for effects of other elements, steels with silicon as the major alloying element appear to be susceptible to crevice corrosion. Steels containing 0.40 to 0.90% Si were worse than those containing from 0.10 to 0.25% Si in this respect.

No effect has been seen on corrosion resistance of HSLA steels strengthened by columbium, titanium, or vanadium. Furthermore, the various methods of sulfide shape control — by means of titanium, zirconium, or rare earth additions, for example — have not shown any effect on corrosion.

In conclusion, we found no alloying element to improve the corrosion resistance of HSLA steels enough to make them superior to low-carbon steel under cars.

Laboratory Simulations — Because under-car road tests are time-consuming, laboratory tests would certainly be desirable to speed up evaluation and provide initial screening prior to lengthy road tests. Unfortunately, they do not reflect under-car tests, which are characterized by pitting (7 to 10 mils [0.18 to 0.25 mm]) exceeding weight loss (1 to 2 mils [0.025 to 0.05 mm]). The standard salt fog test, for example, resulted in corrosion by weight loss greatly exceeding pitting corrosion. Interrupted salt fog tests with dry cycles lowered this ratio, but weight loss still exceeded pitting corrosion.

The Corrodkote test, normally used for plating evaluations, resulted in excessive pitting on these steels with respect to weight loss.

Finally, a dip-and-dry test corrosion, used by one of the automotive companies, resulted in weight loss exceeding pitting corrosion.

Future Work — Protective coatings should be considered because work to date indicates that HSLA steels cannot be modified internally to achieve superior corrosion performance to low-carbon steel, and they will generally be substituted at thinner gages. Our current road tests include various protective coatings, both metallic and nonmetallic.

Another interesting area for future study concerns corrosion fatigue under cars. Data on fatigue properties of HSLA steels are now known, and information on the type and degree of corrosion occurring is appearing. Combining these two properties would provide valuable information to the automotive structure designers.

Compositions, Properties, and Producers of High-Strength Steels

I. Columbium or Vanadium Group

Name	Producer Code	Composition, % C	Mn	P	S	Si	Cu Min	Cb Min	or V Min	Mechanical Properties Yield, 1,000 Psi	Tensile, 1,000 Psi	Elongation in 2 In., %	ASTM or SAE Specification No.
AWHF-45	AW	(0.07)	(0.75)	(0.01)	(0.02)	(0.04)	Opt	(0.035)	—	45	—	—	A572, J410c
50	AW	(0.07)	(0.75)	(0.01)	(0.02)	(0.04)	Opt	(0.035)	—	50	—	—	
AWX-42	AW	(0.12)	(0.45)	(0.02)	(0.03)	(0.04)	Opt	(0.02)	—	42	60	20(a)	
45	AW	(0.12)	(0.45)	(0.02)	(0.03)	(0.04)	Opt	(0.02)	—	45	60	19(a)	A572
50	AW	(0.14)	(0.50)	(0.02)	(0.03)	(0.04)	Opt	(0.02)	—	50	65	18(a)	
55	AW	(0.16)	(0.55)	(0.02)	(0.03)	(0.04)	Opt	(0.02)	—	55	70	17(a)	
AW-Ten	AW	(0.18)	(0.75)	(0.02)	(0.03)	(0.04)	(0.25)	(0.02)	—	50	70	18(a)	
Armco High Strength C (b) (Plates, shapes, bars)													
C-42	ARM	0.21	1.35	0.04	0.05	0.30	Opt	0.005-0.05(c)		42	60	20(a)	
C-45	ARM	0.22	1.35	0.04	0.05	0.30	Opt	0.005-0.05(c)		45	60	19(a)	
C-50	ARM	0.23	1.35	0.04	0.05	0.30	Opt	0.005-0.05(c)		50	65	18(a)	A572, J410c
C-55	ARM	0.25	1.35	0.04	0.05	0.30	Opt	0.005-0.05(c)		55	70	17(a)	
C-60	ARM	0.26	1.35	0.04	0.05	0.30	Opt	0.005-0.05(c)		60	75	16(a)	
C-65	ARM	0.26	1.35	0.04	0.05	0.30	Opt	0.005-0.05(c)		65	80	15(a)	
Armco CT-50(d)	ARM	0.18	1.15-1.60	0.04	0.05	0.15-0.30	Opt	0.01-0.05	—	50	70	17(a)	—
CT-55(d)	ARM	0.18	1.15-1.60	0.04	0.05	0.15-0.30	Opt	0.01-0.05	—	55	70	17(a)	—
CT-60(d)	ARM	0.18	1.15-1.60	0.04	0.05	0.15-0.30	Opt	0.01-0.05	—	60	75	17(a)	—
CT-65(d)	ARM	0.18	1.15-1.60	0.04	0.05	0.15-0.30	Opt	0.01-0.05	—	65	85	17(a)	—
(Sheet)												HR (e)	
Cb/V45	B	0.22	1.35	0.04	0.05	0.30	Opt	0.005	0.01	45	60	25	
Cb/V50	B	0.23	1.35	0.04	0.05	0.30	Opt	0.005	0.01	50	65	22	
Cb/V55	B	0.25	1.35	0.04	0.05	0.30	Opt	0.005	0.01	55	70	20	A607, J410c
Cb/V60	B	0.26	1.50	0.04	0.05	0.30	Opt	0.005	0.01	60	75	18	
Cb/V65	B	0.26	1.50	0.04	0.05	0.30	Opt	0.005	0.01	65	80	16	
(Plates, shapes, bars)													
V42	B(f)	0.22	1.25	0.04	0.05	0.30(g)	Opt	—	0.02	42	63	20(a)	
V45	B	0.22	1.25	0.04	0.05	0.30	Opt	—	0.02	45	65	19(a)	
V50(h)	B	0.22	1.25	0.04	0.05	0.30	Opt	—	0.02	50	70	18(a)	A572
V55(h)	B	0.25	1.35	0.04	0.05	0.30	Opt	—	0.02	55	70	17(a)	
V60(h)	B	0.25	1.35	0.04	0.05	0.30	Opt	—	0.02	60	75	16(a)	
V65(h)	B	0.22	1.25	0.04	0.05	0.30	Opt	—	0.02	65	80	15(a)	
(Sheet)												HR (e) CR	
INX-45	IN	0.22	1.35	0.04	0.05	0.30	Opt	0.005	0.01	45	60	25 22	
50	IN	0.23	1.35	0.04	0.05	0.30	Opt	0.005	0.01	50	65	22 20	
55	IN	0.25	1.35	0.04	0.05	0.30	Opt	0.005	0.01	55	70	20 18	A607, J410c
60	IN	0.26	1.50	0.04	0.05	0.30	Opt	0.005	0.01	60	75	18 16	
65	IN	0.26	1.50	0.04	0.05	0.30	Opt	0.005	0.01	65	80	16 15	
70	IN	0.26	1.65	0.04	0.05	0.30	Opt	0.005	0.01	70	85	14 14	—
(Plates, shapes, bars)													
INX-42	IN	0.21	1.35	0.04	0.05	0.30	Opt	0.01	0.01	42	60	20(a)	
45	IN	0.22	1.35	0.04	0.05	0.30	Opt	0.01	0.01	45	60	19(a)	
50	IN	0.23	1.35	0.04	0.05	0.30	Opt	0.01	0.01	50	65	18(a)	A572, J410c
55	IN	0.25	1.35	0.04	0.05	0.30	Opt	0.01	0.01	55	70	17(a)	
60	IN	0.26	1.35	0.04	0.05	0.30	Opt	0.01	0.01	60	75	16(a)	
65	IN	0.26	1.35	0.04	0.05	0.30	Opt	0.01	0.01	65	80	15(a)	
70	IN	0.26	1.35	0.04	0.05	0.30	Opt	0.01	0.01	70	85	14(a)	J410c
IC-42	I	0.22	1.35	0.04	0.05	0.30	Opt	0.005	—	42	60	25	
45	I	0.22	1.35	0.04	0.05	0.30	Opt	0.005	—	45	60	25	
50	I	0.23	1.35	0.04	0.05	0.30	Opt	0.005	—	50	65	22	
55	I	0.25	1.35	0.04	0.05	0.30	Opt	0.005	—	55	70	20	A572, A607,
60	I	0.26	1.35	0.04	0.05	0.30	Opt	0.005	—	60	75	18	J410c
IV-42	I	0.22	1.35	0.04	0.05	0.30	Opt	—	0.01	42	60	25	
45	I	0.22	1.35	0.04	0.05	0.30	Opt	—	0.01	45	60	25	
50	I	0.23	1.35	0.04	0.05	0.30	Opt	—	0.01	50	65	22	
ICF-42(i)	I	0.22	1.35	0.04	0.05	0.30	—	0.005	—	42	60	25	—
ICF-45(i)	I	0.22	1.35	0.04	0.05	0.30	—	0.005	—	45	60	25	—
ICF-50(i)	I	0.23	1.35	0.04	0.05	0.30	—	0.005	—	50	65	22	—

Maximum values for composition are listed except where ranges, minimum, or typical values are indicated. Typical values are enclosed in parentheses. Mechanical properties are those of sheet or hot rolled plate up to ½ in. thick and are minimums unless typical is indicated by parentheses.

Atmospheric corrosion resistance for these high-strength steels is compared to that of carbon steel. Example: 2 indicates twice the corrosion resistance of carbon steel.

Producer Code: AW = Alan Wood Steel Co.; ALG = Algoma Steel Corp.; ARM = Armco Steel Corp.; B = Bethlehem Steel Corp.; BW = Babcock & Wilcox Co.; DF = Dominion Foundries & Steel Ltd.; IN = Inland Steel Co.; I = Interlake Inc.; JL = Jones & Laughlin Steel Corp.; J = Earle M. Jorgensen Co.; K = Kaiser Steel Corp.; L = Lukens Steel Co.; McL = McLouth Steel Corp.; N = National Steel Corp.; O = Oregon Steel Mills; P = Phoenix Steel Corp.; R = Republic Steel Corp.; SH = Sharon Steel Corp.; SC = Steel Co. of Canada Ltd.; US = United States Steel Corp.; W = Wisconsin Steel Div., International Harvester Co.; WP = Wheeling-Pittsburgh Steel Corp.; Y = Youngstown Sheet & Tube Co.

I. Columbium or Vanadium Group—Continued

Name	Producer Code	Composition, % C	Mn	P	S	Si	Cu Min	Cb Min or	V Min	Mechanical Properties Yield, 1,000 Psi	Tensile, 1,000 Psi	Elongation in 2 In., %	ASTM or SAE Specification No.
IV-55	I	0.25	1.35	0.04	0.05	0.30	Opt	—	0.01	55	70	20	A572, A607,
60	I	0.26	1.35	0.04	0.05	0.30	Opt	—	0.01	60	75	18	J410c
JLX-42(j)	JL	0.20	1.00	0.04	0.05	0.30	Opt	0.01	0.01	42	57	25	A572, J410c
45(k)	JL	0.20	1.10	0.04	0.05	0.30	Opt	0.01	0.01	45	60	24	
50(k)	JL	0.22	1.20	0.04	0.05	0.30	Opt	0.01	0.01	50	65	22	A572, A607,
55(k)	JL	0.24	1.20	0.04	0.05	0.30	Opt	0.01	0.01	55	70	20	J410c
60(k)	JL	0.25	1.35	0.04	0.05	0.30	Opt	0.01	0.01	60	75	18	
65(k)	JL	0.26	1.50	0.04	0.05	0.30	Opt	0.01	0.01	65	80	16	
70(k)	JL	0.26	1.65	0.04	0.05	0.30	Opt	0.01	0.01	70	85	14	A607, J410c
JLX-50 CC(l)	JL	0.12	0.90	0.04	0.05	0.10	Opt	0.01	—	50	65	26	A572, A607, J410c
VAN-50(m)	JL	0.14	1.25	0.03	0.03	0.30	—	0.01	0.02	50	65	25	A572, A607,
VAN-60(m)	JL	0.16	1.40	0.03	0.03	0.30	—	0.01	0.02	60	75	22	J410c
VAN-70(m)	JL	0.18	1.50	0.03	0.03	0.30	—	0.01	0.02	70	85	20	A607, J410c
VAN-80(m)	JL	0.18	1.60	0.03	0.03	0.60	—	—	0.05	80	95	18	A656, J410c
Kaisaloy													
42-CV	K	0.21	1.35	0.04	0.05	0.30	Opt	0.005-0.050	0.01-0.10	42	60	24	
45-CV	K	0.22	1.35	0.04	0.05	0.30	Opt	0.005-0.050	0.01-0.10	45	60	22	
50-CV	K	0.23	1.35	0.04	0.05	0.30	Opt	0.005-0.050	0.01-0.10	50	65	21	A572, A607
55-CV	K	0.25	1.35	0.04	0.05	0.30	Opt	0.005-0.050	0.01-0.10	55	70	20	J410c
60-CV	K	0.26	1.35	0.04	0.05	0.30	Opt	0.005-0.050	0.01-0.10	60	75	18	
65-CV	K	0.26	1.35	0.04	0.05	0.30	Opt	0.005-0.050	0.01-0.10	65	80	15(a)	
UCV-65	L	0.23	1.65	0.04	0.05	0.30	—	—	0.01-0.15	65	80	15(a)	—
MLX-45	McL	0.15	1.00	0.04	0.05	0.10	—	0.005	0.02	45	60	25	
50	McL	0.20	1.00	0.04	0.05	0.10	—	0.005	0.02	50	65	22	
55	McL	0.24	1.20	0.04	0.05	0.30	—	0.005	0.02	55	70	22	A572
GLX-42W (n)	N	0.21	1.35	0.04	0.05	—	Opt	0.01	—	42	60	24	—
45W (n)	N	0.22	1.35	0.04	0.05	—	Opt	0.01	—	45	60	22	
50W (n)	N	0.22	1.35	0.04	0.05	—	Opt	0.01	—	50	65	22	A572, A607,
55W (n)	N	0.25	1.35	0.04	0.05	—	Opt	0.01	—	55	70	20	J410c
60W (n)	N	0.26	1.35	0.04	0.05	—	Opt	0.01	—	60	75	18	
65W (n)	N	0.26	1.35	0.04	0.05	—	Opt	0.01	—	65	80	16	
70W (n)	N	0.26	1.35	0.04	0.05	—	Opt	0.01	—	70	85	14	—
Hi-Yield C-42	N	0.21	0.90	0.04	0.05	—	—	0.005	—	42	60	24	
C-45	N	0.22	1.25	0.04	0.05	—	—	0.005	—	45	60	22	A572, A607,
C-50	N	0.22	1.25	0.04	0.05	—	—	0.005	—	50	65	20	J410c
C-55	N	0.25	1.35	0.04	0.05	—	—	0.005	—	55	70	18	
NAPAC-35	N	0.10	0.55	0.025	0.03	—	—	0.005(o)	—	35	—	—	—
40	N	0.10	0.55	0.025	0.03	—	—	0.005(o)	—	40	—	—	—
45	N	0.10	0.55	0.025	0.03	—	—	0.01(o)	—	45	—	—	—
50	N	0.10	0.55	0.025	0.03	—	—	0.01(o)	—	50	—	—	—
NAPAC-F-40	N	0.10	0.65	0.025	0.03	—	—	0.01(o, p)	—	40	50(q)	30	
F-45	N	0.12	0.75	0.025	0.03	—	—	0.01(o, p)	—	45	55(q)	28	
F-50	N	0.12	0.75	0.025	0.03	—	—	0.015(o, p)	—	50	60(q)	26	
F-55	N	0.12	0.75	0.025	0.03	—	—	0.02(o, p)	—	55	65(q)	24	
F-60	N	0.12	0.75	0.025	0.03	—	—	0.03(o, p)	—	60	70(q)	22	J410c
NAPAC-S-45	N	0.15	0.75	0.025	0.03	—	—	0.01(o)	—	45	60	25	
S-50	N	0.15	0.75	0.025	0.03	—	—	0.01(o)	—	50	65	22	
Orelloy-42	O	0.15	0.75	0.011	0.013	0.18	0.09	—	0.02	42	60	24	
45	O	0.18	0.92	0.006	0.012	0.26	0.16	—	0.02	45	60	22	
50	O	0.18	1.23	0.007	0.006	0.27	0.27	—	0.02	50	65	21	A572
55	O	0.21	1.20	0.009	0.014	0.24	0.10	—	0.02	55	70	20	
60	O	0.23	1.32	0.015	0.017	0.24	0.14	—	0.02	60	75	18	
PX SK42	P	0.27	1.20	0.04	0.05	—	Opt	—	—	42	60	19(a)	
A42	P	0.21	1.35	0.04	0.05	0.30	Opt	—	0.02	42	60	20(a)	
A45	P	0.22	1.25	0.05	0.04	0.30	Opt	—	0.02	45	65	19(a)	
A50	P	0.26	1.30	0.04	0.05	0.30	Opt	—	0.02	50	70	18(a)	A572
A55	P	0.26	1.30	0.04	0.05	0.30	Opt	—	0.02	55	70	17(a)	
A60	P	0.26	1.35	0.04	0.05	0.30	Opt	—	0.02	60	75	18	
A65	P	0.26	1.35	0.04	0.05	0.30	Opt	—	0.02	65	80	15	
X42W	R	0.21	1.25	0.04	0.05	0.30	Opt	0.01	0.01	42	60	24	
X45W	R	0.22	1.25	0.04	0.05	0.30	Opt	0.01	0.01	45	60	22	
X50W	R	0.22	1.35	0.04	0.05	0.30	Opt	0.01	0.01	50	65	22	A572, A607,
X55W	R	0.25	1.35	0.04	0.05	0.30	Opt	0.01	0.01	55	70	20	J410c
X60W	R	0.26	1.35	0.04	0.05	0.30	Opt	0.01	0.01	60	75	18	
X65W	R	0.26	1.35	0.04	0.05	0.30	Opt	0.01	0.01	65	80	16	
X70W	R	0.26	1.65	0.04	0.05	0.30	Opt	0.01	0.01	70	85	14	J410c
Maxi-Form 50	R	0.12	0.90	0.015	0.03	—	Opt	0.01(r)	—	50	60	25	—
80	R	0.09	1.60	0.015	0.03	0.60	Opt	0.06-0.15(r)	0.08	80	90	18	—
Sharalloy 45	SH	0.16	0.90	0.04	0.05	0.30	—	0.01	—	45	60	25	
50	SH	0.18	0.90	0.04	0.05	0.30	—	0.01	—	50	65	22	
55	SH	0.18	0.90	0.04	0.05	0.30	—	0.01	—	55	70	22	J410c
60	SH	0.23	0.90	0.04	0.05	0.30	—	0.01	—	60	75	20	

Name	Producer Code	C	Mn	P	S	Si	Cu Min	Cb Min	V Min	Yield, 1,000 Psi	Tensile, 1,000 Psi	Elongation in 2 In., %	ASTM or SAE Specification No.
		Composition, %								Mechanical Properties			
H45	US	0.20	1.00	0.04	0.05	0.30	Opt	0.01	—	45	60	25	
L45	US	0.13	0.90	0.04	0.05	0.10	Opt	0.01	—	45	55	25	
K45	US	0.13	0.90	0.04	0.05	0.10	Opt	0.01	—	45	55	26	
F45	US	0.13	0.90	0.04	0.05	0.10	Opt	0.01	—	45	55	26	
H50	US	0.20	1.00	0.04	0.05	0.30	Opt	0.01	—	50	65	22	
L50	US	0.13	0.90	0.04	0.05	0.10	Opt	0.01	—	50	60	24	
K50 (sheet)	US	0.13	0.90	0.04	0.05	0.10	Opt	0.01	—	50	60	25	J410c
F50	US	0.13	0.90	0.04	0.05	0.10	Opt	0.01	—	50	60	25	
H55	US	0.20	1.25	0.04	0.05	0.30	Opt	0.01	—	55	70	20	
H60	US	0.22	1.25	0.04	0.05	0.30	Opt	0.01	—	60	75	18	
K60	US	0.13	0.90	0.04	0.05	0.10	Opt	0.01	—	60	70	21	
F60	US	0.13	0.90	0.04	0.05	0.10	Opt	0.01	—	60	70	21	
65	US	0.22	1.35	0.04	0.05	0.30	Opt	0.01	—	65	80	16	
70	US	0.26	1.35	0.04	0.05	0.30	Opt	0.01	—	70	85	14	
IH-50(s)	W	0.22	1.50	0.04	0.05	0.70	Opt	0.005-0.05	0.01-0.15	50	75	20	—
IH-60(s)	W	0.22	1.65	0.04	0.05	0.70	Opt	0.005-0.05	0.01-0.15	60	80	18	—
IHX-42	W	0.21	1.35	0.04	0.05	0.30	Opt	0.005-0.05	0.01-0.15	42	60	24	
IHX-45	W	0.22	1.35	0.04	0.05	0.30	Opt	0.005-0.05	0.01-0.15	45	60	22	
IHX-50	W	0.23	1.35	0.04	0.05	0.30	Opt	0.005-0.05	0.01-0.15	50	65	21	
IHX-55	W	0.25	1.35	0.04	0.05	0.30	Opt	0.005-0.05	0.01-0.15	55	70	20	A572
IHX-60	W	0.26	1.35	0.04	0.05	0.30	Opt	0.005-0.05	0.01-0.15	60	75	18	
IHX-65	W	0.26	1.65	0.04	0.05	0.30	Opt	0.005-0.05	0.01-0.15	65	80	17	
IHX-70	W	0.26	1.65	0.04	0.05	0.30	Opt	0.005-0.05	0.01-0.15	70	85	16	—
Pitt-Ten													
X45W	WP	0.20	1.00	0.04	0.05	0.10	Opt	0.01	0.01	45	60	24	
X50W	WP	0.20	1.00	0.04	0.05	0.10	Opt	0.01	0.01	50	65	22	A572, A607
X55W	WP	0.20	1.00	0.04	0.05	0.10	Opt	0.01	0.01	55	70	20	J410c
X60W	WP	0.20	1.00	0.04	0.05	0.10	Opt	0.01	0.01	60	75	18	
YSW-42	Y	0.20	1.10	0.04	0.05	—	Opt	0.01	0.01	42	62	25	A572, J410c
45	Y	0.20	1.25	0.04	0.05	—	Opt	0.01	0.01	45	60	25	
50	Y	0.22	1.25	0.04	0.05	—	Opt	0.01	0.01	50	65	22	
55	Y	0.25	1.35	0.04	0.05	—	Opt	0.01	0.01	55	70	20	A572, A607,
60	Y	0.26	1.35	0.04	0.05	—	Opt	0.01	0.01	60	75	18	J410c
65	Y	0.26	1.35	—	—	—	Opt	0.01	0.01	65	80	18	
70	Y	0.26	1.50	—	—	—	Opt	0.01	0.01	70	85	16	A607, J410c
Algoform													
45	ALG	0.06	0.35	0.01	0.02	0.10	—	0.01	—	45	55	—	
50	ALG	0.06	0.45	0.01	0.02	0.10	—	0.02	—	50	60	—	J410c
60	ALG	0.06	0.75	0.01	0.02	0.10	—	0.04	—	60	70	—	
80	ALG	0.06	0.95	0.01	0.02	0.10	—	0.10	—	80	90	—	
CB/V45	ALG	0.22	1.35	0.04	0.05	—	—	0.01	0.01	45	60	22	
V50	ALG	0.23	1.35	0.04	0.05	—	—	0.01	0.01	50	65	21	A572, A607,
V55	ALG	0.25	1.35	0.04	0.05	—	—	0.01	0.01	55	70	20	J410c
V60	ALG	0.26	1.35	0.04	0.05	—	—	0.01	0.01	60	75	18	
Dofascoloy													
50F	DF	0.15	1.65	0.025	0.035	0.90	—	0.005-0.10	—	50	60	24	
60F	DF	0.15	1.65	0.025	0.035	0.90	—	0.005-0.10	—	60	70	22	A715
70F	DF	0.15	1.65	0.025	0.035	0.90	—	0.005-0.10	—	70	80	20	
80F	DF	0.15	1.65	0.025	0.035	0.90	—	0.005-0.10	—	80	90	18	
45W	DF	0.22	1.35	0.04	0.05	0.10	—	0.005	—	45	60	19(a)	
50W	DF	0.23	1.35	0.04	0.05	0.10	—	0.005	—	50	65	18(a)	
55W	DF	0.25	1.35	0.04	0.05	0.10	—	0.005	—	55	70	17(a)	A572, A607
60W	DF	0.26	1.35	0.04	0.05	0.10	—	0.005	—	60	75	16(a)	
65W	DF	0.26	1.35	0.04	0.05	0.10	—	0.005	—	65	80	15(a)	
70W	DF	0.26	1.65	0.04	0.05	0.30	Opt	0.005	—	70	85	14(a)	A607
80W	DF	0.26	1.65	0.04	0.05	0.30	Opt	0.005	—	80	90	14(a)	A607
CB/V 42	SC	0.21	1.35	0.04	0.05	0.30	Opt	0.005	0.01	42	60	24	A572
45	SC	0.22	1.35	0.04	0.05	0.30	Opt	0.005	0.01	45	60	22	
50	SC	0.23	1.35	0.04	0.05	0.30	Opt	0.005	0.01	50	65	21	
55	SC	0.25	1.35	0.04	0.05	0.30	Opt	0.005	0.01	55	70	20	A572, A607
60	SC	0.26	1.35	0.04	0.05	0.30	Opt	0.005	0.01	60	75	18	
65	SC	0.26	1.35	0.04	0.05	0.30	Opt	0.005	0.01	65	80	15(a)	A572
Stelmax													
45	SC	0.15	1.50	0.02	0.03	0.30	—	0.005	0.01	45	55	25	
50	SC	0.15	1.50	0.02	0.03	0.30	—	0.005	0.01	50	60	24	
60	SC	0.15	1.50	0.02	0.03	0.30	—	0.005	0.01(t)	60	70	22	A715
70	SC	0.15	1.50	0.02	0.03	0.30	—	0.005	0.01(t, u)	70	80	20	
80	SC	0.15	1.50	0.02	0.03	0.30	—	0.005	0.01(t, u)	80	90	18	

Note: Steels in Group I do not contain more than residual copper; therefore, the atmospheric corrosion resistance is usually equal to that of plain carbon steel. If 0.20% Cu minimum is added, then atmospheric corrosion resistance is up to twice that of plain carbon steel. Usually these are semikilled steels but may be killed, particularly at higher strength levels.

(a) Elongation in 8 in.; (b) C-45, C-50, and C-55 are also available in sheets and strip mill plates; elongations are 25, 22, and 20%, respectively; (c) And/or 0.01-0.10 V, or 0.01-0.10 V + 0.015 N max; (d) Controlled rolled to ¾ in. thickness for certified notch toughness; (e) HR=hot rolled, CR=cold rolled; (f) V42, 45, 50, 55, 60, 65 licensed to Alan Wood Steel Co.; (g) Over ½ to 4 in., 0.15 to 0.30 Si; (h) V50 over ¾ in., V55, 60, 65 contain up to 0.015 N; (i) Al and rare earths added for inclusion shape control; (j) Plate and bar only; (k) Lower C and Mn when made for sheet; (l) CC = controlled cooled; (m) 0.01 Al min and Ce or Zr for inclusion shape control; (n) Lower C and Mn when made for sheet; (o) 0.03 Al; (p) 0.06 Zr; (q) Higher tensile strengths available when specified; (r) 0.02 Al; (s) Also classed as Group III steels; (t) Both Cb and V added; (u) 0.02 N.

II. Low Manganese-Vanadium-Titanium Group

Copper is usually omitted for better formability

Name	Producer Code	Composition, %							Mechanical Properties			Atmos. Corrosion Resistance	ASTM or SAE Specification No.
		C	Mn	P	S	Si	Cu	Other (a)	Yield, 1,000 Psi	Tensile, 1,000 Psi	Elongation in 2 In., %		
VNT-N	ARM	0.22	1.15-1.50	0.035	0.04	0.15-0.50	—	0.04-0.11 V, 0.01- 0.03 N	60	80	23	—	A633
VNT-QT	ARM	0.22	1.15-1.50	0.035	0.04	0.15-0.50	—	0.04-0.11 V, 0.01- 0.03 N	75	90	20	—	—
Ultra-Form 50	B	0.12	0.60	0.025	0.035	—	—	0.10 Ti	50	60	24	—	A715
80	B	0.12	0.60	0.025	0.035	—	—	0.10 Ti	80	90	18	—	A715
Superform 40	JL	0.20	0.90	0.025	0.03	0.20	—	0.05-0.12 Zr, (0.02-0.06 Al)	40	—	—	1	—
ML-F	McL	0.12	0.75	(0.02)	(0.02)	0.25	0.22	0.005 Cb or 0.02 V	(45)	(62)	(28)	1 to 2	—
NAX-Fine Grain (b)	N	0.18	1.05	0.025	0.03	0.90	Opt	0.06 Zr	50	70	22	2	A606, J410c
Republic 35	R	0.12	0.75	0.04	0.05	0.10	—	0.005 Cb or V	(35)	(47)	(30)	1	—
Par-Ten	US	0.13	0.90	0.04	0.05	—	Opt	0.04 Cb or 0.07 V max	(45)	(55)	(30)	1	—
Pitt-Ten #2	WP	0.15	0.75	0.04	0.05	0.10	—	0.035 V	Properties to meet specific forming problems			1	—
YS-T45	Y	(0.10)	(0.40)	0.04	0.05	—	—	0.01 Al, 0.05 Ti	45	60	25	—	J410c
T50	Y	(0.10)	(0.40)	0.04	0.05	—	—	0.01 Al, 0.05 Ti	50	65	22	—	J410c
	Y	(0.10)	(0.40)	0.04	0.05	—	—	0.01 Al, 0.05 Ti	60	75	18	—	J410c
T70	Y	(0.10)	(0.40)	0.04	0.05	—	—	0.01 Al, 0.05 Ti	70	85	18	—	J410c
T80	Y	(0.10)	(0.45)	0.04	0.05	—	—	0.01 Al, 0.05 Ti	80	95	18	—	J410c

(a) Minimum unless otherwise specified. (b) NAX-Fine Grain is licensed to Republic and Sharon Steel.

III. Manganese and Manganese-Copper Groups

Name	Producer Code	Composition, %						Condition	Mechanical Properties			Atmos. Corrosion Resistance	ASTM or SAE Specification No.
		C	Mn	P	S	Si	Cu Min		Yield, 1,000 Psi	Tensile, 1,000 Psi	Elongation in 2 In., %		
A. Manganese Group—Copper not usually indicated													
Armco Lo-Temp(a)	ARM	0.20	0.70-1.35	0.035	0.04	0.15-0.50	—	N	50	70	22	1	A537
Armco Super Lo-Temp(a)	ARM	0.20	0.70-1.35	0.035	0.04	0.15-0.50	—	Q&T	60	80	22	1	A537, A678
Armco LTM-QT	ARM	0.16	0.90-1.50	0.035	0.04	0.15-0.30	—	Q&T	50	70	18(b)	1	A678
Armco LTM-N	ARM	0.14	0.90-1.35	0.035	0.04	0.15-0.30	—	N	42	62	20(b)	1	A633, A662
Armco QTC(c)	ARM	0.20	1.00-1.60	0.035	0.04	0.20-0.50	—	Q&T	75	95	19	1	A678
Armco CT-N(d)	ARM	0.18	1.15-1.60	0.04	0.05	0.15-0.50	Opt	N	50	70	18(b)	—	A633
Armco CT-QT(d)	ARM	0.18	1.15-1.60	0.04	0.05	0.15-0.50	Opt	Q&T	60	80	18(b)	—	—
RQC80	B	0.20	1.35	0.04	0.05	0.15-0.30	—	Q&T	80	95	18	1	—
RQC90(e)	B	0.20	1.35	0.04	0.05	0.15-0.30	—	Q&T	90	100	18	1	—
RQC100(e)	B	0.20	1.50	0.04	0.05	0.15-0.30	—	Q&T	100	110	18	1	—
RQC-60N	B	0.20	0.70-1.60	0.035	0.04	0.15-0.50	—	N	50	70	22	1	A537
RQC 60 Q&T	B	0.20	0.70-1.60	0.035	0.04	0.15-0.50	—	Q&T	60	80	22	1	A537
LT-75HS	L	0.22	1.10-1.60	0.035	0.04	0.20-0.60	Opt	Q&T	75	95-115	17	2	—
LT-75N	L	0.24	0.70-1.35	0.035	0.04	0.15-0.30	—	N	50	70	22	2	A537
LT-75QT	L	0.24	0.70-1.35	0.035	0.04	0.15-0.30	—	Q&T	60	80	22	2	A537
Lukens 45	L	0.20	1.20	0.04	0.05	—	—	—	45	65	24	1	—
50	L	0.20	1.35	0.04	0.05	—	—	—	50	70	24	1	—
55	L	0.22	1.35	0.04	0.05	—	—	—	55	75	23	1	—
60	L	0.22	1.60	0.04	0.05	0.15-0.30	—	—	60	80	23	1	—
PX50	P	0.20	1.00-1.50	0.04	0.05	0.15-0.50	—	N	50	70-90	22	1	—
Char-Pac (Norm.)	US	0.20	0.70-1.35	0.035	0.04	0.15-0.50	0.35 max	N	50	70	22	1	A537
Char-Pac (Q&T)	US	0.20	0.70-1.35	0.035	0.04	0.15-0.50	0.35 max	Q&T	60	80	22	1	A537
Con-Pac 80(f)	US	0.20	1.35	0.04	0.05	0.15-0.40	—	Q&T	80	95	18	1	—
90(f)	US	0.20	1.35	0.04	0.05	0.15-0.40	—	Q&T	90	100	18	1	—
100(f)	US	0.20	1.50	0.04	0.05	0.15-0.40	—	Q&T	100	110	18	1	—
M(f)	US	0.20	1.60	0.035	0.04	0.15-0.50	—	Q&T	75	95	19	1	—
B. Manganese-Copper Group													
AW-440	AW	0.28	1.10-1.60	0.04	0.05	0.30	0.20	—	50	70	18(b)	2	A440
Med. Mn	B	0.28	1.10-1.60	0.04	0.05	0.30	0.20-0.35	—	50	70	20	2	A440, J410c
Hi-Man	IN	0.28	1.10-1.60	0.04	0.05	0.30	0.20	—	50	70	18(b)	2	

III. Manganese and Manganese-Copper Groups (Continued)

Name	Pro-ducer Code	Composition, %						Con-dition	Mechanical Properties			Atmos. Cor-rosion Resist-ance	ASTM or SAE Specifi-cation No.
		C	Mn	P	S	Si	Cu Min		Yield, 1,000 Psi	Tensile, 1,000 Psi	Elonga-tion in 2 In., %		
Jalten-3	JL	0.25	1.60	0.04	0.05	0.30	0.20	—	50	70	22	2	A440, J410c
Kaisaloy 50 MM	K	0.27	1.10-1.60	0.035	0.04	0.30	0.20	—	50	70	18(b)	2	A440, J410c
Lukens A440	L	0.28	1.10-1.60	0.04	0.05	0.30	0.20	—	50	70	18(b)	2	A440
NAX-Hi Mang.	N	0.28	1.10-1.60	0.04	0.05	0.30	0.20	—	50	70	22	2	A440
Orelloy 440	O	(0.19)	(1.28)	(0.020)	(0.030)	(0.23)	(0.35)	—	50	70	18(b)	2	A440
Republic M	R	0.25	1.10-1.60	0.04	0.05	0.30	0.20	—	50	75	20	2	J410c
Yo Man	Y	0.25	1.10-1.60	0.04	0.05	0.30	0.20	—	50	70	20	2	A440, J410c
Dofascoloy M	DF	0.28	1.10-1.60	0.04	0.05	0.30	0.20	—	50	70	18(b)	2	A440, J410c

Note: N, normalized; Q&T, quenched and tempered; (a) 0.25 Ni, 0.25 Cr, 0.08 Mo; (b) Elongation in 8 in.; (c) 0.35 Ni, 0.25 Cr, 0.08 Mo; (d) Contains 0.01-0.05 Cb; (e) 0.0005 B min; (f) Boron-treated.

IV. Manganese-Vanadium-Copper Group

Name	Pro-ducer Code	Composition, %							Mechanical Properties			Atmos. Cor-rosion Resist-ance	ASTM or SAE Specifi-cation No.
		C	Mn	P	S	Si	Cu Min	V Min	Yield, 1,000 Psi	Tensile, 1,000 Psi	Elonga-tion in 2 In., %		
AW-441	AW	0.22	1.25	0.04	0.05	0.30	0.20	0.02	50	70	22	2	A441
Armco High Strength B	ARM	0.22	0.85-1.25	0.04	0.05	0.30	0.20	0.02	50	70	18 (a)	2	A441, J410c
Mn-V	B	0.22	0.85-1.25	0.04	0.05	0.30	0.20	0.02	50	70	18 (a)	2	A441, A606
Tri-Steel	IN	0.22	0.85-1.25	0.04	0.05	0.30	0.20	0.02	50	70	22	2	A441, J410c
Jalten 1	JL	0.15	1.25	0.04	0.05	0.30	0.20	0.02	50	70	22	2	A441, J410c
Kaisaloy 50 MV	K	0.22	0.85-1.25	0.035	0.04	0.30	0.20	0.02	50	70	18 (a)	2	A441, J410c
Lukens A441	L	0.22	1.25	0.04	0.05	0.30	0.20	0.02	50	70	18 (a)	2	A441
GLS-441	N	0.22	1.25	0.04	0.04	0.30	0.20	0.02	50	70	22	2	A441
ML-F (A-441)	McL	0.22	1.25	0.04	0.05	0.30	0.20	0.02	50	70	22	2	A441
Orelloy 441	O	0.17	1.19	0.018	0.031	0.24	0.36	0.04	50	70	22	2	A441
Clay-Loy	P	0.22	1.25	0.04	0.05	0.35	0.20	0.02	50	70	18	2	A441
Republic A-441	R	0.22	0.85-1.25	0.04	0.05	0.30	0.20	0.02	50	70	22	2	A441
Tri-Ten	US	0.22	1.25	0.04	0.05	0.30	0.20	0.02	50	70	18 (a)	2	A441
YSW A441	Y	0.22	1.25	0.04	0.05	0.30	0.20	0.02	50	70	18	2	A441, J410c
Dofascoloy MV	DF	0.22	0.85-1.25	0.04	0.05	0.30	0.20	0.02	50	70	18 (a)	2	A441
Stelco-Vanadium	SC	0.22	1.25	0.04	0.05	0.30	0.20	0.02	50	70	22	2	A441
Wgh.-Pgh. A441	WP	0.22	0.85-1.25	0.04	0.05	0.30	0.20	0.02	50	70	18 (a)	2	A441

(a) Elongation in 8 in.

V. Multiple Alloy and Copper Group

Name	Pro-ducer Code	Composition, %										Mechanical Properties			Atmos. Cor-rosion Resist-ance	ASTM or SAE Specifica-cation No.
		C	Mn	P	S	Si	Cu	Mo	Cr	Ni	Other (a)	Yield, 1,000 Psi	Tensile, 1,000 Psi	Elonga-tion in 2 In., %		
Armco High Strength A	ARM	0.12	0.90	0.04	0.05	0.15-0.70	0.20-0.50	—	0.50-1.00	0.25 min	0.07 Ti max	50	70	18 (b)	4 to 6	A242
Armco High Strength A588	ARM	0.20	0.75-1.25	0.04	0.05	0.15-0.30	0.20-0.40	—	0.40-0.70	0.25-0.50	0.01-0.10 V	50	70	18 (b)	4	A588
Mayari R-50	B	0.20	0.75-1.25	0.04	0.05	0.15-0.30	0.20-0.40	—	0.40-0.70	0.25-0.50	0.01-0.10 V	50	70	21	4	A588, J410c
Mayari R-60	B	0.20	0.75-1.35	0.04	0.05	0.15-0.30	0.20-0.40	—	0.40-0.70	0.25-0.50	0.01-0.10 V	60	80	16 (b)	4	—
Ni-Cu-Ti	JL	0.15	1.00	0.04	0.05	0.50	0.30 min	—	—	0.70	0.05 Ti	50	70	22	4	A588
Kaisaloy 45FG	K	0.12	0.60	0.035	0.04	0.50	0.30	0.10	0.25	0.60	0.02 V, 0.005 Ti	45	60	25	6 max	—
50CR	K	0.20	1.25	0.035	0.04	0.25-0.75	0.20-0.35	0.15	0.10-0.25	0.30-0.60	0.02 V, 0.005 Ti	50	70	22	5 to 8	A242, A588, A606, 410c
60SG	K	0.20	1.25	0.035	0.04	0.35	0.80	0.25	—	0.90	0.005 V	60	80	16(b)	2 to 3	—
70MB	K	0.15	0.75	0.035	0.04	0.35	—	0.40-0.60	—	—	0.001 B	70	85	—	—	—
UCV-60	L	0.25	1.50	0.04	0.05	0.15-0.30	0.25-0.40	—	0.40-0.65	—	0.02-0.10 V	60	70	19	—	—
BCV-42 (c)	L	0.25	1.50	0.04	0.05	0.15-0.30	0.25-0.40	—	0.40-0.65	—	0.02-0.10 V	42	63	19	—	—
46 (c)	L	0.25	1.50	0.04	0.05	0.15-0.30	0.25-0.40	—	0.40-0.65	—	0.02-0.10 V	46	67	19	—	—
50 (c)	L	0.25	1.50	0.04	0.05	0.15-0.30	0.25-0.40	—	0.40-0.65	—	0.02-0.10 V	50	70	19	—	—
55 (c)	L	0.25	1.50	0.04	0.05	0.15-0.30	0.25-0.40	—	0.40-0.65	—	0.02-0.10 V	55	70	19	—	—
60 (c)	L	0.25	1.50	0.04	0.05	0.15-0.30	0.25-0.40	—	0.40-0.65	—	0.02-0.10 V	60	70	19	—	—
70 (c)	L	0.25	1.50	0.04	0.05	0.15-0.30	0.25-0.40	—	0.40-0.65	—	0.02-0.10 V	70	80	16	—	—

V. Multiple Alloy and Copper Group—Continued

Name	Producer Code	Composition, %: C	Mn	P	S	Si	Cu	Mo	Cr	Ni	Other (a)	Mechanical Properties: Yield, 1,000 Psi	Tensile, 1,000 Psi	Elongation in 2 In., %	Atmos. Corrosion Resistance	ASTM or SAE Specification No.
NAX-High Tensile	N(d)	0.16	0.90	0.025	0.03	0.90	Opt	—	0.80	—	0.06 Zr	50	70	22	4 to 6	A242, A588, A606
NAX-80	N	0.12	1.00	0.02	0.03	0.90	0.80	—	—	—	0.01 Cb, 0.06 Zr(e)	80	90	18	4 to 6	J410c
Orelloy 242 #2	O	0.17	1.05	0.013	0.022	0.20	0.25	—	0.50	0.06	—	50	70	18 (b)	4 to 6	A242
	O	0.18	1.16	0.008	0.005	0.20	0.29	—	0.56	0.13	0.03 V	50	70	18 (b)	4 to 6	A588
Republic 50	R	0.15	0.50-1.00	0.04	0.05	—	0.30-1.00	0.10 min	0.30	0.40-1.10	—	50	70	22	4 to 6	A242, A588, A606, J410c
Republic 60	R	0.15	0.50-1.00	0.04	0.05	—	0.30-1.00	0.10 min	0.30	0.40-1.10	0.01 Cb, a/o 0.02 V	60	80	21	4 to 6	
NAX	SH	(0.15)	(0.70)	0.04	0.04	(0.70)	Opt	—	(0.70)	—	0.05 Zr	50	70	22	4 to 6	J410c
Cor-Ten B	US(f)	0.10-0.19	0.90-1.25	0.04	0.05	0.15-0.30	0.25-0.40	—	0.40-0.65	—	0.02-0.10 V	50	70	19 (b)	4	A242, A588
Cor-Ten C	US(f)	0.12-0.19	0.90-1.35	0.04	0.05	0.15-0.30	0.25-0.40	—	0.40-0.70	—	0.04-0.10 V	60	80	16 (b)	4	—
Yoloy HS	Y	0.15	1.00	0.04	0.05	0.30	0.50-1.00	0.25	—	(1.00)	—	50	70	22	4	A242, A588, A606, J410c
HSX	Y	0.15	1.00	0.04	0.05	0.35	0.50-1.00	0.25	—	(1.00)	—	45	62	25	4	
S (g)	Y	0.20	1.00	0.04	0.05	0.30	0.75-1.25	—	—	(1.90)	—	50	70	22	4 to 6	A242, A606, J410c
Yoloy T-50	Y	(0.10)	(0.45)	0.04	0.05	(0.40)	(0.25)	—	(0.65)	—	0.05 Ti, 0.01 Al	50	70	22	4	A242, A606, J410c
T-60	Y	(0.10)	(0.45)	0.04	0.05	(0.40)	(0.25)	—	(0.65)	—	0.05 Ti, 0.01 Al	60	80	20	4	
T-70	Y	(0.10)	(0.45)	0.04	0.05	(0.40)	(0.25)	—	(0.65)	—	0.05 Ti, 0.01 Al	70	85	18	4	
T-80	Y	(0.10)	(0.45)	0.04	0.05	(0.40)	(0.25)	—	(0.65)	—	0.05 Ti, 0.01 Al	80	95	16	4	
Algotuf 50	ALG	0.18	1.25	0.03	0.05	0.30	0.20-0.40	—	0.60	0.50	0.02 V	50	70	22	4 to 5	A588
70	ALG	0.20	1.60	0.025	0.035	0.35	0.50	—	—	0.50	0.06 Cb, 0.10 V	70	90	15	4 to 6	—
Dofascoloy #1	DF	0.25	1.25	0.04	0.05	0.30	0.60	—	—	0.90	—	50	70	18 (b)	4 to 5	A242, A606
Dofascoloy ZR	DF	0.20	1.00	0.04	0.05	0.50-1.00	0.20	—	0.80	—	0.03 Zr	50	70	22	4	A588
Stelcoloy 50	SC	0.15	0.80-1.35	0.04	0.05	0.15-0.30	0.20-0.50	—	0.30-0.50	0.25-0.50	0.01-0.10 V	50	70	21	4 to 5	A242, A588, A606
60	SC	0.20	1.50	0.04	0.05	0.15-0.30	0.20-0.50	—	0.30-0.50	0.25-0.50	0.01-0.12 V	60	80	18	4 to 5	—
70	SC	0.22	1.50	0.03	0.04	0.15-	0.20-0.50	—	—	0.25-0.50	0.02-0.10 V, 0.05 Cb	70	90	14 (b)	2 to 4	—

(a) Minimum, unless otherwise specified; (b) Elongation in 8 in.; (c) Properties are achieved by normalizing BCV-42, 46, 50, and 55; and by quenching and tempering BCV-60 and 70; (d) NAX-High Tensile is licensed to Republic and Sharon Steel; (e) May contain Ti and/or B; (f) Licensed to Alan Wood, Algoma, Colorado Fuel & Iron, Crucible, Granite City (Subs. of National Steel), Inland, Interlake, Jones & Laughlin, Kaiser, Lukens, Republic, and Sharon Steel; (g) Can be precipitation hardened (stress relief annealed) to increase tensile properties.

VI. Multiple Alloy and Copper and Phosphorus Group

Name	Producer Code	Composition, %: C	Mn	P	S	Si	Cu	Ni	Other	Mechanical Properties: Yield, 1,000 Psi	Tensile, 1,000 Psi	Elongation in 2 In., %	Atmos. Corrosion Resistance	ASTM or SAE Specification No.
AW Dynalloy 50	AW	(0.13)	(0.90)	(0.08)	(0.03)	(0.20)	(0.35)	(0.45)	(0.08 Mo)	50	70	22	4 to 5	A242
Mayari R (a)	B	0.12	(0.75)	0.12	0.05	(0.55)	0.50	1.00	0.40-1.00 Cr	50	70	18 (c)	5 to 8	A242, A606, J410c
Orelloy 242	O	(0.10)	(0.50)	(0.09)	(0.03)	(0.48)	(0.43)	(0.43)	(0.86 Cr)	50	70	22	4 to 5	—
Orelloy 242 #1	O	(0.10)	(0.50)	(0.09)	(0.03)	(0.48)	(0.43)	(0.43)	(0.86 Cr)	50	70	22	4 to 6	A242
Cor-Ten A	US(b)	0.12	0.20-0.50	0.07-0.15	0.05	0.25-0.75	0.25-0.55	0.65	0.30-1.25 Cr	50	70	19 (c)	5 to 8	A242, A606, J410c
Pitt Ten #1 (d)	WP	0.20	0.55-0.95	0.07	0.05	0.20	0.30-0.50	0.45-0.95	0.20-0.50 Cr	50	70	22	4 to 6	J410c
Dofascoloy P	DF	0.16	0.90	0.12	0.05	0.15-0.35	0.60	0.90	0.60 Cr	50	70	18	4 to 6	A242, A606
Dofascoloy #2	DF	0.22	1.25	0.12	0.05	0.30	0.60	0.90	0.60 Cr	45	65	22	—	A606
Stelcoloy G	SC	0.12	0.75	0.12	0.04	0.15-0.30	0.30-0.60	0.30-0.60	0.30-0.60 Cr	50	70	22	4 to 6	A242, A606

(a) 0.10 Zr; (b) Cor-Ten A is licensed to Alan Wood, Algoma, Colorado Fuel & Iron, Crucible, Granite City (Subs. of National Steel), Inland, Interlake, Jones & Laughlin, Kaiser, Lukens, Republic, and Sharon Steel; (c) Elongation in 8 in.; (d) Can be precipitation hardened and made with lower phosphorus.

VII. Precipitation-Hardening Alloys

Name	Producer Code	Composition, % C	Mn	P	S	Si	Cu	Mo	Other	Mechanical Properties Yield, 1,000 Psi	Tensile, 1,000 Psi	Elongation in 2 In., %	Atmos. Corrosion Resistance	ASTM or SAE Specification No.
AWP-80(a)	AW	0.08	0.90-1.30	0.04	0.04	0.15-0.35	—	0.20-0.30	0.01-0.07 Cb 0.0005 B min 0.05 Al min	80	90	18	—	—
70	R	0.20	1.00	0.04	0.04	0.15	1.00-1.50	0.20-0.30	1.20-1.75 Ni	70	90	18	4 to 6	—
80	R	0.20	1.00	0.04	0.04	0.15	1.00-1.50	0.20-0.30	1.20-1.75 Ni, 0.01 Cb, a/o 0.02 V	80	100	18	4 to 6	—
Ni-Cu-Cb	US	0.06	0.40-0.65	0.035	0.04	0.20-0.35	1.00-1.30	—	1.20-1.50 Ni, 0.02 Cb	75-85	88-90	18	—	A710B
Yoloy S	Y	0.20	1.00	0.04	0.05	0.30	0.75-1.25	—	1.60-2.20 Ni	65	(90)	20	4 to 6	—
Nicuten 85	ALG	(0.06)	(0.35)	0.04	0.04	—	(1.50)	—	(1.00 Ni), (0.02 V)	85	90	20	—	—

(a) Also classed as a Group V and VIII steel.

VIII. Constructional Alloys (Extra High-Strength Steels)

Name	Producer Code	Composition, % C	Mn	P	S	Si	Ni	Cr	Mo	Other (a)	Mechanical Properties (b) Yield, 1,000 Psi	Tensile, 1,000 Psi	Elongation in 2 In., %	ASTM or SAE Specification No.
SSS-100	ARM	0.12-0.20	0.40-0.70	0.035	0.04	0.20-0.35	—	1.40-2.00	0.40-0.60	0.20-0.40 Cu, 0.04-0.10 V or Ti,0.0015-0.005 B	100	115	16	A514, A517
SSS-100A	ARM	0.13-0.20	0.40-0.70	0.035	0.04	0.20-0.35	—	0.85-1.20	0.15-0.25		100	115	16	
SSS-100B	ARM	0.13-0.20	0.40-0.70	0.035	0.04	0.20-0.35	—	1.15-1.65	0.25-0.40		100	115	16	
HY80	ARM	0.18	0.10-0.40	0.025	0.025	0.15-0.35	2.00-3.25	1.00-1.80	0.20-0.60	—	80	—	20	—
HY100	ARM	0.20	0.10-0.40	0.025	0.025	0.15-0.35	2.25-3.50	1.00-1.80	0.20-0.60	—	100	—	18	—
Stroloy 2A	BW	0.15-0.21	0.70-1.00	0.025	0.025	0.20-0.35	0.40-0.70	0.80-1.10	0.20-0.30	0.001 B	100	115-145	15	—
Stroloy 5C	BW	0.15-0.21	0.70-1.00	0.025	0.025	0.20-0.35	—	0.75-1.10	0.15-0.25	0.001 B	100	115-145	15	—
RQ100A	B	0.12-0.21	0.45-0.70	0.035	0.04	0.20-0.35	—	—	0.50-0.65	0.001-0.005 B	100	115-135	18	A514
RQ100B	B	0.12-0.21	0.45-0.70	0.035	0.04	0.20-0.35	1.20-1.50	—	0.45-0.60	0.001-0.005 B	100	115-135	18	A514
RQ100	B	0.12-0.21	0.45-0.70	0.035	0.04	0.20-0.35	1.20-1.50	0.85-1.20	0.45-0.60	0.001-0.005 B	100	115-135	18	A514
Jalloy S-90	JL	0.10-0.20	1.50	0.04	0.04	0.50	—	1.50	0.30	0.0005 B	90	(100)	(18)	A514
S-100	JL	0.10-0.20	1.50	0.04	0.04	0.50	—	1.50	0.30	0.0005 B	100	(110)	(18)	A514, A517
S-110	JL	0.10-0.20	1.50	0.04	0.04	0.50	—	1.50	0.30	0.0005 B	110	(120)	(17)	A514, A517
J-100	J	0.16-0.20	0.60-0.90	0.025	0.025	0.20-0.35	1.10-1.40	0.50-0.75	0.40-0.60	0.06-0.12 V	100	115	16	—
HY80	J	0.18	0.10-0.40	0.025	0.025	0.15-0.35	2.00-3.25	1.00-1.80	0.20-0.60	—	80	—	20	—
HY100	J	0.20	0.10-0.40	0.025	0.025	0.15-0.35	2.20-3.50	1.00-1.80	0.20-0.60	—	100	—	18	—
HY140	J	0.16	0.60-0.90	0.015	0.015	0.15-0.35	4.75-5.25	0.40-0.70	0.30-0.65	0.05-0.10 V, 0.20 Cu max	140	—	15	—
Ni-Mo	J	0.28	0.15-0.45	0.020	0.020	0.15-0.35	2.75-3.50	—	0.25-0.60	0.08 V max	75	110	20	—
Kaisaloy 100	K	0.15-0.21	0.80-1.10	0.035	0.04	0.40-0.80	—	0.50-0.80	0.28	0.05-0.15 Zr, 0.0025 B max	100	115-135	18	A514
N-A-XTRA-80	N	0.21	0.60-1.10	0.04	0.04	0.40-0.90	—	0.40-0.90	0.28	0.05 Zr, 0.0025 B max	80	95	18	
90	N	0.21	0.60-1.10	0.04	0.04	0.40-0.90	—	0.40-0.90	0.28	0.05 Zr, 0.0025 B max	90	105	18	A514, A517
100	N	0.21	0.60-1.10	0.04	0.04	0.40-0.90	—	0.40-0.90	0.28	0.05 Zr, 0.0025 B max	100	115	18	
110	N	0.21	0.60-1.10	0.04	0.04	0.40-0.90	—	0.40-0.90	0.28	0.05 Zr, 0.0025 B max	110	125	18	
PX80 Plus	P	0.15-0.21	0.80-1.10	0.035	0.04	0.40-0.90	—	0.50-0.90	0.28		80	95	18	A514
PX90 Plus	P	0.15-0.21	0.80-1.10	0.035	0.04	0.40-0.90	—	0.50-0.90	0.28	0.05 Zr, 0.0025 B max	90	105	18	A514
PX100 Plus	P	0.15-0.21	0.80-1.10	0.035	0.04	0.40-0.90	—	0.50-0.90	0.28		100	115	18	A514, A517
PX110 Plus	P	0.15-0.21	0.80-1.10	0.035	0.04	0.40-0.90	—	0.50-0.90	0.28		110	125	18	A514, A517
HY80	US/L	0.18	0.10-0.40	0.025	0.025	0.15-0.35	2.00-3.25	1.00-1.80	0.20-0.60	—	80	—	20	—
HY100	US/L	0.20	0.10-0.40	0.025	0.025	0.15-0.35	2.25-3.50	1.00-1.80	0.20-0.60	—	100	—	18	—
HY130	US/L	0.12	0.60-0.90	0.010	0.015	0.20-0.35	4.75-5.25	0.40-0.70	0.30-0.65	0.05-0.10 V	130	—	15	—
T-1	US/L	0.10-0.20	0.60-1.00	0.035	0.04	0.15-0.35	0.70-1.00	0.40-0.65	0.40-0.60	0.15-0.50 Cu, 0.0005-0.006 B, 0.03-0.08 V	100	115-135	18	A514, A517-F
T-1-A (c)	US/L	0.12-0.21	0.70-1.00	0.035	0.04	0.20-0.35	—	0.40-0.65	0.15-0.25	0.0005-0.005B, 0.01-0.03 Ti, 0.03-0.08 V	100	115-135	18	A514, A517-B
T-1-B (c)	US/L	0.12-0.21	0.95-1.30	0.035	0.04	0.20-0.35	0.30-0.70	0.40-0.65	0.20-0.30	0.0005 B, 0.03-0.08 V	100	115-135	18	A514, A517-H
Yoloy T-80	Y	(0.10)	(0.45)	0.04	0.05	(0.40)	—	(0.65)	—	(0.25 Cu), 0.05 Ti, 0.01 Al	80	95	16	J410c
YS-T80	Y	(0.10)	(0.45)	0.04	0.05	—	—	—	—	0.01 Al, 0.05 Ti	80	95	18	A656, J410c

Note: Bars, structural shapes, and tubing can also be supplied in these alloys. (a) Minimum unless otherwise specified; (b) All alloys quenched and tempered except NAX-80, Yoloy T-80, and YS-T80 which are hot rolled; (c) Copper optional.

IX. Abrasion-Resistant Alloys

Name	Producer Code	Composition, % C	Mn	P	S	Si	Cu	Cr	Mo	Other (a)	Condition	Mechanical Properties Hardness, Bhn	Yield, 1,000 Psi	Tensile, 1,000 Psi
AW-AR	AW	0.35-0.50	1.50-2.00	0.05	0.055	0.15-0.30	—	—	—	—	HR	235	—	—
SSS-AR-321	ARM	0.25	0.40-0.70	0.04	0.05	0.20-0.35	0.20-0.40	0.85-2.0	0.15-0.60	0.04-1.10 Ti or V, 0.0015-0.005 B	Q&T	321	—	—
SSS-AR-360	ARM	0.25	0.40-0.70	0.04	0.05	0.20-0.35	0.20-0.40	0.85-2.0	0.15-0.60	0.04-0.10 Ti or V, 0.0015-0.005 B	Q&T	360	—	—
SSS-AR-400	ARM	0.25	0.40-0.70	0.04	0.05	0.20-0.35	0.20-0.40	0.85-2.0	0.15-0.60	0.04-0.10 Ti or V, 0.0015-0.005 B	Q&T	400	—	—
Armco Abrasion Resisting	ARM	0.33-0.43	1.40-2.00	0.05	0.05	0.15-0.35	—	—	—	—	HR	225	—	—
AR-No. 235	B	0.35-0.50	1.40-2.00	0.05	0.05	0.15-0.30	—	—	—	—	HR	(235)	—	—
RQAR-321	B	0.25-0.32	0.40-0.65	0.035	0.040	0.20-0.35	—	0.80-1.15	0.15-0.25	—	Q&T	321	—	—
RQAR-340	B	0.25-0.32	0.40-0.65	0.035	0.040	0.20-0.35	—	0.80-1.15	0.15-0.25	—	Q&T	340	—	—
RQAR-360	B	0.25-0.32	0.40-0.65	0.035	0.040	0.20-0.35	—	0.80-1.15	0.15-0.25	—	Q&T	360	—	—
RQAR-400	B	0.25-0.32	0.40-0.65	0.035	0.040	0.20-0.35	—	0.80-1.15	0.15-0.25	—	Q&T	400	—	—
RQC-321	B	0.28	1.50	0.040	0.050	0.20-0.60	—	—	—	0.005B	Q&T	321	—	—
RQC-340	B	0.28	1.50	0.040	0.050	0.20-0.60	—	—	—	0.005B	Q&T	340	—	—
RQ321A	B	0.12-0.21	0.45-0.70	0.035	0.04	0.20-0.35	—	—	0.50-0.65	0.001-0.005 B	Q&T	321	—	—
RQ340A	B	0.12-0.21	0.45-0.70	0.035	0.04	0.20-0.35	—	—	0.50-0.65	0.001-0.005 B	Q&T	340	—	—
RQ360A	B	0.12-0.21	0.45-0.70	0.035	0.04	0.20-0.35	—	—	0.50-0.65	0.001-0.005 B	Q&T	360	—	—
RQ321B	B	0.12-0.21	0.45-0.70	0.035	0.04	0.20-0.35	—	—	0.45-0.60	1.20-1.50 Ni, 0.001-0.005 B	Q&T	321	—	—
RQ340B	B	0.12-0.21	0.45-0.70	0.035	0.04	0.20-0.35	—	—	0.45-0.60	1.20-1.50 Ni, 0.001-0.005 B	Q&T	340	—	—
RQ360B	B	0.12-0.21	0.45-0.70	0.035	0.04	0.20-0.35	—	—	0.45-0.60	1.20-1.50 Ni, 0.001-0.005 B	Q&T	360	—	—
RQ-321	B	0.12-0.21	0.45-0.70	0.035	0.04	0.20-0.35	—	0.85-1.20	0.45-0.60	1.20-1.50 Ni, 0.001-0.005 B	Q&T	321	—	—
RQ-340	B	0.12-0.21	0.45-0.70	0.035	0.04	0.20-0.35	—	0.85-1.20	0.45-0.60	1.20-1.50 Ni, 0.001-0.005 B	Q&T	340	—	—
RQ-360	B	0.12-0.21	0.45-0.70	0.035	0.04	0.20-0.35	—	0.85-1.20	0.45-0.60	1.20-1.50 Ni, 0.001-0.005 B	Q&T	360	—	—
Abrasion Resisting, Med. Hard	IN	0.35-0.50	1.20-1.65	0.04	0.05	0.10-0.30	—	—	—	—	HR	210	—	—
Abrasion Resisting, Full Hard	IN	0.70-0.85	0.60-1.00	0.04	0.05	0.30	—	—	—	—	HR	250	—	—
Jalloy AR	JL	0.25-0.31	1.65	0.04	0.04	0.15-0.30	0.20	1.20	0.35	0.0005 B	Q&T	—	90	—
Jalloy AR-280	JL	0.25-0.31	1.65	0.04	0.04	0.15-0.30	0.20	1.20	0.35	0.0005 B	Q&T	260	(130)	—
320	JL	0.25-0.31	1.65	0.04	0.04	0.15-0.30	0.20	1.20	0.35	0.0005 B	Q&T	300	(140)	—
360	JL	0.25-0.31	1.65	0.04	0.04	0.15-0.30	0.20	1.20	0.35	0.0005 B	Q&T	340	(157)	—
400	JL	0.25-0.31	1.65	0.04	0.04	0.15-0.30	0.20	1.20	0.35	0.0005 B	Q&T	400	(184)	—
Jalloy AR Q	JL	0.25-0.31	1.65	0.04	0.04	0.15-0.30	0.20	1.20	0.35	0.0005 B	Q&T	500	(217)	—
Jalloy S-340	JL	0.10-0.20	1.50	0.04	0.04	0.50	—	1.50	0.30	0.0005 B	Q&T	320	(150)	—
Kaisaloy AR	K	0.35-0.50	1.50-2.00	0.05	0.055	0.15-0.35	—	—	—	—	HR	—	—	—
Kaisaloy 360 HI-WR	K	0.17-0.25	0.80-1.20	0.035	0.04	0.40-0.90	—	0.50-0.90	0.18-0.60	0.0025 B max	Q&T	360	—	—
Kaisaloy 400 HI-WR	K	0.20-0.30	0.80-1.20	0.035	0.04	0.40-0.90	—	0.50-0.90	0.18-0.60	0.0025 B max	Q&T	400	—	—
AR-300	L	0.28	1.40	0.04	0.05	0.20-0.50	0.20	—	—	—	Q&T	285-321	—	—
XAR-15	N	0.21	0.60-1.10	0.04	0.04	0.90	—	0.45-0.85	0.30	0.05 Zr, 0.0025 B max	Q&T	360	(165)	(180)
XAR-30 (b)	N	0.30	0.60-1.10	0.04	0.04	0.90	—	0.45-0.85	0.30	0.05 Zr, 0.0025 B max	Q&T	360	(165)	(180)
Orelloy AR	O	0.43	1.50	0.024	0.013	0.29	0.08	—	—	—	HR	235	—	—
PX360 Bhn	P	0.15-0.21	0.80-1.10	0.035	0.04	0.40-0.90	—	0.50-0.90	0.28	0.05 Zr, 0.0025 B max	Q&T	321-360	(165)	(180)
Republic 100-AR	R	0.16	1.40	0.015	0.010	0.25	0.70	1.40	0.45	1.40 Ni 0.04 Al	HR	375-415	100	140
AR	US/L	0.35-0.50	1.50-2.00	0.05	0.055	0.15-0.35	—	—	—	—	HR	(200-250)	—	—
300	US	0.32	1.60	0.04	0.05	0.60	—	—	—	0.0005 B	Q&T	285	—	(136)
350	US	0.32	1.60	0.04	0.05	0.60	—	—	—	0.0005 B	Q&T	321	—	(152)
360	US	0.26-0.33	1.15-1.50	0.035	0.040	0.20-0.35	—	0.40-0.65	0.08-0.15	0.0005 B	Q&T	360	—	(175)
T-1-A 321	US	0.12-0.21	0.70-1.00	0.035	0.040	0.20-0.35	Opt	0.40-0.65	0.15-0.25	0.03-0.08 V,	Q&T	321	—	(152)
340										0.01-0.03 Ti,	Q&T	340	—	(163)
360										0.0005-0.005 B	Q&T	360	—	(175)
T-1-B 321	US/L	0.12-0.21	0.95-1.30	0.035	0.040	0.20-0.35	Opt	0.40-0.65	0.20-0.30	0.30-0.70 Ni,	Q&T	321	—	(152)
340										0.03-0.08 V,	Q&T	340	—	(163)
360										0.0005 B	Q&T	360	—	(175)
T-1 321	US/L	0.10-0.20	0.60-1.00	0.035	0.040	0.15-0.35	0.15-0.50	0.40-0.65	0.40-0.60	0.70-1.00 Ni,	Q&T	321	—	(152)
340										0.03-0.08 V,	Q&T	340	—	(163)
360										0.0005-0.006 B	Q&T	360	—	(175)
Austenitic manganese A	US	0.70-0.90	12.50-14.50	0.07	0.04	0.50-0.80	—	0.50	—	3.00-3.50 Ni	HR	—	—	—
B	US	1.00-1.25	11.50-13.50	0.07	0.04	0.15-0.30	—	0.50	—	—	HR	—	—	—
C	US	0.70-0.90	12.50-14.50	0.07	0.04	0.50-0.80	—	0.50	0.40-0.60	1.75-2.25 Ni	HR	—	—	—

Note: HR, hot rolled; Q&T, quenched and tempered; (a) Minimum unless otherwise specified; (b) Available to Bhn 500.

Guidelines for Selecting a Cold Extruding Steel

BERNARD S. LEVY

In cold extrusion, the characteristics of different fabrication sequences and the metallurgical consequences of different grades of steel provide the information for determining the most economical material-process combination. Fabrication operations affect material requirements in two ways. First, the degree of tensile stress has a predominant effect on the formability which is required to make the part without fracture. Secondly, the amount of deformation affects the required forming forces. In cold extrusion, metal flows in three major ways, as shown in Fig. 1. These modes of deformation are backward extrusion, forward extrusion, and upsetting. More complicated cold extrusions can generally be considered as some combination of these basic metal movements.

EFFECT OF COLD EXTRUSION VARIABLES

While cold extrusion operations generally exhibit substantial average compressive stress, tensile stresses are also generated. In backward extrusion, tensile stresses develop as a result of the pronounced relative shear forces in the extrusion annulus between the corner of the extrusion punch and the die wall. These tensile stresses are increased as the corner radius of the punch is increased in sharpness, and as lubrication effectiveness decreases. Another cause of tensile stresses results from the metal tending to move at a constant mass flow rate. Thus, the metal must decelerate when there is an increase in cross section, and an internal tensile stress is required to produce this effect. If the increase in cross section is outside the extrusion annulus, the average compressive stress is usually lower, which increases the severity of any tensile stresses that are produced. With rounded punch corners and good lubrication, relative shear forces in the extrusion annulus are generally not a problem. However, tensile stresses from deceleration progressively require more ductility in the steel if failure is to be averted.

BERNARD S. LEVY is Supervising Research Engineer, Inland Steel Research Laboratories, East Chicago, Ind. 46312.

In forward extrusion, the primary failure mode is chevron formation. Avitzur[1] has shown that chevrons are caused by periodic tensile stresses that result from nonhomogeneous deformation which requires an abrupt acceleration of the metal in the extrusion annulus. According to Zimmerman and Avitzur,[2] the probability of a chevron failure is increased by decreasing the semicone angle of the extrusion die, by decreasing the percent reduction in area, by increasing friction, and by decreasing work hardening capability in the metal. If the empirical true stress-true strain equation, $\sigma = k\epsilon^n$, is utilized, the Avitzur work hardening parameter, β, can be written as:

$$\beta = \frac{(\phi + \Delta)^n - \phi^n}{1 + \phi^n - (\phi + \Delta)^n}$$

This expression indicates that β decreases as the strain-hardening exponent (n) decreases, as the strain prior to the forward extrusion (ϕ) increases, and as the strain produced by the forward extrusion (Δ) decreases. These observations are consistent with the common observation that applying a light reduction after making heavy reductions increases the likelihood of chevrons.

When metal is upset, the typical formability problem is splitting at the surface. Hoffmanner[3] has shown that tensile stresses rise with increasing barreling, which is the degree to which the sides of a part bulge during upsetting. The effect of various process variables on barreling (and the degree of tensile stress) is summarized below:

1) Increasing reduction increases the degree of barreling.[3,5]

2) Decreasing the aspect ratio increases the degree of barreling for steel,[3,5] but has no effect for aluminum.[3,4] (The aspect ratio is defined, for a cylinder, as the initial height divided by the initial diameter.)

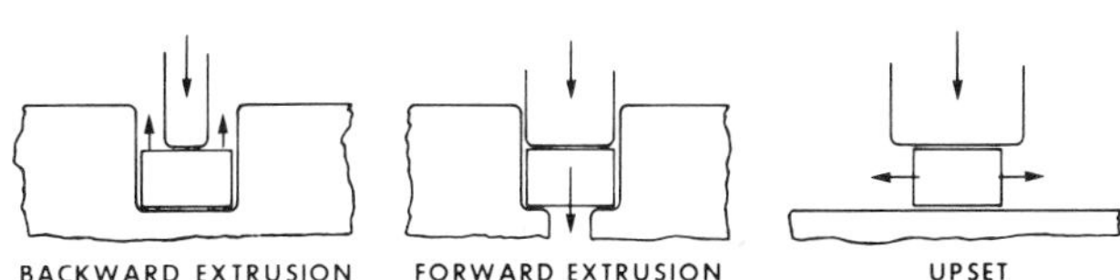

Fig. 1—Metal flow in various types of cold extrusion.

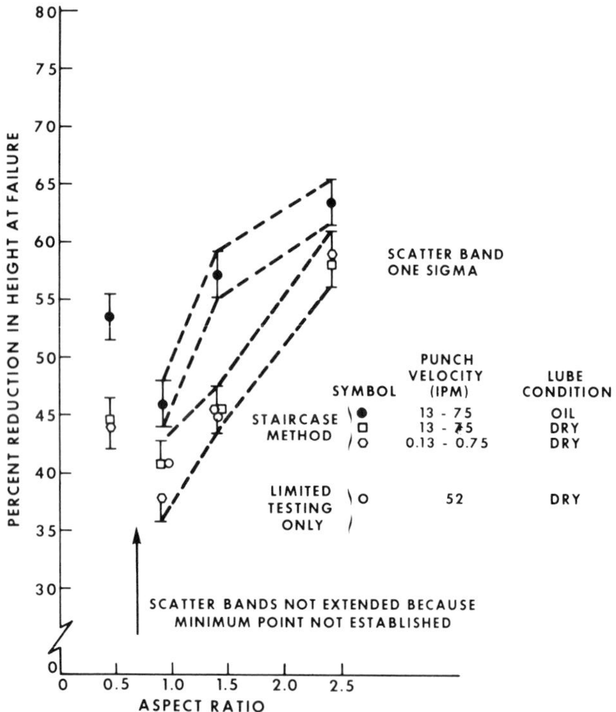

Fig. 2—Effect of aspect ratio and lubrication on reduction in height at fracture during upsetting.[7]

3) Decreasing the lubricant effectiveness in general increases the degree of barreling.[4,6,7]

4) Press speed appears to have no effect on barreling,[4,6] though one would expect it to have an effect when press speed influences lubricant effectiveness.

The effect of lubrication, press speed, and aspect ratio on the reduction in height at failure for samples from the same lot of centerless ground AISI 1045 aluminum killed steel is shown in Fig. 2. These data[7] are consistent with the previous observations. In this work, the failure criteria was the reduction in height at which cracking was observed by careful examination with a 3X eyepiece.

Trap heading is a common type of metal flow by upset in which the expanding metal comes in contact with the die wall so that the diameter can increase only in the portions of the bulged region which previously exhibited lesser diameters. Kuhn *et al.*[8] have shown that this type relieves the tensile stresses in the region in contact with the die wall. Thus, trap heading is an effective method of achieving a uniform increase in diameter.

As has been indicated, the amount of deformation affects the required forming force. In cold extrusion, the required forming force is an important parameter in determining the capacity of required equipment which depends on the reduction in area, the cross sectional area of the slug, and the tensile strength or hardness.[25,26] In backward extrusion, the punch pressure is also important because this stress can exceed the capacity of some punch materials. Punch pressure in backward extrusion depends primarily on the tensile strength, or hardness, of the steel being extruded to reductions-in-area in the 30 to 70 pct range. With respect to selecting steels for cold extrusion, the relative effect of different steels can be assessed by their tensile strength or hardness.

Assessing the formability of steel is less understood, but some basic tensile properties can be used to group different types of steels. The reduction in area at fracture is probably the generally most useful tensile property, though strain-hardening behavior can be more important in forward extrusion. Since chevron failures occur at the center of the bar, this region of the steel is most important in forward extrusion, while the surface is the most critical region when a bar is upset. Ideally, tensile test results should come from the most critical region. Though the effect is hard to rate quantitatively, carbon banding in steel also has a detrimental effect on formability.

EFFECTS OF DIFFERENT STEELS

The metallurgical consequences of different steel grades can be described in terms of the tensile properties which affect performance in cold extrusion. The basic alloying elements in steel are carbon and manganese. To illustrate some of the effects of variations in carbon and manganese content, steels were produced in the laboratory with a semikilled casting practice. forged to flats, normalized or annealed, and tested. Properties of normalized steel are similar to those for hot-rolled stock. Silicon contents were less than 0.05 pct, and aluminum contents were less than 0.01 pct. Other residuals were typical of commercial plain carbon steels.

Fig. 3 illustrates the effect of carbon and manganese content and partial spheroidization on tensile strength. Increasing carbon and manganese contents act to raise tensile strength, the effect of manganese being more pronounced at higher carbon contents. Partial spheroidization reduces significantly the tensile strength, the effect also being more pronounced at higher carbon contents. While these data are generally valid, it is im-

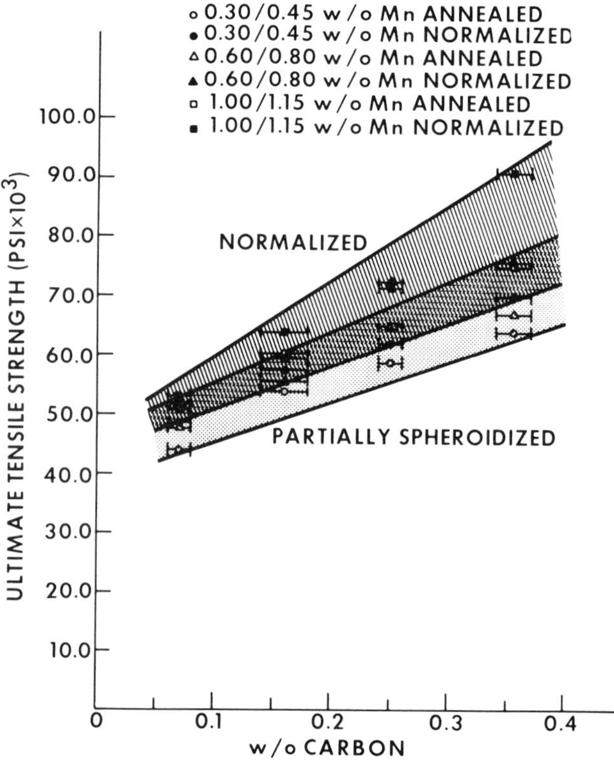

Fig. 3—Effect of manganese content, carbon content, and partial spheroidization on tensile strength.

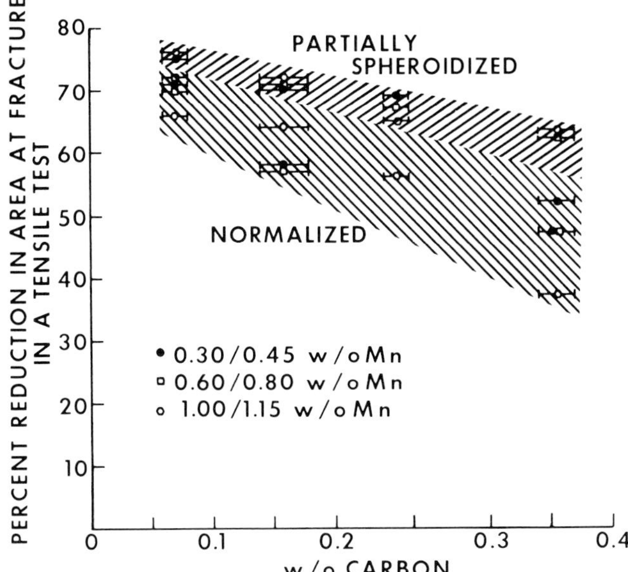

Fig. 4—Effect of manganese content, carbon content, and partial spheroidization on the reduction in area at fracture.

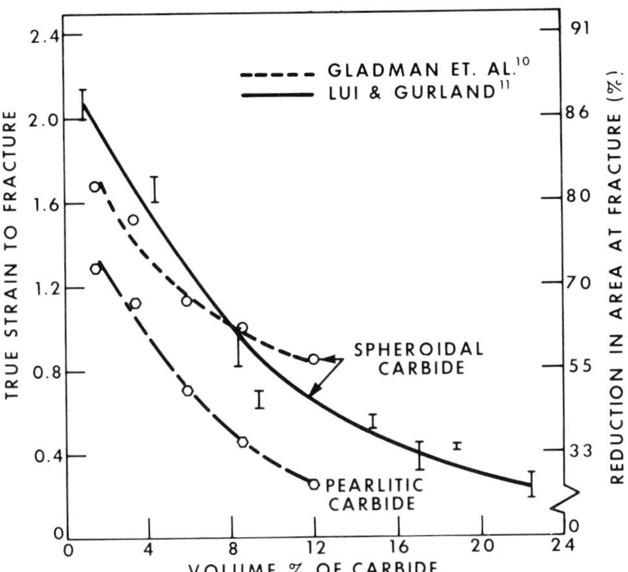

Fig. 5—Effect of volume percent pearlitic and spheroidic cementite on reduction in area at fracture in a tensile test.[10,11]

portant to realize that variations in mill processing can alter properties. Consider, as examples, the effect of controlled cooling or the difference in properties between those of a $\frac{3}{8}$ in. diam rod and a 2 in. diam bar. Furthermore, control cooled steels will be harder than comparably annealed conventional steels even though the degree of spheroidization would be greater for the control cooled steels.[9]

Fig. 4 shows the effect of carbon content, manganese content, and partial spheroidization on reduction in area at fracture. These data indicate that increases in carbon content significantly decrease the reduction in area at fracture, but no consistent effect of manganese is apparent. Increasing either carbon or manganese will increase the volume fraction of pearlite, but only an increase in carbon content will increase the volume fraction of cementite. Therefore, it can be inferred

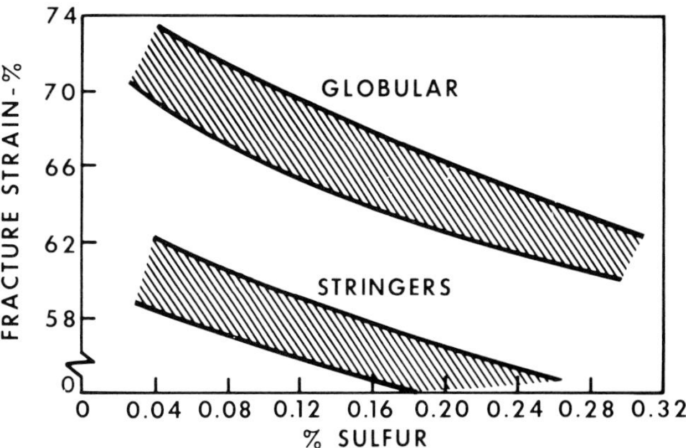

Fig. 6—Effect on volume fraction sulfides and inclusion morphology on fracture strain in upsetting.[21]

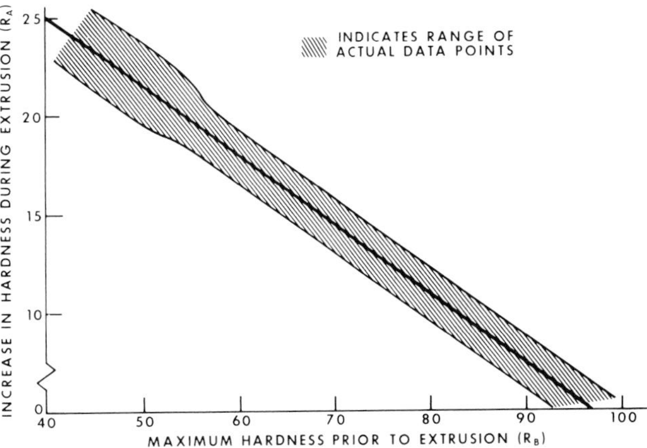

Fig. 7—Effect of hardness prior to extrusion on the hardness after extrusion.[23]

that the increase in volume fraction of cementite is the principal cause of the decrease in reduction in area at fracture. Such an inference is consistent with the work of Gladman *et al.*,[10] and Liu and Gurland,[11] who have shown that the reduction in area at fracture decreases as the volume percent of spheroidized or pearlitic carbide increases, Fig. 5.

The data in Figs. 4 and 5 also show that spheroidizing a normalized steel increases the reduction in area at fracture in a tensile test. As Billigman[12] has shown, fully spheroidizing plain carbon steels containing 0.25 to 0.60 pct C increases the reduction in area at fracture as much as 20 pct in a tensile test. That is, if the reduction in area at fracture was 45 pct with no spheroidization, it would be 65 pct after complete spheroidization. The work of Billigman[12] and the author both indicate that the effect of spheroidization increases as the carbon content increases; these results are generally consistent with the observations of Gladman *et al.*[10] It is also generally accepted that slow cooling and increasing manganese content promote banding, and thus would be expected to reduce the true strain at fracture.

Data on the strain-hardening exponent is more limited. However, James and Kottcamp[13] found that increasing the carbon content decreases the strain-hardening exponent, and that spheroidization increases the strain-hardening exponent. Their data also show that the effect of spheroidization is more pronounced for the higher carbon steels.

Other common alloying elements in steel include Si, B, Mo, Cr, and Ni. In a base composition typical of an AISI 1010 steel, additions of up to 2 pct Si result in an increase in tensile strength of 13,000 psi for each 1 pct Si. However, Grozier *et al.*[14] have shown that, in a base composition typical of AISI 1038 steel, additions of up to 0.8 pct Si increased the tensile strength by about 29,000 psi for each 1 pct Si, plus 800 psi for each 1 pct of pearlite in the microstructure as a result of the silicon addition. Thus, the effect of silicon on tensile strength is apparently similar to that of manganese in that it is more pronounced in higher carbon steels. Silicon also promotes finer carbides, as well as silicate stringers, which can be minimized by aluminum deoxidation.[15] It would be expected that the reduction in area at fracture is adversely affected by both increasing silicon content and numbers of silicate stringers.

Boron, an element which improves hardenability substantially, is believed to have little or no effect on tensile strength or reduction in area at fracture. Therefore, if boron is substituted for other hardenability agents which do increase the tensile strength or reduce the reduction in area at fracture, boron can be said to increase the hardenability of steel with concommittant improved formability and reduced forming loads.[16,17]

Crawford *et al.*[18] have shown clearly, in tests on normalized, annealed commercial steels, that molybdenum increases the volume fraction of ferrite and increases the mean free ferrite path, and that these effects balance or outweigh the solid solution strengthening effect of molybdenum. Cooksey[19] studied the effect of alloying elements on extrusion pressure together with the effect of volume percent pearlite, ferrite grain size, and mean free ferrite spacing. In general, the strain hardening exponent for alloy steels decreases as the stress at 0.2 true strain increases at a rate similar to that of plain carbon steels.[20] It is also important to mention that alloy steels are more likely to be affected by differences in mill processing than are plain carbon steels.

Elements added to increase machinability include S, Pb, and Te. Sulfur markedly reduces the reduction in area in a tensile test, increases the degree of banding, and has no significant effect on the tensile strength. Data taken from Paliwoda and Brown,[21] replotted in Fig. 6, also show the effect of globular and stringer type oxysulfides on fracture strain for the upsetting of cylinders. When the metal movement is upsetting, the detrimental effects of sulfur on the formability of low carbon steels can be avoided by using a sulfur-innoculated steel. Often described as scrapless nut steels, these grades do not exhibit any substantial machinability improvement in the relatively "sulfur free" outer portions of the bar.

At room temperature, lead has no significant effect on the steel's tensile strength or reduction in area at fracture. However, in the vicinity of its melting point (500° to 800°F), the reduction in area at fracture drops significantly due to lead embrittlement. This embrittlement is more severe as the tensile strength increases.[22] As a result of this behavior, leaded steels have fractured in some extrusion jobs. The common conditions for failure seem to be tensile stress components and sufficient deformation rate, or friction, to produce temperatures in the lead embrittlement range. When both of these conditions are not present, lead bearing steels successfully extrude. Also, a reduced lead content can minimize fracture during cold extrusion.

Tellurium has no effect on the tensile strength, reduction in area at fracture, or banding. However, tellurium-containing steels are more costly, and their use is recommended only when the addition of lead or sulfur results in problems during extrusion.

Of significance finally are the properties required in the final part. Since they represent a complex subject, only a few general guidelines are included:

1) Hardenability and tempering characteristics depend on the composition and prior austenitic grain size.

2) Cold work provides substantial static strength increase. Fig. 7 shows the relationship between the maximum hardness prior to extrusion and the increase in hardness during extrusion.[23]

3) Strength from cold work is useful in increasing fatigue strength in long life stress-controlled fatigue, but is not useful in increasing fatigue life for strain control fatigue.[24] If the steel is not heat treated prior to use, residual stresses from the forming operation can also have a significant effect on fatigue life.

CONCLUSIONS

As can be seen, selecting the required steel properties for a particular cold extrusion job can be a difficult, painstaking job. The following is proposed as a guideline for providing the needed information:

1) What steels have worked on the job? What are their properties? What improvements can be made?

2) What steels have worked on similar jobs? What are their properties?

3) Compare the metal movements in the part with the metal movements that can cause tensile stresses.

4) Determine if there are "excessive" loads on the press or tooling? Is it important to try and reduce forming pressures?

Note: If the job or similar jobs have run with no trouble, the steel being used may possess more quality than is required.

REFERENCES

1. B. Avitzur: Study of Flow Through Conical Converging Dies, Dept. of Metallurgy and Materials Science, Lehigh University, Sept., 1970.
2. Z. Zimmerman and B. Avitzur: *Trans. ASME, J. Eng. Ind.*, Feb. 1970.
3. A. L. Hoffmanner: Tech. Rep. AFML-TR-69-174, June, 1969.
4. K. M. Kulkarni and S. Kalpakjian: *Trans. ASME, J. Eng. Ind.*, Aug., 1969.
5. H. A. Kuhn and P. W. Lee: *Met. Trans.*, Nov., 1971, vol. 2.
6. T. E. Fine and B. S. Levy: *Mechanical Working and Steel Processing XI*, AIME, 1973.
7. B. S. Levy: *Mechanical Working and Steel Processing X*, AIME, 1972.
8. H. A. Kugh *et al*: *Modern Developments in Powder Metallurgy*, 1971, vol. 4.
9. T. E. Fine and B. S. Levy: *Mechanical Working and Steel Processing IX*, AIME, 1971.
10. T. Gladman *et al.*: Paper Delivered at BSC/ISI Conference, March 24-25, 1971, Scarborough, England.
11. C. T. Liu and J. Gurland: *Trans. ASM*, 1968, vol. 61.
12. J. Billigman: Brutcher Translation HB-5020, Condensed Translation from *Stahl Eisen*, 1951, vol. 71.
13. C. T. James and E. H. Kottcamp: Paper presented at Regional Technical Meeting of AISI, Chicago, October 20, 1965.
14. J. D. Grozier *et al.*: *Mechanical Working and Steel Processing VII*, AIME, 1969.
15. J. W. Farrell *et al.*: *28th Electric Furnace Conf. AIME*, 1970.

16. J. B. R. Anderson: SME Tech. Paper MF 69-539.
17. F. S. Gezella: SME Tech. Paper MF 69-542.
18. J. D. Crawford *et al.*: *Metals Eng. Quart.*, 1964, vol. 4, no. 4.
19. R. J. Cooksey: *Metal Forming,* April, 1968.
20. J. G. McMullin: Discussion of J. D. Crawford *et al.*: *Metals Eng. Quart.*, May, 1965.
21. E. J. Paliwoda and D. M. Brown: *Mechanical Working and Steel Processing X,* AIME, 1972.
22. D. C. Huffaker: *J. Iron Steel Inst.*, to be published.
23. C. F. Schrader and B. S. Levy: Paper presented at Regional Technical Meeting Of the AISI, Chicago, September 30, 1964.
24. B. S. Levy: SAE, Paper 720017, Jan., 1972.
25. H. L. Pugh *et al.*: Sheet Metal Industries, April, 1966.
26. T. Altan and J. R. Becker: SME Tech. Paper MF 72-142, 1972.

Silicon Steels and Their Applications

F. A. MALAGARI, F. J. MILLER, AND S. L. AMES

A soft magnetic material, silicon steel is used in thousands of motors, generators, and transformers. Though the basic properties of interest are largely present in pure iron, silicon is added to raise resistivity, and to decrease magnetostriction and magnetocrystalline anisotropy. In addition, the α phase is stabilized to higher temperatures, allowing special processing and heat treatments to develop a preferred crystal orientation. The amount of silicon added is limited largely by its effect on cold rollability.

This paper reviews the development of alloys in the iron-silicon family. Factors which affect magnetic properties will be described and related to the magnetic quality of today's steels. Processing, properties, and applications of nonoriented and oriented silicon steels will be discussed.

Silicon steel is undoubtedly the most important soft magnetic material in use today. Applications vary in quantities from the few ounces used in small relays or pulse transformers to tons used in generators, motors, and transformers. Annual consumption of electrical steels in the United States has exceeded 1 million ton, and the consumption rate continually increases. Continued growth in electrical power generation has required development of better steels to decrease wasteful dissipation of energy (as heat) in electrical apparatus and to minimize the physical dimensions of the increasingly powerful equipment now demanded.

The earliest soft magnetic material was iron, which contained many impurites. Researchers found that the addition of silicon increased resistivity (to give a reduction in eddy current loss), decreased hysteresis loss, increased permeability (to allow a smaller exciting current), and virtually eliminated aging. Sometime later, oriented silicon steel was developed—it has a (110)[001] (cube-on-edge) orientation. Strip for electrical equipment is cut so as to maximize passage of the magnetic flux in the easily-magnetized [001] direction.

Substantial quantities of oriented steel are used, mainly in power and distribution transformers. However, it has not supplanted nonoriented silicon steel, which is used extensively where a low-cost, low-loss material is needed, particularly in rotating equipment. Mention should also be made of the relay steels, used widely in relays, armatures, and solenoids. Relay steels contain 1.25 to 2.5 pct Si, and are used in direct current applications because of better permeability, lower coercive force, and freedom from aging.

F. A. MALAGARI, F. J. MILLER, and S. L. AMES are associated with Allegheny Ludlum Industries, Inc., Brackenridge, Pa. 15014.

GENERAL PROPERTIES OF FE-SI STEELS

Important physical properties of silicon steels include resistivity, saturation induction, magnetocrystalline anisotropy, magnetostriction, and Curie temperature, Figs. 1 and 2. Resistivity, quite low in iron, increases markedly with the addition of silicon. Higher resistivity lessens the core loss by reducing the eddy current component.

Magnetic saturation is determined by the magnetic moment per atom of the spining electrons. For iron, the saturation induction at room temperature is about 21,500 G, highest for any element. All elements except cobalt decrease the saturation induction of iron, as is shown for silicon.

Preferential alignment of the magnetic moments along certain crystallographic axes is known as magnetocrystalline anisotropy. As noted earlier, today's grain oriented steels make use of the relative ease of magnetization in the [001] direction, Fig. 3.

Silicon steel changes length when it is magnetized, reaching a limit at saturation—this is referred to as saturation magnetostriction. Because the change in length creates sound, which can be very annoying, many modern power transformers have noise limit specifications. Raising the silicon content will lower magnetostriction, but processing becomes more difficult.

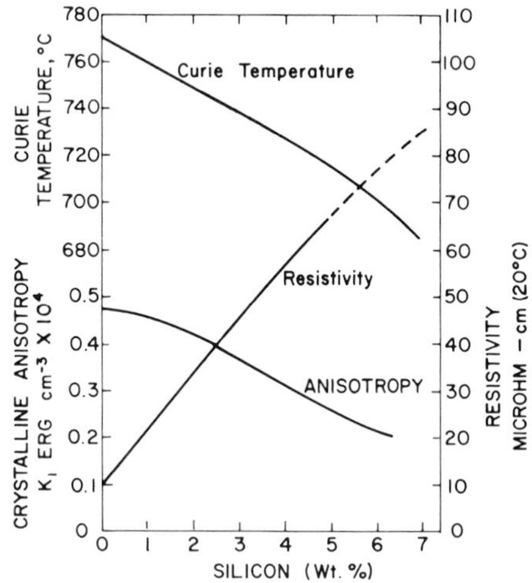

Fig. 1—Curie temperature, resistivity, and anisotropy change with composition.

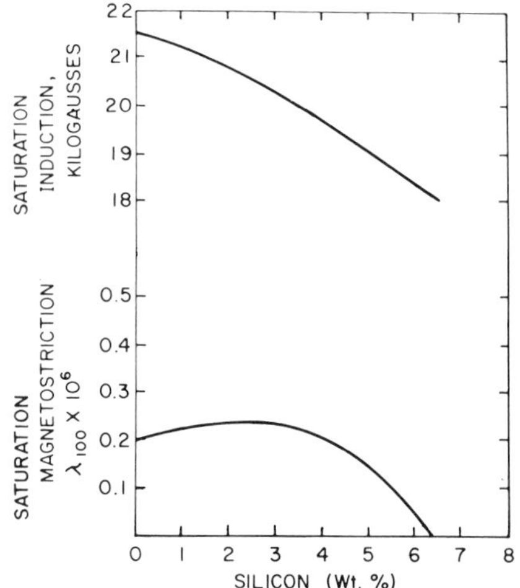

Fig. 2—Silicon content affects saturation induction and magnetostriction.

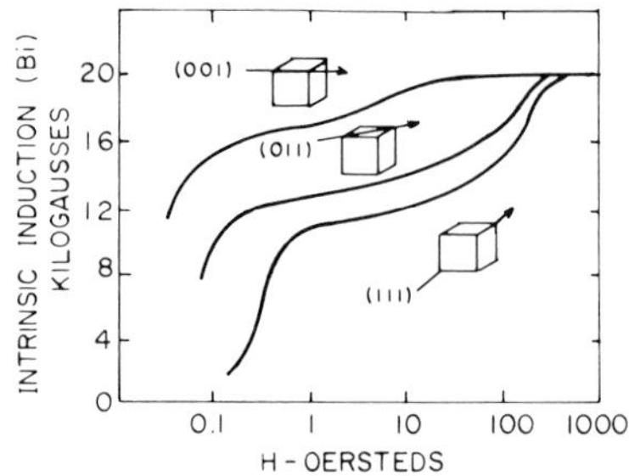

Fig. 3—Crystallographic direction affects magnetization of a Si-Fe crystal.

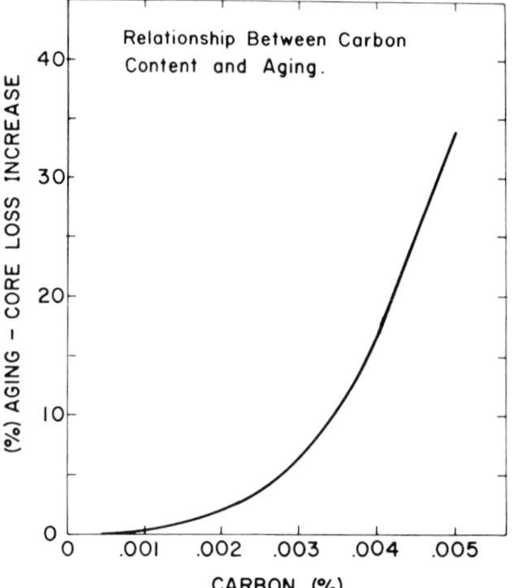

Fig. 4—Core losses increase with carbon content.

The high Curie temperature of iron will be lowered by alloying elements, but the decrease is of little importance to the user of silicon steels.

The magnetization process is influenced by impurities, grain orientation, grain size, strain, strip thickness, and surface smoothness. One of the most important ways to improve soft magnetic materials is to remove impuries which interfere with domain-wall movement; they are least harmful if present in solid solution. Compared with other commercial steels, silicon steel is exceptionally pure. Because carbon, an interstitial impurity, can harm low induction permeability, it must be removed before the steel is annealed to develop the final texture. If the residual carbon is too high, Fe_3C particles will form and grow to cause magnetic aging and an increase in core loss, Fig. 4. Investigations by Yensen and Ziegler,[1] Leak and Leak,[2] and Benford[3] show that C, S, N, and O should be kept to a minimum to assure the best magnetic quality.

To demonstrate the benefits of orientation, 3 pct Si steel at a magnetizing force of 10 oe has an induction of about 14,000 G when randomly oriented, while commercial cube-on-edge oriented stock has a with-grain induction greater than 18,000 G, Fig. 5. Because of this easier magnetization of oriented sheet, transformers can be made lighter, and will have lower core losses.

McCarty,[4] *et al.*, studied the relationship between texture and magnetic properties of oriented silicon steel by defining grain orientation in terms of three

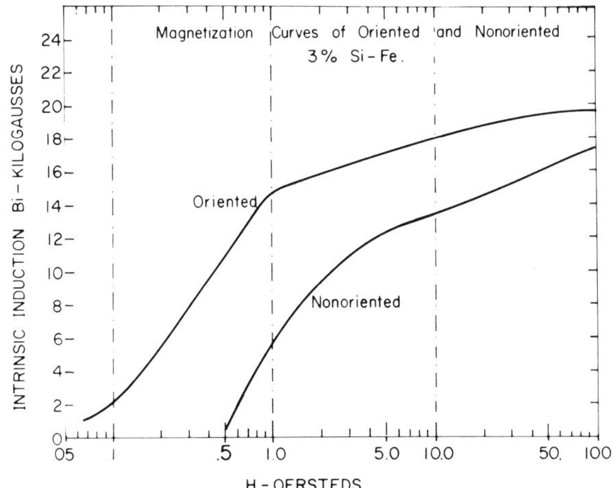

Fig. 5—Magnetization curves of oriented and nonoriented 3% Si-Fe.

angles. In this study, they found that permeability at 10H correlated strongly with the divergence in the plane of the sheet of the [001] direction from the rolling direction. The vertical tilt of the [001] direction from the rolling direction, and the rotation of the (110) plane about the direction of magnetization had less effect on magnetic qualities.

The mechanism for the growth of grains with cube-on-edge orientation during the final anneal is not completely understood. The process involves secondary recrystallization, which, by definition, is characterized by accelerated growth of one set of grains in an already recrystallized matrix, Fig. 6. For secondary recrystallization, normal grain growth must be inhibited in some manner. As the temperature is raised, certain grains break loose from the inhibiting forces, and grow extensively at the expense of their neighbors. Producers know that, on a practical basis, appropriate cold rolling and recrystallization sequences must be carefully followed to obtain the desired secondary recrystallization nuclei and the correct texture. Today's silicon steels use MnS as the grain growth inhibitor, but other compounds, such as carbides, oxides, or nitrides, are also effective. For example, Taguchi *et al.*[5] reported 10H permeabilities greater than 1900 for a cube-on-edge oriented silicon steel containing aluminum nitride.

MAKING AND USING ORIENTED STEEL

Oriented silicon steel is more restricted in composition than nonoriented varieties. The texture is developed by a series of careful working and annealing operations, and the material must remain essentially single-phase throughout processing, particularly during the final anneal because phase transformation destroys the texture. To avoid the γ loop of the Fe-Si phase system, Fig. 7, today's commercial steel has about 3.25 pct Si. Higher silicon varieties, which might be favored on the basis of increased resistivity and lower magnetostriction, are precluded by difficulties in cold rolling.

Silicon steels have been melted in open hearths, electric furnaces, and basic oxygen furnaces. Scrap is carefully selected and furnaces are rigidly controlled for melting and refining, producing a high purity iron; silicon is then added in the ladle. Residual impurities are kept as low as possible with the exception of manganese and sulfur, both necessary for texture development. The ingots are soaked in pits held at 1325°C (2417°F), and either hot rolled direct to band (80 to 100 mil thick) or hot rolled to a slab, which is reheated and finish rolled in a second operation. During the hot-working process, ingot heating rate, maximum temperature attained, and time held at this temperature are carefully controlled to assure good quality in the final product. After the hot-rolled coil is cleaned, it is cold rolled to an intermediate thickness, continuously annealed to recrystallize the structure, and cold rolled to the desired gage, final reduction being carefully controlled. Final normalizing is essential to decarburize the strip to a very low level, preferably less than 0.003 pct C. Temperature, atmosphere composition, and dew point are closely controlled to decarburize the strip without oxidizing the surface. During this treatment, primary recrystallization occurs, forming small, uniform, equiaxed grains. The strip is then coated with magnesium oxide, which keeps the coil convolutions from sticking during the final texture-developing anneal. During this anneal, sulfur is removed, and the magnesium oxide reacts with the film of silicon dioxide on the strip. The coating of magnesium silicate glass which forms will provide electrical insulation between successive laminations when assembled in a transformer core. At this stage, the steel is graded by cutting Epstein samples from the coil; the samples are stress relief annealed and flattened at 1450°F, and tested for core loss.

Applications for oriented silicon steel include transformers (power, distribution, ballast, instrument, audio, and specialty), and generators for steam turbines and water wheels. Because the major application is in

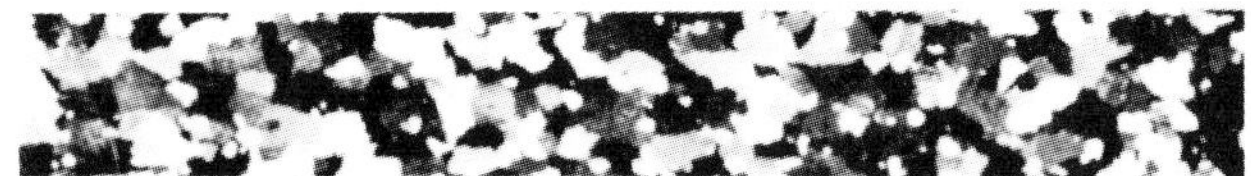
Fig. 6—This grain structure is typical of oriented silicon steel.

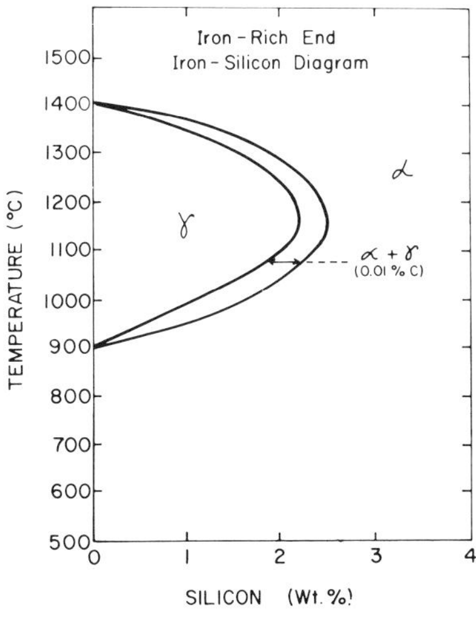

Fig. 7—The Fe-Si diagram indicates that over 2.7 pct Si is needed to eliminate the γ loop.

transformers, three general categories will be considered in detail. The subdivision is based on the type of core—lay-up, wound (or toroidal), or punched lamination. In the order listed, they represent decreasing sizes of transformer.

Lay-up cores, in general, utilize the whole spectrum of grain oriented quality and gages. The gage and grade of material for a given application are determined by economics, transformer rating, noise level requirement, watt loss requirements, density of operation, and even core size. Because the strip must be flat to produce a good core, coils are flattened after the high temperature anneal. Then, the strip is coated with an inorganic phosphate for insulation. Samples from each coil end are graded after a laboratory stress relief anneal, as previously described. The coating thickness, which averages 0.15 mil per side, decreases the stacking factor, but greatly improves the insulation of the strip. Consequently, it decreases the eddy current losses and heat buildup, which is of particular importance in transformers which must withstand an impulse test.

The heat flattening operation may or may not be set up to remove all strain in the material. Slitting induces mechanical strain at the slit edge. Additional strain in the base steel is present because of tension applied during the slitting and subsequent rewinding. Stress-free material is obtained by slitting the heat flattened-coated product, after which the edges are deburred. Then, a thin inorganic coating applied before the strip is continuously annealed under controlled tension to produce a stress-free product. From such strip, the transformer manufacturer cuts his required length (either mitered or square), and stacks his core with maximum assurance that the product has the specified magnetic quality. Material which has been merely heat flattened and slit is also used, but the core loss may be as much as 15 pct poorer. Though the user may stress-relief anneal the laminations, he must exercise extreme care to avoid distorting them.

As noted earlier, an important requirement in the manufacture of lay-up cores is minimizing transformer noise. Noise is a function of manufacturing and core design factors, the core material characteristic being one of the most important. The dependence of magnetostriction on silicon content has already been noted. In addition, magnetostriction is reduced by improving the texture and by introducing tensile stresses through application of glass-type insulative coatings. Because compressive stresses affect magnetostriction adversely, it is important that the lamination remains flat for assembly. Operating induction is also a factor that affects noise, and indeed affects the transformer's general operating characteristics. Operating inductions of lay-up transformers are usually in the 10,000 to 17,000 G range; power ratings extend over the 500 to 1,000,000 kVA range.

Wound cores are wound toroidally with the [100] crystallographic direction around the strip. Processing steps are somewhat different from those used for lay-up transformers though the starting material is the same—large toroidally annealed coil coated with magnesium silicate, which usually provides sufficient insulation.

For wound core application, unreacted MgO powder is removed from the strip surface, and a sample from each coil end is cut into Epstein strips to be tested as before. (Note: The parent coil has not been stress-relief annealed; thus, it contains cooling strains in addition to permanent coil set from the high temperature annealing.) After being graded, the coil is shipped to the transformer manufacturer either as slit multiples or as a full-width coil for subsequent slitting. The slit multiple, wound to the given core dimension, *must* be stress relief annealed at 1450°F in a dry nonoxidizing atmosphere. (If the cores are oxidized in the stress relief anneal, core losses can increase as much as 25 pct.) Annealing trays and plates must be of low carbon steel to eliminate any carbon contamination, which can be very detrimental to quality. After being stress relief annealed, the cores are cut (if they were not cut before being annealed), and the transformer core is assembled by lacing the steel around the copper (or aluminum) current-carrying coils. In the stress relief annealed condition, grain-oriented streel is sensitive to mechanical strain; therefore, cores must be assembled carefully. Regardless of how carefully assembly is accomplished, the final core quality is always poorer than it was in the stress-relief annealed, uncut condition. The difference in quality, commonly referred to as the "destruction factor", is due to the relative strain sensitivity of the grain-oriented steel, the handling procedure in fabrication, and the uniformity and amount of air gap in the core. Being a function of the transformer design and fabrication, the latter two factors are controlled best by the manufacturer. Most wound cores are utilized in distribution transformer applications of 25 to 500 kVA.

Smaller than the other types, punched lamination cores are generally mass produced, so that ease of lamination production becomes very important. Because the base glass on grain oriented stock abrades lamination stamping dies, it must be removed by pickling. To forestall rusting, pickled material is immediately coated with a thin layer of inorganic phosphate, which is cured during heat flattening. As before, samples are graded after a stress relief anneal. After coil slitting, the slit multiple is fed into a punch press, in tandem with a roller leveler which is adjusted to insure flat laminations. The inorganic coating is applied thinly so that, during roller leveling, the coating will not loosen and gum the die lubricant, or cause laminations to be dusty after the customer's 1450°F stress relief anneal. (A higher temperature decomposes the coating, resulting in dusty laminations.) The grain oriented punched laminations are used in specific power transformer applications such as battery chargers, regulated power supplies, and ferroresonant power regulators.

MAKING AND USING NON-ORIENTED SILICON STEELS

Nonoriented silicon steels do not use a secondary recrystallization process to develop their properties, and high temperature annealing is not essential. Therefore, a lower limit on silicon, such as is required for the oriented grades, is not essential. Nonoriented grades contain between 0.5 and 3.25 pct Si plus up to 0.5 pct Al, added to increase resistivity and lower the temperature of primary recrystallization. Grain growth is very desirable in the nonoriented grades, but is generally much smaller than for the oriented grades.

Processing to hot rolled band is similar to that described for the oriented grade. After surface conditioning, the bands are usually cold rolled directly to final gage, and sold to the transformer manufacturer in one of two conditions—fully-processed, or semiprocessed. After final cold rolling, the strip is annealed, decarburizing it to 0.005 pct C or lower and developing the grain structure needed for the magnetic properties. Samples are then taken from each coil end, and tested.

Fully processed nonoriented silicon steels are generally used in applications in which 1) quantities are too small to warrant stress relieving by the consumer, or 2) laminations are so large that good physical shape would be difficult to maintain after a 1550°F stress relief anneal. Nonoriented steels are not as sensitive to strain as the oriented product. Consequently, shearing strains constitute the only strain effects which should degrade the magnetic quality. Because laminations are generally large, these shearing strains can be tolerated. Most of the fully processed grades, are used as stamped laminations in such applications as rotors and stators. When roller leveling fully-processed material, processors must be careful because any significant reduction will deteriorate the magnetic properties by as much as 25 pct.

The nonoriented steels have a random orientation. They are commonly used in large rotating equipment, including motors, power generators, and ac alternators. Fully-processed steels are given a "full" strand anneal (to develop the optimum magnetic quality), making them softer and more difficult to punch than semi-processed products. Grades with a higher alloy content are harder, and thus easier to punch.

Improved punchability can be provided in fully-processed steels by adding an organic coating, which acts as a lubricant during stamping and gives some additional insulation to the base scale. If good interlamination resistance is required, fully-processed material can be purchased with core plate.

Semiprocessed products are generally given a lower-temperature decarburizing anneal after the final cold rolling. Carbon is not necessarily removed to the same low level as in fully-processed material. The transformer manufacturer will subsequently stress relief anneal the material in a wet decarburizing atmosphere to obtain additional decarburization and develop the magnetic properties. Samples are taken after the mill decarburization anneal, cut into specimens, decarburized at 1550°F for at least one hour and tested to grade the coil.

Semiprocessed nonoriented silicon steels are used for applications in which the customer does the stress relief anneal. In general, such products have good punching characteristics, and are used in a variety of applications including small rotors, stators, and small power transformers. Semiprocessed steels can be purchased with a tightly adherent scale, or with an insulating coating over the oxide. The organic coating acts as a lubricant during punching, but it does not withstand stress relief annealing temperatures; therefore, it is not applied to semiprocessed material.

The most common stress relief annealing atmosphere is produced by burning natural gas in a 6.5/1 air-to-gas ratio. The recommended dewpoint is +85°F. During the batch anneal, temperatures greater than 1650°F must be avoided to prevent laminations from sticking together. A minimal amount of welding will also occur at 1550°F, but the incidence of sticking can be reduced if the product has a stippled or matte finish. Such a finish, which can be supplied, will increase surface roughness from about 15/20 to 40/70 μ in.

SELECTION OF MATERIALS

When selecting material for a given application, the engineer must consider many factors. Economics, of course, is the major factor, but there are many instances when space and operating restrictions dictate a specific grade and gage of material. In selecting a material, then, the customer must know his costs for punching, stress relief annealing, and assembly. Also important are his design requirements, including heat restrictions. He must then pick the gage and grade which best suits his needs. Generally speaking:

1) The lower the grade designation the better the core loss,

2) The lighter gages within a grade, the lower the losses,

3) The rougher the surface, the poorer the stacking factor,

4) The lighter the gage, the poorer the stacking factor,

5) The higher the silicon content, the harder the material, and consequently, the better the stamping characteristics,

6) The higher the silicon content the better the low density permeability (below the knee) but the poorer the high density permeability.

Suppliers of electrical steels stand ready to provide the utmost in technical advice and assistance to manufacturers of electrical and electronic equipment. As ardent competitors among themselves, they also engage in research and development aimed at producing more useful materials for new technologies and for better electrical and electronic equipment.

REFERENCES

1. T. D. Yensen and N. A. Ziegler: *Trans. ASM,* 1936, vol. 24, pp. 337-58.
2. D. A. Leak and G. M. Leak: *J. Iron Steel Inst.,* London, 1957, vol. 187, p. 190.
3. J. G. Benford: *J. Appl. Phys.,* 1967, vol. 38, pp. 1100-01.
4. M. McCarty, G. L. Houze, Jr., and F. A. Malagari: *J. Appl. Phys.,* 1967, vol. 38, pp. 1096-98.
5. S. Taguchi and A. Sakakura: *J. Appl. Phys.,* 1969, vol. 40, pp. 1539-41.

Prehardened Strip Steel

By ARTHUR W. MORSE

Processors often blank and form soft, cold-rolled steel strip, then develop maximum strength by piece hardening. But this approach may not produce the part at lowest cost. Coils of high-carbon steel — AISI 1050 to 1095 — now come prehardened and tempered to hardnesses needed for the component.

Makers of blades, knives, springs, saws, and similar items can cut total processing costs by taking advantage of pretempered material. Coils of high-carbon steels such as 1055, 1075, and 1095 now come prehardened and tempered to hardnesses needed for the component. By adjusting their forming and fabricating practices, processors can turn out hardened components without the necessity for in-plant hardening and associated processing steps. Further, scrap is reduced because the possibilities for distortion in heat treatment are eliminated.

Look at Conventional Practices

Products such as blades, knives, and springs are generally made from steel which is hardened and tempered to a high hardness. For such items, the straight high-carbon steels – AISI 1050 to 1095 – are often applied.

Generally, these components are produced from soft cold-rolled strip steel, blanked, formed to shape if necessary, and hardened. Thus, processors take advantage of the good blanking and forming properties of the soft strip, and then develop maximum strength by piece hardening. However, they sometimes overlook the possibility that these methods may not produce the part at the lowest cost.

How Costs Add Up

A blanked, formed part made from annealed stock generally distorts somewhat due to the heat treatment unless jigging fixtures are used. A close-tolerance part will require either the additional cost of heat treating fixtures, or the added operations of inspection and salvage or scrap.

Also, automated assembly machines often require oriented parts for regulated feeding. Baskets of tangled parts from the heat treating furnace can lead to costly sorting and handling. Further, operations that must be performed after the heat treatment – such as grinding, polishing, printing, and plating – call for piece-by-piece handling, adding to processing costs. The actual cost of piece hardening must also be considered.

These drawbacks are not always obvious to the cost-conscious designer or production man. Some of these problems, such as dimensional accuracy and flatness, may have been previously swept aside by considering them unavoidable, and designing corrective measures into other areas. Further, heat treating costs are often hidden in overhead costs, making it difficult to obtain a true cost for processing.

Alternative: Pretempered Stock

In this concept, piece hardening by the fabricator is replaced by a continuous in-line hardening performed by the steel mill. The strip can be processed flat, straight, and well-surfaced to close-hardness tolerances. In coil form, the strip is without coil set, and consequently needs not be leveled ahead of the press.

Uniformity of formed parts depends on the consistency of the material being formed. Because the properties of pretempered strip are more uniform than those of cold-rolled strip, springback is more consistent. Better flatness and dimensional tolerances can act to improve the product quality as well as cut costs due to the elimination of heat treatment by the fabricator.

The cost of handling can be substantially decreased by retaining parts in strip form, as is often feasible with pretempered material. Where parts require subsequent operations or where automated assembly equipment is utilized, this procedure is especially helpful. Further, the use of pretempered materials allows the processor to take advantage of other steel mill operations, such as polishing, coloring, and profiling. Also, a shape can be rolled into a pretempered strip, and the finished part completed by cutting to length.

Some parts are virtually impossible to piece harden and maintain shape or dimensional accuracy. Then, use of pretempered strip is the only way, regardless of cost.

Forming Rules

Pretempered strip is harder than conventional material, but it can be blanked and formed. We are concerned with balancing high hardness needed for wear resistance and spring properties, with low hardness needed for tool life and formability. Because pretempered material wears blanking tools more than cold-rolled stock, certain precautions must be taken to improve tool life.

First, choose the pretempered stock with the lowest practical hardness for the job. Second, use a press with sufficient capacity to withstand high blanking pressures without damaging deflections. (Blanking pressures on pretempered strip can be up to three times as high as on soft-rolled material.) The double-sided press is preferable to the open C-frame press because it is more rigid.

Tool design is important. Whenever possible, avoid punch diameters and slots less than 1.5 times the strip thickness (Fig. 1).

The clearance between punch and die can and should be increased when using pretempered material compared to cold rolled. Higher hardness will prevent edge distortion which could occur because of the increased clearance. For hardnesses in the Rc 35 area, a 10% clearance per side is recommended, while Rc 55 stock calls for a 20% clearance per side.

Such clearances result in a blanked edge of approximately 30% shear and 70% break. Should a greater shear zone be required, use smaller clearances; however, tool wear will be greater. An extreme example of this occurs in fine blanking – the blanked edge is 100% shear, and the clearance is practically zero. Tool wear, however, is so great that fine blanking is not recommended for material harder than Rc 34.

Die life is also influenced by the proper choice of die material. For long runs, carbide tools are the best, but AISI M2 and D4 tool steels suffice for shorter runs.

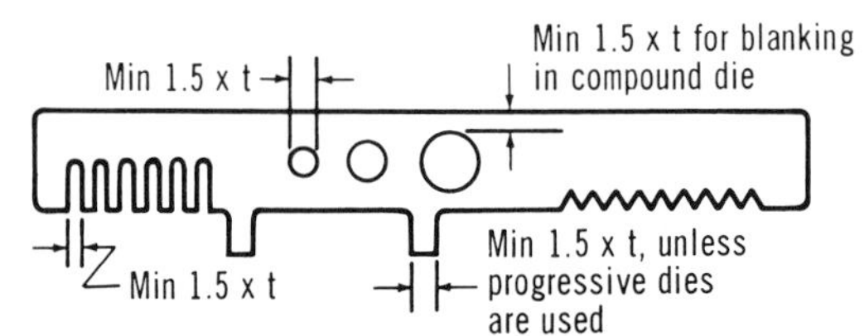

Fig. 1 — To be formed of pretempered strip, parts should be designed with slots and holes which are at least 1.5 times the strip's thicknesses (t). Other design rules are also shown.

Mr. Morse is chief metallurgist, Sandvik Steel Inc., Fair Lawn, N.J.

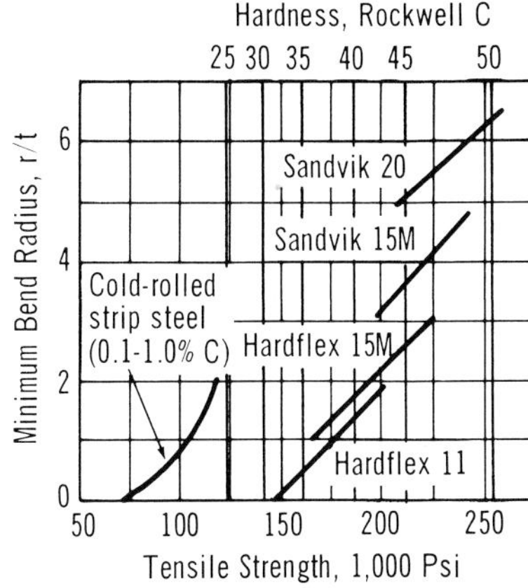

Fig. 2 — Given similar compositions and mechanical properties, bainitic materials (the two Hardflex grades) can be formed to tighter bend radii than martensitic materials (Sandvik 15M and 20).

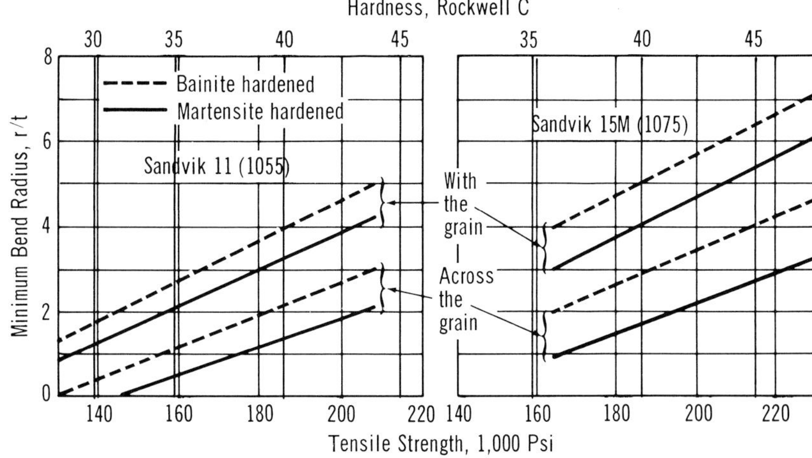

Fig. 3 — Bending tests indicate that bainitic material is more formable than martensitic material when bent with and across the grain.

Importance of Structure

For many years, engineers have known that, given the same grade, a bainitic structure has better formability than a martensitic structure at identical hardnesses. Bainitic structures are produced by heating to the austenitizing temperature, rapid quenching into molten lead maintained at a temperature above the M_S temperature, and holding at this temperature until transformation is completed. Because the holding time is long, it is a problem to produce a wide continuous coil with good flatness.

Sandvik is presently able to produce bainitic strip steel in widths up to 14 in. and thicknesses up to 0.100 in., employing furnaces 450 ft long. The material, marketed under the tradename Hardflex, can be fabricated without breakage. Given the same tensile strength, bainite, compared with martensite, would have a 5 to 10% lower yield strength, a 5 to 10% lower elastic limit, a 20% higher elongation, and a 30% higher impact strength.

Figure 2 shows the effect of hardness, structure, and carbon level on formability when bending across the grain.

Materials are cold-rolled carbon steel, martensitic 1095 (Sandvik 20) and 1075 (Sandvik 15M), and bainitic 1075 (Hardflex 15M) and 1055 (Hardflex 11). The excellent forming properties of the 1055 bainite are quite apparent. Note also the comparison between 1075 in bainitic and martensitic conditions.

Figure 3 illustrates the effect of hardness and bending direction of 1055 and 1075 in martensitic and bainitic structures. Whenever possible, the part should be designed to bend transversely to the rolling direction, preferably not parallel to it. Maximum hardnesses are generally Rc 42 for bainitic 1055 and Rc 47 for bainitic 1075. For parts where formability is not important (flat parts, for example) martensitic 1075 or 1095 is often suitable though tool wear is a drawback.

When wear resistance is needed, the part should be as hard as possible. Blanking a pretempered material cold works the edge, raising its hardness. Conventional blanking, with its large clearances between punch and die, will not develop the high edge hardness that occurs with fine blanking where the amount of cold work on the edge is much greater. This extra hardness on the edge has provided satisfactory wear resistance in many applications formerly calling for case-hardened parts. Though the depth of hardness increase is rather limited, about 0.004 in., it is apparently ample for sliding friction. The cold-worked zone can be removed by grinding or milling, lowering wear resistance.

Impact wear resistance is also improved by increasing the hardness, but the cold-worked zone is generally too thin to withstand impact wear. Here, the designer can specify strip of higher hardness; or, if this is not possible, induction hardening of the wear-affected area is the alternative.

Spring Properties

Spring materials need high elastic limits. Flat springs can be made of a high-carbon, high-hardness martensitic steel. In a formed spring, forming should be severe enough for the properties of the pretempered material to withstand expected stress level. Having better formability than martensitic stock, bainitic material is a candidate for formed springs.

Any forming in a pretempered strip causes internal stresses which reduce the elastic limit in the deformed zone. If "deflect and return" action is essential, the spring should be stress relieved at approximately 480 F for 30 min to restore elastic properties.

Cost Must Be Considered

Pretempered strip generally costs approximately two to three times as much as cold-rolled steel of the same composition and dimensions. To justify the use of pretempered strip, we must make thorough, complete cost studies.

One point is very obvious. Raw material utilization must be effective because the scrap value of pretempered strip is no different from that of cold-rolled. In other words, scrap must be kept low.

The cost of the completed part, the important factor, is determined from the costs of strip material, tooling production, and overhead. If rejects result anywhere in the process, they will be salvaged or scrapped. Either alternative increases costs. When pretempered stock is used, the total cost is somewhat above that for cold-rolled if there are no rejects. For noncritical parts where the reject rate is negligible, use of pretempered stock is thus unjustified, particularly if no other factors warrant its use.

There are some real cost advantages, perhaps more subtle, that come with the use of pretempered stock. For example, production time is shortened because operations are fewer, simplifying production control. Thus, both leadtimes and material-in-process can be reduced. The search to further reduce costs and improve product quality, we feel, will definitely lead to more applications for pretempered strip.

The Selection of Steel for Economy of Manufacture

*By the ASM Committee on Economy of Manufacture**

FOR MAXIMUM ECONOMY, the selection of steel for a production part should be based on processing factors as well as the engineering requirements of service. In this article, the direct processing factors and other considerations relating to fabrication more than to strength are first discussed in general terms for purposes of definition. Later, each selection factor is illustrated by examples of specific parts.

Because the factors that determine selection of steel are usually interdependent, it is unlikely that final selection will be based on any one factor alone. Hence, in any given selection problem, all pertinent factors must be brought together in a single analysis. Cost is the basis to which the various factors are reduced in order to reach a final decision. The controlling selection factors are:†

1. Chemical composition may entail more than the straightforward selection of a particular grade of steel (for example, 4140 vs 4340). It may also involve consideration of buying to a restricted range of chemical composition (a narrow carbon or manganese range, for example, to favor induction hardening) or to a particular specification such as AMS or ASTM.

2. Form. Steel may be bought in the form of bar, rod, tube, wire, plate, sheet, or strip, in addition to a wide variety of structural shapes. The form selected should be the one most compatible with the manufacturing processes required for fabrication, as well as with the design of the part.

3. Size. All forms of steel are made to certain dimensional tolerances in terms of length, width, diameter, or thickness. Economical selection utilizes standard tolerances and a size that is optimum for both fabrication and design.

4. Condition. Steel may be bought in a particular condition, such as cold rolled, cold drawn, hot rolled, annealed, or quenched and tempered. Selection of condition may eliminate additional heat treating, cleaning, or straightening.

5. Surface Finish. The selection of a particular finish may entail extra cost. If it eliminates one or more steps in processing the finished part, or simplifies processing, the higher initial cost of a particular surface finish may be justified.

†Machinability and susceptibility to quench cracking are critical selection factors not covered in detail in this article. Because of their outstanding importance, these two subjects are treated separately in the next two articles in this Handbook.

6. Quality Level. Certain forms of steel can be bought to a particular quality level. Wire, for example, may be specified as "valve spring quality". Sheet may be purchased on the basis of "drawing quality". Bars may be purchased to quality levels designated by terms such as merchant bar, special or cold heading quality. Buying to a particular quality level usually implies extra cost and must be justified.

7. Quantity. The quantity of parts to be made will often influence one or more selection factors.

8. Weldability. If the fabrication of a finished part or assembly requires welding, then the weldability of the steel chosen becomes an inherent factor in selection.

9. Special Requirements. The fact that plating is to be done may also influence steel selection, particularly in terms of surface finish requirements. If the steel is to be used as a conductor of electricity, then special physical (as well as mechanical) property requirements may be involved. Heat treating processes, such as carburizing or nitriding, may also entail selection on the basis of special requirements. A plain carbon steel, for example, is unsuitable for nitriding.

Assuming that both processing and operating requirements have been satisfied, it follows that the predominant factor in steel selection must be cost. To some degree, all of the selection factors described entail cost. Yet, individual cost factors (both higher and lower) may be balanced in such a way as to result in an over-all cost reduction of the finished part.

This balance of factors may lower final product cost by: (*a*) decreasing the cost of the raw material, (*b*) increasing or enhancing adaptability to specific manufacturing operations, (*c*) simplifying or eliminating one or more manufacturing operations, (*d*) simplifying or eliminating one or more inspection operations, and (*e*) reducing or eliminating scrap losses.

Chemical Composition

In planning for economy in manufacture, the type of steel selected is necessarily correlated with mechanical property and performance factors. Some latitude in the selection of a steel on the basis of chemical composition alone is likely. For example, it might be feasible to substitute a comparable alloy, such as 5140 or 8640, for 4140 if the substitution results in a provable cost advantage. Other minor deviations or substitutions are often possible and desirable. Specifically, if parts made from 1045 steel show a high rejection rate because of borderline hardenability, a change to 1046 with slightly higher hardenability might well solve the problem.

However, if 4140 bar is now being used for a part on the basis of the mechanical properties it provides, it would be impractical to substitute a lower-carbon, free-machining grade such as 1137. Although the latter steel is cheaper and easier to machine, its hardenability and performance characteristics are not comparable to those of 4140. If, for any reason, 1137 can be safely substituted, then the initial selection of 4140 was in error.

In rare instances, selection of a standard steel with a restricted range of chemical composition is advantageous. Since such selection involves a chemistry extra, it will invariably increase raw material costs, which must be justified by processing economies. Specifying a particular hardenability range for alloy steels is sounder metallurgically and less costly.

Form

For geometrically simple parts such as bolts or straight shafts, the most economical raw material form and method of manufacture are readily apparent. As the shape of the part becomes more complex, the applicability of two or more forms and methods of fabrication adds complexity to the process of selection. A small gear, for example, may be completely machined from bar stock. On the other hand, it might be more economical to start with a close-tolerance, forged gear blank, depending largely on the total number of parts to be produced.

Among such alternatives, final selection should be based on a comparison of over-all costs. Cost studies are always used to analyze the economic merit of all the various possibilities of form and method

*JOSEPH GURSKI, *Chairman,* Manager, Chemical and Metallurgical Laboratory, Manufacturing Services, Ford Motor Co.; P. H. DAILY, Farm Equipment Research and Engineering, International Harvester Co.; C. R. HANNEWALD, Metallurgical Engineer, Chrysler Corp.; A. L. HARTLEY, Chief Metallurgist, Cincinnati Milling Machine Co.; HENRY HAUSEMAN, Plant Metallurgist, Cedar Rapids Works, Allis-Chalmers Manufacturing Co.

ROBERT N. LIBSCH, Chief Metallurgist, American Bosch Div., American Bosch Arma Corp.; WILLIAM R. MILLER, Chief Metallurgical Engineer, American Steel & Wire Div., United States Steel Corp.; O. G. SAUNDERS, Chief Metallurgist, Hobart Manufacturing Co.; C. H. SHELTON, Director of Metallurgy, White Motor Co.; A. H. SMITH, Chief Metallurgist, Cadillac Motor Car Div., General Motors Corp.

ELMER H. SNYDER, Chief Metallurgist, Austin-Western Div., Baldwin-Lima-Hamilton Corp.; R. L. SPROAT, Chief Metallurgist, Standard Pressed Steel Co.; W. H. SWANK, Chief Metallurgist, Kelsey-Hayes Co.; ARTHUR F. TOROK, Metallurgist, National Acme Co.; G. C. RIEGEL, *Secretary,* retired Chief Metallurgist, Caterpillar Tractor Co.

of manufacture — machining from tube or bar, casting, hot forging, cold forging, extruding or welding.

Production quantity is the factor most likely to determine the selection of form. For a small gear, the total cost of making 100 pieces might favor machining from bar. If the quantity were increased to 10,000 or more, the forged gear blank might show a considerable economic advantage. With a larger gear, if only 500 pieces were to be made, it might be more economical to leave the hub solid and drill the bore. If the required quantity were 5000 pieces, it would probably pay to have the hub pierced in the forging operation. With even larger quantities, forging of minor contours might be justified. As quantity increases, it becomes progressively easier to amortize the initial cost of forging dies. Savings in metal and machining increase.

Size

The design of the part determines the size of steel to be used, to within a fairly narrow range. Nevertheless, there is some flexibility in selection, and savings depend on judicious choice of a final size.

Typifying the problem, consider the manufacture of a simple shaft with a constant diameter that must be held to close tolerance. Such a shaft may be made most economically from bar that is purchased in the turned and centerless-ground condition. This is particularly true if the shaft requires a surface finish that can be achieved only by a secondary operation such as grinding. Neither hot rolled nor cold drawn bars could satisfy this condition without additional machining.

If, however, a similar shaft is made more intricate by requiring a number of different diameters along its length, paying a premium for ground bar of special size would be uneconomical; either hot rolled or cold drawn bar would fit the need better. Since most of the surface will be machined off, there is no justification for a stringent surface requirement on the raw bar.

Ideally, the bar size selected should come as close as possible to finished part size. There are instances, however, where some additional stock removal must be allowed in order to avoid an excessive scrap rate. When bar is slightly out-of-round or otherwise warped, additional stock serves as a safety factor against rejection. Even though extra stock removal is wasteful to some extent, it may be cheaper to purchase slightly larger stock than to pay for scrapped parts.

Restrictions on the amount of surface decarburization and surface seams are sometimes additional reasons for buying oversize stock. The added stock should be adequate to permit complete removal of all decarburized metal. It should also eliminate the need for magnetic particle inspection to detect surface seams and other imperfections.

Conventional machining allowances for hot rolled round bars are ⅛ in. on the diameter for bars of 1½ to 3-in. diam, and ¼ in. for bars over 3-in. diam. This results in added area as follows:

Diam before allowance, in.	Diam after allowance, in.	Increase in area
1.50	1.625	17.4%
2.00	2.125	14.8
2.50	2.625	10.2
3.00	3.125	8.5
3.50	3.75	14.8
4.00	4.25	12.9
4.50	4.75	11.4
5.00	5.25	10.2

Thus, for some applications it may be advantageous for the user to order a smaller bar size and accept a scrap rate a few percentage points above normal in machining or some other operation, rather than to purchase 8 to 17% additional steel on the outside of the bar.

Condition

A number of steels, particularly medium-carbon, plain carbon and medium-carbon alloy bars, are commercially available in a pretreated condition intended to enhance mechanical properties. Steels such as 1050, 1141 and 1144 are specially processed by cold drawing with heavy drafts. The bars are then stress relieved from 700 to 950 F, depending on steel composition and mill practice.

As a result of cold drawing, tensile strength values up to about 150,000 psi can be obtained with hardnesses up to about 300 Bhn. The price of cold drawn stress-relieved bar is only slightly above that of the conventionally cold drawn material. Certain steels (including those mentioned above) are also obtainable in the hot drawn condition. With hardnesses up to about 350 Bhn, these materials have even higher mechanical properties. (Hot drawn bars are passed through dies at about 700 F.)

At higher cost, carbon and alloy steels are also available in the quenched and tempered condition. In this condition, a common hardness range is 250 to 300 Bhn; hardnesses above or below this range can be had on special order.

Whether or not the purchase of steel in the hardened and tempered condition is economical must be determined on the basis of direct costs. The premiums paid for steels in this condition, along with the added cost of machining at higher hardness, must be weighed against the cost of heat treating and cleaning parts in the production line. This is especially true of steels whose hardness is known to have a markedly adverse effect on most machining operations, resulting in decreased machining speeds, decreased tool life, or both.

Part design is most likely to influence the decision to use steel bought in the hardened and tempered condition. Where little machining is required, such pretreatment may have almost no effect on final cost. Where machining operations are extensive, the added time, effort and tool wear involved may rule out the use of pretreated steels.

If a part is likely to distort in heat treatment, the use of steels hardened and tempered before machining is highly attractive. Savings realized through scrap reduction may more than offset the extra costs of raw material and fabrication.

Surface Finish

Although cold drawn steels cost more than their hot rolled counterparts, they provide four advantages: (*a*) better finish, (*b*) closer dimensional tolerance, (*c*) improved machinability, and (*d*) higher strength. Where a number of these qualities can be profitably used in fabrication or in improving the performance of the end product, the higher cost of cold drawn grades may be justified. If finish is not critical and tooling can adjust to the greater size differences encountered, then hot rolled steels may be more economical.

While the cold drawn steels have better machining characteristics, this factor is rarely the sole criterion for selection. The extra strength of cold drawn steel is more important from a cost standpoint, since it will often eliminate the need for heat treatment.

Surface finish is also important when it serves to reduce the number of operations required in processing. To take full advantage of the cold finished surface, therefore, stock size and surface condition must be left virtually intact. For example, a sheet metal stamping may be made of cold finished steel solely to avoid elaborate cleaning prior to plating.

Quality Level

Some forms of steel can be bought to a particular quality level. Manufacturers of household appliances, for example, purchase large quantities of steel sheet of "enameling quality". Without additional processing, this material is suited for enamel coating or porcelainizing. The same material also has excellent formability.

The aircraft industry, as a rule, buys steel to an AMS rather than an AISI or SAE specification. For steels of virtually identical chemical composition, the difference between specifications is likely to be found in special quality considerations not normally included in AISI or SAE specifications. The AMS specifications often call for special bending requirements, more predictable and more uniform response to heat treatment, a maximum content of inclusions, specific mechanical properties at elevated temperatures, and fatigue strength greater than a specified minimum.

All such requirements can be sat-

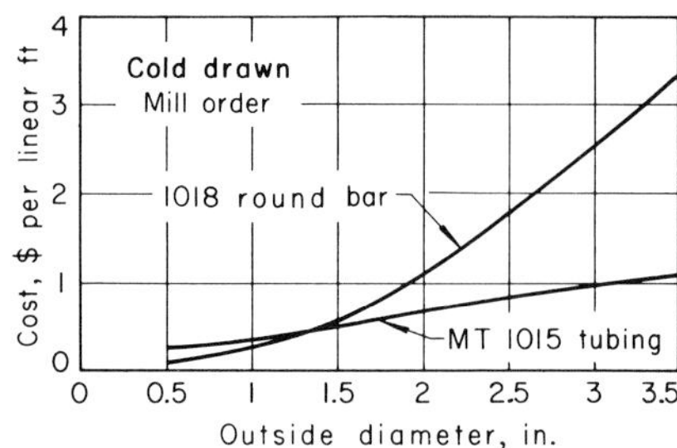

Example 1. Economic break-even point for bar stock compared with tubing with 0.125-in. wall. Minimum quantity of tubing was 600 ft. Tubing and bars were random lengths with no special tolerances or other extras. Material cost only is shown and does not reflect fabricating time.

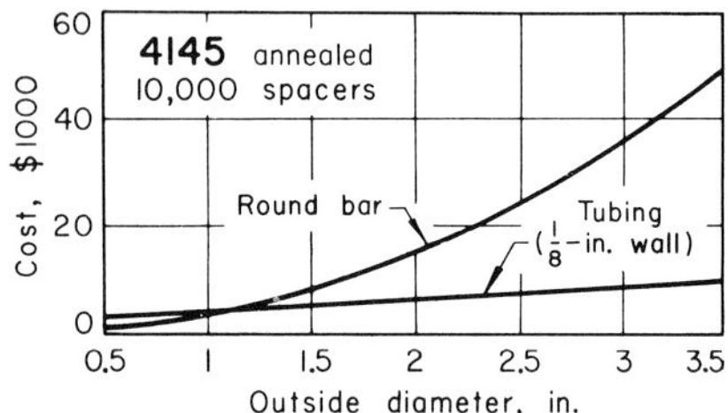

Example 2. Economic break-even point in the production of spacers. Annealed 4145 cold finished bar is compared with tubing. The 10,000 spacers were 12 in. long.

isfied at a premium price. Aside from aircraft applications, most industries restrict their quality level purchases to commercially standard steel products such as merchant bar, special quality, cold heading quality, and others designated in AISI Steel Product Manuals.

Weldability

The weldability of steel is a special factor in its selection and depends primarily on chemical composition. A preferred range of composition is one conducive to welding with maximum ease and speed. Because such a range is slightly more restrictive, it is not to be mistaken for a standard chemical specification, although it is sufficiently close to standard to include many commercial heats. For carbon steel, such a composition range is as follows:

Carbon	0.13 to 0.20%
Manganese	0.30 to 0.60%
Silicon	0.10 to 0.20%
Sulfur	0.05% max
Phosphorus	0.04% max
Cr, Ni, Mo, Cu	0.10% max total

Within the preferred composition range, virtually no restriction need be made as to the type of joint, type of electrode, welding current, or speed of welding. That is, such steel allows the user to take full advantage of large electrodes, high currents, high speeds, and techniques for obtaining deep penetration. Usually steels within this range are available without premium price and can be fabricated easily by the usual forming, shearing, machining and other mechanical processes.

However, it should not be inferred that only steels within the above composition ranges are weldable. At the present time, even the more complex air-hardening 0.35% C alloy steels are regularly welded, but more costly techniques may be required for any steel whose carbon or alloy content exceeds the limits listed above.

Carbon content is the most critical factor affecting the ease with which a steel can be welded without danger of weld cracking. When carbon is increased above about 0.20%, the susceptibility to weld cracking increases; with carbon above about 0.30%, preheating, postheating, or both, are usually required. These requirements are indicated in numerous specifications, such as ASTM A373, establishing composition limits for structural carbon steels that are to be welded. In the ASTM specification, section thickness is also considered. A check analysis of 0.31% max C is allowed for a section thickness of 2 to 4 in., but only 0.30% C for thinner sections.

Compositions of the high-strength low-alloy steels also indicate the importance of carbon content. Despite the fact that these steels may contain appreciable amounts of manganese or other alloying elements, the carbon is seldom higher than 0.28% and is less than 0.20% in a majority of grades that provide minimum yield strengths of 50,000 psi or higher. These high-strength low-alloy steels are regularly welded and seldom require preheating or postheating. High-strength electrodes must be used if the strength of the parent metal is to be matched, thus adding to welding complexity and cost. (See article on high-strength steels, page 87 in this Handbook.)

Special Requirements

Surface conditions required by plating and other metal finishing techniques may sometimes be classified as special requirements. Both carburizing and nitriding frequently involve special requirements in steel selection. A carburized gear that is subject to high loads, for example, will require selection of a low-carbon medium-alloy steel with fairly high hardenability. Hardenability is particularly important to achieve a sufficiently high core hardness to support the compressive loads on the carburized case.

Plain carbon steels are never nitrided, because they build up an excessively brittle "white layer" that cracks and spalls readily. To obtain a satisfactorily nitrided case involves special selection requirements, particularly as regards alloy content. (See Fig. 13, in Selection of Steel for Wear Resistance, page 244.)

Machinability

Selection based on machinability is considered in detail in the next article in this Handbook. In that article, entitled Selection of Steel for Economy in Machining, most of the grades in general use are dealt with, including plain carbon, free-machining, leaded, and carburizing steels, both carbon and alloy. It provides machining data and cost comparisons for the various types of steels, relating them to specific machining operations such as hobbing, turning, drilling, broaching, milling, screw machining, and gear cutting. Cost figures and comparative machining rates are based on actual production parts.

Quench Cracking

Susceptibility to quench cracking, as it applies to steel selection, is discussed in the article on Selection of Steel for Minimum Quench

Examples 3, 4 and 5. Cost Comparisons for Making the Same Parts from Bar or Tubing

	High-cost alternate	Low-cost alternate
Example 3. Spacers(a)	**Bar**	**Tubing**
Material cost	$0.027	$0.049
Labor cost	0.229	0.146
Burden	0.581	0.360
Total cost each	$0.837	$0.555
Cost savings per piece using tubing		$0.282
Example 4. Cylindrical Air Fittings(b)	**Bar**	**Tubing**
Unit weight or size	3.49 lb	7 in.
Work standard, min per cycle	0.351	0.181
Material cost	$9.24 per 100 lb	$19.85 per 100 pc
Material used	8376 lb	1400 ft
Material cost, total	$ 773.94	$ 476.40
Direct labor cost	1566.96	808.08
Tool cost	192.00	96.00
Total cost	$2532.90	$1380.48
Unit savings per piece		$0.48
Total savings using tubing		$1152.42
Example 5. Cylindrical Receivers(c)	**Tubing**	**Bar**
Unit weight and length	$8\frac{7}{16}$ in.	3.235 lb
Material cost	$0.60 per ft	$11.00 per 100 lb
Material used	14,066 ft	64,700 lb
Work standard, hr per 100 pieces	52.79	55.55
Direct labor, hr	10,558	11,110
Material cost	$ 8,440.00	$ 7,117.00
Direct labor cost	15,837.00	16,660.00
Total cost	$24,277.00	$23,777.00
Total savings using cold drawn bar		$500.00
Unit savings per piece		$0.025

(a) Comparative cost for machining spacers 1.93-in. OD by 1.60-in. ID by ⅛ in. thick from 1018 bar and 1015 tubing. (b) Comparative cost for machining 2400 cylindrical air fittings from low-carbon bar and tubing. (c) Comparative cost of machining 20,000 cylindrical receivers for sporting rifles from cold drawn 1137 and from 1118 seamless tubing 1.3125-in. OD by 0.905-in. ID.

Example 6. Cost of Making a Bushing from Bar Stock Compared with Tubing

	Bar stock	Tubing
Material	$0.878	$0.78
Labor	0.25	0.16
	$1.128	$0.94
Reclaim turnings	0.063	....
Net cost	$1.065	$0.94
Savings per bushing		$0.125

Size of bushing: 3-in. OD by 2½-in. ID by 4⅝ in. long. Size of bar stock: 3-in. diam. Size of tubing: 3-in. OD by 5/16-in. wall.

Cracking, page 316 in this Handbook. Detailed consideration is given the various factors that promote cracking, particularly the carbon content, M_s temperature, and maximum hardenability.

Selection of Form

The first 35 examples illustrate the selection of a particular form of raw material, giving cost and other details, where two or more forms have been considered for making the same part.

Example 1 shows the point at which low-carbon steel bar becomes more expensive than tubing of similar composition and wall thickness of 0.125 in. The comparison is based on cost of material per linear foot. Because of the higher cost of producing tubing smaller than about 1¼-in. OD, the cost per foot exceeds bar cost in small sizes.

Example 2 compares cost for making simple spacers from alloy steel tubing and from bar of the same composition. The break-even point in costs is at nearly the same diameter as that in Example 1.

Examples 3, 4 and 5 present data for three different parts made from bar and tubing. In Examples 3 and 4, where similar steels were used for bars and tubing, tubing was the low-cost form. In Example 5, two different steels were used, and although the bar steel was the more expensive, the part made of it was cheaper. There are probably two factors contributing to this reversal — the tube size being close to the break-even point (Example 1), and the heavier wall (0.203 in.). Both steels had the required properties.

Example 6 is another comparison for making parts from low-carbon bars or from tubing of similar composition. These cost figures favor the tubing, but the history of this cost analysis illustrates how variation of a single factor may reverse the selection. The first time this bushing was figured, the parts made from tubing cost $1.30 each, compared with the figure shown ($1.065) for parts made from bar. This was because the cost of tubing was based on purchases of only 220 ft, costing $296.82 per 100 ft. After a standardization program for sizes was initiated, it was possible to purchase tubing in quantities of 700 ft, costing only $202.38 per 100 ft, and the cost of the bushings made from this tubing was decreased from $1.30 to $0.94 each.

Example 7 illustrates the solution of a manufacturing problem and an accompanying reduction in cost by a change of steel form. A type 321 stainless steel tube (shown in sketch) with 1$\frac{15}{16}$-in. OD, $\frac{1}{16}$-in. wall and 21-in. length, was slotted in eight locations. Milling the slots was difficult because the tube was thin and labor cost for removing inside burrs was excessive. The problem was solved by using type 321 sheet, punching the slots, and forming and welding into the required tubular part. Cost of the sheet was slightly less than the tube, and labor was reduced by 50%. However, in such instances availability of fabricating equipment and the quantity of parts to be produced must always be considered.

Example 8 compares seamless with electrically welded tubing for a tractor front axle. Tubing used for this part was of unannealed 1035 steel with a minimum yield strength of 80,000 psi. The same composition and strength in electrically welded tubing had closer tolerances for outside diameter, inside diameter and wall thickness, and 30¢ per tractor was saved. Annual saving amounted to $10,350 for 34,500 tractors.

Example 9. Whether a part should be made by machining it from a solid bar or by forging is often determined by the number of parts to be made. A transmission shaft was made on a screw machine from 4024 steel (see sketch). About 7 lb of chips were generated in machining a 17-lb bar. Cost of this part could be reduced 18% by hot upset forging it from a smaller bar when 100,000 shafts were to be made.

Examples 10, 11 and 12 are additional comparisons of parts made from bar stock or forgings. Forging proved to be the lower-cost method for making main drive axles (Example 10) but bar stock was cheaper for governor gears (see sketch), despite the fact that the quantity being manufactured was relatively large (43,200).

Example 13 gives cost figures for making the gear shown in the sketch from bars or from forgings at three different quantity levels. Cost differential is significantly less as quantity increases, but even in quantities of 250 it is cheaper to machine from bars. Die amortization of $1.30 per piece would be reduced if quantities were larger.

Example 14. Parts such as this

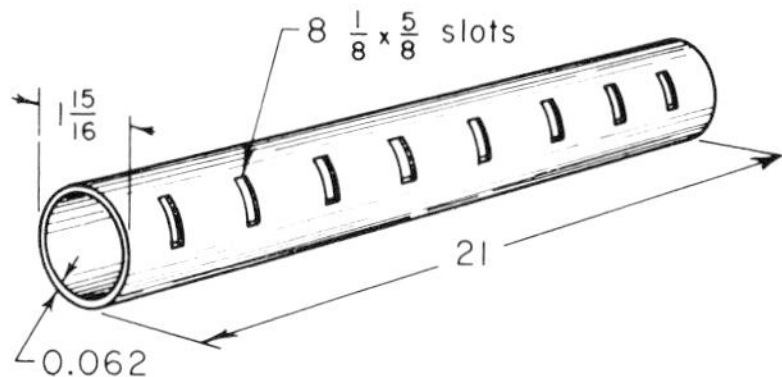

Example 7. Production of slotted tubular part. See text for comparison of making the part from tubing vs sheet.

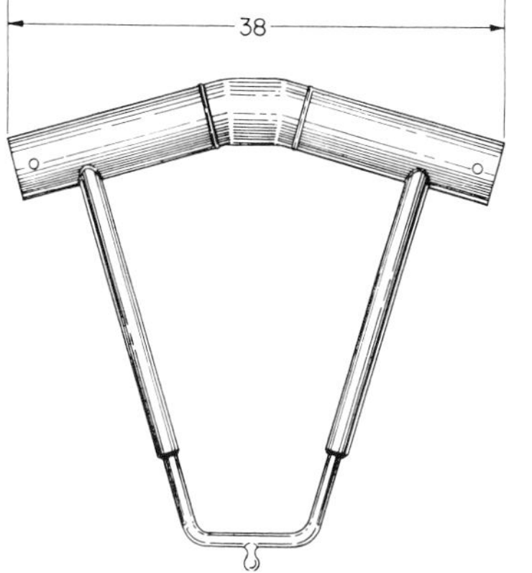

Example 8. Seamless tubing vs welded tubing in a front axle application

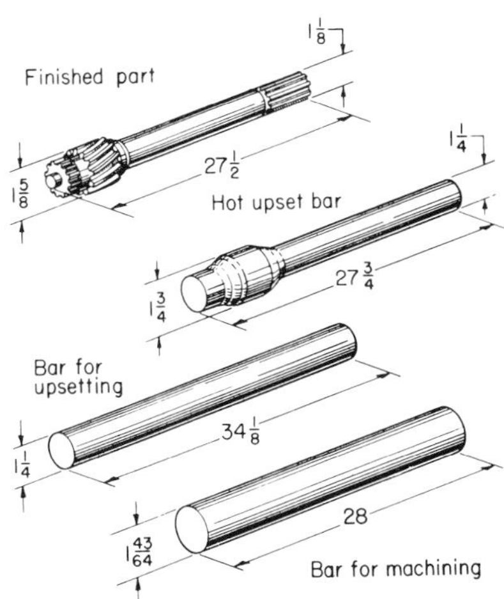

Example 9. The same part made from bar or a hot upset forging

Examples 10, 11 and 12. Cost Comparisons for Making Axles and Gears from 1141 Steel Bar Stock and Forgings

	High-cost alternate	Low-cost alternate
Example 10. Drive Axles(a)	**Bar**	**Forgings**
Weight each, lb	60.21	50.95
Material cost per 100 lb	$7.82	$7.82
Weight of scrap tong hold, lb	....	1.87
Scrap material cost per 100 lb	....	$2.19
Standard, hr per piece	0.401	0.384
Direct material cost, each	$4.71	$4.03
Direct labor cost, each	0.93	0.89
Forge die cost, each	....	0.20
Machine tool cost, each	0.15	0.10
Total direct cost, each	$5.79	$5.22
Savings per unit using forgings		$0.57
Example 11. Governor Gears(b)	**Forging**	**Bar**
Freight	$ 6.42	$ 3.80
Making blank	24.31	23.67
Machining	99.54	40.20
Total cost per 100 pieces	$130.27	$67.67
Cost differential favoring screw machine, per 100 pieces		$62.60
Example 12. Governor Gears(c)	**Forging**	**Bar**
Freight	$ 6.35	$ 3.73
Making blank	24.47	29.28
Machining cost	109.20	103.13
Total cost per 100 pieces	$140.02	$136.14
Cost differential favoring screw machine, per 100 pieces		$3.88

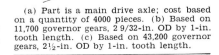

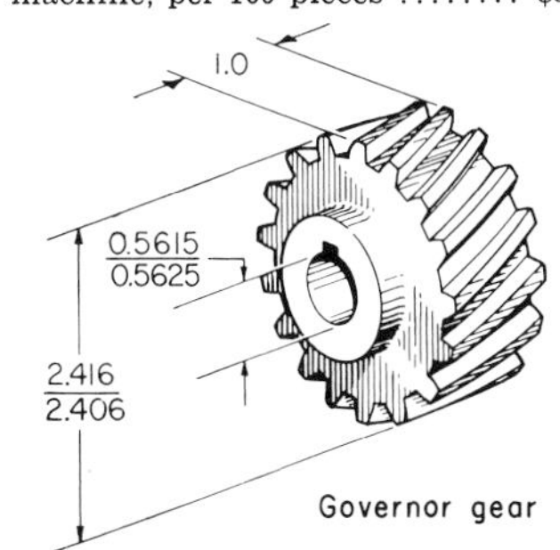

(a) Part is a main drive axle; cost based on a quantity of 4000 pieces. (b) Based on 11,700 governor gears, 2 9/32-in. OD by 1-in. tooth length. (c) Based on 43,200 governor gears, 2½-in. OD by 1-in. tooth length.

Example 13. Cost Comparison for Making Gears from 8650 Steel from Bar Stock Compared with Forgings

	Forgings	Bar
Quantity, 50 Pieces		
Material cost per piece	$10.30	$ 5.00
Machining cost per piece	4.90	6.94
Die amortization(a)	1.30	....
Total cost per piece	$16.50	$11.94
Net saving using bar		$4.56
Quantity, 125 Pieces		
Material cost per piece	$7.85	$ 5.00
Machining cost per piece	4.39	6.50
Die amortization(a)	1.30	....
Total cost per piece	$13.54	$11.50
Net saving using bar		$2.04
Quantity, 250 Pieces		
Material cost per piece	$ 7.20	$ 5.00
Machining cost per piece	4.07	6.34
Die amortization(a)	1.30	....
Total cost per piece	$12.57	$11.34
Net saving using bar		$1.23

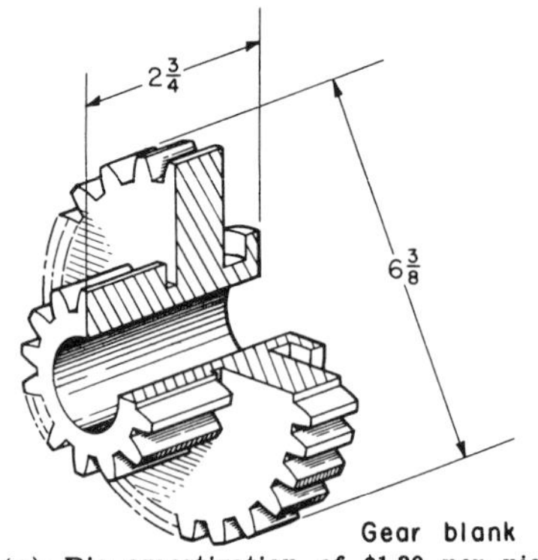

(a) Die amortization of $1.30 per piece at quantity levels up to 250 pieces would be reduced if quantities were higher.

Example 14. Cost of a Solid Bar Compared with an Extrusion(a)

	Round bar	Extruded shape
Unit weight, lb	0.305	0.175
Material cost, per 100 lb	$14.56	$37.46
Direct material, lb	21,960	12,600
Work standard, min per piece	4.04	2.72
Total direct labor, hr	4849.50	3260.40
Total material cost	$3,197.38	$4,719.96
Total direct labor cost	8,695.50	5,896.06
Tool cost	2,942.67	2,535.00
Total cost, 72,000 pieces	$14,835.55	$13,151.02
Unit savings per piece		$0.0234
Total saving		$1,684.53

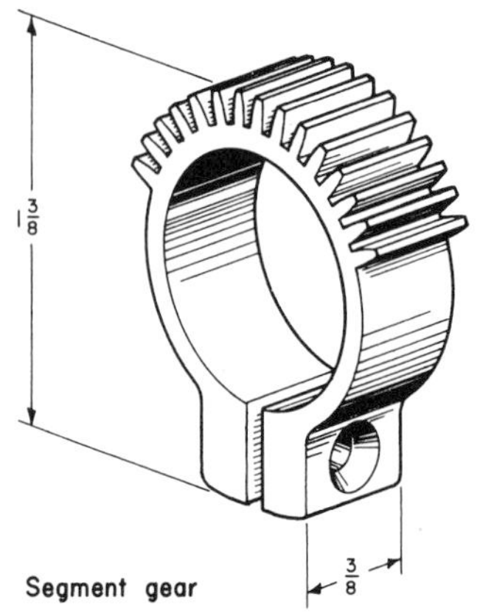

(a) Comparative costs for making 72,000 segment gears (per sketch) from 1118 steel round bar, profile milled and heat treated, compared with 1144 steel, extruded and stress relieved.

segment gear (see sketch) can often be made from an extruded shape with a cost saving. This part was formerly made from round bars of 1118 steel, profile milled and heat treated. It is now made from an extruded bar of 1144 steel and is stress relieved to attain a minimum yield strength of 100,000 psi. This latter method provides an improved finish in addition to the cost saving.

Example 15 illustrates another instance where an extruded shape brought about a cost saving. Three methods for making the retainer rings shown in the sketch were compared. Despite the fact that the ring could be cold drawn in a single pass from either a hot rolled shape or a hot extrusion, the higher cost of dies for the hot rolled shape made a significant difference in the cost per piece.

Examples 16, 17 and 18. Minor alterations in design may permit parts to be made from another form at lower cost, as in Example 16. This portion of a clamp was originally made from cast 1020 steel with rounded corners, as shown in the sketch. A review of the part determined that square corners would be acceptable. The parts were then machined from a flat 1018 bar at a 64% savings, as shown in the tabulation. Example 17 compares a straight bar with a casting for making pivot lugs. No change was made in the design, but a cost differential of about 24% favored machining from a bar.

In other designs, results may be reversed, as shown in Example 18. Cost of the first 8630 steel roller support used in an airplane was considerably more when machined from an investment casting than from bar stock. However, when more than about 60 pieces were made, the casting method became increasingly cheaper until at 4000 pieces the parts cost only $4.00 each made from castings, compared with $10.00 each from bar stock.

Examples 19, 20 and 21 illustrate how cold forming produced substantial savings, compared with other methods for making the same parts from the same steels. The tapered shaft (Example 19) was formerly hot forged from 1035 but is now cold formed. With an annual production of 320,000 pieces, $28,800 is saved.

The automotive engine support (Example 20) was formerly machined from 1212 steel but is now cold formed. A 36% reduction in cost saves $200,000 on an annual production of 2.5 million pieces.

The automotive piston pin (Example 21) was formerly machined from 5015 steel. Much of the bar was converted to chips. Tubing was tried but was too expensive. To save steel the part was cold formed by double backward extrusion of a smaller amount of metal, eliminating rough machining and drilling except for cutoff. Material saving was 19%. Labor costs were about the same for both methods.

Example 15. Cost Comparison for Making Steel Retainer Rings(a)

	Cold drawn from: Hot rolled round	Hot rolled shape	Hot extrusion
Material cost per lb(b)	$0.06	$0.116	$0.236
Die, roll cost	0.033	0.151	0.005
Drawing cost,			
First pass	0.096	0.096	0.096
Second pass	0.096	....	....
Third pass	0.096	....	....
Total cost per lb	$0.381	$0.363	$0.337
Cost, each	0.0342	0.0331	0.0305
Cost per year	547,200	529,600	488,000
Savings	....	$17,600	$59,200

Hot rolled: 0.577 ± 0.031; 0.062 R; 0.463 ± 0.020; 0.062 R

Hot extruded: 0.577 ± 0.031; 0.062 R; 0.463 ± 0.020; 0.062 R

Cold drawn: 0.515 ± 0.005; 3/64 R; 0.421 ± 0.002; Sharp corner

(a) The costs indicated apply to a total annual production of 16 million pieces made from 1050 steel. (b) The material costs for all three alternatives are based on a 30,000-lb order.

Examples 16 and 17. Cost of Machining Clamps from a Solid Bar Compared with a Casting

	Casting	Bar
Example 16. Clamp(a)		
Steel cost per piece	$1.34	$0.22
Machining cost, each	0.95	0.61
Total cost, each	$2.29	$0.83
Net saving per piece from bar		$1.46
Example 17. Pivot Lug(b)		
Steel cost per piece	$7.50	$2.36
Labor cost per piece	1.12	1.96
Burden per piece	2.85	4.57
Total cost per piece	$11.47	$8.89
Net saving per piece from bar		$2.58

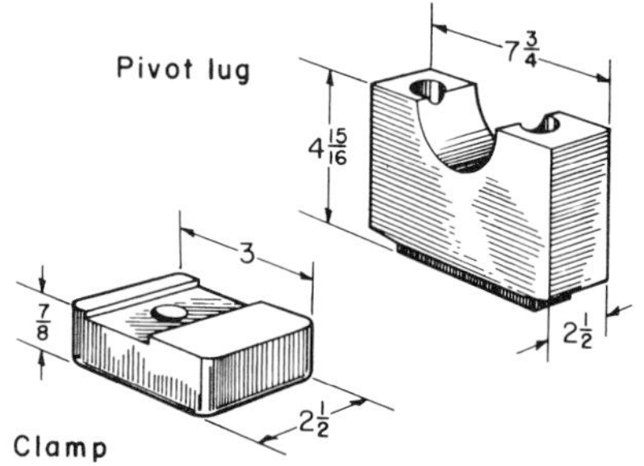

(a) Cost for making clamps (above sketch) from cast 1020 compared with cost for machining from 1018 bar, in four lots of 175 pieces per lot. (b) Comparative costs for making pivot lugs (above sketch) from 1020 steel castings and from 1020 bar.

It should be emphasized that savings in the above three examples depended on a high volume of a single part or on having similar parts that could be made on the same equipment. If high-cost production machinery will remain idle while a part is being made by another method, the purported savings should be reassessed.

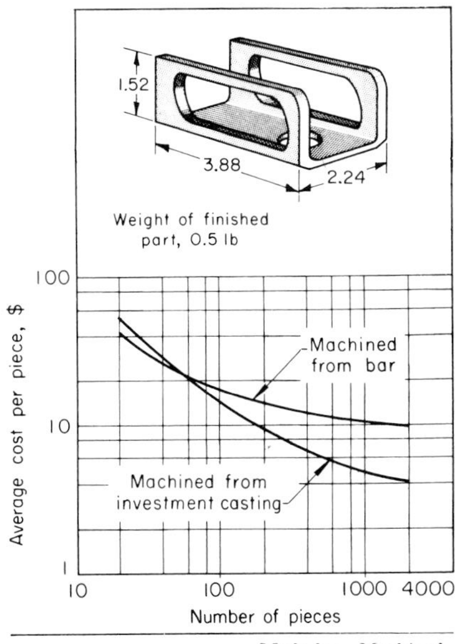

	Made by investment casting	Machined from bar stock
Vendor tooling	$400.00	....
Plant tooling	550.40	$576.00
Plant setup	25.60	108.47
Running time	3.12	7.34
Material	....	0.86
Casting piece charge ..	1.35	....
Total	$980.47	$692.67

Example 18. Unit-one cost breakdown of roller supports made by machining from bar stock compared with investment casting (8630 steel)

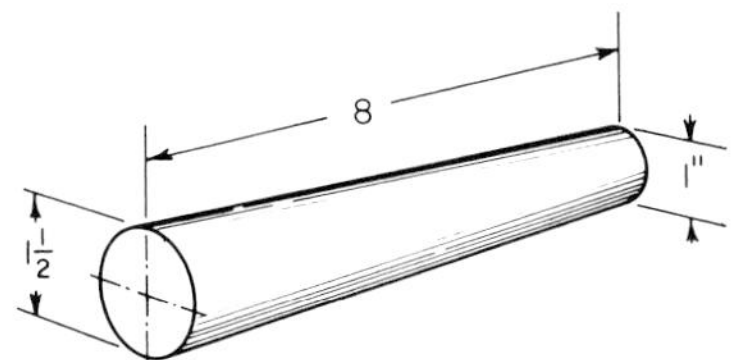

Example 19. Cold forming this tapered shaft cost 9¢ less per piece than hot forging it.

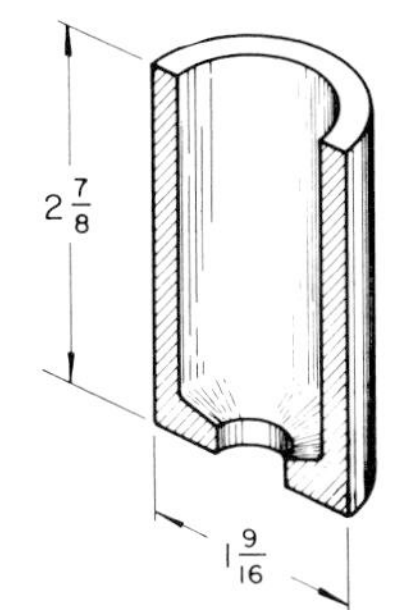

Automotive inner cup engine support

Example 20. Cold forming compared with machining. Cost per piece for machined part was 22¢, for cold formed, 14¢. Cold forming saved $200,000.

Example 22. Quantity is also the deciding factor in selecting a method for producing the special round-headed bolts shown in the sketch. These bolts can be cold headed more economically than they can be machined, for all but extremely small quantities.

Examples 23 to 29 compare the cost of making specific parts (see sketches) by forging and casting. The first two examples illustrate how the cost differential widens for the more complex part, compared with the simple cap. Castings often become more competitive with forgings as complexity increases. In Example 25 castings proved to be lighter and cheaper for connecting rods of 70 to 90 lb weight.

Example 26 compares production costs for steam chests (nozzle boxes) used in steam turbines. In operation, the component is subjected to severe temperature gradients at the time of starting the turbine and during large load changes.

The engineering requirements of the part specify that it must: (*a*) be removable and replaceable, (*b*) have thin walls to reduce thermal stress, (*c*) have no bolted joints, (*d*) retain the smallest possible size to keep outer and inner shell diameters to a minimum and give greater freedom from horizontal leakage, and (*e*) be made to close tolerances.

The part was first designed as a forging (note sketches that accompany tabulation of costs) but developments in technique resulted in castings that satisfied service requirements. The casting method resulted in significant savings through reduced tooling and material costs, as shown in the table.

Example 27. Four gear drive clutch housings, differing in size but otherwise similar, were originally made from bar stock or forged in open dies from 4620 steel. These housings are used on equipment that feeds steel strip to presses or shears.

Manganese-molybdenum steel castings proved a satisfactory replacement, with an attendant reduction in cost. Sketches of the open-die forging, rough casting and finished part are shown with the cost tabulation in Example 27. In all instances but one (the smaller model A), material cost for the casting was less, and in all instances, machining and total costs favored the casting, as indicated by the detailed cost figures for the four castings of different sizes.

Example 28. Heavy winder shafts used on a Sendzimir cold strip mill were originally made from 1045 steel forgings that had a rough weight of about 12,000 lb each. These shafts (see sketch) are now made from castings having a rough weight of only 9000 lb. Details of cost savings gained without sacrifice of required properties are given in the table.

The amount of metal that had to be machined from the outside was about the same for the forging and

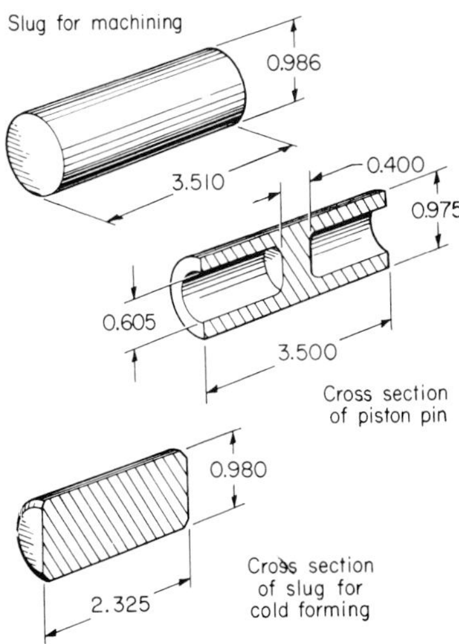

Example 21. Saving in steel by cold forming rather than machining was 19%. Labor costs were not affected.

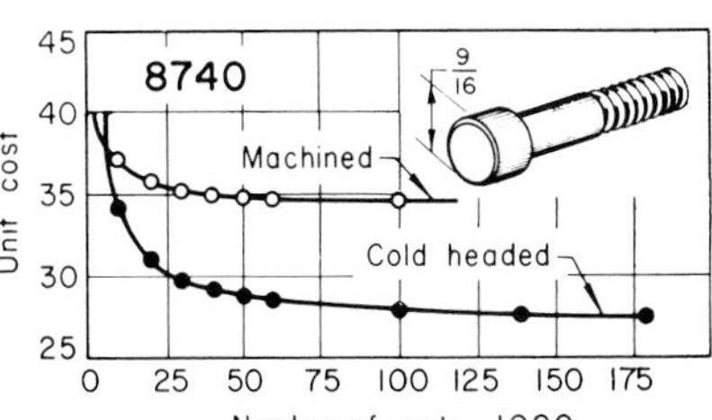

Example 22. Effect of quantity on the unit cost of cold headed bolts and machined bolts

Examples 23 and 24. Cost Comparison of Forgings with Castings

	Casting	Forging
Example 23. Ball Socket(a)		
Cost per 100 pieces	$146.44	$49.80
Freight per 100 pieces .	1.08	2.96
Total per 100 pieces .	$147.52	$52.76
Cost differential favoring forging, per 100 pieces		$94.76
Annual savings		$25,585.20
Example 24. Ball Cap(b)		
Cost per 100 pieces	$18.20	$12.76
Freight per 100 pieces .	0.66	0.41
Cost of increased labor, per 100 pieces	....	0.13
Total cost per 100 ... pieces	$18.86	$13.30
Cost differential favoring forging, per 100 pieces		$5.56
Annual savings		$1501.20

Ball socket

Ball cap

(a) Comparative cost for making stay rod ball sockets (above sketch) from steel castings; cast steel composition was 0.25% C max, 0.65% Mn, 0.60% Si max, 1.00% Mo; 1041 steel was used for forgings. Annual production, 27,000 pieces. (b) Comparative cost for making stay rod ball caps (above sketch) from the same two materials. Annual production, 27,000 pieces.

the casting. However, the cored sections of the casting reduced the metal removed from the inside to 80 lb for the casting, compared with 1600 lb machined from the inside of the solid forging; this resulted in the total saving for metal removal indicated in the tabulation. The savings in metal and cost of metal removal were mainly responsible for a cost reduction of about 37% in favor of the casting.

Example 29. In making a 31½-lb

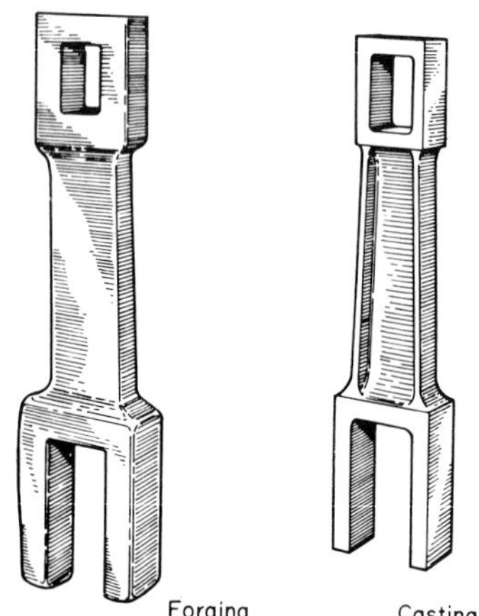

Example 25. Designed originally as a forging, the connecting rod (left) was made of a low-alloy steel. A cast connecting rod (right) of the same material resulted in a 20% reduction in weight from 89 to 70½ lb. The cost of the finished casting was 46% lower, without sacrifice in part performance.

Example 26. Cost Comparison of Forged Versus Cast Steam Chests for Turbines

	Forging	Casting
Die costs	$42,000	....
Pattern costs	....	$ 3,567
Savings in tooling		38,433
Material cost(a)	940	365
Savings per unit	....	575
Savings per year(b)		46,000
Total annual savings using castings(c)		$84,433

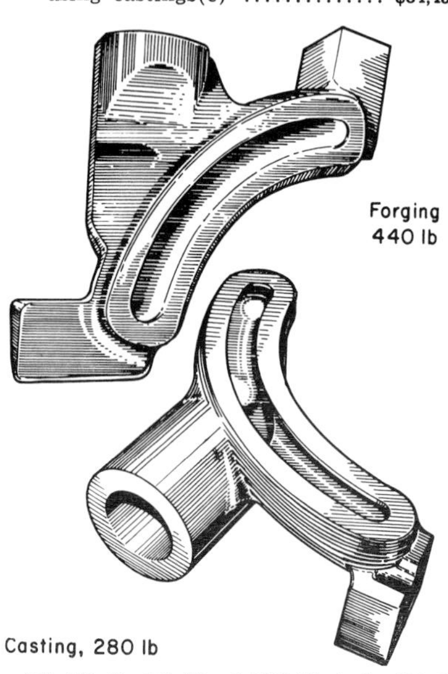

(a) 1% Cr, 1% Mo, 0.25% V steel. (b) 80 castings used per year. (c) There was a 91½% savings in cost of patterns over dies, 61% savings in cost of materials, and a 36% reduction in weight comparing rough forging (440 lb) with rough casting (280 lb).

brace, functional requirements could be fulfilled equally well by castings or forgings. Therefore cost, governed mainly by quantity, was the basis for selection of method.

The higher cost of initial equipment for forging virtually eliminated forgings for low quantities of this specific part. As the quantity of forgings increased, there was a corresponding decrease in price; with steel castings, because of the higher scrap ratio, the decrease in price per piece was not as great as for forgings when the quantity of parts increased.

Two curves for the cost of the casting are given in Example 29 — the first with temporary pattern equipment, having an estimated life of approximately 200 pieces; the second, with permanent equipment for more than 200 pieces. As shown by the graph, a quantity of more than 200 pieces is required to justify the cost of forging dies. The cost of forging dies was approximately 3½ times greater than that of temporary pattern equipment and 15% greater than that of the permanent pattern equipment.

Examples 30 to 35 compare cost of making parts by welding with costs for fabricating from a bar, forging or casting.

Sometimes a combination of product forms decreases cost of the component, as in Example 31, where two forgings were welded together to form a pinion. Welded combinations of castings, or castings with forgings, are also used advantageously in some instances.

Example 33 illustrates how castings proved cheaper because of the complexity of making a garden tractor assembly as a weldment. Originally 15 pieces of bar stock were welded together. A superior component was produced, at a 53% reduction in cost, by casting.

Example 34. The weldment shown in this example did not provide the rigidity required for brake shoe components on large rock-hauling trucks. Steel castings were then tried because pattern costs were negligible compared with the cost of the heavier equipment needed to form thicker, more rigid sections. The cast steel parts proved satisfactory in service and cost about 36% less than the weldments.

Example 35 is a cost comparison for producing an attachment support boot for a large motor grader. In fabricating this part as a weldment, difficulty was encountered because of its complex design (see sketch). The part required eight plates and two angle irons, which had to be cut and scarfed. Material was wasted, and it was difficult to assemble and fabricate the boot because of warpage in welding. It was essential that these boots fit reasonably tight on the grader moldboard and that the shank stock on the attachments fit snugly in the slot where they are inserted into the boot. Any machining would make the cost of the boot prohibitive. The accompanying table shows the cost breakdown in producing the attachment support boot as a mild steel weldment and as a 1040 steel casting.

Stock Size

Examples 36, 37 and 38 point out how cost savings may be effected by selection of a bar size in accordance with finished part requirements. In

Example 27. Cost Comparison of Forged Versus Cast Gear Drive Clutch Housings(a)

	Type or model							
	A		B		C		D	
	Bar stock	Casting	Forging	Casting	Forging	Casting	Forging	Casting
Material cost(b)	$18.00	$22.75	$50.00	$35.40	$74.00	$64.04	$117.00	$98.22
Saving (or loss)	....	(4.75)	....	14.60	....	9.96	....	18.78
Machining cost(c)	12.00	4.50	12.50	4.50	24.00	12.00	30.00	15.00
Saving	....	7.50	....	8.00	....	12.00	....	15.00
Total saving per unit	....	$2.75	....	$22.60	....	$21.96	....	$33.78
Annual saving(d)	....	$550.00	....	$4520.00	....	$2196.00	....	$1689.00
Total annual saving						$8955.00 (on all models)		

Forging Casting Finished part

8 8 7½

(a) This gear drive clutch is used on equipment for feeding coiled steel strip and sheet to presses or shears. (b) Type A housing was machined from bar stock; types B, C and D, from forgings. Cost of rough forging (4620 steel) for type B housing (in 1951) was $50.00 each and weighed 180 lb. Cost of rough steel casting for type B housing (in 1957) was $35.40 each and weighed only 84 lb. Composition of steel casting: 0.20% C, 1.20 to 1.40% Mn, 0.15 to 0.25% Mo. (c) Machine time based on the rate of $6.00 per hour. (d) Based on the following annual production basis: type A, 200 units; type B, 200 units; type C, 100 units; type D, 50 units.

Example 36, although the cost for steel 1/64 in. larger in diameter was slightly higher, scrap caused by surface seams, and the need for inspection, were eliminated, resulting in a total cost saving of more than 37%. In Example 37, a cost saving of 10% was realized by using a larger stock size in manufacturing piston pins from 1046 steel. This cost reduction brought about a sizable saving on an annual production of 1.25 million parts.

Bars purchased closer to finished size were more economical for making spindles, in Example 38. The cold drawn finish of the 1 63/64-in. diam bar was satisfactory for the application. The slightly higher cost for the odd size was insignificant compared with the saving in direct labor for turning the circumference of the long spindle. Savings on 300 parts amounted to about $110.

Example 28. Cost Comparison of Forging Versus Casting of Hollow Shaft for Strip Winder Reducer for a Cold Rolling Mill

	Forging(a)	Casting(b)
Weight before machining, lb	12,000	9000
Weight to be rough machined, lb	2,050	850
Material cost(c)	$4920.00	$3060.00
Pattern cost per piece	....	160.00
Machining cost(d)	615.00	255.00
Total cost of rough machined piece	$5535.00	$3475.00
Savings per piece		$2060.00

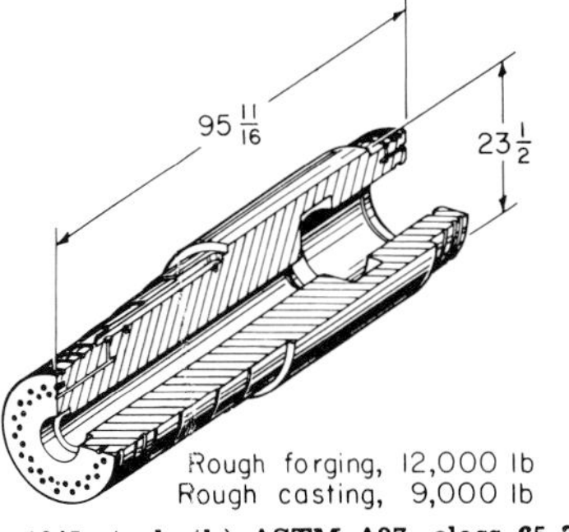

(a) 1045 steel. (b) ASTM A27, class 65-35 steel. (c) Forging cost 41¢ per pound; casting cost 34¢ per pound. (d) Forging, 41 hr at $15.00 per hour; casting, 17 hr; metal removed at rate of 50 lb per hr.

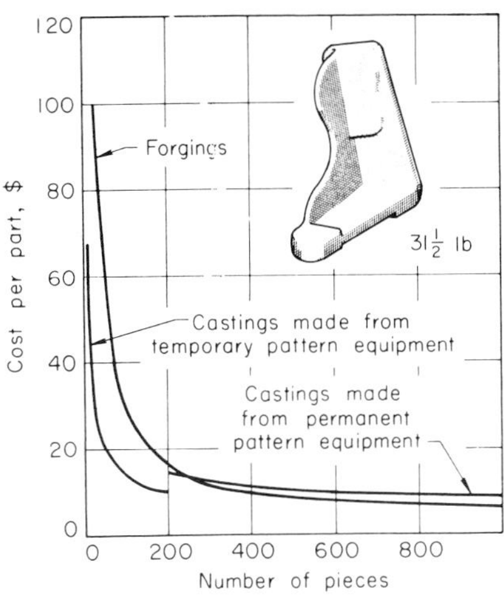

Example 29. Influence of quantity on cost of forging or casting a brace

Condition of Heat Treatment

Examples 39 through 44 compare costs for steels purchased in a pretreated condition to eliminate heat treatment of the parts in an advanced stage of production, with costs for steels in other conditions. In most instances composition has also been changed, and in some there was a sacrifice in mechanical properties. However, for all steels, sufficient testing assured suitability before the alternate steel was adopted on a production basis.

Example 39. Sporting-rifle barrels had been made from 4135 steel, heat treated to about 275 Bhn. Cold drawn stress-relieved 1137 steel, with a minimum yield strength of 90,000 psi, was successfully substituted for the 4135 steel, with a cost saving of almost 30%.

Example 40 shows the cost advantage of stress-relieved 1137 steel, as compared with 4140, for worms.

Examples 30, 31 and 32. Cost of Weldments Compared with Solid Bar, Forgings and Castings

	High-cost alternate	Low-cost alternate
Example 30. Side arm(a)	**Weldment**	**Bar**
Material cost per piece	$19.24	$27.60
Labor cost per piece	5.67	1.06
Burden per piece	12.23	2.07
Total cost per piece	$37.14	$30.73
Net savings per piece using bar		$ 6.41
Example 31. Pinion(b)	**Single forging**	**Two-forging weldment**
Forging cost for both pieces	(b)	$ 8.35
Labor cost	(b)	2.75
Burden	(b)	6.32
Total cost per piece	$32.00	$17.42
Net saving per piece using weldment		$14.58
Annual saving		$23,328.00
Example 32. Support(c)	**Weldment**	**Casting**
Material cost per piece	$ 6.188	$13.72
Labor cost per piece	4.449	(c)
Burden cost per piece	8.360	(c)
Total cost per piece	$18.997	$13.72
Net saving per piece using casting		**$ 5.277**

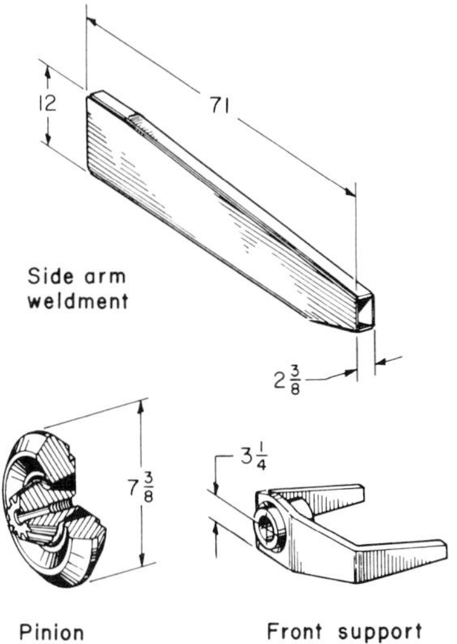

(a) For making 320 parts per year as shown in above sketch from low-carbon steel. (b) For making parts shown in above sketch from low-carbon steel. Cost breakdown figures for parts from single forgings are not available. Figures are based on 1600 pieces. (c) For making parts shown in above sketch from 1030 steel castings or 1020 steel weldments; 207 parts annually. Castings were purchased in finished state.

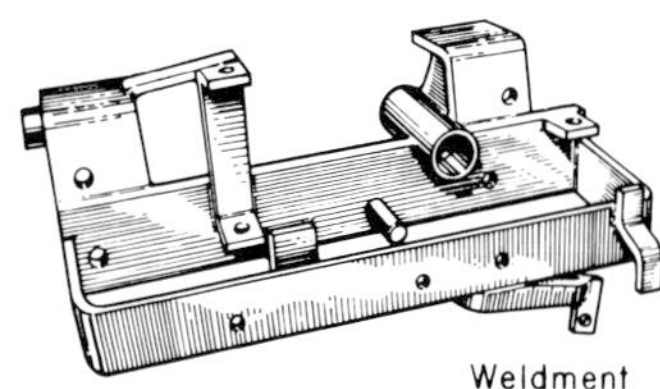

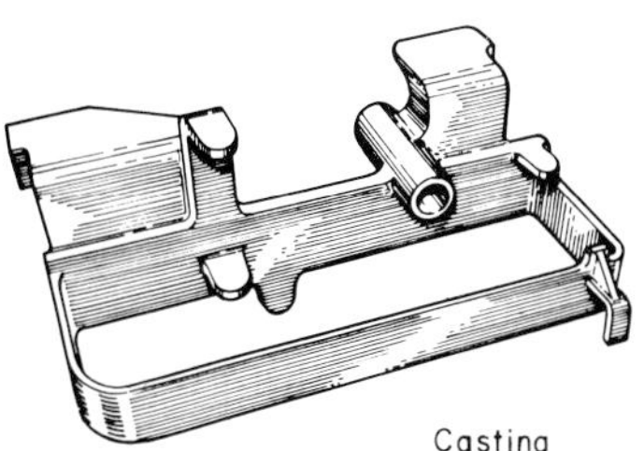

Example 33. Both of the cutter bar and coulter shaft frames were designed for a small riding-type garden tractor. The original frame (top) was fabricated from 15 pieces of bar stock. It required cutting, drilling, bending, and assembly in an elaborate fixture. Straightening operations were often required. Rejections and field failures were frequent. Conversion to a steel casting (bottom) provided greater strength, eliminated rejections and replacements, and reduced manufacturing costs by 53%

Example 34. Cost Comparison for Truck Brake Shoes Produced from Weldments and Castings

	Weldment	Casting
Material cost per piece	$10.72(a)	$20.50(b)
Cutting, forming	18.50	....
Welding	18.40	....
Machining	21.38	21.38
Stress relief	1.80	....
Heat treating	....	2.85
Total	$70.80	$44.73
Savings		$26.07

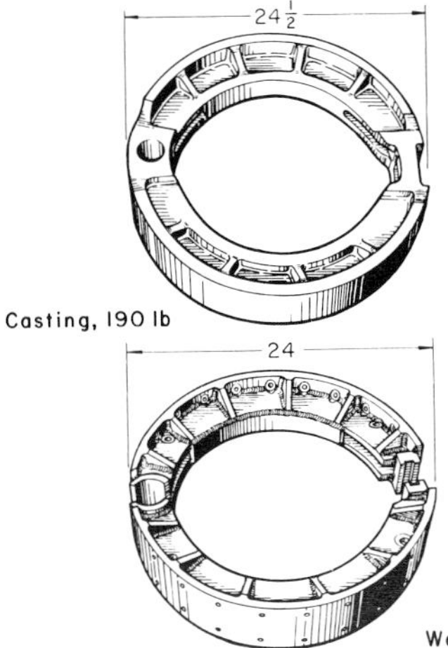

(a) 120 pieces (20 units), mild steel. (b) 600 pieces (200 units). The steel castings complied with ASTM A148, grade 80-50.

The worms made of 1137 (see sketch) received a final heat treatment in an energized salt for surface wear resistance. The improved finish of the stress-relieved 1137 and its lower hardenability (less distortion than 4140) permitted elimination of operations as shown and reduced the cost.

Example 41. Complete cost details were not available for these tie rods (see sketch) but the available data show that parts made from stress-relieved 1144 steel cost less than from 4140H. The lower hardness of the stress-relieved steel (241 to 302 Bhn, compared with 302 to 352 Bhn for 4140H) was satisfactory.

Example 42. Valve caps made from cold drawn stress-relieved 1144 steel proved equal to heat treated 1141. The same mechanical properties could have been obtained with cold drawn stress-relieved 1141 but 1144 was selected because of improved machinability. The higher sulfur content of the 1144 was not detrimental to the engineering properties of this part.

Example 43. Some steels, such as 1144, are available with strength properties higher than can be obtained by ordinary cold drawing and stress relieving. These properties are obtained by heating the steel before reducing it through dies. The cap nuts (see sketch) were made at a substantial saving from 1144 drawn at elevated temperature, compared with 4140 bars heat treated to Rockwell C 30 to 35. The 1144 had a minimum yield strength of 120,000 psi and a hardness of Rockwell C 35. Despite the high hardness, machinability and finish were excellent because of the pearlitic structure and high sulfur content. The same parts, when made of annealed medium-carbon steel and heat treated after machining, distorted excessively, resulting in scrap rates as high as 40%.

Example 44 compares two similar alloy steels for making pinions. Elimination of four manufacturing operations more than offset the increased cost of the pretreated steel.

Example 35. Cost Comparison in Producing an Attachment Support Boot as a Weldment and as a Steel Casting (a)

	Weldment	Casting
Material Cost	$10.00	....
Casting cost	....	$28.84
Welding labor	27.50	....
Freight	....	1.12
Machining cost	7.00	7.00
Total cost	$44.50	$36.96
Savings per unit (b)	....	$ 7.54

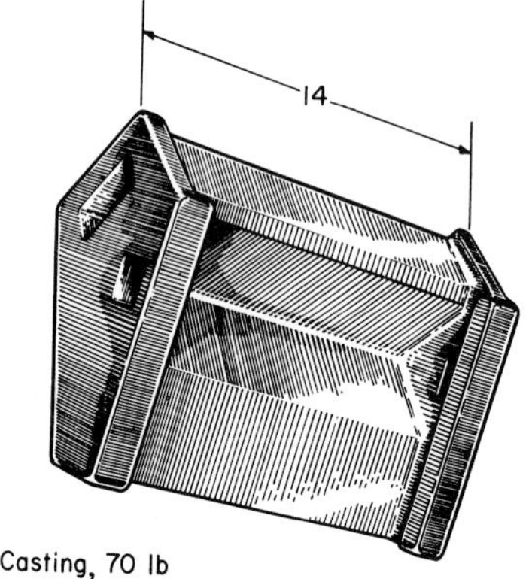

Casting, 70 lb

Support boot

(a) Casting weight was 70 lb; wall thickness, ¾ in.; height, 14 in.; length, 12 in.; width, 3 to 7 in.; 1040 cast steel. (b) 17% less than weldment.

Example 36. Effect of Increased Stock Size on Cost

	Stock diameter ¾ in.	Stock diameter 49/64 in.
Raw material weight. per 1000 pieces	408 lb	424 lb
Raw material cost ... per 1000 pieces	$ 56.30	$ 58.50
Machining cost per... 1000 pieces	187.00	196.00
Magnaflux cost per .. 1000 pieces	49.40	....
Scrap loss	28%	....
Scrap cost per ...... 1000 pieces	81.96	....
Cost per good piece..	0.4065	$ 0.2545
Unit saving per .... piece using 49/64-in. stock		$ 0.1520

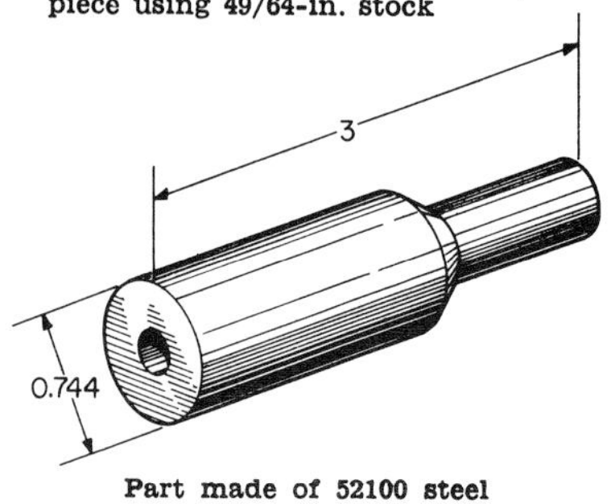

Part made of 52100 steel

Example 37. Selection to Eliminate Surface Seams and Need for Inspection

Original stock size	1.016 in.
Alternate stock size	1.062 in.
Cost saving on finished part... using larger size	10%

For piston pins from 1046 steel. Annual production, 1.25 million parts.

Example 38. Cost Comparison for Fabricating a Spindle from Two Different Sizes of 1113 Steel Stock

	Bar diameter 2 in.	Bar diameter 1 63/64 in.
Unit weight, lb	17.80	17.20
Work standard, ... pieces per hr	10	17
Material, lb .... for 300 pieces	5340	5160
Costs:		
Material	$576.20	$580.50
Labor	261.00	153.75
Tool	29.00	21.75
Total cost	$866.20	$756.00
Savings using $1\frac{63}{64}$-in. diam		$110.20
Savings per piece		$ 0.3673

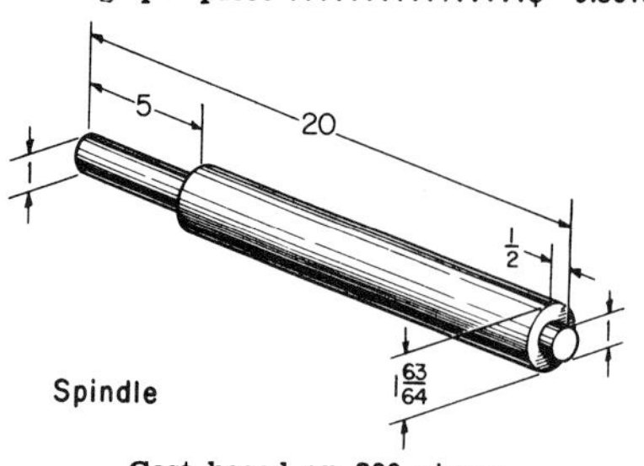

Cost based on 300 pieces

Examples 39 to 42. Costs of Parts Made from Cold Drawn, Stress-Relieved Steels Compared with Heat Treated Steels

	High-cost alternate	Low-cost alternate
Example 39. Rifle Barrels(a)	**4135**	**1137SR**
Unit weight, lb	4.55	4.55
Material cost per 100 lb.	$14.27	$11.90
Material cost per piece.	$ 0.65	$ 0.54
Direct labor, hr .... per piece	0.48	0.36
Direct labor cost ...... per piece	$ 0.79	$ 0.54
Material and .... labor cost	$ 1.44	$ 1.08
Unit savings per piece		$ 0.36
Example 40. Five-Tooth Worm(b)	**4140**	**1137SR**
Material cost .... per 100 pieces	$11.25	$ 6.68
Turn, drill, form, ... bore, ream	12.65	11.50
Face both ends, ... break edges	3.51	3.28
Broach keyway	1.16	1.16
Rream to remove burrs	1.12	....
Remove burrs and ..... sharp edges at ends of keyway	1.96	1.96
Cut teeth	10.08	10.08
File flat on worm ..... teeth, both ends, and burr	5.39	5.39
Heat treat	1.75	1.75
Blast clean	0.75	....
Grind OD	2.50	....
Grind bore	4.00	....
Roto-finish	0.61	0.61
Direct labor .... and material per 100 pieces	$56.73	$42.41
Unit saving		$ 0.143
Example 41. Tie Rods(c)	**4140H**	**1144SR**
Steel cost	$ 0.18	$ 0.19
Heat treating cost	0.05	....
Machining, hr	0.090	0.063
Straightening, hr	0.075	....
Example 42. Valve Cap Screws(d)	**1141HT**	**1144SR**
Unit weight, lb	0.217	0.217
Work standard .... min per piece	7.99	4.00
Material cost .... per 100 lb	$11.60	$12.50
Direct material .... weight, lb	7812	7812
Direct material cost ...$	906.20	$ 976.50
Total direct labor, hr...	4,793.0	2,392.8
Total direct labor cost.$	9,354.11	$4,936.78
Total cost, labor ....$ and material	10,260.31	$5,913.28
Total savings		$4,347.03
Savings per piece ....$		0.12

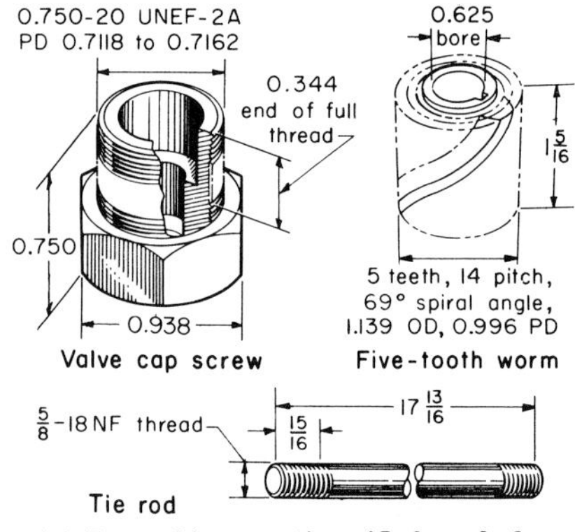

(a) For making sporting rifle barrels from 1⅝-in. diam quenched and tempered 4135, compared with 1⅝-in. diam drawn and stress-relieved 1137. (b) For making five-tooth, 14-pitch worms, as in sketch. 1137 was surface hardened in molten salt; 4140 quenched and tempered. (c) Cost of making tie rods (see sketch) from 4140H, hardened and tempered to 302 to 352 Bhn, compared with 1144 cold drawn and stress relieved, 241 to 302 Bhn. (d) Cost of making valve cap screws from 1141, hardened and tempered to 250 to 300 Bhn, compared with cold drawn and stress-relieved 1144 with minimum yield strength of 100,000 psi. Data based on 36,000 pieces; see sketch.

Eliminating Operations

Examples 45 through 53 deal with selection of the most economical steels for heat treating, based on elimination of operations, reduction of scrap loss, or productivity.

Example 45 shows that the increased cost for the alloy steel is justified for making wear plates (see sketch) because scrap from cracking, and the need for straightening, are eliminated.

Example 46 also shows a higher cost for carbon steel used in making arms (see sketch) because of loss from cracking during heat treating.

Example 47. Elimination of scrap and inspection resulted in a saving in making the shaft shown in the sketch. Excessive hardenability was apparently responsible for cracking of the parts made from 80B42.

Example 48 illustrates how substituting a less costly steel that can be induction hardened can decrease cost. The acme screw was made from a hot rolled nitriding steel. It was necessary to heat treat, rough machine, stress relieve, finish machine, and grind threads before nitriding. After nitriding, extensive straightening was necessary because the portion incorporating the long keyway bowed an appreciable amount in nitriding.

The use of stress-relieved carbon-corrected 1151 steel effected an over-all cost reduction of 28.4%. This saving resulted from a substantial reduction in the number of operations and the use of a lower-cost steel. Field tests proved that the mechanical properties of the in-

Example 43. Cost of a Cap Nut Made from a Carbon Steel Drawn at Elevated Temperature Compared with a Quenched and Tempered Alloy Steel

	4140	1144
Unit weight, lb	0.255	0.255
Work standard	482.15	249.77
Material cost .... per 100 lb	$ 15.07	$ 13.52
Total material cost...	1383.43	1241.14
Direct labor, hr	2893.00	1332.00
Direct labor cost	5980.57	3109.32
Total material and. labor cost	7364.00	$4350.46
Total savings		$3013.54
Savings per piece		$ 0.0837

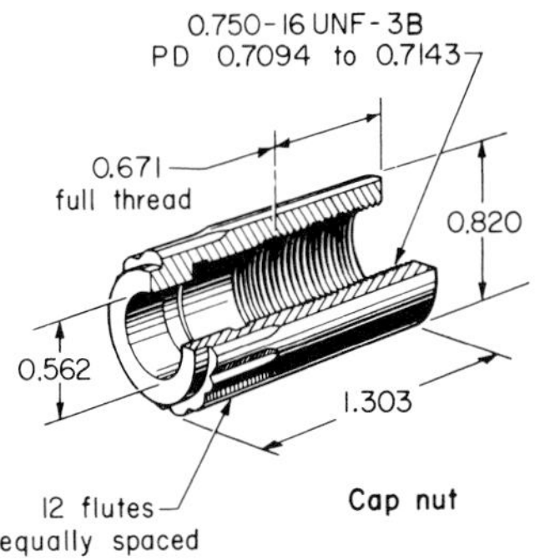

The 4140 steel was a leaded type, quenched and tempered to Rockwell C 30 to 35. The 1144 was drawn at elevated temperature to obtain a minimum yield strength of 120,000 psi. Data based on 36,000 pieces.

Example 44. Cost of a Pinion Made from Heat Treated Bars of 4150 Compared with 4140 Heat Treated after Machining

	4140	4150
Material cost per 100 pieces. (2000-lb purchase)	$ 54.40	$ 61.61
Machining cost	74.10	74.10
Heat treating	6.73	....
Blast cleaning	1.33	....
Grinding OD	2.40	....
Grinding ID	5.67	....
Total for 100 pieces	$144.63	$135.71
Savings per piece		$ 0.0892

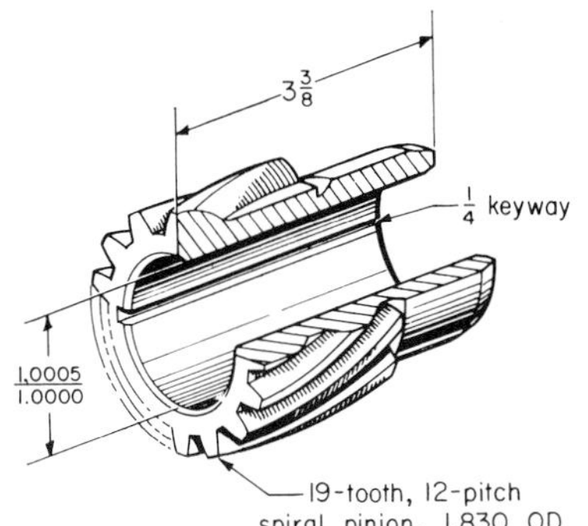

For making pinions, as shown in sketch, from annealed 4140, heat treated in pieces, compared with resulfurized 4150, heat treated in bars to 250 to 300 Bhn.

Examples 45, 46 and 47. Savings by Selecting Steels that Minimize Heat Treating Scrap Losses

	High-cost alternate	Low-cost alternate
Example. 45 Wear Plates(a)	**1045**	**50B50**
Material cost per piece....	$ 0.27	$ 0.37
Shearing and .... straightening	0.79	0.79
Heat treating	0.12	0.12
Straightening	0.42	....
Loss from cracking	0.53	....
Total cost each	$ 2.13	$ 1.28
Savings per piece		$ 0.85
Example 46. Arms(b)	**1045**	**4150**
Weight per piece, lb	31.24	31.24
Material cost per piece ...	$ 2.20	$ 2.93
Machining cost per piece..	34.50	38.60
Loss in heat treating ..... by cracking (30% for 1045)	11.01	....
Total cost per piece	$47.71	$41.53
Savings per piece		$ 6.18
Example 47. Shafts(c)	**80B42**	**8735**
Material cost per piece ...	$ 0.1735	$ 0.1742
Machining cost	0.0813	0.0813
Scrap	15%	....
Cost of scrap	$ 0.0383	....
Cost of inspection	0.0033	....
Total cost each	$ 0.2964	$ 0.2555
Savings per piece	....	$ 0.0409

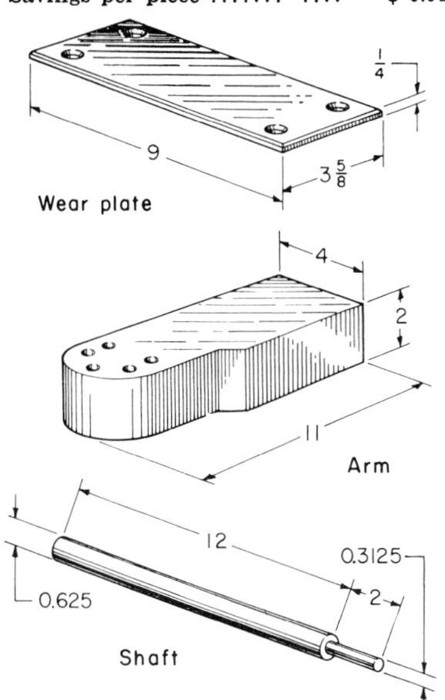

(a) For making wear plates for a fifth-wheel mechanism, per sketch. (b) For making arms, per sketch. (c) For making shafts, per sketch; based on 200,000 pieces.

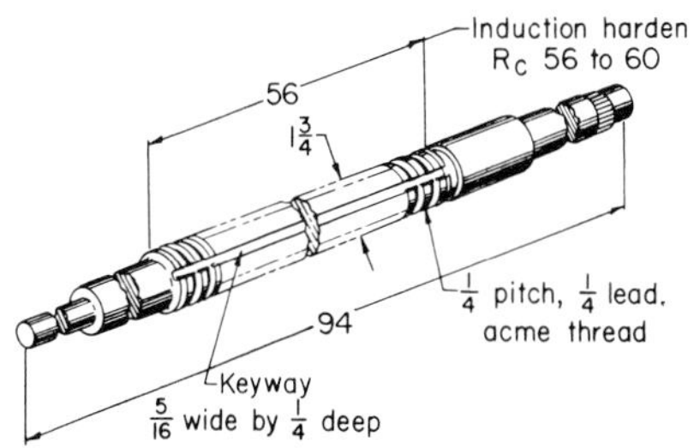

Example 48. Cost reduction by substitution of material and processing. Induction hardened 1151 steel replaced nitrided nitralloy. (See text.)

Example 49. Cost Reduction by Substitution of Material and Processing

When 1041 steel was selected for induction hardened axle shafts in place of the alloy 50B50 steel fully hardened before machining, there were marked savings because of the difference in cost of the two steels.

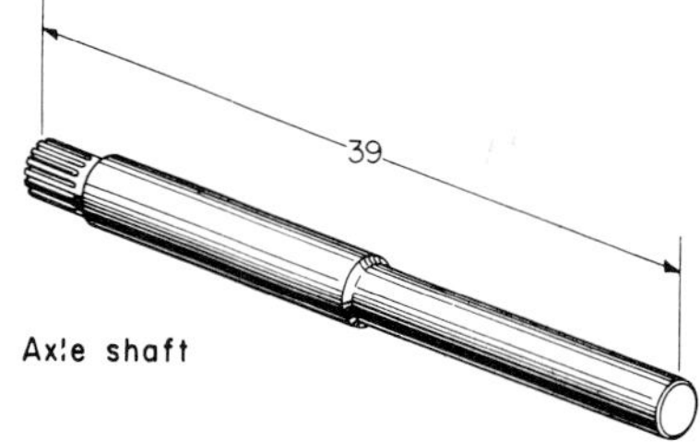

Annual production requirements and cost savings per ton amounted to:

1387.8 tons of $2\frac{7}{8}$-in. diam steel at $30	$41,634
2047.6 tons of $3\frac{1}{8}$-in. diam steel at $28	57,333
Total annual savings	$98,967

Since machining of the 1041 induction hardened axle was done before hardening, there was considerable saving in labor and tool cost; hardness of the 1041 was about 200 Bhn; 50B50 was machined at 321 Bhn. Machining cost was reduced by 26¢ per axle or 52¢ per tractor by changing to 1041 steel.

Example 50. Cost Reduction by Replacing an Alloy Steel with a Carbon Steel for an Induction Hardened Part

	4150	1045
Steel for 1000 .... pieces, lb	2120	2120
Steel cost per 1000 ...... pieces	$ 315.17	$ 285.56
Machining cost per ..... 1000 pieces	2164.00	2034.00
Induction heating ...... cost per 1000 pieces	200.00	150.00
Loss by cracking	22%	....
Cost of scrap loss ...... per 1000 pieces	589.42	....
Total cost per piece ...	$ [illegible].435	$ 2.470
Saving per piece		$ 0.965

Induction harden
R_C 60 to 62
$1\frac{1}{4}$
1
8

High hardenability of 4150 caused cracking.

duction hardened parts were entirely satisfactory.

Example 49 is another part where induction hardening permitted a change from alloy to carbon steel. Axle shafts shown in the sketch were made from 50B50, hardened to 321 Bhn before machining. Induction hardened 1041 proved to be a satisfactory replacement. Annual savings on material alone are tabulated in Example 49. In addition, the change to 1041 steel allowed a decrease in machining cost of 26¢ per axle or 52¢ per tractor.

Example 50 is another illustration of the adaptability of carbon steel to induction heat treating. The high hardenability of the 4150 alloy steel caused excessive cracking in the hardened section (see sketch). Lower initial cost of 1045 and elimination of cracking brought about the savings shown in the tabulation.

Examples 51 and 52 show sketches and tabulated cost data for transmission shafts and hydraulic valves where induction heat treating permitted the replacement of alloy steels with resulfurized carbon steels. Of three steels tried for making the valve plugs, 1132 was the least costly and met requirements.

Example 53. Production can be increased by proper correlation of steel selection with induction heating. Better uniformity resulted when 1039 steel was substituted for 1037 for the axle shafts shown in the sketch. Some heats of 1037 permitted a scan rate of only 0.625 in. per sec, whereas the minimum scan rate for 1039 was 0.700 in. per sec.

Hot Rolled Versus Cold Finished

Examples 54 and 55 give cost analyses showing why hot rolled steels were selected in preference to cold finished steels. Hot finished steels are cheaper if (*a*) finish meets requirements, (*b*) the wider dimensional tolerances can be efficiently adapted to machine tools, and (*c*) the lower strength is sufficient. For the lawn mower tie rods (Example 54), hot rolled and pickled bars had a satisfactory finish and machine-straightened bars could be machined in screw machines. The higher strength of cold drawn steel was of no value in this instance, and the hot rolled grade proved cheaper. Hot rolled steels were also cheaper than cold drawn steels for making transmission pinion gears (see sketch, Example 55).

Example 56. A different situation is presented in the shaft of Example 56. Although it was necessary to heat treat the hot rolled steel, the higher-strength cold drawn steel met mechanical property requirements without heat treatment, thus permitting a marked saving despite the high initial cost.

Other Examples

Examples 57 and 58 illustrate advantages in paying more for carbon-corrected bars. The saving in making the long shafts (sketch, Example 57) is mainly because the use of restricted surface quality and carbon-corrected steel eliminated a major portion of turning. The use of carbon-corrected steel was also advantageous for making track pins in Example 58.

Example 59. Fuel-injection camshafts were made from annealed 1050 steel forgings. Cam designs were similar to the sketch, but varying from the 2½-in. size shown to as small as 1¼ in. across the cam contour. The number of cams integral with the shaft varied from one to eight. Lobe contours were induction hardened to a minimum of Rockwell C 60. Scrap loss from hardening cracks in the steep portions of lobes ran from as high as 5% for some heats to as low as 0.5% for other heats, with an over-all aver-

Examples 51 and 52. Savings Realized by Combination of Steel Selection and Induction Hardening

Example 51. Shafts(a)	4140	1141
Material cost per 100 pieces	$113.23	$ 84.64
Cut off, smooth up ends	6.67	6.07
Automatic lathes, three setups	21.30	19.26
Lathe, form grooves	3.40	3.12
Grind diameter for splining operation	7.62	7.62
Cut spline, two setups	22.84	22.84
Heat treat	5.85	4.95
Temper	1.44	1.44
Blast clean	1.53	...
Power brush splines	6.42	6.42
Straighten to 0.002-in. total indicator	4.27	3.33
Grind eight diameters	23.65	23.65
Form radius	9.60	8.52
Total for 100 pieces, direct labor and material	$227.82	$191.86

Example 52. Valve Plugs(b)	1117	4150	1132
Steel cost	$0.29	$0.41	$0.30
Machining and grinding	1.26	1.40	1.33
Heat treating	0.75	0.15	0.15
Straightening	0.35	0.65	...
Total cost per piece	$2.65	$2.61	$1.78

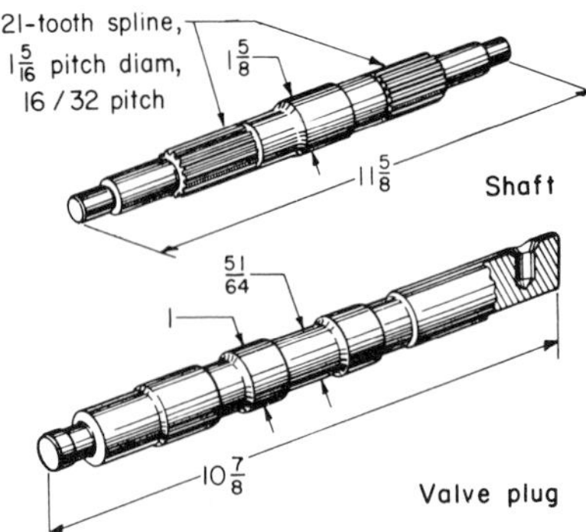

(a) For making splined transmission shafts as per sketch from 4140 conventionally hardened versus 1141 induction hardened. (b) For making hydraulic valve plugs as per sketch from resulfurized 4150, quenched and tempered, 1117, carburized, and 1132, induction hardened. Costs are per piece, based on 4000 pieces annually.

Example 53. Increased Productivity Resulting from Steel Selection for Induction Hardening

Scan rate, in. per sec	No. of heats	% of heats
1037 Steel		
0.800	43	66
0.725	7	11
0.700	12	19
0.625	3	4
1039 Steel		
0.800	26	84
0.700	5	16

~40

Axle shaft

Axle shafts, as sketched, were induction heated. Scanning rate for 65 mill heats of 1037 was 0.625 to 0.800 in. per sec compared with 0.700 to 0.800 in. per sec for 26 mill heats of 1039. Only 66% of the 1037 could be scanned at 0.800 in. per sec compared with 84% for 1039.

Examples 54, 55 and 56. Cost Comparisons for Making Specific Parts from Cold Drawn and Hot Rolled Steels

	High-cost alternate	Low-cost alternate
Example 54. Tie Rods(a)	**1212 CD**	**1212 HR**
Unit weight, lb	1.673	1.673
Work standard, hr per 100	2.5	2.5
Material cost, $ per 100 lb	$10.20	$7.50
Direct material, lb	41,825	41,825
Material cost	$4266.25	$3136.87
Direct labor, hr	625	625
Direct labor cost, $	$905.00	$905.00
Total saving using hot rolled		$1129.38
Unit savings per piece using hot rolled steel		$0.045
Example 55. Transmission Gears(b)	**4024 CD**	**4024 HR**
Material cost, $ per lb.	$0.11425	$0.09625
Weight of raw material per part, lb	0.582	0.606
Cost of material per part	$0.0665	$0.0583
Parts per transmission	6	6
Material costs per transmission	$0.3990	$0.3498
Total material costs (600,000 units)	$239,400	$209,880
Extra tooling costs	....	$13,600
Total material and tooling extra	$239,400	$223,480
Cost reduction using hot rolled		$15,920
Example 56. Shaft(c)	**1045 QT**	**1045 CD**
Material cost per part	$5.93	$11.29
Labor cost per part	5.61	3.00
Burden per part	13.37	6.95
Total cost per part	$24.91	$21.24
Savings per part using cold drawn steel		$ 3.67

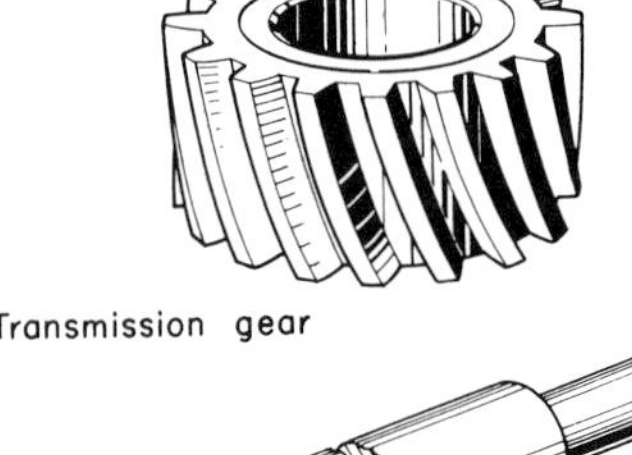

Transmission gear

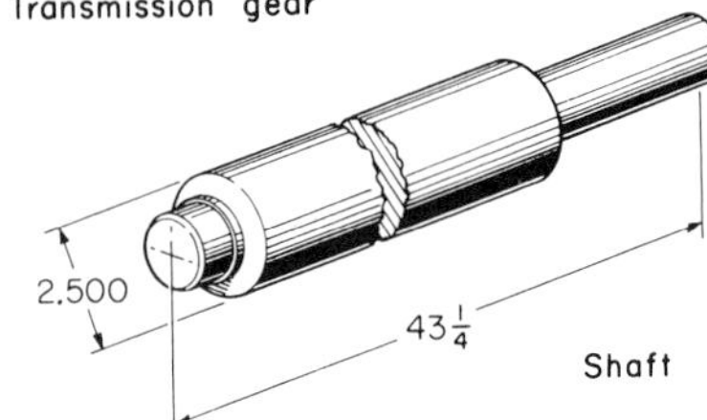

(a) For making lawn mower tie rods from ⅝-in. diam 1212, hot rolled compared with cold drawn, based on an annual production of 25,000 pieces. (b) Cost for making transmission gears, as per sketch from 4024, hot rolled compared with cold drawn. (c) Cost for making parts as per sketch from 1045 hot rolled, quenched and tempered, and from 1045 cold drawn.

age of 1%. The scrap loss was not regarded as serious, but hardening cracks in the cam lobes resulted in fatigue failures early in service life. All shafts were therefore subjected to magnetic particle inspection. Continued examination of cracked shafts proved that high manganese (allowable range, 0.60 to 0.90%), together with excessive residual amounts of chromium and molybdenum, was responsible for cracking. A modified grade of 1050 steel with a maximum manganese content of 0.50% was purchased in five-ton lots at no extra cost.

No harmful cracks appeared for three months after the new steel was specified. Therefore, magnetic particle inspection was discontinued

Example 57. Cost Savings Through Carbon-Corrected Steel

Steel 1151	Cost per shaft
$2\frac{3}{4}$-in. diam HR	$8.56
$2\frac{5}{8}$-in. diam CD	9.40
$2\frac{1}{2}$-in. diam, carbon corrected	8.94
Net cost reduction (from decreased labor)	$4.34

$27\frac{1}{4}$ — $19\frac{1}{4}$ — Induction harden R_C 56 to 60 — $2\frac{1}{2}$ — $52\frac{1}{4}$ — $\frac{1}{4}$ pitch, $\frac{1}{4}$ lead

Induction hardened shaft

Example 58. Comparison of Two Methods for Eliminating Decarburization in Track Pins

Original material	1045 steel, turned and polished
Alternate material	1045 steel, carbon restored
Savings with carbon-restored bars:	
Per 100 lb	$0.77
Per tractor	$2.59

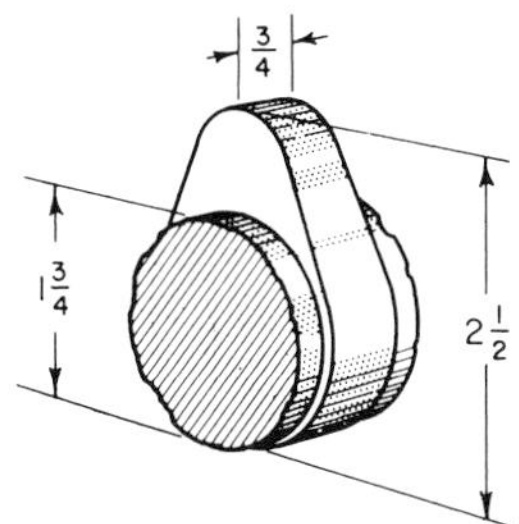

Example 59. Scrap elimination through the use of restricted manganese in 1050 steel forgings. Inspection costs were reduced $7.00 per 100 shafts. See text for discussion.

Example 60. Scrap Reduction in Making Washer and Screw Assemblies

	Original material, 1022	Alternate material, 1024(a)
Scrap loss	40%	Near 0
Cost increase, using restricted manganese content		5%

(a) Restricted manganese

with an average cost saving of $7.00 per 100 shafts. This was in addition to savings from reduced scrap losses and less handling.

Example 60 is another instance where restricted manganese reduced scrap loss because of the erratic hardenability of 1022 steel.

Example 61. By discriminating steel selection, a saving was effected in making the weldment sketched. It was originally made from ASTM A27 (0.35% C max and 0.70% Mn max). Changing to a similar steel (SAE 0030) with a maximum carbon content of 0.30% eliminated the need for preheating, with the cost saving shown in the tabulation.

Example 62 illustrates one instance where a slight extra for killed steel reduced scrap from 7% to less than 1% in making screws. Extra payment for quality level is justified when operations can be eliminated or scrap reduced.

Example 63 is a part where an extra was justified because it permitted the pins shown in the sketch to be made by cold forging instead of the more costly hot forging operation. The cold heading quality 4037 steel was cold drawn from normalized and spheroidized rods.

Example 64. Offset tie rods and torsion shafts (see sketch) made from 4145H showed a 5% scrap loss from cracks initiated by mill seams. Adding an axle shaft quality requirement at $0.30 more per 100 lb eliminated scrap and the necessity

Example 61. Cost Comparison for Weldments Made from Two Different Cast Steels

	ASTM A27 preheated	SAE 0030 no preheat
Labor	$27.38	$22.64
Burden	50.65	41.88
Total	$78.03	$64.52
Net saving per weldment by eliminating preheat		$13.51

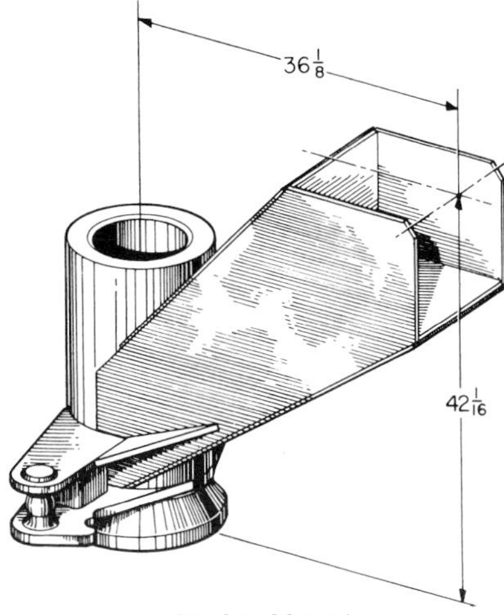

Steel weldment

Example 62. Scrap Reduction in Making Machine Screws

Original material	1010 and 1017, capped
Scrap loss	7%
Alternate material	1010 and 1017, killed
Scrap loss	<1%

for oversize stock. Savings in scrap more than offset the cost extra.

Overspecification

Overspecification frequently occurs from blind copying of specifications of similar parts. Hardenability is a property frequently overspecified.

Example 65 illustrates a side gear formerly made from 4118 steel. A change to 1024 did not sacrifice required properties, and an annual saving of $18,000 was gained.

Example 63. Effect of Steel Selection on Method of Forging and Cost(a)

	Hot forged 4037	Cold forged 4037
Unit weight, lb	0.1088	0.113
Material cost per 100 lb, $	$18.36	$24.00
Direct material cost, $	199.75	271.20
Direct labor cost, $	13,200.00	10,461.00
Total cost, $	13,399.75	10,732.20
Total saving, cold forging		$2,667.55

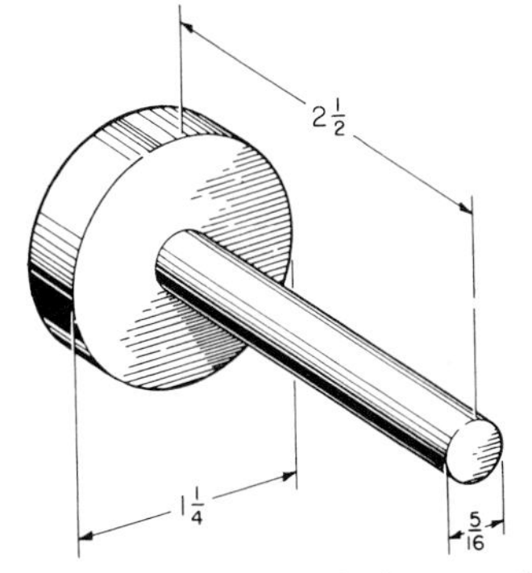

(a) For making headed pins per sketch from 4037 steel, annealed, cold drawn and hot forged compared with cold forging from cold heading quality 4037. Cost data based on 10,000 pieces.

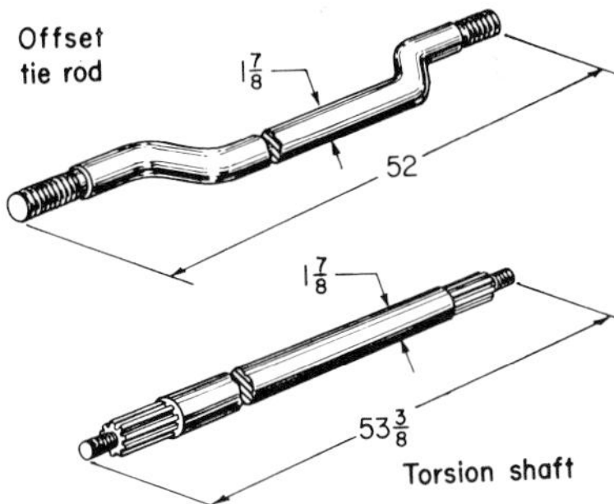

Example 64. Change from 4145H to axle shaft quality eliminated scrap.

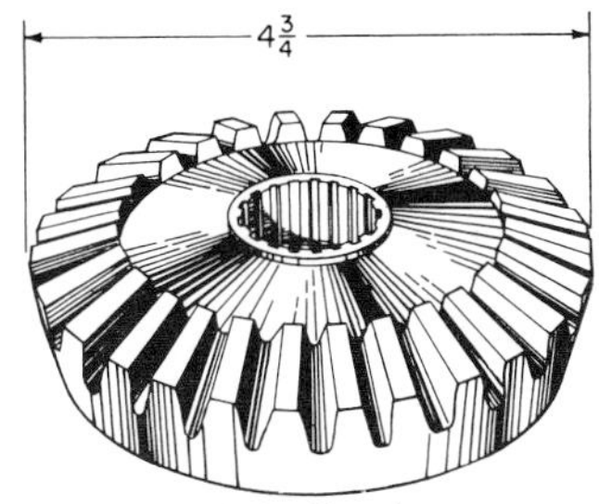

Example 65. Selection for economy by decreasing hardenability. For the side gear shown, a change in steel from 4118 to 1024 resulted in an annual savings of $18,000, based on 750 tons of steel at $24.00 per ton. Heat treatment and machining remained unchanged.

Computer-Based System Selects Optimum Cost Steels-I

By DALE H. BREEN and GORDON H. WALTER

This is the first in a series of articles describing a new approach for the design and selection of alloy and carbon steels for heat treated applications. It is a systems approach incorporating the use of a digital computer. Within International Harvester, it is called the CHAT system, an acronym standing for Computer Harmonized-Application Tailored. These articles will describe, in some detail, the techniques used in the CHAT system. The first article deals mostly with background material, while the remaining articles provide a detailed description of application tailoring (the defining of metallurgical requirements), and computer harmonizing (designing optimum cost steels using a digital computer).

Since the turn of the century, the general nature of steel has continually changed through increased understanding of the metallurgical factors which control reactions in heat treatment and subsequent engineering performance. With increased understanding of the primary influences of microstructure came also the gradual development of quantitative knowledge concerning alloying elements and their influence over microstructures. The importance of the classic works of Shepherd[1] in grain size, Bain and Davenport[2] in transformation rationale, Grossmann[3] in calculated hardenability, and Jominy[4] in hardenability measurement is indisputable. Results of early work, mostly prior to 1959, are essential to current efforts of economical alloy utilization and alloy conservation.

If one would plot the various metallurgical milestones along a time line, he would note that the general trend of alloy usage and selection became more rational with increasing metallurgical insights. Change is healthy, and should be motivated by the desire for efficiency – and, of course, it should be done rationally. The steel to use is that with an alloy combination which will most effectively provide the engineering performance at least cost. In essence, this is the heart of the CHAT procedure.

Mr. Breen is chief, Materials Div., and Mr. Walter is chief engineer, Metallurgical Section, Materials Div., Engineering Research, International Harvester Co., Hinsdale, Ill.

Hardenability Concepts

Since hardenability is essential to this concept of steel design, it is desirable to review the fundamental principles. Hardenability is defined as "a measure of the ease with which one can obtain hardness." Figure 1 illustrates the cross sectional view of steel bars of different alloy content. Their increasing hardenability is shown by the increasing depth of hardness, left to right.

The interrelationship between hardenability, hardness depth, and cooling rate can also be illustrated by the same bars. Assuming that all bars are of the same hardenability, the same pattern of hardness depths could be obtained by increasing the quench severity. That is, if the bars (of equal hardenability) were quenched, left to right, in hot oil, cold oil, water, and brine, the same variation of hardened depth could be obtained. With hardenability and quenching medium held constant, the same variation of hardened depth could also be obtained by decreasing bar diameter, left to right – which, in effect, increases the cooling rate.

Fortunately, the Jominy end-quench test unites the three important parameters into one test, allowing the development of quantitative information. The Jominy bar provides a series of cooling rates, from extremely fast at the quench end to very slow at the air-cooled end. Recording the hardness values obtained along the length of the bar provides a complete hardenability "fingerprint" of a given steel composition.

Figure 2 illustrates Jominy hardenability curves for six different compositions. The curves represent two carbon levels, 0.12 and 0.48%, and three hardenability levels for each of the carbon contents. Three critical features are illustrated. First, the hardness obtained at any given cooling rate, J position, is a function of hardenability and carbon content. Second, the initial hardness (IH, measured at $J = \frac{1}{16}$ in.) is a function of carbon content and independent of hardenability. Third, the Jominy hardenability curve can be characterized, within limits, by a single number known as the "ideal diameter", or D_I[3].

The larger the D_I value, the greater the hardenability. Through standard tables[5], one can determine the hardenability (D_I) required to obtain any specified distance hardness (DH) at any specified J position.

Calculating Hardenability

The next issue concerns the use of alloy multiplying factors to calculate D_I from chemical composition and grain size. This calculation procedure relies on a series of hardenability factors for each alloying element in the composition: multiplied together, they give a D_I value. The D_I value can then be translated into a Jominy curve by using the IH/DH tables[5]. Over the years, values of alloy multiplying factors have been improved. The factors proposed by Doane[6] are utilized here, but sometimes either the Grossmann[3] or Kramer[7] factors can be used advantageously.

Another important concept to be considered in devising new steels is the distinction between case hardenability and base (core) hardenability, designated herein as D_{Ic} and D_{Ib} respectively. The effect of a given alloy addition on hardenability is significantly different for low carbon than for high carbon steel. For carburizing steels, this difference becomes important because the steel will harden as a hypereutectoid steel in the carburized region and as a hypoeutectoid steel in the remaining area. Thus, a steel must possess adequate base and case hardenability.

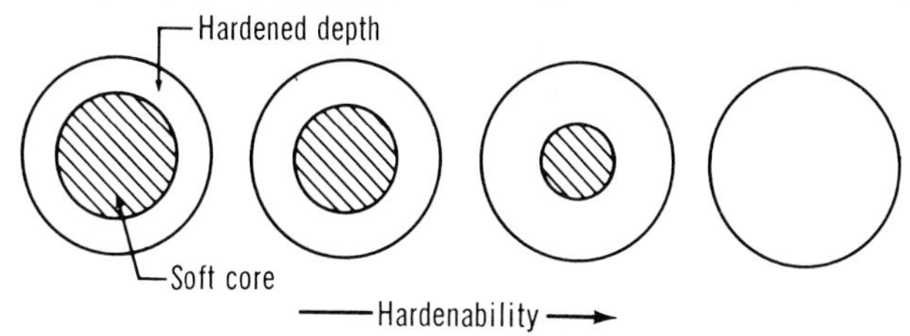

Fig. 1 — Bars of different steels with the same diameter harden to varying depths—hardenability increases from left to right. The same effect is created in bars of the same steel if they are quenched in increasingly effective cooling mediums.

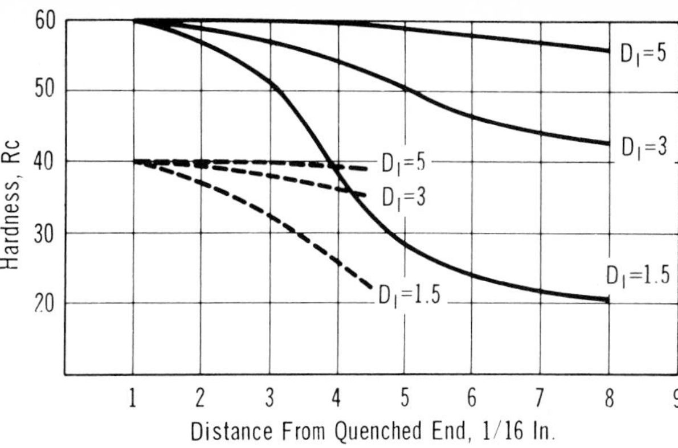

Fig. 2 — Determined by end-quench tests, these curves show hardenabilities of six different steels, three with 0.12% C (dotted lines) and three with 0.48% C. Increasing D_I indicates greater quantities of alloying elements other than carbon.

The case hardenability is calculated in exactly the same manner as D_{Ib} except that the multiplying factors derived by Jatczak and Girardi[8] are used. A major difference, however, is that the Jominy curve for case hardenability cannot be calculated from the D_{Ic} value by use of the IH/DH tables. The D_{Ic} value only locates the Jominy position at which 10% transformation to nonmartensitic products has occurred, approximately Rc 60. Because transformation products are usually not tolerated in quality carburized components, another manipulation is required to assure 100% martensite plus austenite structures, as is discussed in a forthcoming article of the series.

The boron effect on hardenability is dependent upon carbon content. Although a boron multiplying factor[9] has been developed for use with Grossmann factors, a new boron factor is needed for use with the Doane multiplying factors. Based on a limited amount of data, the following equation represents the effect of boron in steel containing 0.20 to 0.40 C:

$$\text{Boron Multiplying Factor} = 1 + 1.76\,(0.74 - \%\text{C})$$

The multiplier approach is not entirely satisfactorily applied to boron. Additional work utilizing regression analysis seems to hold promise for improvements.

Development of Replacement Steels

Now it is appropriate, as an informative exercise, to consider the development of the nickelfree carburizing steels EX 15, 16, 17, and 18, known as Equalloy steels. During the nickel shortage of 1969, the concept and techniques which embody the CHAT system were not fully developed. This example serves to illustrate some of the basic factors and computational techniques to be considered in designing a steel, a replacement steel in this instance. The problem during the nickel crisis involved devising a series of steels containing only residual nickel to replace the International Harvester grades EX 5, 6, 7, and 8, a series of 8600-type carburizing steels with low molybdenum and 0.40 to 0.60 Ni. Other elements in the replacement compositions, however, would have minimum changes so that hardenability, mechanical properties, distortion characteristics, and machinability would be essentially the same as those of the steels replaced. The criterion of "least-change" suggested that carbon, phosphorus, sulfur, and silicon contents employed in the previously used EX series remain unchanged, compositional changes in the respective replacements being limited to manganese, chromium, and molybdenum.

Heat treatable steels have been defined in terms of H-bands. A base hardenability band and a case hardenability band define, within limits, a carburizing steel. Figure 3 shows the standard base hardenability band for 8622H and EX 6. Superimposed are minimum, midrange, and maximum hardenability curves calculated from the minimum, midrange, and maximum composition of the EX 6 grade. Note that the midrange composition gives a hardenability curve which lies in the middle of the base hardenability band. However, computations using the minimum and maximum compositions provide hardenability lines which fall outside of the standard band. The reason for this discrepancy relates to the procedure used to develop H-bands. An empirical procedure, it provides limits in low and high hardenability which assumes remote probabilities for all elements reaching their respective extremes simultaneously. Hence, to compute bands from composition, other means were adopted. This entailed obtaining the statistical distribution of each element in normal practice (for the IH Wisconsin Steel Div.), then utilizing ± 2 standard deviations ($\pm 2\sigma$), as the equivalent high or low compositions.

Use of the $+2\sigma$ and -2σ compositions as the limits provided high and low hardenability curves which

Fig. 3 — Superimposed on the hardenability band for 8622H and EX 6 (0.19 to 0.25 C, 0.70 to 1.05 Mn, 0.20 to 0.35 Si, 0.40 to 0.70 Cr, 0.35 to 0.75 Ni, and 0.08 to 0.15 Mo) are midrange, maximum, and minimum curves. These curves define a very wide band because they were determined through calculations employing midrange, maximum, and minimum compositions; calculated D_{Ib} ranges from 1.4 to 3.7 in.

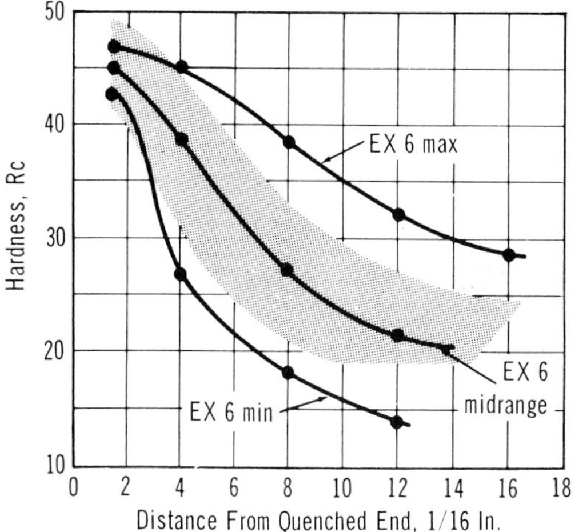

matched the band reasonably well. This is shown in Figure 4, which portrays the calculated hardenability based on the $\pm 2\sigma$ compositions for the EX 6 steel, and the comparison of this calculated hardenability with the published bands for 8622H and EX 6. Though somewhat empirical, the $\pm 2\sigma$ technique has proved to be satisfactory in many instances of steel design. Some compositions may require the use of different multiples of sigma for adequate accuracy.

Matching Case and Base Hardenabilities

Before alloy ranges can be manipulated to produce a steel with a calculated D_{Ib} equivalent to that of the original steel, the case hardenability, D_{Ic}, must be considered. This is a more difficult problem because a replacement designed with equivalent base hardenability may possess insufficient case hardenability, or vice versa. In general, it is difficult to match case and base hardenability simultaneously because the calculated hardenabilities of a given composition are usually unique. However, some latitude exists in attempting to devise a replacement steel with essentially the same case and base hardenabilities as those of the original material.

Base hardenability of the replacement steel should be equal, as closely as possible, to the base hardenability of the steel replaced so that the same hardness occurs at approximately the same J distance. Greater or lesser base hardenability in a replacement steel is undesirable because a part made from the new steel could have significantly different strength properties and possibly different heat treat distortion characteristics.

It is not detrimental for the replacement composition to exhibit greater case hardenability than the original material; however, insufficient case hardenability is unacceptable. In the hypereutectoid region, quenching will yield martensite-retained austenite structures at maximum hardnesses of Rc 60 to 65. Excess case hardenability does not result in higher hardness. However, a lower case hardenability results in the maximum-hardness structure forming to a lesser J-distance than encountered with the original steel. Thus, there is a risk of not hardening the surface of a part or of materially changing the hardness gradient. The

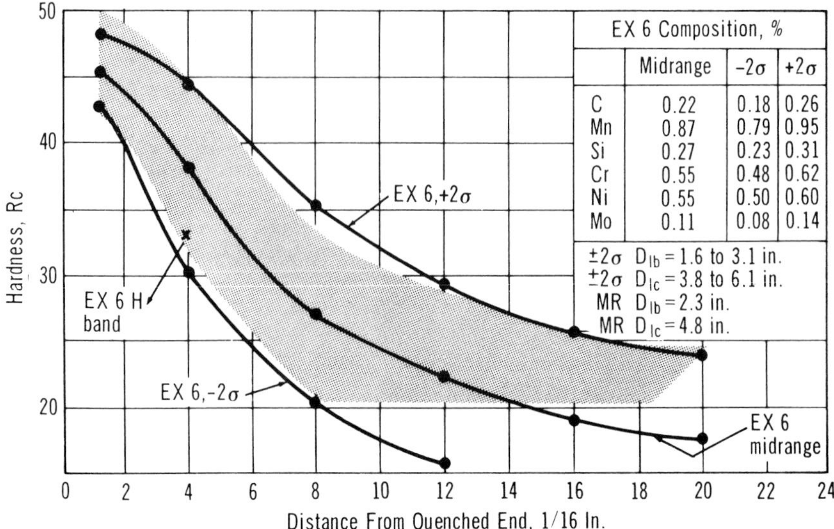

Fig. 4—When hardenability limits for EX 6 are calculated on the basis of the listed $\pm 2\sigma$ limits for the maximum and minimum compositions (rather than with the standard maximum and minimum compositions listed for EX 6 in Fig. 3), maximum and minimum curves coincide closely with the standard band limits for 8622H and EX 6.

replacement composition should possess a calculated D_{Ib} virtually equal to that of the original material, and a calculated D_{Ic} equal to or greater than that of the original material.

When changing an alloy composition, it is important to consider both the specific values of the alloy hardenability multiplying factors and the change in the range of the multiplying factors. This is demonstrated schematically in Fig. 5, which illustrates how changing the mean concentration of an element from a lower to a higher value acted to widen the multiplying factor range, ΔMF. Even though the chemical analysis range remains constant, the nonlinear nature of the multiplying factor function causes a wider spread of the multiplying factor for the new composition. As a result, the hardenability band would be wider for the new steel.

Regarding the nickelfree EX grades described here, analysis of the individual ΔMF ranges indicated a slight reduction in the cumulative ΔMF range for the replacement alloy. Mill experience to date has shown these steels to have hardenability bands which are the same as those of standard 8600 bands.

In designing nickelfree grades to

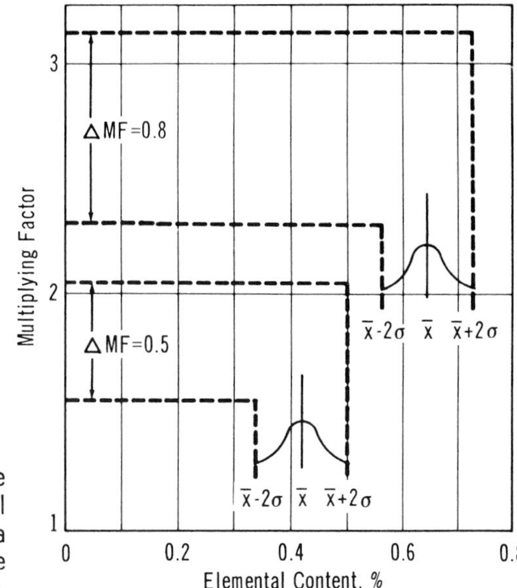

Fig. 5—Changing the amount of an individual alloying element in a steel can increase the multiplying factor range.

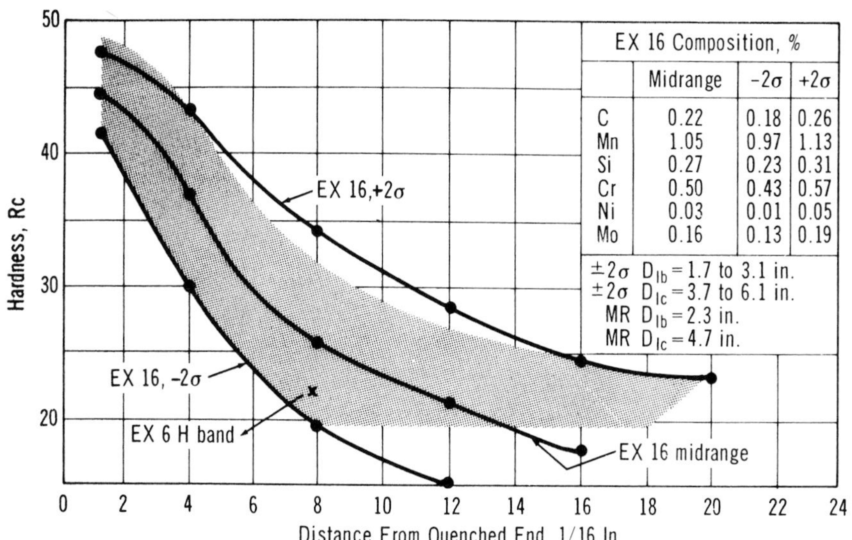

Fig. 6 — Calculated from $\pm 2\sigma$ limits for specific elements of the nickelfree EX 16 composition, maximum and minimum hardenability curves agree well with the hardenability band for EX 6. Respective values for ideal diameters of case and core, also calculated, are similar to those of EX 6 and 8622H (Fig. 4).

replace 8600 types, nickel was removed, and chromium, manganese, and molybdenum contents were adjusted to provide calculated base and case hardenabilities matching those of the previously used EX steels. Figure 6 shows the resultant base hardenability calculated for a $\pm 2\sigma$ analysis for the nickelfree grade, EX 16. Minimum and maximum hardenability curves, calculated, fit the standard hardenability band well, and there is an equivalence between case hardenability values.

Martensite start temperatures (M_S) were not used as criteria in the search for replacement compositions. Once a replacement based on D_{Ib} and D_{Ic} was found, however, the M_S temperatures of this steel at both core carbon and 1% C levels were computed (755 and 297 F, respectively) and compared to those of the steel replaced (752 and 296 F). The M_S values influence the sequence of transformation, which affects residual stress and distortion in heat treatment.

When designing this replacement steel, a basic assumption was applied – metallurgical factors governing the engineering performance of a heat treated part are primarily the carbon content, the microstructure, and the residual stress. This is especially true when dealing with alloy steels of relatively low alloy content. Specific alloy effects, such as the molybdenum effect on temper embrittlement, are recognized. They will be further discussed in the Computer Harmonized section.

It should be noted that, due to the "least change" limitation and the complex requirement of simultaneous matching of D_{Ib} and D_{Ic}, the composition of the replacement is not optimum from a cost viewpoint. The utility of a computer in solving this problem will be shown later.

Evaluation of Performance

Therefore, with equivalent carbon content, microstructure, and residual stress, it was anticipated that these nickelfree steels would give the same engineering performance as previously obtained. This was confirmed by tests conducted on standard 6-pitch test pinions. They showed that the nickelfree steels developed resistance to both bending and pitting fatigue which was equivalent to that of the previously used 8600 steels.

In an additional evaluation, large hypoid rear axle ring gears were run on a chassis test designed to simulate the heavy shock loading endured by rear axles of trucks in severe off-highway service. In all instances, the nickelfree grades performed as well as the previously used EX or 8600 types. Of more significance, however, is three years of successful use with these nickelfree grades of steel, during which no changes have been made in either machining or heat treating practice. Even though this is not an optimum-cost composition, a more efficient utilization of alloys has resulted in substantial savings because of the approximate $20.00 per ton price differential in addition to the conservation of nickel.

Summary

To provide an environment for further discussion, this first article has briefly reviewed some of the fundamental and historical aspects of the hardenability of steel. The discussion concerning the replacement steel, EX 16, demonstrated the utility of computational techniques for designing steels, and showed that engineering performance can be maintained if the design of the alloy is approached judiciously. This discussion also clearly defined a steel in terms of multiple hardenability relationships (range of D_{Ib} and D_{Ic}), and noted the shortcomings of present techniques in arriving at optimum-cost alloys. The next article will present the method used to arrive at and define engineering requirements.

References

1. "The P-F Characteristic of Steel", by B. F. Shepherd, Transactions, American Society for Metals, Vol. 22, 1934.
2. "Transformation of Austenite at Constant Subcritical Temperature", by E. C. Bain and E. S. Davenport, Transactions, AIME, Vol. 90, 1930.
3. "Hardenability Calculated from Chemical Composition," by M. A. Grossmann, Transactions, AIME, Vol. 150, 1942.
4. "A Hardenability Test for Shallow Hardening Steels", by W. E. Jominy, Transactions, ASM, Vol. 27, 1939.
5. "Calculation of End-Quench Hardenability Curve", by Boyd and Field, booklet issued by American Iron & Steel Institute, February 1946.
6. "Predicting Hardenability of Carburizing Steels", by A. F. deRetana and D. V. Doane, Metal Progress, September 1971.
7. "Factors for the Calculation of Hardenability", by I. R. Kramer, S. Siegel and J. G. Brooks, Transactions, AIME, Vol. 167, 1946.
8. "Multiplying Factors for Calculation of Hardenability of Hypereutectoid Steels Hardened from 1700 F", Transactions, ASM, Vol. 51, 1959.
9. "The Effect of Carbon Content on the Hardenability of Boron Steels", Transactions, ASM, Vol. 40, 1948.

Computer-Based System Selects Optimum Cost Steels - II

By DALE H. BREEN, GORDON H. WALTER, and CARL J. KEITH JR.

To define minimum hardenability (D_I) for a steel which will allow a given part to develop a specified strength under production quenching conditions, the Jominy equivalent cooling (Jec) Rate must be determined. Once the D_I is established, standard or special-analysis steels may be selected for the part, depending on such factors as tonnage and inventory requirements.

The first article in this series, which appeared in the December 1972 issue, defined the CHAT (Computer Harmonized Application Tailored) approach to steel design. Briefly, the CHAT system consists of two basic parts (Computer Harmonizing and Application Tailoring), either of which can be executed and applied independently, although they are most effective when used in combination.

In this article, Application Tailoring (AT) will be applied to components requiring the use of quenched and tempered steels – that is, through-hardened steels. Application Tailoring for carburized components will be considered in the next article of this series.

Definition and Basis

Application Tailoring, AT, is the quantitative determination of hardenability requirements, in terms of D_I, for a given application. It is a formalized process in which cooling rates are used to define the hardenability required to meet engineering design criteria such as hardness and microstructure.

A key factor in AT is the determination of the Jominy equivalent cooling (Jec) Rate for a specific part given a production heating and quenching cycle. The Jec Rate at a given location in the material of a part is the cooling rate, expressed as J distance, on a Jominy bar of the same material that produces the same hardening response.

The fundamentals of the AT concept can be illustrated by a simple example of a stepped shaft. It is assumed that design analysis of the shaft shown in Fig. 1 has determined that it should be of a 0.35% C steel. Also, it should have 90% martensite at the center and be tempered to Rc 30 to perform its anticipated load carrying function for an acceptable time.

These requirements (0.35% C and 90% martensite) translate, as shown in Fig. 2, into an as-quenched hardness of Rc 46. AT consists of answering the question, "What minimum D_{Ib} is required to produce Rc 46, at the center of the shaft, with a 0.35% C steel?" For expediency assume that the Jec Rate for the shaft center has been found, by experimental procedures, to equal J4. Shown in Fig. 3 is a series of hardenability curves for 0.35% C steel. This graph reveals that the minimum D_{Ib} required to yield Rc 46 at J4, the shaft center, is 2.7 in. Though any steel with a D_{Ib} greater than 2.7 in. also can be used, such steels have excessive hardenability, and are thus a waste of costly alloying elements.

As Fig. 4 illustrates, the Jec Rate is determined experimentally by comparing hardness values, obtained on the cross section of a part subjected to the production heat and quench cycle, with hardness values obtained on Jominy bars of the same steel. Three critical items must be considered when determining the experimental curves:

1. The Jominy bar should be given the same thermal cycle as the part because hardenability is influenced by thermal history (Fig. 4b).
2. Enough parts should be examined to evaluate effects of load size, position in the load, and changes in quench severity with usage.
3. The steel used for Jec Rate evaluation must have an end-quench curve which drops steeply through the critical cooling rate range so that hardness changes give a sensitive indication of cooling rate.

In some instances, part size, costs, or time prohibit the experimental determination of Jec Rates. Though estimations can be made by using published cooling rate curves[1,2,3] caution should be exercised because accurate quench severity values (H values)[1] are not usually known.

The use of Jec Rates to predict hardness in a quenched part is not a perfect technique. Carney[4] and others[5,6] have clearly indicated a potential shortcoming, presumably related to the dependence of austenite transformation on the state of stress developed during quenching. In most instances, however, the procedure gives accurate, useful results.

Applying AT Analysis

Application Tailoring varies from a very simple procedure which utilizes established design requirements

Fig. 1 – In the text, AT analysis is used to determine the correct hardenability for the 0.35% C steel specified for the diagrammed shaft. Also specified for the cross section indicated by the shaded area is at least 90% martensite at the center. For this section, as-quenched hardness is Rc 46 min, and tempered hardness is Rc 30 min at the center.

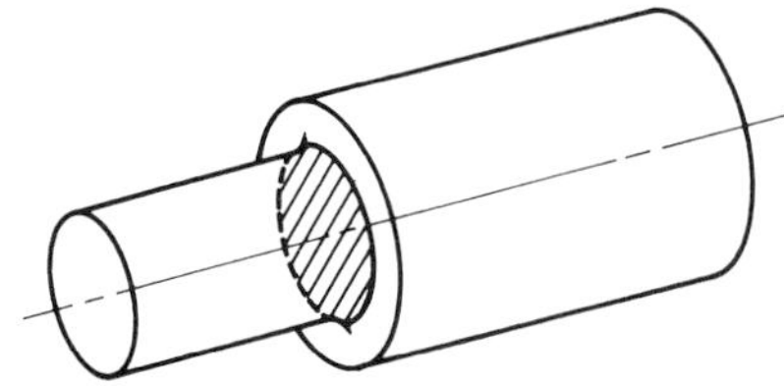

Mr. Breen is chief, Materials Div., Mr. Walter is chief engineer, and Mr. Keith is research metallurgist, Metallurgical Section, Materials Div., Engineering Research, International Harvester Co., Hinsdale, Ill.

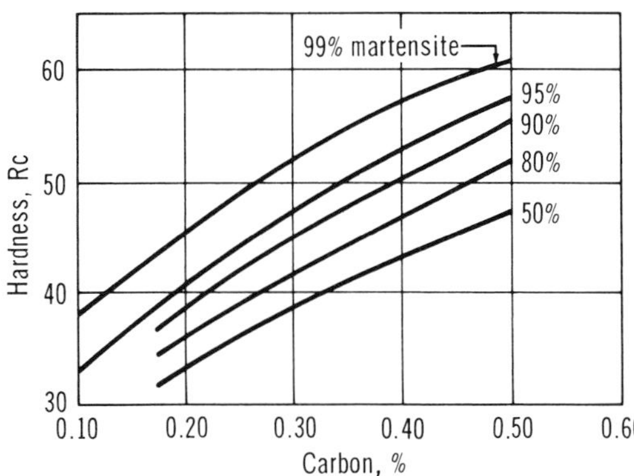

Fig. 2 — As-quenched hardnesses of steel vary with carbon content and the amount of martensite developed by quenching.

in conjunction with Jec Rates, to a more comprehensive procedure which includes a determination of required strength level and microstructure based on an engineering analysis of expected stress spectrum and desired life. The efficiency and reliability of an over-all CHAT analysis is, of course, a direct function of the accuracy of the initial engineering analysis. The AT procedure for quenched and tempered parts with varying degrees of intricacy will be demonstrated by three examples.

Simple Application Tailoring

Shown in Fig. 5 is a wheel spindle, an example of a part for which hardness and heat treatment have previously been established. The procedure for simple Application Tailoring, in this example only, involves determination of Jec Rate and definition of the D_{Ib} required to develop the mechanical properties. The spindle is forged, austenitized, oil quenched, tempered and machined to final dimensions.

Assume the engineering specification requires a 0.40% C steel quenched to 90% martensite at the finish machined surface of the critical section followed by tempering to a hardness range of Rc 30-37. As Fig. 2 shows, the requirement of 90% martensite for a 0.40% C steel necessitates an as-quenched hardness of Rc 50 minimum at the finish machined surface. The Jec Rate is evaluated by determining the cross-sectional hardnesses of a number of production quenched parts, and comparing these values to equal hardnesses on Jominy bars from the same pieces. The resulting minimum Jec Rate curve for the critical cross section is shown in Fig. 6, where the specific Jec Rate for the machined surface is J9.5.

The hardenability may now be determined. One method to do this is by using IH/DH rations [1,8,9], as shown in Table I. The method is as follows:

1. Determine the initial hardness (IH) for a 0.40% C steel. As shown in the second column, the IH, which is equivalent to 99% martensite at the J1 position, is Rc 56.

2. Using distance hardness (DH), which is Rc 50 by definition, calculate the IH/DH Ratio — it is 56/50, or 1.12.

3. Using the defined distance, J9.5, extrapolate between the vertical columns labeled ½ and ¾ in. (J8 and J12). As will be noted, a D_{Ib} of 4.90 in. corresponds to a IH/DH of 1.12 in.

The AT Procedure has thus determined (1) the critical location cooling rate, and (2) that a minimum hardenability of $D_{Ib} = 4.90$ in. is required to develop 90% martensite at that location when the part is processed in the production heat treatment facilities. (A graphical means to determine hardenability, D_{Ib}, will be demonstrated in a subsequent example.)

The values found, carbon level and D_{Ib}, may be used as the input to design a CHAT steel or as the basis for selecting a standard steel. Where the tonnage of steel is sufficient to warrant the purchase of heat lots, a CHAT steel will provide the most efficient and economical steel selection.

When dealing with established heat treat practices, it is expected that the practice itself should be questioned. In this instance, the combination of oil quenching and medium-carbon alloy was acceptable at one time. The advent of new technology has made it appropriate to consider the water quenching of lower-carbon steels as a way to gain additional economies without loss of performance. Because the spindle is machined after heat treatment, distortion does not appear to be a signif-

Fig. 3 — This series of hardenability curves for 0.35% C steels indicates that Rc 46 min at J4 (center of shaft of Fig. 1) corresponds to a D_{Ib} requirement of approximately 2.7 in.

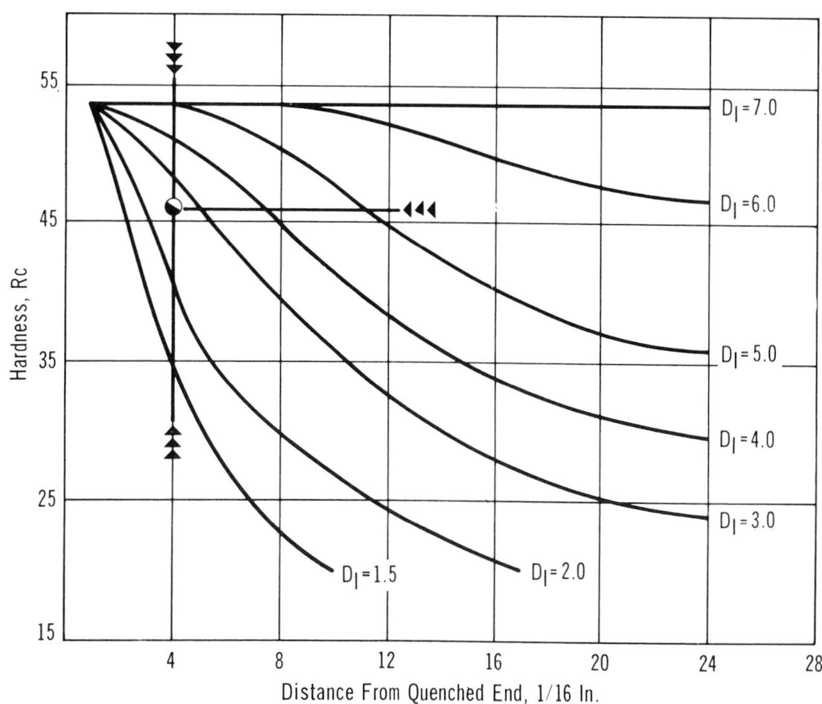

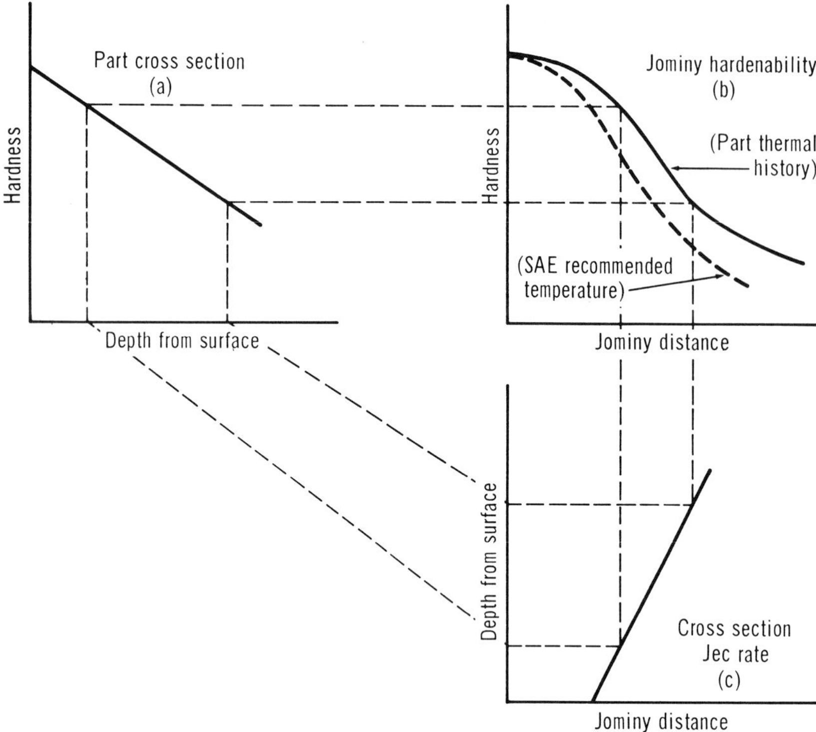

Fig. 4 — Jominy equivalent cooling (Jec) Rates are determined by comparing hardnesses of cross sections of parts receiving the established production heat treatment to hardnesses obtained on end-quenched bars of the same steel.

Fig. 5 — Application Tailoring is used to select the steel for this wheel spindle. Design requirements call for the critical section to harden to Rc 50 min and 90% martensite at the surface when the spindle is quenched. Tempered hardness on the surface should be Rc 30 min.

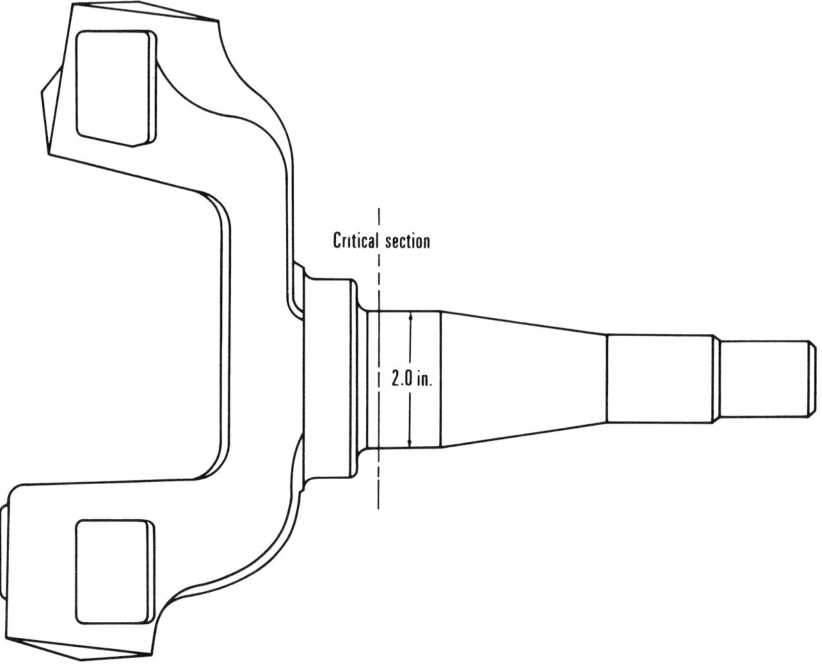

icant factor. Of course, it is generally not desirable to water quench a medium-carbon steel because quench cracks can occur.

However, the wheel spindle is tempered to a Rc 30 minimum, a hardness which is readily attainable in a steel with less carbon. Thus, AT analysis can be reapplied to the wheel spindle with a view toward the use of a water-quenching steel with less carbon and hardenability. The essential design requirements of the wheel spindle are (1) a tempered surface hardness equal to Rc 30 min and (2) a surface microstructure of at least 90% martensite.

As a rule, the as-quenched hardness must be at least 7 Rc points higher than the specified tempered hardness to allow for tempering. (A more exact approach to determine the as-quenched hardness is available from the work of Hollomon and Jaffee[10].) However, in this example, the required as-quenched hardness is taken to be Rc 37.

The first step in the AT procedure is to define the carbon range. Figure 2 indicates that a 0.17% C level is sufficient to develop a hardness of Rc 37 with a microstructure of 90% martensite. However, carbon is the most efficient element for increasing a steel's hardenability. Increasing the carbon content, therefore, reduces the amount of alloy required. Balancing this consideration against the problem of quench cracking, a minimum of 0.25% C is established. A structure of 90% martensite with 0.25% C content will develop an as-quenched hardness of Rc 42 (Fig. 2). The carbon range is established by reference to the applicable AISI Steel Products Manuals,[11,12] which indicate a ladle composition range of 0.06% C.

The next step is to evaluate the production Jec Rate. If the severity of the production quench (Grossmann H Value)[1,13] is known or can be estimated, the Jec Rate can be evaluated from a series of curves such as shown in Fig. 7 and 8. For a well-agitated water quench (H= 1.5), the Jec Rate would be J8.2 at the center of a 2.4 in. round, and the Jec Rate would be J5.3 at a depth equal to 70% of the center-to-surface distance. By this method, the Jec Rate of the cross section is evaluated, as shown in Fig. 6. The machined surface of the section has a Jec Rate of J3.8 (assumed to be

J4 for ease of calculation). The minimum hardenability required may be determined as before.

1. From Table I, the initial hardness, IH, of a 0.25% C steel, is Rc 47.

2. Because the distance hardness is Rc 42 (necessary for 90% martensite), the IH/DH ratio is 47/42, or 1.12, at the J4 position.

3. By extrapolation in the ¼ in. (J4) column, the required D_{Ib} is 2.95 in.

The minimum hardenability may also be determined directly from Fig. 9, which has been prepared from ratios of IH/DH and plots of carbon content against hardness for different percentages of martensite. Figure 9 indicates that 90% martensite at the J4 position can be obtained with a hardenability of D_{Ib} = 3.0 in. The following metallurgical requirements have been determined:

1. Carbon content – 0.25 to 0.31%.

2. Hardenability – D_{Ib} = 2.95 in. min, which will develop a hardness of Rc 42 (90% martensite) at a depth equivalent to the J4 position.

Maximum hardenability allowed is not so significant insofar as the wheel spindle is concerned because a relatively high tempering temperature is used to achieve the final hardness. In some instances where the temperature is very low, however, maximum hardenability can be a factor. In these situations, the maximum D_{Ib} to be permitted would be evaluated in the same manner as the minimum D_{Ib}, based upon maximum carbon content and maximum hardness allowed.

The above requirements are used as the basis for designing a CHAT steel or for selecting a standard steel. If production is low or inventory problems exist, a standard steel should be used. A standard steel should also be evaluated for comparison purposes even when a CHAT steel is being considered.

In the event that a CHAT analysis steel is indicated, the parameters, carbon level, and minimum D_{Ib} are used as input data for Computer Harmonizing (CH) to develop a steel with optimum alloy content. To select an appropriate standard steel, the SAE hardenability bands are consulted. Selection of one steel from

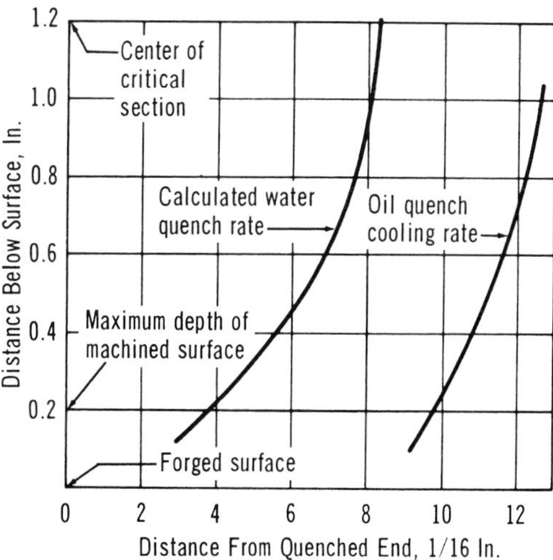

Fig. 6 — Minimum Jec Rates for the wheel spindle shown in Fig. 5 are employed to determine the minimum D_{Ib} required.

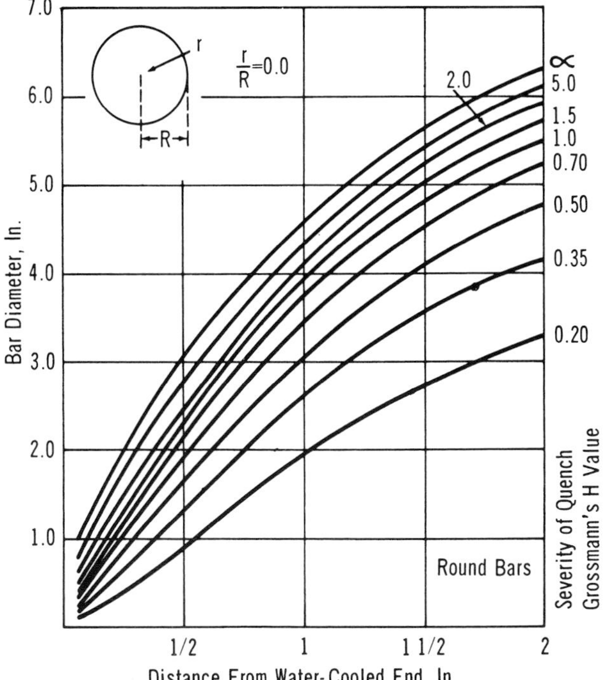

Fig. 7 — This chart shows locations on end-quenched hardenability test bars corresponding to centers of round bars.

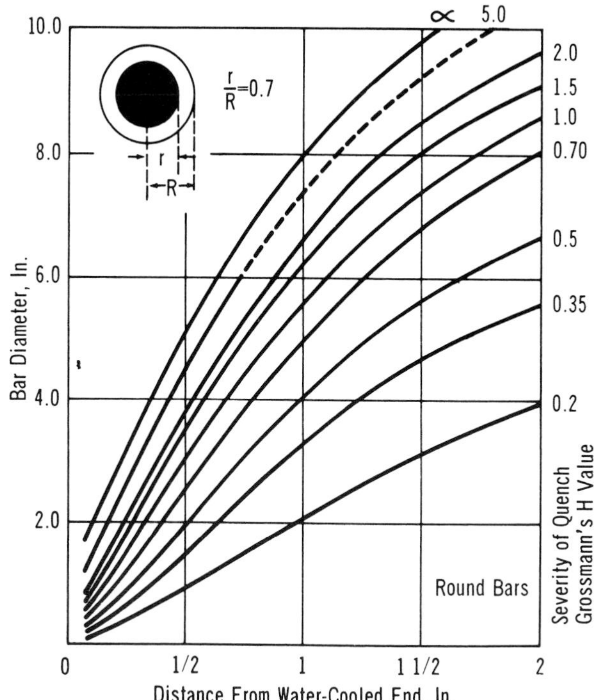

Fig. 8 — Similar to Fig. 7, this chart relates locations on end-quenched hardenability test bars corresponding to 70% from centers of round bars.

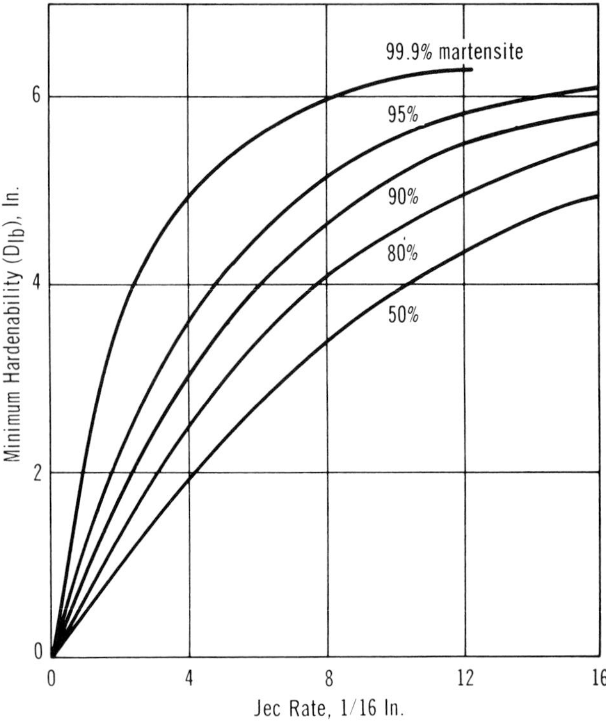

Fig. 9 — Knowing the Jec Rate, it is possible to determine the minimum hardenability (D_{Ib}) needed by a steel to develop the indicated martensitic structures. The graph is suitable for steels containing 0.15 to 0.50% C.

the many available with proper carbon and D_{Ib} values is made on the basis of cost, machinability, availability, and other considerations.

Comprehensive Application Tailoring

This example considers the part in the design stage prior to production. The part (shown in Fig. 10) is an equalizer bar for a crawler tractor. Engineering analysis indicates the tapered beam will be subjected to bending (stress ratio of R=0), with a maximum surface stress of 75,000 psi and zero stress at the neutral axis. A fatigue life of 10 million cycles is desired.

The required surface hardness is determined by reference to a modified Goodman diagram (Fig. 11). As is shown, a tempered surface hardness of Rc 40 will provide a fatigue safety factor of 1.3 for $N = 1 \times 10^7$ cycles. To provide for tempering to Rc 40, a minimum as-quench hardness of Rc 47 is required. In addition, it is desirable to provide 99% martensitic surface structure for optimum fatigue resistance.

The next step, therefore, is to determine the carbon content. Figure 2 indicates that steel with a minimum carbon content of 0.30% will develop Rc 50 when quenched to 99% martensite; this carbon level is thus considered satisfactory for the application. At 0.30% C, a conservative estimate[14] of the tempered center hardness is Rc 15. By comparing the applied stress gradient with the expected strength gradient, Rc 40 to Rc 15, on a fatigue basis, it is determined that a conservative strength gradient will be obtained by requiring Rc 45 as quenched to 1 in. below the surface (Fig. 12).

The metallurgical requirements are:

1. As-quenched surface hardness – Rc 50 min, 99% martensite.
2. Carbon content – 0.30 to 0.36%.
3. As-quenched hardness – Rc 45 min (90% martensite) at a depth equal to 1.0 in. below the surface.
4. Tempered surface hardness – Rc 40 min.

In this preproduction example, prototype parts for direct evaluation of the Jec Rate are not available. It is necessary, therefore, to estimate this rate by using published curves. This is done by assuming that the

Table I — Values for

Carbon, %	Initial (Max.) Hardness (IH), Rc	Ideal Critical Diameter (D_I) In.	Ideal Critical Diameter (D_I) Mm
0.10	38	0.50	12.7
0.11	39	0.55	14.0
0.12	40	0.60	15.2
0.13	40	0.65	16.5
0.14	41	0.70	17.8
0.15	41	0.75	19.1
0.16	42	0.80	20.3
0.17	42	0.85	21.6
0.18	43	0.90	22.9
0.19	44	0.95	24.1
0.20	44	1.00	25.4
0.21	45	1.05	26.7
0.22	45	1.10	27.9
0.23	46	1.15	29.2
0.24	46	1.20	30.5
0.25	47	1.25	31.8
0.26	48	1.30	33.0
0.27	49	1.35	34.3
0.28	49	1.40	35.6
0.29	50	1.45	36.8
0.30	50	1.50	38.1
0.31	51	1.55	39.4
0.32	51	1.60	40.6
0.33	52	1.65	41.9
0.34	53	1.70	43.2
0.35	53	1.75	44.4
0.36	54	1.80	45.7
0.37	55	1.85	47.0
0.38	55	1.90	48.3
0.39	56	1.95	49.8
0.40	56	2.00	50.8
0.41	57	2.10	53.3
0.42	57	2.20	55.9
0.43	58	2.30	58.4
0.44	58	2.40	61.0
0.45	59	2.50	64.0
0.46	59	2.60	66.0
0.47	60	2.70	68.6
0.48	60	2.80	71.1
0.49	60	2.90	73.7
0.50	61	3.00	76.2
0.51	61	3.10	78.7
0.52	61	3.20	81.3
0.53	62	3.30	83.8
0.54	62	3.40	86.4
0.55	63	3.50	88.9
0.56	63	3.60	91.4
0.57	63	3.70	94.0
0.58	64	3.80	96.5
0.59	64	3.90	99.1
0.60	64	4.00	101.6
		4.10	104.1
		4.20	106.7
		4.30	109.2
		4.40	111.8
		4.50	114.3
		4.60	116.8
		4.70	119.4
		4.80	121.9
		4.90	124.5
		5.00	127.0
		5.10	129.5
		5.20	132.1
		5.30	134.6
		5.40	137.2
		5.50	139.7
		5.60	142.2
		5.70	144.8
		5.80	147.3
		5.90	149.9
		6.00	152.4

End-Quench Hardenability Calculation

Distance From Quenched End, In.							
1/4	1/2	3/4	1	1 1/4	1 1/2	1 3/4	2
RATIO: Initial Hardness / Distance Hardness							
4.90							
4.42							
4.03							
3.70	6.00						
3.47	5.15						
3.25	4.50						
3.07	4.18						
2.90	3.88	6.00					
2.75	3.68	5.13					
2.61	3.50	4.70					
2.48	3.33	4.40					
2.33	3.20	4.13	5.28				
2.17	3.08	3.93	4.75	5.70			
2.05	2.96	3.76	4.40	4.95	5.75		
1.96	2.86	3.60	4.15	4.58	5.00	6.00	
1.88	2.76	3.45	3.95	4.32	4.65	5.15	6.00
1.80	2.66	3.32	3.78	4.13	4.40	4.72	5.25
1.73	2.57	3.21	3.65	3.95	4.18	4.45	4.83
1.67	2.49	3.10	3.53	3.77	4.02	4.26	4.53
1.62	2.42	3.01	3.41	3.65	3.87	4.08	4.29
1.57	2.34	2.93	3.30	3.53	3.73	3.91	4.10
1.53	2.27	2.84	3.20	3.43	3.61	3.78	3.96
1.49	2.21	2.75	3.10	3.33	3.51	3.67	3.83
1.46	2.16	2.66	3.01	3.24	3.42	3.57	3.71
1.43	2.11	2.59	2.93	3.16	3.33	3.47	3.59
1.40	2.06	2.52	2.86	3.08	3.25	3.38	3.49
1.38	2.01	2.45	2.80	3.00	3.17	3.29	3.40
1.36	1.96	2.38	2.74	2.94	3.10	3.21	3.32
1.34	1.91	2.33	2.68	2.88	3.04	3.14	3.25
1.32	1.87	2.27	2.63	2.83	2.97	3.08	3.18
1.30	1.83	2.23	2.58	2.78	2.92	3.02	3.11
1.26	1.75	2.13	2.50	2.69	2.82	2.91	3.00
1.24	1.69	2.06	2.42	2.61	2.73	2.83	2.91
1.22	1.64	1.99	2.35	2.53	2.65	2.75	2.83
1.20	1.60	1.93	2.27	2.47	2.58	2.67	2.75
1.18	1.55	1.88	2.22	2.40	2.51	2.60	2.68
1.17	1.52	1.84	2.16	2.34	2.44	2.53	2.61
1.15	1.48	1.80	2.10	2.28	2.38	2.47	2.54
1.14	1.45	1.76	2.05	2.23	2.33	2.41	2.48
1.13	1.42	1.72	2.00	2.18	2.28	2.35	2.42
1.11	1.39	1.68	1.94	2.12	2.22	2.28	2.36
1.10	1.37	1.65	1.90	2.08	2.18	2.24	2.32
1.09	1.35	1.61	1.86	2.04	2.13	2.20	2.27
1.08	1.33	1.58	1.83	2.00	2.08	2.15	2.22
1.07	1.31	1.55	1.80	1.95	2.04	2.11	2.17
1.07	1.29	1.51	1.76	1.91	2.00	2.07	2.13
1.06	1.27	1.48	1.72	1.87	1.96	2.03	2.08
1.06	1.25	1.46	1.68	1.83	1.92	1.98	2.04
1.05	1.23	1.43	1.65	1.80	1.88	1.94	2.00
1.05	1.22	1.41	1.62	1.76	1.84	1.90	1.96
1.04	1.20	1.38	1.59	1.72	1.80	1.86	1.92
1.04	1.18	1.36	1.56	1.68	1.77	1.82	1.88
1.03	1.17	1.34	1.53	1.65	1.73	1.78	1.84
1.03	1.16	1.32	1.50	1.62	1.70	1.75	1.80
1.02	1.15	1.30	1.47	1.58	1.66	1.72	1.76
1.02	1.14	1.28	1.44	1.55	1.63	1.68	1.73
1.02	1.12	1.26	1.41	1.52	1.59	1.64	1.69
1.01	1.11	1.24	1.38	1.49	1.56	1.61	1.65
1.01	1.10	1.22	1.36	1.46	1.53	1.57	1.62
1.00	1.08	1.20	1.33	1.43	1.49	1.53	1.58
1.00	1.07	1.18	1.31	1.40	1.46	1.50	1.54
1.00	1.06	1.17	1.28	1.37	1.43	1.47	1.51
1.00	1.05	1.15	1.25	1.34	1.39	1.43	1.47
1.00	1.04	1.13	1.23	1.31	1.36	1.39	1.43
1.00	1.03	1.12	1.21	1.28	1.33	1.36	1.40
1.00	1.03	1.10	1.18	1.25	1.29	1.33	1.37
1.00	1.02	1.09	1.16	1.22	1.26	1.28	1.33
1.00	1.02	1.07	1.13	1.19	1.23	1.25	1.29
1.00	1.01	1.06	1.11	1.17	1.19	1.22	1.25
1.00	1.01	1.04	1.09	1.13	1.16	1.18	1.21
1.00	1.00	1.03	1.07	1.10	1.13	1.15	1.18

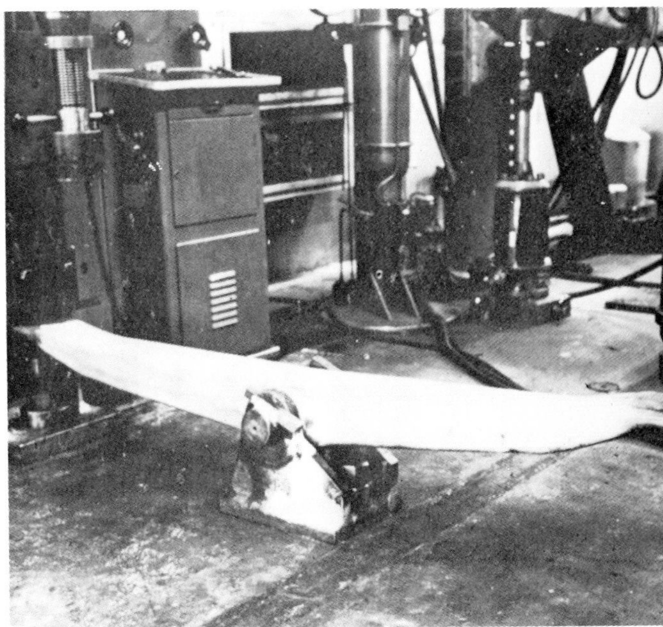

Fig. 10 — This part, an equalizer bar for a crawler tractor, is being tested at International Harvester's laboratories.

Fig. 11 — Allowable diagram is used for determining the strength requirement for the equalizer bar in Fig. 10. This diagram is based on a surface finish between "machine" and "ground".[15]

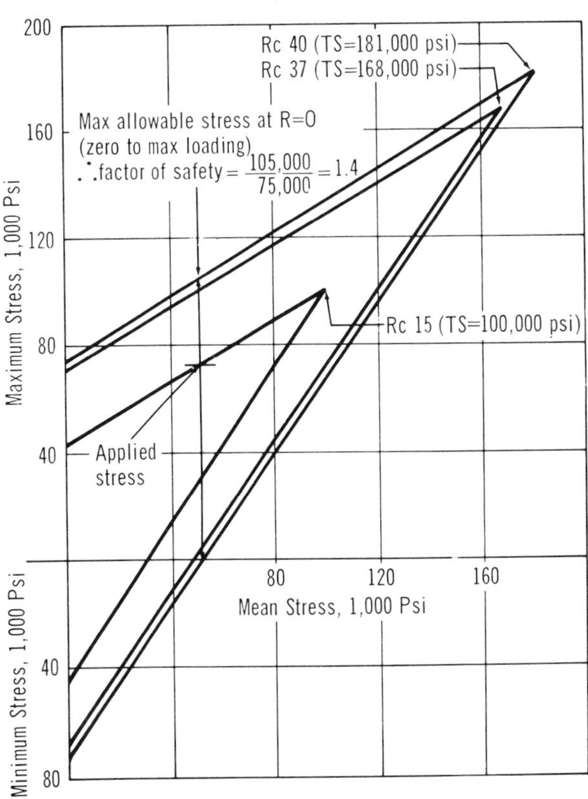

tapered beam may be approximated by a 7 in. round. Figure 8 indicates that in a well-agitated water quench (H Value = 1.5) the round will experience a cooling rate equivalent to J18 to J19 at the 1.0 in. depth.

The hardenability, D_{Ib}, required will be calculated by using the ratio of initial hardness to distance hardness in the manner applied before (Table I).

1. For a 0.30% C steel, the initial hardness, IH, is Rc 50.

2. At the distance, J18, distance hardness (DH) is Rc 45, making the IH/DH ratio = 50/45, or 1.10.

3. By extrapolation between the Table I columns labeled 1 and 1¼ in. (J16 and J20), the value of 1.10 at J18 corresponds to a D_{Ib} of 6.2 in. The requirements of C = 0.30 min and D_{Ib} = 6.2 in. min are used to select a standard alloy steel. If sufficient tonnage is involved, these values may be used for "Computer Harmonizing" to design a more efficient steel.

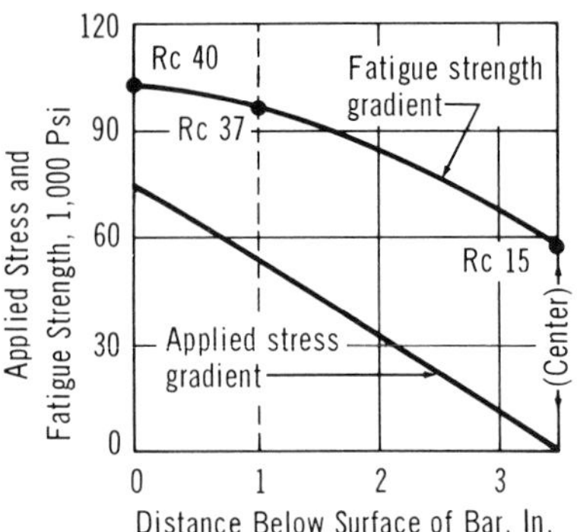

Fig. 12 — Comparisons of gradients for fatigue strength (for R = 0 and N = 10^7) and applied bending stresses are needed for evaluating the durability of a steel part in service.

The AT System in Summary

As has been pointed out, excess hardenability often represents excess cost. Application Tailoring, through use of engineering type analysis and Jec Rates, can be used to determine hardenability requirements accurately. When large tonnages are involved, considerable experimental work can be justified because of the potential of sizable savings. Though Application Tailoring can be quite simple and straightforward, a more comprehensive approach encompassing engineering design and applied mechanics techniques is applicable in some instances.

In the examples described herein, the base hardenability required for particular applications has been the principal concern. Part III of this series will consider the AT procedure as applied to applications utilizing carburized components where core and case hardenability requirements (D_{Ib} and D_{Ic}) are involved.

References

1. "Carilloy Steels," United States Steel Corp., Carnegie-Illinois Steel Corp., 1948.
2. "How to Estimate Hardening Depth in Bars," by J. L. Lamont, Iron Age, 14 October 1943.
3. "The Flow of Heat in Metals," by J. B. Austin, ASM, 1942.
4. "Another Look at Quenchants, Cooling Rate, and Hardenability," D. J. Carney, ASM Transactions, Vol. 46, 1954.
5. "An Estimation of the Quenching Constant, H," by D. J. Carney and A. D. Janulionis, ASM Transactions, Vol. 43, 1951.
6. "Effect of Applied Stress on the Martensitic Transformation," by S. A. Kulin, Morris Cohen, and B. L. Averbach, Journal of Metals, June 1952.
7. "Relationship Between Hardenability and Percentage of Martensite in Some Low Alloy Steels," by J. M. Hodge and M. A. Orehoski, AIME, TP 1800, 1945.
8. "Calculation of the Standard End-Quench Hardenability Curve from Chemical Composition and Grain Size," by L. C. Boyd and J. Field, AISI Contribution to the Metallurgy of Steel, No. 12, 1945.
9. "Republic Alloy Steels," Republic Steel Corp., Handbook Reprint, 1961.
10. "Ferrous Metallurgical Design," by J. H. Hollomon and L. D. Jaffee, John Wiley & Son, New York, 1948.
11. "Alloy Steel: Semi-Finished; Hot Rolled and Cold Finished Bars," AISI Steel Products Manual, AISI, August 1970.
12. "Carbon Steel: Semifinished for Forging; Hot Rolled and Cold Finished Bars; Hot Rolled Reformed Concrete Reinforcing Bars," AISI Steel Products Manual, AISI, May 1964.
13. "Elements of Hardenability," by M. A. Grossmann, ASM, 1952.
14. "Modern Steels and Their Properties," Bethlehem Steel Co., Sixth Edition, Bethlehem Steel Co., Bethlehem, Pa., 1961.
15. "Application of Stress Analysis," by C. Lipson, University of Michigan, Ann Arbor, June 1959.

Computer-Based System Selects Optimum Cost Steels - III

By DALE H. BREEN, GORDON H. WALTER, and CARL J. KEITH JR.

Application Tailoring, a system for determining the exact hardenability requirements necessary for producing specified strength (hardness and microstructure) properties using production heat treat facilities, employs the Jominy equivalent cooling (Jec) rate as an important factor in the analysis. In this article, carburized gears are used as examples. Both the base hardenability (D_{Ib}) and the case hardenability (D_{Ic}) must be considered as well as the carbon gradient. Subsequent articles will tie this information to the Computer Harmonizing aspect of International Harvester's CHAT system, and demonstrate how it can optimize steel costs.

Carburized gears represent a masterpiece in metallurgical and engineering design. Because of the complexities involved, however, they are often made of inappropriate materials from a cost viewpoint.

Many rules of thumb have been substituted for rational engineering approaches. They have had some degree of success, of course. In addition, attempts have been made to set down blanket rules for material selection, but by and large, they too have limitations.

For cost-conscious industries, the best rule is "to use the lowest cost combination of material and heat treatment that will perform the functions for the prescribed time." This entails some detailed analysis, and can result in the use of special compositions.

The analysis should start on a rather elemental level. To be sure, such factors as percentage of retained austenite, intergranular oxide penetration, microcracks, carbides, and inclusion content must be considered in the final analysis. When selecting steels for gear applications, however, the basic problem is to provide a composition which, when carburized and quenched, will produce strength gradients in critical locations (pitch line and root fillet, for example) which will accommodate imposed stresses. This involves, of course, considering carbon level of the base material, the carbon gradient produced in carburizing, the hardenability in the case zone (D_{Ic}) and the core region (D_{Ib}), and the quenching rate.

The procedures used to define the steel characteristics for a given gear application are explained in some detail in the following example. Though some assumptions are made and shortcuts taken in the interest of brevity, the main features related to hardenability are handled.

In this example, the Application Tailoring (AT) procedure will be applied to a six-pitch spur gear (Fig. 1). It is assumed that the design engineer has completed the design analysis, and has found that the specified strength gradient can be obtained with the following metallurgical requirements: surface hardness, Rc 60 to 65; effective depth at pitch line, 0.030 to 0.050 in. (distance to Rc 50); core hardness, Rc 30 to 40. Illustrated graphically in Fig. 2, these three requirements serve as guidelines for the AT procedure.

The intent in this example is to develop a surface microstructure containing only martensite and austenite. It is recognized, however, that intergranular oxidation[1,2] may cause decreased hardenability at the extreme surface due to localized alloy depletion. The development of nonmartensitic transformation products due to this effect will be discussed again in Part V of this series.

Mr. Breen is chief, Materials Div., Mr. Walter is chief engineer, and Mr. Keith is research metallurgist, Metallurgical Section, Materials Div., Engineering Research, International Harvester Co., Hinsdale, Ill.

Determination of Jec Rate

The first, and key, step in the AT procedure is to develop information on the Jominy equivalent cooling (Jec) rate for critical locations on parts subjected to the production heat and quench cycle. This involves machining Jominy bars and gear samples of the required geometry out of a suitable steel – that is, a steel with an end quench curve which drops steeply through the estimated critical cooling rate range so

Fig. 1 – A six-pitch spur gear, such as shown here in two views, is used to demonstrate AT analysis of carburized parts, as described in this article.

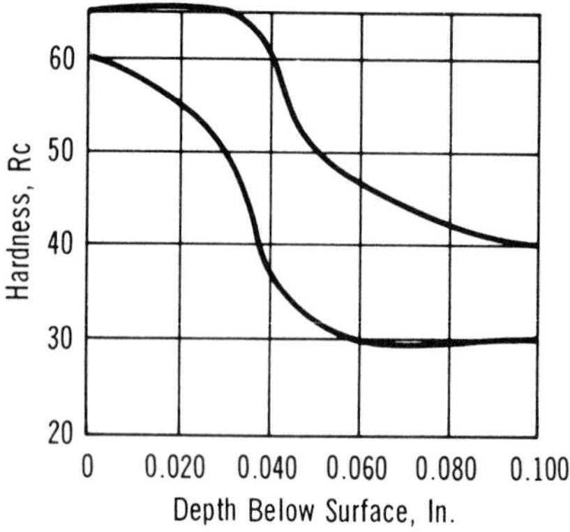

Fig. 2 — The design for the spur gear to be analyzed by AT methods calls for the illustrated hardness-depth limits in the case. Specification of effective case depth applies to the active profile of the gear teeth as measured at the pitch line. However, the root fillets of the gear teeth are required to have a minimum effective case depth of 50% of the minimum pitch line case.

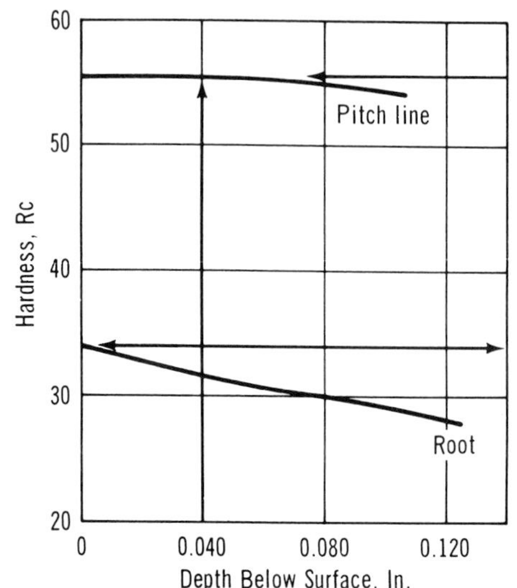

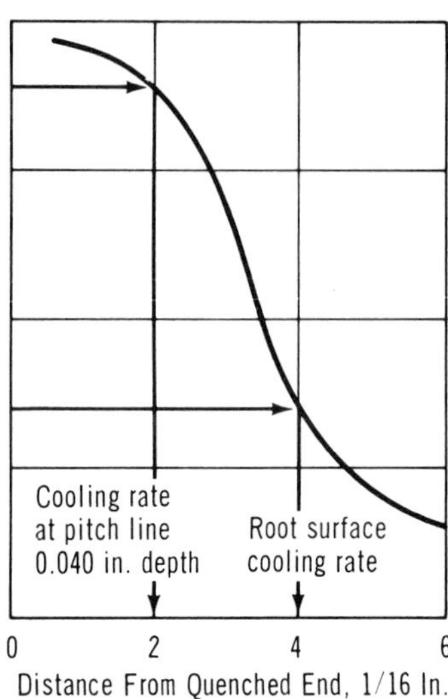

Fig. 4 — Given test gears and Jominy bars representing the same bar of 1040, correlations of hardness gradients in gears with end quench data show that gear hardnesses for the pitch line at 0.040 in. and the root surface correlate with J2 and J4 positions, respectively. These positions give the Jec rates.

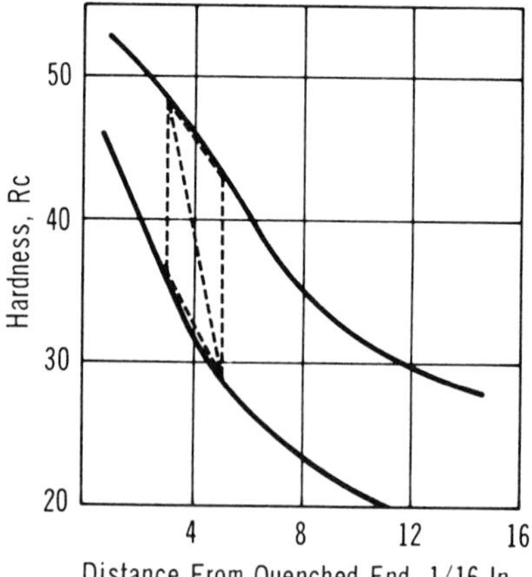

Fig. 3 — Variations in quenching rates contribute to the spread in core hardness. For example, hardness varies from Rc 36 to 48.5 at the J3 position. However, a variation in quenching rate from J3 to J5 will result in a hardness variation of Rc 29 to 48.5 (indicated by diagonal dotted line).

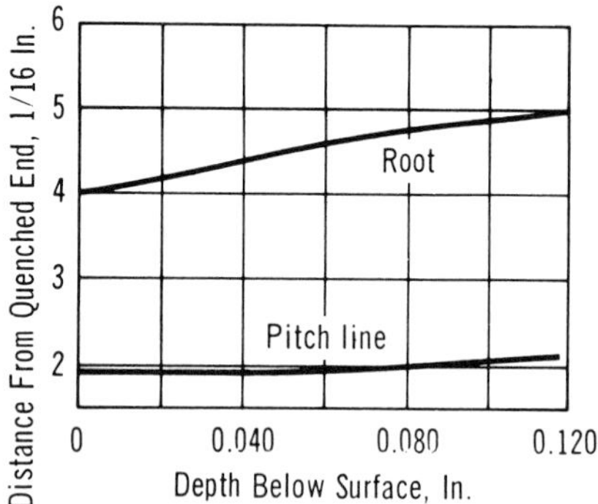

Fig. 5 — The Jec rates, which vary with distances below the surfaces (particularly for the teeth roots), are used to determine minimum and maximum D_{Ib}'s for the gear.

that hardness changes provide a sensitive indication of cooling rate. At this point, the interest is in the quenching rate. Therefore, it is desirable to obtain hardness gradients from parts that have been subjected to the complete heat treatment cycle except for carburization.

To determine the true gradient hardness due to cooling rate variation, then, it is necessary to prevent carbon penetration into the sample part. The simplest method of doing this is to "mock" carburize the part by using copper plating or other stopoff medium which will not interfere with quenching.

Gear samples are copper plated and heat treated at production carburizing and quenching facilities. The location in the furnace tray is judiciously made to determine the range of operational quenching rates. (Figure 3 demonstrates the effect of a typical range.) The Jec rate of the part is then evaluated by comparing hardnesses on the critical section of the part to hardnesses on the Jominy bar. This bar is prepared from the same steel (1040 in this example) from which the gears were machined, and from the same bar if possible. It is subjected to the same thermal cycle (see Part II of this series, February 1973, p. 76) as the test gears, and quenched in a standard Jominy fixture. The test gears are sectioned, and the Jec rates of the gears are determined. For example, Fig. 4 indicates that the Jec rate at the gear tooth root surface is J4, while the Jec rate at the pitch line core is J2.

D_{Ib} Requirements

Having established the Jec rate, the next step is to determine the hardenability levels required. First, the carbon level must be ascertained. Then, using the Jec rate shown in Fig. 5, minimum and maximum D_{Ib}'s are determined.

Selecting the optimum base carbon content involves an engineering compromise. Generally speaking, the lower the base (core) carbon content, the higher will be the residual compressive stress developed in the surface of the carburized case.[3] However, the lower the base carbon

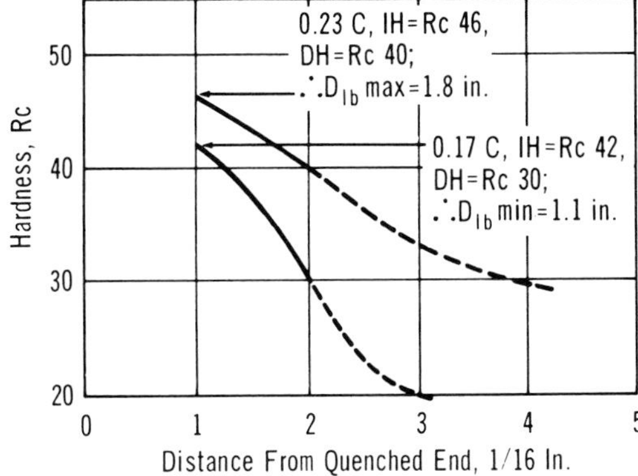

Fig. 6 — The minimum and maximum D_{Ib} values, 1.1 and 1.8 in., are shown as described in the text. At a Jec rate of J2, the minimum D_{Ib} of 1.1 in. will harden to Rc 30, and the maximum D_{Ib} of 1.8 in. will harden to Rc 40 at the pitch line core.

content, the higher the D_{Ib} necessary; thus, greater amounts of alloying elements will be needed to achieve the required core hardness. Also, longer carburizing times are required.

For this particular example, a carbon content typical for a carburizing grade of steel has been selected — 0.17 to 0.23% C. It must be noted that the carbon level selected for this example may not be the optimum carbon level from a cost standpoint; in practice, higher carbon contents should be considered. If the use of higher carbon contents represents a significant departure from previous experience, engineering tests may be required.

In combination with the carbon range selected, the core hardness specified controls the needed base hardenability. Carburized parts are generally martempered, or conventionally oil quenched and tempered at a relatively low temperature which does not significantly change the hardness level. For this reason, it is assumed that the as-quenched and as-tempered hardnesses are approximately equal. In computations, then, use of the final tempered hardnesses (Rc 30 to 40 in this instance) is justified.

The method for computing base hardenability values, D_{Ib} min and D_{Ib} max, will be demonstrated with the aid of Fig. 6. These hardenability values are based upon the core hardness requirements of the gear. At a given cooling rate, the maximum and minimum core hardnesses relate to the extremes of the composition range.

For example, the minimum core hardness will occur in conjunction with the minimum chemical composition, which will contain 0.17% C. A steel with 0.17% C develops an initial hardness, IH, of Rc 42 (from Table I of Part II, columns 1 and 2). Because the distance hardness (DH) required at J2 is taken as the minimum core hardness allowed, Rc 30, the ratio of initial hardness to distance hardness (IH/DH) is 42/30, or 1.42. By extrapolation on Table I (linear extrapolation between J1 and J4), the minimum D_{Ib} value having an IH/DH ratio = 1.42 at the J2 position is 1.10 in.

In a similar manner, the maximum D_{Ib} value is based upon the maximum core hardness (Rc 40) with a composition containing 0.23% C (IH = Rc 46). (IH/DH = 46/40 or 1.15.) The D_{Ib} value having an IH/DH ratio of 1.15 at the J2 position is found by extrapolation, again in Table I of Part II, to be 1.80 in.

Minimum and maximum base hardenability limits (D_{Ib} = 1.1 to 1.8 in.) and the carbon content have now been established. Note: Though these limits (Fig. 6) describe a steel in terms of its hardenability (in other words, they indicate a hardenability band), an actual chemical composition is not defined as yet.

Using these limits, and the Jec rates at a number of locations below the surface at both the pitch line and the root, it is possible to plot the as-quenched hardness of an uncarburized gear made from a steel having the carbon content and hardenability limits described. This is demonstrated in Fig. 7, which shows the minimum hardness gradient expected on the pitch line cross section. In this manner, the minimum and maximum hardness gradients at both the pitch line and root may be ascertained.

The D_{Ic} Requirements

The next major step is the determination of the necessary case hardenability. The other specified hardnesses (Rc 60 to 65 at the surface, and Rc 50 at 0.030 to 0.050 in. below the pitch line surface) will be obtained by carburizing. Recall that a martensite plus austenite structure of Rc 60 to 65 is the desired surface condition.

To determine the D_{Ic} required to yield such a surface structure, the curve shown in Fig. 8 is utilized. This graph is a plot of the thermodynamic relationship[4] between the diameter of an "ideally" quenched round bar having the same cooling rate at its center as a Jominy bar at a given distance along its length. For example, a 2 in. round bar given an "ideal" quench will cool at the same rate at its center as does a Jominy bar 5/16 in. from the quenched end. Jatczak[4,5] used this curve, taken from Carney's work[6], to define D_{Ic}, and to determine the effect of alloying elements on case hardenability. Multiplying factors were developed in a manner similar to the base factors previously discussed. A 10% nonmartensitic transformation criterion was used in this work, making its direct application inadvisable since, as in this instance, many applications require 100% martensite plus austenite structures. The additional step necessary to account for this difference is incorporated into the analysis below.

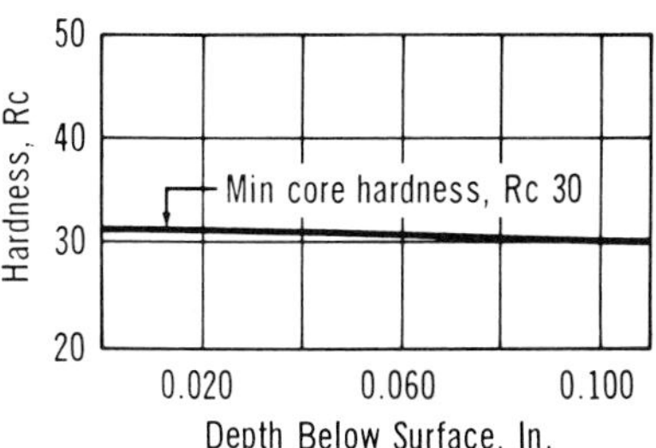

Fig. 7 — Through calculation, the minimum hardness gradient expected at the pitch line of an uncarburized gear can be determined. The minimum core hardness is Rc 30.

To determine the minimum case hardenability (D_{Ic}), refer again to Fig. 5. This figure indicates that

Fig. 8 — This curve relates the ideal diameter to the distances along a Jominy bar. Though J4 indicates the cooling rate at the tooth surface, 3⁄16 in. is added to increase the D_{Ic} to 2.7 in. This increase acts to eliminate the 10% nonmartensitic transformation products, which would develop in the case if the original D_{Ic}, 1.8 in, were used.

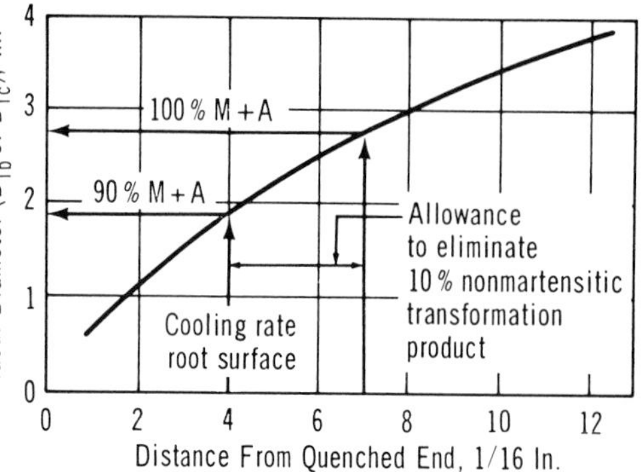

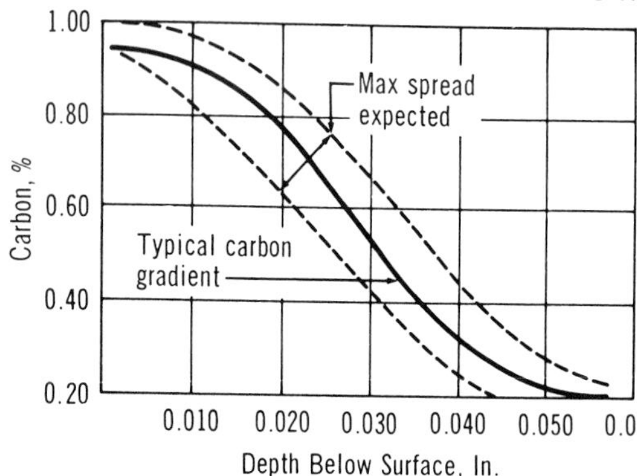

Fig. 9 — In a typical light case, carbon content will vary with the depth below the surface as shown. Base carbon is 0.20%. The carbon gradient spread is approximated from data given in the ASM Metals Handbook, 8th Edition, Vol. 2, p. 101 to 103.

the gear root cools more slowly than the pitch line. Hence, the minimum D_{Ic} value will be selected so that the case microstructure will be only martensite and austenite in the root area. By establishing a case microstructure of 100% martensite plus austenite at the slower-cooling root area, a microstructure of 100% martensite plus austenite will also be developed at the pitch line surface, which cools more rapidly.

In Fig. 8, the minimum D_{Ic} value required is determined by a vertical line drawn from the root surface Jec rate, J4, to the curve. A horizontal line from this point on the curve to the ordinate indicates that a steel with a minimum D_{Ic} of 1.8 in. would develop 90% martensite plus austenite at the root surface.

Because this example does not allow the presence of nonmartensitic transformation products — only a structure of martensite plus austenite is permitted — an additional step is necessary in the sequence. Limited experience gained by the study of carburized hardenability bars has indicated that adding 3⁄16 in. to the experimentally determined Jec rate and using the same thermodynamic relationship (Fig. 8) gives a new D_{Ic} value generally sufficient to produce a surface case structure free from transformation products. (This concept is in need of additional refinement.) When 3⁄16 in. is added to the root Jec rate, J4 + 3, and a vertical line is drawn at the J7 position, the new D_{Ic} value is 2.7 in.

In other words, a steel with a minimum D_{Ic} value of 2.7 in. will develop the desired microstructure of 100% martensite plus austenite. It should be noted that, due to the retained austenite, some gears may develop surface hardness values below Rc 60.

Carbon Gradient Selection

Under production conditions, only a limited number of carburizing cycles are usually available due to practical considerations. Carbon gradients produced by these cycles are potentially fairly controllable. Most cycles are set up to give surface carbon contents in the 0.90 to 1.00% range. The problem then becomes one of selecting a time cycle which will result in a carbon level at the critical depth (in this instance, 0.030 to 0.050 in.) which, in combination with the previously determined D_{Ib}, will give Rc 50 under the expected quenching conditions.

As a first trial, and to demonstrate the computational principles, the typical "light case" carbon gradient shown in Fig. 9 has been selected. Showing the relationship of carbon content to distance below the pitch line surface, this graph enables the calculation of a new D_{Ib} value at each point below the surface. At each point in the cross section, it is possible to consider the increment of hardenability increase (ΔD_{Ib}) due to the increased carbon content. Because the Jec rate, the D_{Ib}, and the carbon content are known at each point below the gear surface, the appropriate hardness values can be calculated. When considering minimum and maximum hardness values, it is necessary to consider the range of carbon, range of quenching rate, and range of hardenability at that location.

For the sake of brevity, a single cooling rate is assumed. For example, consider now the minimum case depth requirement of Rc 50 to 0.030 in. below the pitch line surface.

To say that a steel has a D_{Ib} of 1.1 in. means that the product of the multiplying factors of the alloying elements which make up its composition is also equal to 1.1 — that is, D_{Ib} = MF (carbon) × MF (manganese) × . . . Dividing the D_{Ib} by the multiplying factor (MF) for minimum carbon gives the contribution of the alloys.

Figure 9 shows that 0.41% C is the minimum expected at 0.030 in. The D_{Ib} for the composition at 0.030 in. is found by multiplying the MF for this higher carbon content by the alloy D_{Ib} product, giving (in this

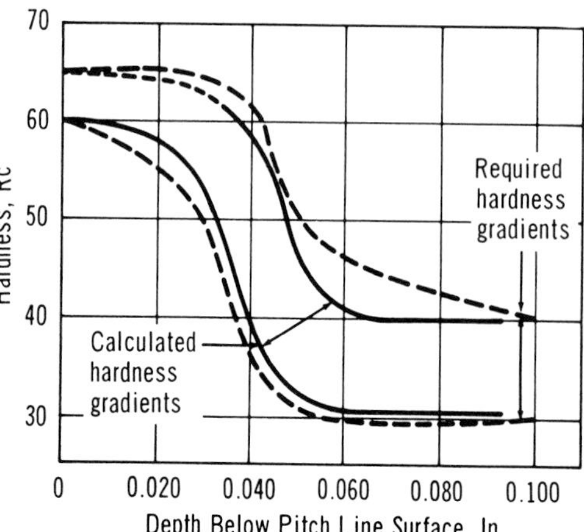

Fig. 10 — Pitch line hardness gradients calculated according to AT precepts correlate with those required for the gear. Minimum and maximum lines are shown.

instance) a D_{Ib} of 2.05 in. To calculate the minimum expected hardness at the 0.030 in. depth, reference is again made to Table I, of Part II. From Column 1 and 2, the initial hardness for 0.41% C is Rc 57. Though the Jec rate for the gear in the pitch line varies from J1.9 to J2.2, it is assumed constant at J2 for ease of calculation. By extrapolation, the IH/DH ratio is 1.075 for a steel with $D_{Ib} = 2.05$ in. at the J2 position. Dividing the IH value (Rc 57) by the IH/DH ratio (1.075) gives a DH value of Rc 53.

This value, Rc 53, represents the minimum hardness expected at the 0.030 in. depth with a minimum composition and the minimum carbon penetration of this cycle. In a similar manner, the maximum hardness level to be expected at the 0.050 in. depth is calculated to be Rc 46. It must be pointed out that, in calculations of this type, the use of D_{Ib} values and the IH/DH tables is restricted to hypoeutectoid compositions.

Figure 10, which shows results of a series of these calculations, compares the required hardness gradients at the pitch line with the calculated hardness gradients, based on the carbon concentration gradient for a typical light case and the base hardenabilities. In this instance, the trial gradient proved satisfactory. Had it not, another trial would have been necessary. It should be noted that, if one had maximum freedom in selecting carbon gradients, an optimum gradient could be selected by manipulating the above principles.

Insofar as the root fillet is concerned, experience has indicated that, when the hardenability of the steel is adequate to satisfy the pitch line strength gradient, it will also harden enough to provide an acceptable strength gradient at the root fillet.

Alternatives for Determining Composition

The Application Tailoring procedure has been used to find the minimum and maximum D_{Ib}'s (1.1 and 1.8 in.) and the minimum D_{Ic} (2.7 in.) required for the steel needed for a particular six-pitch spur gear (Fig. 1). The minimum hardness gradient to be expected in production has also been described. (Note: a max D_{Ic} is not needed; refer to Part I of this series.) Again, attention is called to the fact that the characteristics of the steel have been completely described but the composition has not yet been mentioned.

There are two routes open in this regard. One is to find a standard steel composition that fulfills the requirements. Since standard steels are not designed to have uniform changes in hardenability from grade to grade and are not systematized in terms of D_{Ic}, a compromise of some sort is likely to be necessary. The second route is to design a special steel. Procedures to develop the exact composition fitting multiple requirements comprise part of the CHAT system, and will be developed in future articles.

Selection of a standard steel for the six-pitch gear is, in part, similar to selection of a standard steel for a through-hardened part in that it is based upon the carbon content and base hardenability. Of course, case hardenability must also be considered. A standard steel is selected on the basis of carbon content (0.17 to 0.23% C) and its Jominy curve (required Rc 30 to 40 at J2 for core hardness). Finally, the expected D_{Ic} value calculated for the standard steel must be 2.7 in. minimum.

Summary

Basically, AT is that part of the generalized CHAT procedure which is used to determine the minimum and maximum hardenability levels required to develop the desired hardness and strength levels in a part for a given application. Application Tailoring could include other items such as the M_S temperature or other metallurgical or processing variables. One objective of AT analysis is improved efficiency and accuracy throughout the entire design procedure. Accuracy is emphasized in the design analysis stage because the more accurately that loads and stress levels are known, the more efficient and reliable will be the steel designed for the part.

Experience has shown that AT analysis of a part tends to promote refinements in the design analysis. In addition, AT analysis causes one to look critically at the manufacturing process, particularly the heat treatment. Alterations in quench rates, use of faster quenchants, and selection of optimum heat treating or carburizing cycles frequently result.

Finally, determination of the minimum hardenability requirements leads to a more efficient usage of alloying elements (or combinations of alloying elements) in the selection of a standard steel or the design of a steel by the CHAT approach. In ensuing articles, the logic and some of the algorithmic aspects of Computer Harmonizing (CH) will be discussed.

References

1. Mitsuo Hattori, "Heat Treatment Practice in Japanese Automotive Industry." Private correspondence.

2. "Effect of Surface Condition on the Fatigue Resistance of Hardened Steel," by G. H. Robinson, Fatigue Durability of Carburized Steel, Special ASM Publication, 1957, p. 11-47.

3. "The Distribution of Residual Stresses in Carburized Cases and Their Origin," by D. P. Koistinen, ASM Transactions, Vol. 50, 1958, p. 227-238.

4. "Multiplying Factors for the Calculation of Hardenability of Hypereutectoid Steels Hardened from 1700 F," by C. F. Jatczak and D. J. Girardi, ASM Transactions, Vol. 51, 1959, p. 335-349.

5. "Hardenability in High Carbon Steels," by C. F. Jatczak, preprint of paper presented at Materials Engineering Congress, 17 October 1972, Cleveland. To be printed in Metallurgical Transactions of ASM-AIME, 1973.

6. "Another Look at Quenchants, Cooling Rates, and Hardenability," by D. J. Carney, ASM Transactions, Vol. 46, 1954, p. 882-925.

Computer-Based System Selects Optimum Cost Steels-IV

By DALE H. BREEN, GORDON H. WALTER, and JOHN T. SPONZILLI

This is the fourth in a series of five articles covering a steel design and selection procedure for heat treated components known as CHAT (an acronym for Computer Harmonized Application Tailored). Part IV concerns the metallurgical and mathematical foundation for Computer Harmonizing (CH). Used in designing special optimum cost steels, Computer Harmonizing is the computerized determination of the least costly combination of alloying elements which will meet the property requirements called for by Application Tailoring (AT).

Computer Harmonized (CH) steels are designed with one of three objectives. These are (1) to provide a special steel that meets a component's Application Tailored (AT) requirements; (2) to aid in the selection of the most economical standard steel that meets a component's AT requirements; or (3) to develop "replacement" steels for standard AISI H steels. The distinction between using Computer Harmonizing for the first or second objective lies in whether or not the tonnage or other requirements of the component justify a special steel. For example, if the component under consideration is a low tonnage item, the engineer may lean toward a standard steel that will be readily available from a steel mill or warehouse. Using Computer Harmonizing for the third objective provides a systematic method for devising chemical compositions for "replacement" steels, which (as defined in the first article of this series) are steels which match the base and case hardenability and other characteristics of the original steel.

In each instance, the resulting CH steel has a chemical composition that is optimized with respect to the cost of alloying elements. This usually results in a steel that will also have an optimum price in terms of its "chemistry grade extra", as covered in the steel product price book. The difference between "cost" and "price" in this context must be understood; there is not always a one-to-one correspondence between the two, although a close correlation does exist.

The need for using a computer approach to develop a least-cost steel becomes apparent when the various aspects of the problem are considered. A least-cost steel which only needs to meet a specified D_{Ib} value could be designed fairly easily by a manual method with tables (or nomographs) containing alloy costs and hardenability multiplying factors. When a carburizing grade is designed, however, at least one additional restriction, a minimum D_{Ic} value, is added to the problem. Because hardenability factors for the individual alloying elements are different for the case and the base composition, a steel designed to have a least-cost base composition most likely would not satisfy the D_{Ic} requirement. If further restrictions, such as the martensite start temperatures (M_S) of the case and base analyses are added, it becomes clearly impossible to find, manually, the least-cost combination of alloying elements that satisfies the multiple requirements. Because of the multiple restrictions, as will be shown later, this problem of cost optimization requires the use of separable programming techniques.

Before embarking upon the complexities of linear and separable programming, it is worth while to take a rather simplistic look at the problem of cost optimization. Although the computer system does not use hardenability efficiency directly, the concept of hardenability efficiency is a useful strategem to aid the understanding of the cost optimization procedure.

Mr. Breen is chief, Materials Div., Mr. Walter is chief engineer, and Mr. Sponzilli is materials research stress engineer, Metallurgical Section, Materials Div., Engineering Research, International Harvester Co., Hinsdale, Ill.

Hardenability Efficiency

Although alloying elements' costs and their effects on hardenability have been qualitatively linked to one another almost since steelmaking began, one of the first published quantitative attempts to use this information to devise an optimum steel composition is credited to H. E. Hostetter in the 1940's[1]. At that time Hostetter indicated that the relative costs of the alloying elements chromium, nickel, and molybdenum were about 1:2:5, respectively. Today, the figures indicate that this ratio is about 1:5:9.

A careful and continuing survey of the cost and availability of the various alloying elements used in steelmaking is essential for their efficient utilization in steel design. Quantitative knowledge of the influence of individual alloying elements on such properties as case and base hardenability enables the mathematical coupling of an alloying element's cost with its effect on properties. This is the basis of the procedure for optimizing the chemical composition of a steel.

Ideally, hardenability efficiency should reflect the hardenability contribution of an alloying element with respect to the cost of the element. Because the contributions to base and

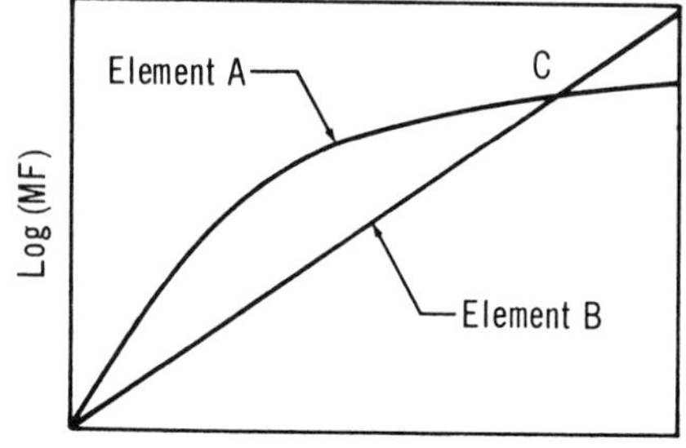

Fig. 1 — These hypothetical curves show the logarithm of the hardenability factor as related to the concentrations of two equal-cost alloying elements, A and B. Point C indicates the point at which the hardenability contribution and cost (relative efficiency) of each is identical.

case hardenability (D_{Ib} and D_{Ic}) are both important in designing a case hardening steel, two efficiencies must be considered for each element. For a through-hardening steel, only base hardenability efficiency is important.

Relative Hardenability Efficiency

Relative hardenability efficiency is defined as the logarithm of the hardenability multiplying factor (MF) divided by the cost (C) of the element:

$$\text{Relative hardenability efficiency} = \frac{\text{Log (MF)}}{\text{C}}$$

Here, the multiplying factor is the factor at any per cent, X, of the element, and C is the cost of X per cent of the element. Logarithms of multiplying factors are used to facilitate computational techniques. The cost is derived from the cost and composition of the addition made to the melt and its recovery level. Cost factors vary somewhat from steel mill to steel mill.

To illustrate the concept of relative hardenability efficiency, suppose that the logarithms of the multiplying factors for two elements, A and B, are as plotted in Fig. 1. Furthermore, assume that the cost per per cent of element A equals that of element B. The relative hardenability efficiency of the two elements would be identical to one another at point C because, at that concentration level, they provide the same contribution toward hardenability at an equal cost. At any percentage less than point C, the relative efficiency of element B is less than that of element A, while at percentages greater than C, the relative efficiency of element B exceeds that of element A.

This concept was used in constructing curves (Fig. 2 and 3) of relative base and case hardenability efficiency for manganese, silicon, nickel, chromium, and molybdenum. (For multiplying factors used, see references 2 and 3.) Note that the relative hardenability efficiencies of manganese and chromium are generally high in the base and case compositions, while those of nickel and molybdenum are generally low. Silicon in the 0 to 0.6% range however, has zero efficiency in the base composition, but has an appreciable efficiency over the same range in the case composition. Thus, silicon can be important when it is necessary to increase case hardenability without affecting base hardenability.

Although carbon and boron are much more effective in low-carbon steels than other alloying elements, curves are not shown for them in Fig. 2 and 3 for several reasons. In "blowing down" a heat of steel, carbon is removed from the melt, so that a cost-vs-hardenability effect is difficult to ascertain. As for boron, its hardenability effect is considered to be largely independent of the quantity of boron present, providing it is maintained within certain limits. Also, the boron effect varies with the carbon level, so that it too does not lend itself to the simple representation of Fig. 2 and 3.

Incremental Hardenability Efficiency

Since the hardenability multiplying factors are generally nonlinear, the relative hardenability efficiencies are not linearly related to the percentage of alloying element in a composition. Use of the efficiency concept in qualitatively understanding the determination of an optimum steel composition is furthered, therefore, by defining an incremental hardenability efficiency. This factor is defined as:

Incremental hardenability efficiency $= \Delta\log\text{ (MF)/C}$

Here, Δlog (MF) represents the incremental change in the logarithm of the multiplying factor, and C is the alloy cost to obtain the Δlog (MF).

This concept is illustrated graphically in Fig. 4 by the slopes of the two multiplying factor curves described previously (Fig. 1). As before, the cost per per cent of element A equals that of element B. The important point to note here is that, beyond X%, the incremental efficiency of element A is significantly lower than that of element B. That is, it requires 2 ΔX of element A to obtain the same hardenability contribution (ΔY) provided by ΔX of element B.

Although both A and B have the same relative hardenability at point C, and steels with a hardenability requirement of Log (MF) $= Y_1$ could be made with either element at the same over-all cost, it is significant that neither steel represents the least-cost analysis. The least-cost analysis would be obtained by adding element A until its incremental efficiency was just less than that which characterized element B. At this point, element B would be added to the analysis until the desired effect was obtained.

One can now visualize the optimization procedure – briefly, the hardenability efficiency factors, relative and incremental, for base and case composition are continuously scanned, and compared with the input requirements of D_{Ib} and D_{Ic}. The system will add elements in the proper amounts according to their hardenability efficiencies in such a way as to meet the D_I requirements and still minimize the total cost of alloying elements.

This analysis of the problem is an oversimplification of the facts concerning cost optimization. To provide a least-cost analysis that satisfies all the requirements involves more than a simple expression of hardenability efficiencies. Instead, linear and separable programming techniques, discussed next, provide the computer means for least-cost alloy design.

Fig. 2 – Given equal amounts of alloy, manganese has the greatest relative base hardenability followed by chromium, molybdenum, silicon, and nickel.

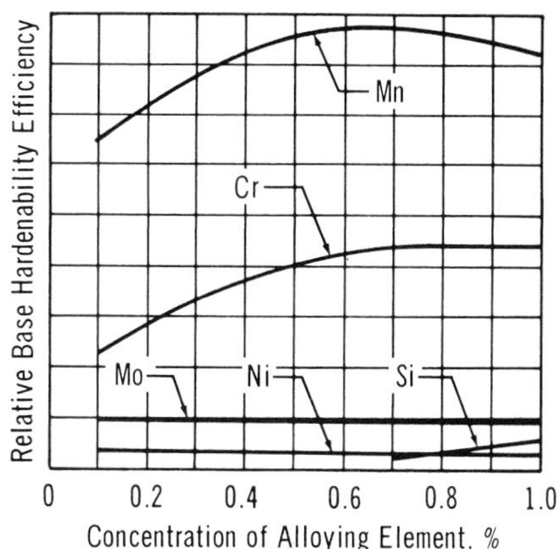

Fig. 3 – In carburized cases, the relative case hardenability efficiency is greatest for silicon, followed by manganese, chromium, molybdenum, and nickel. Compare with Fig. 2.

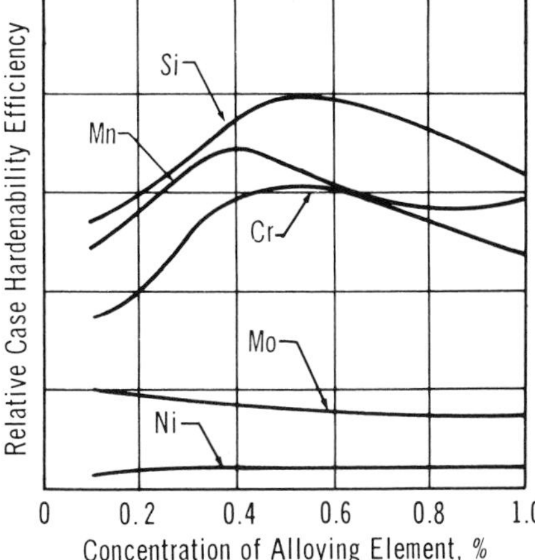

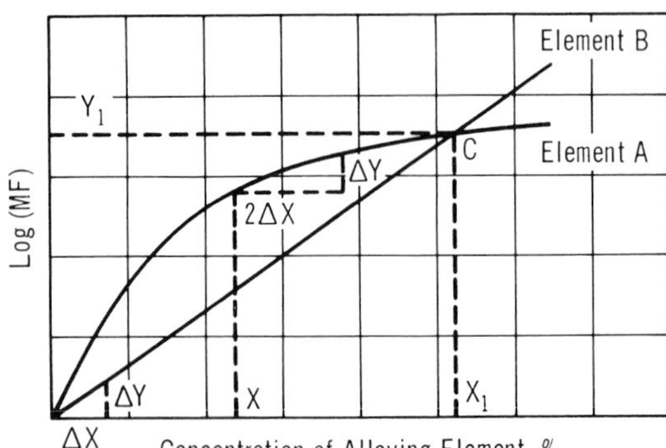

Fig. 4 — Concept of incremental efficiency is illustrated with the curves in Fig. 1. Note how equal increments in log (MF) are obtained through different incremental additions of the equal-cost elements A and B.

Linear and Separable Programming

The concept of cost minimization while simultaneously satisfying a number of other requirements is a familiar one that has been dealt with through the use of linear programming. Basically, the process entails setting up a set of equations that will model the problem accurately, and then solving the system by linear programming.

Two restrictions must be placed on a set of equations for them to be solvable by this method. First, an objective, such as cost, must exist to be optimized, and it must be expressable as a linear function. Second, there must be restrictions on the attainment of the objective, and these restrictions must be expressed as a system of linear equalities[4].

To use a linear programming approach to determine the chemical composition of a steel that will meet AT requirements, it is necessary to quantify the properties under consideration. For discussion purposes, the groundwork for Computer Harmonizing will be developed with only two limited objectives. These are (1) minimizing the cost of alloy additions (objective function), and (2) satisfying the hardenability requirements (restriction functions). The latter are given in the form of ideal critical diameters of the base and the carburized case compositions, D_{Ib} and D_{Ic}.

When accurate equations become available for expressing other properties – nil ductility transition temperature, Charpy V Notch impact energy, fracture toughness, weldability, and others – in terms of chemical composition, they too may be included in the system of equations as restriction equations. For example, M_S temperatures for base and case compositions can be expressed as functions of the chemical composition.

The basic system of equations needed to develop least-cost carburizing steels will enable a resulting composition to fulfill cost, D_{Ib}, D_{Ic}, and other requirements simultaneously. This accomplishment would be virtually impossible without the aid of a computer. The generalized form of some of these equations follows:

1. $\text{Cost} = K_1 \cdot x_1 + K_2 \cdot x_2 + K_3 \cdot x_3 + \ldots$ (Objective function)
2. $D_{Ib} = f_1(x_1) \cdot f_2(x_2) \cdot f_3(x_3) \ldots$ (Restriction function)
3. $D_{Ic} = g_1(x_1) \cdot g_2(x_2) \cdot g_3(x_3) \ldots$ (Restriction function)

Here, cost = total cost of alloy addition; K = cost per per cent of alloying element; x_1, x_2, . . . = per cent carbon, manganese, etc.; and f, g = multiplying factor functions for calculating base and case hardenability respectively – these are generally nonlinear. The optimum solution would be one that minimizes the alloy cost equation while meeting the D_{Ib} and D_{Ic} requirements.

As part of the second requirement, to enable the problem to be solved using linear programming, the terms must be additive. Though Equations 2 and 3 do not meet that requirement, they can be rewritten as:

4. $\log(D_{Ib}) = \log f_1(x_1) + \log f_2(x_2) + \log f_3(x_3) + \ldots$
5. $\log(D_{Ic}) = \log g_1(x_1) + \log g_2(x_2) + \log g_3(x_3) + \ldots$

The system of equations now becomes solvable by a modified linear programming technique known as separable programming. Separable programming requires that some additional conditions must be met. For example, each nonlinear function (multiplying factor function) must be a function of only one variable or a linear combination of such functions. Furthermore, each function must be polygonal, or replaceable by a polygonal representation of it. In other words, it must be capable of representation by a piecewise linear function[5].

The first requirement is met by all of the equations. Dividing each multiplying factor function into a number of linear segments will meet the second requirement. (The utility of using logarithms to define hardenability efficiencies should now become evident.) Of the computer programs available for solving this general type of problem, the one chosen was the IBM Mathematical Programming System/360[5], which can handle several types of linear and separable programming problems.

Computer Program Input

The input data section for the program contains the cost factors, hardenability multiplying factor functions, and M_S temperature equations. The D_{Ib} values are computed with multiplying factors developed by

Fig. 5 — Employing a graphics console, author John T. Sponzilli is shown defining D_{Ib} requirements from Jominy hardness requirements.

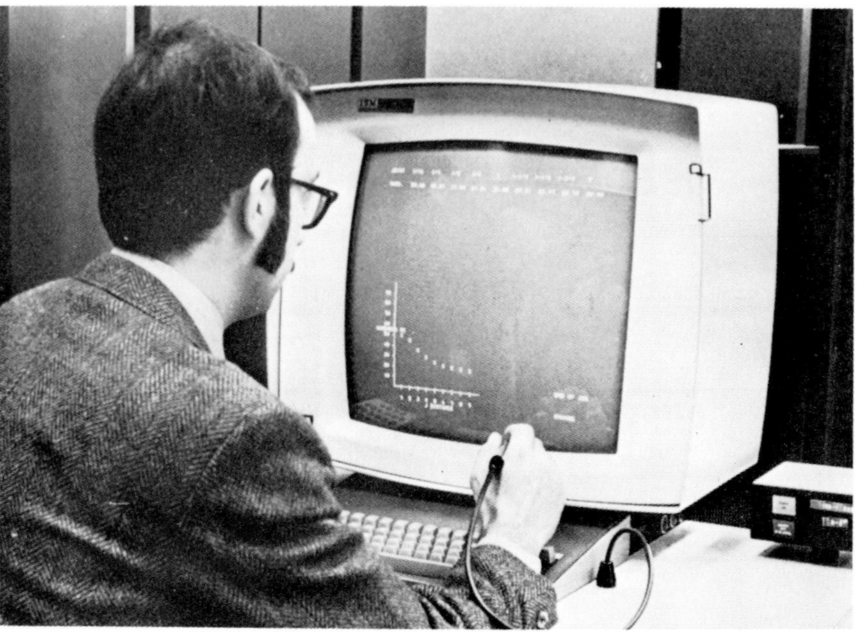

Doane and DeRetana[2]; D_{Ic} values, with factors developed by Jatczak[3]; M_S temperatures of base compositions, with an equation developed by K. W. Andrews[6]; and M_S temperatures of case compositions, with an equation developed by Payson and Savage[7].

To solve the system of equations and obtain a least-cost combination of alloys now only requires that specific requirements be given to D_{Ib}, D_{Ic}, and the M_S temperature equations, and that allowable limits be placed on the chemical elements in the composition when necessary. Changes in the cost of alloy additions can be made simply by revising the necessary terms in the cost equation. If only one requirement, such as D_{Ib}, needs to be satisfied, requirements for D_{Ic} and M_S temperatures can be set equal to zero.

When more than one requirement is placed on a composition, this programming approach provides the only sure way of designing a steel composition that can simultaneously meet all of the requirements with the least-cost combination of alloying elements. As indicated previously, small changes in alloying elements can represent surprisingly sizable savings.

Although they are not yet developed, other computer techniques are being studied for further automating the entire CHAT system. In one such approach, the graphics console shown in Fig. 5 is used. It provides a direct interface with the computer that could enable the operator to carry out both Application Tailoring and Computer Harmonizing functions.

The System in Summary

The CH procedure is designed to optimize one property while simultaneously satisfying multiple restrictions. In the work described here, the CH system determines the least costly alloy combination which will satisfy specified values for D_{Ib}, D_{Ic}, and M_S. Other properties can be optimized and other restrictions added, provided all features can be expressed as quantitative functions of the chemical elements. Working with hardenability and M_S temperatures as a basis, the CH process solves the least-cost alloys problem by using a modified linear programming technique known as separable programming.

In the final article of this series, the selection of standard grades, the designing of special steels for particular applications, and the procedure for developing replacement steels by Computer Harmonizing will be covered in some detail. Specific examples of CHAT components, including test results on the experimental steels, will be presented.

References

1. "Determination of Most Efficient Alloy Combinations for Hardenability," by H. E. Hostetter, AIME, Vol. 167, 1946, p. 643-652.
2. "Predicting Hardenability of Carburizing Steels," by A. F. deRetana and D. V. Doane, Metal Progress, September 1971, p. 65-69.
3. "Determining Hardenability from Composition," by C. F. Jatczak, Metal Progress, September 1971, p. 60-65.
4. "Introduction to Operations Research," by C. West, R. L. Ackoff, and E. L. Arnoff, New York, John Wiley & Sons Inc., 1957, p. 281.
5. "Mathematical Programming System/360, Version 2, Linear and Separable Programming – User's Manual," IBM Application Program, p. 165.
6. "Empirical Formulae for the Calculation of Some Transformation Temperatures," by K. W. Andrews, JISI, July 1965, p. 721-727.
7. Discussion by A. E. Nehrenberg of "The Temperature Range for Martensite Formation," by R. A. Grange and H. M. Stewart, Transactions of AIME, Vol. 167, 1947, p. 467.

Computer-Based System Selects Optimum Cost Steels-V

By DALE H. BREEN, GORDON H. WALTER, and JOHN T. SPONZILLI

This is the final article in a five-part series on procedure for designing and selecting steels for heat treated parts. Part V details the development and application of specific steels.

The first four articles in this series detailed the concepts and applications of the Computer Harmonized Application Tailored (CHAT) steel design and selection system. Included were examples of the Application Tailoring (AT) analyses of through and case-hardened components, as well as a discussion of the metallurgical and mathematical foundation for Computer Harmonizing (CH).

In this article, AT requirements are now used to either select a standard SAE-AISI steel, or serve as input for the computer program that develops an optimum cost steel to meet the requirements of the application. In addition, methods and examples of the development of economical steels for use as replacements to standard H-band steels will be presented.

Alloy Restrictions

Before developing Computer Harmonized steels, it is appropriate to cover the subject of restrictions on the final total quantities of individual alloying elements in a CH composition. Important metallurgical factors that need to be considered when designing a CH steel for through hardening applications include machinability, temperability, forgeability, weldability, and brittle transition temperature. In designing CH steels for case hardening, carbides, retained austenite, intergranular oxidation, case M_S temperature, distortion, machinability, forgeability, and microcracks must be considered. The significance of these factors in governing the final properties of the product will result in certain alloy restrictions, and perhaps the imposition of additional restriction functions, such as martensite start temperatures (M_S).

For example, problems with temper embrittlement have generally been alleviated by adding molybdenum. Accordingly, a minimum molybdenum content might be placed as an input restriction in the computer program to insure against temper embrittlement.

When designing a case-hardening steel, the final alloy balance must be chosen to limit amounts of carbide and retained austenite in the case. The amount of retained austenite can be partially controlled by imposing M_S temperature restrictions, and the tendency to form alloy carbides can be limited by placing maximum values on amounts of strong carbide formers, such as molybdenum and chromium.

Concerning intergranular oxidation which occurs at the surface of carburized steel, it has been recently shown[1] that silicon, manganese, and chromium oxidize, in that order, preferentially at the austenite grain boundaries. For this reason, maximum concentration restrictions may be placed on silicon, manganese, and chromium. These, in turn, could add nickel and molybdenum to the composition, even though the relative base and case hardenability efficiencies of nickel and molybdenum are generally lower than those of silicon, manganese, and chromium.[2]

With respect to distortion, the metallurgical factors most applicable are the relative case and base M_S temperatures, and the relative hardenabilities between the case and base. Since the values of M_S temperatures, D_{Ib}, and D_{Ic} may be specified as computer input requirements, the distortion factor can be substantially controlled.

Another important factor that can influence restrictions on allowable alloying elements, either minimum or maximum, is centered on the problem of price breaks in "chemistry grade extra" sections of steel price books; one such section is shown in Table I.

Continued on p. 44

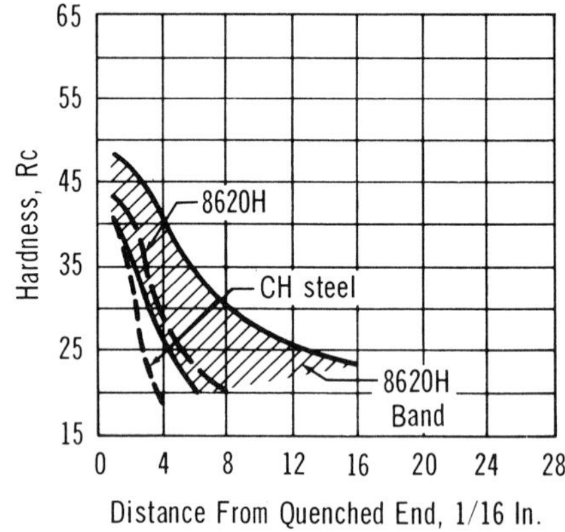

Fig. 1 — The base hardenability of the experimental CH steel designed specifically for a spur gear application is lower than that of a typical 8620H steel, which could also be used. The band for 8620H, a conventional gear steel, is also shown for comparison. The Jec rate at the pitch line core of the gear is J2.

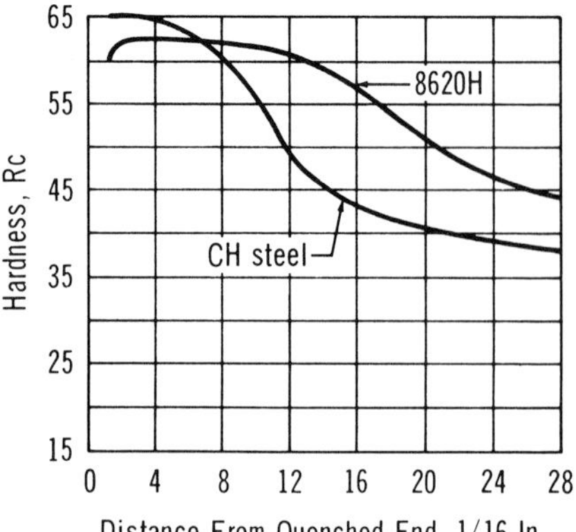

Fig. 2 — Though the case (1% C) hardenability of the experimental CH steel is lower than that of 8620H, it is adequate for the application, a spur gear. The Jec rate at the root fillet surface is J4.

Mr. Breen is chief, Materials Div., Mr. Walter is chief engineer, and Mr. Sponzilli is materials research stress engineer, Metallurgical Section, Materials Div., Engineering Research, International Harvester Co., Hinsdale, Ill.

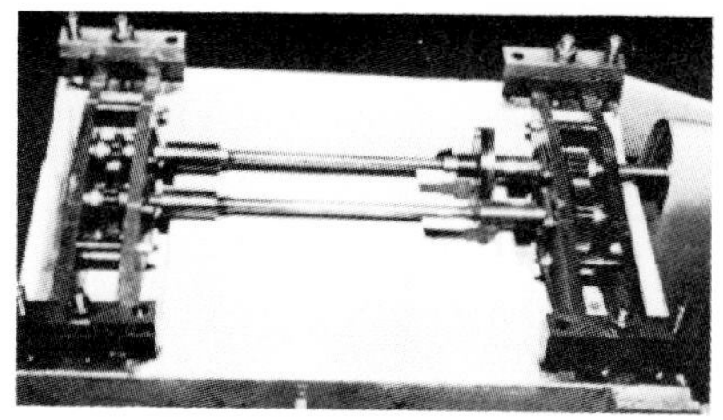

Fig. 3 — Gears made of the experimental CH steel (Fig. 1 and 2) were tested on this dynamometer. See Fig. 4.

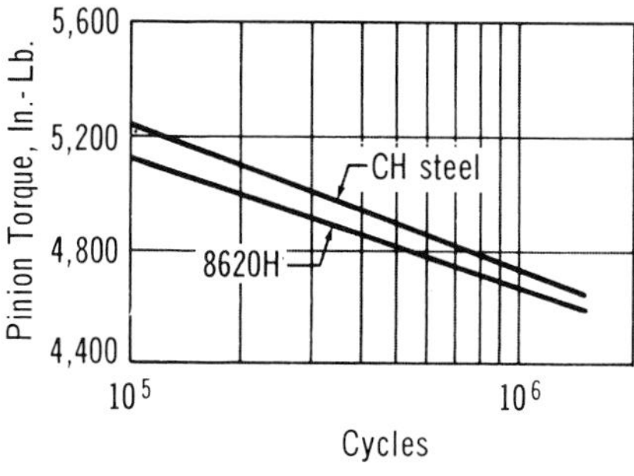

Fig. 4 — Weibull analysis of bending fatigue data from gear tests indicate that gears made from either the experimental CH steel or 8620H have equivalent durability.

Note how chemistry grade extras depend on manganese, chromium, and carbon contents.

The importance of price breaks is better understood through the use of an example. Suppose a nonboron steel with 0.18% C min and a minimum D_{II} requirement of 1.33 in. is needed, and the only special composition restriction is for the resulting composition to contain 0.20% Si min. For explanatory purposes, suppose that the computer program was given these requirements, and produced the following optimum minimum composition: 0.18 C, 1.13 Mn, 0.20 Si, and 0.37 Cr, with nickel and molybdenum being 0.01% minimum residuals.

Residual contents of alloying elements must be considered because they contribute toward hardenability. They are therefore specified as minimum alloy input restrictions for each computer problem. Although present in all heats of steel, residual alloys are not specified in the resulting composition range to which the steel is purchased. If a steel is to be ordered from a particular mill of known high or low residuals, this fact can be taken into account when developing its composition.

Applying standard chemistry ranges to the minimum analysis developed above provides a composition spread to which this steel could be ordered: 0.18 to 0.23 C, 1.10 to 1.40 Mn, 0.20 to 0.35 Si, and 0.35 to 0.55 Cr.

From Table I, the manganese-chromium grade extra would be $1.15 per 100 lb. The CH analysis that resulted from the first computer run could be advantageously revised by placing a maximum restriction of 1.05% on the minimum manganese allowed, and rerunning the problem, leaving other requirements as before.

Restricting manganese results in the following analysis; 0.18 C, 1.05 Mn, 0.20 Si, and 0.44 Cr. This steel could be ordered to the following range: 0.18 to 0.23 C, 1.00 to 1.30 Mn, 0.20 to 0.35 Si, and 0.45 to 0.65 Cr. Again from Table I, the manganese-chromium grade extra is $0.90, which represents a $5.00 per ton savings over the original steel.

In some instances, it is possible to adjust alloy composition to move the final steel from the alloy steel price base to the carbon steel special-quality price base. This results in substantial savings, amounting to a minimum of $19.50 per ton.

In general, restrictions on alloying elements in CH compositions are made and revised carefully, always keeping in view the ultimate effect on the resulting steel or production part. (Many of the alloy restrictions necessary for equivalent "replacement" steels were discussed in the first article of this series.) Whenever possible, adjustment of alloy additions should be considered to insure the design of an economical steel, not only from the alloy additions standpoint but also from the standpoint of consumer grade extra price.

Table I — Typical Grade Extras for Manganese and Chromium in Alloy Steels

Carbon Range	Chromium Range: to 0.40% incl.	0.41 to 0.65% incl.	0.66 to 0.90% incl.	0.91 to 1.20% incl.	1.21 to 1.50% incl.	1.51 to 1.80% incl.
			Manganese, 0.20% max			
to 0.10% incl.	$0.65	$0.85	$1.05	$1.25	$1.55	$1.80
0.11 to 0.20% incl.	0.60	0.75	0.95	1.15	1.40	1.65
0.21 to 0.24% incl.	0.40	0.55	0.70	0.90	1.15	1.40
0.25 to 0.28% incl.	0.40	0.55	0.65	0.80	1.05	1.25
over 0.28%	0.40	0.55	0.65	0.80	1.00	1.20
			Manganese Range, 0.21 to 0.40%			
to 0.10% incl.	0.65	0.85	1.05	1.25	1.55	1.85
0.11 to 0.20% incl.	0.55	0.70	0.90	1.10	1.35	1.65
0.21 to 0.24% incl.	0.40	0.55	0.70	0.90	1.15	1.45
0.25 to 0.28% incl.	0.40	0.50	0.65	0.80	1.05	1.30
over 0.28%	0.40	0.50	0.65	0.80	1.00	1.20
			Manganese Range, 0.41 to 0.70%			
to 0.10% incl.	0.70	0.90	1.10	1.35	1.65	1.95
0.11 to 0.20% incl.	0.55	0.70	0.90	1.15	1.40	1.65
0.21 to 0.24% incl.	0.35	0.50	0.70	0.95	1.25	1.50
0.25 to 0.28% incl.	0.35	0.50	0.65	0.85	1.10	1.30
over 0.28%	0.35	0.50	0.60	0.80	1.00	1.20
			Manganese Range, 0.71 to 1.00%			
to 0.10% incl.	1.00	1.20	1.45	1.70	1.95	2.25
0.11 to 0.20% incl.	0.75	0.95	1.20	1.45	1.65	1.95
0.21 to 0.24% incl.	0.55	0.70	0.95	1.20	1.40	1.70
0.25 to 0.28% incl.	0.50	0.60	0.80	1.05	1.25	1.55
over 0.28%	0.50	0.60	0.80	1.00	1.15	1.45
			Manganese Range, 1.01 to 1.30%			
to 0.10 % incl.	1.25	1.45	1.70	1.95	2.20	2.45
0.11 to 0.20% incl.	1.00	1.20	1.45	1.70	1.90	2.15
0.21 to 0.24% incl.	0.70	0.90	1.15	1.40	1.60	1.85
0.25 to 0.28% incl.	0.65	0.80	1.00	1.25	1.45	1.70
over 0.28%	0.65	0.75	0.95	1.15	1.35	1.60
			Manganese Range, 1.31 to 1.60%			
to 0.10% incl.	1.55	1.80	2.00	2.25	2.50	2.75
0.11 to 0.20% incl.	1.20	1.45	1.65	1.90	2.10	2.35
0.21 to 0.24% incl.	0.90	1.15	1.35	1.60	1.80	2.05
0.25 to 0.28% incl.	0.75	1.00	1.20	1.45	1.65	1.90
over 0.28%	0.75	0.95	1.10	1.35	1.55	1.75

Note: Extras, listed in dollars per 100 lb, apply if manganese and chromium are specified, but only when: (a) max carbon is 0.75% or under and max Cr is 1.80% or under, or (b) max carbon is over 0.75% and max Cr is 0.45% or under, or (c) max Mn does not exceed 2.20%.

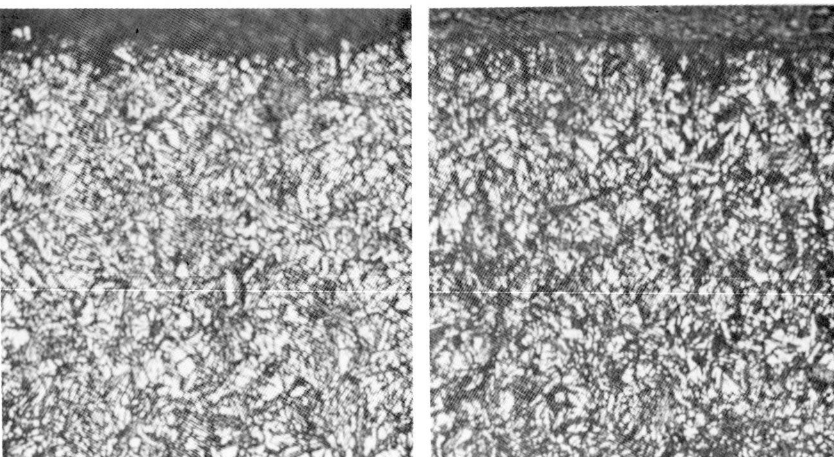

Fig. 5 — Microstructures at roots of 8620H (left) and the experimental CH steel (right) gears indicate both quench to 100% martensite plus austenite at that zone even though the latter has lower hardenability. Etchant, 2% Nital; 500×.

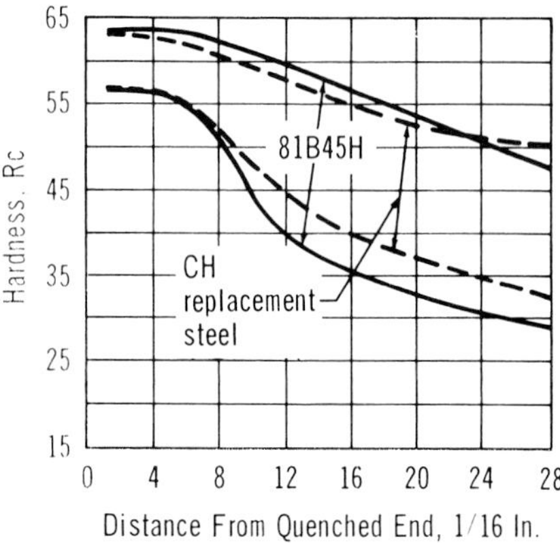

Fig. 6 — The calculated hardenability band for the CH equivalent of 81B45H agrees with that for standard grade.

Selecting and Designing Optimum Steels

Given engineering design requirements, several components were Application Tailored, as described in the second and third articles of this series. Specifying the appropriate steel for two of the components — a wheel spindle and a carburized spur gear — will now be considered in detail.

• **Oil-Quenched Wheel Spindle:** One method of specifying a steel using the hardenability requirements established through AT analysis would be to examine published hardenability bands for the most economical steel that meets the requirements. For comparative purposes, this approach, as well as the Computer Harmonizing approach, will be applied to the development of a steel specification for the wheel spindle.

The engineering requirements for the spindle were a 0.40 to 0.47 C steel, oil-quenched to 90% martensite at the finished machined surface, and tempered at Rc 30 to 37. Through the AT analysis, these requirements translated to a 0.40 C min steel with Jominy hardenability requirements of Rc 56 min at J1 and Rc 50 minimum at J95. The equivalent minimum D_{Ib} requirement was found to be 4.90 in.

Listed in SAE Standard J407c, "Hardenability Bands for Alloy H Steels," are three suitable steels: 4145H, 4147H, and 86B45H (Table II). Of the three, 4145 would be the preferred steel because of its price and carbon content, which is closest to the required 0.40 to 0.47%.

If tonnage requirements are sufficient, a special CH steel could be developed for this component. Because a boron-containing steel is generally the most economical steel for a through-hardening application, a boron steel was Computer Harmonized for the wheel spindle. Its composition follows: 0.40 to 0.46 C, 1.20 to 1.50 Mn, 0.20 to 0.35 Si, 0.55 to 0.75 Cr, and 0.0005 B min. Minimum calculated Jominy hardness values are Rc 50 at J1 and Rc 50 at J10; the grade extra is $1.60 per 100 lb.

This steel has the advantage over 4145H in that it has a carbon content identical to that required, along with a grade extra that represents a $7 per ton savings. Before adopting the new steel for the wheel spindle, laboratory evaluations of its Jominy hardenability and other metallurgical characteristics would be made.

• **Carburized Spur Gear:** In the third article of the series, a spur gear (6 diametral pitch, 3 in. pitch diameter) was Application Tailored. Engineering requirements for this carburized, oil-quenched part were a surface hardness of Rc 60 to 65, an effective case depth of 0.030 to 0.050 in. at the pitch line, and a core hardness range of Rc 30 to 40. The AT procedure indicated that a 0.17 to 0.23 C steel with the following hardenability requirements would be adequate: Rc 30 to 40 at J2 (1.10 to 1.80 in. D_{Ib}). On the 1% C hardenability bar, 100% martensite plus austenite at J4 min was required (2.70 in. D_{Ic} min).

The objective of Computer Harmonizing is to develop an optimum cost steel that would meet the above requirements. Input requirements for the computer program were 0.17 C content, and minimum D_{Ib} and D_{Ic} of 1.1 and 2.7 in. respectively. No other special composition restrictions, except for minimum residuals, were specified. The resulting steel contained 0.17 C, 0.70 Mn, 0.30 Si, 0.50 Cr, plus minimum residuals of 0.01 Ni and Mo; D_{Ib} and D_{Ic} values were 1.1 and 2.8 in., respectively.

A second steel was developed according to the same requirements plus a restriction of 0.02 Cr max — composition was as follows: 0.17 C, 1.10 Mn, 0.55 Si, 0.01 Ni, 0.02 Cr, and 0.01 Mo (the last three are residuals). Though the total cost of alloying additions was increased slightly, the second steel could be priced as a carbon steel rather than as an alloy steel.

Both compositions represent minimum analyses that just meet acceptable hardenability levels called for by the AT procedure. A quantity of the second steel was procured to evaluate for the spur gear. The composition of the experimental steel was 0.16 C, 1.24 Mn, 0.58 Si, 0.07 Ni, 0.05 Cr, and Mo; D_{Ib} and D_{Ic} values were 1.1 and 2.8 in., respectively.
experimental steel is referred to as "CH steel" in the rest of the article, and in Fig. 1, 2, 3, 4, and 5.)

Figure 1 shows base hardenability curves for the experimental steel and a heat of 8620H; the 8620H hardenability band is also given. The base hardenability curves indicate J2 hardnesses of Rc 35 for the experimental CH steel and Rc 41 for the 8620H steel. Gears made from these steels would be expected to have pitch line core hardnesses of Rc 35 and 41. The hardenability comparison is made between 8620H and the experimental steel because 8620H (or one of its EX steel replacements) is commonly used in this type of application.

Figure 2 illustrates case hardenability curves (1% C) for the experimental CH steel and for the heat of 8620H. These curves indicate that gears made from either steel would meet the 100% martensite plus austenite microstructure requirement at the root fillet surface, which has a Jec rate of J4. Note that the experimental CH steel met the AT Jominy hardness requirements for the pitch line core and the root fillet surface, as expected. Although the 8620H steel is acceptable for this part, it actually has excess base and case hardenability. For this gear, therefore, it represents metallurgical overdesign.

Normal machining and heat treating procedures were used for making about 30, 6-pitch test pinions, which were then tested in the 4-square dynamometer test (Fig. 3). As Fig. 4 reveals, the experimental CH steel had fatigue resistance equivalent to that obtained with gears made of 8620H.

Metallurgical data gathered on these gears (including information on microstructure, microhardness, residual stress, surface finish, and fractography) helped establish the adequacy of the new steel. Figure 5 shows microstructures of the root fillet surface of gears made of 8620H and the experimental CH steel. Although the experimental steel had a significantly lower case hardenability, it quenched out to a 100% martensite plus austenite structure at the root fillet surface. Obviously, it had adequate, although not excessive, case hardenability.

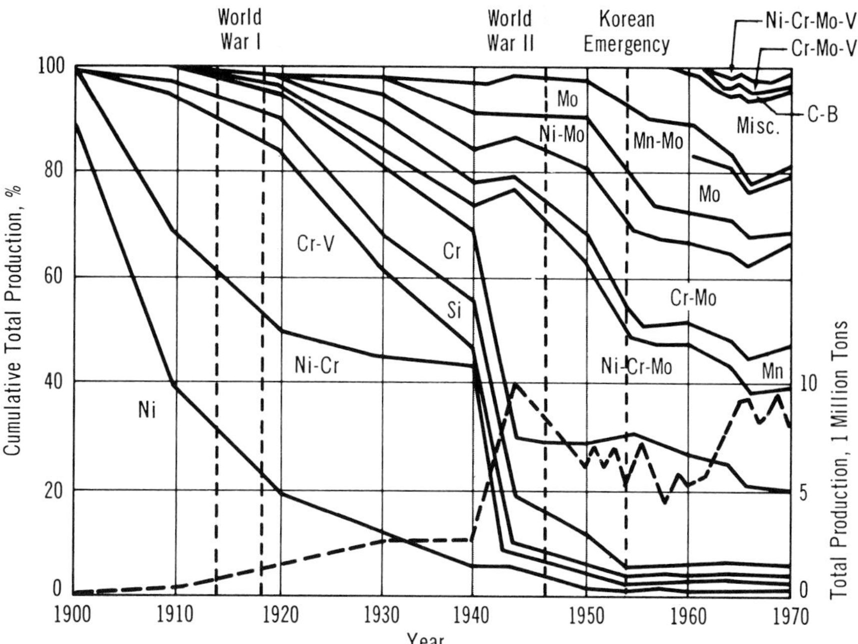

Fig. 7 — Through the years since 1900, the pattern of alloy steel usage has changed greatly. The dotted line indicates alloy steel production.

Designing Replacement Steels

Generally speaking, two steels with equivalent carbon contents, and equivalent case and base hardenability bands as well, will produce end products with equivalent microstructures and hardness gradients. This concept lays the foundation for Computer Harmonizing a standard SAE steel to develop a lower-cost, lower-price replacement with equivalent carbon content and hardenability to the original.

Any standard SAE-AISI steel can be computer-harmonized. That is, it is possible to calculate, for each grade, the average base and case hardenability which is characteristic of that steel, and place these values into the computer program, coupled with whatever restrictions are necessary on minimum and maximum values of certain elements. The result will be a replacement steel of the same carbon content which will develop the same average hardenability, and thus the same hardness gradients and microstructure, in the end product.

Once the average composition is developed, the minimum and maximum expected hardenability and composition can be determined from a statistical analysis of past mill practice in producing many heats of the numerous standard grades. (This technique for calculating hardenability bands — the "plus and minus 2σ method — was discussed in the first article of this series.)

As an example, chemical analyses and midrange base and case hardenabilities for 4817H and its CH equivalent are shown below:

	4817H	CH Equivalent
C	0.14-0.20%	0.14-0.20%
Mn	0.30-0.70	1.25-1.55
Si	0.20-0.35	0.40-0.60
Ni	3.20-3.80	—
Cr	—	0.80-1.00
Mo	0.20-0.30	—
D_{Ib}	2.6 in.	2.6 in.
D_{Ic}	6.4	6.4

Comparative chemistry grade extras are $8.55 per 100 lb for 4817H and $2.45 per 100 lb for the CH equivalent, representing a saving of approximately $122 per ton.

Because this particular example is for illustration purposes, only midrange base and case hardenabilities are shown. The final analysis would be developed by using the $\pm 2\sigma$ analysis technique to assure that the calculated base hardenability band matched the base hardenability band for 4817H. Experimental parts of steel similar to the 4817H replacement are currently undergoing feasibility studies to determine whether they have similar processing and performance characteristics.

This approach was recently applied to design a replacement composition for 81B45H. The chemical analysis range for the original steel and the CH version follow:

Table II — Suitable Wheel Spindle Steels

Steel	Carbon Range	Min. Jominy Hardness, Rc		Grade Extras per 100 Lb
		J1	J10	
4145H	0.42-0.49	56	50	$1.95
4147H	0.44-0.51	57	53	$1.95
86B45H	0.42-0.49	56	51	$2.95

	81B45H	CH Equivalent
C	0.43-0.48%	0.43-0.48%
Mn	0.70-1.05	1.00-1.30
Si	0.20-0.35	0.20-0.35
Ni	0.15-0.45	—
Cr	0.30-0.60	0.45-0.65
Mo	0.08-0.15	—
B	0.0005 min	0.0005 min

Figure 6 shows that hardenability bands for 81B45H and its CH equivalent conform closely. The bands are slightly different because one is calculated while the other is based on Jominy tests of many heats. In practice, the replacement steel could be consistently produced to fall within the published band for 81B45H.

The replacement represents a consumer price saving of $20 per ton. Careful evaluations of these steels similar to those discussed for the wheel spindle example may be necessary, depending on the end use of the replacement steel.

Table III — Equations for Calculating Boron Steel Hardenability Curves

$$J1 = 37.5 \times (\%\ C) + 39.5$$
$$J2 = 37.9 \times (\%\ C) + 38.6$$
$$J3 = 37.8 \times (\%\ C) + 38.1$$
$$J4 = 41.1 \times (\%\ C) + 36.3$$
$$J5 = 44.6 \times (\%\ C) + 8.0 \times (\%\ Mn) + 10.2 \times (\%\ Cr) + 23.6$$
$$J6 = 58.0 \times (\%\ C) + 16.2 \times (\%\ Mn) + 30.4 \times (\%\ Cr) + 5.3$$
$$J7 = 65.5 \times (\%\ C) + 35.1 \times (\%\ Mn) + 66.0 \times (\%\ Cr) - 27.6$$
$$J8 = 54.4 \times (\%\ C) + 42.0 \times (\%\ Mn) + 93.6 \times (\%\ Cr) - 40.7$$
$$J10 = 39.2 \times (\%\ C) + 37.3 \times (\%\ Mn) + 73.9 \times (\%\ Cr) - 37.5$$
$$J12 = 37.5 \times (\%\ C) + 29.9 \times (\%\ Mn) + 54.8 \times (\%\ Cr) - 31.2$$

Note: Calculated values indicate Rc hardnesses. To use the equations, all chemical elements in a composition must fall within the following effective ranges: 0.25 to 0.35 C, 1.20 to 1.38 Mn, 0.23 to 0.31 Si, 0.02 to 0.04 Ni, 0.03 to 0.12 Cr, 0.01 to 0.02 C, and 0.0005 B min.

Concluding Summary

This last article culminates a series which has presented a comprehensive look at a system for arriving at optimum alloy usage for heat-treated components, especially those utilizing large tonnages of steel. Even so, many important issues were only tagged and not fully handled. The contribution to engineering properties by individual alloys per se, for example, was not argued. Neither were residual stress effects, a factor known to be important in metal fatigue. The significant effect of grain size on hardenability was not discussed, all computations being done assuming an ASTM No. 7 grain size to ease calculation. Also, no attempt was made to answer the questions brought forth by the use of a multiplier for boron. Other approaches to this problem, including the development and usage of regression analysis equations such as those of Table III, are being investigated.

We do not believe that the shortcomings noted detract from the utility of the CHAT system, but simply indicate it is not a panacea. As is the situation for many useful tools, the system should be manipulated and applied with discretion, utilizing all available technology as a base.

Two basic concepts have been presented in the CHAT system. One was termed CH – the computer was used to develop an optimum cost steel meeting specific hardenability requirements. The other was termed AT – engineering requirements were translated into quantitive metallurgical requirements.

In addition to these prime engineering and cost aspects, the definition of the CHAT system creates an awareness of the explicit definition of a steel. A steel is not a simple single analysis, as might be implied by the standard numbering system and as it is considered to be by many design engineers. Instead, it is a large, finite number of analyses having quantifiable statistical characteristics in terms of such items as base and case hardenability. This comprehension by designers as to what comprises a steel is an important link in a viable, flexible materials engineering activity.

The principal of "optimum steel usage" has never been completely ignored by the maker of machine parts and assemblies. Many times, however, it has been relegated to a subordinate role, especially when times are good and the climate for change unfavorable. This climate has complex aspects, even to the point of being influenced by the personality of the final decision maker.

As the historical view of alloy steel usage shows (Fig. 7), many metallurgical developments stem from national or international crises.[3] The connotation this carries is not necessarily favorable to the metallurgical community. With the sure knowledge that history repeats itself, an acceptable system to make changes from one alloy system to another will certainly be put to good use in the future.

Maybe the day will come when heat treatable steels will be defined almost solely on the basis of response to heat treating rather than composition. Then, periodic adjustments in alloy content would be made without hesitation, taking immediate advantage of either or both changing alloy costs and availability. The CHAT system should help bring us closer to the ultimate goal – perpetual efficient utilization of alloys in heat treatable steels.

The strategic alloy problem has been the subject of much discussion. In a recent article[4] A. C. Sutton gave an interesting discussion which pointed out that our dependency on other countries for ores could be an important factor in future international relationships. He described a "weak link" principle – an advantage could be gained by a potential enemy who subtly promoted dependency on a strategic alloy import. Then, alternate approaches would be necessary during a war. Having principles established and disseminated can have considerable influence on the reaction time when the need for change arises. This series of articles, we believe, partially serves this purpose.

Finally, the rationale and tools discussed in this series of articles are effective in lowering production costs without sacrificing quality. The concept, though, is important as an ongoing philosophy in terms of efficiency in original designs, preparedness for change, and conservation of strategic materials.

REFERENCES –

1. Mitsuo Hattori, "Heat Treatment Practice in Japanese Automotive Industry," private correspondence.
2. "Computer-Based System Selects Optimum Cost Steels – IV," by D. H. Breen, et al, Metal Progress, June, 1973.
3. "The Sorby Centennial Symposium on the History of Metallurgy," Cyril Stanley Smith, ed., Gordon & Breach Science Publishers, New York, 1963, p. 475.
4. "Soviet Strategy," by A. C. Sutton, Ordnance Magazine, Nov.-Dec., 1969.

The Selection and Purchasing of Steel for Plant Standardization of Grades and Sizes

*By the ASM Committee on Plant Standardization**

IF SELECTION of steel were based entirely on picking the most suitable grade for each part, many grades of steel would be chosen for the many different parts made in each plant. If all parts were produced in equal and large quantities, it might be practical to make each part from a different steel, but since production is seldom in equal and large quantities, selection of a different steel for each part can lead to serious problems, including availability of the grades needed, cost, storage, inventory, distribution, and the heat treating procedures and equipment required by each steel.

The other extreme is to make all parts from the same grade and size of steel. This is obviously impossible except in rare instances. The best policy is to standardize on the smallest number of grades of steel that will satisfy plant-wide needs at the lowest cost.

Some compromises will be necessary; not all requirements will be equally well met. Engineering requirements must be fully satisfied after they have been critically examined with standardization and cost reduction in mind. Fabrication requirements must be reviewed to make sure that cost advantages in eliminating grades and sizes are not more than offset by added manufacturing costs. Effective standardization can be accomplished only by the close cooperation of all interested departments, including materials engineering, design engineering, manufacturing engineering, purchasing and planning, depending on organization of the specific plant.

Likely areas for cost savings through standardization of grades and sizes vary from plant to plant. The following possibilities should be reviewed:

1 Regroup similar bar-stock sizes for reduction of inventory and increase in quantity for each purchase.
2 Regroup similar forging-stock sizes to increase purchase quantity and reduce price extras for sizes.
3 Reduce setups, stock and inventory by making a common set of dies and forging blanks for two or more parts.
4 Change odd gages and thicknesses of sheet and strip metal to standard decimal tolerances, to decrease price extras as well as inventory.
5 Review cold heading practice for the purpose of decreasing the number of similar sizes and grades used for bolts and cap screws.
6 Review all parts made by cold and hot heading to cut down on the number of blanks required. Savings in die expense and inventory charges, as well as steel cost, may be possible.
7 Review new parts for similarity to each other and to parts already in production, to avoid duplication and waste in procurement as well as confusion in production.
8 Compare engineering requirements for similar parts and functions, establishing proof that one grade and one set of mechanical properties may be used for a number of parts.
9 Review plate-stock sizes and grades to increase the number of different parts that can be made from a convenient width and length and from the same grade.
10 Study the importance of finish of stock in relation to final grinding cost and scrap loss. More accurate size, straightness, and surface condition of rough stock may bring about savings.
11 Standardize requirements for surface finishes to permit savings in machin-

*M. F. Garwood, *Chairman*, Chief Engineer – Materials, Chrysler Corp.; W. A. Bachman, Supervisor, Process Development Dept., Ford Motor Co.; D. F. Davis, Metallurgist, Central Steel & Wire Co.; M. L. Frey, Assistant to General Works Manager, Tractor Group, Allis-Chalmers Mfg. Co.; T. A. Frischman, Chief, Metallurgy and Quality Control, Axle Div., Eaton Mfg. Co.; J. R. Hamilton, Manager, Structural, Plate and High-Strength Steel Metallurgy, U. S. Steel Corp.

Lorenz W. Heise, Director of Materials and Process Control, A. O. Smith Corp.; W. F. Hodges, Engineering Laboratory, Medium AC Motor and Generator Dept., General Electric Co.; G. B. Kiner, Works Metallurgist, Milwaukee Works, International Harvester Co.; J. R. LeCron, Metallurgical Engineer, Bethlehem Steel Co.; Peter J. Musante, Jr., Superintendent, Winchester-Western Div., Olin Mathieson Chemical Corp.; W. G. Newnam, Manager, Metallurgical Services, U. S. Steel Supply Div., U. S. Steel Corp.

Homer C. Pratt, Senior Project Engineer, Material Laboratory, Fisher Body Div., General Motors Corp.; Karl Szegedi, Senior Buyer, Airfoil Works Purchasing, Thompson Ramo Wooldridge, Inc.; J. E. Tschopp, Metallurgist, Chicago Screw Co.; Paul K. Zimmerman, Jos. T. Ryerson & Son, Inc.; G. C. Riegel, *Secretary*, retired Chief Metallurgist, Caterpillar Tractor Co.

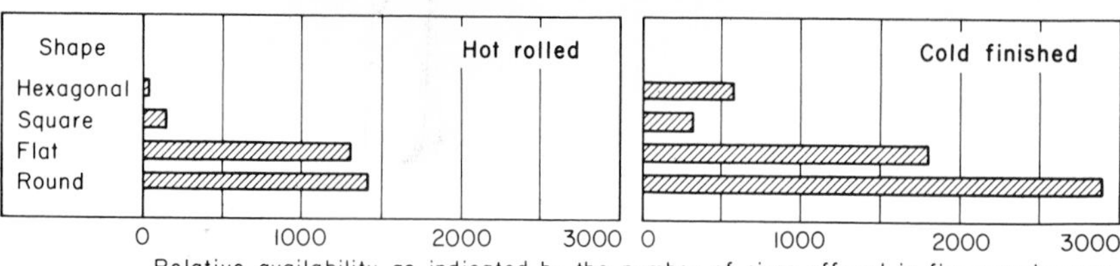

Fig. 1. Relative availability of bar shapes from warehouse stocks

ing and storing several grades used for making similar parts.

12 Reduce the number of grades of steel requiring heat treatment. This will reduce equipment requirements and number of setups, and also increase productivity.

13 Obtain the costs for purchasing, handling, inspecting and storing of all grades and sizes of stock. Simplifying these operations can bring about direct savings and reduce errors.

14 Maintain lists of grades and sizes for possible substitutions or deviations from standard specifications, to reduce the number of steels stored and the number of purchases.

Availability of Steels

The many standard steels are produced in far from equal quantities, and not all grades are made in all steel-producing districts. However, with the exception of the South and the far West, most grades will be available from some mill. If production requirements are such that the grade selected should be purchased in mill-heat quantities or more, there is no problem of availability when the market is free. Since less-than-heat quantities must often be purchased, availability on that basis must be considered. The mill representatives are the best source of information about the grades most often produced in any district so that orders for partial heats have the best chance of being filled quickly. Table 1 lists the 30 grades of nonresulfurized and resulfurized carbon steels that are the most widely available from mills. Their most common product forms are also indicated (AISI Steel Products Manual). Twenty-six grades of alloy steel that are generally the most readily available from mills are listed in Table 2. These grades are listed in order of decreasing tonnage based on a survey for the second quarter of 1957. As noted, these 26 grades comprised about 63% of the alloy steel production.

Table 1. Carbon Steel Grades and Forms Most Readily Available from Mills

Grade	Semi-finished for forging	Hot rolled bars	Cold finished bars
1008	...	X	...
1010	X	X	X
1012	X	X	...
1015	X	X	X
1017	X	X	...
1018	X	X	X
1020	X	X	X
1025	X	X	X
1030	X	X	...
1035	X	X	X
1038	X	X	(a)
1040	X	X	X
1041	(a)	X	...
1042	X	X	X
1045	X	X	X
1046	X	X	X
1050	X	X	X
1095	(a)	X	(a)
1112	...	X	X
1113	...	X	X
1116	...	...	X
1117	...	X	X
1118	...	(a)	X
1137	...	X	X
1141	(a)	X	X
1144	...	X	X
1145	X	(a)	...
1212	...	X	X
1213	...	X	X
12L14	...	X	X

X = frequently specified
(a) = infrequently specified
... = generally not produced

Warehouse Steels. Availability from warehouses is naturally more limited than from producing mills because of the inventory problem and because warehouse trade requirements are a reflection of small-lot rather than large-lot consumption in the area they serve. While the grades available follow a certain general pattern, there will be considerable variation in different parts of the country. Table 3 shows a representative list of grades available from warehouses in the central part of the United States. Whenever warehouse availability is an important consideration in steel selection, inquiry should be made of the local sources for exact information, both because of geographic location and because warehouse demand changes as prices and technology change.

Table 2. Alloy Steels Used in the Greatest Amounts (Second Quarter 1957)

Grade	Tons	% of total
5160	63,558	13.28
8620	31,736	6.61
4028	24,230	5.05
5155	22,699	4.73
4140	21,049	4.39
5132	11,357	2.37
5140	10,594	2.21
4142	9,857	2.06
5147	9,367	1.95
50B60	8,726	1.82
4130	8,678	1.81
8720	8,458	1.76
9260	7,955	1.66
4027	7,931	1.65
4340	6,169	1.29
4037	5,743	1.20
4150	5,561	1.17
E52100	5,111	1.07
8740	4,836	1.01
4145	4,766	0.99
8640	4,575	0.95
8622	4,234	0.88
4047	4,056	0.85
8615	3,997	0.83
4815	3,700	0.77
1340	3,688	0.77
Total(a)	302,631	63.13

(a) Total for the 26 grades listed above. All others comprised 36.87% of total usage.

Table 3. Steels Most Readily Available in Bar Form from Warehouses

Carbon Steels

1008 basic wire, CF	1117 HR and CF	1112 CF
1020 HR under 3 in.	1137 CF	1113 CF
1018 HR and CF	1137 leaded, CF	1212 CF
1035 HR	1141 HR and CF	1213 CF
1042 CF	1144 CF	12L14 CF
1045 HR and CF		
1095 HR		

Alloy Steels

4615 HR and CF	4130(a)
4620 HR and CF	4140 HR and CF, ann and HT(a)
8620 HR and CF	4140 leaded HR and CF, ann and HT
8620 leaded HR and CF	4145 HR and CF, ann and HT
9310 HR and CF, ann(a)	4150 HR and CF, ann and HT
4340 HR ann(a)	4150 leaded HR and CF, ann and HT
6150 HR ann	7140 (nitriding steel) HR and HT(a)
8640 HR	52100 HR and CF, ann
8645 HR	
8740 CF ann(a)	
8745 CF ann	

HR = hot rolled ann = annealed
CF = cold finished HT = heat treated
(a) Also available in aircraft quality

As with grades, sizes are not equally available. Figures 1 and 2 illustrate the general pattern for hot rolled and cold finished bars. A similar chart in the article on low-carbon steel sheet summarizes the availability of various gages and qualities of hot rolled and cold rolled sheets.

Availability listings, such as those in Tables 1, 2 and 3, are completely valid for only a limited time and should be rechecked at least once a year. Costs also change (Fig. 3).

Cost of Purchasing and Storing

Quantity purchases represent the greatest single economy to be realized through grade and size standardization, regardless of whether steel is purchased from mill or warehouse. Figures 4 through 8 indicate the magnitude of savings possible through quantity purchases.

The manner in which steel prices decrease with increasing quantity should be noted. Prices do not follow a curve, but decrease in steps. Economical purchasing always takes advantage of these break-points or price plateaus. For example, if 990 lb of 2-in. diam bars were needed, it would be cheaper to buy more than 1000 lb from a warehouse because of the price change per pound at 1000 lb.

Since the cost of issuing a purchase order is as much for a few hundred pounds of steel as for an entire mill heat, the purchasing cost per pound decreases with increased quantity. Hence, significant savings may be realized by decreasing the number of purchase orders. Elements that are included in the total cost per pound for 100 to 50,000-lb orders are shown in Fig. 8.

Inventory costs make it impractical to purchase all steel in large quantities without thought of when the material is to be used. Few steel users realize how quickly the cost of storing and handling offsets the ini-

Table 4. Cost of Storing $100 Worth of Steel for One Year

Item	Minimum	Maximum
Interest on investment	$ 4.00	$ 6.00
Storage	1.00	5.00
Taxes	0.50	4.00
Insurance	0.15	1.00
Obsolescence	2.00	8.00
Deterioration	1.00	2.00
Inventory taking	0.20	0.50
Record keeping	0.10	0.40
Rehandling	0.20	0.50
Reinspection	0.15	0.30
Total	$ 9.30	$27.70
Average	$18.50	

NOTE: Pilferage was not evaluated.

tial saving in large-quantity purchases. Except in a tight market, prudent managers do not extend steel inventories beyond known schedule requirements. The cost of storing steel for one year is from 10 to 25% of the initial cost (Table 4).

Identification of steels, particularly bars, is a problem in all plants. The magnitude of the problem increases as the number of grades of nearly the same size increases. All the conventional identification methods, including stamping, tape marking and painting, have some disadvantages. Most plants use paint markings, varying from painting bar ends to striping the entire length of the bar. This method is most practicable for identification of about a dozen steels, which can be coded with contrasting colors.

As the number of grades increases, dual and triple color codings are necessary. Such identifying operations are costly and the possibility of mixing steel increases with the number of grades in stock. However, positive steel identification is most desirable and offsets the cost by many fold. One large plant identifies about 90 grades by paint color combinations. Standardization to fewer grades simplifies this problem.

Substitution. Emergency orders and other unforeseen circumstances that result in steel shortage can upset well-planned standardization and necessitate purchase of small quantities with the accompanying cost penalty. One method of handling such situations is to have substitution charts available for immediate use by production control, even though the required substitution may result in some upgrading.

Table 5. Check List of Possible Steel Substitutes

Free-Machining and Carbon Steels

Upgrading ↑					
	1025				
	1020		1025		
	1019	1025	1022		1145
	1018	1022	1120	1030	1050
	1120	1019	1117	1125	1049
	1117	1018	1018	1026	1046
Specified	1112	1120	1020	1025	1045

Alloy Steels

Upgrading ↑				
	4820			
	4817			
	4815			
	4718			
	4620			
	4617			
	4615	8640	8740	
	8720	8637	5145	4340
	8620	8630	8642	9850
	8617	4140	4145	8650
	8615	4137	4142	5155
Specified	4118	4130	4140	4150

Upgrading in necessary substitution brings higher hardenability, usually by higher carbon, alloy or both. Price may or may not increase. Finish and production rate may be questionable and all other processing variables must be evaluated.

An example of such a chart is shown in Table 5. Substitution from stock on hand avoids the necessity for small-quantity purchases during an unexpected steel shortage. A more expensive steel may be used for the specific part, but the added cost is usually much less than that of buying a small quantity of steel in a hurry. In addition, delays in obtaining steel are avoided. If it has already been established that 1117 steel and the others in the same column in the substitution chart are satisfactory substitutes for 1112 steel, and if 1117 or one of the others is in stock in the proper size, it can be used when an unexpected short-

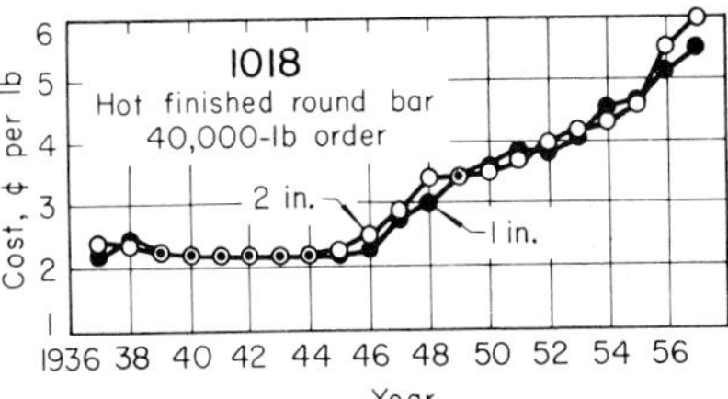

Fig. 3. Variations in the cost of 1 and 2-in. diam 1018 hot finished bar stock purchased in 40,000-lb lots

Hot rolled round bars

Cold finished round bars

Relative availability as indicated by the number of stock sizes offered in five warehouses

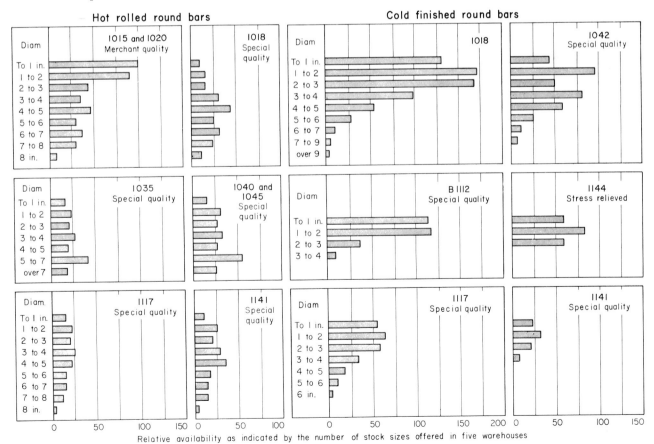

Fig. 2. Typical availability of grades and sizes of cold finished and hot rolled round bars. Survey of five midwestern cities.

age of 1112 occurs. Substitution charts are usually made so that the steel desirability decreases and, so far as possible, price increases from bottom to top.

Table 5 is intended only as an example of a workable system. This specific listing was used in one plant, but the particular steels or their exact order might not be applicable to another plant.

Price of Steel. Prices for the most popular straight-carbon, free-machining, and alloy grades, based on 5-ton lots at the mill, are given in Table 6. Pricing check charts similar to Table 7 are useful for administering standardization through all steps from the requisition to the end product. If standardization is not controlled by some such means, certain requirements may be missed, and future purchases may not resemble the first. A pricing check chart lists metallurgical, physical and other requirements such as size, form, surface treatment, straightness and length, quantity, packaging, marking, and loading to ship.

Similar check charts can be used for steel intended for specific end uses. Table 8 illustrates the use of such a chart in estimating costs of automotive coil springs made from three different steels. Total cost can vary significantly when one extra or more is neglected or purposely omitted. This table also illustrates how extras can build up the total cost of a product.

Table 9 is an example of a check chart specifically applied to an automotive rear output shaft. Successful standardization can be achieved only by study of detailed requirements for each part. For instance, another part using the same grade and size of steel might require more or fewer extras than are listed for this shaft. Therefore, a condensed list of requirements for each part expedites standardization and, at the same time, prevents merging of grades where certain engineering or manufacturing requirements might be lost.

Special qualities of hot rolled, cold finished, free-machining or other types of steel make each more or less suitable for a given operation or product.

If parts are not to be fed through collet chucks, the size variation in hot rolled steel will not be troublesome, but if collet chucks are used, the closer size tolerance of cold finished steels may be necessary. Other advantages of the close sizing of such bars are discussed in the article on Cold Finished Steel.

If carbon steels are to be used and the parts require extensive machining, the appropriate free-machining grade from the 1100 or 1200 series should be specified, as the extra cost will be more than repaid by increased production. These steels should be specified cold drawn because the full machining advantage of free-cutting steels results from a combination of their composition and the cold drawing operation. Obviously, if only small amounts of machining are to be done, such as cutting-off, and drilling a cotter-pin hole at one end of a shaft, free-cutting steels would not be economical. If the piece is to become part of a welded assembly, the high sulfur content of free-cutting steels will be bothersome. If heat treating is required, silicon-killed carbon steels with a minimum of 0.60% Mn and controlled grain size are much to be desired because of more uniform behavior, but in the low-carbon varieties cost is higher than for semikilled steels with lower manganese.

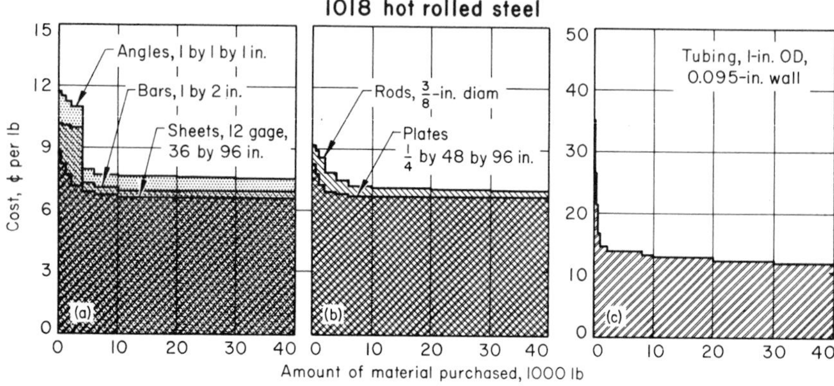

Fig. 4. Relation of cost to quantity for hot rolled products at the mill (1958)

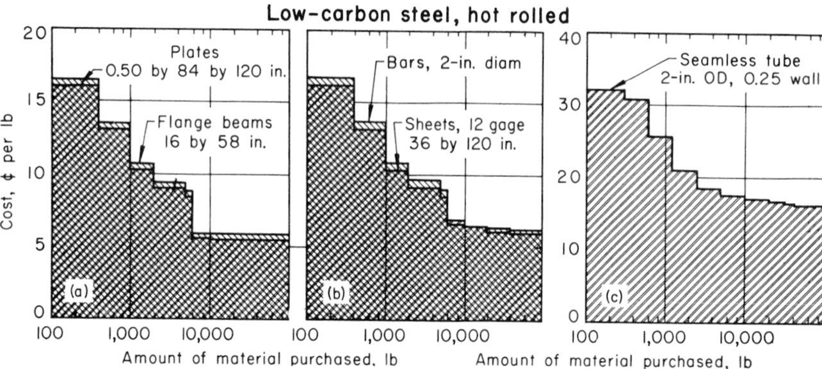

Fig. 5. Relation of cost to quantity for products obtained from warehouses (1958)

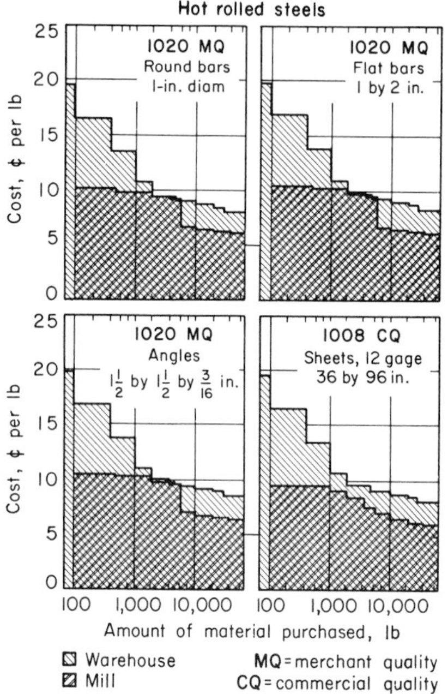

Fig. 6. Relation of cost to quantity purchased for low-carbon steel products. Prices are f.o.b. mill or warehouse plus inbound freight. Lowest cost for size, length, grade, marking, packaging, and tolerance was selected. Prices for shipments to one destination (1958).

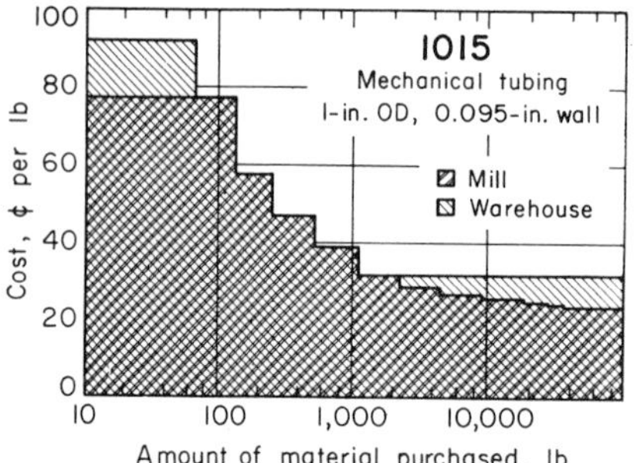

Fig. 7. Effect of quantity purchased on the cost of mechanical tubing (1958).

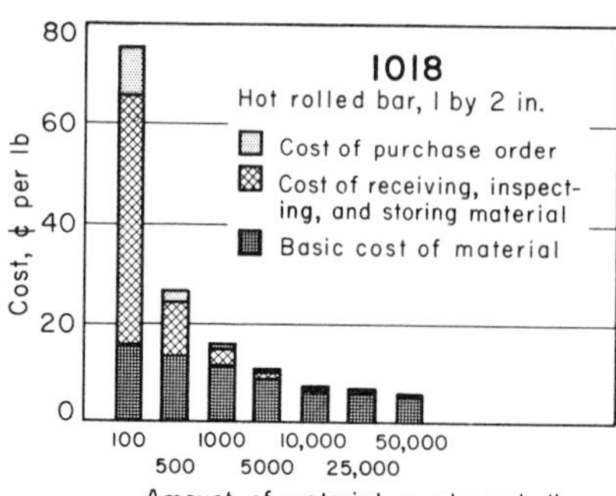

Fig. 8. Cost of steel as influenced by quantity purchased, issuing purchase order, receiving, inspecting and storing

Examples

The range and importance of plant standardization can be illustrated best by citing specific examples from industrial practice. The remainder of this article consists of such examples.

Example 1, Standardization of 1-In. Bars. The following example illustrates what was accomplished in one plant where 31 parts were being made from 1-in. diam bar stock, utilizing ten different steels. It was found that the forged parts listed in the first two groups in Table 10 should be made of hot rolled steel, and the remainder of parts in these groups, not forged, could most economically use hot rolled steel too. Either processing or strength considerations, or both, required eight of the 31 parts to be made of cold drawn steel, listed in group 3.

Design considerations required that three parts in the hot rolled group be made of a case hardened alloy steel. Past production experience with similar parts had shown that a number of alloy steels could be used but some were more acceptable than others from a machining standpoint. **8620H** had not been among the steels previously used for these parts but because of its hardenability and favorable shop experience with this grade for other parts, it was decided that it would be

Table 6. Prices for Special Quality, Hot Rolled Carbon and Alloy Steel Bars, 1-In. Diam (a)

Steel	Price per lb	Steel	Price per lb	Steel	Price per lb	Steel	Price per lb	Steel	Price per lb	Steel	Price per lb
						Carbon Steels					
1008	$0.06825	1018	$0.06825	1038	$0.06825	1046	$0.06975	1116	$0.07825	1144	$0.08025
1010	0.06825	1020	0.06825	1040	0.06825	1050	0.06825	1117	0.07675	1145	0.07275
1012	0.06825	1025	0.06825	1041	0.07475	1095	0.06775	1118	0.07825	1212	0.07825
1015	0.06825	1030	0.06825	1042	0.07475	1112	0.07825	1137	0.07775	1213	0.08025
1017	0.06825	1035	0.06825	1045	0.06825	1113	0.07975	1141	0.07775	12L14	0.08885
						Alloy Steels					
1330	$0.08475	4047	$0.08825	4340	$0.11525	5135	$0.08825	8622	$0.09625	50B50	$0.08675
1335	0.08475	4118	0.08925	4615	0.11175	5140	0.08625	8640	0.09525	50B60	0.08675
1340	0.08475	4130	0.09175	4720	0.10525	5147	0.08825	8642	0.09525	51B60	0.08875
1345	0.08475	4137	0.09375	4815	0.13425	5150	0.08625	8720	0.09875	52100	0.11775
4024	0.09175	4140	0.09375	5046	0.08325	5155	0.08625	8740	0.09675	7140	0.13325
4027	0.08825	4142	0.09375	5120	0.08875	5160	0.08625	9255	0.08825	(nitriding steel)	
4028	0.09075	4145	0.09375	5130	0.08825	8615	0.09975	9260	0.08825		
4037	0.08825	4150	0.09375	5132	0.08825	8620	0.09725				

(a) The above prices, as of January 1959, are quoted for 5-ton lots of as-rolled 1-in. diam bars in random lengths, at the mill.

Table 7. Check Chart for Pricing Billets, Bars, Plates, Shapes, Sheet, Strip and Coils
(Items in italic type are usually extras.)

Base price	Alloy or carbon steel, hot rolled or cold finished
Grade (M)	Standard AISI or *nonstandard;* open hearth or *electric* furnace; rimmed, capped, *semikilled* or *killed.*
Quality (M)	*Aircraft, bearing, cold heading, axle-shaft, gun, shell, rifle-barrel, coiled spring*
Restrictions (M)	Chemistry, *grain size, hardenability,* etc.
Size and form (P)	Semifinished bars, sheet, strip, structurals, tube, wire, *special sections* (spring, flats), *close tolerance*
Treatment (M and P)	*Annealed, box or pipe annealed, normalized, quenched and tempered, stress relieved, spheroidize annealed, pickled, oil, lime, lubrite, hardness*
Straightness (P)	As-rolled or *standard machine straightened,* allowable waviness, twist, camber, edge variation
Preparation (M and P)	*Billet grinding, 100% hot scarfed*
Specifications of Federal agencies and others (M)	*Consult*
Testing (M)	*Extensometer, etch, impact*
Cutting (M and P)	Hot cut, cold shear, *machine cut, torch cut*
Length (P)	*Dead length,* random length, *multiple lengths* and per cent shorts
Quantity (P)	Heats, or less
Marking (M and P)	Heat number stamp, *color code, other identification*
Packaging (P)	Lift weight, coil size, etc., *wrapping, burlap, boxing*
Loading (P)	Gondola or freight cars, *blocking*

(M) indicates consideration of metallurgical factors. (P) indicates attention to physical conditions.

Table 8. Application of Pricing Chart to Automotive Coiled Spring Steels

Item	Cost per 100 lb — Carbon steel 1049		Alloy steel 6150		Alloy steel 5160	
Base price	Special quality	$5.775	Alloy steel	$6.475	Alloy steel	$6.75
Grade	1049, killed, fine grained	0.15	6150	2.65	5160, standard fine grained	1.05
Quality	Open hearth	....	Electric furnace	1.00	Open hearth	....
Restrictions	None	....	Restricted C and Mn	1.00	None, standard steel	....
Size	½-in. diam	1.00	½-in. diam.	1.30	0.649-in. close tolerance (½ standard) Close tolerance	0.95 0.90
Treatment	Hot rolled, as rolled	....	Hot rolled, precision ground	2.55	Hot rolled, pickled and oiled	0.70
Straightness	Standard	....	Special straightness Pickled, oiled	0.75 0.55	½ standard tolerance, machine straightened	0.75
Preparation	None	....	None	....	No surface conditioning allowed	....
Testing	None	....	Restricted hardenability	1.00	Special decarburization, standard chemistry and hardenability	....
Cutting	Shear	....	5 to 8-ft abrasive	0.50	Machine cut	0.35
Length	Random length	....	15-ft dead length	0.20	Dead length 143-in.	0.15
Quantity	Over 40,000 lb	....	Less than 2000 lb	5.50	Heat lots	....
Marking	None	....	Continuous line	2.25	Paint one end of bundle green; attach metal tag; show heat number, weight, part number, grade, length, size	....
Packaging	None	....	Paper wrap 2500-lb or less box with runners, bar 15 to 22 ft	0.10 2.10	3500 to 4000-lb lifts wired with 4 or 5 wires, wrapped in waterproof paper with three steel bands outside for magnet unloading 3500 to 4000-lb lifts Paper wrap	.... 0.10 0.10
Loading	No special blocking, covered trucks	....	Box car loading	0.10	Gondola cars blocked to maintain straightness; not less than 20 tons per car	....
Totals		$6.925		$28.025		$11.80

satisfactory for two of the shafts. However, 8625H was required for the third shaft. This particular alloy combination was high on the availability list and economical. After initial selection, production experience and dynamometer testing of the gears showed that the 8625H could be replaced with 8620H, resulting in standardization on the one hot rolled alloy steel, as shown in group 2.

The strength required of the remainder of the parts made of hot rolled steel indicated that 1045 was necessary for all those not heat treated, except the tie rod; 1045 could also be used to meet the requirements of the heat treated parts. It was decided to make the tie rod also from 1045 in the interest of both standardization and cost reduction. The cost of 1045 purchased from a mill is less than that of a lower-carbon steel purchased in small quantities from a warehouse. This decision eliminated one grade from inventory.

When considering the strength requirements of the cold finished group, it was evident that some parts required a higher minimum strength level than could be obtained from ordinary cold drawn 1045. For these parts (Table 10, group 3) a cold drawn steel with guaranteed minimum mechanical properties was adopted. 1045 was used instead of 1141 because the amount of machining on the parts did not warrant the extra cost of the free-cutting grade.

The parts in this group that required case hardening had core strength requirements compatible with plain carbon steel, silicon-killed, fine grain. A free-cutting steel was needed and 1117 was adopted. To this group were added the remainder of the parts for which free-cutting steel could be used to advantage, thus making up group 4.

Table 9. Check Chart for Steel for a Rear Output Shaft

Base price	Cold drawn alloy	$8.775
Grade	OTS 8622H leaded	2.05
Quality	Open hearth	....
Finish	Cold drawn	....
Restrictions		
Chemistry	0.15 to 0.35% Pb	1.15
Hardenability	Restricted: J30 to 40 at 4/16 in.	1.05
Grain size	Fine	....
Other	None	....
Size	2½-in. diam	1.60
Special tolerance	None	....
Treatment		
Surface	Oiled	....
Heat	Annealed, no special structure	1.25
Straightness	Standard	....
Preparation	Standard practice	....
Testing	Standard	....
Cutting	Machine cut	....
Length	Dead length 12 ft, 2 in.	0.30
Quantity	Heats or less	....
Markings	Identification tags	....
Packaging	Standard wire binding	....
Loading	Covered truck	....
Freight	.....	0.28
Government specifications	None	....
Total cost per 100 lb		$16.455

Table 10. Standardized List of Parts for Stock 1 In. in Diameter (Example 1)

Part	Pounds per piece	Monthly production	Remarks
Group 1. Hot Rolled Carbon Steel (1045, silicon-killed, fine grain)			
Spindle shaft	3.84	750	90,000 psi min yield strength, welded
Tie rod	22.93	Service	25,000 psi min yield strength
4th speed shifter fork	2.3	Service	Forging; 35,000 psi min yield strength
Spindle ext.	3.15	Special	35,000 psi min yield strength, welded
Shaft	3.84	9000	90,000 psi min yield strength
Group 2. Hot Rolled Alloy Steel (8620H)			
Shaft and gear	6.00	4000	Case hardened, part of final drive
Pinion shaft	5.79	500	Case hardened, part of power train
Pinion shaft	21.04	6000	Case hardened, part of final drive
Group 3. Cold Drawn Steel, Guaranteed Minimum Properties (1045, cold drawn, 100,000 psi min tensile strength, 80,000 psi min yield strength, 15% min elongation in 2 in., 40% min reduction of area)			
Stud, special	2.291	Service	80,000 psi min yield strength
Stud, special	2.043	Service	80,000 psi min yield strength
Shaft, clutch drive	6.313	3650	80,000 psi min yield strength
Shaft	5.60	Service	80,000 psi min yield strength
Shaft, special	3.7402	350	80,000 psi min yield strength
Shaft, steering	6.404	5400	80,000 psi min yield strength
Shaft, splined	4.638	900	80,000 psi min yield strength
Shaft, coupling	4.508	Special	80,000 psi min yield strength
Group 4. Free-Cutting Steel (1117, silicon-killed, fine grain)			
Valve lever shaft collar	0.368	Service	No special strength required
Bushing	0.636	Service	No special strength required
Oil seal	0.3157	6000	No special strength required
Roller	0.460	500	Carbonitrided
Roller, cam follower	0.42067	15000	Carburized
Washer	0.23685	3600	No hardness requirement
Shaft, drive gear	2.42	5600	Case hardened, heavily loaded
Shaft, brake cross	21.724	Service	50,000 psi min yield strength
Shifter guide	2.097	Service	No strength requirements
Spacer, wheel guard	0.313	Service	No strength requirements
Bushing	1.25	Service	Carbonitrided
Brake cross shaft brkt	0.375	2450	No special requirements
Group 5. General-Purpose Cold Drawn Steel (1018)			
Shaft, steering	1.29	Service	50,000 psi min yield strength
Shaft, drive	16.83	4300	50,000 psi min yield strength
Shaft, coupling	9.2582	850	50,000 psi min yield strength

The remainder of the cold finished parts, group 5, required neither special strength nor heat treatment; cold drawn 1018 was adopted. This steel costs the same as 1020 but because of its higher manganese content it has slightly better machinability and is carried as a general-purpose, low-carbon, cold drawn steel. It has advantages over 1020 in case hardening. While no parts in this 1-in. size group require this heat treatment, parts in other size groups do, and it is convenient to have one general-purpose steel on hand.

Originally, ten steels were specified in 1-in. round; this came about partly from blind copying of the steels used in older designs but principally from a study of each part separately, without regard for the over-all needs. The previous steels were 4620, 5145, 4140, 1045 HR, 1045 CF, 1020 CF, 1035 CF, 1040 HR, 1120 CF and 1025 HR. Standardization, with the needs of both engineering and manufacturing in mind, and with close attention to costs, reduced the list to the five steels shown in Table 10.

The processing equipment available played an important part in the standardization. The through hardening alloy steels 5145 and 4140 could be replaced with water-quenched carbon steel only because efficient water quenches were available. It is true that the strength specified for the heat treated alloy steel was low, and in that sense the alloy steel was an over-specification, but modern quenching equipment was required to permit an economical replacement.

The availability of carbonitriding equipment made it possible to use 1117 for all the case hardened parts except gears at a considerable saving in both steel and processing cost. For parts requiring low distortion in heat treatment, an oil-quenching alloy steel was unnecessary because the addition of nitrogen in carbonitriding made the case of the carbon steel oil hardening. Processing economy resulted from the fact that finish grinding stock could be considerably reduced because of the low distortion.

The choice of alloy carburizing steel was based on available equipment particularly suitable for direct quenching. Among other things, this equipment permitted close control of the quenching temperature. Had the equipment been less suitable to this type of operation, another steel, perhaps 4620, might have been chosen because its high nickel content provides more leeway in quenching temperature. The result would have been higher steel cost. As it was, a reduction in steel cost was used to help amortize the cost of new equipment. It must be emphasized again that the standardization, particularly as it is affected by processing requirements, would probably be different for a different plant.

All standardization is limited by certain rigid requirements that determine the minimum number of grades which may be specified. The maximum number of grades specified is determined by how many compromises may be made in meeting the less rigid requirements to fit steels to the minimum grouping.

Example 2 is an extension of Example 1, involving ten of the same parts. It is designed to illustrate the kind of analysis that is sometimes necessary in deciding how far the reduction in number of grades may be carried for a specific group of parts, or which three or four grades are the most economical in a balance of steel and processing costs.

In this example (Table 11) it is assumed that a hypothetical plant will soon begin to produce the ten parts listed — in conjunction with other parts currently being made in large volume from steels of higher alloy content, such as 1340, 4118, 4140, 4620, 4640 and 8620. Thus, the example is a test of the area where economical standardization leaves off and uneconomical upgrading begins. For the ten "new" parts, five different groups of steels are assumed and allocated to parts as shown in step 1 of Table 11. Groups are:

Case A — 4140 HR and 8620 CF
Case B — 1340 HR, 4640 HR and 4620 CF
Case C — 1045 CF, 4140 CF, 4118 CF and 1020 CF
Case D — 1045 CF, 1117 CF and 1018 CF
Case E — 1045 CF special, 1117 CF and 1018 CF

Then the unit cost of steel is calculated for each part under each assumption (step 2 of Table 11), the unit processing costs are estimated (step 3), unit steel and processing costs are added together (step 4), and the approximate total monthly costs are computed (step 5) on the basis of the monthly production quantities anticipated.

Monthly totals are as follows:

Case A	$72,013
Case B	74,070
Case C	53,413
Case D	48,246
Case E	47,756

This is the sort of trial calculation that a materials engineer may need to undertake before he can be reasonably sure that standardization has been carried to its most economical conclusion. Necessarily, estimates of processing costs in such a trial balance will be inexact, as they are in this hypothetical example, and a difference of 5% or even 10% in total monthly cost between one case and another may not be significant. But the engineer does gain a reasonable assurance as to how far, in his particular problem, standardization can be carried.

In the example being discussed, it is clear that reduction to two steels (case A — 4140 HR and 8620 CF) would be uneconomical, as would the three-steel combination of case B (1340 HR, 4640 HR, 4620 CF). Depending on actual processing conditions, the other three cases may actually be closer to equality than they appear from the first rough analysis. With an apparent difference of $5000 per month at stake among cases C, D and E, it would be prudent for the engineer to refine his assumptions and recalculate these three cases as precisely as possible before making his final decision.

Example 3. Odd sizes frequently carry a cost penalty. Example 3 (Table 12) shows how cost was reduced for flat rolled merchant bar by specifying a standard thickness (0.250 in.) rather than an odd thickness (0.249). The extra usually applies if the quantity ordered is from 3 to 10 tons and if the item deviates slightly from ⅛-in. increments in thickness or about ¼-in. increments in width. In this example, the cost saving was 7%, despite the fact that 1.4% more steel was used.

Example 4. Usually cold finished steels are specified for parts processed on screw machines because of the need for close dimensional tolerances. However, two conditions have now made it possible to use other grades. First, improvements have been made in the automatic machines so that greater dimensional variation can be tolerated, and second, machine straightening of hot rolled bars means that they can sometimes replace cold drawn bars at a substantial saving. Details of a bevel side gear made on an automatic screw machine are given in Table 13. Tool costs increased when hot rolled bars were used, but not enough to offset the saving for material. In addition, one size of cold finished steel was eliminated from stock.

Forging Blanks

Frequently an unnecessarily large number of forging blank sizes are carried in companies with several plants that buy from multiple suppliers. A survey by one company showed 32 sizes of round-cornered squares between 2½ and 9 in. This was reduced to 18 without affecting manufacturing adversely. By using a slightly longer or shorter blank of stock in upset forging it was possible to take advantage of quantity

Table 11. Trial Calculations to Determine the Most Economical of Five Possible Solutions to a Specific Problem of Standardization (Example 2)

Part	Pounds per piece	Pieces per month	Case A	Case B	Case C	Case D	Case E
Step 1. Identity of Steel Selected							
Spindle shaft	3.84	750	4140 HR	1340 HR	1045 CF	1045 CF	1045(a)
Shaft	3.84	9,000	4140 HR	1340 HR	1045 CF	1045 CF	1045(a)
Shaft, clutch drive	6.3	3,650	4140 HR	4640 HR	4140 CF	1045 CF	1045(a)
Shaft, steering	6.4	5,400	4140 HR	4640 HR	4140 CF	1045 CF	1045(a)
Shaft, splined	4.64	900	4140 HR	4640 HR	4140 CF	1045 CF	1045(a)
Oil seal	0.3157	6,000	8620 CF	4620 CF	4118 CF	1117 CF	1117(b)
Roller	0.460	500	8620 CF	4620 CF	4118 CF	1117 CF	1117(b)
Roller, cam follower	0.42	15,000	8620 CF	4620 CF	4118 CF	1117 CF	1117(b)
Shaft, drive gear	2.42	5,600	8620 CF	4620 CF	4118 CF	1117 CF	1117(b)
Shaft, drive	16.83	4,300	8620 CF	4620 CF	1020 CF	1018 CF	1018(b)
Step 2. Cost of Steel per Piece for Above Selections							
Spindle shaft	3.84	750	$0.34	$0.31	$0.37	$0.37	$0.37
Shaft	3.84	9,000	0.34	0.31	0.37	0.37	0.37
Shaft, clutch drive	6.3	3,650	0.565	0.675	0.77	0.63	0.63
Shaft, steering	6.4	5,400	0.57	0.685	0.78	0.635	0.635
Shaft, splined	4.64	900	0.41	0.495	0.55	0.46	0.46
Oil seal	0.3157	6,000	0.038	0.043	0.034	0.03	0.03
Roller	0.460	500	0.057	0.0625	0.05	0.044	0.044
Roller, cam follower	0.42	15,000	0.043	0.048	0.045	0.04	0.04
Shaft, drive gear	2.42	5,600	0.30	0.335	0.263	0.234	0.234
Shaft, drive	16.83	4,300	2.10	2.32	1.48	1.48	1.48
Step 3. Cost of Processing per Piece for Above Selections							
Spindle shaft	3.84	750	$1.34	$1.34	$0.48	$0.48	$0.48
Shaft	3.84	9,000	0.96	0.96	0.37	0.37	0.37
Shaft, clutch drive	6.3	3,650	2.64	2.64	1.46	1.25	1.28
Shaft, steering	6.4	5,400	2.12	2.12	1.19	1.28	1.28
Shaft, splined	4.64	900	3.07	3.07	2.24	1.97	1.97
Oil seal	0.3157	6,000	0.45	0.45	0.41	0.30	0.20
Roller	0.460	500	0.19	0.19	0.17	0.11	0.11
Roller, cam follower	0.42	15,000	0.14	0.14	0.13	0.09	0.09
Shaft, drive gear	2.42	5,600	0.62	0.62	0.58	0.42	0.42
Shaft, drive	16.83	4,300	2.26	2.26	1.94	1.75	1.75
Step 4. Cost of Steel Plus Processing per Piece for Above Selections							
Spindle shaft	3.84	750	$1.68	$1.65	$0.85	$0.85	$0.85
Shaft	3.84	9,000	1.30	1.27	0.74	0.74	0.74
Shaft, clutch drive	6.3	3,650	3.205	3.315	2.23	1.88	1.91
Shaft, steering	6.4	5,400	2.69	2.805	1.97	1.915	1.915
Shaft, splined	4.64	900	3.48	3.565	2.79	2.43	2.43
Oil seal	0.3157	6,000	0.488	0.493	0.444	0.33	0.23
Roller	0.460	500	0.247	0.2525	0.22	0.154	0.154
Roller, cam follower	0.42	15,000	0.183	0.188	0.175	0.13	0.13
Shaft, drive gear	2.42	5,600	0.92	0.955	0.843	0.654	0.654
Shaft, drive	16.83	4,300	4.36	4.58	3.42	3.23	3.23
Step 5. Total Cost per Month (Steel Plus Processing) for Above Selections							
Spindle shaft	3.84	750	$ 1,260	$ 1,238	$ 638	$ 638	$ 638
Shaft	3.84	9,000	11,700	11,430	6,660	6,660	6,660
Shaft, clutch drive	6.3	3,650	11,698	12,100	8,140	6,862	6,972
Shaft, steering	6.4	5,400	14,526	15,147	10,638	10,341	10,341
Shaft, splined	4.64	900	3,132	3,209	2,511	2,187	2,187
Oil seal	0.3157	6,000	2,928	2,958	2,664	1,980	1,380
Roller	0.460	500	124	126	110	77	77
Roller, cam follower	0.42	15,000	2,745	2,820	2,625	1,950	1,950
Shaft, drive gear	2.42	5,600	5,152	5,348	4,721	3,662	3,662
Shaft, drive	16.83	4,300	18,748	19,694	14,706	13,889	13,889
Total cost per month for ten parts			$72,013	$74,070	$53,413	$48,246	$47,756

(a) Cold finished, guaranteed mechanical properties, as stated in Group 3 of Table 10. (b) Cold finished, normal mechanical properties.

Table 13. Standardization to Eliminate One Stock Size (Example 4)

Cold drawn 2⅝-in. bar, 1.92 lb at $0.121	$0.232
Perishable tools	0.010
Cost per piece	$0.242
Hot rolled machine straightened 2 37/64-in. bar, 1.92 lb at $0.100	$0.192
Perishable tools	0.018
Cost per piece	$0.210
Savings per piece using hot rolled bar	$0.032
Savings per year (739,000 pieces)	$23,648.00

Table 12. Savings from the Use of Standard Rather Than Odd Size Thickness for Door Hinge Anchor Plate (Example 3)

(Hot rolled flat bar, merchant quality)

Item	Specified	Proposed
Size of stock	0.249 by 3 by 91⅞ in.	0.250 by 3 by 91⅞ in.
Makes	68 pieces	68 pieces
Weight of bar	19.9302 lb	20.0102 lb
Unit part weight	0.2931 lb	0.2943 lb
Total weight per year	9617 lb	9752 lb
Base price per pound	$0.05775	$0.05775
Size extra	0.009	0.009
Chemistry extra	0.0015	0.0015
Quantity extra	0.010	0.0035
Total price per pound	0.07825	0.07175
Total material cost	$752.53	$699.70
Savings	....	$52.83

break-points and to reduce procurement and inventory costs by 10 to 20%, depending on stock size.

Example 5. Four similar forgings were consolidated into one blank size of 4118 steel, with cost savings because of quantity procurement. Table 14 compares costs of individual mill purchases with combined mill purchases, and shows savings of 4.9% through combined purchases. Multiple lengths for forgings were adjusted to compensate for differences in diameter so that the same weight of material was used.

Example 6. Sometimes price differences for forging blanks are due to differences in the blanks themselves. An example of this is given in Table 15. Although the blanks with the straight shank from source A cost 1% less than those from source B, the cost of the finished part was 1% greater because of the higher cost of turning and material handling. On a basis of 36,000 pieces per year, $1440 was saved by using blanks from source B.

Example 7 (Table 16) points out savings resulting from standardizing blanks so that one set of dies will make two or more upset forgings. This is particularly advantageous when one part, such as transmission driving shaft A in Table 16, is produced in small quantities. Savings were 3.4%.

Example 8 shows savings realized when two parts, similar in shape but slightly different in size, were standardized to the same steel. In addition to the reduced cost because of using a lower-cost steel and buying in larger quantities, there was a large decrease in the cost of heat treating.

Originally the upper ball joint studs (see sketch in Table 17) were made from 8617 steel, carburized to a case depth of 0.010 to 0.020 in. They were cooled slowly, reheated to 1575 F, oil quenched and tempered at 400 F. The threaded end and tapered section were tempered to a maximum of Rockwell C 37. The shank and head remained at a minimum of Rockwell C 58. The lower ball joint studs were made from 8640 steel, hardened and tempered to Rockwell C 32 to 36 and the head was induction hardened to a minimum of Rockwell C 58.

Satisfactory results from both extensive laboratory and field testing brought about a change to 1041 steel for both parts. The steel was carbonitrided at 1575 F for a case depth of 0.010 to 0.020 in. and tempered at 900 F. Minimum surface hardness was then Rockwell C 50 and core hardness Rockwell C 30 to 35.

Heat treating equipment for the 1041 steel occupied less space and the process was better adapted to automation. Costs of heat treating equipment for the individual and combined processes are shown in Table 17.

Example 9 is a problem growing out of failure to review standards for parts already in production or being produced in other company plants when a new part was specified. Three models of an assembled product specified bolts of the same diameter and thread size, but of different lengths. An analysis of requirements showed that the same bolt could be used on all three models. Cost savings of about 5% are shown in Table 18.

Plate, Sheet and Strip

Sizes of flat rolled steel can frequently be standardized to avoid paying extras for width, thickness and length. With greater widths or lengths there will be an added cost for shearing or slitting. While this cost may vary with the number of smaller widths or lengths that are required for fabrication, it is usually small compared with the savings from the lower cost per pound for the steel. Typical base price and cost extras for cold rolled and hot rolled low-carbon steel sheet are listed in Table 19 on the next page; specific examples are given in Tables 20, 21 and 22.

Example 10. In the upper part of the tabulation of Table 20, cost was reduced $18.00 a ton by purchasing multiple widths of ¾-in. plate. In the lower portion, savings realized by purchasing multiple *lengths* of similar plate are $9.00.

Example 11 shows results of merging specifications for four different parts into a common size of sheet steel. Despite the fact that 8% more steel was required for making the four parts, a 30% saving was realized (Table 21).

Example 12. Narrow cold rolled strips were eliminated by standardizing on the required thickness of multiple-width cold rolled sheet purchased in coils. After deducting the cost of slitting, this specific standardization netted a saving of 12.8% (Table 22).

Design Changes

Minor design changes can often be made to permit standardization. A critical review of existing parts before release of newly designed parts also helps to achieve and maintain standardization. The following two examples are typical:

Example 13 illustrates an instance where a manufacturer was buying two almost identical wire sizes (0.0475 and 0.049-in. diam) for use on two different products. By making a slight design change, the 0.0475-in. wire could be used for both products. Despite the fact that direct saving was only $0.05 per 100 lb (½%), procurement and inventory costs

Table 14. Standardization of Forging Stock Sizes of 4118 Hot Rolled Alloy Steel to Gain Economy in Quantity Purchases (Example 5)

	Individual mill purchase				Combined mill purchase				
					Size, diam				
Part	Ideal size, in. diam	Weight purchased, lb	Cost per 100 lb	Total cost	Permissible, in. Min	Max	Standard, in.	Cost per 100 lb	Total cost
Bevel pinion A	3	87,700	$ 8.825	$7739.53	2¾	3¼	2 15/16	$8.775	$7695.68
Bevel pinion B	2 15/16	2,260	13.185(a)	297.98	2¾	3¼	2 15/16	8.775	198.32
Driving gear	2¾	2,939	13.185(a)	387.51	2⅝	3½	2 15/16	8.775	257.90
Sliding gear	2⅞	3,689	13.185(a)	486.39	2¾	3½	2 15/16	8.775	323.70
		96,588		$8911.41					$8475.60
Savings by combining sizes									$ 435.81
Savings per 100 lb									$ 0.451

(a) Increased cost due to low quantity purchase

Table 15. Standardization of Drive Shaft Forging Blanks (Example 6)

Cost item	Source A	Source B
Material	$ 4.02	$ 4.06
Turning	0.173	0.123
Additional material handling	0.030	...
	$ 4.223	$ 4.183

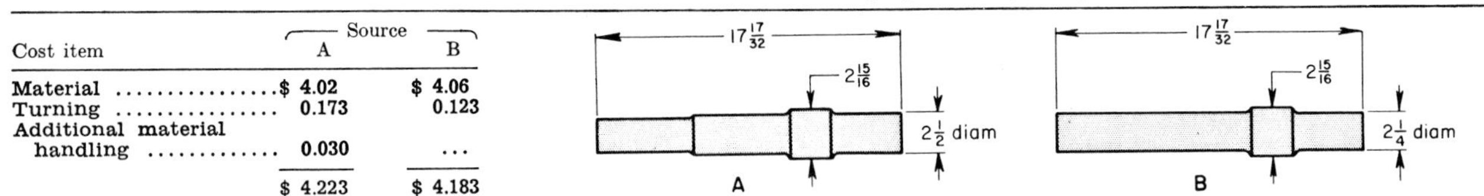

Table 16. Savings Resulting from Combining Two Upset Blanks to Permit Quantity Forging of 4118H Hot Rolled Steel with One Die Setup (Example 7)

		Cost					
Part upset	Quantity	Die	Setup	Die reconditioning	Die and setup, 100 pieces	Forging labor, 100 pieces	Total, 100 pieces
Type A transmission driving shaft	790 pieces (13.5 lb per piece)	$2000.00	6.46 (one setup)	$ 25.00	$2.57	$6.77	$9.34
Type B transmission driving shaft	7160 pieces (13.5 lb per piece)	2000.00	38.76 (six setups)	150.00	0.31	6.77	7.08
Combined total	7950 pieces	4000.00	45.22	175.00	0.53	6.77	7.30
One Die and Setup for Both Blanks							
Type A and B transmission driving shafts	7950 pieces	2000.00	45.22 (seven setups)	175.00	0.28	6.77	7.05
Savings resulting from combining two blanks in one die, per 100 pieces							$0.25

were reduced, and reduction in wire diameter meant that an additional 255 ft was obtained each month at no cost (Table 23).

Example 14. A review of bearing cages already in production revealed that a bearing cage from a previous model would fit the requirements of a new model, producing savings of 34% per cage and totaling $14,760 annually (36,000 pieces). Savings accrued from larger-volume purchases, larger machining quantities, and appreciable decrease in handling costs.

Table 18. Standardization of Cold Heading Steel for Bolts to Gain Economy (Example 9)

Size, in.	Monthly usage	Monthly cost
Original Sizes Used		
7/16 by 14 by ⅝	100,000	$1000.00
7/16 by 14 by ⅞	35,000	385.00
7/16 by 14 by 1	4,500	53.46
Total	139,500	$1438.46
New Standardized Size		
7/16 by 14 by ¾	139,500	$1395.00
Savings in material cost		43.46
Savings in inventory and handling		5.50
Savings in direct labor		35.00
Total savings per month		$ 83.96

Table 20. Standardization of Plate Sizes (Example 10)

Size required	*Width/thickness extra, $ per ton*
Width of Plate	
¾ by 21 by 250 in.	$ 21.00
Order in four multiple widths, ¾ by 84 by 250 in.(a)	3.00
Savings per ton	$ 18.00
Length of Plate	*Length extra, $ per ton*
¾ by 21 by 21 in.	$ 10.00
Order in five multiple lengths, ¾ by 21 by 105 in.(a)	1.00
Savings per ton	$ 9.00

(a) Shearing cost, although small, should be deducted from savings.

Table 22. Standardization of Sash Panel Steel (Example 12)

Weekly requirements, lb. (in coils)	70,000
CR strip 0.043 by 2 in. (cost includes all extras)	$ 6055
CR sheet 0.043 by 36 in. (cost includes all extras and slitting)	5075
Savings per week by standardizing on sheet	$ 980

Table 23. Standardization of Two Wire Sizes into One (Example 13)

Cost of 2500 lb, 0.0475-in. diam	$ 250.00
Cost of 2500 lb, 0.049-in. diam	250.00
Total cost per month	$ 500.00
After standardization by design change:	
Cost of 5000 lb, 0.0475-in. diam	497.50
Direct saving on 5000 lb(a)	$ 2.50

(a) Additional advantages consisted of reduction of procurement and inventory costs and the obtaining of an extra 255 ft of wire in the 5000 lb.

Table 17. Standardization of Upper and Lower Studs to Reduce the Cost of Heat Treating (Example 8)

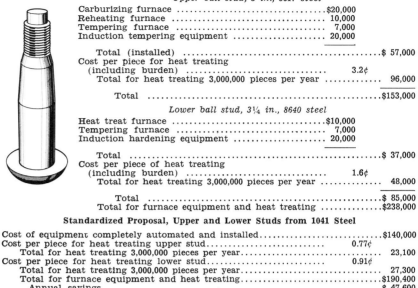

Original Proposal		
Upper ball stud, 3 in., 8617 steel		
Carburizing furnace	$20,000	
Reheating furnace	10,000	
Tempering furnace	7,000	
Induction tempering equipment	20,000	
Total (installed)		$ 57,000
Cost per piece for heat treating (including burden)	3.2¢	
Total for heat treating 3,000,000 pieces per year		96,000
Total		$153,000
Lower ball stud, 3¼ in., 8640 steel		
Heat treat furnace	$10,000	
Tempering furnace	7,000	
Induction hardening equipment	20,000	
Total		$ 37,000
Cost per piece of heat treating (including burden)	1.6¢	
Total for heat treating 3,000,000 pieces per year		48,000
Total		$ 85,000
Total for furnace equipment and heat treating		$238,000
Standardized Proposal, Upper and Lower Studs from 1041 Steel		
Cost of equipment completely automated and installed		$140,000
Cost per piece for heat treating upper stud	0.77¢	
Total for heat treating 3,000,000 pieces per year		23,100
Cost per piece for heat treating lower stud	0.91¢	
Total for heat treating 3,000,000 pieces per year		27,300
Total for furnace equipment and heat treating		$190,400
Annual savings		$ 47,600

Table 19. Range of Typical Costs for Low-Carbon Steel Sheet

Cost item	Cold rolled			Hot rolled		
	26 gage (0.018 in.)	20 gage (0.036 in.)	16 gage (0.0625 in.)	16 gage (0.0625 in.)	12 gage (0.105 in.)	Plate (0.250 in.)
Base price, $ per 100 lb	6.275	6.275	6.275	5.10	5.10	5.30
Extras for Quality Class						
Drawing quality	0.25	0.25	0.25	0.25	0.25	0.35
Drawing quality, killed	0.75	0.75	0.75	0.65	0.65	0.80
Drawing quality, annealed	....	....	....	0.75	0.75	1.35
Extras, Same for All Quality Classes						
Gage and width(a)	0.85	0.45	0.45	0.80	0.80	1.30
Length(a)	0.40	0.30	0.20	0.45	0.35	0.05
Pickling and oiling	....	....	....	0.55	0.45	0.90
Packaging	0.025	0.025	0.025	0.025	0.025	0.10

NOTE: Quality extras are not included and do not apply to mill shipments greater than 40,000 lb. Cost of freight, shearing to specified size, and other extras are not included. Costs are typical as of September 1958.

(a) Mandatory extras, all other optional; width extras for sheet 36 to 48 in. wide. length extras for cold rolled sheet 60 to 180 in. long and hot rolled sheet 60 to 240 in. long.

Table 21. Cost Analysis of Ideal Size Versus Common Size of Cold Rolled Commercial Quality Sheet Steel (Example 11)

Part	Size, in.	Bimonthly requirements, lb	Price per lb	Cost
Ideal				
Door hinge support	0.047 by 36⅛ by 115½	2,532	$0.1055	$267.13
Finishing panel	0.047 by 44 by 199	4,723	0.1055	498.28
Spring support	0.047 by 46 3/16 by 105¼	5,346	0.1045	558.66
Brace	0.047 by 47 by 69⅜	9,059	0.1045	946.66
Total		21,660		$2270.73
Common				
Door hinge support	0.047 by 41 by 122	2,925	....	....
Finishing panel	0.047 by 41 by 122	5,168	....	....
Spring support	0.047 by 41 by 122	5,967	....	....
Brace	0.047 by 41 by 122	9,284	....	....
Total		23,344	0.0675	1575.72
Savings for 2 months				695.01
Savings on yearly production				$4170.06

STEELS FOR LOW-TEMPERATURE SERVICE

By E. GARY MARSHALL

Carbon and alloy grades provide the high strength, ductility, and toughness in vehicles, vessels, and structures which must serve at −50 F and lower.

BECAUSE A NUMBER of steels are engineered specifically for service at low temperature (down to −150 F), selecting the optimum material calls for thorough understanding of the application and knowledge of the mechanical properties that each grade provides. Typically, these steels are used for containment, handling, and transporting of liquefied gas. Other applications include stationary structures and mobile equipment exposed to adverse climates or operating conditions or both. Properties (yield and tensile strengths, fatigue limit, ductility, and notch toughness) must, of course, fit the requirements of the process — pressure, temperature, and any cyclic stresses.

Toughness: an Important Asset

When designing low-temperature systems or equipment, the engineer finds that notch toughness ranks high in importance. Defined as the ability to resist brittle fracture at high stresses, such as can be caused by impact loading, notch toughness is measured by various means. The favored method, recommended by the American Petroleum Institute as well as the ASTM and ASME, is the Charpy V-notch test. It is considered the most appropriate because a part or structure will generally fail due to a notch or other stress concentration. Test results, given in foot-pounds, measure the steel's capacity to absorb energy, and thus signify its ability to resist failure at points of local stress concentration.

Using the Charpy test, we can determine the transition temperature, that at which material becomes brittle rather than ductile. This information helps the designer choose a steel that will remain ductile through the range of temperatures or stresses it will meet in service.

Fatigue limit of a steel also must be considered. At low temperatures, systems are usually subjected to dynamic loads, and structural members, to cyclic stresses. Examples include pumps and vessels that frequently undergo pressure changes, and large structures and mobile equipment that experience extreme stresses imposed by packed snow or high winds. Other considerations include heat conductivity and thermal expansion.

Carbon Steels Grow in Number

Several years ago, only two carbon steels were available for low-temperature service to −50 F, ASTM A201

Mr. Marshall is supervisor, Metallurgical Product Development, Lukens Steel Co., Coatesville, Pa.

Tank cars for low-temperature service must be constructed of steels designed for such use. This one is made of TC-128 Grade B, a fine-grain steel with 81,000 psi min tensile strength.

and A212. (These are now included in ASTM's A516 grades.) Today, four more steels are produced to low-temperature ASTM or proprietary specifications, some of which provide service to −75 F. Less costly than alloy steels, they combine better weldability, greater toughness, and higher strength with low coefficients of thermal expansion and thermal conductivity.

The ASTM A516 grades, probably the most frequently used carbon steels, have tensile strengths ranging from 55,000 to 85,000 psi minimum. These steels listed in Table I are used widely in air liquefaction plants, refrigerating installations, transport equipment, and containment vessels operating down to −50 F. For these applications, the steel is normally made to meet impact test requirements of ASTM A300 Class I specification, which calls for plates to be normalized and to meet a Charpy keyhole minimum of 15 ft-lb at −50 F.

Figure 1 shows how V-notch impact strengths of five normalized low-temperature steels compare. Note that A516 Grade 70, though it offers the least impact resistance, meets requirements for service to −50 F. Figure 2, which relates thickness to the nil-ductility transition temperature (NDTT), shows that resistance to brittle fracture lessens with increasing gage.

How Carbon Grades Compare

The big advantage of A516 steels is their low initial cost. They also feature the lowest ASME allowable stresses, 13,750 to 17,500 psi. (As set by the ASME, Section VIII, Div. 1, allowable stresses equal 25% of the tensile strengths.) Thus, a given design strength requires heavier gages than are needed with high-strength steels.

Compared with A516 steels, those in the A442 class have higher carbon and manganese in plates less than 1 in. thick, and higher carbon and lower manganese beyond 1 in. However, applications for A516 Grades 55 and 60 duplicate those of A442. They are somewhat easier to fabricate than the A442 grades because carbon contents are lower.

Higher strength with good notch toughness is available in carbon steels such as the two listed ASTM A537 grades. Grade B, for example, is a quenched and tempered steel which provides 60,000 min psi yield strength, plus 15 ft-lb of impact strength (Charpy V-notch) at −75 F. Listed in Table I is Lukens LT-75, a steel that meets A537 Grades A and B standards. It can be either normalized or quenched and tempered to raise yield and tensile strengths and impact toughness beyond those of the A516's. Normalized, Lukens LT-75 (A537 Grade A) bridges the toughness gap between that of A516 Grade 70 and those of the two A203 Grades A and B. Furthermore, the typical nil-ductility transition temperature is −60 to −100 F.

Of primary concern to fabricators is the effect of straining – the result of cold forming – on material properties. Tests of A537 Grade B revealed that 7% strain, even without stress relief, reduced the NDTT to −40 F and the impact toughness to

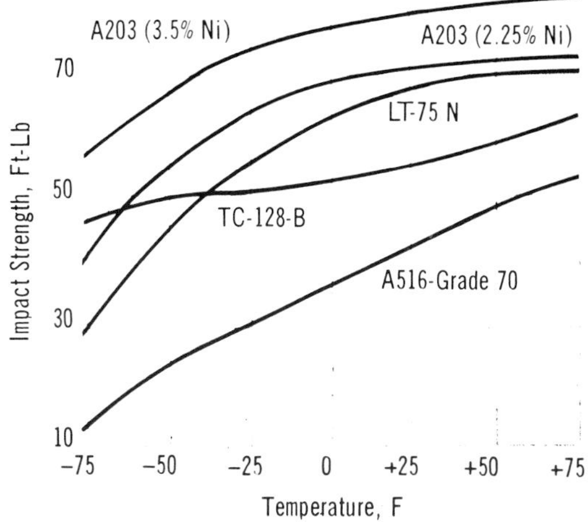

Fig. 1 — High-strength steels vary in their ability to retain toughness to low temperatures. Determinations of impact strength, made by Charpy V-notch tests of normalized stock, indicate that nickel improves toughness appreciably.

Table I — Specifications for Low-Temperature Steels

Designation	Lowest Usual Service Temperature, F	Min Yield Strength, Psi	Tensile Strength, Psi	ASME Allowable Stress, Psi*	Min Elongation, % 2 in.	Min Elongation, % 8 in.	Uses
ASTM A442—Gr. 55	—50	30,000	55,000-65,000	13,750	26	21	Welded pressure vessels and storage tanks; refrigeration; transport equipment.
—Gr. 60	—50	32,000	60,000-72,000	15,000	23	20	
ASTM A516—Gr. 55 †	—50	30,000	55,000-65,000	13,750	27	23	
—Gr. 60 †	—50	32,000	60,000-72,000	15,000	25	21	
—Gr. 65 †	—50	35,000	65,000-77,000	16,250	23	19	
—Gr. 70 †	—50	38,000	70,000-85,000	17,500	21	17	
AAR TC-128—Gr. B †	—50	50,000	81,000-101,000	20,250	21	18	Railroad tank cars.
ASTM A517—Gr. F	—50	100,000	115,000-135,000	28,750	16	—	Highly stressed vessels; tank trucks for LP gases.
Lukens LT-75-HS(≤¾ in.)	—75	75,000	95,000-115,000	23,750	19	—	Offshore drilling platforms; low-temperature environment structures; earthmoving equipment; storage tanks.
ASTM A537—Gr. A (≤ 1¼ in.)	—75	50,000	70,000-90,000	17,500	22	18	
—Gr. B (≤ 1¼ in.)	—75	60,000	80,000-100,000	20,000	22	—	
ASTM A203—Gr. A	—75	37,000	65,000-77,000	16,250	23	19	Tanks, vessels, and piping for liquid propane.
—Gr. B	—75	40,000	70,000-85,000	17,500	21	17	
—Gr. D	—150	37,000	65,000-77,000	16,250	23	19	Land-based storage of liquid propane, carbon dioxide, acetylene, ethane, and ethylene.
—Gr. E	—150	40,000	70,000-85,000	17,500	21	17	
ASTM A533—Class 1	—100	50,000	80,000-100,000	20,000	18	—	Nuclear reactor vessels where low ambient toughness required for hydrostatic testing; some chemical and petroleum equipment.
—Class 2	—100	70,000	90,000-115,000	22,500	16	—	
—Class 3	—100	82,500	100,000-125,000	25,000	16	—	
ASTM A543—Class 1	—160	85,000	105,000-125,000	26,250	14	—	Candidate material with high notch toughness for heavy-wall pressure vessels.
—Class 2	—160	100,000	115,000-135,000	28,750	14	—	

*ASME Section VIII, Div. 1. †To ASTM A300 specification.

30 ft-lb (from a 45 ft-lb unstrained level at —75 F). Both are acceptable for low-temperature service, however.

Other Steels Are Developed

When Lukens engineers evaluated notch-tough steels, they found a need for intermediate-strength weldable plate steels combining good notch toughness with economy. To meet these criteria, they developed two steels: Lukens Penstock Steel and Lukens LT-75 HS. Both are modifications of the ASTM A537 specification, as Table II indicates. The basic approach was to optimize the carbon-manganese balance so as to raise the yield strength without impairing ductility.

Normalized, Lukens Penstock Steel provides 70,000 psi minimum tensile strength in 4 in. gages, with Charpy V-notch impact strength of 30 ft-lb at +10 F. As the name implies, this tough steel is used in hydroelectric applications. The Lukens LT-75 HS plate steel has a minimum yield strength of 75,000 psi in thicknesses up to ¾ in., and a minimum impact strength of 15 ft-lb at −75 F for all thicknesses up to 1½ in. inclusive. Also, this quenched and tempered steel has NDTT's of −80 to −130 F, measured in plate up to 1 in. in thickness.

Large tonnages of Lukens LT-75 HS have already been applied in construction equipment such as crane and shovel booms. It is also useful in other mobile equipment and in stationary structures which have to operate in subzero temperatures under extreme stress and loading conditions. In fact, the steel can be applied where full-alloy strengths do not provide enough rigidity, and the lower-strength carbon steels would call for thicker plates (thus weight would be excessive).

For rail tank cars which may encounter low temperatures, the American Assn. of Railroads has approved TC-128 Grade B, an 81,000 to 101,000 psi tensile strength steel. (On p. 92, a typical application is pictured.) The high forces generated in highballing, starting, stopping, coupling, and uncoupling make it necessary to have more strength than that available in the A516 grade. Use of this steel eliminates the need for complicated extrusions, structural shapes, supports, and stiffeners in tank cars. Because strength is higher, thinner

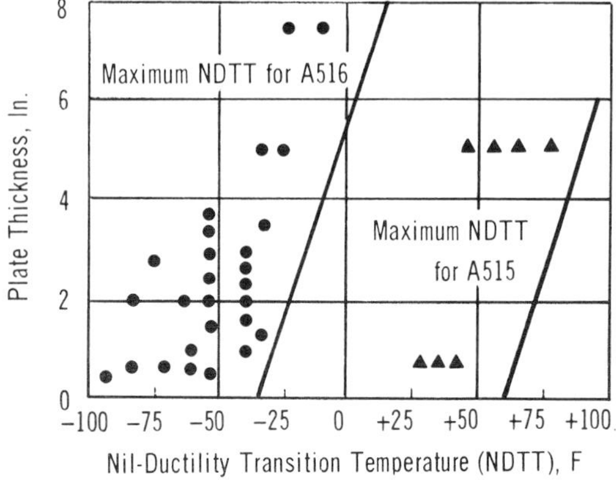

Fig. 2 — In general, the thicker the plate of a given grade, the lower its toughness. Toughnesses and transition temperatures were determined by the drop weight test, as defined by ASTM E208.

Table II — Properties of Low-Temperature Carbon and Alloy Steels

Designation	Composition, % Max C	Mn	Si	Condition*	Min Yield Strength, Psi	Tensile Strength, Psi	Min Elongation, %	Bend Ratio Diam./Thick	Weldability for ½ In. Gage†
Carbon Steel									
ASTM A442—Gr. 55									
1 in. and under	0.22	0.80-1.10‡	‡	AR	30,000	55,000-65,000	26	1	a
Over 1 to 1½ in. incl.	0.24	0.60-0.90	0.15-0.30	N	30,000	55,000-65,000	26	1½	
—Gr. 60									
1 in. and under	0.24	0.80-1.10‡	‡	AR	32,000	60,000-72,000	23	1	a
Over 1 to 1½ in. incl.	0.27	0.60-0.90	0.15-0.30	N	32,000	60,000-72,000	23	1½	
ASTM A516—Gr. 55									
½ in. and under	0.18	0.60-0.90	0.15-0.30	N	30,000	55,000-65,000	27	½§	a
Over ½ to 2 in. incl.	0.20	0.85-1.20	0.15-0.30	N	30,000	55,000-62,000	27	1	
—Gr. 60									
½ in. and under	0.21	0.60-0.90	0.15-0.30	N	32,000	60,000-72,000	25	1§	a
Over ½ to 2 in. incl.	0.23	0.85-1.20	0.15-0.30	N	32,000	60,000-72,000	25	1½	
—Gr. 65									
½ in. and under	0.24	0.85-1.20	0.15-0.30	N	35,000	65,000-77,000	23	1½§	b, c
Over ½ to 2 in. incl.	0.26	0.85-1.20	0.15-0.30	N	35,000	65,000-77,000	23	2	
—Gr. 70									
½ in. and under	0.27	0.85-1.20	0.15-0.30	N	38,000	70,000-85,000	21	2	b, d
Over ½ to 2 in. incl.	0.28	0.85-1.20	0.15-0.30	N	38,000	70,000-85,000	21	2	
AAR TC-128—Gr. B									
To ¾ in. thick	0.25	1.35 max	0.30 max	N	50,000	81,000-101,000	21	2	b, c
Over ¾ to 1 in. incl.	0.25	1.50 max	0.50 max	N	50,000	81,000-101,000	21	2	
Lukens LT-75-HS									
To ¾ in. thick	0.22	1.10-1.60	0.20-0.60	QT	75,000	95,000-115,000	19	2	e
Over ¾ to 1½ in. incl.	0.22	1.10-1.60	0.20-0.60	QT	70,000	90,000-110,000	19	3	
ASTM A537—Gr. A									
To 1¼ in. thick	0.24	0.70-1.35	0.15-0.50	N	50,000	70,000-90,000	22	1½	b, c
Over 1¼ to 2 in. incl.	0.24	0.70-1.35	0.15-0.50	N	46,000	65,000-85,000	22	2	
—Gr. B									
To 1¼ in. thick	0.24	0.70-1.35	0.15-0.50	QT	60,000	80,000-100,000	22	2	b, c
Over 1¼ to 2 in. incl.	0.24	0.70-1.35	0.15-0.50	QT	56,000	75,000-95,000	22	2½	
Lukens Penstock									
To ¾ in.	0.25	0.90-1.35	0.15-0.30	N	50,000	70,000 min	23	1	
¾ to 2 in.	0.25	0.90-1.35	0.15-0.30	N	50,000	70,000 min	24	1½-2½	b, c
2 to 4 in.	0.25	0.90-1.35	0.15-0.30	N	42,000	70,000 min	24	3	

Alloy Steels (for plates up to 2 in. thick)	C	Mn	Si	Ni	Cr	Mo	B	V	Cu	Condition*	Min Yield Strength, Psi	Tensile Strength, Psi	Min Elongation, %	Bend Ratio Diam./Thick	Weldability for ½ In. Gage†
ASTM A517—Gr. F	0.10-0.20	0.60-1.00	0.15-0.35	0.70-1.00	0.40-0.65	0.40-0.60	0.002-0.006	0.03-0.08	0.15-0.50	QT	100,000	115,000-135,000	16	2-3	e
ASTM A203—Gr. A	0.17 max	0.70 max	0.15-0.30	2.10-2.50	—	—	—	—	—	N	37,000	65,000-77,000	23	1-1½	e
—Gr. B	0.21 max	0.70 max	0.15-0.30	2.10-2.50	—	—	—	—	—	N	40,000	70,000-85,000	21	1½-2	e
—Gr. D	0.17 max	0.70 max	0.15-0.30	3.25-3.75	—	—	—	—	—	N	37,000	65,000-77,000	23	1-1½	e
—Gr. E	0.20 max	0.70 max	0.15-0.30	3.25-3.75	—	—	—	—	—	N	40,000	70,000-85,000	21	1½-2	e
ASTM A533—Gr. A	0.25 max	1.15-1.50	0.15-0.30	—	—	0.45-0.50	—	—	—	QT	Class 1 50,000	80,000-100,000	18	3½	f
—Gr. B	0.25 max	1.15-1.50	0.15-0.30	0.40-0.70	—	0.45-0.60	—	—	—	QT	Class 2 70,000	90,000-115,000	16	3½	f
—Gr. C	0.25 max	1.15-1.50	0.15-0.30	0.70-1.00	—	0.45-0.60	—	—	—	QT	Class 3 82,500	100,000-125,000	16	3½	f
—Gr. D	0.25 max	1.15-1.50	0.15-0.30	0.20-0.40	—	0.45-0.60	—	—	—	QT					
ASTM A543	0.23 max	0.40 max	0.20-0.35	2.60-3.25	1.50-2.00	0.45-0.60	—	0.03 max	—	QT	Class 1 85,000	105,000-125,000	14	3½	f
											Class 2 100,000	115,000-135,000	14	3½	f

*AR, as rolled; N, normalized; QT, quenched and tempered.
†a—No preheat or special electrodes; no welding below +32 F.
b—Low hydrogen electrodes; no welding below +10 F.
c—100 F minimum preheat with other than low hydrogen electrodes.
d—200 F minimum preheat with low hydrogen electrodes.
e—200 F minimum preheat with low hydrogen electrodes.
f—Special considerations to strength level, design, gage.
‡May be same as for over 1 in., if steel is made to fine grain practice.
§For 1 in. and under.

plates can be used so that tanks hold larger volumes for their weight.

The American Assn. of Railroads allows use of as-rolled or normalized TC-128 steel, since both forms provide adequate toughness at 81,000 psi. Normalizing, however, gives greater uniformity in tensile and yield strength, and impact strength is much higher at −50 F (Fig. 3).

Alloy Steels Are Even Stronger

Of the low-temperature alloy steels listed in Table II, A517 Grade F has the highest allowable stresses. A low-carbon alloy steel, it combines minimum yield strength of 100,000 psi with toughness and weldability. At −50 F, its impact strength (Charpy V-notch) is 40 ft-lb, and its notch and crack resistance are sufficient to encourage wide usage. Since its NDTT is −30 to −90 F, this steel performs excellently at low temperatures. Typical applications include earthmoving and transportation equipment subject to seasonal low temperatures; missile and aircraft ground handling equipment; tank cars and trucks for LP gas and propane.

When service requirements call for a lower transition temperature, as needed in the storage, piping, and handling of liquid propane at −44 F, designers apply ASTM A203 Grades A and B, for example. Though both have 2.25% Ni, Grade B is more popular because it has a higher ASME maximum allowable design stress, 17,500 psi compared with Grade A's 16,250 psi. Determinations of NDTT on normalized 2.25% Ni steels read from −70 F downward. Quenching and tempering lower the NDTT to below −90 F.

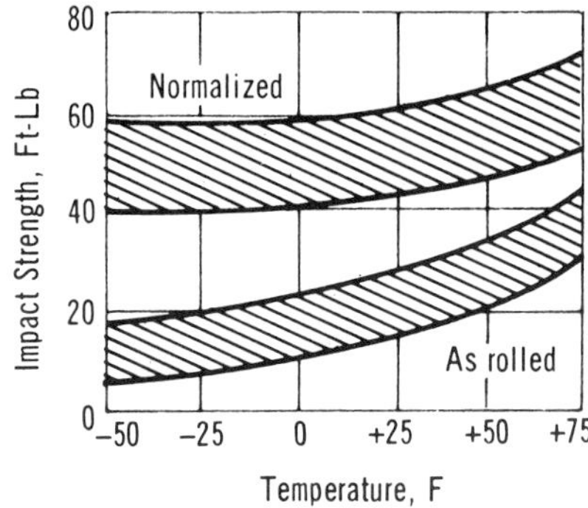

Fig. 3 — In comparison with as-rolled stock, normalized TC-128 Grade B steel has much higher impact strength (Charpy V-notch) down to −50 F. The steel has been accepted by the American Assn. of Railroads for tank cars and similar applications.

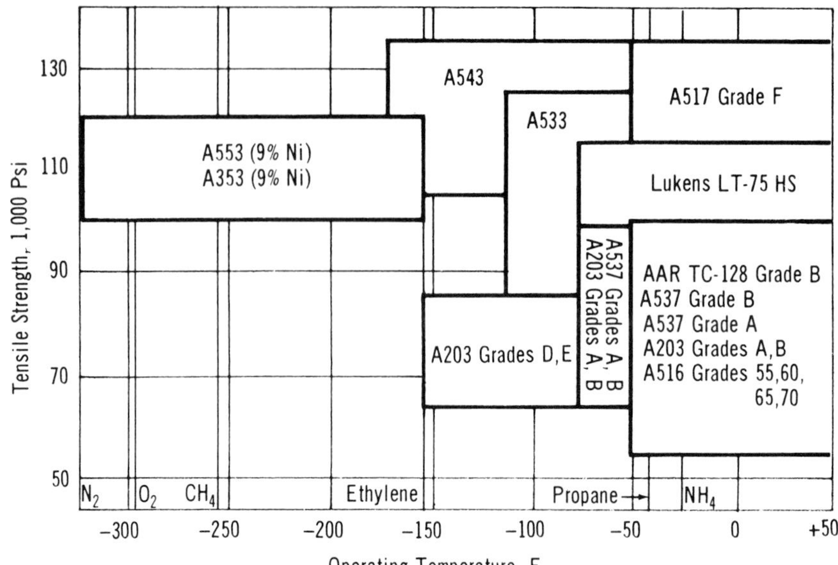

Fig. 4 — Steels listed in the blocks can be used at the indicated tensile strengths and temperatures. Points along bottom scale indicate liquefying temperatures of gases. Once the engineer determines the operating temperature of the part in question and converts the design stresses to tensile strengths required of the construction material, he can find, on this chart, those grades of steel meeting the requirements of the application. From among those steels, then, he chooses that particular grade which offers the best economics.

Grades D and E of ASTM A203 contain 3.5% Ni, an amount which lowers the transition temperature to below −150 F, the so-called upper level of cryogenic applications. For these two grades, NDTT values start at −100 F normalized, and drop to under −165 F in quenched and tempered stock.

Each of the A203 grades provides high impact strength at low temperature. Their formability, weldability, fatigue, and other design characteristics are well-known through experience. Of course, they cost more than carbon steels, but the premium buys increased corrosion resistance along with improved toughness.

Because they provide extra high tensile strengths and very low transition temperatures, ASTM A533 and A543 are popular for nuclear pressure vessels. A Mn-Mo-Ni steel, A533 has an impact strength of 15 ft-lb to −100 F for tensile strengths of 80,000 to 125,000 psi. Offering high yield strengths at those tensile strengths is the Ni-Cr-Mo steel, A543. This steel provides better than 35 ft-lb of impact strength (Charpy V-notch) at −50 F or 15 ft-lb at −180 F.

Choosing a Steel

Since a variety of low-temperature steels are available, the engineer must consider the advantages each has to offer according to the application. The cost-strength ratio is but one factor; others, such as welding and fabricating costs, have equal or greater bearing on final costs. Where extra high strength and good impact properties to below −75 F are called for, an alloy steel should be chosen. However, the chief low-temperature steels today are the heat-treated carbon grades. Besides offering excellent low-temperature toughness plus fabricability, these grades are lower in initial cost.

Using A516 Grade 70 as a base equal to 100, the relative cost per unit of tensile strength for A516 Grade 55 is 136; of A537 Grade A, 98; of A537 Grade B, 98; of LT 75-HS, 95; and of AAR TC-128 Grade B, 94. With this information and that in the selection guide (Fig. 4), the engineer can better relate a steel's properties to the product performance requirements.

Pipeline Steel for Low Temperature Uses

By JOHN L. MIHELICH and J. H. SMITH

Combining strength, toughness, and weldability, Arctic Grade steel is a hot-rolled material which features 70,000 to 100,000 psi yield strengths. Applications: automotive, railroad, heavy equipment, construction and ship building.

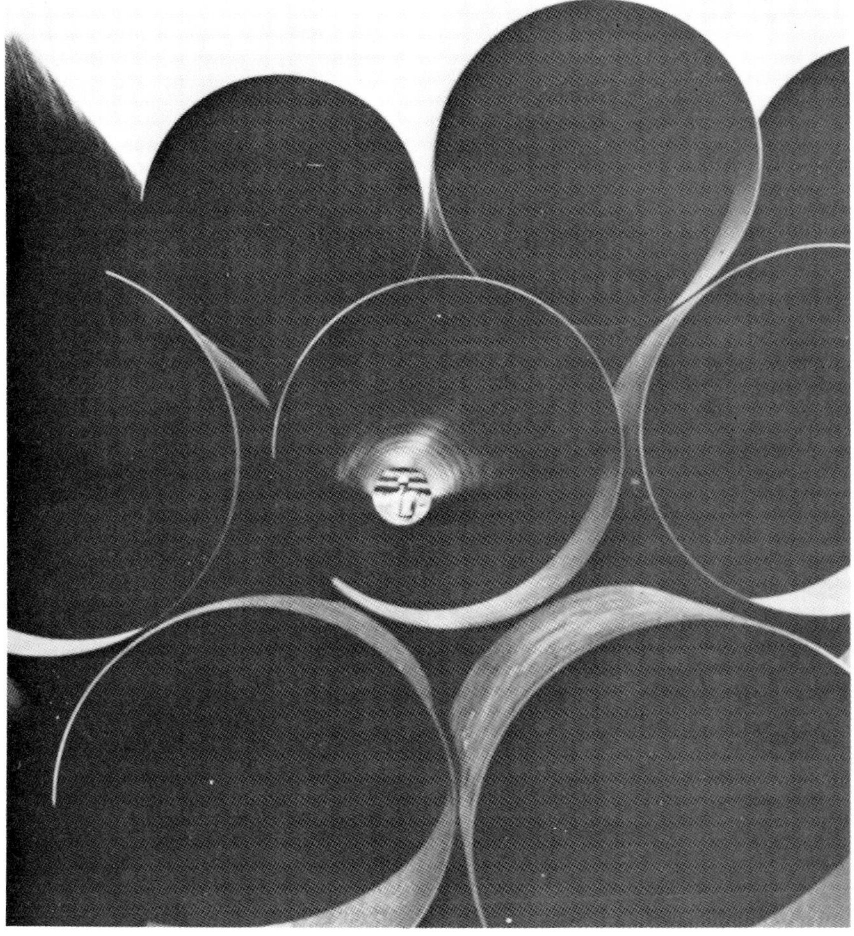

Spirally welded X-70 pipe of Arctic Grade steel is used in a natural gas pipeline in Canada. The steel is alloyed with molybdenum and columbium.

Pipelines 42 in. in diameter, now carrying natural gas through 240 miles of central Canada, may be forerunners of a giant, 2,500-mile pipeline system connecting Alaska's vast natural gas resources with industrial markets in the eastern United States.

These pipelines are made of a steel developed to meet the requirements of subzero arctic conditions.

Mr. Mihelich is manager, High Strength Steel Development, Climax Molybdenum Co., Pittsburgh; and Mr. Smith is chief metallurgist, Interprovincial Steel & Pipe Corp., Regina, Saskatchewan.

Called Arctic Grade steel, it features high strength, exceptional toughness, and excellent weldability. The first commercial application of the manganese-molybdenum-columbium acicular ferrite concept, the steel contains 0.05 C, 1.60 Mn, 0.05 Si, 0.25 Mo, and 0.06 Cb.

Some Development Problems

Steel for natural gas pipelines must meet more exacting requirements than that used for oil. For example, they carry compressed gas at −10 F to +25 F, making crack growth and brittleness a problem in the severe arctic environment. Achieving low-temperature notch toughness, grain size control, and low sulfur content were among major problems in developing the steel, particularly since economic feasibility had to be considered.

Because hot-rolled steels presented a good opportunity to cut both cost and weight if the cost per unit strength could be reduced, the research team began with them. (Substantial savings in transportation and welding costs can be achieved by increasing the allowable stress per unit of pipe weight.)

Knowing that molybdenum exerts a powerful influence in delaying the transformation of austenite to polygonal ferrite and pearlite, the researchers developed the concept of a manganese-molybdenum-columbium formulation. While all three of these major alloying elements contribute to this transformation control, molybdenum and manganese are most effective in producing a stronger microstructure of acicular ferrite. Because it forms at relatively low temperature, acicular ferrite is fine-grained and highly substructured. A low transformation temperature also precludes the rapid precipitation of columbium carbonitride, which occurs during transformation at high temperatures.

Transformation to acicular ferrite at low temperature involves cooling through the lower part of the cooling curve for columbium carbonitride precipitation. Under these conditions, precipitation is only partially completed. Enough columbium remains in solution so that the Mn-Mo-Cb steel can be reheated and age hardened, if necessary, under controlled conditions.

Moreover, low temperature transformation leads to homogeneity of properties throughout large plates or coils, which is a significant advantage. The net result is a fine-grained steel of high dislocation density, capable of high yield strength and toughness with excellent weldability

How the Composition Works

A yield strength of 80,000 psi in thick plate and coiled steel had been set as an objective. To achieve this aim, along with good toughness and weldability, the carbon level was kept low – to a maximum of 0.06%. Such a carbon level also permits the dis-

Pipeline sections, up to 80 ft long, are inspected carefully before being shipped.

solution of a greater amount of columbium, increasing the potential for precipitation strengthening.

Additions of about 1.90 Mn with varying amounts of molybdenum were made to produce the appropriate transformation from austenite to acicular ferrite in the required plate thickness or in coils.

By delaying the transformation of austenite to polygonal ferrite and pearlite, molybdenum, in amounts up to at least 0.50% produces a corresponding increase in the amount of acicular ferrite. At the highest molybdenum content, the structure is completely acicular. Addition of 0.25 Mo increases yield strength from 70,000 to 88,000 psi, and the microstructure consists largely of fine-grained acicular ferrite with a high dislocation density.

As strengths of high-strength, low-alloy steels rise, toughness usually drops. In this steel, however, molybdenum raises both strength and toughness. Impact values obtained in the laboratory are about 100 ft-lb, even with a yield strength of 90,000 psi.

The effect is even more striking in terms of the FATT (Fracture Appearance Transition Temperature) test, which employs full-size Charpy-V notch specimens. An addition of about 0.50 Mo to 0.05C-1.9Mn-0.09Cb composition shifts the FATT (rated at 50% shear on the fracture surface) from ambient to below 0 F. A yield strength of 80,000 psi with a 50% shear FATT below 0 F can be produced with consistency.

Carbon Is Low

The traditional role of iron carbide as the primary strengthener of hot-rolled steel is not applicable in manganese-molybdenum-columbium steels. Carbon is reduced to make columbium more soluble, and to improve weldability and impact strength. While only 0.01 to 0.02 C is required, carbon was set at 0.06% max to satisfy the requirements of commercial melting practice.

Columbium Adds Strength

Columbium combines with carbon and nitrogen to form columbium carbonitride, the precipitation strengthening phase. Yield strength can be increased by 2,600 psi for each 0.01 Cb increment, up to 0.09 Cb.

Precipitation strengthening is determined by a time-temperature relationship. Steels with small and large amounts of columbium have similar precipitation kinetics; higher strengths are produced by larger quantities of columbium, however.

The effects of columbium on the secondary strengthening component are well known. In addition to this, columbium also promotes hardenability, which is needed to develop an acicular-ferrite microstructure.

Manganese and Sulfur

Manganese, along with molybdenum, helps to inhibit transformation to polygonal ferrite in the steel. Sulfur is kept to a low level because of its tendency to form harmful sulfide stringers—low sulfur insures good transverse toughness. Where sulfur cannot be kept low, however, rare earth additions will control the shape of the sulfide inclusions.

Influence of Processing Variables

The way the steel is processed determines, to a great extent, its ability to respond properly to transformation strengthening and precipitation hardening. Our engineers have found that adjustments in processing schedules could produce considerable variation in the strength-toughness relationship.

During hot working, grain refinement is enhanced because columbium has a grain-boundary pinning effect. This effect makes it possible to produce a highly substructured austenite prior to transformation, which helps in assuring transformation to fine-grained acicular ferrite.

Researchers also found that lower finishing temperatures usually produce the best combinations of strength and toughness in acicular-ferrite steels. And the amount of intermediate temperature deformation that precedes finishing also has a strong influence on these properties.

Carrying the processing investigation into strip, they determined that the Arctic Grade steel composition lends itself to the production of coiled products. Data on the effect of coiling temperatures and rates of cooling before coiling show that these factors have relatively minor effect on the favorable strength/toughness combination. This characteristic is important because commercial processing of strip usually subjects the material to significant variations in coiling temperatures and cooling rates.

Making a Commercial Product

Though it is often difficult to scale up to commercial production on the basis of laboratory data, Arctic Grade steel was another matter. The first commercial heat exceeded expectations, and later heats confirmed that no major obstacles exist in the commercial production of this composition.

Specifications were for a minimum yield of 65,000 psi and tensile strength of 80,000 psi, with a minimum of 35% shear fracture and a minimum all-heat average of 75% shear fracture at 25 F transverse to the pipe axis, as determined by the Battelle Drop Weight

Tear Test.

After the first heats proved to be successful, IPSCO began full-scale production of about 40,000 tons of Arctic Grade X-65 linepipe in a 42 in. diameter and 0.371 in. wall thickness for TransCanada Pipeline Ltd. This amounted to 80 miles of pipe. Subsequently, more than 160 miles of X-65 (0.371 in. wall thickness) and X-70 (0.390 and 0.468 in. wall thickness) pipe was produced in the same diameter for TransCanada and Alberta Gas Truck Line Co. Ltd.

Yield strength of the Arctic Grade X-70 pipe averaged about 75,000 psi transverse to the pipe axis. In the same orientation, impact specimens (⅔ size) tested to ASTM A-370 showed a mean energy of 57 ft-lb at 0 F, and 90% shear in a Battelle Drop Weight Tear Test at this same temperature.

Spiral welding, a notable feature of the pipe's fabrication, was accomplished by a two-pass, submerged-arc process at a maximum heat input of about 35 kJ per in. The steel exhibited good weldability – the heat-affected zone of the weld softened slightly, but there was no hardened region.

The Long-Range Outlook

Adding up the advantages of this new steel, IPSCO and Climax realized they had made a useful advance, not only for pipelines but possibly for applications in many other areas. Steels of this type can be rolled into either flat plate products that are air cooled from the finish-rolling temperature, or into coiled strip. In ½ in. plates, yield strengths range from about 70,000 psi (for the as-rolled condition) to about 100,000 psi for the controlled rolled-plus-aged condition.

Toughness, as measured by Charpy-V notch impact tests, is nominally over 100 ft-lb at room temperatures. High toughness is maintained well below 0 F. Contributing to high strength and good impact resistance is the transformation mechanism – austenite changes to fine-grained acicular ferrite, which is further strengthened by the precipitation of columbium carbonitride. Other advantages include good formability and most important, excellent weldability.

Aside from pipeline, the new steel can be used in the automotive, railroad, heavy equipment, construction, and shipbuilding industries, application areas in which the keynote is low cost per unit of strength. Because of their inherently good strength-toughness relationship, the manganese-molybdenum-columbium steels may well satisfy this requirement.

Strong, Tough Molybdenum Steels for the Arctic

G. TITHER AND J. W. MORROW

A series of molybdenum-containing high strength steels has been developed to comply with the stringent property requirements necessary for Arctic application. These steels exhibit either an acicular ferrite or pearlite-reduced microstructure, dependent on both composition and controlled processing techniques. In the as-rolled condition, the steels show excellent combinations of high strength and toughness, which satisfy mainline piping specifications. They also respond to quenching and tempering to achieve yield strengths in excess of 65 ksi (450 N/mm^2) with 50 pct shear fracture appearance transition temperatures below −110°F (−80°C) when quenched to simulate sections up to 3 in. (75 mm) thick.

This paper highlights the effects, on mechanical properties and microstructure, of controlled rolling with water spray cooling between hot rolling passes, and simulated hot-strip mill practice, in addition to quenching and tempering. Results show that these steels will meet proposed Arctic specifications for both mainline piping [up to 3/4 in. (19 mm) thick] and linepipe fittings; the latter in the quenched and tempered condition. Property changes are brought about by both longitudinal-welded and spiral-welded pipe fabrication. In this respect, the acicular ferrite steel exhibits advantages, which are substantiated by recent commercial data. Commercial production by several steelmakers illustrates that molybdenum-containing high strength low alloy steels are prime contenders for Arctic application.

ALTHOUGH the method for transporting natural gas from Alaska to the rest of continental United States has not yet been finalized, there is no doubt that a large tonnage of high strength, low alloy steel will be required for pipeline fabrication. The steel (or steels) must be of high strength [70 ksi (480 N/mm^2) minimum yield strength] and possess good toughness at subzero temperatures [85 pct shear Battelle Drop Weight Tear Test (BDWTT) at −10°F (−25°C)][1] in plate thicknesses up to 0.72 in. (18 mm). Based on recent developments[2,3] it may be said that the prime contender for Arctic application is an acicular ferrite, high strength, low alloy steel containing manganese, molybdenum, and columbium. This steel is already being supplied by Italsider[4] for large diameter gas transmission pipelines for service in extremely cold climatic regions of the Soviet Union. More recently, data presented by The Steel Company of Canada (Stelco)[3] strongly favored the Mn-Mo-Cb steel approach for the production of skelp for Arctic pipelines. It is worthy of note that pipe skelp supplied by Italsider is fabricated by longitudinally welding (U-O-E process) while Stelco would produce spiral-welded pipe.

Pearlite-reduced Mo-Cb steels have also been developed, and have found application in pipelines requiring somewhat less stringent property specifications.[5] Both USINOR (France) and United States Steel Corporation have provided Mo-Cb steel pipe for the North Sea gas and oil fields. Current research at Climax suggests that this type of steel can be upgraded to meet more demanding applications, and results applicable to spiral-welded pipe are included in the present report.

The purpose of this paper is to report more recent results pertaining to process control of plate and coiled skelp for both U-O-E and spiral-welded pipe fabrication. In addition to pipe, stringent Arctic specifications also exist for pipeline fittings, and it will be shown that these requirements are easily met by quenched and tempered Mn-Mo-Cb steel in section sizes up to 3 in. (75 mm).

G. TITHER is Research Supervisor and J. W. MORROW is Senior Research Assistant, Climax Molybdenum Company of Michigan, a subsidiary of AMAX Inc., 1600 Huron Parkway, Ann Arbor, Mich.

BACKGROUND

Development of acicular ferrite steels was based on sound metallurgical principles. To improve toughness, reduction or elimination of carbide aggregates was necessary. This was accomplished by reducing the steel's carbon content to about 0.06 pct. Controlled rolling plus addition of columbium maintained a fine-grained austenite prior to transformation. One of the distinctions of acicular ferrite is a high dislocation density, which is achieved by transformation control. Additions of manganese (1.5 to 2.0 pct) and molybdenum (up to 0.50 pct, but generally around 0.30 pct) effect this control by reducing the gamma-to-alpha transformation temperature. The matrix is further strengthened by precipitation of fine Cb(C,N) particles in the ferrite during cooling from the finish rolling temperature. Low carbon content provides a readily weldable product.

Since development of pearlite-reduced steels is well known and well documented,[6] it will suffice to say that a combination of low carbon content (0.08 to 0.15 pct), controlled rolling, and an austenite grain refining addition produce a relatively high-strength, tough, pearlite-reduced polygonal ferrite structure. The more common grain refining additions made to this type of steel are columbium or vanadium, or both. Improvements in toughness of both the acicular ferrite and pearlite-reduced steels, transverse to the rolling direction, may be achieved by cross-rolling, desulfurization, or inclusion shape control.

ACICULAR FERRITE STEELS

Much of the data published to date on acicular ferrite steels have been concerned with the mechanical properties obtained after subjecting these steels to a typical controlled rolling schedule.[7,8] It has been shown, in both the laboratory[9] and commercially,[2,10] that acicular ferrite Mn-Mo-Cb steel exhibits a yield strength of approximately 65 ksi (450 N/mm^2) in 3/4 in. (19 mm) thick plate. One notable exception is the work by McCutcheon *et al.*,[3] who reported results pertaining to spiral-welded pipe fabricated from plate skelp with a yield strength in excess of 80 ksi (550 N/mm^2). The purpose of the present work, therefore, was to examine in more detail the effect of soaking temperature, rolling deformation, and water spray cooling during rolling on mechanical properties of a Mn-Mo-Cb acicular ferrite steel.

The steel was prepared as 75 lb (34 kg) heats, melted and cast under an argon atmosphere to prevent excessive nitrogen contents. Composition: 0.05 C, 1.87 Mn, 0.09 Si, 0.31 Mo, 0.06 Cb, 0.012 P, 0.006 S, 0.006 N, 0.04 Al. Ingots were press forged at 2250°F (1230°C) to slabs 3 1/2 in. thick × 5 in. wide × 10 in. long (90 mm × 125 mm × 250 mm) ready for rolling to 3/4 in. (19 mm) plate, according to rolling schedules presented in Table I. Water spray cooling (cooling rate approximately 6°F/s, 3.3°C/s) was employed between all rolling passes made in the 1900 to 1500°F (1040 to 815°C) range.

Mechanical Properties and Microstructure

Mechanical properties of 3/4 in. (19 mm) thick plates of the Mn-Mo-Cb steel after austenitizing at the higher soaking temperature of 2150°F (1175°C) and finish rolling at 1400°F (760°C), are presented in Table II (Schedules A through D). Yield strength varied notably with the amount of total deformation below 1750°F (955°C), as illustrated in Fig. 1. Decreasing the percentage deformation from 65 to 30 pct resulted in an incremental increase in yield strength from 63.9 ksi (440 N/mm^2) to 71.5 ksi (492 N/mm^2). This increase, however, was generally accompanied by only a small sacrifice in the 50 pct shear fracture appearance transition temperature (FATT), Fig. 2, with the FATT value of the higher strength steel being −70°F (−55°C). The effect of increased deformation on impact properties is also illustrated in Fig. 3, which shows the relationship between yield strength and impact energy absorbed at −10°F (−25°C) as a function of rolling deformation. As can be seen, both yield strength and energy absorbed are impaired by

Table I. Rolling Schedules for Mn-Mo-Cb Steel Plates

Rolling Schedule*	Slab Reheat Temperature, °F (°C)	Finish Rolling Temperature, °F (°C)	Total Amount of Deformation Below 1750°F (955°C), Pct
A	2150 (1175)	1400 (760)	30
B	2150 (1175)	1400 (760)	40
C	2150 (1175)	1400 (760)	50
D	2150 (1175)	1400 (760)	65
E	2050 (1120)	1400 (760)	40
F	2050 (1120)	1400 (760)	65

*Plates were water spray cooled between rolling passes in the temperature range 1900 to 1500°F (1040 to 815°C). Thickness: 3/4 in. (19 mm).

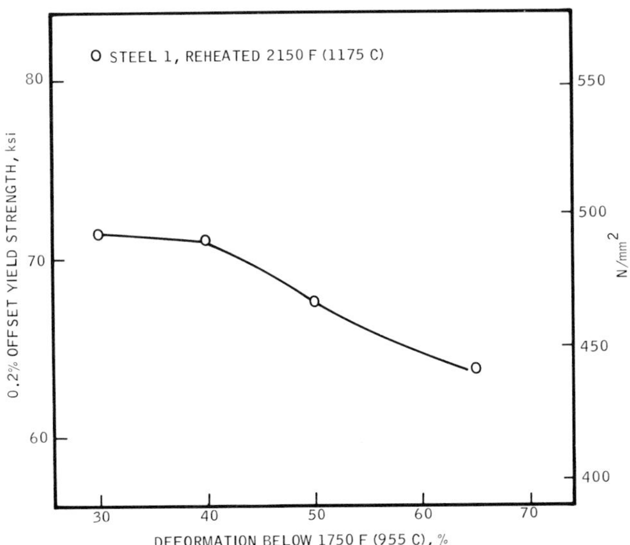

Fig. 1—Effect of total rolling deformation below 1750°F (955°C) on yield strength of 3/4 in. (19 mm) thick Mn-Mo-Cb steel plates (Steel 1), spray cooled during rolling between 1900°F (1040°C) and 1500°F (815°C).

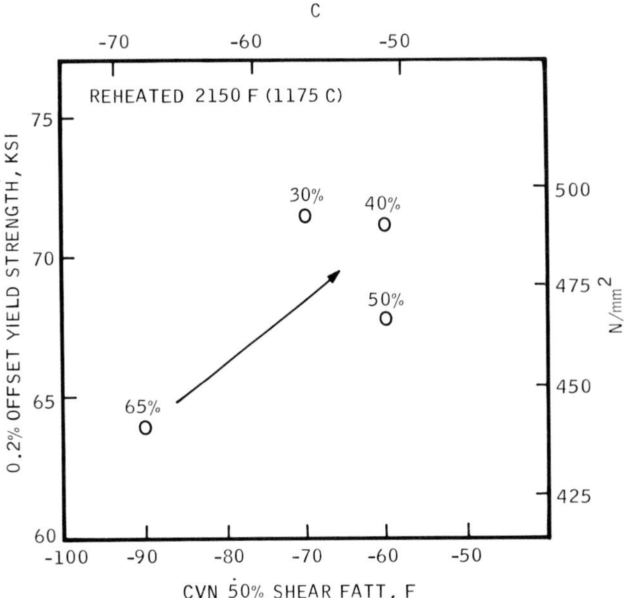

Fig. 2—Effect of rolling deformation below 1750°F (955°C) on strength-toughness balance of 3/4 in. (19 mm) thick Mn-Mo-Cb steel plates (Steel 1) spray cooled during rolling between 1900°F (1040°C) and 1500°F (815°C).

Table II. Longitudinal Tensile Properties and Impact Data of Mn-Mo-Cb Steel Plates*

Reheat Temperature, °F (°C)	Rolling Schedule (Deformation) Pct†	0.2 Pct Offset Yield Strength, ksi (N/mm²)	1.5 Pct Offset Yield Strength, ksi (N/mm²)	Tensile Strength, ksi (N/mm²)	Yield/Tensile Ratio	Eongation 1 in. (25 mm), Pct	Reduction in Area, Pct	Charpy V-Notch Impact Data: Energy Absorbed, ft lb (J)				Transition Temperature, °F (°C)	
								-80°F (-62°C)	-10°F‡ (-23°C)	+25°F (-4°C)	C_V 100¶	50 ft lb (68 J)	50 Pct Shear FATT
2150 (1175)	A (30)	71.5 (493)	82.5 (569)	98.8 (681)	0.72	30	74	30 (41)	170 (231)	198 (268)	190 (258)	-70 (-57)	-70 (-57)
	B (40)	71.2 (491)	85.0 (586)	98.8 (681)	0.72	29	73	20 (27)	166 (225)	178 (241)	178 (241)	-55 (-48)	-60 (-51)
	C (50)	67.8 (467)	77.7 (536)	94.5 (651)	0.72	30	75	20 (27)	160 (217)	200 (271)	220 (298)	-50 (-46)	-60 (-51)
	D (65)	63.9 (441)	75.5 (520)	92.4 (637)	0.69	29	73	62 (84)	144 (195)	174 (236)	185 (251)	-90 (-68)	-90 (-68)
2050 (1120)	E (40)	64.5 (445)	75.2 (518)	94.4 (651)	0.68	29	77	108 (146)	194 (263)	206 (279)	208 (282)	-95 (-71)	-90 (-68)
	F (65)	68.5 (472)	78.9 (544)	94.6 (652)	0.72	29	73	138 (187)	200 (271)	202 (274)	198 (268)	-135 (-93)	-120 (-84)

*Plates were water spray cooled between passes in the 1900 to 1500°F (1040 to 815°C) range. Thickness: 3/4 in. (19 mm).
†See Table I.
‡All specimens exhibited ≥85 pct shear.
¶Energy at lowest temperature at which 100 pct shear fracture was observed.

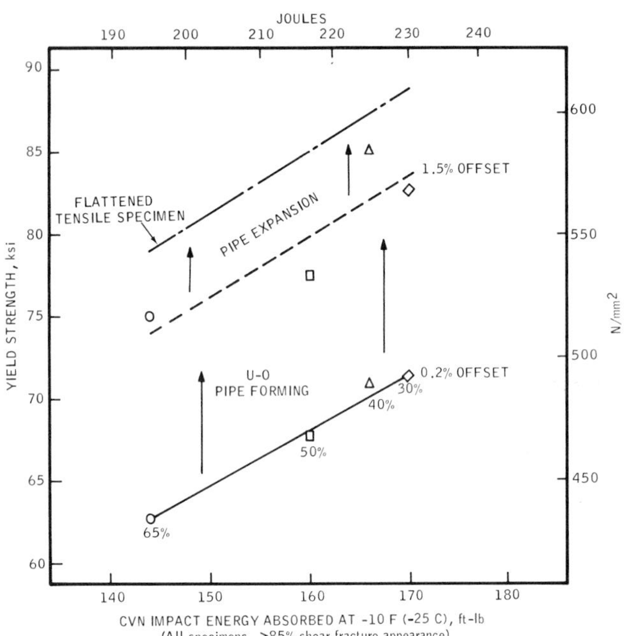

Fig. 3—Effect of rolling deformation below 1750°F (955°C) on yield strength and impact energy absorbed at −10°F (−23°C) of 3/4 in. (19 mm) thick Mn-Mo-Cb steel plates (Steel 1), austenitized at 2150°F (1175°C) and spray cooled during rolling between 1900°F (1040°C) and 1500°F (815°C).

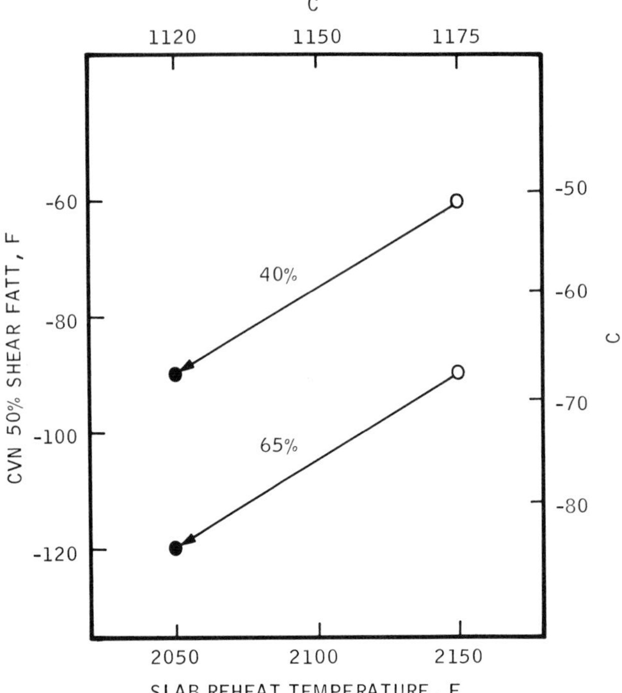

Fig. 4—Effect of slab reheat temperature on the 50 pct shear FATT of 3/4 in. (19 mm) thick Mn-Mo-Cb steel plates (Steel 1), spray cooled during rolling between 1900°F (1040°C) and 1500°F (815°C).

increasing deformation (Fig. 3) even though the 50 pct shear FATT is slightly improved (Fig. 2). Impact energy values, however, would meet the requirements for Arctic service at a design temperature of −10°F (−25°C).

It is well known[3,11] that lowering the reheating temperature generally improves impact properties in the final plate. Results of the present work on Mn-Mo-Cb steels are no exception, as is shown in Fig. 4. The detailed mechanical properties of Steel 1, reheated to 2050°F (1120°C) and finish rolled at 1400°F (760°C), are given in Table II. As can be seen, decreasing the soaking temperature improved the 50 pct shear FATT by 30°F (15°C) and significantly raised impact energy absorbed at −10°F (−25°C). For the plates rolled from slabs soaked at 2050°F (1120°C), the energy at −80°F (−60°C) was more than 100 ft lb (135 J). The effect on yield strength, of greater rolling deformation at the lower soaking temperature, was opposite to that observed after reheating to 2150°F (1175°C), producing an increase of 4 ksi (30 N/mm²) to a value of 68.5 ksi (472 N/mm²).

Microstructures of these steels explain variations in mechanical properties. In the steels that were reheated at 2150°F (1175°C), increasing the percentage deformation from 30 to 50 pct produced little or no change in basic optical microstructure. All specimens showed a mixed structure of acicular ferrite, fine-grained polygonal ferrite, and isolated martensitic islands; Fig. 5. In the material deformed 65 pct below 1750°F (955°C), the structure was somewhat similar except for the presence of a larger percentage of fine-grained polygonal ferrite (Fig. 6).

The decrease in yield strength with increasing deformation may be attributed to two factors. First, greater deformation in the lower austenite temperature range would promote precipitation of Cb(C,N) in the austenite. Fig. 7 shows that increased deformation produced a greater volume fraction of Cb(C,N) particles. It has been shown[7] that Cb(C,N) particles of this size (≈100Å) can form in the austenite. While this effect would retard austenite recrystallization and sub-

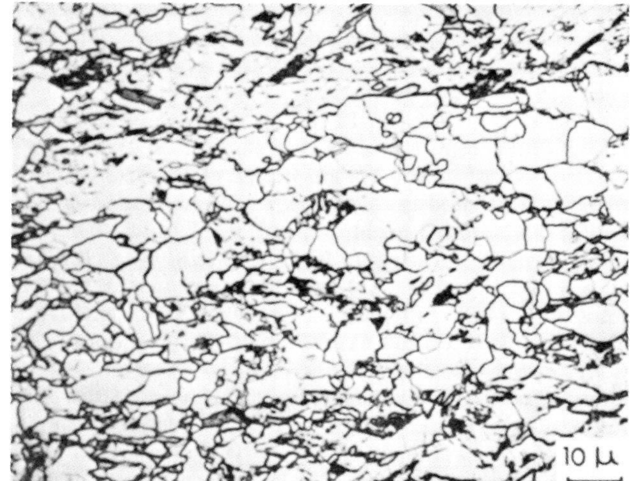

Fig. 5—Microstructure of Steel 1 rolled from 2150°F (1175°C) according to schedule B [40 pct total deformation below 1750°F (955°C)]. 2 pct Nital; Magnification 750 times.

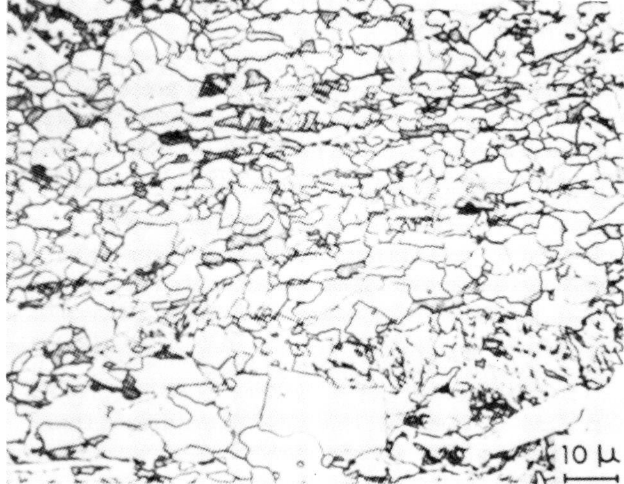

Fig. 6—Microstructure of Steel 1 rolled from 2150°F (1175°C) according to schedule D [65 pct total deformation below 1750°F (955°C)]. 2 pct Nital; Magnification 750 times.

sequent grain growth, it would offer little or no direct contribution to strengthening. Further, more precipitation in austenite would leave less columbium in solution available for subsequent precipitation as Cb(C,N) during gamma-to-alpha transformation and in the ferrite matrix. (It is the precipitation of fine Cb(C,N) particles in the ferrite which imparts a significant strengthening effect.)

The second factor contributing to the loss in yield strength was the presence of more polygonal ferrite. Because the polygonal ferrite, in all instances, showed no sign of being "cold-worked", transformation must have taken place after finish rolling. This "low-temperature" polygonal ferrite would not have the strength of the high dislocation density acicular ferrite, nor would it have the same Cb(C,N) strengthening potential as that of polygonal ferrite formed at a relatively higher temperature.[3,12] It is possible that laboratory rolling did not provide for sufficient deformation at the higher temperatures and a mixture of fine and coarse grained austenite resulted. Retention of the mixed austenite structure would give two different hardenability levels with the lower hardenability austenite transforming to polygonal ferrite.

Improvement in impact transition temperature observed by increasing the deformation to 65 pct below 1750°F (955°C) may be attributed to 1) the presence of more fine-grained polygonal ferrite, and 2) less precipitation of Cb(C,N) particles in the ferrite.

In both specimens controlled rolled after reheating to 2050°F (1120°C), the basic structure was somewhat similar to that observed in material austenitized at 2150°F (1175°C). The main difference, however, was that the samples rolled from the lower reheat temperature exhibited more fine-grained polygonal ferrite. Also, more fine-grained polygonal ferrite was evident when comparing samples deformed 65 pct with those deformed 40 pct below 1750°F (955°C) during rolling from a reheat temperature of 2050°F (1120°C).

Reheating to the lower austenitizing temperature would reduce the amount of dissolved Cb(C,N) particles, which in turn would prevent austenite grain growth. The finish rolling sequence, therefore, would be started with a material having an overall finer austenite grain size which would be retained

(a)

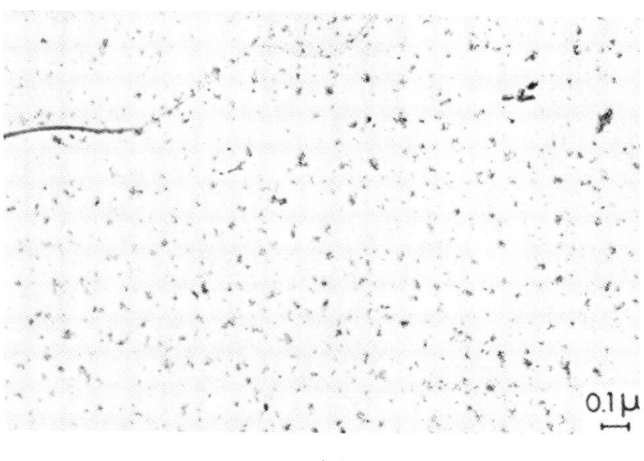

(b)

Fig. 7—Effect of total deformation below 1750°F (955°C) on Cb(C, N) precipitation in Steel 1 austenitized at 2150°F (1175°C) (carbon extraction replicas). Top, 30 pct deformation; bottom, 65 pct deformation; Magnification 37,500 times.

throughout rolling. This effect would tend to reduce hardenability of the steel, even in the presence of manganese and molybdenum, with the resultant transformation product being mainly fine-grained polygonal ferrite. Sufficient hardenability is retained, however, to produce a reasonable amount of acicular ferrite. The combined factors of extremely fine polygonal ferrite and less precipitation of Cb(C,N) particles in ferrite—by virtue of the fact that more Cb(C,N) remained out of solution in the austenite because of the lower reheat temperature—would account for the superior toughness properties of these specimens.

Arctic Linepipe Application

Although these results suggest that the acicular ferrite Mn-Mo-Cb steel may be marginal for X70 pipe application, consideration of the stress-strain behavior of the steel refutes this argument. Unlike other candidate steels for Arctic application, which produce a typical discontinuous (*i.e.*, yield-point) stress-strain curve, the Mn-Mo-Cb steel exhibits continuous yielding.[9] The rapid work hardening potential of this steel is more than sufficient to offset the Bauschinger effect observed in testing of flattened tensile specimens from pipe. The result is that the yield strength of the flattened tensile specimen is much greater (15 to 20 pct) than the yield strength of the plate-skelp. Commercial data from Italsider[2] substantiate this phenomenon, and Fig. 8 shows how the yield strength varies during U-O-E pipe fabrication for both air-cooled acicular ferrite and pearlite-reduced steels. It can be seen that the yield strength after U-O pipe forming is increased significantly for the acicular ferrite steel, whereas a large drop (about 15 pct) in strength occurs with the pearlite-reduced steel. The latter steel, however, does recover a percentage of this strength loss after expansion (1.5 pct) of the pipe, but not enough to equal the yield strength of the skelp. The acicular ferrite steel, on the other hand, continues to increase in yield strength on pipe expansion (because of the shape of the stress-strain curve) to approximately 76 ksi (525 N/mm^2), which is about 10 ksi (70 N/mm^2) greater than that of the skelp. Referring to Fig. 3, the 1.5 pct offset yield strength of the steels studied in this work was approximately 10 ksi (70 N/mm^2) greater than the 0.2 pct offset yield strength. A strain of 1.5 pct would be a typical strain that the plate skelp would undergo when formed into 48 in. (1220 mm) diam pipe by the U-O process. The additional 1.5 pct expansion of the pipe would further increase the yield strength of a flattened tensile specimen by about 5 ksi (35 N/mm^2) (Ref. 2). Reduction in yield strength of pearlite-reduced steels, due to the Bauschinger effect, has been reported to range between 5 to 10 ksi (35 to 70 N/mm^2) for 65 ksi (450 N/mm^2) and 80 ksi (550 N/mm^2) yield strength pipe, respectively.[13] However, based on commercial data[14] it is thought that this decrease may be much less in acicular ferrite steels. Even so, if the Bauschinger effect does reduce the yield strength of acicular ferrite steels by up to 10 ksi (70 N/mm^2), all of the Mn-Mo-Cb steels still meet the CAGPL X70 specification for U-O-E pipe, and would usually satisfy X80 requirements.

Although most spiral-welded pipe is produced from coiled skelp, Stelco fabricates this type of pipe from plate at the present time. Data (Fig. 9, after McCutcheon *et al.*[3]) show that the yield strength of spiral-welded pipe is similar to the yield strength of the plate-skelp for acicular ferrite steels of 70 ksi (490 N/mm^2) yield strength. This would make the steels studied in the present work marginal for spiral-welded pipe for Arctic application. However, Stelco's successful approach of slightly increasing the manganese and molybdenum content of the Mn-Mo-Cb steel to comply with their processing technique ensured that this steel transformed completely to acicular ferrite, and it easily met the CAGPL X70 Arctic specification.

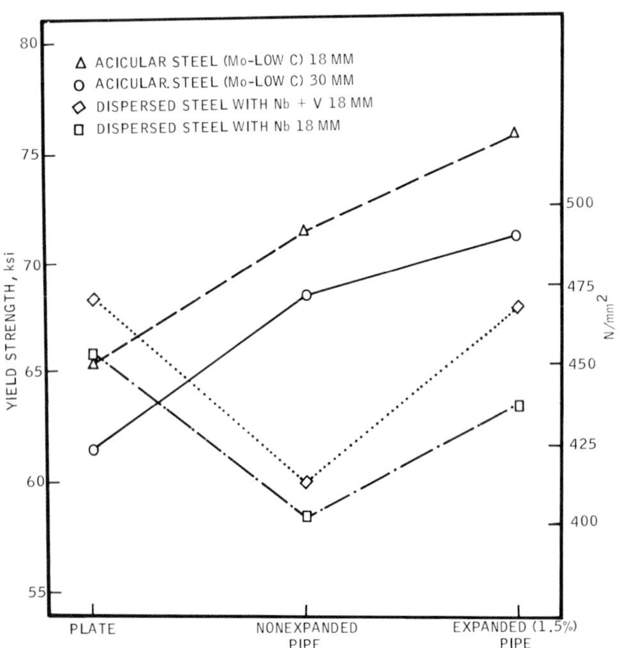

Fig. 8—Effect of pipe fabrication on yield strength of molybdenum-containing and molybdenum-free steels (after Civallero and Parrini[2]).

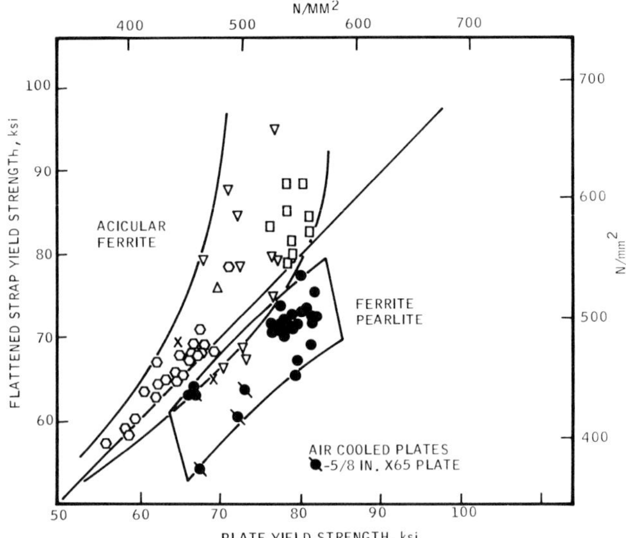

Fig. 9—Flattened specimen (pipe) yield strength *vs* as-rolled yield strength for ferrite pearlite and acicular ferrite steels (after McCutcheon, *et al.*[3]).

In summary, acicular ferrite Mn-Mo-Cb steel of 65 ksi (450 N/mm^2) yield strength (which can be achieved by a typical air-cooled controlled rolling schedule) meets the stringent X70 Arctic specification when fabricated to U-O-E pipe. If a slightly faster cooling rate of 6 F/sec (3.3 C/s) is effected by water spraying during rolling, properties can be upgraded to satisfy X80 requirements. The more rapid cooling through the austenite range retains more columbium in solid solution, thereby allowing an increase in subsequent Cb(C,N) precipitation in the ferrite. A slight modification of the steel composition has shown that spiral-welded pipe, produced from plate, will also comply with Arctic requirements.

PEARLITE-REDUCED STEELS

To determine the potential of pearlite-reduced Mo-Cb steels for large-diameter, spiral-welded pipe fabricated from coiled skelp, several steel compositions were rolled and simulated-coiled in the laboratory. Effects of variation in carbon and molybdenum content were studied. Steels were prepared as described in the previous section; compositions are given in Table III. Slabs were austenitized at 2300°F (1260°C), and rolled to 1/2 in. (13 mm) plate according to the schedule presented in Table IV. After finish rolling at 1550°F (845°C), the plates were spray cooled at a rate of approximately 20°F/s (11°C/s). Plates from Steel 2 were also spray cooled at 10°F/s (6°C/s) to one of three preselected coiling temperatures, and then slow cooled at a rate of 50°F/h (30°C/h) to below 600°F (315°C). Accelerated cooling immediately after rolling was used to simulate cooling on a hot strip mill run-out-table, while the subsequent slow cooling simulated cooling of a commercial coil of strip steel. Coiling temperatures were 1075, 1150, and 1225°F (580, 620, and 665°C).

Table III. Composition of Pearlite-Reduced Mo-Cb Steels

Steel No.	Element, Pct								
	C	Mn	Si	Mo	Cb	P	S	N	Al
2	0.09	1.40	0.10	0.26	0.05	0.016	0.011	0.006	0.03
3	0.09	1.40	0.09	0.19	0.05	0.016	0.009	0.006	0.06
4	0.05	1.39	0.09	0.18	0.05	0.016	0.009	0.006	0.07

Table IV. Rolling Schedule for Simulated Coiled Mo-Cb Steel Plates

Pass No.*	Entry Temperature, °F (°C)	Exit Thickness, In. (mm)	Reduction Per Pass, Pct
0	2300 (1260)	3.25 (83)	(Initial)
1	2100 (1150)	2.80 (71)	13.8
2	2000 (1095)	2.30 (58)	17.9
3	1950 (1065)	1.80 (46)	21.7
4	1850 (1010)	1.35 (34)	25.0
5	1670 (910)	1.12 (28)	17.0
6	1640 (895)	0.85 (22)	24.1
7	1610 (875)	0.75 (19)	11.8
8	1580 (860)	0.62 (16)	17.4
9	1550 (845)	0.50 (13)	19.3

*Thickness: 1/2 in. (13 mm).

Mechanical Properties and Microstructures

Detailed results of tensile tests are given in Table V, while a summary of the yield strength and impact data is presented in Table VI. Yield strengths in the 67.5 ksi to 92.6 ksi (465 to 940 N/mm^2) range were obtained with 50 pct shear FATT between −160 to −55°F (−105 to −50°C). All steels simulated coiled at 1150°F (620°C), after cooling at 20°F/s from the finish rolling temperature, exhibited yield strengths greater than 80 ksi (550 N/mm^2) with 50 pct shear FATT values down to −80°F (−60°C). An exceptionally low transition temperature of −160°F (−105°C) was shown by Steel 4 [the lower carbon steel simulated coiled at 1075°F (580°C)] while still maintaining a yield strength in excess of 70 ksi (485 N/mm^2).

The effect of coiling temperature on the yield strength can be seen in Fig. 10. Steels 2 and 4 both showed a maximum yield strength at the 1150°F (620°C) coiling temperature, and, although the curve is not complete, Steel 3 would appear to be following a similar pattern. With the exception of Steel 4, the 50 pct shear FATT value did not increase as yield strength increased, and actually decreased 20°F (11°C) in Steel 2. The strength-toughness relationship for the Mo-Cb steels simulated-coiled at 1150°F (620°C) is shown in Fig. 11.

Fig. 10 indicates how the cooling rate from the finish rolling temperature to the coiling temperature affected yield strength. Although the basic shape of the curve remained similar, decreasing the cooling rate from 20°F/s (11°C/s) to 10°F/s (6°C/s) for Steel 2 resulted in a corresponding lower yield strength after simulated-coiling at the "optimum" temperature of 1150°F (620°C).

In general, the microstructures produced (Fig. 12) were somewhat dissimilar from typical controlled rolled plate material of similar chemistry. In all but one instance, the gamma-to-alpha transformation was completed during spray cooling to the coiling

Table V. Longitudinal Tensile Test Data for Simulated Coiled Mo-Cb Steel Plates

Steel No.*	Cooling Rate from Finish Rolling Temperature to Coiling Temperature, °F/s (°C/s)	Simulated Coiled at 1075°F (580°C)					Simulated Coiled at 1150°F (620°C)					Simulated Coiled at 1225°F (665°C)				
		0.2 Pct Offset Yield Strength, ksi (N/mm^2)	Tensile Strength, ksi (N/mm^2)	Elongation, Pct	Reduction in Area, Pct	Yield/Tensile Ratio	0.2 Pct Offset Yield Strength, ksi (N/mm^2)	Tensile Strength, ksi (N/mm^2)	Elongation, Pct	Reduction in Area, Pct	Yield/Tensile Ratio	0.2 Pct Offset Yield Strength ksi (N/mm^2)	Tensile Strength, ksi (N/mm^2)	Elongation, Pct	Reduction in Area, Pct	Yield/Tensile Ratio
2	~20 (11)	75.0 (517)	88.4 (609)	27	77	0.85	92.6 (638)	104.9 (723)	21	74	0.88	67.5 (465)	79.0 (545)	33	80	0.85
	10 (6)	75.8 (523)	90.0 (620)	29	66	0.84	78.4 (540)	92.5 (638)	30	69	0.85	72.9 (503)	82.0 (565)	33	69	0.89
3	~20 (11)	–	–	–	–	–	84.3 (581)	96.4 (665)	24	78	0.87	71.0 (489)	80.3 (554)	31	80	0.88
4	~20 (11)	70.4 (485)	77.2 (532)	32	74	0.91	81.2 (560)	91.4 (630)	29	84	0.89	75.4 (520)	83.1 (573)	29	83	0.91

*Thickness: 1/2 in. (13 mm).

Table VI. Yield Strength and Impact Data for Simulated Coiled Mo-Cb Steel Plates

		Simulated Coiled at 1075°F (580°C)					Simulated Coiled at 1150°F (620°C)					Simulated Coiled at 1225°F (665°C)				
			Full-Size Charpy V-Notch Impact Data					Full-Size Charpy V-Notch Impact Data					Full-Size Charpy V-Notch Impact Data			
			Energy Absorbed, ft lb (J)		Transition Temperature, °F (°C)			Energy Absorbed, ft lb (J)		Transition Temperature, °F (°C)			Energy Absorbed, ft lb (J)		Transition Temperature, °F (°C)	
Steel No.*	Cooling Rate from Finish Rolling Temperature to Coiling Temperature, °F/s (°C/s)	0.2 Pct Offset Yield Strength, ksi (N/mm^2)	−40°F (−40°C)	C_V 100†	20 ft lb (27 J)	50 Pct Shear	0.2 Pct Offset Yield Strength ksi (N/mm^2)	−40°F (−40°C)	C_V 100	20 ft lb (27 J)	50 Pct Shear	0.2 Pct Offset Yield Strength ksi (N/mm^2)	−40°F (−40°C)	C_V 100	20 ft lb (27 J)	50 Pct Shear
2	~20 (11)	75.0 (517)	70 (95)	82 (111)	−65 (−54)	−55 (−48)	92.6 (638)	63 (85)	65 (88)	−75 (−59)	−75 (−59)	67.5 (465)	75 (102)	110 (149)	−65 (−54)	−55 (−48)
	10 (6)	75.8 (523)	57 (77)	108 (146)	−55 (−48)	−50 (−46)	78.4 (540)	77 (104)	95 (129)	−70 (−57)	−65 (−54)	72.9 (503)	51 (69)	89 (121)	−55 (−48)	−45 (−43)
3	~20 (11)	–	–	–	–	–	84.3 (581)	52 (71)	115 (156)	−80 (−62)	−55 (−48)	71.0 (489)	53 (72)	90 (122)	−55 (−48)	−55 (−48)
4	~20 (11)	70.4 (485)	170 (231)	125 (170)	−175 (−115)	−160 (−107)	81.2 (560)	96 (130)	139 (188)	−90 (−68)	−80 (−62)	75.4 (520)	104 (141)	153 (207)	−85 (−65)	−75 (−59)

*Test specimens taken from the longitudinal direction. Thickness: 1/2 in. (13 mm).
†Energy at lowest temperature at which 100 pct shear fracture was obtained.

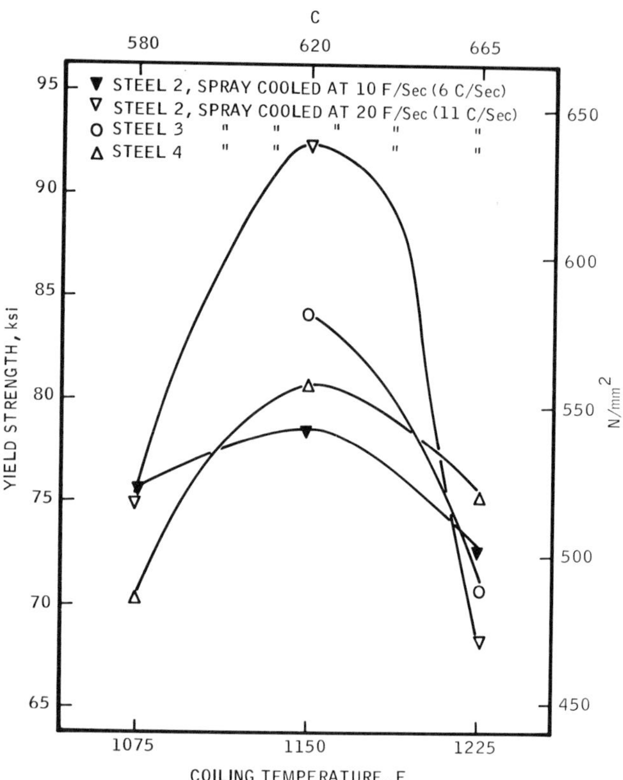

Fig. 10—Effect of coiling temperature and spray cooling down to the coiling temperature on yield strength of 1/2 in. (13 mm) simulated coiled Mo-Cb steel plates.

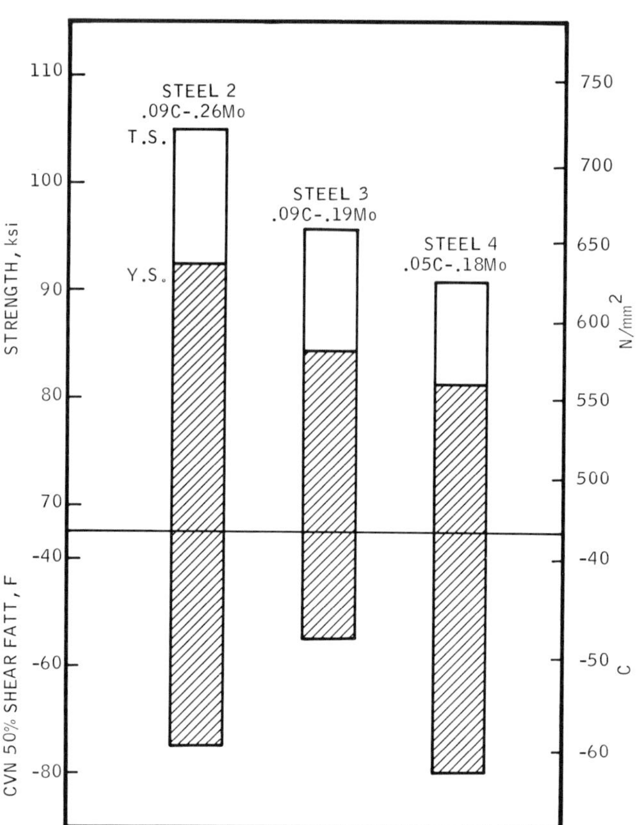

Fig. 11—Strength-toughness relationship for 1/2 in. (13 mm) thick Mo-Cb steel plates simulated coiled at 1150°F (620°C). All steels spray cooled at 20°F/s (11°C/s) down to the coiling temperature.

temperature, forming a fine, irregular shaped ferrite structure plus degenerate pearlite (or bainite). The exception was the more-rapidly spray cooled strip of Steel 2 [coiled at 1150°F (620°C)], which exhibited a mixed structure of acicular ferrite, polygonal ferrite, and degenerate pearlite (or bainite). As might be expected, this steel contained the higher molybdenum (0.26 pct), which would assist in promoting acicular ferrite. Ferrite grain size of all strip samples was in the ASTM 11-12 range.

Rapid cooling of these low alloy steels directly after finish rolling would be expected to produce a somewhat highly dislocated substructure, which, apart from itself being a strengthening mechanism, would also provide numerous nucleation sites for subsequent Cb(C,N) precipitation. Also, cooling would be fast enough to retain additional columbium in solid solution in the austenite, working together with the simulated-coiling process (*i.e.*, an aging treatment) to produce a fine strengthening dispersion of Cb(C,N) particles in the ferrite. A thin foil study tended to support this hypothesis; Fig. 13 (top) shows the dislocation structure of Steel 2, simulated-coiled at 1150°F (620°C). The typical high dislocation density of the acicular ferrite areas can be observed together with the somewhat high dislocated substructure of the polygonal ferrite. Fig. 13 (bottom), a higher magnification of the same area, shows a fine dispersion of Cb(C,N) particles precipitated both on dislocations and within the ferrite matrix. The majority of these particles ranged between 40 and 80Å in size. Because the optical microstructure of the lower-carbon steel (Steel 4) was more typically polygonal ferrite plus pearlite, the dislocation density of this steel after coiling at 1150°F (620°C) was somewhat lower, Fig. 14. Also, the lower carbon content would reduce the amount of Cb(C,N) precipitation during cooling from the coiling temperature. These two factors would account for the lower yield strength [81.2 ksi, (560 N/mm^2)] observed.

Simulated-coiling at 1225°F (665°C) produced a more typical polygonal ferrite-pearlite optical structure, suggesting that the gamma-to-alpha transformation occurred at a higher temperature. The effect of coiling temperature on precipitation can be seen by comparing carbon extraction replicas of Steel 2, shown in Fig. 15. In the material simulated-coiled at the higher temperature, *i.e.*, 1225°F (665°C), a more coarse Cb(C,N) particle size, mainly between 100 and 200Å, was observed (Fig. 15 (bottom)). The very coarse particles observed in this material were apparently Cb(C,N) particles precipitated in the austenite which coarsened during slow cooling from the coiling temperature. These noncoherent particles would not contribute to strengthening. In addition to the basic microstructure having inherently lower strength, it appears therefore that coiling at the higher temperature promotes an over-aged precipitate. At the lower simulated-coiling temperature of 1075°F (580° C), very little fine precipitation of Cb(C, N) was observed, indicating that precipitation and particle growth were suppressed by the rapid cooling. This would result in an underaged condition and hence account for the low yield strength obtained.

SUMMARY

From the present work, therefore, it is suggested that the maximum in the yield strength level exhibited after coiling at 1150°F (620°C) was due to a relatively high dislocation substructure coupled with optimum

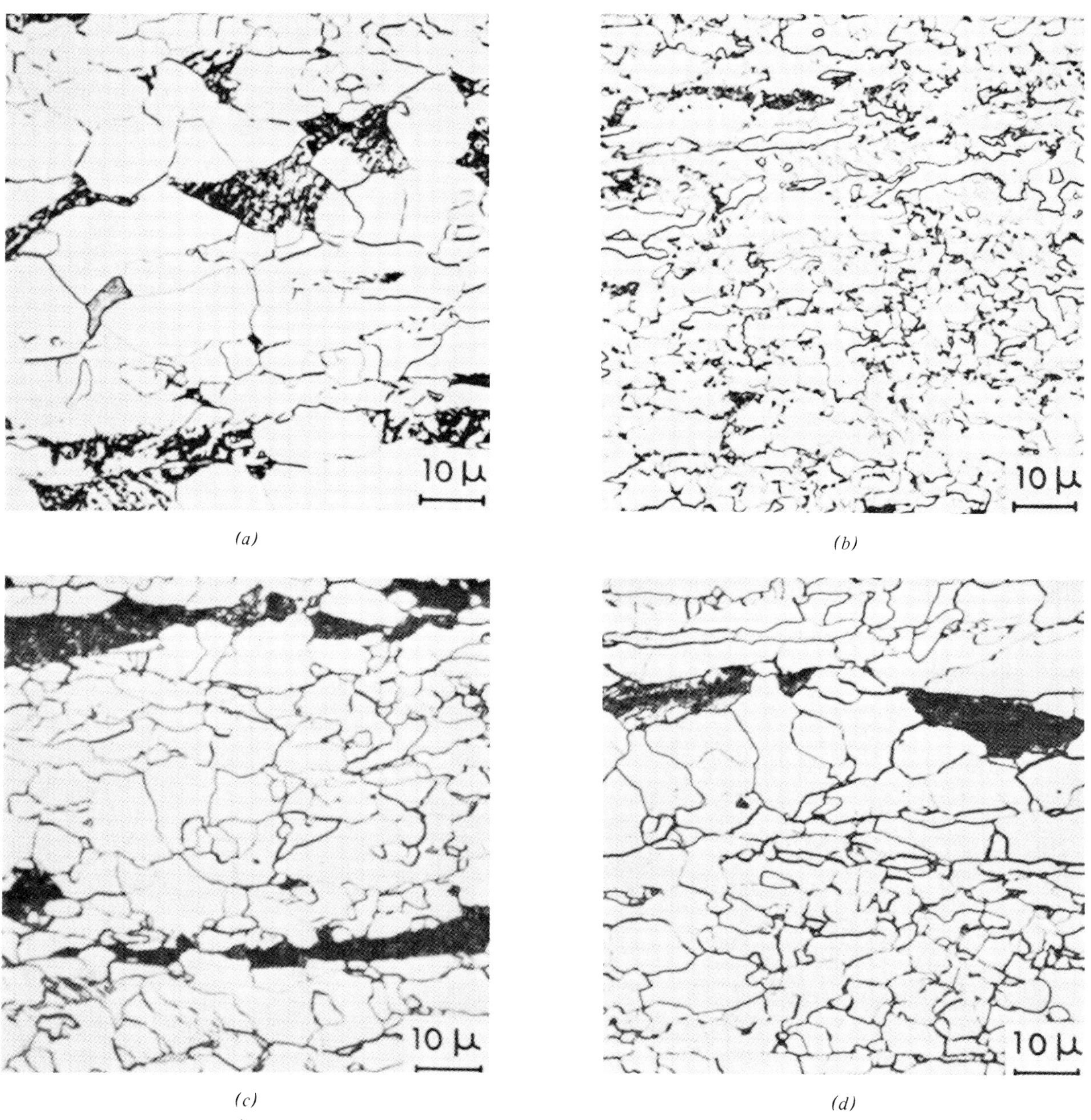

Fig. 12—Microstructures of 1/2 in. (13 mm) thick Mo-Cb steel plates simulated coiled at 1150°F (620°C) compared to a typical as-rolled, pearlite-reduced structure (upper left). Steels 2 (upper right), 3 (lower left), and 4 spray cooled at 20°F/s (11°C/s) down to the coiling temperature. 2 pct Nital; Magnification 1000 times.

precipitation strengthening in the coil. The highest yield strength of 92.6 ksi (640 N/mm^2), Steel 2, may be attributed, in part, to the presence of acicular ferrite areas promoted by depression of the gamma-to-alpha transformation temperature due to spray cooling and increased molybdenum content. Reducing the cooling rate of Steel 2 resulted in a more typical polygonal ferrite-plus-pearlite structure with less Cb(C,N) precipitation in the ferrite, and hence a lower yield strength, 78.4 ksi (540 N/mm^2). The good toughness exhibited by these Mo-Cb steels may be related directly to the fine grain size attained together with a low percentage of detrimental pearlite.

QUENCHED AND TEMPERED STEELS

In addition to the high strength-toughness requirements of linepipe skelp, it is obvious that linepipe fittings and valves required for Arctic application should also possess similar properties. Therefore, in an attempt to meet the CAGPL requirements, the Mn-Mo-Cb steel was subjected to a quench and temper treatment. Heat treatment is necessitated because these parts are formed at high temperature, eliminating any effect of prior working on mechanical properties.

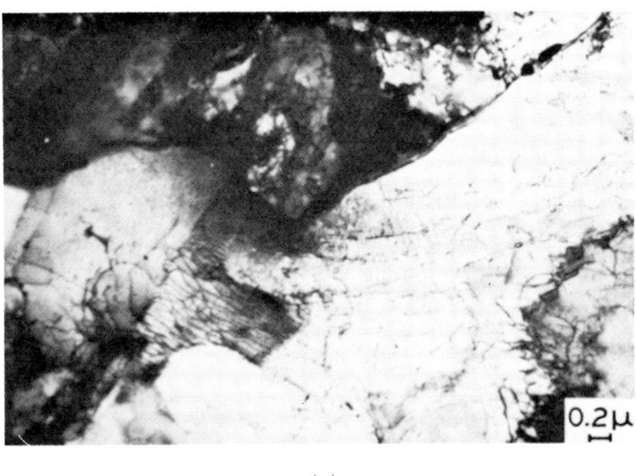

(a)

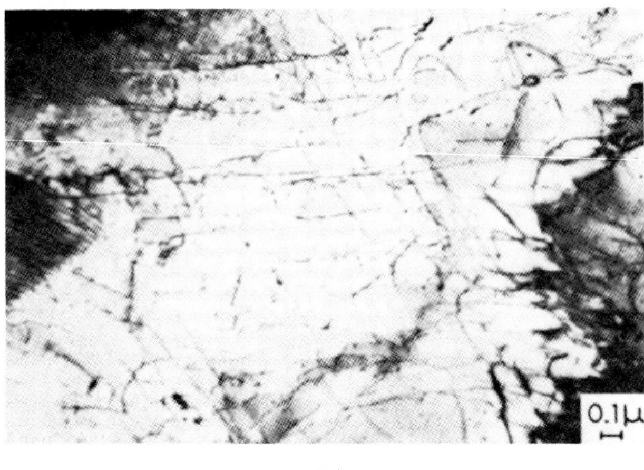

(b)

Fig. 13—Dislocation substructure and Cb(C, N) precipitation in simulated-coiled Mo-Cb steel [Steel 2, spray cooled at 20°F/s (11°C/s) down to the coiling temperature of 1150°F (620°C)]. Top; magnification 15,000 times: Bottom; magnification 30,000 times.

Composition of the steel investigated initially, and hardenability data obtained using standard end-quench specimens quenched from the optimum austenitizing temperature [1700°F (925°C)] are presented in Fig. 16. The range of austenitizing temperatures studied was

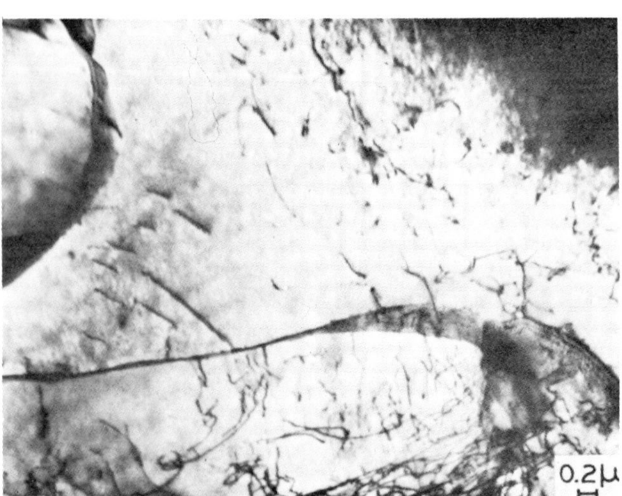

Fig. 14—Low dislocation density of polygonal ferrite in simulated-coiled Mo-Cb steel [Steel 4, coiled at 1150°F (620°C)]. Magnification 15,000 times.

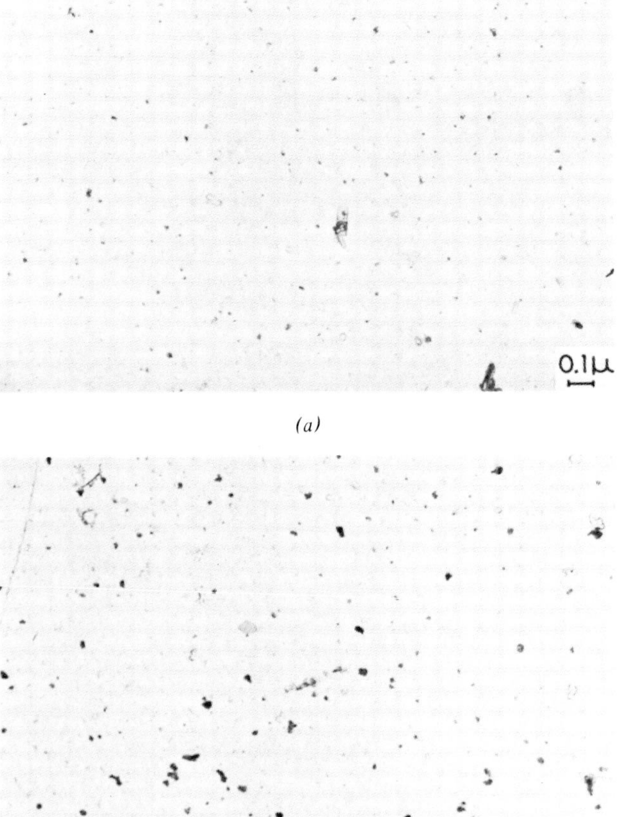

(b)

Fig. 15—Effect of coiling temperature on Cb(C, N) particle size of simulated-coiled Mo-Cb steel (Steel 2), spray cooled at 20°F/s (11°C/s) down to the coiling temperature—1150°F (620°C), top; 1225°F (665°C), bottom: Magnification 37,500 times.

1600°F (870°C) to 1800°F (980°C). An austenitizing temperature of 1700°F (925°C) was chosen because 1) 1600°F (875°C) provided incomplete austenitization, and 2) 1800°F (980°C) did not offer a significant increase in hardness over 1700°F (925°C), and would probably have resulted in reduced toughness due to an increased grain size. Hardness tests also predicted that 1200°F (650°C) was the optimum tempering temperature. As can be seen from Fig. 16, there was a small initial drop in hardness over the first 1/4 in. (6 mm), after which the hardness level was constant along the remaining length of the end-quench specimen.

The Mn-Mo-Cb steel was rolled to 1 in. (25 mm) plate, then reaustenitized and water spray quenched to simulate the cooling rate of immersion-quenched thicker sections. The cooling rates, measured at 1300°F (705°C), were 35°F/s (20°C/s), 20°F/s (11°C/s), 11°F/s (6°C/s), and 7°F/s (4°C/s), to simulate cooling rates at centers of 1 1/2 in. (38 mm), 2 in. (50 mm), 2 1/2 in. (65 mm), and 3 in. (75 mm) thick sections, respectively.[15] After cooling each plate was tempered at 1200°F (650°C) for 1 h.

Tensile and impact data for the quenched and tempered plates are summarized in Table VII, and are illustrated as a function of plate thickness in Fig. 17. A minimum yield strength of 65 ksi (450 N/mm^2) was achieved in all instances, with a yield strength of 73.4 ksi (505 N/mm^2) being obtained in the simulated-quenched 2 1/2 in. (65 mm) thick plate. These high strengths were coupled with excellent toughness, with 50 pct shear FATT well below −100°F (−75°C) for each of the plates. Specimens tested at −20°F (−30°C) and −80°F (−60°C) all exhibited a 100 pct ductile fracture appearance and impact energy values in excess of 190 ft lb (260 J).

The microstructure observed at the center of each of the four plates was a mixed structure of acicular ferrite with few polygonal ferrite grains. A typical structure [*e.g.*, the simulated-quenched 2 in. (50 mm) plate] is shown in Fig. 18.

From these results, it is evident that in plate thicknesses up to 3 in. (75 mm), the quenched and tempered Mn-Mo-Cb steel can meet a specification of 65 ksi

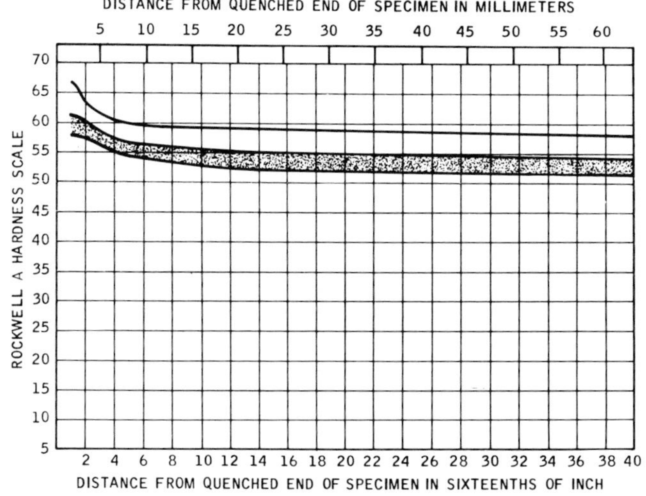

Fig. 16—Rockwell A hardenability data on as-quenched and quenched-and-tempered bars from Steel 5 austenitized at 1700°F (925°C). Temper band represents hardness data obtained after tempering for 1 h at 1100, 1150, 1200, and 1250°F (595, 620, 650, and 675°C). Composition: 0.06 C, 1.89 Mn, 0.15 Si, 0.33 Mo, 0.03 Cb, 0.06 Al, 0.006 N, 0.005 P, 0.004 S.

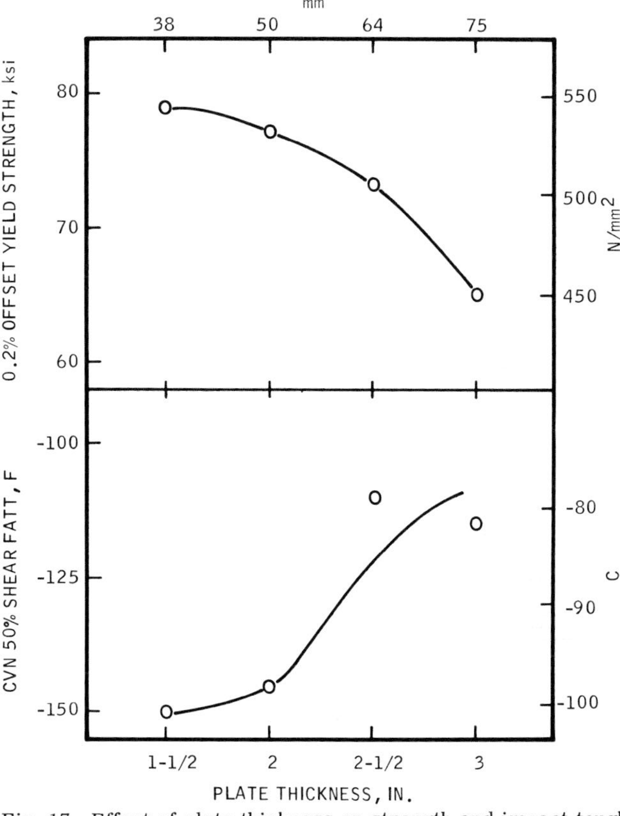

Fig. 17—Effect of plate thickness on strength and impact toughness of quenched and tempered Mn-Mo-Cb steel plates (Steel 5). Specimens were quenched from 1700°F (925°C) and tempered at 1200°F (650°C) for 1 h.

Table VII. Properties of Mn-Mo-Cb Steel Plates Spray Cooled to Simulate Immersion-Quenched Plates*

Simulated Plate Thickness, in. (mm)	0.2 Pct Offset Yield Strength, ksi (N/mm^2)	Tensile Strength, ksi (N/mm^2)	Elongation in 1 in. (25 mm), Pct	Reduction in Area, Pct	Full-Size Charpy V-Notch Impact Data: Energy Absorbed at −80°F (−62°C) ft lb (J)†	Full-Size Charpy V-Notch Impact Data: 50 Pct Shear Transition Temperature, °F (°C)
1 1/2 (38)	79.0 (545)	90.0 (620)	27	80	191 (259)	−150 (−100)
2 (50)	77.3 (533)	88.4 (609)	28	80	218 (296)	−145 (−100)
2 1/2 (65)	73.4 (506)	85.5 (589)	29	81	215 (292)	−110 (−80)
3 (75)	65.3 (450)	79.2 (546)	32	82	217 (294)	−115 (−80)

*Steel 5: Austenitized at 1700°F (925°C) and tempered for 1 h at 1200°F (650°C).
†Also equal to the upper shelf energy.

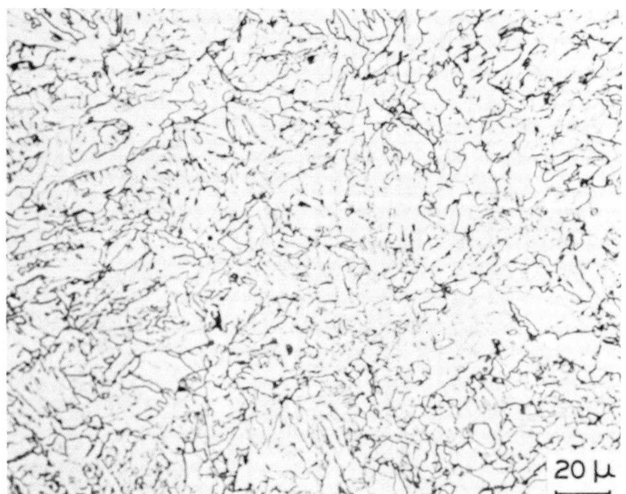

Fig. 18—Microstructure of simulated 2 in. (50 mm) thick Mn-Mo-Cb steel plate (Steel 5) quenched from 1700°F (925°C) and tempered at 1200°F (650°C) for 1 h. 2 pct Nital; Magnification 375 times.

(450 N/mm^2) minimum yield strength [with excellent toughness at a design temperature of −80°F (−60°C)] for linepipe fittings and valves. More recent data[16] have shown that raising the carbon content from 0.06 to 0.09 pct increased the yield strength of a simulated quenched and tempered 3 in. (75 mm) section to greater than 70 ksi (480 N/mm^2).

While only the "as heat treated" mechanical properties have been determined in this work, it is realized that many specifications require strength and toughness properties after a 3 pct plastic straining and aging treatment [550°F (290°C) for 1/2 h]. However, a recent study[17] compared three candidate steels (including the Mn-Mo-Cb steel) after subjecting them to the specified treatment. Although tests were performed on thinner sections than those simulated in the present report, it was concluded that the properties of the Mn-Mo-Cb steel, in the quenched and tempered condition, complied with the most stringent contemplated Arctic fittings requirements.

SUMMARY AND CONCLUSIONS

While commercial data have already proven that as-rolled Mn-Mo-Cb steels comply with the harsh requirements of Arctic linepipe application, the current laboratory program has shown that these acicular ferrite steels can be upgraded to meet higher yield strength-toughness property specifications. This aim was accomplished by utilizing moderate water spray cooling, such as that possible by descaling sprays, during the rolling sequence. Yield strength in excess of 70 ksi (480 N/mm^2) was produced in 3/4 in. (19 mm) thick plate by spray cooling at a rate of 6°F/s (3.3°C/s) between 1900°F and 1500°F (1040°C and 815°C). Because of the stress-strain behavior of these steels—continuous yielding, with a high rate of work hardening during the initial stages of plastic deformation—forming into U-O pipe, and then expanding, would produce a pipe of very high yield strength. In the present report, it is estimated that the yield strength of an expanded pipe, as measured by a flattened tensile specimen, could be as high as 88 ksi (605 N/mm^2). These steels also maintain good toughness. Increasing the amount of rolling deformation to 65 pct below 1750°F (955°C) lowered yield strength by about 8 ksi (55 N/mm^2), an effect attributed to increased precipitation of Cb(C,N) in austenite and the presence of more polygonal ferrite. It is suggested, therefore, that high strength can be developed by heavy deformation only when the steel possesses sufficient hardenability to transform completely to acicular ferrite. Lowering the reheat temperature to 2050°F (1120°C) improved toughness effectively without significantly affecting the yield strength, and produced impact transition temperatures surpassing the most stringent −80°F (−60°C) requirement for Arctic service.

Results on simulated-coiled pearlite-reduced type Mo-Cb steels as skelp for spiral-welded pipe fabrication indicated yield strengths in excess of 80 ksi (550 N/mm^2), plus 50 pct shear FATT values down to −80°F (−60°C). Properties were attributed to 1) a fine-grained ferrite matrix containing only a small amount of pearlite, 2) a relatively high dislocation density substructure, and 3) a fine dispersion of Cb(C,N) particles. Optimum coiling temperature was found to be 1150°F (620°C).

Potential application of quenched and tempered Mn-Mo-Cb steels for pipe-line fittings and valves under Arctic conditions was substantiated by strength and toughness properties reported for sections up to 3 in. (75 mm) in thickness. Yield strengths over 65 ksi (450 N/mm^2) were coupled with excellent toughness. A small increase in carbon content (to 0.09°C) successfully raised the yield strength of a 3 in. (75 mm) section to greater than 70 ksi (480 N/mm^2).

REFERENCES

1. *Material Specification,* Canadian Arctic Gas Pipeline Limited, Specification No. 2950-6-6, August 1, 1973.
2. M. Civallero and C. Parrini: *Proceedings of 16th Mechanical Working and Steel Processing Conference,* p. 413, Dolton, Illinois, January 23-24, 1974.
3. D. B. McCutcheon, T. W. Trumper, and J. D. Embury: Paper presented at 13th Annual Conference of Metallurgists, Toronto, August 25-28, 1974.
4. M. Civallero and C. Parrini: *Pipeline and Gas J.,* July, 1974, p. 35.
5. J. LeClerc: Metallurgical Director—USINOR (France); private communication.
6. *Strong Tough Structural Steels,* ISI P104, London, 1967.
7. Y. E. Smith, A. P. Coldren, and R. L. Cryderman: *Toward Improved Ductility and Toughness,* Symposium, Climax Molybdenum Company, Kyoto, Japan, October 25-26, 1971.
8. A. P. Coldren, Y. E. Smith, and R. L. Cryderman: *Proceedings, AIME Symposium on Processing and Properties of Low-Carbon Steel,* Cleveland, Ohio, October 1972.
9. G. Tither, A. P. Coldren, and J. L. Mihelich: Paper presented at 13th Annual Conference of Metallurgists, Toronto, August 25-28, 1974.
10. C. Parrini and T. Badino: Paper presented at "Journee Internationale de Siderurgie", Paris, France, October 4, 1974.
11. J. Malcolm Gray: *Proceedings, AIME Symposium on Processing and Properties of Low-Carbon Steel,* Cleveland, Ohio, October 1972.
12. J. Malcolm Gray: Paper presented to the Metallurgical Task Force on the Northwest Project, Dallas, Texas, January, 1972.
13. T. Taira, T. Osuka, and Y. Ishida: *Proceedings of 15th Mechanical Working and Steel Processing Conference,* p. 33, Pittsburgh, Pa., January 24, 1973.
14. C. Parrini: Taranto Works, Italsider, private communication.
15. A. L. Boegehold: *Metal Progr.,* May 1948, vol. 53, pp. 697-700.
16. G. T. Eldis: Climax Molybdenum Company, unpublished data.
17. R. J. Jesseman and R. C. Smith: ASME Publication 74-Pet-9, Paper presented at the Petroleum Mechanical Engineering Conference, Dallas, Texas, September 15-18, 1974.

A Constructional Alloy Steel for Arctic Service

B. G. REISDORF

An 0.1C-0.6Mn-1Ni-1Cr-0.3Mo heat-treated alloy steel has been developed for use in Arctic regions in applications requiring a minimum yield strength of 65 ksi and high toughness. This new steel should be suitable for welded applications for critical bridge members, line pipe fittings, and other structural members for which a high degree of toughness is required at subzero temperatures.

In a study to develop a constructional steel having a minimum yield strength of 65 ksi (448 MN/m^2) and having high toughness at as low as −80°F (−62°C), nine heats of steel with variation in C, Mn, Ni, Cu, Cr, V and N were melted in the laboratory, rolled to plate, quenched, and tempered. Tests of the plates showed that a 0.1C-0.6Mn-1.0Ni-1.0Cr-0.3Mo steel met the desired strength level and had excellent transverse Charpy V-notch (CVN) impact properties at as low as −80°F in plate thicknesses to at least 1 1/2 in. (38 mm). The 50 pct shear-fracture-appearance transition temperature was below −80°F, and the energy absorption exceeded 50 ft-lb at −80°F. An 80 ton electric-furnace heat of this composition was melted and processed to plate up to 2 in. (51 mm) thick and to 24 in. OD by 0.969 in. wall (610 by 24.6 mm) seamless pipe. Mechanical property tests of the heat-treated plate samples and pipe showed that the product met the desired strength and had excellent toughness at −80°F. Weld underbead cracking tests showed that the steel has negligible susceptibility to hydrogen-induced cracking.

DEVELOPING the oil and gas fields in the Arctic regions and transporting the product to market will require immense quantities of steel, some of which must have a higher-than-usual level of toughness at as low as −80°F (−62°C). To date, major applications for which these low-temperature, high-toughness steels have been specified have been line pipe and bridges, but other applications are sure to arise as development of Arctic resources progresses.

The purpose of the investigation discussed here was to develop an economical, weldable, steel in plate and tubular form suitable for producing hot-formed and heat-treated line-pipe fittings with a high level of toughness down to at least −80°F. For line pipe fitting applications, the steel would be provided in the hot-rolled condition; for other low-temperature applications not involving hot forming, steel plate would be supplied in the heat-treated condition.

STEEL COMPOSITIONS AND PROCESSING

In the initial part of this investigation, eight steels were melted in the laboratory as 500 lb (227 kg) heats, and processed to 1 in. (25 mm) thick plate with a cross-rolling ratio of 1.0. The aim base composition of these steels, in percent, was 0.08 C, 0.60 Mn, 1.0 Ni, 0.50 Cr, and 0.30 Mo; elements varied were C, Mn, Cu, Ni, Cr, V, and N. Chemical compositions of the eight steels (*A* through *H*) are given in Table I; elements varied are underlined.

Two plate samples from each heat were used to fabricate a simulated 2 in. (51 mm) thick plate. This was done by surface grinding a major surface of each of two 1 in. thick plates, placing the ground surfaces together, and welding the plates together at the faying edges. Both the 1 in. and the simulated 2 in. thick plates were heated in a 1650°F (900°C) furnace for 1 h per in. of thickness (time in furnace), and quenched in mildly agitated water. The plates were then tempered at 1250°F (677°C) for 1 h per in. of thickness (time in furnace), and air-cooled.

From each of these plates, two longitudinal 0.505 inch (12.8 mm) diameter tension-test specimens, and 12 longitudinal and 12 transverse Charpy V-notch (CVN) impact-test specimens were obtained. The test specimens were machined from the midthickness of the 1 in. thick plate, and from the quarterthickness of the 2 in. thick plate. The CVN specimens were tested over a range of temperatures to obtain the ductile-to-brittle transition temperature.

Results of tension tests and Charpy V-notch im-

B. G. REISDORF is associated with U.S. Steel Corp. Research Laboratory, Monroeville, PA 15146.

pact tests in combination with alloy addition costs were used to select the "best" steel (Steel *D*). An additional steel, Steel *J* (Table I) was melted to the aim composition of Steel *D* to assess the reproducibility of mechanical properties, and to provide material for weldability tests and strain-aging tests. Melting and rolling of Steel *J* was the same as that for the initial eight steels, but only 1 in. thick plate was heat-treated and examined. Plate samples of Steel *J* were tempered at 1200°F (649°C) for 1 h, and air cooled.

Strain-aging studies were conducted on a plate sample of Steel *J*; its dimensions were 1 × 4 × 18 in. (25 × 102 × 457 mm). The sample was cold-rolled in the 18 in. (transverse) direction to a total elongation of 3 pct in four passes, and then aged at 550°F (288°C) for 30 min. Weld underbead-cracking tests were conducted on Steel *J* by using the procedure[1] illustrated in Fig. 1.

Mechanical-test results from Steel *J*, which will be discussed subsequently, confirmed the good strength and toughness values obtained with Steel *D*, and encouraged planning of a production heat of the Ni-Cr-Mo steel. Accordingly, an 80 ton (73 metric ton) electric-furnace heat was melted, and processed to four 5/8 in. (15.9 mm) thick plates, four 1 in. thick plates, three 2 in. thick plates, and four lengths of 24 in. OD × 0.969 in. wall (610 × 24.6 mm) seamless pipe. Steel composition° 0.09C, 0.58Mn, 0.007P, 0.010S, 0.3Si, 1.05Ni, 0.98Cr, 0.30Mo, 0.03Al. A plate sample from one plate of each thickness and a pipe sample were heat-treated and mechanical properties were determined.

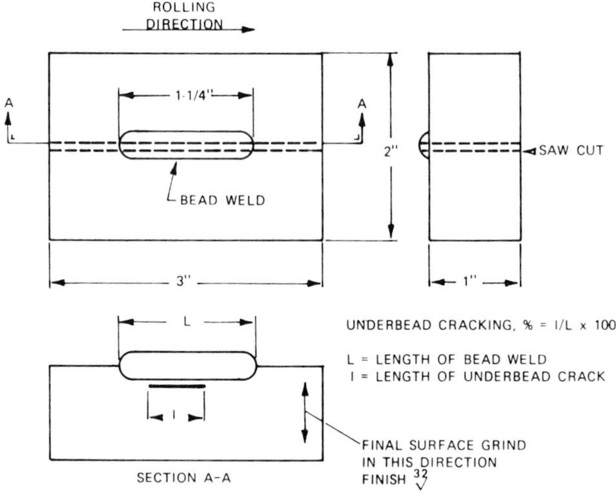

WELDING CONDITIONS

ELECTRODE	E6010
ELECTRODE DIAMETER, inch	1/8
CURRENT, amperes	100
ARC VOLTAGE, volts	25
SPEED OF TRAVEL, ipm	10.0
TYPE OF CURRENT	DC-RP
HEAT INPUT, joules per inch	15,000

* DIRECT CURRENT - REVERSE POLARITY

Fig. 1—Preparation of underbead-cracking specimen and conditions used for welding 1 in. thick plates.

RESULTS

Eight Laboratory-Melted Steels

Except for the 2 in. thick sample of Steel *A*, all the plates had a yield strength in excess of 65.0 ksi (448 MN/m^2). Strength and toughness values of these steels showed only minor differences, except for Steels *E* and *F*. These two steels contained vanadium, and were stronger and poorer in toughness than the other steels. Among the remaining five steels, Steel *D* with a transverse 50 pct shear-fracture-appearance transition temperature (FATT) of −125°F (−87°C) for the 1 in. thick plate and −90°F (−68°C) for the 2 in. thick plate, generally had the best toughness. At the FATT, the CVN energy absorption of all the plates exceeded 50 ft-lb (68 J) in both longitudinal and transverse directions.

Laboratory-Melted Steel *J*

Test results for Steel *J*, which was made to the aim composition of Steel *D*, are given in Table III. Yield and tensile strengths of 1-in.-thick plate of Steel *J* are about 10 ksi (69 MN/m^2) higher than those of Steel *D*, probably because of the lower tempering temperature used to treat Steel *J*. Transverse CVN specimens tested at −80°F (−62°C) exhibited 100 pct shear fracture, and the FATT was about −140°F (−95°C).

Straining 3 pct and aging at 550°F (258°C) for 30 min raised the yield strength about 20 ksi (138 MN/m^2) and the tensile strength about 10 ksi. Transverse CVN specimens from the strained and aged plate sample had 100 pct shear when tested at −80°F, and the FATT was about −110°F (−79°C) or about 30°F (17°C) higher than that of the unstrained plate.

Table I. Compositions of Laboratory Steels

Steel*	C	Mn	P	S	Si	Cu	Ni	Cr	Mo	V	Al	N
A	0.083	0.60	0.010	0.010	0.24	–	0.98	0.50	0.30	–	0.024	0.007
B	0.093	0.60	0.010	0.011	0.24	–	1.50	0.51	0.30	–	0.026	0.007
C	0.086	0.56	0.010	0.011	0.23	0.49	0.98	0.50	0.30	–	0.023	0.007
D	0.084	0.59	0.010	0.010	0.24	–	1.00	0.99	0.30	–	0.026	0.007
E	0.076	0.59	0.009	0.010	0.24	–	0.98	0.50	0.30	0.081	0.023	0.007
F	0.078	0.60	0.010	0.010	0.23	–	0.99	0.50	0.30	0.081	0.024	0.011
G	0.076	0.98	0.010	0.010	0.24	–	0.98	0.49	0.31	–	0.023	0.007
H	0.12	0.60	0.010	0.011	0.23	–	0.98	0.50	0.30	–	0.025	0.007
J	0.10	0.59	0.007	0.008	0.23	–	0.99	0.99	0.29	–	0.035	0.006

*Amounts are given in percentage; elements varied are underlined.

Table II. Mechanical Properties of Production Heat (Steel *J*)

Tensile Properties							
Heat-Treated Plate*				Strain-Aged Plate†			
Yield Strength, ksi	Tensile Strength, ksi	Elongation in 1.4 In., Pct	Reduction of of Area, Pct	Yield Strength, ksi	Tensile Strength, ksi	Elongation in 2 In., Pct	Reduction of Area, Pct
78.4	93.0	‡	80.4	98.4	103.2	20.0	75.9

Transverse CVN Impact Properties						
	Heat-Treated Plate			Strain-Aged Plate		
Test Temperature, °F	Energy Absorbed, ft-lb	Shear-Fracture Appearance, Pct	Lateral Expansion, mils	Energy Absorbed, ft-lb	Shear-Fracture Appearance, Pct	Lateral Expansion, mils
−80	192, 197, 206	100, 100, 100	96, 95, 95	167, 187	100, 100	98, 102
−100	–	–	–	114, 149	55, 100	74, 90
−120	179, 140, 169	100, 65, 85	95, 83, 90	74, 121	40, 55	50, 78
−140	107, 96, 130	50, 50, 60	68, 61, 74	–	–	–

*Austenitized at 1650°F for 1 h, water quenched, tempered at 1200°F for 1 h, and air cooled. Longitudinal tests of 0.357 in. diam specimens.
†Transverse tests of 0.505 in. diam. specimens.
‡Both test specimens broke at gage marks.

Table III. Mechanical Properties of Production Heat

	Tensile Properties				CVN Tests at −80°F (−62°C)			
					Longitudinal		Transverse	
Product*	Yield Strength, ksi	Tensile Strength, ksi	Elongation in 2 In., Pct	Reduction of Area, Pct	Energy Absorbed, ft-lb	Shear Fracture Appearance, Pct	Energy Absorbed, ft-lb	Shear-Fracture Appearance, Pct
5/8 in. plate	78.8	91.4	26.0†	70.2	120, 114, 106	100, 90, 90	74, 59, 86	70, 55, 90
1 in. plate	73.6	88.1	25.5	73.5	148, 134, 157	65, 60, 100	126, 125, 156	70, 65, 100
2 in. plate	65.9	82.4	28.5	76.4	201, 191, 208	100, 100, 100	140, 100, 116	60, 50, 55
0.969 in. wall pipe	73.9	88.0	26.5	77.8			176, 180, 175	100, 100, 100

*Plates were tempered in a 1275°F furnace for 1 1/2 h per in. of thickness, and the pipe was tempered in a 1200°F furnace for 2 h.
†Elongation in 1.4 in. (35.6 mm).

Weld-underbead-cracking tests made at 75°F (24°C) resulted in an average cracking of only 5 pct with a range of zero to 12 pct on eight tests. This low value of cracking indicates the steel has negligible susceptibility to hydrogen-induced cracking.

80 Ton Production Heat

As noted, the composition of an 80 ton electric-furnace heat of the steel is very similar to that of the laboratory-melted Steels *D* and *J*. Tension-test results (Table III) showed that all the product tested met a 65 ksi (448 MN/m^2) minimum yield strength and an 80 ksi (552 MN/m^2) minimum tensile strength, although the 2 in. thick plate was marginal with respect to a minimum yield strength of 65 ksi. Transverse CVN tests showed that all product tested met or exceeded 50 pct shear fracture at −80°F. At this temperature, the minimum average CVN energy absorption was 73 ft-lb (99 J) and the minimum energy absorption for any single specimen was 59 ft-lb (80 J). Thus the 0.1C-0.6Mn-1.0Ni-1.0Cr-0.3Mo steel meets minimum requirements of 65 ksi (448 MN/m^2) yield strength, and 50 pct shear-fracture appearance, and 50 ft-lb (68 J) energy absorption from transverse CVN specimens in section thicknesses through at least 1 1/2 in. (38 mm).

CONCLUSIONS

1) As the result of a series of laboratory studies and production trials, a silicon-aluminum-killed, fine-grain steel with the nominal composition, in percent, of 0.08C, 0.60Mn, 1.0Ni, 1.0Cr and 0.30Mo has been developed for use in hot-formed and heat-treated line-pipe fittings for Arctic use.

2) The steel exhibits 65 ksi minimum yield strength in sections to at least 1 1/2 in., high shelf energy (generally over 100 ft-lb), and a 50 pct shear-fracture-appearance transition temperature (FATT) of less than −80°F. These properties have been obtained in plate and seamless pipe from an 80 ton electric-furnace production heat.

3) Strain-aging studies showed that when the heat-treated plate was strained 3 pct and aged at 550°F for 30 min, the yield strength was raised about 20 ksi and the tensile strength about 10 ksi. The 50 pct shear FATT was raised about 30°F.

4) Weld underbead-cracking tests indicated that the steel has negligible susceptibility to hydrogen-induced cracking.

REFERENCE

1. R. D. Stout and W. D. Doty: Weldability of Steels, 2nd ed. p. 252, Welding Research Council, New York, N.Y., 1971.

The Selection of Mild Steel for Corrosion Service

*By the ASM Committee on Mild Steel Corrosion**

DURING the last half century, the corrosion of iron and steel has become accepted as a phenomenon that is essentially electrolytic, especially where attack depends on the simultaneous presence of moisture and oxygen. Because corrosion is generally electrolytic, dissolved salts, in the absence of an inhibiting effect, increase the rate of corrosion by increasing the electrical conductivity.

It has been demonstrated, notably by W. R. Whitney, that ferrous metals, in the absence of oxygen, will go into solution in water until a state of equilibrium exists. Ferrous and hydroxyl ions result from the displacement of hydrogen, which is "plated out" on the surface of the metal:

$$Fe + 2H_2O = Fe^{++} + 2OH^- + H_2 \quad (1)$$

Inasmuch as the reactants are in their standard states, the equilibrium constant for Reaction (1) may be given as:

$$K = [Fe^{++}][OH^-]^2 P_{H_2}$$

where the terms in brackets are usually molar concentrations. If the equilibrium constant is low enough and the vapor space small enough, the reaction will stop before there is saturation with respect to ferrous hydroxide. If the constant is high enough and the vapor space large enough, ferrous hydroxide will precipitate until the reaction builds up the pressure of hydrogen to a level which will stop the reaction. Whitney's experiment suggests that a condition can exist that will stop the reaction before precipitation occurs.

In the event the iron or steel is in electrical contact with another metal, lower in the galvanic series, the hydrogen will be plated out preferentially on the contacting metal even though the ferrous ions are still formed at the surface of the ferrous metal.

Where there is no electrolytic coupling, the film of hydrogen will be on the surface of the iron or steel and, in the absence of oxygen, will hinder further corrosion by insulating the metal from the water and by its own tendency toward a back reaction. The formation of such a gas film on metal is called "polarization". Oxygen, if present, reacts with the hydrogen film, removes it from the surface of the metal in the formation of water and also reacts with the ferrous ions to form ferric ions in such concentration as to exceed the solubility product of ferric hydroxide and, hence, a precipitate of ferric hydroxide is formed. Both reactions are involved in the corrosion process; the removal of the hydrogen film allows the solution to contact the iron or steel and the precipitation of ferric hydroxide reduces the concentration of dissolved iron, thus promoting further dissolution. Since iron or steel will go into solution in water in the absence of air or oxygen by an ionic displacement reaction, the phenomenon is electrolytic.

In the absence of air, the dissolved iron exists as ferrous ions. Oxygen, in the presence of water, reacts with ferrous ions to produce ferric ions:

$$4Fe^{++} + O_2 + 2H_2O = 4Fe^{+++} + 4OH^- \quad (2)$$

As this reaction proceeds and ferrous ions are removed, Reaction (1) will go to the right, giving still more ferrous ions for Reaction (2).

Because of the very low solubility product of ferric hydroxide, this compound will be precipitated as a reddish brown solid almost as soon as Reaction (2) commences.

None of these reactions takes place in dry air, and Reaction (2) will not take place in the absence of air or oxygen, but since air is usually present if water is, atmospheric corrosion of ordinary steel can be expected if it is subjected to an environment where there is moisture.

Rate of Corrosion. The rate of corrosion of every piece of iron or steel depends on the individual conditions existing in the particular locality in which the metal is in use, as well as on the precautions that are taken to prevent corrosion.

The two substances besides iron necessary for ferrous corrosion in most natural environments are liquid water and oxygen. Iron samples that are wholly immersed corrode faster if uninhibited water is moving rapidly than if it is in a state of rest, but less rapidly where the velocity is high if an inhibitor is present. The rate of corrosion of iron tanks or pipes holding water is much slower if they are kept full than if they are alternately wet and dry. Atmospheric corrosion proceeds much more rapidly in districts where the atmosphere is chemically polluted by vapors of sulfur oxides and the like than it does in districts where the atmosphere is free from contamination. Steel will corrode much faster in the vicinity of naturally occurring sea water, because of the sodium chloride in the atmosphere, than it will in ordinary rural areas.

The corrosion of iron or steel, where an electrolyte is present, is accelerated or retarded by contact with other metallic materials, de-

*WILLIAM A. PENNINGTON, *Chairman*, Professor of Metallurgy, University of Maryland; CHESTER ANDERSON, Assistant Directing Engineer, Crane Co.; W. G. ASHBAUGH, Group Leader, Corrosion and Materials Group, Union Carbide Chemicals Co.; J. M. BIALOSKY, Metallurgical Laboratory, Koppers Co., Inc.; H. M. CANAVAN, Chief Inspector, Boiler Div., Mutual Boiler and Machinery Insurance Co.; GRAHAM B. COOPER, Graham Research Laboratory, Jones & Laughlin Steel Corp.

G. A. ELLINGER, Chief, Corrosion Section, National Bureau of Standards; O. B. ELLIS, Senior Research Engineer, Armco Steel Corp.; E. R. HELMER, Corrosion Engineer, Chain Belt Co.; H. F. HINST, Chief Metallurgist, Keystone Street Plant, Tubular Products Div., Babcock & Wilcox Co.; RICHARD B. KESLER, Chemical Engineering Group, Institue of Paper Chemistry; C. P. LARRABEE, Applied Research Laboratory, U. S. Steel Corp.

T. E. LARSON, Head, Chemistry Section, State Water Survey Div., State of Illinois; JOHN R. LECRON, Metallurgical Engineer, Bethlehem Steel Co.; LOGAN MAIR, Senior Metallurgist, Service Dept., Inland Steel Co.; JOHN MIKULAK, Assistant to Vice President, Worthington Corp.; HAROLD MYERS, Research Metallurgist, Edward Valves, Inc.; N. A. NIELSEN, Senior Research Associate, Materials Technology Section, E. I. du Pont de Nemours & Co., Inc.

pending on whether the metal is electropositive or electronegative with respect to iron. The corrosion rate, like the rate of most chemical reactions, usually increases with rising temperature.

The rate of corrosion depends also on the character of the steel as determined by its chemical composition and heat treatment. For instance, the addition of 0.2% Cu improves the resistance of steels to atmospheric corrosion but generally not to underwater corrosion.

Scope. This article deals with low-carbon and structural steels in which the alloy content is no higher than in the group of materials known as high-strength low-alloy steel. A few data are also included for steel containing 5% Cr. The article is limited to the behavior of uncoated steels even though in many applications the metal is coated before or after it is put in service.

Atmospheric Corrosion

Dry rural atmospheres produce the least corrosion. High humidity or salt in the air or both accelerate corrosion. Certain industrial atmospheres give the highest corrosion rates.

ASTM Perforation Tests. In 1916 Committee A-5 of the American Society for Testing Materials exposed groups of corrugated black (pickled) sheets from several manufacturers at three test locations. The sheets, about 24 by 96 in., were placed at an angle of 30° to the horizontal, facing south. Periodic inspections were made by a subcommittee representing producers, consumers and general-interest members. Failure consisted of a visible perforation. The results of these life tests are shown in Fig. 1. The tests at Pittsburgh (5.8 years) and Ft. Sheridan (11 years) had to be discontinued before the data were complete, but at Annapolis the exposure continued for 34.5 years until all 214 sheets had failed. Pittsburgh has an industrial, Ft. Sheridan an urban and Annapolis a rural atmosphere.

In 1926 a test was started at Sandy Hook, N.J., in an industrial-marine atmosphere. The specimens were 28 by 30 in., and the means of exposure was the same as in the earlier tests. Results are plotted in the right-hand chart of Fig. 1, as a function of the copper content of the steels.

The 1954 report of the ASTM committee contains the following summarizing statements:

1. Sheets of open-hearth iron, open-hearth steel, Bessemer steel, and wrought iron (hand puddled), with copper contents of about 0.2%, remained unperforated longer than similar grades with residual copper contents below 0.04%.
2. The beneficial effect of both copper and phosphorus on the atmospheric corrosion resistance of steels was demonstrated by the longer life of the higher-phosphorus steels (Bessemer copper steel and acid open-hearth steel).
3. The benefit of copper does not cease at comparatively low values (0.1%). Steels containing about 0.2% Cu (or more) had longer life at each location than similar steels with 0.1% Cu.
4. Considering open-hearth iron, open-hearth steel, and Bessemer steel, the copperized materials had 1.7, 2.5, and 2.9 times the life of the residual-copper (less than 0.05%) materials at Annapolis, Ft. Sheridan, and Pittsburgh, respectively. For the two wrought irons (copper-bearing and residual-copper), the similar factor was 1.1 at both Annapolis and Pittsburgh. Too few of the wrought iron sheets perforated during the 11-year exposure at Ft. Sheridan to make a comparison.
5. The ratios of corrosivity of the three atmospheres varied greatly when the years to failure of different steels were used as a means of comparison. The Pittsburgh and Ft. Sheridan locations were, on the average, 8.1 and 2.6 times as corrosive as Annapolis, respectively.
6. The corrosiveness of the several atmospheres varies more than the differences in corrosion resistance of the steels or irons tested. The greatest ratio found for the average life of the most resistant steel to the average life of the least resistant is about 4 to 1, whereas the greatest ratio, for the corrosiveness of the Pittsburgh atmosphere to that of Annapolis, is 12 to 1.

Other Tests, Rural Atmospheres. Plain carbon, copper-bearing, and high-strength low-alloy steels were tested in a rural atmosphere in Nebraska where the relative humidity is predominantly below 60% and yet averages somewhat higher than 50%. The specimens, sheets 8 by 10 in., were exposed for periods of ½, one, two, three and six years. The corrosion data and ranges in composition are shown in Fig. 2.

Marine, 80 Ft from Shoreline. A similar test using the same materials was made at Kure Beach, N.C., where the steels were exposed about 80 ft from the shoreline on the land side. Curves of weight loss are shown in Fig. 2.

Marine, 800 Ft from Shoreline. Another test was conducted at Kure Beach, N.C., where 4 by 6-in. sheet specimens were exposed 800 ft from the sea. Compositions are given in Table 1 and corrosion results in Fig. 2. Again plain carbon, copper-bearing, and low-alloy steels were tested. The results are much the same as for the samples 80 ft from the sea. Wrought iron (0.12% Cu) was included in this test. It corroded linearly with time, being completely destroyed in 7.5 years, in which time about 900 g per sq ft of surface would have wasted away if the specimen had been sufficiently thick. This amount is equivalent to an average penetration of 0.050 in. for this period.

Mine Shaft. The same three types of steel, in the form of 5 by 7-in. specimens, were exposed for periods of one, two and three years in a coal mine hoisting shaft where the temperature range was 65 to 78 F and the air was saturated with water vapor. The points from which the curves of Fig. 2 were constructed are the averages of two specimens of each steel.

Industrial Atmosphere, Relative Humidity Greater Than 60%. The same three types of steel were exposed in an industrial atmosphere at East Chicago, Ind., where the relative humidity was predominantly higher than 60%. The specimens, 18 by 2 in., were exposed vertically. Time-corrosion curves are shown in Fig. 2 (right).

In another series of atmospheric corrosion tests, the same three types of steel were exposed for ten years (1947 to 1957), the test panels being 4 by 6 in., hot rolled to approximately 0.125 in. thick. The specimens were exposed at 30° to the horizontal, facing south. Rust was removed by scrubbing the speci-

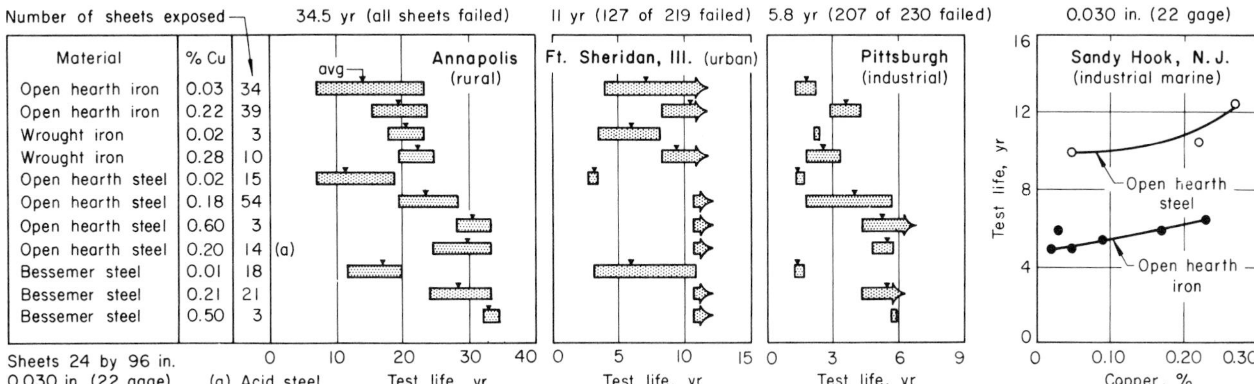

Fig. 1. Comparison of inspection results of atmospheric exposure tests at four locations. Number of sheets refers to the Annapolis site; the same or a larger number of sheets were exposed at the other locations, except that in five instances, one less sheet was tested. (Proc ASTM, 54, 110, 1954)

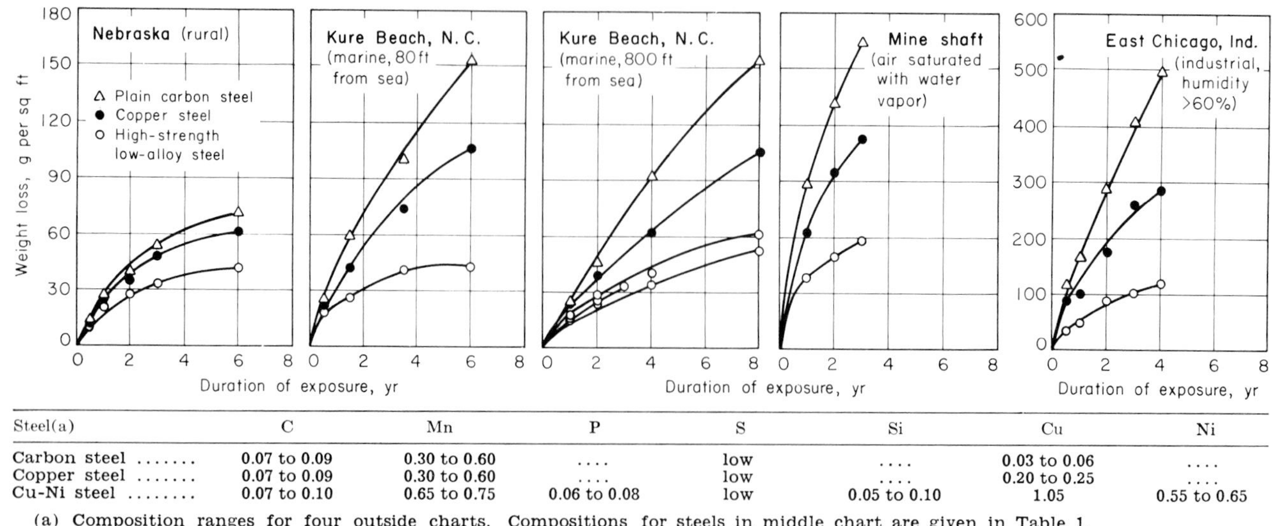

Steel(a)	C	Mn	P	S	Si	Cu	Ni
Carbon steel	0.07 to 0.09	0.30 to 0.60	....	low	....	0.03 to 0.06	....
Copper steel	0.07 to 0.09	0.30 to 0.60	....	low	....	0.20 to 0.25	....
Cu-Ni steel	0.07 to 0.10	0.65 to 0.75	0.06 to 0.08	low	0.05 to 0.10	1.05	0.55 to 0.65

(a) Composition ranges for four outside charts. Compositions for steels in middle chart are given in Table 1.

Fig. 2. Atmospheric corrosion data for three types of steel in five locations

mens with a wire brush and then pickling in inhibited acid. Corrosion results and chemical analyses are given in Fig. 3, corrosion being expressed as weight loss per sq ft of surface.

Four testing sites were employed: Kansas City, Mo., New Orleans, Pittsburgh and New York City. The atmospheric conditions of these sites are as follows, the humidity figures being taken from records of the U.S. Weather Bureau for the ten-year period of test exposure:

Kansas City—nonindustrial, average humidity 66%

New Orleans—light industrial, semitropical, average humidity 70%

Pittsburgh—moderate industrial, average humidity 70%

New York City—heavy industrial, semimarine, average humidity 66%

On comparing the results obtained in these tests with those shown in Fig. 2, 4 and 5, there may be apparent discrepancies. For example, the weight loss in Kansas City for carbon steel is only about half that obtained for the Nebraska, South Bend or Bethlehem sites in ten years. This difference can be explained, in part, by the higher phosphorus content of the carbon steel exposed in Kansas City. Phosphorus, even in small concentrations, is most effective in enhancing atmospheric corrosion resistance of ferrous alloys, especially those of low alloy content.

Semirural, Industrial and Marine. Figure 4 and caption give test results and compositions of steels exposed

Table 1. Compositions of Steels Tested at Kure Beach, N. C., 800 Ft from the Sea

Steel	C	Mn	P	Cu	Ni
1	0.06	0.36	0.007	0.08	0.02
2	0.06	0.35	0.008	0.30	0.13
3	0.05	0.37	0.007	0.91	1.65
4	0.16	0.46	0.073	0.98	1.93
5	0.10	0.82	0.016	0.42	0.53

Corrosion-time curves are shown in the center chart of Fig. 2 (above).

at South Bend, Pa., Kearny, N.J., and Kure Beach, N.C., representing conditions classified as semirural, industrial and marine, respectively. Data for the corrosion-time curves were obtained from weighed scale-free (pickled) specimens (4 by 6-in. sheet), supported by porcelain insulators mounted on steel racks attached to galvanized pipe frames. The specimens were at an angle of 30° to the horizontal, facing south. Specimens were removed at the times shown, cleaned of rust by sodium hydride, and reweighed. After cleaning, specimens were not re-exposed.

The atmosphere at South Bend, Pa., is clean, but sulfur compounds from burning coal slack dumps a few miles away render it intermediate in corrosiveness between that of Kearny, N.J., and a rural atmosphere such as the ASTM test site at State College, Pa. The industrial atmosphere at Kearny has virtually the same corrosiveness as that of New York City.

The corrosiveness to steel specimens on test racks at the 800-ft lot at Kure Beach depends on the length of exposure. Sea salts deposited by winds are largely removed by rain from skyward surfaces but accumulate on the groundward surfaces so that the corrosion rates of carbon steels frequently increase with time and those of alloy steels do not decrease with time as they do in industrial atmospheres.

Rural and Industrial. Thirteen steels were exposed for ten years at Rankin, Pa., Columbus, Ohio, and in two locations in and near Bethlehem, Pa. The specimens, 4 by 5⅛ in. by 16 gage, were exposed in a vertical position. Amount of corrosion was determined by weight loss, the specimens being cleaned by the Bullard-Dunn process, which involves cathodic cleaning in sulfuric acid containing a small amount of a tin salt. Tin plated on the specimen is removed anodically in a strong alkaline bath. The weight loss data for the ten-year period are given in Fig. 5, each horizontal bar representing the average of two specimens.

Industrial. A test was conducted where machined cylindrical specimens (0.5-in. diam by 3.25 in. long) were exposed to West Milwaukee industrial atmosphere. The specimens were suspended vertically and rotated each week to insure a more nearly uniform exposure of the entire surface. Among the various metals exposed was 1020 steel containing 0.09% Cu, there being three specimens for each test period. The

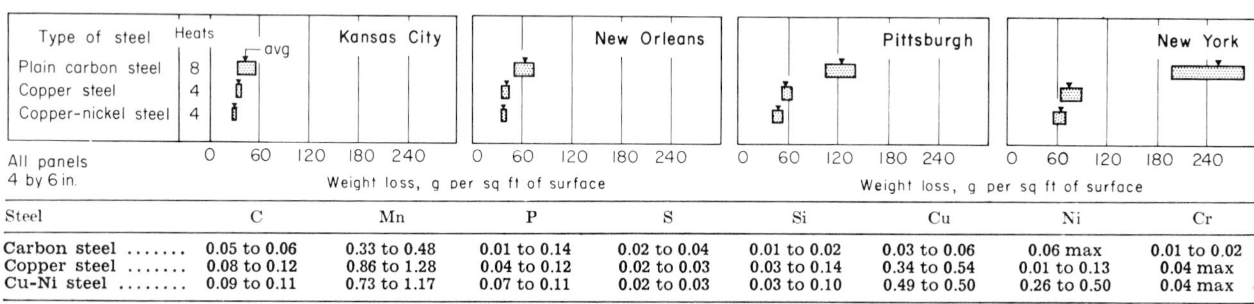

Steel	C	Mn	P	S	Si	Cu	Ni	Cr
Carbon steel	0.05 to 0.06	0.33 to 0.48	0.01 to 0.14	0.02 to 0.04	0.01 to 0.02	0.03 to 0.06	0.06 max	0.01 to 0.02
Copper steel	0.08 to 0.12	0.86 to 1.28	0.04 to 0.12	0.02 to 0.03	0.03 to 0.14	0.34 to 0.54	0.01 to 0.13	0.04 max
Cu-Ni steel	0.09 to 0.11	0.73 to 1.17	0.07 to 0.11	0.02 to 0.03	0.03 to 0.10	0.49 to 0.50	0.26 to 0.50	0.04 max

Fig. 3. Atmospheric corrosion data for three types of steel in four locations

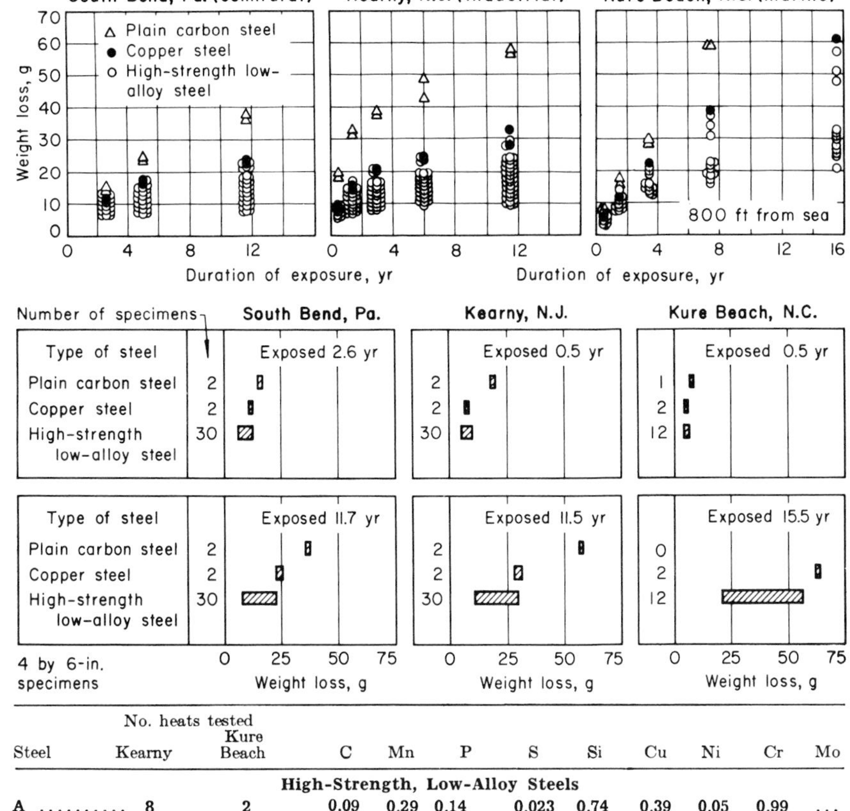

Steel	No. heats tested Kearny	Kure Beach	C	Mn	P	S	Si	Cu	Ni	Cr	Mo
			High-Strength, Low-Alloy Steels								
A	8	2	0.09	0.29	0.14	0.023	0.74	0.39	0.05	0.99	...
B	5	2	0.13	0.48	0.10	0.025	0.008	0.69	1.5	0.11	0.12
C	2	2	0.08	0.90	0.009	0.016	0.02	1.5	1.0	0.03	0.09
D	1	1	0.08	0.49	0.11	0.020	0.14	0.95	0.57	0.01	...
E	3	1	0.09	0.40	0.10	0.021	0.064	0.55	0.64	0.04	0.10
F	3	.	0.06	0.37	0.02	0.018	0.054	1.01	2.0	0.01	...
G	2	1	0.19	0.65	0.022	0.020	0.88	0.23	0.11	0.65	...
H	1	.	0.09	0.56	0.08	0.027	0.06	0.52	0.30	0.31	...
I	3	2	0.11	0.52	0.08	0.027	0.004	0.40	0.10	0.05	...
J	2	1	0.20	1.0	0.012	0.028	0.24	0.21	0.05	0.03	...
			Structural Copper Steel								
K	2	1	0.25	0.48	0.019	0.038	0.020	0.24	0.01	0.08	...
			Structural Carbon Steel								
L	2	2	0.21	0.53	0.016	0.034	0.023	0.025	0.01	0.03	...

Fig. 4. Atmospheric corrosion data for three types of steel in three locations (C. P. Larrabee, Corrosion, 9, 259, 1953)

average weight loss was 1.192, 1.513 and 1.950 g per specimen for one, two and three-year exposure, respectively, the corresponding ranges being 0.250, 0.206 and 0.304 g. As would be expected, the time-corrosion curve is nearly parabolic.

Heavy Industrial. Reproducible data are difficult to obtain in an industrial atmosphere near some kinds of manufacturing plants. At one time a statistical analysis of the nature of the industrial atmosphere at a large chemical plant was attempted. The analysis was defeated by the large short-range (hourly) fluctuations in the composition of the atmosphere at a selected location. This short-range variation occurred because of the changes in plant operations and shifts in wind velocity. Gross seasonal changes were also encountered. It is standard plant practice to cover exposed steel surfaces with protective coatings.

Fresh Water

Data on the useful life of mild steel in contact with fresh water are incomplete without information on: (1) mineral quality of the water, (2) acidity, (3) presence or absence of dissolved oxygen, (4) velocity of flow, (5) temperature, and (6) environmental conditions such as contact with copper-bearing metals and irregularly covered surfaces causing localized corrosion. The corrosiveness of water depends especially on the mineral content, pH (acidity) and dissolved oxygen content.

Steel pipe is not used for domestic water, and other steel containers are seldom used for household or industrial water or for water distribution systems without protective coatings of zinc, cement, enamel, paint, coal tar or a coal-tar-base product. Such coatings are necessary not only to reduce corrosion but also to avoid "red water" problems. Steels exposed in holidays or joints will not resist aggressive water, but will be protected by nonaggressive water or by water that has been treated properly.

No Minerals — Dissolved Oxygen Present. In the absence of minerals, increasing pH decreases corrosion rates by water containing dissolved oxygen. However, if the pH is near, but not above, that required for complete protection, pitting occurs, which rapidly decreases the useful life of mild steel. Crevices at joints and welds which do not permit oxygen to be maintained at the surface, are subject to local corrosive attack.

Pitting takes place at local unprotected points of corrosion where the corrosion products prevent the diffusion of oxygen to the metal surface and thereby permit differences in oxygen concentration at the metal surface. Similar pitting can occur under deposits of debris. Also, differences in oxygen concentration, as at the water line of surfaces exposed partly to air and partly to water, will cause pitting. Corrosion rate increases with temperature.

No Minerals — Dissolved Oxygen Absent. In the absence of dissolved oxygen, the pH of "high-purity" water, in mild steel containers, adjusts itself to about 8.4 and corrosion becomes negligible. However, all natural waters are mineralized to some degree.

Noncarbonate Minerals — Dissolved Oxygen Present. In the absence of carbonate minerals, increasing concentrations of other common minerals, as chloride and sulfate salts, increase the corrosion rate at all pH values below the pitting range of pH, and increase pitting when the pH is just below that required for protection in the presence of dissolved oxygen. Increasing temperature accelerates both general corrosion and pitting.

Carbonate Minerals — Dissolved Oxygen Present. Carbonate minerals, indicated by the alkalinity determination for bicarbonates, inhibit corrosion, acting contrary to the accelerating salts of chloride and sulfate in waters containing dissolved oxygen. In the absence of calcium, this inhibition is maximum at a pH of 6.5 to 7.0 in concentrations above five to ten times the chloride and sulfate salt concentration and is minimum at pH 8 to 9. At concentrations decreasing below five times the chloride and sulfate salts, corrosion rates increase. Since nearly all natural domestic waters contain carbonate minerals and in addition usually contain chloride and sulfate salts, this too is a criterion in classification.

Two examples involving carbonate minerals may be cited: (1) A high-capacity hot water heater using zeolite-softened water, containing virtually no chloride or sulfate salts but 300 ppm (as $CaCO_3$) alkalinity and 6 ppm dissolved oxygen, was still in service at a hotel in Urbana, Ill., after 11 years. For the first two years, 13 ppm dissolved oxygen was present by deliberate addition for experimental purposes. This is an unusual example of prolonged useful life of mild steel resulting from the unusual quality of water. (2) In steam condensate waters that are corrosive because of their content of carbonic acid, neutralization with some such compound as ammonia or an amine is necessary for inhibition with or without dissolved oxygen present.

Minerals — Dissolved Oxygen Absent. In the absence of dissolved oxygen, the types of mineralization are less important with respect to useful life of mild steel as indicated by the following two examples: (1) Plumbing experience has indicated that properly designed hot water heating systems can be made of steel, provided no water (domestic) is lost from the system; avoiding the addition of make-up water containing dissolved oxygen also prevents loss of corrosion products that inhibit the corrosion. (2) The useful life of steel used for feedwater heaters, boilers and piping for power plants may be extended considerably by maintaining an oxygen-free boiler feedwater.

Calcium Salts — Dissolved Oxygen Present. From the standpoint of corrosivity, stability as indicated by saturation with calcium carbonate is the most widely accepted criterion in classification. However, for bare mild steel a very significant supersaturation must exist to form a visible white deposit of calcium carbonate; thus this criterion is not infallible. It has specific limitations:

1 A minimum alkalinity of 50 to 100 ppm (calculated as $CaCO_3$) and a minimum of about 50 ppm (as $CaCO_3$) calcium must be present at normal temperatures (32 to 160 F) for even a small degree of extended life.
2 The greater the concentrations of calcium and alkalinity, the greater is the protective action of the water. However, such increasingly high concentrations are responsible for an increasing tendency to deposit objectionable quantities of scale at temperatures above that at which saturation stability is established.
3 The protective action is enhanced by movement of the water and decreased at near-stagnant conditions.
4 The protective action may be nullified at higher temperatures when the pH is high enough to deposit nonprotective magnesium hydroxide.
5 Pitting and tuberculation will occur in the presence of dissolved oxygen if the stability and velocity are near, but still below, that required for complete protection.
6 The protective action is decreased by increasing proportions of chloride and sulfate salts above a ratio of about 0.1 or 0.2 to 1 with respect to alkalinity. This limitation becomes less significant in the absence of dissolved oxygen.

Experience at state institutions in Illinois has shown that:

1 The useful life of unprotected steel hot water tanks (500 to 1500 gal) may be from two to more than 20 years, depending on the quality of the water, temperature, use, and thickness of steel.
2 Elevated and ground steel storage tanks offer no standard of criterion since the maintenance by paint, and cathodic protection and repair work are as variable as the quality of water. With adequate cathodic protection, the useful life depends on the protection and maintenance provided to the exterior and to the interior above the water line.
3 Copper-bearing steel has no advantage over mild steel under conditions of fresh-water exposure.

The qualities of water are too numerous to classify, except in a general way. The mineral quality, pH (acidity) and dissolved oxygen all have their influence in a supernumerary of combinations and permutations, making it very probable that experiences involving life will sometimes seem contradictory.

Gatineau River (Canada). A test was made in the headwaters of the Gatineau River ("clear river water"), using the same steels described in connection with the atmospheric corrosion tests in rural Nebraska (Fig. 2 and associated text). Specimens were sheets 5 by 7 in. Data are given in the top row of Fig. 6. Copper-bearing steel has better corrosion resistance in this fresh water than does plain carbon steel. In the bar graph, the rate represented by the left end of the bar is the average for 2½ years; the right end, for six months; and "the average", for a period of 1¼ years.

Monongahela, Allegheny and Mississippi Rivers. A summary of the results of immersion tests made on steels in the Monongahela, Allegheny and Mississippi Rivers is given in the bottom part of Fig. 6. The data show that in each river the average corrosion rates decrease with time. The Monongahela River water has a pH of 3.5 to 4.0, which explains the high rate of attack. The contaminant is largely ferric sulfate from coal-mine drainage.

In the Allegheny River the corrosion losses of the various steels were less dependent on composition than in the Monongahela. The decrease in average corrosion rate with time, while less than in the Monongahela, is still significant. The maximum depth of pits (measured with a pit-depth gage on the corroded surfaces) is somewhat greater in the Allegheny than in the Monongahela, possibly because the lower turbulence of the Allegheny was more conducive to the formation of oxygen-concentration cells.

In the Mississippi River, specimens were immersed in a compartment of a dam where the water was nearly stagnant. The low oxygen content of the water is responsible for the much lower corrosion rate here than in the other two rivers. Special tests showed that the amount of corrosion in a given time would have been about doubled by placing the specimens in the free-flowing water; because of ice conditions at the site it was not practicable to expose the specimens in the free-flowing water.

Sea Water

In tests made at four widely separated locations (Plymouth, England; Halifax, Nova Scotia; Colombo, Ceylon; and Aukland, New Zealand) specimens were removed from sea water after 5, 10, and 15 years; attack on these structural steels was essentially linear.

In a test of several steels immersed in sea water at Kure Beach, N.C., a slight decrease in corrosion rate with time was found for most grades when the period of uninterrupted exposure was as short as 1.5, 2.5, and 4.5 years (left chart in Fig. 7). From the above two tests it appears that a slight decrease in corrosion rate occurs during the first few years, after which the average corrosion rates of most steels in sea water become constant at about 3 to 5 mils per year; however, for the 2.6% Cr steel, the rate is virtually constant at 1.5 mils per year. I-beams in sea water at

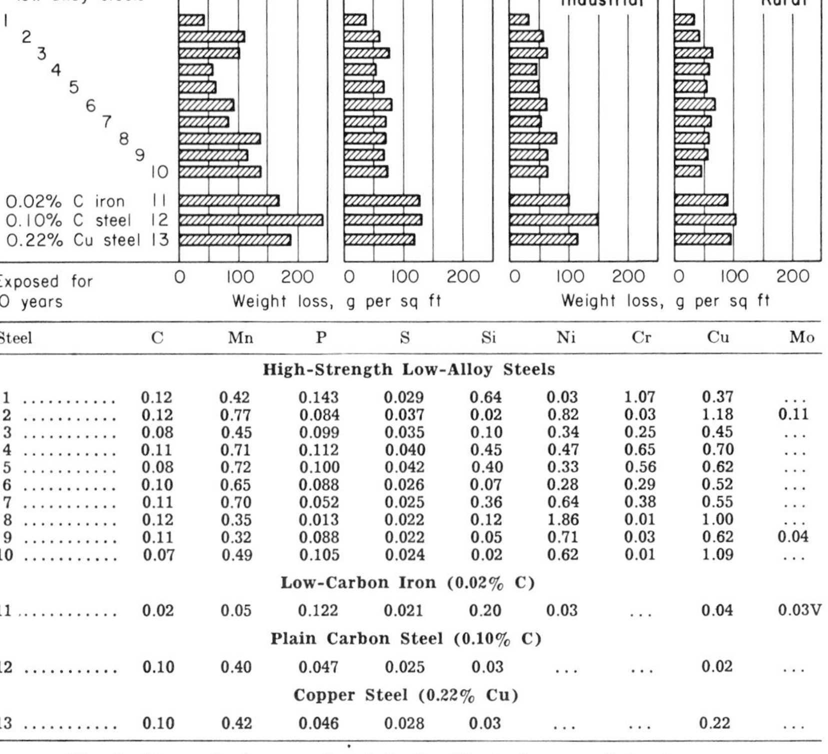

Steel	C	Mn	P	S	Si	Ni	Cr	Cu	Mo
High-Strength Low-Alloy Steels									
1	0.12	0.42	0.143	0.029	0.64	0.03	1.07	0.37	...
2	0.12	0.77	0.084	0.037	0.02	0.82	0.03	1.18	0.11
3	0.08	0.45	0.099	0.035	0.10	0.34	0.25	0.45	...
4	0.11	0.71	0.112	0.040	0.45	0.47	0.65	0.70	...
5	0.08	0.72	0.100	0.042	0.40	0.33	0.56	0.62	...
6	0.10	0.65	0.088	0.026	0.07	0.28	0.29	0.52	...
7	0.11	0.70	0.052	0.025	0.36	0.64	0.38	0.55	...
8	0.12	0.35	0.013	0.022	0.12	1.86	0.01	1.00	...
9	0.11	0.32	0.088	0.022	0.05	0.71	0.03	0.62	0.04
10	0.07	0.49	0.105	0.024	0.02	0.62	0.01	1.09	...
Low-Carbon Iron (0.02% C)									
11	0.02	0.05	0.122	0.021	0.20	0.03	...	0.04	0.03V
Plain Carbon Steel (0.10% C)									
12	0.10	0.40	0.047	0.025	0.03	...	...	0.02	...
Copper Steel (0.22% Cu)									
13	0.10	0.42	0.046	0.028	0.03	...	...	0.22	...

Fig. 5. Atmospheric corrosion data for 13 steels exposed for ten years

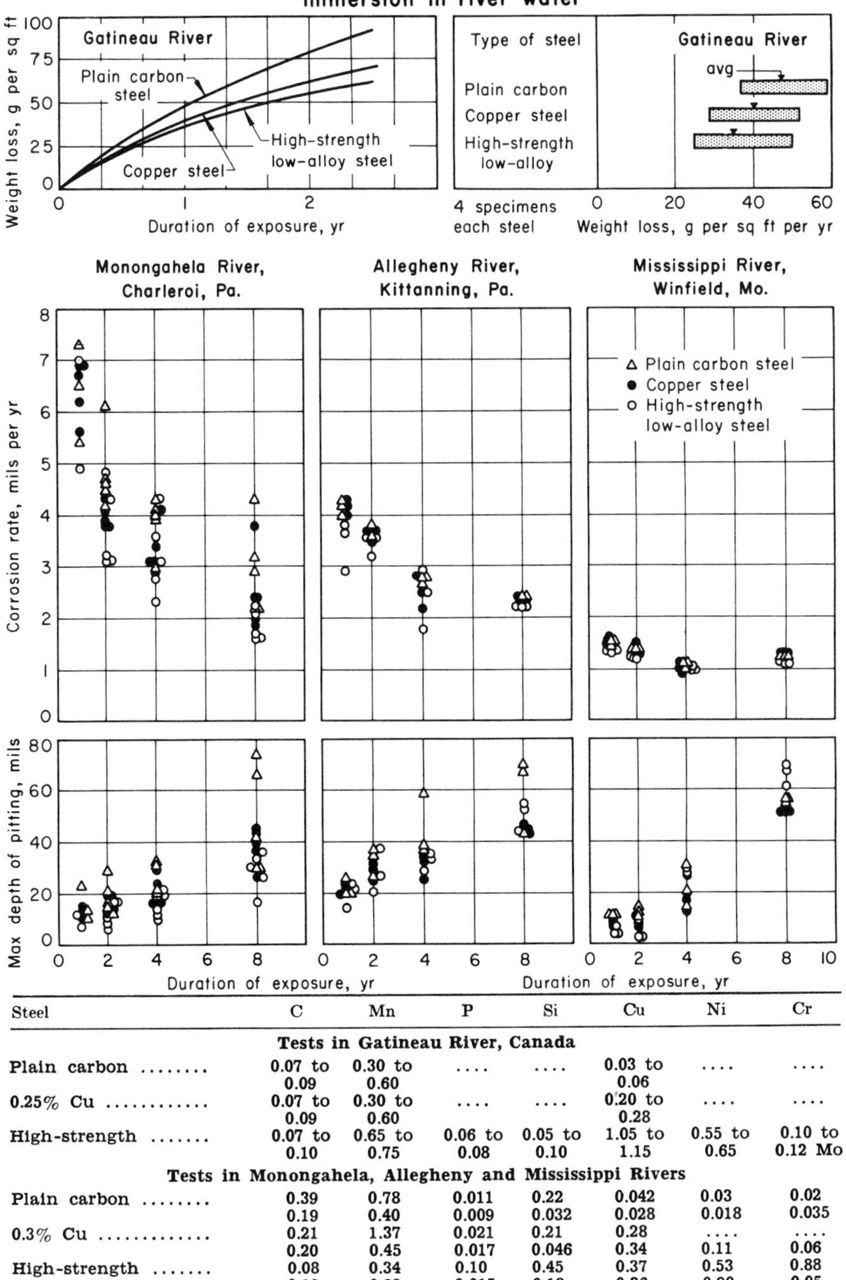

Steel	C	Mn	P	Si	Cu	Ni	Cr
Tests in Gatineau River, Canada							
Plain carbon	0.07 to 0.09	0.30 to 0.60	....	....	0.03 to 0.06	....	....
0.25% Cu	0.07 to 0.09	0.30 to 0.60	....	....	0.20 to 0.28	....	....
High-strength	0.07 to 0.10	0.65 to 0.75	0.06 to 0.08	0.05 to 0.10	1.05 to 1.15	0.55 to 0.65	0.10 to 0.12 Mo
Tests in Monongahela, Allegheny and Mississippi Rivers							
Plain carbon	0.39 0.19	0.78 0.40	0.011 0.009	0.22 0.032	0.042 0.028	0.03 0.018	0.02 0.035
0.3% Cu	0.21 0.20	1.37 0.45	0.021 0.017	0.21 0.046	0.28 0.34	.... 0.11	.... 0.06
High-strength	0.08 0.19	0.34 0.92	0.10 0.015	0.45 0.18	0.37 0.36	0.53 0.90	0.88 0.05

Fig. 6. Corrosion of three types of low-carbon steel in four rivers

Kure Beach, N.C., have been examined just above high tide level for loss in thickness after exposure. The steels were of compositions noted in the table below **Fig. 7**, and data are shown in the two center charts of **Fig. 7**.

Extensive immersion tests of many steels have been made in the Panama Canal. Test specimens 12 by 12 by ¼ in. were exposed in the tidal zone of sea water and were also continuously immersed in brackish water and in fresh water. Specimens were removed after exposure of one, three and five years. Some results, including those obtained on specimens of several high-strength low-alloy steels, are given in Fig. 8, which shows that during five years of exposure in brackish water, the average corrosion varied from 6 to 9 mils per year. In fresh water the calculated rates were only 3 or 4 mils per year. In sea water at the mean-tide position, the corrosion rates were 5 or 6 mils per year.

Boiler Water, Steam and Condensate

Experience has shown that the useful life of steel in steam service is unlimited if the proper steel is selected or if the service condition is controlled properly. Some steels have been in continuous service in steam for 40 to 50 years.

If the proper steel is selected and failure occurs, impurities in the steam are responsible. The solution, in this event, is correction of the service conditions rather than the use of more corrosion-resistant steel. Occasionally a change in operating conditions may be impossible and stainless may be required.

The following is a conservative account of the service life of steels in steam service based on field experiences with valves:

1 Carbon steel (ASTM A216, grade WCB) has been used in steam service with temperatures up to 750 F for 25 to 30 years.
2 Carbon-molybdenum steel (ASTM A217, grade WC1) has been available since 1934 and has been entirely satisfactory during that period in steam service with temperatures up to 850 F. However, the steel is not recommended for welded structures.
3 Chromium-molybdenum steel (ASTM A217, grade WC6) has been available since 1946 and has been entirely satisfactory during that period in steam service with temperatures up to 1000 F. This steel is recommended for welded structures.
4 Chromium-molybdenum steel (ASTM A217, grade WC9) has been available since 1948 and has been entirely satisfactory during that period in steam service with temperatures up to 1050 F. This also is a weldable steel and very few flanged connections are used.

Return condensate lines in boiler systems frequently deteriorate along their bottom inside surfaces because of the formation of carbonic acid by the carbon dioxide gas in the condensing steam. The carbon dioxide is present in the steam mainly through the breakdown of the bicarbonate and carbonate alkalinity of the make-up water to the boiler.

Where an open feedwater heater precedes the boiler much of the bicarbonate alkalinity breaks down to carbonate plus CO_2 and water, and some of the CO_2 is lost by venting at this point. However, in the boiler the carbonate breaks down further to carbon dioxide plus hydroxide, and this last carbon dioxide joins the steam, eventually forming carbonic acid as the steam condenses.

Various materials may be employed for combatting return-line corrosion caused by carbon dioxide. These include ammonia, alkalies, polyphosphate, neutralizing amines and filming amines. Ammonia neutralizes the carbon dioxide and raises the pH of the condensate. It is generally undesirable in this type of application because it can seriously corrode copper and zinc-bearing metals. Alkalies and polyphosphate are also undesirable because they increase solid content of the steam, and control is difficult.

Volatile amines such as cyclohexylamine and morpholine can be used to combat this corrosion. These compounds, fed directly to the boiler, volatilize with the steam and neutralize the carbon dioxide in the condensate. They are not corrosive to copper and zinc-bearing metals. They are limited to applications where the attack is strictly by carbon dioxide since they do not prevent oxygen corrosion. Because they neutralize the carbon dioxide, the feed rate depends on the CO_2 content and may be high when the CO_2 content is high. They can also be prohibitively expensive when a large percentage of the steam is lost.

Exposed at Kure Beach, N.C.

Immersion of sheet

Steel 4 (2.6% Cr)

Penetration, mils per yr (avg for each total test period)

Duration of immersion, yr

Corrosion rate of I-beams just above the high tide level

Plain carbon steel

Copper steel

High-strength low-alloy steel

Avg thickness loss, mils

Avg weight loss, g per sq ft

Duration of exposure, yr

Salt water, splash area

Type of steel

avg

Plain carbon

Copper steel

High-strength low-alloy

2 specimens each steel

Weight loss, g per sq ft per yr

Steel	C	Mn	P	S	Si	Cu	Ni	Cr
Immersion of Sheet (Left Chart)								
1 High-strength low-alloy	0.08	0.36	0.08	0.026	0.31	0.41	0.49	0.73
2 High-strength low-alloy	0.17	1.2	0.026	0.025	0.18	0.31	0.53	0.07
3 High-strength low-alloy	0.19	0.44	0.030	0.032	0.025	0.81	1.8	0.22
4 High-strength low-alloy(a)	0.07	0.39	0.014	0.020	0.09	0.020	0.04	2.6
5 Plain carbon	0.25	0.42	0.020	0.032	0.025	0.012	0.02	0.06
Other Three Charts								
High-strength low-alloy (b)	0.07 to 0.10	0.65 to 0.75	0.06 to 0.08	....	0.05 to 0.10	1.05 to 1.15	0.55 to 0.65	...
0.25% Cu	0.07 to 0.09	0.30 to 0.60	....	....	....	0.20 to 0.28	....	...
Plain carbon	0.07 to 0.09	0.30 to 0.60	....	....	....	0.03 to 0.06	....	...

(a) Molybdenum, 0.52%. (b) Molybdenum, 0.10 to 0.12%

Fig. 7. Results of corrosion tests of specimens exposed to sea water at Kure Beach, N.C.

The filming amines, among the newer developments in this field, have proved successful in combatting return-line corrosion. They function by laying down an impervious film on the metal surfaces. They afford protection against CO_2 attack and also against oxygen. The feed rate does not depend on the amount of carbon dioxide or oxygen since it is necessary to feed only enough to form and maintain the film.

There is ample evidence that the process of manufacture has no bearing on the ability of mild steel to withstand corrosion where it is in contact with boiler water. Pitting by dissolved oxygen in the boiler water proceeds regardless of whether the steel is rimmed or killed, and whether the tubes are seamless or welded. In a number of actual instances, tubes installed in boilers at the same time, made by both processes of both types of steel, have failed by oxygen pitting at about the same time.

A recent series of tests in small atmospheric-pressure boilers indicates some difference in depth and frequency of pits using carbon steels and steels of the low-alloy grades.

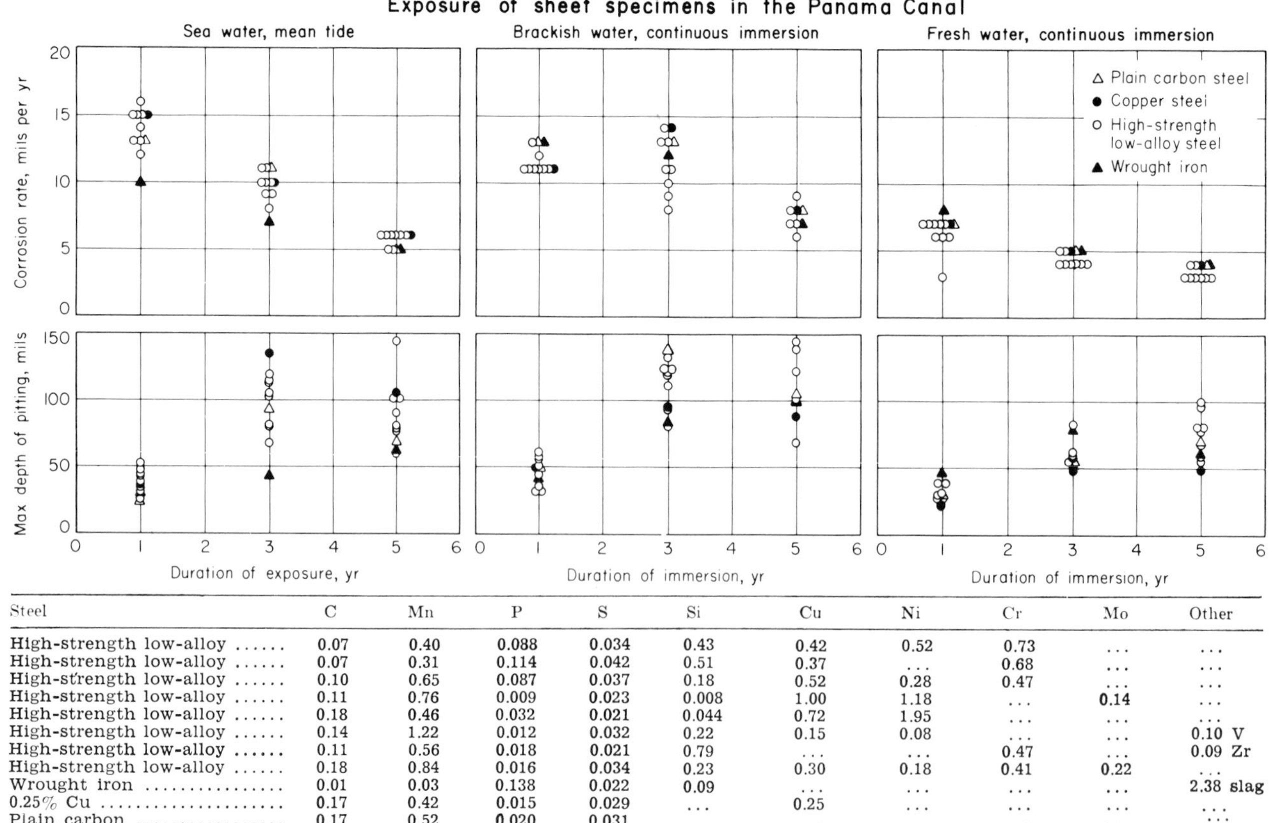

Steel	C	Mn	P	S	Si	Cu	Ni	Cr	Mo	Other
High-strength low-alloy	0.07	0.40	0.088	0.034	0.43	0.42	0.52	0.73	...	...
High-strength low-alloy	0.07	0.31	0.114	0.042	0.51	0.37	...	0.68	...	...
High-strength low-alloy	0.10	0.65	0.087	0.037	0.18	0.52	0.28	0.47	...	...
High-strength low-alloy	0.11	0.76	0.009	0.023	0.008	1.00	1.18	...	0.14	...
High-strength low-alloy	0.18	0.46	0.032	0.021	0.044	0.72	1.95	...	...	...
High-strength low-alloy	0.14	1.22	0.012	0.032	0.22	0.15	0.08	...	...	0.10 V
High-strength low-alloy	0.11	0.56	0.018	0.021	0.79	...	...	0.47	...	0.09 Zr
High-strength low-alloy	0.18	0.84	0.016	0.034	0.23	0.30	0.18	0.41	0.22	...
Wrought iron	0.01	0.03	0.138	0.022	0.09	...	...	...	...	2.38 slag
0.25% Cu	0.17	0.42	0.015	0.029	...	0.25	...	...	...	...
Plain carbon	0.17	0.52	0.020	0.031	...	...	...	...	...	...

Fig. 8. Sea water corrosion results obtained in the Panama Canal. There was one 12 by 12-in. specimen of each steel for each removal period. Corrosion rates were calculated from the weight loss of each specimen.

These tests were run without treatment in the boiler water which was fed make-up water containing approximately 5 ml per liter of dissolved oxygen. The boiler water was maintained at approximately 100 ppm chloride. Pit depth and frequency were as plotted in Fig. 9, for a condition involving 2500 to 3000 hr of operation and 1000 to 1500 hr of downtime.

It is obvious from the data that operating boiler tubes in this manner is not economical nor good practice. The need for elimination of dissolved oxygen and maintenance of suitable alkalinities, plus water treatment to prevent scale deposits, have all been recognized for many years as necessary for the prevention of corrosion of such materials. The above tests serve only to show the relative corrosion rates of four steels under the most undesirable operating conditions that could be devised.

For the experiences related below, either ASTM A70 copper-bearing steel or ASTM A285, grade B, with or without copper, were used. Examples follow:

Condensers (Steam Power). On condensers pH values of 6½ to 7 are typical, and the service is on the line with highly saturated steam for continuous duty. There has not been a failure from corrosion during the 25 years that steel has been used.

Deaerators. The pH value is somewhat below that in condenser service, and in addition, the operating temperature is relatively low. The water is deaerated to hold dissolved oxygen down to 0.03 ml per liter. These conditions provide relatively low corrosion rates and there has been only one failure in 20 years of operating experience. In this instance, the deaerating equipment was heavily overloaded with highly saturated steam with actual reversal flow, and failure occurred in less than one year. Recent advancement in design of deaerating equipment has made it possible to provide a guarantee that the oxygen content will be reduced to 0.005 ml per liter or less.

Pitting in small atmospheric-pressure boilers

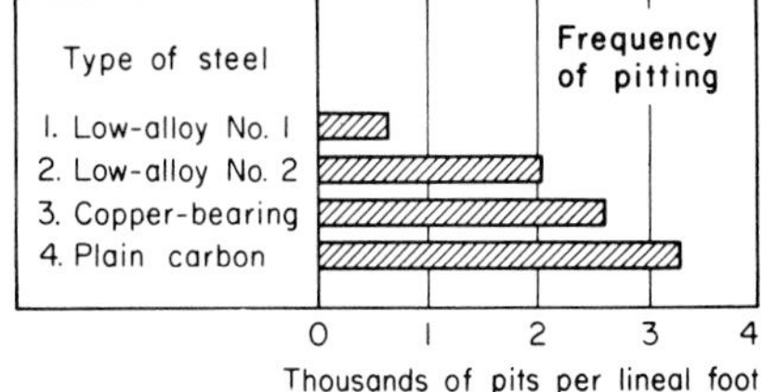

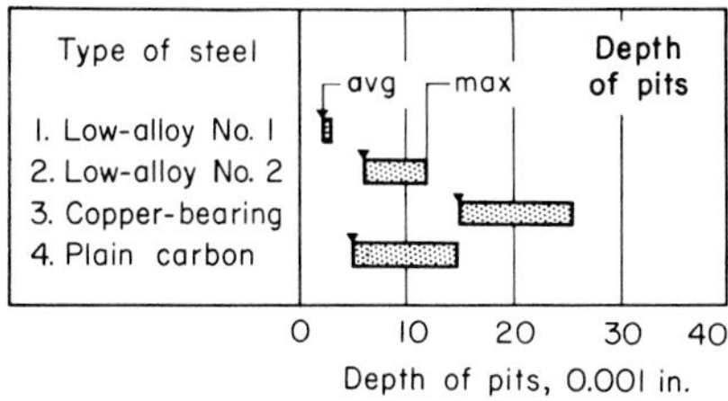

Steel	C	Mn	P	Cu	Ni
1(a)	0.08	0.34	0.10	0.37	0.53
2(b)	0.10	0.75	0.03	0.75	1.00
3	0.11	0.42	0.008	0.37	0.09
4	0.17	0.35	0.008	0.06	0.06

(a) Also, 0.88% Cr. (b) Also, 0.12% Mo.

Fig. 9. Pitting of boiler tubes in untreated boiler water (100 ppm chloride) that was fed make-up water containing 5 ml per liter of dissolved oxygen. Service period involved 2500 to 3000 hr of operation and 1000 to 1500 hr of downtime. This is uneconomical practice (see text in col 1).

Feedwater Heaters. The pH values are higher than in condenser service — from 8 to 8.3. However, with continuous service corrosion has been negligible. One job was followed in service to determine whether corrosion was a problem requiring statistical attention; results indicated no significant corrosion. The shell of this unit was sandblasted before shipment to obtain a silky gray color and the shell was installed without applying paint or primer. After three years of service, the condition of the shell was virtually unchanged. Sparsely located minor pinhole attack had occurred in which the depth was approximately 0.001 in. in areas the size of a common pinhead.

Plain carbon and low-alloy steels commonly used for valve bodies, flanges, valve bonnets, valve disks and gates, steam piping, feedwater piping, and other pressure-containing parts regularly encountered in steam generating plants are listed in Table 2. Valve seat areas in carbon or low-alloy steel valves are generally hard faced with cobalt-chromium-base alloy or are made from hardened martensitic stainless steel to resist erosion by high-pressure steam flow. Pressure-containing parts such as valve bodies, valve disks, steam piping and feedwater piping are designed to the allowable stress established by codes of the American Society of Mechanical Engineers and other code-writing organizations. Such allowable stresses are based on long-time high-temperature properties of the steels. Below the creep range, allowable stresses are established at the lowest value of stress obtained from using 25% of the specified minimum tensile strength at room temperature, or 25% of the minimum expected tensile strength at temperature, or 62.5% of the minimum expected yield strength for 0.2% offset, at temperature. (See Sec VIII, Unfired Pressure Vessels, ASME Boiler and Pressure Vessel Code.)

In addition to the above, designers are required by the Boiler Code to provide a corrosion allowance in all pressure-containing vessels exposed to steam, water, air or any combination thereof. This allowance of ad-

Table 2. Plain Carbon and Low-Alloy Steels Commonly Used for Pressure-Containing Components in Steam Generating Plants

Common name	ASTM designation	Pertinent nominal composition, %	Usual maximum service temperature, F
Cast carbon steel	A216-53T, grade WCB	0.35 C max	850
Cast carbon steel	A216-53T, grade WCA	0.25 C max	850
Forged carbon steel	A105-55T, grades 1 and 11	0.35 C max	850
Cast chromium-molybdenum steel	A217-55, grade WC6	0.20 C max, 1.25 Cr, 0.50 Mo	1050
Cast chromium-molybdenum steel	A217-55, grade WC9	0.18 C max, 2.50 Cr, 1.00 Mo	1050
Cast carbon-molybdenum steel(a)	A217-55, grade WC1	0.25 C max, 0.50 Mo	850
Forged carbon-molybdenum steel(a)	A182-55T, grade F1	0.25 C, 0.50 Mo	850
Forged 1.25% chromium-molybdenum steel	A182-55T, grade F11	0.15 C, 1.25 Cr, 0.50 Mo	1050
Forged 1% chromium-molybdenum steel ..	A182-55T, grade F12	0.15 C, 1.00 Cr, 0.50 Mo	1050
Forged 2.25% chromium-molybdenum steel.	A182-55T, grade F22	0.15 C max, 2.25 Cr, 1.0 Mo	1050
Seamless carbon steel pipe	A106-55T, grade A	0.25 C max	850
Seamless carbon steel pipe	A106-55T, grade B	0.30 C max	850
Seamless carbon steel pipe	A106-55T, grade C	0.35 C max	850
Carbon-molybdenum steel pipe	A335-55T, grade P1	0.15 C, 0.50 Mo	850
1.25% chromium-molybdenum steel pipe...	A335-55T, grade P11	0.15 C max, 1.25 Cr, 0.50 Mo	1050
1% chromium-molybdenum steel pipe	A335-55T, grade P12	0.15 C max, 1.0 Cr, 0.50 Mo	1000
2.25% chromium-molybdenum steel pipe...	A335-55T, grade P22	0.15 C max, 2.25 Cr, 1.0 Mo	1050
Boiler tubes	A83, grade A	0.06 to 0.18 C	850
Boiler tubes for high pressure	A192	0.06 to 0.18 C, 0.35 Si max	850
Boiler and superheater tubes	A210	0.27 C max	850
Boiler tubes	A178, grade A	0.06 to 0.18 C	850
Boiler tubes	A178, grade C	0.35 C max	850
Boiler tubes for high pressure	A226	0.06 to 0.18 C, 0.25 Si max	850
Carbon-molybdenum boiler and superheater tubes	A209, grade T1a	0.20 C, 0.50 Mo	950
Alloy steel boiler and superheater tubes ...	A213, grade T11	1.25 Cr, 0.5 Mo	1100
	A213, grade T3b	2.00 Cr, 0.5 Mo	1200
	A213, grade T22	2.25 Cr, 1.0 Mo	1200
	A213, grade T21	3.00 Cr, 1.0 Mo	1200
	A213, grade T5	5.00 Cr, 0.5 Mo	1200
	A213, grade T9	9.00 Cr, 1.0 Mo	1200
	A213, grade TP304	18.00 Cr, 8.0 Ni	1500
	A213, grade TP321	18.00 Cr, 8.0 Ni, Ti	1500

(a) Carbon-molybdenum steel may graphitize in high-temperature service, particularly if it has been welded.

ditional wall section is required so that the reduced metal section produced by corrosion will still be stressed within allowable limits throughout the life of the part. The reader is referred to the appropriate section of the ASME Boiler Code for selection of corrosion allowances.

Oxidation Data. Table 3 shows expected oxidation rates in steam and air for a number of steels commonly used in the construction of steam-containing components. All tests were run on specimens 0.7-in. diam by 1.0156 in. long. Test pieces for oxidation tests in air were supported upright on Monel screens in a laboratory muffle furnace. Oxidation tests in steam were run in a closed tube furnace with steam under one atmosphere pressure. Tests were conducted for 10,000 hr with specimens being removed at 50, 100, 500, 1000, 5000 and 10,000 hr. Results were extrapolated to 100,000 hr.

The specimens were cleaned cathodically in 10% sulfuric acid to which 0.1% quinoline ethiodide had been added as an inhibitor, the current density being 1 amp per sq in. Unoxidized specimens treated by this procedure gave a weight loss of 3 mg in 6 hr, low enough not to influence test results significantly. (See "Corrosion of Unstressed Steel Specimens and Various Alloys by High-Temperature Steam", H. L. Solberg, G. A. Hawkins and A. A. Potter, Trans ASME, **64**, 303, 1942.)

A steam valve body or pipe is oxidized by steam on the inside and by air on the outside. A valve stem is decreased in diameter because of steam oxidation only and a flange bolt in a steam line is decreased in root diameter because of air oxidation only. Table 4 lists the total loss in section that would be expected under the indicated conditions, and the resulting increase in stress. These tables indicate that only slight allowance for extra metal section is required for many of these steels for service at 1000 F for 100,000 hr, but at 1150 F many of them oxidize too rapidly to be useful.

In general these test results are confirmed by service experience where plain carbon steels are performing quite satisfactorily in steam and boiler feedwater service at 850 F and lower, while 1050 F seems to be near the upper limit for the low-alloy chromium-molybdenum steels.

Products of Combustion

Trouble from fireside deposits and corrosion in boiler furnaces has greatly increased during recent years. This has been caused by at least four factors: (1) increased rating on existing units, (2) economic pressure to burn lower-cost fuel, (3) the critical operating ranges of gas temperatures, and (4) higher sulfur content in fuel oil. The first three factors are not difficult to control, but the higher sulfur content in oil must be tolerated because of the high cost of low-sulfur oil.

There have been two principal causes for the increasing deterioration in the quality of heavy fuel oils used for steam generation and for the consequent increase in corrosion rates caused by deposits in furnaces, superheater, economizer and air-heater sections. First, the quality of the crude oil from certain new oil fields, particularly the foreign fields, is lower than from the older domestic fields that supplied most of the United States demand in the past. This decrease in quality has included a change in the nature of the ash and an increase in sulfur content. Also, there has been an increase in ash content as well as a change in ash quality.

Early records show oil ash contents ranging from 0.01 to 0.10%, while today ash contents as high as 0.2% are encountered and 0.10 to 0.15% ash is common. Without the parallel change in oil ash composition, this increase in ash content would not be significant. However,

Table 3. Oxidation of Common Steam Plant Steels in Air and Steam

Oxidizing medium	Temperature, F	Penetration, mils — Time, hr. 100	1,000	10,000	100,000
ASTM A216, Grade WCB					
Air	850	0.12	0.22	0.4	0.74
	1000	0.18	0.52	1.6	4.90
	1150	0.88	3.70	15.5	66.0
Steam	850	0.16	0.2	0.34	0.6
	1000	0.29	0.86	2.6	8.0
	1150	3.1	11.6	46.0	150
ASTM A105, Grade 11 (AISI 1029)					
Air	850	0.12	0.22	0.57	1.55
	1000	0.38	0.68	2	6
	1150	0.68	1.61	14.0	105
Steam	850	0.10	0.21	0.43	0.90
	1000	0.22	0.50	1.2	2.7
	1150	1.25	5.8	28	102
ASTM A108, Grade 1035 CD (AISI 1035)					
Air	850	1.81	0.32	0.58	1.1
	1000	0.26	0.64	1.5	3.7
	1150	0.88	3.3	11	40.1
Steam	850	...	....	....	....
	1000	0.19	0.84	3.8	17
	1150	4.3	13	39	101
AISI 1116 CD					
Air	850	0.15	0.42	1.3	3.8
	1000	0.35	1.0	2.8	8.6
	1150	0.70	3.2	14.8	68.0
Steam	850	...	....	....	....
	1000	0.23	0.72	2.2	7.0
	1150	2.7	8.8	29	96.8
ASTM A193, Grade B7 (AISI 4140)					
Air	850	0.09	0.17	0.44	1.1
	1000	0.2	0.54	1.5	4.1
	1150	0.9	2.9	9.5	30
Steam	850	...	....	....	....
	1000	0.3	0.96	3.2	10.2
	1150	6.8	13.8	27	55
ASTM A217, Grade WC1					
Air	850	0.12	0.23	0.44	0.82
	1000	0.216	0.58	1.58	4.4
	1150	1.4	3.5	10.4	40.1
Steam	850	....	....	....	....
	1000	0.175	0.57	1.9	6.6
	1150	1.3	6.2	30	124
ASTM A182, Grade F6 (410 Stainless)					
Air	850	0.04	0.027	0.052	0.15
	1000	0.037	0.068	0.125	0.23
	1150	0.074	0.18	0.42	1
Steam	850	0.036	0.16	0.62	0.95
	1000	....	....	....	....
	1150	0.019	0.29	8	1.3
ASTM A217, Grade WC6					
Air	850	0.095	0.19	0.53	1.9
	1000	0.15	0.45	1.4	4.3
	1150	0.55	2.5	11.5	53
Steam	850	0.099	0.18	0.36	0.62
	1000	0.2	0.62	1.88	5.8
	1150	1.2	3.6	9.4	26.5
ASTM A217, Grade WC9					
Air	850	0.07	0.22	0.58	1.6
	1000	0.1	0.41	1.75	7.4
	1150	0.27	1.09	4.1	16
Steam	850	...	....	....	....
	1000	0.8	2.1	5.4	14
	1150	1.3	3.6	8.2	21

For compositions of ASTM grades, see Table 2. For test conditions, see text under **Oxidation Data** (col 1, this page). 100,000-hr data extrapolated from 10,000 hr.

Table 4. Effect of Steam and Air Oxidation on Design

ASTM designation of steel	Estimated section loss in 100,000 hr, mils: 850 F	1000 F	1150 F	% increase in stress resulting from section loss in 100,000 hr — 1000 F, Section size: ½ in.	1 in.	2 in.	1150 F, Section size: ½ in.	1 in.	2 in.
Exposed to Steam on One Surface, Air on the Other Surface									
A216, grade WCB	1.4	13	240	2.7	1.3	0.65	92.0	31.6	13.6
A105, grade 11	2.4	8.6	250	1.7	0.9	0.4	100	33.0	14.3
A108, grade 1035	..	20	160	4.2	2.0	1.0	47.0	19.0	8.7
AISI 1020	..	14	160	2.9	1.4	0.7	47.0	19.0	8.7
A217, grade WC1	..	11	190	2.3	1.0	0.6	61.0	23.5	10.5
A217, grade WC6	2.1	10	79	2.0	1.0	0.5	18.8	8.6	4.1
A217, grade WC9	..	20	38	4.2	2.0	1.0	8.2	3.9	1.9
Exposed to Steam on Both Surfaces									
A182, grade F6 (300 Bhn)	1.9	....	2.6	..	..	...	0.1	0.025	0.010
A182, grade F6 (400 Bhn)	..	1.9	3.3	..	..	...	0.13	0.032	0.016
Exposed to Air on Both Surfaces									
A193, grade B7	2.0	8.2	60	0.4	0.2	0.1	58.5	16.0	6.7

the composition of the ash of present-day troublesome oils is somewhat different from the early-day trouble-free oils. Ash from trouble-free oils consisted chiefly of iron oxide, silica, lime and magnesia compounds, with small amounts of other materials being present. The ash from troublesome oils carries large quantities of alkali sulfates and in addition often contains considerable vanadium. Vanadium is more prevalent today and in much larger quantities than in the past. The source of the vanadium is the crude oil itself; the Venezuelan crudes are notoriously high in vanadium, and several other fields also yield crudes high in vanadium. Typical ash analyses are shown in Table 5.

In the past, with some exceptions, the sulfur contents of the heavy fuel oils burned in the United States were about 1% or less. Today sulfur contents of 3.0 to 3.5% are encountered and 2% is common. The percentage of total sulfur in extreme cases is as much as 7% by weight. Although neither the increase in sulfur content nor the change in ash quality has been correlated precisely with the increase in deposit difficulties, the high sulfur content probably contributes to the slagging and corrosion problems.

The corrosivity of sulfur depends on the kind of compound, concentration and temperature. In the absence of water, sulfur is most corrosive in the form of hydrogen sulfide. In the presence of water, sulfur dioxide corrosion occurring below the dew point of the corrosive medium must be considered.

As would be expected, corrosion increases gradually with increasing concentration of active sulfur compounds. Sulfur corrosion of mild steel becomes apparent at temperatures above 390 F but is not disturbing up to about 625 F. In the temperature range between 705 and 795 F, the corrosion rates are more than 390 mils per year. Both below and above these temperature limits, the corrosion is less severe. (These relations are shown quantitatively in the graphs of Fig. 21, in the article on selection of steel for service in petroleum refineries.) Measurements were taken to determine the metal being lost on tubes located in the air heater section of a boiler, and the rate of corrosion was found to be 14 to 24 mils per 1000 hr. Simulated corrosion tests indicate that

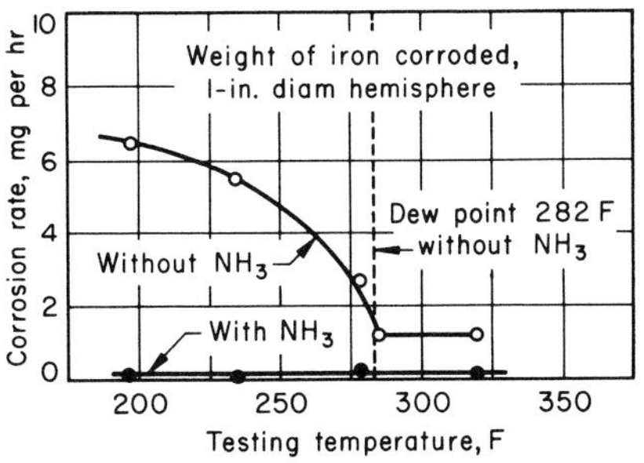

Fig. 10. Effect of ammonia on corrosion of iron in an experimental furnace, the extent of corrosion being measured by the iron content of the probe washings

Table 5. Typical Compositions of Ash from Heavy Fuel Oil

Constituent	Trouble-free	Trouble-some
Ferric oxide	54%	6%
Silica	24	5
Alumina	6	1
Lime	3	1
Magnesia	9	2
Vanadium pentoxide	..	35
Alkali	1	37

corrosion rates in the neighborhood of 250 mils per year are possible.

Since the deposits obtained from many installations indicate that alkali sulfate and compounds of vanadium are the principal offenders, an obvious remedy would be to remove such metallic ions from the fuel oil before combustion. However, the ions are present in the crude oil and in fuel oil in very small amounts, and in such chemical combinations that their removal has not yet proven economically feasible.

Since the ash slag deposits from boilers having superheater slag troubles had relatively low fusing temperatures, the possibility was suggested of changing the characteristics of the deposits favorably by adding to the oil some material that would raise the fusing temperature of the ash. Experiments were conducted on various materials considered promising. Some of these materials were mixed with the fuel oil, while others were blown into the combustion areas. Some of the materials used were silica powders, oil-soluble metallic soaps, magnesium oxide, zinc dust, dolomite and ammonia. The most successful results were obtained with materials that combined directly with the SO_3 in the combustion gases to form noncorrosive products which largely remained suspended in the flue gases and could be vented to atmosphere without excessive condensation or settling.

The use of ammonia in an experimental furnace was the most successful (Fig. 10) and indicated that the corrosion rate could be reduced from 284 to 8.7 mils per year.

The products of combustion of oils high in sulfur may have dew points as high as 340 F. Whenever metal temperatures in contact with these

Table 6. Compositions of Steel Used in Corrosion Tests in Regenerative Air Preheaters (See Fig. 11 for Corrosion Data.)

Steel	Corrosion	C	Mn	P	Si	Cu	Ni	Cr	Mo
1 High-strength low-alloy steel	92	0.14	0.99	0.029	0.47	0.30	0.90	0.50	0.47
2 High-strength low-alloy steel	93	0.11	0.65	0.123	0.23	0.70	0.42	0.61	...
3 High-strength low-alloy steel	100	0.10	0.40	0.105	0.54	0.42	0.45	0.89	...
4 High-strength low-alloy steel	125	0.055	0.40	0.089	0.05	0.47	0.88	0	...
5 High-strength low-alloy steel	140	0.08	0.42	0.007	...	0.77	1.51	...	...
6 High-strength low-alloy steel	150	0.13	0.70	0.020	0.80	...	...	0.60	0.11 Zr
7 High-strength low-alloy steel	195	0.055	0.33	....	0.42	0.08	0.32	5.24	0.59
8 High-strength low-alloy steel	200	0.080	0.72	0.074	0.04	1.12	0.55	0.31 Al	0.13
9 High-strength low-alloy steel	280	0.25	0.26	0.19	0.01	0.15	0.71	0.09	...
Copper steel	130	0.030	0.13	0.006	0.005	0.48	0.04	0	...
Plain carbon steel	175	0.06	0.43	0.046	0.10	0.028	0	0	...

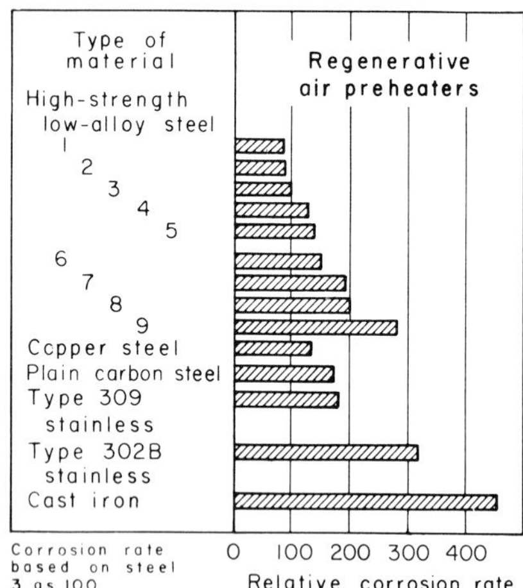

Fig. 11. Relative corrosion rates in Harding Street preheaters. Each rate shown is the average of at least four specimens. For compositions of steels, see Table 6, bottom of preceding page.

gases are lower than the dew point, sulfuric acid condensation occurs, resulting in corrosion. This becomes a problem in air preheaters, both the regenerative and tubular types. The obvious solution is to reduce the sulfur content of the fuel. However, this is impractical considering the many sources of oil and the improved refinery methods, which have a tendency to leave more of the sulfur behind in the oil used as fuel. Much effort has been expended in the form of field and exposure tests to find a material more serviceable than carbon steel.

In tubular air heaters, a high-strength low-alloy steel (0.10 C, 0.40 Mn, 0.10 P, 0.55 Si, 0.40 Cu, 0.45 Ni, 0.90 Cr) gave from two to three times the life of 1020 steel. This improvement in service life is gained with an increased material cost of approximately 1.7 to 1. Total cost including installation has a lower ratio of approximately 1.25 to 1. Thus it is apparent that with outage time reduced, a considerable saving can be accomplished. This can be a major item in marine boiler usage if cost is not the only consideration.

Steels of compositions given in Table 6 have been tested in regenerative air preheaters by the U.S. Bureau of Mines, with the results shown in Fig. 11.

Locomotive Smoke. The Iowa State Highway Commission ran a ten-year test of several kinds of steels used as blast plates. The uncoated specimens 6 by 24 by 0.250 in. were placed about 23 ft above the rails, where they were subjected to about 140,000 locomotive movements. The location was on a steep grade and sharp curve; consequently, the stack blast was severe. Chemical compositions and corrosion results are given in Table 7.

Kraft Digesters

Information on the service life of Kraft digesters made of mild steel and used in the manufacture of paper is presented in Fig. 12. Attention should be called to the fact that the manganese content of A70 steels depends on the year in which it was manufactured. In one of these sets of results (bottom chart), service life in calendar years is given for nine digesters made of steel ASME S1. This is an old designation, identical with ASTM A70-27 and presently comparable to ASTM A285. The service life ranged from 13.2 to 16.3 years. It is almost universal practice to build these digesters from 2-in. plate, 1 in. being for corrosion allowance.

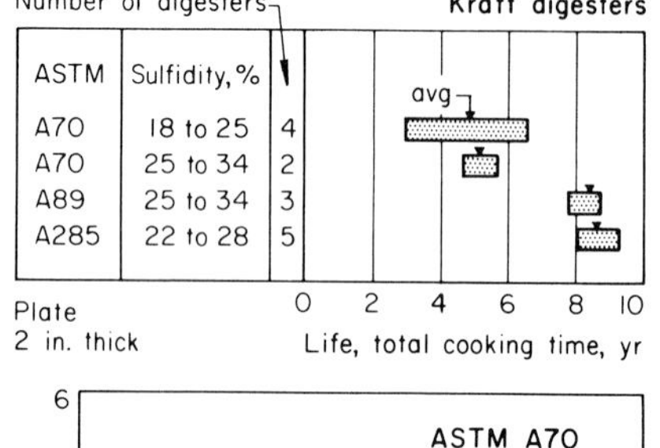

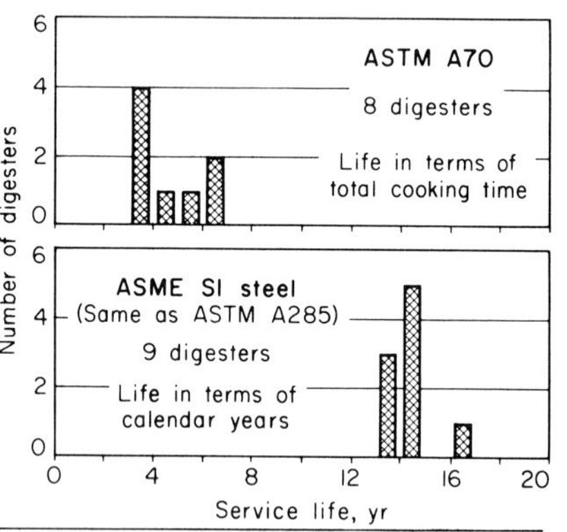

Spec	Grade	C	Mn
A70-33	....	0.3	0.3 to 0.5
A70-42	....	0.25 to 0.30	0.8
A89-43	A.........	0.12	0.8
A89-43	B.........	0.22	0.8
A285-46	A.........	0.15	0.8
A285-46	B.........	0.20	0.8
A285-46	C.........	0.25	0.8

All steels: 0.035 to 0.040% P, 0.040% S max

Fig. 12. Service life of Kraft digesters. (J. C. Hair and A. W. Duskin, Paper Mill News, 74, 88, 1951; R. S. Peoples and G. L. Ericson, Tappi, 35, 403, 1952; A. Ungar and T. E. Caywood, Tappi, 37, 177, 1954)

Acids

Hydrofluoric. Corrosion rates for mild steel in various concentrations of hydrofluoric acid are given in Fig. 13. The stock solution contained 69.9% H_2F_2, 0.12% H_2SO_4, and 0.08% H_2SiF_6 and was diluted with distilled water to get the concentrations below 69.9% H_2F_2. Specimens were exposed in sealed polyethylene bottles on a roof for 26 days. Note the large change in corrosion rate in going from 63 to 62% H_2F_2. Although the data in Fig. 13 indicate a comparatively low rate of corrosion where the concentration is as high as 63%, the safe limit is set at 70%. It has been found that killed steel is more resistant to hydrofluoric acid than is semikilled steel, but it is not expected that the difference would be more than 10%.

Sulfuric. Mild steel is widely used for the handling of sulfuric acid in concentrations over 70%. Storage tanks, pipelines, tank cars and shipping drums made of steel are very common for 78% (60° Bé), 93% (66° Bé), 98% and stronger acids such as oleum. (Pumps and valves are often made of high-alloy material such as type 316 stainless because of erosion-corrosion of steel.) Much of the equipment in plants manufacturing sulfuric acid by the contact process is made of mild steel. Steel is attacked rapidly by the more dilute sulfuric acids. Figure 14 shows corrosion of carbon steel by concentrated sulfuric acid as a function of temperature. Most of the tests were made on ordinary steel of 0.20% C, with exposure periods of 48 hr. Specimens were

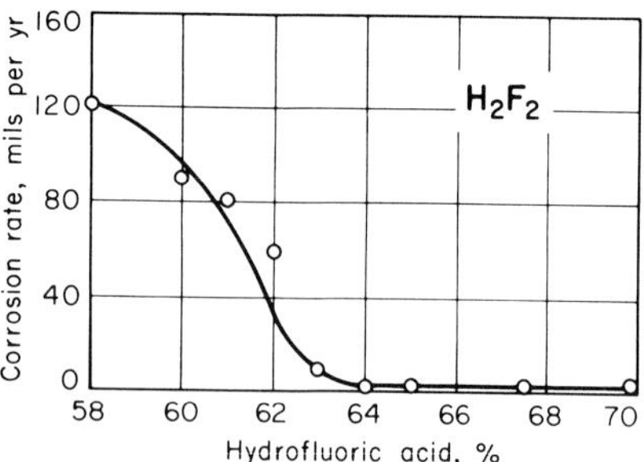

Fig. 13. Corrosion of mild steel in hydrofluoric acid at room temperature

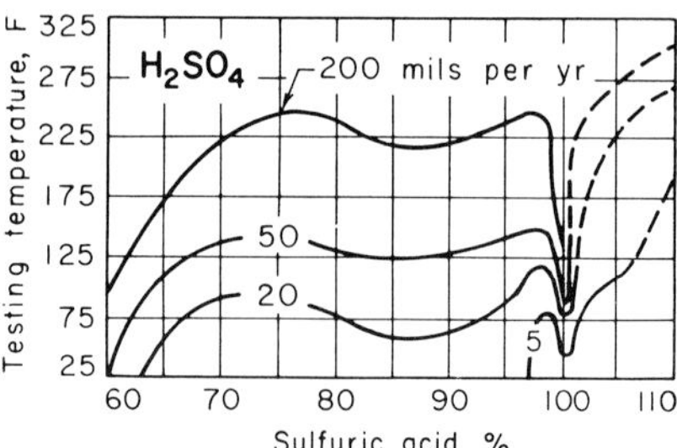

Fig. 14. Corrosion of mild steel in concentrated sulfuric acid as a function of temperature (M. G. Fontana, Ind Eng Chem, August 1951)

Table 7. Ten-Year Corrosion Data for Steels Subjected to Locomotive Smoke

Steel	Composition, %: C	Mn	P	S	Si	Cr	Ni	Cu	Average corrosion loss: Wt %	Mils
1..........	0.12	0.75	0.100	0.025	0.05	0.20	0.50	0.60	51	128
2..........	0.02	0.05	0.120	0.025	0.10	2.50 slag	...	...	55	138
3..........	0.09	0.36	0.133	0.025	0.71	0.93	0.04	0.39	59	148
4..........	0.25	0.25	0.015	0.140	...	0.05 Mo	...	0.40	63	158
5..........	0.16	0.45	0.029	0.032	0.08	...	...	0.27	72	180
6..........	0.16	0.44	0.010	0.024	0.04	...	...	0.04	94	235

Specimens 23 ft above rails at a location where stack blast was severe because of the combination of a steep grade and sharp curve. About 140,000 locomotive movements occurred during the ten-year test, sponsored by the Iowa State Highway Comm.

prepared by abrading the surface with 120 emery cloth. The curves in Fig. 14 represent corrosion rates of 5, 20, 50 and 200 mils per year. That is, the outlined areas represent regions where corrosion rates of 0 to 5, 5 to 20, 50 to 200, and over 200 mils per year would be expected. The corrosion of steel by strong sulfuric acid is complicated because of the peculiar dips in the curves or the rapid increase in corrosion in the neighborhood of 100 to 101% acid. The narrowness of this range means that the acids must be carefully analyzed in order to obtain reliable results. The dip or increased attack around 85% is more gradual and less difficult to establish.

Figure 14 indicates that mild steel would not be generally suitable in concentrations below about 65% at any temperature. Above 70% concentration this type of steel can be used, depending on the temperature. It is generally unsuitable above 175 F and concentrations up to about 100%.

Mild steel is also generally unsatisfactory for handling sulfuric acid in concentrations from 100 to 101% within the oleum range. The corrosion rates are fairly low below 125 F for concentrations from 65 to 98%. The rates are also low for concentrations above 104%.*

Hydrosulfuric (H_2S). The results of a laboratory test involving the corrosion of steels containing 0 to 5% Cr in hydrogen sulfide gases under a pressure of 190 to 515 psia hydrogen are given in Fig. 15. Below about 1000 F the corrosion rate for a given concentration increases with temperature, but at higher temperatures the corrosion rate decreases with temperature. In addition to corroding mild steel, hydrogen sulfide also causes it to develop blisters, in an aqueous environment, as shown in Fig. 16.

Nitric. Mild steel reacts very rapidly in dilute HNO_3 and in concentrations up to 65% at room temperature. At concentrations above 65%, steel (iron) becomes passive and the corrosion rate decreases to

*A percentage of 104 signifies an oleum solution (SO_3 in H_2SO_4), 100 lb of which will require an addition of 4 lb H_2O to give 104 lb of 100% H_2SO_4.

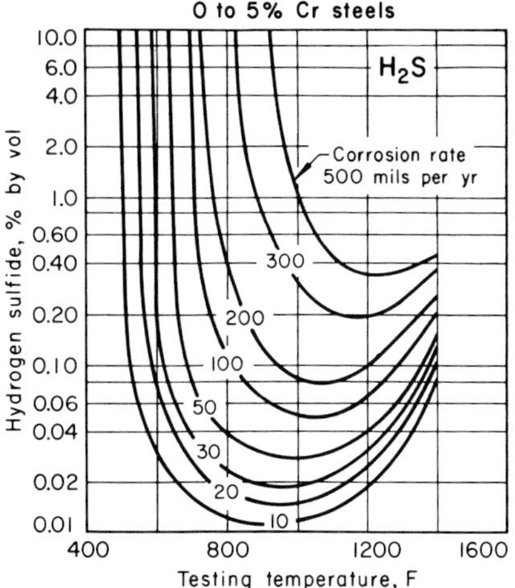

Fig. 15. Effect of temperature and hydrogen sulfide concentration on the corrosion of steel containing up to 5% Cr (F. B. Backensto, R. D. Drew and C. Stapleford, Corrosion, 12, 22, 1956)

a very low value. The occurrence of passivity is temperature-dependent — at 170 F the concentration must be 90% and at 190 F passivity cannot be induced.

Hydrochloric. The corrosion of mild steel is prohibitive in hydrochloric acid of all concentrations.

Liquid Metals

Quantitative information is not available for the attack of liquid metals on solid metals. Table 8 indicates the degree of suitability of mild steel for containers to hold certain liquid metals.

Liquid Fertilizers

A partial immersion test of mild steel and 5% Cr steel in liquid fertilizers at temperatures from 100 to 135 F for 14 and 28 days developed the results shown in Table 9.

Hydrogen

The chart shown as Fig. 17 has been developed to reveal the operating limits for carbon and low-alloy steels in contact with hydrogen at high temperature and pressure. The line for each steel represents the maximum temperatures at which the material should be used. (See, also, the article on Selection of Steel for Petroleum Refinery Applications, in this Handbook.)

Other Substances

Some substances not discussed elsewhere in the article are listed in Table 10 with indications of conditions that will permit the satisfactory use of mild steel. Service may be assumed to be satisfactory at pressures and temperatures lower than those in the table.

Soil

Disturbed Soil. During the 30 years (1922 to 1952) of the National Bureau of Standards soil-corrosion exposure program, data were obtained on approximately 37,000 specimens of 330 varieties of material for exposures up to 17 years in

Table 8. Suitability of Mild Steel as a Container for Liquid Metals

Good	Limited	Poor
Temperature, 600 F		
Sodium, Potassium, Na-K, Lithium, Indium	Lead, Bi-Pb, Pb-Bi-Sn, Bismuth	Tin, Gallium
Temperature, 1150 F		
Cadmium, Mercury, Lead, Bismuth, Magnesium	Sodium, Potassium, Na-K, Lithium, Bi-Pb	Zinc, Aluminum, Gallium, Indium, Tin, Antimony

Table 9. Corrosion Rates for 1020 and 5% Cr Steels in Liquid Fertilizers

Fertilizer	Corrosion rate, mils per month 1020	5% Cr
Nitrana	1.6	1.5
Feran	0.11	<0.01
6-6-6 with NH_4NO_3	0.66	0.49
10-10-5 with NH_4NO_3	0.31	0.68
6-6-6 with uran	0.21	0.08
10-10-5 with uran	0.12	0.04
6-6-6 with urea	0.28	0.05
10-10-5 with urea	0.22	0.04
3-9-9	0.31	1.19
8-24-0	0.14	<0.01

Temperature, 100 to 135 F; duration of test, 14 or 28 days. D. C. Vreeland and S. H. Kalin, Corrosion, 12, 569t (1956).

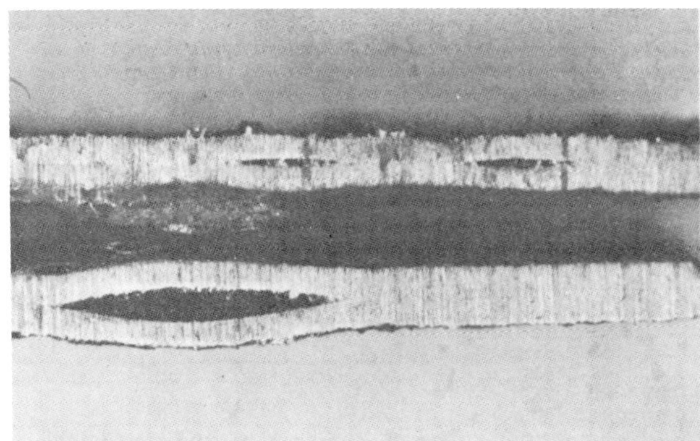

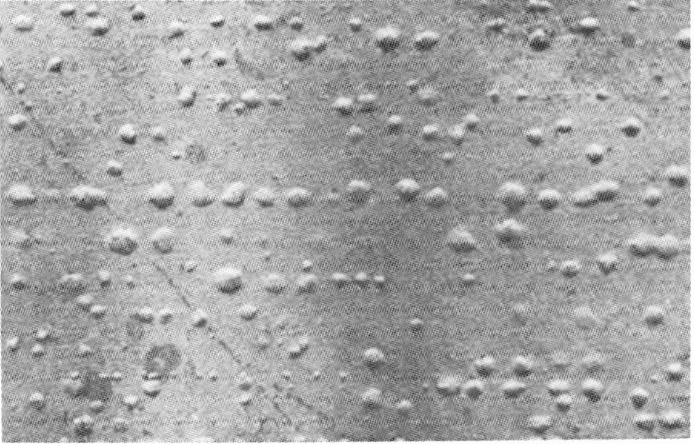

Fig. 16. (Left) Hydrogen blisters formed in ⅛-in. plate by H_2S corrosion. 2.5×. (Right) Hydrogen blisters in rimmed sheet steel exposed to H_2S-H_2O. 8×.

128 test sites representing 95 different types of soil. The soils included as wide a range of chemical and physical properties as are found in soils throughout the United States. The plain ferrous metals included in the tests were Bessemer steel, open-hearth steel, open-hearth iron, and hand-puddled and mechanically puddled wrought iron. The specimens were pipe of 1½-in. OD, 0.145-in. wall thickness, 12 in. long.

All of the plain steels were affected by corrosion in all soils, and in the same soil environment all of the plain ferrous metals corroded at nearly the same rate. This is shown in the top pair of graphs in Fig. 18, in which the losses in weight and maximum pit depths for carbon steel, wrought iron manufactured by two different processes and open-hearth iron are plotted against the duration of exposure in a poorly drained soil (Sharkey clay). If the curves for each material were drawn, the result would be four curves crossing and recrossing each other between the two extreme curves shown. No one material was superior to any of the others, and the different materials corroded at approximately the same rate in the same soil environment. The two upper charts in Fig. 18 are typical of the behavior of plain ferrous metals in any one environment.

Data obtained on plain steels in the different soil environments show that the initial rate of corrosion (either loss in weight or pitting, may be maintained in one soil, whereas in another soil the corrosion may decrease with time. This is shown in the bottom pair of graphs in Fig. 18, where the loss in weight and maximum pit depth are plotted against the duration of exposure for carbon steel in soils typical of each of six environments.

For comparison of the behavior of low-alloy sheet specimens (2.5 by 12 by 0.188 in.) under different soil conditions, the corrosion data for the soils classified according to environment were calculated for each material after 13 years of exposure (Fig. 19). Each value in Fig. 19 represents the average for two specimens removed from the same trench. The data for the earlier removals (after 4, 9 and 11 years) of the specimens show trends similar to that for the 13-year removals except that the 5% Cr steels had higher initial rates of pitting than plain steel. But these rates decreased considerably with time in all but the most corrosive soils.

The soils in Fig. 19 have been classified according to soil environment. However, because of the many factors that affect the corrosion of metals underground, other factors such as electrical resistivity, aeration and pH of the soil must be considered in determining the cor-

Table 10. Conditions Permitting the Satisfactory Use of Mild Steel

Service	Pressure, psig	Temperature F
Acetone	150	700
Acetylene(a)	150	300
Air (compressed)	150	680
Air (compressed)	300	amb
Alcohol	300	390
Ammonia (anhydrous gas)	600	930
Ammonia (anhydrous liquid)	600	930
Ammonia (aqueous)	600	930
Benzene(a)	470	840
Brine (calcium chloride)(b)	50	212
Butanol(a)	150	725
Carbon dioxide(a)	450	300
Carbon disulfide (anhydrous)(a)	300	930
Carbon tetrachloride(a)	300	930
Caustic (concentration under 5%)(a)	400	250
Caustic (concentration 0 to 10%)	150	355
Caustic (concentration 11 to 50%)(c)	150	250 to 300
Chlorine (anhydrous gas)	50	300
Chloroform(a)	300	930
Dowtherm "A"	150	1380
Gas (city)(a)	20	212
Gas (inert)	150	660
Gas (natural)(a)	50	285
Gas (natural)	100	600
Gas (natural)	600	175
Hydrogen	150	840
Hydrogen(a)	600	930
Hydrogen chloride (anhydrous gas)(a)	150	930
Kerosine(a)	125	660
Methanol(a)	150	555
Nitrogen	700	930
Phosphate (boiler feed)(a)	400	250
Sodium	150	930
Sodium cyanide (26% solution)(a)	25	212
Sodium polysulfide solution	150	930
Sulfuric acid (commercial grade)(e)		
60° Bé	100	220
66° Bé	100	250
109% (40% oleum)	100	220
Xylene(a)	75	300

(a) Copper-free steel. (b) Economical life of steel, normal maintenance, minimum temperature −15 F. (c) Stress-relieved welds and cold bands, if steam traced. (d) Hardened (high-carbon) steel is subject to sulfide or hydrogen embrittlement under intermediate operating conditions. (e) Limit velocity to 3 per sec, 6 to 8-yr life at temperatures shown.

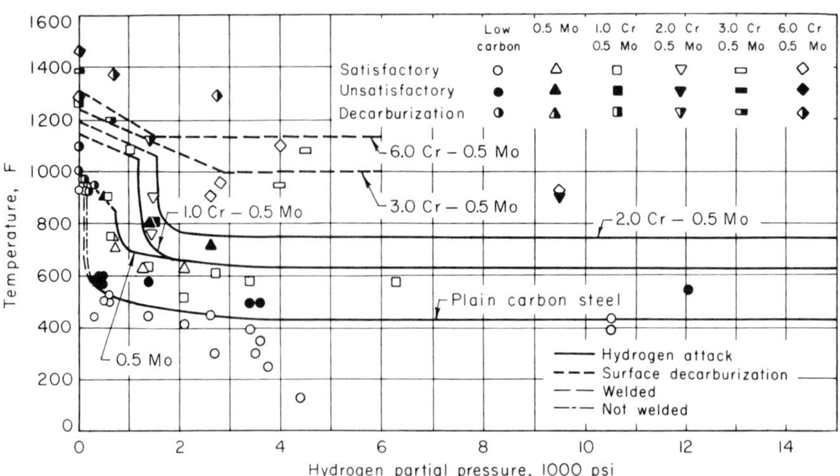

Fig. 17. Safe operating limits for carbon and low-alloy steels in contact with hydrogen at high temperature and pressure (G. A. Nelson and R. T. Effinger, Welding Journal, Jan 1956). See also article on Petroleum Refinery Applications.

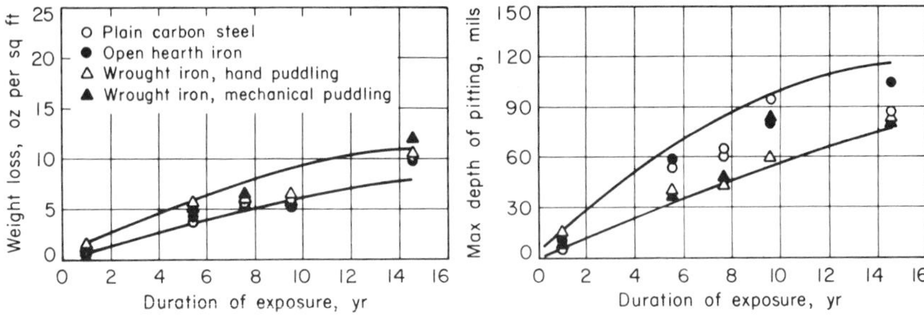

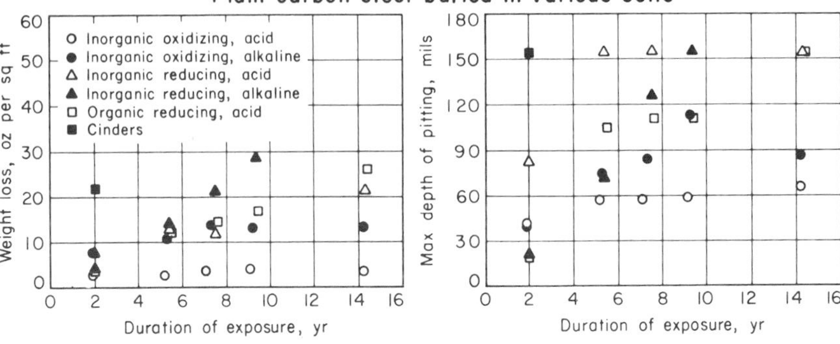

Fig. 18. Corrosion of ferrous metal buried in Sharkey clay and other soils

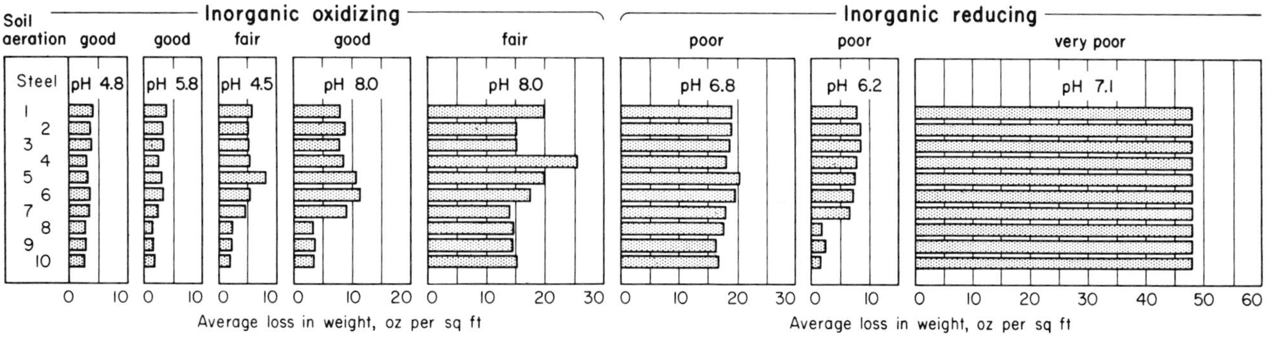

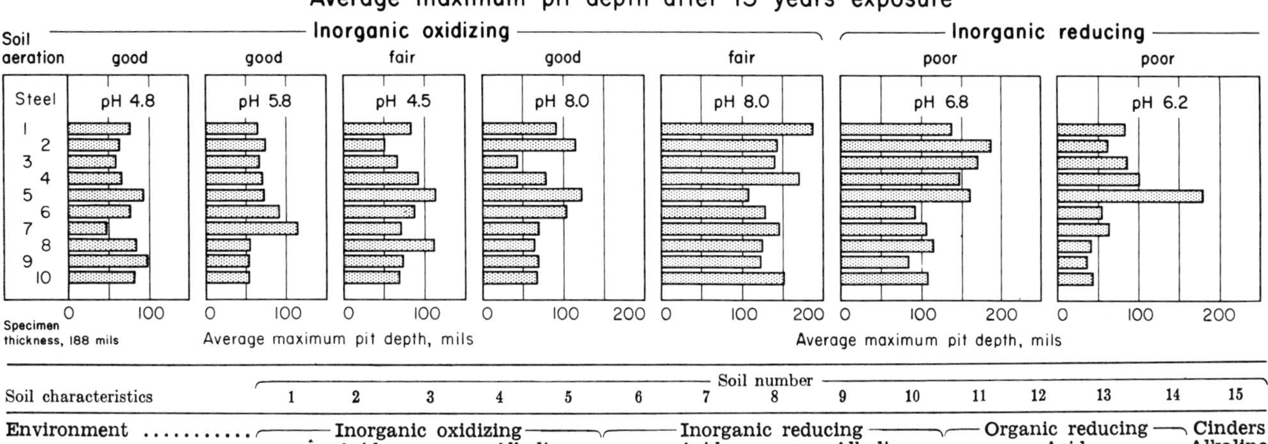

Soil characteristics	1	2	3	4	5	6	7	8	9	10	11	12	13	14	15
Environment	Inorganic oxidizing					Inorganic reducing					Organic reducing				Cinders
	Acid			Alkaline		Acid			Alkaline		Acid				Alkaline
Aeration	Good	Good	Fair	Good	Fair	Poor	Poor	Very poor	Fair	Fair	Poor	Poor	Very poor	Very poor	Very poor
Resistivity, ohm-cm	17,790	5210	6920	148	232	190	943	406	62	278	712	218	1660	84	455
pH	4.8	5.8	4.5	8.0	8.0	6.8	6.2	7.1	7.5	9.4	4.8	2.6	5.6	6.9	7.6

Element	Steel number 1	2	3	4	5	6	7	8	9	10
Cr	0.049	0.02	0.02	...	...	1.02	2.01	5.02	4.67	5.76
Ni	0.034	0.15	0.14	0.52	1.96	0.22	0.07	0.09	0.09	0.17
Cu	0.052	0.45	0.54	0.95	1.01	0.428	0.004	0.008	0.004	0.004
Mo	....	0.07	0.13	...	...	....	0.57	....	0.51	0.43

Fig. 19. Effect of composition on corrosion of low-alloy ferrous materials in various soils. (Charts continued on next page)

rosiveness of a particular environment. These soil properties are tabulated below Fig. 19.

The inorganic-oxidizing-acid soils, which are always characterized by high electrical resistivity, indicating the absence of soluble salts in the soil, and good to fair aeration, show no appreciable difference in behavior between plain and low-alloy steels, except for the lower weight losses of the 5% Cr steels. In soils of this environment, the rate of corrosion for the plain ferrous metals may be high initially, but decreases after a few years of exposure to almost complete cessation of pitting because the initial corrosion products are favorable for the formation of tubercles over the pits, which retards further corrosion.

In the other environments, corrosiveness varied from moderate to severe. In the severly corrosive soils, especially those in which failure of the reference metal (steel 1) occurred at an early exposure period, none of the low-alloy materials was more resistant to weight loss and pitting than plain steel. An exception occurs in the cinder fill, where the copper-molybdenum open-hearth iron and the 5% Cr steels containing molybdenum were consistently more corrosion-resistant than plain steel.

In the less corrosive soils of the inorganic and organic reducing environments, the 5% Cr steels consistently show lower weight loss than plain steel. This greater resistance to corrosion, however, is less marked with respect to pitting. In two of the organic reducing soils (soils 12 and 14) the 5% Cr steels pitted more deeply than the control specimens (steel 1). These two soils are characterized by high concentrations of sulfates and sulfides.

Included in the National Bureau of Standards soil-corrosion tests were specimens of the same lot of Bessemer steel pipe, with and without mill scale. These were exposed to 47 different soils for periods up to 17 years. The scale-free specimens corroded slightly less during the early exposure periods than the mill-scale coated specimens. However, with longer periods of exposure, no appreciable differences in corrosion could be observed between the two specimens, either in weight loss or pitting. (NBS Circular 579, "Underground Corrosion", U.S. Government Printing Office, 1957)

Undisturbed Soil. A soil which may be very corrosive if it has been disturbed, as in the laying of a pipe line, may be rather mild in its behavior if it has not been disturbed. Tests have been conducted where ordinary steel H-piles were driven in two locations (G. G. Greulich, Engineering News-Record, p 41, August 24, 1950). One pile was withdrawn and examined at each site.

One of the piles examined was in a group of eight that had been driven in a swamp beside a highway near the Bonnet Carré Floodway in Louisiana and the other was at the turning basin in the harbor at Houston, Texas.

The Lousiana pile, a 12-in. 65-lb steel section, had been driven 17 years before to a depth of 122 ft 3 in. below ground level, which was 116 ft below sea level. The soil penetrated ranged from a blue sandy clay at the surface through fine sands, humus with some sand, and then through clays with some sand to stiff blue clay for the last 38 ft.

When this pile was withdrawn, the looser soils fell out of the space be-

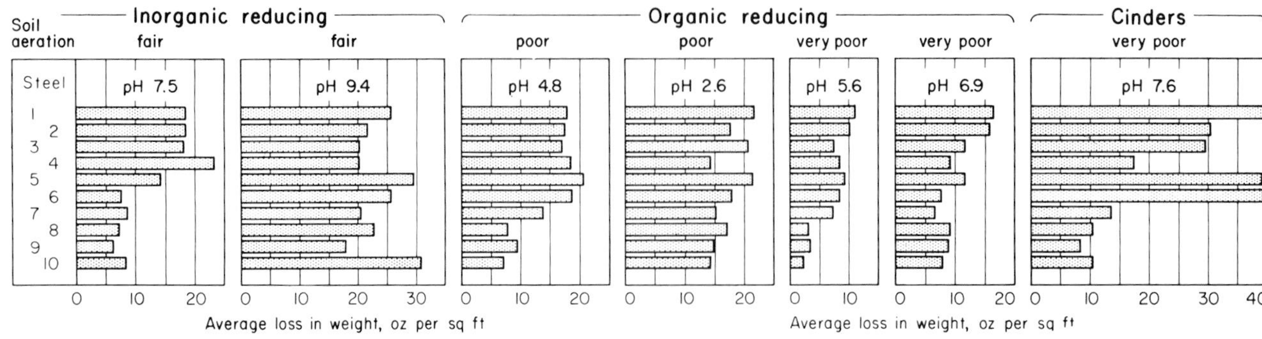

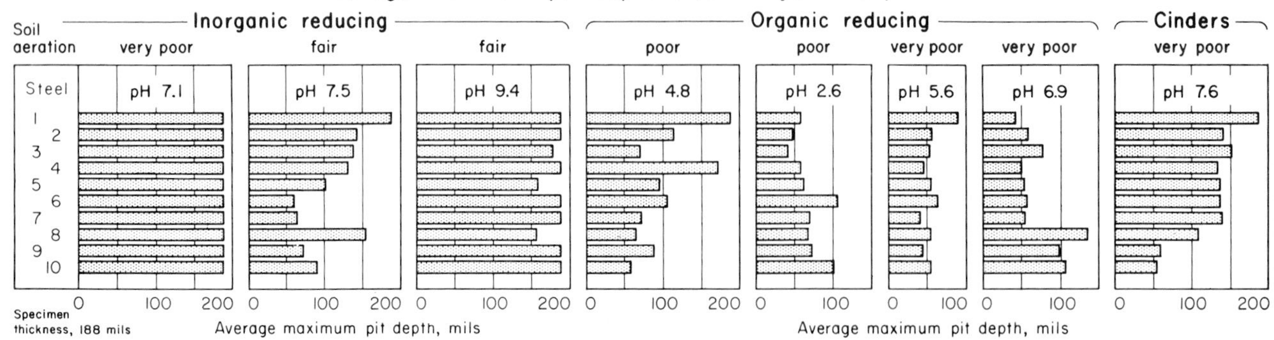

Fig. 19 (continued). See preceding page for identification of steels and soils.

tween the flanges, but the cohesive soils remained in place, taking the form of a roughly circular plug. When this material was checked against the log of test borings made nearby, there was close agreement, showing that the pile, during driving, had cut through the layers and did not carry with it a hard plug from near the surface as happens when such piles penetrate hard layers near the surface.

No measurable corrosion was found in the pile although two of the soil strata, selected at random, were proved by a National Bureau of Standards test method to be corrosive to steel when in the presence of oxygen. Red-lead paint on the tip of the pile was as fresh as new. White painted footage numerals on the flanges also had a new appearance. What appeared to be a roughened zone for a length of 30 in. over the range of the fluctuating water table (about 5 ft below the surface) developed on laboratory examination to be a thin hard mineral crust.

The second pile examined was driven at the site of the proposed Wharf No. 9 in the harbor turning basin at Houston in August 1937. It was a 12-in. 72-lb pile driven to a depth of 72 ft through various layers of sand and clay.

After driving, the pile was braced and a sand box was erected on steel brackets to permit a test load of 100 tons to be applied. This load was maintained until May 1938 at which time the total settlement, including elastic compression in the pile, was ⅛ in.

Shortly thereafter the bracing of the pile was removed to permit it to sway slightly, as it would in a wharf, to determine whether such movement would accelerate settlement. When no further settlement had occurred by June 1939, the test load was removed.

When first driven, this pile was at the edge of the shoreline, just away from the water, but in subsequent years the earth was removed to a depth of 6 ft and the pile used as a mooring for barges. This resulted in its being bent back and forth at a point 3 to 4 ft below the mud line, and finally it was bent down out of sight below water level by a barge. From this position the upper part was recovered by breaking it off with a clamshell bucket at the point where initial fracture had started from the bending.

Examination of the straight part of the recovered piece, which was between 1 and 2 ft below the mud line and which seemed to be corroded as much as any part of the pile above the point of recovery, revealed the condition shown in Table 11 as a result of over 12 years in mud and water. The average corrosion rates of the upper end of the sample, calculated from the original nominal thicknesses, were 2.6 mils per year on each surface of the flanges.

It was estimated that the life of such a pile in wharf service would be about 75 years and hence steel piles were used in the wharf constructed in 1949. Apparently the undisturbed soil had long since been depleted of oxygen by the organic matter present. Disturbing the soil by removing it and putting it back furnished a supply of oxygen which led to such results as the National Bureau of Standards found with New Orleans muck.

Refrigerants

Data are given in Fig. 20 for the corrosion rates of mild steel in Refrigerants 11 and 113 (ASRE designations, Standard 34). Two kinds of test were conducted: (1) under reflux conditions where the metal boiler was sitting on a hot plate in the laboratory air for 480 days, and (2) specimens in the refrigerants at room temperature for 786 days.

The following conclusions were drawn from the tests: (1) Regardless of oil or water content, neither Refrigerant 11 nor 113 is corrosive to mild steel in the liquid, vapor or air space above the vapor. (2) The corrosion rate decreased markedly from the eighth week to the end of the test. (3) Oil accelerates corrosion in the vapor space. (4) The corrosion rates were not significantly different in the vapor and air space in any one test. (5) The low corrosion rates where there was far more water (2%) than enough to saturate the refrigerant indicates that the corrosion is caused almost entirely by the the air present.

In general the halocarbon refrig-

Table 11. Corrosion of Steel H-Pile During 12-Year Period in Houston Harbor

Section	Original nominal thickness, in.	Thickness after exposure, in. — Max	Min	Avg of five	Loss in thickness, in.
Right flange	0.688	0.631	0.605	0.617	0.071
Left flange	0.688	0.640	0.625	0.633	0.055
Web	0.438	0.404	0.401	0.403	0.035

erants are not corrosive to metals. This is not true of some of the older refrigerants. Ammonia, for example, particularly in the presence of water and some air, is devastating in its effect on copper or the yellow copper alloys. Refrigerant 11 centrifugal machines have suffered severe corrosion where the cooling brine containing fairly low amounts of ammonia as an impurity has leaked into the evaporator. Tests indicate that the refrigerant accelerated the destruction. It is imperative that ammonia be kept out of machines containing copper or copper alloys which are yellow or reddish bronze. Also, methyl chloride must be kept out of machines where it will come in contact with aluminum. It reacts directly with this metal, forming aluminum trimethyl, a spontaneously combustible gas when in contact with air.

Galvanic Corrosion

Galvanic corrosion is associated with the current of a galvanic cell consisting of two dissimilar conductors in an electrolyte or two similar conductors in dissimilar electrolytes. Where the two dissimilar metals are in contact, the resulting reaction is referred to as bimetal or "couple action". The couple action of steel is regulated by: (1) the metal to which the steel is coupled, (2) the conductivity of the solution in which it is in service, (3) the area relationship between the steel and the other metal, and (4) the presence or absence of oxygen or other depolarizing agents. A specific unfavorable combination of all of the above factors is necessary for accelerated galvanic corrosion to occur.

Steel is affected adversely by galvanic corrosion only when in contact with a metal below it in the electrochemical series for a specific environment. A practical version of the galvanic relationship of metals in sea water is shown in Table 12. Steel is protected when coupled to any of the metals above it in this table and it may be corroded at an accelerated rate when in contact with metals below it. For example, zinc, as in galvanizing, will protect steel in sea water, while copper will cause severe attack of steel in the same environment. However, the coupling of steel and copper does not always mean a prohibitive increase in the corrosion rate of steel. If the conductivity of the solution is very low, the current flow may be so restricted as to prevent any appreciable galvanic corrosion. The satisfactory use of stainless steel or bronze-trimmed steel valves in steam condensate and many organic chemical environments illustrates this relation.

Table 12. Galvanic Series in Sea Water

1	Magnesium (most active)	18	Inconel (active)
2	Magnesium alloys	19	Hastelloy B
3	Zinc	20	Brasses
4	Aluminum 1100	21	Copper
5	Cadmium	22	Bronzes
6	Aluminum 2017	23	Copper-nickel alloys
7	Steel (plain)	24	Titanium
8	Cast iron	25	Monel
9	Chromium iron (active)	26	Silver solder
10	Nickel cast iron	27	Nickel (passive)
11	304 stainless (active)	28	Inconel (passive)
12	316 stainless (active)	29	Chromium iron (passive)
13	Hastelloy C	30	304 stainless (passive)
14	Lead-tin solders	31	316 stainless (passive)
15	Lead	32	Silver
16	Tin	33	Graphite (least active)
17	Nickel (active)		

In some services where conductivity is measurable but still low, corrosive attack may occur on the steel in a narrow band adjacent to the more noble metal to which it is joined. This has been noticed as undercutting of stainless steel welds on mild steel pipe in dilute organic acid services.

Another example of severe localized attack is that on steel in contact with 18-8 stainless in aerated concentrated water solutions of lithium bromide (50 to 65% LiBr) at about 250 F. The attack is very much localized even though the area of steel is large compared to that of the stainless steel.

The area relationship between the anodic or corroding metal and the more noble metal is important as may be illustrated by an example of a stainless steel weld on mild steel pipe in a solution of high conductivity, such as sea water. Under these conditions the steel is subject to accelerated corrosion; however, the current flow is distributed over such a large area that no significant wastage of steel will be noticed. If, on the other hand, carbon steel trim were put in a stainless steel valve the large area of stainless steel compared with the carbon steel would cause rapid attack of the carbon steel trim. Steel rivets used in a railroad coal-hopper car made of 12% Cr steel were quite unsatisfactory because of the concentrated attack on the rivets' small area.

The presence of a depolarizing agent such as oxygen is also necessary to permit accelerated attack in a bimetal system. Increased corrosion will not occur between dissimilar metals in sea water in the complete absence of oxygen. However, the presence of oxygen is not always necessary in other environments for the accelerated corrosion of steel. In certain media such as dilute organic acid, steel may be adversely affected by the presence of stainless, and the expected straight acid attack may be accelerated.

Two situations involving failures

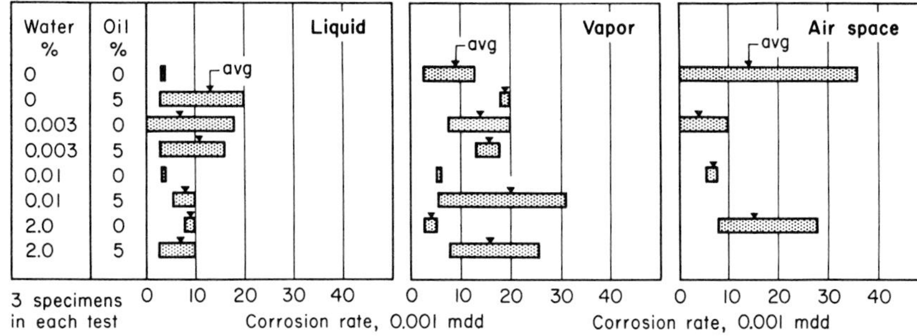

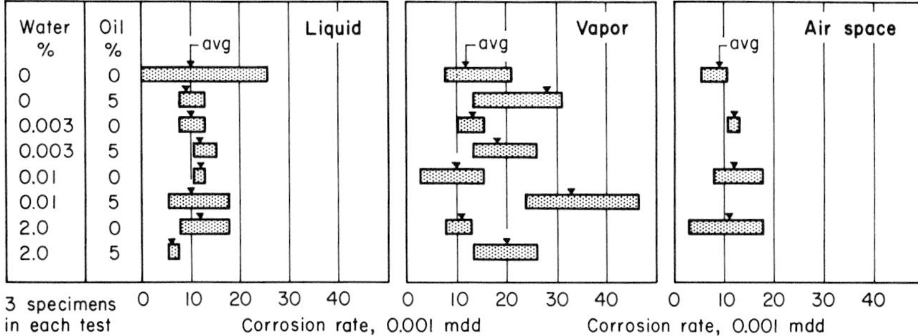

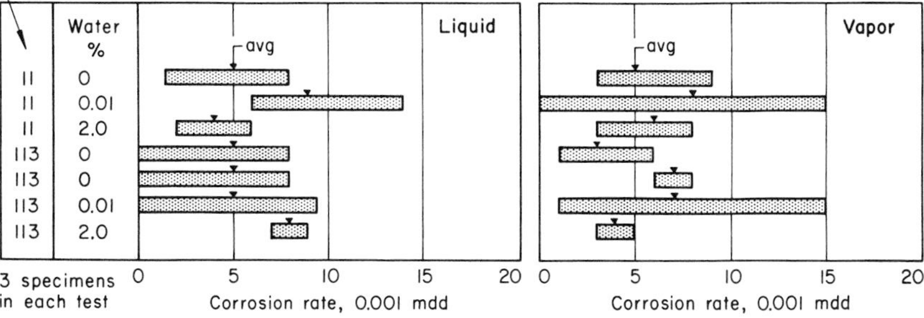

Fig. 20. Corrosion rates of mild steel in two commercial refrigerants

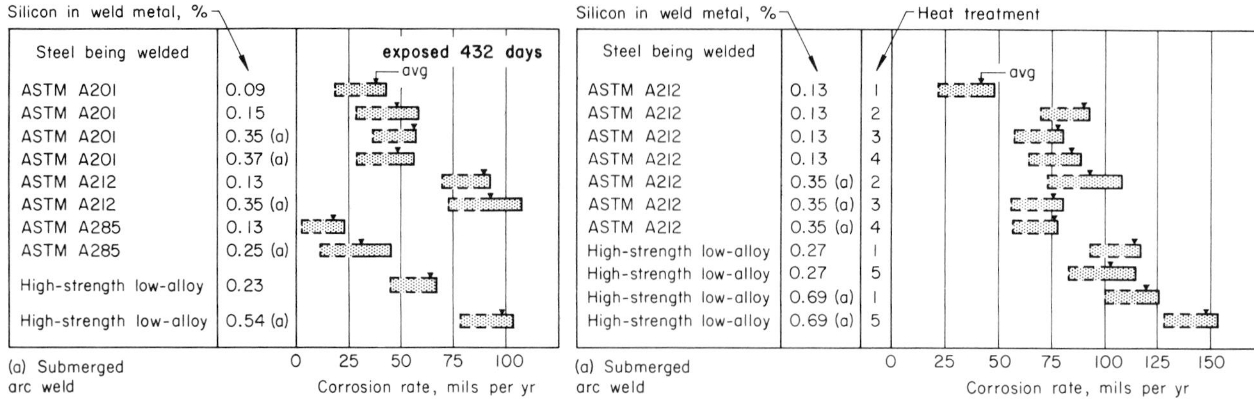

Fig. 21. Corrosion rates at welded seams in Kraft digesters. Measurements were made along a single 15-in. length of welded seam. Average and maximum rates were recorded; minimum rates are unknown. Heat treatment code is as follows: (1) stress relieved at 1150 to 1200 F (2) annealed 2 hr at 1300 F, furnace cooled; (3) annealed 2 hr at 1450 F, furnace cooled; (4) annealed 1 hr at 1600 F, furnace cooled; (5) normalized at 1650 F. (R. A. Huseby and M. A. Scheil, Tappi, 34, 202, 1951)

of mild steel coupled with stainless steel in sea water service may be cited: (1) plain carbon steel Alemite grease fittings inserted into type 410 stainless steel chain pins failed within four to six months in ocean water in the Los Angeles area, and (2) type 1040 bolts and nuts used on a type 304 stainless steel structure failed after five weeks in ocean water at Redondo Beach, Calif.

Welded Specimens in Sea Water. A seven-year test was conducted at Kure Beach, N.C., involving couples of plain and low-alloy steel, the amount of corrosion being measured by weight loss and converted into loss of thickness (Table 13). Equal areas of the two steels were used, the specimens before welding being 6 by 12-in. plate. The plain steel would have corroded even more had the ratio of the area of the low-alloy steel to the plain steel been higher. The low-alloy steel caused the plain steel to corrode faster than if it were not coupled. Apparently the scale had little effect except where coupled to a pickled specimen of plain steel.

Welded Specimens in Kraft Digesters. In one test in Kraft digesters composite specimens were made of two 15 by 5 by 0.75-in. pieces of the same steel welded, from specimen to specimen, with different electrodes. All specimens were sandblasted and stress relieved at 1150 F; except for two tests the material was annealed 2 hr at 1300 F and furnace cooled.

The specimens were bolted to the lower third of the middle section of a Kraft digester and submerged in sulfate "white liquor" for 432 days. The corrosion rates are given in Fig. 21. This liquor consists of approximately 0.2 lb per gal Na_2S with 0.7 lb per gal NaOH and small amounts of impurities, mainly, Na_2CO_3, Na_2SO_3, Na_2SO_4, $Na_2S_2O_3$ and polysulfides. The liquor is used in a cyclic batch process in a steam-heated pressure vessel to produce pulp from wood chips.

Further results were obtained by the same investigators concerning the effect of heat treatment on the corrosion of welded specimens, using the same conditions of exposure. Results are given in the right-hand chart of Fig. 21 for the five different heat treatments, which are defined in the caption.

Various Swedish electrode materials were exposed during the course of 20 Kraft cooking periods, each requiring 3 to 3½ hr to reach 338 F and 30 min at temperature. The results are shown in Table 14.

Mild steel is often coupled with stainless steel in Kraft digesters that are partially lined or repaired with stainless steel overlays or weld deposits. Table 15 gives the results of tests to define quantitatively the galvanic effects of such lining and repair procedures. The tests were made by welding together a mild steel (0.20% C, 1.51% Mn, 0.10% Si) and a stainless steel. The couples were then bolted to, and electrically insulated from, the inside of a mild steel tumbling Kraft digester and exposed for 432 days in 100% white liquor. After welding, the specimens were about 30 by 15 by ¾ in.

In general, the coupling of mild steel to various stainless steels does not greatly accelerate corrosion of the mild steel. In the corrosive environment of a Kraft digester, the galvanic effect is usually nullified rather quickly by film formation, which opens the galvanic circuit be-

Table 13. Galvanic Effect of a Low-Alloy Steel on Plain Steel in Sea Water

Surface condition of plain steel	Loss in thickness of plain steel, mils — Uncoupled	Coupled to low-alloy steel which was — Pickled	Scaled
Pickled	44	56	62
Scaled	49	59	56

Table 14. Corrosion Rates of Various Electrode Materials in Kraft Digesters (a)

Electrode type	Typical composition, % — C	Si	Mn	P max	S max	N max	Corrosion rate, mpy
A, decarburizing	0.05	0.01	0.10	0.040	0.040	0.040	352
B, organic rutile	0.10	0.30	0.50	0.040	0.040	0.020	641
C, rutile	0.10	0.30	0.60	0.040	0.040	0.007	325
D, basic No. 1	0.10	0.50	0.60	0.040	0.040	0.007	439
E, acid No. 1	0.10	0.15	0.55	0.040	0.040	0.015	325
F, acid	0.10	0.15	0.45	0.040	0.040	0.010	303
G, acid No. 2	0.15	0.05	0.70	0.040	0.040	0.013	290
H, basic No. 2	0.10	0.70	0.60	0.040	0.040	0.007	400
I, basic No. 3(b)	0.15	0.70	0.80	0.040	0.040	0.007	466

(a) L. Ruus and L. Stockman, Svensk Papperstidn, 56, 857 (1953). (b) Also contains 1.15% Cr and 0.5% Mo.

Table 15. Galvanic Corrosion of Mild Steel Welded to Various Stainless Alloys

Alloy type	Ratio of alloy to mild steel	Weld metal	Heat treatment	Average corrosion rate, mpy
430	1	430	None	50
430	1	430	1550 F (1 hr)	53
430	8	430	None	77
430	1/6	430	None	35
Inconel	1	80 Cr – 20 Ni	None	55
Inconel	1	80 Cr – 20 Ni	1550 F (1 hr)	44
Inconel	8	80 Cr – 20 Ni	None	29
Inconel	1/6	80 Cr – 20 Ni	None	32
347	1	310 + Cb	None	31
347	1	310 + Cb	1550 F (1 hr)	26
347	8	310 + Cb	None	37
347	1/6	310 + Cb	None	33

Exposed 432 days in 100% Kraft white liquor. R. A. Huseby and M. A. Scheil, Tappi, 34, 202 (1951)

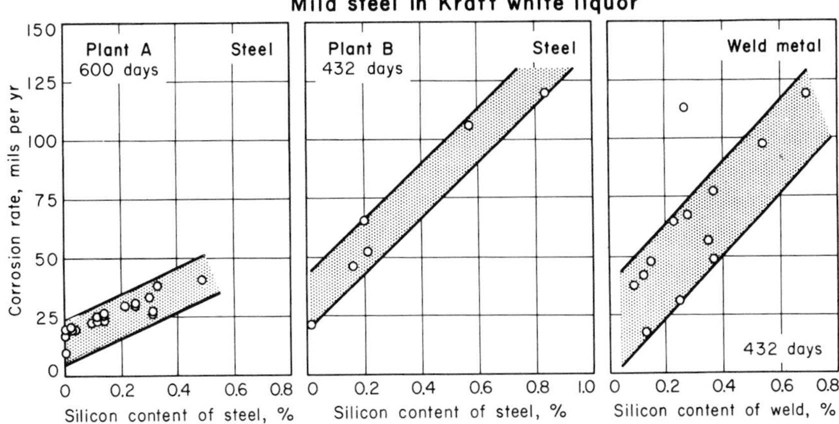

Fig. 22. Effect of silicon content on the corrosion rate of mild steel in Kraft white liquor (B. Roald, Norsk Skoglind, 11, 446, 1957; and R. A. Huseby and M. A. Scheil, Tappi, 34, 202, 1951)

tween metal and electrolyte. Galvanic corrosion of mild steel welded to Inconel or type 347 stainless is not dangerously severe, but type 430 considerably accelerates corrosion if its area is as large as that of the plain steel.

The successful coupling of stainless steel and mild steel is explained on the basis of potential measurements of the same couples placed in boiling white and black liquor. In both solutions, the potential difference was 0.23 volt, with the mild steel anodic to the stainless; however, in less than 9 min the mild steel was covered with a black film of iron sulfide, and there was no potential difference between the two metals.

Effect of Alloying Elements

In the first part of this article, many data were presented to show the effect of copper in atmospheric corrosion. Copper-bearing steel has better atmospheric corrosion resistance than plain steel. High-strength low-alloy steels have still greater resistance.

Chromium, nickel, manganese and phosphorus in steel enhance the atmospheric corrosion resistance; phosphorus is particularly effective in small amounts — up to 0.15%. The life of a chromium-copper steel is several times that of plain carbon steel. Chromium also gives greater corrosion resistance in 50% H_2SO_4.

Carbon, either in the dissolved or carbide form, has little effect on corrosion in the atmosphere, fresh water or sea water.

Nitrogen in iron or steel increases the susceptibility to stress-corrosion cracking in alkaline nitrate solutions. Steels containing aluminum are more resistant to stress-corrosion failure because the dissolved nitrogen has been tied up as aluminum nitride. Manganese, sulfur, phosphorus and silicon ordinarily present in constructional steels have no effect on corrosion in fresh or salt water. In acids, the presence of manganese, sulfur and phosphorus in the steel may increase corrosion. Some data indicate that the addition of about 3% Cr to mild steel decreases the maximum depth of pitting in steels exposed to fresh water.

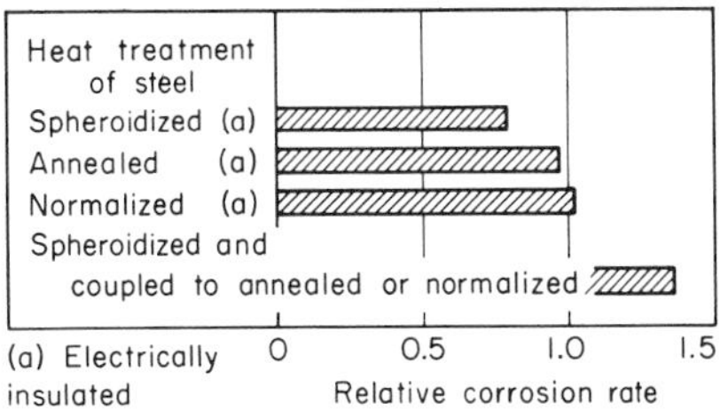

Fig. 23. Effect of heat treatment on corrosion of mild steel in sulfuric acid

Silicon in steel is detrimental in alkaline liquors. Steel specimens 4 by 8 in., of thickness from 13/32 to 15/16 in. were exposed in Kraft sulfate digesters in a scale-free area. The specimens were machined to size and sandblasted from unannealed rolled plate. (The contents of this white liquor have been previously described.) The corrosion rates for the 600-day test are given in Table 16 and the left-hand chart of Fig. 22; the other charts in Fig. 22 are from the same data as Fig. 21.

Fabrication and Processing

The service life of any pressure vessel fabricated from mild steel depends on the precautions taken to insure good workmanship in fabrication, and rigid shop inspection and testing.

Cold working increases the corrosion rate of steel in acid solutions; the amount of increase depends on the degree of cold work as well as the carbon and nitrogen contents. Stress relieving will reduce the corrosion rate, but normalizing or annealing results in the lowest rate.

In hardenable carbon and low-alloy steels, tempering of martensite between 700 and 800 F will increase the corrosion rate to a maximum. Heating to higher temperatures results in a reduction of the corrosion rate. So far as is known, this effect of tempering temperature on corrosion rate is observed only in acids.

The formation of spheroidal carbides in pearlite areas by local heating also results in high or excessive corrosion in these areas in acid environments. Full annealing or normalizing is the only effective means of reducing the rate. Stress relieving, where carbides are present, is unsatisfactory. Figure 23 shows the relative corrosion rates of mild steel, in three heat treated conditions, in 60 to 67% sulfuric acid at 76 F. The spheroidized material had a lower corrosion rate than either of the other two where it was electrically insulated. However, its corrosion rate was almost doubled where it was coupled to either the annealed or normalized material. Localized corrosion of steel known as ringworm attack in oil-well service has been encountered in steels having a spheroidized carbide structure. Steel possessing well formed pearlitic lamellae shows measurably better resistance to this type of corrosion, apparently because the plate-like carbides permit more adherent corrosion products.

F. N. Speller, in his book "Corrosion, Causes and Prevention" (McGraw-Hill), states that a 0.38% C steel quenched from 1560 F and tested in distilled water for 9 months corroded at a rate of 4.3 mils per year. Tempered specimens (570 to 1470 F) corroded at 3.3 mils per

Table 16. Effect of Silicon Content on Corrosion Rates of Steel in Kraft White Liquor

Steel	Corrosion rate, mpy	Composition, % C	Si	Mn	P	S	Cu	N	Al
1	27.2	0.20	0.31	0.52	0.016	0.040	0.16	0.007	....
2	27.6	0.20	0.31	0.60	0.020	0.037	0.13	0.006	0.041
3	38.2	0.22	0.33	1.12	0.024	0.032	0.17	0.007	....
4	22.4	0.18	0.10	1.14	0.019	0.025	0.12	....	....
5	33.1	0.20	0.30	1.10	0.018	0.032	0.13	0.006	0.041
6	17.4	0.17	trace	0.53	0.031	0.028	0.11	0.002	....
7	20.1	0.20	trace	0.55	0.035	0.032	0.15	0.003	....
8	20.1	0.11	0.03	0.79	0.015	0.022	...	0.005	....
9	20.1	0.14	0.03	0.90	0.016	0.019	...	....	0.016
10	9.9	0.04	0.00	0.38	0.056	0.016	...	0.015	....
11	29.2	0.19	0.25	0.73	0.024	0.034	0.15	0.007	0.060
12	21.2	0.14	0.02	1.12	0.019	0.023	...	0.007	0.006
13	30.8	0.17	0.25	0.75	0.033	0.026	0.16	0.006	0.070
14	23.2	0.14	0.14	0.50	0.017	0.026	0.08	0.007	0.030
15	26.8	0.13	0.14	0.47	0.026	0.023	0.08	0.007	0.045
16	40.6	0.17	0.49	1.11	0.045	0.025	0.10	0.007	0.025
17	26.4	0.18	0.14	0.66	0.035	0.029	0.10	0.007	0.040
18	29.2	0.14	0.21	0.58	0.031	0.030	0.08	0.008	0.025
19	23.6	0.09	0.12	0.52	0.032	0.036	0.14	0.007	0.025
20	25.6	0.14	0.12	0.57	0.020	0.029	0.14	0.007	0.025

Corrosion test period was 600 days. B. Roald, Norsk Skoglind, 11, 446 (1957)

year. There is some indication that this is true for exposure to seawater. However, heat treatment is probably a minor factor affecting steel corrosion in natural water.

The corrodibility of metals used in boiler construction has been determined at various temperatures in inhibited hydrochloric acid solutions. Twenty-two metals representing 14 different ASME code specifications as well as 40 boiler handhole plates were used in this investigation. (P. H. Cardwell and S. J. Martinez, Ind Eng Chem, **40,** p 1956, 1948.) Considerable variation in the corrodibility of the metals was attributed in part to their carbon and silicon contents. The metals with the lowest silicon contents corroded the least, as shown in Fig. 24. In 10% acid, corrodibility was also found to increase above about 0.2% C except that some metals in the low-carbon range, but with high silicon contents, corroded at higher rates than metals with more than 0.2% C.

Corrodibility of certain fabricated metals with abnormally high rates of corrosion was lowered substantially by annealing.

In the immediately previous section of this article it was shown that silicon in mild steel adversely affected the corrosion resistance in Kraft white liquor. It seems, therefore, if mild steel is to be used in an alkaline corrosive environment, rimmed steel should be selected rather than one deoxidized with silicon (semikilled).

In one of the tests involving sulfate white liquor (Fig. 21), exposures were made to determine the effect of surface condition and heat treatment. Steel A285, grade C, in the hot rolled (scale intact) stress annealed condition had a corrosion rate of 11 mils per year; the same steel with a sandblasted surface, 21 mils per year. The respective values for A212, grade B, were 45 and 52. After holding at 1600 F for 1 hr and furnace cooling, the corrosion rate of A285 was increased from 21 to 82 mils per year for the sandblasted surface; A212, from 52 to 94.

A test has been conducted to determine the effect of normalizing hot rolled stock. Specimens of mild steel of varying composition were exposed in a laboratory digester for a large number of Kraft cooking cycles. Duplicate specimens of each type of steel were exposed, one in the hot rolled condition and the other normalized at 1740 F. The results are given in Fig. 25.

The units used to express corrosion rates, grams per square foot per cooking cycle, are more meaningful than mils per year in this application. At the beginning of a cooking cycle, after the digester has been charged with wood chips and liquor, the temperature is raised linearly with time to the desired cooking temperature, which is usually in the range from 325 to 350 F. During this period, usually 1 to 3 hr, reaction of the liquor with the wood

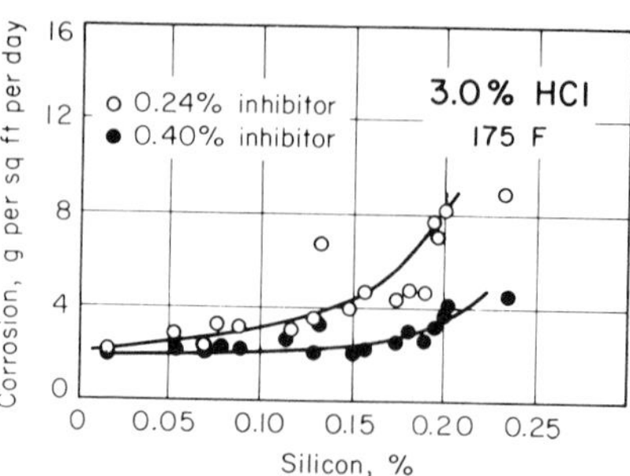

Fig. 24. Corrosion of mild steel in inhibited hydrochloric acid (P. H. Cardwell and S. J. Martinez, Ind Eng Chem, 40, 1956, 1949)

proceeds at an ever-increasing rate, and the nature of the liquor changes greatly as a result. Also, during the heating-up period, the corrosion rate is directly proportional to the temperature up to about 250 F, at which point corrosion nearly ceases for the remainder of the cycle.

For these reasons, it is not quite appropriate to express corrosion rates of Kraft digesters on a time basis, although it is commonly done; an understanding of corrosion rates on the common basis of mils per year would require knowledge of the rate of temperature rise for a given digester plus the number of cooking cycles per year with that digester. These two variables differ greatly from mill to mill and even from one digester to another.

A number of years ago a condition was investigated where grinding of the surface had introduced stresses into the surface of metal that subsequently developed stress-corrosion cracking. A series of laboratory tests indicated conclusively that the cracking was from stress corrosion, and the condition could be duplicated in the laboratory. Temperatures of 3000 F can be reached on the surface of steel while it is being ground, and consequent surface stresses may exceed 100,000 psi. Thus, grinding can be a major contributing source of accelerated stress corrosion.

Such variations in surface finish are apparently much more important and have much more influence on the life of the metal than any of the ordinary variations in composition which occur incidental to manufacture. The corrosion environment has a lot to do with the type and rate of attack. In some services a highly polished surface often shows less initial attack than a rough surface. Frequently it withstands exposure for a long time before showing any corrosion, while an unpolished surface may be badly attacked in a short period.

Embrittlement

Hydrogen embrittlement of mild steel may occur under at least three different sets of conditions:

1 Hydrogen attack — at high temperature and (usually) low pressure — resulting from the reaction of hydrogen with some nonmetallics in the metal.
2 Electrolysis and pickling embrittlement — occurring in contact with electrolytes either with an impressed voltage or without an impressed voltage as in ordinary acid pickling.
3 High-pressure hydrogen embrittlement — usually encountered at low temperatures where reaction with nonmetallics is not involved.

In petroleum refinery applications carbon steel is badly attacked at temperatures as low as 625 F. A 1% Cr-V steel shows improved hydrogen resistance. Cr-Al-Mo and Ti-Cr-Mo steels show even better resistance, while a 5% Cr-Mo steel is unaffected by these conditions.

In 100-hr tests of resistance to hydrogen attack plain carbon steels were damaged starting at 930 F under 50 atm, at 750 F under 100 atm, and below 660 F at 600 atm pressure. Silicon, nickel and copper are not beneficial as alloying elements. Manganese slightly increases resistance to hydrogen attack. A 3% Cr addition markedly increases resistance. Tungsten is more effective than chromium, and molybdenum better than tungsten. At critical alloy conditions, vanadium, titanium, columbium, tantalum, zirconium and thorium all give high resistance to attack. In general, alloying elements that form stable carbides confer high resistance to hydrogen damage. The historical upgrading of steels to counteract hydrogen attack has been from carbon steel to chromium-nickel and chromium-tungsten, to chromium-vanadium, to chromium-molybdenum and chromium-molybdenum with vanadium or tungsten additions.

At high temperature and pressure,

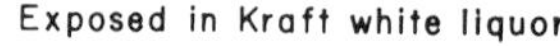

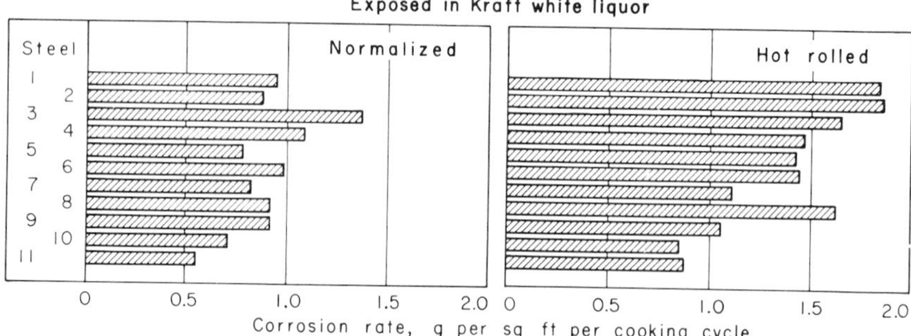

Fig. 25. Effect of heat treatment on corrosion of carbon and low-alloy steel in Kraft white liquor. All steels had 0.02 to 0.17% C, 0.01 to 0.24% Si and 0.02 to 0.63% Mn, except steel 9 (1.42% Mn). Steel 5 had 1.06% Ni; steel 10, 2.09% Ni; and steel 11, 5.02% Ni. (L. Stockman and L. Ruus, Svensk Papperstidn, 57, 831, 1954)

atomic hydrogen from thermal dissociation causes decarburization of mild steel and general intergranular cracking, sometimes referred to as hydrous embrittlement. At low temperature, diffusion of atomic hydrogen produced by corrosion leads to blisters and isolated cracks without decarburization, through recombination within the steel to molecular hydrogen, which does not readily diffuse through the metal and accordingly builds up high internal pressures. For damaging quantities of hydrogen to enter steel, the hydrogen must pass through the metal unabated for a considerable period of time. Thus, the corrosion or other electrolytic process supplying the hydrogen must be stifled if hydrogen damage is to be avoided. For instance, corrosion reactions that start on a clean metal surface at an appreciable rate but gradually become stifled, cause little or no hydrogen damage if the duration of hydrogen passage is short.

All grades of mild carbon and alloy steel can be permeated by the dissociated nascent hydrogen. Differences in permeability among various mild carbon steels are slight. However, the damage will be influenced by the cleanliness of the steel. For example, a conventional rimmed carbon steel is more susceptible to blistering, because of its heterogeneity and the likelihood of blowholes, slag inclusions and phosphide stringers, than a fully killed steel.

Electrolysis and pickling embrittlement have been associated with faults in steels such as blisters, cracks, flakes and "fish-eyes". Metals with greater residual or applied stress are more easily affected by dissolved hydrogen than similar metal with lower stress. Electrolysis embrittlement becomes dangerous in carbon steels, particularly those harder than Rockwell C 40.

Caustic and Boiler Embrittlement. Embrittlement in conjunction with steam boilers (sometimes called caustic embrittlement) may give rise to cracks at riveted joints or other areas where contact between metal surfaces permits the accumulation of concentrated solution and where the stresses are high. Such embrittlement has been found also in rolled tube ends, tube ligaments,

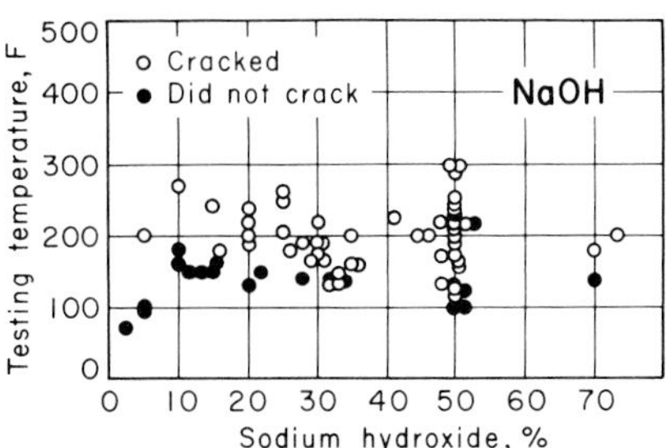

Fig. 26. Effect of temperature and concentration of caustic on cracking of mild steel (H. W. Schmidt and others, Corrosion, 7, 295, 1951)

headers and threaded pipe connections. There is scant likelihood of embrittlement in mild steel in boilers unless it has been stressed beyond the yield point; the stress created by steam pressure or uniformly distributed structural loads has slight effect; on the other hand, stresses left from the roll forming of plate into a shell or drum, distortion during riveting, or any cold work that causes permanent deformation, can provide the stress condition necessary for cracking. There are no data definitely showing any grade of mild steel to be more or less susceptible to this type of cracking than another grade.

Although mild steels differ in silicon and manganese content, and low-residual steels may have less carbon than others that contain copper, any difference in susceptibility to cracking is negligible. Cracks from intercrystalline corrosion may occur in any type of pearlitic steel. Composition of the steel is less important than heat treatment or design, and much less important than the basic requirements for cracking, which are hydroxide in water, conditions favorable to hydroxide concentration, high stress in the metal in contact with the concentrated hydroxide, and absence or insufficiency of inhibitors.

Cracks in unfired and fired pressure vessels (caustic embrittlement) should be distinguished from those in steam boilers (boiler embrittlement). Caustic embrittlement cracks in pressure vessels are usually restricted to vessels handling caustic, basic waters and (to a minor extent) nitrates and phosphates, and can be identified by three characteristics: location and occurrence of the cracks, the obviously brittle nature, and, especially, the characteristic appearance of polished specimens under the microscope. The typical embrittlement crack is not a single line of penetration, but a honeycomb of fine cracks, predominantly intercrystalline, following grain boundaries. Actually, boiler embrittlement cracks and caustic embrittlement cracks in pressure vessels differ only in that boiler embrittlement requires some mechanism by which an alkaline solution can be concentrated.

In pressure vessels the caustic strength must be sufficient to cause cracking of the low-carbon steel. Temperature is important; it speeds up the chemical reaction between caustic and steel. Here again one type of mild steel appears to be no more resistant than another. The environment, stress and temperature are far more important than the type of mild steel selected.

The effect of temperature and caustic concentration on the cracking of mild steel is shown in Fig. 26; a micrograph of a stress-corrosion crack in mild steel is presented as Fig. 27; and a fatigue crack on the fire side of a mild steel heater tube is illustrated in Fig. 28.

Erosion and Velocity Corrosion

External erosion of boiler tubes, superheater tubes and economizer tubes is one of the outstanding causes of forced shutdowns with high-pressure watertube boilers. Low-pressure firetube and watertube boilers are usually not exposed to severe external wasting other than by corrosion. In recent years, fly ash has been recirculated in an increasing number of boiler installations using spreader stokers and pulverized coal units. Unburned combustible is reclaimed and reburned in the furnace.

The feasibility of fly ash recirculation without accelerated erosion depends on the design of the installation and the operating conditions. A boiler tube failed from erosion after less than a year of service during which fly ash had been recirculated. In another furnace, tubes were eroded at the rate of 1/64 in. per month from impingement of recirculated fly ash.

In boilers that do not recirculate fly ash, tubes may erode from operation at increased boiler ratings and accompanying higher gas velocities through the boiler and other heat recovery surfaces. Erosion has not been confined to one fuel or type of firing. When solid fuel is used, either with pulverized-coal burners or spreader-type stokers, there is always danger of erosion.

Types of Erosion. Tube erosion has been associated with one or more of the following conditions:

1. Streams of ash concentrated on baffles by centrifugal force caused by a change in direction of gas flow.
2. Ash particles well distributed in a

Table 17. Gas Velocity Limits in Fuel Erosion

Fuel	Maximum velocity, ft per sec — Condition(a)	Condition(b)
Pulverized coal	75	100
Spreader stoker, Anthracite chain grate, Coke breeze chain grate	60	75
Cyclone furnace	No erosion limit	
Underfeed stoker, Blast furnace gas	75	100
Bituminous chain grate	100(c)	120(c)
Wood or other waste fuels containing sand	50	60
Cement dust	....	45

(a) Limit at, or anywhere downstream from, a baffle that will concentrate ash and then permit the concentrated stream of ash to impinge on the boiler tubes. If the concentrating type of baffle is equipped with a "cinder catcher" or slot arranged to prevent the concentrated ash stream from striking the tubes, condition (b) will apply. (b) Single-pass boilers or those with baffle arrangements that do not concentrate the fly ash. (c) Arbitrary limit; no erosion ever experienced.

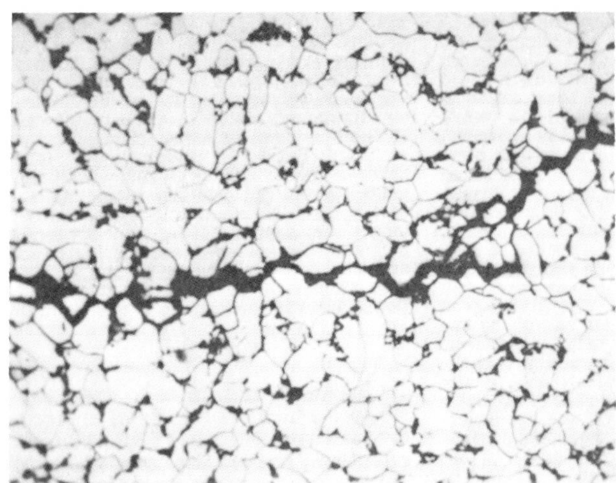

Fig. 27. Stress-corrosion crack in mild steel exposed to 70% NaOH at 175 F. The crack occurred in a welded area. The specimen was etched in nital. Magnification is 80×.

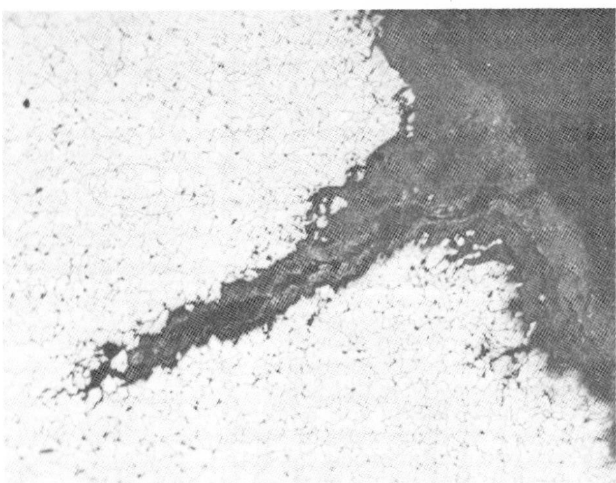

Fig. 28. Fatigue crack on fire side of a mild steel heater tube. Note oxide layer on the outer surface and on the sides of the crevice. Specimen etched in nital. Magnification, 80×.

high-velocity gas stream that erodes an entire face of a tube bank.

3 Fly ash particles leaking through baffle joints and voids and between drums and baffles.

4 High-velocity ash flowing between wall tubes and wall.

5 Ash caught in eddy currents around U-bolts (or other projections) and eroding the surface close to the projection.

6 Extremely turbulent conditions at the entrance to the tubes of a firetube boiler before the gas straightens out for long flow through tubes.

Causes of Erosion. Erosion is a function of the weight and velocity of individual ash particles, the quantity of ash that hits a particular spot, the abrasive qualities of the ash, and the angle at which the particles strike the surface. For any combination of ash type and concentration there is some critical range of velocity below which no trouble from erosion will be experienced and above which serious erosion will occur. In an intermediate zone, metal surfaces will be polished but not seriously eroded.

Location of Erosion. Tube erosion may occur anywhere from the furnace tubes back to the last pass and economizer tubes. However, in critical locations tube wear develops at an accelerated rate. Lanes or open spaces along the walls at the outside of the tube bank should be held to a minimum, as gases passing at high velocities along either the edge of the tube bank or the waterwalls can cause appreciable damage.

Serious erosion is common on superheater tubes and on first and second pass tubes in multipass boilers. This area of wasting is closely associated with, and influenced by, the relative location of the primary gas baffles. A change in the direction of gas flow at the baffle extremities causes wear of tubes in the succeeding pass. Exposed U-bolts or other projections will cause eddy currents resulting in wear close to the projections.

Cracks in the baffles or open refractory joints will allow bypassing of gases at high velocity, which will result in cutting of tube or shell surfaces. Voids between baffles and drums will allow bypassing and cause eventual damage to the drum and tubes. Slag deposits between generating or superheater tubes can raise gas velocities above the critical range, thereby causing erosion that normally would not develop. Since 1956, erosion in the economizer region of stoker and pulverized-fired boilers has developed to a point at which major changes are necessary for handling the gases at the economizer. Consideration is being given to the use of deflector plates and fly ash traps at the gas inlet to the economizer.

Erosion defects develop from the use of solid fuels with spreader-type stokers and with pulverized-coal units mainly in multipass boilers, as indicated by the higher tolerable gas velocities for condition (b) than for condition (a) in Table 17.

Increased gas velocities, which have accompanied increased boiler ratings over the years, and the use of solid fuels have combined to cause considerable damage to many conventional multipass boilers. Single-pass boilers have now been designed that eliminate concentration of ash on the surfaces and also lower the gas velocities. With existing multipass units provision can be made to level off the flow of gases so as to retard velocities and eliminate abrupt changes in the direction of gas flow.

Effect of Velocity on Corrosion. The usual effect of increased velocity is an increase in corrosion rate because the protective film is removed from the metal surface. However, after an initial increase there may be a decrease in the corrosion rate if oxygen availability increases enough to form a passive film on the steel. This phenomenon is undependable in design or operation because other variables enter, such as film permeability and turbulent or streamlined flow.

In dilute sulfuric acid exposed to air the corrosion rate of steel at first diminishes when velocity is increased but then rapidly increases. At velocities around 12 ft per sec all concentrations of sulfuric acid from 0.0043 to 5N corrode iron at the same rate. Figure 29 shows the effect of velocity and oxygen concentration on the corrosion rate of steel in tap

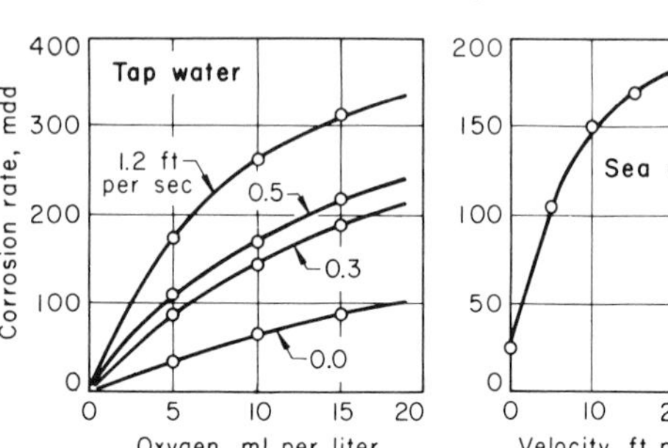

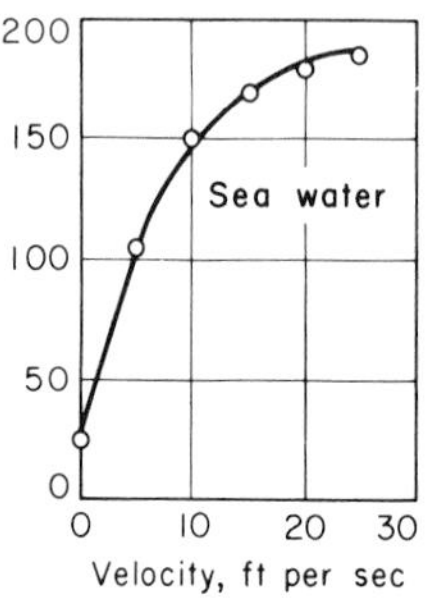

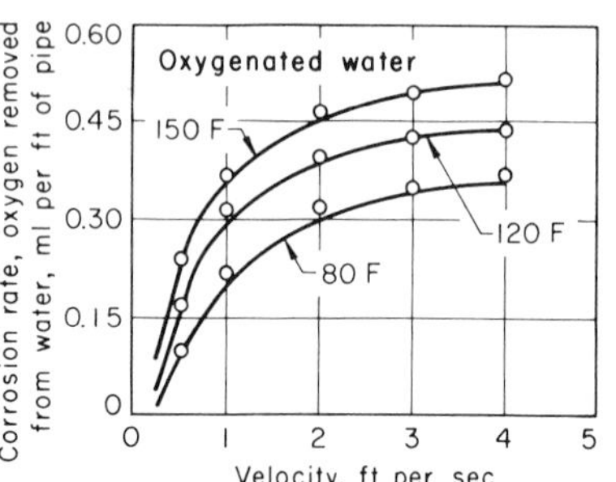

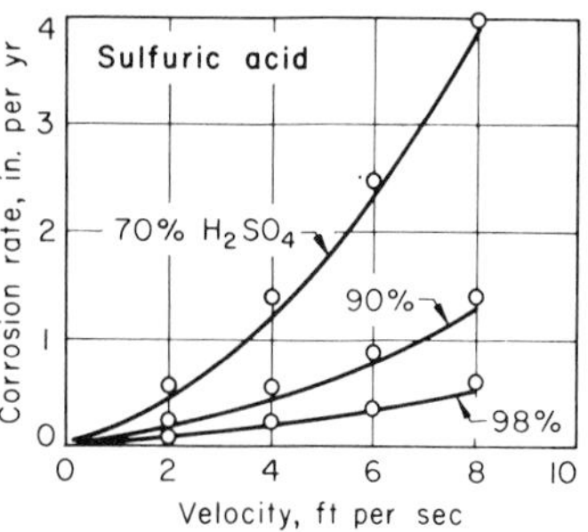

Fig. 29. Effect of velocity on corrosion rates of plain carbon steel (two charts at left) and steel pipe (two at right) (F. L. LaQue, Corrosion Handbook, 391, 1948, John Wiley and Sons; G. L. Cox and B. E. Roetheli, Ind Eng Chem, 23, 1012, 1931; F. N. Speller and V. V. Kendall, Ind Eng Chem, 15,134, 1923)

water and sea water, and also (right-hand graph) illustrates the effect of velocity of strong sulfuric acid on steel pipe and the greater effects of velocity in somewhat weaker acid solutions. This is a clear example of the destruction of the natural iron sulfate film. Below 1 ft per sec the steel will last almost indefinitely in 70 to 100% acid.

Localized Corrosion and Pit Depth

Steel equipment may fail by perforation before 10% of the metal has been removed by corrosion. A thickness survey of metal in service may be undertaken for a safety check, for a measurement of corrosion rate or simultaneously for both. The purpose of a safety check is to locate areas too thin for the operating pressure or approaching the minimum thickness. The purpose of a corrosion-rate measurement is the exact determination of the annual metal loss in various locations of a vessel or piping. In boilers and pressure vessels, a determination of the size of the affected area is more important than the detection of small pit holes. Widely scattered pits usually have little or no effect on the life of the object or its safe working pressure, as they can be easily repaired by welding.

Maximum penetration or maximum depth of pitting is usually determined by reference to the original surface, some of which usually remains intact. The average penetration is compared with the maximum penetration, and a pitting factor established to provide an index. The pitting factor is obtained by dividing the maximum penetration in inches per year by the average penetration. Thus, if the maximum penetration is 0.05 in. per year and the average penetration 0.01 in. per year, the pitting factor is 5. Complete penetration would then be expected when the average penetration is only 1/5 of the thickness. However, the maximum depth of penetration usually limits the life of the vessel from a safety standpoint.

Example. In evaluating the corrosion rate in a digester used in a pulping process in a paper mill, the extent of corroded areas is of primary concern and isolated pits are ignored, provided their depth is less than ½ the thickness of the vessel wall and their total area does not exceed 7 sq in. within any 8-in. diam circular area.

Thickness measurements must be taken at exactly the same locations over a predetermined period of time or on an annual basis. The points of measurement must be located accurately so that results of successive inspections may be compared and accurate records of corrosion rates and trends may be prepared. The first step in the location of test points should be the accurate orientation of some conspicuous point within the digester, such as a vertical seam or nozzle, with reference to the building. A popular method is to lay out a grid system by marking off the circumference of the digester into eight equal parts. At each of these eight points and 2 in. below the top girth a stainless steel stud or hook of ¼-in. diam and about 2 in. long is welded to the side wall. From these studs chains are hung, with each stud being assigned a letter or number in a clockwise direction. The chains used for the spacing of the readings taken on the side of the digester are hung plumb from each stud or hook. The chains are made up of a washer with a 1¾-in. diam hole and joined to another washer with a distance of 24 in. between centers. The total length of the chain and washers reaches the bottom of the digester.

Thickness measurements are then taken in the center of each washer in each chain. Enough readings are taken to give a clear picture of the condition of the vessel. Additional readings may be taken around the washer if a serious corrosion condition is observed at a washer. By taking several readings around a washer, the corrosion area can be outlined and any critical areas established. Each time readings are taken the critical area is determined for the section where the corrosion rate is the highest. From the original thickness, a rate of corrosion can be established, permitting an estimate of the remaining useful life of a vessel. The life expectancy can be plotted and the curve extrapolated, assuming a uniform rate of corrosion, to give a fairly accurate prediction of the time when a pressure reduction will be required in the process, based on the safe working pressure of the vessel.

In the application of this method to an actual digester an internal corrosion and scale pattern map was constructed, as shown on the left in Fig. 30. Thickness measurements were then recorded on another copy of the map (Fig. 30, right). An area of active corrosion (circled on the map) was found at the 10.5-ft level in the WSW section where the thickness (1.12 in.) was much less than elsewhere. A straight line drawn through the thickness points y and z, representing the thinnest metal found during the last two inspections, and extrapolated to the level representing the minimum thickness permissible at the operating pressure, gives an intersection at the end of the ninth year, indicating a life expectancy of three additional years. At the next-to-last inspection the extrapolated line x-y had indicated a much longer life, demonstrating that the strain in the badly corroded area is increasing and the corrosion is being greatly accelerated. Hence the prediction of three additional years of service life is questionable.

Inhibitors

Corrosion inhibitors are substances which, when added to corrosive environments in small amounts, effectively reduce the corrosion rate. Inhibitors function in three ways, under the categories of anodic, cathodic and adsorption.

Anodic inhibitors function by restraining the anodic reaction in the corrosion process. These inhibitors form an invisible, very thin protective film over the anodic areas of the metal surface. The most important of these for commercial use are the chromates, which are used extensively in treating cooling water and brines. In insufficient concentrations, anodic inhibitors cause accelerated attack by pitting, thus leading to premature failure of equipment. When used in sufficient concentrations, these inhibitors are the most effective of the three types, and are capable of arresting the corrosion almost completely.

Cathodic inhibitors function by laying down a thick visible film on the cathodic surface, thus polarizing the cathode and reducing the corrosion potential. These inhibitors are less effective than the anodic type, but are safe in virtually any concentration. Zinc and calcium compounds, for the most part, are cathodic inhibitors. Calcium bicarbonate is an example of a cathodic inhibitor that precipitates calcium carbonate at the cathode in aqueous systems. Glassy polyphosphates are probably the most common type of cathodic inhibitor used in industry.

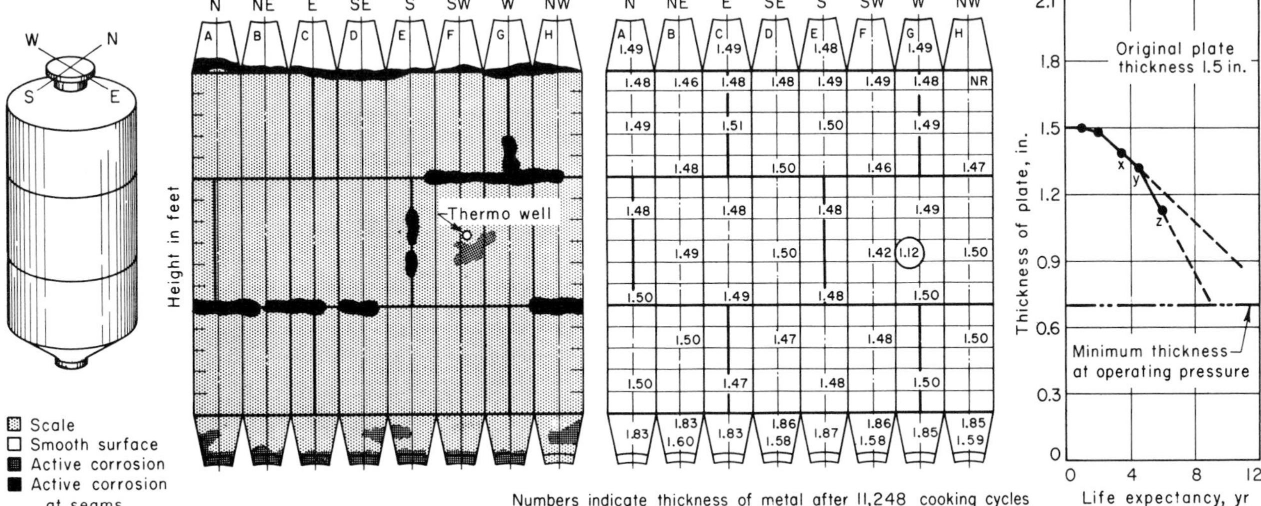

Fig. 30. Corrosion maps and life expectancy chart for a paper-mill digester. See text above for discussion of this example.

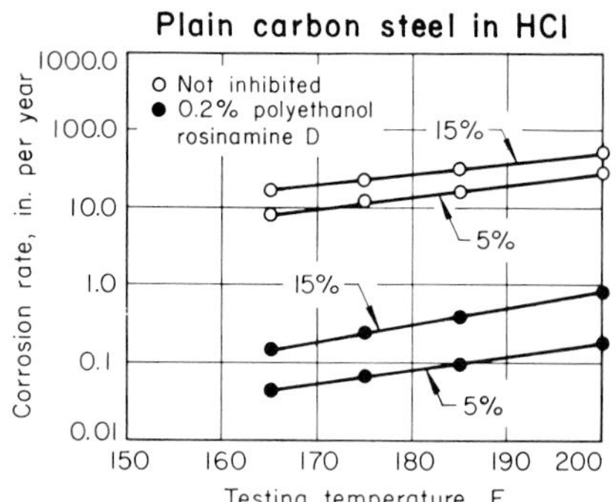

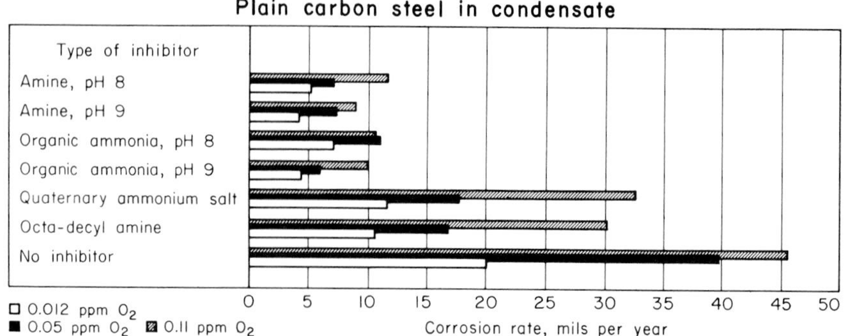

Fig. 31. Effect of inhibitors on corrosion of plain carbon steel (E. D. Bried and H. M. Winn, Corrosion, 7, 180, 1951; R. C. Ulmer and J. W. Wood, Ind Eng Chem, 44, 1761, 1952)

The adsorption type of inhibitors protect the metal surface by forming a barrier between the metal and the corrosive constituents. An example is cyclohexylamine for use in steam condensate. The cyclic, heterocyclic and aliphatic amines in this class are important in inhibiting acid used for cleaning industrial equipment. The left-hand graph in Fig. 31 is an example of the protection afforded in hydrochloric acid by the addition of one of the amine-type inhibitors, and the right-hand graph shows the effect of amine-type inhibitors in preventing corrosion by steam condensate.

Many proprietary inhibitors are formulated as combinations of the three types of inhibitor. Definite synergistic effects from some of these combinations give much better protection at a lower concentration than do any of the materials used alone.

Liquid Phase Inhibitors. In waters with pH above 7.0, dissolved oxygen is the chief cause of corrosion. In the absence of dissolved oxygen the corrosion rate is very slow. By controlling the dissolved oxygen in boilers, for example, tubes have been made to last 25 or 30 years. Without such control, tubes have been known to fail in six months. Users of high-pressure boilers eliminate dissolved oxygen from their feedwater and internal boiler water by mechanical deaeration followed by the addition of oxygen-scavenging materials to the boiler water itself. All but a few tenths of a part per million of dissolved oxygen can be removed by heating the feedwater in an open heater under a back pressure of 7 to 10 psi so that the water attains a temperature of 220 F min. These last traces of oxygen are generally removed by the addition of sodium sulfite and an excess of this chemical is kept in the boiler water as a precautionary measure.

Hydrazine has been proposed as an alternate to sodium sulfite with the advantage that neither the hydrazine nor its reaction product with oxygen add to the concentration of dissolved solids in the boiler water. This is an important feature in large utility boilers where it is desirable to keep the dissolved solids as low as possible. Sodium sulfite also is impractical at pressures above about 1800 psi because it decomposes.

The hydrazine reaction produces only water and nitrogen as end products. Since hydrazine may break down to ammonia above 400 F and since ammonia may be corrosive to other parts of the system, the residual quantities kept in the boiler water must be maintained at a low level, and close chemical control is necessary. For this reason hydrazine has been less widely used than test results and experience would suggest. It has been especially effective when residuals are properly controlled.

Corrosion during idle time must not be ignored in boiler equipment. Even though the boiler is drained, cleaned, and refilled with deaerated water, it can readily pick up dissolved oxygen cascading down over boiler surfaces, and much scab formation can result during such periods of idleness. Boilers should be either laid up dry or, if laid wet, the water should be alkalized to a minimum pH of 11.0, and as a precaution against dissolved oxygen, at least 100 ppm sodium sulfite should be maintained in the water to scavenge any oxygen that is picked up.

Special problems exist in many low-pressure heating boilers since the means and the knowledge for controlling water composition are usually unavailable. A series of tests on small heating boilers under atmosphere boiling conditions and using distilled water make-up saturated with dissolved air have shown that pitting of the tubes can be eliminated by the addition and maintenance of sodium chromate at 2200 ppm, even with 100 ppm chloride ion present.

Experimental work and actual operating installations have shown that sodium silicate is effective in preventing corrosion in closed hot water recirculating systems, even of the high-temperature type.

Chromates and sodium nitrate-borate combinations are also used in diesel engine cooling systems and in hot water recirculating systems of the low-pressure type.

In oil-well operations some types of inhibitors are effective in preventing corrosion and can reduce to a large extent the maintenance costs of a unit. As an example of this, the use of 9% Ni steel tubing in oil wells seems to have been progressively discontinued and replaced by the injection of inhibitors. In one American company, 35 oil wells were originally equipped with 9% Ni steel tubing, the price of which is about five times the price of ordinary steel tubing. Moreover, nickel is sometimes difficult to obtain and the behavior of 9% Ni steel was not found to be fully satisfactory in every installation. This same oil company started the injection of inhibitors, and in three production divisions their use has saved about $350,000 per year.

As another example of the value of inhibitors in the petroleum field, the use of inhibitors in four oil wells has reduced the cost of corrosion by the following amounts:

Location	Estimated annual cost of corrosion per well before use of inhibitors	Cost reduced to
Arkansas	$1200	$100
Kansas	2000	225
New Mexico	270	220

This experience was reported in "Corrosion Problems and Prevention in the Chemical and Petro-Chemical Industries in the USA", OEEC Document TAR/130(54)1.

Vapor Phase Inhibitor (VPI). In discussing boiler water, steam, and condensate earlier in this article attention was called to the use of amines such as cyclohexylamine and morpholine in combatting corrosion in condensate because of the presence of carbon dioxide. These amines, while they can be used as liquid-phase inhibitors, are vapor-phase inhibitors. Dicyclohexylammonium nitrite, called Dichan, is a vapor-phase inhibitor used to prevent rusting of packaged steel.

A test was conducted in an industrial-marine atmosphere, the test site being on the roof of a two-story building on the banks of the Oakland estuary on San Francisco Bay. Average summer temperature varied daily from about 53 to 71 F; average winter temperature, from 41 to 58 F, and humidity from 65 to 86%. Annual rainfall averages about 17 in. Wind is nearly continuous and mostly greater than about 9 mph.

The data show the increased shelf life of packaged steel articles through the use of VPI papers. Clean, machined 1020 steel was

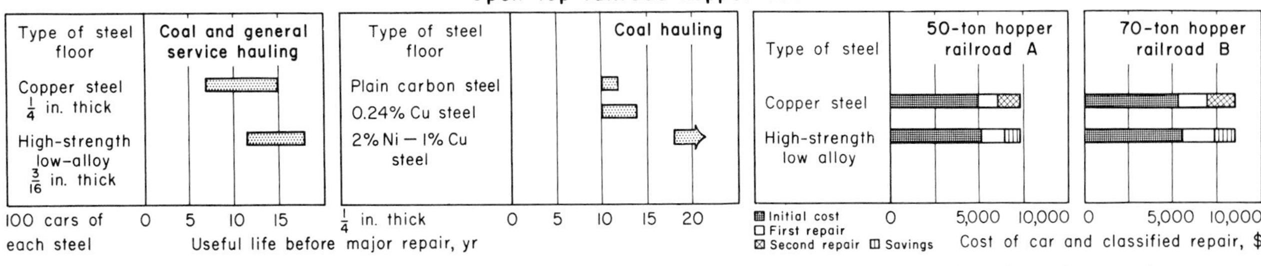

Fig. 32. (Left) Useful life of open-top railroad hopper cars in hauling coal and in general service before repairs. (Right) Cost of hopper cars made of copper steel and high-strength low-alloy steel (1955).

wrapped in various kinds of paper, placed in a box made of waxed cardboard, and stored at the site for four years under partial shelter to protect the boxes from direct rain. About 50% of the surface of the steel wrapped in untreated Kraft paper was rusted, while only 1% of the surface of the steel wrapped in grease-proof paper after the steel had been coated with rust-preventive oil, was marred with corrosion products. As little as 1 g of dicyclohexylammonium nitrite per square foot within the paper was sufficient to avoid virtually all corrosion.

Another more involved test was made at the San Francisco Bay site. Machined bars of 1020 steel, 5 in. by 1-in. diam, were cleaned with a solvent and stored, with and without oil treatment, in packages with an inner wrap of plain Kraft or of Kraft impregnated with 2 to 3 g of dicyclohexylammonium nitrite per square foot and with an outer wrap of waterproof barrier laminates. The edges of the packages were stapled and sealed with wax. These packages were exposed without shelter of any kind for five years. Results are in Table 18.

Results are shown in Table 19 for the corrosion of mild steel and cast iron in the vapor space over solutions containing 5% inhibitor. The mild steel specimens were 2 by 1 by 1/16-in., the cast iron 1½ by 1 by ¼-in. They were degreased in benzene and hung from a hook in the upper part of a test tube 10 in. long by 1 ¼-in. diam. The test tube containing 50 ml of solution was placed in a water bath at 104 F for 14 days. The top of the tube was sealed so as to get a refluxing action with condensation on the surface of the specimens. The corrosion products were removed cathodically in 10% sodium cyanide solution. The weight losses of the cast iron, where inhibitors were used, are ascribed to loss of graphite during cleaning.

Another test was carried out by the same investigators in the same manner to find the effect of concentration of still other inhibitors. The results are given in Table 20, three inhibitors and a control being used.

Service Life of Railroad Hopper Cars

The data presented in Fig. 32 (left) represent the useful life as determined in one set of tests, of open-top railroad hopper test cars in coal hauling and general service before major replacement of body sheets became necessary. The structural copper steel test cars were operated on nine railroad lines and the others on five railroad lines. The high-strength low-alloy (Cr-Cu-P) steel has the following composition; 0.09 C, 0.38 Mn, 0.09 P, 0.48 Si, 0.41 Cu, 0.84 Cr and 0.28 Ni.

In a wholly different set of tests, (Fig. 32, center) a low-alloy steel containing 2% Ni — 1% Cu required no major repairs until after 18 years. Some cars made of plain carbon and some made of copper-bearing steel required major repairs at the end of 10 years. All plain steel cars required repairs by 12 years; all copper-bearing by 14 years. All cars made of low-alloy steel had not been repaired just after 20 years when the test was discontinued.

The normal life of a standard AAR hopper car is about 36 years. Service performance tests have indicated that cars constructed with structural copper steel will need major repairs every 12 years or two major repairs during the normal life of 36 years. Cars constructed with a high-strength low-alloy steel with atmospheric corrosion resistance equal to that of the Cr-Cu-P steel of Fig. 32 (left) need only one major repair in 36 years.

The data presented in the right-hand portion of Fig. 32 represent the cost (1955 dollars) of hopper cars over a period of 36 years, including repairs, constructed from structural copper steel and from the same Cr-Cu-P high-strength steel.

The repair costs of hopper cars built from the low-alloy steel do not include the savings effected by avoiding the time out of service for the additional shop work required by the structural copper steel car nor the savings realized from the greater paint life found on the high-strength low-alloy steel cars.

Table 18. Storage Test Involving Dicyclohexylammonium Nitrite Inhibitor

Treatment of steel	Nature of wrap: Inner	Outer(a)	Surface rusted, %: 3 yr	5 yr	Residual VPI, g per sq ft
None	Kraft	K-A-K	55	85	...
None	VPI paper	K-A-K	0	0	0.7
Oil X	Kraft	K-A-K	5	10	...
Oil X	VPI paper	K-A-K	0+	0+	0.07
Oil Y	Kraft	K-A-K	6	14	...
Oil Y	VPI paper	K-A-K	2	2	0.23
Oil X	Kraft	Al foil	3	60	...
Oil X	VPI paper	Al foil	0	0	1.1
Oil Y	Kraft	Al foil	15	85	...
Oil Y	VPI paper	Al foil	0+	0+	0.6

(a) K-A-K means 60-lb Kraft - 60-lb asphalt - 60-lb Kraft laminated paper. A. Wachter, T. Skei and N. Stillman, Corrosion, 7, 284 (1951)

Table 19. Inhibition of Corrosion of Steel Where Various Amine Carbonate Inhibitors (VPI) Were Used

Solution	Weight loss (mg per specimen): Mild steel	Cast iron
Control (no VPI)	31.0	199.4
Cychlohexylamine carbonate	0.1	1.1
α-Phenylethylamine carbonate	0.4	1.2
β-Phenylethylamine carbonate	0.1	2.4
Dibutylamine carbonate	0.4	2.3
Diethylenetriamine carbonate	0.5	2.8
Phenyl-p-phenylenediamine carbonate	0.7	1.2
Benzylamine carbonate	0	1.5

E. G. Stroud and W. H. J. Vernon, J Appl Chem, 2, 178 (1952)

Table 20. Inhibition of the Corrosion of Steel with Different Concentration of Inhibitors

Conc, %	Weight loss (mg per specimen): Mild steel	Cast iron
Ammonium Carbonate Inhibitor		
0.5	9.9	15.6
1.0	0.5	5.8
2.5	0	2.1
5.0	0.1	1.9
Ammonium Carbamate Inhibitor		
0.5	0	3.2
1.0	0	1.2
2.5	0	1.8
5.0	0	1.4
Hydrazine Carbonate Syrup Inhibitor		
2.5	0.7	9.3
5.0	0.1	1.8
No Inhibitor		
0.0	30.7	202.5

E. G. Stroud and W. H. J. Vernon, J Appl Chem, 2, 178 (1952)

Corrosion Performance of Constructional Steels in Marine Applications

by R. J. Schmitt and E. H. Phelps

Corrosion is one of the most important factors to be considered in the design of offshore steel structures. In many instances, relatively simple design changes may prevent the occurrence of serious corrosion. Also, careful consideration should be given to the selection of methods for preventing corrosion to ensure that long term, economical protection is obtained.

High-strength low-alloy and alloy steels are available that have improved corrosion resistance over that of structural carbon steel in marine atmospheres and in the splash and tidal zones. Under total immersion conditions, however, the low-alloy and the alloy steels offer no advantage over carbon steel from a corrosion standpoint. Long-term corrosion data on various constructional steels under different exposure conditions will be reviewed. Galvanic corrosion, stress-corrosion cracking, and corrosion fatigue are discussed with emphasis on means for minimizing these forms of attack. Cathodic protection, metal sheathing, concrete encasement, and organic and metallic coatings are methods to be considered for preventing corrosion in marine applications.

INTRODUCTION

Exploration of the ocean both by industry and government has required either the temporary or permanent installation of many types of structures and vessels in marine environments. Carbon steel and the high-strength low-alloy and alloy steels have been the basic construction materials for these structures and vessels and will continue to be used in the future because of their relatively low cost, high strength, good fabricability, and good corrosion resistance.

The purpose of this presentation is to review the corrosion behavior of the constructional steels in marine environments and to discuss the specific types of corrosion that these steels will encounter. Important factors to be considered in the selection and design of steel structures are emphasized. The information given here will be primarily concerned with the performance of the standard grades of constructional steels in marine atmospheres and in seawater. The properties of the various constructional steels discussed are shown in Table I. The stainless steels will not be discussed; however, it should be recognized that they also have wide application in marine environments.

R. J. SCHMITT and E. H. PHELPS are affiliated with the Research Laboratory of U. S. Steel, Monroeville, Pa. **Paper presented at the 1969 Offshore Technology Conference, Houston, Texas, May 18–21, 1969.**

CORROSION PERFORMANCE

Marine atmospheres

After World War II considerable emphasis was placed on developing quantitative corrosion data from tests in marine atmospheres.

It has been learned from these studies that the rate of atmospheric corrosion of steel is dependent on 1) the length of time that moisture is in contact with the surface, 2) the extent of contamination or pollution of the atmosphere, and 3) the chemical composition of the steel. The results of atmospheric-corrosion tests conducted by various American Society for Testing Materials (ASTM) technical committees[1,2] and other societies and associations over the years have shown that the time of wetness of the corroding surface and the amount of chloride contamination of the atmosphere are the critical factors in controlling the corrosivity of marine environment.

Data from tests in marine atmospheres reveal that large differences in corrosivity can exist at locations only a few hundred feet apart. The results of tests

Table I—Properties of various constructional steels evaluated.

ASTM Specification	Yield Strength (Minimum), psi	Tensile Strength Minimum or Range, psi	Brand Name* of Steel Evaluated
A242 Type 1**	50,000	70,000	USS COR-TEN A steel
A588 Grade A	50,000	70,000	USS COR-TEN B steel
A572 Grade 50	50,000	65,000	USS EX-TEN 50 steel
A441	50,000	70,000	USS TRI-TEN steel
A517 Grade F	100,000	115,000–135,000	USS "T-1" Constructional alloy steel

* COR-TEN, EX-TEN, TRI-TEN, and "T-1" are registered trademarks of United States Steel Corp.
** Requirements for plates ½-in. and under.

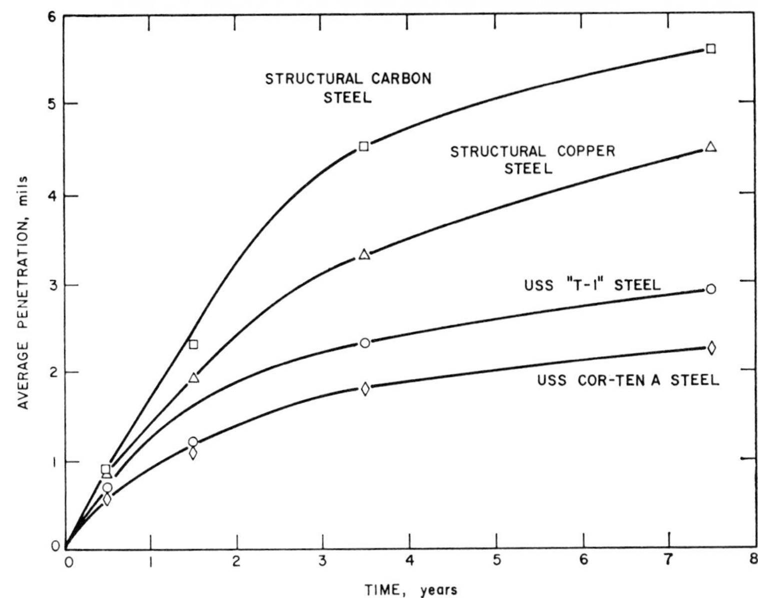

Fig. 1—Comparative corrosion performance of constructional steels exposed to moderate marine atmosphere, Kure Beach, N. C.

conducted both 80 and 800 ft from the ocean are given in Table II.[3] They show that for iron the corrosion rate 80 ft from the ocean was 10 times that obtained 800 feet from the ocean. Substantial differences in corrosivity were also observed for stainless steel and aluminum.

Further evidence of the wide variations in atmospheric corrosivity that can exist along the seashore can be found in tests conducted by the ASTM at Cape Kennedy, Florida.[4] The atmospheric corrosivity of five locations at Cape Kennedy was determined by exposing test specimens of carbon steel 60 yds from the ocean at ground level and at elevations of 60 ft and 30 ft, as well as on the beach and one-half mile from the ocean. The results of these tests are given in Table III. It is evident that the beach site is the most corrosive location, followed by the ground elevation 60 yds from the ocean. The corrosion rates decrease with increasing elevations. The corrosivity one-half mile from the ocean appears to be slightly less than that obtained at the 60-ft elevation. It is evident that corrosion and maintenance costs for outdoor structures can be appreciably reduced by locating the structures some distance back from the shore.

The chemical composition of the steel is as important a factor in determining the corrosion rate of steel as are the environmental factors. Since the pioneering work of Buck,[5] and the early publications and interpretations of the results of ASTM tests,[6] it has been generally acknowledged that carbon steel containing 0.2% or more copper has twice the atmospheric-corrosion resistance of a similar steel containing only 0.01–0.02% residual copper. Larrabee[7,8] confirmed these findings and also showed that nickel, chromium, silicon, and phosphorus singly are beneficial in improving the corrosion resistance of steel. He further showed that the greatest improvements in corrosion resistance are obtained by the addition of specific combinations of these alloying elements.

As a result of these and other corrosion studies and concurrent programs to improve strength and weldability, there are several proprietary corrosion-resistant high-strength low-alloy and alloy steels available today. To illustrate the advantages to be obtained by using these types of steels, let us compare the losses in thickness for four commercial steels exposed for 7.5 yrs in a moderate marine atmosphere (800 ft lot, Kure Beach) and 5 yrs in a severe marine atmosphere (80 ft lot, Kure Beach) (see Table IV). A plot of the average penetration versus time for these steels in the moderate marine atmosphere is presented in Fig. 1. Table IV shows that structural copper steel is more corrosion resistant than structural carbon steel in both environments. Further, the A242 Type 1 (Cr-Si-Cu-Ni-P) steel and the A517 Grade F steel were appreciably more corrosion resistant than either copper steel or carbon steel in both atmospheres. The differences in the corrosion performance of the four steels are caused by the

Table II—Comparison of corrosion rates 80 and 800 ft from the ocean.*

Metal	Corrosion Rate, mils per year: 80 ft from Shore	800 ft from Shore	Ratio of Rates
Iron	29.0	3.0	10
AISI Type 302 stainless steel	0.006	0.002	3
Aluminum (1100-H14)	0.085	0.004	21

* All specimens exposed for two years at International Nickel Co.'s test site at Kure Beach, N. C.

Table III—Corrosion rate of carbon steel at various locations at Cape Kennedy, Fla.

Corrosion Rate, mils per year*: 60 Yards from Ocean: Ground Level	30-ft Elevation	60-ft Elevation	½ Mile from Ocean	Beach
17.4	6.5	5.3	3.5	**21.0**

* Based on a two-year exposure except for the beach test, which was only a one-year exposure.

differences in the protective qualities of the oxide films that form on the steels. The slopes of the curves in Fig. 1 show that the corrosion-resistant low-alloy and alloy steels are developing a more protective oxide film with time than are either the copper or carbon steels.

Copson and Larrabee[9] report that both field tests and service experiences have shown that paint coatings are more durable on low-alloy steel than on carbon steel. Any rust that forms at breaks or holidays or underneath the paint film is less voluminous on low-alloy steels. Thus, because of the smaller volume of rust, there is less rupturing of the paint film and less moisture reaches the steel to promote further corrosion. It follows, therefore, that the extra durability of the paint is due to better atmospheric-corrosion behavior of the low-alloy steel.

Splash and tidal zones

Corrosion of carbon-steel piling installed in seawater is most severe in the splash zone, the area from mean tide to the upper limit of wave action.[10] The attack on steels immersed in seawater is a function of the availability of the dissolved oxygen at the metal-water interface. Inasmuch as the splash zone of piling is alternately wet and dry there is a plentiful supply of oxygen available for the corrosion process. Between mean tide and low tide, the steel is generally subject to less severe corrosion than in the splash zone. This is due to the decreased oxygen supply resulting from periodic submergence of the steel, and apparently to the fact that the submerged steel provides some cathodic protection to the steel in the tidal zone. As will be discussed later in the paper, it has been shown that in water which is relatively uncontaminated, the rate of corrosion in the completely submerged zone is low and, in many cases, the desired life may be attained without protection.

Cathodic protection, which is an effective method of controlling corrosion of steel piling immersed in seawater, will not extend to the splash zone since this area does not continually stay wet. In the past the only means available for obtaining longer life from steel in the splash zone was to use reinforced concrete jacketing or coating systems. However, neither of these methods has provided a completely satisfactory answer to the splash-zone problem, mainly because of their high cost. Recognizing the need for a piling steel with improved splash-zone corrosion resistance, U.S. Steel Corporation conducted a series of corrosion tests with certain experimental steels. For this study ¼ in. by 6 in. wide by 20 ft long specimens of carbon steel and experimental steels were exposed on a dock at Wrightsville Beach, North Carolina, so that their tops extended above the "splash zone" and their bottoms were in the mud. The results obtained after 5 and 9 yr exposures[11] showed that a steel* containing 0.5% Ni, 0.5% Cu, and 0.12% P was appreciably more resistant than carbon steel to corrosion in the splash and tidal zones (Fig. 2).

To confirm the earlier findings and to obtain longer-term data on Ni-Cu-P steel, tests were conducted at 14 different coastal sites in the United States. The corrosion data obtained from these studies, which ranged from 1–10 years duration, confirm the results of the previous tests that showed Ni-Cu-P steel to be at least twice as corrosion resistant as carbon steel.

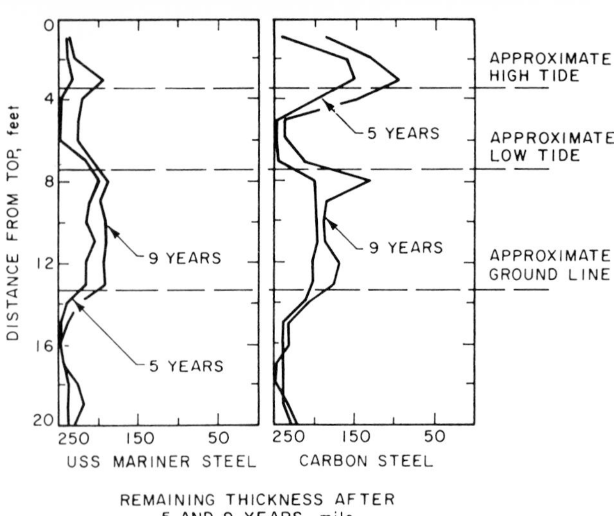

Fig. 2—Comparative corrosion of Ni-Cu-P steel and carbon steel in marine environment. Test strips exposed for 5 and 9 years at Wrightsville Beach, N. C.

Table IV—Corrosion performance of constructional steels evaluated in moderate and severe marine atmospheres, Kure Beach, N. C.

		Average Penetration, mils			
Type of Atmosphere	Time, yr	Structural Carbon Steel	Structural Copper Steel	A242 Type 1 (Cr-Si-Cu-Ni-P) Steel	A517 Grade F Steel
Moderate Marine (800-ft lot)	0.5	0.9	0.8	0.6	0.7
	1.5	2.3	1.9	1.1	1.2
	3.5	4.5	3.3	1.8	2.3
	7.5	5.6	4.5	2.3	2.9
Severe Marine (80-ft lot)	0.5	7.2	4.3	2.2	1.1
	2.0	36.0	19.0	6.4	(2)
	3.5	57.0	38.0	(1)	3.9
	5.0	(3)	(3)	19.4	5.0

(1) Specimen not removed for evaluation.
(2) Specimen lost.
(3) Specimen corroded so severely it was lost from test rack.

Included in the 14 tests were two shoreline bulkhead installations made from carbon steel and Ni-Cu-P steel-sheet piling sections. One installation was made at Wrightsville Beach, North Carolina, and the other at Norfolk, Virginia. The performance of the piling was determined by periodically measuring the thickness of the steel ultrasonically. The results of the measurements made after 8 yrs at the Wrightsville installation are shown graphically in Fig. 3 and the results of measurements made after 6 yrs at the Norfolk installation are shown graphically in Fig. 4. The superior performance of the Ni-Cu-P steel over that of carbon steel is evident in these figures.

Seawater immersion

The corrosion rate most commonly used for carbon steel in quiescent seawater is 5 mils per year (mpy), which is generally considered to be linear with time.[12] However, these earlier data were based on relatively short-term tests, and longer-term data are now avail-

* U. S. Steel's USS MARINER Steel.

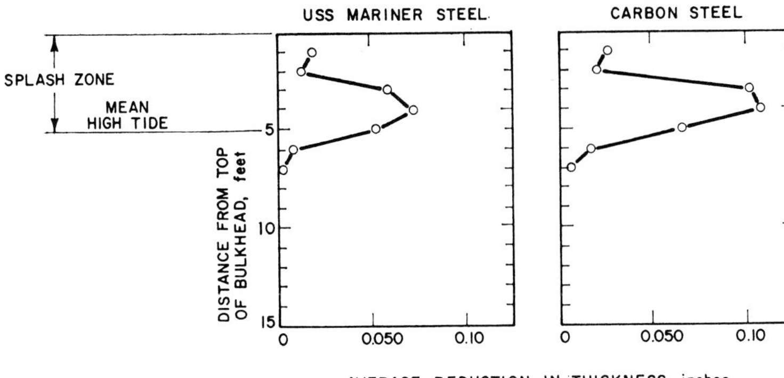

Fig. 3—Corrosion performance of Ni-Cu-P steel and carbon steel piles exposed 6 years in a shoreline bulkhead, Wrightsville Beach, N. C.

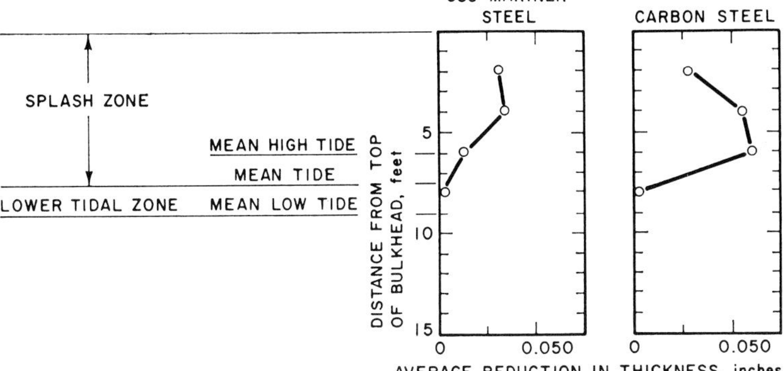

Fig. 4—Corrosion performance of Ni-Cu-P steel and carbon steel piles exposed for 6 years in a shoreline bulkhead, Norfolk, Va.

Table V—Corrosion rates of steels in seawater.

Material	Corrosion Rate, mils per year Years Exposure 1.5	2.5	4.5	8.5
ASTM A242 Type 1 (Cr-Si-Cu-Ni-P) steel	4.2	4.3	3.8	3.1
ASTM A588, Grade A steel	4.4	3.8	3.0	2.6
1.8 Ni-0.8 Cu steel	5.3	4.5	3.5	3.2
2.6 Cr-0.5 Mo steel	1.4	1.6	1.6	0.3
Structural Carbon steel	4.8	4.1	3.3	2.7

able which show that the corrosion rate of steel actually decreases to values below 5 mpy with time. Carefully controlled tests conducted over a 16 yr period by Southwell and Alexander[13] showed a decrease of corrosion rate with time: 5.8 mpy after a one yr exposure and 2.7 mpy after 16 yrs exposure (See Fig. 5). Larrabee[14] in conducting an investigation of carbon-steel piles immersed in seawater for 23.6 yrs, concluded that steel has a low corrosion rate during long seawater exposure. From a study of some 20 piles, he reported that the corrosion rate for carbon steel in seawater averages 2 mpy for the first 20 yrs, then drops to 1 mpy.

As mentioned earlier, the corrosion of steels immersed in seawater is a function of the availability of the dissolved oxygen at the metal-water interface. Therefore, as might be anticipated, increasing flow velocity raises the corrosion rate because it increases the amount of oxygen diffusing to the steel surface. As an example of the effect of flow rate on corrosion rate, Kirk, et al.[15] reported corrosion rates for mild steel after a one yr exposure of 4 mpy under quiescent conditions and 9 mpy at a flow rate of 2 ft per second. Another important factor is that differences in oxygen concentration at different parts of the steel surface can lead to localized corrosion. For example, if there are small areas where the soluble oxygen is deterred from reaching the metal as a result of slow diffusion through rust, and large areas exist where the oxygen remains in ready contact with the steel, pitting will result at the areas where there is a deficiency of oxygen. The rate of pitting of steel in seawater is usually found to be several times higher than the average rate of penetration of steel. However, long-term studies of 16 yrs duration[16] show that after 2–4 yrs exposure there is an appreciable decrease in the pitting rate of steel, see Fig. 5. In fact, after about 8 yrs, the pitting curve in Fig. 5 parallels the weight-loss curve and the average pitting penetration over long periods is about 3 mpy.

The corrosion rates obtained from test specimens of carbon steel and low-alloy and alloy steels immersed in seawater at Wrightsville Beach, North Carolina, for periods up to 8.5 yrs are given is Table

Table VI—Solution potentials of steels in aerated 3.5% sodium chloride solution.

Steel	Solution Potential, Volts*	Standard Deviation, Volts**
AISI C1005	−0.692	0.002
0.04% C, 0.24% Cu	−0.686	0.003
AISI C1010	−0.686	0.004
0.25% C, 1.5% Mn	−0.684	0.002
AISI 5140	−0.680	0.006
ASTM A242 (Cr-Si-Cu-Ni-P)	−0.671	0.009
AISI 4140	−0.671	0.004
AISI 1020	−0.670	0.005
AISI 8640	−0.669	0.003
AISI Type 501	−0.653	0.001
0.06% C, 2.5% Ni	−0.650	0.010
AISI 4340	−0.633	0.005
0.6% C, 3% Cr, 0.25% Ti	−0.627	0.002
0.6% C, 8.5% Ni	−0.615	0.002
AISI 3340	−0.601	0.002

* Measured against saturated calomel electrode.
** Determined with at least six specimens.

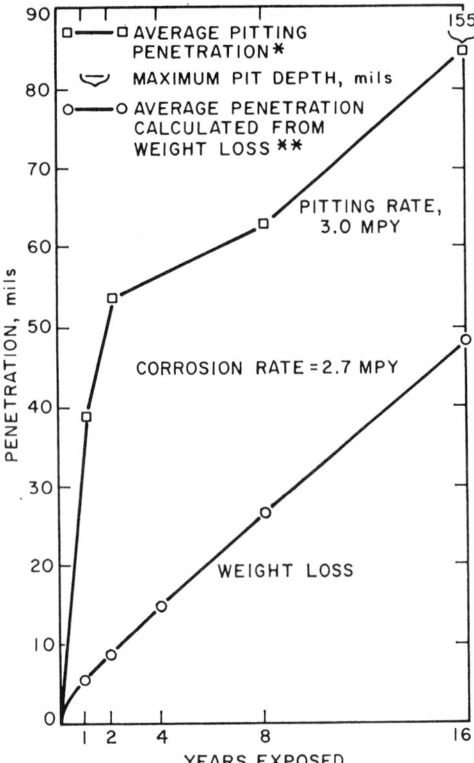

* EACH "□" REPRESENTS THE MEAN OF 60 MEASURED PITS - 5 DEEPEST ON EACH SURFACE OF 6 PANELS

** EACH "○" REPRESENTS THE MEAN LOSS FOR 6 PANELS

Fig. 5—Corrosion of carbon steel continuously immersed in seawater. (Courtesy of U.S. Naval Research Laboratory).

V. With the exception of the Cr-Mo-steel, the low-alloy and alloy steels showed no significant improvement over carbon steel when exposed to seawater.

Although this investigation and those of others[17,18,19] suggest that chromium in the range of 2–5 percent effects an appreciable reduction in the corrosion rate of steel immersed in seawater, Southwell and Alexander[20] have shown that steels containing 3 and 5% chromium corroded at rates appreciably higher than those for carbon steel. Further, they showed that the pit depths after 8 yrs exposure were about the same for the chromium steels and carbon steel. The 16 yr data also showed 2 and 5% nickel steels to corrode at the same rate as carbon steel, 2.7 mpy. Resistance to pitting, however, was appreciably less for the nickel steels as compared with carbon steel.

GALVANIC CORROSION

When two dissimilar metals are in electrical contact with each other and immersed in an electrolyte, the difference in solution potential between the metals tends to drive electric current through the solution between the two metals and back through the metallic contact. The more active metal acts as the anode and the corrosion rate of this electrode is increased. The metal that is more noble acts as the cathode and corrosion of this electrode is decreased or prevented altogether.

The solution potentials of several different types of steel in aerated 3.5% sodium chloride solution are given in Table VI. The measurements were made at 30° C in an apparatus in which the specimens were moved at a constant linear velocity of 31 ft per min through the solution.

Table VII—Corrosion rates of coupled and uncoupled steels in quiescent seawater. Exposure period, 174 days.

Couple		Corrosion Rate, MDD					
		Area Ratio					
		1:1		1:8		8:1	
1	2	1	2	1	2	1	2
C1020	A242*	82	20	170	32	67	27
C1020	Type 410	130	0.3	470	0.4	70	0.0
C1020	Type 446	100	0.2	340	0.1	63	0.0
C1020	Type 304	91	0.0	370	0.0	65	0.0
C1020	Type 316	120	0.0	410	0.0	68	0.0
C1020	(Control)	55	—	—	—	—	—
A242*	Type 410	95	0.3	350	0.2	62	0.4
A242	Type 304	82	0.0	390	0.0	51	0.0
A242	(Control)	45	—	—	—	—	—

NOTE: Separation between coupled specimens was about four inches.
* Cr-Si-Cu-Ni-P type.

In Table VI, it is evident that the solution potentials of the various steels range from approximately 0.69 volts (versus a saturated calomel electrode) for unalloyed carbon steel to a range of about − 0.60 to − 0.65 volts for steels with alloy contents of 3–8%. Fig. 6 compares these potentials measured for steels with those of several different metals in the aerated 3.5% sodium chloride solution. The values shown in Table VI and Fig. 6 are somewhat more negative than those measured by LaQue and Cox[21] in flowing seawater. However, the ratings for the different metals were generally similar in the two investigations.

The relative areas of the two electrodes can be extremely important in galvanic corrosion. If a large anodic area is coupled to a relatively small cathodic area, the galvanic attack on the anode normally will be diffuse and moderate. However, if a large area of cathode metal is coupled to a relatively small area of anode metal, attack can be very severe.

The importance of the area ratio is illustrated by the data in Table VII. The coupling of carbon steel to a stainless-steel specimen of equal area approximately doubles the corrosion rate of the carbon steel. Moreover, coupling to a stainless area 8 times that of the carbon steel results in a corrosion rate for the carbon steel that is about 8 or 9 times that of the uncoupled steel. The reason for these increases is that the rate of corrosion of the carbon steel is determined by the rate of diffusion of oxygen to its cathode areas. Coupling to stainless steel increases the effective cathode area by the area ratio of the two metals, and hence the corrosion rate would be expected to increase proportionately.

It is also noteworthy in Table VII that the corrosion rate of carbon steel is significantly increased by coupling with A242 Type 1 (Cr-Si-Cu-Ni-P). This effect is somewhat surprising because the solution potentials for the two steels, measured over a 24 hr time period, are practically the same (Table VII). However, over a long exposure period, as in the couple tests, it is apparent that the higher alloy content of the A242 steel has resulted in this steel becoming more noble than the carbon steel.

The importance of relatively minor potential differences on the corrosion behavior of a couple is il-

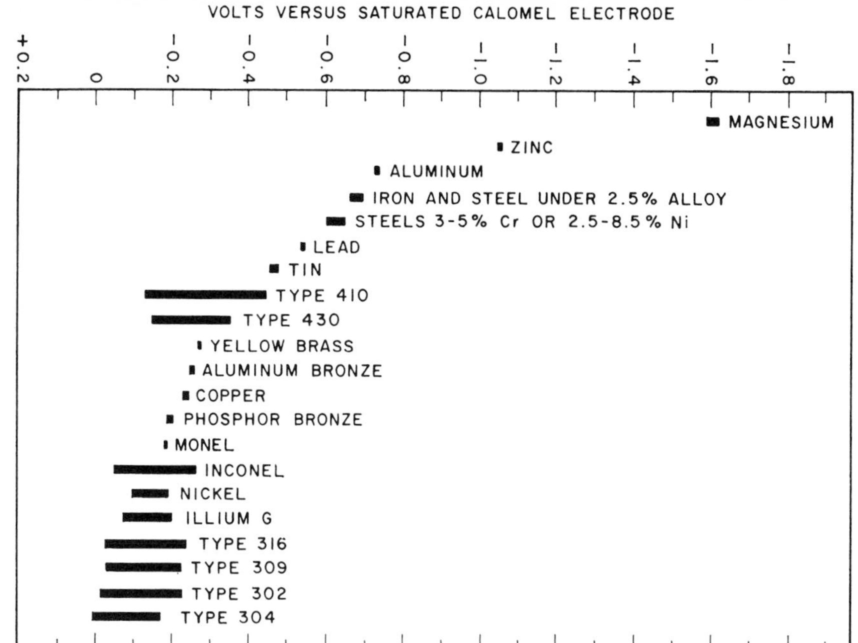

Fig. 6—Solution potentials of metals and alloys in 3.5% NaCl solution at 30°C.

lustrated by the data on HY-80 weldments given in Table VIII. In this example, a very moderate difference in solution potential has resulted in a significant difference in corrosion rate for the different weld zones. For welds that are to be exposed to conducting solutions, such as seawater, it is very important that the weld metal be chosen so that its solution potential will be cathodic to the base metal. As a general rule, this can be accomplished for carbon, low-alloy, and alloy steels by selecting weld metals with alloy contents at least equal to that of the base metal.

When it is necessary to use constructional steels in contact with other dissimilar metals, one or more of the following precedures should be considered for preventing galvanic corrosion:

1. Electrically insulating the dissimilar metals.
2. Adding a "waster" section of a third metal that is anodic to the other two and which will cathodically protect both of the original dissimilar metals.
3. Covering the surface of the cathodic member of the couple with an insulating coating.
4. Making the anodic member of the couple extra thick and designing it so that easy replacement is possible.

STRESS-CORROSION CRACKING

One of the present authors has reviewed the stress-corrosion behavior of high-yield-strength steels in a previous paper.[22] It was shown that constructional alloy and ultra-high-strength steels at yield strengths up to about 180,000 psi are resistant to stress-corrosion cracking in environments containing chlorides. From this yield strength to about 210,000 psi these types of steel may be resistant, depending on the specific steel and heat treatment. At yield strengths over about 210,000 psi, these steels are generally susceptible to stress-corrosion cracking. Fig. 7 shows the stress-corrosion failure time as a function of yield strength for five types of steel, all of which were heat-treated to obtain high strength. The tests were conducted with unwelded sheet specimens of steels stressed by bending to 75% of their room-temperature yield strength.

Table VIII—Electrochemical and corrosion measurements on HY-80 weldments.

Portion of Weldment	Potential, mv*	Average Corrosion Rate in Seawater, mpy**
Base metal	660	4.3
Heat-affected zone	650	1.8
Weld metal***	700	6.1

* Potential vs saturated calomel electrode in synthetic seawater.
** One-year exposure period in quiescent seawater.
***AWS Class 110-18 (.06% C, 1.5% Mn, 0.3% Cr, 1.7% N.)

Table IX—Results of stress-corrosion tests on constructional steels in marine atmospheres 80-ft lot, Kure Beach, N. C.

ASTM Specification	Type of Specimen	Exposure Time, Days
A441	Bent Beam	NF 750
A441	Tensile	NF 750
A572 Grade 50	Bent Beam	NF 425
A572 Grade 50	Tensile	NF 750
A242 Type 1 (Cr-Si-Cu-Ni-P)	Bent Beam	NF 750
A242 Type 1 (Cr-Si-Cu-Ni-P)	Tensile	NF 750
A588 Grade A (Mn-Cr-Cu-V)	Bent Beam	NF 425
A588 Grade A (Mn-Cr-Cu-V)	Tensile	NF 750

NOTE: Five specimens of each steel were exposed at a stress level of 75% of yield strength. See reference 30 for description of test specimens.
NF—No failure in time indicated, test terminated.

Mechanical Properties of Steels Tested

Steel	Yield Strength, ksi	Tensile Strength, ksi	Elongation in 1 inch, %
A441	52	82	30
A572 Grade 50	53	71	33
A242 Type 1	50	72	37
A588 Grade A	72	101	27

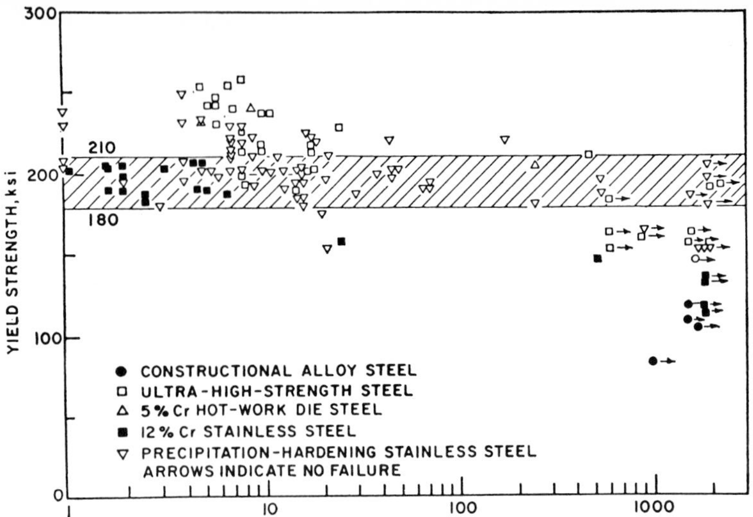

Fig. 7—Stress-corrosion behavior of steels exposed to marine atmosphere at 75% of the yield strength.[22]

The presence of hydrogen sulfide in the environment can also be very detrimental from the standpoint of stress corrosion of the higher-strength steels. As a general rule, steels with hardness values over Rockwell C22 can be expected to be susceptible to cracking if sulfide is present. Furthermore, steels with hardness values below Rockwell C22 may be susceptible under severe conditions of stress and/or cold work and in the presence of high hydrogen sulfide concentrations. It should also be recognized that the use of welding to join steels having a base-metal hardness below Rockwell C22 may cause high hardness values in the heat-affected zones near the welds. Although hydrogen sulfide is not a normal constituent in seawater, this substance is generated by sulfate-reducing bacteria under anaerobic conditions. Since such conditions can occur at and below the mud line on offshore structures, the possibility of sulfide corrosion cracking should be considered during the design of such structures.

Constructional alloy and low-alloy steels have been used in many applications involving exposure to marine atmospheres and seawater. To the knowledge of the writers, stress-corrosion cracking of these steels has not been encountered in any of these applications. The results of stress-corrosion tests on several different low-alloy and alloy steels exposed to marine atmospheres and to seawater are presented in Tables IX, X, and XI. A review of these tables will show that none of the steels were susceptible to cracking in exposure times ranging up to 2200 days. These data in conjunction with field experience furnish solid support for the contention that low-alloy and alloy constructional steels are resistant to stress-corrosion cracking in chloride environments.

CORROSION FATIGUE

Corrosion fatigue is defined as the reduction in the fatigue resistance of a steel as a result of exposure to a corrosive environment. This phenomenon is complicated because it involves the various factors involved in fatigue, such as stress range and cycle frequency, as well as environmental factors, such as dissolved oxygen content, conductivity, and flow conditions. The effect of the combination of corrosion and fatigue for steels is to essentially eliminate the fatigue limit that is otherwise obtained. Hence, in marine applications that involve fatigue loading, this phenomenon has to be given careful consideration in the design stage so that preventive measures can be taken to avoid future difficulties.

Table X—Results of stress-corrosion tests on alloy steels in marine atmosphere, exposure at 80-ft lot, Kure Beach, N. C.

Steel	Type of Specimen	Stress Level Percent of Yield Strength	Exposure Time, Days
HY-80	Bent Beam	75	NF 1700
HY-80	Bent Beam	90	NF 1700
STS	Bent Beam	75	NF 1700
STS	Bent Beam	90	NF 1700
A517 Grade F	Bent Beam	75	NF 1700
A517 Grade F	Tuning Fork	75	NF 425
A517 Grade F	Tuning Fork	Cold-worked, welded, & stressed	NF 425

NOTE: Five specimens were exposed for each test condition. Plate materials were machined to permit use of bent-beam specimens (see reference 30 for description of test specimen).
NF—No failure in time indicated, tests terminated.

Mechanical Properties of Steels Tested

Steel	Yield Strength, ksi	Tensile Strength, ksi	Elongation in 1 inch, %
HY-80	100	108	11.2
STS	106	121	12.5
A517 Grade F	115	122	14.2

Table XI—Results of stress-corrosion tests on alloy steels in seawater. Specimens exposed 1.5 ft below surface in quiescent seawater, Harbor Island, N. C.

Steel	Type of Specimen	Stress-Level Percent of Yield Strength	Exposure Time, Days
HY-80	Bent Beam	75	NF 250
HY-80	Tensile	75	NF 2200
STS	Bent Beam	75	NF 70
STS	Tensile	75	NF 2200
A517 Grade F	Bent Beam	75	NF 250
A517 Grade G	Tensile	75	NF 2200

Mechanical Properties

Steel	Yield Strength, ksi	Tensile Strength, ksi	Elongation in 1 inch, %
STS	163	175	6.2

Other steels tested have the same properties as in Table X.

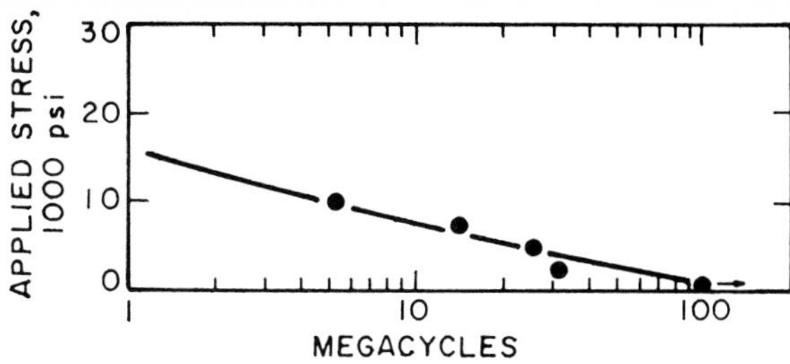

Fig. 8—Corrosion fatigue of mild steel rotated at 1400 rpm in seawater.[13]

A typical example of the corrosion-fatigue behavior of steel is shown in Fig. 8, which was taken from the work of Kirk et al.[23] To obtain the data shown, cantilever-beam specimens were rotated at 1400 rpm in seawater. It is apparent that decreasing the stress results in a longer time to failure (greater number of cycles); however, a definite fatigue limit is not observed.

Corrosion fatigue can be prevented by several methods. One approach is to use protective coatings; the effectiveness of this method will depend on how well the coating prevents corrosion. It has been found that metallic zinc coatings are very effective, most likely because the zinc furnishes cathodic protection at any discontinuities in the coatings. The application of cathodic protection has been effective in laboratory experiments and can be expected to be similarly effective in field installations. Shot peening to introduce residual compressive stresses in the surface is another method of preventing corrosion fatigue. However, with this method steps should also be taken to prevent corrosion, otherwise general corrosion and pitting will gradually penetrate the layer containing the residual compressive stresses.

PROTECTIVE MEASURES

Cathodic protection

Cathodic protection is an effective method of controlling the corrosion of steel piling immersed in seawater. Depending on the type and location of the structure, cathodic protection may be achieved either by a galvanic system using zinc or magnesium anodes or by an impressed current system employing a rectifier and perhaps graphite, lead-silver, or platinum anodes. The need for cathodic protection should be based on the actual performance of the structure, which can be determined by periodic inspection. It is always desirable in the construction of steel structures to tie the unit together electrically so that if cathodic protection is needed sometime during the life of the structure it can be applied quickly and at low cost. In some highly aggressive warm water or polluted seawater, it may be necessary to apply cathodic protection at the time the steel structure is installed in order to obtain long-term service.

When a cathodic-protection system is installed, the location of the anodes should be considered carefully to obtain maximum performance. Anodes for a galvanic system should be installed prior to installation of the structure, with such considerations as service life of the structure, predicted life of the anodes, and current requirements taken into account. In all cases the system should be designed to resist damage by boats, debris, and heavy wave action. Generally speaking, a current density in the range of about 5–10 milliamperes per square foot is generally required to protect a stationary steel structure in seawater. On coated structures where the coating remains fairly well intact, the amount of current applied to protect the structure will be less than that required for an uncoated structure. This will result in a savings in power cost and anode-replacement costs. However, on structures where the coating deteriorates relatively rapidly the initial current requirements will increase. An excellent series of papers reviewing field experience in the application of cathodic protection to offshore structures is presented in the January 1967 issue of *Materials Protection.*

One aspect of cathodic-protection application that has received attention in recent years is the question of whether this protective method can lead to hydrogen embrittlement. In laboratory experiments with very high strength steels (yield strengths of over 200,000 psi), for example, several investigators have shown[25] that the application of cathodic current can cause failure as a result of hydrogen entering the steel. The situation is not as clear, however, with steels of lower strength. Brown[26] has indicated that heat-treated steels begin to exhibit susceptibility to this type of damage at strength levels on the order of 150,000–175,000 psi, and that cathodic protection should not be used to prevent stress-corrosion cracking in sea water with steels of this strength range or higher. He further points out that the cathodic-protection systems now in use for underwater structures made of low-strength steels may mitigate against corrosion fatigue by preventing the formation of pits which otherwise could nucleate corrosion-fatigue cracks; with such steels there is little chance of any adverse effects from hydrogen. With AISI 4330 steel heat-treated to a yield strength of 160,000 psi, however, he indicates that the application of relatively high levels of cathodic protection may cause an adverse effect on fatigue performance as a result of hydrogen entry into the steel, even though corrosion is effectively prevented.

In stress-corrosion tests with precracked cantilever-beam specimens, Leckie and Loginow[27] found no deterioration in the load-carrying capacity of HY-80 steel cathodically charged at 1.2 volts (saturated calomel electrode) in 3% NaCl solution. Although the load-carrying capacity of HY-130 steel was reduced by the application of cathodic protection, the normal stress for failure was higher than the yield strength of the steel. These investigators also found that predeformation (up to 5%) had a detrimental effect on the performance of HY-130 but not on that of HY-80. Furthermore, cathodic polarization reduced the load-carrying ability of maraging steels at yield-strength levels of 180,000 psi and 250,000 psi.

In commenting on the possible effects of cathodic protection on line-pipe steels buried in soil, Elsea[28] concludes from laboratory tests and field experience that hydrogen is not a problem with steels having yield strengths below 60,000 psi. He further states that although steels with yield strengths in the range of from 70,000–110,000 psi are susceptible to hydrogen stress-cracking under severe laboratory test conditions, it is doubtful that enough hydrogen would ever be absorbed under field conditions to cause cracking.

Unfortunately the information obtained in these investigations is not sufficient to delineate in a quantitative manner the effect of cathodic protection on the properties of steel. More studies are definitely needed with steels in the intermediate strength range exposed to charging conditions that might be encountered in service. Furthermore, it is important that the future studies encompass all the metallurgical conditions used in the field, including base metal, weld metal, and heat-affected zone.

METAL SHEATHING AND CONCRETE ENCASEMENT

As mentioned earlier, corrosion of steel is most severe in the splash zone, and cathodic protection is ineffective in controlling corrosion in this area. Metal sheathing to protect new and existing structures is an effective method of combatting splash-zone corrosion. The U.S. Coast Guard has used heavy iron plates to protect the legs of fixed offshore towers along the Atlantic seaboard. One popular protection method is the use of Monel sheathing because of its excellent resistance to seawater corrosion. Stainless steel also has received attention as sheathing. When either Monel or stainless are used as sheathing, a sheet of insulating material should be placed between the metallic sheath and the carbon steel to prevent galvanic corrosion.

Concrete provides good protection to steel and has been successfully used for many years in protecting dock structures. However, cracking and spalling problems have been encountered with concrete. To minimize the deterioration of concrete marine structures, the construction procedures outlined by the American Concrete Institute should be followed. These include using a 3-in.-thick concrete cover on the steel, selecting the proper portland cement and aggregates, conducting necessary ASTM material tests at the job site, and obtaining high-quality workmanship.

ORGANIC AND METALLIC COATINGS

Organic coatings are probably the most widely used means of protecting steel against corrosion in seawater. The tremendous growth of the chemical industry, along with the need for improved coatings, has led to some outstanding developments in the organic-coating field. The various types of coatings are too numerous to discuss in a paper of this type. The generic classes of coatings that have performed well in protecting steel in seawater include vinyls, epoxies, urethanes, and coal-tar epoxies. Of course, maximum performance from coatings depends to a large extent on good surface preparation; preferably, blastcleaning the steel surfaces to remove mill scale, dirt, and rust before applying the coatings. Heavy-bodied amine-terminated epoxies can be applied in place to coat corroded steel in splash-zone areas. Reportedly,[28] installation only requires the removal of loose scale, rust, dirt, grease, and marine growth. The epoxy displaces water and forms a firm bond to the steel.

Metallized aluminum coatings to protect steel in seawater have been in use for over a decade and have established an excellent performance record. About 6–12 mils of aluminum are sprayed on the steel surface, which is first blast-cleaned to white metal. The American Welding Society reports[29] that steel panels metallized with 6 mils or more of aluminum and sealed with a clear vinyl coating are in excellent condition after 12 yrs exposure to seawater.

SUMMARY

The constructional steels have a wide range of application in marine environments and when properly used can be expected to give long service life. However, decisions concerning the anticipated performance of carbon steel and the high-strength low-alloy and alloy steels are not always simple to make because in many cases the conditions governing the corrosion of these materials are very complex. Therefore, a detailed study of the conditions is usually necessary before a material or protective method can be selected. In a review paper of this kind, it is impossible to include complete and detailed information on the performance of these steels under all possible conditions. However, it is hoped that the information presented here will promote a better understanding of the use of the constructional steel in marine environments.

REFERENCES

1 *Proceedings of ASTM*, Vol. 59, 1959, p. 183.

2 Guttman, H., and Sereda, P. J., "Measurement of Atmospheric Factors Affecting the Corrosion of Metals", *Metal Corrosion in the Atmosphere, ASTM STP 435*, ASTM 1968, pp. 326–359.

3 Mears, R. B.: *Australasian Corrosion Engineering*, Vol. 6, No. 12, December.

4 A report of ASTM Committee G-1, Subcommittee IV, Section 1, *Metal Corrosion in the Atmosphere, ASTM STP 435*, ASTM, 1968, p. 373.

5 Buck, D. C.: *Proceedings of ASTM*, Vol. 19, 1919, Part 2, p. 224.

6 Reports of Committee A-5, *Yearly Proceedings of ASTM*, 1916–1951.

7 Larrabee, C. P.: *Corrosion*, Vol. 9, 1953, No. 8, pp. 259–271.

8 Larrabee, C. P., and Coburn, S. K.: *Proceedings of the First International Congress on Metallic Corrosion*, London, 1961, Butterworth, London, 1962, p. 283.

9 Copson, H. R., and Larrabee, C. P.: "Extra Durability of Paint on Low-Alloy Steels", *ASTM Bulletin No. 242*, December 1959.

10 Humble, H. A.: "The Cathodic Protection of Steel Piling in Seawater", *Corrosion*, Vol. 5, 1949, No. 9, p. 292.

11 Larrabee, C. P.: "Corrosion-Resistant Experimental Steels for Marine Applications", *Corrosion*, Vol. 14, 1958, No. 11, pp. 501t–504t.

12 Uhlig, H. H.: *Corrosion Handbook*, pp. 383–388, John Wiley and Sons, Inc., New York City, 1948.

13 Southwell, C. R., Jr., and Alexander, A. L.: "Corrosion of Structural Ferrous Metals in Tropical Environments—Sixteen Years' Exposure to Sea and Fresh Water", U.S. Naval Research Laboratory Report No. 6862, 1969.

14 Larrabee, C. P.: *Materials Protection*, Vol. 1, 1962, No. 12, December, pp. 95–96.

15 Kirk, W. W., Covert, R. A., and May, T. P.: "Corrosion Behavior of High-Strength Steels in Marine Environments", *Metals Engineering Quarterly*, American Society for Metals, November 1968, pp. 31–38.

16 Southwell, C. R., Jr., and Alexander, A. L.: *op. cit.*

17 Uhlig, H. H.: *op. cit.*, p. 392.

18 Hudson, J. C.: "Corrosion of Bare Iron or Steel in Seawater" *Journal of the Iron and Steel Institute*, October 1950.

19 Larrabee, C. P.: "Corrosion Resistance of High-Strength Low-Alloy Steels as Influenced by Composition and Environment", *Corrosion*, Vol. 9, 1953, No. 8, August, pp. 259–271.

20 Southwell, C. R., Jr., and Alexander, A. L.: *op. cit.*

21 LaQue, F. L., and Cox, C. L.: *Proceedings of ASTM*, Vol. 40, 1940, p. 670.

22 Phelps, E. H.: "Stress-Corrosion Behavior of High-Yield-Strength Steels", *Proceedings of the Seventh World Petroleum Congress*, 1967.

23 Kirk, W. W., Covert, R. A., and May, T. P.: *op. cit.*

24 Phelps, E. H.: "A Review of the Stress-Corrosion Behavior of Steels with High-Yield-Strength", To be published in the *Proceedings of the Ohio State Conference on Stress-Corrosion Cracking*, September 11-15, 1967.

25 Brown, B. F.: "Stress-Corrosion Cracking and Corrosion Fatigue of High-Strength Steels", *DMIC Report 210*, Battelle Memorial Institute, Columbus, Ohio, pp. 91–102.

26 Leckie, H. P., and Loginow, A. W.: "Stress-Corrosion Behavior of High-Strength Steels", *Corrosion*, Vol. 24, 1968, pp. 291–297.

27 Elsea, A. R.: "Studies on Hydrogen Stress Cracking", Symposium on Line Pipe Research, *American Gas Association Catalog No. L30000.*

28 "Splash Zone Coating Protects Corroded Steel Piling", *Materials Protection*, Vol. 2, 1963, October, pp. 81–82.

29 *American Welding Society Report No. C2.11-67*, "Corrosion Tests of Metallized Coated Steel—12 year Report", 1967.

30 Loginow, A. W.: "Stress-Corrosion Testing of Alloys", *Materials Protection*, Vol. 5, 1966, pp. 33–39.

Acknowledgement

The authors wish to point out that many of the results presented in this paper were obtained in programs conducted by their present and former co-workers. Of particular value were studies conducted by C. P. Larrabee, C. X. Mullen, and A. W. Loginow.

Pressure-Vessel Steels: Promise and Problem

JOHN H. GROSS

During the past 25 years, numerous pressure-vessel problems have been solved, but in many instances the solutions have led to other problems. Currently, promising developments in pressure-vessel steels are providing such solutions and such additional problems with respect to fabrication and various failure modes. With respect to fabrication, lamellar tearing is being minimized by special melting and solidification practices. However, continuous casting and electroslag-remelting of slabs are currently limited in plate size that can be produced. With respect to bursting, recent studies indicate that high-yield-strength steels have higher burst-strength indices than lower-strength steels even for vessels with nozzles and notches up to 25 pct of the wall thickness. Strength, and therefore resistance to bursting, can be increased without loss in toughness through new control-rolling practices, except that these practices are limited to plates up to 3/4 in. (19 mm) thick. New high-toughness line-pipe steels are now available that should be very attractive for pressure vessels that require very high resistance to shear tearing. These low-sulfur steels may be somewhat impaired by sensitivity to splitting as a result of the control-rolling practice. To date, steels have not been developed with improved resistance to fatigue failure. Fortunately, pressure vessels have rarely failed by fatigue. Similarly, resistance to failure by environmental effects is not basically improved by changing steel composition, usually because of difficulty in defining the effects of the numerous environments that may be involved.

THE many changes that have occurred in pressure-vessel technology have provided solutions to many problems that had temporarily deferred advances in pressure-vessel applications, and indicate the promise for continued advancement. In almost all instances, however, the promise of these advancements has merely whetted the appetite for further advancements. In many respects, this is a nonphysical example of Newton's Third Law—For every action there is an equal and opposite reaction. One of the best illustrations of this reaction in the materials area is the continuing push-pull between increased strength and increased thickness of pressure-vessel steels. When the continued increase in operating pressure resulted in an increase in plate thickness that prohibitively complicated fabrication practices, new high-strength steels were developed. However, the continued increase in operating pressure soon raised the thickness requirements to those previously required of carbon steels. This created new, difficult problems because the design, fabrication, and inspection requirements for high-strength steels are much more sophisticated than those for carbon steels. Thus, the promise of high-strength steels became a complex problem.

It is in this context that I will discuss some current problems in pressure-vessel materials, developments that promise to solve the problem, and problems that may result from the solution. My comments will include steels for piping as well as those for pressure vessels because the previously mentioned promise and problem apply to piping as well as to pressure vessels. As in previous related papers, these steels will be discussed with respect to fabrication problems and to the mode of failure—bursting, brittle fracture, fatigue, and environmental effects.

JOHN H. GROSS is Director-Research, Research Laboratory, United States Steel Corp., Monroeville, Pa. Reprinted from Part III-Discussions Proceedings of the Second International Conference on Pressure Vessel Technology, San Antonio, Texas—Oct. 1-4, 1973. Published by American Society of Mechanical Engineers, 345 E. 47th St., New York, N. Y. 10017. Printed in the U.S.A.

FABRICATION

Fabrication problems arise continuously as new materials are introduced, and for various other reasons. One such problem is lamellar tearing, which arises when a high stress is imposed in the through-thickness direction, as can occur when restrained attachments are welded to another plate (Fig. 1). The plate locally separates internally in a plane parallel to the plate surface where the through-thickness stress is high. The internal separation, or lamellar tear, occurs along planes where segregates and nonmetallic inclusions act to reduce through-thickness strength and ductility. The problem can be avoided by designing the structure to eliminate this type of configuration, but this is not always possible. Therefore, producers

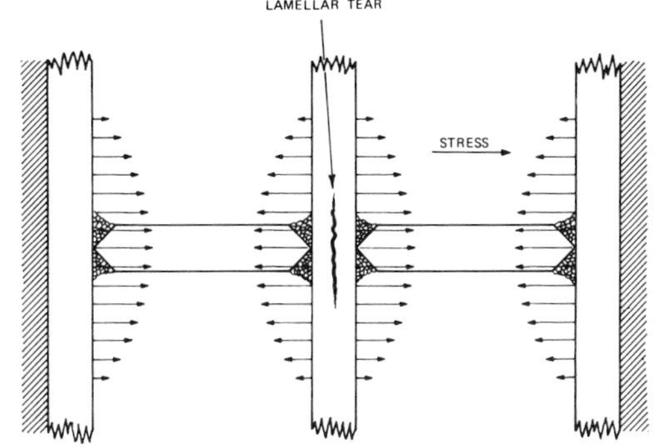

Fig. 1—Lamellar tearing can occur in welded structures.

of steel plates have developed practices to minimize segregation and nonmetallic inclusions in plates that will be used in applications where the through-thickness stress is high.

The solution involves special melting and solidification practices that reduce the area of discontinuous phases in a plane parallel to the plate surface. One obvious practice is to melt cleaner steels, particularly to minimize sulfide inclusions which spread into platelets during hot rolling. In one such experiment,[1] a heat of A537-B was melted to a low-sulfur practice. A comparison of the properties for the low-sulfur and regular-sulfur steel (Table I) shows that the tensile and impact properties for all orientations were better for the low-sulfur steel but that the marked improvement was for the through-thickness properties. In a similar experiment for a low-sulfur A517-B steel (Table I), the through-thickness properties proved to be much better than those for a regular-sulfur steel.

Another approach is to "ball-up" the sulfides into refractory-type spheres that do not spread out into relatively large platelets during rolling. A number of special additives have been developed that effectively control sulfide morphology.

Another solution involves solidifying the metal so rapidly that segregates cannot agglomerate, dispersing instead as small particles throughout the steel. In this form, they are much less deleterious. When slabs are continuously cast rather than being rolled from ingots, the area of the discontinuous phase parallel to the plate surface is reduced, and through-thickness properties are improved.

One attractive solution is based on electroslag remelting (ESR) of ingots. In this process, some nonmetallics are volatilized, some are fluxed off into the slag, some are rejected to the sidewalls where they are removed during conditioning, and the remainder are rather uniformly dispersed as fine particles because of the rapid solidification rate. By far the largest tonnage of ESR steels is produced in Russia, and they report properties in the through-thickness direction that are essentially equivalent to the transverse properties. As shown in Table II, properties are reasonably similar in all three directions. The method requires different melting facilities than conventional steelmaking; at present, ESR capacity in this country is very small.

Several of the solutions to lamellar tearing also create problems. In continuous casting, slab width and thickness limit the plate size that can be satisfactorily rolled. This is also true, but to a lesser degree, for ESR slabs. Nevertheless, advancements in continuous casting and ESR remelting hold promise for significantly reducing the problem of lamellar tearing.

Table I. Effect of Sulfur Content on Directional Properties

Orientation	Tensile Properties: Tensile Strength, ksi (MPa)	Tensile Properties: Red. of Area, Pct	Charpy V-Notch Properties at 75 F (22C): Energy Absorbed, ft-lb (J)	Charpy V-Notch Properties at 75 F (22C): Lateral Expansion, mil (mm)
A. ASTM A537-B Steel				
	0.024 Pct Sulfur			
Longitudinal	80.9 (558)	69	86 (117)	70 (1.8)
Transverse	80.2 (553)	69	69 (94)	61 (1.6)
Through-Thickness	75.3 (519)	18	23 (31)	30 (0.76)
	0.005 Pct Sulfur			
Longitudinal	89.8 (619)	72	139 (188)	90 (2.3)
Transverse	89.6 (617)	72	115 (156)	81 (2.1)
Through-Thickness	89.0 (614)	37	54 (73)	50 (1.3)
B. ASTM A517-B Steel				
	0.005 Pct Sulfur			
Longitudinal	128 (882)	66	80 (108)	58 (1.5)
Transverse	128 (882)	65	70 (95)	52 (1.3)
Through-Thickness	127 (876)	38	35 (47)	26 (0.66)

FAILURE MODES

Bursting Strength

Resistance to bursting depends primarily on the strength of the material. The relative importance of yield strength versus tensile strength in resisting bursting is not fully established. Nevertheless, the ASME Code essentially ignores yield strength. Currently, the Pressure Vessel Research Committee (PVRC) is sponsoring studies at the University of Kansas to confirm the observation by Langer[2] and Gross[3,4] that credit should be given for yield strength. Studies to data have shown that the Svensson equation, which indicates that bursting strength rises with yield strength, is applicable to conventional thin-walled pressure vessels. Royer, Rolfe, and Easley[5] show that vessels containing weld seams and nozzles exhibit sufficient plasticity to behave in accordance with the Svensson equation (Fig. 2). Work which will be published in 1974 shows that sharp notches up to 25 pct of the wall thickness can be tolerated before the plasticity of 100 ksi (690 MPa) yield-strength steels decreases to the point that the Svennson equation is no longer applicable (Fig. 3). Although the results do not establish the highest level at which an increased yield-to-tensile strength helps resist bursting, they clearly indicate that it applies to pressure-vessel steels with minimum specified yield strengths up to at least 100 ksi (690 MPa).

Strength to resist bursting can be raised in many ways. For unhardened steels, small amounts of va-

Table II. Directional Properties of an Open-Hearth and an Electroslag Remelted Steel

Orientation	Tensile Properties: Yield Strength, ksi (MPa)	Tensile Properties: Tensile Strength, ksi (MPa)	Tensile Properties: Elongation, Pct	Tensile Properties: Red. of Area, Pct	Impact Energy Absorption, ft-lb (J)
		Open-Hearth Steel			
Longitudinal	50.2 (346)	65.8 (454)	22.9	61.0	15.8 (21.4)
Transverse	49.7 (343)	64.2 (443)	24.4	61.1	12.7 (17.3)
Through-Thickness	47.6 (328)	62.6 (430)	11.7	32.7	4.1 (5.56)
		Electroslag-Remelted Steel			
Longitudinal	47.8 (330)	63.4 (437)	25.8	68.2	19.5 (26.5)
Transverse	47.7 (329)	63.9 (441)	26.1	66.0	13.1 (17.8)
Through-Thickness	45.7 (316)	62.1 (428)	23.4	63.6	12.6 (17.1)

nadium or columbium can be added to form strengthening precipitates. However, as shown in Fig. 4, strengthening by precipitation is accompanied by an undesirable increase in the transition temperature.[6] When strengthening is obtained by grain refinement, the transition temperature is also improved. As shown in Fig. 5, this can be accomplished by special rolling of carbon steels. This figure also shows that precipitation and grain refinement can be combined

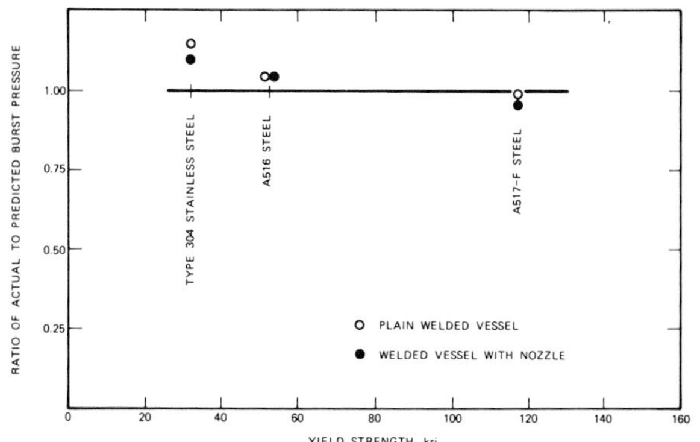

Fig. 2—Variation in resistance to bursting with yield strength.

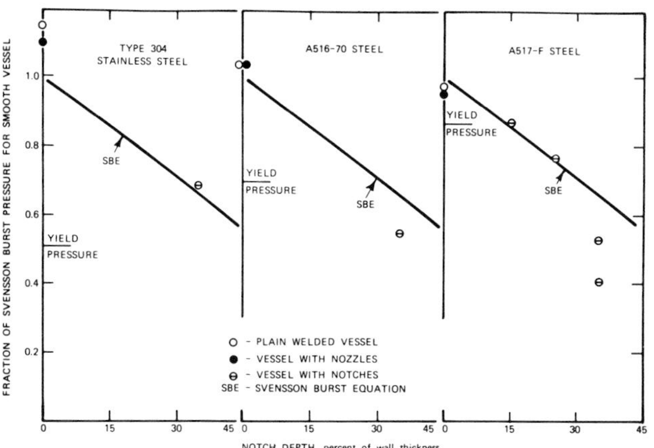

Fig. 3—Effect of welding, nozzles, and notches on bursting of pressure vessels.

to increase strength without raising the transition temperature, and that special control-rolling significantly improves strength and transition temperature.

The limitation on the use of control-rolled high-strength columbium steels is that even the largest plate mills lack the power necessary to control-roll wide plates over about 3/4 in. (19 mm) thick. For line pipe, this thickness limitation is not serious because the wall thickness for line pipe is still generally less than 3/4 in. (19 mm). Thus, most plate for line pipe is control-rolled. For pressure vessels, however, the thickness limitation is a serious deterrent. Moreover, control-rolling slows production, and consequently limits availability. Improved control-rolling schedules and better control of mill loading promises to increase the thickness limit on control-rolling, making these steels generally applicable to pressure vessels. Of course, increased resistance to bursting is of limited value if resistance to other modes of failure is not correspondingly increased.

Brittle Fracture

When pressure vessels contain compressible fluids at high pressures, the stored energy can cause cracks to extend for long distances. Agencies concerned with pipeline transmission of gas are particularly aware of this problem because of pipeline failures that have been miles in length. This problem has been studied by Battelle in full-scale tests for the American Gas Association, and by U. S. Steel in full-scale tests, which were subsequently supported by AISI. The results of one of the early U. S. Steel tests are summarized in Fig. 6.

The failure began catastrophically in the 14 ft-lb (19 J) initiation section where the crack grew at almost 900 ft (270 m) per second. To the right, the crack velocity decreased shortly after it entered the 18 ft-lb (24.4 J) material, but reached an equilibrium velocity of about 750 feet (230 m) per second, which indicates that the crack would probably have grown for a great distance. When the crack entered the 48 ft-lb (65 J)

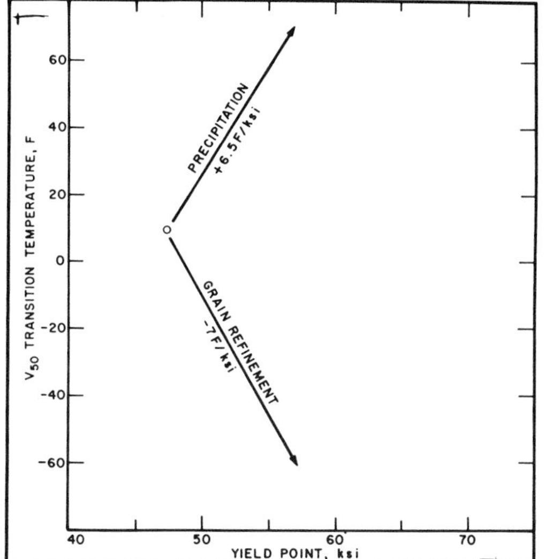

Fig. 4—Vectors for hardening by precipitation and grain refinement.

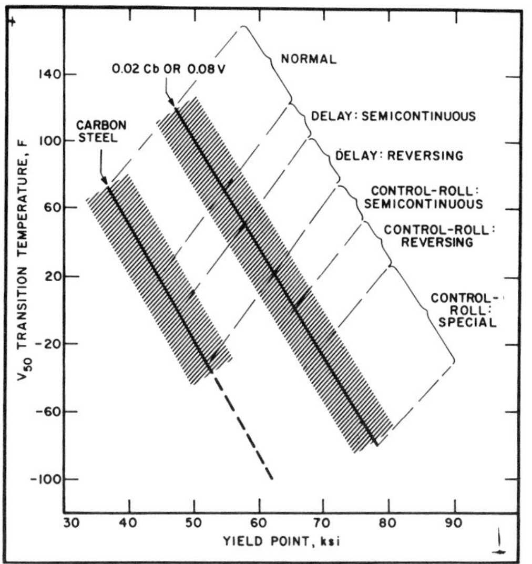

Fig. 5—Relation between rolling mill processing and mechanical properties.

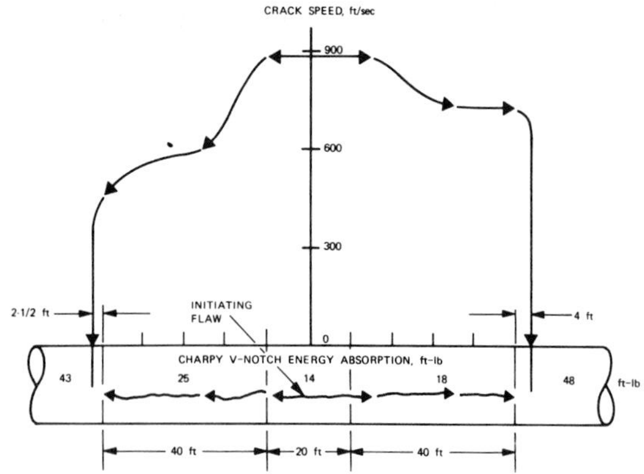

EFFECT OF TOUGHNESS ON CRACK PROPAGATION IN LINE PIPE BURST TEST

Fig. 6—Toughness inhibits crack growth in line pipe burst test.

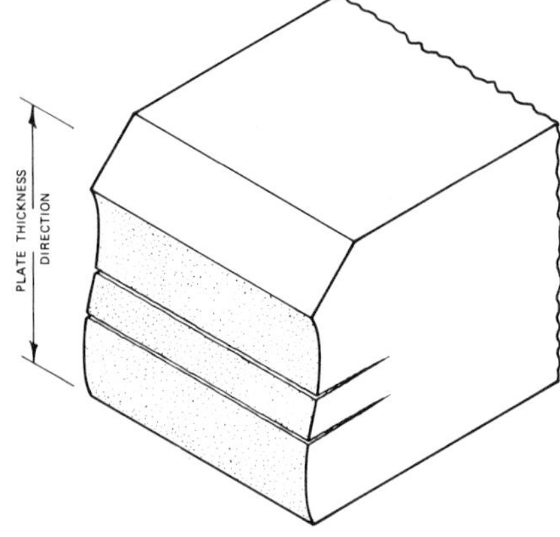

Fig. 7—Schematic illustration of fissuring or splitting in Charpy V-notch specimen.

pipe, it decelerated very rapidly, and stopped in 4 feet (1.22 m).

To the left, the crack decelerated rapidly to about 600 feet (180 m) per second upon entering the 25 ft-lb (24 J) pipe, and then continued to decelerate to about 500 feet (150 m) per second when it reached the end of this section. We believe that the crack would have been arrested in the 25 ft-lb (34 J) pipe if it had been several sections longer. Consequently, for these test conditions, 25 ft-lb (34 J) appears to be close to the minimum shelf energy required to arrest a long-running fracture within several pipe lengths. The crack stopped growing after traveling only 2½ feet (0.76 m) in the 43 ft-lb (58 J) pipe.

These results clearly established that crack growth could be arrested only in high notch-ductility (high notch-toughness) material. To obtain this notch ductility required the development of a steel with a transition temperature low enough so that the steel would be on the shelf at the operating temperature, and that the shelf energy or ductility be much higher than in conventional line-pipe steels. These improvements were obtained by developing a new composition, and by optimizing the control-rolling practice. This innovation, which should reduce susceptibility to long-running fractures, has permitted the necessary increase in strength and operating pressure.

Impairing this promising development is the occurrence of splitting or fissuring in the transition-temperature range when the steel is fractured at a high strain-rate (Fig. 7). Although splitting is obviously related to some weakness in the through-thickness direction, probably induced by the control-rolling practice, its exact cause has not been established. However, Fig. 8 shows that the transition temperature in the through-thickness direction is much higher than in the transverse direction.

The extent, if any, to which splitting reduces the ability of a steel to arrest growing fractures has also not been established. However, the fracture appearance is disconcerting even if not actually deleterious. The importance of the foregoing comments on long-running fractures to pressure-vessel performance is currently being studied by the ASME Boiler and Pressure Vessel Committee Subgroup on Toughness.

Another problem associated with toughness criteria is related to the effect of thickness on susceptibility to brittle fracture. Current codes, including the ASME

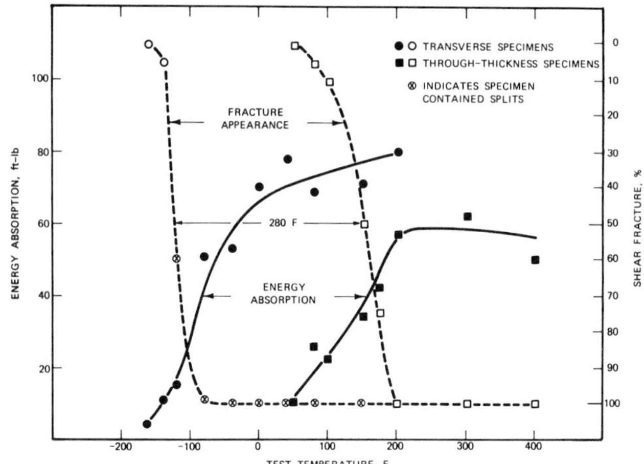

Fig. 8—Transition-temperature characteristics of transverse and through-thickness Charpy V-notch specimens of control-rolled steel.

Boiler and Pressure Vessel Code, do not include any effect of thickness on toughness requirements for non-nuclear vessels. On the basis of fracture-mechanics considerations, however, requirements should increase with plate thickness, as illustrated in Fig. 9 for plane strain behavior: $[(K_{Ic}/\sigma_y)^2 = 0.4]$. The curves indicate the problems that would arise if fracture-mechanics concepts were imposed, particularly for high-strength steels. Fortunately, testing of full-thickness specimens for plates of A533-B steel indicates that plane-strain behavior is not extended indefinitely with plate thickness, and that a metallurgical transition temperature is reached beyond which the steel behaves ductilely regardless of plate thickness. The cut-off point in the effect of increased thickness on increased toughness for other steels has not been established. Consequently, the magnitude of this potential problem has not been really assessed. This problem is also being considered by the ASME Subgroup on Toughness.

Fatigue Failure

The incidence of failures of any type in pressure vessels is extremely small, and is almost infinitesimally small for failure by fatigue. This is understand-

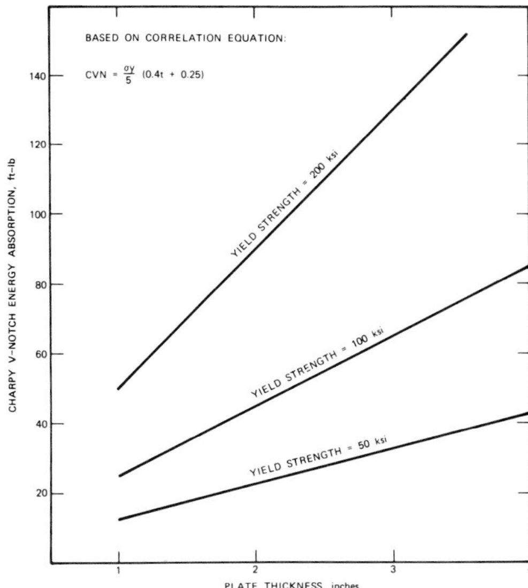

Fig. 9—Effect of plate thickness on toughness requirements, according to fracture-mechanics calculations.

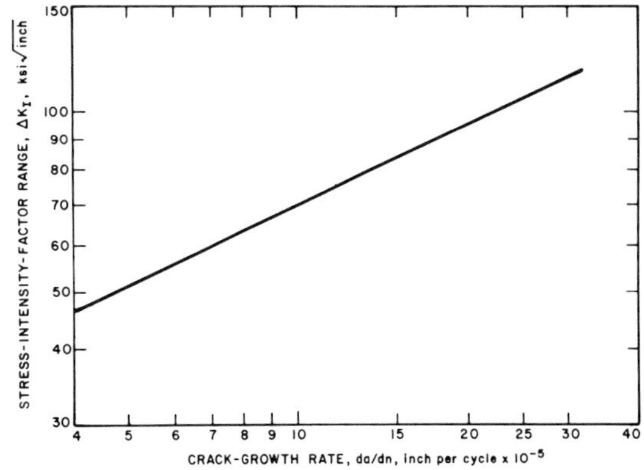

Fig. 10—Idealized fatigue crack-growth rate behavior for steels of various yield strengths.[7]

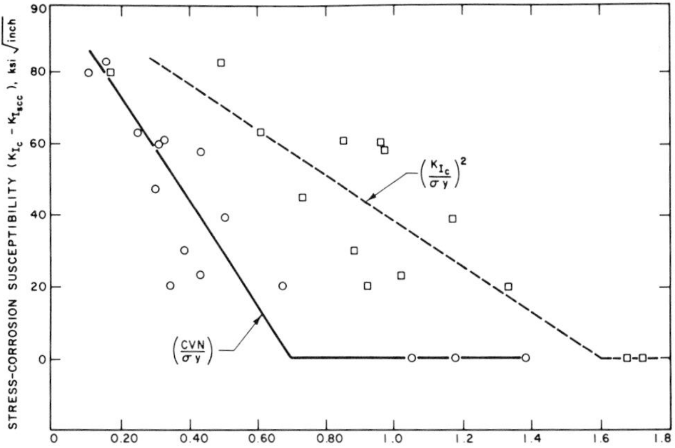

Fig. 11—Effect of fracture-toughness–strength ratio on susceptibility to stress corrosion in sea water.

able when one considers that a daily pressure cycle over a twenty-year life amounts only to 7300 cycles. Even hourly pressurization amounts only to 175,000 cycles. Thus, in the cycle range of interest (5,000 to 200,000 cycles), crack initiation in the membrane region is highly unlikely; safety factors of 2 or 3 to 1 safeguard against the initiation of fatigue cracks.

In regions where stress raisers reduce the factor of safety, fatigue cracks can conceivably start if the discontinuity associated with notches, nozzles, weld reinforcements, etc., is sharp enough. For nuclear vessels and for Division 2, Section VIII vessels which utilize higher design stresses, fatigue analyses safeguard against fatigue failure by controlling the magnitude of stress raisers.

It is indeed fortunate that pressure vessels are not particularly susceptible to fatigue failure because measures for improving the fatigue strength of steels are not currently available. Problems could arise, particularly for high-strength steels because the resistance to fatigue-crack initiation does not scale up with tensile strength. Likewise, as shown in Fig. 10, crack growth is essentially independent of steel strength. Consequently, control of fatigue in pressure vessels relies almost entirely on design and fabrication practices that maintain stress raisers at an acceptable level.

Detailed discussions of fatigue in pressure vessels is contained in Chapter 2 by Langer in the book, "Pressure-Vessel Engineering Technology" by R. W. Nichols, and a report is currently being prepared by Barsom[7] under PVRC sponsorship. This information is most useful in understanding the fatigue of pressure vessels. However, as Dolan[8] noted with respect to pressure vessels, "if peak stresses are significantly reduced, the question of exact fatigue life becomes more academic than of real interest." Thus, although there is no current promise for improving the basic fatigue strength of pressure-vessel steels, there is no significantly identifiable problem of fatigue failure of pressure vessels.

Environmental Effects

The effect of environment on pressure-vessel service is even more difficult to analyze than that of fatigue because pressure vessels may operate in so many different environments. Hydrogen sulfide, high-pressure hydrogen, and various media that cause stress corrosion can lead to failure. As with fatigue, high-strength steels tend to be more susceptible than lower-strength steels to these forms of failure. In general, steel composition is not a significant factor in resisting failure in various environments. As shown in Fig. 11, some evidence indicates that high-toughness steels may have increased resistance to stress corrosion in sea water. Coatings, inhibitors, cathodic protection, and similar approaches, however, appear more reliable in protecting against environmental failure. Thus the question of promise and problem with respect to environmental effects on pressure vessels is not readily defined.

SUMMARY

This discussion covers several of the current steel developments that promise to solve certain existing pressure-vessel problems, but that also create new problems in many instances. This promise and problem may be summarized as follows:

With respect to fabrication, lamellar tearing is being minimized by special melting and solidification practices. However, continuous casting and electro-slag-remelting of slabs are currently limited in plate size and quantity that can be produced.

With respect to bursting, recent studies indicate that high-yield-strength steels have higher burst-strength indices than lower-strength steels even for vessels with unground weld reinforcements, with nozzles, and with notches up to 25 pct of the wall thickness. Strength, and therefore resistance to bursting, can be increased without loss in toughness through new control-rolling practices, except that these practices are limited to plates up to about 3/4 in. (19 mm) thick.

New high-toughness line-pipe steels are now available that should be very attractive for pressure vessels that require very high resistance to shear tearing. These low-sulfur steels may be somewhat impaired by sensitivity to splitting as a result of the control-rolling practice.

To date, steels have not been developed with improved resistance to fatigue failure. Fortunately, pressure vessels have rarely failed by fatigue. Similarly, resistance to failure by environmental effects is not basically improved by changing steel composition, the main difficulty being in defining the effects of the numerous environments that may be involved.

REFERENCES

1. Unpublished investigation by U. S. Steel.
2. B. F. Langer: *Design-Stress Basis for Pressure Vessels,* presented at 1970 SESA Fall Meeting, Boston, Mass., Oct. 18-28, 1970.
3. J. H. Gross: *PVRC Interpretive Report of Pressure Vessel Research: Section 2-Material Considerations,* Welding Research Council Bulletin No. 101, November 1964.
4. J. H. Gross: *J. Eng. Ind.,* vol. 93, no. 4, *Trans. ASME,* November 1971, pp. 962-68.
5. C. P. Royer, S. T. Rolfe, and J. T. Easley: *Effect of Strain Hardening on Bursting Behavior of Pressure Vessels,* presented at the Second International Conference on Pressure-Vessel Technology.
6. J. H. Gross: *Transformation Characteristics of Low-Carbon Cb-Containing Steels,* Symposium on Low-Alloy High-Strength Steels, May 1970.
7. J. M. Barsom: *Fatigue Behavior of Pressure Vessels,* PVRC Interpretive Report, to be published.
8. T. J. Dolan: *Weld. J.,* 1954, vol. 33, no. 6.

Guide to ASTM Specifications for Pressure Vessel Forging Steels

ASTM A 336 — Seamless Drums, Heads, and Other Components

Class	Composition, % (a)								Mechanical Properties (b)				Typical Applications
	C	Mn	P	S	Si	Ni	Cr	Mo	Tensile Strength, Min, 10^3 psi (MPa)	Yield Strength, Min, 10^3 psi (MPa)	Elongation (Tangential), Min, %	Reduction in Area (Tangential) Min, %	
Ferritic Steels													
F1	0.20-0.30	0.60-0.80	0.040	0.040	0.20-0.35	—	—	0.40-0.60	70 (480)	40 (280)	20	30	Drums, headers, barrels, nozzles, covers, plugs.
F12	0.10-0.20	0.30-0.80	0.040	0.040	0.10-0.60	—	0.80-1.10	0.45-0.65	70 (480)	40 (280)	18	25	
F5	0.15	0.30-0.60	0.030	0.030	0.50	0.50	4.0-6.0	0.45-0.65	60 (410)	36 (250)	19	35	
F5a	0.25	0.60	0.040	0.030	0.50	0.50	4.0-6.0	0.45-0.65	80 (550)	50 (340)	19	35	
F6	0.12	1.00	0.040	0.030	1.00	0.50	11.5-13.5	—	85 (590)	55 (380)	18	35	
F21	0.15	0.30-0.60	0.030	0.030	0.50	—	2.65-3.25	0.80-1.06	75 (520)	45 (310)	18	35	
F21a	0.15	0.30-0.60	0.030	0.030	0.50	—	2.65-3.25	0.80-1.06	60 (410)	30 (210)	20	35	
F22	0.15	0.30-0.60	0.030	0.030	0.50	—	2.00-2.50	0.90-1.10	75 (520)	45 (310)	18	25	
F22a	0.015	0.30-0.60	0.030	0.030	0.50	—	2.00-2.50	0.90-1.10	60 (410)	30 (210)	20	35	
F30(c)	0.45	0.50-0.90	0.040	0.040	0.15-0.45	—	—	0.30-0.60	80 (550)	50 (340)	21	35	
F31(d)	0.35	0.50-0.90	0.040	0.040	0.10-0.40	2.25-3.00	—	0.20-0.50	95 (650)	70 (480)	18	35	
F32(e)	0.35	0.50-0.90	0.040	0.040	0.15-0.45	0.50-1.00	3.00-3.60	0.30 0.50	100 (690)	60 (410)	18	35	
Austenitic Steels													
F8	0.08	2.00	0.040	0.030	1.00	8.00-11.00	18.00-20.00	—	70 (480)	30 (210)	30	35	
F8m	0.08	2.00	0.040	0.030	1.00	10.00-14.00	16.00-18.00	2.00-3.00	70 (480)	30 (210)	30	35	
F8c(f)	0.08	2.00	0.040	0.030	0.85	9.00-12.00	17.00-19.00	—	70 (480)	30 (210)	30	35	
F8t(g)	0.08	2.50	0.035	0.030	0.85	9.00 min	17.00	—	70 (480)	30 (210)	30	35	
F10	0.10-0.20	0.50-0.80	0.030	0.030	1.00-1.40	19.00-22.00	7.00-9.00	—	80 (550)	30 (210)	25	35	
F25	0.15	2.00	0.040	0.030	1.00	19.00-22.00	24.00-26.00	—	75 (520)	30 (210)	30	35	

(a) Maximum unless otherwise noted; (b) Minimum requirements for heat treated forgings; (c) Also 0.10-0.25 V; (d) Also 0.15 max V; (e) Also 0.05-0.15 V; (f) Cb is not less than 10 x C content, but not more than 1.0%; (g) Ti is not less than 5 x C content, but not more than 0.60%.

ASTM A 372 — Thin-Walled Pressure Vessel Steels

Class	Composition, % (a)								Mechanical Properties (b)			Typical Applications
	C	Mn	P	S	Si	Ni	Cr	Mo	Tensile Strength, Min, 10^3 psi (MPa)	Yield Strength, Min, 10^3 psi (MPa)	Elongation, Min, %	
I	0.30	1.00	0.04	0.05	0.15-0.30	—	—	—	60 (410)	35 (240)	20	Thin-walled gas bottles and vessels.
II	0.40	1.29	0.04	0.05	0.15-0.30	—	—	—	75 (520)	45 (310)	18	
III	0.48	1.65	0.04	0.05	0.15-0.30	—	—	—	90 (620)	55 (380)	15	
IV	0.40-0.50	1.40-1.80	0.035	0.04	0.15-0.35	—	—	0.17-0.27	105 (720)	65 (450)	15	
V-A	0.26-0.34	0.40-0.70	0.035	0.04	0.15-0.35	—	0.80-1.15	0.15-0.25	120 (830)	70 (480)	18	
V-B	0.31-0.39	0.70-1.00	0.035	0.04	0.15-0.35	—	0.80-1.15	0.15-0.25	120 (830)	70 (480)	18	
V-C	0.26-0.34	0.70-1.00	0.035	0.04	0.15-0.35	0.40-0.70	0.40-0.65	0.15-0.25	120 (830)	70 (480)	18	
V-D	0.31-0.39	0.75-1.05	0.035	0.04	0.15-0.35	0.40-0.70	0.40-0.65	0.15-0.25	120 (830)	70 (480)	18	
V-E	0.35-0.50	0.75-1.05	0.035	0.04	0.15-0.35	—	0.80-1.15	0.15-0.25	120 (830)	70 (480)	18	
VI	0.18	0.10-0.40	0.025	0.025	0.15-0.35	2.00-3.25	1.00-1.80	0.20-0.60	100 (690)	80 (550)	20	
VII	0.38-0.43	0.60-0.80	0.035	0.04	0.20-0.35	1.65-2.00	0.70-0.90	0.20-0.30	155 (1070)	135 (930)	12	
VIII	0.35-0.50	0.75-1.05	0.035	0.04	0.15-0.35	—	0.80-1.15	0.15-0.25	135 (930)	110 (760)	15	

(a) Maximum unless otherwise noted; (b) Minimum requirements for heat treated forgings.

ASTM A 508 — Vacuum-Treated Steels

Class	Compositon, % (a)									Mechanical Properties (b)					Typical Applications
	C	Mn	P	S	Si	Ni	Cr	Mo	V	Tensile Strength, Min, 10^3 psi (MPa)	Yield Strength, Min, 10^3 psi (MPa)	Elongation, %	Reduction in Area, %	Impact Energy ft-lbf(J) (c)	
1	0.35	0.40-0.90	0.025	0.025	0.15-0.35	0.40	0.25	0.10	0.05	70-95 (480-655)	35 (240)	20	38	15 (20)	Reactor systems — vessel, closures, shells, flanges, tube.
2	0.27	0.50-0.90	0.025	0.025	0.15-0.35	0.50-0.90	0.25-0.45	0.55-0.70	0.05	80-105 (550-720)	50 (340)	18	38	30 (41)	
2a	0.27	0.50-0.90	0.025	0.025	0.15-0.35	0.50-0.90	0.25-0.45	0.55-0.70	0.05	90-115 (620-790)	65 (450)	16	35	35 (48)	
3	0.15-0.25	1.20-1.50	0.025	0.025	0.15-0.35	0.40-0.80	0.25	0.45-0.60	0.05	80-105 (550-720)	50 (340)	18	38	30 (41)	
4	0.23	0.20-0.40	0.020	0.020	0.30	2.75-3.90	1.50-2.00	0.40-0.60	0.03	105-130 (720-895)	85 (590)	18	45	35 (48)	
4a	0.23	0.20-0.40	0.020	0.020	0.30	2.75-3.90	1.50-2.00	0.40-0.60	0.03	115-140 (790-965)	100 (690)	16	45	35 (48)	
4b	0.23	0.20-0.40	0.020	0.020	0.30	2.75-3.90	1.50-2.00	0.40-0.60	0.03	90-115 (620-790)	70 (480)	20	48	35 (48)	
5	0.23	0.20-0.40	0.020	0.020	0.30	2.75-3.90	1.50-2.00	0.40-0.60	0.08	105-130 (720-895)	85 (590)	18	45	35 (48)	
5a	0.23	0.20-0.40	0.020	0.020	0.30	2.75-3.90	1.50-2.00	0.40-0.60	0.08	115-140 (790-965)	100 (690)	16	45	35 (48)	

(a) Maximum unless otherwise noted; (b) Minimum requirements for heat treated forgings; (c) Minimum average values for three Charpy V-notch specimens tested at 40 F (4 C), Classes 1, 2, and 3; 70 F (20 C), Class 2a; or −20 F (−29 C), Classes 4, 4a, 4b, 5, and 5a.

ASTM A 541 — Quenched and Tempered Forgings

Class	Compositon, % (a)									Mechanical Properties (b)					Typical Applications
	C	Mn	P	S	Si	Ni	Cr	Mo	V	Tensile Strength, Min, 10^3 psi (MPa)	Yield Strength, Min, 10^3 psi (MPa)	Elongation, %	Reduction in Area, %	Impact Energy ft-lbf(J) (c)	
1	0.35	0.40-0.90	0.050	0.050	0.15-0.35	—	—	—	0.06	10-95 (480-655)	35 (240)	20	38	15 (20)	Nozzles, manways, manholes, rings.
2	0.27	0.50-0.90	0.035	0.040	0.15-0.35	0.50-0.90	0.25-0.45	0.55-0.70	0.06	80-105 (550-720)	50 (340)	18	38	30 (41)	
2A	0.27	0.50-0.90	0.035	0.040	0.15-0.35	0.50-0.90	0.25-0.45	0.55-0.70	0.05	90-115 (620-790)	65 (450)	16	35	35 (48)	
3	0.15-0.25	1.20-1.50	0.035	0.040	0.15-0.35	0.40-0.80	—	0.55-0.70	0.06	80-105 (550-720)	50 (340)	18	38	30 (41)	
4	0.18	1.30	0.035	0.040	0.15-0.35	0.25	0.15	0.05	0.02-0.12	80-105 (550-720)	50 (340)	18	38	30 (41)	
5	0.10-0.20	0.30-0.80	0.035	0.040	0.50-1.00	—	1.00-1.50	0.45-0.65	—	80-105 (550-720)	50 (340)	18	38	15 (20)	
6	0.15	0.30-0.60	0.035	0.040	0.50	—	2.00-2.50	0.90-1.10	—	80-105 (550-720)	60 (410)	20	50	35 (48)	
6A	0.15	0.30-0.60	0.035	0.040	0.50	—	2.00-2.50	0.90-1.10	—	115-140 (790-965)	100 (690)	15	40	25 (34)	
7	0.23	0.20-0.40	0.035	0.040	0.30	2.75-3.90	1.25-2.00	0.40-0.60	0.03	105-130 (720-895)	85 (590)	18	48	35 (48)	
7A	0.23	0.20-0.40	0.035	0.040	0.30	2.75-3.90	1.25-2.00	0.40-0.60	0.03	115-140 (790-965)	100 (690)	16	45	35 (48)	
7B	0.23	0.20-0.40	0.035	0.040	0.30	2.75-3.90	1.25-2.00	0.40-0.60	0.03	90-115 (620-790)	70 (480)	20	48	35 (48)	
8	0.23	0.20-0.40	0.035	0.040	0.30	2.75-3.90	1.25-2.00	0.40-0.60	0.05-0.15	105-130 (720-895)	85 (590)	18	48	35 (48)	
8A	0.23	0.20-0.40	0.035	0.040	0.30	2.75-3.90	1.25-2.00	0.40-0.60	0.05-0.15	115-140 (790-965)	100 (690)	16	35	35 (48)	

(a) Maximum unless otherwise noted; (b) Minimum requirements for heat treated forgings; (c) Minimum averages for three Charpy V-notch specimens tested at 40 F (4 C); 2A is tested at 70 F (20 C).

ASTM A 592 — Forged Fittings and Parts

Class	Composition, % (a)										Mechanical Properties (b)				Typical Applications
	C	Mn	P	S	Si	Ni	Cr	Mo	B	Other	Tensile Strength, 10^3 psi (MPa)	Yield Strength, 10^3 psi (MPa)	Elongation, %	Reducton in Area, %	
A	0.15-0.21	0.80-1.10	0.035	0.040	0.40-0.80	—	0.50-0.80	0.18-0.28	0.0025	0.05-0.15 Zr	To 2½ in. (64 mm) incl.	100 (690)	18	45	Fittings, miscellaneous parts.
E	0.12-0.20	0.40-0.70	0.035	0.040	0.20-0.35	—	1.40-2.00	0.40-0.60	0.0015-0.005	0.04-0.10 Ti 0.20-0.40 Cu	115-135 (790-930) Over 2½ in. (64 mm)	90 (620)	17	40	
F	0.10-0.20	0.60-1.00	0.035	0.040	0.15-0.35	0.70-1.00	0.40-0.65	0.40-0.60	0.002-0.006	0.03-0.08 V 0.15-0.50 Cu	to 4 in. (102 mm) incl. 105-135 (720-930)				

(a) Maximum unless otherwise noted; (b) Minimum requirements (unless otherwise noted) for heat treated forgings.

Source: ASTM A 336, A 372, A 508, A 541, and A 592. Printed by permission of the American Society for

AISI-SAE Standard Carbon Steels

Free-Machining Grades

AISI No.	Composition*, %				SAE No.
	C	Mn	P	S	
Resulfurized					
1108	0.08 to 0.13	0.50 to 0.80	0.040 max	0.08 to 0.13	1108
1109	0.08 to 0.13	0.60 to 0.90	0.040 max	0.08 to 0.13	1109
1110	0.08 to 0.13	0.30 to 0.60	0.040 max	0.08 to 0.13	1110
1116	0.14 to 0.20	1.10 to 1.40	0.040 max	0.16 to 0.23	1116
1117	0.14 to 0.20	1.00 to 1.30	0.040 max	0.08 to 0.13	1117
1118	0.14 to 0.20	1.30 to 1.60	0.040 max	0.08 to 0.13	1118
1119	0.14 to 0.20	1.00 to 1.30	0.040 max	0.24 to 0.33	1119
1132	0.27 to 0.34	1.35 to 1.65	0.040 max	0.08 to 0.13	1132
1137	0.32 to 0.39	1.35 to 1.65	0.040 max	0.08 to 0.13	1137
1139	0.35 to 0.43	1.35 to 1.65	0.040 max	0.13 to 0.20	1139
1140	0.37 to 0.44	0.70 to 1.00	0.040 max	0.08 to 0.13	1140
1141	0.37 to 0.45	1.35 to 1.65	0.040 max	0.08 to 0.13	1141
1144	0.40 to 0.48	1.35 to 1.65	0.040 max	0.24 to 0.33	1144
1145	0.42 to 0.49	0.70 to 1.00	0.040 max	0.04 to 0.07	1145
1146	0.42 to 0.49	0.70 to 1.00	0.040 max	0.08 to 0.13	1146
1151	0.48 to 0.55	0.70 to 1.00	0.040 max	0.08 to 0.13	1151
Resulfurized and Rephosphorized					
1211	0.13 max	0.60 to 0.90	0.07 to 0.12	0.10 to 0.15	1211
1212	0.13 max	0.70 to 1.00	0.07 to 0.12	0.16 to 0.23	1212
1213	0.13 max	0.70 to 1.00	0.07 to 0.12	0.24 to 0.33	1213
1215	0.09 max	0.75 to 1.05	0.04 to 0.09	0.26 to 0.35	1215
12L14†	0.15 max	0.85 to 1.15	0.04 to 0.09	0.26 to 0.35	12L14

*When silicon is required, the following ranges and limits are commonly used: for nonresulfurized steels: up to 1015, 0.10% max; 1015 to 1025, 0.10% max, or 0.10 to 0.20%, or 0.15 to 0.30%, or 0.20 to 0.40%, or 0.30 to 0.60%; over 1025, 0.10 to 0.20%, or 0.15 to 0.30%, or 0.20 to 0.40%, or 0.30 to 0.60%; 1513 to 1524, 0.10% max, or 0.10 to 0.20%, or 0.15 to 0.30%, or 0.20 to 0.40%, or 0.30 to 0.60%; 1525 and over, 0.10 to 0.20%, or 0.15 to 0.30%, or 0.20 to 0.40%, or 0.30 to 0.60%. For resulfurized steels: to 1110, 0.10% max; 1116 and over, 0.10% max, or 0.10 to 0.20%, or 0.15 to 0.30%, or 0.20 to 0.40%, or 0.30 to 0.60%. It is not common practice to produce resulfurized and rephosphorized steels to specified limits for silicon because of its adverse effect on machinability.

Copper can be added to a standard steel.

Standard killed carbon steels, which are generally fine grain, may be produced with a boron addition to improve hardenability. Such steels can be expected to contain 0.0005% B min. These steels are identified by inserting the letter "B" between the second and third numerals of the AISI number — for example, 10B46.

†0.15 to 0.35% Pb. When lead is required as an added element to a standard steel, a range of 0.15 to 0.35%, inclusive, is generally used. Such a steel is identified by inserting the letter "L" between the second and third numeral of the AISI number.

Sources: American Iron & Steel Institute, New York; SAE Standard J403f.

Nonresulfurized Grades

AISI No.	Composition*, %				SAE No.
	C	Mn	P Max	S Max	
1005	0.06 max	0.35 max	0.040	0.050	1005
1006	0.08 max	0.25 to 0.40	0.040	0.050	1006
1008	0.10 max	0.30 to 0.50	0.040	0.050	1008
1010	0.08 to 0.13	0.30 to 0.60	0.040	0.050	1010
1011	0.08 to 0.13	0.60 to 0.90	0.040	0.050	1011
1012	0.10 to 0.15	0.30 to 0.60	0.040	0.050	1012
1013	0.11 to 0.16	0.50 to 0.80	0.040	0.050	1013
1513	0.10 to 0.16	1.10 to 1.40	0.040	0.050	1513
1015	0.13 to 0.18	0.30 to 0.60	0.040	0.050	1015
1016	0.13 to 0.18	0.60 to 0.90	0.040	0.050	1016
1017	0.15 to 0.20	0.30 to 0.60	0.040	0.050	1017
1018	0.15 to 0.20	0.60 to 0.90	0.040	0.050	1018
1518	0.15 to 0.21	1.10 to 1.40	0.040	0.050	1518
1019	0.15 to 0.20	0.70 to 1.00	0.040	0.050	1019
1020	0.18 to 0.23	0.30 to 0.60	0.040	0.050	1020
1021	0.18 to 0.23	0.60 to 0.90	0.040	0.050	1021
1022	0.18 to 0.23	0.70 to 1.00	0.040	0.050	1022
1522	0.18 to 0.24	1.10 to 1.40	0.040	0.050	1522
1023	0.20 to 0.25	0.30 to 0.60	0.040	0.050	1023
1524	0.19 to 0.25	1.35 to 1.65	0.040	0.050	1524
1025	0.22 to 0.28	0.30 to 0.60	0.040	0.050	1025
1525	0.23 to 0.29	0.80 to 1.10	0.040	0.050	1525
1026	0.22 to 0.28	0.60 to 0.90	0.040	0.050	1026
1526	0.22 to 0.29	1.10 to 1.40	0.040	0.050	1526
1527	0.22 to 0.29	1.20 to 1.50	0.040	0.050	1527
1029	0.25 to 0.31	0.60 to 0.90	0.040	0.050	1029
1030	0.28 to 0.34	0.60 to 0.90	0.040	0.050	1030
1033	0.29 to 0.36	0.70 to 1.00	0.040	0.050	1033
1034	0.32 to 0.38	0.50 to 0.80	0.040	0.050	—
1035	0.32 to 0.38	0.60 to 0.90	0.040	0.050	1035
1536	0.30 to 0.37	1.20 to 1.50	0.040	0.050	1536
1037	0.32 to 0.38	0.70 to 1.00	0.040	0.050	1037
1038	0.35 to 0.42	0.60 to 0.90	0.040	0.050	1038
1039	0.37 to 0.44	0.70 to 1.00	0.040	0.050	1039
1040	0.37 to 0.44	0.60 to 0.90	0.040	0.050	1040
1541	0.36 to 0.44	1.35 to 1.65	0.040	0.050	1541
1042	0.40 to 0.47	0.60 to 0.90	0.040	0.050	1042
1043	0.40 to 0.47	0.70 to 1.00	0.040	0.050	1043
1044	0.43 to 0.50	0.30 to 0.60	0.040	0.050	1044
1045	0.43 to 0.50	0.60 to 0.90	0.040	0.050	1045
1046	0.43 to 0.50	0.70 to 1.00	0.040	0.050	1046
1547	0.45 to 0.51	1.35 to 1.65	0.040	0.050	1547
1548	0.44 to 0.52	1.10 to 1.40	0.040	0.050	1548
1049	0.46 to 0.53	0.60 to 0.90	0.040	0.050	1049
1050	0.48 to 0.55	0.60 to 0.90	0.040	0.050	1050
1551	0.45 to 0.56	0.85 to 1.15	0.040	0.050	1551
1552	0.47 to 0.55	1.20 to 1.50	0.040	0.050	1552
1053	0.48 to 0.55	0.70 to 1.00	0.040	0.050	1053
1055	0.50 to 0.60	0.60 to 0.90	0.040	0.050	1055
1059	0.55 to 0.65	0.50 to 0.80	0.040	0.050	—
1060	0.55 to 0.65	0.60 to 0.90	0.040	0.050	1060
1561	0.55 to 0.65	0.75 to 1.05	0.040	0.050	1561
1065	0.60 to 0.70	0.60 to 0.90	0.040	0.050	1065
1566	0.60 to 0.71	0.85 to 1.15	0.040	0.050	1566
1069	0.65 to 0.75	0.40 to 0.70	0.040	0.050	1069
1070	0.65 to 0.75	0.60 to 0.90	0.040	0.050	1070
1572	0.65 to 0.76	1.00 to 1.30	0.040	0.050	1572
1074	0.70 to 0.80	0.50 to 0.80	0.040	0.050	1074
1075	0.70 to 0.80	0.40 to 0.70	0.040	0.050	1075
1078	0.72 to 0.85	0.30 to 0.60	0.040	0.050	1078
1080	0.75 to 0.88	0.60 to 0.90	0.040	0.050	1080
1084	0.80 to 0.93	0.60 to 0.90	0.040	0.050	1084
—	0.80 to 0.93	0.70 to 1.00	0.040	0.050	1085
1086	0.80 to 0.93	0.30 to 0.50	0.040	0.050	1086
1090	0.85 to 0.98	0.60 to 0.90	0.040	0.050	1090
1095	0.90 to 1.03	0.30 to 0.50	0.040	0.050	1095

Selecting Plate Steels for Nuclear Steam Supply Systems

By ROBERT H. STERNE JR., WILLIAM H. GORMAN JR., and ROBERT L. BROOKS

Carbon, alloy, and clad steel plates play vital roles in the construction of nuclear powerplants. Applications include heavy-wall reactor vessels, steam generators, pressurizers, containment structures, heat exchangers, and piping.

Two types of nuclear steam supply systems (NSSS) are prevalent in the United States: the light-water-cooled, pressurized water reactor (PWR) and boiling water reactor (BWR).

In both, steam is produced from water heated under pressure by the neutron flux generated in the core. In PWR systems, steam is generated in a secondary heat exchanger, and then piped to a turbine generator. In BWR systems, steam generated inside the primary reactor vessel is piped directly into the turbine.

Other system designs in commercial operation in the Western Hemisphere include the high-temperature, gas-cooled reactor (HTGR) and the Canadian-built, heavy-water moderated reactor.

This article covers applications for plate steels in these four nuclear steam supply systems.

Primary Reactor Vessels

PWR System – Current PWR pressure vessels must operate under pressures of about 2,000 psi at 550 to 600 F.

Historically, these vessels have been fabricated from Mn-Mo-Ni alloy steel plates, usually compositions described by ASTM A302 and A533. Thicknesses vary from 4 in. to nearly 12 in. at large nozzle openings; diameters range from 11 to 15 ft.

Other materials approved for construction of these Class 1 nuclear vessels include ASTM A542 and and A543 (see Table I). These more highly alloyed steels exhibit increased strength and develop specified properties in extremely heavy plate gages.

Because PWR reactors presently weigh from 300 to 450 tons, the use of higher strength grades is preferred for reduced vessel weight and wall thickness, making for lower fabrication costs, easier transportation, and less expensive foundations.

Other factors affecting this economic relationship include base and weld metal costs, notch toughness requirements, and potential irradiation effects.

BWR System – Boiling water reactor vessels can use plate steels less thick than PWR systems because of lower operating pressures.

Mr. Sterne is market development manager; Mr. Gorman is manager, application engineering; and Mr. Brooks is application engineer, Lukens Steel Co., Coatesville, Pa.

Virtually the entire steam supply system for this pressurized water reactor uses nuclear-quality plate steels. The system is a major element in Consumers Power Co.'s 710 Mwe Palisades plant.

Table I — Steels for PWR and BWR Primary Reactor Vessels, to ASME Section III Rules

Specification (Pressure Vessel Quality)	ASTM A533 Grade B	ASTM A533 Grade C	ASTM A542*	ASTM A543† Grade B
Class 1				
Min yield strength, psi	50,000	50,000	85,000	85,000
Tensile strength, psi	80,000-100,000	80,000-100,000	105,000-125,000	105,000-125,000
Elongation, %	18	18	14	14
Class 2				
Min yield strength, psi	70,000	70,000	100,000	100,000
Tensile strength, psi	90,000-115,000	90,000-115,000	115,000-135,000	115,000-135,000
Elongation, %	16	16	13	14
Composition, %				
Carbon	0.25 max	0.25 max	0.15 max	0.23 max
Manganese	1.15-1.50	1.15-1.50	0.30-0.60	0.40 max
Silicon	0.15-0.30	0.15-0.30	0.50 max	0.20-0.35
Nickel	0.40-0.70	0.70-1.00	—	2.60-3.25 (to 4 in.) 3.00-4.00 (over 4 in.)
Chromium	—	—	2.00-2.50	1.50-2.00
Molybdenum	0.45-0.60	0.45-0.60	0.90-1.10	0.45-0.60
Vanadium	—	—	—	0.03 (0.05 max)
Required heat treatment, F	Quench 1,550-1,800, temper 1,100 min	Quench 1,550-1,800, temper 1,100 min	Quench 1,650-1,850, temper 1,050 min	Quench or double quench 1,650-1,850, temper 1,100 min
Max specified gage	Not specified	Not specified	Not specified	Not specified
Charpy V-notch impact requirement, ft-lb	30	30	35	35

*ASME Case Interpretation 1414 approves Class 1 only.
†See ASME Case Interpretation 1358 for Class 1 vessels and 1408 for Class MC vessels.

Here thicknesses range from 3 to 8 in.

Specifications governing these plate thicknesses include: 1,250 psi pressure, diameters from 18 to 21 ft, and heights from 55 to 70 ft. Weights range from 350 to 850 tons for 500 and 1,100 megawatt systems.

Most BWR vessels manufactured since 1965 use Mn-Mo-Ni plate steels covered by ASTM A533 Grade B, Class 1. A recent example is shown in Fig. 1. Those manufactured prior to 1965 use this specification's predecessor – A302 Grade B (Mn-Mo steels).

Specifications for nuclear-quality A533 call for vacuum degassing to provide internal soundness. Vacuum degassing also allows flexibility in adjusting final chemical analysis to close tolerances by adding small quantities of alloying elements through the degasser.

Proposed BWR vessels require plates up to 12 in. thick. Such vessels could be as large as 36 ft in diameter and 100 ft high. Current pressure vessel grades approved for construction by ASME Section III (Table I) will be capable of meeting these requirements.

Because these future behemoths

Table II — Steels for HTGR and Heavy-Water Primary Reactor Vessels, to ASME Section III Rules

Specification	ASTM A537 Grade B	ASTM A387 Grade D	ASTM A516 Grade 55
Min yield strength, psi	60,000	45,000	30,000
Tensile strength, psi	80,000-100,000	75,000-100,000 (to 8 in.)	55,000-65,000 (to 12 in.)
Composition, %			
Carbon	0.24 max	0.15 max	0.26 max*
Manganese	0.70-1.35 (to 1½ in. incl.) 1.00-1.60 (over 1½ in. to 2½ in.)	0.30-0.60	0.60-1.20†
Silicon	0.15-0.50	0.50 max	0.15-0.30
Nickel	0.25 max‡	—	—
Chromium	0.25 max‡	2.00-2.50	—
Molybdenum	0.08 max‡	0.90-1.10	—
Required heat treatment, F	Quench and temper (1,100 min)	Normalize (§) and temper (1,200 min)	Normalize
Max specified gage, in.	2½	Not specified	12
Charpy V-notch impact requirement, ft-lb	30	30	15

*Over 8-12 in., less for lighter gages. †0.60-0.90, ½ in. and under. ‡Incidental elements.
§Liquid quenching may be used for plates 4 in. and over.

overreach today's transport facilities, their construction most likely will require on-site assembly.

For on-site assembly, use of plates heat treated at the mill will insure the steel's compliance to the ASME Nuclear Code before fabrication. In addition, use of such plates and subsequent cold forming operations can save fabricators the time-consuming heat treatments required after hot forming.

HTGR System — The high-temperature, gas-cooled reactor, in its latest commercial design, uses a reinforced, prestressed concrete reactor vessel. To insure gas tightness, it's lined with A537 Grade B, a quenched and tempered carbon steel plate with the notch toughness required by ASME Section III to provide resistance to degradation resulting from low-temperature irradiation.

In addition, A387 Grade D, a heat-resistant 2.25Cr-1Mo alloy plate, is used in the system's integral heat exchangers. Data on these and other approved grades are shown in Table II.

Heavy Water System — Atomic Energy of Canada Ltd. builds natural uranium-fueled, heavy-water moderated reactor systems. One unusual characteristic of this design is on-load refueling — fuel elements terminate in a calandria consisting of a massive circular steel plate, 13

Fig. 1 — As BWR primary reactor vessels become larger, transportation difficulties will most likely force on-site assembly. For example, it took three weeks by barge and crawler-type transporter to deliver this 8 in. thick, 62 ft long vessel to Philadelphia Electric Co.'s Peach Bottom Atomic Power Station No. 3. It's fabricated from ASTM A533 Grade B, Class 1 plate steel.

Table III — Steels for PWR Steam Generators and Pressurizers, to ASME Section III Rules

Specification	ASTM A533 Grade A	ASTM A533 Grade B	ASTM A516 Grade 70
Class 1			
Min yield strength, psi	50,000	50,000	—
Tensile strength, psi	80,000-100,000	80,000-100,000	—
Class 2			
Min yield strength, psi	70,000	70,000	—
Tensile strength, psi	90,000-115,000	90,000-115,000	—
Min yield strength, psi	—	—	38,000
Tensile strength, psi	—	—	70,000-85,000
Composition, %			
Carbon	0.25 max	0.25 max	0.31 max*
Manganese	1.15-1.50	1.15-1.50	0.85-1.20
Silicon	0.15-0.30	0.15-0.30	0.15-0.30
Nickel	—	0.40-0.70	—
Molybdenum	0.45-0.60	0.45-0.60	—
Required heat treatment, F	Quench from 1,550-1,800, temper 1,100 min†	Quench from 1,550-1,800, temper 1,100 min†	Normalize‡
Max specified gage, in.	12	8	8
Charpy V-notch impact requirement, ft-lb	30	35	20

*4-8 in., lower for lighter gages. †Double quench in gages over 8-9 in. ‡Quenching and tempering may be used (paragraph NB2170).

Table IV — Steels for Containment Structures, to ASME Section III Rules

Specification	ASTM A442		ASTM A516				ASTM A537	
	Grade 55	Grade 60	Grade 55	Grade 60	Grade 65	Grade 70	Grade A	Grade B
Min yield strength, psi	30,000	32,000	30,000	32,000	35,000	38,000	50,000	60,000
Tensile strength, psi	55,000-65,000	60,000-72,000	55,000-65,000	60,000-72,000	65,000-77,000	70,000-85,000	70,000-90,000	80,000-100,000
Ladle composition, %								
Carbon max								
to ½ in.	0.22*	0.24*	0.18	0.21	0.24	0.27	0.24	0.24
over ½ in.-2 in.	0.24†	0.27†	0.20	0.23	0.26	0.28	0.24†	0.24†
over 2 in.-4 in.	—	—	0.22	0.25	0.28	0.30	—	—
over 4 in.-8 in.	—	—	0.24	0.27	0.29	0.31	—	—
over 8 in.-12 in.	—	—	0.26	—	—	—	—	—
Manganese								
to ½ in.	0.80-1.10	0.80-1.10	0.60-0.90	0.60-0.90‡	0.85-1.20	0.85-1.20	0.70-1.35	0.70-1.35
over ½ in.-1 in.	0.80-1.10	0.80-1.10	0.60-1.20	0.85-1.20	0.85-1.20	0.85-1.20	0.70-1.35	0.70-1.35
over 1 in.-1½ in.	0.60-0.90	0.60-0.90	0.60-1.20	0.85-1.20	0.85-1.20	0.85-1.20	0.70-1.35	0.70-1.35
over 1½ in.-2½ in.	—	—	0.60-1.20	0.85-1.20	0.85-1.20	0.85-1.20	1.00-1.60	1.00-1.60
over 2½ in.-12 in.	—	—	0.60-1.20	0.85-1.20§	0.85-1.20§	0.85-1.20§	—	—
Phosphorus max	0.04	0.04	0.035	0.035	0.035	0.035	0.035	0.035
Sulfur max	0.05	0.05	0.040	0.040	0.040	0.040	0.040	0.040
Silicon								
to 1 in.	—	—	0.15-0.30	0.15-0.30	0.15-0.30	0.15-0.30	0.15-0.50	0.15-0.50
over 1 in.	0.15-0.30	0.15-0.30	0.15-0.30	0.15-0.30	0.15-0.30	0.15-0.30	0.15-0.50	0.15-0.50

*To 1 in. †Over 1 in.-1½ in., A442; to 2½ in. incl., A537. ‡0.85-1.20 Mn permissible. §8 in. max.

in. thick and about 20 ft. in diameter.

Plates are supplied in half circles, two of which are joined together and machined with hundreds of through-gage apertures. The specification used is A516 Grade 55, a notch-tough, vacuum-degassed carbon steel.

Steam generators use A533, while A543 has been used for tube sheets in secondary heat exchangers.

Additional Components in the NSSS

BWR systems, generating steam internally in the reactor pressure vessel, are the only systems not requiring external generators or pressurizers. This section is confined to components in PWR systems.

Steam Generators — PWR steam generators are of two kinds: conventional U-tube exchangers and once-through steam generators that superheat the steam slightly.

Materials most commonly used for generators are A533 Grades A and B, and A516 Grade 70 (see Table III), depending on vessel geometry and configuration.

Plate thicknesses for once-through-types have reached 9 in., for U-tube generators, greater than 8 in.

The method of construction for steam generators varies. Sometimes unheat-treated plates are hot formed to the desired radius and then heat treated prior to, or after, welding.

Other times plates are fabricated, fully heat treated and tested, by cold forming and welding. This procedure requires only stress relieving of completed welds.

PWR steam generators (Fig. 2, top) are huge — often larger than the reactors. Diameters range from 10 to 18 ft and heights can soar to 60 ft. Weights from 300 to 600 tons are typical.

Pressurizer — In a PWR nuclear system, the pressurizer maintains pressure at start-up and during operation. At the time of plant start-up, electrically heated water is used to maintain system pressure as temperature increases.

During operation, the pressurizer keeps the pressure constant, acting somewhat like a reserve tank. During this phase, the pressurizer is about half filled with water, half with steam. As electrical output of the plant increases, water flows into the PWR loops from the pressurizer; the reverse occurs when output is reduced.

Pressurizers are typically 7 to 9 ft in diameter (Fig. 2, bottom) and use plates from 4 to 6 inches thick. A533 Grades A and B are often specified.

Nuclear Piping

Steel piping is used in all water-cooled nuclear systems to carry steam or water from the primary reactor to either a secondary steam generator or directly into a turbine. Internal pressures up to 2,000 psi and corrosion-inducing chemicals in steam or water call for strong, corrosion-resistant alloys.

An economical choice for this application is roll-bonded stainless-clad steel. Here, a single plate combines the corrosion resistance of stainless steel with the economy of carbon or alloy steel backing.

Specifications usually call for Charpy V-notch tested carbon steel backing such as A516 Grade 70 for pressure-retaining components, and about ¼ in. thick type 304 or 304L cladding, for corrosion resistance. Total thickness of clad plate for piping is usually 2 in. to more than 4 in.

An additional safety advantage to using clad steel relates to stress-corrosion cracking. If this phenomenon occurs in solid stainless steel pipes, cracks can propagate entirely through the wall, resulting in failure. When clad piping is used, stress-corrosion cracks are usually arrested at the bond line. This characteristic also applies to vessels such as storage tanks and nuclear fuel pools.

Another benefit is found in welding — 4 in. thick joints can be made

Fig. 2 — The shell of the 426-ton, 60 ft high PWR steam generator at top is fabricated from pressure-vessel-quality carbon steel. Compared to steam generators and reactor vessels, pressurizers (bottom) are relatively simple to fabricate.

using stainless steel filler metal only on the cladding; lower cost carbon steel wires can be used on the backing.

Furthermore, ASME Codes permit higher allowable stresses for A516 backing over solid stainless steel. The result is a substantial reduction in the pipe's wall thickness. For example, at 650 F, type 304L has an allowable stress of 12,700 psi while A516 Grade 70 has an allowable stress of 18,400 psi — a 45% improvement.

Cost too favors clad steel plate construction. The difference between solid stainless and stainless-clad steel will vary as a function of size, cladding thickness, and backing steel specification. In general, however, savings of at least several hundred dollars per ton of plate are possible by using clad.

Auxiliary Vessels

A number of auxiliary items complement nuclear systems. Examples include accumulators or flooding tanks and chemical storage and make-up tanks. To maintain purity, many of these vessels are constructed of clad steels.

Accumulators — Accumulator tanks are designed to inject borated water into the primary reactor vessel, forestalling core meltdown in the event of coolant loss.

These pressure vessels operate under static pressures of several hundred psi and must be constructed of materials that can withstand this pressure and maintain the borated water's purity.

Type 304 or 304L austenitic stainless steel cladding, with carbon steel backing, is commonly used. The cladding prevents water contamination and corrosion of the steel backing. Plate thicknesses from ⅞ in. to nearly 3 in. have been used.

As with piping, clad steel in these applications can offer significant cost savings. Available cladding materials include 300 and 400 series stainless steels, copper, cupronickel, Inconel, Monel, and nickel; backing materials include a variety of carbon and low-alloy steels.

Fuel Storage Tanks — Fuel elements (spent and new) must be stored under water for safe handling. When reactors are initially fueled, and during subsequent reloads, the cavity surrounding the reactor vessel must also be flooded with water.

These storage tanks and cavities require linings that remain uncontaminated and facilitate cleaning.

Steel clad with an austenitic stainless steel can provide the necessary corrosion resistance, at lower cost than solid austenitic stainless. And because clad steel is available in wider plates than solid cold-rolled stainless, fabricating costs can be reduced. The most widely used combination is type 304 or 304L stainless on carbon steel backing.

Clad steel is normally provided annealed with a blasted finish, but when optimum smoothness is required, special polished finishes can be ordered.

Containment

To protect against mechanical failure in nuclear power plants, leak-tight outer containment structures are specified.

This additional safeguard provides an outer barrier to fission products created during normal operation and guards against accidental release of larger amounts of radioactivity.

Because a pipe or reactor rupture would release large quantities of high-pressure steam, containment vessels are designed with large volumes to allow for steam expansion to lower pressures. Depending on containment type, normal pressures may range from 15 psi to about 60 psi. Compositions and properties of steels approved for containment structures are given in Table IV.

BWR Containment — One type of containment for a boiling water reactor system consists of a lightbulb-shaped central structure, containing the reactor vessel, surrounded by a torus. Steam escaping from the reactor enters the torus through headers and condenses, filling its bottom

Specification
Min yield strength, psi
Tensile strength, psi
Composition, % Carbon max
Manganese
Silicon
Nickel
Chromium
Molybdenum
Vanadium
Copper
Boron
Required heat treatment
Max available gage, in.

portion and reducing the pressure.

Nearly all of this type of containment has been constructed from A516 Grade 70 steel plates furnished heat treated for improved notch toughness.

Plates under about 2 in. thick are normalized, while heavier plates are often quenched and tempered to maintain toughness levels comparable to those in the lighter, normalized material. A516 steel can be specified to provide 20 ft-lb Charpy V-notch values at temperatures as low as —30 to —50 F, depending on thickness and specimen orientation.

PWR Containment — In contrast to the proprietary design described above, there is a variety of PWR containment structures; they can, however, be divided into two basic types: full-pressure steel containment, and steel-lined, reinforced concrete.

One manufacturer of PWR steam supply systems has developed an ice condenser containment. Here, pressure is reduced by condensing steam with tons of ice in a lattice-type structure inside the containment vessel.

Reported advantages of such structures are reduced design pressures and associated lower costs of the vessel's pressure-containing portion. And the ice condenser can be adapted readily to reinforced-concrete designs with steel plate linings.

PWR containment vessels were originally spherical and fabricated from steel plate. As systems grew larger, cylindrical containment vessels and concrete-steel designs gained in favor. "Concrete" containment actually uses steel to withstand pressures caused by expansion in the event of an accident — reinforcing bars and post-tensioned tendons normally carry the load. Relatively thin steel plate linings, ¼ to ½ in. thick, are typical.

Specifications for lining plate vary. But as a general rule, low-strength plates with high ductility and good notch toughness are most desirable.

ASTM A442 Grade 55 or 60, and A516 Grade 55 or 60, are commonly used. Charpy impact testing and normalized fine-grain structures are often specified for A442. Additionally, these carbon steels have good weldability because of their low carbon contents.

Steel plates also are used to anchor supports for concrete post-tensioning systems. Both A516 Grade 70 and higher strength, proprietary grades such as T-1 (A517) have been considered here.

In 1969-71, a number of PWR plants were sold using all-steel, free-standing containments, reversing the trend toward concrete construction. Reported advantages of such designs include faster containment construction, known cost for the structure, and one-contractor responsibility.

These steel structures are typically 140 ft in diameter and over 200 ft tall, cylindrically shaped with elliptical heads and surrounded by an ordinary reinforced-concrete radiation shield.

Plate steels with higher strength and improved toughness such as ASTM A537 Grades A and B will be used.

Structural Steel Applications

Miscellaneous components may require structural steel plates. Examples are steam generator supports, reactor vessel supports, neutron shielding vessel support skirts, and even skids or carriages used to transport reactor vessels.

Reinforcing steel fabrications used for equipment and personnel hatches have utilized T-1 high-strength steels. Other specifications (Table V) seeing structural use include A36, A515, A516, A588, A533, A542, and A543.

Table V — Plate Steels for Structural Uses in Nuclear Powerplants

ASTM A36	ASTM A515 Grade 70	ASTM A516 Grade 70	ASTM A517 Grade F	ASTM A537 Grade A	ASTM A588 Grade A
36,000 (to 8 in.) 32,000 (over 8 in.-16 in.)	38,000	38,000	100,000 min	50,000	50,000 (to 4 in.) 46,000 (over 4 in.-5 in.)
58,000-80,000 (all gages)	70,000-85,000	70,000-85,000	115,000-135,000	70,000-90,000	70,000 min (to 4 in.) 67,000 min (over 4 in.-5 in.)
Varies to 0.29	0.31 (to 1 in.) 0.33 (1 in.-2 in.) 0.35 (2 in.-8 in.)	0.27 (to ½ in.) 0.28 (over ½ in.-2 in.) 0.30 (over 2 in.-4 in.) 0.31 (over 4 in.-8 in.)	0.10-0.20	0.24	0.10-0.19
1.20 max	0.90 max	0.85-1.20	0.60-1.00	0.70-1.35 (up to 1½ in.) 1.00-1.60 (over 1½ in.-2½ in.)	0.90-1.25
0.15-0.30 (over 1½ in.)	0.15-0.30	0.15-0.30	0.15-0.35	0.15-0.50	0.15-0.30
—	—	—	0.70-1.00	0.25 max	—
—	—	—	0.40-0.65	0.25 max	0.40-0.65
—	—	—	0.40-0.60	0.08 max	—
—	—	—	0.03-0.08	—	0.02-0.10
—	—	—	0.15-0.50	0.35 max	0.25-0.40
—	—	—	0.002-0.006	—	—
Optional	Normalize (over 2 in.)	Normalize (over 1½ in.)	Quench and temper	Normalize	Optional
16	8	8 (fine grain practice)	2½	2½	8

Section III: Cast Irons and Steels

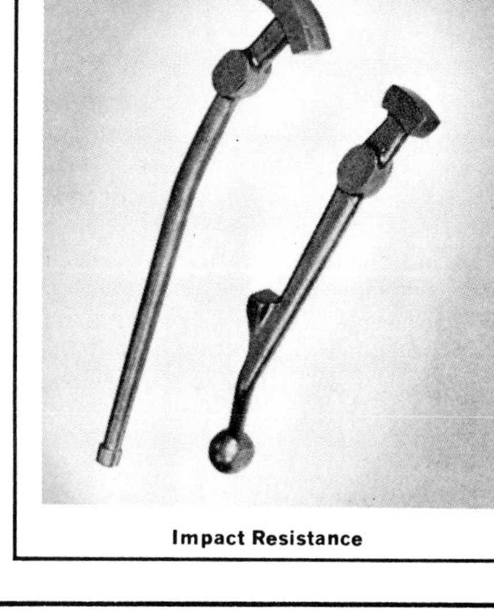
Impact Resistance

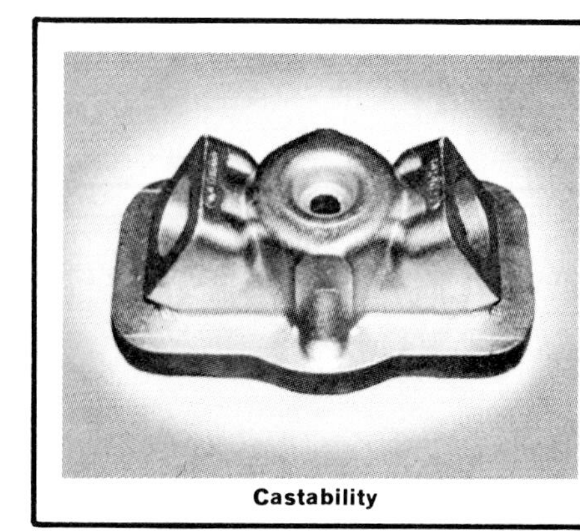
Castability

Iron Castings

By HARRY E. CHANDLER

In the current revolution in materials, change is both natural and constant. And we generally don't "put the pieces together" in a given field until we have some reason to step back and take a considered look at the past, present, and future of individual developments. Such is the case in iron casting.

A statement, made recently, prompted a look at this technology. It ran, "Over the next 40 years, cast iron, the first cast composite made by man, will still be the cheapest structural material and still maintain its position in industry."

The ongoing revolution implied in that prediction, we discovered, was undoubtedly triggered in the late forties by the introduction of ductile iron as a commercial material. It and ensuing developments in processing technology, such as shell molding, combined to spur the general resurgence of interest in castings so evident today.

In the following article, the spotlight is on ductile iron, the prime mover in this revolution. By any standard, its record has been sensational. Industry production, for example, was 173,000 tons in 1959. This year, the total is expected to cross the 1-million-ton mark. Such progress, as indicated by the adjoining graphic case histories, is largely explained by ductile's unusually broad utility.

When you talk with producers and users of ductile iron, they are sure to advise you sooner or later, "Don't oversell this material." You can see the wisdom in their admonition as they go on to explain, "There is a natural tendency to oversell because ductile has earned what must be termed a unique position among engineering materials. It is not a cure-all, but it is generally agreed that no other ferrous material can match its combination of castability and mechanical properties."

Mr. Chandler is managing editor, Metal Progress.

"Versatility" is the key word used by Harvey E. Henderson, technical director, Lynchburg Foundry Co. Div., Woodward Co., a division of Mead Corp., Lynchburg, Va. He adds, "Ductile has the processing advantages of cast iron, engineering properties resembling those of steel, plus some of the desirable properties of gray iron."

Properties – Ductile, like gray and malleable iron, can be described as steel plus graphite. The presence of graphite contributes directly to the lubrication of rubbing surfaces and provides reservoirs to accommodate and hold lubricants. This means good resistance to mechanical wear. Graphite also contributes to machinability because it acts as a lubricant during cutting and also tends to break up chips. Like gray iron,

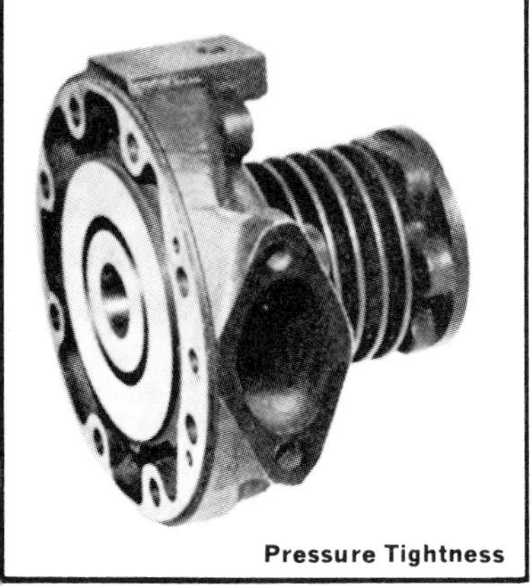
Pressure Tightness

Surface Hardness

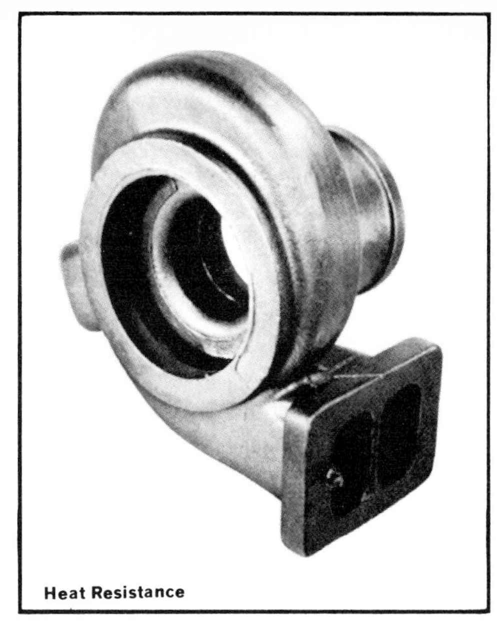
Heat Resistance

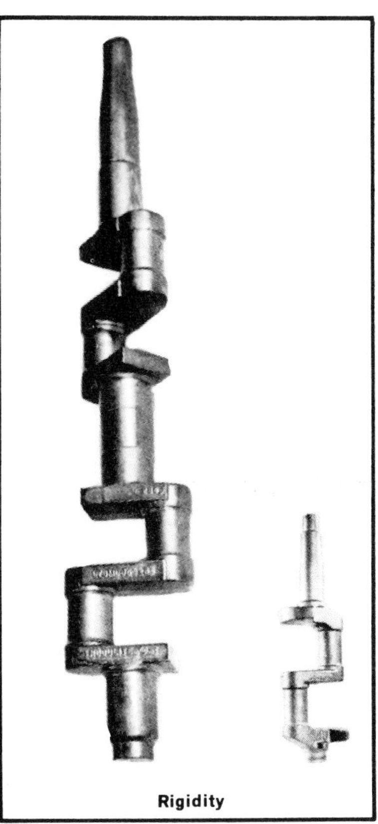
Rigidity

Surface Finish

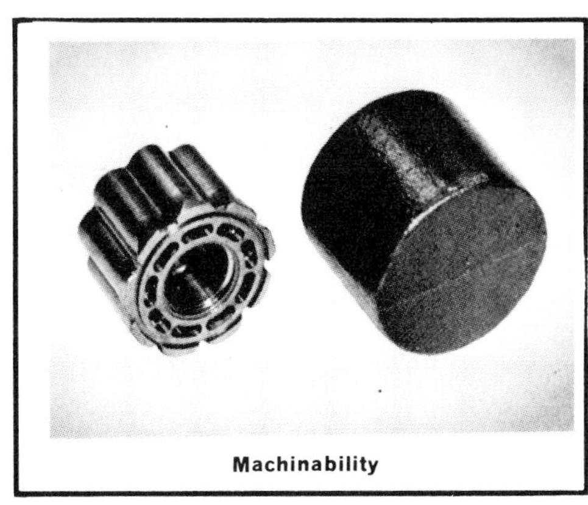
Machinability

Impact Resistance—Machine tool control lever, 15 in. long, 2 lb, annealed for 20 to 25% elongation, grade 60-40-18.

Pressure Tightness—Air compressor top head, 127 lb, gas-pressure tested to 2,400 psi, grade 80-60-03.

Surface Hardness—Compacting press ejection cam, 170 lb, flame hardened to Rc 52-58, grade 80-55-06.

Heat Resistance—Turbocharger housing for service at 1,500 F, 31 lb, Ni-Resist ductile iron.

Castability—Helicopter blade control, 3.5 lb, grade 100-70-03.

Rigidity—Crankshafts: 18 in. long, 9.2 lb, grade 80-55-06; 4 ft 2 in. long, 125 lb, grade 120-90-02.

Surface Finish—High-pressure valve body cast to close tolerances, 30 lb, grade 60-45-15.

Machinability—Hydraulic pump cylinder, 9 oz, grade 100-70-03.

All of these castings are products of Hamilton Foundry Inc., Hamilton, Ohio. Its Ni-Resist is an alloy containing nickel and chromium.

ductile also has inherent corrosion resistance.

Important engineering properties include high strength, toughness, ductility, hardenability, damping characteristics, and pressure tightness.

"Ductile," Mr. Henderson explains, "is a eutectic alloy, which means it has a low melting point, good fluidity, and castability. The most intricate, complex castings with light sections can be made with it."

Within this context, it becomes abundantly clear why the word "versatile" is chosen to typify ductile iron, and you begin to see why it is so interesting to materials engineers. As the products pictured on these two pages indicate, the material can be tailored to fit an unusually broad diversity of needs.

Range – There are five basic ASTM grades: 120-90-02, 100-70-03, 80-55-06, 65-45-12, and 60-40-18. (All available grades are listed in the Data Sheet on p. 74). Generally, composition is not specified. Specifications are based on mechanical properties, in this order: tensile strength, 10^3; yield strength, 10^3; and percentage of elongation in 2 in. The 60-40-18 grade, for example, has a minimum tensile strength of 60,000 psi, a minimum yield strength of 40,000 psi, and 18% elongation.

Typical applications of these grades cited by James D. Voss, process controls manager, Hamilton

Foundry Inc., Hamilton, Ohio, add further documentation of ductile's utility:

65-45-12: pressure castings; compressor bodies; pipe; pipe fittings; valves; cylinders; pump bodies; connecting rods (nonautomotive); shock-resistant parts for automotive, agricultural, electrical, railroad, machine, marine, and general use; plus high-temperature applications requiring toughness and ductility.

80-55-06: gears, cams, bearings, pistons, crankshafts, sheaves, sprockets; wear and strength applications in the automotive, aeronautical, diesel, agricultural, heavy machinery, mining, paper, textile, and related industries.

100-70-03: gears, crankshafts, camshafts, pistons, and agricultural implement parts such as bolsters, bolster forks, ratchets, governor weights, track shoes, and reactor brake drums.

120-90-02: pinions, gears, cams, dies, machine guides, track rollers, idlers, tractor steering gear arms, drill columns, and pumps.

60-40-18: recommended for same applications as 65-45-12, where resistance to severe thermal shock and maximum toughness are required.

Next Plateau – Even broader versatility is being obtained with alloys, special compositions, and special heat treatments.

There is only one essential difference between the compositions of gray and ductile irons: the presence of 0.035 to 0.050% Mg in ductile (see Table I). Manganese, nickel, molybdenum, and copper are common alloying elements. For example, one austenitic alloy grade, D2, contains 3.00 total carbon max, 1.50-3.00 Si, 0.70-1.25 Mn, 0.08 P max, 18.00-22.00 Ni, and 1.75-2.75 Cr. Brinell hardness ranges from 139 min to 202 max. Applications include valve stem bushings; valve and pump bodies in petroleum, salt-water, and caustic service; manifolds; turbocharger housings; and air compressor parts.

Table I — Typical Analyses of Irons Poured at Lynchburg

Element	Unalloyed Gray Iron	Ductile Iron
Total carbon	3.25-3.45	3.50-3.80
Sulfur	0.10 max	Not run *
Silicon	1.80-2.30	2.30-2.60
Nickel	0.05-0.15	0.35-0.65
Copper	0.20-0.40	0.10-0.15
Magnesium	None	0.035-0.050
Aluminum	0.012-0.018	0.017-0.023
Phosphorus	0.12 max	0.08 max
Molybdenum	0.10-0.20	Trace
Chromium	0.05-0.20	0.08 max
Manganese	0.60-0.90	0.20-0.40
Vanadium	0.005-0.020	0.005-0.015
Titanium	0.020-0.040	0.02-0.03
Tin	0.006-0.010	Trace

* Base analysis was 0.02% S max.

The effects of various amounts of molybdenum in D2 are shown in Fig. 1.

Here is an example of a special heat treatment. The part, an agricultural idler wheel, must have adequate wear resistance because it guides the track chain. Impact strength is needed because the tractor may drop several feet suddenly. The starting material, a pearlitic ductile, was treated in this manner: induction heat 5 min, oil quench, and temper for 1 hr at 950 F. The result: a minimum hardness of Rc 35 to a depth of ⅛ in. beneath the wearing surface of the tread. Considerable savings were realized through the combination of castability, machinability, and hardenability.

Where Ductile Gets Its Properties

The essential difference between gray and ductile iron is the magnesium in ductile. There are, of course, minor differences in other elements, such as sulfur and manganese.

"In gray cast iron," Mr. Henderson explains," the graphite form is flake-shaped; and these flakes, which have practically no strength, act as notches or sharp cracks and extensively weaken the material. In the case of ductile, the small amount of magnesium present causes the

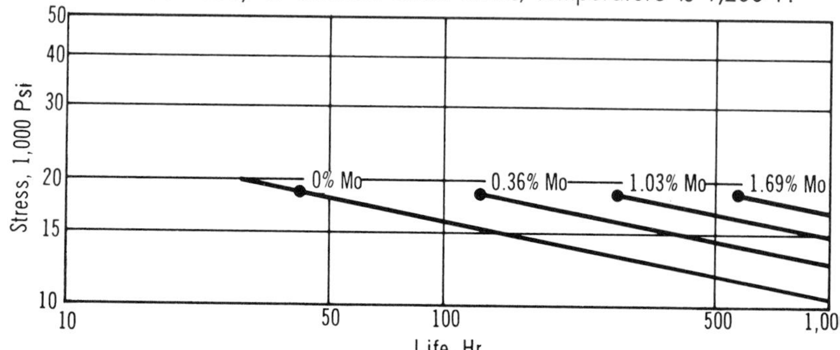

Fig. 1 — Various amounts of molybdenum affect the life of D2 ductile iron alloy at different stress levels; temperature is 1,200 F.

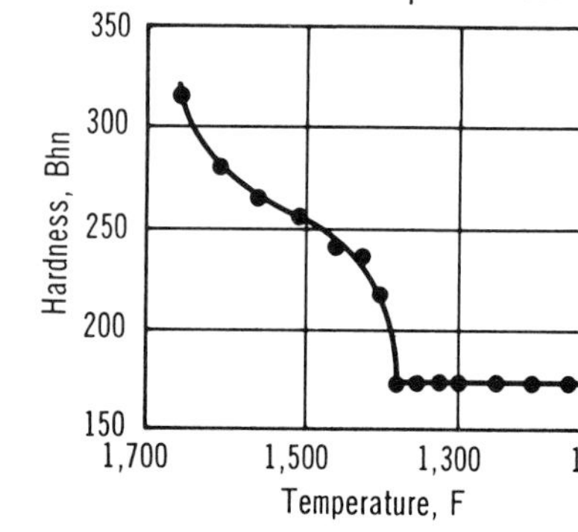

Fig. 2 — This curve gives you an idea of the hardnesses that can be obtained with ductile iron by austenitizing at 1,650 F and air quenching from several temperatures.

graphite to precipitate in the form of near spheres or spheroids. They give it its high strength and ductility."

Ductile iron, like steel, is an elastic material; "and stress is proportional to strain under loads up to the proportional limit," states Mr. Henderson. "Gray iron, on the other hand, does not follow Hooke's law because the graphite, in addition to interrupting the matrix, causes internal notches which act to concentrate stress locally when a load is applied."

The microstructure of the matrix (percentage of ferrite, pearlite, or martensite) depends upon the heat treatment and mechanical properties desired.

Acicular ductile iron is used in some instances. Steel mill rolls weighing up to 68,000 lb are an example. Requirements include resistance to wear at high temperatures, bending stresses, and shock. Here is a typical composition: 3.00-3.50 total carbon, 1.0-2.50 Si, 0.3-0.9 Mn, residual-1.50 Cr, 0.25-1.0 Mo, and 1.5-4.75 Ni.

Ductile need not be heat treated to produce a specific grade, but heat treatment reduces variations in microstructure and hardness. The material's response to heat treatment is shown in Fig. 2.

In ductile iron with 90% types I and II graphite, and no carbides, says Mr. Henderson, the Brinell hardness of a casting is a good indicator of mechanical properties. (In fact, SAE has proposed specifications based on hardness and microstructure.) Here is the hardness range for the five general grades:

Grade	Bhn Range
60-40-18	149 to 187
65-45-12	170 to 207
80-55-06	179 to 248
100-70-03	217 to 269
120-90-02	240 to 300

Initially, control of nodularity was a problem in the production of ductile iron. However, the desired graphite structures are now obtained routinely with established industry practices. (Procedures used by GM's Central Foundry Div. in the production of crankshafts are described on p. 293.)

Fig. 3 — Lynchburg Foundry has used the shaking ladle method to desulfurize its ductile iron. Calcium carbide is added to the melt.

As stated previously, ductile iron is not a cure-all. If the user wants to avoid heat treatment, he may encounter variations in properties. For example, the properties of as-cast 80-55-06 can vary in tensile strength from 107,000 to 80,000 psi; in yield strength, from 73,000 to 55,000 psi; in elongation, from 6.0 to 10.5%; and in hardness, from Bhn 243 to 180. Carbides can be a problem in thin sections of as-cast pearlitic ductile castings.

Tiny pits can be a disadvantage where surfaces are critical.

Several Engineering Case Histories

Ductile iron is competing with cast steel, steel forgings, high-strength gray iron, and steel fabrications, including weldments, in parts ranging from a fraction of a pound (see the pump cylinder on p. 61) to steel mill rolls (previously cited) weighing up to 68,000 lb.

Pressure pipe (it doesn't break when it is dropped) accounts for more than a third of total production. In 1967, for example, the product took 300,000 tons of the 850,000 tons produced in this country. Other major markets include motors, vehicles, and parts; farm machinery and equipment; paper industry machinery; internal-combustion engines; metalworking machinery; pumps and compressors; construction machinery; electric motors and generators; and power-transmission equipment.

The following case histories will give you an idea of where ductile iron castings are being used to advantage. (Other applications are illustrated on p. 64-66.)

Seven in One – Ductile iron castings are serving a range of functions in a heavy-duty pump and motor. Mr. Henderson names the parts and their requirements:

- Pump and motor end caps: they must be pressure tight and have high fatigue strength in the kidney area. The grade: 80-60-03. Hardness is Bhn 200-260.
- Servosleeve: it must be pressure tight and have good wear resistance on its inside diameter. The grade: 80-60-03. Hardness is Bhn 200-260.
- Servopiston: it must have good wear properties on its outside diameter, plus good toughness, strength, and wear properties in the linkage area. The grade: 60-45-15. Hardness is Bhn 143-207.
- Thrust plate: it is heat treated to Rc 50-56, ground and lapped to a high surface finish. Grade: 80-60-03.
- Pump swashplate: this is mainly a strength requirement. Grade: 80-55-06. Hardness is Bhn 179-255.
- Trunnion: this is mainly a strength requirement. Grade 80-55-06. Hardness is Bhn 179-255.
- Motor swashplate: its angle face is selectively hardened to Rc 50-56, ground and lapped to a high surface finish. Grade: 80-60-03. Hardness is Bhn 20-260.

Autos–Detroit is the largest single user of ductile iron castings. Major applications are crankshafts (weighing 80 to 100 lb) and differential carriers (35 to 40 lb). Housings for disc brake calipers are among the newer applications, and a number of other parts are candidates for conversions. Normally, the auto industry prefers to avoid heat treatment.

Here is an exception: a pin-type rocker arm. As engine performance has increased, the geometry of the rocker arm has been changed to the pin type. The weight of the component has decreased, and a major design consideration—fatigue resistance—has been imposed. This application involves a combination of casting and forming. The as-cast part is first annealed to essentially ferrite and graphite. Then it is coined and drilled in this relatively soft condition. In the last step, it is hardened by quenching from 1,650 F and drawing at 400 F.

Carriers – As stated, differential carriers are a major application in the auto industry. Comparable parts are also used in earth-moving equipment. Planetary carrier castings made at Lynchburg, for example, range from 58 to 118 lb. Engineering requirements include good machinability, freedom from cracking, a high modulus of elasticity, and good wear resistance in the gear pin bore.

A manufacturer of road-building equipment comments on his experience with power-shift transmission carriers: "Because of strength considerations, the part could not be made in gray iron. At engineering's request, we first went to malleable, but hot tears, cracks, and dimensional control were problems.

"We tried steel castings next, and found that they could not be cast in thin sections. Cold shuts and shrinks were encountered. Eventually, the 80-60-03 grade of ductile was chosen. Strength levels were met, and the material was castable in all sections."

"In the early days," the user continues, "ductile was subject to carbide and shrink problems. On occasion, there was trouble with flotation or wormy graphite. These problems, however, have been practically eliminated by controlled melting practices, proper inoculation, pouring times, gating and risering, and other related foundry practices."

Conversions – Cost reduction is a common motive for switching to ductile castings. For example, one of the winners in the annual Design Contest sponsored by the Gray & Ductile Iron Founders' Society Inc. (GDIFS), Cleveland, reported that manufacturing costs were slashed 69% in converting a two-piece fabrication – a steam turbine governor case – to a one-piece, 60-40-10 casting.

William Straslicka, Elliott Co., Jeannette, Pa., stated that the cost of the part, which contained fly-weights and related speed-control parts, was trimmed from $40.12 to $12.51. Formerly, steel bar stock 4 in. in diameter and 3 in. steel tubing were joined by welding and machined in eight separate operations. The one-piece casting exceeds maximum engineering requirements.

Another winning entry, from C. W. Posey Jr., Materials Handling Systems, Birmingham, involves a conversion from a steel forging. As a forging, the part, a wire rope

Where Ductile Iron Castings Are Being Applied

Ductile iron has been tailored to hundreds of different applications in this country and abroad. The diversity is indicated by uses of special compositions and alloys cited in a book, *The Uses of Molybdenum in Nodular Irons*, put out by Climax Molybdenum Co., New York. They include: bull rings, sinter pallets, cams, cylinder liners, burring discs, drawing dies, drawing wheels, mining knives, asphalt mixer liners, hot-forming dies, rock jaw crushers, coal pulverizer rings, bridge bearings, impellers for outboard diesel engines, heavy-section sheet metal dies, pulp press rolls, hot air valves, pump casings that carry coal-containing mud, wire straightening dies, drop balls, compressor valves, bearing cages, auto-body dies, explosive-forming dies, sprockets, bushings, piston-rings, Hydrospin mandrels, grinder rolls, oil-well pump beams, sheet and bar mill rolls, and grooved rolls for finishing stands.

Other applications – of all grades –are illustrated here and on the following page.

Oil manifold for earth-moving equipment was made of ductile iron because the application called for a combination of high strength and good ductility.

Piston for diesel locomotive engines is stronger and has more wear resistance than previously used types. Reduction in weight (to 50.5 lb) also means less stress on other parts, such as bearings.

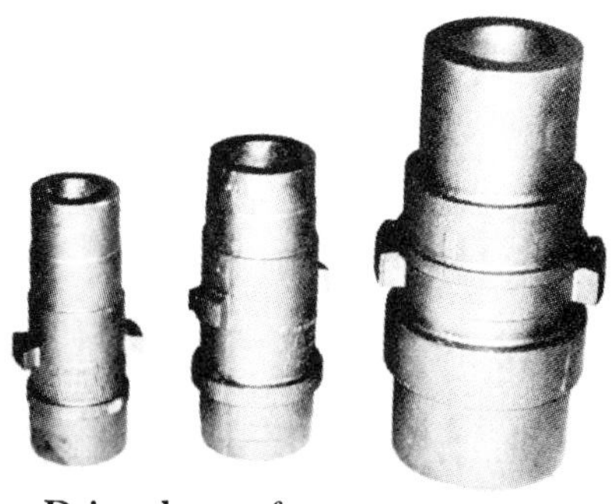

Drive sleeves for automatic torque control devices were formerly steel castings or made from solid steel

billets. Service conditions range up to 150,000 lb of thrust and 4,100 ft-lb of torque.

Driveshaft for farm equipment was formerly gray iron. Full-length, heavy-duty steel pipe replaces sand core for hub, which simplifies machining and assembly. High-strength part weighs 72.4 lb.

Planetary carriers for earth-moving equipment combine high strength, dimensional stability, and resistance to wear, shock, and sudden loading. Weights range from 58 to 117.9 lb.

Suspension link for railroad track scale transmits 70,000 lb of force from platform to a lever. Bending movements in lever system were reduced by 50% because a smaller link is possible.

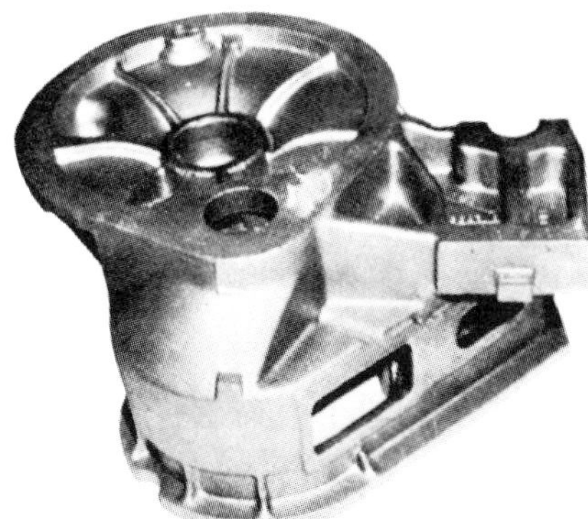

Final drive housing for earth-moving equipment (261 lb) was switched from gray iron to get more strength to handle unusual shock loads. Existing pattern for gray iron casting was used.

Base for rubber impeller in centrifugal slurry pumps was formerly a six-part steel weldment. The casting trimmed 12% from component weight and reduced per-part cost from $69.40 to $32.35.

Front wheel casting for heavy machinery was switched from steel to ductile iron primarily to bring about a reduction in costs. This complex part weighs 160 lb.

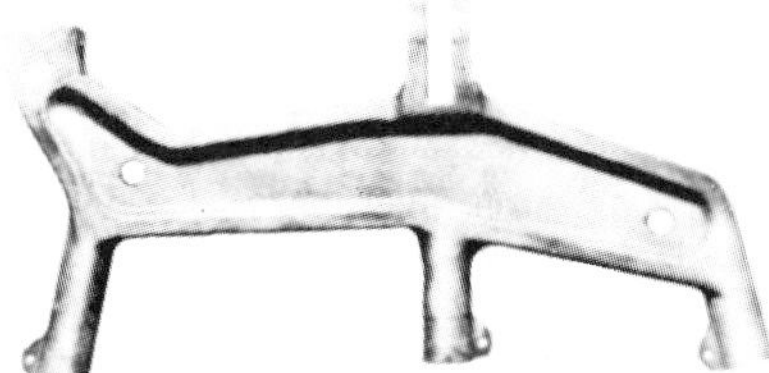

Door mechanism for bomber aircraft, formerly an eight-part steel weldment, was converted because it had a tendency to warp. Casting reduced part cost from $182 to $41.

Tool storage index for a numerically controlled machining center formerly required 2,115 manufacturing operations. Conversion to one-piece casting cut number to 24.

clamp for conveyor idler and support stands, cost $2.31 per unit to manufacture. By going to a sand-cast clamp of 80-55-06, the company was able to shave part cost by 91%. The castings, which cost 20¢ each, have more than adequate strength and toughness; and machining was eliminated because functional surfaces had sufficient accuracy in the as-cast condition.

A third example from GDIFS has special interest because the 65-45-12 casting is lined with Teflon. The complex plug valve, which controls the flow of extremely corrosive liquids in chemical processing plants, was formerly made of a variety of exotic materials, including type 316 stainless steel, Monel, Hastelloy B, titanium, and tantalum. In some cases, reports Russell G. Smith, director of engineering, Continental Mfg. Co., Cincinnati, unit costs ran as high as $229. The figure is now down to $79. Cast recesses and machined grooves in the body and plug lock the Teflon in place. It will not collapse in a high vacuum or during thermal cycling.

How Ductile Iron Is Made

Cupola melting is the most common method, but a number of foundries now use electric induction furnaces. Where both ductile and gray iron are melted in the same cupola, the acid-type prevails. When separate melting facilities for ductile

Where Ductile Iron Castings Are Being Applied

Drive gear for tire changer was formerly machined from steel. Smooth finish on casting cut per-part machining cost from $4.55 to 23¢. Part cost dropped from $9 to $3.50.

Compressor discharge casing was switched from a cast-weld structure that called for 200 manufacturing operations to a one-piece casting that requires only ten.

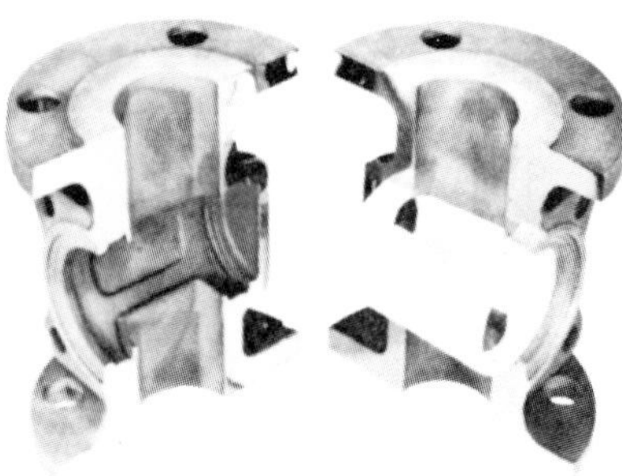

Bidirectional valve bodies were made from carbon steel. Now, tapered body bore does not require machining or lapping. Per-unit cost of machining is now $2. It was $14.

are provided, the basic cupola is preferred. It's estimated that as much as 85% of all tonnage is produced in basic cupolas.

The sulfur content of acid-cupola iron is higher than that made in basic cupolas—it generally runs 0.06 to 0.12% vs 0.025 to 0.035% in basic cupolas. And if the iron is not desulfurized prior to the spheroidizing treatment, an appreciable amount of high-cost magnesium (usually in the form of magnesium-nickel, magnesium-ferrosilicon, or magnesium-silicon alloys) is consumed. Lynchburg, for example, has acid cupolas, and it desulfurizes its iron to 0.010% S by the shaking ladle method (Fig. 3). The company adds 0.5 to 0.6% calcium carbide to the melt, which is shaken for 10 min. Soda ash is also used in desulfurization.

The magnesium treatment develops the desired spheroidal graphite structure. But it is of great importance to keep the added and retained amounts of magnesium at low levels. There are several reasons for this. One is that excessive magnesium content can cause dross defects. Another is that high magnesium can promote carbides in castings because the element is a carbide stabilizer.

Generally, minimum retained magnesium of 0.015 to 0.050% is considered adequate. Several techniques are employed to add magnesium. Lynchburg, for example, has used what is known as the plunging method, and it reports that magnesium recovery runs 50 to 75%. Central Foundry uses the sandwich method.

Inoculation with ferrosilicon alloys (those containing 85% Si are usually preferred) increases the number of graphite spheroids formed during solidification. The aim is to eliminate eutectic carbides that raise hardness and tend to lower mechanical properties.

Silicon content is also carefully controlled. "Ductile iron with a high silicon (above 2.8%) may have low impact resistance, particularly at temperatures of −40 F or below," Mr. Henderson explains. "On the other hand, a low silicon content may increase the chilling tendency enough to cause the formation of excess carbides in thin sections. Such castings may be unsuitable for machining and unresponsive to heat treatment."

Ductile iron foundries have established rigid controls. Emission spectrometers are commonly used to quickly determine the chemical composition of both the base and final ductile iron. To check on nodularity and matrix structure, it is common to pour microspecimens from the last part of each ladle of iron treated with magnesium alloy and examine it under a microscope. Results are obtained in minutes. At Lynchburg, if the microstructure contains less than 90% of types I and II graphite, the castings are scrapped at shakeout. If the matrix structure shows centerline carbides or massive carbides, the castings are salvaged by heat treatment or scrapped.

What's on the Horizon

Over the last few years, the yearly growth rate in the production of ductile iron in this country has been bounding along at a solid 15%. Some expert observers, including Charles F. Walton, technical director, GDIFS, foresee an easing to an 8% growth rate, starting this year.

The reasoning runs this way: production has reached a high level—the estimate for 1968 is 960,000 tons and the 1969 forecast is 1,037,000 tons. For one thing, it would probably be unrealistic to expect a continuation of the 15% rate at this level. Also, people in the auto industry will tell you, "the bulk of the rapid growth in recent years has been brought about by mass conversions of crankshafts and differential carriers to ductile iron. Now this process is about completed. At least 7 million of the 9 million or so crankshafts produced for passenger cars each year, for example, are ductile."

This slowdown, however, is not interpreted as a sign of weakness. At the 1-million-ton level, yearly growth of 8% is respectable by any standard. Further, it is generally agreed that a plateau has been reached following several years of sensational growth; and it is not improbable that a new surge is in the making.

Ford Motor Co., for example, recently announced that it was reviewing all forgings and steel castings in its truck lines. Its goal is to replace at least 75% of these parts with ductile iron castings. "I would assume that other manufacturers will follow this lead," comments Mr. Walton. He adds that many companies in other industries are also taking this route.

Developments in this technology are also expected to make this material more attractive to users. In the words of P. S. Cowen, assistant technical director, GDIFS, "The potential versatility of ductile iron's metallurgical resources has not been fully utilized in mass production. Special heat treatments, basic composition alterations, and alloy refinements provide surprising properties that can only be classified in terms of those obtained by superalloys.

"Tensile strengths of 250,000 psi show up in martensitic ductile irons processed through conventional quench hardening. Impact properties are radically improved by diminishing silicon content and the use of ferritic matrices. A stable austenitic matrix has found uses in pressure parts for cryogenic service through the manipulation of maganese contents against nickel and heat-treating techniques.

"It is likely that very substantial progress could be made in extracting greater versatility from iron. No real attempt has been made to push research deep into its metallurgical mysteries because its elementary makeup has produced astounding results."

Hamilton's Mr. Voss adds, "Ductile iron has come a long way. In earlier years, casting users had some valid reasons to question reliability and integrity. Implementation of refined process controls by the producer has developed a very high level of quality assurance, resulting in an equally high level of confidence and acceptance by the casting user."

There's no doubt that this material helped to bring about the revolution in iron castings.

How GM's Central Foundry Div. Makes . . .

Ductile Iron Crankshafts

The man facing the line is sonically testing ductile iron crankshafts for internal soundness and metal structure. In preceding steps on this line, the blast-cleaned parts are end milled and center drilled. The workman at the right is paddle gaging dimensional tolerances. In the final step here, parts are coded for shipment—with this information the time a crankshaft was poured can be pinpointed to the hour. Other quality control checks at this stage include Magnaglo inspection, cobalt-60 radiography, and Brinell hardness tests.

These operations reflect the care that go into volume production of pearlitic ductile crankshafts at Central Foundry's No. 2 plant, Defiance, Ohio. The General Motors Corp. facility can turn out more than 12,000 ductile crankshafts per day. (See p. 98 for further information on this unusual foundry.)

Beginning—Close control of each heat of ductile iron starts at the cupola's spout. Samples taken from it, the 6,000 lb treatment ladle, and the 3,000 lb pouring ladles are transported by vacuum tube to the metallurgical laboratory, where the iron undergoes a metallographic test and spectrographic analysis.

Specimens are cut, polished, and examined under a microscope for nodularity, percentage of ferrite, percentage of carbide, and carbon flotation. Results are obtained within 30 min, and people at the shakeout operation are advised whether the heat should be accepted or rejected.

Results of the spectrographic test are relayed to the melting floor within 8.5 min. Control limits on pearlitic are 3.65-3.85 C, 2.30-2.50 Si, 0.65-0.70 Mn, 0.03-0.05 Mg, and 0.05 Cr max.

Crankshafts are made four per mold. The casting cooling rate is carefully controlled to maintain the desired hardness. The journey through enclosed cooling conveyors—from pouring to shakeout—takes 45 to 60 min, depending on the casting.

At the shakeout area, castings which are acceptable are unloaded mechanically onto a main shakeout. Rejected castings go to a second shakeout. Shotblast cleaning and the operations described above follow. The crankshafts are not heat treated.

Melting—The cupola, which is 114 in. in diameter, melts into a 60-ton Asea holding furnace at a rate of 50 to 55 tons per hour. The cupola uses a 1,000 F hot blast through copper-coil, water-cooled tuyeres.

Magnesium is added (by the sandwich method) in the treatment ladle previously mentioned. The pouring temperature is about 2,600 F.

The No. 2 plant has two, 1,230 ft lines. One is for crankshafts exclusively. The other one which is served by four, 33-ton induction furnaces and one on standby, turns out ferritic ductile, gray iron, Armasteel, and malleable iron castings. Castings are heat treated in two, 125 ft long malleabilizing kilns and two conveyorized draw furnaces. The automated malleabilizing kilns feature air-blast cooling.

Ductile Iron: Where and How It's Used

By JOHN SCHUYTEN

In specification, yield strength is particularly important. It establishes the limit of tensile loading and indicates minimum compressive strength, torsional strength, and shear strength.

LIKE OTHER cast irons, ductile iron is specified by its tensile properties. A three-number symbol is used to designate the various grades. Numbers refer, respectively, to minimum tensile strength in thousands of psi, minimum yield strength in thousands of psi, and minimum percentage of elongation in 2 in. (50.9 mm).

The more common specifications range from the high ductility grade 60-40-18, which is generally annealed for a ferritic matrix, to the 120-90-02 high strength grade, which has been heat treated to a high but machinable hardness.

The 60-40-18 grade is generally annealed; the 65-45-12 grade is usually as-cast; the 80-55-06 grade is generally as-cast; the 100-70-03 grade is normalized; and the 120-90-02 grade is either air quenched or oil quenched and tempered.

The remarkable fact is that one material without alloy additions can produce such a wide range of properties by simple heat treatment. This versatility is illustrated by the stress strain curves of annealed, normalized, and oil quenched and tempered grades shown in Fig. 1a, which also shows the corresponding hardness ranges and a high elastic modulus of 23 to 25 million (15 860 to 17 240 MPa). Fig. 1b illustrates the direct relationship between Brinell hardness and tensile strength and also shows the high yield to tensile ratio.

From a structural design standpoint, the yield strength figure is especially important. It not only establishes the limit of tensile loading but also indicates minimum compressive strength, torsional strength, and shear strength. It should be noted that yield strengths in ductile iron are consistently higher than those of comparable tensile strength grades of unalloyed cast steel, an important advantage to the material specifier.

The unique property of ductile iron is its ductility, which is evidenced by the percentage of elongation in the tensile test. In the ferritic, fully annealed grade, it is possible to produce elongations comparable to those of cast steel (20 to 25%) consistently. Ductility is also of importance as an indicator of impact resistance. Although ductile iron does not have as high impact resistance as comparable grades of steel, it is entirely satisfactory for many severe applications involving impact, such

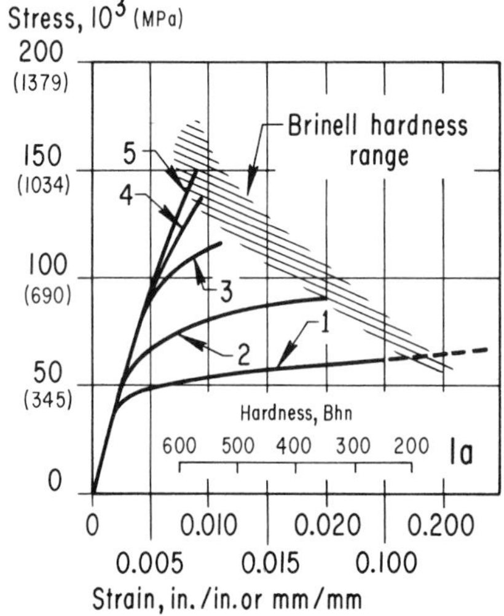

1. Fully annealed
2. Normalized from 1650 F (900 C)
3. Oil quenched from 1650 F (900 C), drawn 2 h at 1000 F (540 C)
4. Oil quenched from 1650 F (900 C), drawn 2 h at 800 F (425 C)
5. Oil quenched from 1650 C (900 C), not drawn

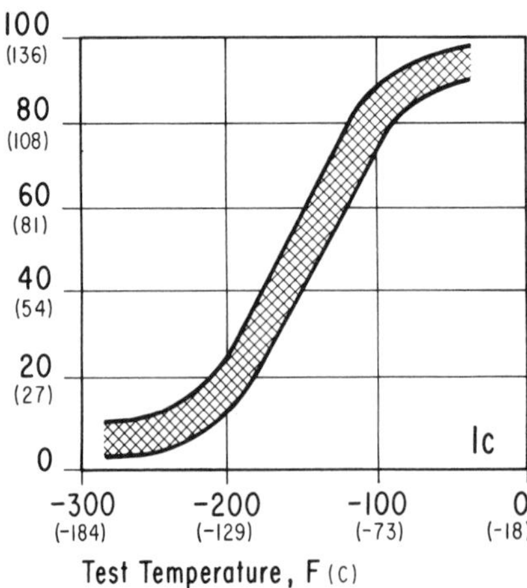

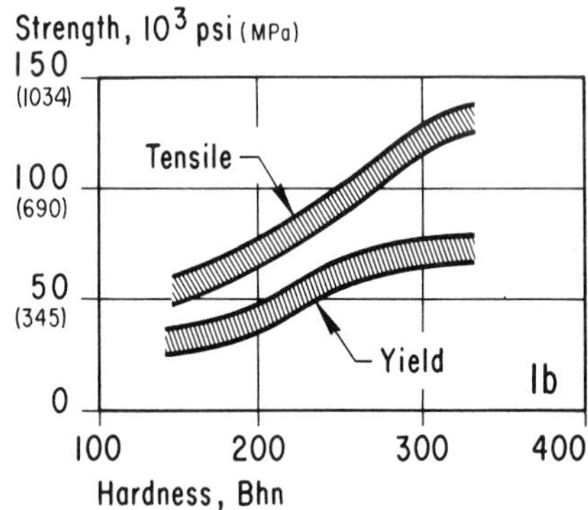

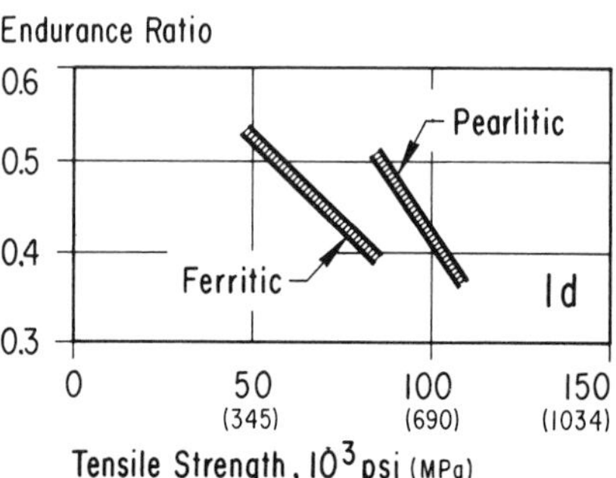

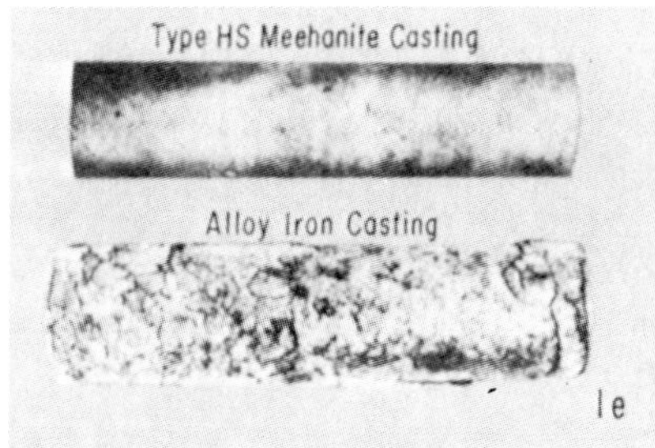

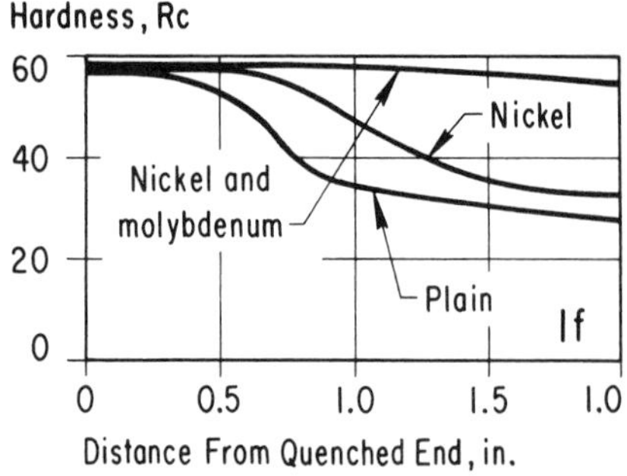

Fig. 1 — (a) Stress-strain curves for annealed, normalized, and quenched and tempered grades. (b) Relationship between hardness and tensile strength. (c) Impact energy of ferritic ductile iron at low temperatures. (d) Ratio of endurance limit to tensile strength. (e) Comparative processing properties of ductile and alloyed cast irons heated to 1600 F (870 C), cycled for 300 h. (f) Contributions of nickel and molybdenum to hardenability.

Mr. Schuyten is president of Vulcan Foundry Co., Oakland, Calif. This article is adapted from a paper, "Ductile Iron: Growth Metal of the Century," which he presented at WESTEC '75.

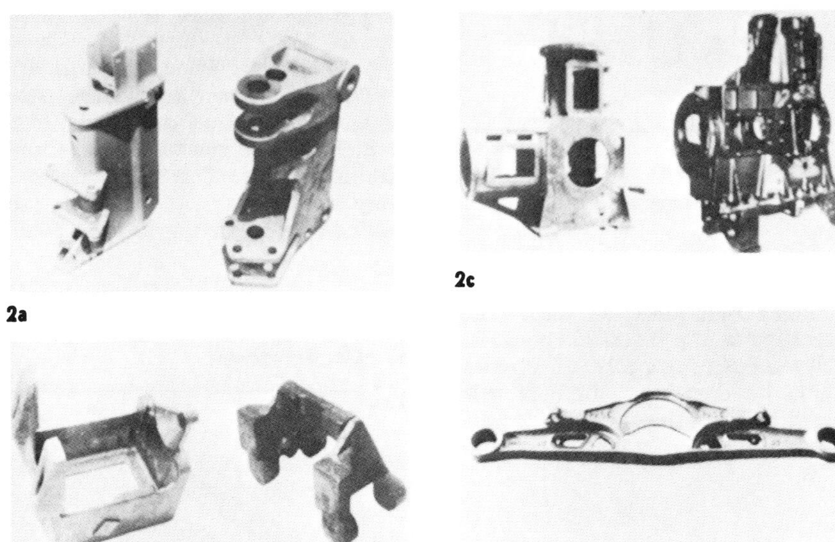

2a 2b 2c 2d

Fig. 2 — (a) Ductile iron casting replaced steel fabrication at the left. (b) Castings replaced cold-rolled steel plate fabrications. (c) Castings replaced steel weldment on left. (d) Ductile iron rocker arm replaced forged and welded component.

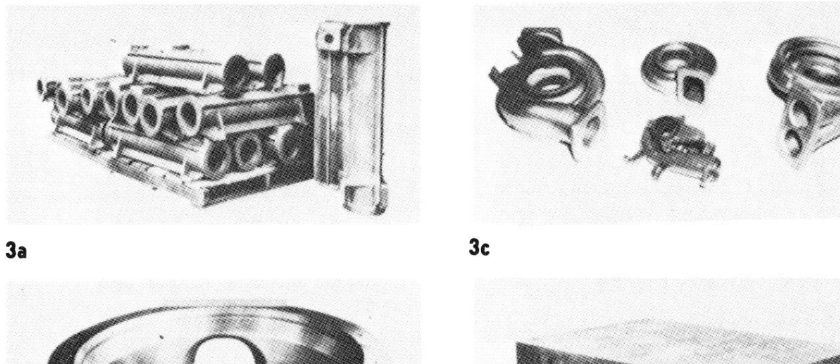

3a 3c

3b

3d

Fig. 3 — (a) Pump cases required a variety of material and processing properties. (b) Combination of strength and toughness was provided by nickel-molybdenum alloy. (c) High-temperature strength and resistance to growth are key properties for this part. (d) Special processing techniques were used to meet requirements for high-pressure hydraulic casting.

as gears, housings, steel mill rolls, and valves.

However, like steel, ductile iron is both notch and temperature sensitive in impact. Fig. 1c shows the impact energy of ferritic ductile iron at low temperatures and indicates a transition temperature of about −100 F (−73 C). This transition temperature of ductile to brittle fracture is raised by the presence of notches and by increasing pearlite content in the matrix. This means that well-designed ductile iron castings are giving excellent service in low-temperature and shock-loaded applications. For example, ductile iron has been accepted for shipboard use by the U.S. Navy and is used in vital applications such as valves and fittings aboard tankers.

Another excellent property of ductile iron is its resistance to fatigue which makes it an excellent choice for critical parts on high-speed machinery such as crankshafts on compressors. This property is expressed as the ratio of endurance limit to tensile strength. As shown in Fig. 1d, for commercial ductile irons, this ratio is between 0.4 and 0.5.

Ductile irons also exhibit excellent high-temperature strength as well as resistance to oxidation and growth. High-temperature strength and ability to resist oxidation and growth make ductile iron ideal for elevated-temperature applications such as frames and doors of coke ovens, exhaust manifolds, turbocharger and gas turbine housings. Ductile iron is also a superior material for hot forming and sizing dies such as those used to make supersonic aircraft parts.

For more severe temperature applications, ductile iron's ability to withstand growth is enhanced by increasing silicon content from a normal average of 2.5 up to 4%. Fig. 1e illustrates the superior growth and scaling resistance of high-silicon ductile iron versus alloyed cast iron after 300 h cycling to 1600 F (870 C). For higher short-time tensile strengths at elevated temperatures small additions can be made of vanadium and/or molybdenum.

The principles of alloying applicable to steel are, with some modifications, also applicable to ductile irons. Nickel, molybdenum, and chromium are useful alloying elements, although chromium should be used sparingly because of the tendency to form a brittle carbide network. Fig. 1f shows the improved hardenability imparted by nickel and molybdenum as determined by the Jominy hardenability test.

Higher nickel contents with molybdenum will produce a bainitic structure as-cast, which after a tempering treatment can develop unusual combinations of high strength, ductility, and toughness.

A group of manganese-molybdenum alloyed ductile irons (tradenamed Almanite) has been developed by Meehanite Metal Corp. They are air hardening and develop excellent abrasion resistance in a number of applications. Heat-treated chromium-molybdenum alloyed ductile irons also develop very high hardness and abrasion resistance.

Applications — Automotive castings represent almost 50% of total ductile iron consumption. The most common applications are crankshafts, cam shafts, rocker arms, differential housings, steering knuckles, disc brake caliper housings, bearing caps, and, more recently, exhaust manifolds.

The first and most famous application is the crankshaft. The combination of high fatigue strength and the castability and soundness of ductile iron coupled with low cost made this a natural application.

In the farm equipment industry, ductile iron has achieved wide utilization for such parts as tractor axles, axle housings, clutch plates, lift arms, and steering knuckles.

The following examples were prize winners in the Gray & Ductile Iron Founder's Society Annual Design Contest.

The casting of 65-45-12 ductile iron on the right in Fig. 2a is a 177 lb (81 kg) swing pivot used on a heavy-duty backhoe boom. It replaced the steel fabrication shown on the left in the figure. Ductile iron was used for its strength, ductility, machinability, and exceptional wear resistance. In addition, it had a better appearance, and costs were reduced from $90 to $63 per piece.

The housing and carriage in Fig. 2b are used to support and contain components in a Du Pont automatic photometer chemical analyzer. The castings were designed to 65-45-12 ductile iron to supplant fabrications from cold-rolled steel plate and resulted in a shape with appreciably greater strength at a weight of 30 lb (14 kg). The fabrication weighed 100 lb (46 kg). Total cost for the two parts was reduced by more than 40% — from about $550 to about $300, most of which was in machining cost.

In Fig. 2c, the casting on the right is a compressor block which functions as the main structural component in a multistage compressor. The 60-45-15 ductile iron casting replaced the steel weldment shown on the left with no sacrifice in mechanical properties. Damping characteristics and appearance were improved. The total cost per ductile iron part is $1435, or 53% of the $2675 weldment cost. Material costs dropped only $15 — from $195 to $180; but machining costs were reduced from $1680 to $1255, and $790 in welding and fabricating costs were eliminated.

The rocker arm in Fig. 2d replaced one made of forged and welded steel. It is said to be the most complicated part of a truck axle suspension system and is subjected to vertical twist and horizontal-plane loading as well as high weight and shock loads. The ductile iron is heat treated to provide a minimum of 52 000 psi (359 MPa) yield strength and is 25% lighter than the forging. At a cost of $32 as against the forging price of $85, the manufacturer saved more than $1 million in one year.

Vulcan Foundry Co. has pioneered a number of ductile iron applications in cooperation with its customers. The following were primarily replacements of steel castings, forgings, weldments, or nonferrous castings where the properties of ductile iron were superior or where the end product could be produced at a lower cost or both. These examples were selected to show the wide diversity of applications.

Fig. 3a shows high pressure pump cases of fully annealed ductile iron for nuclear submarines. Strength, ductility, castability, and machinability were all important. Fig. 3b shows an explosive forming die for aircraft and missile parts. The nickel-molybdenum ductile iron provided the required high strength and toughness. Fig. 3c illustrates a group of turbocharger housings for aircraft and diesel engines where the high-temperature strength and growth resistance as well as ductility and castability of ductile iron in complex configurations are utilized.

Fig. 3d shows a high-pressure hydraulic casting. It's essentially a 12 by 24 by 36 in. (305 by 610 by 915 mm) block weighing 3000 lb (1361 kg) which was made as a riserless casting in a furan sand mold utilizing the expansion during the solidification process. The casting was absolutely flat with no pull-in, and the drilled holes through the center were completely free of porosity.

Quality Control — The phenomenal rate of growth of ductile iron castings over the last 25 years could not have been possible without the cooperative technological and research efforts of many individual foundries, foundrymen, suppliers, and the various trade associations and technical societies of the world. The common denominator in all this effort was a new emphasis on metallurgical and foundry quality control from the base metal through the treating process and all the familiar foundry variables.

In all methods of manufacture, it is absolutely essential to insure proper metallurgical control by the examination of metallographic pins from the end of each ladle to determine nodularity of the graphite and insure against magnesium fade as well as to determine the matrix structure.

It is also necessary to pour keel bars for tensile test bars and pins for determining complete chemistry. In this connection it is very convenient to use the new BCIRA carbon calculator utilizing special Tectips for determining both liquidus and solidus arrests correlating with carbon equivalent and carbon content from which silicon can then be determined. Chemical analysis should also include retained magnesium for best control of the treating operation since magnesium recoveries are very dependent on type of alloy, treating temperatures, and treating techniques used.

Emission spectrometers are also being increasingly used in ductile iron foundries for the basic elements as well as for various alloying elements. Techniques and equipment have also been developed for ultrasonic measurement of graphite nodularity.

Correct pouring temperatures, properly designed gating and risering systems are also a part of foundry quality control procedures.

It is a tribute to the industry and the individual producers that a very high reliability factor has been established for ductile iron castings. Without this increased emphasis on quality control procedures, this success story would not have been possible.

Ductile Iron for High-Temperature Service

By JOHN E. BEVAN

Adding molybdenum to ductile iron with 4% Si increases high temperature mechanical properties without changes in foundry practice. Based on cost and creep-rupture properties, optimum addition is 1% Mo.

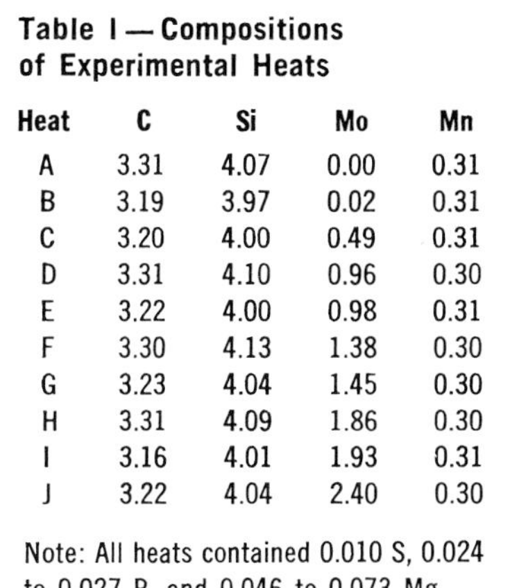

Table I — Compositions of Experimental Heats

Heat	C	Si	Mo	Mn
A	3.31	4.07	0.00	0.31
B	3.19	3.97	0.02	0.31
C	3.20	4.00	0.49	0.31
D	3.31	4.10	0.96	0.30
E	3.22	4.00	0.98	0.31
F	3.30	4.13	1.38	0.30
G	3.23	4.04	1.45	0.30
H	3.31	4.09	1.86	0.30
I	3.16	4.01	1.93	0.31
J	3.22	4.04	2.40	0.30

Note: All heats contained 0.010 S, 0.024 to 0.027 P, and 0.046 to 0.073 Mg.

A HIGH-TEMPERATURE ductile cast iron that could be made at most foundries with little change in normal foundry practice has been developed, as indicated by mechanical property data recently released on a series of 4% Si molybdenum irons. In a ductile iron containing nominally 3.2 C, 4.0 Si, and 0.3 Mn, the optimum molybdenum content appears to be 1%.

Molybdenum is useful in raising high-temperature strength because it produces the highest increase in hot strength per unit addition. And being neither a graphitizer nor a strong carbide former, molybdenum can be added without altering normal foundry practices.

Experimental Heats — Ten 125 lb (66.7 kg) heats of ductile iron were prepared for this investigation; compositions are given in Table I. Charges consisted of Armco iron, graphite, and part of the total ferrosilicon requirement. Melts were held at 2850 F (1565 C) for 5 min, and then cooled to 2800 F (1540 C) at which temperature alloy additions were made. Each melt was tapped into a nodulizing ladle containing magnesium ferrosilicon, and transferred to a pouring ladle where 0.5% Si (as 85% ferrosilicon) was added as an inoculant. The heats were cast into ½ in. (12.7 mm) Y-block molds, and analyzed.

Test Data — A high-temperature dilatometer was used to determine Ac_1 temperatures for Heats B, C, E, G, I, and J. The transformation temperature for the molybdenum-free iron (Heat A) was 1525 F (830 C); those for the molybdenum-containing irons ranged from 1510 to 1535 F (820 to 835 C). Hardness of as-cast and annealed iron (treated above and below the critical temperatures) rose with molybdenum content (0 to 2.40% Mo). Ranges: as-cast, Bhn 199 to 246; annealed (1450 to 1500 F [790 to 815 C]), Bhn 199 to 234; annealed (1650 F [900 C]), Bhn 196 to 221.

The number and shape of graphite spheroids in each heat appeared suitable, few nonspheroidal graphite particles being observed in any of them.

All as-cast irons contained some pearlite in the cell boundary regions, and subcritical annealing spheroidized the pearlitic regions of the molybdenum - containing irons. Amounts of primary carbides and pearlite at cell boundaries increased progressively with molybdenum content.

Tensile Properties — Samples from the ductile irons were tensile tested at room temperature and up to 1500 F (815 C). At all temperatures, strength increased at an essentially uniform rate with molybdenum content. Figure 1 illustrates the effect of annealing on tensile strength at 1200 F (650 C). Annealing at 1650 F (900 C), above the critical temperature, resulted in a significantly lower high-temperature strength than that of iron an-

Mr. Bevan is senior research associate, Climax Molybdenum Co. of Michigan, Subs. AMAX Inc., Ann Arbor, Mich.

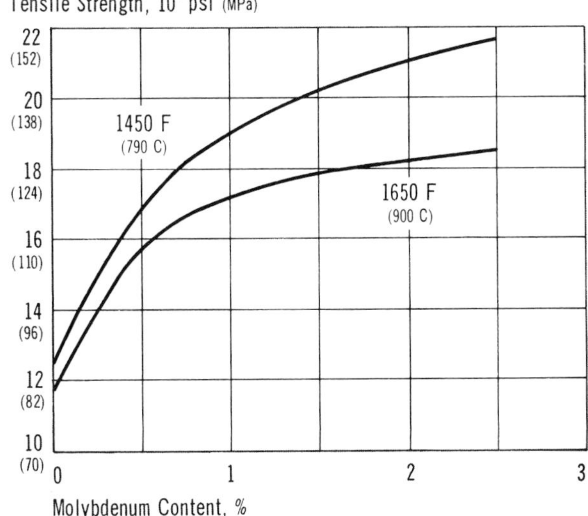

Fig. 1 — High-temperature tensile strength of ductile iron containing 4% Si rises with molybdenum content. Tests were made at 1200 F (650 C) on stock annealed at the indicated temperatures.

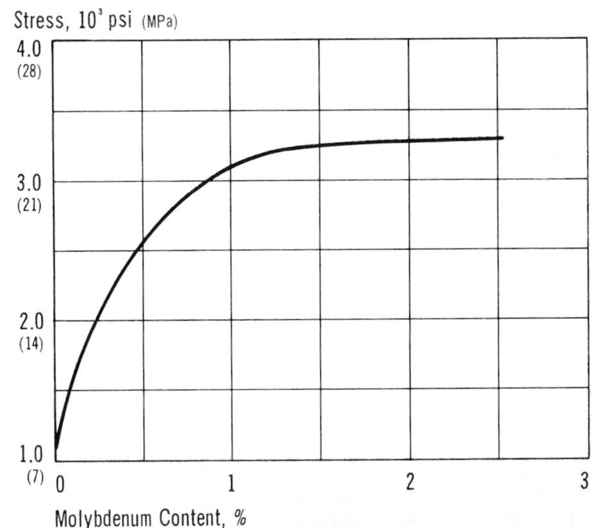

Fig. 2 — Creep strength rises with molybdenum content in ductile irons containing 4% Si. Graph indicates stress to produce 1% creep in 1000 hr at 1300 F (700 C) in specimens annealed at 1450 F (790 C).

nealed at 1450 F (790 C).

Tensile ductility tended to drop with increasing molybdenum contents, but was considered adequate even at the highest molybdenum level that was evaluated.

Creep Rates — Resistance to creep increased dramatically up to about 1% Mo, then tended to level off (Fig. 2). From a practical standpoint, creep-rate data are quite useful. Unlike tensile-test results, they can be used to determine accurate design stresses that will produce a known (allowable) amount of strain over the life of the part.

Apparently, molybdenum additions raise long-time creep-rupture properties because they promote stable fine-matrix carbides. Metallographic examinations of samples that had been tested for short and long times indicate that 1% or more molybdenum stabilized matrix carbides. In the molybdenum-free irons, matrix carbides dissolved (or dissociated), and the carbon migrated to the nodules during long-time exposures at elevated temperatures — that is, secondary graphitization took place.

Oxidation — Several of the experimental ductile irons were subjected to oxide penetration tests at 1500 F (815 C). Samples were given 50 exposures, each lasting 2 hr, and the depth of oxide penetration was determined by metallographic examination. Results indicate that molybdenum additions slightly decrease oxide penetration, as shown:

Heat	Mo	Oxide Depth, in. (mm)
A	—	0.0030 (0.076)
D	0.96%	0.0015 (0.038)
F	1.38	0.0020 (0.051)
H	1.86	0.0020 (0.051)

Of perhaps more importance, oxide layers were much more adherent in the molybdenum-containing irons. Oxide layers on irons containing at least 1% Mo were, in fact, difficult to remove, and did not flake off during thermal cycling tests.

Bainitic Ductile Irons Offer Higher Strength

Nickel and small molybdenum additions make it possible to obtain significant improvements in mechanical properties in a variety of section thicknesses. High temperature heat treatment isn't necessary, it's reported.

Bainitic ductile irons providing yield strengths of 100,000 to 170,000 psi, tensile strengths of 140,000 to 200,000 psi, and adequate ductility for many applications can now be produced by alloying. High-temperature heat treatments are not necessary, reports International Nickel Co. Inc.

Excellent properties are made possible in sections up to 4 in. thick or even larger by nickel alloying, which causes the bainitic structures to be obtained without stabilizing deleterious intercellular carbide networks. Molybdenum plays an important part in providing the desired structure.

These irons should replace steel castings in some applications by taking advantage of their superior castability, coupled with the advantage of not needing an austenitizing heat treatment. They should also replace some pearlitic ductile iron, particularly in cases where the latter's properties are marginal, or where redesign of a part permits a lighter casting to take advantage of increased strength capability.

Background – Use of ductile irons as a replacement for steel has been somewhat limited because the highest strength specification is for a material with 90,000 psi minimum yield strength and 120,000 psi tensile strength with at least 2% elongation.

In addition, these properties are usually obtained by quenching and tempering. Cracking and distortion problems can limit these heat treatments to relatively simple shapes. Pearlitic structures, which can be obtained as-cast or by normalizing, provide a maximum of about 80,000 psi yield strength, but ductility and especially impact toughness are poor.

Improvement – Significantly better mechanical properties cannot be obtained with a ferritic or pearlitic structure. The iron must develop a structure of upper or lower bainite or martensite. The desired structure and properties should be obtained as-cast or after a low-temperature temper to avoid heat treatment problems. Further, the bainitic or martensitic structures will provide unnecessarily high strength with inadequate ductility if the iron has a eutectoid carbon content of 0.8%.

An alloying element is required which will lower the eutectoid carbon content, increase the hardenability (by decreasing the pearlite forming tendency), and promote the formation of bainite – all without promoting deleterious intercellular carbides or adversely affecting graphite shape.

Nickel will do all of these things. Because molybdenum increases hardenability and encourages bainite formation when present in small quantities, it also can be used without significantly increasing carbide stability.

Greatest interest is in upper bainitic irons. They're being used in rolling mill rolls and have excellent potential in large gears, crankshafts, cams, diesel blocks and heads, tractor treads, cultivating spiders, hoist brake drums, and clutch plates.

Characteristics – Bainitic ductile irons encompass a range of compositions, microstructures, and mechanical

Table I — Tensile Properties of Upper Bainitic Ductle Irons

Composition						Section Thickness, In.	Yield Strength, Psi	Tensile Strength, Psi	Elongation %	Reduction in Area, %	Hardness, Bhn
C	Mn	Si	Ni	Mo	Mg						
3.65	0.39	2.23	2.67	0.49	0.064	1	103,600	138,800	4.8	6.0	321
3.78	0.41	2.25	4.38	0.25	0.077	3	110,700	138,500	2.5	3.5	334
3.67	0.40	2.00	4.90	0.24	0.070	6	109,300	135,400	1.8	2.8	334

Note: Tempered 4 hr at 600 F, air cooled.

properties. Irons containing sufficient alloy will avoid pearlite formation on mold cooling. Relatively slow cooling, particularly below the temperature range for pearlite formation, will then favor transformation to bainite. The leanest alloy which will transform fully to bainite will have the coarsest (and the weakest) structure.

Further alloying will depress the

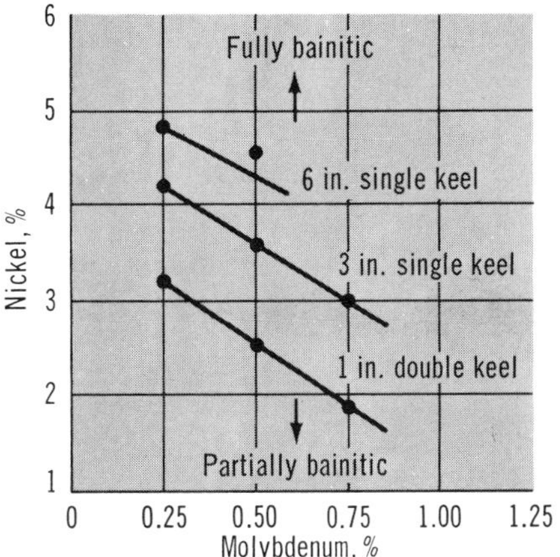

Fig. 1 — The effect of nickel, molybdenum, and section size on the as-cast microstructure of ductile iron.

bainite transformation temperature and produce a finer structure. If transformation is suppressed too far, however, some martensite transformation results.

The irons described here attain a range of structures from a coarse acicular structure termed upper bainite to a fine structure termed lower bainite. Mixed bainites with intermediate properties can result at certain compositions, which makes possible a range of mechanical properties.

Under certain conditions martensite will form, but since these irons are always tempered, the martensite seems to have little effect on the strength, ductility, or toughness of a predominantly lower bainitic iron.

Compositions – The composition ranges most satisfactory in providing predominantly upper bainitic structures are shown in Table I. The particular composition chosen to provide upper bainite depends on the nickel-molybdenum balance and section thickness as shown in Fig. 1. At 0.25% Mo, the highest nickel levels are required. Nickel requirements are reduced as molybdenum is increased. However, it's reported that 0.75% Mo is embrittling. Molybdenum must be restricted if good ductility is to be obtained.

The effect of nickel on yield strength and microstructure as a function of section thickness up to 4 in. is shown in Fig. 2. The effect of silicon and manganese variations and the removal of molybdenum on the matrix structure and strength of an iron containing 5.2% Ni is also shown.

Increasing silicon from 1.86 to 2.33% decreases yield strength about 20,000 psi in the 1 in. section thickness, probably by reducing carbon content of the matrix and the amount of lower bainite, but has little effect in the coarser, lower strength upper bainite of the 2 and 4 in. sections.

Increasing manganese from nil to 0.39% boosts strength significantly in all section sizes studied. It is likely, however, that a further increase in manganese would promote intercellular carbide formation, particularly in heavier sections. Adding 0.19% Mo to a molybdenum-free iron has a dramatic influence on structure (Fig. 2), changing a predominantly pearlitic structure to one that is fully bainitic. This change increases strength and improves toughness considerably.

Properties – Optimum properties are obtained by heat treatment for 4 hr at 600 F and air cooling. This low temperature heat treatment avoids distortion and stresses always possible in quench and temper heat treatments, especially in heavy section castings.

Tensile properties obtained in properly alloyed upper bainitic irons are given in Table I. Because it is most economical to alloy upper bainitic irons to match section size, different compositions are used in the 1, 3, and 6 in. sections. The matrix is predominantly upper bainitic in each case. All of the irons attain 100,000 to 110,000 psi yield strength and nearly 140,000 psi tensile strength characteristic of this structure.

Ductility is somewhat lower in larger sections; this probably results from several factors, including the larger graphite spheroids and larger cell size, the enhanced probability of irregular graphite shapes, the increased incidence of segregation-induced intercellular carbides, and an increase in microshrinkage.

Notch tensile properties have been determined for pearlitic, upper bainitic and lower bainitic structures. The pearlitic iron was notch sensitive in a 1 in. section; the notched:unnotched tensile ratio (NTS/UTS) was 0.74. The greater ductility of the upper bainitic structure resulted in relatively notch tough behavior. This iron has a NTS/UTS of 1.12. The higher strength lower bainitic iron was notch sensitive, having a ratio of 0.76 in a 1 in. section.

The impact toughness of irons can also be important. Data for three compositions of upper bainite are given in Table II.

Table II — Toughness of Upper Bainitic Ductile Irons

Composition						Section Thickness, In.	Charpy V-Notch Impact Strength at 70 F, Ft-Lb
C	Mn	Si	Ni	Mo	Mg		
3.53	0.42	2.00	2.87	0.50	0.07	1	5
3.63	0.41	2.15	4.65	0.25	0.08	3	6.2
3.67	0.40	2.00	4.90	0.24	0.07	6	6.8

Note: Tempered 4 hr at 600 F, air cooled.

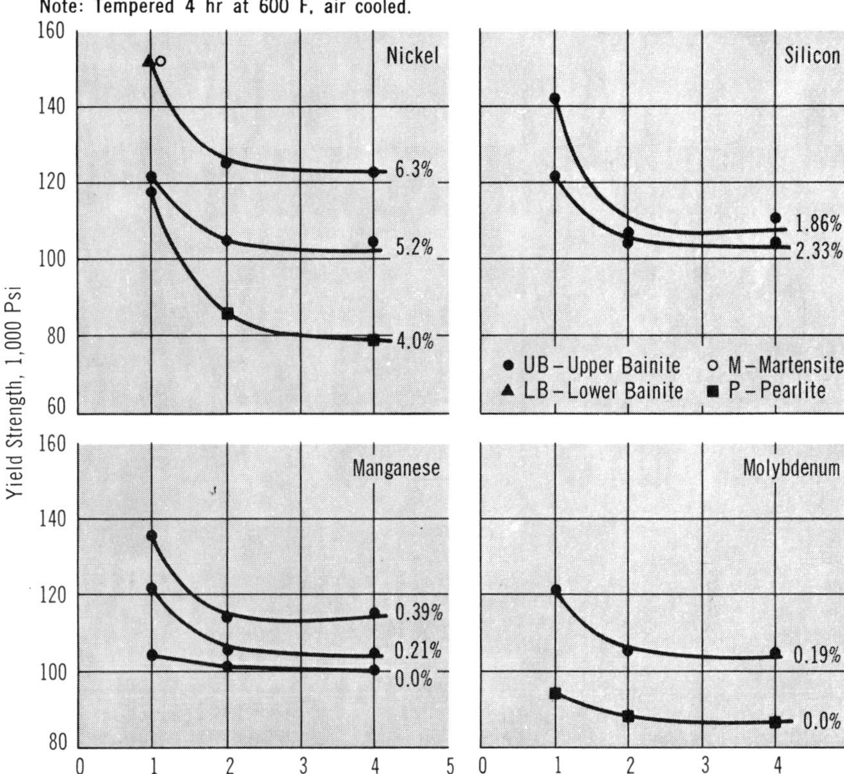

Fig. 2 — The relationship between nickel, silicon, manganese, and molybdenum contents and the yield strength of upper bainitic castings of various thicknesses. Nominal composition is: 3 C, 0.2 Mn, 2.3 Si, 5.2 Ni, 0.2 Mo, 0.05 Mg.

Specifications and Properties of

Gray Iron Castings

Specification No. (a)	Grade or Class	Min Tensile Strength, Psi	Hardness, Brinell	Other Requirements	Typical Applications
ASTM A126-73	A B C	21,000 31,000 41,000		Composition, %: P Max 0.75; S Max 0.15 1.2 in. dia. x 12 in. transverse test is optional.	Stock valves, flanges, and pipe fittings, and castings not requiring critical tensile test evaluation.
ASTM A48-74 ANSI G25.1 QQ-1-652c	20 (b) 25 (b)	20,000 (c) 25,000 (c)		Test bar size shall be related in cooling rate to the critical section of the casting and so specified. At least two test bars shall be cast and prepared for each casting lot, the lot size being designated. Test bars shall be cast in dry silica sand molds similar to that in which the castings are poured. Tension test shall be under true axial loading. Hardness, chemical composition, microstructure, pressure tightness, radiographic soundness, dimension, surface finish, etc., can be established as requirements upon written agreement between manufacturer and purchaser.	Small or thin-sectioned castings requiring good appearance, good machinability, and close dimensions.
	30 (b) 35 (b)	30,000 (c) 35,000 (c)			General machinery, municipal and water works, light compressors.
	40 (b) 45 (b)	40,000 (c) 45,000 (c)			Machine tools, medium gear blanks, heavy compressors.
	50 (b) 55 (b) 60 (b)	50,000 (c) 55,000 (c) 60,000 (c)			Dies, crankshafts, high pressure cylinders, heavy-duty machine tool parts, large gears, press frames.

Specification No. (a)	Grade or Class	Min Tensile Strength, Psi	Hardness, Brinell	Total Carbon, %	Microstructure	Typical Applications
ASTM A159-72 (d) SAE J431b (d)	G1800		187 Max (e)		Ferritic-pearlitic	For machinability where higher strength is not necessary.
	G2500		170-229 (e)		Ferritic-pearlitic	Small cylinder blocks and heads, pistons, clutch plates, pump bodies, gear boxes, housings, light-duty brake drums.
	G2500a		170-229 (e)	3.40 min mandatory	"A" graphite size 2-4, 15% max ferrite	Brake drums and clutch plates to minimize heat checking.
	G3000		187-241 (e)		Pearlitic	Cylinder blocks, heads, liners, fly wheels, pistons, medium-duty brake drums, clutch plates.
	G3500		207-255 (e)		Pearlitic	Truck cylinder blocks and heads, heavy flywheels and transmission cases, differential carriers.
	G3500b		207-255 (e)	3.40 min mandatory	"A" graphite size 3-5, 5% max ferrite or carbide	Brake drums and clutch plates for heavy service requiring heat resistance and higher strength.
	G3500c		207-255 (e)	3.50 min mandatory	"A" graphite size 3-5, 5% max ferrite or carbide	Extra-heavy-duty service brake drums.
	G4000		217-269 (e)		Pearlitic	Diesel engine castings, liners, cylinders, pistons, heavy parts in general.

Specification No. (a)	Grade or Class	Min Tensile Strength, Psi	Hardness, Brinell	Carbon Equivalent Max	P Max	S Max	Other	Typical Applications
ASTM A278-75 ASME SA278	40 (f) 50 60 70 (g) 80 (g)	40,000 (h) 50,000 (h) 60,000 (h) 70,000 (h) 80,000 (h)		3.8 [CE = %C + 0.3 (%Si + %P)]	0.25	0.12	Castings and test bars must be stress relieved by prescribed methods.	Pressure-containing parts for use to 650 F — valve bodies, paper-mill drier rolls, chemical process equipment, pressure vessel castings.

Specification No. (a)	Grade or Class	Min Tensile Strength, Psi	Hardness, Brinell	Carbon Equivalent	Carbon Min	Other	Typical Applications
ASTM A319-71	I	Low strength (i)	Maximum hardness at casting locations to be machined shall be agreed on by manufacturer and purchaser	3.81-4.40	3.50	When chromium is present as an alloying element, each class shall be subdivided as: (j) Type / Cr: A 0.20-0.40 B 0.41-0.65 C 0.66-0.95 D 0.96-1.20	Nonpressure-containing parts at elevated temperatures — stoker and fire box parts, grate bars, process furnace parts, ingot molds, glass molds, caustic pots, metal melting pots. Class I — superior thermal shock resistance, low strength: Class II — average thermal shock resistance: Class III — high strength at temperature.
	II	Above 30,000 may be expected (i)		3.51-4.10	3.20		
	III	As high as 40,000 may be expected		3.20-3.80	2.80		

High Alloy Gray and White Iron Castings

Specification No. (a)	Grade or Class	Min Tensile Strength, Psi	Hardness, Brinell	Composition, %	TC	Si	Mn	Ni	Cr	Cu	S	Other	Typical Applications
MIL-G-858B	1	25,000	120-180	Min	2.60	1.25	1.0	13.5	1.8	5.5			Resists corrosion, scaling, warpage, and growth. For use at elevated temperatures, galley range tops to resist acid, caustic, and salt solutions. (For galley range tops, maximums shall be 0.20 S, 0.70 P.)
				Max	3.00	2.20	1.5	17.5	3.5	7.5	0.10	0.20 P	
	2	25,000	120-180	Min	2.60	1.25	0.80	18.0	1.75				
				Max	3.00	2.20	1.30	22.0	3.50	0.50	0.10	0..20P	

Gray and White Iron Castings

Specification No. (a)	Grade or Class	Min Tensile Strength, Psi	Hardness, Brinell	Other Requirements									Typical Applications
ASTM A436-72a (k)	1	25,000 (l)	131-183	Min		1.00	0.50	13.50	1.50	5.50			Valve guides, insecticide pumps, flood gates, piston ring bands.
				Max	3.00	2.80	1.50	17.50	2.50	7.50	0.12		
	1b	30,000 (l)	149-212	Min		1.00	0.50	13.50	2.50	5.50			Sea water valve and pump bodies, pump section belt.
				Max	3.00	2.80	1.50	17.50	3.50	7.50	0.12		
	2	25,000 (l)	118-174	Min		1.00	0.50	18.00	1.50				Fertilizer applicator parts, pump impellers, pump casings, plug valves.
				Max	3.00	2.80	1.50	22.00	2.50	0.50	0.12		
	2b	30,000 (l)	171-248	Min		1.00	0.50	18.00	3.00(m)				Caustic pump casings, valve, pumps impellers.
				Max	3.00	2.80	1.50	22.00	6.00	0.50	0.12		
	3	25,000 (l)	118-159	Min		1.00	0.50	28.00	2.50				Turbocharger housings, pumps and liners, stove tops, steam piston valve rings, caustic pumps and valves.
				Max	2.60	2.00	1.50	32.00	3.50	0.50	0.12		
	4	25,000 (l)	149-212	Min		5.00	0.50	29.00	4.50				Range tops.
				Max	2.60	6.00	1.50	32.00	5.50	0.50	0.12		
	5	20,000 (l)	99-124	Min		1.00	0.50	34.0					Glass rolls and molds, machine tool, gauges, optical parts requiring minimum expansion and good damping qualities, solder rails and pots.
				Max	2.40	2.00	1.50	36.00	0.10	0.50	0.12		
	6	25,000 (l)	124-174	Min		1.50	0.50	18.00	1.00	3.50			Valves.
				Max	3.00	2.50	1.50	22.00	2.00	5.50	0.12	1.00 Mo	
					TC	Si	Mn	Ni	Cr	Mo	S	Other	
ASTM A-532-750 (m)	I Grade 1		500 Sand Cast	Min	3.0	0.3	0.3	3.3	1.4				
			600 Chill Cast	Max	3.6	0.8	0.8	5.0	2.5	0.75	0.15	0.30 P	
	I Grade 2		450 Sand Cast	Min	2.5	0.3	0.3	3.3	1.4				
			500 Chill Cast	Max	3.0	0.8	0.8	5.0	2.5	0.75	0.15	0.30 P	
	I Grade 3		525 Sand Cast	Min	2.9	0.3	0.2	2.7	1.1				
			600 Chill Cast	Max	3.7	0.6	0.5	4.0	1.5	0.75	0.15	0.30P	Grinding balls.
	II Grade 1		500 As Cast	Min	3.1	0.3	0.4		14.0	2.50			
			600 Hardened	Max	3.6	0.8	0.9	0.5	18.0	3.50	0.06	0.10 P	
	II Grade 2		500 As Cast	Min	2.4	0.3	0.4		14.0	2.50			Can be annealed to 330 max. Bhn to improve toughness.
			550 Hardened	Max	3.1	0.8	0.9	0.5	18.0	3.50	0.06	0.10 P	
	III Grade 1		400 As Cast	Min	2.3	0.2			24.0				
			550 Hardened	Max	3.0	1.5	1.5	1.2	28.0	0.60	0.06	0.10 P	
	III Grade 2		400 As Cast	Min	2.3	0.2			24.0				Can be annealed to 380 max. Bhn to improve toughness.
			550 Hardened	Max	3.0	1.5	1.5	0.5	28.0	0.06	0.06	0.10 P	
ASTM A518-64		(o)	(p)	Min	0.7	14.50			0.5	0.5		0.5 Cu	Pumps and piping for corrosive liquids.
				Max	1.1	14.75	1.5						

Table I — ASTM Tensile Bar Dimensions

Controlling Section of Castings, In.	Test Bar	Cast Bar Length, In. Min	Cast Bar Length, In. Max	Cast Bar Average Diameter, In.	Machined Bar Diameter, In.
0.25-0.50	A	5.0	6.0	0.88	0.50
0.51-1.00	B	6.0	9.0	1.20	0.75
1.00-2.00	C	7.0	10.0	2.00	1.25
Under 0.25 Over 2.0	S	Intended for use when standard bars are not satisfactory. All dimensions shall be agreed upon by manufacturer and purchaser.			

(a) ANSI = American National Standards Institute; ASME = American Society of Mechanical Engineers; ASTM = American Society for Testing and Materials; QQ-1-652c is a Federal specification, and MIL-G-858B is a U. S. Military specification.

(b) Each class number is followed by a letter, either A, B, C, or S, indicating the test bar size required for the class. All test bars must be separately cast and machined, and the tension test result is required for casting qualification. At least two test bars are required for each lot of castings intended to conform to this specification.

(c) The test bar size shall be determined by the controlling casting section if a test bar is not specified. Recommended dimensions are in Table I.

(d) Automotive castings cast in sand molds.

(e) Hardness of the casting in properly prepared area or areas established by agreement and shown on drawings.

(f) Classes 20, 25, 30, and 35 are also covered but limited to use below 450 F.

(g) Not in ASME SA278.

(h) Required test bar size is similar to Note d (except size A bar used for all sections under 0.50 in.).

(i) Low strength is desired for thermal shock resistance. Strength may be specified where essential (up to strength prescribed for Class 40, ASTM A48-74).

(j) Other alloys to increase strength and to improve and stabilize structure for elevated temperature service may be used in all classes.

(k) Austenitic gray iron castings with heat, wear, and corrosion resistance. Austenitic matrix contains uniformly distributed graphite flakes plus some carbides.

(l) Test bars machined from 1 in. keel block, or a "y" block in ½, 1, or 3 in. size by option of purchaser.

(m) When some machining is required, the 3.00-4.00% Cr range is recommended.

(n) Abrasion resisting white irons for mining, milling, and earth handling uses.

(o) Transverse strength shall be 930 lb min with 0.026 in. min deflection on special test bar.

(p) Heat treatment required.

Specifications and Properties of Ductile (Nodular) Iron Castings

Specification No.	Class or Grade	Min Tensile Strength, Psi(a)	Min Yield Strength, Psi(b)	Elongation in 2 In., %	Heat Treatment	Other Requirements		Typical Applications
ASTM A536-71	60-40-18	60,000	40,000	18	May be annealed	Chemical composition is subordinate to mechanical properties; however, the content of any chemical element may be specified by mutual agreement.		Pressure castings, such as valve and pump bodies.
	65-45-12	65,000	45,000	12				Machinery castings subject to shock and fatigue loading.
	80-55-06	80,000	55,000	6				Crankshafts, gears and rollers.
	100-70-03	100,000	70,000	3	Usually normalized			High strength gears, automotive and machine components.
	120-90-02	120,000	90,000	2	Quenched and tempered			Pinions, gears, rollers and slides.
						Hardness, Bhn(c)	Microstructure(d)	
SAE J434b	D-4018				May be annealed	170 max	Ferritic	Steering knuckles
	D-4512					156-217	Ferritic-pearlitic	Disc brake calipers
	D-5506					187-255	Ferritic-pearlitic	Crankshafts
	D-7003				May be normalized	241-302	Pearlitic	Gears
	DQ & T				Quenched and tempered	Range specified	Martensitic	Rocker arms
11466A (Ordnance)	D-4018	60,000	40,000	18	Ferritized by annealing			Components of motor vehicles, agricultural equipment, and general machinery.
	D-4512	65,000	45,000	12				
	D-5506	80,000	55,000	6				
	D-7003	100,000	70,000	3	Normalized			
	DQ & T	By agreement			Quenched and tempered			

Specification No.	Class or Grade	Min Tensile Strength, Psi(a)	Min Yield Strength, Psi(b)	Elongation in 2 In., %	Heat Treatment	Composition, %						Bhn	Typical Applications
							TC	Si	P	Other	CE (f)		
ASTM A395-71 ASME SA395	60-40-18	60,000(e)	40,000	18	Ferritized by annealing, Bhn 143-187	Min	3.0						Valves and fittings for steam and chemical plant equipment.
						Max		2.50	0.08				
ASTM A476-70	80-60-03(g)	80,000	60,000	3	To be used in as-cast condition. Hardness shall be minimum Bhn 201	Min	3.0				3.8		Paper mill dryer rolls; used up to 450 F.
						Max		3.0	0.08	0.05 S	4.5		
MIL-I-24137 (Ships) amended	Class A	60,000	45,000	15	Shall be ferritized by annealing to Bhn 190 max (h)	Min	3.0	(i)	(i)				Shipboard electric equipment, engine blocks, pumps, compressors, gears, valves, clamps, and hydraulic equipment.
						Max		2.50	0.08		4.3(j)	190	
MIL-I-22243 amended	10 ft-lb min Charpy V-notch at 20 F	55,000	37,000	20	Ferritized. See details in specifications	Max		2.20	0.05	0.05 Ti		175	For use where maximum notch toughness is needed.
ASTM A445-71 API 604	60-40-18(k)	60,000	40,000	18	Heat treated to ferritic structure	Min	3.0					143	Piping components — valves, flanges, fittings, etc.
						Max		2.50	0.08			187	
AMS 5315		60,000	45,000	15	Annealed	Min	3.2	1.7					
						Max	4.0	2.5	0.08	0.8 Mn		190	
AMS 5316		80,000	60,000	3(l)		Min	3.2	1.7				202	Parts, requiring wear resistance and strength.
						Max	4.0	2.5	0.08	0.8 Mn		269	
AGMA 244.02	165(m)	65,000	45,000	10	Annealed								
	180(m)	70,000	55,000	7	Normalized and tempered								
	210(m)	85,000	70,000	5									
	225(m)	89,000	75,000	4									
	255(m)	103,000	87,000	3									
	265(m)	107,000	92,000	2									
	285(m)	115,000	100,000	1.5									
	300(m)	123,000	105,000	1									
	350(m)	143,000	123,000	0.5									
	180(m)	98,000	75,000	7	Quenched and tempered								
	210(m)	105,000	82,000	6									
	255(m)	115,000	90,000	4									
	265(m)	120,000	95,000	3.5									
	285(m)	130,000	105,000	3									
	300(m)	135,000	110,000	2.5									
	350(m)	158,000	130,000	1									

High Alloy Ductile (Nodular) Iron Castings

Specification No.	Class or Grade	Min Tensile Strength, Psi(a)	Min Yield Strength, Psi(b)	Elongation in 2 In., %	Heat Treatment		TC	Si	Mn	P	Ni	Cr	Bhn	Typical Applications
ASTM A439-71	D-2(n)	58,000	30,000	8		Min		1.50	0.70		18.00	1.75	139	Valve stem bushings, valve and pump bodies in petroleum, salt water, and caustic service, manifolds; turbocharger housings; air compressor parts.
						Max	3.00	3.00	1.25	0.08	22.00	2.75	202	
	D-2B	58,000	30,000	7		Min		1.50	0.70		18.00	2.75	148	Turbocharger housings, rolls.
						Max	3.00	3.00	1.25	0.08	22.00	4.00	211	
	D-2C	58,000	28,000	20		Min		1.00	1.80		21.00		121	Electrode guide rings, steam turbine dubbing rings.
						Max	2.90	3.00	2.40	0.08	24.00	0.50(o)	171	
	D-3(n)	55,000	30,000	6		Min		1.00			28.00	2.50	139	Turbocharger nozzles and housings, steam turbine diaphragms, gas compressor diffusers.
						Max	2.60	2.80	1.00(o)	0.08	32.00	3.50	202	
	D-3A	55,000	30,000	10		Min		1.00			28.00	1.00	131	High-temperature bearing rings requiring gall resistance.
						Max	2.60	2.80	1.00(o)	0.08	32.00	1.50	193	
	D-4	60,000				Min		5.00			28.00	4.50	202	Diesel engine manifolds, manifold joints.
						Max	2.60	6.00	1.00(o)	0.08	32.00	5.50	273	
	D-5	55,000	30,000	20		Min		1.00			34.00		131	Guidance system housings, gas turbine shroud rings, glass rolls.
						Max	2.40	2.80	1.00(o)	0.08	36.00	0.10	185	
	D-5B	55,000	30,000	6		Min		1.00			34.00	2.00	139	Optical system mirrors and parts for dimensional stability, compressor stators.
						Max	2.40	2.80	1.00(o)	0.08	36.00	3.00	193	
ASTM A571-71 (t)	D-2M	65,000	30,000	30	Annealed	Min	2.20	1.50	3.75		21.00		121	Compressors, expanders, pumps and other presure containing parts requiring a stable austenitic matrix at −423 F.
						Max	2.70	2.50	4.50	0.08	24.00	0.20	171	
AMS 5394		55,000	32,000	7(p)	Stress relieved	Min	2.40	2.00	0.80	(q)	18.00	1.70	140	Parts requiring strength to 1,200 F. Austenitic.
						Max	3.00	3.20	1.60	0.25	22.00	2.40	180	
AMS 5395		50,000	25,000	20(s)		Min	2.50	2.00	1.90		20.00		125	Good castability and corrosion resistance. Can be fabricated by welding.
						Max	3.00	3.00	2.50	0.15	24.00	0.50(r)	175	
MIL-I-24137 (Ships) amended	Class B	55,000	30,000	7	Stress relief, 1,200 F (carbide solution at 1,750 F if necessary)	Min	2.40	1.80	0.80		18.00	1.70		Resistance to heat, corrosion, shock; nonmagnetic; shipboard use and propellers.
						Max	3.00	3.20	1.50	0.20	22.00	2.40	190	
	Class C	50,000	25,000	20		Min	2.70	2.00	1.90		20.00			
						Max	3.10	3.00	2.50	0.15	23.00	0.50	175	

Footnotes:

(a) Test bars are machined from 1 in. keel block or a Y block in ½, 1, or 3 in. size by option of purchaser.

(b) As determined by "offset method" at 0.2%.

(c) Hardness in properly prepared area is established by agreement and shown on drawings.

(d) Graphite shall be at least 80% spheroidal conforming to Types I and II in ASTM A247.

(e) Test specimen shall be machined from a test coupon based on size of controlling section of casting. Recommended dimensions are: 1 in. and under controlling section of casting — 1 in. coupon (may be Y or keel block); 1 to 3 in. — 3 in. coupon (must be Y block); over 3 in. — larger coupons may be used by agreement.

(f) % Carbon Equivalent = % TC + 0.3 (% Si + % P).

(g) Tensile strength, yield strength, and elongation properties obtained from 1 in. thick test coupon. For 3 in. thick coupon, tensile and yield strength specifications are identical; elongation not required.

(h) One metallographic test shall be made for each lot after annealing. Microstructure at 50 × shall show a matrix of 90% ferrite min with no primary carbides; all graphite spheroidal.

(i) For castings with ½ in. sections and smaller 2.75 Si max and 0.08 P are allowed or 3.00 Si max with 0.05 P max.

(j) Applies to castings with sections 2 in. and over. CE = TC + 1/3 Si.

(k) Microscopical examination may be substituted for some tension tests.

(l) 2% on bars cut from castings.

(m) Recommended tooth hardness (Bhn min).

(n) Additions of 0.7 to 1.0 Mo will increase the mechanical properties above 800 F.

(o) Not intentionally added.

(p) 5% on bars cut from castings.

(q) 0.003 Pb max; 0.50 Cu max.

(r) 0.30 Mo max.

(s) 15% on bars cut from castings.

(t) Notched impact of 15 ft-lb at — 320 F.

Source: Gray & Ductile Iron Founders' Society Inc., Rocky River, Ohio.

Properties, Applications, and Cutting Conditions for Malleable Iron Castings

Properties and Applications

Designation	Mechanical Properties: Tensile Strength*, Psi	Yield Strength*, Psi	Elongation in 2 In.*, %	Brinell Hardness†	Other Properties	Typical Applications
Ferritic						
35018	53,000	35,000	18	110-156	Modulus of elasticity in tension: 25 x 10^6 psi; fatigue strength: 40 to 60% of tensile strength; damping capacity (energy lost in one vibration cycle): 4.2% at 30,000 psi; compressive strength: 200,000 psi; torsional strength: equals tensile strength; shear strength: 80 to 90% of tensile strength.	Iron grillework, railroad car hardware, hand tools, high-pressure parts, hardware for oil industry.
32510	50,000	32,500	10	110-156		Gear cases and housings, chain links, auto hinges, brackets, mounting pads, brake shoes, wheel hubs.
Pearlitic						
40010	60,000	40,000	10	149-197	Modulus of elasticity (tension): 25.5 x 10^6 to 28 x 10^6 psi; modulus of elasticity (compression): 23.2 x 10^6 psi; modulus of elasticity (torsion): approx. 9.97 x 10^6 psi (range, 8.79 to 10.71 x 10^6 psi); damping capacity (energy lost in one vibration cycle): 1.9% at 5,000 psi, 5.3% at 35,000 psi; compressive strength: 197,000 to 290,000 psi; compressive yield strength: 63,000 to 122,000 psi; torsional strength: 72,000 to 160,000 psi; yield strength (torsion): 33,000 to 110,000 psi; shear strength: 50,000 to 77,000 psi.	C-clamps, diesel engine brackets, levers, transmission cases, artillery shells, gears, farm implement parts.
45008	65,000	45,000	8	156-197		
45006	65,000	45,000	6	156-207		
50005	70,000	50,000	5	179-229		
60004	80,000	60,000	4	197-241		Pistons for diesel engines, differential axle cases, rocker arms, clutch hubs, transmission gears, universal joint yokes, crankshafts, idler gears and shafts.
70003	85,000	70,000	3	217-269		
80002	95,000	80,000	2	241-285		
90001	105,000	90,000	1	269-321		

* ASTM minimum (A220-68).
† Hardnesses are listed for informational purposes only; they are typical but not part of the ASTM specification.

Recommended Cutting Conditions for Turning

Grade	Type of Cut‡	Feed, In. per Rev.	Depth of Cut, In.	Cutting Speed, Ft per Min (for Indicated Tool Life): Dry, 20 Min	Dry, 30 Min	Dry, 40 Min	Soluble Oil, 20 Min	Soluble Oil, 30 Min	Soluble Oil, 40 Min
32510 (Bhn 109)	Roughing skin	0.015	0.100	440	390	340	500	410	340
		0.030	0.100	350	300	250	410	360	600
	Coarse underskin	0.015	0.060	600	510	430	840	720	630
		0.030	0.060	440	380	350	600	530	460
	Finish	0.003	0.010	1,380	1,240	1,150	1,660	1,520	1,400
		0.007	0.010	1,210	950	700	1,380	1,220	1,080
48004 (Bhn 179)	Roughing skin	0.015	0.100	230	180	130	270	220	130
		0.030	0.100	160	120	70	200	150	110
	Coarse underskin	0.015	0.060	300	260	230	355	330	280
		0.030	0.060	230	185	150	260	230	210
	Finish	0.003	0.010	590	515	460	700	650	615
		0.007	0.010	510	470	450	670	610	550
60003 (Bhn 230)	Roughing skin	0.015	0.100	175	140	115	200	165	140
		0.030	0.100	130	115	100	150	120	100
	Coarse underskin	0.015	0.060	245	220	200	285	240	225
		0.030	0.060	170	145	125	210	165	135
	Finish	0.003	0.010	525	495	465	540	510	480
		0.007	0.010	470	430	400	480	440	415
80002 (Bhn 250)	Roughing skin	0.015	0.100	—	—	—	195	165	135
		0.030	0.100	—	—	—	145	115	90
	Coarse underskin	0.015	0.060	185	160	145	190	170	155
		0.030	0.060	160	125	100	160	125	100
	Finish	0.003	0.010	—	—	—	470	420	385
		0.007	0.010	—	—	—	365	330	305

‡Tool geometry for roughing skin and coarse underskin cuts: —5° BR; 15° SCEA; —5° SR; 15° ECEA; 5° relief. Tool geometry for finish cuts: 0° BR; 15° SCEA; 5° SR; 15° ECEA; 5° relief.

All tests were made under the direction of the Machining Subcommittee of the Malleable Founders Society, Cleveland. Grade C2 cutting tools were used dry or with soluble-oil cutting fluid. Tests were discontinued at a uniform wearland of 0.015 in. Discontinued grades 48004 and 60003 approximate grades 50005 and 60004, respectively.

Types of Graphite Flakes in Gray Iron

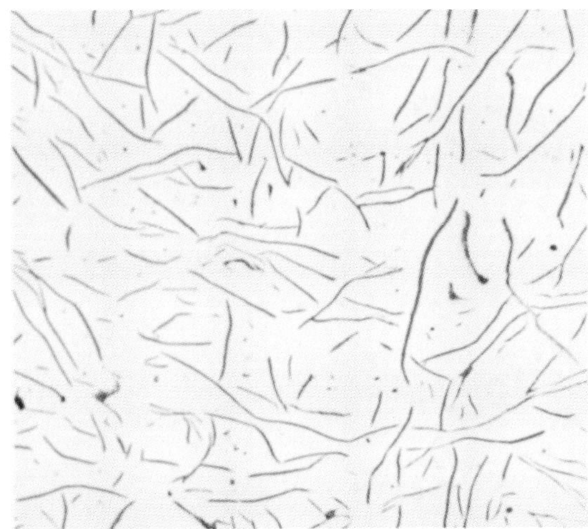

As polished (not etched) 100×

645 Type A distribution of graphite flakes in gray iron, characterized by uniform distribution and random orientation.

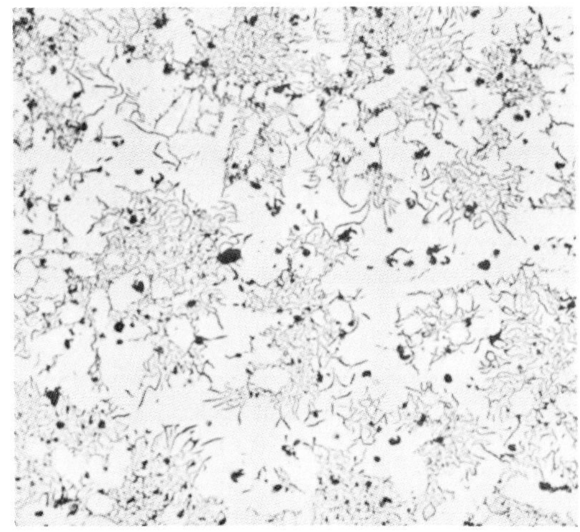

As polished (not etched) 100×

646 Type B distribution of graphite flakes in gray iron, characterized by rosette grouping and random orientation.

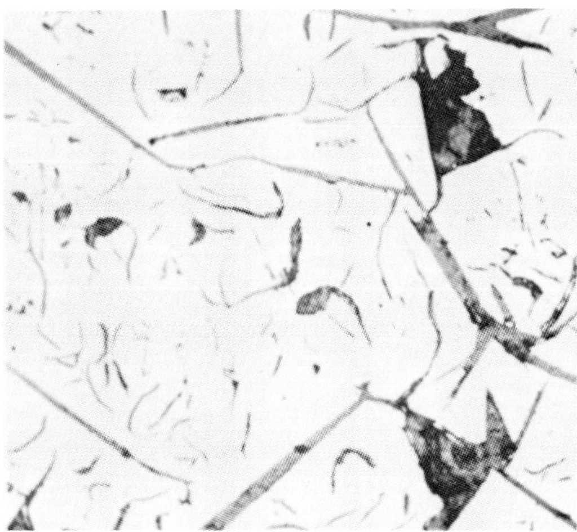

As polished (not etched) 100×

647 Type C distribution of graphite flakes in gray iron, characterized by superimposed flake size and random orientation.

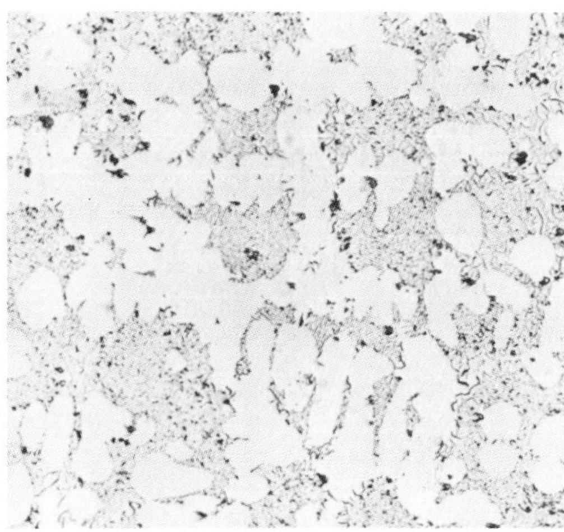

As polished (not etched) 100×

648 Type D distribution of graphite flakes in gray iron, characterized by interdendritic segregation and random orientation.

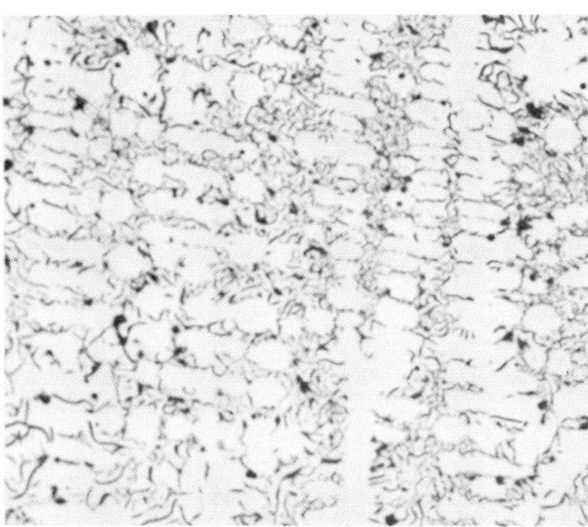

As polished (not etched) 100×

649 Type E distribution of graphite flakes in gray iron, characterized by interdendritic segregation and preferred orientation.

3:1 methyl acetate – liquid bromine 130×

650 Scanning electron micrograph of hypereutectic gray iron with matrix etched out to show position of type B graphite in space.

Steel Castings

*By the ASM Committee on Steel Castings**

THE CARBON AND LOW-ALLOY STEEL CASTINGS dealt with in this article are classified in four general groups according to their carbon or alloy contents. Carbon steel castings account for three of these groups: (*a*) low-carbon steel castings with less than 0.20% C, (*b*) medium-carbon with 0.20 to 0.50% C, and (*c*) high-carbon with more than 0.50% C. The fourth group, low-alloy steel castings, is limited to grades with a total alloy content of less than 8%.

Other types of steel castings are discussed in separate articles on heat-resistant castings, stainless (corrosion-resistant) steel castings, and austenitic manganese steel castings.

The many types of carbon and low-alloy steel produced in wrought form can also be made as steel castings. Such castings are produced by pouring molten steel of the desired composition into a mold of the desired configuration and allowing the steel to solidify. The mold material may be silica, zircon or olivine sand, graphite, metal or ceramic. Choice of mold material depends on the size, intricacy and accuracy of the casting, and on cost. While the producible size, surface finish and dimensional accuracy of castings vary widely with the type of mold, the properties of the cast steel are not affected significantly. Steel castings produced in any of the various types of molds and wrought steel of equivalent chemical composition respond similarly to heat treatment, have the same weldability and similar physical and mechanical properties. Cast steels do not exhibit the effects of directionality on mechanical properties that are typical of wrought steels.

Specifications

Steel castings are usually purchased to meet specified mechanical properties. Table 1 lists some standard ASTM, SAE and Government specifications. In the low-strength ranges, some specifications limit carbon and manganese content, usually to insure satisfactory weldability. In the SAE specifications, carbon and manganese are specified to insure that the minimum desired hardness and strength are obtained after heat treatment. For special applications other elements may be specified, either as maximum or minimum, depending on the characteristics desired.

Other ASTM specifications for carbon and low-alloy castings are A216, A217, A352, A356, A389 and A426. Chemical composition is often left to the discretion of the casting supplier. However, limits may be established to facilitate welding, uniform response to heat treatment, or other requirements. Hardness is specified in some SAE specifications to insure machinability, ease of inspection for high-production-rate items, or certain characteristics pertaining to wear.

SAE steel castings specifications include three grades, HA, HB and HC, with specified hardenability requirements. Figure 1 plots hardenability requirements, both minimum and maximum, for these steels. Hardenability is determined by the end-quench hardenability test described in the article on Selection of Steel for Hardenability. Other specifications require minimum hardness at one or two locations on the end-quench specimen. In general, hardenability is specified to insure a predetermined degree of transformation from austenite to martensite during quenching, in the thickness required. This is important in critical parts requiring optimum resistance to fatigue and shock.

Among the most commonly selected grades of steel castings are (*a*) low-carbon annealed steel corresponding to QQ-S-681 class 2, ASTM 65-35, or SAE 0030, and (*b*) a higher-strength steel, often alloyed or fully heat treated, or both, similar

*J. F. WALLACE, *Chairman,* Associate Professor, Metallurgical Engineering, Case Institute of Technology; JESS L. BELSER, Assistant Chief, Operations, Watertown Arsenal; T. A. BENTON, Chief Metallurgist, Engineering Materials Research, International Harvester Co.; M. R. BOWERMAN, Director of Research, Alliance Machine Co.; CHARLES W. BRIGGS, Technical and Research Director, Steel Founders' Society of America; W. FRED CARN, Chief Metallurgist, Precision Metalsmiths, Inc.; F. B. HERLIHY, Director of Metallurgical Research, American Brake Shoe Co.

W. T. HUNT, Manager, Laboratory and Services, Defense Engineering, Chrysler Corp.; CLYDE B. JENNI, Chief Metallurgist, General Steel Castings Corp.; SIDNEY LOW, Director of Research, Chapman Valve Mfg. Co.; R. H. MARSHALL, Chief Materials Engineer, Frank G. Hough Co.; L. H. MCCREERY, Boeing Airplane Co.; GEORGE P. MESSENGER, Materials Branch, Ordnance Tank-Automotive Command, Detroit Arsenal; JOHN D. NIELSEN, Design Engineer, Thew Shovel Co.

CLYDE O. PENNEY, Metallurgist, Denver and Rio Grande Western Railroad Co.; JACK H. PENROSE, Chief Metallurgist, Watervliet Arsenal; JOHN A. RASSENFOSS, Manager, Manufacturing Research Laboratory, American Steel Foundries; JOHN E. READ, General Supervisor of Laboratories, Cadillac Motor Car Div., General Motors Corp.; A. T. WESTBROOK, Department of Research and Development, Canadian National Railways; C. H. WYMAN, Chief Metallurgist, LFM Manufacturing Co., Inc., Rockwell Mfg. Co.; R. S. ZENO, Large Steam Turbine-Generator Dept., General Electric Co.

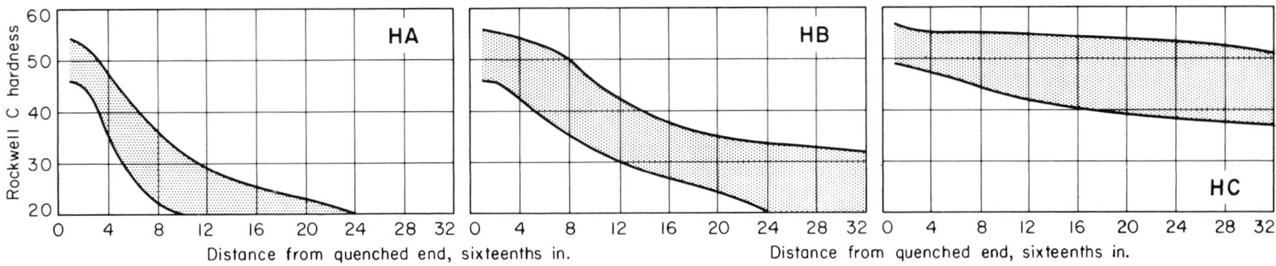

Fig. 1. End-quench hardenability limits (SAE) for three cast steels with nominal carbon content of 0.30% C (see Table 1). Manganese and other alloying elements are added as required to produce castings that meet these limits.

to QQ-S-681 class 4C2, ASTM 105-85, or SAE 0105.

Particularly where the purchaser heat treats the part after other processing, a casting will be ordered to composition limits closely equivalent to the SAE-AISI wrought steel compositions with somewhat higher silicon permitted. As in other steel castings, it is best not to specify silicon to a range, but to permit the foundry to utilize the silicon and manganese combination needed to achieve required soundness in the shape being cast. The silicon content is frequently higher in cast steels than for the same nominal composition in wrought steel. The effect of silicon on mechanical properties is not significant up to about 0.70%. Silicon above 0.80% is considered an alloy addition, since it contributes significantly to resistance to tempering.

Railroad equipment manufacturers and other major users of steel castings may prefer their own or industry association specifications. Users of steel castings for extremely critical applications such as aircraft may use their own, industry association, or special-purpose military specifications. Foundries frequently make nonstandard grades for special applications or have their own specification system to meet the needs of the purchaser. Savings may be realized by using a grade that is standard with a foundry, especially when small quantities are needed.

Mechanical Properties

If ferritic steels are compared at a given level of hardness and hardenability, the tensile and yield strengths of cast, rolled, forged and welded metal are virtually identical regardless of alloy content. Consequently, where tensile and yield properties are controlling criteria, the designer can interchange rolled, forged, welded and cast steel.

Ductility. If ductility properties are compared at a prescribed hardness, cast, forged, rolled or welded steels are nearly the same. The longitudinal properties of rolled or forged steel are somewhat higher than the properties of cast steel or weld metal. However, the transverse properties are lower by an amount that depends on the amount of working. When service conditions involve multidirectional loading, the nondirectional characteristic of cast steels is advantageous.

Impact. The notched-bar impact test is often used as a measure of the toughness of materials and is particularly useful in determining the transition temperature from ductile to brittle fracture. The impact properties of wrought steels are usually listed for the longitudinal direction and the values shown are higher than those for cast steels of equivalent composition and thermal treatment. The transverse impact properties of wrought steels are usually 50 to 75% of those in the longitudinal direction above the transition temperature and, in some conditions of composition and degree of working, even lower. Since cast steels are nondirectional, their impact properties usually fall somewhere between the longitudinal and transverse properties of wrought steel of similar composition.

Specimens. Mechanical properties are determined from test bars machined from coupons cast as a part of the casting, from test coupons separately cast from the same melt

Table 1. Summary of Specification Requirements for Steel Castings

Class or grade	Tensile strength, min psi	Yield strength, min psi	Elongation in 2 in., min %	Reduction of area, min %	Chemical restrictions, max(a) C	Mn
Federal Specification QQ-S-681						
X	....	....	....	....	0.30	1.00
0	....	....	....	....	0.45	1.00
	60,000	30,000	24	35	0.30	0.60
2	65,000	35,000	20	30	0.35	0.70
3	80,000	40,000	17	25	0.50	....
4A1	75,000	40,000	24	35	....	....
4A2	85,000	53,000	22	35	....	....
4B1	85,000	55,000	22	40	....	....
4B2	90,000	60,000	22	45	....	....
4B3	100,000	65,000	17	30	....	....
4C1	90,000	65,000	20	45	....	....
4C2	105,000	85,000	15	30	....	....
4C3	120,000	100,000	12	30	....	....
4C4	150,000	125,000	10	25	....	....
ASTM Specification A27-58(b)						
N-1	....	....	....	....	0.25	0.75
N-2	....	....	....	....	0.35	0.60
N-3	....	....	....	....	....	1.00
U-60-30	60,000	30,000	22	30	0.25	0.75
60-30	60,000	30,000	24	35	0.30	0.60
65-30	65,000	30,000	20	30	....	....
65-35	65,000	35,000	24	35	0.30	0.70
70-36	70,000	36,000	22	30	0.35	0.70
70-40	70,000	40,000	22	30	0.25	1.20
ASTM Specification A148-58						
80-40	80,000	40,000	18	30	....	....
80-50	80,000	50,000	22	35	....	....
90-60	90,000	60,000	20	40	....	....
105-85	105,000	85,000	17	35	....	....
120-95	120,000	95,000	14	30	....	....
150-125	150,000	125,000	9	22	....	....
175-145	175,000	145,000	6	12	....	....
SAE Specifications(c,d)						
0022	....	....	....	....	0.12 to 0.22	0.50 to 0.90
0030(g)	65,000	35,000	24	35	0.30	0.70
0050(h)	85,000	45,000	16	24	0.40 to 0.50	0.50 to 0.90(e)
0050(i)	100,000	70,000	10	15	0.40 to 0.50	0.50 to 0.90(f)
080(j)	80,000	40,000	18	30	....	....
090(k)	90,000	60,000	20	40	....	....
0105(l)	105,000	85,000	17	35	....	....
0120(m)	120,000	100,000	14	30	....	....
0150(n)	150,000	125,000	9	22	....	....
0175(o)	175,000	145,000	6	12	....	....
HA(p)	....	....	....	....	0.25 to 0.34	....
HB(p)	....	....	....	....	0.25 to 0.34	....
HC(p)	....	....	....	....	0.25 to 0.34	....

(a) Carbon and manganese are maximum limits unless a range is given. All specifications restrict phosphorus to 0.05% and sulfur to 0.06% max. Silicon and alloying elements are restricted on some grades. (b) For each reduction of 0.01% C below the maximum specified, an increase of 0.04% Mn above the maximum specified will be permitted to a maximum of 1.40% for grade 70-40 and 1.00% for the other grades. (c) For each reduction of 0.01% C below the maximum specified, an increase of 0.04% Mn above the maximum specified will be permitted to a maximum of 1% Mn. (d) Hardness values given in footnotes (g) through (o) are nominal and applicable to casting sections not over 3 in.

(e) Normalized or normalized and tempered. (f) Quenched and tempered. (g) 131 Bhn. (h) 170 Bhn. (i) 207 Bhn. (j) 163 Bhn. (k) 187 Bhn. (l) 217 Bhn. (m) 248 Bhn. (n) 311 Bhn. (o) 363 Bhn. (p) Purchased on the basis of hardenability. Manganese and other elements added as required (Fig. 1).

of steel used to produce the castings, or from test bars prepared from specified areas of the casting. Even though test specimens of the first two types are heat treated in the same operation as the castings they represent, their mechanical properties could differ significantly from those of similar specimens taken from the actual casting. Coupons machined from sections of a casting may show properties ranging from 100 to 75% as great as those exhibited by separately cast or cast-on test bars. These factors contribute to the disparity: (*a*) pouring and solidification conditions for separately cast or cast-on test bars can be made practically ideal, and (*b*) response to heat treatment may be vastly different because of differences in cross section or thickness at the time of heat treatment unless the composition has been adjusted adequately to insure hardening throughout the casting thickness.

For these reasons, separately cast test bars and cast-on test bars have been replaced in some instances by test bars prepared from specified areas, for those castings that require a minimum weight or minimum design factor, as in aeronautical applications. The preparation of test specimens from actual castings is costly, both from the standpoint of the casting destroyed and from the fact that much work and expense are involved in preparing a specimen. However, this type of specimen, used on a statistical basis, provides a high degree of reliability.

When cast steel is used in place of a wrought product, increasing the strength requirements or design safety factor to compensate for variations in the casting is neither necessary nor desirable. This increases cost and weight of the part and may even decrease service life or fatigue life, particularly at higher strength levels. In general, when tensile strength is 200,000 psi or more, the ratio of yield strength to tensile strength may decrease to some extent. Fatigue limits of cast steel are in the same range as for wrought steel.

Heat treatment can provide the wide range of mechanical properties required for different applications. Lowest strength and hardness are obtained with low-carbon steel in

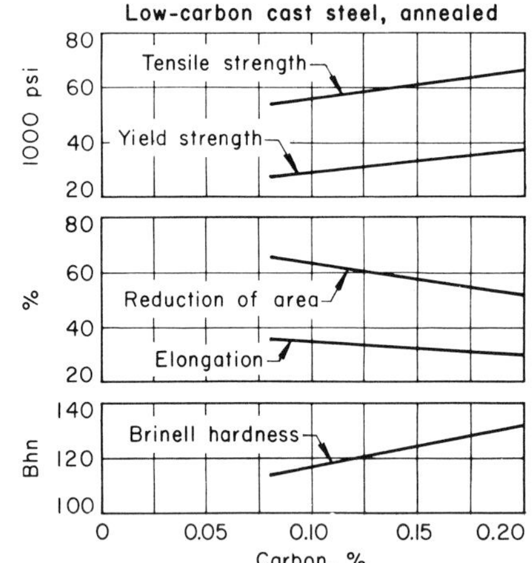

Fig. 2. Effect of carbon content on mechanical properties of annealed low-carbon cast steel. Normalizing these steels has negligible effect on mechanical properties. Properties of the steels as cast are also closely similar to those of the annealed material.

the annealed condition and highest strength with liquid-quenched alloy steel. Chemical composition and mechanical properties are usually specified as a range. Mechanical properties may be specified as minimums, since there is a variation in composition and mechanical properties within any one grade and even within any one heat of steel. Variation in mechanical properties will also depend on specific response to heat treatment. Table 2 shows the variations in composition and mechanical properties of 12 heats of closely similar carbon-molybdenum steel, some annealed and some in the normalized and tempered condition.

Low-Carbon Cast Steels

Low-carbon cast steels are those with a carbon content less than 0.20%. Most of the tonnage produced in the low-carbon classification contains between 0.16 and 0.19% C, with 0.50 to 0.80% Mn, 0.35 to 0.70% Si, 0.05% max P and 0.06% max S. For electrical equipment, in order to obtain high magnetic properties, the manganese content is usually held between 0.10 and 0.20%.

The mechanical properties at room temperature of annealed cast steels containing from 0.08 to 0.20% C are illustrated in Fig. 2. The properties of dynamo steel may be slightly below those shown because of its low manganese content. There is very little difference in the properties of the low-carbon steels resulting from the use of normalizing or full annealing heat treatments; hence the data are presented as one set of curves. In cast steels, as in rolled steels of this composition, increasing carbon increases strength and decreases ductility. Although the mechanical properties of low-carbon cast steels are nearly the same in the as-cast condition as they are after annealing, low-carbon steel castings are often annealed for stress relief.

Low-carbon steel castings are made in two important classes. One may be termed "railroad castings" and the other "miscellaneous jobbing castings". The railroad castings consist mainly of comparatively symmetrical and well-designed castings where possible adverse stress conditions have been carefully studied and avoided. Miscellaneous jobbing castings present a wide variation in design and frequently involve the joining of light and heavy sections. Varying sections make it more difficult to avoid high residual stress in the as-cast shape. Since in many service applications, residual stresses of large magnitude cannot be tolerated, stress relieving becomes necessary. The annealing of those castings is, therefore, decidedly beneficial even though it may cause little improvement of mechanical properties.

The composition and properties of low-carbon cast steel in the as-cast condition, averaged from a study of over 2000 consecutive heats, are:

Carbon	0.189%
Manganese	0.740%
Silicon	0.370%
Phosphorus	0.013%
Sulfur	0.026%
Tensile strength	64,465 psi
Yield point	34,690 psi
Elongation in 2 in.	32.9%
Reduction of area	53.0%

A comparison of these data and those presented in Fig. 2 will show that very little change is produced by annealing this grade of steel. There is, however, an increase in the impact value as the result of heat treatment. The Charpy V-notch impact value at room temperature for

Table 2. Composition and Typical Mechanical Properties of 12 Heats of Closely Similar Cast Carbon-Molybdenum Steel (a)

Item	Composition, %						Tensile strength, psi	Yield strength, psi	Elongation, %	Reduction of area, %	Hardness, Bhn	Charpy V-notch, ft-lb at 70 F
	C	Mn	Si	Ni	Cr	Mo						
1 (b)...	0.24	0.54	0.41	0.70	0.11	0.15	73,500	40,000	28.0	46.0	143	44, 40, 34, 39
2 (b)...	0.26	0.59	0.45	0.09	0.14	0.14	76,600	42,500	27.0	46.3	149	34, 36, 34, 35
3 (c)...	0.23	0.55	0.43	0.18	0.10	0.11	70,500	41,500	27.5	43.5	140	42, 40
4 (b)...	0.23	0.55	0.43	0.18	0.10	0.11	67,800	37,000	28.0	39.5	138	30, 28
5 (c)...	0.21	0.56	0.50	0.16	0.10	0.14	68,250	43,000	35.0	53.3	140	47, 56
6 (b)...	0.21	0.56	0.50	0.16	0.10	0.14	65,500	36,000	32.0	48.5	138	30, 28
7 (c)...	0.22	0.52	0.50	0.15	0.10	0.12	72,000	43,500	33.0	49.0	141	64, 61
8 (b)...	0.22	0.52	0.50	0.15	0.10	0.12	68,750	37,250	27.5	40.0	138	34, 30
9 (c)...	0.21	0.57	0.34	0.14	0.16	0.13	69,000	42,300	33.0	53.0	138	58, 46
10 (b)...	0.21	0.57	0.34	0.14	0.16	0.13	67,000	37,250	35.0	46.3	136	35, 32
11 (b)...	0.20	0.56	0.42	0.07	0.11	0.07	66,000	37,500	28.0	54.1	...	40
12 (b)...	0.20	0.58	0.49	0.06	0.05	0.06	70,000	40,000	33.0	55.2	...	41, 40

(a) Equivalent to ASTM A27, grade 65-35; properties from a 3 by 5 by 7-in. keel block; basic electric steel, deoxidized with 2½ lb of aluminum per ton of molten metal. (b) Annealed at 1700 F for 12 hr, furnace cooled. Microstructure, 30% fine pearlite, 70% primary ferrite. (c) Normalized 1700 F, 8 hr, air cooled, tempered 1240 F, 2 hr, air cooled. Structure, 25% fine pearlite, 75% primary ferrite.

an annealed 0.18% C steel will be about 40 ft-lb, while in the as-cast condition, it is about 25 ft-lb.

Since the as-cast properties are very similar to annealed properties, it is unnecessary to heat treat a considerable tonnage of commercial steel castings, especially of the railroad type, which contain less than 0.20% C. Castings for electrical or magnetic equipment are usually fully annealed as this improves the electrical and magnetic properties.

An increase in mechanical properties above those reported in Fig. 2 may be obtained by quenching and tempering, provided the design of the casting is such that it can be liquid quenched without cracking. The impact resistance is improved by quenching and tempering, especially if a high tempering temperature is employed.

Fatigue Strength. The ratio of fatigue limit to tensile strength for the low-carbon cast steels varies somewhat but is approximately 45%. This approximate ratio is also maintained at low and high temperatures, and is not much affected by the various types of heat treatment which the steel may receive.

In designing cast steel structures based on the fatigue ratio, it is advisable to use 40% of the tensile strength for a smooth bar when actual fatigue test values cannot be obtained. To this approximate figure is added a factor of safety.

Uses. As has been mentioned, important castings for the railroads are produced from low-carbon cast steel. Some castings for the automotive industry are produced from this class of steel, as are annealing boxes, annealing bottoms and hot metal ladles. Low-carbon steel castings are also produced for case carburizing, by which process the castings are given a hard, wear-resistant exterior while retaining a tough, ductile core. The magnetic properties of this class of steel make it useful in the manufacture of electrical equipment. Free-machining cast steels containing 0.08 to 0.30% S are also produced in low-carbon grades.

Medium-Carbon Cast Steels

The medium-carbon grades of cast steel contain 0.20 to 0.50% C and represent the bulk of the steel casting production. In addition to carbon, they contain 0.50 to 1.00% Mn, 0.35 to 0.80% Si, 0.05% max P and 0.06% max S. The mechanical properties at room temperature of as-cast steels containing from 0.20 to 0.50% C are shown in Fig. 3. Steels in this carbon range are seldom used in the as-cast condition but are usually heat treated, which relieves casting strains, refines the as-cast structure, and improves the ductility of the steel.

Figure 3 also shows that by fully annealing medium-carbon steel castings it is possible to increase the yield strength, the reduction of area and the elongation over the entire range, compared with as-cast properties. This increase is pronounced for steel with carbon content between 0.25 and 0.50%. The hardness and tensile strength of the as-cast steel fall off slightly following full annealing. A very large proportion of steel castings of this grade are given a normalizing treatment, which sometimes is followed by a tempering treatment.

The improvement in mechanical properties of medium-carbon cast steel that may be expected after normalizing or normalizing and tempering are shown in the right-hand part of Fig. 3.

If the design of a casting is suitable for liquid quenching, further improvements are possible in the mechanical properties. In fact, to develop mechanical properties to the fullest degree, steel castings should be heat treated by liquid quenching and tempering. Commercial procedure calls for tempering to obtain the desired strength level. Tempering temperatures of 1200 to 1300 F are usually used to obtain higher ductility and impact properties.

Figure 4 illustrates the range of mechanical properties and carbon and manganese contents for more than 2000 heats of low-carbon and medium-carbon steel castings in the as-cast, annealed, normalized, and normalized and tempered conditions. For the 0.17% C, 0.75% Mn steel, the beneficial effect of the normalizing and tempering treatment on yield strength, elongation, and reduction of area is apparent. Comparing this steel with the 0.20% C, 1.25% Mn normalized steel indicates the strengthening effect of higher carbon and higher manganese. The elongation and reduction of area of the latter steel can be improved by tempering.

Effect of Mass. The effect of increasing mass on the mechanical properties of a medium-carbon cast steel in the as-cast and annealed conditions is illustrated in Fig. 5. In sizes up to 8 in. square, as size increases tensile strength decreases. Most of the total effect occurs as size is increased up to 4 in. square. The difference between the as-cast and annealed tensile strength is minor. Differences in elongation and reduction of area are much more significant. Specimens in the annealed condition have twice the elongation and more than twice the reduction of area of as-cast material.

High-Carbon Cast Steels

Cast steels containing more than 0.50% C are classified as high-carbon steels. This grade also contains 0.50 to 1.00% Mn, 0.35 to 0.70% Si, 0.05% max P and S. The mechanical properties of high-carbon steels at room temperature are shown in Fig. 6. These test results were obtained on cast steel in the fully annealed condition. Occasionally a normaliz-

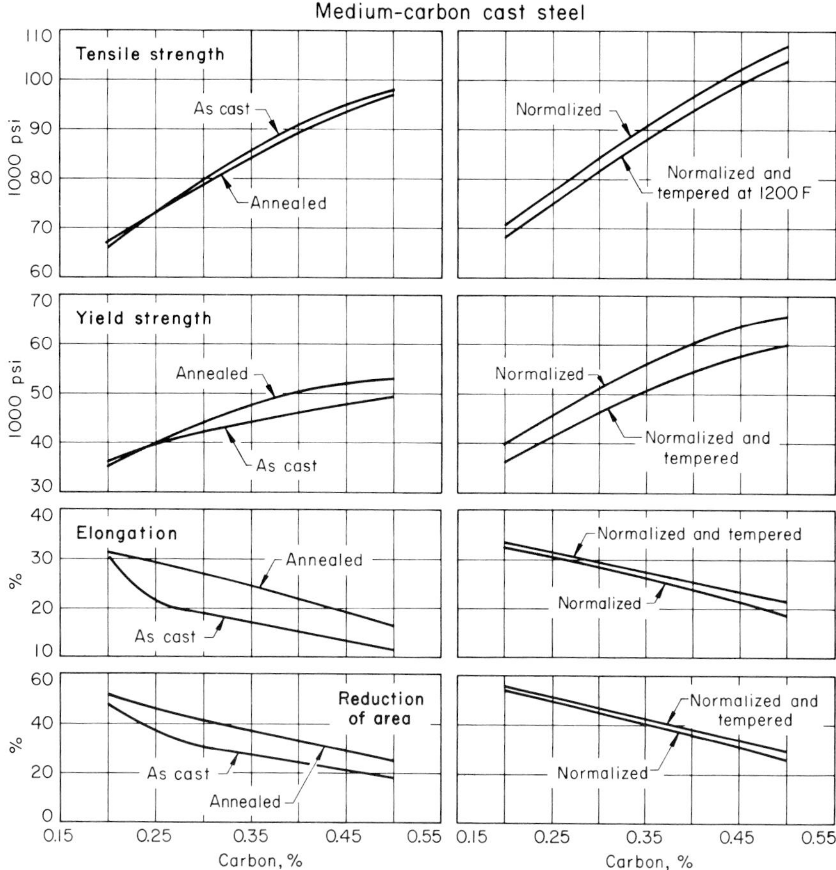

Fig. 3. Effect of carbon on mechanical properties of medium-carbon steel castings in as-cast, annealed, normalized, and normalized and tempered conditions

ing and tempering treatment is given, and for certain applications, an oil quenching and tempering treatment may be used.

Low-Alloy Cast Steels

Low-alloy cast steels have been developed with properties to satisfy special requirements such as higher and lower operating temperatures, ability to withstand greater pressures, wear resistance, higher impact resistance, higher strength with increased toughness, and higher hardenability. These materials are produced to meet specification requirements of from 70,000 to 200,000 psi tensile strength with a total alloy content less than 8%.

Alloy cast steels are used in machine tools, high-speed transportation units, steam turbines, valves and fittings, railway, automotive, excavating and chemical processing equipment, pulp and paper machinery, refinery equipment, rayon machinery and various types of marine equipment. They are also used in the aeronautical field.

Low-alloy cast steels may be divided into two classes according to their use: (*a*) those for structural parts of increased strength, hardenability and toughness and (*b*) those resistant to wear, abrasion or to heat and corrosive attack. There can be no sharp distinction between the two classes as many steels serve in both fields.

The present trend toward decreasing weight through the use of high-strength materials in lighter sections has had a marked effect on the development of low-alloy cast steels. Efforts are now being directed to producing steels with a yield strength 50% higher and a tensile strength 40% higher than carbon steels, with a ductility and impact resistance at least equal to unalloyed steels. Some 75 to 100 combinations of the available alloying materials have been regularly or occasionally used. It is doubtful that this many variations in composition are necessary or economical.

Manganese is the cheapest of the alloying elements and has an important effect in increasing the hardenability of steel. For this reason many of the low-alloy cast steels now contain between 1 and 2% Mn.

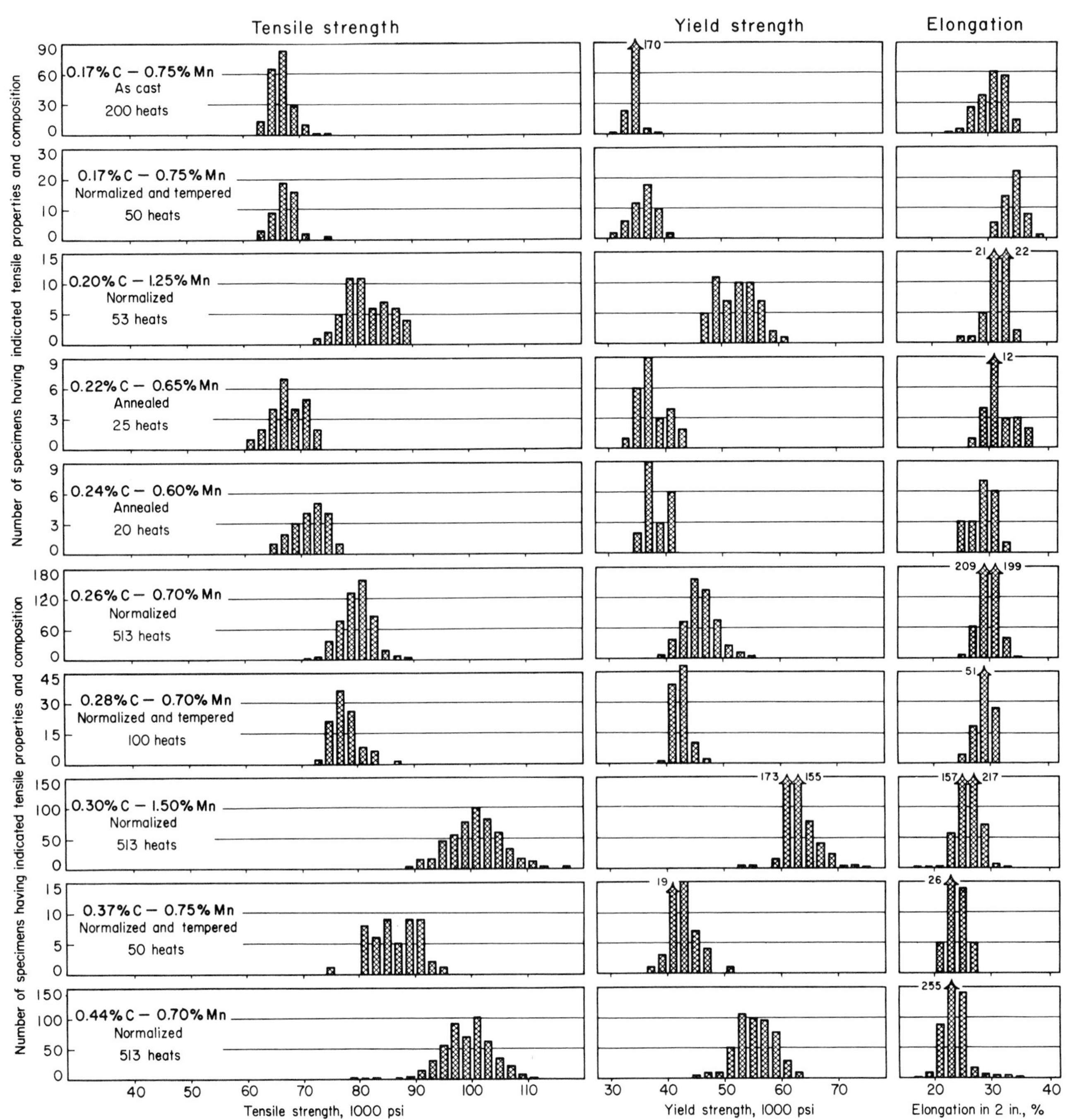

Fig. 4. Distribution of mechanical properties, carbon and

In the normalized steels where grain refinement also is needed, vanadium, titanium or aluminum is often added. Nickel or molybdenum with manganese refines the grain structure to a lesser extent, but each is important for increasing the ability of the steel to air harden. Copper and chromium impart considerable hardenability. Copper-containing steels are sometimes precipitation hardening, and therefore have increased tensile and yield strengths.

Cast steels containing chromium, molybdenum, vanadium and tungsten have given good service in valves, fittings, turbines and oil-refinery parts, which are subjected to steam temperatures up to 1200 F. Nickel and nickel-vanadium steels are being used for parts exposed to subzero conditions, such as return headers, valves and pump castings in oil-refinery dewaxing processes, because of good notch toughness at the lower temperatures. Chromium leads the field as an alloying element for wear-resistant steels. However, it is seldom used alone. Substantial quantities of steels with chromium and molybdenum, nickel, vanadium or manganese are being produced. Nickel-vanadium, manganese-molybdenum and nickel-manganese cast steels are used for numerous wear-resistance and high-strength structural purposes.

A massive 900-lb test casting made from a silicon-deoxidized Cr-Mo-V steel comparable to ASTM A389, grade C24, is shown in Table 3. Standard tension test specimens were machined from normalized and tempered sections A and B; mechanical properties obtained are tabulated with the figure and show the similarity of properties obtained within a given section. Although there is some variation, all tests met the specified requirements.

Two 500-lb test castings (approximately 1-ft cubes) were also poured and mechanical properties were determined from test bars machined from a slab cut horizontally through the middle. The steel was basic electric, silicon-deoxidized, with the following composition: 0.16 C, 0.83 Mn, 0.30 Si, 0.007 S, 0.012 P, 1.12 Cr, 0.11 Ni, 0.97 Mo, and 0.25 V. The castings were homogenized at 1925 F for 12 hr, air cooled, tempered at 1275 F (casting 1, for 40 hr; casting 2, for

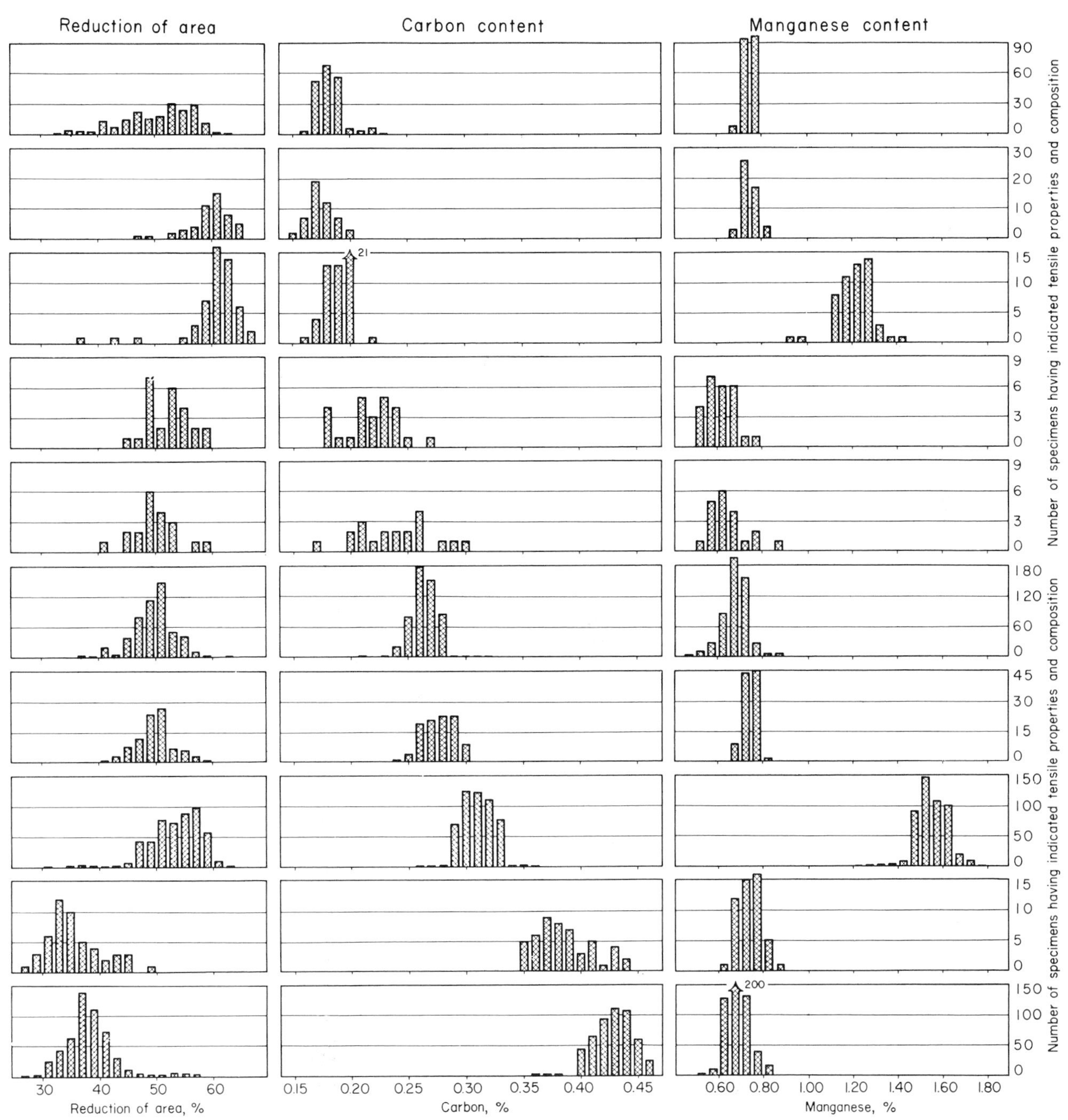

manganese for low-carbon and medium-carbon steel castings

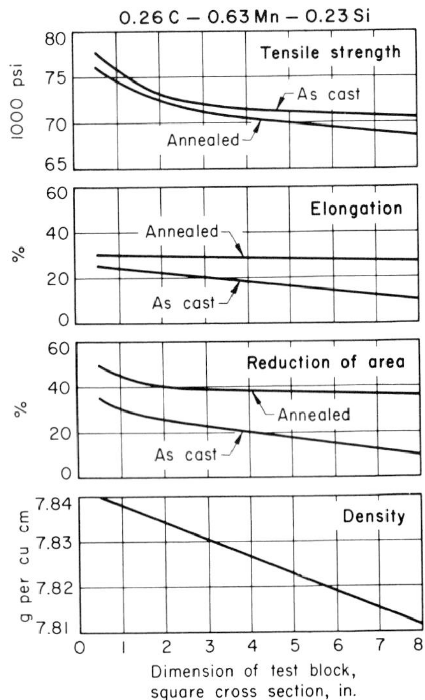

Fig. 5. Decrease in properties with increase in mass of steel castings in the as-cast and annealed conditions

48 hr), and furnace cooled. ASTM grain size was 2 to 3, and the microstructure was upper bainite. Properties obtained were as follows: tensile strength, 104,800 to 110,000 psi; yield strength, 83,000 to 88,000 psi; elongation in 2 in., 19.5 to 21.0%; and reduction of area, 61.8 to 63.8%. These values were well above the minimum requirements which were: tensile strength, 95,000 psi; yield strength, 60,000 psi; elongation in 2 in., 15%; and reduction of area, 45%.

These data show that acceptable mechanical properties can be developed in massive low-carbon alloy steel castings by normalizing and tempering.

Figure 7 shows the wide range of properties obtainable through changes in carbon and alloy content and heat treatment—note the properties for 0.30% C, 1.50% Mn, 0.35% Mo steel (bottom three rows of charts).

In Fig. 8, the relations between tensile strength and other properties of a 4330 steel in the quenched and tempered condition are presented. The ratio of yield strength to tensile strength is uniform at 90% ± 5% for tensile strengths up to 200,000 psi. Above 200,000 psi, the ratio decreases at a rather rapid rate until, at tensile strengths around 275,000 psi, the ratio has dropped to 75%. There is a decrease in the scatter of yield strength as the tensile strength is increased to about 200,000 psi and then an apparent increase in scatter as the tensile strength is further increased.

The fatigue limit, or endurance limit, for smooth-machined specimens is generally about one-half the tensile strength, but is reduced considerably by notches or a rough cast surface. Fatigue test data for carbon-manganese steel are shown in the lower portion of Fig. 9. The effect of notches can be seen by comparing the tests results on notched specimens in the center row of the figure with those on un-

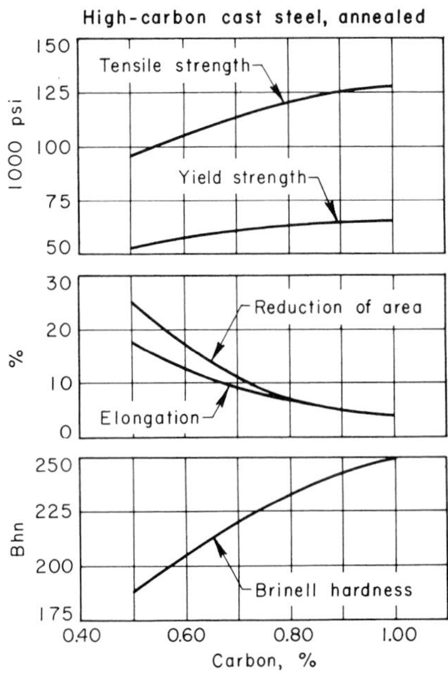

Fig. 6. Effect of carbon content on mechanical properties of fully annealed high-carbon cast steel

notched specimens below. The higher fatigue limit for the higher-hardenability, higher-carbon manganese steel is also indicated.

The two S-N curves at the upper left and upper center of Fig. 9 compare wrought 8640 and cast 8630 steel in two different conditions of heat treatment. In both of these comparisons the wrought 8640 is superior, but in the notched fatigue test the two steels are practically identical. This is significant because

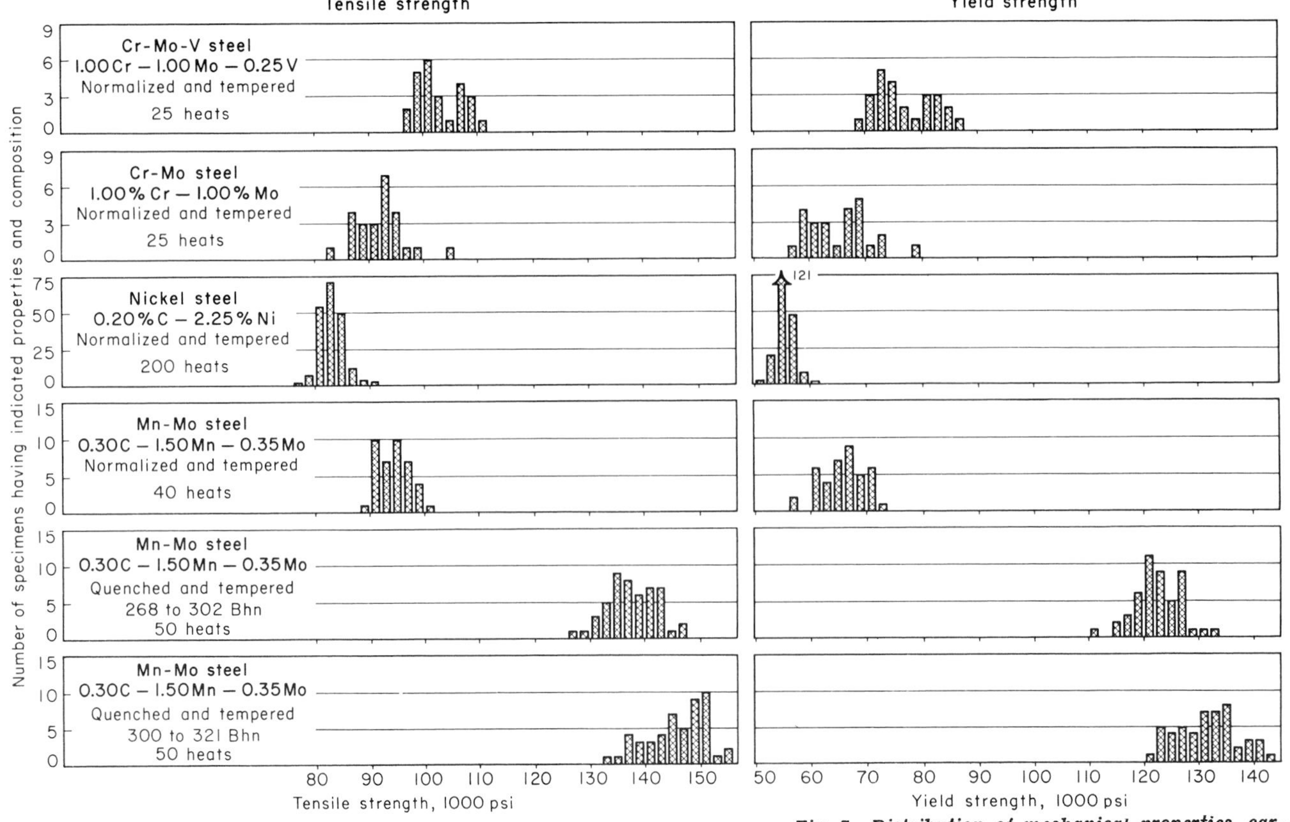

Fig. 7. Distribution of mechanical properties, car-

most articles fabricated from either wrought or cast steel contain more than one notch and more than one type of notch. Again in the upper-right chart of Fig. 9, the effect of notches is apparent in comparing similar cast and wrought steels with regard to available fatigue limit at selected static tensile strengths.

Impact resistance of steel depends on microstructure, chemical composition, strength, hardness and soundness. This is true for both cast and wrought steels. Gross inclusions, segregation and high gas content, singly or in combination, will cause large variations in impact resistance and may decrease it to a dangerously low level.

Table 4 lists some Charpy V-notch impact test results for two different steels with the same hardness and given approximately the same heat treatment, but with slightly different microstructures. The 2.25% Cr, 1% Mo steel with a smaller variation in grain size and no primary ferrite shows consistently better impact results than does the 1% Cr, 1% Mo steel with only a small amount of primary ferrite (up to 10%) and greater variation in grain size.

In the upper half of Fig. 10, results of Charp V-notch tests on two different quenched and tempered steels tested at temperatures from −80 F to +80 F are shown. The Mn-Cr-Mo steel quenched and tempered to 307 to 337 Bhn shows a gradual decrease in impact resistance with decreasing temperature. This is an indication of the transition-temperature behavior of ferritic steels

Table 3. Properties of Test Castings of Chromium-Molybdenum-Vanadium Steel

Casting No.	Location of bar in casting	Tensile strength, psi	Yield strength, psi(a)	Elongation in 2 in., %	Reduction of area, %	Brinell hardness (avg)
Minimum requirements(b)		95,000	60,000	15.0	45.0	200 to 240
1	A	107,500	81,300	23.0	60.0	210 to 215
1	A	106,500	83,000	21.0	60.5	210 to 215
1	A	105,500	81,500	20.0	60.5	210 to 215
2	A	104,500	81,300	17.0	59.4	210 to 215
2	A	103,500	78,800	20.0	60.7	210 to 215
2	A	105,500	80,000	20.0	60.8	210 to 215
3	A	98,500	73,800	21.0	61.7	210 to 215
3	A	98,000	73,800	22.0	63.3	210 to 215
3	A	97,000	73,500	21.0	58.7	210 to 215
4	A	107,500	81,300	23.0	60.0	229
4	A	106,500	83,000	21.0	60.5	229
4	A	105,500	81,500	20.0	60.5	229
5	A	102,500	78,800	18.5	58.8	223
5	A	104,000	80,000	21.0	57.0	223
5	A	104,000	80,000	21.5	61.1	223
6	A	107,500	81,500	20.5	59.4	241
6	A	107,500	76,300	21.5	60.1	241
6	A	107,800	80,500	20.5	58.3	241
7	B	111,500	91,500	18.0	51.4	234 to 239
7	B	111,500	91,500	18.0	50.8	234 to 239
8	B	109,000	85,500	17.0	46.9	230 to 235
8	B	109,000	86,300	19.0	55.5	230 to 235

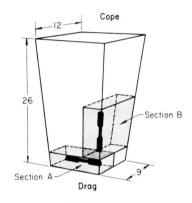

A 900-lb test casting from which test bars were machined at sections A and B for determination of properties is shown in the sketch at the left. Test castings were made from a silicon-deoxidized basic electric chromium-molybdenum-vanadium steel equivalent to ASTM specification A389, grade C24, with the following composition: 0.20 C, 0.38 Si, 0.67 Mn, 0.011 to 0.017 S, 0.011 P, 1.05 Cr, 0.15 Ni, 1.10 Mo, 0.17 V, and 0.008 N. The castings were normalized at 1925 F for 10 hr, air cooled, and tempered at 1275 F to the indicated hardness levels. The ASTM grain size was 2 to 3, and the microstructure was upper bainite with up to 5% primary ferrite. There was some variation in properties, but all tests met the specifications.

(a) Determined by the 0.2% offset method. (b) Hardness values are for castings in the normalized and tempered condition only.

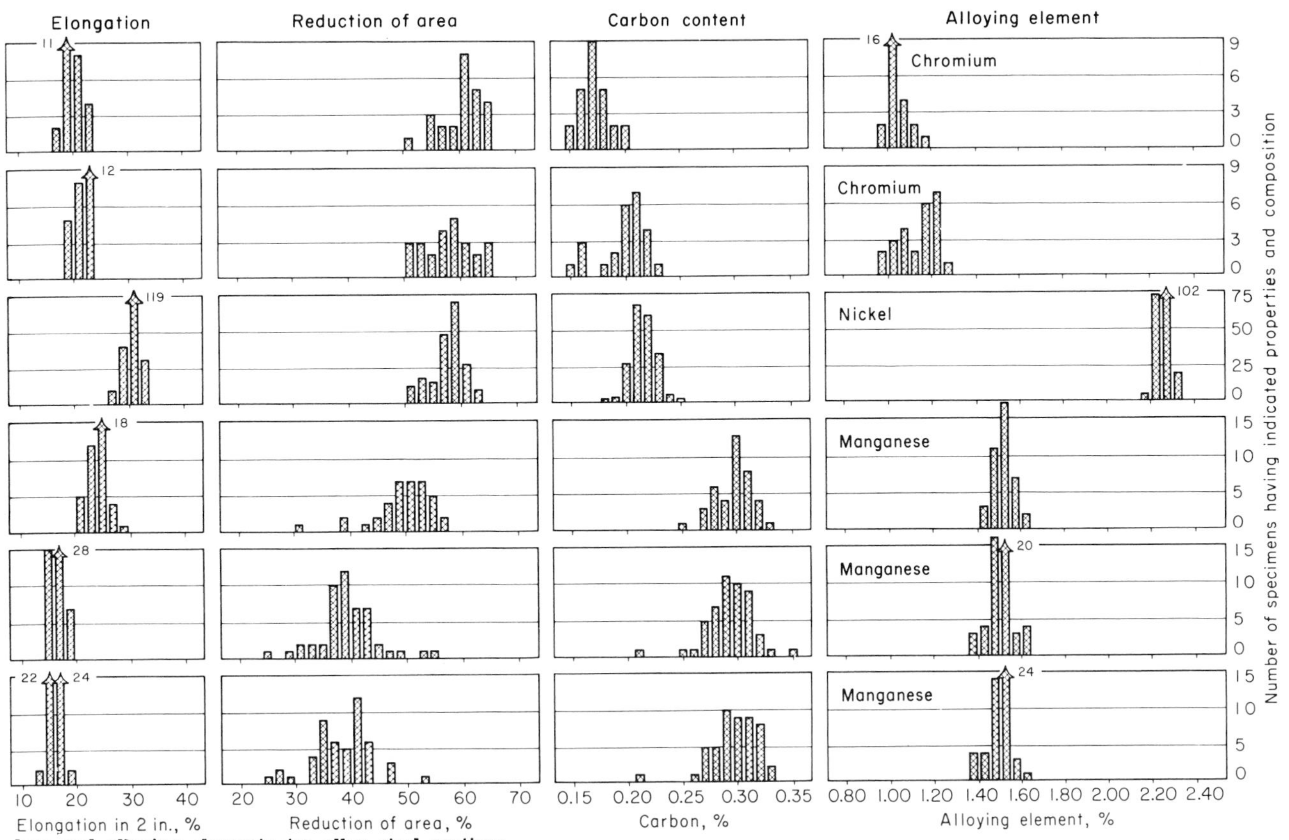

bon and alloying elements for alloy steel castings

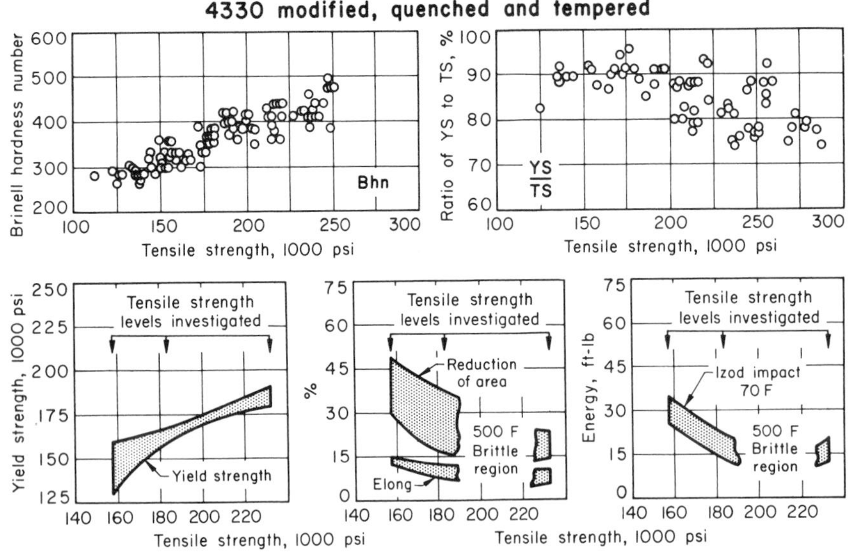

Fig. 8. Range of mechanical properties obtained in modified 4330 steel castings in the quenched and tempered condition. Composition of the steel was as follows: 0.27 C, 1.00 Mn, 0.45 Si, 0.015 P, 0.025 S, 1.85 Ni, 0.90 Cr and 0.40 Mo.

when the type of fracture changes from ductile to brittle as the testing temperature is lowered. The impact test results for cast and wrought steels are similar. When tested at room temperature, the Mn-Ni-Cr-Mo steel, quenched and tempered to a lower hardness of 207 to 229 Bhn, had a significantly higher impact resistance than the other steel.

The lower half of Fig. 10 shows the effect of heat treatment on the impact resistance of a carbon steel. Here, the more homogeneous microstructure that results from normalizing and tempering, as compared to annealing, is evident in the decrease in scatter of results. These tests, conducted over a range of temperature from slightly below zero to 450 F, also show the decrease in impact resistance with decreasing temperature, characteristic of ferritic steel.

The effect of hardness, microstructure and composition on the Charpy V-notch impact resistance at -40 F for modified 8630 steel and a molybdenum-boron steel is shown in Fig. 11. At the same hardness, both steels exhibit better impact resistance with a martensitic microstructure than with a pearlitic. The difference is of considerable magnitude. With either microstructure or steel composition, impact resistance decreases with higher hardness. Increasing hardness, however, has less effect with a pearlitic microstructure. For applications requiring good impact resistance, not only chemical composition and hardness but also microstructure must be considered in selection.

Figure 12 also shows the effect of hardness and testing temperature on the impact resistance of a molybdenum-boron steel in the quenched and tempered condition. Again, at any selected temperature the effect of increasing hardness is to lower impact resistance, although the scatter in results is large.

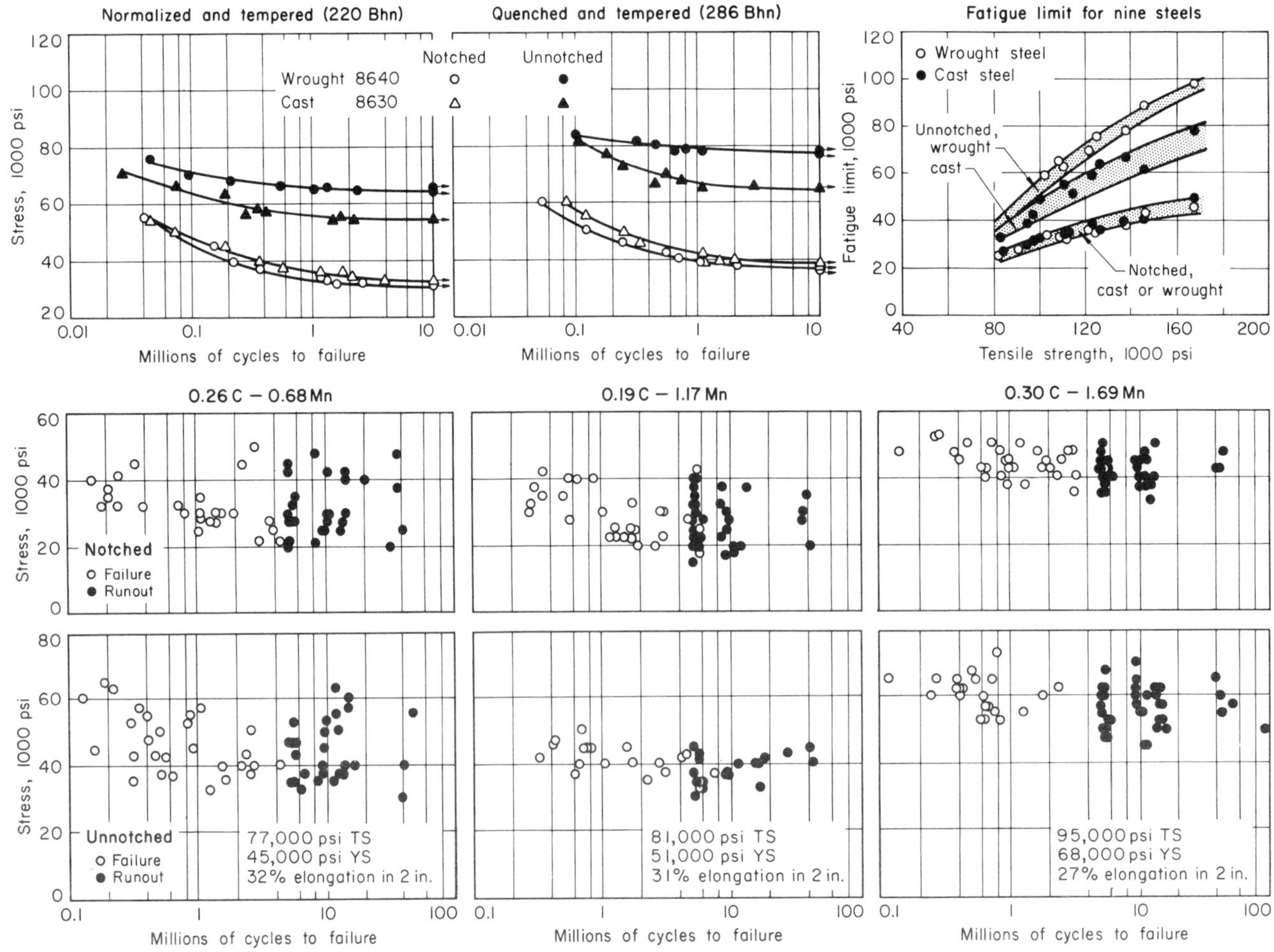

Fig. 9. (Top row) *Fatigue properties of cast 8630 steel compared with those of wrought 8640 steel in the normalized and tempered, quenched and tempered, notched and unnotched conditions.* (Bottom two rows) *Results from three production heats each of three grades of cast carbon-manganese steel.*

Effect of Mass. The size of a cast coupon or casting can have a marked effect on its mechanical properties. This effect reflects the influence of hardenability and microstructure, as modified by heat treatment, on mechanical properties. Results of tests on specimens taken from very heavy sections (Fig. 5) and from large castings are helpful in predicting minimum properties in cast steel parts.

Figure 13 shows the variations in tensile properties and impact resistance of specimens taken from quenched and tempered high-hardenability cast steel blocks of various sizes. The specimens were taken from the centers of a 3 by 3 by 9-in. block and a 4 by 4 by 11-in. block, and from several locations in a ¾ by 7 by 16-in. plate, parts A, B and C of Fig. 13. The castings were quenched and tempered to the same nominal hardness or tensile strength.

An increase in effective thickness results in a decrease in average elongation and reduction of area and a marked decrease in impact resistance. Essentially, these results denote less response to heat treatment for thicker test blocks. The specimen from part D was quenched and tempered to a higher tensile strength. As a result, average elongation and reduction of area are reduced. Specimens from part D had a lower average impact resistance at 70 F than specimens from parts A, B, or C tested at −20 F. The low impact resistance in these tests is influenced more by microstructure than by testing temperature.

The effect of mass or effective thickness is further illustrated in Fig. 14. Sections were taken from tubes of different wall thickness, but the same chemical composition and similar heat treatment.

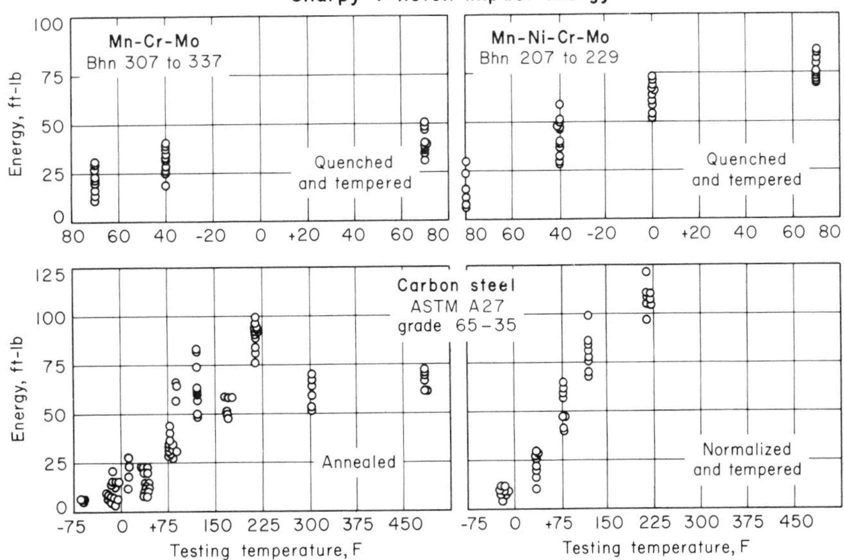

Fig. 10. Effect of temperature on impact energy for carbon and low-alloy steels

Table 4. Scatter in Charpy V-Notch Impact Properties of Chromium-Molybdenum Cast Steels at Room Temperature

Charpy V-notch, ft-lb(a)					
1% Cr, 1% Mo steel(b)			2.25% Cr, 1% Mo steel(c)		
8, 3	11, 12	21, 25	44, 39	56, 65	62, 60(d)
7, 10	10, 9	9, 10	66, 68	38, 53	55, 61(d)
9, 9	7, 8	8, 10	24, 23	33, 41(d)	66, 85(d)
8, 11	9, 10	12, 8	37, 34	54, 50(d)	70, 90(d)
5, 6	11, 9	17, 8	29, 31	71, 78(d)	78, 102(d)
10, 12	19, 12	16, 12		68, 86(d)	
9, 9	11, 17	7, 8			
10, 14	8, 9	11, 12			

(a) Each pair of values represents one keel block.
(b) Equivalent to ASTM A356, grade 8. Properties from 3 by 5 by 7-in. keel blocks of basic electric steel, silicon deoxidized, normalized at 1925 F for 12 hr, air cooled, tempered at 1275 F for 20 hr, furnace cooled; average grain size, 1 to 3; average mcirostructure, upper bainite with 0 to 10% primary ferrite; minimum mechanical property specifications for this steel: tensile strength, 80,000 psi; yield strength, 0.02% offset, 50,000 psi; elongation in 2 in., 18%; reduction of area, 45%; and hardness, 180 to 220 Bhn.
(c) Equivalent to ASTM A356, grade 10. Properties from 3 by 5 by 7-in. keel blocks, basic electric steel, deoxidized with 2 lb of aluminum per ton; normalized at 1925 F for 12 hr, air cooled, tempered at 1275 F for 28 hr, furnace cooled; average grain size, 2 to 3; average microstructure, upper bainite; minimum mechanical property specifications for this steel: tensile strength, 70,000 psi; yield strength, 0.02% offset, 40,000 psi; elongation in 2 in., 20%; reduction of area, 45%; and hardness, 180 to 220 Bhn.
(d) Basic induction melted.

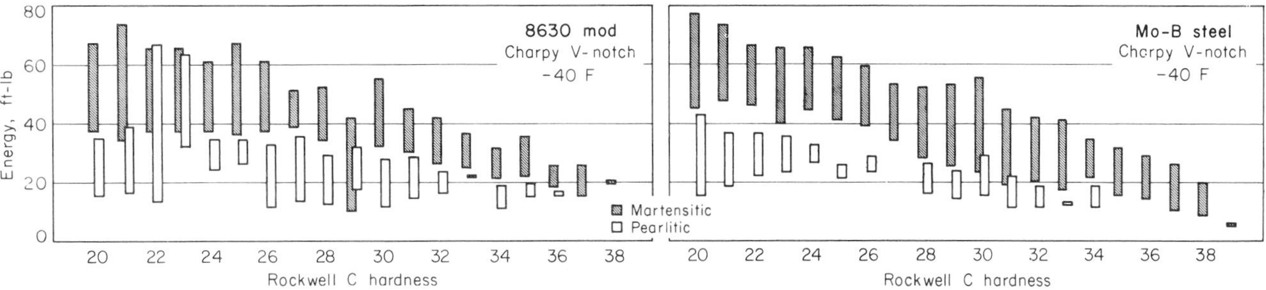

Fig. 11. Effect of hardness and microstructure on Charpy V-notch properties of cast alloy steels at −40 F

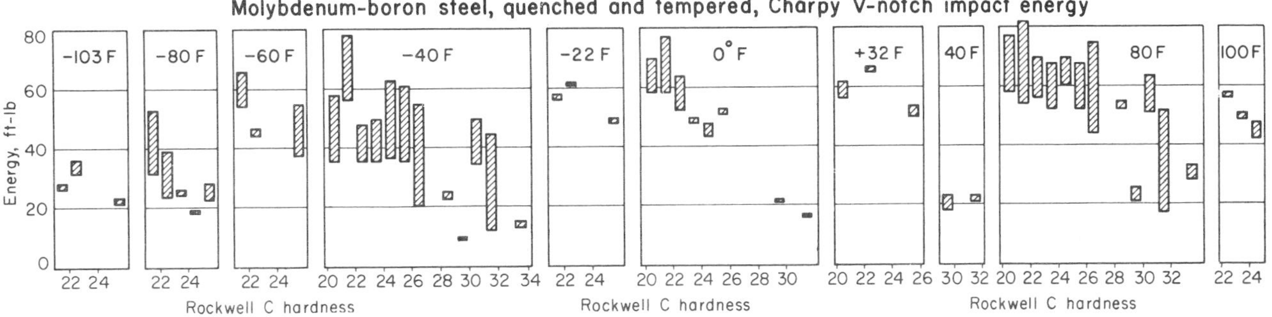

Fig. 12. Effect of hardness and testing temperature on Charpy V-notch properties of cast alloy steel

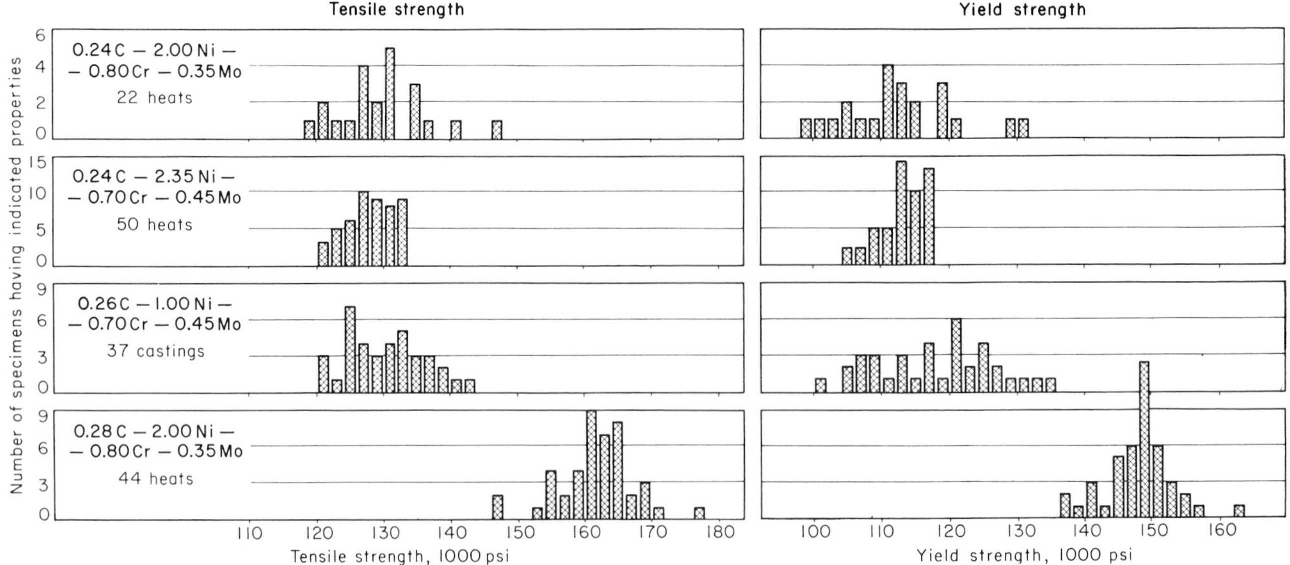

Fig. 13. Effect of mass on distribution of mechanical prop-

The castings with 3 and 4-in. wall have a slight decrease in average tensile strength and a more pronounced decrease in average yield strength. In these tests there was practically no difference in average elongation, but decidedly lower reduction of area for the heavy-wall low-strength specimens. It might be assumed that at lower strength levels the average ductility would increase but lower strength does not always imply higher ductility. The correlation depends on a number of factors, such as chemical composition, effective thickness, and conditions or effect of heat treatment.

Impact resistance shows a more predictable response. Specimens taken from the heavy section, when tested at room and low temperatures, have lower resistance to impact than those from the lighter section. This reflects the effect of microstructure. Heat treatment of the light-wall castings produces a more homogeneous microstructure of tempered martensite than that produced in the heavy-wall castings.

Figure 15 presents the results of tests made on bars cut from several locations in a casting of moderate complexity, quenched and tempered to a relatively high strength level. Uniformity of properties in the three selected areas indicates what can be done when castings of moderately complex configuration are properly gated, risered and heat treated.

Physical Properties

The physical properties of cast steel are generally similar to those of wrought steel.

Elastic constants of carbon and low-alloy cast steels as determined at room temperature are only slightly affected by changes in composition and structure. The modulus of elasticity, E, is about 30 million psi, Poisson's ratio is 0.3, and the modulus of rigidity is 11.2 million psi. Increasing temperature has a marked effect on the modulus of elasticity and the modulus of rigidity. The modulus of elasticity at some elevated temperatures is: at 400 F, 28 million psi; at 680 F, 26 million; at 830 F, 24 million; at 910 F, 22 million. Above 900 F, the value of the modulus of elasticity is rapidly reduced.

Density of cast steel is sensitive to changes in composition, structure and temperature. The density of medium-carbon cast steel is in the range from 7.825 to 7.830 g per cu cm. Steel castings have a weight of 490 lb per cu ft or 0.283 lb per cu in. The density of cast steel is also affected somewhat by mass or size of section (Fig. 5).

Electrical properties of carbon and low-alloy steel castings do not account for any significant usage. The only electrical property that may be regarded as of any importance is resistivity, which, for various annealed carbon steel castings with 0.07 to 0.20% C, is 13 to 14 microhm-cm. Resistivity increases with carbon content and at 1.0% C is about 20 microhm-cm.

Magnetic Properties. Steel castings form the housings for electrical machinery and magnetic equipment, and carry only stray fluxes around the machines; hence the magnetic properties of steel castings are less important than they were formerly when core material was manufactured from commercial cast iron and steel. Low-carbon cast dynamo steel has supplanted other cast metals for housings and frames for magnetic circuits.

The carbon content of the steel casting is very important in determining the magnetic properties. This point is illustrated in Fig. 16, which shows a decrease in maximum permeability and saturation magnetization and an increase in the coercive force as the carbon content increases. Manganese, phosphorus, sulfur and silicon also increase the magnetic hysteresis loss in cast steels. This loss is equal to about 100 ergs per cu cm per cycle for B = 10,000 gausses for each 0.10% Mn, 0.01% S and 0.01% P. Other factors being unchanged, the magnetic hysteresis loss is unaffected by more than 0.02% P. Magnetic properties change considerably, depending on the mechanical treatment and heat treatment of the steel.

Cast dynamo steels contain about 0.10% C with other alloying elements held to a minimum; the castings are furnished in the annealed condition. Specifications require 0.05 to 0.15% C, 0.20% Mn, 0.35 to 0.60 or 1.50 to 2.00% Si.

The magnetic properties of an-

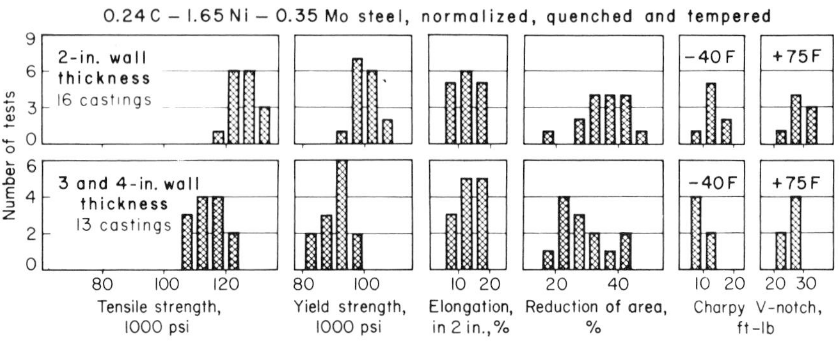

Fig. 14. Effect of wall thickness of tubes on distribution of mechanical properties of normalized, quenched and tempered alloy steel

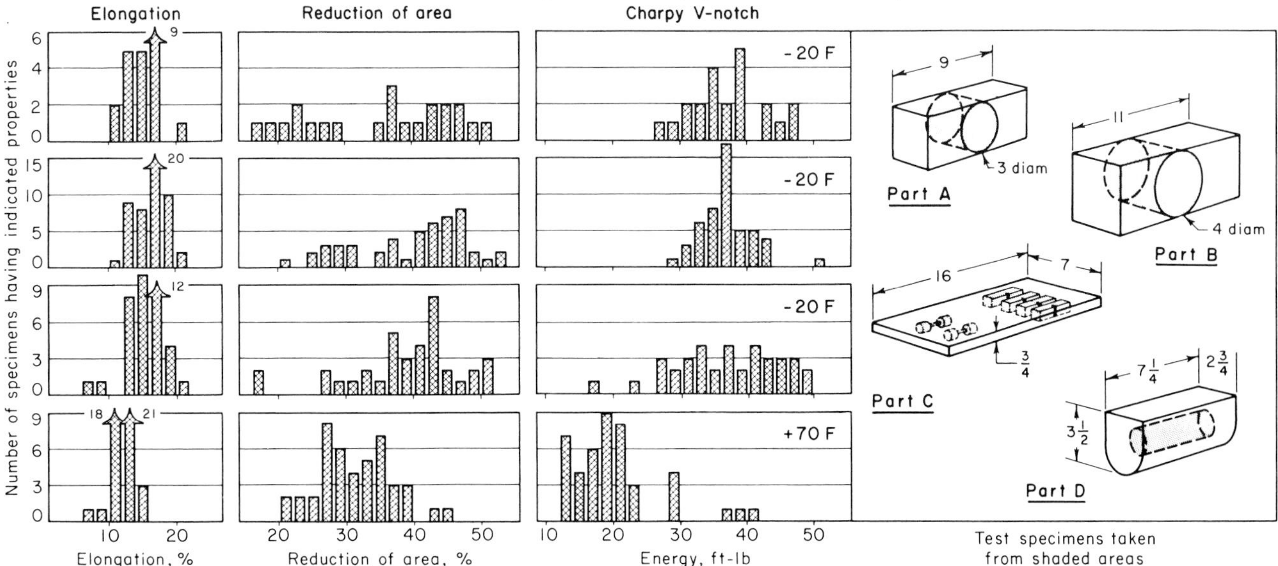

erties of normalized, quenched and tempered cast alloy steels

nealed cast dynamo steel that may normally be expected are as follows:

Maximum permeability	14,800
Hysteresis loss (induction for H = 150)	19,100
Saturation magnetization, gausses	21,420
Residual induction, gausses	11,000
Coercive force, oersteds	0.37

As the carbon content is increased, maximum permeability and saturation magnetization decrease and coercive force increases. Also, an increase in manganese and sulfur content increases the magnetic hysteresis loss.

Silicon and aluminum eliminate the allotropic transformations in iron and permit annealing at high temperature without recrystallization during cooling; thus large grains can be obtained. These elements can be added in large quantities without affecting magnetic properties, but they do reduce the saturation value and increase the brittleness of the metal. Grain size is directly proportional to hysteresis loss; therefore the larger the grain size, the better the properties.

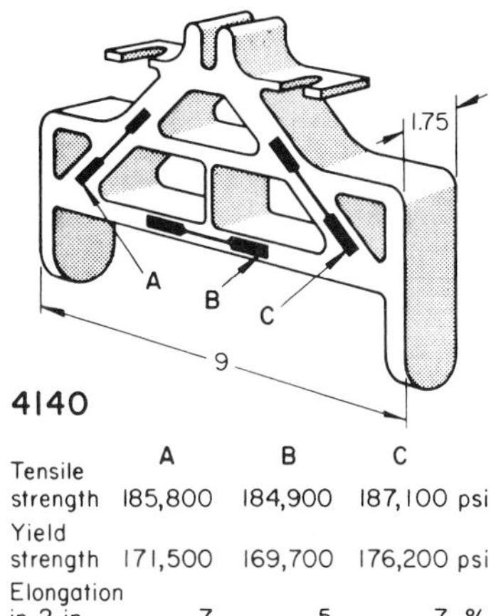

	A	B	C
Tensile strength	185,800	184,900	187,100 psi
Yield strength	171,500	169,700	176,200 psi
Elongation in 2 in.	7	5	7 %

Fig. 15. Consistency of mechanical properties in three locations of a 7.3-lb casting of 4140 steel hardened and tempered to Rockwell C 37 to 40

Residual alloy content should be low, as it lowers saturation value.

The factors that improve machinability of dynamo steel decrease the magnetic properties. A disadvantage in the use of pure iron for dynamo steel is low resistivity. The iron must be rolled thin to keep eddy currents down; otherwise the magnetic properties are poor.

Volumetric Changes. In the foundry, all volume changes of the metal are pertinent, whether occurring in the liquid state, during solidification, or in the solid state. Of particular interest is the contraction that results when molten steel solidifies.

Volume changes that occur in the liquid state as the cast metal cools affect the planning for adequate metal to fill the mold. Contraction is of the order of 0.9% per 100 F for a 0.30% C steel. The exact amount of contraction will vary with the chemical composition, but it is usually within the range from 0.8 to 1.0% per 180 F for carbon and low-alloy steels. A larger contraction occurs on solidification (2.2% for nearly pure iron to 4% for a 1.00% C steel). For cast carbon and low-alloy steels a solidification contraction of 3.0% is generally assumed.

The greatest amount of contraction occurs as the solidified metal cools to room temperature. Solid-state contraction from the solidus to room temperature varies between 6.9 and 7.4% as a function of carbon content. Alloying elements have no significant effect on the amount of this contraction. The rigid form of the mold hinders contraction and results in the formation of stresses within the cooling casting which may be great enough to cause fracture or hot tears in the casting. The hot metal has low strength just after solidification. The rigidity of the mold makes the proper relation of casting configuration to accommodate this contraction one of the most important factors in successfully producing a casting.

In commercial production, a combination of all three elements of contraction may operate simultaneously. Molten metal in contact with the mold wall solidifies quickly and proceeds to solidify toward the center of the casting. The solid envelope undergoes contraction in the solid state, while a portion of the still-molten metal is solidifying. The remaining molten metal contracts as its temperature decreases toward the freezing point. Because of contraction factors, many casting designs require considerable development to produce a sound casting.

Engineering Properties

Wear Resistance. Cast steels have wear resistance comparable to wrought steels of similar composition and condition. (See the article on Selection of Steel for Wear Resistance, in this Handbook.)

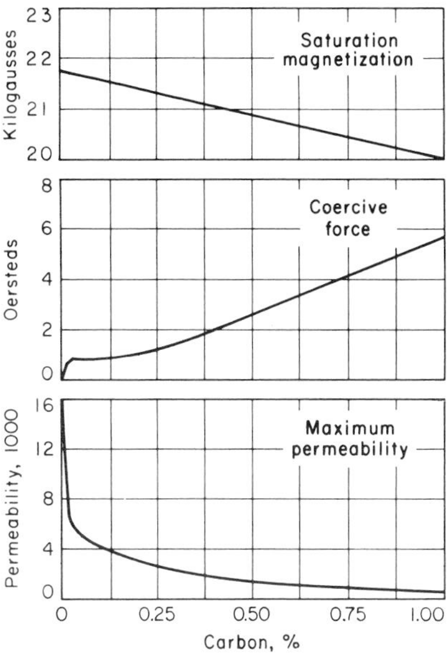

Fig. 16. Effect of carbon on magnetic properties of annealed carbon steel (T. D. Yensen)

Corrosion resistance of cast steel is similar to that of wrought steel of equivalent composition. Data published on corrosion resistance of wrought carbon and low-alloy steels under various conditions may be applied to cast steels. A full discussion of this subject is contained in the article on Mild Steel Corrosion.

A study was conducted at Kure Beach, N. C., and at East Chicago, Ind., to determine corrosion resistance of carbon and low-alloy cast steels exposed to marine and industrial atmospheres. The surfaces of some specimens were machined before exposure, while others were tested with as-cast surfaces. The tests covered periods of one year in the marine atmosphere and three years in the industrial atmosphere. Figure 17 presents the results.

The following conclusions were reached: (*a*) machining the surfaces of the castings had no significant effect on corrosion resistance in these atmospheres; (*b*) cast steels containing nickel, manganese and chromium have superior resistance to corrosion in these atmospheres to cast steels with only a manganese addition; (*c*) all cast steels corroded most rapidly in the marine atmosphere 80 ft from the ocean, at an intermediate rate in the industrial atmosphere, and at the slowest rate in the marine atmosphere 800 ft from the ocean.

Soil Corrosion. Cast steel pipe has been tested for various periods up to 14 years in different types of soil. The results of these tests were compared directly with results from wrought steel pipe of similar composition, and no significant difference in the corrosion of the two materials could be detected. The actual corrosion rate and rate of pitting of the cast pipe, however, varied widely depending on the soil and aeration conditions. Data on soil corrosion of pipe is found in National Bureau of Standards Circular 579, April 1957, and is summarized in graphs in the article on Corrosion of Mild Steel, page 257.

Heat Resistance. At temperatures above 900 to 1000 F, carbon and low-alloy steels oxidize rapidly, and the oxide formed does not protect the underlying metal from further oxidation. When these steels are held for a long time at these high temperatures, they are gradually converted to oxide. For effective resistance to oxidation at high temperatures, more highly alloyed steels must be used (see the article on Heat-Resistant Castings).

Machinability. Extensive lathe and drilling tests on steel castings have brought out no significant differences in machinability of steel made by different melting processes nor between wrought and cast steel, provided strength, hardness and microstructure are equivalent. The skin or surface on a sand mold casting often wears down cutting tools rapidly, possibly because of adherence of abrasive mold materials to the casting. Therefore, the initial cut should be deep enough to penetrate below the skin, or the cutting speed may be reduced to 50% of that recommended for the base metal.

Microstructure has considerable effect on machinability of cast steels. It is sometimes possible to improve machining characteristics of a steel casting by 100% through normalizing, normalizing and tempering, or annealing.

Weldability. Steel castings have welding characteristics comparable to wrought steel of the same composition, and welding them involves the same considerations.

Cast steels with less than 0.25% C and less than 0.50% Mn present no problem from the standpoint of loss of ductility next to the weld. Steels with more than 0.25% C and slightly over 0.50% Mn are borderline for service in the as-welded condition, but a stress-relief treatment produces sufficient ductility. Steels having more than 0.35% C are not considered weldable unless preheated.

To prevent cracking in carbon and low-alloy cast steels, the hardness of the weld bead should not exceed 350 Vickers, except where conditions are such that only compressive forces result from the welding. This value may not be low enough in configurations where extreme restraint is involved.

The following suggestions are offered for preheating and stress-relieving heat treatments:

Average Vickers hardness of weld bead	Preheat	Stress relief
Under 200	None	None
200 to 250	None	Advisable
250 to 300......	300 F	Necessary
300 to 350......	400 F	Without cooling from preheat

Virtually all castings receive a stress-relief heat treatment after welding, even composite fabrications where steel castings are welded to wrought steel shapes.

Since stress-relief heat treatment is universally used in the steel casting industry, the maximum limits that have been placed on readily weldable grades of castings by the industry are 0.35% C, 0.70% Mn, 0.30% Cr and 0.25% max Mo plus W, with a total of the undesirable elements of 1.00%. For each 0.01% below the specified maximum carbon content, most specifications permit an increase of 0.04% Mn above the maximum specified, up to a maximum of 1.00% Mn. Three specifications for weldable grades of cast steel are ASTM A27, A216, and A217. Specifications covering control of weld quality are ASTM E99 and E52.

Many welds that fail do not fail in the weld but in the zone immediately adjacent to the weld. While the weld is being made this zone is heated momentarily to a melting temperature. Temperature ranges downward as distance from the weld increases. Such heating induces structural changes with the development of hard, brittle areas adjacent to the weld deposits, which reduce the toughness of the area and frequently cause cracking during and after cooling. Certain of the alloying elements other than carbon, such as nickel, molybdenum and chromium, likewise bring about air hardening of the base metal. For these reasons the quantity of alloying elements to be used must be limited unless special precautions are taken, such as the preheating of the base metal to 300 to 600 F. Increased hardness in the heat-affected zone of the base metal can be removed by annealing the welded casting or by heating it for definite periods at 1200 to 1250 F. This also relieves stresses from welding.

For the arc welding of steel castings, a high-grade heavy-coated electrode (AWS E6016 type) or granular flux or CO_2 atmosphere is generally desirable. These coatings contain little or no combustible material. Mineral coatings are often

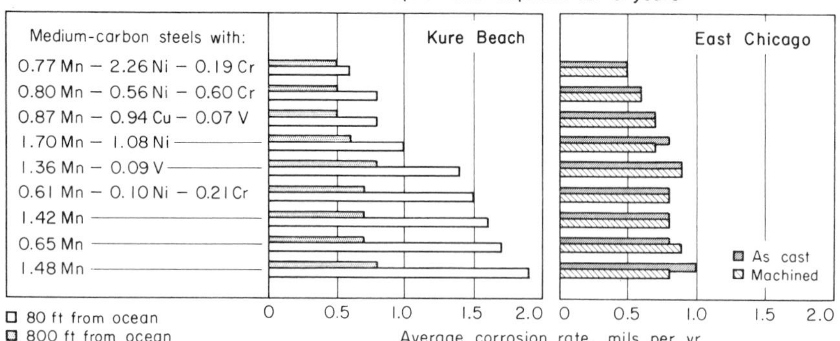

C	Mn	Si	Ni	Cr	Other
0.17	0.77	0.65	2.26	0.19	...
0.26	0.80	0.44	0.56	0.60	0.15 Mo
0.28	0.87	0.42	0.94 Cu	...	0.07 V
0.27	1.70	0.42	1.08	...	...
0.37	1.36	0.34	...	...	0.09 V
0.14	0.61	0.41	0.10	0.21	...
0.37	1.42	0.38	...	...	...
0.25	0.65	0.51	...	...	...
0.33	1.48	0.40	...	...	0.05 Ti

Fig. 17. Corrosion data. Steels are in the same order in both graph and table.

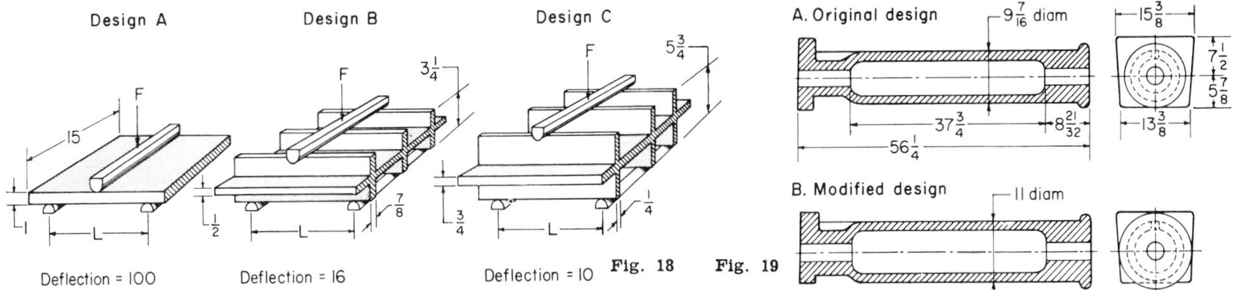

Fig. 18. Under a uniform bending load F exerted midway between two supports of span L, deflections are in the ratio shown. Fig. 19. Charging arm for open-hearth furnace, showing original and modified designs.

used to keep hydrogen absorption at a minimum and thereby to limit underbead cracking. Selection of number of passes and welding conditions is similar to welding practice for wrought steels.

Welds in castings may be radiographed by gamma-ray or x-ray methods to ascertain the degree of homogeneity of the welded section. Most common defects are incomplete fusion, slag inclusions and gas bubbles. Magnetic particle inspection is also useful in the detection of small cracks.

The mechanical properties of welds joining cast steel to cast steel and of welds joining cast steel to wrought steel are of the same order as similar welds joining wrought steel to wrought steel. Most specimens machined across the weld will break outside the weld, in the heat-affected zone, in a tension test. This does not mean that the weld is stronger than the casting base metal. Closely controlled welding technique and stress relieving are necessary to prevent brittleness in the heat-affected zone.

Nondestructive Inspection

Highly stressed steel castings for aircraft, high-pressure or high-temperature service must pass rigid radiographic inspection. ASTM E71 covers radiographic standards for steel castings. Radiographic acceptance standards should be agreed upon by the user and producer before production begins. Critical areas to be radiographed may be identified on the casting drawing.

Magnetic particle inspection is used on highly stressed steel castings to detect surface discontinuities or imperfections just below the surface. ASTM E109, "Dry Powder Magnetic Particle Inspection", should be consulted in establishing cause for rejection before the castings are produced. Also, the critical areas of the casting must be denoted on the blueprint.

The fluorescent penetrant test can be used on steel castings, but it is used primarily to inspect nonmagnetic materials such as nonferrous metals and austenitic steels for possible surface discontinuities.

Ultrasonic testing is sometimes used on steel castings to detect imperfections below the surface in heavy sections that extend from 1 to 28 ft. For proper operation, the casting must be ground smooth. This technique is used on rolling mill rolls, for example, which are too massive for radiographic inspection.

Hydrostatic testing, or pressure testing, is used on valves and castings intended to contain steam or fluids, such as those made to ASTM specifications A216, A217, A352, A356 and A389. The hydrostatic test generally follows radiographic inspection. If a casting must pass a pressure test, essential factors must be noted on the blueprint, and the details of the test should be understood by the buyer and producer.

Engineering Factors in Design*

One of the first properties required of steel castings is an assurance of mechanical strength commensurate with the stresses to which they will be subjected in service. The resistance of steel castings to the stresses in question depends on: (*a*) the actual strength of the metal as a function of the thickness; and (*b*) the shape of the casting, which is determined by the stress system it must withstand.

Considerations of Shape. In general, all castings consist of a "body" onto which are built various additions (flanges, bosses, stringers, ribs). When designing a casting, the following points should be taken into account: (*a*) similarity with another casting; (*b*) position occupied in mechanical assembly; (*c*) service conditions such as mechanical properties, resistance to oxidation or corrosion, pressure tightness, thermal or electrical conductivity, and magnetic permeability.

When choosing shapes giving the required mechanical strength, the following principles should be observed:

1 From a structural standpoint, in castings subject to flexural stresses, minimize the amount of material on the neutral axis (at the position where there is no stress). Considering the three equal sections A, B and C of Fig. 18, under the same bending load *F* applied between two points at a distance *L*, the deflections vary as **100 : 16 : 10**, the increased resistance to bending being at-

*This section is based on, and in part is an abridgment of, "A Practical Guide to the Design of Steel Castings", prepared by the Commission Technique de la Metallurgie des Aciers, of the Centre Technique des Industries de la Fonderie, Paris. An English translation of the original document was prepared by the British Steel Castings Research Association and is made available in the United States through the Steel Founders' Society of America, Cleveland 15, Ohio.

Figures 18 through 47 and Tables 5, 6 and 7 are from the English translation.

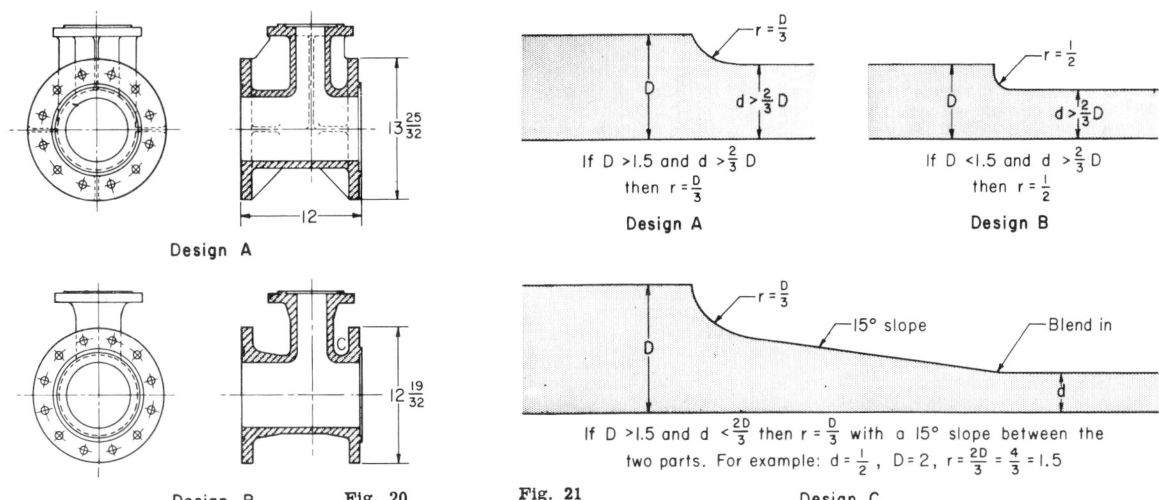

Fig. 20. T-piece for steam-pressure service, showing defective (A) and correct (B) designs. Fig. 21. Section changes on one side.

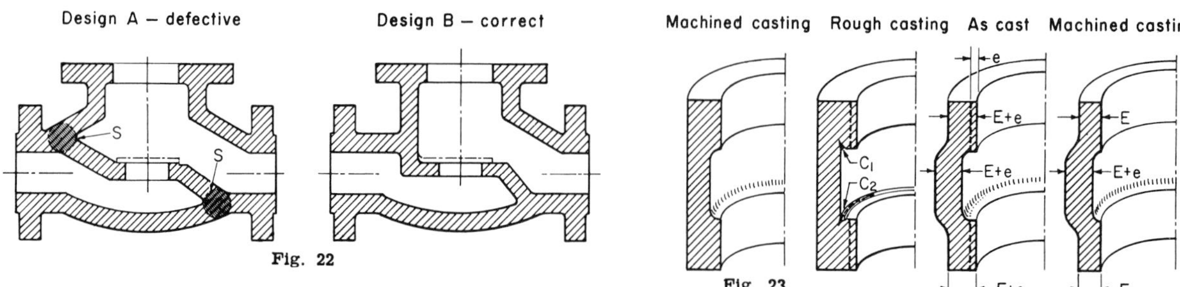

Fig. 22. Valve body, showing defective and correct designs. Fig. 23. Influence of extra machining thicknesses, illustrating defective design (sections paired at left) and improved design providing uniform thickness of the rough casting (pair at right).

tributable to the provision of ribs of different heights and thicknesses. A shape compatible with the requirements for correct production of the casting in a foundry involves the adoption of web-carrying alternating ribs which may be one third of the plate thickness.

2 In castings subject to torsional stresses, closed sections should be considered. Increased resistance to torsion is secured by the use of closed profiles. However, they are difficult to produce in the foundry, and it is wise to make use of them only when it is essential to obtain high torsional resistance or when the casting must be of closed section (wheels, frames). Certain precautions must be taken in design in order to minimize the risk of cracking.

3 Sharp angles should be avoided where stresses can become concentrated and be the cause of fractures. A notched testpiece fractures more easily than one that is not notched. The effect of notching is harmful both from mechanical and foundry viewpoints. It can be the cause of the formation of shrinkage cavities and of cracks or tears in the skin of the metal. These risks can be reduced to a minimum by appropriately shaping the junctions of walls, webs and ribs.

Selection of Wall Thicknesses. Having determined the shape of castings, the first guiding principle is to make all wall thicknesses as uniform as possible, compatible with foundry requirements. If they cannot be properly risered, localized masses of metal attributable to sudden change of thicknesses are the causes of foundry defects, notably tears and surface cracks. An increase in some local thicknesses in the hope of securing only a slight increase in the strength, often introduces the risk of compromising the soundness of the casting and causing defects.

The strength and soundness of a casting depend more on uniformity of thickness than on the local reinforcement of certain parts. For example, casting shape may be modified instead of reinforcing certain thicknesses. Thus, in the charging arm for an open-hearth furnace (Fig. 19) that had broken in service, it became necessary to modify the design in order to strengthen it. Instead of increasing the thickness of the arm, the same result was reached simply by making the internal diameter 8 instead of 6.4 in., which, with the same thickness, gave higher bending and torsional moments. The use of cast steel instead of cast iron permits reductions in section, as well as equalizing the sections by doing away with some massive parts.

An important application of steel castings involves the production of complicated shapes, which cannot always be given a uniform thickness. In order to approach as closely as possible to the ideal of having a uniform section thickness, the number of different thicknesses must be reduced to a minimum. In the steam T-piece shown in Fig. 20, the design A is faulty because the nonuniformity of the thickness of the cylindrical parts interferes with the satisfactory behavior of the fitting under steam pressure. In order to correct this defect, the thickness was made uniform at 0.055 in., as shown in design B.

In addition, the first design of the cylindrical body and the flanges of the fitting was defective. The outside diameter of the flanges was too great with respect to that of the cylindrical parts, making it necessary to provide ribs to brace the flanges. The ribs adopted, being much too large, were the cause of tears during casting, which gave rise to leakage during testing. In the corrected design, the flange diameter was reduced, which made it possible to do away with the ribs, and the number of different thicknesses in the whole assembly was reduced to two.

It is necessary to provide for progressive and uniform variations of thickness. Having reduced to a minimum the number of different thicknesses, the transitions from one to the other should be progressive toward an area that can be risered, and should avoid sharp changes of section, which always give rise to defects.

Considering flat walls having different thicknesses, the progressive variations should be made according to the recommendations in Fig. 21.

Thicknesses of Walls, Webs and Ribs. The walls constitute the external contour of the casting; their thickness is governed by the strength required. The webs unite several walls, and their object is usually that of increasing the rigidity of the casting. The ribs serve particularly to strengthen certain parts such as joints between webs and walls.

The webs are not usually in direct contact with the risers and consequently are less well fed with metal. It is therefore necessary to reduce their thickness and it is usual to adopt a value of four fifths of the thickness of the walls, which *are* connected with the risers. A lower limit to the thickness of the webs is set by the fluidity of the steel. Thus, for two vertical walls

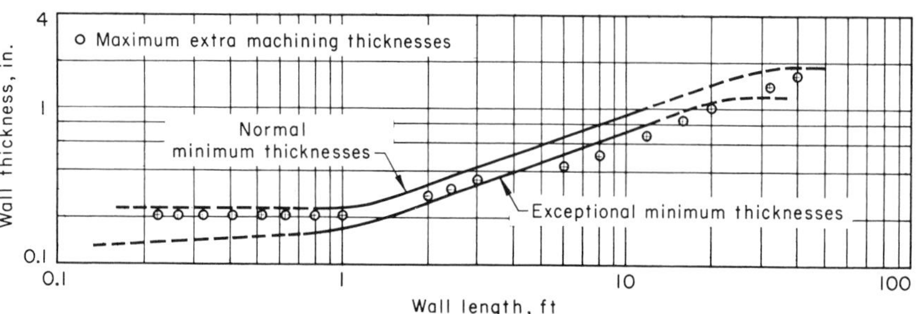

Fig. 24. Minimum thickness as a function of length

Table 5. Effect of Size of Steel Castings on Machining Allowances

	Largest dimension of casting		
	Less than 9 in.	9 to 36 in.	More than 36 in.
Minimum extra thickness, in. ..	0.120	0.120	0.120
Maximum extra thickness, in. ..	0.160 + 6D/1000	0.180 + 5D/1000	0.200 + 4D/1000

D = Longest dimension of the casting in inches

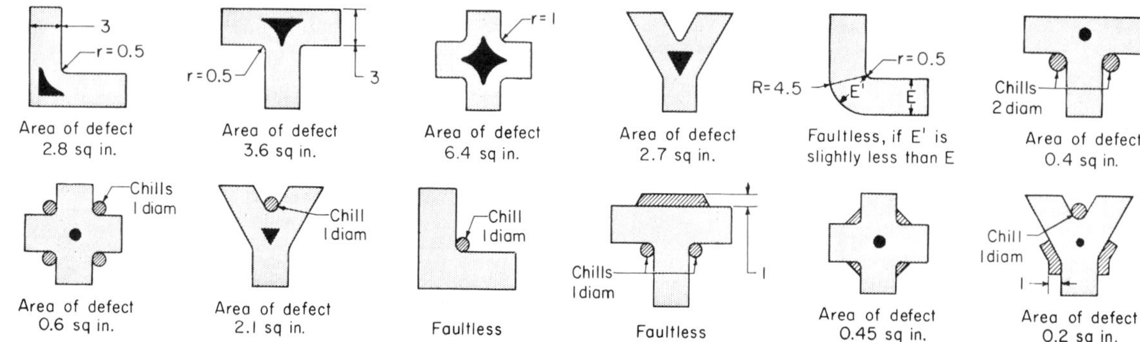

Fig. 25. Defects in the principal types of web intersection. L and T-junctions are best. (Briggs, Gezelius and Donaldson)

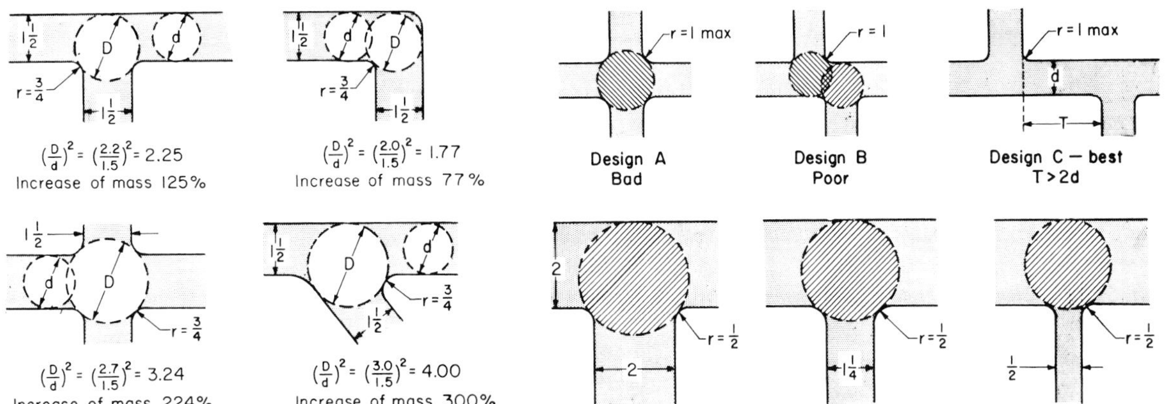

Fig. 26. Use of inscribed circles to determine effect of mass. *Fig. 27. Design of X-sections (top) and T-sections (bottom).*

1 in. thick, fed by runners and joined by a horizontal web, the latter will be given a thickness of about $\frac{25}{32}$ in. Ribs are often made two thirds the wall thickness.

Consider the valve body shown in Fig. 22. If the thickness of the seating is made equal to that of the external walls, there is a risk of having tears at the points *S*, since solidification is slower at these points. To avoid this defect, the thickness of the seating should be reduced by one fifth to one third.

Machining Allowances. Assuming that the minimum extra thickness called for by a manufacturer for machining is of the order of 0.120 in., Table 5 indicates the limits within which the maximums of the extra thickness for machining may vary as a fraction of the dimension *D*, on which the allowance is based.

Tables 5 and 6 indicate that the extra thicknesses for machining may sometimes equal that of the walls themselves. It is difficult to form an idea of the relative importance of these extra thicknesses merely by considering the drawing of the finished casting after machining, without studying the design of the rough casting. It is for this reason that it is advisable for designers to show on separate drawings, where necessary, various sections indicating actual wall thicknesses of the casting and extra thickness for machining.

In Fig. 23 the design A of a finished casting may seem correct in that it shows no excessively sharp differences in thickness. In effect, the addition of extra thicknesses for machining, provided by the patternmaker, may make the casting incorrect from the foundry viewpoint. There will be a risk of tears occurring at the points C_1 and C_2. On the other hand, by showing the extra thickness on a supplementary drawing as well as on the drawing of the finished machine casting (design B), the design of the rough casting is provided so that one can deduce from it what precautions to take in order to avoid sharp differences in thickness.

Minimum Thickness as a Function of Dimensions. To profit by reduction in weight often afforded by cast steel, specified wall thicknesses should be sufficient to avoid the risk of casting defects.

The minimum thicknesses obtainable depend on the casting involved and the position of the particular wall with respect to the runners and the risers. They are made greater for external horizontal

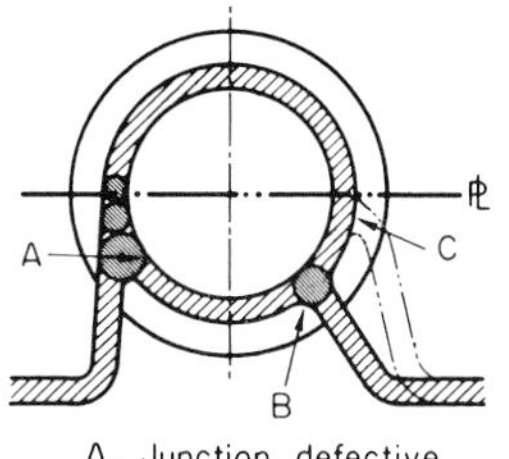

Fig. 28. Cross section of a motor frame casting, showing defective and improved junctions

Table 6. Machine Finish Allowances for Radii in Steel Castings

Diameter, in.	Allowance on radius, in.
Outside Radii on Circular Castings(a)	
Up to 18	1/4
18 to 36	5/16
36 to 48	3/8
48 to 72	1/2
72 to 108	5/8
Over 108	3/4
Bore Radii on Holes	
Up to 1	Solid
2 to 7	1/4
7 to 12	3/8
12 to 20	1/2

(a) Rings, spoked wheels, spoked gears and other circular castings

webs than for vertical walls. They also depend on the width of the wall, since for a given thickness the steel flows better in a narrow web than it does in a wide one. The minimum wall thicknesses as a function of wall length, shown in Fig. 24, are, therefore, not absolute but are indicative of common practice.

The normal minimum thickness shown should be adopted initially when replacing other metals by steel. The exceptional minimum thicknesses are those below which one does not usually go. However, it is possible that certain highly specialized founders can produce minimum thicknesses lower than those indicated here. Their success, however, requires special and costly precautions.

Avoidance of Hot Spots

Hot spots are those parts of castings that cool more slowly than the rest. Unless adequately fed by an adjacent riser, they are the points where shrinkage cavities and tears are likely to occur (Fig. 25). The risk of defects attributable to variations in thickness increases with the massiveness of the part concerned with respect to the casting as a whole.

The designer should evaluate the relative importance of the massive parts caused by variations in thickness. It can be done by simple means. The "inscribed circle" method enables the mass effect in any particular section of a casting to be represented by the surface of the inscribed circle in this section. It is assumed that the relationship of the metal masses at two different points in the same section is equal to that of the surfaces of the corresponding inscribed circles (that is, the relationship of the squares of the radii). This method is not strictly accurate, although it provides an approximation that is more than sufficient and can be of great service to designers (see Fig. 26 to 29).

The application of the inscribed circle method enables an evaluation of the effect of accumulations of metal resulting particularly from the intersections of walls, webs and ribs, and to deduce the shapes these points should have in order to reduce to a minimum the risk of the formation of objectionable hot spots.

The problems of defects caused by the accumulations of metal have been investigated experimentally. It has been found that the increase in mass at the right angle of a junction by the inscribed circle method may be evaluated as a percentage with respect to the mass of the wall by taking the quotient of the squares of the radii of the circles inscribed in the junction and the wall. If the figures thus obtained for the four kinds of junction shown in Fig. 26 are compared with the corresponding defective areas determined experimentally (Fig. 25), it will be seen that the size of the defect is largely proportional to the increase in mass, which proves the efficacy of the inscribed circle method.

Figure 25 also shows that the use of external chills tends to lessen the seriousness of the defects and even to overcome them in the case of L and T-pieces, which are the least bad.

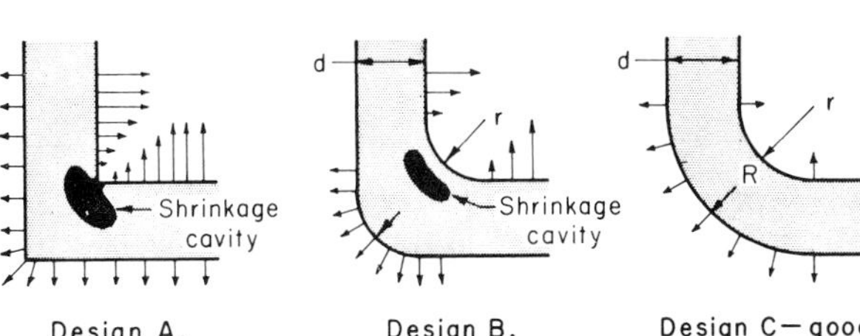

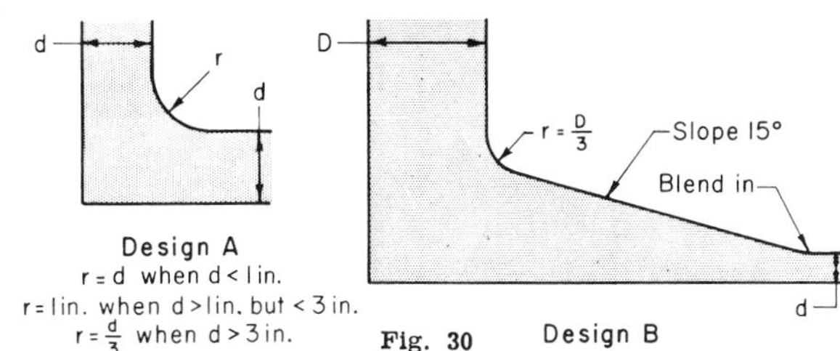

Fig. 29. Evolution of the L-design, eliminating shrinkage. *Fig. 30. Dimensional relationships of L-design with exterior corners.*

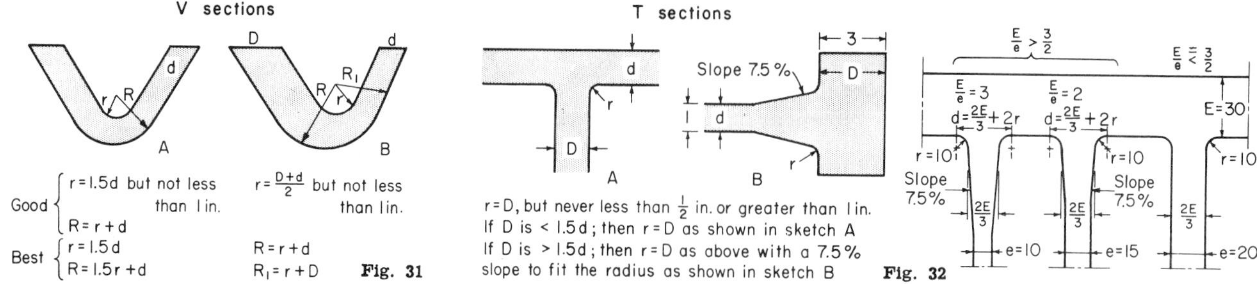

Fig. 31. Recommended V-design relationships. Fig. 32. Recommended T-design (left) and T-junction joining unequal walls.

By using the graphic method (measuring the radii of the inscribed circles) the designer can easily evaluate the increase in the mass with sufficient accuracy. Figure 26 shows the results obtained for a wall thickness of 1½ in. and fillets having a radius of ¾ in. It will be seen that the crosses and acute angle branches should be avoided and that T and L-junctions are to be recommended.

To reduce the accumulation of metal at the intersections of walls, webs and ribs, it is advisable, where possible, to modify the design as shown in the following examples. Replacement of a cross by an offset crosspiece is illustrated in Fig. 27. If the intersections of walls and webs are examined, it will be seen from Fig. 27 that the accumulations of metal are not so great when the webs are thinner than the walls. The sharp-angled union shown in Fig. 28 on the left is defective, while that on the right is correct. The dotted union has the added advantage of facilitating molding.

The fillets to be provided at junctions between walls and webs should be wide enough to avoid a notching effect, but they should not be so wide as to involve a significant increase in the mass of metal. Junctions to be adopted for the different types of walls are as follows:

Junction of Two Walls (Right-Angle Intersection). In a wall the rate of solidification of the metal is a function of the external perimeter of its section. In the case of a flat wall (one having an infinite curvature) the rates of solidification of the surfaces in contact with the sand are equal. If the radius of curvature decreases, the outer surface, having an area greater than that of the internal part, cools more rapidly, since the heat liberated by the metal during solidification can be absorbed by a greater volume of sand.

With a radius of zero (Fig. 29, left) this difference attains its maximum and the process of solidification is as follows: the outer area of sand being very large, the molten metal solidifies rapidly on coming into contact with it; the inner area of sand being very small, the sand becomes overheated and within a short time is raised to the temperature of the molten metal itself. This results in a hot spot within the solidifying metal section, so that after solidification there is a shrinkage cavity (often complicated by cracks) at the angle. If the internal radius r is not equal to zero, the size of the shrinkage cavity decreases; it becomes smaller as the outer radius increases — that is, as the ratio of the inscribed circles becomes closer to unity. The case of a 90° angle is shown in Fig. 30 (left), indicating the preferred relationships for r and d under three different conditions. With certain L-junctions, one arm of the L is thicker than the other (Fig. 30, right), as in a flanged fitting. The radius must then be based on the thickness of the largest section and may be connected by a tangent of 15° slope to the lighter section. With a junction angle less than 90° (V-junction), it is advisable to provide for equal wall thicknesses when r is less than 1½ in.; recommended design relationships are shown in Fig. 31.

Junction of Three Walls (T-Piece). For T-junctions the inscribed circle method shows that the accumulation of metal increases with the radius of the fillets. It is therefore necessary to reduce this radius to a minimum, but this is limited by the drawbacks attributable to sharp angles (effects of notches, hot spots). If the thicknesses to be joined differ widely, the thin parts cool more rapidly and exert a pull on the thick parts. To avoid tears, or even fractures, it is advisable to provide tapered junctions. Recommended design relationships for simple T-junctions and tie-in members in the form of T-junctions are given in Fig. 32.

The inscribed circle method shows that Y-joints are very defective. They can be replaced by a combination of a T and an L-piece, which is definitely more advantageous. This can be effected for two equal thicknesses, bearing in mind the above rules for L and T-junctions. For unequal thicknesses, the necessary information is given in Fig. 32.

Junction of Four Walls. In Fig. 25 it was shown that the action of external chills is not sufficient to prevent the formation of shrinkage cavities in such a joint (at least for thicknesses of 3 in.). If soundness of the casting is to be guaranteed, this kind of junction should not be used unless its length is short and it is possible for it to be cast with direct risers.

However, in a crosspiece, the central part is the least stressed, so that a casting is just as strong and much easier to produce sound if made either by inserting a central chill or by providing a central cavity. The designer should know that, when designing a crosspiece, the founder usually will be compelled to make use of an internal chill at the center part with the object of improving the soundness of the casting and, in consequence, its strength, because the accumulation of metal is such that it cannot be fed completely unless it is placed directly under a riser. Alternatively, to avoid the use of an internal chill, a central core may be used, provided it is of sufficient diameter (see dimensioning of holes, below) to avoid fusing of the core. This is equivalent to replacing the crosspiece by four T-pieces having uniform thicknesses around the central cavity.

It is often possible, while still retaining the same moments of inertia, to replace the cross by an L or a Z-piece and even by a closed section in the form of an O. Figure 33 indicates the various methods of joining two webs without using X-junctions. The first two should be avoided. They can be substituted by No. 3, 4 and 5, which are better from both mechanical and foundry viewpoints. No. 6 could be replaced with advantage by No. 7, 8 and 9, which possess better mechanical properties and lessen the risk of tears. Numbers 10 and 11, which call for the use of a core, have, on the other hand, the advantage of offering high resistance to bending and torsional stresses. For the internal ribbing of a plate or frame it is advantageous to replace the checkerwork consisting of four walls at an angle of 90° by junctions of three walls at 120°.

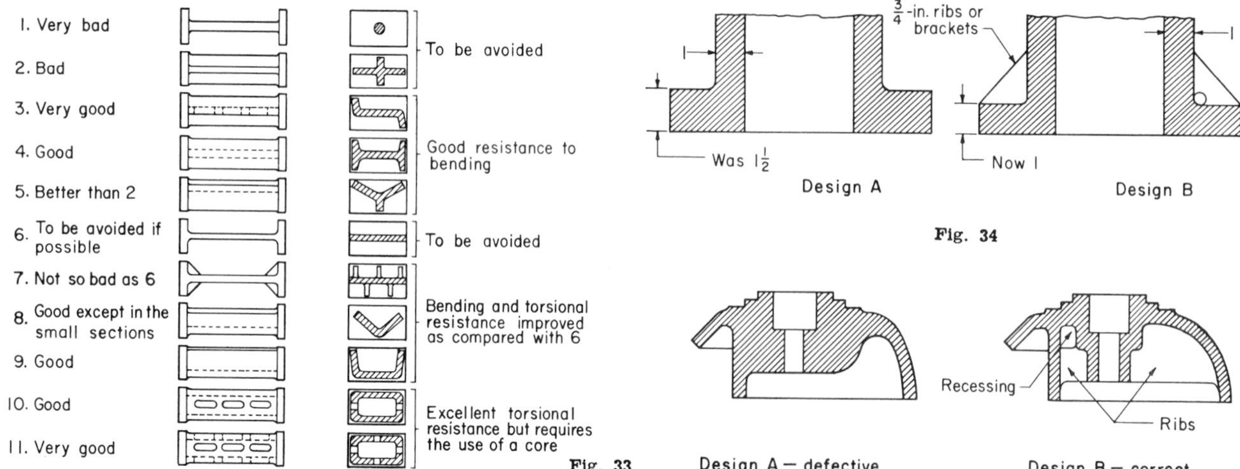

Fig. 33. Methods for joining two walls by means of open or closed sections. Fig. 34. Use of brackets in steel casting design (top). Lightening of a casting that is too massive by means of ribs and recessing (bottom).

Junction of Five or More Walls. Where the junction of five or more walls can be envisaged it is advisable to place a central clearance in such a way as to avoid the meeting of more than three. The solution to this important problem of joining walls, webs and ribs can be summarized by the two following rules:

1. Utilize L and T-pieces as much as possible and avoid crosses or acute angles.
2. Arrange for ordinary junctions if the ratio of the thicknesses D/d is less than 1.5 and for tapered junctions if the ratio D/d is higher than 1.5.

Strengthening Ribs and Recessing. The addition of ribs and the use of recessing often permit the avoidance of accumulations of metal at certain points. These two means, when judiciously adopted, help the founder economically to obtain sound castings that are free from tears.

Increasing the thickness of a casting to obtain higher strength often results in the production of a massive part which may not be sound. By adding ribs rather than increasing wall thickness, the casting is strengthened without adding to possible defects.

The double-flanged sleeve of Fig. 34 (top) is a common design. The flanges are often thicker than the barrel, resulting in accumulations of metal likely to cause cracks if certain preventive measures are not taken (horizontal pouring, separate feeding heads for each flange). When this sleeve must be cast upright, an obvious step is to make the lower flange of the same thickness as the barrel and to strengthen it with ribs, which by quick solidification serve to resist the shrinkage forces that may distort the flange, cause hot tears and interfere with the mechanical strength of the flange.

In order to guard against distortion of the flanges or the formation of tears, the founder is often compelled to make use of strengthening ribs that must be removed subsequently, thus adding to the cost of the casting. It would be more economical for the designer to provide strengthening ribs having the double role of replacing objectionable increased thicknesses, and strengthening the profile in its resistance to contraction forces.

In the example examined above, the addition of ribs ¾ in. thick will enable the thickness of the lower flange to be reduced from 1½ to 1 in.

Recessing enables massive parts to be lightened and at the same time lessens the risks of defects resulting from the formation of hot spots in such parts. However, the recessing should be done with great care. The placing of small cores complicates molding; they are likely to shift and, because of their low density, compared with that of the molten metal, they tend to become entrapped in the casting under the effect of ferrostatic pressure. Moreover, small cores can fuse and be very difficult to remove. The use of recesses, although sometimes essential, may entail a considerable increase in the cost of the casting. The following examples show the possibilities and the limits governing their use.

To lighten the casting (Fig. 34, bottom), the massive body has been recessed and strengthened by ribs; however, to prevent a crack at the top of the rib on the left under a massive part, this rib has been recessed. Additional coring costs are more than offset by the elimination of casting defects. Where the area in question can be fed directly by a riser, recessing is usually unnecessary, since it does not contribute to metal soundness and it complicates production.

Hot Spots in Sand. Consideration of the L-piece showed that the overheating of thin parts of sand enclosed between two walls gave rise to the formation of hot spots. The same applies when a part of the mold is enclosed on several sides by the walls of the casting; the sand there heats up much more than in the rest of the mold, causing the formation of a hot spot and the occurrence of defects.

Regions of sand, belonging to either the mold or the cores, which are not sufficiently voluminous must be avoided. In this respect, it is advisable to arrange for cores and mold parts to be at least twice the thickness of the metal surrounding them. For example, the fitting shown in Fig. 20, which had already been modified by making the thicknesses uniform, can undergo a second improvement involving an increase in the sand thickness at the point C. This can be done either by increasing the distance between the two vertical flanges or by displacement of the horizontal barrel. In general, recesses and holes of too-small dimensions should be avoided.

Dimensioning of Holes. As-cast holes are made by inserting sand cores in the mold. The holes may be straight through or blind (Fig. 35). A core becomes fused when surrounded by a mass of metal too great in relation to its dimensions. This causes great difficulties when the casting is being cleaned. In some instances, it is almost impossible to remove a fused core from the casting.

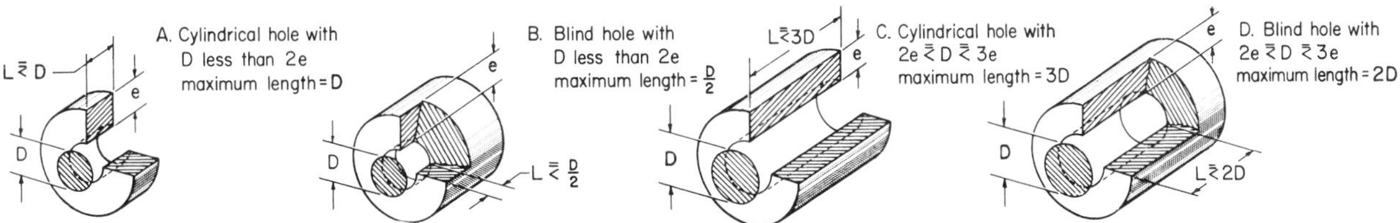

Fig. 35. Dimensional relationships for cylindrical through holes and blind holes

Careful choice of the diameter and length of holes is closely related to the thickness of the surrounding metal. In effect, the process of cleaning, which in foundry practice is an operation just as costly as molding, is largely dependent on the dimensional ratio between the diameter and length of the holes. Three cases serve to illustrate this principle:

1. If the diameter D of the hole is less than twice the thickness e surrounding the core, its length L should be no more than equal to the diameter for a cylindrical hole (Fig. 35, A); thus, for holes in walls and webs the diameter of the hole D may equal the thickness of the wall L. But for a blind hole the length should not exceed half the diameter Fig. 35, B).
2. If the diameter D of the hole is between twice and three times the thickness e surrounding the core, its length L should not exceed three times the diameter for a cylindrical hole (Fig. 35, C), and twice the diameter for a blind hole (Fig. 35, D).
3. If the diameter D of the hole exceeds three times the thickness, the length of the core is no longer limited by fusing, but by the fluidity of the metal being poured and the ability of the cores to withstand the ferrostatic forces to which they are subjected when the mold is being filled. Under these conditions, the graph in Fig. 24 should be consulted in order to ascertain the permissible length of the core as a function of the surrounding thickness.

The difficulty in obtaining clean surfaces free from sand usually requires considerable reduction in the machining speeds of cast steel when compared with other metals. It is, therefore, seldom advisable to provide holes in castings that can be more rapidly and economically made by drilling. It is preferable to provide holes in castings only when their machined diameter is greater than 1¼ in., which corresponds to an as-cast hole of about ¾ in.

On the other hand, in order to avoid breakage of drills, it is necessary that the bosses be cast sound. The absence of cores or chills in the bosses often gives rise to shrinkage cavities. For this reason, the founder frequently adopts expedients such as internal chills made of iron rods, soft iron wire spirals, or merely the use of large-headed nails. This practice can greatly simplify the work of the founder without affecting the mechanical strength of the casting, especially if the chills used are removed during machining.

The designer should leave to the founder the responsibility of choosing the chills and should prohibit their use only where, for reasons of mechanical strength, the use of internal chills in a massive part of the casting cannot be permitted. This should be clearly indicated on the drawing, and its effect on the cost should be understood.

Closed Sections. Shrinkage can vary considerably from one part of a casting to another. These shrinkage variations give rise to residual stresses which may show themselves as soon as solidification takes place, when the metal has little strength. They can cause cracks and even fractures if the profile involved cannot withstand a certain amount of deformation. Two typical closed sections, a pulley and a flywheel, illustrate the problem. If the thicknesses of the spokes and rim are different, stresses are set up during cooling, and the stresses can be the cause of cracks. Thus, a pulley or a flywheel with straight spokes can develop cracks beside the hub or the rim. On the other hand, when a design with curved spokes is adopted, the risk of cracks is overcome since the stresses are accommodated by the variations in the curvature of the spokes.

Generally speaking, it is best in closed sections to replace some straight elements by curved ones, to accommodate stresses set up during the transition through the critical hot tearing temperatures without affecting the strength of the whole assembly when cold. For the same reason, wheels should be given an odd number of spokes.

The principle of wave construction is useful in providing for the relief of stresses in cast structures. It makes use of members slightly waved or curved. This principle has been utilized in the design of the wheel shown in Fig. 36 (left). A certain elasticity is imparted by the curvature of the spokes and their inclination relative to the axle. Drawing upon the same principle, Fig. 36 (right) compares a rigid and an elastic design for a flywheel, the desired elasticity resulting from the use of an offset section.

The susceptibility of closed sections to cracking increases with any variation in thicknesses and large accumulations of metal. In a frame heavily ribbed by webs intersecting at right angles, accumulations of metal occurring at these intersections will give rise to hot spots and the risk of tears. By reducing these accumulations of metal, the tendency to tearing is diminished. This can be attained by coring out heavy sections, provided the cores are given a diameter at least twice the thickness of surrounding metal and have uniform thicknesses of metal around them.

If, in the same design, the use of cores is to be avoided, the webs can be stag-

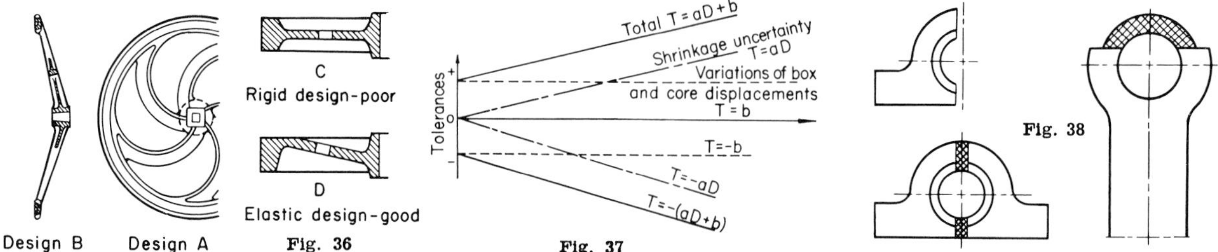

Fig. 36. Wave construction, showing rigid and elastic design. Fig. 37. Variations in tolerance, shown as a function of dimensions. Fig. 38. Grouping of two castings to facilitate machining (left). Providing additional metal to facilitate machining (right).

gered, allowing the central bar to take a slight deformation instead of breaking. This procedure also has the advantage of lessening the accumulations of metal by replacing crossings with T-junctions.

Accuracy of Dimensions. The considerable contraction of steel during cooling causes the patternmaker to calculate the dimensions of his pattern by increasing their values, which may differ in different parts of the casting, in order to compensate for the contraction. However, owing to variations in contraction and the molding operations, it is impossible to be sure of obtaining castings having exactly the dimensions given in the drawing. Therefore, it is permissible that the as-cast dimensions of a casting may differ from the exact dimensions laid down, provided that the plus or minus differences do not exceed certain limits or tolerances. The tolerances in dimensions allowed by standards or specifications are of the form:

$$t = \pm(aD + b)$$

when D denotes the length involved, while a and b are constants.

The difference of a casting dimension from its exact value is obviously the sum of two terms, one being a constant arising from molding and coremaking procedures, the other being proportional to the length and corresponding to the uncertainty of the contraction (Fig. 37).

Table 7 indicates the tolerances calculated on this principle for unalloyed steel castings. These tolerances apply to the parts of castings destined to remain as cast. The system of tolerances adopted, however, can differ for various parts of the same casting. It is often found that certain as-cast parts are indicated as having a dimensional accuracy unattainable within the limits of the tolerances. It should be borne in mind that the responsibility of the founder concerning the accuracy of the dimensions of parts to be machined is limited to making this machining possible by providing the necessary extra thickness. Machining allowances are shown in Tables 5 and 6; they may vary, taking into account the tolerances, so that the maximum extra thickness equals the minimum extra thickness with the plus tolerance added.

Setting Up for Machining. When castings are to be machined under mass-production conditions, the machining is carried out on jigs, which determine the position of the surface to be machined with respect to as-cast surfaces. Therefore, the as-cast surfaces coming into contact with the jigs should vary as little as possible from one casting to another — that is, they should be obtained by direct molding. For this reason it is essential to indicate jig location points to the founder, to enable him to select an appropriate method of molding.

Continuity of Machining Operations. Boring or thread-cutting operations can be correctly carried out only on continuous cylindrical surfaces. To insure continuous working of the tool, it is advisable to complete the segments of the cylindrical surfaces by grouping the castings. Thus, a divided nut can be made from a single pattern providing two castings (Fig. 38, left). If grouping is not feasible, the cylindrical shape is completed with supplementary material (Fig. 38, right). This is also true where two boring operations are to be carried out.

Tool clearances are often shown on drawings to provide for machining. However, they are usually insufficient because the designer has not taken machining allowances into account, with the result that the castings contain layers of sand practically surrounded by metal (Fig. 39). It is impossible to remove the sand, and the tools are quickly blunted.

It is necessary to provide very wide tool clearances. In this connection, it is well to show on the drawings the tool clearance in full-scale transverse section, together with the machining allowance and the tolerance, and to insure that the gaps provided satisfactorily comply with the conditions regarding core sizes (Fig. 39).

Table 7. Dimensional Tolerances for Steel Castings Not To Be Machined

	Dimension to which tolerance applies		
	Under 12 in.(a)	12 to 36 in.	36 to 120 in.
Average	0.060 + 0.006D	0.060 + 0.006D	0.080 + 0.006D
Concise	0.040 + 0.005D	0.050 + 0.005D	0.070 + 0.005D
Minimum	0.030 + 0.004D	0.040 + 0.004D	0.060 + 0.004D

(a) Tolerance, $\frac{1}{16}$-in. min. D is the longest dimension of the casting in inches.

Foundry Factors in Design

An understanding of the engineering factors in design is not in itself sufficient for the production of sound and economical castings. In most cases, only knowledge of the pouring and molding techniques enables drawings to be prepared which insure obtaining castings that are both sound and economical. Soundness is achieved by modifying the design as a function of the direction of pouring, while the economic factor involves changes in design to suit ease of molding.

Pouring Direction. Steel founders generally find it necessary to make their molds and cores from established natural materials and to bottom run the castings — that is, to introduce the liquid metal into the lower part of the molds. Generally, by adopting this recognized method, defects such as sand inclusions and displaced cores, and turbulent pouring can be avoided when filling the mold. The adoption of this procedure insures that the casting has the best possible external appearance.

Because of the considerable shrinkage that occurs in steel as it solidifies, shrinkage cavities may form in those parts of castings insufficiently fed with liquid metal as solidification progresses. Once the mold is filled, extra metal cannot be provided except by means of the feeding heads. Thus, the feeding of castings is essentially a function of the direction of solidification. It is important that the designer have a knowledge of the principal factors on which the founder bases his choice of gating and risering practice.

The principal factors that determine the rigging (or gating and risering) of a casting are, in their order of importance, as follows:

1 The position of the massive parts
2 The position of the surfaces to be machined
3 The possibility of venting the core gases

Positioning Massive Parts. Since the feeding of a casting takes place by gravity, it is important that the most massive

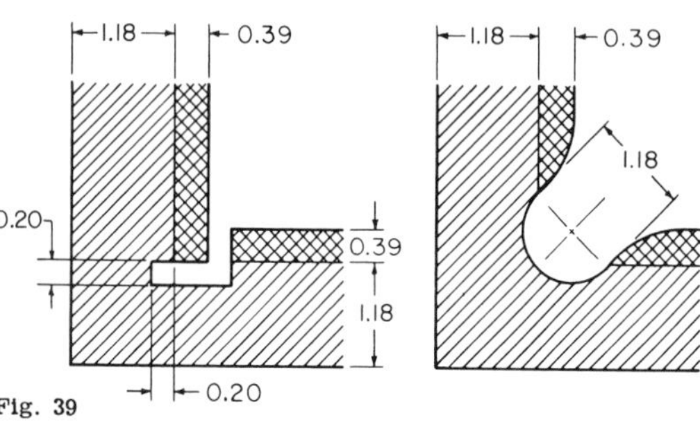

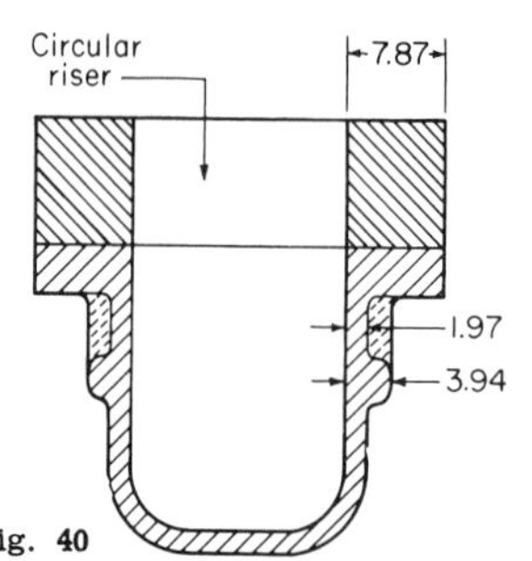

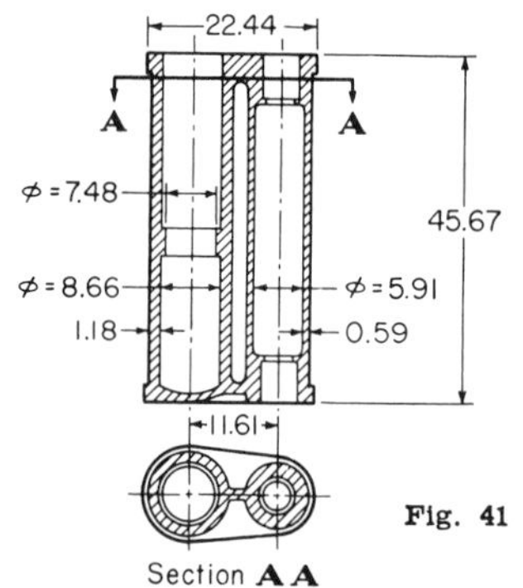

Fig. 39. Design for tool clearance. Fig. 40. Pressure vessel cast upright. Fig. 41. Double cylinder cast upright, large diameter at top.

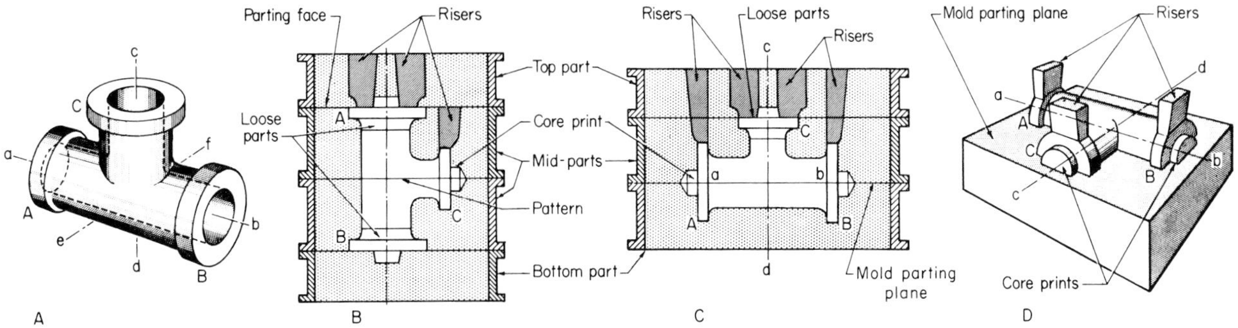

Fig. 42. The triple-flanged T-piece (A) may be poured in complicated molds with poor pouring directions (B and C), or in a simple, well-directed mold (D).

parts be arranged uppermost in the mold. Since these are the last to cool they can feed the lower part of the casting with molten metal and draw on the feeding heads for the extra amount of metal needed to compensate for this supply and for their own shrinkage.

This condition is all the more essential the greater the proportion of massive parts present in the casting as a whole. It is for this reason that in some castings the pouring direction is imposed on both the designer and founder alike. This is what happens with the pressure vessel shown in Fig. 40, the flange of which must be uppermost so that it is properly fed by the riser. Similarly, to obtain proper feeding of the rib, the change in the design indicated by the broken hatching is necessary in practice.

Surfaces to Be Machined. Dense and sound surfaces for machining should be arranged vertically or at the bottom of the mold. The upper parts can, in effect, contain inclusions (gas bubbles or particles of slag), which have a tendency to rise to the top of the casting because of their low density compared with the metal; these impurities are discovered during machining and affect the condition of the surface obtained.

This second condition is less essential than the first; it can, however, predominate and decide the pouring direction if the casting has no major differences in thickness. For this reason, cylinders that are to be bored are usually cast vertically. Casting horizontally would give a less clean surface along the upper half circle. The vital part of the double-cylinder casting shown in Fig. 41 is the 8-in. bore. The casting would be poured in the position shown, the critical section being in the upper part of the mold, where the required machining allowance will give it an extra thickness in relation to the lower cylinder. The feeding of the bottom portion of this casting would require special precautions, such as blind risers or chills.

Similarly, it is always essential to avoid arranging machined parts in such a way that the metal reaches them cool, slowly and without pressure after having traversed a considerable distance in the mold.

The casting shown in Fig. 42,A is a flanged T-piece to be machined. There are three possible molding and pouring positions, depending on which of the axes, *ab*, *cd*, *ef*, is vertical:

Choice of Pouring Direction. *First position: ab* vertical (Fig. 42, B). — The flange *B* cannot be fed, and the machined part of the flange *A* is at the top of the mold. This position need not be considered.

Second position: cd vertical (Fig. 42, C). — The three flanges may be fed (separate runners), but the machined part of the flange *C* is at the top of the mold; better position.

Third position: ef vertical (Fig. 42, D). — The three flanges may be fed (separate runners), and their machined parts are vertical; excellent position.

Choice of Molding Direction. *First position: ab* vertical (Fig. 42, B). — The pattern has to be divided at three points (very complicated); the molding has to be done in four boxes (very expensive); assembly is difficult; cleaning is normal. A very complicated and expensive solution.

Second position: cd vertical (Fig. 42, C). — The pattern has to be divided at two points (complicated); the molding has to be done in three boxes (expensive); assembly is difficult; cleaning is normal. A better solution, but still costly.

Third position: ef vertical (Fig. 42, D). — The pattern can be divided advantageously in the plane *ab*, *cd*, which simplifies molding. Molding in two boxes will also be simplified, thanks to the use of the natural taper of the pattern. Assembly is easy and cleaning is normal. This is the solution to be adopted because it is the least complicated and the cheapest.

The third position provides simultaneously the pouring direction enabling a sound casting to be obtained and the cheapest molding position. It sometimes happens that the most suitable pouring and molding directions do not emerge as clearly as in this well-known example. It is often necessary to compromise between several solutions, choosing the one embodying the fewest drawbacks and with the maximum advantages.

Directional Solidification

The principle of directional solidification refers to the solidification of molten metal that takes place progressively inward from the mold surface and upward to the reservoirs of liquid steel located in feeder heads or risers. Thus, the direction of solidification corresponds to the direction of increasing temperature gradients. A casting design conforms with this desirable condition when the radii of inscribed circles are uniformly increasing both from the bottom of the casting upward and horizontally toward the risers. Section thicknesses decrease gradually from the top to the bottom.

When applying the inscribed circle method to determine that each part of a casting is suitably fed, the rule of diminishing masses should be observed equally in both the horizontal and vertical sections. In order to make the design comply with these principles, it is possible either to provide padding at certain points and remove such extra metal subsequently by cleaning or machining, or to modify the design so as to make the casting nonsymmetrical.

Padding. The wheel casting (Fig. 43) demonstrates the difficulties encountered in the production of even simple steel castings. Circles corresponding to the application of the inscribed circle method have been drawn. The amount of padding to be allowed for by the founder to obtain a sound casting and the influence of the padding on the size of the risers are also indicated. Such padding is difficult and expensive to remove.

Nonsymmetry. In many instances it is possible to avoid the increased costs that result from padding by a simple modification of the design of the casting. Specifically, the casting may be given a nonsymmetrical shape that eliminates padding and promotes directional solidification. As a rule, a casting design that is correct from the foundry viewpoint is generally nonsymmetrical in its section thicknesses.

For example, a casting that would be difficult to produce sound if kept symmetrical is shown in Fig. 44. In this instance, the preferred design also demonstrates how decreasing thickness from top to bottom can be achieved simply.

In applying the principle of nonsymmetry, it is not necessary or desirable to increase section thickness excessively. Massive sections are prone to shrinkage cavities and tears, since the feeding of these masses is difficult unless they are directly under the risers. In this connection, flanges and bosses should not be placed so that the extra thickness created at their junction with the wall of the casting cannot be properly fed.

Molding. There are a number of ways

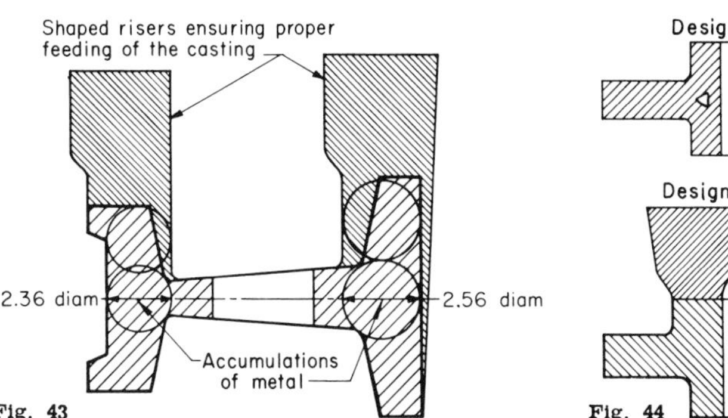

Fig. 43

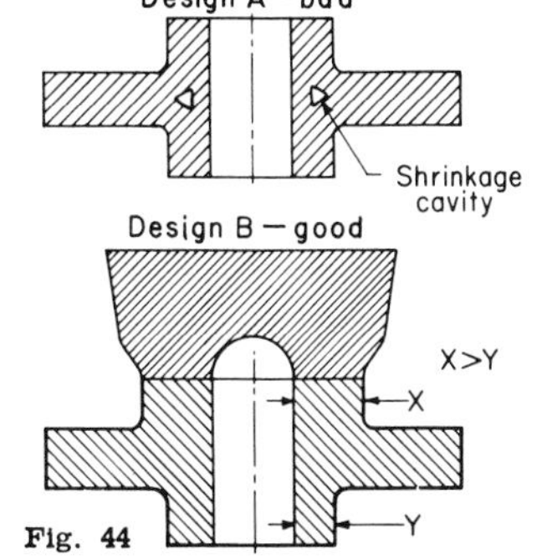

Fig. 44

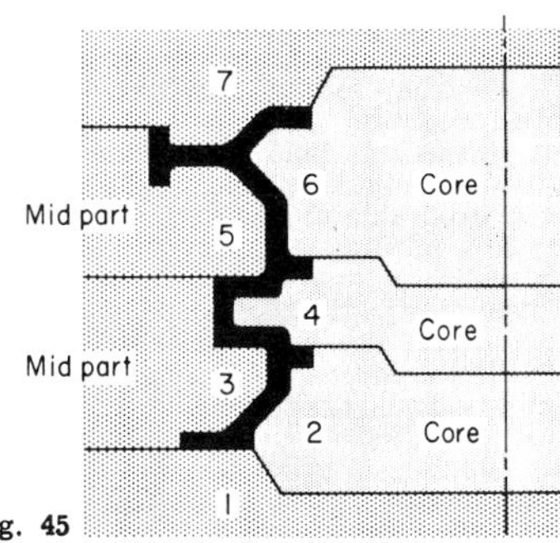

Fig. 45

Fig. 43. Pouring technique for a wheel casting. Fig. 44. Principle of nonsymmetry. Fig. 45. Molding a complicated shape.

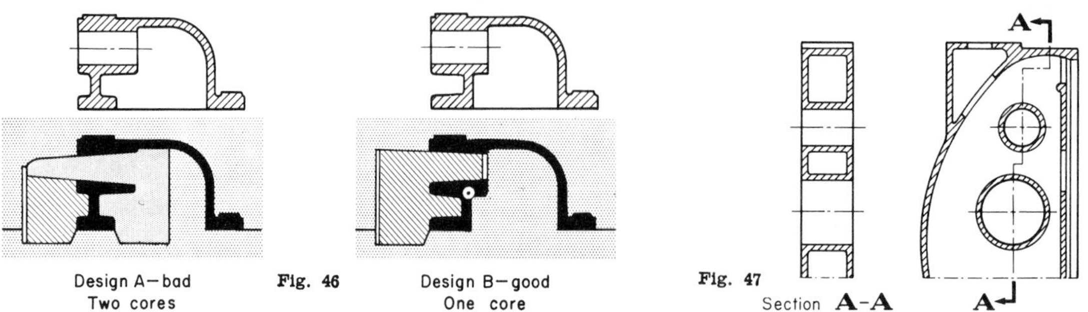

Fig. 46. Design illustrating reduction in number of cores. Fig. 47. Housing showing accessibility of cored parts for cleaning.

in which casting design can help to simplify and reduce the cost of molding. Making provision for suitable taper is very important in this respect. Taper is applied to the pattern so that it can be withdrawn without spoiling the mold. To accomplish this objective, the vertical parts of the pattern are given a slight slope, depending on molding direction and the position of the molding joint line. In addition to facilitating pattern removal, proper taper will sometimes permit the elimination of an internal core.

A taper of about 2.5° is usually sufficient for the easy withdrawal of the pattern. With special equipment, it is possible to use patterns with even less taper. If the amount of taper is not indicated on the drawing, it is still advisable to note whether taper is to be made plus or minus. A casting may be provided with natural taper by giving sufficient slope to the vertical walls.

Parting and Sectioning. A mold is made by dividing the pattern into a certain number of parts, which are molded separately in boxes and then superimposed at the parting lines following withdrawal of the pattern.

The more complicated the casting, the greater are the number of parting surfaces and sections of the pattern needed. Figure 45 illustrates the molding of a complicated profile with two midparts and three cores. The mold is divided at the maximum external cross sections, whereas the core is divided at the minimum internal cross sections. Withdrawal takes place in reverse numerical order from 7 to 1, and reassembly of the cores and mold takes place in numerical order.

In general, the parting surfaces of patterns are always located at maximum cross sections. When cores have to be sectioned, this is usually done at the minimum cross sections. These joints (inner or outer) give rise to fins, which have to be trimmed off, and to displacements, which may cause rejection. For this reason, it is well to keep their number down to a minimum. This can be done by minimizing the changes in cross section relative to the direction of molding.

A simplified design not only makes possible a reduction in the number of joints, but also enables them to be arranged at positions on the casting where they are more accessible for chipping and grinding. This is a point of great importance because of the contribution of cleaning and rejects to final costs.

Reducing the Number of Cores. Cores are assembled into a mold to form the cavities required in the castings. They must be securely held in position by ample bearing surfaces (core prints). It is usually advisable to reduce the number of cores to a minimum since difficult assemblies result in loss of time and increase the risk of rejection. However, the molds for some castings are made almost exclusively from cores. It is often advantageous to adopt core assemblies in order to attain less rigidity in the mold and improve drying. Figure 46 shows how a change in design can affect the shape of cores and reduce the number of cores required.

To retain adequately a core that overhangs, it is sometimes necessary to provide metallic supports (chaplets) that become embedded in the solidified metal. This practice should be avoided whenever possible because it adds to the difficulty of assembling the mold. Moreover, the metallic supports may act as chills at points where chilling is undesirable. Chaplets can often be dispensed with by providing a second bearing surface for the core.

With vertical cores, it is advisable to provide openings in the upper part of the casting through which core gases may be vented. To avoid entrapment of gases, the escape route should follow the shortest possible path to the vent.

When bosses or ribs are poorly located, they can greatly complicate molding. This is particularly true when such details are counter-tapered. In such castings, the pattern must be divided, or removable loose pieces must be used in the mold. If this condition is unavoidable, it is better, particularly for mass-produced castings, to facilitate molding by making more liberal use of cores.

Trademarks and lettering are also details that should never be counter-tapered. Their position should be fixed only after the molding direction has been determined. When the pouring direction has been established, it is advisable to avoid placing bosses and ribs on the upper, more massive parts of the casting to which risers will be applied. This will minimize machining costs.

Cleaning Considerations. Manufacturers frequently complain about castings that are supplied with the internal parts contaminated with adhering core sand. These parts cannot be cleaned unless the openings provided are sufficiently large to permit ready access for the removal of the sand and reinforcing core wires. Therefore, when castings are required to be clean internally, this should be specified and access for cleaning should be provided by enlarging existing openings or providing others. In most instances these measures will also improve venting and, consequently, the soundness of the casting. Such openings may or may not have to be reclosed. Figure 47 shows openings which have been provided in a casting for making the cored parts accessible for cleaning.

Certain features of design may greatly simplify cleaning and thereby lower costs markedly. Proper design of sections located near risers, for example, can eliminate the need for removal of excess metal after the risers have been cut off. Elimination of unwanted fins will also help to simplify final dressing.

The cost of cleaning may become prohibitive when the areas to be cleaned are virtually inaccessible and it is not feasible to provide openings for the purpose. This is true of cylinders having a large ratio of length to diameter. In such parts, it is better to cast the end separately and fabricate it afterward. The casting can also be split, thus enhancing its soundness and simplifying its production.

Tolerances

The dimensional tolerances discussed in the preceding section on casting design and summarized in Table 8 are applicable to sand castings. Other molding materials have been developed to provide improved dimensional accuracy, and other advantages not usually obtainable with conventional sand molds. Consequently, the dimensional tolerances applicable to these mediums will differ from those for sand castings.

Shell mold casting is particularly adapted to the production of large quantities of relatively small, high-quality components. Because of reduced draft, greater dimensional accuracy of pattern, and superior mold stability, this technique usually permits closer dimensional tolerances than sand mold casting. Minimum dimensional tolerances for shell mold steel castings that are not to be machined are shown in Table 9. Recommended minimum section thicknesses for shell mold castings as a function of section length are shown in Table 10 and Fig. 48. A statistical presentation of actual dimensions obtained with sand mold and shell mold components is also shown in Fig. 48. With shell molded parts A and B, for example, measurements were taken between the lugs and across the parting line, repectively.

Part C, a kingpin latch, was also made in a shell mold; measurements were taken of the diameter of the hole on both the

Table 8. Suggested Minimum Tolerances for Steel Sand Castings Not To Be Machined (a)

Length, in.(b)	Tolerance, in.(c)	Length, ft(d)	Tolerance, in.(c)
Up to 4	+ 0.078, − 0.063	Up to 8	+ 3/8, − 3/8
4 to 8	+ 0.125, − 0.063	8 to 15	+ 1/2, − 3/8
8 to 16	+ 0.188, − 0.063	15 to 30	+ 3/4, − 1/2
16 to 32	+ 0.250, − 0.125	30 to 50	+ 1 1/4, − 3/4
		Over 50	+ 1 1/2, − 1
Thickness, in.	Tolerance, in.	Thickness, in.	Tolerance, in.
1/4 to 1	+ 1/8, − 1/16	1/2 to 1	+ 1/8, − 1/16
1 to 1 1/2	+ 3/16, − 1/8	1 to 1 1/2	+ 3/16, − 1/8
1 1/2 to 3	+ 1/4, − 3/16	1 1/2 to 3	+ 1/4, − 3/16

(a) From a design and economic standpoint, tolerance should be as liberal as application warrants, with due regard for assembly, service and replacement. (b) Transverse dimensions from established longitudinal centerline ±1/8 in. (c) For longitudinal dimensions between unmachined surfaces. (d) Transverse dimensions from established longitudinal centerline ±3/8 in.

cope side and the drag side of the mold. Part D was produced in both sand and shell molds, thus providing a basis for comparing the two molding mediums. The range of dimensions for the sand mold castings is greater than for the shell mold castings. Finally, part E is a cap nut casting made from 1045 steel in a shell mold. Four different dimensions were measured on 108 of these hexagonal-top nuts, with variations of ±0.008 in. for four dimensions less than 3 in.

The investment mold casting process is best suited to production of fine detail and close dimensional tolerances. Close control of dimensions in the wax, plastic or frozen mercury pattern, smooth surface and rigidity of the ceramic investment or shell mold, and melting and drainage of the pattern all contribute to accuracy. The accuracy obtained on various dimensions of several castings is shown in Fig. 49. Suggested minimum tolerances for investment castings are shown in Table 11 and Fig. 50.

The minimum thickness that can be produced by the investment mold process is considerably less than by either sand or shell mold casting, primarily because of the higher temperature of the investment

Table 9. Minimum Dimensional Tolerances for Shell Mold Steel Castings Not To Be Machined

Dimension	Tolerance(a)
Length, in.	
Less than 4	±0.010 in.
4 to 8	±0.015 in.
8 to 16	±0.035 in.
16 to 32	±0.10 in.
Across the parting line	±0.010 in.
Draft	0.002 in. per in. min
Hole diameter	±0.005 in. per in.
Concentricity	0.002 in. per in.
Angular tolerance	±0°, 45′

(a) Between unmachined surfaces

Table 10. Recommended Minimum Section Thickness for Steel Castings Produced in Shell Molds

Length, in.	Minimum section thickness, in.
Less than 4	3/16
4 to 12	¼
12 to 18	5/16
18 to 48	⅜
48 to 80	½
Over 80	⅝

mold during pouring. The pouring methods used also contribute to soundness in thin sections. Table 12 shows minimum thickness of sections of investment castings as related to length.

The casting designer should carefully consider all specified tolerances. The closer the tolerance, the higher the final cost of the casting. For greater economy, tolerances should be only as close as the application actually requires. Table 13 shows how rejection rate increases with decreasing dimensional tolerance. In the example in Fig. 51, no casting was rejected because of failure to meet a ±0.010-in. tolerance for the 0.750-in. dimension, although 398 castings were checked. However, had the tolerance been narrowed to ±0.005 in., 35 of the castings (9%) would have been rejected. A still closer tolerance would result in a larger number of rejections.

This example is representative of small dimensions. As dimensions increase, the tolerance range increases. In the foregoing example, only one dimension was considered. If there were more dimensions with the same close tolerances, recovery would be proportionately lower.

Many steels can be cast by the investment mold process; hence most requirements for mechanical properties can be

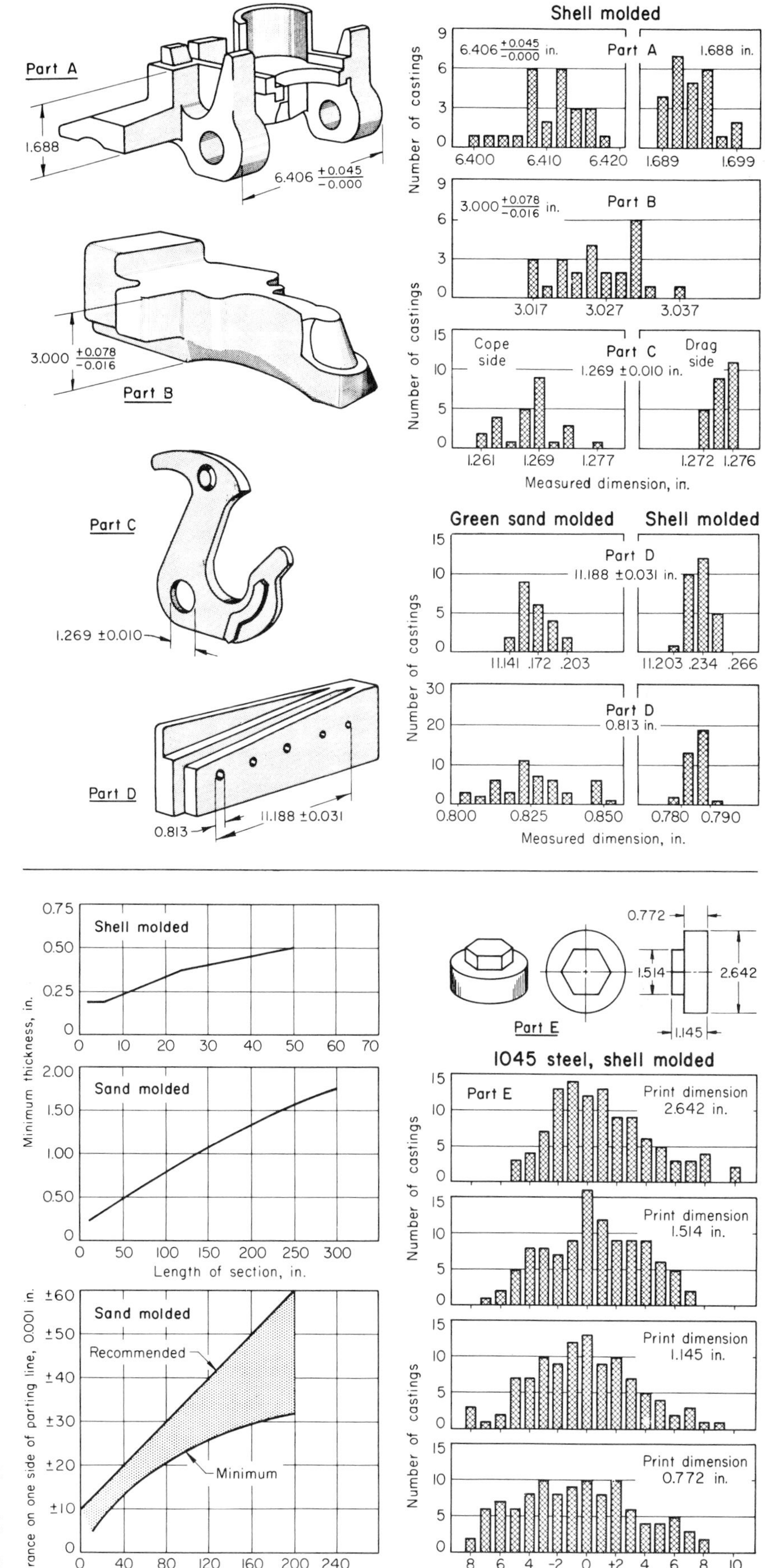

Fig. 48. Dimensional and tolerance data on sand and shell mold castings. Shell mold casting is adapted to the production of large quantities of relatively small, high-quality components and normally permits the use of closer dimensional tolerances than sand molding because of improved mold stability and other factors.

met. It is desirable to allow a hardness range of six points on the Rockwell C scale and a range for tensile strength of approximately 25,000 psi.

Graphite molds permit casting of steel parts with smooth surfaces and close tolerances in a re-usable mold. Dimensional variations were recorded on a production run of railroad freight car wheels 33 in. in diameter produced by this process. Table 14 shows the effect of location on the tolerances for these dimensions.

Centrifugal Castings. Carbon and low-alloy steels are used in making castings by the centrifugal casting process, in which metal and sand molds, and sometimes combinations of both, are used. Metal is poured into a central gate and is distributed to the exterior of the mold cavities by centrifugal force. Molds are revolved vertically or horizontally. Machining of the surface may be required to remove surface laps and discontinuities from castings made in metal molds. Centrifugal casting in sand molds usually prevents formation of these laps, permitting use of the as-cast surfaces. Dimensional accuracy of the as-cast outside diameters of 16-ft lengths of carbon and low-alloy steel tubes centrifugally cast in sand molds is shown in the next column.

Outside diameter, in.	Tolerance, in.
3.00 to 13.55	±0.10
13.56 to 26.75	±0.12
26.76 to 37.70	±0.16
37.71 to 50.25	±0.20

Wall thicknesses are subject to a tolerance of 8%, except where the tubing is to be bored. Then the requirements are more exacting, and outside diameter tolerances must be held on both inside and outside diameters.

Size of Cores

Another dimensional limitation in making steel castings is the minimum diameter of cores that can be used without core failure. In green sand molds, cores of dry sand, oil-bonded baked sand, and shell molded sand have similar limitations.

The minimum diameter core that can be used successfully in steel castings depends on the thickness of the metal section surrounding the core, the length of the core, and the special precautions and procedures used by the foundry. The thermal conditions that the core must withstand increase in severity as metal thickness increases and core diameter decreases. More heat must be dissipated in heavier sections, and the core is less able to absorb and dissipate this heat as the diameter decreases. As heat conditions become more severe, the possibilities for metal penetration and sand burn-on increase, making the cleaning of the casting and removal of the core more difficult.

The thickness of the metal section surrounding the core and the length of the core affect the bending stresses induced in the core by flotation forces. Thus they affect the ability of the foundry to obtain the required tolerances in cored cavities. The core must be large enough so that it can be reinforced with wire or rod to withstand these stresses. As metal thickness and core length increase, the amount of rodding required increases. The minimum core diameter must also increase to accommodate the extra rodding.

Figure 52(a) shows the recommended minimum diameter of cores to be used in cylindrical or boss sections in steel castings, as a function of section thickness and core length. These curves are for cores assembled in a mold in a horizontal position and supported at the ends only. Minimum core diameters determined from these curves can be reduced by 25% if the core is to be used in a vertical position. A similar curve for cores in plate sections of steel castings is presented in Fig. 52(b). Smaller core diameters may be used but they involve special practices, such as the use of stronger core materials, which increase the cost of the castings.

Because stronger molding materials are used in the investment casting process, thinner and longer cores may be used. It is feasible to use cores of considerable length (up to 2½ in.) that are supported at one end only. The minimum diameters of such cores are shown in Table 15.

Core Tolerance

One of the influences on casting tolerances that is often overlooked is the dimensional variation in the mold and the cores, themselves. Before the metal is poured into the mold, variations have been introduced. While it is difficult to measure the mold to record these variations, it is comparatively easy to do this with a core.

The data in Fig. 53 show the diametral dimensional variations for two blown shell cores. One set of dimensions was taken parallel to the parting line of the core box and the other set perpendicular to the parting line. Core A was approximately 19 in. long and 2¾ in. in diameter.

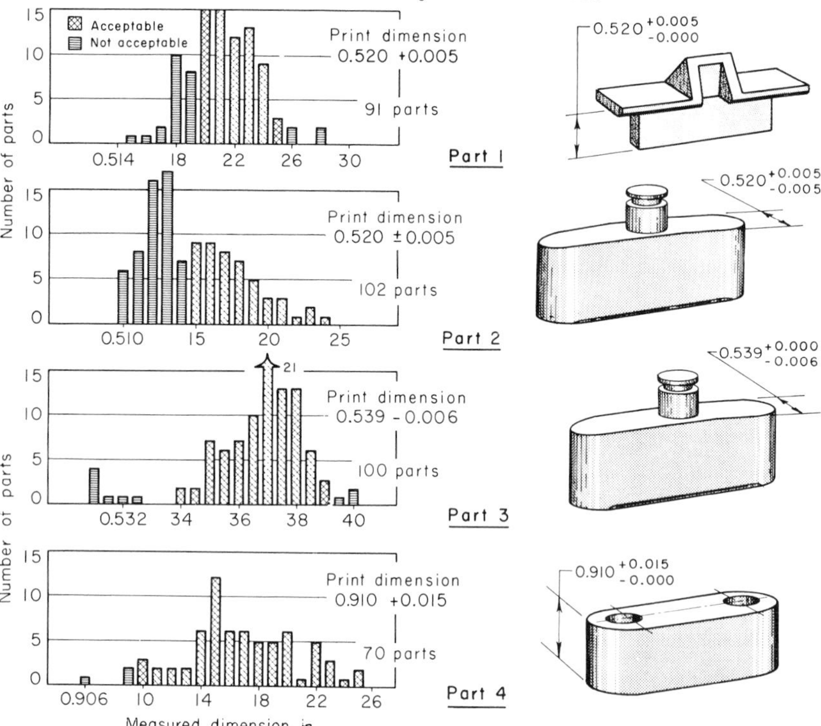

Fig. 49. Acceptability limits and variation in dimensions for castings of 8620 steel produced by the investment method

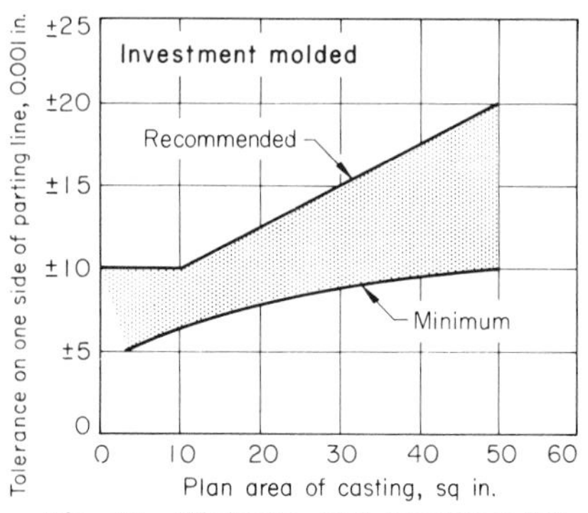

Fig. 50. Minimum and recommended tolerances for steel investment castings on one side of parting line, as a function of the plan area of the casting

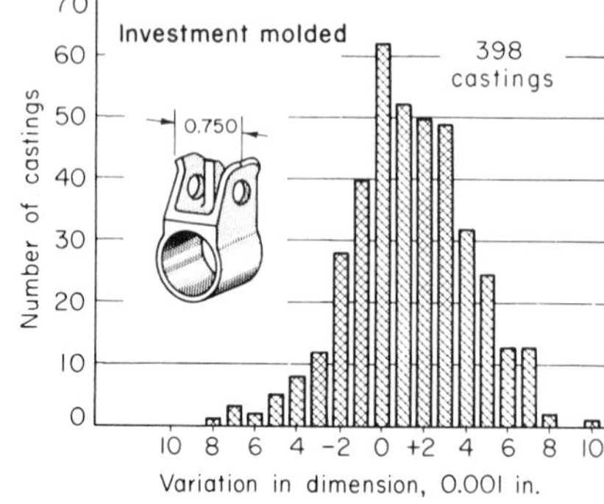

Fig. 51. Variation of the 0.750-in. dimension within the specified ±0.010-in. tolerance for the gun sight casting shown, investment molded from steel

Table 11. Minimum Tolerances for Investment Mold Steel Castings

Dimension, in.(a)	Minimum tolerance, ± in.	Dimension, in.(a)	Minimum tolerance, ± in.
½	0.005	8	0.040
1	0.005	9	0.045
1½	0.009	10	0.050
2	0.010	15	0.075
2½	0.015	16	0.080
4	0.020	20	0.100
5	0.025	25	0.125
6	0.030	30	0.150
7	0.035	32	0.160

(a) As, for example, the dimension between centerlines of the large bore and the small bore of a connecting rod

Table 12. Minimum Thickness of Investment Cast Steel Sections

Length, in.	Minimum thickness, in.	Length, in.	Minimum thickness, in.
¼	0.030	1¼	0.060
½	0.040	1½	0.060
¾	0.050	2	0.060
1	0.060	2½	0.060

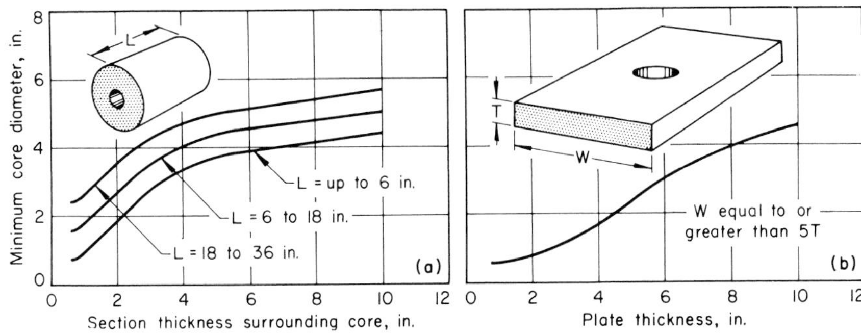

Fig. 52. (a) Recommended minimum diameters for horizontal cores supported at ends only, in cylindrical or boss sections of steel castings. Minimum core diameters shown in the three curves can be reduced by 25% if the core is vertical in the mold. (b) Recommended minimum core diameter for plate sections in steel castings.

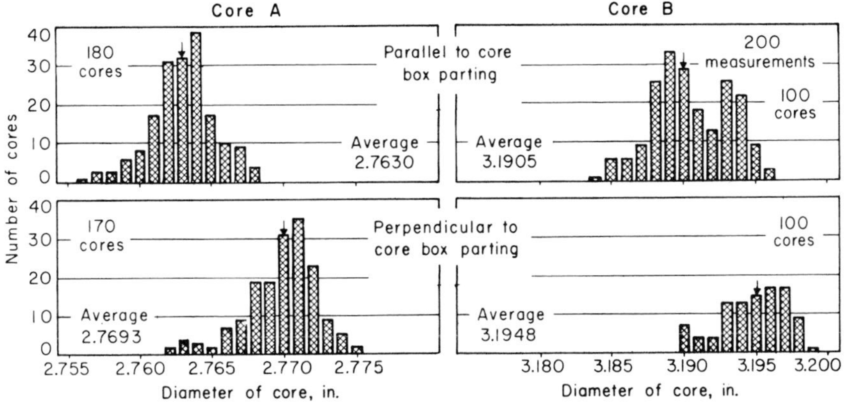

Fig. 53. Frequency distributions of diameters of shell cores (W. C. Truckenmiller, C. R. Baker and G. H. Bascom, Modern Castings, March 1958)

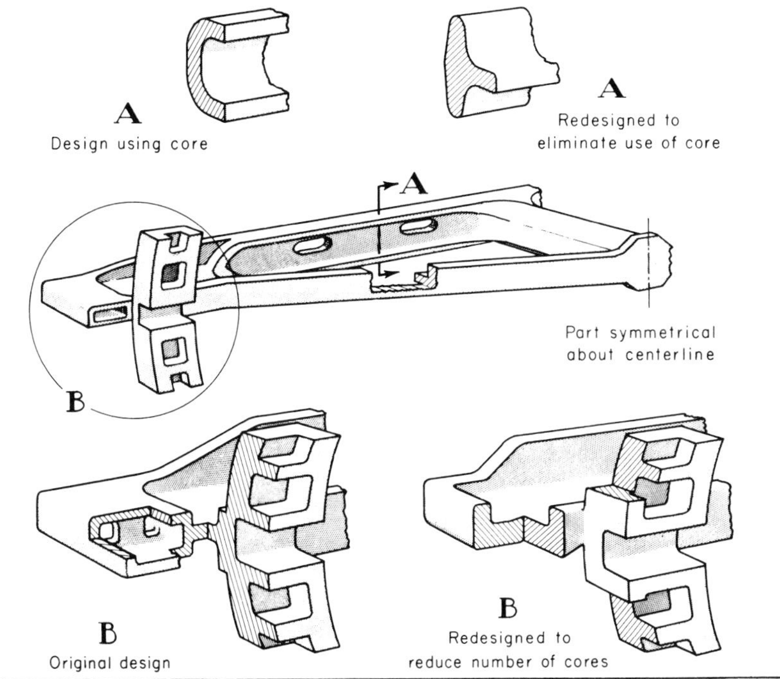

Item	Relative cost — With original number of cores (129 lb)	Relative cost — With reduced number of cores (135 lb)
Metal(a)	11	12
Melting	11	11
Coremaking(a)	7	4
Molding(a)	20	19
Cleaning, grinding, finishing(a)	24	20
Inspection	2	2
Overhead(b)	25	24
Total per casting	100%	92%

(a) Includes departmental overhead. (b) General plant overhead.

Fig 54. Freight car brake beam casting redesigned so that fewer cores were necessary, thereby reducing production costs by 8%

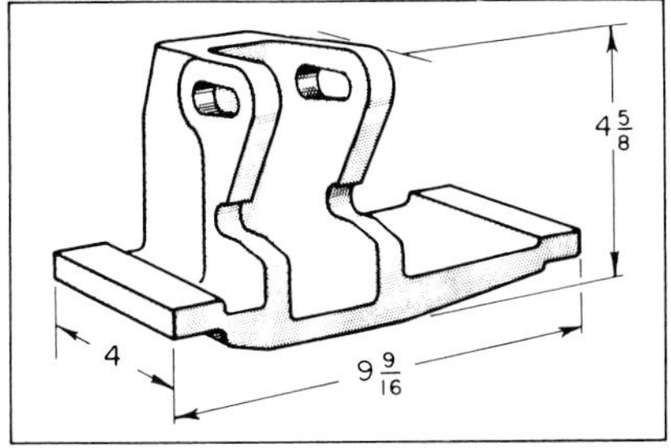

Item	Cost — With three cores (9 lb)	Cost — With one core (9 lb)
Metal and melting	$0.85	$0.85
Coremaking	0.52	0.37
Molding	0.57	0.62
Cleaning, grinding, finishing	1.08	1.52
Inspection	0.09	0.09
Overhead	0.49	0.55
Total per casting	$3.60	$4.00
Total per pound	$0.40	$0.44

Fig. 55. Steel casting originally produced with three cores was redesigned for production with one core. Although coremaking costs were reduced appreciably, total cost with fewer cores was 10% greater. The increase resulted from higher molding, finishing and overhead costs.

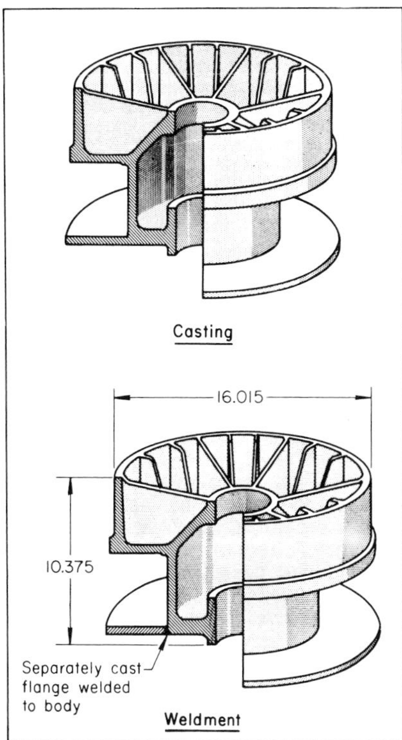

Cost item	With original number of cores (193 lb)	With reduced number of cores (193 lb)
Metal	$ 8.69	$ 7.24
Melting, pouring	17.04	13.28
Coremaking	4.94	2.67
Molding	5.43	6.18
Cleaning, grinding, finishing	16.86	6.79(a)
Inspection and overhead	25.11	14.52
Total per casting	$78.07	$50.68
Total per pound	$ 0.405	$ 0.263

(a) Includes additional welding cost of combining the two parts.

Fig. 56. Drum-type casting redesigned for production with a reduced number of cores. The bottom flange was cast separately and welded as indicated. This change in method reduced cost about 35%.

Table 13. Estimated Effect of Tolerances on Rejection Rate for Investment Castings

Tolerance, in.	Rejection rate, %	Tolerance, in.	Rejection rate, %	Tolerance, in.	Rejection rate, %
Up to 0.250-In. Dimension		**0.250-In. to 0.500-In. Dimension**		**0.500-In. to 0.750-In. Dimension**	
0.002	70	0.002	80	0.002	90
0.003	40	0.003	50	0.003	80
0.004	3	0.004	30	0.004	50
0.005	0	0.005	15	0.005	35
		0.006	3	0.006	15
		0.007	0	0.007	3
				0.008	0

Table 14. Effect of Dimension Location on Tolerances of Steel Castings Made in Graphite Molds

Dimension location	Tolerance, ± in.	Castings measured, %
Circumference	0.125	100
Radius	0.020	100
Rotundity	0.015	100
Plate thickness	0.020	2.5
Flange thickness	0.020	10
Tread width	0.010	2.5

For railroad car wheels 33-in. diam

Table 15. Minimum Diameter of Cylindrical Cores for Investment Castings

Core length, in.	Minimum diameter, in.: Supported at both ends	Minimum diameter, in.: Supported at one end only
¼	3/32	3/16
½	⅛	¼
¾	3/16	½
1	¼	⅝
1¼	5/16	¾
1½	⅜	¾
2	7/16	1
2½	½	1

Core B was approximately 12 in. long and 3¼ in. in diameter. Each core was made in a single opening of a two-opening box.

The histogram for core A shows almost the same spread (0.012 in.) for the measurements parallel to the parting line as for those taken perpendicular to the parting line (0.013 in.). However, the mean dimensions differ by approximately 0.007 in., with the mean dimension perpendicular to the parting line being larger.

In the study of core B, somewhat more elaborate work went into the preparation of the cores. A rigid carriage for handling the core box in and out of the core blower was employed, and the opening of the box was mechanized so as not to deform the cores in any manner. One hundred cores were produced. Two readings of diameter were taken on each core, approximately parallel to the parting, one on each side and both immediately adjacent to the parting.

The histogram for these dimensions shows an unusual twin-peaked curve. This is due to a slight mismatch caused by slightly different diameters of the core box halves at the parting, and assignable to different temperatures of the core box halves during molding. If the diameter parallel to the parting line had been measured only once, as for core A, the total spread would probably have been 0.009 or 0.010 in., and a normal distribution curve would have been plotted. The dimensional stability of core B is due to greater rigidity of equipment and mechanization of core box opening.

From this study, it is evident that a certain amount of tolerance control is sacrificed when cores are used. In this example, a small amount of metal could be removed from the faces of the core boxes to bring the mean dimensions into agreement. This would not, however, decrease the amount of spread for the dimension taken perpendicular to the parting, but would only serve to move the spread to an area more compatible with concentricity or accuracy.

Designers should recognize that cores exert a detrimental influence on dimensional stability, and should design to eliminate coring, where practical, or to require the most simple cored shapes.

Cost of Cored Cavities

The use of cores markedly increases cost. Making the cores, assembling them in the mold cavity and removing them after pouring add considerably to the expense of production. While designing for production of a casting without cores may not always reduce the total cost, the possibility of its leading to cost saving should always be considered.

To demonstrate the effect of elimination of cores on cost, three separate castings with and without cores are evaluated in Fig. 54, 55 and 56. The parts shown in Fig. 54 and 56 were redesigned for production with fewer cores, and in both castings production costs were reduced as shown in the tabulation. However, producing the casting shown in Fig. 55 with one core instead of three cores increased the cost because of increased cost of making the mold and cleaning and finishing the casting.

Properties and Applications of Carbon and Low-Alloy★ Steel Castings

Typical Specifications	Mechanical Properties, Min. Spec. Values: Tensile Strength, 1000 Psi.	Yield Strength, 1000 Psi.	Elong. in 2 in., %	Red. in Area, %	Hardness, Bhn.†	Charpy Impact, Ft-Lb.†: 70°F. Key-hole	70°F. V-Notch	−40°F. Key-hole	−40°F. V-Notch	Endurance Limit 1000 Psi.†: Un-notched	Notched	Machinability Speed Index†: HSS‡	Carbide Tool	Heat Treatment†	Other Current Specifications	Application and Outstanding Characteristics
Carbon Steels																
ASTM A27-65 Grade 60-30	60	30	24	35	131	30	12	8	5	30	19	160	400	Anneal	ASTM A216-60T, WCA; AAR M201-53, Grade AU, Grade AA; MIL-S-15083B, Class B; ABS Class 1, Hull	Excellent weldability; can be case hardened or carburized, low electrical resistivity, desirable magnetic properties.
ASTM A27-65 Grade 65-35	65	35	24	35	131	30	35	15	12	30	19	135	230	Normalized	SAE Grade 0030; Federal QQ-S-681d, Class 65-35; Lloyds Class A; ASTM A352-60 LCB; MIL-S-15083B, Grade 65-35	Excellent machinability and weldability; combine moderate strength, good ductility and machinability.
MIL-S-15083B Class 70-36	70	36	22	30	143	30	30	13	12	35	22	135	230	Normalized	ASTM A27-60, Grade 70-36; ASTM A216-60T, WCB; AAR M201-53, Grade B; ABS Class 2	
MIL-S-15083B Class 80-40	80	40	17	25	163	25	35	12	10	37	26	135	400	Normalized and tempered	SAE Grade 080; Federal QQ-S-681d, Class 80-40	High strength, good machinability, toughness and excellent fatigue resistance; readily weldable.
SAE Automotive Grade 0050A	85	45	16	24	179	20	26	10	10	39	28	120	325	Normalized and tempered		
SAE Automotive Grade 0050B	100	70	10	15	212	30	40	15	12	45	31	80	310	Quenched and tempered	Federal QQ-S-681d, Class 0050	High hardness, wear resistance.
Low-Alloy Steels★																
ASTM A352-68a Grade LC1	65	35	24	35	137	50	60	18	20	32	20	130	400	Normalized and tempered	ASTM A352-60T, LC2, LC3; ASTM A217-60T, Grade WC1; MIL-S-870B	Suitable for high and low-temperature service; excellent weldability.
ASTM-A217-69 Grade WC4	70	40	20	35	143	48	55	25	22	35	23	120	230	Normalized and tempered	ASTM A217-60T, WC4, WC5, WC6, WC9; MIL-S-15464B, Class 1, 2 and 3	Excellent weldability; combine moderate strength, high toughness and machinability; suitable for high-temperature service.
ASTM A148-65 Grade 80-50	80	50	22	35	170	45	48	25	18	39	25	110	240	Normalized and tempered	ASTM A148-60, Grade 80-40; SAE Grade 080; Fed. QQ-S-681d, 80-50; MIL-S-15083B, 80-50	
ASTM A148-65 Grade 90-60	90	60	20	40	192	40	40	20	16	42	31	95	290	Normalized and tempered	ASTM A-217-60T, C5; SAE Grade 090; QQ-S-681d, Class 90-60; AAR M201-47, Class C; MIL-S-15083B, 90-60	Excellent combination of strength and toughness. Certain steels in this group are deep-hardening grades and are suitable for high and low-temperature service. High resistance to impact. Readily weldable.
ASTM A148-65 Grade 105-85	105	85	17	35	217	50	58	40	40	53	34	90	310	Quenched and tempered	QQ-S-681d, 105-85; MIL-S-15083B, 105-85; SAE Grade 0105	
ASTM A148-65 Grade 120-95	120	95	14	30	262	43	45	35	31	62	37	75	180	Quenched and tempered	SAE Grade 0120; QQ-S-681d, 120-54; MIL-S-15083B, 120-95	
ASTM A148-65 Grade 150-125	150	125	9	22	311	28	30	20	17	74	44	45	200	Quenched and tempered	SAE Grade 0150; QQ-S-681d, 150-125; MIL-S-15083B, 150-125	Deep hardening, high strength, resistance to wear and fatigue.
ASTM A148-65 Grade 175-145	175	145	6	12	352	22	24	15	12	84	48	35	180	Quenched and tempered	SAE Grade 0175; QQ-S-681d, 175-145	High strength and hardness, resistance to wear and fatigue.
No Specification	200	170	6	17†	401	—	14	—	8	88	50	—	—	Quenched and tempered	None specified.	

★Below 8% total alloy content.
†Typical properties; should not be used as design or specification limit.
‡Machinability speed index for standard 18-4-1 high speed steel based on cutting speed which gives a tool life of 1 hr. For carbide tools, cutting speed for tool life of 1 hr. based on 0.015-in. wearland.
Source: Steel Founders' Society of America, Cleveland.

Section IV: Wrought Stainless Steels

Ferritic Stainless Steels

By Remus A. Lula

GENERALLY, FERRITIC STAINLESS STEELS provide about the same corrosion resistance as their austenitic counterparts at less cost — lower amounts of alloying elements are needed. Further, these grades have certain useful corrosion properties in their own right, such as resistance to chloride stress-corrosion cracking, corrosion in oxidizing aqueous media, oxidation at high temperature, and pitting and crevice corrosion in chloride media.

The newer ferritics, especially those with high chromium content, have become possible through vacuum and argon-oxygen decarburization, electron-beam melting, and large-volume vacuum induction melting. Compositions are listed below:

Type 409	0.05 C, 11 Cr, 0.5 Ti
Type 439	0.05 C, 18 Cr, 0.5 Ti
18SR (Armco)	0.05 C, 18 Cr, 0.4 Ti, 0.5 Ni, 1 Si, 2 Al
20-Mo (J&L)	0.02 C, 20 Cr, 0.5 Cb, 1.6 Mo
18-2	0.02 C, 18 Cr, 0.4 Ti, 2 Mo
26-1S	0.03 C, 26 Cr, 0.5 Ti, 1 Mo
E-Brite 26-1 (Airco)	0.002 C, 0.010 N, 26 Cr, 1 Mo
29Cr-4Mo (Du Pont)	0.004 C, 0.01 N, 29 Cr, 4 Mo
29Cr-4Mo-2Ni (Du Pont)	0.004 C, 0.01 N, 29 Cr, 4 Mo, 2 Ni

A discussion of present and potential applications can be simplified by grouping the alloys according to chromium level:

The 12% Cr steels, including type 409 and various proprietary modifications, are low in cost; formability and weldability are good. Usage thickness is limited to approximately 0.150 in. (3.8 mm) max if ductile-to-brittle transition temperature (DBTT) at room temperature or lower is needed. Atmospheric corrosion resistance is adequate for functional uses, but not for decorative applications.

Applications include automobile exhaust equipment, radiator tanks, catalytic reactors, containerization, culverts, dry fertilizer tanks, animal containment housings. Type 409 and its proprietary modifications can replace coated carbon steel and brass. Example: a radiator cap made of a low-residuals-modified type 409. Outstanding

ESCOA Fintube Corp., Pryor, Okla., uses MF-1 (AISI type 409) for fin strips of heat exchanger tubes.

Corrosion Resistance + Economy

formability permits replacement of brass in this instance.

The 18-20% Cr steels — such as type 439, 18Cr-2Mo, 18SR, and 20-Mo — resist chloride stress-corrosion cracking. Resistance to general and pitting corrosion is approximately equivalent to that of austenitic types 304 and 316; type 18SR has oxidation resistance equivalent to that of austenitic stainless steels. Limitations: sheet cannot exceed approximately 0.125 in. (3.2 mm) thickness if DBTT at room temperature or lower is needed. Embrittlement at 885 F (470 C) and low strength at high temperatures are also problems.

These grades are suitable for equipment exposed to aqueous chloride environments, heat transfer applications, condenser tubing for fresh water powerplants, food handling uses, and water tubing for domestic and industrial buildings. Type 439 can be used where corrosion resistance must be equivalent to that of type 304; in this respect, 18-2 and 20-Mo are closer to type 316. The ferritic steels add the benefit of stress corrosion resistance.

These steels can replace brass, bronze, cupronickels, and austenitic stainless steels. Example: gas-fired hot water tanks made out of welded type 439, selected because of its resistance to stress corrosion. Another example: flexible hose for connecting appliances to gas supply lines. It too is made of type 439, which demonstrates the excellent formability of this alloy in light gages.

26-1 steels are typified by the low-residual version, E-Brite 26-1, and the titanium-stabilized steel, 26-1S. They feature resistance to stress-corrosion cracking, oxidizing corrosion conditions, and chloride solutions. E-Brite 26-1 has good toughness in heavier sections. The 26-1S grade is limited to thicknesses of less than about 0.100 in. (2.5 mm) if DBTT at room temperature or lower is needed. It also embrittles at 885 F (470 C) and has low strength at elevated temperatures.

Applications: condenser tubing, heat exchangers, equipment for handling acids (organic acids in general) in the chemical and petrochemical industries, synthesis of urea, the pulp and paper industry.

29 Cr steels include 29Cr-4Mo and 29Cr-4Mo-2Ni. They resist stress-corrosion cracking, corrosion in acids, and chlorides better than do existing stainless steels. Another feature is excellent pitting and crevice corrosion resistance in seawater. The 29-4-2 type has good corrosion resistance in sulfuric acids, and both alloys are ductile in heavy sec-

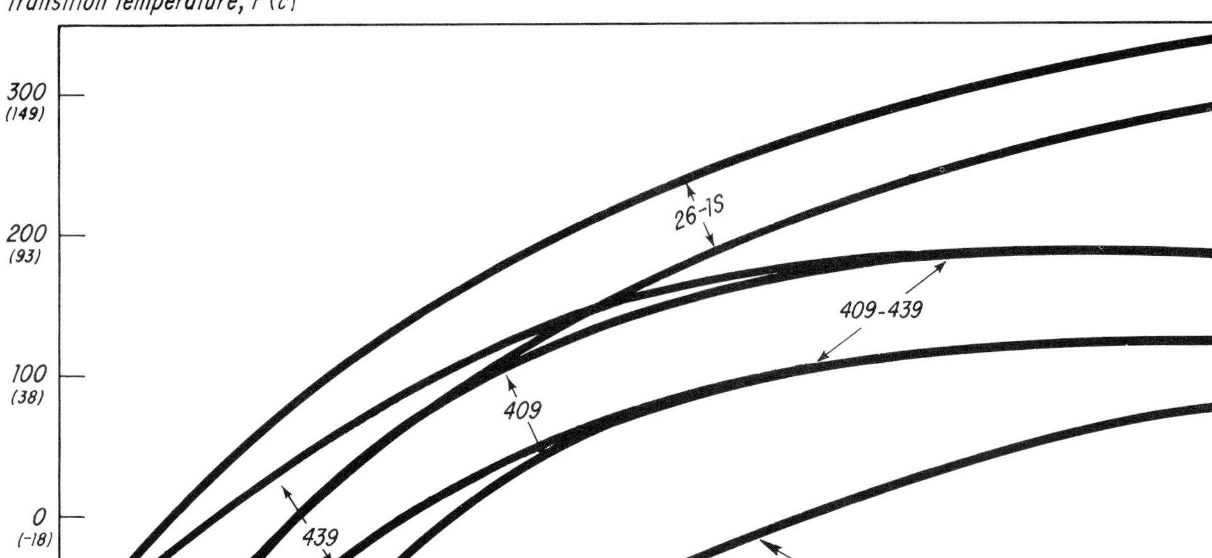

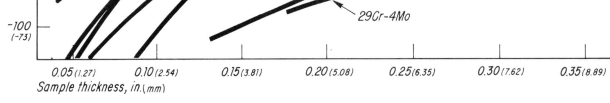

Fig. 1 — Ductile-to-brittle transition temperatures (DBTT) for ferritic stainless steels rise with section thickness. Bands for 26-1S, 409, and 439 indicate data scatter.

tions. Like others, they embrittle at 885 F (470 C) and have low strength at elevated temperatures.

They are used in equipment exposed to seawater, and in chemical industry equipment for highly corrosive acids or pitting media.

A Closer Look at the Alloys

Most of the new ferritics are either available commercially or in an advanced stage of development. Type 409 is an AISI standard grade, while types 439, 26-1S, and E-Brite 26-1 are being standardized by the ASTM.

Structures are completely ferritic at room and high temperature — via additions of titanium or columbium, or by melting to very low levels of carbon and nitrogen, or both. Such microstructures provide ductility and corrosion resistance in weldments.

Molybdenum improves pitting corrosion resistance, while silicon and aluminum add resistance to high-temperature oxidation. Type 18-2 is also available in a resulfurized version for machinability purposes; type 409 comes in several proprietary modifications, primarily developed for forming.

Ferritic steels containing above approximately 13% Cr precipitate alpha prime phase in the 650 to 1000 F (340 to 540 C) range — the maximum effect is at about 885 F (470 C). Precipitation rates vary with chromium content — the higher the chromium content, the faster the rate. Because precipitation hardening lowers room temperature ductility, it must be taken into account in both processing and usage of ferritic stainless steels, especially those with higher chromium content.

Sigma phase also forms above 1050 F (570 C) in types which contain chromium above approximately 20%. Chi phase has also been detected in molybdenum-containing types. The tendency to form sigma, chi, and alpha prime precipitates should always be considered when planning to use these alloys for long-time service at high temperature.

Transition Temperatures: Toughness or ductile-to-brittle transition temperature (DBTT), as measured by impact testing, is important in applying these steels. Impact strength is often low at ambient temperature, and the DBTT is above room temperature. Problems with the DBTT, however, are generally limited to heavier section sizes; in lighter gages, all the ferritic steels behave in a ductile manner at room temperature.

Figure 1 shows the DBTT, as determined with Charpy V-notch specimens. Note that it rises with thickness of the material for the titanium-bearing types (409, 439, and 26-1S) and the low-residual grades (E-Brite 26-1 and 29Cr-4Mo). The other grades are expected to follow a corresponding trend.

Transition temperatures of titanium-bearing steels are below room temperature up to approximately 0.1 to 0.2 in. (2.5 to 5.1 mm) thickness; consequently, these steels are ductile at ambient temperature only in light gages. This thickness dependence of the DBTT is caused primarily by the mechanical constraint associated with increasing thickness, which inhibits through-thickness yielding. Because heavier gages of titanium-bearing steels have transition temperatures above room temperature, they should be used selectively and with caution.

Titanium additions raise transition temperatures of ferritic steels, but are important for producing a ferritic structure, needed especially to prevent intergranular corrosion. Lowering carbon and nitrogen contents increases toughness of ferritic steels, even in titanium-bearing types.

The pronounced scatter band in the impact data of Fig. 1 is attributed to several metallurgical factors. The first is grain size — the smaller the size, the lower the transition temperature. Lacking an austenite-ferrite or martensite phase transformation, these steels can only be grain-refined by cold rolling followed by a recrystallization anneal. For this reason, thinner gage material, which receives more cold reduction, generally has smaller grains and is tougher.

Another factor is the cooling rate from the annealing temperature. Steels lower in chromium, such as type 409, are relatively insensitive to variations in cooling rate. In types containing 18, 26, and 29% Cr, fast cooling improves toughness because even short-time exposure to the alpha prime precipitation range caused by a slower cooling rate will cause some increase in the DBTT.

Low Residuals Important: Stainless steels with very low carbon and nitrogen contents have low ductile-to-brittle transition temperatures, and are thus more ductile, even in heavier sections. Examples are E-Brite 26-1 and 29Cr-4Mo, which are produced by electron-beam hearth refining process and by vacuum induction melting, respectively. Because their DBTT's are below room temperature even in heavier gages (Fig. 1), these low interstitial alloys are preferred in applications requiring ductile, heavy-gage materials.

Argon-oxygen decarburizing (AOD) melting and refining permit effective lowering of carbon and nitrogen to under 200 ppm each without the use of pure raw materials. These values are intermediate between those produced by electric arc melting and vacuum induction melting. Because they lower the DBTT, these levels of carbon and nitrogen prove useful in production of titanium-bearing steels.

Room temperature tensile properties of ferritic steels are shown in Table I. Variations in yield strength are mainly due to the solid-solution strengthening effect of chromium and molybdenum. These steels drop drastically in strength above 1000 F (540 C), indicating that they should be used with caution when load bearing at high temperature is important.

Weldability of ferritic steels is good; tensile properties, including elongation, are very similar to those of the base metal (Table I). The DBTT of weld metal shows only a small increase compared with that of the base metal in titanium-bearing steels, and is unchanged in the titanium-free low-carbon and nitrogen grades, providing proper shielding prevents absorption of interstitial elements.

The high-temperature oxidation resistance of ferritic stainless steels is equivalent to that of austenitic grades, while the thermal expansion is lower. Oxidation resistance can be enhanced by aluminum and silicon additions, as illustrated by 18SR. Experience shows that ferritic stainless steels perform very well in high-temperature applications where applied stresses are low.

Corrosion Resistance of Ferritics

Resistance to stress-corrosion cracking is the most obvious advantage of the ferritic stainless steels. In contrast with austenitic types, they resist chloride and caustic-stress corrosion cracking very well. Results of tests in boiling 42% $MgCl_2$ are shown below, indicating time to fracture:

Types 304, 305, 316, 321, and 347	<10 h
Type 310	<50
Type 439	>200 (no cracks)
26-1S	>200 (no cracks)
E-Brite 26-1	>1200 (no cracks)
29Cr-4Mo	200 (no cracks)

Nickel and copper residuals lower resistance of ferritic steels to stress corrosion, as determined in boiling 42%

Table I — Tensile Properties of Ferritic Stainless Steels

Base Alloy	Yield strength, psi (MPa)	Tensile strength, psi (MPa)	Elongation, %
Type 409	30 000-40 000 (210-280)	60 000-70 000 (410-480)	25-35
Type 439, 18-2	40 000-60 000 (280-410)	65 000-85 000 (450-590)	25-35
26-1S, E-Brite 26-1	50 000-55 000 (340-380)	75 000-80 000 (520-550)	25-35
29Cr-4Mo, 29Cr-4Mo-2Ni	70 000-80 000 (480-550)	75 000-85 000 (520-590)	25-35
Weld Metal			
Type 409	30 000-40 000 (210-280)	50 000-60 000 (340-410)	30-35
Type 439	45 000-55 000 (310-380)	70 000-75 000 (480-520)	25-35
18SR	61 000-64 000 (420-440)	84 000-88 000 (580-610)	19-28
E-Brite 26-1	55 000-60 000 (380-410)	65 000-70 000 (450-480)	10-25
26-1S	60 000-65 000 (410-450)	70 000-80 000 (480-550)	15-25
29Cr-4Mo	70 000-75 000 (480-520)	90 000-95 000 (620-650)	20-27

$MgCl_2$; also, the tolerable amount of these two elements is lower for the higher chromium steels. For instance, the Ni+Cu content of 26-1S has to be kept below approximately 0.30%, but can be slightly higher in type 439.

Boiling $MgCl_2$, however, is a very severe test; and it does not accurately represent conditions encountered in chemical plant exposure. The wick test is more realistic — the stressed sample is heated by an electric current while it is in contact with glass wool immersed in a NaCl solution. Even high-chromium ferritic steels with up to 2% Ni can survive the wick test.

Field experience with ferritic stainless steels confirms their resistance to cracking. A typical example is type 439, as used in domestic hot water tanks. No stress-corrosion failures have been reported in spite of the high chloride content in many areas. Similar results are claimed for type 439 heat exchanger tubing.

Intergranular Corrosion Resistance: Susceptibility of the ferritics to intergranular corrosion is due to chromium depletion, caused by precipitation of chromium carbides and nitrides at grain boundaries. Because of the lower solubility for carbon and nitrogen and higher diffusion rates in ferrite, the thermal cycle for sensitization is, however, different from that for austenitic steels. For this reason, the sensitized zone of welds in ferritic steels is in the weld and adjacent to the weld; in austenitic steels, it is located at some distance from the weld.

To eliminate intergranular corrosion, either reduce carbon to very low levels, or add titanium and columbium to tie up the carbon and nitrogen. In ferritic types, carbon and nitrogen have to be considerably lower than in austenitic steels. For example, total C+N of about 0.01% max is needed to prevent intergranular corrosion, with the exception of 29Cr-4Mo, which can tolerate about 0.025% max. This low level of residuals cannot be consistently attained by electric arc melting or argon-oxygen refining. Vacuum induction or electron beam melting have to be used. Low interstitial residuals are also essential for DBTT control, while titanium or columbium are needed to insure a ferritic structure. Thus, all the steels listed on p. 24 contain the features needed to resist intergranular corrosion.

The amount of titanium or columbium required is usually expressed as ratios to carbon or carbon plus nitrogen. These quantities should be at least stoichiometric, but preferably higher, to satisfy the various testing methods used for checking corrosion. Though steels with a Ti:C ratio of 6:1 resist nitric tests, type 439 requires a Ti:C ratio of 12:1 minimum for long-time exposure in boiling water.

Pitting and Crevice Corrosion: Pitting, an insidious localized type of corrosion occurring in halide media (particularly chlorides), can put installations out of operation in a relatively short time. Resistance to this type of corrosion is affected by such factors as chloride concentration, exposure time, temperature, and oxygen content. In general terms, resistance to pitting increases with chromium content. Molybdenum also plays an important role. It is considered equivalent to several percentages of chromium.

Relative pitting corrosion resistance can be determined in the laboratory by electrochemical techniques or by simple exposure in selected severe pitting solutions. Accelerated tests are more useful in determining comparative properties of various alloys (including those with known field performance) than they are in predicting field performance with any great accuracy. For instance, the minimum critical pitting potentials of types 439 and 304 in various chloride solutions are very close, indicating that they have similar pitting corrosion resistance, a fact confirmed by field experience.

The pitting potential of several ferritic stainlesses in 1 M NaCl at 25 C (70 F) was determined by investigators at Climax Molybdenum Co. Types 304 and 439 are very close (0.22 and 0.28 V [vs SCE]), as are 18-2 (0.43 V) and type 316 (0.48 V). No pitting could be produced in the higher-alloyed ferritic steels such as 26-1S, E-Brite 26-1, and 29Cr-4Mo.

A more severe test for pitting and crevice corrosion consists of exposing rubber-banded samples in a solution of 10% $FeCl_3 \cdot 6H_2O$ for 72 h at room temperature, and gradually at higher temperatures. Per cent weight loss from original is shown below:

	70 F	95 F
Type 409	14.5	—
Type 304	12.8	—
Type 439	10.9	—
18-2	4.67	—
Type 316	1.4	4.08
26-1S	0.030	13.0
E-Brite 26-1	0.001	5.05
29Cr-4Mo	0	0
Hastelloy C	0	0

A material free of pitting or crevice attack at room temperature in this test is considered to have good corrosion resistance in seawater. Any weight loss (expressed in percentage of sample weight) indicates that the material shows crevice corrosion, but the difference between the loss of various materials has only approximate comparative value.

A limited amount of seawater exposure confirms that 29Cr-4Mo is not attacked. Thus, this alloy belongs to a select group consisting of Hastelloy alloy C, Inconel 625, and titanium, which are considered to be immune to corrosion in seawater. Chloride-contaminated waters frequently encountered in industrial conditions can be handled by lower-alloyed materials, including ferritic steels discussed here (other than type 409).

General Corrosion Resistance: The atmospheric corrosion resistance of the ferritic steels discussed equals — and in many instances is superior to — that of type 304, again excepting type 409. Table II shows that the ferritic steels, starting with type 439, have good corrosion resistance to strongly oxidizing acids such as nitric acid. In organic acids, all the listed ferritic steels are superior to type 304. Type 439 and 18-2 are about equivalent to type 316, while the higher-chromium grades are superior to type 316.

The situation is different in reducing media; as illustrated by the 10% H_2SO_4 data, losses for ferritic steels are higher than for type 304 or type 316. A notable exception is 29Cr-4Mo-2Ni, which has been passivated by the 2% Ni addition.

Ray Lula is assistant to the vice president-technical director, Advanced Technology Planning, Allegheny Ludlum Steel Corp., Div. of Allegheny Ludlum Industries, Research Center, Brackenridge, Pa. 15014.

Table II — General Corrosion Resistance of Ferritic Stainless Steels

	Rate, in./mo. (mm/yr)			
Alloy	**65% nitric acid**	**20% acetic acid**	**45% formic acid**	**10% sulfuric acid**
Type 439	0.00261 (0.795)	0.00027 (0.082)	0.03961 (12.07)	19.44236 (5926)
18-2	—	0.0002 (0.061)	0.0340 (10.36)	—
26-1S	0.00033 (0.101)	0.00001 (0.003)	0.00011 (0.034)	7.94675 (2422)
E-Brite 26-1	0.0005 (0.152)	—	—	—
29Cr-4Mo	0.00052 (0.158)	0.00001 (0.003)	0.00013 (0.0396)	Dissolved
29Cr-4Mo-2Ni	—	—	—	0.0004 (0.122)
Type 304	0.00067 (0.204)	0.0250 (7.62)	0.14292 (43.56)	1.36833 (417.1)
Type 316	0.00092 (0.280)	0.00017 (0.052)	0.04333 (13.21)	0.07125 (21.72)

A High-Chromium Ferritic Stainless Steel

K. E. PINNOW, J. P. BRESSANELLI,
AND A. MOSKOWITZ

HIGH-chromium ferritic stainless steels such as Types 442 and 446 have excellent resistance to corrosion and to oxidation in many industrial environments. Until recently, however, these steels were only considered for high temperature service because their relatively low toughness (as-welded) and susceptibility to intergranular corrosion after welding precluded use in most applications at ambient temperature. It has long been recognized that carbon and nitrogen are the key elements in such steels in relation to weld toughness and corrosion resistance after welding.[1-4] Lowering the total carbon and nitrogen content of these steels to about 0.02 pct substantially improves weld toughness, whereas still lower carbon plus nitrogen content (on the order of 0.006 pct) is needed to eliminate susceptibility to intergranular corrosion or weld decay.[5-11] Production of the high chromium ferritic stainless steels to a maximum carbon plus nitrogen content of 0.006 pct is extremely difficult and expensive, and has not yet been accomplished on a tonnage basis.

As an alternate to reducing carbon and nitrogen to extremely low levels, titanium can be added– it combines with these elements, reducing their effective concentration to near zero. Titanium stabilization is highly effective in improving toughness and reducing susceptibility of high chromium ferritic stainless steels to intergranular attack after welding, particularly for alloys with relatively low levels of carbon and nitrogen. Recent advances in steelmaking, such as the development of argon-oxygen-decarburization (AOD), have made it possible to achieve low carbon and nitrogen contents and to control alloy composition more closely. Economic production of titanium-stabilized high chromium ferritic stainless steels with low carbon and nitrogen contents is therefore now possible on a tonnage basis.

Because the Crucible Stainless Steel Division has an AOD unit of 100 ton capacity, it is in a unique position to produce high chromium ferritic alloys with low carbon and nitrogen contents. Crucible 26-1 was developed with this capability in mind. Several commercial heats of the new alloy have been made, and both strip and tubing from these heats are in field service in a variety of applications. The alloy has high resistance to weld decay and chloride stress corrosion, and outstanding resistance to pitting and general corrosion in a wide variety of acid and chloride-containing environments. Detailed information on its composition and properties are presented in this paper, together with comparative data for several standard ferritic and austenitic stainless steels. Information on applications for the new steel is also given.

K. E. PINNOW is Manager, New Stainless Steels, and J. P. BRESSANELLI is Technical Director, Stainless Steels, Crucible Materials Research Laboratory, Colt Industries, Pittsburgh, Pa. A. MOSKOWITZ is Vice President, Technical Services, Crucible Stainless Steel Division, Colt Industries, Midland, Pa.

I. COMPOSITION

The composition of Crucible 26-1 is shown along with the compositions of several ferritic and austenitic stainless steels in Table I. Carbon-plus-nitrogen content of Crucible 26-1 is quite low, typically about 0.05 pct or less. Titanium in the alloy combines with the small amount of carbon and nitrogen present, thereby preventing intergranular chromium carbide and nitride precipitation during welding or processing. Chromium at the 26 pct level together with molybdenum at 1 pct provides for excellent resistance to pitting and general corrosion. The remaining elements are kept low to maintain good toughness and corrosion resistance.

II. CORROSION RESISTANCE

The new alloy is characterized by outstanding corrosion resistance in a wide variety of environments. It is especially resistant to general attack in organic acids, and to pitting and crevice corrosion in basic and slightly acid chloride-containing environments. Because of its ferritic structure and controlled composition, the alloy exhibits good resistance to chloride stress corrosion cracking.

A. Intergranular Corrosion

Welds in conventional ferritic stainless steels, such as Types 430 and 446, are susceptible to sensitization and intergranular corrosion. Recent investigations in-

Table I. Compositions of Crucible 26-1 and Other Stainless Steels

Grade	C	Mn	Si	Ni	Cr	Other
Crucible 26-1*	0.04 max	0.75 max	0.75 max	0.50 max	25.00/27.00	0.75/1.25 Mo
Type 304	0.08 max	2.00 max	1.00 max	8.00/12.00	18.00/20.00	–
Type 316	0.08 max	2.00 max	1.00 max	10.00/14.00	16.00/18.00	2.00/3.00 Mo
Type 321	0.08 max	2.00 max	1.00 max	9.00/12.00	17.00/19.00	Ti [5XC] min
Type 430	0.12 max	1.00 max	1.00 max	–	14.00/18.00	–
Type 446	0.20 max	1.50 max	1.00 max	–	23.00/27.00	0.25 max N

*Also contains 0.20 max Cu, 0.04 max N, and 0.20 to 1.00 Ti [7(C + N) min].

dicate that the cause is a chromium-depleted zone along the grain boundaries, which results from precipitation of chromium carbides or nitrides at the boundaries.[12,13]

Because of the susceptibility of standard high chromium stainless steels to intergranular corrosion after welding, particular attention was given to the composition of Crucible 26-1 so as to control this problem. Our early results confirmed those of other investigations,[14] indicating that the carbon-plus-nitrogen content of *unstabilized* 26 pct Cr-1 pct Mo alloys must be below 0.01 pct to obtain good resistance to intergranular corrosion after welding. Since this interstitial level is difficult to achieve, even by vacuum melting, development of titanium-stabilized 26 Cr-1 Mo alloys was undertaken at carbon and nitrogen levels attainable by AOD refining methods. Furthermore, our initial results showed that the isothermal sensitization treatments commonly used to evaluate austenitic stainless steels for welding applications—*e.g.*, 1 h at 675°C (1250°F)—were not reliable for screening ferritic 26 Cr-1 Mo alloys that might become susceptible to intergranular corrosion after welding. For this reason, our studies of sensitization resistance were largely conducted on TIG-welded materials.

Several intergranular corrosion tests (Huey, Streicher, Strauss, Warren, etc.) included in ASTM A262 were employed during the development of the alloy to insure that it would be resistant to intergranular chromium carbide and nitride precipitation, and therefore to preferential intergranular corrosion. The Warren test (10 pct nitric acid-3 pct hydrofluoric acid, 70°C) when conducted for two or four 2 h test periods was found to be exceptionally sensitive to chromium carbide or nitride precipitation; it was capable of causing intergranular attack in TIG weldments of unstabilized 26 Cr-1 Mo alloys containing as little as 0.006 pct carbon-plus-nitrogen. The Warren test was therefore used extensively in the development of the alloy.

Table II presents data on the corrosion resistance of 0.060 in. thick autogenous TIG welds of several stainless steels in the Warren intergranular corrosion test. Results clearly show that the new alloy is more resistant to seld decay than the other steels. Fig. 1 illustrates the degree of attack occurring on the weldments of some of these materials in the Warren test. As can be seen, the Type 430 sample showed severe attack immediately adjacent to the weld after only 5 min exposure, whereas Types 304, 316, and the non-stabilized 26 Cr-1 Mo samples showed varying degrees of intergranular attack slightly away from the welds in the heat-affected zone after 4 h of testing. The new alloy appears unaffected.

Table II. Weld Corrosion Resistance of Stainless Steels in the Warren Test*

Material†	Weld Metal	Weld Line	Heat Affected Zone
Crucible 26-1	Nil	Nil	Nil
Unstabilized 26-1‡	Nil	Trace	Light
Type 304	Nil	Nil	Light
Type 316	Nil	Nil	Light
Type 321	Nil	Nil	Nil
Type 430	Severe	Severe	Nil
Type 446	Light	Severe	Nil

*ASTM Standard A262; corrosion resistance rated microscopically according to severity of attack.

†TIG welded at 0.060 in. without filler metal.

‡0.014 pct (C + N).

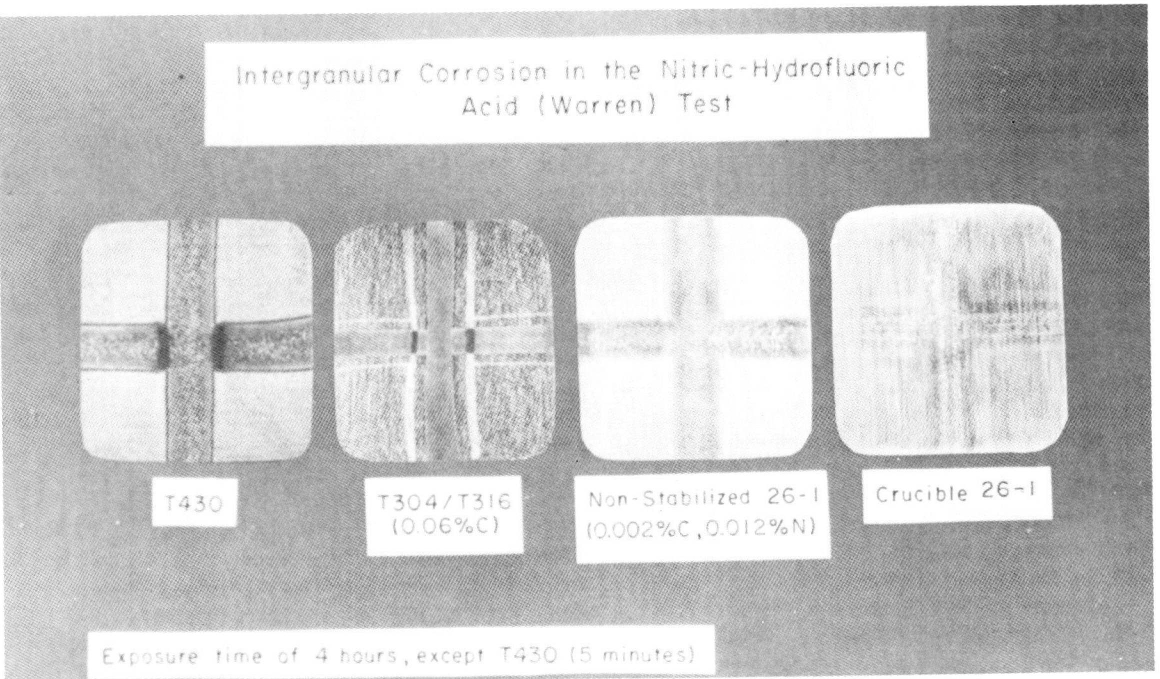

Fig. 1—Intergranular corrosion resistance of Crucible 26-1 and other stainless steels in the cross-welded condition; 0.060 in. thick TIG welds. Nitric-hydrofluoric acid (Warren) test.

Crucible 26-1 is not completely resistant to weld decay in the highly oxidizing environments employed in the sulfuric acid-ferric sulfate (Streicher) and boiling nitric acid (Huey) tests. In this respect, the alloy's behavior parallels that of the titanium-stabilized austenitic stainless steel Type 321, which when welded is also subject to intergranular corrosion in the Streicher and Huey tests. This effect is attributed to selective attack on the intergranular titanium carbide or nitride precipitates by the highly oxidizing media used in these tests. Accordingly, as is the situation for Type 321, the Streicher and Huey tests are not recommended for evaluating weld corrosion resistance of Crucible 26-1 except when the alloy is to be used under strongly oxidizing conditions, which are rarely encountered in practical situations.

B. Pitting and Crevice Corrosion

Remarkable resistance to pitting and crevice corrosion in many chloride-containing environments is a key characteristic of the high-chromium, molybdenum-bearing ferritic stainless steels. Therefore, a number of tests were conducted to evaluate this property of Crucible 26-1. Evaluations involved laboratory anodic polarization studies, laboratory immersion tests, and field exposures in seawater.

Anodic Polarization Studies. Though stainless steels exposed to seawater and many chloride-containing environments normally have a low rate of general corrosion, they may become susceptible in these environments to corrosion by pitting. In neutral chloride-containing media such as seawater there is a specific electrochemical potential at which the corrosion current rises, sharply reflecting local loss of the passive film on stainless and the initiation of pitting attack. Known as the breakthrough or pitting potential, this factor is a valuable index of pitting susceptibility.[15,16] The specific pitting potential of a sample in a given solution is influenced by chemical composition, metallurgical condition, and surface finish. Factors associated with the corrosive media such as temperature, chloride concentration, and pH also affect the pitting potential.[17] Questions have been raised concerning the practical significance of the pitting potential,[18,19] but it has proven to be useful for comparing alloy performance in our laboratory.

Pitting potential can be determined by measuring the change of current density as the potential of a specimen (relative to a standard electrode) is made to increase in either small steps or at a slow uniform rate, as is possible with an electronic potentiostat. In the present work, a uniform scanning rate of 10 mV/min was used. This rate conforms to that used in recent studies for the Office of Saline Water (OSW) on the pitting of stainless steels in hot seawater.[20]

In this work, anodic polarization tests were conducted in nitrogen-deaerated synthetic seawater at 25, 60, and 90°C (77, 140, and 194°F). Synthetic seawater was prepared by dissolving sufficient "sea salt"* in distilled water to approximate the nominal composition of seawater (3.5 pct NaCl). Prior to the test, the synthetic seawater was boiled to lower oxygen and carbon dioxide content, and then adjusted to a pH of about 7 with sulfuric acid. This practice

*"Sea Salt," a product of Lake Product Co., Inc., has a formula based on ASTM-1141-52. Composition: 58.49 pct NaCl, 26.46 pct $MgCl_2 \cdot 6H_2O$, 9.75 pct Na_2SO_4, 2.765 pct $CaCl_2$, 1.645 KCl, 0.477 pct $NaHCO_3$, 0.238 pct KBr, 0.071 pct H_3BO_3, 0.095 pct $SrCl_3 \cdot 6H_2O$, and 0.007 pct NaF.

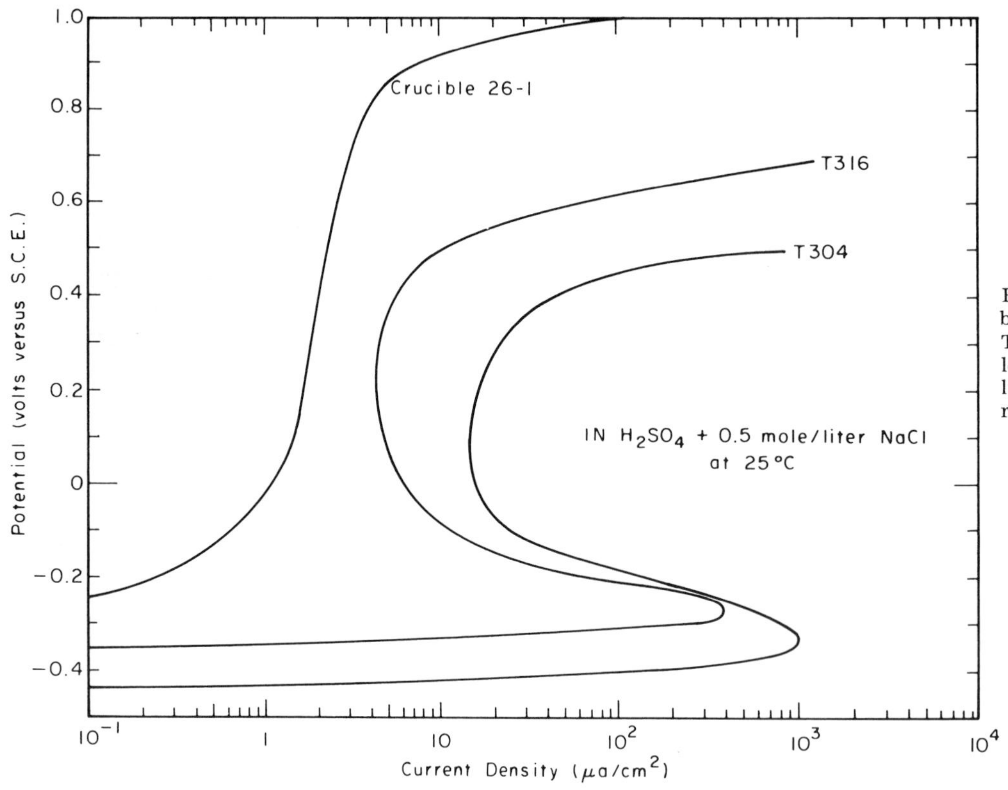

Fig. 2—Anodic polarization behavior of Crucible 26-1, Type 304, and Type 316 stainless in 1 N H_2SO_4 + 0.5 mole/liter at 25°C (Hydrogen-deaerated).

was used to approximate the feedwater treatment employed in flash distillation plants to minimize the formation of carbonate scale. In addition to the tests in seawater, anodic polarization tests were made at 25°C (77°F) in a hydrogen-deaerated 1.0N sulfuric acid solution containing 0.5M of sodium chloride per liter.

Samples for the polarization tests were smoothed with 600-grit silicon carbide abrasive, and then cleaned in chloroform prior to mounting in the specimen holder, which exposed a surface area of about 1 cm^2. Before being anodically polarized (at a constant rate of 10 mV/min), specimens were held for approximately 1 h in the test solution to allow them to reach an equilibrium rest potential. All potentials were measured relative to a saturated calomel electrode (SCE). Details on the electronic potentiostat and the cell used for these studies are given in the literature.[21]

Table III compares the pitting potentials of Crucible 26-1 with those obtained for Types 304 and 316 stainless in hot deaerated synthetic seawater. Data show that the new alloy was immune to pitting in this environment at 25°C (77°F), and that its pitting resistance was significantly superior to that of Types 304 and 316 stainless at higher temperatures.

Fig. 2 shows the polarization behavior of Crucible 26-1, Type 304, and Type 316 at 25°C (77°F) in a solution of 1 N H_2SO_4 containing 0.5 mole/liter of sodium chloride. The new alloy passivated instantaneously in this environment, and was completely resistant to pitting up to the breakdown potential of the test solution. Conversely, Types 304 and 316 were initially active in this solution, and pitted at potentials below the breakdown potential of the solution.

Laboratory Immersion Tests. Pitting and crevice corrosion resistance of Crucible 26-1 was further evaluated in laboratory immersion tests conducted in 0.1 N HCl solution containing 10 pct $FeCl_3 \cdot 6H_2O$ (acid ferric chloride) and in modified synthetic seawater. In the acid ferric chloride test, commonly used for evaluating pitting resistance of highly corrosion resistant stainless steels,[22-24] specimens are immersed for 24 h in stagnant solutions at room temperature. The modified synthetic seawater test, developed more recently, has been used by others[25,26] to evaluate pitting resistance of stainless steels in marine and saline environments. In this latter test, samples are immersed for 24 h in neutral synthetic seawater to which has been added 10 grams per liter of potassium ferricyanide, which considerably increases the corrosivity of seawater by raising its oxidation-reduction (redox) potential. In effect, the test solution acts as a natural potentiostat. It theoretically differentiates between materials having a pitting potential above or below about +460 mV (SCE). Test severity can be increased further by testing above room temperature. Corrosion on ferrous materials is readily discernible in that corrosion products are blue due to formation of a ferro-ferricyanide complex (Turnbull's Blue).

Specimens for immersion tests, generally prepared with 240 grit finishes, were suspended vertically in the test solutions, each by a glass hook inserted in a hole located near the upper edge. Contact between the hook and the specimen formed a natural crevice at which corrosion initiated on some materials. All samples were visually rated for both crevice and pitting attack.

Fig. 3 shows samples of Crucible 26-1, Type 304,

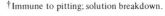

Table III. Pitting Resistance of Stainless Steels in Synthetic Sea Water

Grade	Pitting Potential*		
	25°C (77°F)	60°C (140°F)	90°C (194°F)
Crucible 26-1	>+1000†	+395	+330
Type 316	+250	+180	+50
Type 304	+130	+40	–

*Rated in mV relative to saturated calomel electrode.
†Immune to pitting; solution breakdown.

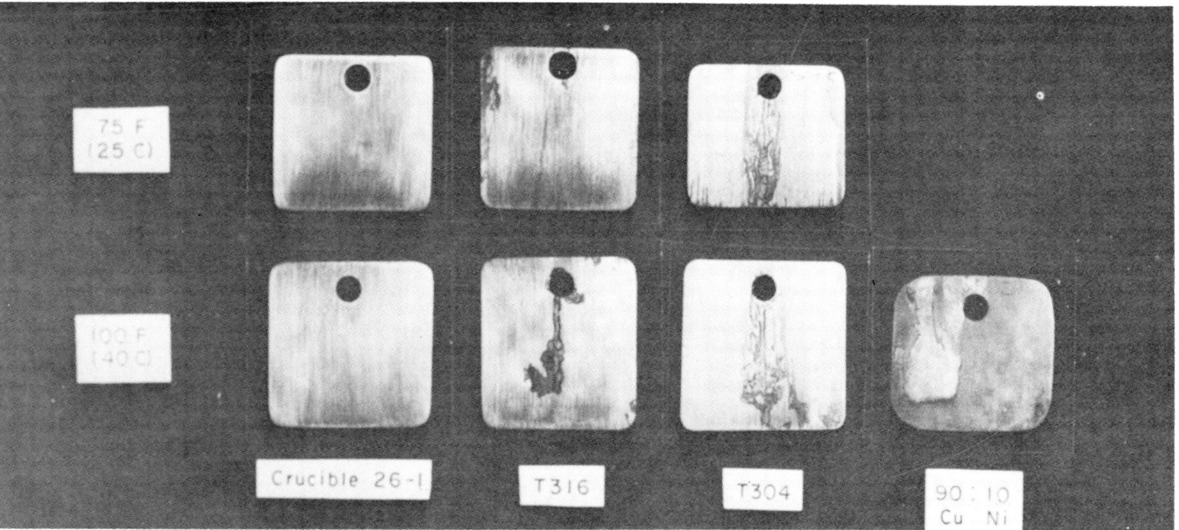

Fig. 3—Corrosion resistance of Crucible 26-1 and other corrosion resistant materials in synthetic seawater containing 10 g/liter of potassium ferricyanide (24 h exposure).

Type 316, and 90/10 cupronickel after exposure in modified synthetic seawater for 24 h at 25°C (77°F) and at 40°C (104°F). In this figure, dark portions of stainless steel samples represent areas that had been stained blue by the potassium ferricyanide addition in the seawater. In this test, the cupronickel showed severe general attack with considerable thinning of the cross section. As expected for this environment, Types 304 and 316 showed the typical pit corrosion with rundown of corrosion products at both test temperatures. In contrast, the new alloy showed no attack at either test temperature.

Fig. 4 shows samples of Crucible 26-1, Type 304, and Type 316 after one day exposure in acid ferric chloride at room temperature. In this severe test, the new alloy showed no evidence of attack, whereas Type 304 and 316 showed some pitting at or near the edges of the specimen.

To further evaluate crevice corrosion resistance, panels of Crucible 26-1, Type 316, and Type 304 fitted with artificial crevices were immersed at room temperature in acid ferric chloride. The artificial crevices on the panels were created by stretching rubber bands over the surface of the samples. After 30 days of exposure in the acid ferric chloride, attack beneath the rubber band on the Type 304 created a notch that extended for about 1 in. from the most severely attacked edge, and Type 316 exhibited a notch of about 1/2 in. The new alloy, however, showed no attack whatsoever beneath the rubber band, again indicating it to possess excellent resistance to pitting and crevice corrosion.

Seawater Immersion Tests. Seawater tests comparing the corrosion resistance of Crucible 26-1, Type 316, and Type 321 stainless were conducted for one year at ambient temperature by R. Baboian, Texas Instruments, at Battelle's William F. Clapp Laboratory located in Duxbury, Massachusetts. Location for these tests was an outdoor aquarium, 12 foot by 6 foot by 4 foot deep, through which seawater from Duxbury Harbor was continuously pumped.

Samples were fastened to plexiglass racks with nylon nuts and bolts, and thus contained crevices in areas around the bolt holes. Completely immersed in the seawater, all specimens were exposed in areas where flow rates were negligible. Considerable fouling developed on surfaces of samples during the year of exposure.

Fig. 5 shows the samples of Crucible 26-1, Type 316, and Type 321 stainless after cleaning to remove fouling. Note that considerable attack with rust rundown occurred around the bolt hole on the Type 321 sample. Similar, but less severe, attack was present on the Type 316 sample. Some light, scattered pitting of the Type 321 and Type 316 samples (indicated by arrow) also occurred under the fouling deposits. The sample of Crucible 26-1, however, showed essentially no attack, even in crevices around bolt holes.

Results of seawater immersion tests support those of the accelerated laboratory corrosion tests, and indicate that the new alloy has significantly better resistance than Type 316 to crevice and pitting corrosion in marine environments. These results also support those of other investigators indicating that crevice corrosion is probably the most important form of attack on metals in seawater.[18]

C. General Corrosion Resistance

Conventional chromium ferritic stainless steels such as Types 430 and 446 have insufficient resistance to general corrosion or pitting in many organic acid environments. Since molybdenum is known to improve the performance of ferritic stainless steels in these environments,[7] tests were made in boiling 50 pct acetic, 10 pct formic, and 10 pct oxalic acids to compare resistance of Crucible 26-1 to that of the austenitic stainless steels widely used in such service. Table IV shows the results of these tests, and indicates that the corrosion resistance of the

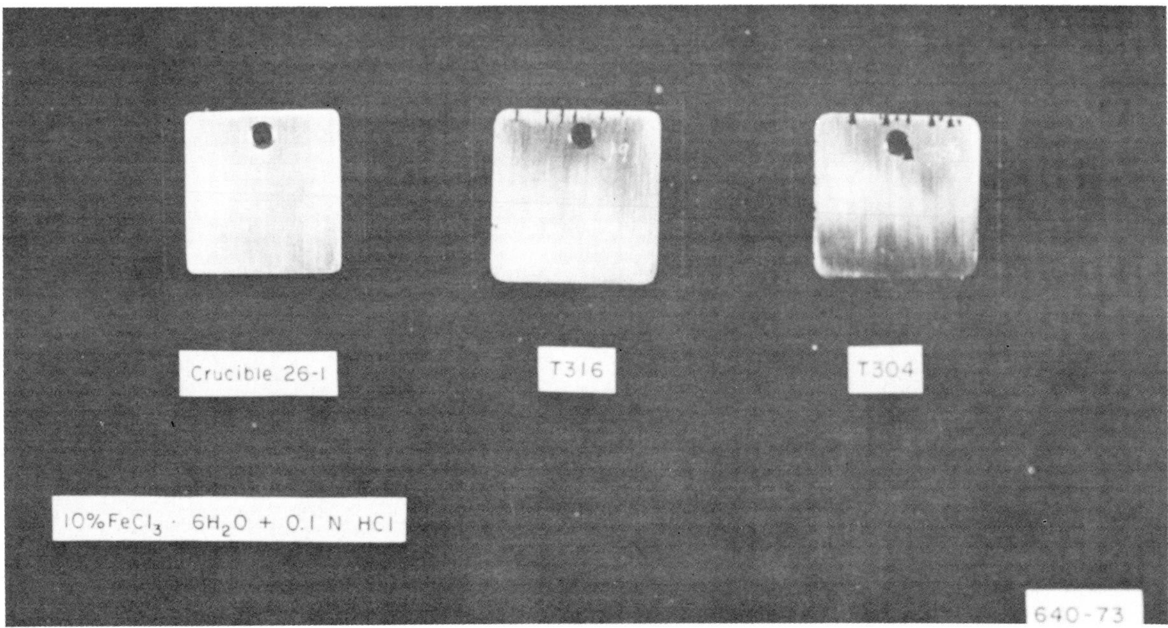

Fig. 4—Pitting resistance of Crucible 26-1, Type 316, and Type 304 stainless in acid ferric chloride (75°F, 24 h).

new alloy in these media is excellent, being superior to that of Types 304 and 316 stainless.

D. Stress Corrosion Cracking

The straight chromium ferritic stainless have long been considered to have excellent resistance to chloride stress corrosion cracking.[27,28] However, use of these stainless steels in equipment operated with chloride-bearing cooling water is not widespread owing to problems with low toughness and poor corrosion resistance after welding and to insufficient pitting resistance. These potential disadvantages have largely been overcome by the compositional features included in the new alloy.

Susceptibility of Crucible 26-1 to chloride stress corrosion was first evaluated in boiling solutions of 45 pct magnesium chloride at 155°C (309°F) and of 62 pct calcium chloride at 150°C (302°F) using conventional U-bend specimens. These solutions will crack susceptible alloys in short times, although there are important questions regarding their severity and applicability to service in dilute chloride media or where sodium chloride is involved.[29,30]

Our results in boiling magnesium chloride and in boiling calcium chloride paralleled those of other investigators, indicating that resistance to stress corrosion cracking of the molybdenum-bearing high chromium ferritic stainless steels was excellent in these solutions, providing nickel and copper residuals were controlled.[31] For this reason, the nickel-plus-copper content of Crucible 26-1 is kept relatively low. Stress corrosion resistance of Crucible 26-1, Type 304, and Type 316 in boiling 45 pct magnesium

Table IV. Corrosion Resistance of Stainless Steels (Annealed) in Organic Acids

	Corrosion Rate (Microns/Year)*		
Grade	50 Pct Boiling Acetic Acid	10 Pct Boiling Formic Acid	10 Pct Boiling Oxalic Acid
Crucible 26-1	1	1	9
Type 430	>500	>1200	>1200
Type 304	185	950	490
Type 316	1	85	88
Type 316L	1	115	64

*Specimens were exposed seven days.

Table V. Stress Corrosion Cracking Resistance of Stainless Steels in Accelerated Test Media

	Hours to Failure			
Material*	45 Pct $MgCl_2$ (155°C)	62 Pct $CaCl_2$ (150°C)	5N NaCl (106°C)	Polythionic Acid (25°C)†
Crucible 26-1	>500	>308	>1440 (60 days)	>720 (30 days)
Type 304	2	21	120 (5 days)	–
Type 316	16	21	720 (30 days)	–

*Test conducted with U-bend specimens prepared by bending 1/2 in. wide by 5 in. long by 0.060 in. thick strips of annealed material over a 1/2 in. diam mandrel. Springback was prevented by mounting the specimens in a holder block.

†ASTM G35-73: Tests conducted on mill-annealed and on TIG-welded materials—a sensitized Type 302 control sample failed within one hour.

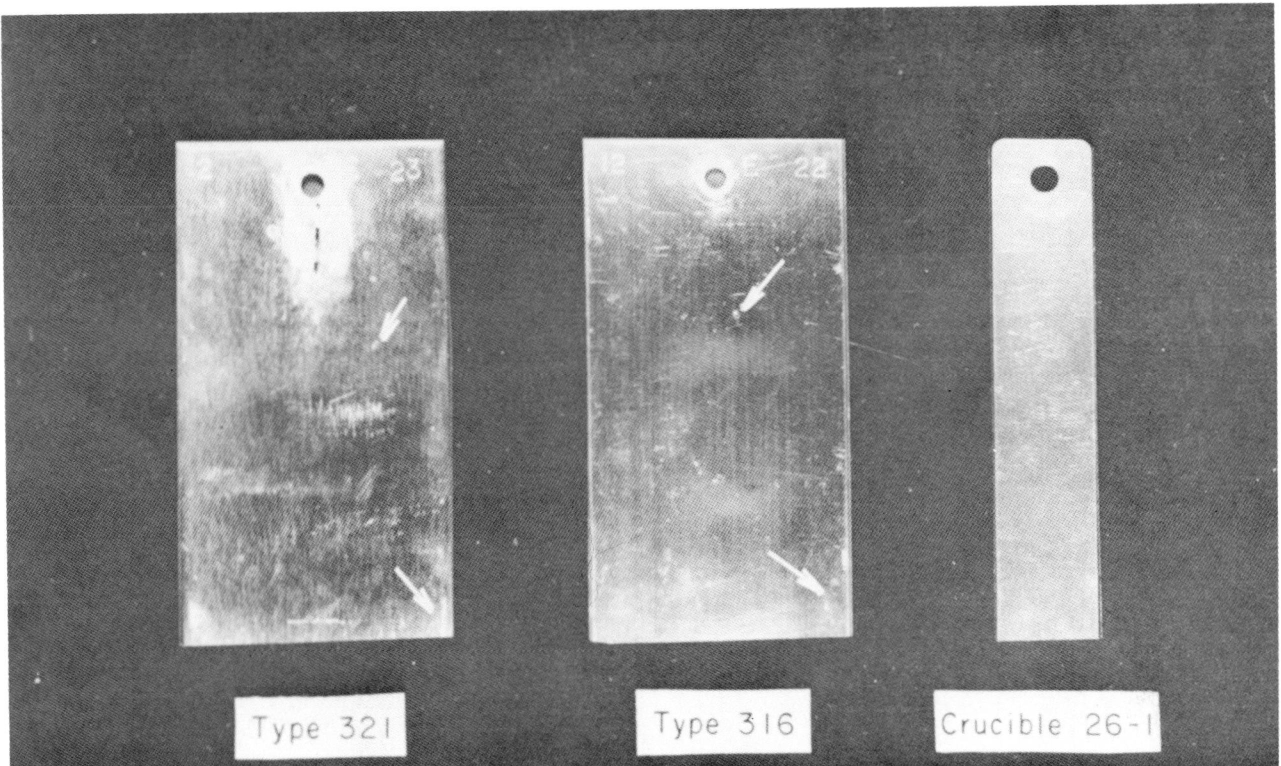

Fig. 5—Corrosion of Type 321, Type 316, and Crucible 26-1 after total immersion for 1 yr in seawater at ambient temperature at William F. Clapp Laboratory, Duxbury, Mass. (Arrows denote pitting attack.)

chloride and in boiling 62 pct calcium chloride is compared in Table V. Data show that the new alloy is considerably more resistant to stress corrosion in these test media than are Types 304 and 316 stainless.

Stress corrosion tests were also conducted on Crucible 26-1 in boiling 5 N sodium chloride solutions at 106°C (220°F) using U-bend specimens. The alloy has shown excellent resistance to stress corrosion in this test, and has been immune to cracking in sodium chloride test media in all tests. Fig. 6 shows the comparative performance of Crucible 26-1, Type 304, and Type 316 U-bends in the boiling sodium chloride test. Here, the new alloy shows no signs of pitting or stress corrosion after 60 days exposure. Types 304 and 316, as expected, showed cracking and some rust formation after respective exposures of 5 and 30 days.

Because of interest in the new steel by the petrochemical industry, annealed as well as as-welded samples were exposed in polythionic acid according to procedures described in ASTM G35-73. None of the samples cracked during the 30 day period the test was run, whereas the sensitized Type 302 control sample (specified as part of the test procedure) failed within 1 h.

III. OXIDATION RESISTANCE

Similar to other high chromium stainless steels, Crucible 26-1 has excellent oxidation resistance at elevated temperatures. Fig. 7 illustrates the oxidation resistance of Crucible 26-1 during continuous exposure in air at 2000°F, and reveals the oxidation resistance of the alloy to be superior to that of Type 309S and comparable to that of Types 310S and 446 stainless.

IV. WELDABILITY

Crucible 26-1 is weldable by inert gas welding processes such as TIG welding. Autogenous TIG welds made in stock 0.060 in. thick have good ductility, and are much more formable than those produced in conventional ferritic stainless such as Type 430. Fig. 8 illustrates this advantage, showing that Crucible 26-1 welds in material of this thickness have good Olsen cup ductility and can be bent through 180 deg without cracking. Commercially produced TIG-welded tubing shows good ductility in various crush, flare, flattening, and bend tests, as shown in Fig. 9.

V. PHYSICAL PROPERTIES

Table VI compares the physical properties of Crucible 26-1 and several other stainless steels. As would be expected, the physical properties are similar to those of Type 446 and other high chromium ferritic stainless steels. In contrast to the austenitic stainless steels such as Type 304 and 316, Crucible 26-1 is ferromagnetic, and has significantly lower density and expansivity. Compared to austenitic alloys, the new alloy also has higher thermal conductivity, a characteristic which can be advantageous in heat exchangers and condensers.

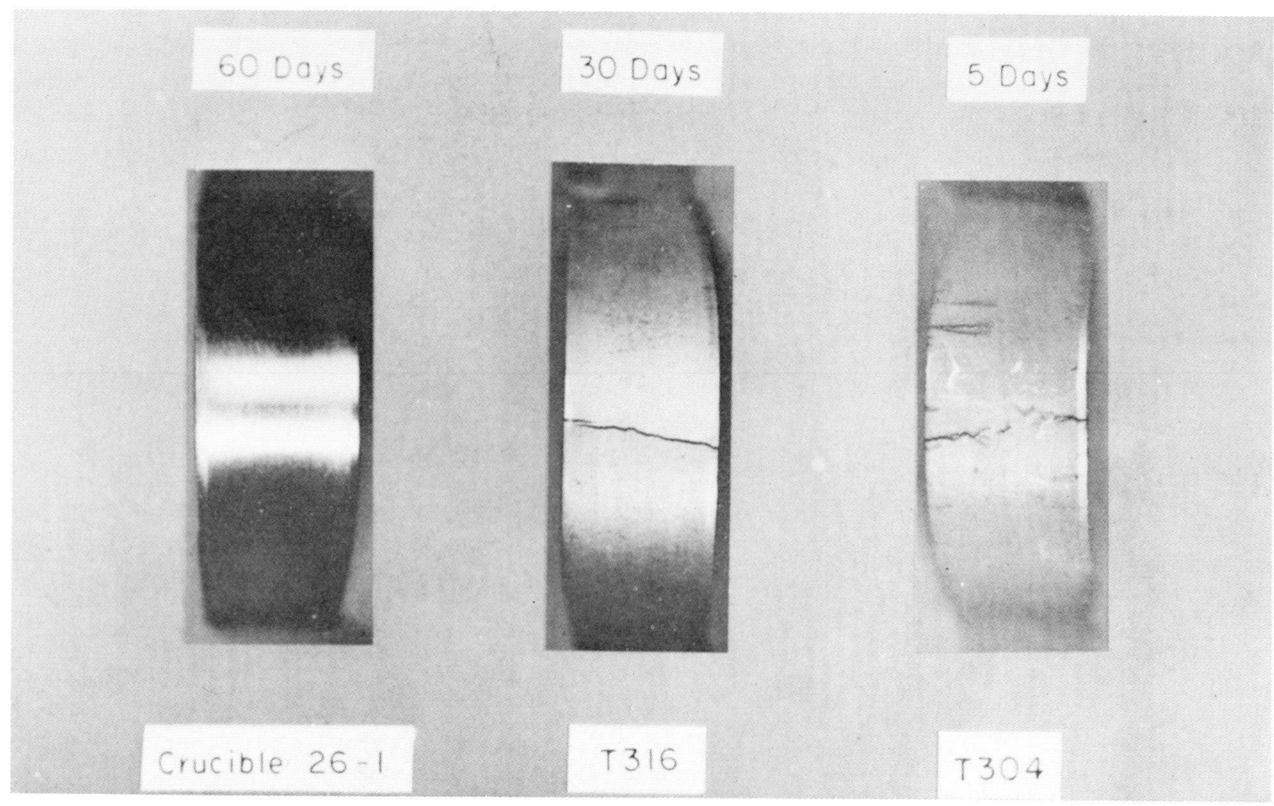

Fig. 6–Stress corrosion resistance of Crucible 26-1, Type 316, and Type 304 stainless (U-Bends) in boiling saturated (28 pct) sodium chloride (220°F, pH 7).

VI. MECHANICAL PROPERTIES

Table VII presents data on mechanical properties of Crucible 26-1 and several other corrosion resistant alloys at ambient temperature. In general, tensile properties and formability of the new alloy are similar to those of other ferritic stainless steels such as Type 430 or Type 434. Compared with austenitic stainless steels such as Type 304 and 316, it has significantly higher yield strength, but work hardening rate and tensile ductility are lower. The work hardening behavior of Crucible 26-1 is compared with that of Type 304 in Fig. 10.

Charpy V-notch impact properties of Crucible 26-1 at a thickness of 0.075 in. are compared with those of Type 430 and 446 stainless in the unwelded and/or welded conditions in Table VIII. Data for the welded condition were developed using subsize

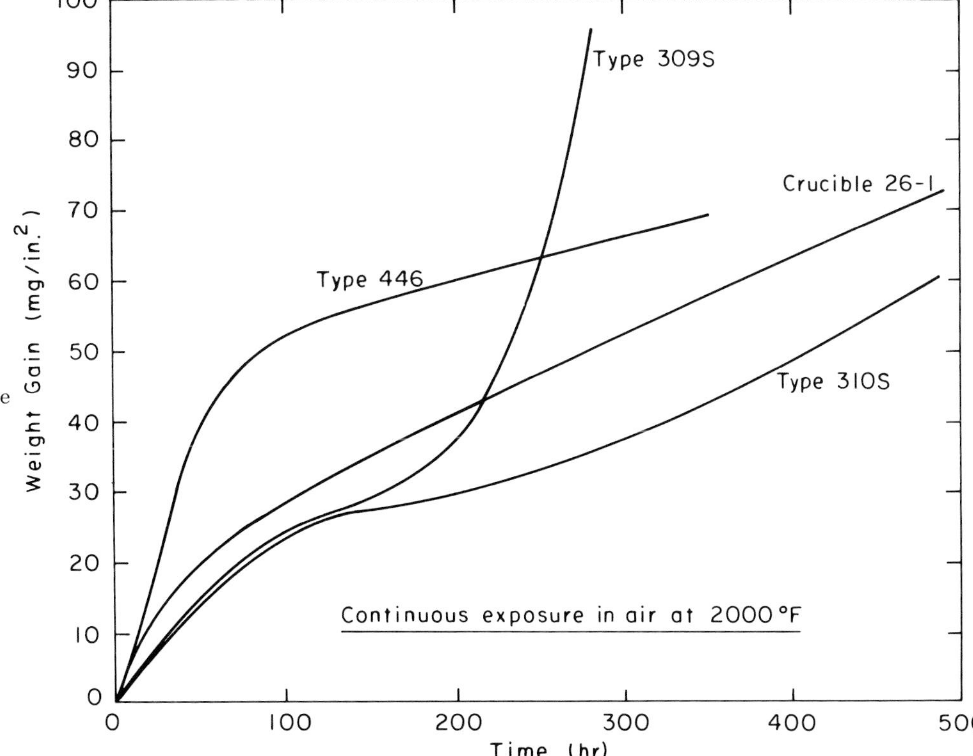

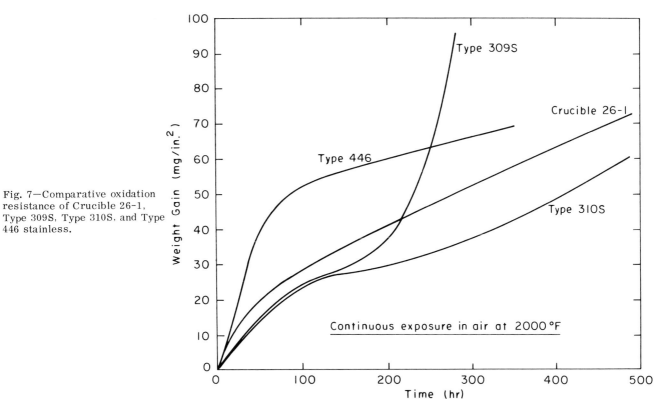

Fig. 7—Comparative oxidation resistance of Crucible 26-1, Type 309S, Type 310S, and Type 446 stainless.

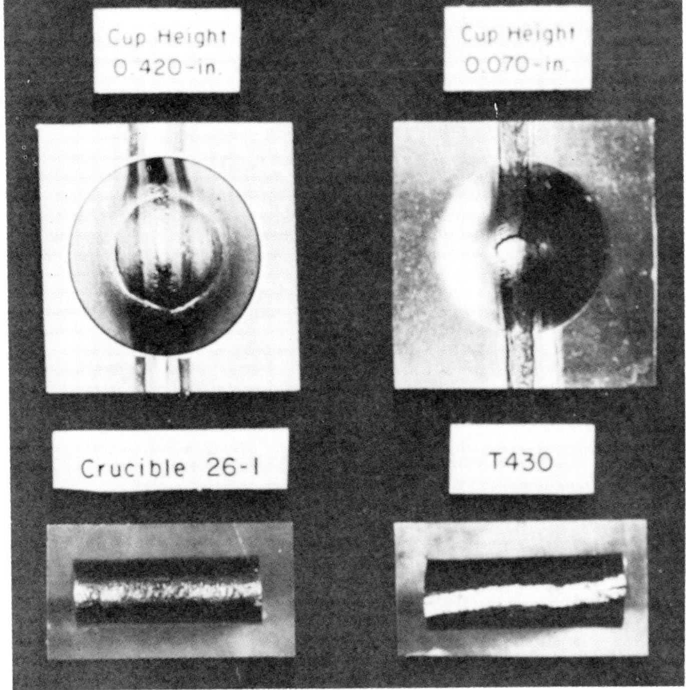

Fig. 8—Ductility of Crucible 26-1 and Type 430 TIG welds (0.060 in. thick) in Olsen Cup and 180° bend tests.

Fig. 9—Ductility of Crucible 26-1 tubing (1 in. O. D. by 0.062 in. wall) in various bend, crush, flattening, and flare tests.

Table VI. Physical Properties of Corrosion Resistant Alloys

Material	Crucible 26-1	Type 446*	Type 304*	Type 316*	Alloy 600†	Alloy 800†	Alloy 825†
Density							
(ibs/ cu in.)	0.28	0.27	0.29	0.29	0.30	0.29	0.294
(grams/cc)	7.63	7.45	8.02	8.02	8.43	8.02	8.14
Electrical resistivity (70°F)							
(microhm ft)	2.18	2.19	2.36	2.43	3.38	3.25	3.71
(microhm cm)	66.3	66.8	72.0	74.0	103.0	99.0	113.0
Magnetic permeability (70°F)							
(H = 200 Oersteds)	Ferromagnetic	Ferromagnetic	1.004	1.004	1.008	1.009	1.005
Thermal conductivity (212°F) (Btu/h/sq ft/ft °F)	10.4‡	10.4¶	9.4	9.3	9.0	7.4	7.1
Average coefficient of expansion (per °F)	5.6×10^{-6} (70 to 500°F)	5.8×10^{-6} (32 to 212°F)	9.6×10^{-6} (32 to 212°F)	8.9×10^{-6} (32 to 212°F)	7.4×10^{-6} (32 to 212°F)	7.9×10^{-6} (70 to 200°F)	7.8×10^{-6} (70 to 200°F)

**Properties of Some Metals and Alloys*, International Nickel Company, New York, New York, 1968.
†*Huntington Alloys*, Huntington Alloy Products Division, International Nickel Company, Huntington, West Virginia.
‡Estimated value based on electrical resistivity and comparative data for Type 446.
¶*Thermophysical Properties Research Center Data Book*, vol. 1, Table 1086, Purdue University, Lafayette, Indiana, 1964.

Table VII. Mechanical Properties of Stainless Steels

Grade	Tensile Strength (ksi)	0.2 Pct Yield Strength (ksi)	Elongation (Pct)	Olsen Cup Height (in.)
Crucible 26-1	75	52	30	0.400
Type 430	74	50	28	0.340
Type 304	88	42	55	0.500
Type 316	90	43	50	0.480

Charpy V-notch specimens prepared from 0.100 in. thick autogenous TIG weldments which were ground after welding to remove excess weld metal or weld undercut. Notches in welded specimens were located in the weld metal. Transition temperature data show that toughness of Crucible 26-1 is substantially superior to that of Type 430 and 446 in the welded condition and to that of T-446 in the cold-rolled and annealed condition.

VII. POTENTIAL APPLICATIONS

The excellent resistance of Crucible 26-1 to chlorides, organic acids, and chloride stress corrosion indicates that it should be suitable for a wide range of applications in which conventional stainless steels or other materials are either inadequate or uneconomical. The new alloy would appear particularly useful in heat exchanger tubing, feedwater tubing, condenser tubing, and in equipment operated with chloride-bearing or brackish cooling waters. Its excellent corrosion resistance also suggests use in commercial cooking and food handling equipment, beverage storage, and chemical processing equipment. Crucible 26-1 is presently moving into usage for a number of such applications in the electrical power, desalination, petrochemical, chemical, food processing, and pulp and paper industries. The alloy is included in ASTM specifications A176-74 (Chromium Stainless Flat Products), A240-74 (Stainless for Unfired Pressure Vessels),

Table VIII. Charpy V-Notch Impact Properties of Stainless Steels

Material	Metallurgical Condition*	Transition Temperature†
Type 430	Welded	185 to 212°F
Type 446	Annealed	32 to 50°F
Type 446	Welded	120 to 140°F
Crucible 26-1	Annealed	−10 to 15°F
Crucible 26-1	Welded	0 to 32°F

*Weld tests conducted on 0.075 in. thick specimens prepared from butt-welded 0.100 in. cold rolled and annealed strip. Welds made in one pass, without filler metal, at 10 to 12 volts, 160 amps and 9.5 ipm using a 3/32 in. diam thoriated tungsten electrode. Argon was used for shielding (15 cfh) and back-up (10 cfh). Weld specimens notched in weld metal. Base metal tests conducted on 0.075 in. thick specimens notched in the rolling direction.

†Based on energy absorption (3.8 ft lb/0.1 in. width) or lateral expansion (0.015 in.).

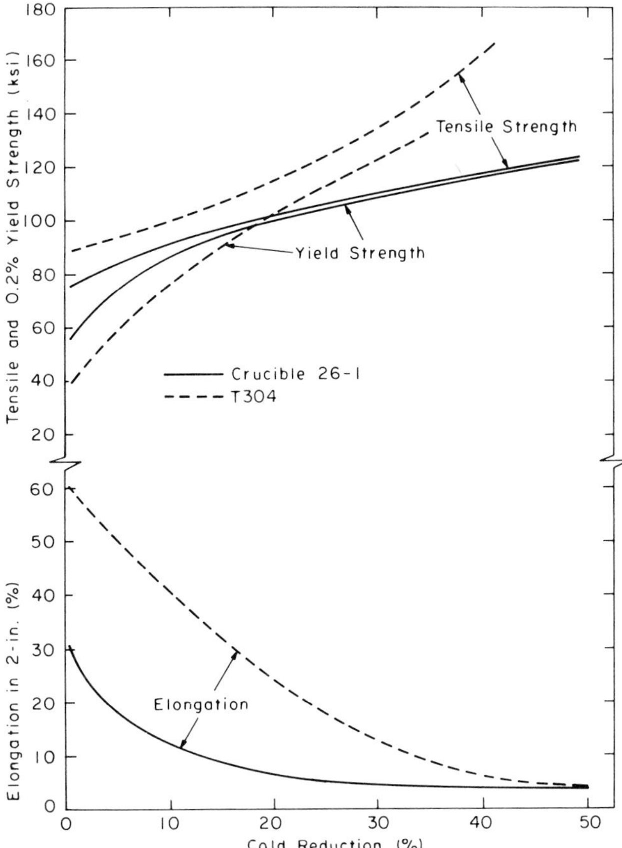

Fig. 10—Comparative work hardening behavior of Crucible 26-1 and Type 304 stainless.

and A268-74 (Ferritic Stainless Steel Tubing for General Service). It is also covered by ASME Code Case 1673.

VIII. SUMMARY

Crucible 26-1 has been developed to provide an outstanding combination of resistance to chloride stress corrosion cracking, resistance to general, pitting, and crevice corrosion, and good fabricability. Available in sheet, strip, tubing, and welding wire, the alloy is finding substantial application in replacing brass and cupronickel, corrosion-resistant high-nickel alloys, and other materials in the food processing, power, chemical, petrochemical, marine, and pulp and paper industries.

REFERENCES

1. J. Hochmann: *Mem. Sci. Rev. Met.*, 1951, vol. 48, no. 10, pp. 734-58 (French).
2. W. O. Binder and H. R. Spendelow, Jr.: *Trans. ASM*, 1951, vol. 43, pp. 759-77.
3. J. L. Ham and F. L. Carr: *Impact Properties of Vacuum Melted Iron-Chromium Alloys*, Symposium on Electrothermics and Metallurgy, Electrochemical Society, Inc., Boston, Mass., October 1954.
4. R. A. Lula, A. J. Lena, and G. C. Kiefer: *Trans. ASM*, 1954, vol. 46, pp. 197-230.
5. E. Baerlecken, W. A. Fischer, and K. Lorenz: *Investigations on the Transformation Behavior, Impact Strength, and Susceptibility to Intergranular Corrosion of Iron-Chromium Alloys with Chromium Contents Up to 30 Pct*, Stahl u. Eisen, vol. 81, pp. 768-78, 1961 (German).
6. R. Mayoud, P. Coulomb, and J. Hochmann: *Rev. Met.*, 1964, vol. 61, no. 2, pp. 199-203 (French).
7. L. Colombier and J. Hochmann: *Stainless and Heat Resisting Steels*, St. Martin's Press, New York, 1968.
8. M. A. Streicher: *The Effect of Heat Treatment on the Corrosion of 16 Percent and 25 Percent Chromium Stainless Steels*, NACE Corrosion Conference, Cleveland, March 1968.
9. R. J. Hodges and E. Gregory: *Some Corrosion Properties of Iron-Base Alloys Refined in the Electron Beam Cold Hearth Furnace*, American Vacuum Society, Vacuum Metallurgy Conference, Pittsburgh, June 1969.
10. M. Semchyshen, A. P. Bond, and H. J. Dundas: *Effects of Composition on Ductility and Toughness of Ferritic Stainless Steels*, Proceedings, Symposium on Improved Ductility and Toughness, Kyoto, Japan, October 1971.
11. B. Pollard: *Metals Technol.*, January 1974, vol. 1, no. 1, pp. 31-36.
12. A. P. Bond: *Trans. AIME*, 1969, vol. 245, pp. 2127-134.
13. R. P. Frankenthal and H. W. Pickering: *J. Electrochem. Soc.*, 1973, vol. 120, no. 1, pp. 23-26.
14. R. F. Steigerwald: *Low Interstitial Fe-Cr-Mo Ferritic Stainless*, Soviet-American Symposium on New Developments in the Field of Molybdenum-Alloyed Cast Iron and Steel, Moscow, January 1973.
15. J. M. Defranoux: *Corros. Sci.*, 1963, vol. 3, pp. 75-86 (French).
16. V. Hospadaruk and J. V. Petrocelli: *J. Electrochem. Soc.*, 1966, vol. 113, pp. 878-83.
17. O. Steensland: *Corros. Prev. and Contr.*, May/June 1968, pp. 25-29.
18. B. E. Wilde and E. Williams: *J. Electrochem. Soc.*, 1971, vol. 118, no. 7, pp. pp. 1057-62.
19. J. Degerbeck: "On Accelerated Pitting and Crevice Corrosion Tests," *J. Electrochem. Soc.*, 1973, vol. 120, no. 2, pp. 175-82.
20. N. Pessall, F. C. Hull, and C. Lim: *Development of a Low-Cost Iron Base Alloy to Resist Corrosion in Hot Seawater*, Research and Development Progress Report No. 478, U.S. Dept. of Interior, September 1969.
21. H. D. Greene: *Experimental Electrode Kinetics*, Rensselaer Polytechnic Institute, Troy, New York, 1965.
22. E. A. Lizlovs and A. P. Bond: *J. Electrochem. Soc.*, January 1971, vol. 118, no. 1, pp. 22-28.
23. R. F. Steigerwald: *Tappi*, April 1973, vol. 56, no. 4, pp. 129-33.
24. C. R. Rarey and A. H. Aronson: *Corrosion*, July 1972, vol. 28, no. 7, pp. 255-58.
25. H. Stoffels and W. Schwenk: "Investigations of the Pitting Corrosion of Chemically-Resistant Steels with Aid of the Turnbull-Blue-Color Reaction," *Werkst. Korros.*, 1961, vol. 8, pp. 493-500 (German).
26. K. Lorenz, H. Fabritius, and G. Medawar: *Mem. Sci. Rev. Met.*, 1969, vol. 66, pp. 779-93 (French).
27. J. E. Truman and H. W. Kirby: *Metallurgia*, August 1965, vol. 72, pp. 67-71.
28. A. P. Bond, J. D. Marshall, and H. J. Dundas: "Resistance of Ferritic Stainless Steels to Stress Corrosion Cracking," *Stress Corrosion Testing*, pp. 116-26, ASTM STP 425, Am. Soc. Testing Mats., 1967.
29. D. Warren: *Chloride-Bearing Cooling Water and the Stress-Corrosion Cracking of Austenitic Stainless Steel*, Proceedings of the Fifteenth Annual Industrial Waste Conference, pp. 420-38, Purdue University, May 1960.
30. M. A. Streicher: *Stress Corrosion Cracking of Ferritic Stainless Steels: The Effect of Environment, Alloy Composition, and Heat Treatment*, NACE Corrosion Research Conference, Chicago, 1974.
31. A. P. Bond and H. J. Dundas: *Corrosion*, October 1968, vol. 24, pp. 344-52.

Nitronic Family of Nitroge

By *JOSEPH A. DOUTHETT*

Five types are specifically formulated to give unique combinations of strength, formability, weldability, and resistance to corrosion, oxidation, and wear. They are alternatives to standard grades.

THE NITRONICS are nitrogen-strengthened chromium-manganese stainless alloys with alloying elements such as nickel, molybdenum, silicon, columbium, and vanadium. The family comprises five austenitic steels having higher strength at both room and elevated temperatures than the standard type 300 stainlesses. Exceptionally stable, they resist martensitic transformation due to cold work, therefore retaining very low magnetic permeability — it is less than 1.02 even in material cold worked to 70%. They also feature exceptional strength and toughness at cryogenic temperatures.

Compositions of alloys, together with their new and former names, are shown in Table I. The system includes weld designations. For example, Nitronic 40W would designate the wire for welding Nitronic 40.

Nitronic 32 and 33 — Compositions are similar, the major differences being in carbon and nickel as Table I shows. Room-temperature tensile strengths of annealed stock are approximately the same — 120 000 psi (830 MPa) for Nitronic 32 and 115 000 psi (790 MPa) for Nitronic 33. Both grades can be cold drawn to high strengths, with Nitronic 32 requiring less cold work to reach a given strength; more ductility also remains. Nitronic 32 has been used primarily for bars and wire, while Nitronic 33 is made in sheets, strip, pipe, and tubing.

Mr. Douthett is senior research engineer, Special Metals Research, Armco Steel Corp., Middletown, Ohio.

The two grades have similar corrosion resistance in most media, but the lower carbon content of Nitronic 33 helps its resistance to more closely approximate that of AISI type 304. This is particularly important for maintaining corrosion resistance of heat-affected zones in welds, and for cryogenic properties.

As for conventional mechanical properties, Nitronic 33, annealed, has almost double the yield strength of type 304 — 68 000 psi (470 MPa) as opposed to 35 000 psi (240 MPa) for the other, annealed. This superiority holds with increasing temperature so that Nitronic 33 is approximately 15 000 psi (100 MPa) higher in yield strength at 1000 F (540 C). At cryogenic temperatures, Nitronic 33 also demonstrates superior strength and less notch sensitivity, while maintaining a respectable level of toughness.

Wear resistance, another important property, has been determined by metal-to-metal sliding wear tests. Tests were conducted on an LFW-1 machine; specimens were abraded by a water-lubricated carburized ring (Rc 64) with a 30 lb (14 kg) load, running at 300 rpm for a travel distance of 3300 ft (1040 m). Results indicated Nitronic 33 was about three times more wear resistant than type 304. It was even superior to several much harder martensitic stainlesses.

This alloy also has good corrosion resistance, remaining unaffected after being exposed for 240 h to either 5% NaCl fog at 95 F (35 C) or 100% relative humidity at 120 F (50 C). Eighteen months of exposure to marine atmosphere at Kure Beach, N.C., 800 ft (240 m) from the Atlantic Ocean caused only light superficial corrosion on both Nitronic 33 and type 304, with the former corroding less severely.

Corrosion resistance of Nitronic 33 to mild acids and pitting media approaches that of type 304, but more severe environments cause greater corrosion on the former. Resistance of Nitronic 33 to intergranular attack is good — weldments did not display accelerated corrosion when tested in boiling 65% HNO_3 or in the boiling $CuSO_4$ test (ASTM A393).

As for resistance to stress-corrosion cracking (SCC), Nitronic 33,

Fig. 1 — Skid wire of Nitronic 32, an austenitic stainless steel, is being wound around high-voltage electrical cable. The wire protects the cable from being damaged while being dragged through underground line pipe during installation.

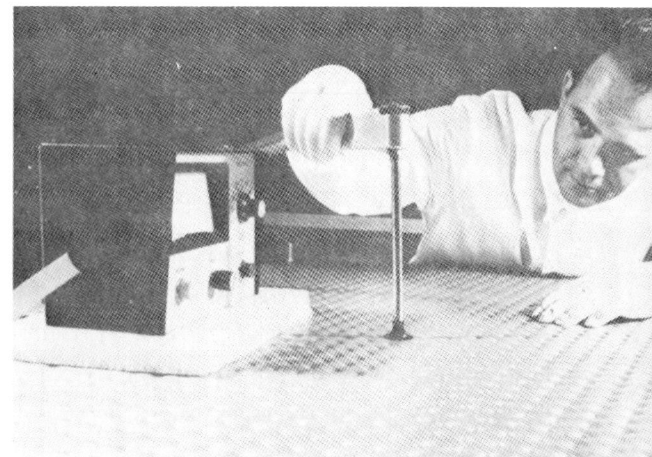

Fig. 2 — Technician checks welds in meteoroid detection panels, made of Nitronic 40 sheets, for spacecraft. The alloy was chosen for its strength (40 000 psi [280 MPa] annealed) and weldability.

...aring Stainless Steels

like most austenitic stainlesses, may crack in hot chloride environments. Based on most design considerations, however, the alloy exhibits improved SCC resistance when compared with type 304 at low stress levels, and SCC resistances are equal in alloys tested at high stress levels in boiling 42% $MgCl_2$.

Stress-corrosion resistance superiority is also demonstrated in welded structures — as-welded samples of Nitronic 33 remained crackfree for 264 h in $MgCl_2$ while type 304 tested similarly failed in less than a day. In room-temperature polythionic acid, Nitronic 33 showed no trace of cracking after 500 h of exposure.

Nitronic 33 is readily welded in all forms; welds are autogeneous with strengths equivalent to the base metal. As for fabricability, Nitronic 33 sheet exhibits the same type of forming characteristics as type 300 stainlesses, but its higher strength calls for more power.

Used primarily as a bar and wire grade, Nitronic 32 is applied for fasteners, clips, cables, industrial screens, and spring-type applications such as whip antenna bases. Here its lower cost and density, compared with those of type 302, provide more steel for the same cost.

Considered an economical replacement for type 304, in strip and sheet form, Nitronic 33 is suited for heat exchangers, process vessels, pipe, and tubing where type 304 is borderline with regard to stress-corrosion cracking. For example, one textile manufacturer had type 316 dye vats which failed by stress corrosion within a few months of service. He switched to more economical Nitronic 33 tanks and has had no sign of a failure in over a year of service.

Table I — Typical Compositions of Nitronic Stainless Steels

Name*	C	Mn	Cr	Ni	N
Nitronic 32 (18-2 Mn)	0.10	12.0	18.0	1.6	0.32
Nitronic 33 (18-3 Mn)	0.05	12.0	18.0	3.2	0.32
Nitronic 40 (21-6-9)	0.03	9.0	21.0	7.0	0.30
Nitronic 50† (22-13-5)	0.04	5.0	21.2	12.5	0.30
Nitronic 60‡	0.07	8.0	17.0	8.5	0.14

*Former name in parentheses
†Also contains 2.2 Mo, 0.20 Cb, and 0.20 V
‡Also contains 4.0 Si

Other applications are cryogenic tanks, valves, piping, and components for electrical transmission. In above-ground transmission, Nitronic 33 is popular for crossarms on transmission poles because of its high strength and corrosion resistance, and Nitronic 32 is being used for pole line hardware. An important use for Nitronic 33 is high-voltage underground transmission pipe because of the steel's low magnetic permeability and strength. Nitronic 32 wire has found extensive underground usage as skid wire, which is wrapped around expensive transmission cable to protect it while being pulled through the pipe (Fig. 1).

Nitronic 40 — Formerly called 21-6-9, this stainless steel is the oldest of the Nitronic family. This alloy has about twice the room-temperature yield strength of comparable austenitic stainlesses, such as types 304 and 347, as well as excellent oxidation and corrosion resistance. Tests indicate that Nitronic 40 has high strengths and good toughnesses down to −423 F (−253 C), suggesting use in highway trailers for transporting liquefied gases. At elevated temperatures, Nitronic 40 is superior in strength to type 304 as was Nitronic 33.

Nitronic 40's corrosion resistance in industrial and marine atmospheres falls between those for type 304 and type 316. Carbon is 0.04% at the most, giving the alloy immunity to harmful carbide precipitation due to welding, but it will

Fig. 3 — Hooks of Nitronic 50 suspend coils of stainless steel wire for pickling in acid. Composed with type 316, the alloy is said to be twice as strong and more corrosion resistant.

Fig. 4 — Built to process 36 000 wieners per hour, this machine has roller chains with link side bars of Nitronic 33 and link pins of Nitronic 60. The alloy provide both corrosion and wear resistance at comparatively low cost.

sensitize after long exposures in the carbide precipitation temperature range, 1000 to 1500 F (540 to 820 C). Welded specimens exposed to marine atmospheres have shown excellent resistance to stress-corrosion cracking.

In the development of this alloy, a cooperative program was initiated with Trent Tube, which resulted in aircraft-quality hydraulic tubing with a minimum tensile strength of 140 000 psi (965 MPa), a minimum yield of 120 000 psi (830 MPa), and at least 20% elongation. Today, Nitronic 40 is the standard hydraulic tubing for such aircraft as the 737, 747, DC-10, and L1011. It has also become popular for bleed air ducting systems for aircraft, including the bellows and clamping devices found in those systems. And it has served in refrigeration tankage and tanks for 747 and F-15 fire control systems.

In addition, Nitronic 40 has excellent resistance to embrittlement by hydrogen under very high pressures, such as those being used in the space shuttle (Fig. 2). And it finds a great deal of usage in both the ground support as well as the flying systems.

Nitronic 50 — The most highly alloyed corrosion-resistant member of the Nitronic family, this alloy has more corrosion resistance than provided by type 316 and 316L plus approximately twice the room-temperature yield strength. As Table I shows, it is a low-carbon stainless steel strengthened with nitrogen; molybdenum, columbium, and vanadium are added to increase its strength and corrosion resistance. Like other members of the Nitronic family, it can be strengthened only by cold work. The alloy is available in bar, wire, rod, and billet form, and sheet, strip, pipe, and tubing products are being developed.

Yield strength is about 60 000 psi (410 MPa) and tensile strength is about 120 000 psi (830 MPa) in stock annealed at 2050 F (1120 C). With a 1950 F (1070 C) anneal, strengths are about 5000 psi (35 MPa) higher. A very respectable yield strength of 40 000 psi (280 MPa) is retained to essentially 1350 F (730 C). At cryogenic temperatures, strengths and toughnesses are also quite good, as is typical of these steels.

This alloy is very stable, and high tensile properties can be developed by cold work — 245 000 psi (1690 MPa) tensile strength after 75% reduction in area, together with 8% elongation, is typical. Nitronic 50 retains its low permeability even after severe cold working, making it a natural for applications requiring a combination of high strength, excellent corrosion resistance, and low magnetic permeability. The endurance limit in reverse bending is about 42 000 psi (290 MPa) in air and about 22 000 psi (150 MPa) in seawater.

Nitronic 50 has excellent corrosion resistance, as we can demonstrate with some data comparing it with type 316 or 316L. The materials show excellent performance in the sodium chloride fog test, but exposure to 10% ferric chloride for 50 h pits the standard stainlesses while leaving Nitronic 50 unaffected.

Researchers also tested specimens in sulfuric acid at 175 F (80 C) for 48 h. Judging by averages of five exposure periods, they found essentially no corrosion on Nitronic 50 in 1, 2, and 5% sulfuric acid, but the type 316 and 316L samples showed significant weight loss. Though boiling sulfuric acid had more effect on Nitronic 50, it was still significantly better than type 316.

Tested against other standard corrosion-resistant materials, Nitronic 50 often exhibits superior corrosion resistance. We compared its resistance with that of alloy 400 (Monel), type 316, and type 304. Four pieces, each with a rubber band around it, were exposed to 10% ferric chloride for 50 h at room temperature.

Only the Nitronic 50 sample remained bright and shiny, while Monel showed general attack (except in the area protected by the rubber band), and type 316 and 304 specimens were badly pitted, showing severe corrosion in the area where the rubber bands were placed. The Huey Test in boiling nitric acid, commonly used to measure intergranular corrosion resistance, produced minimal corrosion on both annealed and sensitized Nitronic 50.

As for resistance to stress-corrosion cracking, Nitronic 50 appears equivalent to type 316 in hot chloride solutions. However, U-bend samples of Nitronic 50 sensitized ½ h at 1250 F (680 C) showed no stress cracks in room-temperature polythionic acid in 500 h exposure.

In an effort to evaluate stress cracking and corrosion resistance in the hydrogen sulfide environments of sour wells, we exposed stressed samples of various austenitics including Nitronic 50 to solutions of 6% NaCl-0.5% acetic acid saturated with H_2S. Under constant stress levels of 50 000 and 100 000 psi (345 and 690 MPa) neither Nitronic 50, in a high-strength, hot-rolled condition, nor A286 cracked in 1000 h of exposure. When tested as U-bends in the annealed and in the sensitized (1250 F [680 C], 1 h) conditions, Nitronic 50, type 304, and type 304N all passed the 1000 h endurance test as-annealed but not as-sensitized. Only type 304L exceeded 1000 h in both conditions.

Of the materials tested, however, only high-strength Nitronic 50 and aged A286 met the strength (100 000 psi [690 MPa] minimum yield), general corrosion, and stress-cracking resistance requirements necessary for sour well applications. Of these two alloys, Nitronic 50 exhibited the better resistance to pitting and general corrosion.

Major market areas for Nitronic 50 are the chemical processing and marine-oriented fields. In the chemical processing industry, Nitronic 50 has been used for pumps, valves, fittings, fasteners, cables, chains, screens, springs, and photographic equipment. Specific applications include seal rings for high-performance industrial butterfly valves, and pickling hooks (Fig. 3). Comparative tests in ammonium carbamate indicated that type 304 became severely etched in two weeks, type 316 showed severe corrosion in six weeks, and Nitronic 50 was unaffected. Because of tests like these, Nitronic 50 has replaced type 316 for pump blocks in fertilizer plants producing ammonium carbamate.

In marine environments, Nitronic 50 has found use for fittings, fasteners, chains, and other hardware. As an example, the alloy is employed in the high pressure microbial sampler built by the Woods Hole Oceanographic Institute. Weighing 200 to 250 lb (90 to 115 kg) and approximately 3 ft (0.94 m) high, this device is lowered as deep as 19 000 ft (5800 m) into the ocean to gather water samples. At these great depths, it becomes a bomb, containing water at 8800 psi (60 MPa).

Another marine application: pleasure boat shafts. In this instance, Nitronic 50 is marketed under the name Aquamet 22, which has 105 000 psi (720 MPa) yield strength, and 135 000 psi (930 MPa) tensile strength as hot rolled. In quiet seawater tests, its corrosion resistance has proved to be superior to that of type 316, long considered the most corrosion resistant of the stainlesses. Though both alloys were covered with barnacles and other marine organisms in nine months, cleaning revealed the Aquamet 22 to be unaffected, while type 316 had suffered random pitting and crevice corrosion.

Nitronic 60 — This alloy, which just recently became available in bar, rod, and wire forms, provides wear resistance in unlubricated metal-to-metal sliding. Looking at the typical composition of Nitronic 60 (Table I) we see that it contains nitrogen, manganese, carbon (0.07% to assure strength), and 4% Si, which acts to aid wear resistance through its tendency to improve oxidation resistance.

Room temperature mechanical properties of Nitronic 60 are 103 000 psi (710 MPa) tensile strength; 60 000 psi (410 MPa) yield strength; and 62% elongation. At elevated temperatures, short-time tensile tests indicated that the alloy is significantly stronger than type 304 up to 1500 F (815 C). Longer-time stress rupture tests, however, reveal the two alloys to be approximately equivalent in strength, particularly at higher temperatures. At cryogenic temperatures, Charpy V-Notch impact energy is in the neighborhood of 150 ft-lbf (200 J) at −320 F (−195 C).

Corrosion resistance of Nitronic 60 in the annealed condition appears as good as that of type 304, but is not quite equivalent to that of type 316. In $FeCl_3$ pitting tests, however, Nitronic 60 outperformed both standard types, while both type 304 and Nitronic 60 exhibited light scattered corrosion as tested in 5% NaCl fog.

High silicon content also provides Nitronic 60 with better-than-average oxidation resistance. Static oxidation tests between 2000 and 2200 F (1090 and 1200 C) indicated Nitronic 60 to be superior to type 304 and essentially equivalent to type 310 in this property. However, cyclic oxidation tests in the 1600 to 1700 F (870 to 925 C) range show the alloy to fall between the two standard types in oxidation resistance.

Metal-to-metal sliding wear tests were conducted on an LFW-1 machine under the same conditions outlined earlier for Nitronic 33. They revealed that Nitronic 60 was inferior only to hardened type 440C.

A comparative test devised to simulate severe wear and galling employed two polished flat specimens placed against each other. As a load was applied, by a modified Brinell Hardness Tester, one block was slowly rotated 360°. Then the two samples were examined for galling.

If no galling occurred, two additional samples were tested at a higher load, the procedure continuing either until galling occurred or until the galling stress rose above 50 000 psi (345 MPa). In this test, the Nitronic 60 (Bhn 205) sample did not gall when run against the following alloys: Nitronic 60 (Bhn 216), type 304 (Bhn 140), type 316 (Bhn 150), Nitronic 50 (Bhn 205), type 440C (Bhn 560).

Diesel Equipment Div. of General Motors has chosen Nitronic 60 for the shaft of the EFE (Early Fuel Evaporation) valve in 1975 models. This valve must be capable of operating at as high as 1500 F (815 C) for 50 000 miles.

Oscar Meyer & Co. uses the alloy for link pins in the conveyor chain of a continuous "wiener processing" machine (Fig. 4). These units produce 36 000 skinless wieners per hour for two to three shifts a day. These 1250 ft (380 m) long chains are estimated to last nine years, as contrasted with four years for 17-4 PH pins previously used.

PROCESSING INFLUENCES SELECTION of STAINLESS for CORROSIVE ENVIRONMENTS

By A. MOSKOWITZ

Susceptibility to intergranular corrosion was studied by exposing a large number of austenitic stainless steel samples in numerous process environments. Over 2,500 specimens of nine stainless types were exposed in the unwelded, welded, and heat-treated conditions.

THE GENERAL CONCLUSION resulting from an extensive field testing program by the Welding Research Council is that certain stainless steels that have been welded or subjected to thermal treatment may show intergranular corrosion in some environments.

Welding is less likely to cause corrosion than thermal treatment for an hour or longer at temperatures of 1,100 to 1,250 F. With an appropriate choice of alloy, problems associated with weldments should not be encountered in service. Relatively few cases of weld-line attack developed on as-welded samples in this program, and these were mostly with the higher-carbon grades (types 302 and 317).

Carbon content is the most critical factor influencing sensitization susceptibility of the nonstabilized stainless steels.

The environment which most often produced intergranular attack were strong acids and seawater. Many, some quite severe, did not produce intergranular attack, even on relatively high-carbon (0.11%) type 302 given a severe sensitizing treatment (1 hr at 1,250 F). Extra-low-carbon and stabilized stainless types resisted attack even in the most severe environments.

Field Testing Program

The austenitic stainless steels are usually among the top candidates when alloys are being considered for highly corrosive environments. Types 302 and 304, containing 17 to 20 Cr and 8 to 12 Ni, are the basic members of the group. Types 316 and 317 incorporate molybdenum for increased general corrosion resistance.

After exposure in the temperature range 800 to 1,500 F, all of the above alloys can exhibit lower corrosion resistance in specific environments. The steel has become "sensitized," and corrosive attack under these circumstances is intergranular in nature. Sensitization is generally attributed to the precipitation of chromium carbides at the grain boundaries which lowers the effective chromium content (and corrosion resistance) near the boundaries. Welding is the most common source of sensitization; the affected region is parallel to and at a slight distance from the weld. Sensitization can also result if the steel is exposed to temperatures within the 800 to 1,500 F range during stress relief after fabrication — or in service.

A high-temperature solution heat treatment will redissolve chromium carbides and prevent intergranular corrosion, but this treatment often is not practical in service. Stainless steels resistant to sensitization include extra-low-carbon stainless steels (types 304L and 316L) and alloys containing columbium (types 347 and 318) or titanium (type 321) to tie up the carbon.

The deteremental effect of sensitization on corrosion resistance in boiling 65% nitric acid (Huey Test) is well known, but there is not much information available on many aspects of practical importance.

How serious is sensitization caused by welding compared to that caused by longer isothermal heat treatments? How wide a range of environmental conditions might cause intergranular attack of sensitized stainless and what would be the worst environment? Would use of stabilized or extra-low-carbon grades completely eliminate intergranular attack in the worst environments?

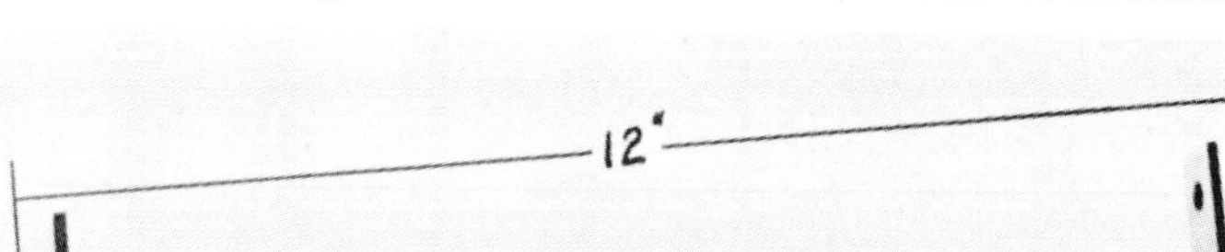

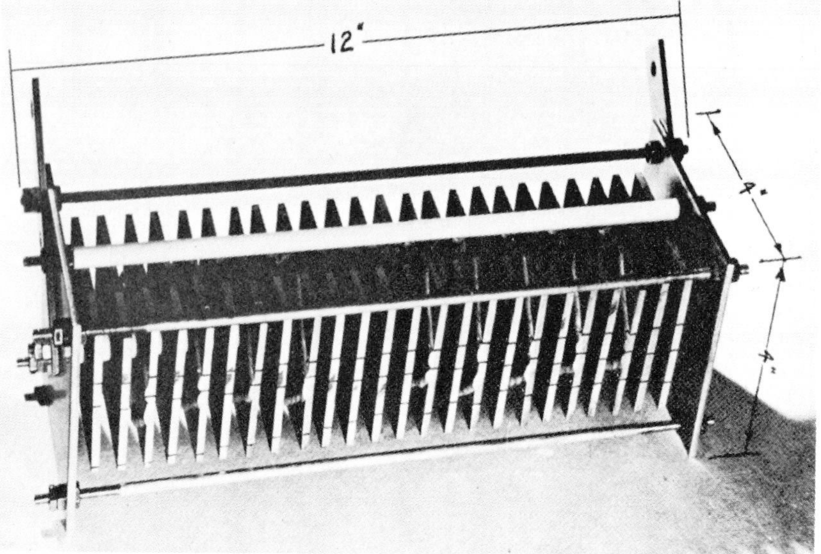
Fig. 1 — Test rack is used for plant corrosion tests.

Mr. Moskowitz, director, technical services, Crucible Stainless Steel Div., Colt Industries, Midland, Pa.

To obtain answers to such questions with particular regard to chemical processing media, an extensive field testing program (see box) was conducted under the auspices of the Welding Research Council, and many companies cooperated by providing plant locations for exposure of test racks. Table I lists the composition of the steels tested.

Results of the Program

Specific media that did or did not cause intergranular attack are listed in Tables II and III. Any single specimen in one rack showing some intergranular attack (IGA) or weld-line attacks (WLA), or even one that had been subjected to artificially severe sensitizing conditions, resulted in the environment involved being classified as causing intergranular attack. Forty-six of the racks had one or more specimens showing some evidence of intergranular attack.

It is difficult to generalize because the variety of environments was somewhat limited, and the exact conditions such as temperature and concentration strongly influence results. Over-all, however, intergranular attack was encountered most often in strongly acid environments and in seawater:

- **Nitric Acid** – All hot nitric acid solutions in concentrations from 5 to 60% caused some intergranular attack in all racks exposed and in most stainless types. In one instance, there was accelerated attack of grain boundaries in nonsensitized (an-

Fig. 2 — Corrosion test specimen illustrates severe weld-line attack (WLA).

Fig. 3 — Cross sectional view shows severe weld line attack. 140×.

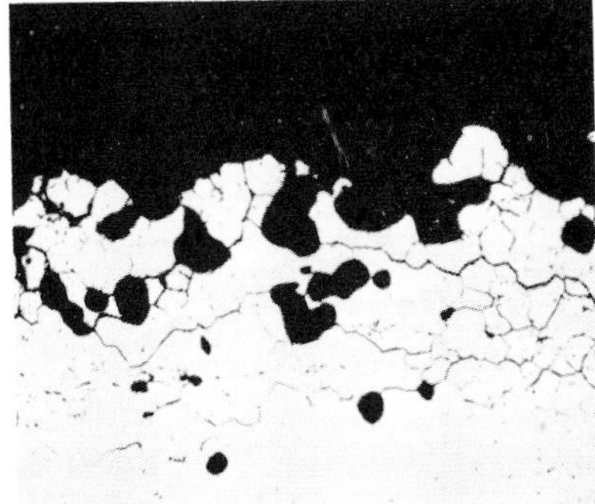

Table I — Compositions of Stainless Steels in WRC Field Testing Program

Type	Composition, % C	Mn	P	S	Si	Cr	Ni	Mo	Cb	N	B
302	0.11	0.72	0.029	0.007	0.50	17.15	9.57	—	—	0.056	< 0.0005
304	0.05	1.34	0.025	0.018	0.56	19.18	9.04	—	—	0.052	< 0.0005
304L	0.028	1.20	0.030	0.022	0.45	18.82	10.0	—	—	—	< 0.0005
321*	0.06	1.54	0.032	0.011	0.68	17.76	10.22	0.19	—	—	< 0.0002
347†	0.060	1.58	0.018	0.025	0.63	17.69	9.64	—	0.48	—	< 0.0005
316	0.042	1.60	0.025	0.018	0.72	17.85	13.38	2.38	—	—	< 0.0002
316L	0.025	1.39	0.021	0.017	0.46	18.54	13.76	2.75	—	0.038	< 0.0002
317	0.072	1.96	0.027	0.025	0.50	18.33	13.69	3.25	—	0.038	< 0.0002
318	0.056	1.84	0.020	0.009	0.40	19.00	14.81	2.40	0.85	0.042	< 0.0002

* Type 321 also contains 0.44 Ti. † Type 347 also contains 0.19 Ta.

Table II — Environments Producing Intergranular Attack in Austenitic Stainless Steels

Acetic acid, glacial, turbulent vapor, 293 F
Acetic acid (100%), 242 F
Acetic acid-butanol (esterification reaction), 257 F
Acetic acid (35%) + formic acid (1%), 268 F
Acetic anhydride-acetic acid, 212 F and 230 F
Chlorinated kraft pulp, pH 2.4, 75 F
Cornstarch slurry, pH 1.5 due to HCl, 120 F
Crude fatty acids (tall oil), 472 F
Maleic acid-xylene-water, 68 F
Maleic acid-maleic anhydride-fumaric acid-xylene-water, 302 F and 374 F
Maleic acid (20-30%), saturated with pthalic anhydride, plus chlorides, 160 F
Maleic anhydride, 140 F
Monochlorobenzene (64%)-DDT (35%), organic phase + sodium hydroxide (20%)-NaCl (4%), aqueous phase, 212 F
Monochlorobenzene, plus steam, phenol, HCl, air, 840 F
Nitric acid (60%) plus chlorides and fluorides, 190 F
Nitric acid (50%), plus chlorides and fluorides, 180 F
Nitric acid (48%) plus chlorides and fluorides, 175 F
Nitric acid (47% to < 1%) plus chlorides and fluorides 176 F
Nitric acid (12% max) plus chlorides and fluorides, 155 F
Nitric acid (20%) + metal nitrates (6-9%) and sulfate (2%), 190 F
Nitric acid (5%), 214 F
Phosphoric acid (75%), 140 F and 203 F
Phosphoric, nitric, and sulfuric acids (dilute), pH 1.6, 140 F
Phosphoric acid ("phossy" water), pH 6.2, 131 F
Phthalic anhydride (crude), 450 F
Seawater, 1 to 2 ft/sec flow, ambient temperature
Seawater, tidal flow conditions, ambient temperature
Seawater, turbuent flow, ambient temperature
Sugar liquor, 66 to 67% sugar solids, pH 7, 167 F
Sulfuric acid (98%), 110 F
Sulfuric acid (78%), ambient temperature
Sulfuric acid (13%), 113 F
Sulfuric acid (4%), 190 F
Sulfuric acid (~1%), pH 2.0, 150 F
Sulfuric acid (0.1%) + ammonium sulfate (1%) and chlorides, 220 F
Starch, Milo, pH 1.6 due to H_2SO_4, 120 F

Table III — Environments Which Did Not Produce Intergranular Attack in Austenitic Stainless Steels

Acetic acid (99%), plus traces of acetaldehyde, acetic anhydride, and esters, 257 F
Acetic acid (85 to 97%), 75 F
Acetic acid (30%), 77 F
Acetic acid vapor — methylene chloride-steam-air, plus HCl traces, 260 F
Ammonium bisulfite (5% total SO_2), 75 F
Aqueous waste water, plus $NaCO_3$, $NaHCO_3$, HCl, traces of paranitrophenol, pH 2 to 9, 122 F
Boron trichloride (99%), plus $CoCl_2$, 86 F
Cane sugar juice, pH 5 to 8, 130 F and 165 F
Cane sugar juice (limed), pH 7, 150 F
Fumaric acid vapor, 392 F
Hydrochloric acid — Cl_2-H_3BO_3-chloride salts, 356 F
Lactic acid (20%) in corn steepwater liquor, pH 4, 165 F
Maleic acid-maleic anhydride, 153 F
Paranitrophenol (PNP)-NaPNP-Na_2CO_3-$NaCO_3$-acetone-NaCl, 131 F
Phenol-acetone-cresol-Na_2SO_4-water, 140 F
Phthalic anhydride-phthalic acid-maleic anhydride-benzoic acid-water vapor, 400 F
Phosphoric acid (75%), 90 F
Sodium chloride (25%)+lime Na_20 (1.5%), 180 F
Sodium orthophosphate solution (42% $NaHPO_4$+ 18% NaH_2PO_4), pH 6.5, 203 F
Sodium sulfite (25%), pH10, 115 F
Sugar liquor, 10 to 20% sugar solids, 167 F
Sulfuric acid (105%), 115 F
Sulfuric acid (0.3%) in ammonium sulfate liquor, pH 3, 158 F
Sulfuric acid (0.1%)-acetic acid (0.1%)-acetaldehyde, 257 F
Sulfurous acid (1% SO_2), pH 1.5 to 2, calcium bisulfite liquor, 74 F
Sulfurous acid (0.1% SO_2), pH 4.5, ground corn slurry, 110 F
Sulfurous acid ($<1\%$ SO_2), pH 5 to 6.5, washed sulfite pump-weak calcium bisulfite liquor, 75 F
Tall oil rosin, 293 F and 302 F and 311 F
Water (NYC), hardness 21.7 ppm $CaCO_3$, pH 6.8, 47 F
Xylene (80%)-maleic anhydride (13%) + malic acid (2%) and water, 221 F
Xylene (75%) + maleic acid (0.5%) and water, 221 F
Xylene-maleic anhydride, 140 F

nealed) specimens; this can occur when corrosion products formed in nitric acid (dichromates) are allowed to accumulate.

• **Sulfuric Acid** – Neither dilute (0.1 to 1.0%) nor highly concentrated (78 to 105%) sulfuric acid was particularly severe in causing intergranular attack. Conversely, concentrations of 4 and 13% caused attack in exposed racks.

• **Acetic Acid** – Acetic acid caused no intergranular attack at ambient temperatures but did produce corrosion at temperatures above 200 F. However, higher concentrations of acid (about 100%) were less aggressive. There was little WLA in the many racks exposed to acetic acid; and in the as-welded condition, only one sample (type 317) showed WLA.

• **Phosphoric Acid** – Temperature is important in 75% phosphoric acid. Several specimens showed intergranular attack at 203 F, only one at 140 F, and none at 90 F. Only one stainless steel (type 317) showed any WLA in the phosphoric acid environments.

• **Maleic Acid** – There was little WLA in the racks exposed to maleic acid and only one steel (type 302) was attacked.

• **Seawater** – Some sensitized steels (types 302 and 317) developed severe "internal IGA" not evident on the surface in sea water. The attack was apparently initiated in the crevice area of the mounting hole and it then spread internally throughout the specimen. As-welded specimens also showed severe internal WLA; the attack intiated at pits and then progressed under the surface. Other stainless types showed no WLA and little IGA.

Stainless Types

Types 302 and 317 were by far the most susceptible to intergranular attack after isothermal sensitization or welding. Carbon content is responsible. Type 302 contained 0.11% C, which is within AISI specifications but above the usual percentage for this steel as produced today. Type 317 had the next highest carbon content (0.072%) which is close to the maximum (0.08%) specified for this steel.

Table IV — How Type of Heat Treatment Affects Susceptibility to Intergranular Attack

Type	Heat Treatment	No. of Specimens Showing Surface IGA			
		Mild Attack	Mod. Attack	Severe Attack	Total
304	1,100 F, 2 hr	3	2	2	7
304	1,250 F, 1 hr	2	3	5	10
316	1,150 F, 2 hr	3	0	0	3
316	1,250 F, 1 hr	2	0	1	3
317	1,100 F, 4 hr	5	2	1	8
317	1,250 F, 1 hr	4	6	12	22
317	1,300 F, 1 hr	6	5	8	19

The low-carbon and stabilized types were generally highly resistant to intergranular attack. Types 304L and 316L did not show WLA in any environment. Type 347 showed WLA only once; in this instance, sample was sensitized at 1,300 F after welding and the environment was hot concentrated nitric acid.

Thermal Treatments

Of the various heat treatments evaluated, some were applied to each stainless type tested, while others were given to only certain types. Thus it is difficult to generalize, but pertinent observations can be made:

• **Sensitizing Heat Treatments** – Table IV summarizes data on the severity of the sensitizing heat treatments on types 304, 316, and 317. Results for types 304 and 316 indicate that 1 hr at 1,250 F was more detrimental than 2 hr at 1,100 or 1,150 F. The differences were more marked for type 317 with its higher carbon content. The 1,250 F treatment was the most severe, the 1,300 F somewhat less so, and the longer-time 1,100 F treatment produced even less sensitization.

• **Stress Relief - Stabilizing Heat Treatments** – Treatments at 1,600 F are sometimes specified for "stabilizing" types 347 and 321 and may be used for stress relieving various stainless types. With type 347, a 4 hr, 1,600 F "stabilization" treatment given prior to a 1 hr, 1,250 F treatment resulted in IGA rather than preventing it as has sometimes been postulated. Treatments at 1,600 F were also detrimental to the molybdenum-containing steels, resulting in IGA in several environments. Type 316 was adversely affected by a 1 hr, 1,600 F "stress relief" treatment. This treatment was also detrimental to type 316L,

Field Testing Program of the Welding Research Council

The work on which this article is based was done by the Subcommittee on Field Corrosion Tests (E. A. Tice, chairman) of the High Alloys Committee of the Welding Research Council. A detailed report has been issued as WRC Bulletin No. 138, "Intergranular Corrosion of Chromium-Nickel Stainless Steels," February 1969.

Typical AISI stainless steels were evaluated in the following conditions: (a) solution annealed, (b) heat treated, (c) as-welded, (d) welded and heat treated.

A total of 106 test racks each containing 24 specimens were evaluated by field exposure. Forty-seven racks contained the nonmolybdenum steels (types 302, 304, 304L, 321, 347); and 56 racks contained the molybdenum steels (types 316, 316L, 317, 318). Exposures were in service environments for times ranging from 30 to 1,600 days.

Although the program evaluated susceptibility to intergranular attack, corrosion rates based on weight losses were also determined. The program was not designed to obtain information on susceptibility to stress-corrosion cracking or pitting, although pitting was noted when it occurred.

There is no accepted quantitative rating system for intergranular attack. Weight loss is not adequate because there can be intergranular attack with only a slight loss in weight; and a weight loss due to intergranular attack along a relatively small heat-affected-zone caused by welding could be obscured by overall corrosion of the specimen. It was to rate specimens by microscopic examination of surfaces, the specimens being rated against standards chosen to represent different degrees of intergranular attack. Ratings were given both for general surface intergranular attack (IGA) and for weld-line attack (WLA) in the heat-affected base metal in a localized zone adjacent to the weld deposit. Specimens for which surface examination was inconclusive were cross sectioned and metallographically examined for intergranular penetration.

and even worse was a 4 hr treatment at 1,600 F prior to 1 hr at 1,250 F (compared with the 1,250 F treatment alone). With this extra-low-carbon type, there is the possibility of sigma phase formation at 1,600 F.

• **High-Temperature Heat Treatment** – Both of the columbium-stabilized steels were susceptible to IGA after being heated for 1 hr at 2,400 F (thus dissolving columbium carbides) and then for 1 hr at 1,250 F. Thus the mere presence of columbium in the steel is not necessarily sufficient; it must be combined with the carbon to be effective. Related to this is the "knifeline attack" sometimes encountered immediately adjacent to the weld deposit in columbium or titanium-stabilized stainless steels.

Welding vs Heat Treatment

Welding was much less severe than sensitizing heat treatments in causing susceptibility to intergranular attack. Table V shows that in many environments, surface IGA (excluding the weld-affected areas) developed on heat-treated specimens, but no WLA developed in the weld-heat-affected areas of welded specimens. Thus weldments might be quite satisfactory in environments wherein sensitization-treated materials may show IGA.

Almost as many racks of type 302 showed WLA as IGA, but the severity of the WLA was generally less. With types 347 and 318, most cases of IGA resulted from heating at 2,400 F prior to sensitizing at 1,250 F. In these types, "knifeline attack" might rate more attention than ordinary weld line attack.

Table V — Corrosion Results in Welding and Various Heat Treatments

Type	Number of Racks Showing Intergranular Attack: WLA (Welded Specimens)	General Surface IGA (Heat-Treated Specimens)
316L	0	11
316	1	15
317	6	22
318	1	21
302	15	19
304	1	10
304L	0	5
321	6	12
347	1	13

A High Strength Stainless Steel for Rivets

DONALD WEBSTER

Compositional modifications to existing Cr-Mo-Co stainless steels have enabled ductility to be increased by means of a stress-induced martensitic transformation.

WHAT are the incentives for improving ductility of high strength stainless steels? In a limited number of applications, plastic deformation has to be applied to the fully heat treated part to develop a combination of strength and high ductility. A good example of this is a rivet, which has to be bucked without cracking and yet cannot usually be further heat treated *in situ*.

For this type of application, a sacrifice in strength is generally accepted to obtain the necessary ductility. In aerospace structures, the weight penalty that this step involves provides an incentive for improving the strength level at which steels can maintain sufficient ductility for bucking.

Some of the most ductile stainless steels are the 300 series austenitic steels, which owe much of their ductility to a stress-induced transformation to martensite that occurs during plastic deformation. Because these steels do not have the strength for many applications, it was decided to use the same mechanism to improve the ductility of one of the stronger martensitic stainless steels, AFC 77. This steel was modified to produce a small amount of fine-grained austenite, which transforms readily to martensite during plastic deformation. Composition of the test alloy, designated AFC 77B, is 0.16 C, 0.20 Mn, 0.018 P, 0.017 S, 0.06 Si, 13.94 Cr, 13.67 Co, 5.22 Mo, 103 Ni, 0.22 Cb, and 0.032 N.

During a tensile test of this material, areas which attempt to neck raise the local stress at that portion of the gage length, causing stress-induced martensite to form. This effect hardens the local area sufficiently to remove any further tendency to neck, and the process occurs continuously along the entire gage length, ensuring a high uniform elongation.

The effect of this mechanism on the shape of the stress-strain curve of AFC 77B is shown in Fig. 1 for two heat treated conditions. To produce Condition I, material is austenitized at 2100°F, cooled to room temperature, and then tempered at 800°F without cooling to subzero temperatures. In this condition, it contains substantial amounts of retained austenite, which produces a low yield point (about 40 ksi); work hardening capacity is sufficient to produce an tensile strength of about 240 ksi. Condition II material is also austenitized at 2100°F, but is tempered at 800°F after a subzero (1 h at − 100°F) treatment, which removes a large amount of retained austenite. This results in a substantial increase in yield strength with only a small decrease in ductility. For comparison, the stress-strain curve of one of the best medium strength martensitic stainless steels PH 13-8 Mo is also shown. The low *uniform* elongation of PH 13-8 Mo is typical of that obtained with low carbon martensitic stainless steels and 18 pct nickel maraging steels; it is a result of their low work hardening rates.

Table I. Properties of AFC 77B

Material	Condition	Yield Strength ksi	Tensile Strength ksi	Shear Strength Before Driving	Shear Strength After Driving	Impact Energy, Charpy V-Notch ft-lb
AFC77 B	T800A	40	238	126	172	96
AFC77 B	T800	200	258	155	–	34

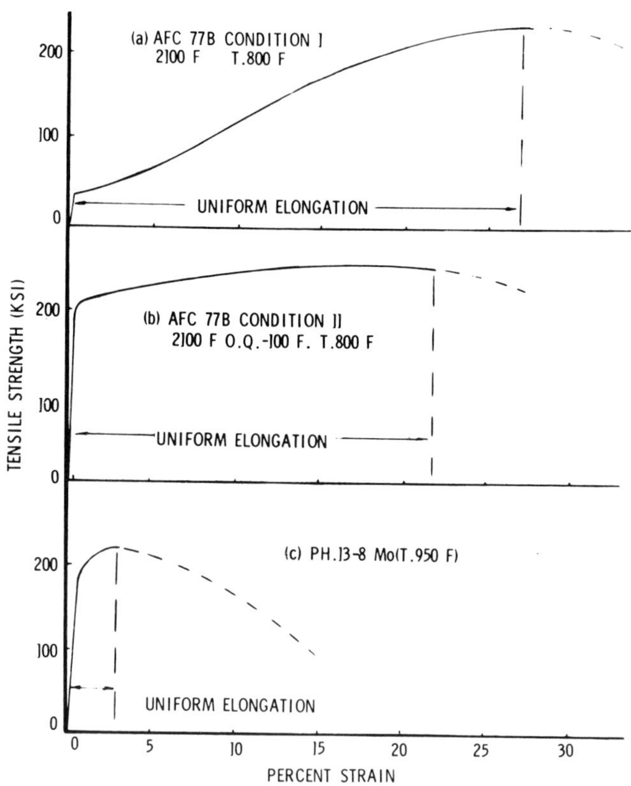

Fig. 1—Stress strain curves: (*a*) AFC77B, Condition I; (*b*) AFC77B, Condition II; (*c*) PH 13-8 Mo (T 950).

Metallographic changes that occur during the plastic deformation of AFC 77B can be studied by observing the martensite transformation around a Rockwell hardness impression in Condition I material (Fig. 2). Density of stress-induced martensite (dark areas) is high close to the hole, but diminishes to zero about two hole diams away. At higher magnifications, the individual martensite plates can be resolved (Fig. 3).

The practical effects of high ductility are apparent from an examination of driven rivets. Rivets of AFC 77B which have been gun driven or squeezed show no

DONALD WEBSTER is Research Scientist, Lockheed Missiles & Space Co., Palo Alto, Calif.

Fig. 2—Stress-induced martensite around a Brinell hardness impression in AFC77B (condition I). Dark area shows extent of the newly formed martensite. Magnification 76 times.

Fig. 3—In the zone shown in Fig. 2, individual martensite needles in the prior austenite grain and across a twin band can be resolved. Magnification 790 times.

head cracking. In contrast, rivets of many other high strength steels with lower ductility cracked in heads even when installed by squeezing.

As mentioned above, AFC 77B has the properties desired in a high strength rivet. For rivets that are to be installed by squeezing, material in either Condition I or Condition II can be used. Where rivets are to be installed by gun driving however, the lower yield strength of Condition I should make driving easier.

Table I shows that AFC 77B offers a high tensile and shear strength while maintaining high toughness. The high work hardening rate of AFC 77B in Condition I causes an increase in shear strength during installation.

To summarize, a high strength stainless steel of high ductility has been developed which will probably find its first application as a rivet material. It should also find application wherever extensive plastic deformation in the fully heat treated condition is required.

Mechanical Properties of AFC77 Stainless Steel Bolts

DONALD WEBSTER

The mechanical properties of AFC77 bolts have been measured for different heat treated conditions; tensile strength ranged from 236 to 277 ksi. With rising tempering temperatures to 900°F, stress corrosion resistance of the bolts decreased down to a minimum, which corresponded to the point of maximum crevice corrosion susceptibility. A large difference between tensile strengths measured by uniaxial tension and torque-tension was found; it is attributed to combined torsional and tensile stresses during torquing.

BOLT manufacturers continue to search for high-strength stainless bolts which can withstand environmental attack. Stainless bolts with minimum tensile strengths above 220 ksi are available only in the heavily cold worked condition, which limits their size and form and renders them uneconomical for many applications. Bolts of stainless steels presently used for applications requiring 220 ksi minimum can fail prematurely when in contact with both humidity and aluminum.[1]

In this program, AFC77 bolts were investigated for two reasons. First, they can be heat treated to a tensile strength of 280 ksi without thermomechanical processings, thereby expanding the present strength range of stainless bolts. Second, the superior stress corrosion resistance of AFC77[2] should reduce the incidence of environmental failure in contact with aluminum.

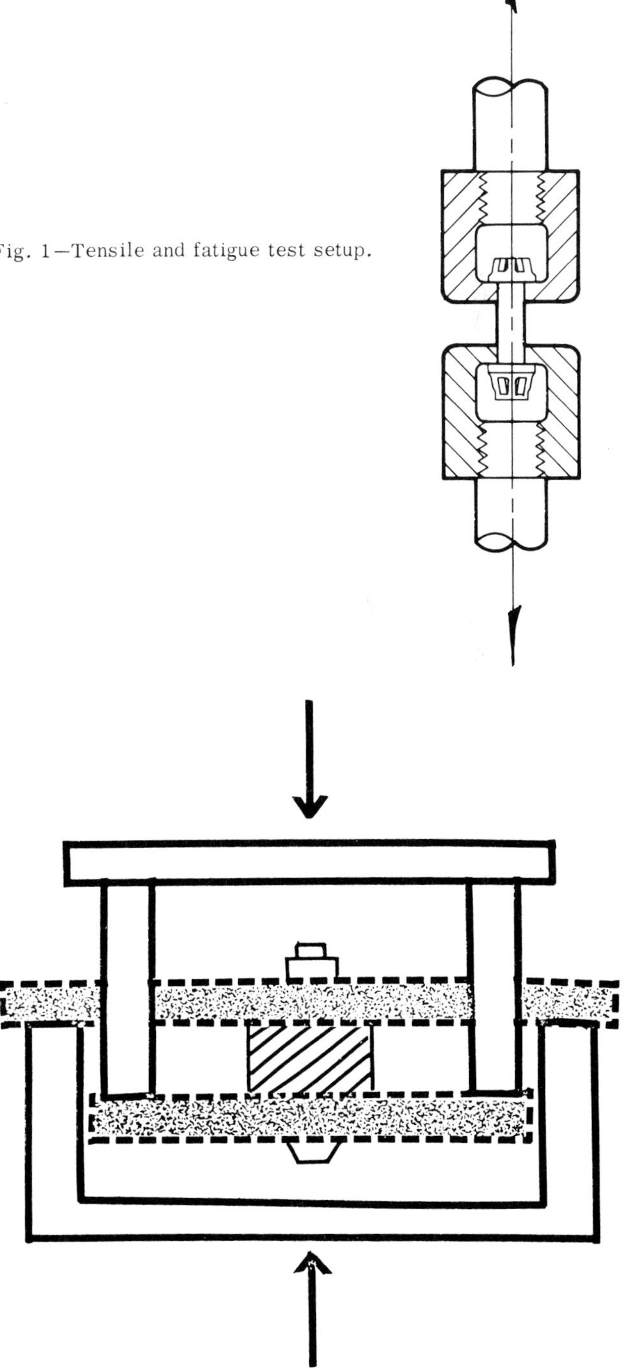

Fig. 1—Tensile and fatigue test setup.

Fig. 2—Jig for determining torque tension stress-strain curve for high strength bolts.

MATERIAL AND HEAT TREATMENT

The 0.25 in. diam bolts used for this program were manufactured by the Hi Shear Corporation from 0.275 in. diam rod supplied by Crucible Steel Co. The composition follows: 0.16 C, 0.24 Mn, 0.20 Si, 14.66 Cr, 0.58 Ni, 14.07 Co, 4.76 Mo, 0.20 B, and 0.031 N.

To forestall the formation of delta ferrite,[3] the bolts were hot headed at 2000°F, austenitized at 2100°F for 1 h, and held 1 h at 1900°F. Then, they were cooled to −100°F for 1 h, and tempered for 4 h in the 500° to 900°F range. Threads were then rolled at 750°F.

Experimental Techniques

Tensile, double shear, and fatigue strength tests were performed on the finished fasteners in accordance with MIL-STD-1312. Fig. 1 shows the test fixture for the tensile and tension-tension fatigue testing. In fatigue tests, the load fluctuated between a maximum (equal to 45 pct of the minimum tensile strength required for bolts), and a minimum, equal to 10 pct of the maximum load (R = 0.1).

The tension developed during torquing of the bolt was measured in the following manner. Bolts were placed in a jig consisting of two steel plates, $\frac{3}{8}$ by 3 by 3 in., between which was sandwiched a hollow steel cylinder, Fig. 2. The nut was torqued to compress the steel plates onto the central cylinder. As each bolt was torqued, a small amount of its elongation was accurately measured with an accurate micrometer. The

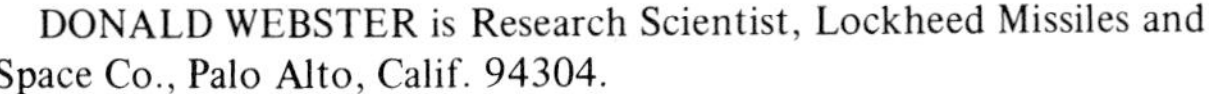

DONALD WEBSTER is Research Scientist, Lockheed Missiles and Space Co., Palo Alto, Calif. 94304.

tensile load on the central cylinder was then determined by applying a known tensile load in an Instron tensile testing machine to the fixture shown in Fig. 2 (dashed lines). The tensile load was slowly increased until the cylinder was just free to rotate. At this stage, the applied tensile load was equal to the compressive force applied by the bolt to the cylinder, which in turn was equal to the tensile load on the bolt. This procedure was repeated until either the bolt failed, or (as occurred in most instances) until the threads were too extended to allow further rotation of the nut. In this way, a torque-tension stress strain curve could be determined for bolts representing each heat treated condition.

Stress corrosion tests were conducted on bolts inside cylinders of stainless steel (A286) or Ti-6 Al-4 V. The cylinders were 1 in. in diam with an axial hole slightly larger than the start diameter of the bolts. Four $\frac{1}{4}$ in. holes were spaced around the cylinder at right angles to the axial hole to allow easy access of the environment. In many of the stress corrosion tests, a sheet of 2024 aluminum (2 by 2 by 0.125 in. thick) was placed either under the head of the bolt or under the nut, Fig. 3. The stress corrosion jigs were tested by alternately immersing them in $3\frac{1}{2}$ pct NaCl solution for $2\frac{1}{2}$ day, and then drying in air for $2\frac{1}{2}$ day. In some instances, the bolts were swabbed with salt solution before assembly, but this procedure was discontinued when results were found to be similar to those for alternate immersion.

TENSILE TEST RESULTS

Basic data for comparison with competitive high strength steels were provided by stress strain curves on PH13-8 Mo and H11, both heat-treated to 220 ksi minimum tensile strength, Fig. 4. Both alloys show greater elongation and lower maximum stress when the load is applied by torquing. The H11 bolt fractured at an extension of 0.017 in., while the PH13-8 Mo bolt remained unbroken at 0.035 in. extension, although the load it could maintain had decreased slightly at this stage.

Corresponding stress-strain curves for AFC77 bolts tempered at 500°, 800°, 850°, and 900°F are shown in Fig. 5, 6, 7, and 8. These curves also indicate stress corrosion results, which will be described later. Elongation at failure in uniaxial tension ranged from 0.015 in. for T500°F to 0.012 in. for T900°F.

In all heat treatment conditions, the AFC77 bolts could not be broken by torquing, although elongations up to 0.029 in. were achieved before thread seizure. As Table I shows, tensile strength increases with tempering temperature, in the manner described previously for smooth tensile specimens of AFC77.[4] Table II also indicates that, as expected, shear strength increases with tempering temperature.

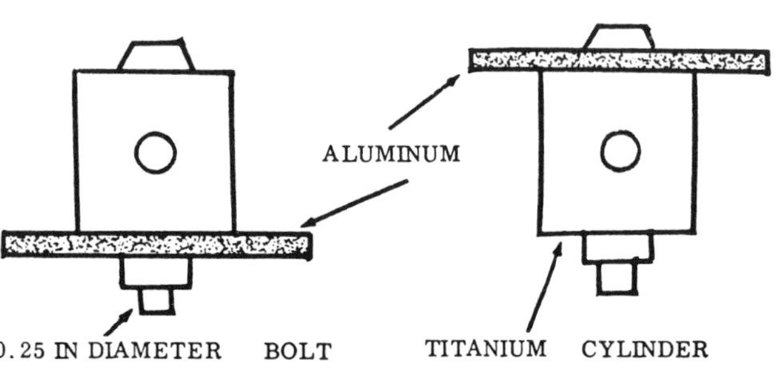

Fig. 3—Location of aluminum couple in stress corrosion tests.

Fatigue Tests

Fig. 9 shows tension-tension fatigue results for bolts of AFC77 (tempered at 500°F), H11, and PH13-8 Mo; the latter two were heat treated to a minimum tensile

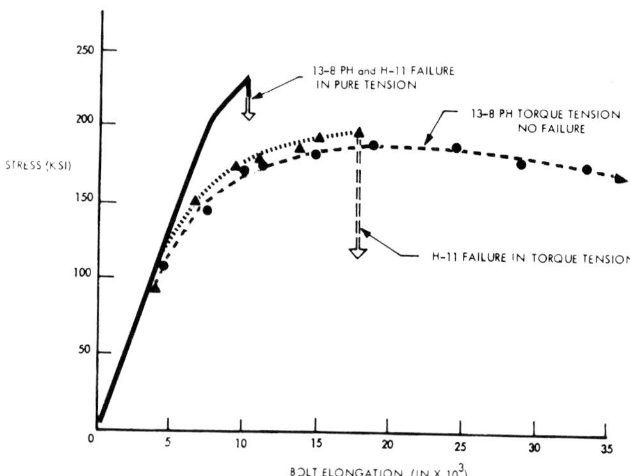

Fig. 4—Stress strain curves for uniaxial tension and torque-tension for PH 13-8 Mo and H11.

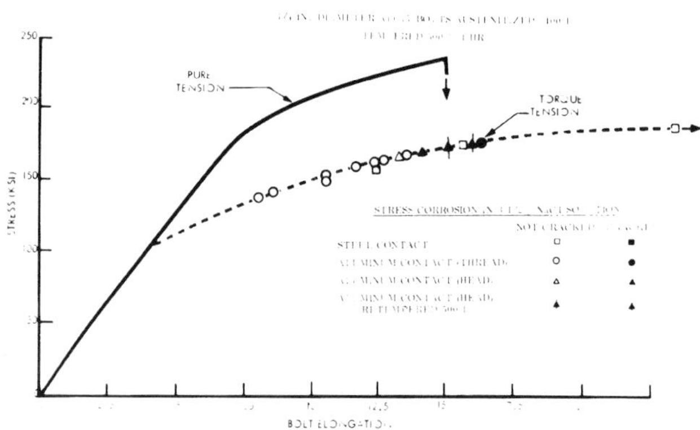

Fig. 5—Stress strain curves for uniaxial tension and torque-tension for AFC77 tempered at 500°F (T500).

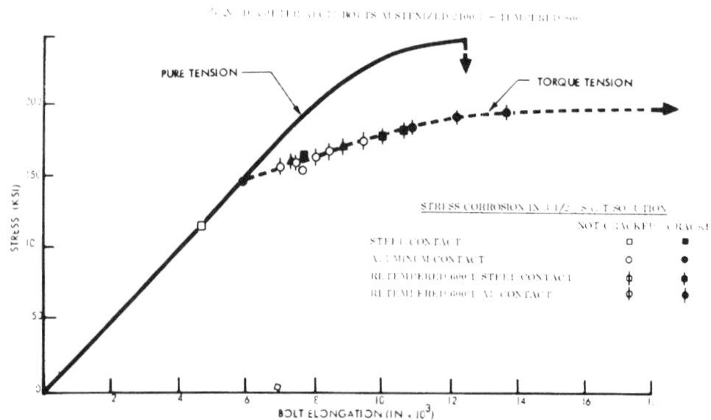

Fig. 6—Stress strain curves for uniaxial tension and torque-tension for AFC77 tempered at 800°F (T800).

Table I. Tensile Strengths of AFC 77 Bolts

Tempering Temperature, °F	500	800	850	900
Average Load at Failure, lb.	9700	10400	10550	10900
Average Tensile Strength, ksi	240	258	261	270

Table II. Shear Strengths of AFC 77 Bolts

Tempering Temperature, °F	500	800	850	900
Average Load at Failure, lb.	14100	15280	15750	16400
Average Shear Strength, ksi	144	156	161	167

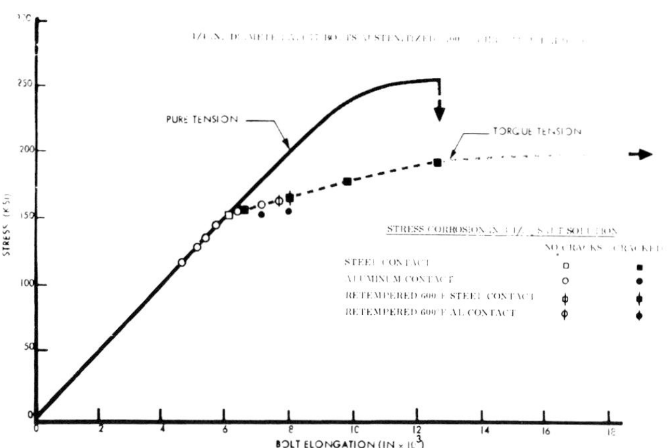

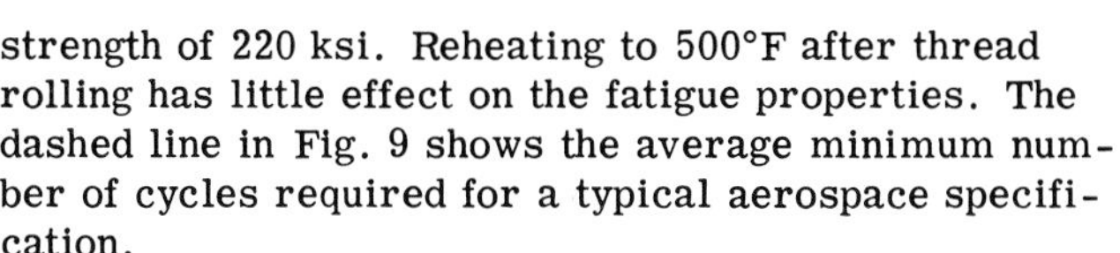

Fig. 7—Stress strain curves for uniaxial tension and torque-tension for AFC77 tempered at 850°F (T850).

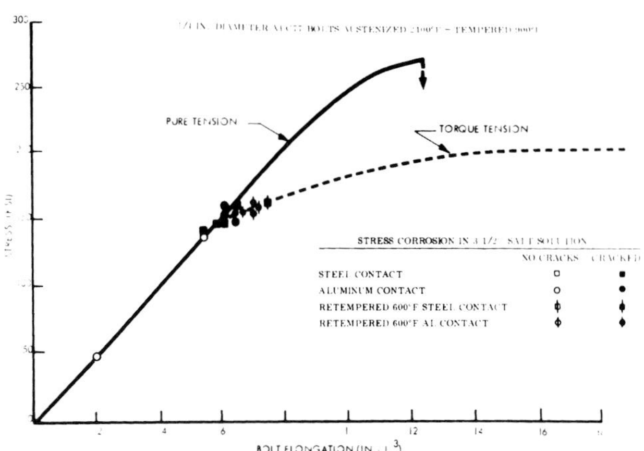

Fig. 8—Stress strain curves for uniaxial tension and torque-tension for AFC77 tempered at 900°F (T900).

strength of 220 ksi. Reheating to 500°F after thread rolling has little effect on the fatigue properties. The dashed line in Fig. 9 shows the average minimum number of cycles required for a typical aerospace specification.

Fatigue results for AFC77 bolts tempered between 500° and 900°F show that average life increases rapidly with tempering temperature, Fig. 10. Retempering at 600°F for 2 h, which simulates elevated temperature excursions, cuts the lifetime of the bolts by approximately 50 pct.

Corrosion Tests

Although most of the stainless steels used for high strength bolts suffer only slight pitting in moist environments, they are often subject to severe local corrosion in the presence of a crevice which prevents free access of oxygen to the corroding surface. In bolts, this localized corrosion frequently occurs in the crevice formed in the vicinity of the head-to-shank fillet. Stress corrosion failures in this region are common in spite of the higher stress existing at the root of the threads.

In understanding the stress corrosion behavior of stainless steel bolts, therefore, it is of interest to compare the crevice corrosion susceptibility of AFC77 and other stainless steels in various heat treated conditions. A crevice was formed by binding a portion of the bolt shank with to 0.25 in. wide adhesive tape, in the manner described by Patel and Taylor.[5] Because no crevice corrosion of AFC77 bolts was observed after three month in $3\frac{1}{2}$ pct NaCl solution, the tests were repeated in 5 pct ferric chloride solution, a much more severe corrodent. Results of these tests are shown in Fig. 11.

No crevice corrosion occurred in AFC77 bolts tempered below 800°F. Corrosion was visible on bolts tempered at 800°F, reached a measurable amount on

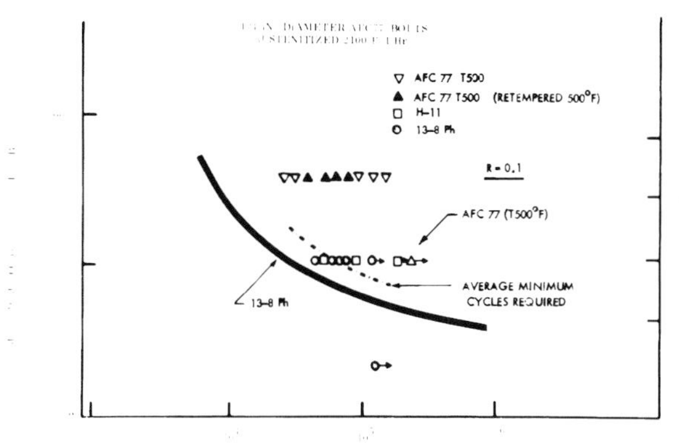

Fig. 9—Tension-tension fatigue results for bolts of AFC77 T500, PH13-8 Mo, and H11. Broken line indicates average minimum cycles required for an aerospace application.

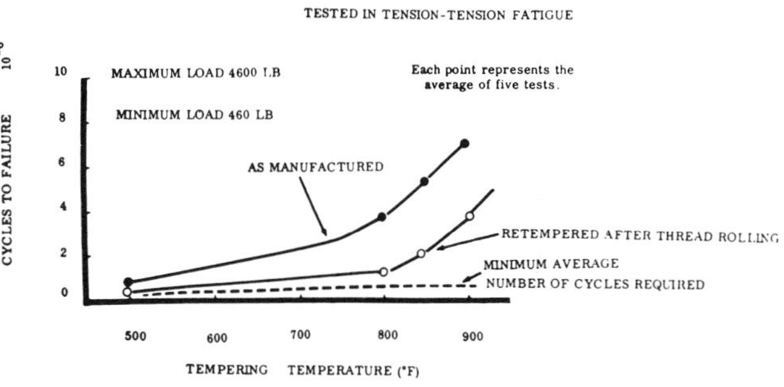

Fig. 10—Tension-tension fatigue results for AFC77 bolts tempered at indicated temperatures show that fatigue life rises with tempering temperature.

bolts tempered at 900°F, and then decreased rapidly to zero for bolts tempered at 1100°F. The other stainless steels, particularly the austenitic types, were considerably more susceptible to crevice corrosion. When the tests were repeated with the steels coupled to aluminum sheet, crevice corrosion did not occur.

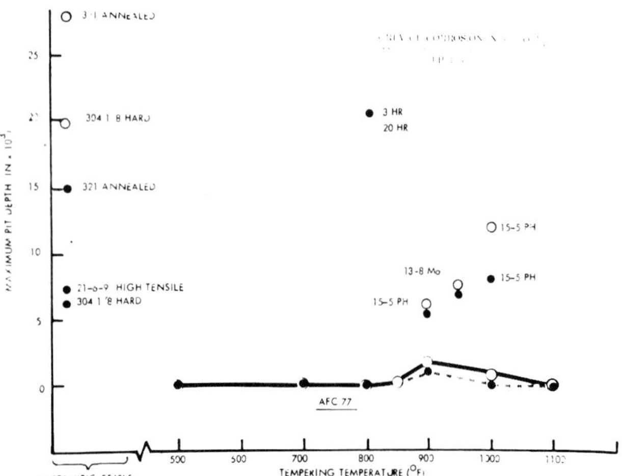

Fig. 11—Depth of crevice corrosion pits in 5 pct $FeCl_3$ for a variety of stainless steels.

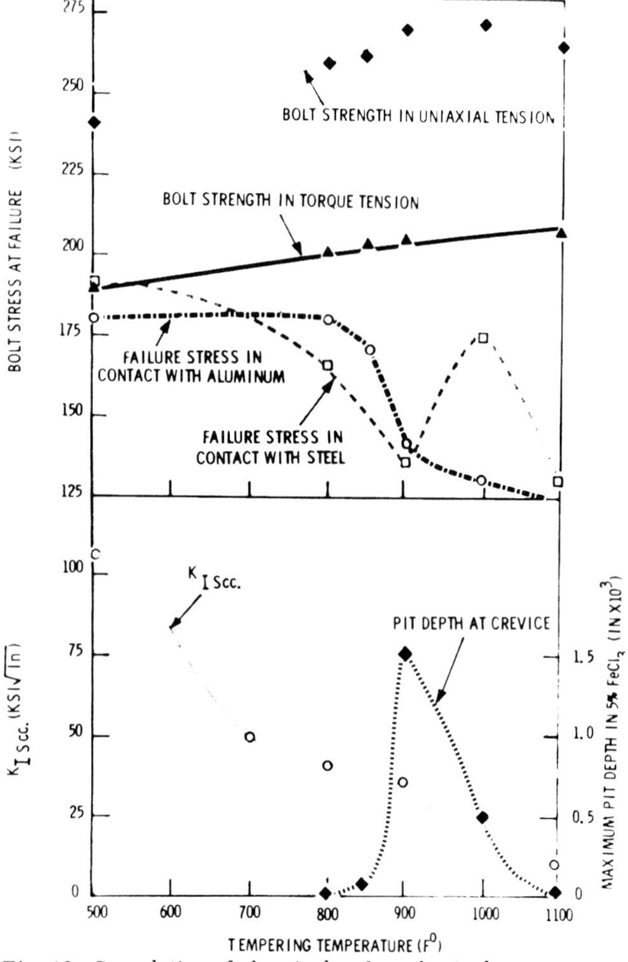

Fig. 12—Correlation of chemical and mechanical properties with stress for environmental failure.

Stress Corrosion Tests

Figs. 5, 6, 7, and 8 show results of stress corrosion tests or AFC77 bolts. In the T500 condition, the uncoupled bolts showed no failure at any load, Fig. 5. Bolts coupled to aluminum, either at the threads or at the head, showed failure in the head-to-shank fillet at an elongation (in 1 in.) of about 0.015 in., which is about the same strain required to cause tensile failure in an uniaxial tension test.

In the T800 condition, uncoupled bolts failed at an elongation (in 1 in.) of about 0.007 in., while coupled bolts failed at 0.006 in. extension, Fig. 6. T800 bolts which were retempered at 600°F were more resistant to stress corrosion, and required an extension of 0.010 in. to cause failure in contact with steel, and 0.011 in. when coupled to aluminum. Bolts in the T850 condition failed about 0.007 in. extension in both coupled and uncoupled conditions, Fig. 7. Retempering after thread rolling had a slight beneficial influence.

In the T900 condition, both coupled and uncoupled bolts failed at 0.006 in. extension, which corresponds to the onset of plastic flow in torque-tension, Fig. 8. Retempering had no measurable effect on stress corrosion resistance.

DISCUSSION OF RESULTS

The wide difference between tensile strength measured in uniaxial tension and torque-tension was unexpected, although Hansen[6] noted similar behavior in medium carbon steel bolts with lower strength. Hansen attributed the lower strength of torqued bolts to the additional torsional stresses that were present during torquing. These carbon steel bolts decreased in elongation to failure during torque testing, while the high strength bolts tested here all showed marked increases in elongation.

The difference in achievable stress levels in the two types of loading should be a cause for concern because the clamping stress produced by torquing is generally the stress of practical interest in design, while the simple tensile test is used both for quality control and as a basis for introducing new bolt materials. Uniaxial tensile testing shows identical stress strain curves for H11 and PH 13-8 Mo, and gives no indication of the higher elongation to fracture available with the stainless bolt.

The fatigue resistance of AFC77 bolts is somewhat higher than other steel of similar strength, Fig. 9, and is only marginally lowered by retempering at 500°F. Retempering at 600°F halves the number of cycles to failure, probably because the treatment relieves the residual compressive stresses produced at thread roots during thread rolling. The retempered condition is nevertheless of practical importance as the fatigue properties indicate the effects to be expected in a structure undergoing intermittent operation at elevated temperatures.

The stress corrosion resistance of AFC77 bolts seems to be a combination of at least two factors—the stress corrosion and crevice corrosion resistance of the material itself. After being tempered at 500°F, the alloy has a high resistance to stress corrosion cracking (K_{Iscc}) and a high resistance to crevice corrosion, Fig. 12. It is therefore not possible to cause a

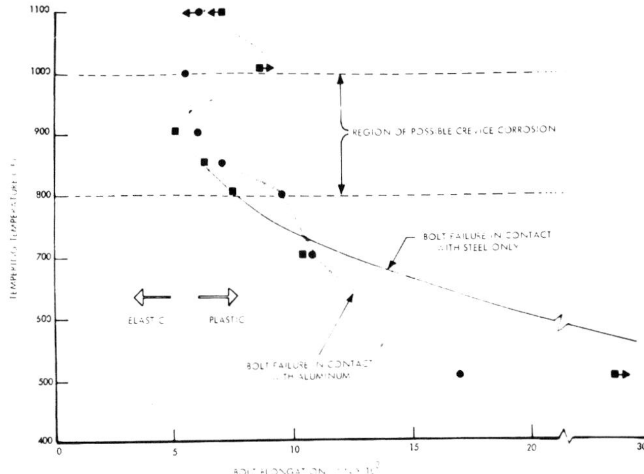

Fig. 13—Correlation of bolt elongation to failure with tempering temperature.

bolt in this condition to fail by environmental attack, unless it is coupled to aluminum. (Aluminum causes a potential difference between it and the bolt, which liberates hydrogen at the bolt surface.) Results indicate that the actual point of coupling is not important in determining either the failure stress or its origin. As the tempering temperature is increased to 800°F, the K_{Iscc} drops and the failure stress in contact with steel decreases. Because there is no drop in the failure stress in contact with aluminum, it is giving some protection against environmental attack. The minimum failure stress in contact with steel occurs at 900°F, and corresponds to the condition of maximum crevice corrosion. No minimum occurs for bolts coupled to aluminum, the bolts being protected by the aluminum from crevice corrosion. Fig. 13 shows the same features in terms of elongation to failure, indicating that the strain to failure is greater for uncoupled bolts, except for the area of crevice corrosion susceptibility where the protection given by the aluminum reverses this trend. In all instances failure is at or above the elastic-plastic limit, as determined by torque-tension.

CONCLUSIONS

1) AFC77 bolts austenitized at 2100°F and tempered at 500° to 900°F have tensile strengths of 236 to 277 ksi.

2) Tempered at 500°F, AFC77 bolts have better stress corrosion, crevice corrosion, and fatigue resistance than other stainless bolts of the same strength.

3) In the higher strength conditions AFC77 bolts tempered at 800°, 850°, and 900°F show a reduction in stress corrosion resistance, which reaches a minimum with 900°F tempering.

4) The failure stress of bolts tempered at 500° or 1000°F under environmental attack is lowered by coupling the bolts to aluminum. Aluminum coupling increases the failure stress for bolts tempered at 800°, 850°, and 900°F by protecting against crevice corrosion.

ACKNOWLEDGMENTS

The author would like to express his appreciation of the cooperation of Hi-Shear Corporation of Torrance, California, and in particular of their Chief Metallurgist, Pete Rush.

REFERENCES

1. R. P. Stewart: private communication, Wright-Patterson Air Force Base, Ohio.
2. D. Webster: *Trans. TMS-AIME,* 1970, vol. 1, pp. 2919-25.
3. D. Webster: *Trans. ASM Quart.,* 1968, vol. 61, pp. 817-28.
4. D. Webster: *Trans. ASM Quart.,* 1968, vol. 61, pp. 817-28.
5. S. Patel and E. Taylor: *Metal Progr.,* Sept. 1971, vol. 100, p. 98.
6. R. Hansen: *Assembly Eng.,* Aug. 1967, p. 18.

HIGH-STRENGTH FASTENER MATERIALS RESIST CORROSION

By SUMANT PATEL and EDWARD TAYLOR

Stress-corrosion cracking is minimized by such alloys as PH 13-8 Mo, PH 12-9 Mo, Custom 455, Inconel 718, and MP35N. However, some have problems with hydrogen embrittlement.

Several high-strength bolting materials are available to industry as alternatives for alloys that are susceptible to stress-corrosion cracking and hydrogen embrittlement. In addition, the new alloys are capable of meeting the 220,000 psi tensile strength requirement of the aerospace industry.

Candidate materials include precipitation-hardenable stainless steels, such as PH 13-8 Mo, PH 12-9 Mo, and Custom 455. All are martensitic as-solution-treated and develop the desired 220,000 psi minimum tensile strength as a result of an aging cycle at around 1,000 F. Inconel 718, a nickel-base, precipitation-hardenable alloy, is also capable of reaching the 220,000 psi level by a combination of cold work and aging.

A truly corrosion-resistant material capable of attaining 260,000 psi tensile strength while maintaining a good balance of other fastener properties is Multiphase MP35N. This material derives its tensile strength from a combination of cold work, transformation, and aging.

Limitations – These new fastener materials can provide better mechanical properties than several alloys previously used, but they are not completely free of problems. The galvanic compatibility of these bare alloys with aluminum structures does not compare well with that of cadmium-plated steel. However, platings and coatings compatible with aluminum can be used to overcome the problem.

The ability of these steels to resist stress-corrosion cracking permits their use in many applications formerly denied to high-strength alloys. However, hydrogen embrittlement may be a possibility if the aluminum structure corrodes and generates hydrogen at the surface of the bare bolt. If hydrogen embrittlement does occur and the applied stress is high enough, catastrophic cracking could result. To prevent this, coatings may be required on the bolts.

All possibilities considered, the new corrosion-resistant fasteners offer considerable improvement over other types, either as a substitute for them in existing applications or in new applications where high strength is required.

Mr. Patel is metallurgical engineer and Mr. Taylor corrosion engineer, Standard Pressed Steel Co., Jenkintown, Pa.

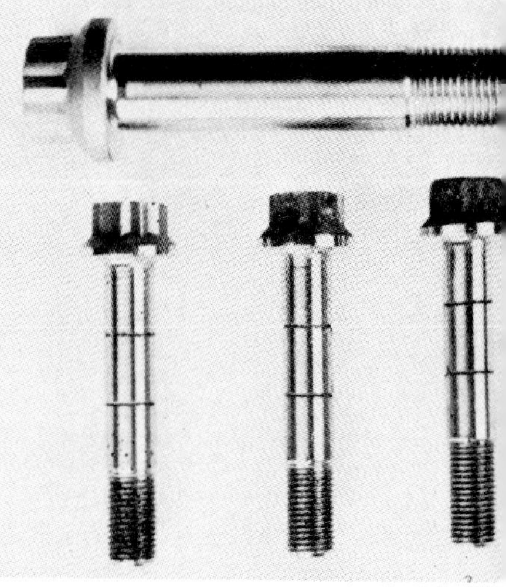

Fig. 1 – Rings show crevice corrosion on PH 13-8 Mo (left), Custom 455 (center), and PH 12-9 Mo (right). MP35N (top) was not affected.

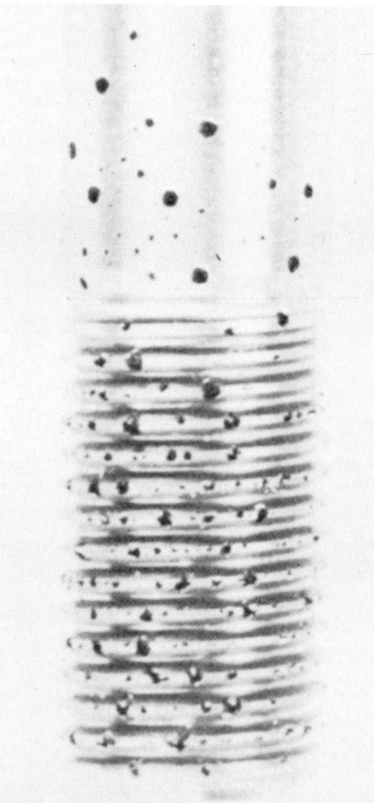

Fig. 2 – PH 13-8 Mo bolt shows severe grooving from crevice corrosion and wormhole attack after ferric chloride test.

Materials Evaluated

Standard Pressed Steel Co. evaluated several corrosion-resistant materials for their suitability in high-strength fasteners. Both mechanical properties and properties related to corrosion resistance were checked. Test results, given in the accompanying tables, are for 220,000 psi bolts (made from Custom 455, PH 13-8 Mo, PH 12-9 Mo, Inconel 718, and H11) and for 260,000 psi bolts (made from H11, MP35N, and Marage-300).

Tests were conducted on 12-point external wrenching bolts with MIL-S-8879 threads.

The manufacturing sequence was as follows: forge, heat treat, grind shank and thread roll diameters, roll threads, and roll head-to-shank fillet radius.

Heat treatments, given in Table I, were selected to optimize mechanical properties, while maintaining nominal tensile requirements.

Following is a description of tests:

Tensile and Shear Strength – Tensile tests were performed on bolts using appropriate companion nuts, following MIL-STD 1312 procedures. Shear tests were those involving double shear of the bolt

Table I — Heat Treatment of Test Bolts

Alloy	Heat Treatment	Min Tensile Strength, Psi
Custom 455	Solution treat and 950 F age harden	220,000
PH 13-8 Mo	Solution treat and 950 F age harden	220,000
PH 12-9 Mo	Solution treat and 1,000 F age harden	220,000
Inconel 718	Cold work and age harden	220,000
H11	Harden and temper	220,000
H11	Harden and temper	260,000
MP35N	Cold work and age	260,000
Marage-300	Solution treat and 900 F age harden	260,000

Table II — Room-Temperature Tensile and Double Shear Strength of High-Strength Bolts

Alloy	Average Tensile Strength, Psi*	Average Shear Strength, Psi†
220,000 Psi Tensile Strength		
Custom 455	240,100	139,100
PH 13-8 Mo	243,300	141,600
PH 12-9 Mo	248,500	142,900
Inconel 718	232,600	137,800
H11	238,000	135,800
260,000 Psi Tensile Strength		
H11	269,000	159,000
MP35N	283,000	161,500
Marage-300	276,500	148,800

*Based on the basic pitch diameter.
†Based on twice the area at nominal shank diameter.

shanks in accordance with accepted fastener industry procedures. Results are set forth in Table II.

Fatigue Properties – All bolts were tested in tension-tension fatigue at room temperature with the minimum load equal to 10% of the maximum load ($R = 0.1$). Cadmium-plated H11 nuts heat treated to Rc 38-42 were used for all tests.

Fatigue test results for 220,000 psi and 260,000 psi bolts are presented in Table III. The maximum load for all the tests was 45% of tensile strength. Results indicate that the high-strength, corrosion-resistant alloys have fatigue strength in excess of the aerospace requirements. All fatigue failures were in the threads.

Crevice Corrosion – Fastener applications create situations where crevice corrosion can be a problem. In a 10% by weight solution of ferric chloride, severe crevice corrosion was observed on three corrosion-resistant alloys after only 6 hr immersion.

A band of tape ½ in. wide was placed around the bolt shank before total immersion in the solution at room temperature. After 6 hr, the tape was removed, and two grooves were observed where the edge of the tape had been. As shown in Fig. 1, no attack was seen between the grooves. Wormhole-like attack was observed primarily on crests, one of the locations where infoliation sometimes occurs as shown in Fig. 2.

Results show that the PH 12-9 Mo, PH 13-8 Mo, and Custom 455 are highly susceptible to crevice corrosion which is likely to be found in the thread crests or wherever else a differential aeration cell is produced. Inconel 718 and the MP35N alloy showed no detrimental effects in this test.

Stress Corrosion – Testing followed the procedure in MIL-STD 1312, where the bolt is loaded to

Table III — Tension-Tension Fatigue Properties of High-Strength Bolts

Normal Cycle Requirements of Specification: 45,000 cycles minimum; 65,000 cycles average

220,000 psi bolts:
Maximum load — 99,000 psi*, 45% of tensile strength
Minimum load — 9,900 psi, 4.5% of tensile strength

Type Material	Average Cycles to Failure
Custom 455	500,000
PH 13-8 Mo	5,000,000
PH 12-9 Mo	2,000,000
Inconel 718	2,000,000
H11	5,000,000

260,000 psi bolts:
Maximum load — 117,000 psi*, 45% of tensile strength
Minimum load — 11,700 psi, 4.5% of tensile strength

Type Material	Average Cycles to Failure
H11	300,000
MP35N	5,000,000
Marage-300	156,000

*Based on the basic pitch diameter.

Table IV — Stress-Corrosion Cracking

Alloy	Time to Failure, Hr (Cause)
PH 13-8 Mo	3,099 (SF), 6,534 (NF)
	2,615 (NF, 4 specimens)
	2,600 (NF, 5 specimens)
Custom 455	5,020 (NF, 2 specimens)
	6,534 (NF, 2 specimens)
	2,615 (NF, 4 specimens)
	2,735 (SF)
	2,338 (NF, 2 specimens)
	2,316 (SF)
	2,318 (SF)
MP35N	1,535 (NF, 5 specimens)
	2,500 (NF, 5 specimens)
H11 (260,000 psi)	138 (SF), 144 (SF), 457 (SF), 100 (SF), 164 (SF), 149 (SF)
H11 (220,000 psi)	1,356 (TRO), 1,074 (TRO), 660 (TRO), 4,274 (NF)

SF—shank failure; TRO—thread runout failure; NF—no failure

Table V — Simulated Service Test Results

Alloy	Coating	Time to Failure, Hr (Cause)
PH 13-8 Mo	Bare	66, 137, 137, 137, 217 (HF)
Custom 455	Bare	1,800 (NF, 3 pieces)
PH 12-9 Mo	Bare	65 (HF), 990 (NF, 5 pieces)
Inconel 718	Bare	2,900 (NF, 3 pieces)
H11 (220,000 psi)	Bare	2,200 (NF, 3 pieces)
PH 13-8 Mo	Copper	1,000 (NF, 2 pieces)
PH 13-8 Mo	Nickel	1,000 (NF, 2 pieces)

Loaded by elongation to 75% tensile strength. NF — no failure; HF — head failure.

Table VI — Hydrogen Embrittlement Test (Sustained Load, Hr*)

Alloy	Cadmium Plate		Bare	
	No Bake	375 F, 23 Hr, Bake	Acid Charged	Alkaline Charged
PH 13-8 Mo	23.3	95.7 (NF)		(FL)
K_t 3.5	215.1 (NF)	240 (NF)		(FL)
	72.6			
PH 13-8 Mo	5.7	200.3 (NF)	4.8	11.3
K_t 6.0	27.1	259.5 (NF)	FL	
Custom 455	215.7 (NF)	235.6 (NF)	30.0	100.5 (NF)
K_t 3.5	172.9 (NF)			
Custom 455	215 (NF)		38.5	
K_t 6.0				

*75% notched tensile strength. NF — no failure; FL — failed loading.

75% tensile strength and exposed to alternate immersion in a 3.5% NaCl solution for 10 min followed by 50 min of drying in forced air. This cycle is repeated until catastrophic cracking of the bolt occurs, or until cessation of the test. H11 steel or Ti-6Al-4V cylinders (or both) were used in lieu of the similar metal designated in the standard test method.

Results in Table IV clearly show that these high-strength alloys are not very susceptible to stress-corrosion cracking. In general, they have lasted at least 3,000 hr when failures were recorded. Some alloys have not failed after 5,000 hr even though they did appear rusty. Bare H11 steel bolts, on the other hand, usually fail after several hundred hours.

Galvanic Compatibility – In today's aircraft, fasteners of these alloys will probably be used in aluminum, rather than titanium, structures. The open circuit potential (voltage) between the noble metal in the corrosion-resistant fasteners and the active metal in the structures is about 0.5 v in the presence of an aqueous electrolyte containing chlorides. With a potential this high, galvanic corrosion of the aluminum is inevitable.

In a one-year exposure of an MP35N bolt in a 7075-T6 aluminum panel at Kure Beach, N. C., the panel was attacked around the fastener head, a classical case of galvanic or dissimilar metal corrosion. When plated with cadmium, the same bolt provided complete protection to the aluminum panel. Thus a sacrificial coating was employed not to protect the bolt, as with low-alloy steels, but to protect the structure around the bolt.

Simulated Service Test – When swabbed with 3.5% NaCl solution and loaded to 75% tensile strength in 7075-T651 aluminum blocks, the PH 13-8 Mo bolts catastrophically cracked after very short periods, as shown in Table V. Subsequent macro and micro investigation indicated that stress-corrosion cracking was not present, but that the alloy was being embrittled by hydrogen. The source of the hydrogen was probably the intense galvanic reaction between the bare corrosion-resistant bolt and the aluminum block. Because the aluminum was the anode and the bolt the cathode, atomic hydrogen evolved on the latter, some of it diffusing into the surface and the remainder recombining to form molecular hydrogen.

When sacrificial coatings were applied to the bolts, no failures occurred under the same conditions. This shows that either the reduction of galvanic corrosion did not permit enough hydrogen to be generated to meet the threshold condition or that the coating itself was enough of a diffusion barrier to the hydrogen to severely reduce the amount getting through to the bolt surface.

When nickel or copper diffusion barrier coatings were applied to the bolts, no cracking was observed even though the corrosion reaction between nickel or copper and the aluminum structure was intense and produced large quantities of corrosion products. The impenetrability of the barrier coatings provided sufficient protection to the bolts. An improvement of this system would be to overcoat the copper or the nickel with either cadmium or aluminum to provide galvanic compatibility with the aluminum structure.

Bolts made of Custom 455 did not crack under the same conditions as those under which the PH 13-8 Mo did. Although the exact reason for this is not known, it appears that Custom 455 had considerably more of an immunity to hydrogen embrittlement than PH 13-8 Mo.

Hydrogen Embrittlement – Specimens were prepared from bolts by notching the shank and producing a K_t of either 3.5 or 6.0. These were either charged or plated to produce hydrogen and then hung at sustained load until failure or runout. An acid charging solution consisting of 0.1 N H_2SO_4 with 0.1N As_2O_3 and an alkaline solution consisting of 0.1N NaOH with 0.1N NaCN were used for charging at 1 amp (about 20 amp per sq ft) for 30 min. Plating was done in production cyanide cadmium or sulfamate nickel baths at normal current densities.

Table VI shows the results of tests where baking after plating had a definite effect on the ability of the PH 13-8 Mo to sustain the imposed load. These results clearly show that hydrogen embrittlement of this alloy does occur, but threshold conditions must first be met. The sharper notch provides more susceptibility to hydrogen embrittlement and the acid charging provides much more hydrogen than necessary to produce a delayed failure.

Custom 455 was only affected when acid charged, showing that it, too, is susceptible to hydrogen embrittlement, but not nearly to the degree of the PH 13-8 Mo alloy. This condition indicates a need to insure safe applications so that this threat is minimized, if not eliminated.

Ultrahigh-Strength Stainless for Light-Gage Spring Applications

By SETH R. THOMAS
and ERIC C. SHARPLESS

Type 301 is suggested for light-gage applications. Two strip grades combine the corrosion-resistant properties of stainless steel with ultrahigh strength.

Fig. 2—These Hunter Neg'ator constant force springs have more uses when made from the Rodflex grades of type 301.

WHILE INCREASING numbers of ultrahigh-strength alloys are being developed and introduced, many applications are well served by special processing of more familiar grades, such as type 301 stainless steel.

Rodflex 270 and Rodflex 290 are cold-rolled strip tempers of type 301 with guaranteed minimum tensile strengths of 270,000 psi and 290,000 psi, respectively. Rockwell hardness values range from Rc 49 to 54.

Finished rolled in thicknesses of 0.001 to 0.031 in., the two grades are popular in several types of springs and other mechanical elements such as collapsible and extendible structures, diaphragms, and retainer rings.

Properties – Rodflex tensile test data are compared at given hardness values with a typical commercial type 301 grade in Fig. 1.

Stress-relief cycles are usually employed by fabricators of Rodflex strip. Test data for 2 hr exposures suggest an optimum cycle of 2 hr at 700 F. Tensile strength increases of 15,000 to 20,000 psi are common.

Tensile strengths of each Rodflex grade furnished by the mill generally range within 20,000 psi above the minimum.

Improved Spring – One spring, the Hunter constant force Neg'ator (Fig. 2), is available in a wide variety of sizes and lengths. By producing it from Rodflex, its application range has been considerably broadened. The grades impart the corrosion-resistant properties of stainless steel combined with the mechanical properties normally associated with springs made from high-carbon steel.

Fig. 1 — Here's how tensile properties of Rodflex compare with those of type 301.

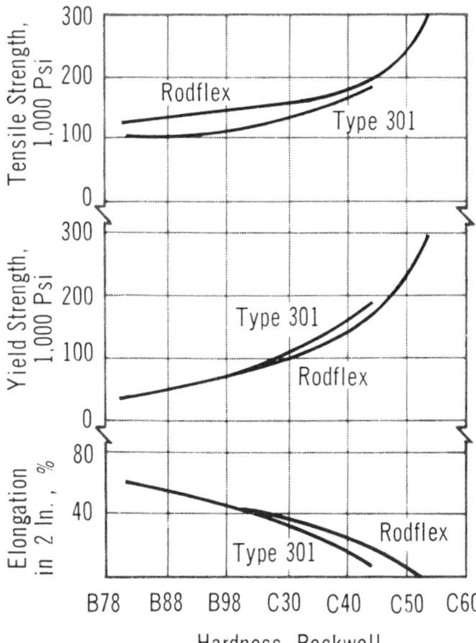

Satellite Antennas – The Hunter Stacer self-erecting tube is a new application. One version of this free-standing tube extends from a 6 in. compressed state to a full 21 ft boom. Deployment takes less than 1 sec. The tube is being used in a satellite.

Process – Type 301 is specially processed to produce the two Rodflex grades. Type 301 is cold-rolled at temperatures substantially below the M_d temperature, causing large percentages of metastable austenite to transform to the high-strength martensite phase.

Controlled rolling practices insure extra high strength levels above the usual minimums.

Cold-rolled and stress-relieved tensile strengths as high as 333,000 psi have been achieved in production lots.

Newly introduced computerized gage controls on Sendzimir rolling mills provide tighter control of thickness tolerances. This improvement is expected to be of considerable value because in the case of the Neg'ator, its energy is directly proportional to the cube of its thickness. Thus, closer control of this parameter should enable the spring manufacturer to produce a more uniform and closely controlled product.

Mr. Thomas is manager of metallurgy and quality control, Teledyne Rodney Metals, New Bedford, Mass., and Mr. Sharpless is engineering manager, Hunter Spring Div., Ametek Inc., Hatfield, Pa.

Section V: Cast Stainless Steels

Corrosion-Resistant Steel Castings

*By the ASM Committee on Corrosion-Resistant Castings**

CORROSION-RESISTANT STEEL CASTINGS are distinguished by their ability to serve where carbon and low-alloy steels would be destroyed by the corrosive action of the environment, rather than by the mechanical conditions of loading. Corrosion-resistant steel castings are used to resist corrosion by aqueous solutions at or near room temperature, and by hot gases and high-boiling-point liquids at elevated temperatures up to 1200 F.

All of the corrosion-resistant steel castings contain more than 11% Cr and most of them from 1 to 30% Ni (a few contain less than 1% nickel). Carbon content, especially important in its influence on both corrosion resistance and strength, is usually under 0.20% and is sometimes as low as 0.03%.

Chemical compositions of these steels are given in Table 1. Physical and mechanical properties, melting temperature, microstructure, heat treatment and other important fabricating considerations are shown in the data compilations immediately following this article, for alloys standardized by the Alloy Casting Institute (ACI).

Corrosion-resistant nickel-base castings are discussed in the Nickel Section of this Handbook.

Although twelve of the ACI casting grades (Table 1) have counterparts among the AISI types of wrought stainless steel, the chemical composition ranges for the cast and wrought alloys differ. Therefore, the casting alloys should be referred to by their ACI designations, a system of nomenclature that is used in ASTM specifications and by many individual producers.

The Aircraft Material Specification group of SAE has adopted these chemical ranges in a number of specifications, and they are also used for several specifications of the Navy Department.

About two thirds of the corrosion-resistant steel castings contain from 18 to 22% Cr and 8 to 12% Ni; the straight chromium compositions are also produced in considerable quantity, particularly the steel with 11.5 to 14.0% Cr (Fig. 1).

Chromium imparts passivity to ferrous alloys when present in amounts of more than about 11%, particularly if conditions are strongly oxidizing. Corrosion resistance improves as chromium content increases beyond the minimum required for passivity. In general the addition of nickel to iron-chromium alloys improves ductility and impact strength although there are some exceptions where nickel does not impart this effect, especially in the region of the ferrite-plus-austenite to austenite phase boundary. Resistance to corrosion by neutral chloride solutions and weakly oxidizing acids increases in proportion to nickel content. Addition of molybdenum increases resistance to attack by pitting chloride solutions. It also extends the range of passivity in solutions of low oxidizing characteristics. In all iron-chromium-nickel stainless alloys, the resistance to corrosion, especially in environments that cause intergranular type of attack, can be improved by lowering the carbon content.

The type CA-15 alloy, containing 11.5 to 14% Cr, can be quenched and tempered to achieve hardness where wear is involved. It is used widely for trim in carbon steel valves, for pumps to handle acid mine water containing abrasives, and in paper mill applications where wear is as important as corrosion resistance. Type CB-30 alloy, containing 18 to 22% Cr, is ferritic at all temperatures, and thus cannot be heat treated to gain hardness. However, its higher chromium content allows a slightly higher range of acid concentration and temperature.

The iron-chromium-nickel alloys have found wide acceptance and constitute about 60% of the total production of cast high alloys. They are generally referred to as being austenitic and by this virtue they are generally nonmagnetic. However, the balance here is fairly narrow; they may be slightly magnetic depending on the composition of the heat or compensating effects of various alloying elements. The most popular alloys of this type are CF-8 and CF-8M. Their nominal composition is 18% Cr and 8% Ni, with or without 2.5% Mo. Carbon is maintained at 0.08% max.

*J. H. JACKSON, *Chairman*, Manager, Department of Metallurgy, Battelle Memorial Institute; ERROLL V. BLACK, Chief Metallurgist, Lebanon Steel Foundry; W. T. BRYAN, Duriron Co., Inc.; E. L. HAILE, Plastics Div., Monsanto Chemical Co.; A. W. LEMMON, Research Coordinator, The Jeffrey Mfg. Co.; SIDNEY LOW, Director of Research, Chapman Valve Mfg. Co.; E. A. MACHA, Manager, Pump Development Section, Atomic Equipment Dept., Westinghouse Electric Corp.; N. S. MOTT, Chief Metallurgist, Cooper Alloy Corp.; E. A. SCHOEFER, Executive Vice President, Alloy Casting Institute; HAROLD C. TEMPLETON, Chief Metallurgist, Alloy Steel Products Co.

Table 1. Standard Designations and Composition Ranges for Corrosion-Resistant Steel Castings

ACI type(a)	Wrought alloy type(b)	Composition					
		C (max)	Mn (max)	Si (max)	Cr	Ni	Other(c)
CA-15	410	0.15	1.00	1.50	11.5 to 14	1 max	Mo 0.5 max(d)
CA-40	420	0.40	1.00	1.50	11.5 to 14	1 max	Mo 0.5 max(d)
CB-30	431	0.30	1.00	1.00	18 to 22	2 max	....
CC-50	446	0.50	1.00	1.00	26 to 30	4 max	....
CE-30	....	0.30	1.50	2.00	26 to 30	8 to 11	....
CF-3	....	0.03	1.50	2.00	18 to 21	8 to 11	....
CF-3M	....	0.03	1.50	1.50	18 to 21	9 to 12	Mo 2.0 to 3.0
CF-8	304	0.08	1.50	2.00	18 to 21	8 to 11	....
CF-20	302	0.20	1.50	2.00	18 to 21	8 to 11	....
CF-8M	316	0.08	1.50	1.50	18 to 21	9 to 12	Mo 2.0 to 3.0
CF-12M	316	0.12	1.50	1.50	18 to 21	9 to 12	Mo 2.0 to 3.0
CF-8C	347	0.08	1.50	2.00	18 to 21	9 to 12	Cb(e)
CF-16F	303	0.16	1.50	2.00	18 to 21	9 to 12	Mo, Se(f)
CG-8M	....	0.08	1.50	1.50	18 to 21	8 to 11	Mo 3.00 min
CH-20	309	0.20	1.50	2.00	22 to 26	12 to 15	....
CK-20	310	0.20	1.50	2.00	23 to 27	19 to 22	....
CN-7M	....	0.07	1.50	(g)	18 to 22	21 to 31	Mo-Cu(g)

(a) Most of these standard grades are covered by ASTM A296-55. (b) Type numbers of wrought alloys are listed only for nominal identification of corresponding wrought and cast grades. Composition ranges of the cast alloys are not the same as for the corresponding wrought alloys; cast alloy designations should be used for castings. (c) Phosphorus is 0.04% max except in CF-16F, which has 0.17% max; sulfur is 0.04% max in all grades. (d) Molybdenum not intentionally added. (e) Cb, 8 × C min, 1.0% max; or Cb-Ta, 10 × C min, 1.35% max. (f) Mo, 1.5% max; Se, 0.2 to 0.35%. (g) Several proprietary alloy compositions within the stated chromium and nickel ranges contain varying amounts of silicon, molybdenum and copper.

Heat Treatment

The optimum corrosion resistance of these 18-8 grades of stainless steel, which are essentially austenitic, is obtained by heating in the range from 1950 to 2050 F, and then quenching in water, oil or air to insure complete solution of carbides. Water is generally the accepted quenching medium; air is used only on relatively thin sections. Holding time at temperature will vary with thickness of the casting section, but should be long enough to heat all sections uniformly throughout. This process is known as solution quenching, solution annealing or quench annealing. If this process is omitted or is not done properly, or if the alloy is exposed in the range from 800 to 1600 F after solution quenching, complex chromium carbides may reform preferentially at grain boundaries. These carbides are attacked selectively in oxidizing solutions and will in time lead to failure by intergranular corrosion. These alloys are vulnerable to formation of both sigma and chi phases above 1200 F.

The extra-low-carbon grades, CF-3 and CF-3M, which contain 0.03% C max, are made in quantity for applications where heat treatment is impractical or fabrication by welding is required after machining. In these grades as cast, because of the low carbon content, the amount of chromium carbide present is insufficient for selective corrosive attack to occur. Therefore, these grades are relatively immune to intergranular corrosion failure.

The columbium-modified grade of 18-8, known as CF-8C, is produced for similar applications where heat treatment is impractical. This alloy in the as-cast condition has most of the carbon in the form of columbium carbide, thus precluding the precipitation of chromium carbide in the critical temperature range of 800 to 1600 F (particularly 1050 to 1200 F). For maximum resistance to intergranular attack, this alloy can be solution treated at 2050 F, followed by quenching to room temperature and then reheating to 1600 to 1700 F, where precipitation of columbium carbide occurs. An alternate method of treating is to solution treat at 2050 F, cool to the 1600 to 1700 F range, and then hold at this temperature before cooling to room temperature. For maximum corrosion resistance, it is not recommended that this alloy be solution treated without using the stabilizing treatment at 1600 to 1700 F as described.

The accepted test for evaluation of proper heat treatment is ten days in boiling 65% nitric acid. Maximum corrosion rate permitted in the test is 30 mils per year. Recently, a rapid qualitative etching test using oxalic acid has been developed for screening purposes. These tests are described in ASTM specifications A262-55T and A393-55T, respectively.

As with wrought grades, steels are available that contain selenium or sulfur (about 0.30%) to improve machinability. These are not widely used. The ability to machine the conventional CF-8 alloy has advanced to the point where the slight assistance from selenium or sulfur is generally not needed.

Alloys Containing Molybdenum

Alloys of the CF-8M type are modifications of CF-8, to which molybdenum has been added in amounts from 2 to 3%, to enhance the general corrosion resistance. Their passivity is more stable under weakly oxidizing conditions. They have good resistance to corrosive mediums such as sulfurous and acetic acids, and they are more resistant to pitting when exposed to mild chlorides. These alloys are widely used in sea-water service.

Structure of these alloys is essentially austenitic, with some ferrite (5 to 20%) distributed throughout the matrix in the form of discontinuous pools. In ordinary service where this steel may be heated in the range of 800 to 1200 F, there is an indication of carbide precipitation at the edges of the ferrite pools in preference to the austenite grain boundaries. It is claimed that this condition results in improved resistance of the steel to intergranular attack. In general, these alloys are not recommended for service temperatures above 1200 F. However, when they are heated in this temperature range, the ferrite pools can transform to chi or sigma phase. If these pools are distributed in such a way that some continuous network is formed, embrittlement of the steel or a network of corrosion penetration may result.

The amount of ferrite decreases as the carbon content increases with other elements unchanged. Thus, the low-carbon grade may be more strongly magnetic than the higher-carbon grade. Chromium, molybdenum, and silicon promote the formation of ferrite (magnetic), whereas carbon, nickel, nitrogen, and manganese favor the formation of austenite (nonmagnetic). It is inconsistent to specify the extra-low-carbon grade (0.03% C max) and at the same time request that the castings be completely nonmagnetic. This can be done, but it requires a minimum of 12 to 15% Ni.

The equivalent wrought grade (type 304L) normally contains about 13% Ni. It is intentionally made fully austenitic to get improved rolling and forging characteristics. However, the partially ferritic cast structure may have increased corrosion resistance under certain conditions.

Alloys With Higher Nickel

For some types of service iron-nickel-chromium grades in which the nickel content is higher than the chromium find extensive applications. Most important of this group is alloy CN-7M, with a nominal composition of 28 Ni, 18 Cr, 3.5 Cu, 2.5 Mo and 0.07 C max. In effect, 20% Ni and 3.5% Cu have been added to the CF-8M type alloy. This gives greatly improved resistance to hot concentrated, weakly oxidizing solutions such as sulfuric acid and also improves its resistance to severe oxidizing media. Alloys of this type will withstand all concentrations of sulfuric acid up to 150 F, and many up to 175 F. They are widely used in nitric-hydrofluoric pickling solutions, phosphoric acid, cold dilute hydrochloric acid, hot acetic acid, strong hot caustic solutions, brines, and many complex plating solutions and rayon spin baths.

It is easier to define the conditions for which these higher-nickel alloys are not satisfactory—generally those involving unstable chlorides. The influence of contaminants is one of the most important considerations in selecting an alloy for a particular process application. Ferric chloride in relatively small amounts, for example, will cause concentration cell corrosion and pitting even though this salt is not added intentionally. A buildup of corrosion products in a chloride solution may increase the iron concentration enough to be destructive. Thus, chlorine salts, unstable chlorinated organic compounds, or wet chlorine gas cannot be handled by any of the iron-base alloys. This creates a need for the nickel-base alloys dealt with in the nickel section of this Handbook.

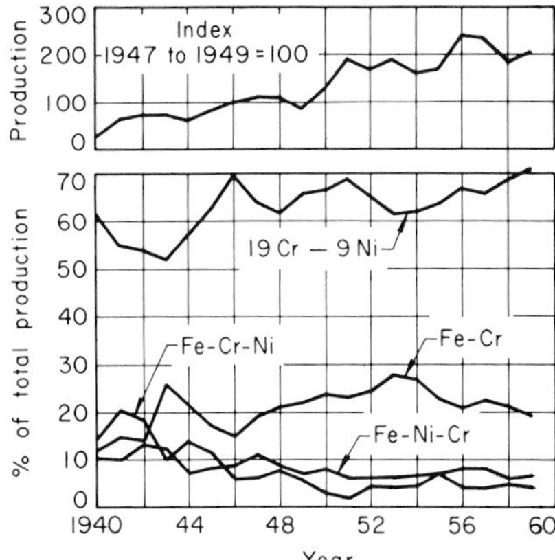

Fig. 1. Production of corrosion-resistant steel castings from 1940 to 1959

Figure 2 shows tensile data for 490 heats of four alloys, based on test specimens from separately cast test blocks representing metal poured into production castings. Tensile properties of the metal in the thicker and thinner sections of an intricate casting will be affected by the cooling rate in the mold, and the properties may differ from those produced in the test block.

A comparison of mechanical properties after heat treatment, for test bars taken from different locations in castings, is shown in Fig. 3 for two different stainless steels.

Corrosion Variables

In alloys of the CF type, the effects of composition on the rates of corrosion caused by *general attack* have been studied, and certain definite relationships have been established. Through use of the Huey test (5 to 48 hr in boiling 65% nitric acid), it has been shown that in this standardized environment, carbide-free quench-annealed alloys of varying nickel, chromium, silicon, carbon and manganese contents have corrosion rates directly re-

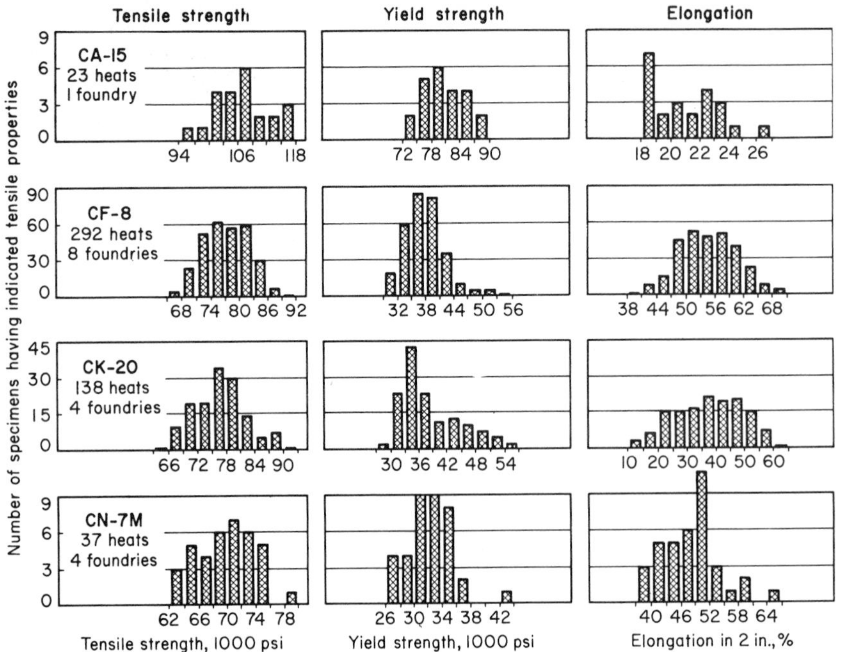

Fig. 2. Distribution of tension test results at room temperature for four types of corrosion-resistant casting alloys. Obtained from test blocks, these data indicate a range above and below the nominal values given for the heat treated alloys.

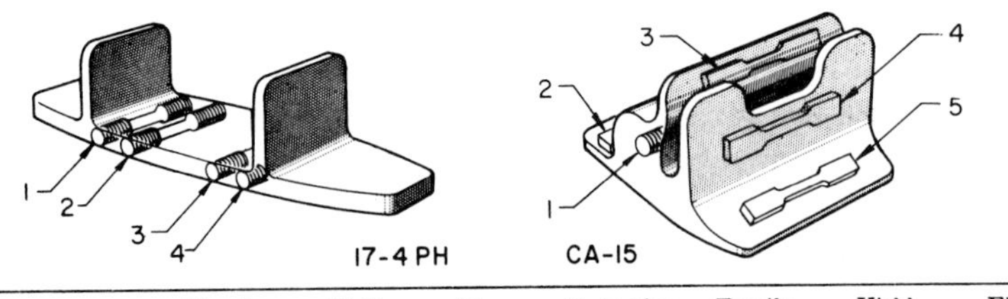

Test bar	Tensile strength, psi	Yield strength, psi	Elongation, %	Reduction of area, %	Tensile strength, psi	Yield strength, psi	Elongation, %
	17-4 PH(a)				CA-15(b)		
1	193,600	161,500	17	48.1	147,500	126,100	2
2	194,700	164,800	16	50.0	188,500	147,800	6
3	192,200	158,800	15	41.6	187,500	148,700	8
4	194,800	163,600	12	34.9	194,700	153,600	6
5	....	....	..	...	188,600	145,900	9

(a) Solution heat treated and aged to a hardness of Rockwell C 43. (b) Hardened and tempered to a hardness of Rockwell C 42.

Fig. 3. Variation in mechanical properties for different locations in a type CA-15 steel casting after heat treatment, and mechanical properties of four bars taken from the same section and same direction of a 17-4 PH casting solution treated and aged. Cast 17-4 PH is used extensively in the aircraft industry and has the following nominal composition: 0.07 max C, 1.00 max Mn, 1.00 max Si, 0.04 max P, 0.04 max S, 15.50 to 17.50 Cr, 3.0 to 5.0 Ni, 3.0 to 5.0 Cu, rem Fe.

lated to the amounts of the elements present in the composition.

Figure 4 shows the influence of each of the elements on corrosion rate. Variations of nickel, manganese and nitrogen contents for the ranges shown have relatively slight influence on the rate, but chromium, carbon and silicon have marked effects. The relationship of composition to corrosion rates for properly heat treated CF alloys in boiling 65% nitric acid is summarized in the nomograph of Fig. 5.

Intergranular attack may be avoided in the CF alloys *(a)* by the addition of the stabilizing element columbium, *(b)* by the use of extra-low-carbon grades such as CF-3 or CF-3M or *(c)* possibly by the formation of small amounts of ferrite that can be induced in the structure by reduction in amount of some of the austenite-stabilizing elements such as nickel or carbon, or by increase of such ferrite-stabilizing elements as molybdenum, silicon, or chromium. The addition of columbium to molybdenum-containing alloys of the CF compositions has been found unsatisfactory for castings.

In general, intergranular corrosion is of less concern in the straight chromium alloys, especially those containing 25% or more of chromium.

Simulated Service Tests. Results of plant tests on CF-8, CF-8M and CN-7M alloys that were made to simulate service tests are shown in Table 2. These tests illustrate the specific effect of molybdenum on 19 Cr – 9 Ni alloys in reducing selective attack and pitting, and the over-all corrosion rate computed from loss in weight. The effect of the higher nickel plus copper and molybdenum in the CN-7M alloy is to reduce the rate of corrosion as compared with the CF-8M alloy.

The austenitic steel used in equipment handling sulfite pulp requires a controlled molybdenum content for optimum serviceability. It varies somewhat with the composition of the acid cooking solution, but a

Table 2. Plant Corrosion Tests on CF-8, CF-8M and CN-7M Alloys

Type and composition of corroding solution	Temperature of solution, F	Alloy	Metal loss on surface, mpy	Surface condition by visual examination	Remarks
Neutralizer after formation of ammonium sulfate Ammonium sulfate plus small excess of sulfuric acid, ammonia vapor and steam.	212	CF-8 CF-8M CN-7M	26.2 1.1 0.7	Very heavy etch (a) Light tarnish (b) Bright	CF-8M was installed for low corrosion tolerance equipment in this service and performed satisfactorily.
Settling tank after neutralizer................. Ammonium sulfate plus excess of sulfuric acid.	122	CF-8 CF-8M CN-7M	15.2 0.4 0.1	Very heavy etch (a) Slight tarnish Bright (b)	CF-8 in service showed excessive corrosion rate plus heavy solution cell attack.
Ammonium sulfate processing................. Ammonium sulfate; pH is 8.0	122	CF-8 CF-8M CN-7M	27.0 6.8 2.0	Heavy etch Moderate etch Light etch	CF-8M had too high a corrosion rate in service for good valve life, although suitable for equipment of greater corrosion tolerance. CN-7M was installed in this service.
99 to 100% fuming nitric acid................	68	CF-8 CN-7M CF-8M	9.6 3.1 13.5	Moderate etch Light etch Moderate etch	CF-8 was satisfactory except for low tolerance equipment such as valves. CN-7M valves performed satisfactorily in service.
Saturated. Sodium chloride plus............... 15% sodium sulfate; pH is 4.5	140	CF-8M CF-8	0.1 9.5	Bright Concentration cell corrosion at various small areas of specimen	CF-8M was installed for valves in service.

(a) Solution cell attack under insulating washer. (b) Slight solution cell attack under insulating washer.

molybdenum content within the range of 2.25 to 2.5% is desirable. In one application where the chloride content of the cooking solution was higher than normal because the logs had been floated on salt water, pitting was encountered in valves made of CF-8M alloy containing 2.15% Mo. For the same installation, it was found that the wrought evaporator heater tubes of type 316 required a minimum molybdenum content of 2.75%.

Molybdenum may produce a detrimental catalytic reaction. For example, the residual molybdenum in CF-8 alloy must be held below 0.5% in the presence of hydrazine.

Influence of Heat Treatment. Stainless steel castings are almost always heat treated. For the hardenable ferritic-martensitic straight chromium compositions the heat treatment is primarily for the purpose of obtaining desired mechanical properties. The 12% Cr alloy castings are air cooled or oil quenched from about 1800 F and tempered between 1000 and 1400 F, depending on the specified properties. Tempering between 700 and 1000 F causes a marked loss of impact resistance and should be avoided.

Austenitic alloys such as 19 Cr – 9 Ni are solution heat treated for corrosion resistance, especially the intergranular type. However, the beneficial effects of solution heat treatment are often removed by welding, with resulting impairment of corrosion resistance in the region of the weld. This difficulty can be avoided by using the 0.03% max C or columbium-stabilized grades.

The relationship between rates of corrosion in 65% boiling nitric acid, the quenching temperatures and the carbon, silicon, and molybdenum contents, respectively, of a series of 19 Cr – 9 Ni alloys is shown in Fig. 6. The solid curves on these charts form the boundary between alloys with constant rates of corrosion and those with rates that increase with time. All alloys to the left and above the broken curves have average rates of corrosion slower than 2.5 mils per month, as measured by five 48-hr test periods in boiling nitric acid. It is advisable to select a quenching temperature at least 50 °F above the solid line

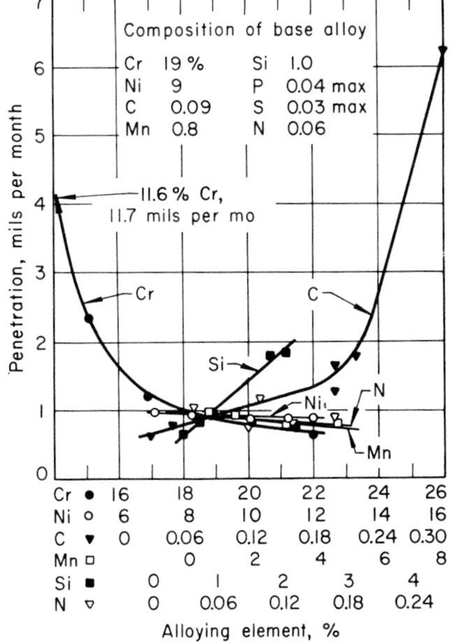

Fig. 4. Effect of each element in a 19 Cr – 9 Ni casting alloy on rate of corrosion penetration as determined from quench-annealed specimens in boiling 65% nitric acid

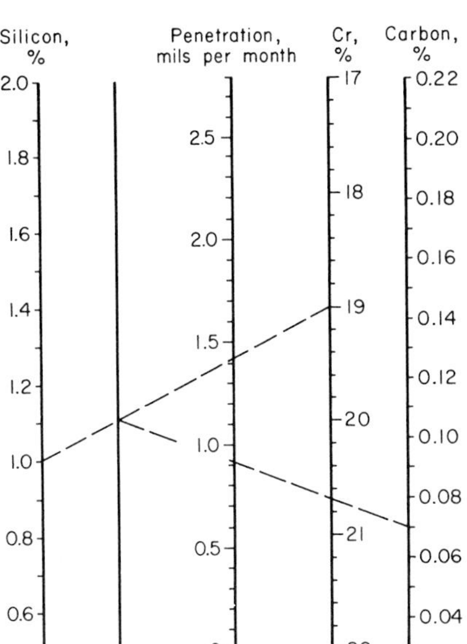

Fig. 5. Nomograph for determining rate of corrosion penetration in boiling 65% nitric acid for quench-annealed type CF alloys

Base composition: 19 Cr, 9 Ni, 0.80 Mn, 1.0 Si, 0.02 S, 0.02 P, 0.06 N

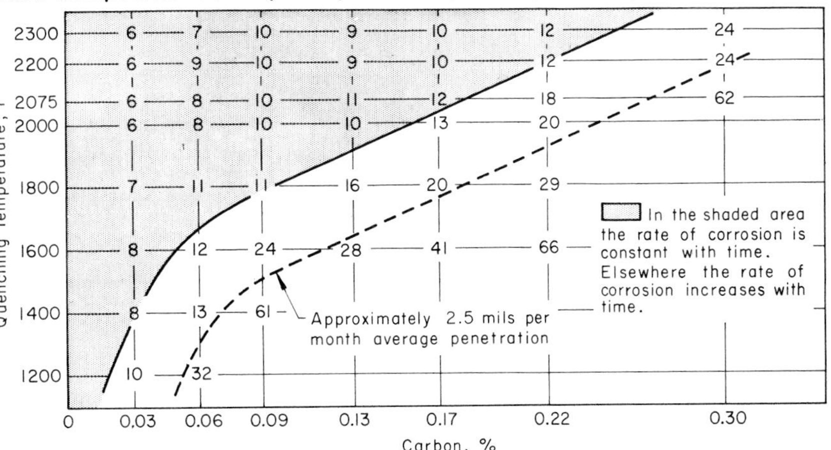

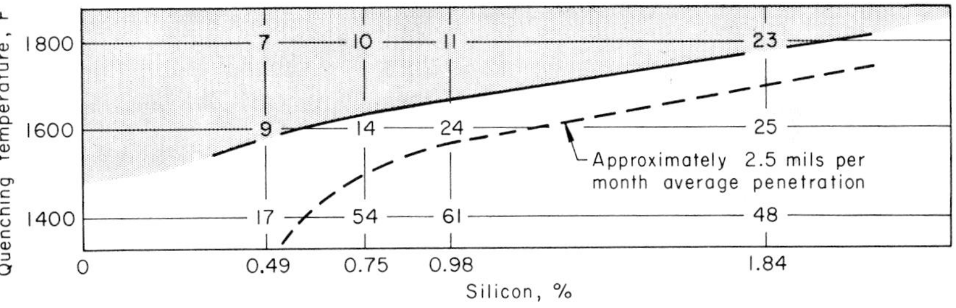

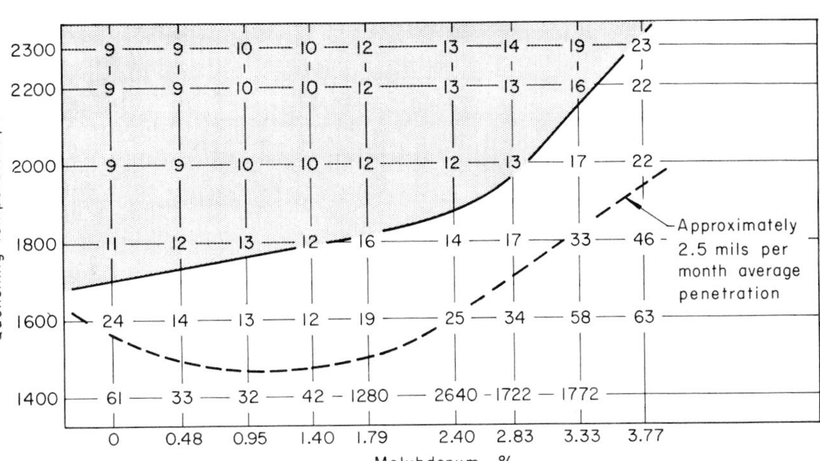

Fig. 6. Effect of carbon, silicon and molybdenum contents on resistance of 19 Cr – 9 Ni steel castings to corrosion in boiling 65% nitric acid. Specimens were quenched in water after ½ hr at the temperatures indicated. Numbers on the graphs indicate the penetration in ten-thousandths of an inch per month, calculated from the average loss in weight during five 48-hr periods in boiling nitric acid. In the shaded areas the rate of corrosion is constant with time; elsewhere the rate of corrosion increases with time.

for the composition under consideration. In no instance should the heat treating temperature be below 1950 F for maximum resistance to intergranular corrosion.

Where the usual quench-annealing practice is difficult or impossible, holding for 24 to 48 hr at 1600 to 1800 F and air cooling is helpful in improving resistance of castings to intergranular corrosion. However, except for alloys of very low carbon content and castings with thin sections, this treatment fails to produce material with as good resistance to intergranular corrosion as properly quench-annealed material.

As noted earlier, resistance to intergranular penetration for the 19 Cr – 9 Ni alloys can be increased by having small amounts of ferrite in the structure. Control of the ferrite content is achieved by balancing the chemical composition.

Corrosion resistance of the 19 Cr – 9 Ni alloys is also influenced by the ferrite content, which is controlled by properly balancing the chemical composition. The Schaeffler diagram (Fig. 7) is generally used for calculating the ferrite content from the chemical composition. The calculated amount of ferrite can be in error by 4%. Since the alloy becomes increasingly ferromagnetic with increase in concentration of ferrite, the calculations may be supplemented by a measurement of the magnetic properties and comparison with standards.

Magnetic properties of high-alloy castings depend on their microstructure. The straight chromium types, CA, CB and CC, are ferritic and ferromagnetic. All the other grades are mainly austenitic, with or without minor amounts of ferrite, and are either weakly magnetic or wholly nonmagnetic.

The tensile properties of 19 Cr – 9 Ni steels are affected appreciably by the percentage of ferrite in the alloy. The completely austenitic alloys have the lowest strength; strength increases with an increase in the ferrite content (Table 3) and is accompanied by a loss of ductility. On the other hand, strength and ductility of these alloys can be affected by the amount of carbon and the distribution of the carbides. As a result, alloys with carbon content up to 0.08% and low ferrite content can, after heat treatment, have higher strength or lower ductility, or both, than alloys with similar carbon content but higher ferrite content.

At temperatures above 1000 F, austenite has better creep resistance than ferrite. The weaker ferrite phase may give better plasticity to the alloy but after long exposure it transforms to sigma or chi phase, which reduces resistance to impact, although in some instances sigma or chi phase is deliberately formed to increase the strength of the austenite. Austenite can transform directly to sigma or chi without going through the ferrite phase.

The criterion for low-temperature service is a minimum Charpy keyhole impact value of 15 ft-lb. As the ferrite in the cast structure increases, the impact strength decreases (Table 4).

Test data on the effect of ferrite in the 19 Cr – 9 Ni alloy on the resistance to general corrosion attack are meager, although it appears from isolated observations that the presence of ferrite neither impairs nor benefits corrosion resistance. It is indicated that in most environments corrosion resistance is determined by the alloy content.

Calcium chloride solutions attack the austenite phase, whereas a cornstarch solution of 10° Baumé and pH of 1.8, with sulfuric acid at 275 F attacks the ferrite in heat treated alloys.

Intergranular corrosion may occur in corrosive mediums because of carbide precipitation at the austenite grain boundaries. This could result from detrimental heat produced in welding operations or from any form of heating in the temperature range from 1000 to 1600 F. Since carbides precipitate preferentially in ferrite pools rather than at the grain boundaries, this form of corrosion may be controlled through the inclusion of 5 to 20% of ferrite phase in the austenitic matrix. Corrosion now takes place at the edges of these separated pools of ferrite, which is less detrimental than at grain boundaries. If the amount of ferrite is too great, it may form continuous stringers along which corrosion can take place, producing a condition similar to grain boundary attack.

Where castings are to be welded without subsequent heat treatment, the injurious effect of carbide precipitation that results from the heat of welding, can be eliminated by addition of columbium to the alloy. Carbon in the alloy combines preferentially with the columbium, so

Table 3. Effect of Ferrite on Mechanical Properties of 19 Cr – 9 Ni Alloy

Property	Ferrite 3%	10%	20%	41%
Tested at Room Temperature				
Tensile strength, psi	67,400	72,200	84,700	91,880
Yield strength, 0.2% offset, psi	31,300	34,000	43,000	48,000
Elongation in 2 in., %	60.5	61	53.5	45.5
Reduction of area, %	64.2	73	58.5	47.9
Tested at 670 F				
Tensile strength, psi	49,130	50,750	66,300	70,750
Yield strength, 0.2% offset, psi	15,130	15,840	26,500	27,250
Elongation in 2 in., %	45.5	43	36.5	33.8
Reduction of area, %	63.2	69.7	47.5	49.4

Table 4. Effect of Ferrite Content on Charpy Impact Strength at –320 F

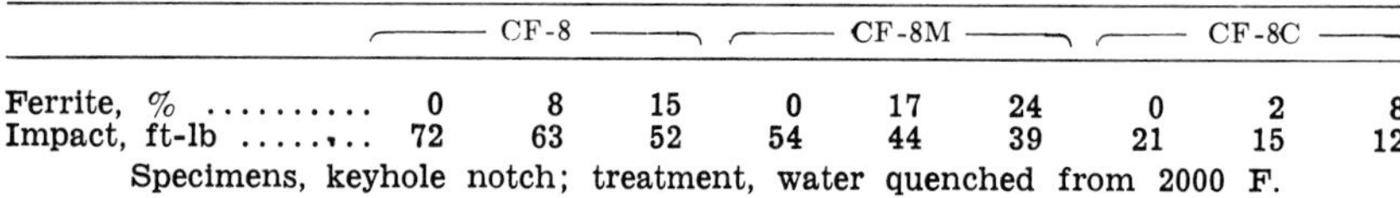

	CF-8			CF-8M			CF-8C		
Ferrite, %	0	8	15	0	17	24	0	2	8
Impact, ft-lb	72	63	52	54	44	39	21	15	12

Specimens, keyhole notch; treatment, water quenched from 2000 F.

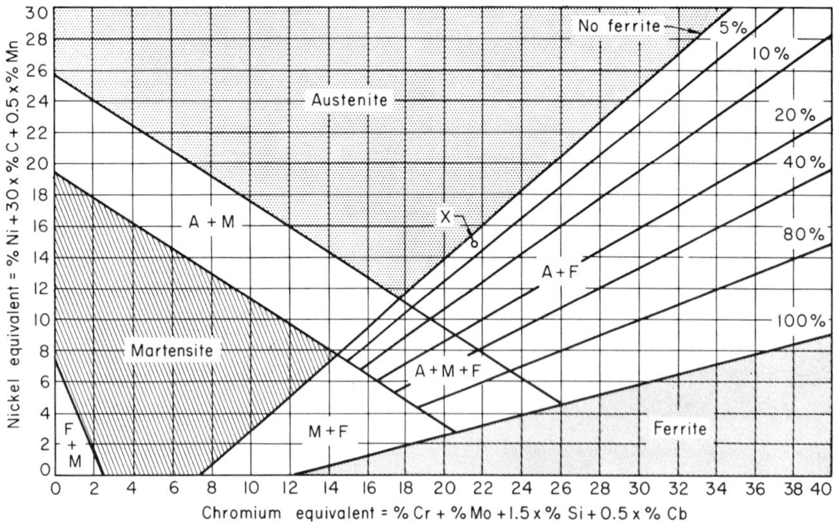

Example: Point X on the diagram indicates the equivalent composition of a type 318 (type 316 with columbium) weld deposit containing 0.07 C, 1.55 Mn, 0.57 Si, 18.02 Cr, 11.87 Ni, 2.16 Mo and 0.80 Cb. Each of these percentages was multiplied by the "potency factor" indicated for the element in question along the axes of the diagram, in order to determine the chromium equivalent and the nickel equivalent. When these were plotted, as point X, the constitution of the weld was indicated as austenite plus from 0 to 5% ferrite. Magnetic analysis of the actual sample revealed an average ferrite content of 2%. For austenite-plus-ferrite structures, the diagram predicts the percentage ferrite within 4% for the following corrosion-resistant steels: CH-20, CH-20C, CK-20, CF-12M, and CF-8MC.

Fig. 7. Schaeffler diagram for estimating ferrite content from composition

Table 5. Magnetic Permeability of Thick and Thin Sections of CF-8 Castings Correlated with Ferrite Content

Thick section (3 in.)		Thin section (¾ in.)	
Magnetic permeability	Ferrite, %	Magnetic permeability	Ferrite, %
1.005	0	1.005	0
1.025	0.4	1.010	0
1.105	2.3	1.084	0.7
1.615	7.5	1.150	1.8
1.665	7	1.316	3.2
2.315	10	1.659	8
3.753	19	2.829	16.2

columbium carbides rather than chromium carbides are precipitated. Depletion of chromium at the grain boundaries is prevented by this action; therefore, intergranular corrosion does not occur. Another means of minimizing carbide precipitation is to use an extra-low carbon grade such as CF-3.

Columbium-containing alloys that have been heated to sensitizing temperatures around 1200 F are not susceptible to intergranular corrosion. They are more susceptible to over-all corrosion when tested in nitric acid, compared with the columbium-free, quench-annealed alloys of the same nickel, chromium and carbon content.

Weld crack sensitivity of CF alloy containing columbium (CF-8C) is more pronounced in the fully austenitic grade, and may be alleviated through the introduction into the weld deposit of a small amount of ferrite, usually between 4 and 10%. However, appreciable amounts of ferrite in columbium-bearing corrosion-resistant steels will transform, at least partially, to the sigma or chi phase on heating between 1000 and 1700 F.

In weld deposits the presence of sigma or chi phase is extremely detrimental to ductility. When welding for service at room temperature or up to 1000 F, 4 to 10% of ferrite may be present and will greatly reduce the tendency to weld cracking. However, when service conditions are raised to between 1000 and 1500 F, the amount of ferrite in the weld must be reduced to less than 5% to avoid embrittlement from excessive sigma or chi phase.

Magnetic Permeability. Corrosion-resistant steel castings for nonmagnetic parts for such applications as radar and mine sweepers require close control of the ferrite content. Thicker sections of the casting have a higher permeability than the thinner sections, and magnetic permeability checks should be made on the former. For data on thick and thin sections, see Table 5.

Design

Corrosion-resistant steel castings have a somewhat lower modulus of elasticity in tension than carbon or low-alloy steel, varying from about 24 million psi for the high-nickel grades to 29 million psi for the high-chromium low-nickel alloys. This compares with 30 million psi for carbon and low-alloy steel. Density of the alloys is very similar to that of steel. The specific heat of steel, 0.12 Btu per lb per °F, is bracketed by the casting alloy values of 0.11 to 0.14.

Galling. Stainless steel castings are not resistant to galling and seizing. The surfaces of the castings can be nitrided so that they are hard and wear-resistant. Tensile properties are not impaired. The 19 Cr – 9 Ni and straight chromium alloys, after nitriding, are resistant to superheated steam (1500 psi at 930 F), saturated steam, boiler feed water and petroleum-base fuels. They are not resistant to halogen acids or salts, nor to any corrosive medium that will attack the untreated alloy. Nitriding reduces resistance to corrosion by concentrated nitric or mixed acids.

Table 6. Patternmakers' Shrinkage for Cast Corrosion-Resistant Alloys(a)

ACI type	Shrinkage allowance, in. per ft
CC-50	7/32
CA-15, CA-40, CB-30	1/4
CE-30, CF-8, CF-8M, CF-12M, CF-16F	5/16
CH-20, CK-20, CN-7M, CF-8C, CF-20	11/32

(a) These values are for unhindered contraction; considerable variation may occur, depending on casting shape.

Parts such as gate disks for gate valves and plugs for plug valves are usually furnished in the solution-treated condition but may be nitrided to reduce susceptibility to seizure in service. Similar results are obtained by hard facing with Co-Cr-W alloys.

Design considerations for corrosion-resistant steel castings do not differ greatly from those for low-alloy and carbon steel castings. (See section on design in the article on Steel Castings, page 122.)

Pattern design is an essential consideration. Loose wood or metal patterns are subjected to damage from warping and mechanical abuse and should be considered only for large castings or when a few small or experimental parts are being made. Mounted patterns, matchplate patterns, or separate cope and drag patterns are essential to produce dimensionally accurate castings of high quality. The surface to be machined should be in the drag half of the mold if possible.

Minimum section thickness should be held to 3/16 in. or greater if possible. Somewhat lighter sections are feasible for most of the alloys, depending on pattern equipment and casting design, but some difficulty is experienced in running thin sections in the straight-chromium alloys of the CA, CB and CC types. Where intricate designs are involved, the greater fluidity of austenitic chromium-nickel grades is advantageous. Designs requiring appreciable changes in section should be avoided.

Dimensional tolerances depend on pattern equipment and configuration. In general, over-all dimensions and location of cored holes can be held to 1/16 in. per ft. Solidification shrinkage is compensated for as noted in Table 6; the figures apply only to unhindered contraction. Finish allowances of ⅛ in. or more on surfaces to be machined should be considered in the casting design.

Because varying amounts of resistance to free contraction of the castings are offered by the shape of the mold, a single shrinkage allowance is necessarily a compromise. Extreme demand for dimensional accuracy may require that several different shrinkage allowances be used on a single pattern. Furthermore, an entirely different pattern may be needed to produce the same casting design if different molding methods are employed.

Areas of low density, called shrinkage, from inadequate feeding during solidification, will show accelerated attack in a corrosive environment. This is particularly true in rotating parts such as pump impeller castings or pump casings subjected to erosion.

Performance of valves for corrosion service has been improved by

Table 7. Speeds and Feeds for Machining Corrosion-Resistant Steel Castings

Operation	Approximate feed, ipr	Machining speeds, sfm								
		CF-20 CF-8	CF-16F	CE-30 CF-8M CF-12M CH-20 CK-20	CF-8C	CN-7M	CA-15	CA-40	CB-30	CC-50
Broaching	0.001 to 0.005	8 to 15	10 to 20	8 to 15	8 to 15	8 to 15	10 to 20	8 to 15	10 to 20	10 to 20
Tapping	0.003 to 0.007	10 to 20	15 to 30	10 to 25	10 to 25	12 to 20	10 to 25	10 to 20	10 to 25	10 to 25
Threading	0.003 to 0.008	10 to 20	10 to 25	10 to 25	10 to 25	10 to 20	10 to 20	10 to 20	10 to 25	10 to 25
Reaming	0.003 to 0.008	20 to 60	30 to 100	40 to 80	40 to 80	20 to 60	20 to 60	20 to 60	40 to 120	40 to 120
Drilling	0.003 to 0.007	15 to 40	35 to 85	30 to 50	30 to 50	30 to 60	35 to 75	30 to 60	40 to 60	40 to 60
Turret lathe	0.003 to 0.008	60 to 90	90 to 130	60 to 80	60 to 90	60 to 80	80 to 110	60 to 100	70 to 100	60 to 100
Milling	0.003 to 0.008	35 to 65	75 to 110	30 to 50	40 to 60	35 to 70	70 to 105	35 to 70	40 to 60	40 to 60
Turning, boring	0.003 to 0.008	40 to 85	85 to 120	60 to 120	60 to 120	60 to 80	80 to 115	40 to 80	60 to 100	60 to 120
Screw machine	0.003 to 0.008	60 to 90	90 to 130	60 to 80	60 to 90	60 to 80	80 to 110	60 to 100	70 to 100	60 to 100
Hack sawing	Use a coarse-tooth blade (not over 10 teeth per in.) at about 50 strokes per minute with positive pressure.									

Table 8. Welding of Corrosion-Resistant Alloys (a)

ACI designation	Type of electrodes used (b)	Preheat	Post-treatment
CA-15	410	400 to 600 F	1125 to 1400 F, air cool
CA-40	420	400 to 600 F	1125 to 1400 F, air cool
CB-30	431	600 to 800 F	1450 F min, air cool
CC-50	446	350 to 400 F	1650 F, air cool
CE-30	Same composition	Not required	Quench from 1950 to 2050 F
CF-8	304	Not required	Quench from 1950 to 2050 F
CF-8C	347	Not required	Quench from 1950 to 2050 F
CF-8M	316	Not required	Quench from 1950 to 2100 F
CF-12M	316	Not required	Quench from 1950 to 2100 F
CF-16F	304 or 303	Not required	Quench from 2000 to 2100 F
CF-20	302	Not required	Quench from 2000 to 2100 F
CH-20	309	Not required	Quench from 2000 to 2100 F
CK-20	310	Not required	Quench from 2000 to 2150 F
CN-7M	Same composition	400 F	Quench from 2000 F

(a) Metal-arc, inert-gas arc and oxyacetylene gas welding methods can be used, but oxyacetylene gas welding is not recommended because of possible impairment of corrosion resistance caused by carbon pickup. The following table lists suggested electrical settings and electrode sizes for various section thicknesses:

Section thickness, in.	Electrode diam, in.	Current, amp	Max arc volts
⅛ to ¼	3/32	45 to 70	24
⅛ to ¼	⅛	70 to 105	25
⅛ to ¼	5/32	100 to 140	25
¼ to ½	3/16	130 to 180	26
½ and over	¼	210 to 290	27

(b) Lime-coated electrodes are recommended.

changes in design which reduce the probability of solution cell or crevice attack. Threaded connections produce a crevice between the pipe line and the valve body which is subject to crevice corrosion. Tight threaded joints are difficult to obtain with austenitic steels because they have a tendency to gall on tightening. The flanged type of connection avoids threads and increases the life of the joint. The valve seat is another source of crevice corrosion when a renewable seat ring is used in the valve body. The normal valve for handling corrosive mediums is made with an integral seat ring cast with the body (Fig. 8).

Machining. The machinability of straight chromium alloys is as good or better than that of annealed 19% Cr – 9% Ni type alloys. The machining characteristics of all the iron-chromium-nickel type alloys (chromium in excess of nickel) are about on a par with the quench-annealed CF composition. The CE type and alloys that contain columbium are somewhat easier to machine; CH is slightly less machinable than CF. More detailed information for several alloys and operations is given in Table 7.

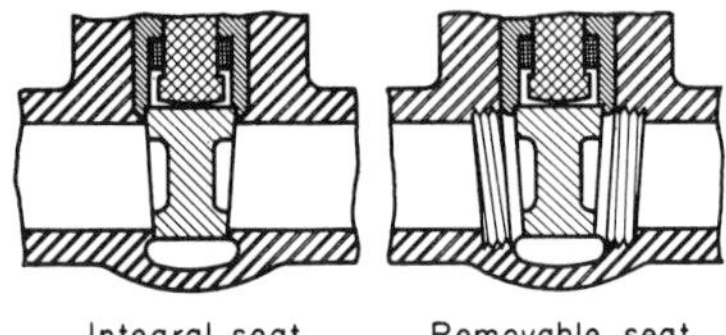

Fig. 8. Valve with integral seat (left) *satisfactorily handles corrosive mediums with less probability of crevice corrosion than valve with removable seat* (right)

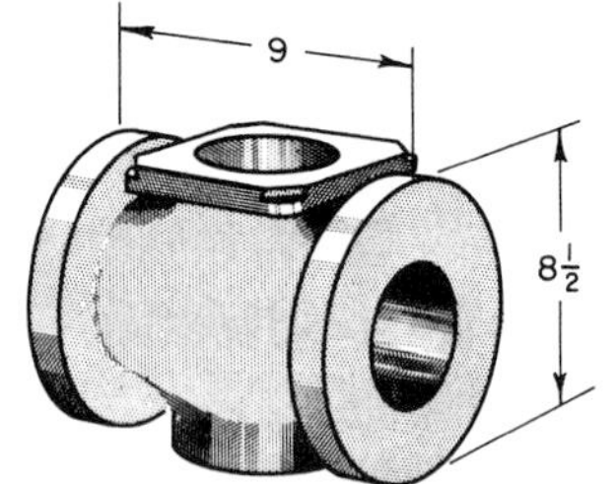

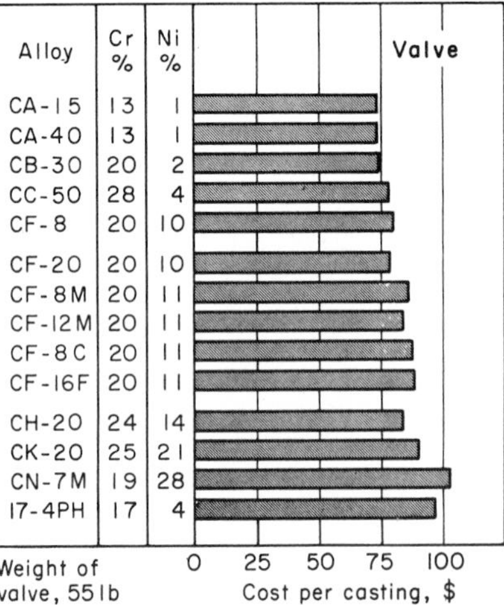

Fig. 9. Influence of grade of alloy on cost of a specific valve body casting in quantities of 50. Pattern cost is not included in cost per casting.

Welding. Corrosion-resistant steel castings can be welded by metal-arc, inert-gas and oxyacetylene methods. Oxyacetylene welding causes carbon pickup with impairment of corrosion resistance. Austenitic castings are normally welded without preheat, and are solution annealed after welding. Martensitic castings require preheating to avoid cracking during welding, and are given an appropriate post-heat treatment. Specific considerations for welding the individual alloys are dealt with in Table 8.

When the welds are properly made, tensile and yield strengths of the welded joint are similar to those of the castings. The elongation is generally lower for specimens taken perpendicular to the seam. The tests recorded in Table 9 were from a peripheral weld in a cylinder approximately 1½ in. thick. The specimens were machined with longitudinal axes perpendicular to the welded seam and with the seam at the middle of the gage length.

Cost of corrosion-resistant steel castings generally increases as the alloy content increases, ranging from CA-15, which contains a total of about 14% Cr and Ni, to CN-7M, which contains a total of about 50% Cr and Ni. However, many operations in making a specific casting will cost about the same; thus the cost of producing castings of the same size and weight from different alloys will vary less than the basic alloy costs, which may differ by 100% or more. This is illustrated by the data of Fig. 9, which show the influence of grade and alloy content on the cost per casting for a specific 55-lb valve body purchased in lots of 50. This casting was not actually produced in all of the 14 alloys listed, but the costs shown are quotations from a single producer. Here it is seen that the difference in casting cost is only about 30% between the lowest and highest alloy. For larger quantities (250 or more) the cost per casting would decrease by about 3%.

Table 9. Short-Time Tensile Properties of Welded Cylinders Fabricated from CF-8 Cast Alloy

	Base metal						Welded joint				
		Testing temperature									
Property	Keel block (a)	Room	600 F	800 F	1000 F	1100 F	Room	600 F	800 F	1000 F	1100 F
Tensile strength, psi	72,500	72,500	47,750	49,250	42,250	40.400	71,000	49,500	51,500	47,250	39,375
Yield strength, 0.2% offset, psi	34,500	37,750	24,500	24,250	20,250	18,750	35,750	28,750	24,750	27,250	19,500
Proportional limit, psi (b)	...	26,000	13,000	8,500	8,000	6,500	21,500	10,500	10,000	9,000	8,000
Reduction of area, %	59	62.1	54.9	58.6	60.8	59.1	70.8	58.3	46.3	62.8	70.4
Elongation in 2 in., %	49	58.0	33.5	37.5	32.5	38.0	42.0	15.5	24.5	23.5	31.0
Modulus of elasticity, million psi (b)	...	27.0	22.0	19.5	17.0	16.0	27.0	22.0	19.0	16.5	15.5
Location	...	...	...	...	...	...	Base metal	Base metal	Base metal	Base metal	Base metal

(a) Separately cast from the same heat as cylinder. (b) Values for proportional limit and modulus of elasticity at elevated temperatures are apparent values since creep occurs.

Stainless Pumps and Valves Play Vital Part in BWR Plants

Corrosion resistance is a critical requirement. Particulate matter due to corrosion in a system must be at an absolute minimum when a plant is shut down. Also, smooth contours are needed for complex shapes.

STAINLESS STEEL castings for critical fuel-handling components have been a vital factor in the impressive growth of boiling water reactor (BWR) nuclear power plants as economically competitive power sources.

Since the first BWR plant went into operation in 1957, 29 others have gone on line all over the world (Fig. 1). An additional 68 are now planned or under construction.

Requirements — Stringent water clarity specifications are of vital importance for critical BWR components such as those handling demineralized water in the recirculation system. The amount of particulate matter due to corrosion in the system must be at an absolute minimum when the plant is shut down.

Stainless steel provides the corrosion resistance as well as the strength required, but for complex-shaped fluid-handling components such as recirculation system pumps, the production of smoother contours becomes a major consideration. Here the use of stainless alloys in the cast form allows the economical production of complex parts without the disadvantages of wrought-assembled construction.

By producing such shapes as one-piece units, the possibility of corrosion due to "nooks and crannies" is avoided. In addition, stainless castings give assurance of high reliability and long, troublefree service life.

Materials — Several cast stainless alloys are used: types CA-15 (nominal composition 12 Cr, 1 Ni) and CA-6NM (13 Cr, 4 Ni) are hardenable high-strength alloys containing sufficient chromium to prevent significant corrosion. Types CF-8 (19 Cr, 9 Ni) and CF-8M (19 Cr, 9 Ni, 2.5 Mo) offer superior corrosion resistance.

The ASME Nuclear Power Code specifies higher allowable stresses for the CF-8M alloy; this grade is most often specified for pumps and valves in nuclear power plants since it permits the use of somewhat thinner walls.

Pumps — Numerous pumps are utilized in the reactor plant. Feedwater, which has been condensed and demineralized after passing through the turbine, is heated and pumped back to the reactor vessel. Feedwater pumps contain cast stainless parts.

Burns & Roe Inc. reports that the Oyster Creek installation (Toms River, N.J.) has feedwater pumps able to supply 0.315 m^3/s (5000 gpm) against an 853 m (2800 ft) pressure head formed by the

Fig. 1 — Cast stainless steel volute weighing about 6804 kg (15 000 lb) is being finished machined. It's for a double-suction recirculation pump with a capacity of 2.1 m^3/s (34 200 gpm) used in a 322 MW BWR plant designed by General Electric Co. for the Japan Atomic Power Co., Tsuruga, Japan. Credit: Byron Jackson Pump Div., Borg-Warner.

Fig. 2 — Jet pump, which is enclosed within a BWR reactor vessel, recycles water through a demineralizer. Critical parts are cast of type CF-8 alloy. Credit: Atomic Power Equipment Dept., General Electric Co.

boiling water in the reactor. Impellers, shaft sleeves, and stuffing box parts are cast of type CA-15, heat treated to Rc max.

At Oyster Creek, the systems supplying cooling water to the condensers contain four 27.4 mm (81 in.) water pumps, each of which delivers 7.245 m^3/s (115 000 gpm). These pumps contain impellers cast of CF-8M, and wear rings cast of CA-15.

With the construction of the Dresden 2 plant, operated by Commonwealth Edison of Illinois, reactor vessels began to include an innovative jet pump which has no moving parts (Fig. 2). The pump receives input energy in the form of a fluid stream from outside the reactor and utilizes it to force circulation of the water inside the reactor vessel. A complex weld-assembled unit, the jet pump has critical parts cast of type CF-8 stainless. Elbows, tees, and tapered fittings and sleeves were also cast because of their economic advantages.

Recirculation pumps, which keep water within the reactor circulating by supplying water to the jet pumps and which recycle reactor through a demineralizer, are also stainless. Type CF-8M is used here for all wetted parts. Because recirculation pumps must operate indefinitely without maintenance, pump castings are given rigorous tests that simulate performance conditions (Fig. 3), as well as radiographic and other nondestructive testing procedures.

Valves — Valves are used throughout the plant to provide control over water and steam routing. Valve types used include gate (Fig. 4), globe, check and plus. All have type CF-8M castings for critical parts such as body, bonnet, cap, disc, plug and seat.

The stainless valves used vary in size from 12.7 to 711 mm (½ to 28 in.). They are installed in the cleanup demineralizer system, in the shutdown cooling system, and in the core spray system. The control-rod drive provides reactor-control maneuvers during normal operation and rapid insertion of the neutron-absorbing control rods when a potentially unsafe condition arises; because the valves in this system must function perfectly at all times, stainless is a must.

Similarly, the shutdown cooling system removes decay heat from the reactor water during shutdown, so perfect performance of these valves is essential. No less critical, valves in the core spray system must be certain to open to prevent reactor core meltdown under postulated accident conditions involving loss of coolant.

Many of these valves, designed to operate up to 9 MPa (1250 psig) and 283 C (575 F), are made of cast stainless to insure reliability and long life.

Valve parts are extensively tested prior to installation. Radiographic and liquid penetrant examination of all castings larger than 21.7 mm (2½ in.) insure that they will meet ASME Nuclear Code requirements. For valves rated at 272 kg (600 lb) and higher, seat leakage tests make sure they do not permit more than 2 cc per h per 25.4 mm (1 in.) of diameter to pass the fully closed valve.

Fig. 3 — Cast CF-8M recirculating pump undergoing simulated temperature and pressure tests. It'll be used by Caroline Power & Light Co. in one of its Brunswick BWR plants. Credit: Bingham-Willamette Co.

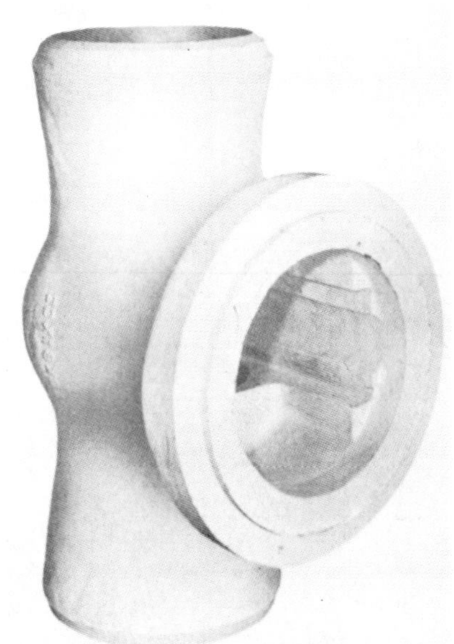

Fig. 4 — Cast CF-8M alloy gate valve performs vital function in BWR reactor. Castings must pass stringent radiography and dye penetrant inspections. Credit: Darling Valve & Mfg. Co.

CA-6NM Cast Stainless

By WILLIAM H. RICE and LEROY E. FINCH

An alternative to CA-15 solves problems in casting, reduces tendency toward cracking, makes it possible to cast high-strength complex shapes, and often improves mechanical, corrosion, and other properties.

An improved cast martensitic stainless steel designated CA-6NM promises to provide the solution to a costly problem faced by producers of the widely used CA-15 alloy included in ASTM A-296 and A-351. Because of its metallurgical nature and composition balance, CA-15 has tended to present, in some instances, problems in casting; and its tendency toward cracking frequently requires that castings be repair welded. CA-6NM, a 13Cr-4Ni alloy, provides better casting behavior, improved weldability, and mechanical, corrosion, and cavitation resistance properties equal to or better than those of CA-15, CA-6NM is also included in ASTM A-296 and A-351.

Chemical composition of the alloy is: 0.06 C max, 1.50 Mn max, 1.00 Si max, 11.50-14.00 Cr, 3.50-4.50 Ni, 0.40-1.00 Mo, 0.04 S max, 0.04 P max.

Fig. 1 — Pump housing made of CA-6NM.

Fig. 2 — Several impellers cast in CA-6NM for the Edmonston Plant, Los Angeles Water System.

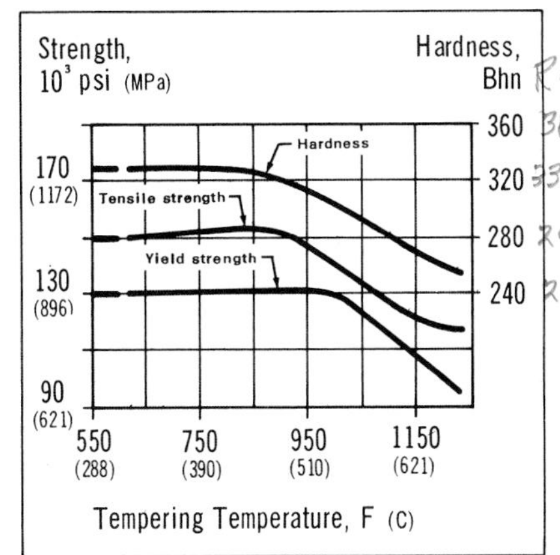
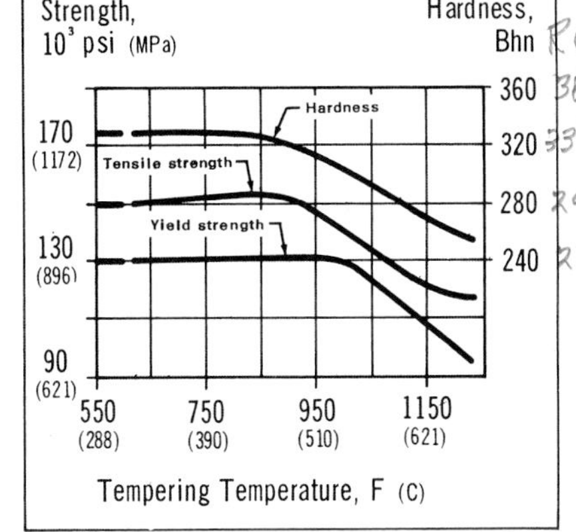

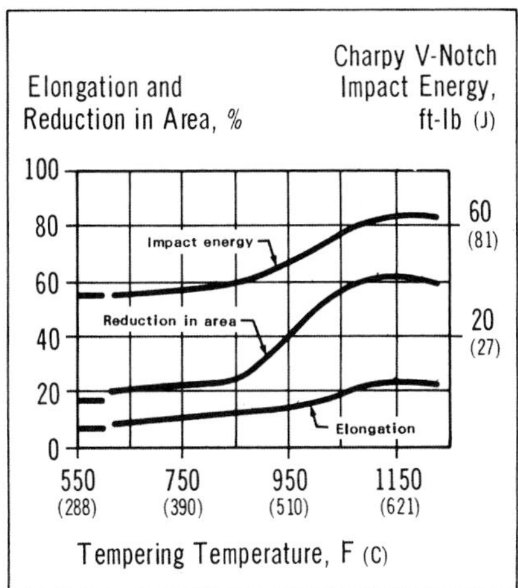

Fig. 3 — Effect of tempering temperature on the mechanical properties of a CA-6NM standard keel block. Source: Alloy Notebook No. 13, May 1968, ESCO Corp.

Minimum mechanical properties are specified as follows: tensile strength, 758 MPa (110 000 psi); yield strength, 552 MPa (80 000 psi); elongation, 15%; reduction in area, 15%.

Nickel content is the key to improved performance. Its presence expands the gamma loop of the iron-chromium system so that at 4% Ni, carbon content can be as low as 0.06% to permit development of properties equal to or greater than those obtainable with CA-15.

Plus — Castings of complex shapes, which need high strength, hardness, and corrosion resistance for good service life, were often difficult or almost impossible to produce with CA-15. Use of CA-6NM makes it possible and feasible to produce such castings.

A complex pump housing cast in CA-6NM is shown in Fig. 1. Impellers cast for the Edmonston plant of the California Water System, are shown in Fig. 2. The Edmonston plant is designed to lift 7.57 million l (2 million gal) of water per min to a height of 600 m (2000 ft) from the San Joaquin Valley over the Tehachapi Mountains into the Los Angeles basin. Use of CA-6NM for these castings resulted in a 58% reduction in the time and effort needed to finish them, with a consequent reduction in cost. Before the development of the alloy, it is doubtful that these impellers would have been cast in martensitic stainless steel. One foundry in Canada is casting very large water wheels and Francis runners in CA-6NM. Here too it is doubtful that CA-15 would have been specified for these large intricate castings.

Heat Treatment — The optimum austenitizing temperature range for CA-6NM is 1010 to 1065 C (1850-1950 F). Figure 3 shows the effect of tempering temperature on the mechanical properties of the alloy when quenched from this temperature range. Re-austenitizing will occur upon tempering above 620 C (1150 F), the amount increasing with increasing temperature.

Depending on the amount of this

Dr. Rice is welding consultant and Mr. Finch is senior metallurgist, ESCO Corp., Portland, Oreg.

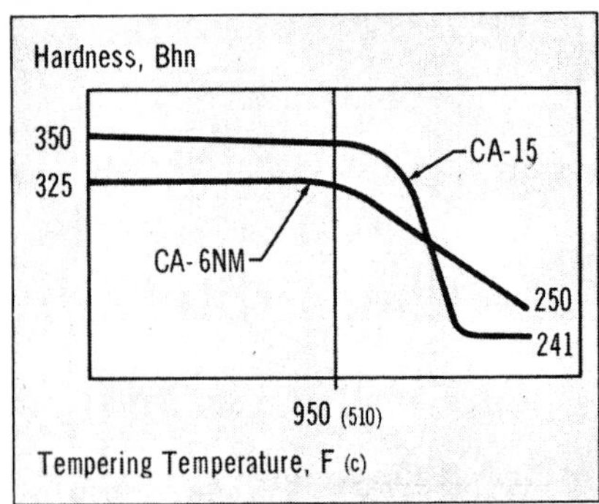

Fig. 4 — Influence of tempering temperature on the hardness of CA-6NM and CA-15. Source: Alloy Notebook No. 13, 1968, ESCO Corp.

Table I — Comparing Minimum and Typical Mechanical Properties of CA-6NM and CA-15

Alloy	Tensile Strength, MPa (10^3 psi)	Yield Strength, MPa (10^3 psi)	Elongation, %	Reduction in Area, %
CA-6NM				
Minimum*	758.4 (110)	551.6 (80)	15	35
Typical†	827.4 (120)	689.5 (100)	24	60
CA-15				
Minimum*	620.5 (90)	448.2 (65)	18	30
Typical‡	792.9 (115)	689.5 (100)	22	60

*955 C (1750 F) min, air cool, plus 595 C (1100 F) min temper (ASTM 2-296 and A-351). †1050 C (1925 F), air cool, plus 620 C (1150 F) temper. ‡980 C (1800 F), air cool, plus 650 C (1200 F) temper.

transformation, cooling from tempering temperatures above 620 C (1150 F) may adversely affect both ductility and toughness through the transformation to untempered martensite. As noted in Fig. 3, a wide range of properties is available. The best combination of strength and ductility is obtained by tempering above 510 C (950 F). The highest strength impact is obtained by tempering above 593 C (1100 F).

The rapid hardness drop on tempering CA-15 above 510 C (950 F) is not exhibited by CA-6NM, as shown in Figure 4. From a practical point of view, the gradual decrease in hardness of CA-6NM makes heat treatment easier and cheaper compared with CA-15, and decreases the frequency of rejects and/or the necessity for reheat treatment.

Mechanical Properties — CA-6NM has no major strength advantages over CA-15. Table I compares the two alloys on the basis of minimums specified in ASTM A-296 and A-351, as well as on the basis of typical properties developed in practice. Although the specifications differ, the properties obtained in practice are very similar.

Elevated temperature properties: Although CA-6NM castings are most commonly used at room temperature, typical short-time elevated tensile properties are shown in Table II. The data, based on both European and ESCO tests, show that CA-6NM will maintain the equivalent of room-temperature properties up to 400 C (750 F).

Table II — Short-Time Elevated-Temperature Properties

Test Temperature, C (F)	Tensile Strength, MPa (psi)	Yield Strength, MPa (psi)	Elongation, %	Reduction in Area, %
200 (390)	772.2 (112 000)	730.8 (106 000)	14	59
300 (570)	737.8 (107 000)	689.5 (100 000)	12	50
400 (750)	758.5 (110 000)	655 (95 000)	11	52
482 (900*)	579.2 (84 000)	524 (76 000)	18	53
500 (930)	641.2 (93 000)	586.1 (85 000)	12	53
538 (1000*)	475.8 (69 000)	455.1 (66 000)	23	58
593 (1100*)	441.3 (64 000)	317.2 (46 000)	39	77
600 (1110)	493 (71 500)	424 (61 500)	17	74

*ESCO data.

Impact strength: Figure 5 shows the ductile-to-brittle transition behavior of CA-6NM normalized at 996 C (1825 F) and tempered at 593 C (1100 F). Impact strength exceeds 20.4 J (15 ft-lb) at a temperature as low as −171 C (−275 F). Work done in Europe shows that CA-6NM is less susceptible to brittle fracture than a 13% Cr steel containing 1.4% Ni. A comparison of the two steels is shown in Fig. 6.

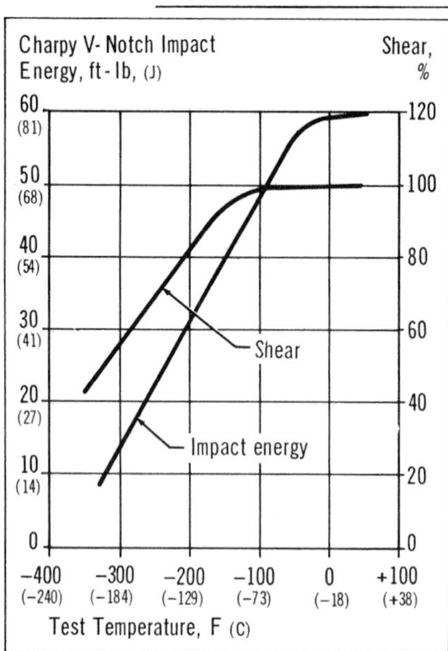

Fig. 5 — Ductile-brittle transition behavior of CA-6NM 6 by 6 by 6½ in. (150 by 150 by 165 mm) test block. Heat treatment: 1825 F (996 C), air cooled, plus 1100 F (593 C). Source: Canadian Dept. of Mines, Energy & Resources, Mines and Branch Investigation Report, IR 10-22.

The impact strength of CA-6NM is significantly higher than that of CA-15, which has typical room-temperature Charpy V-notch impact energy of 13.6 to 27.2 J (10 to 20 ft-lb).

Fatigue strength: Results of some bending fatigue tests performed in Europe show that a steel with 13 Cr-1.2Ni had a life of 10^6 cycles at a stress of 275.8 MPa (40 000 psi). An alloy of CA-6NM had a life of 10^6 cycles at a stress of 324.1 MPa (47 000 psi). Samples were air-cooled. When they were sprayed with water during the test, the stress for 13Cr-1.2Ni was reduced to 255.1 MPa (37 000 psi), while that of CA-6NM was unchanged.

Tests in artificial seawater also

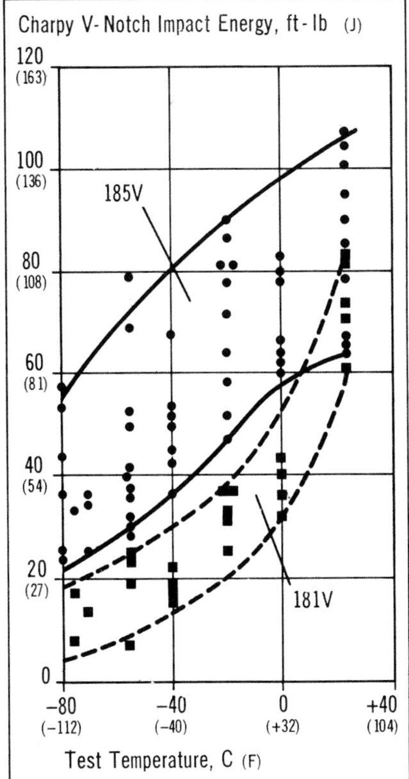

Fig. 6 — Influence of test temperature on notch impact strength of 13Cr-1.4Ni (181V) and CA-6NM (185V) steels.

illustrated the superior fatigue resistance of CA-6NM when compared with a 13Cr-1.4Ni analysis. Results that follow show MPa first, with psi in parentheses:

	CA-6NM	13Cr-1.4Ni
10^6 cycles	289.6 (42)	NA
5×10^6 cycles	220.6 (32)	186.2 (27)
10^7 cycles	193.1 (28)	172.4 (25)

Resistance to corrosion: Corrosion tests in artificial seawater show the following corrosion rates after 18 months:

	Corrosion Rate ($g/m^2/day$)	
	Corrosion Immersion	Alternate Immersion*
CA-6NM	0.005	0.006
CA-15	0.10	0.006

* 4 h in air.

Data obtained in artificial seawater after 1000 hr of rotation at 36.6 m/s (120 ft/s) indicated a weight loss of 0.20 $g/m^2/day$ for CA-6NM and 0.30 $g/m^2/day$ for CA-15.

Resistance to cavitation: Because of its application in large intricate systems involving water and steam, the cavitation resistance of CA-6NM is an important property to consider. The data in Fig. 7 show the marked increase in cavitation resistance exhibited by CA-6NM compared with CA-15.

Other work has indicated the same advantage. Tests in Mersey Estuary seawater, followed by a measurement of the cavity formed, showed that CA-6NM had a cavity with a volume of 0.0041 cm^3. A cavity in CA-15 had a volume of 0.0122 cm^3.

Resistance to erosion: Erosion resistance of CA-6NM was compared with that of a 13Cr-1.2Ni alloy by rotating a disc holding test pins through a slurry of two parts sand and one part water. Weight loss was expressed as a ratio relative to carbon steels. CA-6NM, with a tensile strength of 861.9 MPa (125 000 psi) had an erosion index of 1.26. The 13Cr-1.2Ni composition, with a tensile strength of 737.8 MPa (107 000 psi) had an index of 1.18. Erosion resistance is directly related to hardness. Since hardness and tensile strength are directly related, the increased erosion index of CA-6NM is readily explained.

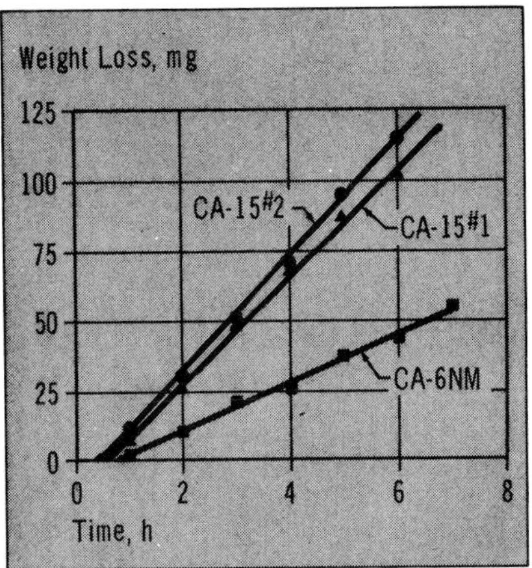

Fig. 7 — Weight loss by cavitation in distilled water at 120 F (50 C) of CA-6NM and CA-15. Weight loss ratio:

$$\frac{\text{CA-15}}{\text{CA-6NM}} = \frac{0.352\ \text{mg/min}}{0.1415\ \text{mg/min}} = 2.49.$$

Sources: Birdsboro Corp. and University of Michigan, November 1969.

Weldability — CA-6NM is superior to CA-15 in this respect. A minimum preheat of 200 C (400 F) is normally required for the latter, while the former may be welded if it is warm and dry. A very slight preheat of 40 C (100 F) will assure that there is no surface condensation that could be a source of hydrogen embrittlement. Castings up to 50 mm (2 in.) and, in some instances, up to 125 mm (5 in.) thick have been welded without preheating.

Welding should consist essentially of stringer beads with weaving not to exceed four times the diameter of the core wire of coated electrodes, or twice the diameter of the gas cup orifice when using the inert-gas-shielded metallic arc process. Each weld pass should be thoroughly cleaned before proceeding to the next.

Filler metal should be modified to a lower Cr-Ni ratio than the base metal. This will expand the gamma loop upward to prevent formation of delta ferrite on solidification. AWS Specifications A5.4 and A5.9 for corrosion-resistant chromium and chromium-nickel-steel-covered electrodes and bare electrodes, designated E-410NM and ER-410NM respectively, have the following composition: 0.06 C max, 1.00 Mn max, 0.03-0.9 Si, 11.0-12.5 Cr, 4.0-5.0 Ni, 0.4-0.7 Mo.

The postweld stress-relief heat treatment should not exceed 620 C (1150 F), since the increased nickel also lowers the gamma loop, permitting some re-austenitization to occur above this temperature. The ideal stress relieving temperature for optimum notch toughness is 550 C (1250 F).

When coated electrodes are used, it is most imperative that storage practices be used which minimize the possibility of moisture absorption by the coating.

Properties and Applications of Corrosion Resistant Stainless and High Alloy Steel Castings

ACI Cast Alloy Designation	AISI Wrought Alloy Type	Nominal Composition, %				Hardenable	Thermal Conductivity at 212 F, Btu/Hr/Sq Ft/Ft/F	Thermal Expansion, 70 to 212 F In./In./F, 10^{-6}	Magnetic Permeability	Yield Strength at 0.2% Offset, Psi (a)	Elongation in 2 In., % (a)	Brinell Hardness (a)	Charpy Key Hole Impact, Ft-Lb (a)	Machinability	Weldability (b)	Characteristics and Applications
		Ni	Cr	C Max	Other											
CA-15	410	1 max	12	0.15	—	Yes	14.5	5.5	Ferromagnetic	150,000	7	390	15	Good	Fair	Hardenable alloys having wide range of mechanical properties (up to 220,000 psi tensile strength); tough, good resistance to abrasion and to mildly corrosive environment. Used for valve bodies, disks and seats, steam equipment and pumps in chemical and petroleum plants.
CA-15M	—	1 max	12	0.15	0.6 Mo	Yes	—	—	—	150,000	7	390	—	Good	Fair	
CA-6NM	—	4	12	0.06	0.40-1.0 Mo	Yes	—	—	—	106,000	14	300	—	Good	Good	
CA-40	420	1 max	12	0.40	—	Yes	14.5	5.5	Ferromagnetic	165,000	1	470	1	Fair	Fair	
CB-30	431	2 max	20	0.30	—	Yes	12.8	5.7	Ferromagnetic	60,000	15	195	2	Good	Fair	More corrosion resistant than the CA alloys, but nonhardenable, except for a modified version of CB-30. High chromium content of CE-30 allows welding without subsequent heat treatment; used in sulfite digester fittings. CB-30 resists nitric acid, alkali and many organic chemicals; used for pump and valve parts. CC-50 resists attack by nitric, sulfuric and oxidizing acids; used to handle acid mine waters.
CC-50	446	4 max	28	0.50	—	No	12.6	5.9	Ferromagnetic	65,000	18	210	—	Good	Fair	
CE-30	312	9	29	0.30	—	No	—	9.6	>1.5	63,000	18	170	—	Good	Good	
CB-7Cu	17-4 PH	4	17	0.07	Cu	Yes	10.3	6.5	Ferromagnetic	156,000	3	418	23	Good	Fair	Ductile and tough, good corrosion resistance. Used in marine applications for propellers and pumps; in food and chemical industries as valves, pumps and other handling equipment.
CD-4MCu	—	5	26	0.040	Mo, Cu	Yes (c)	8.8	6.5	Ferromagnetic	85,000	25	260	37	Good	Good	High tensile strength and superior resistance to stress-corrosion cracking. Highly resistant to brine, steam, sodium carbonate, sulfuric acid, nitric acid and other strongly oxidizing media. Used in marine, chemical processing, power plant and pulp and paper industries as pumps, valves, digester liners and paper-making rolls.
CF-3	304L	10	19	0.03	—	No	9.2	9.0	1.0 to 1.3	37,000	55	140	75	Good	Good	Most widely used alloys. Good corrosion resistance in variety of environments, including bleaching compounds; caustic salts; nitric, sulfuric or phosphoric acid; organic acids and compounds; halogen acids and salts; and wet chlorinated hydrocarbons. Used in cryogenic applications, since they retain high impact strength at very low temperatures, and, as other CF alloys, in chemical, textile, petroleum, pharmaceutical and food industries.
CF-8	304	9	19	0.08	—	No	9.2	9.0	1.0 to 1.3	37,000	55	140	75	Good	Good	
CF-20	302	9	19	0.20	—	No	9.2	9.6	1.01	36,000	50	163	60	Good	Good	
CF-3M	316L	11	19	0.03	Mo	No	9.4	8.9	1.5 to 2.5	42,000	50	156 to 170	70	Good	Good	Similar in many respects to CF types, but molybdenum added to enhance corrosion resistance (especially to halogen ion solutions and reducing acids) and to increase strength. Used in severe environments, such as sulfite pulp production, and in milder environments where continuous, maintenance-free operation is required.
CF-8M	D319 (316)	9	19	0.08	Mo	No	9.4	8.9	1.5 to 2.5	42,000	50	156 to 170	70	Good	Good	
CG-8M	317	11	19	0.08	Mo	No	9.4	9.7	1.5 to 2.5	43,000	50	170	70	Good	Good	
CF-8C	347	9	19	0.08	Cb	No	9.3	9.3	1.2 to 1.8	38,000	39	149	30	Good	Good	Further modifications of basic CF types. Columbium in CF-8C makes it useful where field welding does not permit postweld heat treatment. CF-16F contains selenium to improve machinability.
CF-16F	303	9	19	0.16	—	No	9.4	9.0	1.0 to 2.0	40,000	52	150	75	Excellent	Good	
CG-12	—	12	22	0.12	—	No	—	—	—	28,000	35	—	—	—	—	
CH-20	309	12	24	0.20	—	No	8.2	8.3	1.71	50,000	38	190	30	Fair	Good	Higher nickel and chromium contents than CF types for better resistance to many corrosive environments and good performance at elevated temperatures. Used for fittings, pump parts, ore roasting equipment and valves in chemical, pulp processing, power plant and oil refining industries.
CK-20	310	20	25	0.20	—	No	8.2	8.0	1.02	38,000	37	144	50 (d)	Good	Good	
CN-7M	—	30	20	0.07	Mo, Cu	No	12.1	8.6	1.01 to 1.10	31,000	48	130	70	Excellent	Fair (e)	Contains more nickel than chromium. Excellent resistance to various concentrations of hot sulfuric acid, dilute hydrochloric acid and many reducing chemicals, as well as nitric-hydrofluoric pickling solutions.
CW-12M	—	40	18	0.12	18 Mo	No	—	—	—	46,000	4.0	—	—	—	—	
CY-40	—	67	16	0.40	—	No	—	—	—	28,000	30	—	—	—	—	
CZ-100	—	90	—	1.00	—	No	—	—	—	18,000	10	—	—	—	—	

Note: All standard grades listed are covered for general applications by ASTM Specification A296. ASTM Specifications A217, A351, A362, A451 and A452 also apply to some grades.

(a) Properties are for alloys in the solution-annealed condition, except types CA-15, CA-40, CB-30, CB-7Cu and CC-50 which are in the heat treated condition providing maximum yield strength; (b) All should be heat treated after welding to restore maximum corrosion resistance; CE-30, CF-3 and CF-8C may be field welded without post-treatment; (c) Properties for alloy in solution-annealed condition. Under special conditions alloy can be hardened, but foundry must be consulted; (d) Izod V-notch; (e) Preheat required.

Data supplied by Alloy Casting Institute, New York.

Section VI: Tool and Die Materials

Tool Steels

*By the ASM Committee on Tool Steel**

TOOL STEELS are characterized by high hardness and resistance to abrasion coupled in many instances with resistance to softening at elevated temperature. Generally these attributes are attained with high carbon and alloy content.

In melting, scrap and raw material must be carefully selected not only for alloy content but also to insure cleanliness and homogeneity of the finished product. Tool steels are almost always melted in electric furnaces because of greater cleanliness, more precise control of melting conditions, and economy with small tonnages. Because many tool steels are highly alloyed, forging and rolling practices are complicated and frequently result in a large amount of process scrap. The rigorous inspection procedures generally involve examination, in the finished or semifinished form, of both ends of each bar for center quality, cleanliness, hardness, grain size, annealed structure and hardenability, while the entire bar may be subject to magnetic particle and sonic inspection for surface and internal defects. It is also important that the finished tool steel bars be as free from decarburization as possible, or at least have decarburization within carefully controlled limits. This requires special annealing procedures and thorough inspection.

Such precise production requirements and quality control are reasons for the high cost of tool steels, apart from the extra cost of alloying elements. The insistence on quality in the manufacture of these steels is justified, however, since many bars of tool steel are made into complicated cutting and forming tools worth many times the cost of the steel itself. Although several grades of standard alloy constructional steel (for instance, SAE 9260, 6150, 4340 and 52100) are similar in composition to some tool steels, the standard alloy steels are seldom used for expensive tools because in general they are not manufactured to tool steel quality requirements.

Types of Tool Steel

To simplify the classification and selection of tool steels, a system has been developed by the American Iron and Steel Institute and the Society of Automotive Engineers which groups grades of similar properties as shown in Table 1. A brief description of this classification with regard to alloy content, properties and uses of the various groups follows.

Water-Hardening Tool Steels, Group W. Carbon is the principal "alloying element" of this group of tool steels, with small additions of chromium and vanadium in most steels of the group. Chromium is added to increase the hardenability and wear resistance; vanadium, to refine the grain for added toughness. Although various carbon contents are available in 0.10% ranges from 0.60 to 1.40%, the most popular grades contain approximately 1.00% C. These grades are also available in different hardenabilities, as will be described later. Basically, group W steels are shallow hardening, and when heat treated in sections over ½ in. in diameter, possess a hard case with a strong, tough, resilient core. This combination of properties coupled with the low resistance to heat softening or tempering makes

*G. D. Dolch, *Chairman*, Materials Engineering Manager, Tapco Group, Thompson Ramo Wooldridge, Inc.; Edward A. Dolega, Metallurgical Engineer, Bell Aircraft Corp.; Ural H. Gillett, Metallurgist, Small Tool Div., Barber-Colman Co.; C. W. Hangosky, Chief Metallurgist and Materials Engineer, Reo Div., White Motor Co.

Elmer B. Hauser, Metallurgist, Weldon Tool Co.; John J. Hoffer, Apparatus and Optical Div., Eastman Kodak Co.; David P. Hughes, Manager, Technical Services, (Peter Leckie-Ewing, alternate), Latrobe Steel Co.; Ralph G. Kennedy, Director of Laboratories, Cleveland Twist Drill Co.; J. P. Long, Metallurgical Section Supervisor, Materials Engineering Dept., Tapco Group, Thompson Ramo Wooldridge, Inc.

S. R. Prance, Chief Metallurgist, Inland Manufacturing Div., General Motors Corp.; D. A. Stewart, Machinability Testing Engineer, Aircraft Gas Turbine Div., General Electric Co.; Rex F. Supernaw, Metallurgist, National Twist Drill & Tool Co.; A. L. Trueax, Jr., Project Engineer, Ternstedt Div., General Motors Corp.; E. D. Wilson, Tool & Die Materials Div., Allegheny Ludlum Steel Corp.

Table 1. Classification and Compositions of Principal Types of Tool Steels

(AISI-SAE except for last group of steels)

Designation	C	Mn	Si or Ni	Cr	V	W	Mo	Co
Water-Hardening Tool Steels(a)								
W1*	0.60 to 1.40(a)	....	....	....	....	....	....	....
W2*	0.60 to 1.40(a)	....	....	....	0.25	....	....	....
W3	0.60 to 1.40(a)	....	....	....	0.50	....	....	....
W4	0.60 to 1.40(a)	....	....	0.25	....	....	....	....
W5	0.60 to 1.40(a)	....	....	0.50	....	....	....	....
W6	0.60 to 1.40(a)	....	....	0.25	0.25	....	....	....
W7	0.60 to 1.40(a)	....	....	0.50	0.20	....	....	....
Shock-Resisting Tool Steels								
S1*	0.50	....	....	1.50	....	2.50	....	....
S2	0.50	....	1.00 Si	....	....	....	0.50	....
S3	0.50	....	....	0.75	....	1.00	....	....
S4	0.50	0.80	2.00 Si	....	....	....	....	....
S5*	0.50	0.80	2.00 Si	....	....	....	0.40	....
Oil-Hardening Cold Work Tool Steels								
O1*	0.90	1.00	....	0.50	....	0.50	....	....
O2	0.90	1.60	....	....	....	....	....	....
O6	1.45	....	1.00 Si	....	....	....	0.25	....
O7	1.20	....	....	0.75	....	1.75	0.25 opt	....
Air-Hardening Medium-Alloy Cold Work Tool Steels								
A2*(b)	1.00	....	....	5.00	....	....	1.00	....
A4	1.00	2.00	....	1.00	....	....	1.00	....
A5	1.00	3.00	....	1.00	....	....	1.00	....
A6	0.70	2.00	....	1.00	....	....	1.00	....
A7	2.25	....	....	5.25	4.50	....	1.00	....
High-Carbon High-Chromium Cold Work Steels								
D1	1.00	....	....	12.00	....	....	1.00	....
D2*(b)	1.50	....	....	12.00	....	....	1.00	....
D3*(b)	2.25	....	....	12.00	....	....	....	....
D4*	2.25	....	....	12.00	....	....	1.00	....
D5(b)	1.50	....	....	12.00	....	....	1.00	3.00
D6	2.25	....	1.00 Si	12.00	....	1.00	....	....
D7(b)	2.35	....	....	12.00	4.00	....	1.00	....
Chromium Hot Work Tool Steels								
H11	0.35	....	....	5.00	0.40	....	1.50	....
H12*	0.35	....	....	5.00	0.40	1.50	1.50	....
H13*(b)	0.35	....	....	5.00	1.00	....	1.50	....
H14	0.40	....	....	5.00	....	5.00	....	....
H15	0.40	....	....	5.00	....	....	5.00	....
H16	0.55	....	....	7.00	....	7.00	....	....
Tungsten Hot Work Tool Steels								
H20	0.35	....	....	2.00	....	9.00	....	....
H21*	0.35	....	....	3.50	....	9.50	....	....
H22	0.35	....	....	2.00	....	11.00	....	....
H23	0.30	....	....	12.00	....	12.00	....	....
H24	0.45	....	....	3.00	....	15.00	....	....
H25	0.25	....	....	4.00	....	15.00	....	....
H26	0.50	....	....	4.00	1.00	18.00	....	....
Molybdenum Hot Work Tool Steels								
H41	0.65	....	....	4.00	1.00	1.50	8.00	....
H42	0.60	....	....	4.00	2.00	6.00	5.00	....
H43	0.55	....	....	4.00	2.00	....	8.00	....
Tungsten High Speed Tool Steels								
T1*(b)	0.70	....	....	4.00	1.00	18.00	....	....
T2(b)	0.85	....	....	4.00	2.00	18.00	....	....
T3(b)	1.05	....	....	4.00	3.00	18.00	....	....
T4	0.75	....	....	4.00	1.00	18.00	....	5.00
T5	0.80	....	....	4.00	2.00	18.00	....	8.00
T7	0.75	....	....	4.00	2.00	14.00	....	....
T8	0.80	....	....	4.00	2.00	14.00	....	5.00
T15(b)	1.50	....	....	4.00	5.00	12.00	....	5.00
Molybdenum High Speed Tool Steels								
M1*(b)	0.80	....	....	4.00	1.00	1.50	8.50	....
M2*(b)	0.85	....	....	4.00	2.00	6.25	5.00	....
M3*(b)(c)	1.00	....	....	4.00	2.40	6.00	5.00	....
M4	1.30	....	....	4.00	4.00	5.50	4.50	....
M6	0.80	....	....	4.00	1.50	4.00	5.00	12.00
M7	1.00	....	....	4.00	2.00	1.75	8.75	....
M10*(b)	0.85	....	....	4.00	2.00	....	8.00	....
M15	1.50	....	....	4.00	5.00	6.50	3.50	5.00
M30	0.80	....	....	4.00	1.25	2.00	8.00	5.00
M33	0.90	....	....	3.75	1.15	1.75	9.50	8.25
M34	0.90	....	....	4.00	2.00	2.00	8.00	8.00
M35	0.80	....	....	4.00	2.00	6.00	5.00	5.00
M36	0.80	....	....	4.00	2.00	6.00	5.00	8.00
Low-Alloy Special-Purpose Tool Steels								
L1	1.00	....	....	1.25	....	....	....	....
L2	0.50 to 1.10(a)	....	....	1.00	0.20	....	....	....
L3	1.00	....	....	1.50	0.20	....	....	....
L4	1.00	0.60	....	1.50	0.20	....	....	....
L5	1.00	1.00	....	1.00	....	....	0.25	....
L6	0.70	....	1.50 Ni	0.75	....	....	0.25 opt	....
L7	1.00	0.35	....	1.40	....	....	0.40	....
Carbon-Tungsten Tool Steels								
F1	1.00	....	....	....	....	1.25	....	....
F2	1.25	....	....	....	....	3.50	....	....
F3	1.25	....	....	0.75	....	3.50	....	....
Low-Carbon Mold Steels								
P1	0.10 max	....	....	....	....	....	....	....
P2	0.07 max	....	0.50 Ni	1.25	....	....	0.20	....
P3	0.10 max	....	1.25 Ni	0.60	....	....	....	....
P4	0.07 max	....	....	5.00	....	....	....	....
P5	0.10 max	....	....	2.25	....	....	....	....
P6	0.10	....	3.50 Ni	1.50	0.20	....	....	....
P20	0.30	....	....	0.75	....	....	0.25	....
PPT	0.20	1.20 Al	4.00 Ni	....	....	....	....	....
Other Alloy Tool Steels(d)								
6G	0.55	0.80	0.25 Si	1.00	0.10	....	0.45	....
6F2	0.55	0.75	0.25 Si, 1.00 Ni	1.00	0.10 opt	....	0.30	....
6F3	0.55	0.60	0.85 Si, 1.80 Ni	1.00	0.10 opt	....	0.75	....
6F4	0.20	0.70	0.25 Si, 3.00 Ni	....	....	....	3.35	....
6F5	0.55	1.00	1.00 Si, 2.70 Ni	0.50	0.10	....	0.50	....
6F6	0.50	....	1.50 Si	1.50	....	....	0.20	....
6F7	0.40	0.35	4.25 Ni	1.50	....	....	0.75	....
6H1	0.55	....	....	4.00	0.85	....	0.45	....
6H2	0.55	0.40	1.10 Si	5.00	1.00	....	1.50	....

*Stocked in almost every warehousing district and made by the majority of tool steel producers.

(a) Various carbon contents are available in 0.10% ranges. (b) Available as free-cutting grade. (c) Available with vanadium contents of 2.40 or 3.00%. (d) The designations of these steels are similar to those used in the 1948 Metals Handbook, except they were previously written with Roman numerals (VI F2, etc). Neither AISI nor SAE has assigned type numbers to these steels.

these grades suitable for cold heading, striking, coining, and embossing applications and for woodworking tools, hand metal-cutting tools such as taps and reamers, wear-resistant machine-tool uses, and cutlery.

Steels of the W group are available in as many as four different grades or quality levels for the same nominal composition. These quality levels have been given various names by different manufacturers. These include such designations as extra carbon, special, standard and commercial, and range from a very clean carbon tool steel with precisely controlled hardenability, grain size, annealed hardness and microstructure to a grade that is less carefully controlled but is satisfactory for a noncritical low-production application.

The Society of Automotive Engineers now defines four grades of carbon tool steels as follows:

Special (grade 1) is the highest-quality water-hardening carbon tool steel, controlled for hardenability, with composition held to close limits and bars subject to the most rigid tests to insure maximum uniformity in performance.

Extra (grade 2) is a high-quality water-hardening carbon tool steel, controlled for hardenability, subject to tests that will insure good service for general applications.

Standard (grade 3) is a good-quality water-hardening carbon tool steel, not controlled for hardenability, recommended for application where some latitude in uniformity can be tolerated.

Commercial (grade 4) is a commercial grade of water-hardening tool steel not controlled for hardenability nor subject to special tests.

Limits on manganese, silicon and chromium contents are not generally required on "special" and "extra" grades because of the following Shepherd hardenability limits:

	Hardenability penetration, 1/64-in.	Minimum fracture grain size
0.70 to 0.95% C		
Shallow	10 max	8
Regular	9 to 13	8
Deep	12 min	8
0.95 to 1.30% C		
Shallow	8 max	9
Regular	7 to 11	9
Deep	10 to 16	8

Standard and commercial grades generally require a maximum of 0.35% for both manganese and silicon and a maximum of 0.15% Cr in the standard grade and 0.20% in the commercial grade. The total of manganese, silicon and chromium should not exceed 0.75% in the standard and commercial grades.

Shock-Resisting Tool Steels, Group S. The principal alloying elements in these steels are silicon, chromium, tungsten and sometimes molybdenum or nickel. (Although a nickel content is not indicated for any of the shock-resisting steels listed in Table 1, it is present in many.) Silicon and nickel strengthen the ferrite and increase hardenability, while chromium improves hardenability, provides some heat resistance, and contributes slightly to abrasion resistance. Molybdenum is also important in increasing hardenability. With carbon content maintained at about 0.50%, these steels have high strength with moderate wear resistance. Tools made from these steels have measurable ductility even at Rockwell C 60. Principal uses are for chisels, rivet sets, hammers and other tools where repetitive high-impact loading is developed. The high-tungsten types are sometimes used for hot shearing or heading where heat resistance is important.

Oil-Hardening Cold Work Steels, Group O. These steels contain tungsten, manganese, chromium and small amounts of molybdenum. Alloy additions increase the hardenability, permitting oil quenching, with much less distortion and less cracking hazard than with the group W steels. Group O steels are relatively inexpensive and their high carbon content produces adequate wear resistance for short-run applications at or near room temperature. The high silicon in O6 steel is only partly for increasing the hardenability; its main function is to induce graphitization of part of the carbon (carbide) thereby improving machinability in the annealed condition and wear resistance in the hardened condition. Typical uses for group O steels are short-run cold forming dies, blanking dies and gages where distortion is unimportant, and cutting tools where no high temperatures are generated.

Air-Hardening Medium-Alloy Cold Work Tool Steels, Group A. Manganese, chromium, molybdenum and vanadium are the principal alloying elements in this group of tool steels. The main function of the alloy additions is to promote deep hardening thereby inducing air-hardening characteristics with consequent low distortion. In addition, the high carbon content promotes high wear resistance but, except for A2 and A7, these steels have slight resistance to heat softening. The high-manganese grades can be hardened from temperatures as low as some plain carbon steels, thereby further reducing dimensional change from hardening and scaling loss caused by oxidation. The low dimensional change and high wear resistance make these steels useful for intricate die shapes, thread-rolling dies, and slitters. Although in general their wear resistance is less than that of the group D steels and their susceptibility to distortion is about the same, the lower cost, lower hardening temperatures and better toughness of the group A steels contribute to their widespread use. It should be mentioned that A7 has extremely high wear resistance, exceeding that of many of the group D steels.

High-Carbon High-Chromium Cold Work Steels, Group D. The principal alloying elements in these steels are chromium and carbon, but they may also contain tungsten, molybdenum, cobalt and vanadium. Group D steels are highly wear resistant with deep hardening promoted by the high carbon and chromium contents, and with the hardenability accentuated by minor additions of tungsten and molybdenum. These grades, especially the D7 grade, contain large amounts of hard carbides. A careful balance of alloying elements and air-hardening properties results in extremely low dimensional change in hardening. Medium resistance to heat softening, however, limits the use of group D tool steels to applications below 900 F. Typical uses are long-run blanking and forming dies, thread-rolling dies, brick molds, gages and abrasion-resistant liners. Unfortunately these steels are susceptible to edge brittleness, which, in combination with relatively low heat resistance, makes them unsuitable for cutting tools.

Chromium Hot Work Tool Steels, Group H11 to H16. These steels contain chromium and tungsten with additions of molybdenum and vanadium. They have good resistance to heat softening because of their medium chromium content, supplemented by the addition of carbide-forming elements such as molybdenum, tungsten or vanadium. The low carbon and relatively low total alloy content promote toughness at the normal working hardnesses of Rockwell C 40 to 55. Higher tungsten and molybdenum contents increase red hardness and hot strength, but slightly reduce toughness. Vanadium is added to increase resistance to washing at high temperature.

All types are extremely deep hardening and may be air hardened to full working hardness in sections up to 12 in. The air-hardening qualities and balanced alloy content are responsible for low distortion in hardening. These grades are especially adapted to hot die work of all kinds, particularly white metal extrusion dies and die-casting dies, forging dies, mandrels and hot shears. The alloy and carbon contents are low enough so that the steels may be water cooled in service without cracking and with consequent increase in service life.

An interesting application of the H11 grade is its use in highly stressed structural parts, particularly for supersonic aircraft (see article on wrought heat-resisting alloys).

The chief advantage of this grade over conventional high-strength steels is its ability to resist softening during continued exposure to temperatures up to 1000 F and at the same time to provide moderate toughness and ductility at tensile strength levels of 250,000 to 300,000 psi. In addition, the high tempering temperature permitted with H11 because of its secondary hardening characteristic affords the nearly complete relief of residual hardening stresses necessary for maximum toughness at high strength levels.

Other important advantages of H11 for these applications include its exceptional ease of forming and working, good weldability, relatively low coefficient of thermal expansion, above-average resistance to corrosion and oxidation, and low content of strategic elements.

Tungsten Hot Work Steels, Group H20 to H26. Principal alloying elements of these steels are carbon, tungsten and chromium with some vanadium. The higher alloy content increases resistance to high-temperature softening and washing, compared with the H11 to H16 hot work steels, but also makes them more susceptible to brittleness at the normal working hardnesses of Rockwell C 45 to 55. The high alloy content also makes it impossible for these die steels to be water cooled safely in service. They can be air hardened but are usually quenched in oil or hot salt to minimize scaling. When air hardened, they exhibit low distortion.

Inasmuch as they require somewhat higher hardening temperatures than the chromium hot work steels, they have a greater tendency to scale under typical hardening conditions. Although they have much greater toughness, these steels have many characteristics of the high speed steel grades; in fact, type H26 is a low-carbon version of T1 high speed steel. If these steels are preheated to operating temperature before use, breakage can be avoided and they may be used for mandrels and extrusion dies for high-temperature applications such as the extrusion of brass, nickel alloys and steel. They are also suitable for use in hot forging dies of rugged design.

Molybdenum Hot Work Steels, Group H41 to H43. Molybdenum chromium, vanadium, and carbon, with varying amounts of tungsten are used in these steels, which are similar to the tungsten hot work steels, having almost identical characteristics and uses. As may be seen from their compositions, they resemble the various types of molybdenum high speed steels although they have lower carbon content and greater toughness. Their principal advantage over the tungsten hot work steels is lower initial cost. These steels are more resistant to heat checking than the tungsten hot work grades, but in common with all high-molybdenum steels, require greater care in heat treatment, particularly with regard to decarburization and control of hardening temperatures.

Tungsten High Speed Steels, Group T. Tungsten, chromium, vanadium, cobalt and carbon are the principal alloying elements in these grades.

Grade T1 was developed partly as the result of the work of Taylor and White who, in the early 1900's, found that certain steels with over 14% W, 4% Cr and about 0.3% V developed red hardness. In its earlier form T1 contained about 0.68%

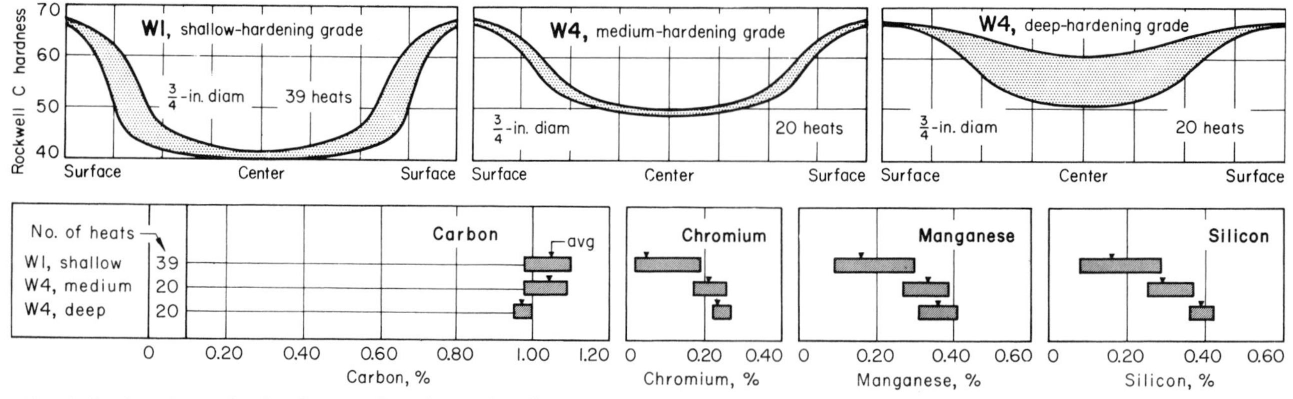

Fig. 1. Surface-to-center hardness values for ¾-in. diam water hardening tool steels ordered as shallow, medium and deep-hardening grades. All specimens were 3 in. long, heated for 40 min at 1600 F, oil quenched, reheated to 1450 F for 15 min, and quenched in agitated 10% brine. Grain size, 8¾ to 9¼. Lower charts indicate variation in composition.

C, 18% W, 4% Cr and about 0.3% V. By 1920 the vanadium had been increased to about 1.0% and the carbon was slightly increased over a 30-yr period to its present level of 0.70% or slightly higher.

Grade T1 and the other grades of this group are characterized by high red hardness and wear resistance. They are also deep hardening and can in general be hardened to Rockwell C 65 or over in oil or hot salt in sections up to 3 in. The high alloy and high carbon contents produce a large number of hard wear-resistant carbides in the microstructure, particularly in those grades containing more than 1.5% V and higher carbon contents. T15 is the most wear-resistant grade of this group, although it is somewhat lacking in toughness. Cobalt additions increase red hardness at a sacrifice in toughness. The presence of many wear-resistant carbides in a hard heat-resistant matrix makes these steels suitable for cutting tool applications, while their toughness allows them to outperform the sintered carbides in delicate tools and interrupted-cut applications. Applications are almost entirely for such cutting tools as tool bits, drills, reamers, taps, broaches, milling cutters and hobs, although some usage is found for dies and punches and for high-load, high-temperature structural components such as aircraft bearings and pump parts.

Molybdenum High Speed Steels, Group M. Principal alloying elements are molybdenum, tungsten, chromium, vanadium, cobalt and carbon. These steels are similar in properties to the tungsten high speed tool steels, but are generally considered to have slightly greater toughness at the same hardness. Their main advantage, considering their equivalent performance, lies in their lower initial cost, almost 30% less than similar grades of the group T steels. Increasing carbon and vanadium content increases wear resistance; increasing cobalt raises red hardness as described for the group T high speed steels. M15 is the most wear resistant steel of the M group.

The M steels are somewhat more sensitive than the T steels to hardening conditions, particularly to temperature and atmosphere, since they will decarburize and overheat easily under adverse treating conditions. This is especially true of the high-molybdenum grades. Typical uses are for cutting tools of all kinds. At present approximately 85% of all high speed tool steels produced in the United States are in this group.

Low-Alloy Special-Purpose Steels, Group L. Principal alloying elements in this group are chromium and manganese, with additions of vanadium, molybdenum and nickel. These group L steels are similar to the W group of carbon tool steels with the addition of chromium and other elements for greater wear resistance and hardenability. Nickel is added to the L6 grade for increased toughness as well as hardenability. The high chromium contents promote wear resistance through the formation of hard complex iron-chromium carbides and together with molybdenum and manganese increase hardenability; vanadium serves to refine the grain.

These steels are oil hardening and thus only fair in resisting dimensional change. They may be hardened from the temperature range of 1500 to 1600 F, which, along with their good resistance to decarburization, simplifies hardening problems. Typical uses are various machine-tool applications where high wear resistance with good toughness is required, as in bearings, rollers, clutch plates, high-wear springs, feed fingers and chuck parts. L1, L3, L4, and L7 are similar to SAE 52100 steel and are used for similar applications.

Carbon-Tungsten Tool Steels, Group F. Principal alloying element is tungsten with some chromium. These steels are generally shallow-hardening, water-quenching steels with high carbon and tungsten contents to promote high wear resistance. Type F3 possesses the greatest hardenability and because of its chromium content may be oil hardened with lower distortion. Under some conditions of operation these steels possess four to ten times the wear resistance of the plain carbon group W tool steels. These steels were formerly known as the "fast finishing steels" because of their great hardness and ability to retain a sharp cutting edge and because they are used for finish machine cuts on soft materials. Since they are low in red hardness, however, feeds must be light. They are relatively brittle so that in general they are used for high-wear, low-temperature and low-shock applications. Typical uses are paper-cutting knives, wiredrawing dies, plug gages, forming tools and brass-cutting tools.

The high distortion of the water-hardening grades is occasionally put to advantage in rehardening used wiredrawing dies, since it is possible to shrink the bore of these dies considerably by the proper spout quenching method and thus restore the original internal diametral size.

Low-Carbon Mold Steels, Group P. Chromium and nickel are the principal alloying elements of this group, with vanadium, molybdenum and aluminum as additives. Most of these steels are alloy carburizing steels produced to tool steel quality. They are generally characterized by very low hardness in the annealed state and resistance to work hardening, both factors suiting them for hubbing operations. After the impression has been formed or cut, the steels are generally carburized for wear resistance before they are hardened. The addition of nickel and chromium increases hardenability though at the expense of annealed hardness so that impressions in the higher-alloy P steels are generally cut rather than pressed. This is particularly true of P6, P20 and PPT.

With the exception of P4, these steels possess poor red hardness and low wear resistance (unless carburized) and thus are used almost entirely for low-temperature die-casting dies and for molds for injection or compression molding of plastics. Types PPT and P20 normally are supplied in a heat treated condition to a hardness of Rockwell C 30 to 35, in which condition they can be readily machined into large intricate dies and molds. Since they are prehardened, no subsequent heat treatment is required and distortion and size change are avoided.

Other Alloy Tool Steels, Groups 6G, 6F and 6H. Principal alloying elements are nickel, molybdenum and chromium with vanadium and silicon additions. These are multi-alloy deep-hardening steels of medium carbon content, characterized by high toughness and in some instances (6F4) good heat resistance. The high nickel and silicon contents promote strengthening of the ferrite and assist hardenability, which is intensified by the additions of molybdenum and chromium. This group is used extensively for hot upset forging dies and dies for hammer and press forging. They are also used for applications requiring high toughness at working hardnesses up to approximately Rockwell C 58.

They have low wear resistance, as might be expected from their toughness qualities and carbon content. They are oil hardening, and resistance to dimensional change on hardening is relatively good. This combination of qualities renders them suitable for shear blades, forming dies, master hubs for cold hubbing, swaging dies, and stamps. The more heat-resistant grades are often used for mandrels for hot and cold work applications.

Hardenability

Water-Hardening Tool Steels, Group W. Controlled hardenability is probably the most important property of these grades. Carbon tool steel owes its popularity in large extent to the fact that it will harden with a hard case (Rockwell C 65) and a soft core (Rockwell C 40) in sizes over approximately ½ in. in diameter, thereby providing in these larger sizes a wear-resistant surface and a tough core. Controlling the depth of hardness controls the steel's basic properties with respect to the balance between case and core. The hardenability, or depth of hard case measured to the 50% martensite level (approximately Rockwell C 55), is controlled in these steels mainly by the austenitic grain size, melting technique, alloy content, amount of excess carbide present at the quenching temperature, and to a lesser degree by the initial structure of the steel prior to heating for quenching.

Figure 1 shows hardness traverse results on three types of water-hardening tool steels in ¾-in. bars hardened as noted. The plain carbon steel, W1, contains no intentional alloying elements such as chromium, molybdenum or vanadium and is therefore a shallow-hardening steel compared with the W4 grade, which contains 0.25% Cr. The deep-hardening W4 is a controlled variation resulting from melting techniques and higher residual elements. In general, chromium is added to increase the hardenability and vanadium to refine the grain size. Hardenability can also be controlled within one type by varying the manganese or silicon content within narrow limits or by different deoxidation methods in melting. For example, deoxidizing with heavy aluminum additions will produce a shallower-hardening steel in all grades.

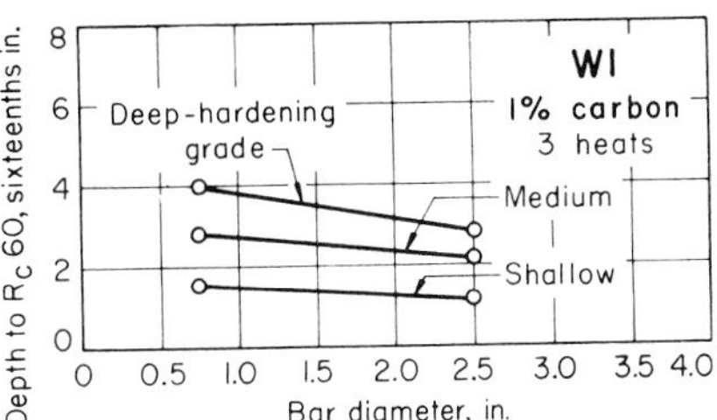

Fig. 2. Relation of bar diameter and depth of zone hardened to Rockwell C 60 for shallow, medium and deep-hardening grades of water-hardening tool steel (Shepherd PF test data)

The hardenability of carbon tool steel has been the subject of considerable research and is measured in many ways, possibly the most common being the Shepherd PF test. This test measures the penetration (P) in 64ths of an inch and fracture grain size (F) as compared with standard samples, on a ¾-in. round sample hardened under controlled conditions of prior microstructure and hardening temperature. Results are given numerically; for example, a PF value of 6-8 obtained from 1450 F means a case depth or penetration of 6/64 in. from the surface of the sample and a fracture grain size of 8, measured in the case zone, when the ¾-in. bar, after normalizing, is quenched from 1450 F into brine. Fine-grained steels are those with grain sizes (F values) of 8 or more. Deep-hardening steels have P values of 12 or more, medium hardening 9 to 11, and shallow hardening 6 to 8.

Increased fracture grain size indicates an increased austenitic grain size and results in increased hardenability. Typical results indicate an increase in P value of 2/64 in. for every one-point increase in grain size number for the same grade. Increased amounts of undissolved carbides at the hardening temperature will reduce hardenability. This is doubly important in the hypereutectoid grades, which are deliberately quenched to contain carbides undissolved at the quenching temperature, in order to provide increased wear resistance. A fine lamellar prior microstructure (such as obtained from normalizing) will contain fewer undissolved carbides than a prior spheroidized microstructure at the normal quenching temperature. Fewer carbides result in a deeper hardening reaction because carbides act as nucleation centers for high-temperature transformation products on cooling; thus normalized bars will show deeper hardenability than will spheroidize-annealed bars of the same grade.

The addition of vanadium frequently decreases hardenability under normal hardening conditions because of the formation of many very fine carbides, which act not only as nucleation centers for transformation but also refine the austenitic grain size. However, abnormal hardening conditions, such as overheating, will dissolve these excess carbides so that the normal effect of vanadium in increasing hardenability will be observed.

Group W steels of lower than eutectoid carbon often have greater hardenability than the hypereutectoid grades. This follows partly from grain coarsening resulting from the higher quenching temperatures used for these grades, but mainly from the lack of nucleating carbides present at the hardening temperature.

Figure 2 shows a typical relationship between bar diameter and depth of hard case (measuring Rockwell C 60 or above) for three group W tool steels with 1% C and different hardenability. Hardenability in this instance is varied by the amount of manganese and silicon and the deoxidation procedure. This chart indicates the need for precise specification of hardenability in the selection of these grades, since group W grades purchased without hardenability requirements could vary widely enough in this property to cause severe processing difficulties or actual tool failures.

Shock-Resisting Tool Steels, Group S. Hardenability varies in these steels from the deep-harden-

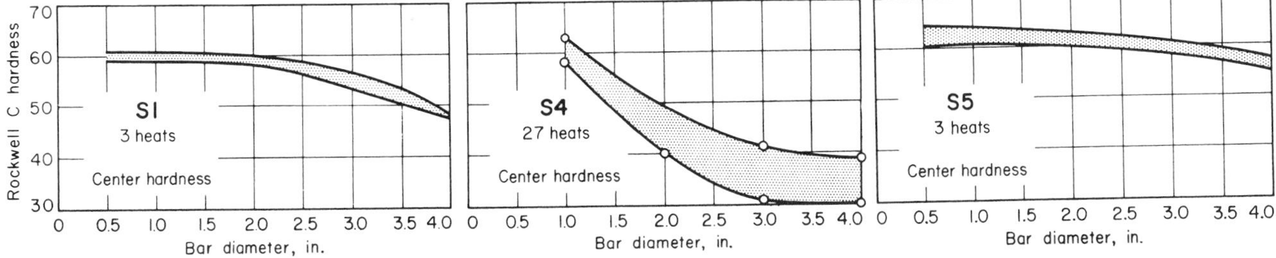

Fig. 3. Effect of bar diameter, grade and heat treatment on center hardness of shock-resisting tool steels. Grade S1 was quenched in agitated oil from 1625 F, S4 in agitated water from 1575 F, and S5 in agitated oil from 1575 F.

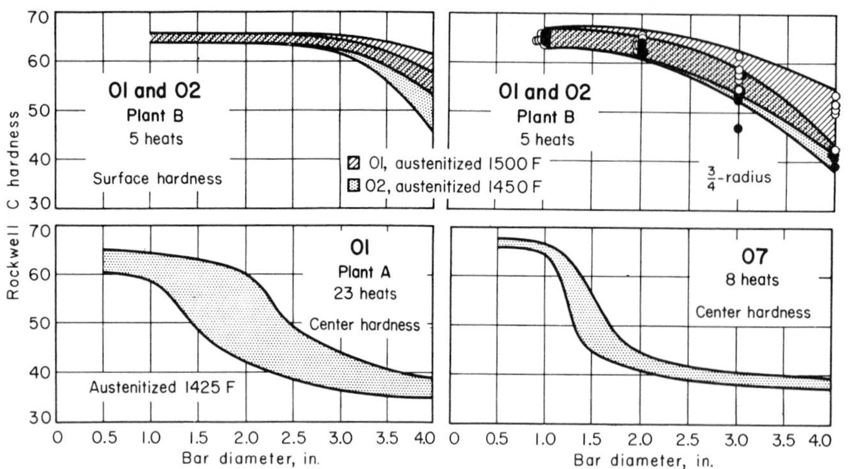

Fig. 4. Effect of bar diameter and grade on oil-quenched hardness at center, surface and ¾-radius for oil-hardening tool steels

ing S1 and S5 grades to the relatively shallow-hardening S4 grade (Fig. 3). These are steels of intermediate alloy content whose hardenability is controlled to a much greater extent by the actual composition of the steel than by the incidental effects of grain size and melting practice, which are so important to the group W steels. The S steels require relatively high hardening temperatures to achieve good hardness; consequently undissolved carbides are not a factor in the controlling of hardenability.

These points are brought out in Fig. 3, where the S1 and S5 steels are much higher in alloy content than the lower-hardenability S4 steel. The S1 and S5 steels normally are oil hardening while S4 is strictly a water-hardening grade; in fact S4 will harden with a case-core relationship basically similar to the group W grades in sizes larger than 1 in. round. Steels S2 and S3 are either water or oil hardening; sections of ½ in. and under are satisfactorily hardened in oil. In water, sections ¾ in. and smaller in these two grades will harden throughout.

Oil-Hardening Cold Work Steels, Group O. Typical relationships between bar diameter and Rockwell hardness at the surface, ¾-radius, and center for some of these grades are shown in Fig. 4. Examples of using the end-quench test for measuring hardenability of tool steels are shown in Fig. 5. Generally only the group O, L and S steels are adaptable to end-quench hardenability testing.

Since wear resistance of cold work die steels is an important factor, carbon contents of group O steels are high. This results in considerable undissolved carbide at the hardening temperature so that despite their relatively high alloy contents, some of these grades do not harden as deeply as the lower-alloy S grades (for instance S1 in Fig. 3, compared with O7 in Fig. 4). The higher carbon makes these steels higher in surface hardness than any S grade.

Raising the hardening temperature has several effects on hardenability: It increases grain size, increases solution of alloying elements and dissolves more of the excess carbide so that, as shown in the results for O1 and O2, plant B, in Fig. 4, hardenability is increased. Based on end-quench hardenability (the majority of tool steel metallurgists consider this the more accurate test for evaluating hardenability of oil-hardening steels), all of the O grades will harden to Rockwell C 60 in the center in sizes up to 1¼-in. diam; the higher-alloy O1 normally hardens to Rockwell C 60 in 1¾-in. rounds. The surface hardness of all grades is Rockwell C 65 to 66 in sizes up to 2-in. diam (Fig. 4).

Comparing the O grades with W grades of similar carbon content, it is noted that W1 will harden to Rockwell C 55 to a depth of only about 3/32 in. while O7 will harden to a *center* hardness of Rockwell C 55 in a 1½-in. round and O1 to a center hardness of Rockwell C 55 in a 2-in. round.

Air-Hardening Die Steels, Group A. It is important to note for all air-hardening steels that cooling conditions are such that the surface and center cool at nearly the same rate. Thus center hardnesses of sections of all sizes of air-cooled steels will be the same as the surface hardnesses.

By definition these air-hardening steels must harden under the relatively slow cooling rate of air quenching. Therefore they are high in alloy content. Undissolved carbides, grain size and melting techniques are unimportant in their influence on hardenability. The hardenability of the A2 grade, which is the most popular of all the A steels, is such that in 3½-in. rounds it will harden to Rockwell C 62 in the center (Fig. 6). Hardening temperature and slight variations in the manganese content have a pronounced effect on the hardenability of this grade (Fig. 7). Hardenability may be increased by raising the hardening temperature slightly, so that some of the excess alloy carbides are dissolved. It should be emphasized, however, that hardening temperature can be increased only slightly for the A2 type; too great an increase will cause considerable austenite to be retained, with consequent serious deterioration of mechanical properties and loss in dimensional accuracy. This is particularly true of the high-manganese varieties of A2.

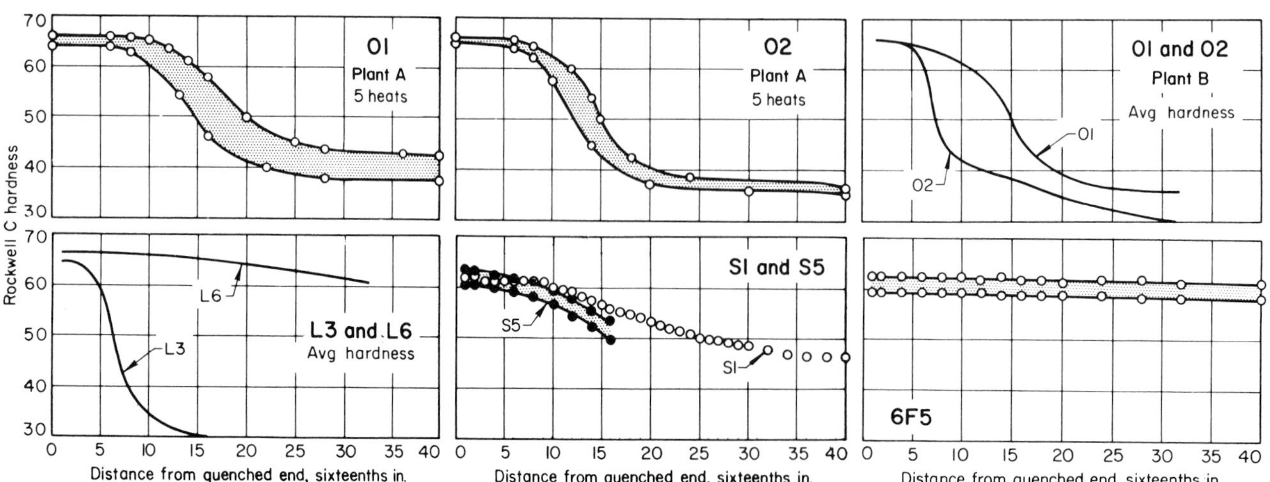

Fig. 5. End-quench hardenability values for seven tool steels. Steel O1 was austenitized at 1500 F, O2 at 1450, L3 at 1500, L6 at 1475, S1 at 1725, S5 at 1600 and 6F5 at 1600 F.

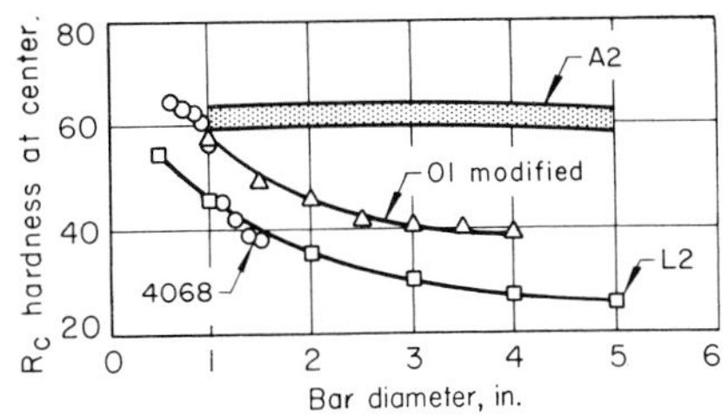

	A2	O1 mod	L2	4068
C	1.01	1.00	0.55	0.70
Mn	0.36	1.17	0.70	0.86
Si	0.70	0.25	0.22	0.27
Cr	5.28	0.56	1.00	...
Mo	1.15	0.09	...	0.24
W	...	0.53	...	...
V	0.24	0.27	0.20	...

Sizes 1 to 3-in. diam of A2 were air cooled from 1750 F; sizes 4 to 6-in. diam were air cooled from 1775 F. Steel O1 modified was oil quenched from 1450 F, and L2 and 4068 from 1525 F.

Fig. 6. Effect of bar diameter on as-quenched center hardness

Although the A4, A5 and A6 grades are less popular than A2, they have the advantage of lower hardening temperatures with consequent heat treating simplification. Sections of A6 up to 4 in. square will harden to Rockwell C 61 to 63 throughout and an 8-in. square will harden to Rockwell C 60 at the center; in this respect A6 is similar to A2. The A4 and A5 steels have similar hardenability but in general show a one-point lower surface hardness. The A7 grade is a direct counterpart of A2 in hardenability but has considerably higher wear resistance.

High - Carbon High - Chromium Cold Work Steels, Group D. Tools made from air-hardening steels D2, D5 and D7 have been hardened above Rockwell C 60 in minimum section sizes of 12 in., and therefore hardenability is not an important consideration. In D3 and D6, so much of the chromium is tied up with the high carbon content that the actual hardenability is less than in D2 and similar grades. For this reason D3 and D6 are frequently oil hardened in sections of more than 1½ in. where full hardness is required. Because of the molybdenum addition in D4, this grade can be air hardened satisfactorily.

All of these grades except D7 provide surface hardnesses of approximately Rockwell C 62 to 64 under normal hardening conditions. D7, because of its particularly high alloy content and numerous hard vanadium carbides, will quench out to Rockwell C 65 to 67.

Hot Work Tool Steels, Group H. Relative hardenability of the chromium hot work steels is not an important or limiting factor in their selection, and within the limits of presently commercially available sizes, steels H11 to H16 possess the ability to be air hardened to Rockwell C 50 or higher. Tungsten hot work steels H21, H24 and H26 will readily develop hardnesses greater than Rockwell C 50 in all commercially available section sizes. However, steels H23 and H25 will not harden to Rockwell C 50 in any section size because of their low carbon content. For hardnesses of Rockwell C 55 the selection of tungsten hot work tool steels is limited to H24 or H26. H21 normally will not attain Rockwell C 55 without an increase in the carbon content. All molybdenum hot work steels, H41, H42 and H43, are capable of developing a minimum hardness of Rockwell C 55 in all available section sizes.

Tungsten and molybdenum hot work steels require high quenching temperatures to develop full hardenability and must be hardened in the same type of furnace required for high speed steel.

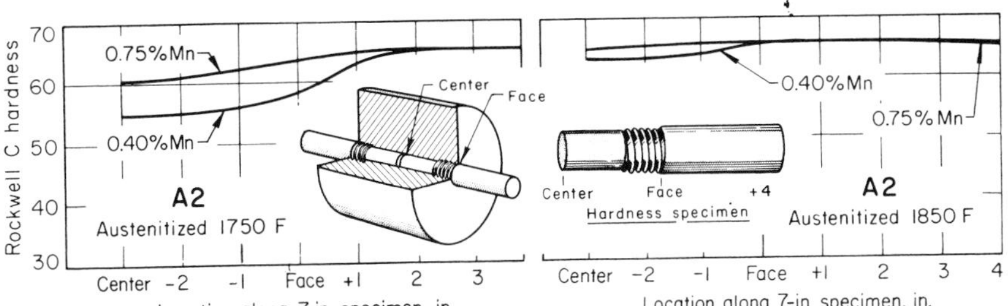

Fig. 7. Effect of manganese and austenitizing temperature on the hardenability of A2 air-hardening tool steel. (See text for description of test.) Approximate relationship of specimen location and corresponding bar diameter is as follows:

Location on specimen, in.(a)	Center to −1	Face	+1	+2	+3	+4
Corresponding bar diameter, in.(b) ..	6	5	3½	3¼	1¼	1

(a) refers to location along 7-in. test specimen in inches; (b) refers to corresponding value for approximate bar diameter in inches.

Tungsten High Speed Steels, Group T. These steels are all deep hardening when quenched from their recommended hardening temperatures (2150 to 2350 F). However, except in unusual circumstances, they are seldom used in hardened sections of over 3 in. Even very large cutting tools such as 3 and 4-in. diam drills have a relatively small effective section for hardening because of the removal of metal for flutes. Some very large diameter solid tools are made from high speed steel — for example, broaches and cold extrusion punches up to 4 and 5 in. in diameter. However, with such tools only the surface hardness is important. Figure 8 indicates a possible drop in center and surface hardness with increase of bar size. These values are more an indication than a specific result, and the foregoing comments with regard to tool section should be held in mind. For tools of extremely large diameter and heavy section, it is common to use an accelerated oil quench that is effective in gaining one or two Rockwell C points over the more common hot salt quenching and air-cooling procedure. This direct oil quench is preferred for full hardness in sizes over 3 in.

Frequently the word "hardenability" when used in connection with high speed steel does not refer to the depth of hardness so much as it refers to the specific hardness that can be obtained under standard procedures of quenching and tempering. Used in this sense it is a common and useful criterion of high speed steel quality.

The maximum hardness of the tungsten high speed steels varies somewhat with the alloy content and especially the carbon content. Rockwell C 64.5 is obtained on all high speed steels. Some grades such as T15, because of their very high content of carbon and hard carbides, will harden to Rockwell C 67.

Molybdenum High Speed Steels, Group M. The great similarity between the tungsten and molybde-

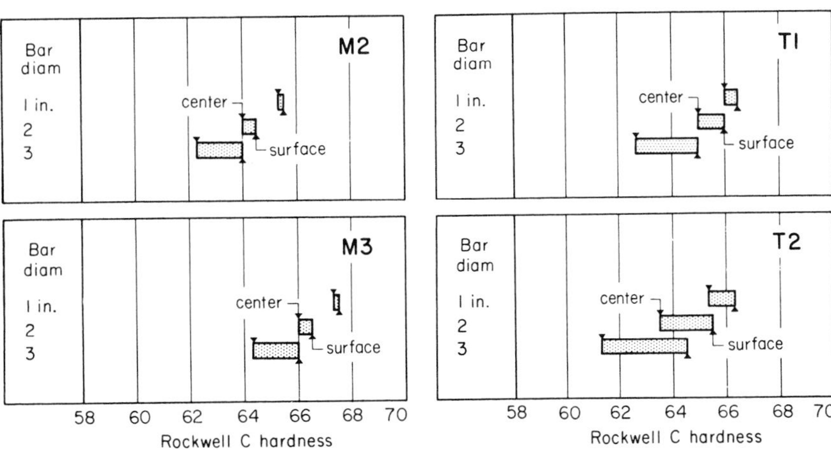

Fig. 8. Effect of bar diameter on range of surface-to-center hardness of high speed steels. Steels M2 and M3 were oil quenched from 2200 and 2250 F respectively. Steels T1 and T2 were oil quenched from 2350 F.

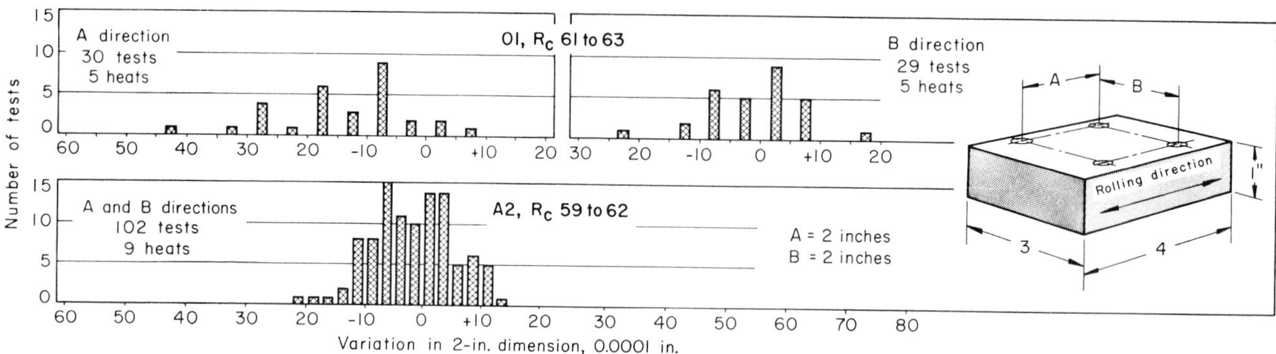

Fig. 9. Dimensional changes in heat treatment of oil and air-hardening tool steels. O1 was oil quenched from 1460 F and tempered at 350 F. A2 was air cooled from 1750 F and tempered at 350 F.

num high speed grades makes the comments for the group T steels applicable to these steels also. A possible exception is the M10 grade, which usually has slightly lower hardenability than other molybdenum grades, necessitating the use of an oil quench in sections larger than about 1½ or 2 in.

Again the actual composition will determine the maximum hardness possible; for example, the M3, M4, M7, M15 and M33 grades will harden to at least one or two points Rockwell C higher (Rockwell C 66 or slightly higher) than other lower-carbon group M steels. When properly hardened all group M steels will test at least one-half to one point Rockwell C harder than comparable group T steels.

Low-Alloy Special-Purpose Tool Steels, Group L. Considerations developed for the group O steels with regard to hardenability apply to the group L steels except that their greater variation in composition causes the group L steels to show a wider range in hardenability. This is particularly evident when hardenability characteristics of the L2, L3 and L6 steels are examined, as in Fig. 5 and 6. These figures show that the L2 grade will harden to Rockwell C 55 at the center only in sizes of ½ in. or smaller (Fig. 6) while L3 will harden to Rockwell C 60 at the center of a ¾-in. round. On the other hand, L6 will harden to Rockwell C 65 in the center of 3-in. rounds when oil quenched (Fig. 5). The relatively high alloy content of these steels makes them oil hardening, except that L1 and L3 are usually water hardened in the larger sections.

Carbon-Tungsten Tool Steels, Group F, are in general low-hardenability steels. The high tungsten and carbon contents combine to form complex iron-tungsten carbide, which both removes the tungsten from the matrix where it could slow down the high-temperature cooling transformations, and produces numerous particles of iron-tungsten carbide, which is present at the hardening temperature, further reducing hardenability. Chromium in the **F3** grade, however, increases the hardenability to the extent that this material is an oil-quenching steel. Actually the **F1** and **F2** steels will show a shallower penetration than typical grades of W steels of the same carbon content. Some increase in core hardness of group **F** steels over group W steels is apparent, however. For example, Rockwell C 45 to 50 is a typical core hardness in the 1-in. round for an **F** steel as compared with Rockwell C 40 to 45 in the W steels.

Low-Carbon Mold Steels, Group P. Hardenability of the group P steels varies from the extremely low results typical of P1 to the high hardenability of the P4, P6 and PPT steels. The low hardenability of carburized P1 means that a special quench must be used, which may offset the advantages of the extremely low annealed hardness. On the other hand, P4 is an air-hardening steel after carburizing. The PPT steel is precipitation hardening, attaining maximum hardness of Rockwell C 38 to 40 in sections as large as can be practically produced.

Other Alloy Steels, Groups 6G, 6F and 6H. The alloy content of these materials indicates that they are all extremely deep-hardening steels, so much so in fact that only 6F2 and 6G are strictly oil quenching, the remainder being either air and oil-quenching steels or air-quenching alone. Hardenability results measured by end-quench tests on 6F5 steel show that it is essentially through hardening to Rockwell C 60 in sizes up to at least 3½-in. when air hardened (Fig. 5).

Hardenability Tests

Hardenability testing of tool steels presents many difficulties not normally encountered in similar testing of the standard alloy steels. A wide range of hardenability is encountered in tool steels, from the very shallow-hardening plain carbon steels to the extremely deep-hardening die and high speed steels. Some common hardenability testing techniques, specifically adapted for tool steels, follow.

Hardness traverse is probably the oldest and most obvious method of testing hardenability. A sample of any desired size is hardened and then sectioned roughly in the center of its length by nicking and fracturing or abrasive cutting. The cross section is then ground and Rockwell hardness readings are taken from surface to center at specified distances apart. Although this method is time-consuming, expensive and relatively inaccurate, particularly for the shallow-hardening steels, it is still widely employed as a spot-check method.

Shepherd PF Test. In this test a ¾-in. round sample, 3 in. long, of regular mill-annealed stock, is first normalized by oil quenching from 1600 F, reheated to various hardening temperatures as specified in the particular test involved, then quenched in 8% brine. The sample is nicked with a cutoff wheel in the center of its length and fractured. Polishing and etching the fractured surface delineates the case-core areas and allows a visual determination of case depth. At the same time the opposite half of the sample is evaluated for fracture grain size in the case region by comparison with a set of standard specimens. This test determines the hardenability as well as the grain-coarsening characteristics of the material at various temperatures.

The Shepherd PV test is specifically adapted to the shallow-hardening carbon steels, which are difficult to rate by the hardness traverse or Shepherd PF test. In this test a sample is sawed or otherwise machined to produce a 90° V-point on the end of the bar to be tested. Bar diameter may vary from 1¼ to 1¾ in. Sizes larger than 1¾ in. should be machined to 1½ in.; sizes smaller than 1¼ in. may be quenched in a special fixture. Samples are cut off to 1⅛ in. long and quenched from an originally spheroidized condition at specified temperatures in a jet fixture which rapidly quenches the V-point.

The hardened samples are then sectioned to remove a disk with parallel sides from the center of the testpiece at 90° to the V-point. After proper preparation, the samples are etched and tested to determine the depth of hardness penetration below the point. This enables a much wider hardness gradient to be observed than is possible on round samples and is particularly suited to the shallow-hardening steels. Results are reported in inches penetration to Rockwell C 55 as converted from superficial or Rockwell A hardness values. Results are also coded with the quenching temperature.

The taper cone test, which is extremely sensitive, was suggested for the very shallow-hardening steels by C. B. Post and O. V. Greene (Trans ASM, **30**, 1202, 1942). Samples are prepared from bars by machining a tapered section 5 in. long, 1¼ in. in diameter at the large end and ¼ in. in diameter at the small end. These samples are then quenched from specified austenitizing temperatures into brine and sectioned axially. After metallographic polishing and etching, the depth of hardened zone, accentuated by the specimen taper, is clearly shown. The chief disadvantage of this test is its relative difficulty in execution, but in certain instances this difficulty is justified.

Standard end-quench hardenability testing is frequently used to determine the hardenability of the medium-hardenability or oil-hardening types of tool steels. This is a standard test and is described in the article on Selection of Steel for Hardenability. Data in Fig. 5 for some of the oil-hardening tool steels were secured by use of the end-quench test.

Air-Hardening Steels. Because of the deep hardenability of the air-hardening steels, end-quench tests and the various tests described above are not applicable since specimens would harden uniformly throughout the entire section. For this reason a special test has been developed by C. B. Post, M. C. Fetzer and W. H. Fenstermacher (Trans ASM, **35**, 85, 1945) and has been previously sketched in Fig. 7. This test allows the hardenability evaluation of sizes from a 6-in. round to a 1-in. round. It is a useful test for large sizes of deep-hardening steels.

The 6-in. diam test block used in this test is a 30% Ni alloy that has the same thermal characteristics as most air-hardening steels whose hardenability is to be determined. The small testpiece of the air-hardening steel is convenient and inexpensive to produce. To perform the test, the samples are screwed into the large 6-in. test block and heated to the hardening temperature. The entire assembly is then removed from the furnace and air cooled. At room temperature, the testpieces are removed from the test block and ground for standard Rockwell hardness testing along the entire length.

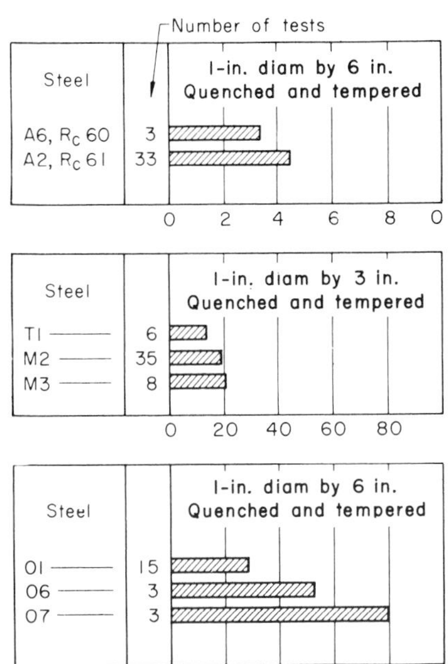

(Upper) Steel A2 was air cooled from 1750 F and tempered at 350 F. Steel A6 was air cooled from 1550 F and tempered at 350 F.

(Center) All three steels were preheated at 1300 and 1600 F and salt quenched at 1125 F. Steel T1 was austenitized at 2305 F and tempered at 1050 F. Steel M2 was austenitized at 2195 F and tempered at 1025 F. Steel M3 was austenitized at 2210 F and tempered at 1000 F.

(Lower) Steel O1 and O6 were oil quenched from 1460 F and tempered at 350 F. Steel O7 was oil quenched from 1525 F and tempered at 400 F.

Fig. 10. Growth of seven tool steels during heat treatment

Dimensional Changes in Heat Treatment*

Dimensional changes in heat treatment are affected by stresses induced by machining or other fabrication, thermal stress from heating or cooling, transformation stresses caused by the difference in volume of austenite and martensite or other transformation products, and finally by the shape of the piece being heat treated. Rapid heating and cooling can intensify thermal stresses to the extent that the tool steel piece may actually be cracked by this alone. The most rapid cooling is that developed in brine quenching, so that the deformation during heat treatment of the water-hardening steels is much greater than that of the oil and air-hardening types. Transformation stresses are intensified by cooling conditions that allow different parts of the same piece to pass through the austenite-to-martensite transformation range at different times. Such cooling rates naturally exist in large tools, particularly if small sections adjoin heavy sections.

The shape of the piece influences the rate of cooling and thereby the thermal stress gradients as well as transformation stress gradients. Although shrinkage on cooling causes thermal stresses, frequently in different amounts depending on the specimen shape, transformation stress will cause an expansion since martensite occupies a greater volume than austenite. Total deformation of a quenched tool will also be influenced by the transformation involved in tempering, which generally involves shrinkage followed by slight expansion, finally followed by further shrinkage.

*For additional data and comparisons, the reader may refer to the article "Distortion in Tool Steels", immediately following this article, on page 654.

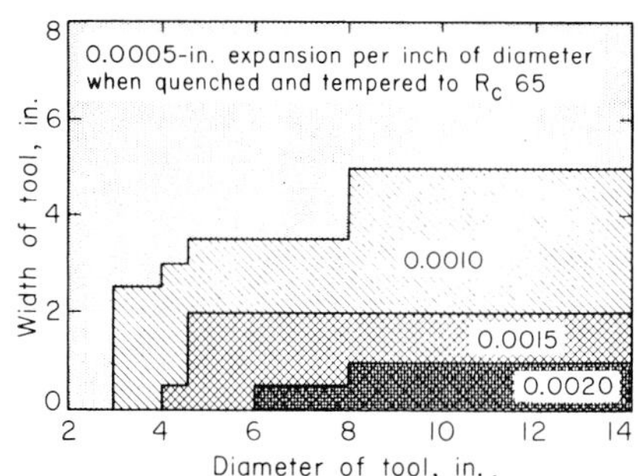

Fig. 11. Typical expansion allowances for heat treatment of M2 high speed steel tools of various sizes, indicating effect of size on deformation.

Figure 9 gives some experimental results obtained in treating oil and air-hardening steels of a specific shape. It will be noted that the oil-hardening steel deforms almost twice as much as the air-hardening steel. This is a typical ratio of deformation between the two types of tool steel. The spread of dimensional change from contraction to expansion is probably influenced by warping of the testpiece as well as by uniform deformation.

The deformation of the water-hardening steels is diffcult to predict because of the pronounced effect of section size and hardenability. In general the shallow-hardening water-quenching steels will deform less than the deeper-hardening grades of this group of steels. Large sections of these grades, which harden only superficially, will deform much less than even the oil or air-hardening grades in the same section. Long thin sections or tools of varying cross section will warp and distort to a high degree because of the high thermal stresses as well as the nonuniform transformation stresses developed by water quenching. However, this type of distortion is not truly an inherent quality of the steel but depends on factors of size and shape.

Figure 10 gives comparative data for hardening deformation of various grades of oil and air-hardening steels, including high speed steels. This shows the greater deformation of the shallower-hardening O6 and O7 steels as compared with O1 steel, and the effect of lower thermal stress on the deformation of steel A6 compared with A2.

The results for the tungsten and molybdenum high speed steels (Fig. 10) are useful in estimating hardening deformation to be compensated for in machining allowances. Note in Fig. 11 that different sections of M2 distort by different amounts. This is an example of the effect of tool shape on deformation.

In general the hardening deformation of the lower-alloy tool steels is uniform in all directions. Higher-alloy steels in groups D, T and M are anisotropic in this respect. For example, D2 steels will expand about twice as much in the direction of rolling as across the direction of rolling. Over-all volumetric deformation of the D2 steels, however, is

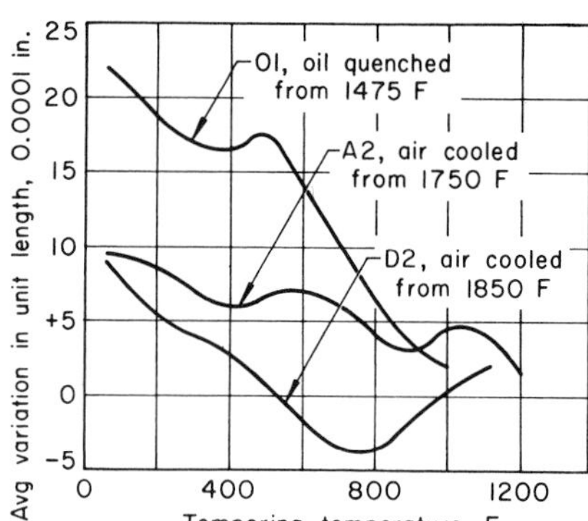

Fig. 12. Effect of tempering temperature on size changes during heat treatment of three tool steels. Length changes are averages of three directions for 1 by 2 by 6-in. testpieces.

less than half that of the A2 grades, and for this reason nonuniform deformation of D2 is not ordinarily significant.

Relative size changes for three grades of air and oil-hardening steels as an effect of tempering temperature are presented in Fig. 12. The O1 steel shows greater over-all deformation at normal tempering temperatures principally because of the oil quenching required and the high ratio of martensite to austenite in the as-quenched structure. Tempering first reduces expansion by the transformation of the tetragonal martensite to cubic martensite, with consequent shrinkage. The slight expansion that occurs after this original contraction is a result of the transformation of retained austenite. Continued tempering contracts the martensite with gradual shrinkage from all dimensions. Also, the transformation temperatures for the retained austenite vary greatly depending on the steel composition, and the A2 grades show a double range of austenite transformation.

The effect of increased austenitizing temperature on hardening deformation is extremely important for all grades. In the lower-alloy grades its main effect is in increased hardenability and higher thermal stresses, but in the high-carbon high-chromium die steels its effect is amplified by the fact that these grades will retain a large amount of austenite if quenching temperatures are much higher than those recommended. This increased amount of retained austenite causes severe shrinkage, and if high-temperature tempering or subzero treating is used to transform the retained austenite, the attendant expansion will also be severe and may cause cracking or excessive deformation. Under normal conditions tempering of the high-carbon high-chromium steels at 400 to 800 F gives the best results for over-all deformation of as little as 0.0005 in. per in.

The effect of section size and method of quenching is important even in the high speed steels. This is shown in Fig. 13, in which the pronounced barreling of the section is apparent from the measurements given. Salt quenching of such a section would have substantially eliminated this effect because lower thermal shock and more uniform cooling would allow the cross section to shrink more or less uniformly regardless of the section size.

Figure 14 gives some comparative results in determining the relative tendency of W1 tool steel to crack in a delicate section. As noted from the shape of the special testpiece pictured in Fig. 14, the 0.25-in. hole was varied in location so that the wall thickness between the hole and the surface varied from about 0.094 to 0.250 in. Even the 0.93% C grade showed total cracking when the wall thickness was 0.094 in. The incidence of cracking decreased as the section thickness increased until no cracking was found in a section thickness of about 0.188 in. for the 0.93% C steel. However, as the carbon was increased the section thickness required for no cracking increased to about 0.203 and 0.234 in., respectively, for grades containing 1.03 and 1.09% C. Under practical conditions such a high incidence of cracking would warrant a change to oil or air-hardening steel.

Quench cracking is intensified by low martensitic transformation temperatures. If, for example, martensitic transformation temperatures are near room temperature, the expansion caused by the austenite-martensite transformation is resisted by hard, relatively nonductile material. In large sections of high-alloy steels, particularly the high speed steels, there is considerable probability of cracking during cooling from the austenitizing temperature, even during air cooling if the cooling is continued to room temperature. This tendency toward cracking is also influenced by the amount of austenite (a relatively ductile constituent) that is retained in cooling. The amount of austenite is influenced by the amount and type of alloying elements and the hardening temperature. However, even though higher percentages of austenite may lower the probability of cracking in cooling from the austenitizing temperature, the probability of cracking during tempering is increased. In any instance it is advisable to temper such large sections as soon as they are down to approximately 100 to 150 F, but not before they can be picked up comfortably with bare hands. Double tempering is a recommended practice for the higher-alloy steels.

Heat-Resisting Properties

Tool steels that must operate at high temperature must first resist softening or tempering under the effects of heat. For the hot work steels, this resistance to softening must be coupled with resistance to erosion or "washing" at the high temperatures and also with resistance to heat checking or fire checking, which is the formation of shallow "craze cracks" on the heat-affected surface.

For cutting tools, the most important property in this respect is red hardness, or resistance to softening under the temperatures generated at the tip of the tool. Although many cutting tools are flooded with oil or other coolant to the extent that the development of high temperatures in their operation would appear unlikely, nevertheless wear and eventual failure of cutting tools is largely connected with the red hardness properties of the steel.

Softening of hardened tool steels is the result of the precipitation and agglomeration of carbides from the hardened martensitic matrix. Steels that resist this precipitation effect are high in alloying elements that resist diffusion, thereby preventing or slowing down the formation and agglomeration of the alloy carbide particles. Resistance to the softening effect of heat is also improved by the presence of high-alloy carbides that retain their hardness at elevated temperatures. Finally a steel that has high hardness at elevated temperatures will also have high hardness at room temperature.

For extremely high red hardness, therefore, a tool steel with high room-temperature hardness, a microstructure containing numerous high-alloy carbides, and a composition heavily alloyed with diffusion-slowing elements must be selected. Alloying elements that promote all the above requirements are molybdenum, chromium, tungsten and

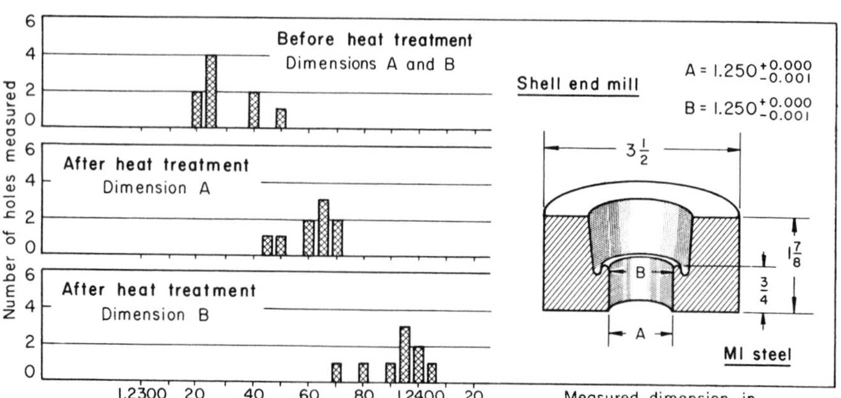

Fig. 13. Change in dimensions in heat treating tools made from M1 high speed steel. Tools were oil quenched from 2200 F and tempered at 1050 F.

vanadium, with supplemental additions of cobalt being particularly effective in the high speed steels.

Hot work steels, since they require good toughness in service and good but not necessarily highest red hardness, are lower in carbon and alloying elements than the high speed steels. Here again the high-temperature properties are obtained by the content of chromium, molybdenum, tungsten and vanadium. Vanadium also produces hard carbides that resist erosion or washing, while the lower-alloy steels resist heat checking because of their greater toughness. The ability to withstand water cooling is inherent in the lower-alloy hot work steels. Some die steels with lower red hardness will be more useful than the more brittle materials of high red hardness such as H26 and H41 because of their ability to be water cooled in service. Cobalt rarely is used in increasing the red hardness of the hot work die steels because of its cost and tendency to promote brittleness for these applications involving high-shock loads.

Since high room-temperature hardness and high red hardness are necessary for the good performance of cutting tools, high speed steels for this application are highly alloyed with the metallic alloying elements and carbon. This produces a material that can be hardened to the highest operating hardness of all steels and will retain most of this hardness up to approximately 1100 F and regain it completely upon cooling to room temperature, provided the original tempering temperature has not been exceeded. The many hard wear-resistant carbides in these steels not only contribute to red hardness but also are most important in increasing wear resistance and basic hardness.

Because of the great range in operating conditions and requirements for red hardness and wear resistance for both the hot work and high speed steels, it is difficult to formulate definite criteria for selection of a particular grade for a special operation. In general, however, it will be satisfactory to select a grade of good ductility and preferably lowest cost to determine actual operating conditions. Depending on the performance of this first tool, steps can be taken either in the direction of improving wear resistance or providing greater toughness in order to secure optimum results.

Effect of Composition on Tool Performance

Chemical compositions of tool steels are held to ranges considerably closer than those generally obtainable in constructional alloy steels. This is because of the pronounced effect of slight variations in composition on tool performance; for example, note in Fig. 7 the effect of manganese on hardening response of the A2 steels. Actual composition ranges vary considerably,

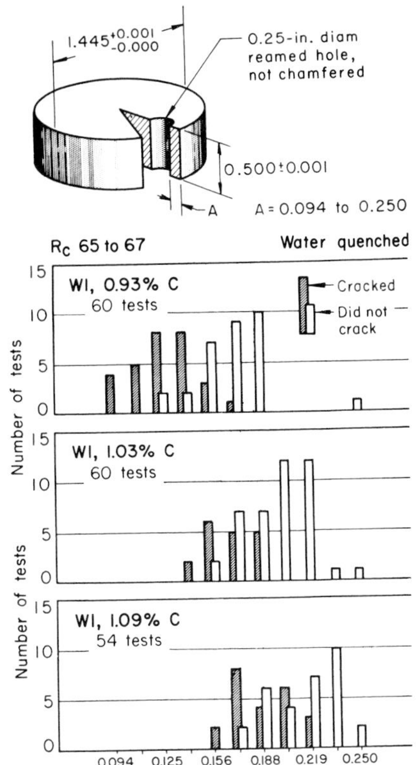

Fig. 14. Relative susceptibility to quench cracking in the thin section (purposely varied from 0.094 to 0.250 in.) of three W1 steels normally used for the same tools. The test specimens, water quenched from 1420 F, were not tempered.

depending on the alloying elements involved and the type of steel produced. Manufacturers maintain as close limits as practical, considering both the economics of steelmaking and the resultant alloy properties.

Carbon in high speed steel is commonly held to ±0.02% because of its pronounced effect on hardness and other properties. On the other hand, tungsten in high speed steel

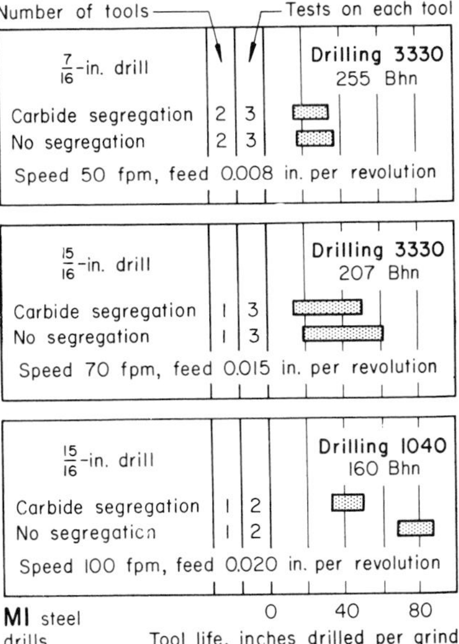

Fig. 15. Effect of carbide segregation on the life of M1 steel drills

need be held to only ±0.75 or ±1.00% since variations of this amount do not affect eventual tool performance. In water-hardening tool steels, however, a rather unusual situation exists in that from three to four different grades or qualities of these steels are available with exactly the same nominal composition. (These qualities were discussed earlier in this article, under water-hardening tool steels.) Therefore it follows that there may be a difference in performance among steels of the same nominal composition because of the differences in grain size, microstructure and hardenability.

All tool steels are alloyed (even the W grades may be included in this generalization if carbon is considered an alloying element), and the probability of segregation is directly proportional to alloy and carbon content. Much work has been done by various producers of tool steel to minimize segregation since its effect is usually very deleterious to the production and ultimate life of the finished tool. In the high-alloy tool steels, segregation causes nonuniform deformation, hardness variations from zones of retained austenite, and microstructural inhomogeneity responsible for planes of weakness or actual breakage after heat treatment. Melting and forging techniques are the main factors in controlling segregation, with melting technique paramount.

In recent years with more advanced processing techniques, the effect of segregation on the high-alloy steels has become much less severe. Some results obtained in drill testing high speed steel drills made from segregated and nonsegregated bars are shown in Fig. 15. As might be expected, segregation generally reduces drill life. This is particularly important for drills since the most highly segregated area is in the web, which is also the most highly stressed area of the tool.

Some of the higher-alloy steels are also available with controlled additions of sulfur to improve machinability. The advantage of easier machining is generally recognized in the softened state, where the tool life or production rate in the manufacture of tools can be increased about 25%. An added advantage is the improvement in surface finish; such improvement in finish is, of course, important only on those tools or portions of tools that are not to be ground after heat treatment.

The addition of sulfur can also enable die steels to be machined after certain heat treatments. Hot work die steels, for example, have been machined at hardnesses in the range from Rockwell C 45 to 46. Figure 16 indicates some results obtained in the life-testing of tools made from sulfurized and nonsulfurized steels, including prehardened hot work tool steels. Machining of prehardened sulfurized steel results in a die with freshly ma-

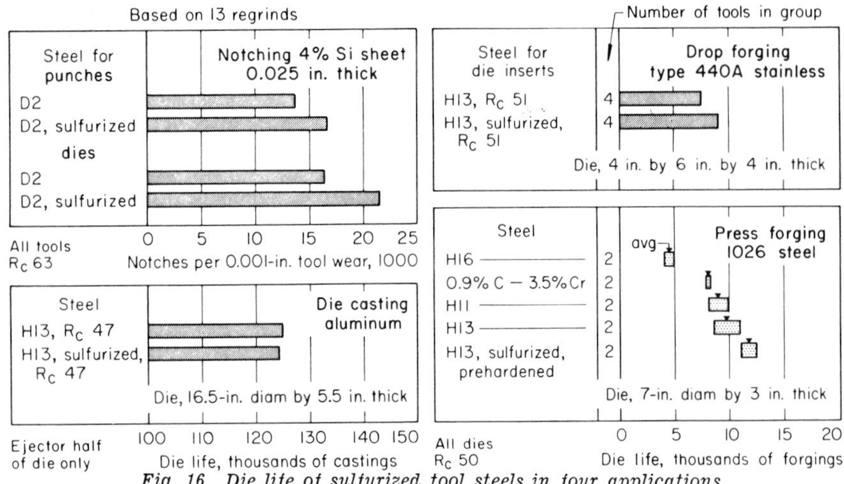

Fig. 16. Die life of sulfurized tool steels in four applications

chined surfaces free from detrimental conditions that might have developed during heat treatment. Machining tool steel at Rockwell C 45 or higher is generally impractical unless the steel is free machining. Recent work has also indicated an improvement in grindability with an addition of sulfur, particularly for the higher-alloy tool steels that are difficult to grind.

The effect of sulfur on the life of M1 high speed steel drills is shown in Fig. 17. In general, no improvement in tool performance can be expected from the use of the sulfurized tool steels. Their advantage must be realized during the manufacture of the tool, as a result of easier machinability.

Sulfur additions must be carefully controlled to prevent gross inclusions that would adversely affect tool performance. As it is presently added, sulfur for machinability is visible in the microstructure only under high magnification, as thin elongated stringers dispersed so that mechanical properties of the material are not affected to any significant extent.

Table 1 on page 638 notes those types of tool steel generally available in a free-cutting grade.

Modern melting techniques, careful scrap selection, and quality control of the tool steel mill products have virtually eliminated incidental residual elements such as arsenic, tin, copper and selenium from tool steel. These elements are usually detrimental; Fig. 18 gives some indication of the influence of tin and arsenic on the life of T1 drills.

Defects in Tool Steels

The alloy additions to tool steels add considerably to the complexities of subsequent hot working of the steel. Steels in the A, D, T and M categories are likely to develop carbide segregation at their centers during normal solidification of the ingots, and an ingot or billet containing carbide concentrates in the center will not flow smoothly or uniformly during hot fabrication. All of these factors make it very difficult to fabricate tool steel by hot pressing, forging and rolling, and all of these fabrication operations create the likelihood of developing internal fissures or bursts, even with good hot working practices. All of these grades must be heated and

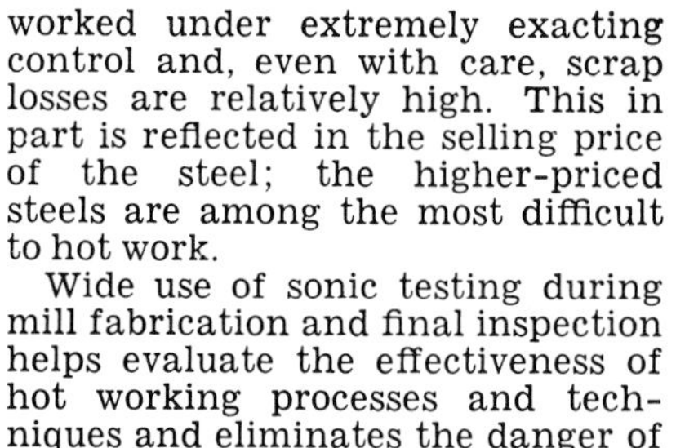

worked under extremely exacting control and, even with care, scrap losses are relatively high. This in part is reflected in the selling price of the steel; the higher-priced steels are among the most difficult to hot work.

Wide use of sonic testing during mill fabrication and final inspection helps evaluate the effectiveness of hot working processes and techniques and eliminates the danger of shipping material with gross internal flaws from fabrication.

Directionality. Tool steels in the form of rolled or forged bar exhibit fiber directionality or anisotropy because generally the hot working is confined to two axes. This elongates the grains, alloy segregate when present, and undissolved carbide constituents in the direction of working. As a result, properties of the material vary depending on orientation across the fiber or with the fiber of the steel. For example, size change in heat treatment of D2 steel is greater when measured with the fiber, compared with measurements at right angles to the fiber. Strength is influenced little if any by direc-

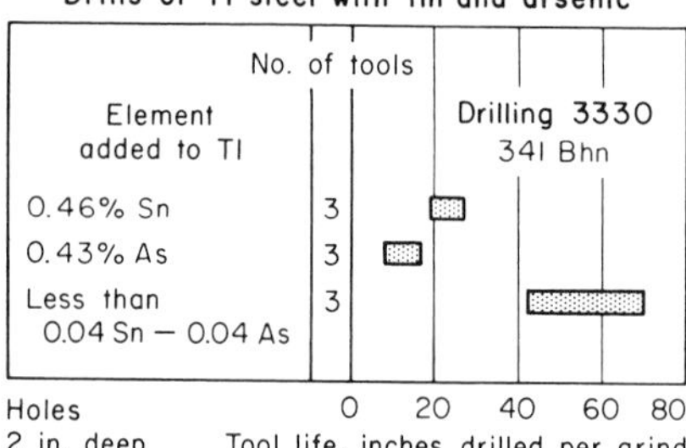

Fig. 18. Effect of tin and arsenic on the life of T1 high speed steel drills

tionality, but directionality has a strong influence on ductility and toughness.

Figure 19 illustrates the influence of directionality on toughness for several of the popular cold work die steels. Here the steels with the minimum amount of undissolved carbides in their heat treated structure exhibit the least anisotropy. Types O1, A2 and A6, all steels with few excess carbides, have a longitudinal-transverse toughness ratio of 0.5 or higher. D2 and D3, containing a large volume of excess carbides, show directional toughness ratios of 0.3 or less. This is one of the drawbacks to the use of the higher-alloy steels and one of the strong points favoring the lower-alloy varieties.

Effects of directionality can be reduced by orienting dimensionally weak or highly stressed areas in a tool or die across the fibers of the steel. When the tool design is such that layout for preferred orientation is impossible, as in die-casting dies, extrusion dies, large milling cutters, or hobs, the use of individually upset-forged pieces is recommended. Upset forging permits hot working along three axes, which deforms the material more or less equally in all directions, minimizing directionality.

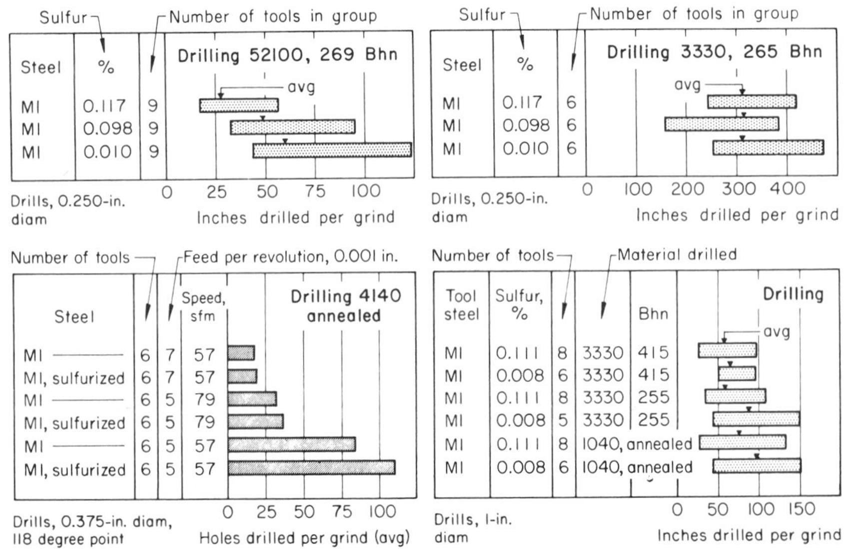

Fig. 17. Sulfurized tool steels compared with nonsulfurized grades for drilling

Decarburization. Tool steels decarburize in oxidizing atmospheres above 1400 F; those with significant molybdenum, silicon and cobalt contents are considered more susceptible. [The effect of silicon (and probably cobalt), however, may be confused with effects of these elements in lowering the hardenability of low-carbon areas, as discussed by P. Payson in *Metal Progress* for December 1960.]

Steel manufacturers work to a table of decarburization allowances common to all of the grades (Table 2). Individual suppliers or specific products often are held to more restrictive limits than shown in the table. For example, high speed steel rounds more than 5/8 in. in diameter are usually supplied free from decarburization even though the material is ordered as hot rolled.

Unless mill bark is entirely removed it will adversely affect performance of many tools. A notable example is the outside diameter of a drill. Mill bark is often more harmful than loss of a few points of carbon from the surface layers during heat treatment.

In addition to the elements mentioned above that promote decarburization, those steels requiring austenitizing temperatures higher than about 1700 F, those requiring long soaking periods because of the sluggish absorption of carbon in the austenite, and those grades with sufficient hardenability to be air hardening usually exhibit the greatest tendency to decarburize under adverse heat treating conditions. The M series of steels, the cobalt-bearing T group, the entire D series and the molybdenum-base hot work steels, as well as A2, S2, S4 and S5, are rated as generally poor in resisting decarburization and must be protected in some way during the heat treating cycle. The entire W group is particularly good in resisting decarburization and no special protection is needed to obtain satisfactory results. The group O steels are also good in this regard.

No inflexible rules can be established for the amount of material that should be removed after heat treatment to insure complete freedom from decarburization, as this is a function not only of the steel composition but more importantly of the efficiency of the protective mediums used and the hardening cycle. With the proper protection of the surfaces, such as that afforded by rectified salt baths or controlled atmosphere furnaces, any tool steel can be heat treated so that no decarburization results, and unground tools produced from all grades of tool steel are satisfactorily treated regularly. Where suitable surface protection during heat treatment is not available and little or no stock is to be removed in finishing, the tendency to decarburize should be another important consideration influencing the selection of tool steel.

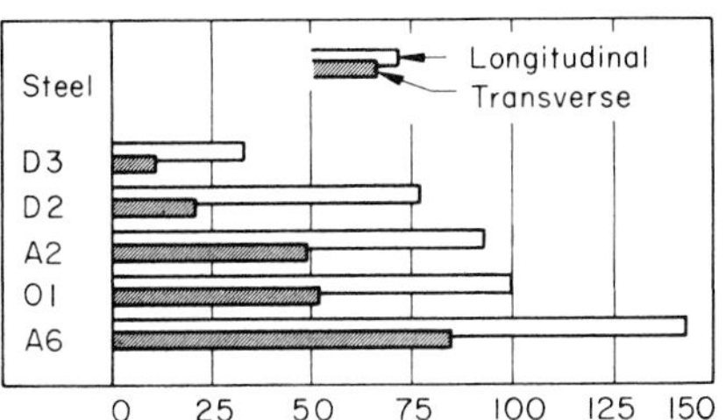

Fig. 19. Relative toughness of unnotched Izod specimens of five tool steels; O1 longitudinal specimens rated at 100. Evaluations were made with substandard-size specimens 1/8 by 5/16 by 1 5/8 in. long. All steels Rockwell C 60 to 61.

Table 2. Decarburization Allowances for Tool Steel Bars (AISI)

Ordered size, in.	Hot rolled	Hammered	Rough turned	Cold drawn	Rough ground
Allowance per Side from Ordered Size for Rounds, Hexagons and Octagons, In.					
Up to ½, incl.	0.016	....	....	0.016	0.004
Over ½ to 1, incl.	0.031	....	....	0.031	0.008
Over 1 to 2, incl.	0.048	0.072	....	0.046	0.012
Over 2 to 3, incl.	0.063	0.094	0.020	0.063	0.016
Over 3 to 4, incl.	0.088	0.120	0.024	0.088	0.020
Over 4 to 5, incl.	0.112	0.145	0.032	....	....
Over 5 to 6, incl.	0.150	0.170	0.040	....	....
Over 6 to 8, incl.	0.200	0.200	0.048	....	....
Over 8	....	0.200	0.072	....	....

Ordered thickness, in.	Side (see cut)	Ordered width, in.: 0 to ½, incl	Over ½ to 1, incl	Over 1 to 2, incl	Over 2 to 3, incl	Over 3 to 4, incl	Over 4 to 5, incl	Over 5 to 6, incl	Over 6 to 7, incl	Over 7 to 8, incl	Over 8 to 9, incl	Over 9 to 10, incl
Hot Rolled Square and Flat Bars												
0 to ½, incl.	A	0.020	0.020	0.024	0.028	0.032	0.036	0.040	0.044	0.048	0.048	0.048
	B	0.020	0.036	0.044	0.056	0.068	0.092	0.104	0.120	0.136	0.144	0.152
Over ½ to 1, incl.	A	....	0.036	0.036	0.040	0.044	0.048	0.056	0.056	0.060	0.060	0.060
	B	....	0.036	0.052	0.064	0.080	0.104	0.120	0.136	0.160	0.160	0.160
Over 1 to 2, incl.	A	....	....	0.052	0.052	0.056	0.056	0.060	0.060	0.072	0.076	0.080
	B	....	....	0.052	0.068	0.084	0.112	0.124	0.144	0.168	0.180	0.180
Over 2 to 3, incl.	A	....	....	....	0.068	0.068	0.068	0.068	0.072	0.080	0.080	0.080
	B	....	....	....	0.068	0.092	0.120	0.136	0.160	0.180	0.200	0.200
Over 3 to 4, incl.	A	....	....	....	....	0.092	0.092	0.092	0.092	0.100	0.100	0.100
	B	....	....	....	....	0.092	0.120	0.140	0.180	0.200	0.200	0.200
Over 4 to 5, incl.	A						0.120	0.120	0.120	0.120	0.120	0.120
	B						0.120	0.152	0.180	0.200	0.200	0.200
Over 5 to 6, incl.	A						....	0.152	0.152	0.152	0.152	0.152
	B						....	0.152	0.200	0.200	0.200	0.200
Over 6	A						....	....	0.200	0.200	0.200	0.200
	B						....	....	0.200	0.200	0.200	0.200
Hammered Square and Flat Bars												
0 to ½, incl.	A	0.024	0.024	0.028	0.032	0.036	0.044	0.052	0.056	0.060	....	....
	B	0.024	0.048	0.064	0.080	0.100	0.120	0.144	0.168	0.200	....	....
Over ½ to 1, incl.	A	....	0.048	0.048	0.052	0.052	0.060	0.064	0.068	0.072	0.080	0.088
	B	....	0.048	0.072	0.084	0.100	0.120	0.144	0.168	0.200	0.200	0.200
Over 1 to 2, incl.	A	....	....	0.072	0.072	0.072	0.080	0.088	0.092	0.100	0.112	0.120
	B	....	....	0.072	0.096	0.108	0.124	0.148	0.172	0.200	0.200	0.200
Over 2 to 3, incl.	A	....	....	....	0.096	0.096	0.100	0.104	0.108	0.120	0.128	0.140
	B	....	....	....	0.096	0.120	0.128	0.148	0.172	0.200	0.200	0.200
Over 3 to 4, incl.	A	....	....	....	....	0.120	0.120	0.128	0.144	0.152	0.168	0.180
	B	....	....	....	....	0.120	0.144	0.152	0.180	0.200	0.200	0.200
Over 4 to 5, incl.	A						0.144	0.144	0.152	0.168	0.180	0.200
	B						0.144	0.168	0.200	0.200	0.200	0.200
Over 5 to 6, incl.	A						....	0.168	0.180	0.180	0.200	0.200
	B						....	0.168	0.200	0.200	0.200	0.200
Over 6	A						....	....	0.200	0.200	0.200	0.200
	B						....	....	0.200	0.200	0.200	0.200

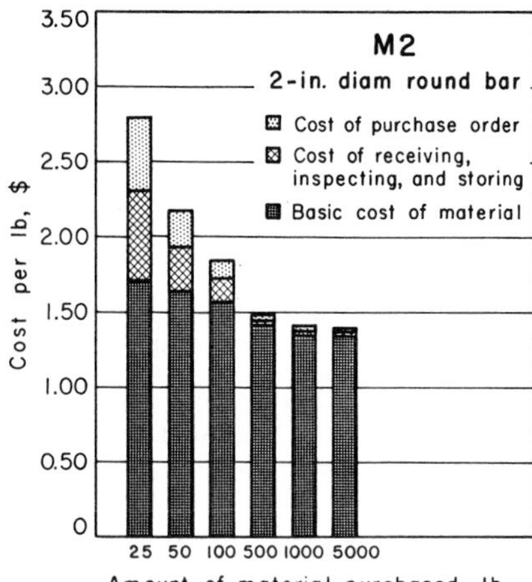

Fig. 20. Effect of quantity purchased on basic material cost and total cost of M2 high speed steel. Inspection costs are based on statistical quality control methods.

Specification, Inspection and Storage

Slight deviations from standard tool steel quality may seriously impair the service life of tools ultimately made from such material.

Specifications. Large consumers of tool steel generally establish their own standards and specifications, and the tool steel supplier makes each order to these requirements. The consumer then spot checks each shipment on some pre-established percentage schedule to insure conformance to the specification. This usually is feasible and economical only with large-quantity purchases, as illustrated in Fig. 20. The producer usually must begin to meet the specifications at the melting stage of manufacturing and therefore considerable lead time (usually five to eight weeks for appreciable-quantity purchases) is required for delivery of the tool steel.

Smaller consumers normally have neither the quantity nor facilities to warrant the added cost of incoming inspection and therefore rely on the general standards of the tool steel industry or those established specifically for them by individual suppliers. The greatest quantity of tool steel purchases falls into this latter category.

Mill Inspection. The initial problem in providing high-quality material is control of the chemical composition. This control must start with the raw materials used in melting the steel. Scrap and ferroalloys are chemically sampled to determine not only the percentage of desirable elements, but also to detect the presence and extent of undesirable elements such as arsenic, tin, antimony, copper, and sometimes nickel and cobalt. Spectrographic analysis renders this analytical load somewhat lighter than with the conventional wet method of analysis, and also maintains a more precise chemical quality.

Consumer specifications normally require the supplier to submit a certified report of chemical analysis for each lot of material. This heat analysis may be supplemented by qualitative spark testing of each individual bar to insure that lots have not been mixed. It is also common practice to spark test the prepared tools before heat treatment, again to insure that no material mixing has occurred during tool fabrication or preparation.

Control of internal quality has been greatly extended in recent years by ultrasonic inspection, which helps to assure internal soundness. Only rarely are gross internal defects such as pipe and hammer bursts evident in modern tool steels, although until recently these defects occasionally occurred because no means were available for inspecting the material adequately without destroying it in the process.

Extremely critical applications, such as die-casting dies, forging die blocks, large extrusion tool blanks, and high-speed cutter blanks, may be 100% sonically inspected by the consumer as a phase of raw material approval. Figure 21 shows the results of sonic testing hot work die blocks before sinking the die impressions. Although the percentage of rejects is low and the defects minor, the consumer found it economically advantageous to employ sonic testing to supplement that of the suppliers.

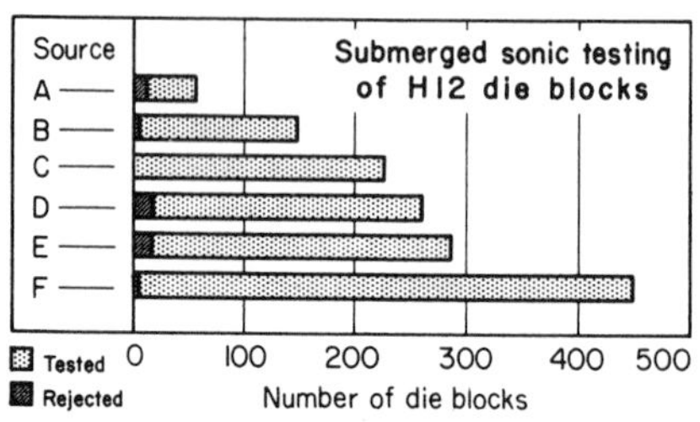

Fig. 21. Results of sonic testing hot work steel die blocks to detect defects such as inclusions and forging bursts. This method of inspection saved $2800 in one plant during an eight-month period.

Another popular procedure to determine internal quality is a disk-inspection test. From a representative number of bars, disks are cut, ground and hot acid etched in a 1:1 hydrochloric acid solution for a period of time suitable for making macroscopic examinations. Such defects as excessive porosity, inclusions and weak or spongy centers can be detected by hot acid etch testing. Although these tests are performed by the manufacturer, consumers often find their own hot acid etch test invaluable for correlating tool performance with hot acid etch standards. Where material is destined to be made into cold heading dies, coining dies and other tools employing the geometric center of the bar as a critical working surface, the added cost of this type of inspection usually can be justified.

The hot acid etched disks of high-alloy tool steels in the D, T and M categories are also hardened, tempered, polished and etched in a 4% nital solution to determine the extent of carbide segregation and carbide formation in the steel. Disks prepared in this manner are examined both macroscopically and microscopically to detect the presence of harmful carbide clusters, undue envelope structure or unusually large carbide particles.

The control of surface quality in tool steels is maintained by both visual and magnetic particle inspection. These testing techniques detect such mechanical surface imperfections as fins, seams, pinches, laps, scabs, strain cracks and air checks. Common causes of such defects are improper mold coatings, fire checks on the mold walls, improper forging techniques, ineffective billet preparation, improper roll pass design or roll setup, or improper cooling rates from hot working operations.

A more difficult surface quality control that affects all producers is that of decarburization on the surface of hot finished products. These carbon gradients at the surface develop when the steel is heated above its critical temperature of approximately 1500 F, which occurs at practically every forming step in the manufacturing cycle. High-alloy steels, particularly those containing large percentages of cobalt, silicon or molybdenum, are especially susceptible to decarburization during forging, rolling and annealing. These more expensive steels often are supplied rough machined or ground to remove decarburization.

Steel Identification. There is no standard means of identifying various grades of tool steel by color coding or other uniform identification. Color coding along the entire length of the bar is popularly employed to insure ready and accurate identification of the grade and to prevent mixing of grades during use or storage. Other popular techniques include stamping the steel or stenciling with paint every few feet along the length of the bar so that its identity is known regardless of how much material may have been cut from it during processing. Consumers generally specify the means for identifying steel. Where no preferred identification system exists, each supplier normally employs his own particular system.

Availability of Tool Steels

About 16 grades of steel are stocked in most tool steel warehousing areas (Table 3). In addition, at least 30 grades can be obtained from

Table 3. Tool Steels Stocked in Most Warehousing Areas

W1, W2	D2*	H12 or H13*
O1	D3* or D4	H21
	M2*	S1
A2*	M3*	S5
	M10	
	T1*	
	T2	

*Available in free-machining types.

stock in certain areas or from the main warehouses of producers.

Grade Popularity. The molybdenum-bearing types now make up approximately 85% of all high speed steels produced. Molybdenum high speed steels received great impetus during the tungsten shortage of the second World War and the Korean War. In the 1950 to 1952 period, major increases in the price of tungsten brought about changes in high speed steel pricing (particularly for those grades containing high percentages of tungsten), reflected in the relative usage of tungsten to molybdenum high speed steel shown in Fig. 22. This difference in price of 50¢ a pound, applied to 20,000 tons of high speed steel per year, amounts to more than $20 million per year difference in total cost — a large economic factor in the selection of steel.

A further trend has been noted toward greater use of vanadium in high speed steels because of the better wear resistance and longer tool life imparted by the hard vanadium carbides. Use of these types (such as M3, type 1; M3, type 2; M4 and T15) has increased in recent years at about three times the rate of all other high speed steels. (Steel M3, type 1 contains about 2.40% V compared with a vanadium content of about 3.00% in type 2, other elements being the same.)

In the cold work die steels, there is a similar trend toward greater use of the high-carbon high-chromium types to obtain longer die life. Over-all, the trend is as follows:

From	To
W1, W2	Oil-hardening O1
O1	Air-hardening 5% Cr A2
A2	Air-hardening 12% Cr D2

Tool steel consumption fluctuates widely with business conditions, varying from approximately 130,000 tons annually in a busy year down to 70,000 tons in a recession year. Among the various types this breaks down approximately as follows:

Type	Percentage of total
High speed steels	18
Hot work steels	13
Chromium die steels	13
Oil-hardening tool steels	18
Shock-resisting and chisel steels	8
Special-purpose tool steels	19
Carbon tool steels	11
	100%

Standardization

Standardization of grades can be practiced by large or small consumers of tool steel, whereas standardization of sizes and quantities to achieve economies in purchasing, handling, inspection and storage usually can be adopted successfully only by large-volume consumers.

The following are likely areas for savings through standardization:

1 Regrouping of sizes to permit quantity purchases.
2 Reduction of ordering, checking, handling and storage costs by purchasing in larger quantities.

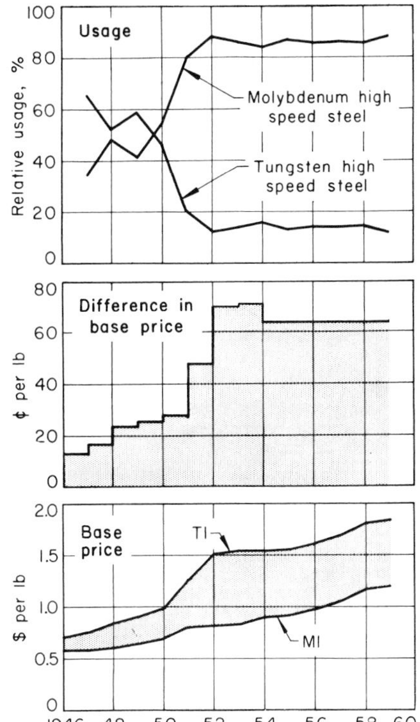

Fig. 22. Relative usage and difference in base price of molybdenum and tungsten high speed steels

3 Standardization of hot rolled sizes into groupings carrying the lowest possible size extras.
4 Standardization of specified steels by review of hardenability requirements because of improvements or alterations in heat treating procedures, particularly quenching rates.
5 Standardization of grades to permit establishment of uniform heat treating cycles without loss of furnace time for grade change-overs.
6 Combining grades of steel and elimination of more expensive ones, in certain instances, by review of engineering and application requirements.
7 Review of engineering specifications before release to production with the aim of standardizing materials and processing procedures.
8 Study of relative machinability of different grades of tool steel used, with standardization on one grade for all similar parts.
9 Review of engineering specifications to eliminate nonstandard-size material that is difficult to procure.

It is perhaps unfortunate that so many individual types of tool steel are available, since many of these compositions are similar to one another in properties. In most instances the choice of suitable steels for a particular tooling problem is not limited to a single type or even a single group of steels, and in this regard the long list of steels in the classification table is more confusing than helpful. For example, the general workability, properties, heat treatment and ultimate performace of O1 and O2 steels are very similar if not identical, yet it is not unusual to find both types being purchased by a given plant. Similarly, S4 and S5 are practically identical in performance and workability, and many other similar comparisons are readily evident in the classification table. Table 4 illustrates how one plant successfully reduced the number of steels purchased in 2-in. bar stock sizes from seven grades to two.

Grade standardization provides direct benefits and cost savings by reducing unit purchasing charges, materials handling and simplification of inventory records, steel identification and storage. The decreasing cost per pound as the quantity of individual orders is increased is shown in Fig. 20.

The potential savings offered by standardization must always be balanced against the added costs of overstandardization. Table 5 shows the penalty of combining too many sizes. For each set of circumstances a break-even point must be established for maximum economy.

Table 4. Example of Plant Standardization of Tool Steel Grades
(Seven grades of 2-in. bars were reduced to two.)

Part name	Tool steel used: Original	Replacement	Result of change
Thread roll	D2	A2	Savings in material and grinding
Thread roll	M2	A2	Savings in material and grinding
Number die	O1	A2	Longer life obtained
Marking die	O1	A2	Longer life obtained
Engraving	O2	A2	Longer life obtained
Gear	SAE 6150	S2	Improved toughness
Ratchet	S1	S2	Reduced cost
Spindle	SAE 4150	S2	Improved toughness
Spindle	S1	S2	Reduced cost

Table 5. Cost of Machining Tool Steels(a)

Bar diam, in.	Cost: Machining (b)	Material removed	Total	Cost differential over 2-in. size
$2\frac{1}{8}$	$0.23	$1.41	$1.64	(c)
$2\frac{1}{4}$	0.25	2.89	3.14	$1.50
$2\frac{1}{2}$	0.52	6.25	6.77	5.13
$2\frac{5}{8}$	0.80	8.05	8.85	7.21

(a) Machining and material cost involved in turning various stock sizes to 2-in. diam. Material is priced at $1.155 per pound. (b) One turning cut is required to reduce $2\frac{1}{8}$ and $2\frac{1}{4}$-in. sizes to 2 in. Two turning cuts are required to reduce $2\frac{1}{2}$-in. sizes to 2 in. Three turning cuts are required to reduce $2\frac{5}{8}$-in. sizes to 2 in. Cutting speed, 80 fpm; feed, 0.015 in. per revolution; depth of cut, variable, $\frac{1}{8}$ in. max; 6-in. length; labor rate, $3.00 per hr, including setup. (c) It is assumed that $2\frac{1}{8}$-in. stock is ordinarily used to produce a 2-in. diam part, insuring removal of any deleterious surface conditions that might be present in the bars.

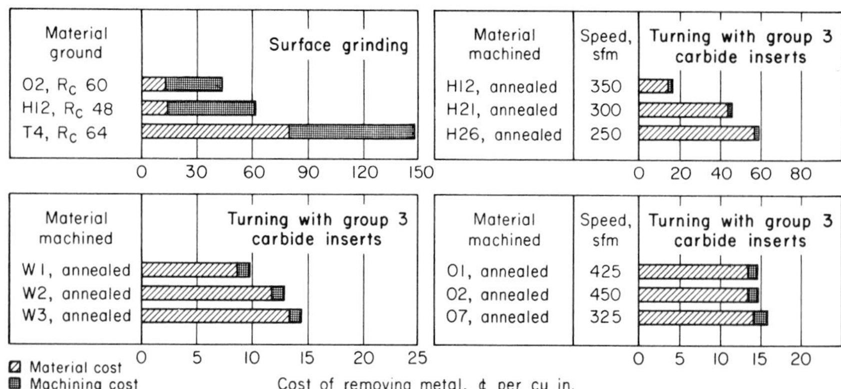

Fig. 23. Effect of composition on cost per cubic inch of tool steel removed, compared with the cost of the grinding and turning operations. No credit is included for the scrap value of chips.

Cost of Tool Steels

Table 6 lists typical base prices and price extra classifications (from a survey by one user among seven tool steel companies in February 1960) for a group of popular grades of tool steels. Previous discussion has indicated the effect of higher alloy content on cost of tool steels with regard to manufacturing losses and fabrication problems. In addition, there is the cost involved in the actual price per pound of the various alloying elements used. This is particularly important in tool steels containing tungsten and cobalt. The high content (18%) of expensive tungsten in the T1 grade, for example, has made this grade one of the most expensive tool steels.

Tool steels are usually sold on a per pound basis. The industry practice is to add fabricating costs to a base price to arrive at a net price for any specific size, shape, condition or quantity. These fabricating costs are commonly known as extras and are defined as follows.

Base price reflects the cost of raw materials and expenses incurred to melt and refine to the required chemical composition, as well as the yield of usable tool steel.

Quantity extra reflects the cost of producing and handling small and large quantities, much the same as any other item sold on a per pound or per piece basis.

Annealing extra covers relieving rolling or forging strains, refinement of the grain, and making the steel machinable.

Heat treating extras cover the cost of thermal treatments other than annealing.

Size extras reflect the cost of hot rolling or forging rounds, squares, rectangulars and shapes in bar form.

Rough-turning extra covers the machining of round bars to tolerances closer than hot rolled tolerances, and improving the surface finish and removing decarburization.

Rough-machining extra covers planing, shaping or milling of blocks, disks or other special shapes to improve finish, remove decarburization and facilitate testing.

Grinding extra is for flat grinding of disks to improve finish, remove decarburization and facilitate testing.

Cold drawing extra covers the additional costs to process hot rolled material to closer tolerances and a better finish by cold drawing.

Centerless ground extra covers the cost to process round bars that are usually free from decarburization to specified tolerances and finish.

Cutting and wastage extras charge for cutting to a dead length or multiple length, reflecting the cost of work done and the scrap produced in cutting these lengths.

Shape extras cover special rolls, roll changes and special handling to produce the shape required.

Forging extra covers the cost of hammering or pressing shapes and sections, such as disks, blocks, rings and special die forgings, that cannot be rolled or otherwise produced.

Since manufacturing techniques involve different problems with different groups of tool steels there are actually three different schedules of extras (Table 6).

Table 7 illustrates typical magnitudes of price extras expressed as a percentage of the base price for a group of blanks in various diameters and grades. It assumes that ten pieces of tool steel, each 12 in. long, are purchased from a warehouse, and that such an order would include annealing, hot rolled size, warehouse quantity and cutting extras. For W1 steel these would amount to an additional 93 or 60% on the base price for 1 and 4 in. diam material (Table 7).

Possibly the most important extra is that pertaining to quantity. Base prices usually apply to a quantity of 1000 lb or more, and therefore it is most economical to buy tool steel in lots of this amount. Where production involves more than one size, it may become necessary to buy base quantities of large-size bars and use these larger bars for a variety of smaller-sized parts. It is important to note here that the extra machining involved in turning down larger sizes is sometimes more costly than buying smaller quantities of the correct size even though a quantity penalty must be paid. As shown in Table 5, the cost of the metal reduced to chips is from five to ten times the cost of the turning operation to remove the unwanted stock.

Calculation of the proper break-even point with regard to machining costs and quantity extra is complicated and varies with conditions in each plant as well as with the type of tool steel involved. For example, if 50 lb of 2-in. diam A2 die steel is required as a final product it will be less expensive, providing no 2⅛-in. round is available, to turn down a 2¼-in. round that has been purchased in any amount greater than a 400-lb lot, than it will be to buy the 50 lb in the 2⅛-in. round. However, turning down 2⅜ or 2½-in. rounds to a 2-in. round in this quantity would be much more expensive than buying the 2⅛-in. size in 50-lb lots. Similar calculations can be made for different weights.

Consumer Processing Costs

The ability to machine or otherwise fabricate tool steels in the annealed or soft state usually is a relatively minor consideration influencing grade selection, since the perfection of the finished product is of much greater importance than the ease or low cost of machining.

Table 6. Typical Base Price and Price Extra Classification of 17 Common Tool Steel Grades (a)

(See also Table 1 on page 672.)

Grade	Base price per lb, $	Price extra classification
W1	0.33 to 0.46 (b)	Tool steel (c)
W2	0.35 to 0.48 (b)	Tool steel
S1	0.64	Tool steel
S5	0.305 to 0.33	Tool steel
O1	0.505	Tool steel
A2	0.54	Tool steel
H12	0.53	Tool steel
D2	0.955	High-alloy tool steel (d)
D3	0.955	High-alloy tool steel
H21	1.425	High-alloy tool steel
M1	1.20	High speed steel (e)
M2	1.345	High speed steel
M3	1.59	High speed steel
M10	1.20	High speed steel
T1	1.84	High speed steel
T2	2.005	High speed steel
T5	2.915	High speed steel

(a) Based on a survey of seven tool steel companies, March 1960. (b) This price depends on quality as discussed in the text. (c) For tool steels, except those containing 3% or more of tungsten or molybdenum or both, or 8% Cr or more, with or without additional alloying elements. (d) For all tool steels containing 3% or more tungsten or molybdenum or both, or 8% Cr or more, with or without additional alloying elements, except those classified as high speed steel round bars. (e) For all tool steels containing 3% or more tungsten or molybdenum or both, or 8% Cr or more, with or without additional alloying elements.

Table 7. Typical Tool Steel Extras Expressed as Percentage of Base Price (a)

Grade	Bar diameter, in. ½	1	2	4	8
W1	103	93	65	60	63
A2	73	67	46	43	45
D2	51	47	30	28	32
M1	51	39	25	22	26
T1	33	26	17	14	17

(a) Based on a survey of seven tool steel companies, March 1960. Extras are based on order from warehouse of ten pieces, each 12 in. long, and includes annealing (as applicable), size, quantity and cutting.

Table 8. Machinability Ratings of Tool Steels

(Water-hardening grades rated at 100)

Tool steel group	Machinability rating
W	100
S	85
O	90
A	85
D	40 to 50
H(Cr)	75
H(W or Mo)	50 to 60
T	40 to 55
M	45 to 60
L	90
F	75
P	75 to 100

Where several available grades have equivalent properties in the finished tools, it is, of course, economical to select the grade exhibiting the best machinability and general workability.

Machining. The mechanical properties of steel generally regarded as producing good machinability are low strength, low ductility and low abrasive hardness. This combination of properties is not inherent in tool steels, and the properties are even incompatible with each other; for example, high ductility normally is associated with low hardness. Tool steels as a class possess high strength, low ductility and usually high abrasive hardness.

When compared with the conventional production alloy steels, tool steels are considerably more difficult to machine and, as a class, at best (W types) have a machinability rating of about 30% that of B1112 screw stock. Because of the low ratings, it is usual to compare the machinability of tool steels with W1 at an arbitrary rating of 100. On this basis, machinability for each of the different types of tool steel is rated in Table 8. The free-machining varieties have a machinability rating about 25% higher than equivalent non-free-machining grades.

The machinability and general workability of tool steels decreases with increase of carbon and alloy content. Carbon, particularly in combination with strong carbide-forming elements such as chromium, tungsten, molybdenum and vanadium, adversely affects machinability through the formation of a great number of hard, abrasion-resistant carbide particles. Also, as the carbon and alloy content increases, it usually becomes more difficult to obtain low annealed hardnesses. For example, W1 steel can be annealed to a hardness of about 200 Bhn whereas the typical annealed hardness for T15 and M15 high speed steels is from 255 to 269 Bhn. Tool steels containing high percentages of nickel and cobalt, which strengthen the ferrite, also have lower machinability because of the generally higher annealed hardnesses. T5 and T6 cobalt high speed steels can be annealed no softer than about 285 to 300 Bhn.

Probably the greatest factor influencing consumer processing costs in forming, working or machining the various tool steels is the value of the material discarded in the form of chips. In any operation producing a large amount of chips, such as turning or milling, the value of the discarded chips is the major cost of machining. As shown in Table 5 and Fig. 23, the turning costs for a wide variety of tool steels are insignificant in relation to the value of the lost material. This is true even for the relatively low-cost type W and O tool steels and is applicable to heavy roughing operations. For finishing work involving low rates of metal removal, it is not true.

The data in Fig. 23 and Table 5 take no account of income from the sale of the scrap produced. Careful identification, segregation and classification of the scrap makes it possible to recover as much as 10% of the original cost of the material through the sale of such scrap. This recovery can result in a significant dollar saving with the more expensive higher-alloy steels.

Table 9. Surface Grinding of Regular and Free-Machining Grades of Tool Steel(a)

Steel being ground	Grinding ratio: Regular grade	Grinding ratio: F-M grade(b)	Increase, %
Dry Grinding			
M2	17	34	100
M3(c)	5.5	15.4	180
M4	2.8	9.5	240
D7	...	...	...
Wet Grinding			
M2	6.4	9.6	50
M3(c)	2.7	4.2	60
M4	0.9	2.1	130
D7	0.9	3.1	240

(a) L. P. Tarasov, "Grinding Sulfurized Tool Steels", Tool Engineer, Feb. 1959. (b) Free-machining grade. (c) Type 1.

Grinding. In the production of intricate tools and dies where considerable finish grinding after heat treatment is involved, the costs of finish grinding can be an important consideration. Usually this is of greater significance than is the comparative machinability or workability of the steel in the annealed state. Finish grinding operations involving form grinding, thread grinding or the internal grinding of small holes sort the difficult-to-grind tool steels from the easy-to-grind grades, more than mere surface grinding of flats. However, in surface grinding of the more abrasion-resisting longer-wearing high speed and die steels, the rate of wheel wear often equals or exceeds the rate of metal removal (L. P. Tarasov, Trans ASM, **43**, 1144, 1951). This necessitates frequent wheel dressing and in part accounts for the increased costs.

Steel composition affects the ease of grinding after heat treatment, which is usually inversely proportional to the base hardness, volume of carbide phase present, and hardness of individual carbide particles. Steels containing high percentages of carbon and carbide-forming alloying elements are all relatively difficult to grind. Sometimes tool steels that are relatively high in content of carbide-forming elements are more difficult to grind than a carbon or low-alloy grade even though the carbon content and Rockwell hardness are higher for the low-alloy type. An example of such is indicated in the data at the upper left of Fig. 23. The higher cost of grinding the H12 compared with O2 is attributed to the more rapid loading of the grinding wheels in grinding H12 at Rockwell C 48 compared with O2 at Rockwell C 60.

Vanadium additions in particular exert a strong influence on grindability and all steels containing more than 1% V are difficult to grind. Vanadium is soluble in the matrices of most tool steel compositions up to about 1%, and little if any undissolved vanadium carbide remains in the normal heat treated structure. Beyond 1% V, however, excess undissolved vanadium carbides begin to appear in the microstructure, and these increase rapidly in size and number at higher vanadium concentrations. T1 high speed steel contains only about 0.5% undissolved vanadium-rich carbides after conventional heat treatment, whereas the T15 type contains approximately 8% (Francis Kayser and Morris Cohen, Carbides in High Speed Steel — Their Nature and Quantity, Metal Progress, p 79, June 1952).

Vanadium carbides are relatively pure carbides formed during solidification of the steel, and are contaminated very little by iron or other metallic elements. These carbides have a hardness of 2300 to 2700 Knoop, averaging about 2520. Aluminum oxide grinding wheel abrasive has a hardness of 1900 to 2900 Knoop, averaging 2440. This overlapping of hardness ranges partly explains the difficulty in grinding high-vanadium steels.

Figure 24 shows the relative length of time for grinding high speed steels of different vanadium contents. The same relative time to grind T1 and M2 steels is reported even though M2 has a higher vanadium content. However, wheel wear is greater when grinding M2 than when grinding T1, and this cost factor has not been taken into consideration in the presentation of data on grinding time.

Table 10. Effect of Hardening Temperature of M3 High Speed Steel on Grindability (a)

Temp, F	Dry grinding: Regular grade	Dry grinding: F-M grade(b)	Wet grinding: Regular grade	Wet grinding: F-M grade(b)
Grinding Ratio				
2210	5.7	18.7	2.9	4.4
2240	5.5	15.4	2.7	4.2
2270	5.5	14.6	2.5	3.2
Net Power Input (c)				
2210	8.7	8.3	10.8	10.3
2240	9.2	8.9	10.6	10.2
2270	10.2	9.2	10.7	9.4

(a) L. P. Tarasov, "Grinding Sulfurized Tool Steels", Tool Engineer, Feb. 1959. (b) Free-machining grade. (c) Net power input is expressed in arbitrary units.

In the A and D series of cold work die steels, the same effects of vanadium on grindability have been observed. A7 and D7 are difficult to grind not only because of the high volume of vanadium carbides present but also because of the great number of complex iron-chromium carbides. As the wear resistance of steels is increased, grindability is sacrificed. The abrasion-resistant high-vanadium high speed and die steels, when properly applied, offer considerably longer tool life than the lower-alloy easier-to-grind materials and this helps to compensate for the extra cost in fabricating the material. Use of the difficult-to-grind materials for intricate tools and dies, such as taps, small twist drills, compacting dies, and spline broaches, adds to fabrication costs and sometimes makes their use prohibitive, particularly when such tools are made and sold on a competitive basis.

Sulfurized tool steels show improved grindability in the heat treated condition. Recent work in determining the grindability of sulfurized and nonsulfurized tool steels is summarized in Tables 9 and 10. The sulfurized grades show the greatest improvement in grindability under adverse conditions of dry grinding of the higher-alloy more difficult-to-grind materials. These experimental findings have been confirmed under production grinding conditions.

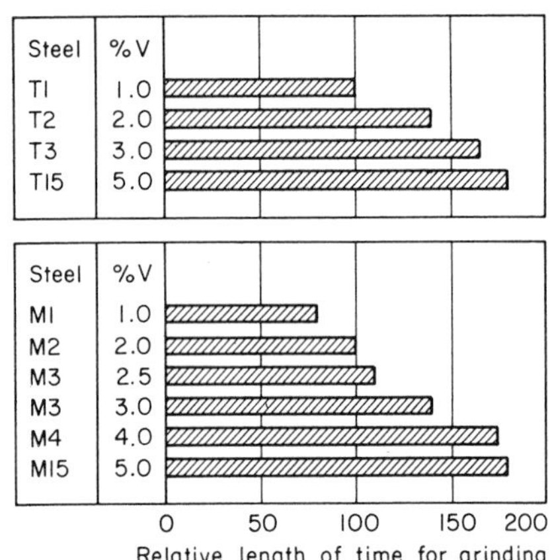

Fig. 24. Effect of vanadium content on grindability of high speed steels, based on typical experience in toolroom grinding with the same type of grinding wheels for all of the steels

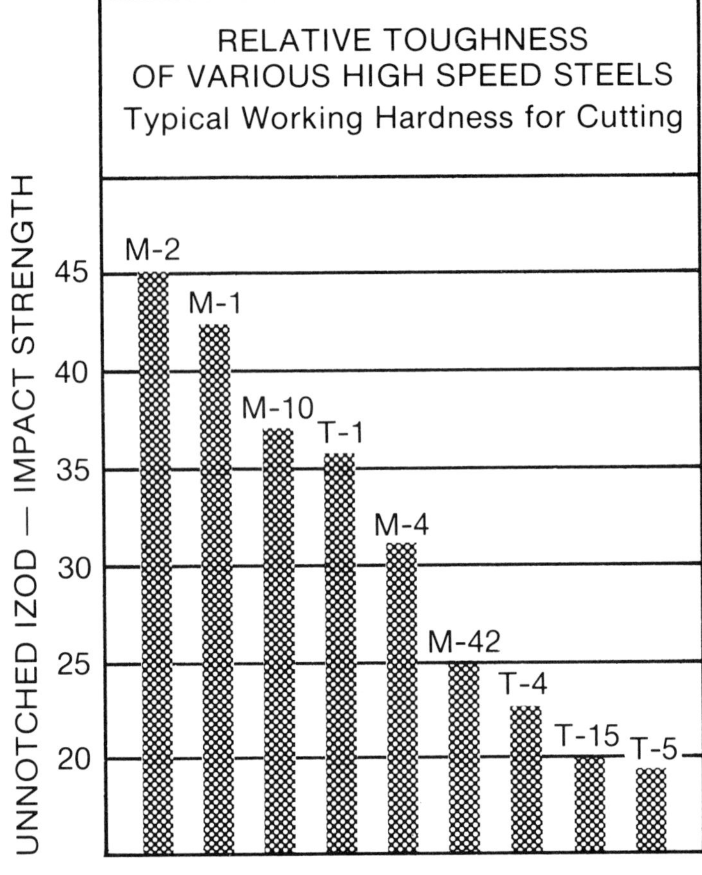

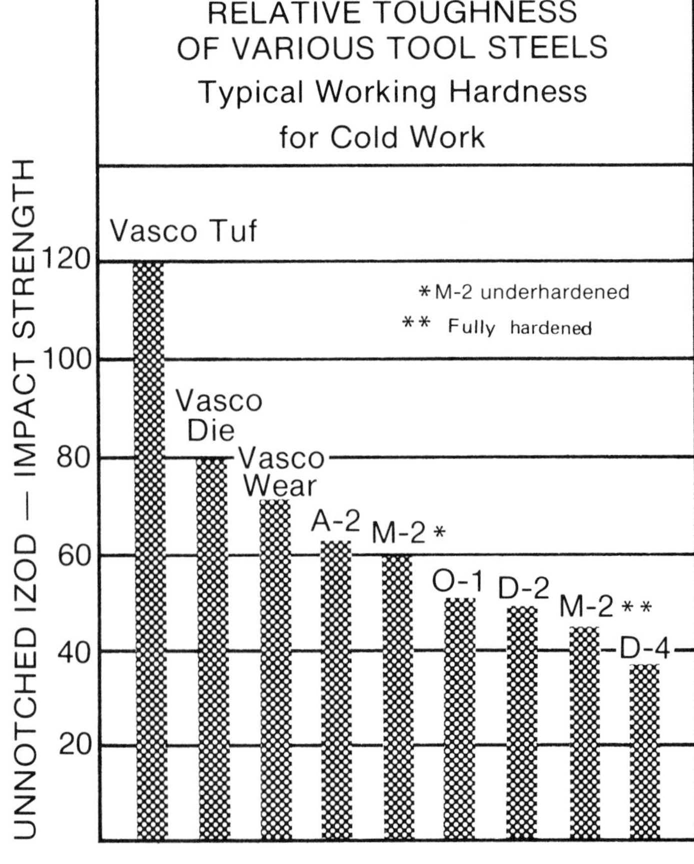

Classification and Selection of Tool Steels

AISI Type	Composition, %						Typical Applications
	C	W	Mo	Cr	V	Other	
							Water Hardening
W1	0.60 to 1.40*	—	—	—	—	—	Low carbon: blacksmith tools, blanking tools, caulking tools, cold chisels, forging dies, rammers, rivet sets, shear blades, punches, sledges. Medium carbon: arbors, beading tools, blanking dies, reamers, bushings, cold heading dies, chisels, coining dies, countersinks, drills, forming dies, jeweler dies, mandrels, punches, shear blades, woodworking tools. High carbon: glass cutters, jeweler dies, lathe tools, reamers, taps and dies, twist drills, woodworking tools. Vanadium content of W2 imparts finer grain, greater toughness and shallow hardenability.
W2	0.60 to 1.40*	—	—	—	0.25	—	
W5	1.10	—	—	0.50	—	—	Heavy stamping and draw dies, tube drawing mandrels, large punches, reamers, razor blades, cold forming rolls and dies, wear plates.
							Shock Resisting
S1	0.50	2.50	—	1.50	—	—	Bolt header dies, chipping and caulking chisels, pipe cutters, concrete drills, expander rolls, forging dies, forming dies, grippers, mandrels, punches, pneumatic tools, scarfing tools, swaging dies, shear blades, track tools, master hobs.
S2	0.50	—	0.50	—	—	1.00 Si	Hand and pneumatic chisels, drift pins, forming tools, knock-out pins, mandrels, machine parts, nail sets, pipe cutters, rivet sets, screw driver bits, shear blades, spindles, stamps, tool shanks, track tools.
S5	0.55	—	0.40	—	—	0.80 Mn, 2.00 Si	Hand and pneumatic chisels, drift pins, forming tools, knock-out pins, mandrels, nail sets, pipe cutters, rivet sets and busters, screw driver bits, shear blades, spindles, stamps, tool shanks, track tools, lathe and screw machine collets, bending dies, punches, rotary shears.
S7	0.50	—	1.40	3.25	—	—	Shear blades, punches, slitters, chisels, forming dies, hot header dies, blanking dies, rivet sets, gripper dies, engraving dies, plastic molds, die casting dies, master hobs, beading tools, caulking tools, chuck jaws, clutches, pipe cutters, swaging dies.
							Oil Hardening
O1	0.90	0.50	—	0.50	—	1.00 Mn	Blanking dies, plastic mold dies, drawing dies, trim dies, paper knives, shear blades, taps, reamers, tools, gages, bending and forming dies, bushings, punches.
O2	0.90	—	—	—	—	1.60 Mn	Blanking, stamping, trimming, cold forming dies and punches, cold forming rolls, threading dies and taps, reamers, gages, plugs and master tools, broaches, circular cutters and saws, thread roller dies, bushings, plastic molding dies.
O6‡	1.45	—	0.25	—	—	1.00 Si	Blanking dies, forming dies, mandrels, punches, cams, brake dies, deep drawing dies, cold forming rollers, bushings, gages, blanking and forming punches, piercing and perforating dies, taps, paper cutting dies, wear plates, tool shanks, jigs, machine spindles, arbors, guides in grinders and straighteners.
O7	1.20	1.75	—	0.75	—	—	Mandrels, slitters, skiving knives, taps, reamers, drills, blanking and forming dies, gages, chasers, brass finishing tools, dental burrs, paper knives, roll turning tools, burnishing dies, pipe threading dies, rubber cutting knives, woodworking tools, hand reamers, scrapers, spinning tools, broaches, blanking and cold forming punches.
							Cold Work (Medium Alloy, Air Hardening)
A2	1.00	—	1.00	5.00	—	—	Thread rolling dies, extrusion dies, trimming dies, blanking dies, coining dies, mandrels,

*Other carbon contents may be available. †Optional. ‡Contains free graphite in microstructure to improve machinability.
Some of these grades can be made with added sulfur to improve machinability.

AISI Type	Composition, % C	W	Mo	Cr	V	Other	Typical Applications
A3	1.25	—	1.00	5.00	1.00	—	shear blades, slitters, spinning rolls, forming rolls, gages, beading dies, burnishing tools, ceramic tools, embossing dies, plastic molds, stamping dies, bushings, punches, liners for brick molds.
A4	1.00	—	1.00	1.00	—	2.00 Mn	Blanking dies, forming dies, trimming dies, punches, shear blades, mandrels, bending dies, forming rolls, broaches, knurling tools, gages, arbors, bushings, slitting cutters, cold treading rollers, drill bushing, master hobs, cloth cutting knives, pilot pins, punches, engraver rolls.
A6	0.70	—	1.25	1.00	—	2.00 Mn	Blanking dies, forming dies, coining dies, trimming dies, punches, shear blades, spindles, master hubs, retaining rings, mandrels, plastic dies.
A7	2.25	1.00†	1.00	5.25	4.75	—	Brick mold liners. drawing dies, briquetting dies, liners for shot blasting equipment and sand slingers, burnishing tools, gages, forming dies.
A8	0.55	1.25	1.25	5.00	—	—	Cold slitters, shear blades, hot pressing dies, blanking dies, beading tools, cold forming dies, punches, coining dies, trimming dies, master hobs, rolls, forging die inserts, compression molds, notching dies, slitter knives.
A9	0.50	—	1.40	5.00	1.00	1.50 Ni	Solid cold heading dies, die inserts, heading hammers, coining dies, forming dies and rolls, die casings, gripper dies. Hot work applications: punches, piercing tools, mandrels, extrusion tooling, forging dies, gripper dies, die casings, heading dies, hammers, coining and forming dies.
A10‡	1.35	—	1.50	—	—	1.80 Mn, 1.25 Si, 1.80 Ni	Blanking dies, forming dies, gages, trimming shears, punches, forming rolls, wear plates, spindle arbors, master cams and shafts, stripper plates, retaining rings.
							Cold Work (High Carbon, High Chromium)
D2	1.50	—	1.00	12.00	1.00	—	Blanking dies, cold forming dies, drawing dies, lamination dies, thread rolling dies, shear blades, slitter knives, forming rolls, burnishing tools, punches, gages, knurling tools, lathe centers, broaches, cold extrusion dies, mandrels, swaging dies, cutlery.
D3	2.25	—	—	12.00	—	—	Blanking dies, cold forming dies, drawing dies, lamination dies, thread rolling dies, shear blades, slitter knives, forming rolls, seaming rolls, burnishing tools, punches, gages, crimping dies, swaging dies.
D4	2.25	—	1.00	12.00	—	—	Blanking dies, brick molds, burnishing tools, thread rolling dies, hot swaging dies, wiredrawing dies, forming tools and rolls, gages, punches, trimmer dies, dies for deep drawing.
D5	1.50	—	1.00	12.00	—	3.00 Co	Cold forming dies, thread rolling dies, blanking dies, coining dies, trimming dies, draw dies, shear blades, punches, quality cutlery, rolls.
D7	2.35	—	1.00	12.00	4.00	—	Brick mold liners and die plates, briquetting dies, grinding wheel molds, dies for deep drawing, flattening rolls, shot and sandblasting liners, slitter knives, wear plates, wiredrawing dies, Sendzimir mill rolls, ceramic tools and dies, lamination dies.
							Hot Work (Chromium)
H10	0.40	—	2.50	3.25	0.40	—	Mandrels, extrusion and forging dies, die holders, bolsters and dummy blocks, punches, die inserts, gripper and header dies, hot shears, aluminum die casting dies, inserts for forging dies and upsetters, shell piercing tools.
H11	0.35	—	1.50	5.00	0.40	—	Die casting dies, punches, piercing tools, mandrels, extrusion tooling, forging dies, high strength structural components.
H12	0.35	1.50	1.50	5.00	0.40	—	Extrusion dies, dummy blocks, holders, gripper and header dies, forging die inserts, punches, mandrels, sleeves for cold heading dies.
H13	0.35	—	1.50	5.00	1.00	—	Die casting dies and inserts, dummy blocks, cores, ejector pins, plungers, sleeves, slides, extrusion dies, forging dies and inserts.
H14	0.40	5.00	—	5.00	—	—	Backer blocks, die holders, aluminum and brass extrusion dies, press liners, dummy blocks, forging dies and inserts, gripper dies, shell forging points and mandrels, hot punches, pushout rings, dies and inserts for brass forging.
H19	0.40	4.25	—	4.25	2.00	4.25 Co	Extrusion dies and die inserts, dummy blocks, punches, forging dies and die inserts, mandrels, hot punch tools.
							Hot Work (Tungsten)
H21	0.35	9.00	—	3.50	—	—	Mandrels, hot blanking dies, hot punches, blades for flying shear, hot trimming dies, extrusion and die casting dies for brass, dummy blocks, piercer points, gripper dies, hot nut tools (crowners, cutoffs, side dies, piercers), hot headers.
H22	0.35	11.00	—	2.00	—	—	Mandrels, hot blanking dies, hot punches, blades for flying shear, hot trim dies, extrusion dies, dummy blocks, piercer points, gripper dies.
H23	0.30	12.00	—	12.00	—	—	Extrusion and die casting dies for brass, brass and bronze permanent molds.
H24	0.45	15.00	—	3.00	—	—	Punches and shear blades for brass, hot blanking and drawing dies, trimming dies, dummy blocks, hot press dies, hot punches, gripper dies, hot forming rolls, hot shear blades, swaging dies, hot heading dies, extrusion dies.
H25	0.25	15.00	—	4.00	—	—	Hot forming dies, die casting and forging dies, die inserts, extrusion dies and liners, shear blades, blanking dies, gripper dies, punches, hot swaging dies, nut piercers, piercer points, mandrels, high temperature springs.
H26	0.50	18.00	—	4.00	1.00	—	Mandrels, hot blanking dies, hot punches, blades for flying shear, hot trimming dies, extrusion dies, dummy blocks, piercer points, gripper dies, pipe threading dies, nut chisels, forging press inserts, extrusion dies for brass and copper.
							Hot Work (Molybdenum)
H41	0.65	1.50	8.00	4.00	1.00	—	Cold forming and hot forming dies, dummy blocks, extrusion dies for copper base alloys, hot or cold heading dies, hot or cold punches, gripper dies, thread rolling dies, hot shear blades, trimming dies, spike cutters.
H42	0.60	6.00	5.00	4.00	2.00	—	Cold trimming dies, hot upsetting dies, dummy blocks, header dies, hot extrusion dies, cold header and extrusion dies and die inserts, hot forming and swaging dies, nut piercers, hot punches, mandrels, chipping chisels.

of Tool Steels (continued)

AISI Type	Composition, % C	W	Mo	Cr	V	Other	Typical Applications
H43	0.55	—	8.00	4.00	2.00	—	Cold forming punches, hot forming dies and punches, swaging dies, hot gripper dies, hot heading dies, hot and cold nut dies, hot spike dies, grippers and headers, hot shear knives, hot working punches, insert punches.
							High Speed (Tungsten)
T1	0.75*	18.00	—	4.00	1.00	—	Drills, taps, reamers, hobs, lathe and planer tools, broaches, crowners, burnishing dies, cold extrusion dies, cold heading die inserts, lamination dies, chasers, cutters, taps, end mills, milling cutters.
T2	0.80	18.00	—	4.00	2.00	—	Lathe and planer tools, milling cutters, form tools, broaches, reamers, chasers.
T4	0.75	18.00	—	4.00	1.00	5.00 Co	Lathe and planer tools, drills, boring tools, broaches, roll turning tools, milling cutters, shaper tools, form tools, hobs, single point cutting tools.
T5	0.80	18.00	—	4.00	2.00	8.00 Co	Lathe and planer tools, form tools, cutoff tools, heavy duty tools requiring high red hardness.
T6	0.80	20.00	—	4.50	1.50	12.00 Co	Heavy duty lathe and planer tools, drills, checking tools, cutoff tools, milling cutters, hobs.
T8	0.75	14.00	—	4.00	2.00	5.00 Co	Boring tools, lathe tools, heavy duty planer tools, tool bits, single point cutting tools for stainless steel.
T15	1.50	12.00	—	4.00	5.00	5.00 Co	Form tools, lathe and planer tools, broaches, milling cutters, blanking dies, punches, heavy duty tools requiring good wear resistance.
							High Speed (Molybdenum)
M1	0.80*	1.50	8.00	4.00	1.00	—	Drills, taps, end mills, reamers, milling cutters, hobs, punches, lathe and planer tools, form tools, saws, chasers, broaches, routers, woodworking tools.
M2	0.85; 1.00*	6.00	5.00	4.00	2.00	—	Drills, taps, end mills, reamers, milling cutters, hobs, form tools, saws, lathe and planer tools, chasers, broaches and boring tools.
M3-1	1.05	6.00	5.00	4.00	2.40	—	Drills, taps, end mills, reamers and counterbores, broaches, hobs, form tools, lathe and planer tools, cheeking tools, milling cutters, slitting saws, punches, drawing dies, routers, woodworking tools.
M3-2	1.20	6.00	5.00	4.00	3.00	—	Drills, taps, end mills, reamers and counterbores, broaches, hobs, form tools, lathe and planer tools, cheeking tools, slitting saws, punches, drawing dies, woodworking tools.
M4	1.30	5.50	4.50	4.00	4.00	—	Broaches, reamers, milling cutters, chasers, form tools, lathe and planer tools, cheeking tools, blanking dies and punches for abrasive materials, swaging dies.
M6	0.80	4.00	5.00	4.00	1.50	12.00 Co	Lathe tools, boring tools, planer tools, form tools, milling cutters.
M7	1.00	1.75	8.75	4.00	2.00	—	Drills, taps, end mills, reamers, routers, saws, milling cutters, lathe and planer tools, chasers, borers, woodworking tools, hobs, form tools, punches.
M10	0.85; 1.00*	—	8.00	4.00	2.00	—	Drills, taps, reamers, chasers, end mills, lathe and planer tools, woodworking tools, routers, saws, milling cutters, hobs, form tools, punches, broaches.
M30	0.80	2.00	8.00	4.00	1.25	5.00 Co	Lathe tools, form tools, milling cutters, chasers.
M33	0.90	1.50	9.50	4.00	1.15	8.00 Co	Drills, taps, end mills, lathe tools, milling cutters, form tools, chasers.
M34	0.90	2.00	8.00	4.00	2.00	8.00 Co	Drills, taps, end mills, lathe tools, milling cutters, form tools, chasers.
M36	0.80	6.00	5.00	4.00	2.00	8.00 Co	Heavy duty lathe and planer tools, boring tools, milling cutters, drills, cutoff tools, tool holder bits.
M41	1.10	6.75	3.75	4.25	2.00	5.00 Co	Drills, end mills, reamers, form cutters, lathe tools, hobs, broaches, milling cutters, twist drills, end mills. Hardenable to Rockwell C 67 to 70.
M42	1.10	1.50	9.50	3.75	1.15	8.00 Co	
M43	1.20	2.75	8.00	3.75	1.60	8.25 Co	
M44	1.15	5.25	6.25	4.25	2.25	12.00 Co	
M46	1.25	2.00	8.25	4.00	3.20	8.25 Co	
M47	1.10	1.50	9.50	3.75	1.25	5.00 Co	
							Special Purpose (Low Alloy)
L2	0.50 to 1.10*	—	—	1.00	0.20	—	Automotive gears and forgings, arbors, crank pins, chuck jaws and liners, chain feed sprockets, dogs, drift pins, die rings, friction feed disks, gun barrels, gun hoops, jack screws, lead and feed screws, machine shafts, pinions, rivet sets, shear blades, spindles, wrenches, die insert holders.
L3	1.00	—	—	1.50	0.20	—	Bearing races, mandrels, cold rolling rolls, ball bearings, arbors, precision gages, broaches, taps, dies, drills, thread rolling dies, knife edges, cutlery, files, knurls.
L6	0.70*	—	0.25	0.75	—	1.50 Ni	Arbors, blanking dies, forming dies, disk saws, drift pins, brake dies, hand stamps, hubs, lead and feed screws, machine parts, punches, pawls, pinions, shear blades, spindles, spring collets, swages, tool shanks, metal slitters, wood cutting saws.
							Special Purpose (Carbon-Tungsten)
F1	1.00	1.25	—	—	—	—	Taps, broaches, reamers.
F2	1.25	3.50	—	—	—	—	Burnishing tools, tube and wiredrawing dies, extruding dies, reamers, taps, paper knives, eyelet dies, inserts for heading or sizing dies, forming dies, blanking dies, piercing punches, cold heading dies, sizing plugs and gages, roll turning tools, deep drawing dies.
							Mold
P2	0.07	—	0.20	2.00	—	0.50 Ni	Plastic molding dies (hobbed and carburized).
P3	0.10	—	—	0.60	—	1.25 Ni	Plastic molding dies (hobbed and carburized).
P4	0.07	—	0.75	5.00	—	—	Plastic molding dies, die casting dies (hobbed and carburized).
P5	0.10	—	—	2.25	—	—	Plastic molding dies (hobbed and carburized).
P6	0.10	—	—	1.50	—	3.50 Ni	Heavy duty gears, shafts, bearings, plastic molding dies (hobbed and carburized).
P20	0.35	—	0.40	1.25	—	—	Molds for zinc and plastic articles, holding blocks for die casting dies.
P21	0.20	—	—	—	—	4.00 Ni, 1.20 Al	Thermoplastic injection molds, zinc die casting dies, holding blocks for plastic and die casting dies.

*Other carbon contents may be available. †Optional. ‡Contains free graphite in microstructure to improve machinability. Some of these grades can be made with added sulfur to improve machinability. **Source:** AISI Tool Steel Products Manual and producers of tool steels.

Selected Tooling Materials

Grades recommended where extrusion and forming temperatures exceed 1,150 F . . . a new family of high-speed steels that can be heat treated to Rc 70 . . . urethanes for foundry patterns and core boxes. . . and a stainless that is gaining popularity in the molding of corrosive plastics . . . Those are among the significant developments covered in this comprehensive Metal Progress Forum Report. Specific subjects are listed below:

Superalloys

At extrusion and forming temperatures above 1,150 F, high-temperature alloys carry on where hot-work tool and die steels leave off. Typical of these are Pyrotools, seven alloys which have found increasing acceptance for a wide variety of tooling applications. While most production experience has been in the hot extrusion of brass, engineers in other fields such as extrusion and hot press-forming of titanium also report success.

The key to these alloys is their retention of high hardness at elevated temperatures. Rams, dummy blocks, mandrels, liners, and dies made from them offer high strength and great resistance to softening above 1,200 F. (Table I lists compositions and properties of these grades.)

Using the Alloys

During brass extrusion, usually performed between 1,350 and 1,850 F, liners (which normally average 1,500 F) may approach billet temperatures as high as 1,850 F. And some titanium forming dies are maintained constantly at 1,400 to 1,500 F. In such applications, where tooling actually operates above 1,150 F for substantial periods, conventional hot-work steels such as AISI H11 soften. In fact, the hot-work steels (H11 through H26) will, if run red hot, eventually return to an annealed condition. Result: erosion, swelling, or tearing of tool parts.

On the other hand, Pyrotool alloys never become as soft as the hot-work steels when exposed to high temperatures in service. Even if exposed beyond their temperature limitations, they continually maintain hardnesses of at least Rc 25 to 30.

We should emphasize that these alloys are not intended to replace conventional hot-work steels. Instead, they are offered for those high-temperature applications where the drop in strength and hardness of the hot-work steels interferes with production schedules.

Mr. Lane is metallurgist, tooling, Carpenter Technology Corp., Reading, Pa.

r Supertools

By E. J. LANE

Grades Have Varying Properties

Selection of a grade depends upon desired life expectancy compared with over-all cost. For example, Grades A and V are adequate for liners. The first is satisfactory for short-cycle operations in which copper and brass alloys are extruded at temperatures to 1,400 and 1,500 F. Grade V handles longer cycles, and can be applied for working cupronickel alloys at temperatures of 1,800 F and above. (Within the past few years, many brass extrusion companies have changed to Pyrotool A for liners, and a number of extrusion companies recently upgraded further to Pyrotool V.) Grade V is also used for dummy blocks where long life is required. For extrusion dies, Grade EX is suggested because it is easier to work than the other alloys. Where extrusion dies need the longest possible life expectancy, Grade M is a suitable alternate. However, if extrusion dies are susceptible to thermal shock, Grade 7 should be used instead. And Grade 15 is recommended for die insert holders because they must retain maximum strength without expansion or relaxation.

Pyrotools give impressive results. Liners made from Pyrotool A can last twice as long as those fabricated from H11 and H12 steels, and use of Pyrotool V may quadruple liner life. Also, dummy block life could be extended as much as eight times with the use of Pyrotool V.

The greatest savings lie in reduction of downtime and toolroom maintenance work. Sometimes in changing a liner the entire container must be removed from the press before the liner can be reached. Following this, the new liner must be inserted into a second preheated container before the press can be reassembled. Thus, quadrupling of liner life and consequent elimination of downtime often justify the initial cost of a high-temperature alloy.

Examples of Usage

One customer found it impossible to maintain production schedules for hot extrusion of cupronickel alloys which were high in nickel. Attempts were made to extrude seamless tubing on a 2,400-ton hydraulic press, but dummy blocks failed often, cutting down on production. The terrific heat and compression caused outer edges of the dummy to fold over and the inner holes to fill in so that the block no longer cleared the mandrel. In fact, blocks would mushroom and expand, eventually scoring the liner and pushing impurities into the extruded product.

When our metallurgists suggested Pyrotool V for this application, the extruder ordered two sets of dummies. After completing 804 pushes with the first set, the processors sent them to the machine shop for inspection; checks confirmed the original hardness, Rc 40 to 42. To date, there has been no need to use the second set; and the first, which has completed over 10,000 extrusions, is still in service.

Such performance is not unusual. In many instances, these alloys actually improve with time and use, seeming to grow stronger and better. In effect, cycling at elevated temperature tends to improve the surface interface. This effect, in turn, progressively reduces galling so that no galling problems occur during extrusion after the break-in period.

Some Cautions

Because these alloys are unable to withstand severe thermal shock, most should not be water cooled.

In hot extruding, the pushing dummy (right) acts as a buffer between the press ram and hot billet. Dummies may crack, roll over along the outer edges, or fill in at center holes.

Table I — High-Temperature Alloys for Tools and Dies

Pyrotool Type	Composition, %										Hardness, Rc	
	C	Cr	Mo	Ni	Fe	Ti	Mn	Si	Co	Other	Room Temperature	Test Temperature
A	0.04	14.50	1.50	25.00	55.06	2.20	1.20	0.50	—	—	30 to 35	28 to 29 (1,200 F)
EX	0.05	14.00	6.00	42.50	28.85	3.00	0.20	0.20	4.00	1.20 Al	40 to 44	26 to 27 (1,500 F)
M	0.12	19.00	10.00	54.98	2.00	2.50	0.20	0.20	10.00	1.00 Al	38 to 42	27 to 28 (1,500 F)
V	0.04	14.50	1.25	27.00	53.51	3.00	0.25	0.25	—	0.20 V	35 to 39	28 to 29 (1,300 F)
W	0.05	19.50	4.25	57.50	1.00	3.10	0.20	0.20	13.00	1.20 Al	36 to 40	32 to 33 (1,500 F)
7	0.05	19.00	3.00	51.50	18.60	1.00	—	—	1.00	0.60 Al, 5.25 Cb+Ta	40 to 44	34 to 35 (1,300 F)
15	0.03	15.00	2.90	0.20	61.87	—	—	—	20.00	—	46 to 50	34 to 35 (1,050 F)

(Alloys 7 and 15 have some resistance to thermal shock, we should note.) All tooling should be preheated to about 600 F before being placed into operation (as with the hot-work steels) and then air cooled during cycling to try to maintain that temperature as far as practicable. In some press cycles where hot material (above 1,500 F) contacts dies and dummy blocks for up to 1 min, two sets should be available so that those components can be rotated to allow time for normal or air-blast cooling back to 600 to 800 F.

Some extruders of tubing use internally cooled mandrels of H11 hot-work steel tipped with Pyrotool M. Instead of cooling the outer surface of the bimetallic mandrel by spraying with water, however, only lubricants are used to avoid thermal shock. Result: much longer mandrel life.

The high-temperature alloys are not as sensitive to crack growth as are the hot-work tool steels. While the martensitic hot-work grades will break apart once a crack appears, parts of Pyrotool alloys can be operated with numerous cracks because crack growth is limited.

In general, shop practices determine the best grade for the individual application. With the current degree of knowledge in choosing tooling material, empirical testing seems to be the only effective method in making a selection.

Maraging Steels for Die Casting

By ALEXANDER NAGY

Maraging steels are paying off as tooling material in aluminum die casting, the extrusion of aluminum, and plastic molding. Three varieties are being used; and their compositions are shown below:

Grade	Ni	Co	Mo	Ti
18250	18.00	8.00	4.80	0.40
18300	18.50	9.00	4.80	0.60
18350	17.50	12.00	4.80	1.50

All grades also contain 0.01 C, 0.05 Mn, 0.05 Si, and 0.10 Al. Ideally, these steels should be annealed (solution treated) by air cooling from 1,500 F, then aged at 900 F to develop the properties shown in Table I.

For tooling applications, the 18300 grade has become most popular. It is used for aluminum die-casting dies and cores, plastic molding dies, and various support tooling in the extrusion of aluminum. The grade features a tensile strength of 280,000 psi coupled with excellent ductility (12% elongation and 55.0% reduction in area). Aging at 900 F for 3 to 10 hr produces a nominal hardness of Rc 52 to 54. A wider range of hardnesses (from Rc 51 to 46) can be obtained by overaging at 1,000 to 1,075 F.

Mr. Nagy is associated with Universal-Cyclops Specialty Steel Div., Cyclops Corp., Pittsburgh.

Used in Dies

For die casting aluminum, AISI H13 is standard. Years of experience have shown that a range of Rc 42 to 48 provides the optimum combination of strength and toughness. Normally, dies fail by heat checking or erosion of the die, or both. Resistance to these effects would be improved by increasing the hardness, but raising the hardness of H13 out of the prescribed range leads to premature cracking.

Die hardness of the 18300 maraging steel can be Rc 52 to 54, however. The accompanying increase in resistance to heat checking and erosion substantially improves die life. The photo (opposite page) shows a casting (a basket hub for a washing machine) taken from a two-cavity mold wherein one cavity was made of H13 and the other of 18300. This casting was taken after 20,000 shots had been made. Note that the casting obtained from the H13 insert contains numerous surface irregularities due to heat checking on the die surface. The identical surface on the casting taken from the maraging steel insert, however, is quite clean and free of such irregularities, indicating that the insert is unharmed. At the latest count, 125,000 shots have been made without affecting the maraging steel insert.

Although excellent results have been obtained with maraging steel, these alloys need to be further improved and refined to assure consistently satisfactory results. Just as much experimentation was involved in establishing Rc 42 to 48 as the ideal range for H13 for aluminum die-casting dies, we must establish proper hardness ranges for maraging steel. Obviously, some comparatively small, simple dies work well at Rc 54. However, larger, more complex dies might work better in the Rc 48 to 50 range.

Table I — Mechanical Properties of Maraging Steels

Grade	Tensile Strength, Psi	Yield Strength, Psi	Elongation, %	Reduction in Area, %	Charpy V-Notch Impact Strength, Ft-Lb	Hardness, Rc
			Annealed Stock*			
18250	140,000	95,000	18.0	75.0	—	30
18300	150,000	110,000	17.0	72.0	—	32
18350	165,000	130,000	16.0	70.0	—	35
			Aged Stock*			
18250	260,000	250,000	12.0	56.0	23.0	50.0
18300	280,000	270,000	12.0	55.0	20.0	54.0
18350	360,000	355,000	6.0	30.0	10.0	58.0

* Specimens, 1 in. round, were annealed at 1,500 F for 1 hr, cooled in air, and aged at 900 F for 3 hr.

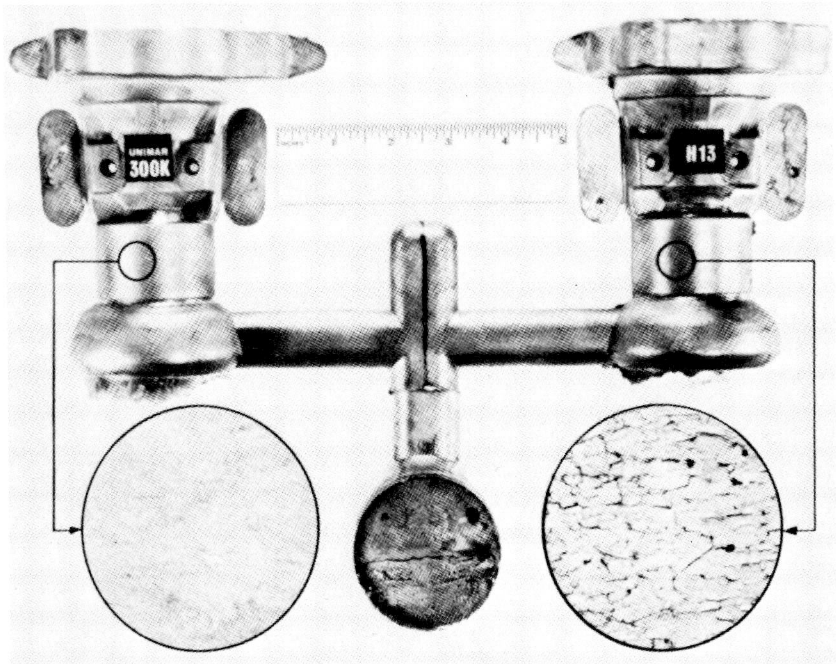

Casting taken after 20,000 shots shows effects of heat checking on right (H13-cast section). Section cast in maraging steel insert of die is smooth and defectfree. Insets: 2.3×.

Another useful feature of maraging steel is the relative ease of welding. In contrast with H13, which must be very carefully preheated and postheated, maraging steel can be welded without any preheat. In fact, the welder should make sure that heat input is as low as possible by making very light passes when laying down a deposit. Another indication of excellent results obtained in welding: maraging steel welding rods have been used to repair defective H13 dies.

Plastic Molds Another Application

Maraging steels find wide use in the plastic molding industry where the negligible size change that occurs on heat treatment is of value. For plastic molding applications, 18300 is the most popular. A minimum hardness of Rc 50 – the usual requirement – is obtained by aging at 900 F for 3 hr. Since short aging times result in comparatively little size change, intricate molds can be machined from steel in the annealed condition down to final dimensions, then aged.

Unlike other tool steels which expand when heat treated, maraging steel shrinks. The amount of contraction is a function of aging temperature and time as follows:

Aging Treatment	Shrinkage
900 F, 3 hr	0.00025 in./in.
900 F, 6 hr	0.00050
900 F, 10 hr	0.00100
1,025 F, 6 hr	0.00125

With the development of extremely abrasive plastics, a need for stable mold steels of higher hardness has arisen. This need is fulfilled by using an 18350 maraging steel which develops a nominal hardness of Rc 58. Complex, intricate molds can be machined from this alloy, and then heat treated to the desired hardness with practically no distortion or size change.

A number of other applications are being evaluated. These include cold impact extrusion dies; high-strength, thin-webbed forging dies; and other highly stressed, complex tools. Though it will not solve all problems, maraging steel is a strong candidate for a number of seemingly unsolvable die problems.

A Stainless Grade for Plastic Molds

By WILLIAM YOUNG

AISI TYPE 420 STAINLESS has been used for plastic molds to a limited extent for many years. Now it is gaining in popularity because of the increasing use of corrosive plastics, such as polyvinyl chloride. The government has approved this plastic for plumbing tubing and fittings, and many molders are specifying type 420 (0.30 C, 0.80 Mn, 0.50 Si, 13.0 Cr) for their molds.

Another reason for the alloy's increasing use as a mold material is its resistance to rusting. Shorter cycling times needed to raise production call for molds to be cooled with refrigerated water, which causes condensation on the mold surface. Since those made from type 420 will not corrode in storage, and water lines will not clog with rust, this grade is ideal for injection, compression, transfer, and glass molds – particularly when the environment or plastic (or both) are corrosive.

Importance of Heat Treatment

Molds of type 420 can be hardened from an austenitizing temperature of 1,850 F. Full hardness is obtained in air when cooling molds up to 4 in. thickness. For thicker ones, an interrupted oil-quench is recommended. Quenching should be stopped at 1,100 F, followed by cooling in air. Larger molds with complicated designs can be fan-cooled in air with little sacrifice in hardness.

Molds should be double tempered at 750 F for at least 4 hr each time; this treatment will result in hardnesses of Rc 48 to 54. Molds heat treated in this manner will have the optimum corrosion resistance, toughness, and impact strength. Tempering between 800 and 1,100

Mr. Young is technical consultant, Crucible Specialty Metals Div., Colt Industries, Syracuse, N.Y.

F should be avoided because it will reduce corrosion resistance and impact strength. And tempering below 750 F will produce a structure which lacks toughness and may result in early fatigue and cracking.

Molds Are Getting Bigger

During the last year, molds have been made from blocks weighing as much as 3,000 lb. Blocks of this size have to be sound to obtain the needed high polish and luster. Our densifying process for type 420 molds has been used for the past eight years in manufacturing tool steels for molds and dies. The method imparts more uniform properties and a more homogeneous structure. Maximum center soundness is of prime importance because the bottom of the mold cavity, where the mold is polished to high luster, quite often coincides with the middle of the block.

During the process, forging forces are essentially directed into the center of the ingot, where the greatest amount of porosity exists. First, the hot top of the ingot is forged to fit into the bottom die when the ingot is set up vertically in the press (see drawing). As the top die moves down, force is transmitted first to the hot top, and then into the middle of the ingot. During initial pressing, very little compression work is done on the ingot surface. As the ingot is further squeezed down, its shoulders come in contact with the die so that more work is directed into the center portion. Upsetting continues until the ingot is half its original length. During this operation, the ingot's girth increases. Thus, more reduction in cross section can take place while the steel block is being conventionally reduced on flat dies to its original cross sections. For very large pieces, upsetting and drawing out are repeated two or three times until center is as dense as the outside.

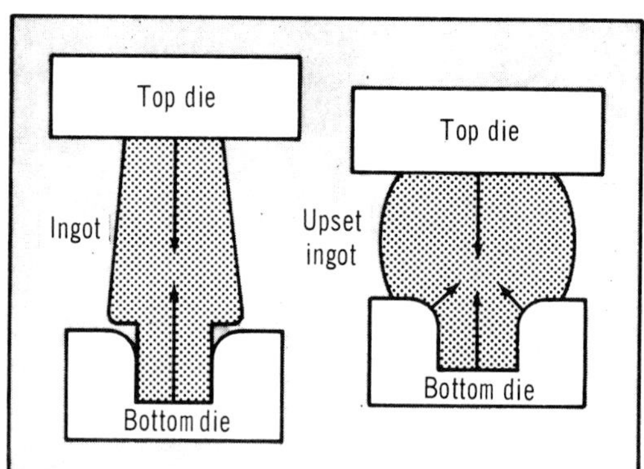

Inserted upside down into the forging press (left), ingots of type 420 stainless are compressed to increase their girth (right). This extensive working helps to densify interiors of die blocks.

Carvable Plastics for Models, Patterns

By RICHARD B. PETERSON

IN THE CONSTRUCTION of prototypes and models, polyester plastics were first used as a patching material, and later to face the entire surface of the model. As time passed, epoxy plastics appeared. And they are being accepted more and more by tool and design engineers. They consider plastics to be dependable materials, and use them for many applications. Models constructed from carvable epoxies, for example, give excellent performance, and can be constructed economically.

Epoxies have several advantages when compared with natural materials. First, they are grainless, and will absorb little, if any, moisture when cured. In addition, engineering changes and corrections can be made rapidly, and adhesion of epoxy to epoxy is excellent. Since material can be added and removed quickly, leadtime is reduced. Also, the final model holds tolerances well because it is made of like materials. Rough models man be built close to final dimensions, and finished surfaces can be developed without wasting material or labor.

Plastic models may be used as intermediate tooling. Because the surface has strength and hardness, the tool can be employed as a duplicating model, or as a master model for constructing further intermediate tooling. Further, surfaces will not break down during use.

Also available are paste plastic materials which can be used to incorporate engineering changes. These pastes can be applied to create an easily worked surface on the model. Other carvable epoxies have been formulated for application by splining, casting, and spraying. Finished dimensions are developed by the use of finished templates and traditional carving techniques, or by machining with equipment which can be programmed by numerical control.

Constructing a Model

A typical model buildup starts with the construction of a rough female mold. This rough mold may be constructed of plywood, sheet metal, or any other inexpensive material that can be roughly contoured to create a cavity oversized by ¼ to ½ in.

After the mold is coated with a parting agent, epoxy paste is applied to the surface, and built up to approximately 1 in. in thickness. A laminate is then applied to the back of the paste to add strength, if required.

Honeycomb bulkheads are attached to develop the back-up structure, and epoxy tubing with cast base is applied to establish the base plane. Then, the model is left in the mold and cured. After oversized grid slots for locating templates are machined in the carvable surface, additional carvable material is wiped into them. Templates coated with a parting agent are then pressed into the slots to establish the final contour grid. It is important, when the templates are set in the wet plastic, that the model be in body position and the templates be aligned on all three axes. After the paste has cured, templates are removed. When the

---

Mr. Peterson is associated with Ren Plastics Inc., Lansing, Mich.

final surface is developed (by removing the excess stock and correlating flow and section templates), the model is ready for inspection and painting.

With some variations, the same basic techniques may be used to fabricate models from epoxies formulated for casting and spraying. When using casting materials, a rough core is suspended over the rough female mold which has been prepared with a release agent. The epoxy is then cast to fill the area between the mold and the suspended core to develop the surface which will ultimately be worked to finished dimensions.

As for spray-up materials, spraying equipment is used to build up an oversized epoxy surface on an undersized armature. A rough core is constructed of aluminum honeycomb panels, rough cut and bonded together. Then, carvable epoxy is sprayed over the armature to develop the surface to be machined. This armature is next mounted in machining position, and the final contour is cut by tape-controlled equipment.

In fact, as numerical control becomes more universally available, carvable epoxies should play a greater role in tooling programs. Not only can a mathematically computed tape be verified for accuracy, but a permanent master can be quickly constructed for use as a reference tool. A laminate shell can be quickly taken directly from a clay styling model, reinforced for rigidity, and coated on the inside with carvable epoxy paste, casting material, or spray material. The final contour is developed by numerical control tape which is programmed by scanning the clay model and establishing key engineering reference points. Surfaces may be resprayed or recast if there are engineering changes or errors to be corrected.

New Developments Boost Popularity of S7

By JAMES J. McCARTHY

FIRST INTRODUCED in the 50's, AISI S7 has evolved into a tool steel delivering performance expected in the 70's. Containing 0.50 C, plus 3.25 Cr and 1.40 Mo, this air-hardening grade was originally used where extreme shock resistance was required and to a lesser extent for hot-work tooling involving moderate temperatures. Features include safety and dimensional stability in hardening, low levels of residual stresses, and consistency from part to part.

Over the years, case histories ranging from cold-impact tools to dies for plastic molding of the injection, compression, and transfer types have attested to the breadth of applications for S7. As an example, the accompanying photo shows two-cavity molds of S7 which have produced several million camera frames of styrene. (The steel is used at Rc 55.) Its relative ease of machining and minimum dimensional change in hardening contributed to performance in this application.

Mr. McCarthy is assistant metallurgical engineer, staff of vice president-operations, Bethlehem Steel Corp., Bethlehem, Pa.

Hardening the Steel

To measure the hardenability of a deep-hardening tool steel such as S7, we use a specimen which is 2½ to 3 times as long as its diametral measurement. Our tests show that the following maximum section sizes harden to centers at specimen mid-lengths: AISI S7 – 2½ to 3 in. round; AISI A2 and A4 – 4 to 5 in. round; and AISI D2 – 6 to 7 in. round.

Regardless of the type of tool steel, a key factor in performance has been control of the surface composition. This can be done by either providing sufficient stock allowance for removal by grinding after hardening, or by neutral hardening. The latter method is preferable on air-hardening steels since their inherently lower distortion reduces the need for grinding to adjust dimensions. Such work is handled by controlled atmosphere and vacuum furnaces.

Demonstrating the need for careful processing, we investigated a cracked S7 die measuring 12 by 12 by 8 in. which had been preheated at 1,425 F for 1 hr in endothermic atmosphere and then austenitized at 1,725 F for 4½ hr with enriching gas added the last hour. The resulting carbon-enriched surface or case, more than 0.050 in. in thickness, had an appreciable quantity of unstable retaining austenite, creating differential quenching stresses which were sufficient to cause cracking.

To help heat treaters, we have studied the effect of various atmospheres on S7 tool steel when heated at 1,725 F, recommended for austenitizing. Our results indicate the following atmospheres can be used without danger of carburizing or decarburizing:

- Rich endothermic (AGA Clsas 302) with dew points in the 21 to 29 F range.
- Rich combusted ammonia (AGA Class 622 – 18% H_2, balance N_2) at a dew point of –70 F or lower.
- Lean combusted ammonia (AGA Class 621 – 100% N_2 with a dew point of –52 F or lower.
- Prepared nitrogen-base (AGA Class 223 – 4% H_2, balance N_2) with a dew point of –44 F or lower.
- Argon with a dew point of –42 F or lower.

The steel can also be vacuum treated where the pressure is maintained below 1×10^{-3} torr during

Dies of S7 turn out complex plastic forms for cameras.

the cycle.

These results highlight the importance of strict attention to dew point control regardless of type of atmosphere. We should also note that vacuum hardening does not guarantee freedom from surface decarburization even through scaling may not occur.

Wrapping in Foil Helps

Our engineers have run tests to establish the effectiveness of wrapping tools of S7 in type 321 stainless steel foil, 0.002 in. thick by 24 in wide. Judging by results, airtight wrapping effectively prevents decarburization. Also air-hardening tool steels will be fully hardened on cooling as long as the austenitizing temperature does not exceed 2,000 F and the holding time at high heat does not exceed 4 hr. For steels with 0.50 C or lower, the wrap is effective for austenitizing temperatures below 2,000 F for up to 8 hr. (Such wrapping also works for oil-hardening grades, but should be removed just prior to quenching. Otherwise, it appears to interfere with attainment of expected hardness during quenching.)

Thus, dies or mold components of S7 in sizes small enough to harden in air effectively can be austenitized and air-quenched in airtight foil wrap without carburizing, decarburizing, or scaling.

It is also possible to obtain prefinished flats and squares of S7. Normally furnished in mill-length bars free of decarburization with good surface finish (100 to 125 rms), squareness, and size accuracy, this product has undoubtedly contributed to better toolmaking practice. Users have saved money not only by cutting direct material expense but also by reducing time in shop preparation.

Die Blocks of Forged H13 Are Tough

By ARNE OMSÉN and BENGT E. SKOOG

WELL KNOWN for its high-temperature hardness, wear resistance, and excellent resistance to thermal shock and fatigue, H13 is generally considered the standard die steel for die casting, hot pressing, and extrusion of aluminum and copper alloys. However, this grade and related steels have one inherent weakness – a tendency to precipitate grain-boundary carbides during cooling. This effect has proved hazardous because it embrittles heavy sections and jeopardizes uniform properties throughout tool bodies.

Modern steelmaking technology has, however, given us ways to overcome these difficulties. Our research has enabled us to use advanced melting practice, including electroflux refining, to process our H13 steel – UHB ORVAR – to excellent cleanliness and structural homogeneity even in heavy ingots. Subsequent forging is handled by our fully integrated forging press system. The press, with its 3,000 tons of pressing force, produces rapid deformation of ingots. Auxiliary equipment for heating, transport, and manipulation is automated to give a maximum of accuracy in the forging cycle.

Treatment includes preforging the ingot to oversized dimension, followed by reheating for carbide solution.

By inducing a high degree of deformation energy at a high temperature during the subsequent final forging, we achieve immediate, complete recrystallization, which transforms the large austenitic grains to small ones, and impedes carbide nucleation. After being cooled and annealed, the structure is characterized by mechanical properities, ductile fracture, and homogeneity as good as the properities in small-sectioned material.

Effects of Heating

The UHB ORVAR (5.3 Cr, 1.4 Mo, 1 V) contains three types of carbides in annealed stock: $(FeMo)_6C$, $(FeCr)_7C_3$, and VC, represented in amounts of 2:2:1 in the structure. When heated in the austenitic region, carbides dissolve – chromium carbide at about 1,650 F, and molybdenum and vanadium carbides at about 1,850 F. All carbide phases are almost completely dissolved over 2,000 F.

Large Grains Lower Properties

During such treatments, austenitic grains grow spontaneously unless un-

The 3,000-ton forging press at the Uddeholm plant in Sweden handles large ingots of tool steel.

Dr. Omsén is supervisor of metal physics research and Mr. Skoog is manager of product and material development, Steel Research Dept., Uddeholm AB, Hagfors, Sweden.

dissolved carbides in the matrix effectively restrain growth. It is a well known fact that larger grains result in decreased toughness. We find conclusively that a thorough recrystallization must be introduced in the material.

As the carbides dissolve, the austenite will be enriched in alloying elements. Above 2,000 F, the composition of the austenite equals that of the steel. When such an austenite cools slowly, proeutectoid carbides precipitate predominantly in the grain boundaries. They will develop into very heavy aggregates with time, further impairing the toughness.

Practical Consequences

Our investigations further indicate that large forged blocks of UHB ORVAR can have substandard properties because holding for long times at high temperatures promotes austenite grain growth. Frequently, the forging time for heavy blocks is extended so that the temperature of the block drops; then, recovery of introduced deformation energy becomes more difficult. This effect causes problems because unrelaxed deformation energy activates nucleation of austenite decomposition and carbide precipitation. Finally, restricted heat dissipation from heavy forgings may retard cooling through the critical temperature range, allowing harmful grain-boundary carbides to agglomerate.

Based on this briefly presented structural evidence, we realized that UHB ORVAR – as an example among similar grades – can be processed to high-quality standards in big blocks. The two-step forging cylce, as has been noted, practically elimintes the detrimental effects caused by grain growth and precipitation of grain-boundary carbides. Dwell times at higher temperatures are kept to a minimum. Thorough grain refining by recrystallization is achieved by final forging at a crosshead speed of 3 to 3½ in. per sec at maximum pressing forces. The efficient system guarantees strict timing and temperature control, excluding hazards of critical treatment conditions.

Machinable Carbides Solve Problems

By STUART E. TARKAN

More and more manufacturing companies are turning to machinable carbides for long-wearing tools and components because these steel-bonded carbide materials provide solutions to many different problems. It is not unusual for users to raise productivity 100 times above that of tool steel and increase the quality of the finished part at the same time. Further savings accrue due to reduction of costs for downtime (for tool changes), for standby production equipment, and for tool maintenance.

Machinable carbides have an unusual combination of properties. For example, the Ferro-Tic family comprises a group of five materials that machine like tool steel and, when hardened, wear like carbides. Briefly, they are sintered composites of hard grains of titanium carbide in a matrix of tool steel or alloy. They can be made in round or rectangular blanks, rings, or special shapes.

Mr. Tarkan is vice president, Sintercast Div., Chromalloy American Corp., West Nyack, N.Y.

As Table I shows, M-6 and HT-6 have high resistance to heat and corrosion resistance, while C and CM have better machinability. All, of course, have satisfactory abrasion resistance.

For general-purpose tooling, Ferro-Tic C – it has a tool steel matrix – is widely accepted. Annealed blanks, at a hardness of Rc 40, can be turned, milled, bored, or otherwise machined as easily as tool steel. Hardening by oil quenching from 1,750 F gives it a wear resistance equivalent to that of conventional carbides.

Preparing Machinable Carbides

Any processor who can work with D2 steel can handle machinable carbides just as easily. There are, however, a few recommendations to follow to assure good results. In machining, for example, cutting speeds are slow – from 20 to 35 surface ft per min for turning to 10 to 15 surface ft per min for drilling. But feeds are heavy (5 to 8 mils per rev for rough turning, for example), so that total machining time is not excessive. Cutting is done dry, and at least a 3 mil cut should be taken to prevent glazing. Either annealed or hardened material can be ground, with open-structured wheels of aluminum oxide doing the best job. Dry grinding is suggested for annealed material; after hardening, mist cool or flood with conventional coolant when grinding.

Hardening techniques are similar to those for tool steel. The material, however, must be protected against decarburization and overtempering by processing in a controlled-atmosphere furnace, a salt bath, pack hardening box, or a stainless steel envelope which excludes all air. Since Ferro-Tic C will grow about 0.0003 in. per in. during hardening, a die generally can be furnished to size in the annealed state. If grinding is necessary to hold tight tolerances, only one or two thousandths

Table I — Properties of Machinable Carbides

Property*	Ferro-Tic Types				
	C	CM	J	M-6	HT-6
Machinability	G	G	F	F	F-P
Hardness, Rc					
Annealed	40	45	45	49	46
Hardened	68-72	67-69	68-70	63	54
Wear resistance	E	E	E	G	G
Heat resistance	P	F	F-G	G	E
Corrosion resistance	P	P-F	P	G	E
Max operation temperature, F	400	900	1,100	900	2,000

*E, excellent; G, good; F, fair; P, poor.

need be left for this operation. Warpage is minimized.

When building composite tooling, the steel-bonded carbide can be fusion-bonded to back-up steel. In fact, standard sections are available; they consist of 90% tool steel with a fusion-bonded facing of machinable carbide.

Because machining and heat treating procedures can be followed readily in most plants, steel-bonded carbides have great appeal. With a supply of blanks on hand, the toolroom can supply a die within a matter of days – sometimes hours – using available equipment.

Applications Are Diverse

Machinable carbides are used for stamping, drawing, and forming tools, gages, form rolls, guides, drill bushings, and wear parts.

- At Boonton Molding Co. Inc., a machinable carbide die has compression molded more than 120,000 pot handles of asbestos-filled phenolic resin without maintenance. In a typical gang die, six 1/4 in. pins shape the hollows within the handles. When the die is closed, these pins are constantly subjected to abrasion by the resin flowing over them.
- Kwikset Powdered Metal Products Co. uses Ferro-Tic C to make compacting and coining dies for high-density parts. High pressures greatly increase friction on die walls, severely testing the dies's abrasion resistance. With Ferro-Tic, some runs go over a million pieces without die maintenance.
- American Can Co. has used machinable carbides for a variety of tooling applications in several plants. They include body hammers to flatten side seams of cans; segmented curling rings which have been run over four years without maintenance; and body blanking dies which have withstood 5 million hits between regrinds.
- Wix Corp. has substantially reduced costs per piece by employing machinable carbide for draw, pinch trim, and perforating dies. Production of 2½ million to 3 million pieces between regrinds equals that for dies of tungsten carbide. However, die cost ran 75% less because the plant's shop could make its own dies.

Heat, corrosion, and wear-resistant grades are finding use in heavy cold heading and hot forging and in such parts as valves and nozzles, rotating seals, and gas bearings.

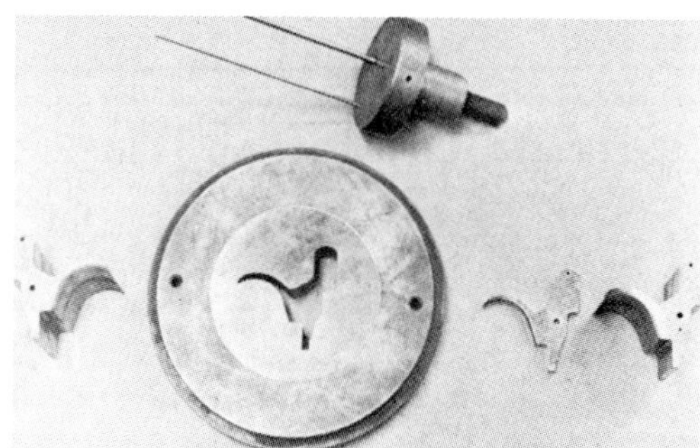

Die and core rods are made of machinable carbide. Note compressed powder metal parts at sides.

Carbide Tools Ease Cold Extrusion

By W. L. KENNICOTT

For most managers of metalworking plants, processors, and tooling experts, the mention of carbides evokes thoughts of cutting tools for shaping parts by turning, facing, boring, or milling. These basic tooling materials, however, can be modified to give equally good results in the forming of metals. Generally somewhat softer than compositions used for cutting tools, die grades with 10 to 25% Co – the balance is essentially WC – are applied efficiently in heading, swaging, or drawing operations.

Developed recently, techniques for controlling the grain size and dispersion have enabled engineers to raise shock resistance of carbide tools without sacrificing wear resistance or stiffness. With this grain size control, the same shock resistance is retained with a lower binder content. For example, WC with 12% Co will have a hardness of Ra 88 and a modulus of elasticity of over 82 million psi. (A conventional composition of equal shock resistance having 20% Co reaches only Ra 86 with a modulus under 70 million psi.)

The most common use for these newer carbides is in punches for cold extruding steel parts. It is not unusual for a back extrusion punch to have an average stress of more than 400,000 psi in compression at the end of its stroke. In fact, with normal variations in alignment, work material dimensions and properties, localized stresses in excess of 600,000 psi may be reached. Under such conditions, only a tool material with a high modulus can resist bending or upsetting. At the same time, the punch is subjected to the shock of contact on each stroke, plus wear which occurs as steel flows between the punch and the die wall.

Given the same deflection, mechanical hysteresis of carbides is only about one-fifth that of hardened steel. Since these carbides are also almost three times as stiff as steel, mechanical hysteresis is actually about one-fifteenth that of steel for a given load. Because fatigue is closely related to hysteresis – which indicates the extent of permanent setting each time there is a reversal of stresses – these carbides are very resistant to fatigue under the cyclic stresses involved.

Cup-shaped parts made by cold back extrusion of steel include roller bearing cups, wrist pins, and valve lifters. (One much larger job involves a steel slug weighing 13 lb; the 11 in. long punch has a working diameter of over 2 in.) Forward extrusion is used to produce small shafts, pins, and worm gear blanks, while combination back and forward

---

Mr. Kennicott is vice president, engineering, Kennametal Inc., Latrobe, Pa.

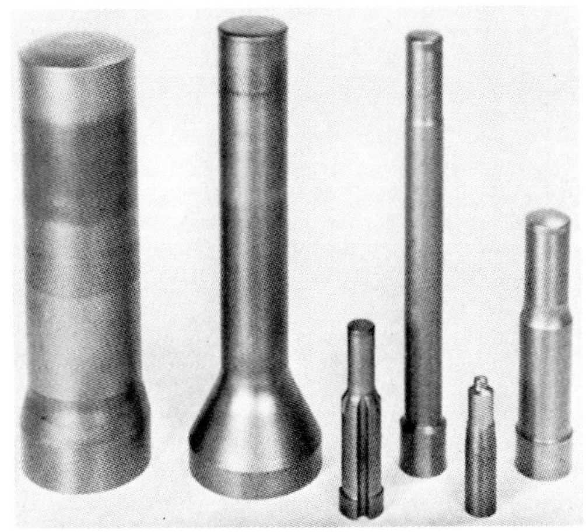

Punches of high-strength, shock-resistant tungsten carbide are used for cold back extrusion of parts ranging from a few ounces to 13 lb.

extrusion is entailed in making spark plug shells. In all instances, the punch is a highly stressed component in the tooling setup.

Carbides in the same family are in widespread use as rolls in new high-speed rod mills, built to deliver hot steel rod at over 10,000 ft per min. Here, again, the need is for mechanical shock, wear, and gall resistance and freedom from fatigue as well as resistance to heat shock.

Other hot-working tools are in limited use. To date, the limitation is length of time in high-pressure contact with hot steel. Carbide surfaces can develop heat checks. Thus, hot punching is successful due to the very short time of contact per cycle. Some hot forging operations look feasible, but only in special high-speed machines where the contact between work and tooling lasts less than 1 sec.

P20: A Familiar Grade Updated for Molders

By E. E. LULL

CONCURRENT WITH NEW tool steel developments, producers have been active in upgrading traditional grades — those most commonly used in day-to-day tooling applications. One outstanding example is the continuing progress which has been made in improving the quality of AISI P20, the fastest growing tool steel of the past decade.

Since the growth in usage of P20 can be attributed to the rapid increase of zinc die casting and injection plastic molding, we can examine the changes which have been made to enable this popular grade to keep pace with this fast-growing industry — particularly as they pertain to plastic molds. A decade ago, popular injection molding machines handled shots of 16 to 24 oz, utilizing a clamping pressure of 300 to 450 tons. A large machine could inject up to 60 oz with clamping pressures of 600 to 750 tons. Tools for this machine were made of P20 in thicknesses up to 10 to 12 in., widths of 26 to 30 in., and weights of 4,000 to 5,000 lb.

Today, big injection molding machines can handle shots to 1,000 oz, and have a melt capacity of several thousand pounds per hour. Machines with clamping pressures of 4,000 tons are not uncommon, and 10,000-ton machines are being built. Plastic molds require steel blocks up to 30 in. thick and more than 80 in. wide — many weigh over 20,000 lb. Further, tools are not only larger, but are taking greater abuse because cycle times have been reduced by automatic injection to 5-25 sec.

Steels Are Improved

As molding machines and their tools grew larger, tool steels kept pace. Problems attended this progress because prehardened blocks required greater hardenability to obtain the necessary center hardness, but machinability or weldability could not be sacrificed. Also, center soundness had to be maintained in the larger blocks to assure good polishability from top to bottom of cavities.

These aims are being realized by consistently improving quality. For example, the composition of P20 has changed. Back in 1955, it was 0.30 C, 0.75 Cr, and 0.25 Mo. By 1963, contents of the alloys had risen to 0.35 C, 1.25 Cr, and 0.40 Mo to boost hardenability. Currently, melting ranges of P20 are 0.30 to 0.35 C, 0.65 to 0.85 Mn, 1.55 to 1.85 Cr, and 0.35 to 0.45 Mo. Some manufacturers even add boron to increase hardenability, and others modify sulfur to improve machinability — but not to the extent of impairing polishability and weldability.

More Efficient Processing

To fulfill demands for cleaner, sounder material for larger tools, producers also devised new melting and forging techniques. Since the early 60's, at least 13 variations of vacuum processes have been introduced. The most recent is vacuum arc degassing (VAD), a patented process developed by our organiza-

Huge press forges ingots of P20 die steel.

Mr. Lull is product manager, specialty steels, A. Finkl & Sons Co., Subs. of Republic Steel Corp., Chicago.

tion, and discussed in *Metal Progress* (December 1969, p. 12). Degassing is of interest to builders of plastic molds because it reduces hydrogen (eliminating flakes) and oxygen, cutting down on nonmetallic inclusions. Prevention of reoxidation also helps to make a cleaner steel, capable of being polished to a finer finish with a higher luster.

In another advance, press sizes have been increased and presses have become more versatile in operation. Typical of the newer equipment is the 5,000-ton computer-controlled press which went into operation in our plant late last year. It can operate at a penetration rate ranging from 0.07 to 2.54 in. per sec. Positive control of penetration rates helps to close and permanently weld small internal voids that could not, if they remained, be eradicated by polishing.

Through computerization, the press is also capable of forging to a tolerance of $\pm 1/16$ in. Our forgers can obtain maximum reduction in area from a given ingot size to a given finished size for a plastic mold. By this means, blocks for large plastic molds are upset from smaller ingots. Since small ingots are inherently cleaner than large ones, the forgings which result are also cleaner. And a clean steel is essential to good polishability – regardless of the size of the mold or insert.

Graphitic Tool Steel for Precision Parts

By PHILIP A. NAGY

WITH ITS COMPLICATED computers and electronic equipment, the Space Age demands much of tool steels. To answer needs for better grades, we offer our air-hardening graphitic tool steel – AISI A10. Features of this steel, which we call Graph-Air, are dimensional stability and good hardenability. Typical composition: 1.35 C, 1.80 Mn, 0.025 P max, 0.025 S max, 1.20 Si, 1.85 Ni, and 1.50 Mo.

The steel is considered suited for applications where varying cross sections are involved or close dimensional tolerances are required. They include gages, draw dies, blanking dies, forming dies, spindles, cams, punches, drill plates, print-out rolls for computers, and dies for printed circuits boards.

The structure contains controlled amounts of graphite achieved by adjusting the composition, melting practices, and subsequent hot working operations. Free graphite is, in many ways, beneficial. It acts as an antiseizing agent, and provides built-in discontinuities, preventing rapid buildup of surface stresses during grinding which, in turn, inhibits crack formation.

Mr. Nagy is service metallurgist, Timken Roller Bearing Co., Canton, Ohio.

In annealed material (Bhn 269), graphite particles control the size of machining chips, and act as a solid lubricant, providing ease of machining and excellent surface finish. Thus, in spite of its deep hardening characteristics, the grade machines as well or better than the conventional oil-hardening tool steels. Graphite pockets also act as reservoirs for externally applied lubricants.

Processing the Grade

Graph-Air is austenitized between 1,450 and 1,500 F. Since the hardening temperature is fairly low, the thermal gradient is less than those associated with the higher hardening temperatures recommended for other air-hardening tool steels. This effect, in turn, reduces distortion during heating and cooling; scaling is also less. Parts which are 2 in. and under in cross section are heated at 1,450 F; those 2 to 4 in. in cross section are heated at 1,475 F; and those 4 in. and greater in cross section are heated at 1,500 F. Allow parts to reach furnace temperature and hold 30 min. Then the material should be cooled in still air to room temperature, an essential operation because of the low M_S temperature (325 F). As-quenched hardnesses in the Rc 60.0 to 62.0 range can be expected in section sizes up to 3 in., and Rc 59 to 61 in section sizes over 3 in.

Tempering after hardening is required to obtain the desired hard-

Extreme accuracy is required in dies for making printed circuits. A graphitic tool steel, A10, gives ease of machining, plus durability in service.

When tempered at the indicated temperatures, the hardened steel changes little in dimensions. Samples, 4 in. long by 0.750 in. in diameter, were austenitized at 1,450 F and quenched in air to under 100 F before being tempered.

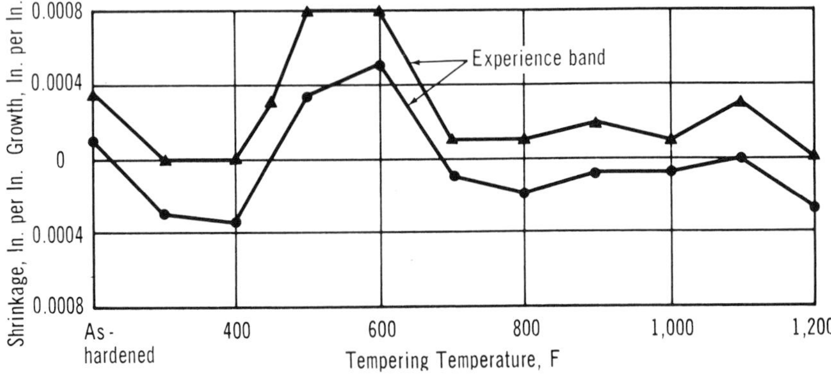

ness, toughness, and dimensional stability. The steel has good resistance to tempering, and hardness remains in the range of Rc 55.0 to 57.0 after a 700 F temper. Dimensional changes experienced in hardening and tempering of Graph-Air are minute (as shown in the graph on p. 80) and are essentially equal in both the longitudinal and transverse directions. The graph also shows that tempering between 300 and 400 F will return the section to its original size. As a result, grinding after hardening can be held to a minimum; in many instances, hardened sections only require polishing. Original dimensions are also maintained extremely well at tempering temperatures of 700 F and above.

Hardened and tempered, the steel has excellent dimensional stability because the retained austenite present is stabilized. For optimum long-time stability, we recommend cold treating, however.

Carbon is ample for both free graphite particles and evenly distributed, wear-resisting carbides. Tests show that the combination of the graphite particles and carbides provides resistance to wear, galling, and pickup. Therefore, it is possible to obtain both toughness and wear resistance. Finally, under conditions of inadequate lubrication, debris formed between two surfaces acts as a lubricant rather than an abrasive third body.

Cemented Titanium Carbide Has Big Future

By HERBERT S. KALISH

CEMENTED TITANIUM CARBIDE is becoming an important tool material for machining steel. In fact, we feel that the end of the 1970's will see 75% of all steel machined with solid tools being handled by cemented titanium carbide. Though available in 1929, this type of material really began to draw attention after 1959 when Ford Motor Co. announced development of a high-strength, high-hardness composition.

Properties of our Titan 60 and Titan 80 (C6 and C8 basic grades) are shown below:

	C6	C8
Hardness, Ra	91.9	93.0
Transverse rupture strength, psi	250,000	200,000
Density, g per cu cm	5.72	5.63

The major constituent of both is TiC with a $Ni\text{-}Mo_2C$ binder. The C6 grade is for general-purpose machining of steel, while the C8 grade is for semifinish and finish machining.

Titanium carbide tools have several advantages. First, they have resistance to edge buildup and cratering much superior to that of conventional WC-base grades, and even to that of ceramic materials. With high resistance to edge wear, tools hold keen edges longer, and give closer tolerances on long finishing cuts, or where part size is critical. Surface finishes as low as 10 rms are easily attained on a high production basis.

Tools of TiC tolerate wider variations in surface cutting speed than either WC-base carbides or ceramics. In fact, they can be run at cutting speeds approaching those of the ceramics. However, they have superior strength so that heavier cuts can be taken at high speed. Such tools last longer than WC tools at equal speed, or allow higher speeds for the same tool life. The graph compares use of tools of the Titan 80 type titanium carbide with those of the regular C8 grade of cemented tungsten carbide for machining 1045 steel. Note that, for any machining speed, tool life is increased almost five times. Also, it is practical to use machining speeds of 700 to 1,200 surface ft per min with TiC tools.

In other tests with AISI 1045, we compared flank wear on tools of C8 cemented titanium carbide and a regular C7 cemented tungsten carbide. For the C7 type, flank wear was more than three times as much under equal conditions. Again, the effect becomes more pronounced at the higher machining speed. From 800 to 1,200 surface ft per min, flank wear of the cemented tungsten carbide tool increased drastically, while that of the cemented titanium carbide rose gradually.

Showing what can be done, a machinist used an LNU 5464 steel insert in Titan 60 grade on steel rolls 13 ft long by 27 in. in diameter. As the tool took off 1¼ in. diameter at 375 surface ft per min with a 0.040 in. feed, it wore but little, and the tip blued just slightly from the heat.

Wear Resistance Is High

Great potential for cemented titanium carbide also lies in special structural and wear parts. For example, titanium carbide was used for wire feeding jaws in a high-production repetitive operation in an electronics plant. The titanium carbide jaws lasted for six 8 hr shifts. The best performance of tungsten carbide jaws was one 8 hr shift.

Another wear application involved an induction heat-treating operation. Six wear pads of Titan 60 were copper brazed by the vacuum method into pockets of stainless steel holders. These pads held truck crankshafts measuring up to 6 ft in length and weighing several hundred pounds. Prior to the use of the titanium carbide, hardfaced wear pads had to be replaced every day. The Titan 60 types lasted more than two months.

Tools of C8 cemented TiC are more durable than those of cemented WC. Material: AISI 1045, Bhn 170. Condition: feed, 0.011 in. per rev; depth of cut, 0.060 in.; wear land, 0.010 in.; coolant, soluble oil.

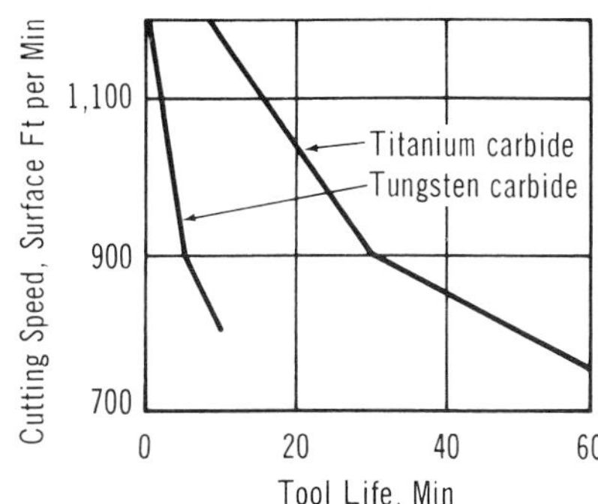

Mr. Kalish is director, research and development, Adamas Carbide Corp., Kenilworth, N.J.

Where Urethanes Are Used in the Foundry

By JAMES KOSMALA

In the foundry, patterns and core boxes are subjected to drastic abrasion. Extensive deterioration occurs because sand slinger molding machinery and core shooters drive abrasive materials against them at high rates. Urethane elastomers have a combination of tensile strength (5,000 to 8,000 psi) and good abrasion resistance that makes them a natural for patterns and core boxes. When used with conventional sand slingers, such items have outlasted their counterparts by a factor of 5 to 1. Of course, this means significantly less downtime and more cores per mold.

Due to their impact resistance (>10 ft-lb Izod), urethane components withstand punishment without degradation or breakage. Another advantage: their high load-bearing capacity helps them resist the high pressures and temperatures needed to produce finely detailed, true reproductions in cores and molds. Foundry urethanes also exhibit 200 to 700% elongation, plus excellent tear strength (200 to 600 psi; ASTM D624), impact resistance, hardness (Shore D, 60 to 70), and good load-bearing properties. Other foundry applications for which urethanes could be used include squeeze boards, sprues, and hopper liners.

Mr. Kosmala is project leader, Hysol Div., Dexter Corp., Olean, N. Y.

Handling Urethanes

For the most part, urethane compounds are supplied as two-package systems. The two components are merely blended together and poured into a mold. Curing at room temperature, the mix reproduces the exact pattern or core box in a relatively short period. Handling commercially available urethanes requires no extensive capital investment. For the most part, a blender and, at times, a degassing chamber are generally the only equipment required.

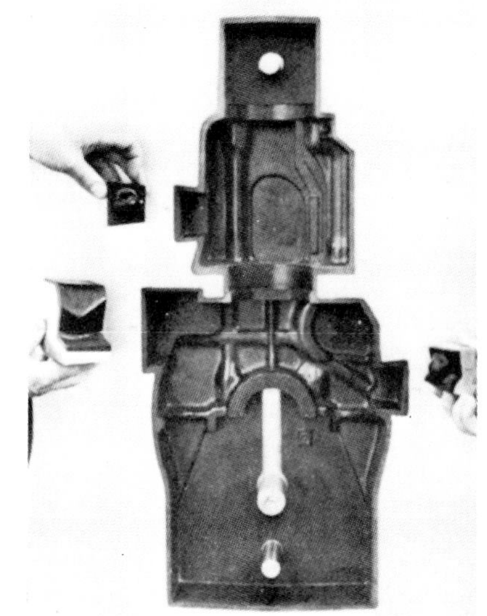

Made of urethane, this core box produces cores in an engine block for a farm tractor. Note the accurate detail and the small inserts.

Tools are light, and have high impact and shock resistance, so that they may be stacked without fear of breaking or being damaged by accident. Finally, urethanes adhere well to most substrates, allowing use of low-cost material for cores.

An Ultrahard Steel for Machining Jet Age Materials

By HARRY H. CORNELL

With the advent of modern jet aircraft and rapid innovations in space technology, greater use is being made of alloys which are difficult to machine and form. To meet demands for better tools, our engineers have developed an ultrahard, high-speed steel which we call Braetuf. A modification of our Braecut (AISI M44), it contains 1.15 C, 4.00 Cr, 2.00 V, 5.00 W, 6.75 Mo, and 7.25 Co. Though capable of reaching a hardness of Rc 69 to 70, it is normally used in the Rc 66 to 68 range, where it possesses more toughness. The steel also has enough hot hardness to assure adequate cutting ability for long times under adverse conditions. Compressive strength is over 600,000 psi, and wear resistance far exceeds that of conventional high-speed steels.

Mr. Cornell is technical service metallurgist, Braeburn Alloy Steel Div., Continental Copper & Steel Industries Inc., Braeburn, Pa.

Some Appropriate Uses

Braetuf has proved itself many times. In broaches formerly made from T15, it has performed well in machining aerospace alloys such as Inconel 718, Incoloy 901, René 41, L605, 17-4 PH, and Waspaloy. Added savings are realized because the alloy is easier to grind than T15 at comparative hardness levels. Dovetail form tools made from Braetuf have also shown excellent results on high-temperature alloys and other metals. For example, during the rough machining of type 301 austenitic stainless steel, it outperformed M3 and T15 by better than 3 to 1.

As another example, Braetuf excelled as a material for internal thread chasers. Threads on SAE 1031 and 1061 components had become surface hardened as a result of prior machining with carbide tools. (When surface hardnesses exceed Rc 45, threading can cause excessive tool wear.) After trying chasers made of such grades as M2, M4, M7, M33, and M43 (with the best results being 80 to 100 pieces before failure), the producer switched to Braetuf at Rc 67 to 67.5 and production rose to 450-500 pieces.

The steel is also considered excellent for tool bits, and for other tooling where cutting ability and high wear resistance are prerequisites. It has replaced T1 for end mills, drills, and reamers for machining bearing races, and has been specified for inserted milling cutter blades used on hard nodular iron; taps for high-temperature alloys; spade drills for hardened alloy steels; and roll-turning tools. Another potential application: tools for gear rolling of automatic transmission parts.

Handling Is Simple

When hardening Braetuf, two preheats are preferred (approximately 1,000 and 1,500 F). To assure a minimum of distortion, stress relieve the tools prior to heat treating. Hardening in salt minimizes decarburization and distortion. The recommended austenitizing range is 2,160 to 2,200 F, the maximum temperature being used for single-point cutting tools. When hardening in an atmosphere, austenitize between 2,185 and 2,225 F. Quenching may be in air, oil, or salt maintained at a maximum of 1,050 F. Braetuf should be tempered between 1,000 and 1,100 F. Triple tempering is required, and a fourth tempering is recommended to minimize residual stresses. Austenitizing at 2,170 F in salt and triple tempering at 1,025 F typically hardens the steel to Rc 67.5.

Since Braetuf has high wear resistance, it is naturally more difficult to machine and grind than less wear-resistant steels. Its machinability rating lies between that of M1 and T15, and hard surface grinding may be done with a wheel similar to Carborundum's Type PA46-D8-V40. Running at 40 surface ft per min and turning at 3,450 rpm, such a wheel will normally grind the steel at the rate of 0.001 to 0.002 in per min. For form grinding, a wheel such as GA80-J6-V10 has given very good results.

Ceramic Inserts Expedite Machining

Said to provide hard, strong tools, an alumina (99.9% Al_2O_3) ceramic produced by Coors Porcelain Co. has these properties: modulus of elasticity of 56 million psi, a compressive strength of 650,000 psi, a transverse rupture strength of over 100,000 psi, a tensile strength of more than 60,000 psi, and a hardness of R45N 90 to 91. Supplied through an agreement with Kennametal Inc., inserts of this material (called AD-999 Alumina Ceramic by Coors and CO6 by Kennametal) offer many benefits, including excellent resistance to breakage, edge strength, and wear resistance – all of which aid in production.

- For example, tool and production costs were both lowered when Owatonna Tool Co., Owatonna, Minn., switched to Kendex CO6 inserts, the Kennametal designation for these tools. Previously, the plant had used positive-rake triangular inserts of titanium carbide to face 5¼ in. diameter discs of cold-rolled 1215 steel. Because of the low feeds and high speeds required to maintain the necessary finish (125 rms max), production was limited to 15 pieces per insert (five pieces per edge) at an average of 21.14 pieces per hour.

The ceramic inserts turned out ten pieces per edge and 60 pieces per insert. A surface finish of 55 rms was maintained, and flatness was 0.0005 in., well within the ±0.0015 in. tolerance. Production jumped to 38.5 pieces per hour, while tool costs reportedly dropped from $0.102 for titanium carbide inserts to $0.028 for ceramic inserts. With ceramic inserts, the speed was 1,800 rpm at a 6 in. feed to the halfway point. Then, holding the same feed, the lathe was run at 2,300 rpm for the rest of the cut. Depth of the cut remained 0.030 to 0.040 in.

- In another example, ceramic inserts are being used to machine shafts of hardened steel bar stock at Dodge Mfg. Corp., Div. of Reliance Electric Co., Mishawaka, Ind. The plant orders quenched and tempered 4140 at a hardness of Bhn 265 to 302. The operation consists of two roughing cuts and one finishing cut on a lathe. Roughing cuts are made with a Kendex TNG-443 insert, and the finishing cut with a Kendex TNG-432 insert, both of which are of the CO6 alumina ceramic. Because breakage is insignificant, all six corners of the triangular inserts are generally utilized. Other styles of shafts are also machined with CO6 inserts. Roughing cuts have been as deep as 5/16 in. at 0.015 in. feed rates.

Further, the company has eliminated grinding operations on some shafts by finish turning with ceramic inserts. The nominal ±0.002 in. tolerance on the diameters can be held with finishes as low as 25 rms.

- By substituting ceramic inserts for tungsten carbide types, a manufacturer of mining equipment hiked production while lowering tool costs. Previously, when profiling 4320 steel at Rc 40, machinists could turn 36 pieces with one triangular insert of carbide. Feed was 0.014 in., depth

Use of CO6 ceramic inserts cuts machining costs for a manufacturer of mining equipment.

This article is based on information provided by Kennametal Inc., Latrobe, Pa.

of cut was 0.030 in., and speed was 670 surface ft per min.

Switching to the CO6 ceramic, the machinists maintained the same depth of cut, but feed became 0.009 in., and speed was increased to 2,200 surface ft per min. The insert was also able to cut 25 pieces per edge (150 pieces per insert). Tool costs were cut from $0.036 to $0.013 per piece, and machine costs, from $0.585 to $0.462 per piece. Savings, as a result, amounted to $0.146 per piece.

Matrix Grades Offer Strength, Toughness

By ALAN M. BAYER

THE MATRIX STEELS combine the high strength of high-speed steels with a degree of toughness which is usually associated with tool steels appreciably lower in alloy content. Since chemical compositions are the same as those of matrix compositions of high-speed steel, microstructures are free of excess carbides.

Table I lists compositions and typical heat treatments of Vasco MA CVM and Matrix II CVM, two steels developed according to this concept. Both have performed well in punching, gear rolling, cold heading, tubing cutter extrusion tooling, and fastener applications. Other potential uses: torsion bars, gears, shafting, bearings, die-cast components, and thread rolling dies.

Applications Are Many

Matrix steels are suitable in applications calling for a combination of very high toughness, good strength, and good fatigue resistance. Since Matrix II can be hardened to a slightly higher level, it is better for applications demanding higher wear resistance, while Vasco MA CVM is better for parts which require greater toughness.

Mr. Bayer is manager, Metallurgical Services, Teledyne Vasco, Latrobe, Pa.

Here are examples of typical applications:

- A hot forging punch of Vasco MA CVM produced an average of 20,000 pieces before failure by wear. The best production that had been obtained previously was 7,900 pieces with a low-alloy die steel. And AISI H21 produced 4,000 pieces, while H12 produced only 2,500. All materials except the Vasco alloy failed by heat checking, confirming our laboratory tests which suggested that matrix alloys have far better resistance to thermal fatigue.
- One producer of semicircular gear sections had a particularly difficult problem with splined punches (1¼ in. in diameter by 7 in. long) for powder compacting. Those of A2, D2, M2, and M4 always split longitudinally in a short time. Since the application demanded high compressive strength in combination with good toughness, Matrix II was selected. Hardened to Rc 63 to 64, it worked beyond all expectations.
- Matrix II CVM heat treated to a hardness of Rc 63 was recently tried as a cold heading die insert 2¼ in. in diameter. In comparison with inserts made of a standard tooling grade, M2 high-speed steel, tool life of the M2 types averaged approximately 10,000 pieces, while that of the Matrix II CVM inserts averaged almost 15,000 pieces.
- In the automotive industry, some gears are rolled – as bars pass through a gear rolling machine, teeth are roll formed. With M1 dies, the best average life was 160 ft of bar before failure occurred by chipping. But dies of Vasco MA CVM heat treated to Rc 59 to 60 (350,000 to 365,000 psi compressive yield strength) rolled 5,025 ft of bar and could still be redressed.
- A large tubing manufacturer is enthusiastic about rotary tubing cutters of Matrix II CVM. The lock-seam tubing has 0.050 in. thick cores of cold-rolled steel and sheaths of 0.020 in. sheet of type 430 stainless. Cutters of Matrix II CVM hardened to Rc 63 have worked far better than those of the next best alloy – and many were tried.

Some Promising Uses

In warm and cold extrusion, the matrix alloys have been very successful. Extruders now have a tool material with the compressive strength of a high-speed steel, which is desirable, plus the toughness of a low-alloy tool steel, which is necessary.

Because of their secondary hardening characteristics, the matrix alloys have excellent temper resistance and more usable fatigue strength than high-speed steels. We have received a number of reports suggesting the punch life is four to ten times better than that of the next best tooling for extruding materials ranging from copper to alloy steels. Where extrusion punch failures resulted from breakage or fatigue, Vasco MA CVM or Matrix II CVM has been the answer.

Because of its combination of strength and toughness, Vasco MA

Table I — Compositions and Heat Treatments of Matrix Alloys

	Composition, %						Heat Treatment*			
Alloy	C	W	Cr	V	Mo	Co	Preheating Temperature, F	Austenitizing Temperature, F	Tempering Temperature, F	Hardness, Rc
MA CVM	0.50	2.00	4.50	1.00	2.75	—	1,500 to 1,600	2,025 to 2,050	950 to 1,150	50 to 60
Matrix II CVM	0.55	1.00	4.00	1.00	5.00	8.00	1,500 to 1,600	2,025 to 2,050	950 to 1,150	52 to 64

* MA CVM can be quenched in salt, oil, or air; Matrix II CVM in salt or oil. Triple tempering is recommended.

CVM has been widely used for ultrahigh-strength aircraft fasteners, which are generally heat treated to a minimum tensile strength of 300,000 psi.

Some Words on Development

Matrix steels were developed on the theory that we could keep the high strength of a high-speed steel while raising toughness by minimizing the residual carbides. (Although hard carbides add wear resistance, they act as stress raisers from which failures can initiate.) Therefore, our engineers analyzed matrices of several high-speed steels and developed steels containing just enough alloying additions to go into solution.

To take the greatest advantage of the potential of the matrix steels we developed (MA CVM and Matrix II CVM, U. S. Patent No. 3,117,863) and to obtain the best combination of strength and toughness, both grades are manufactured only by consumable vacuum melting. Experience indicates that this method removes dissolved gases and reduces inclusions to improve the transverse toughness 2 or 3 to 1.

The two grades are relatively simple to heat treat as Table I shows. Best results are obtained by hardening in a salt bath or a controlled-atmosphere furnace. Both steels possess secondary hardening properties, eliminating possibilities of strength and hardness reductions due to overheating in service. Multiple tempering is needed to complete the transformation of retained austenite and assure maximum stress relief.

A summary of properties, as heat treated, is given below:

	MA CVM	Matrix II CVM
Hardness, Rc	61	63
Compressive yield strength, psi	375,000	403,000
Impact strength, ft-lb	12	4
Wear resistance index	4.5	5.5

For wear resistance measurements, the higher the value, the greater the resistance. (AISI M4 is the basis with an index of 1.00.) Index values are determined by laboratory grindability tests.

High strength is not the only feature of matrix alloys. Vasco MA CVM is significantly tougher than A2 or D2, and Matrix II CVM is tougher than D2. The gain in toughness results from the absence of residual carbides.

Titanium Carbide Coating Raises Wear Resistance of Inserts

By GERHARD PERSSON

A CUTTING TOOL needs to retain its shape as long as possible without decreasing in efficiency. Thus, it should wear slowly if at all, and withstand deformation (both elastic and plastic), breaking, and rapid temperature changes. Unfortunately, these needs are contradictory. Since increased hardness generally leads to less toughness, the most wear-resistant grades are also the least tough.

Composite Inserts Appear

The introduction of indexable inserts, however, eliminated the need for regrinding. As a result, it is now possible to use heterogeneous materials – processors can get wear resistance and toughness by combining materials with the desired properties. We apply this concept by coating tungsten carbide cores of inserts with titanium carbide.

Our tests show that a 0.0002 in. layer of pure, fine-grained titanium carbide will give a tough core of tungsten carbide much more wear resistance. In fact, results proved so successful that Sandvik is now producing GC (Gamma Coated) carbide inserts for commercial use. Three grades are available: Sandvik GC 125 and GC 135 are for cutting steel, and GC 315 is for cast iron.

Our findings show that an operator can take advantage of GC inserts in three ways. He may elect to retain the normal machine operating conditions, lengthening tool life. Or he may raise cutting speed so that the Gamma Coated insert and conventional insert have the same life, thus increasing production substantially. Third, he can balance the two factors (speed and tool life) to yield maximum reduction in total machining cost.

Fracture surface through hard titanium carbide layer (top) and tough core of tungsten carbide shows the tight interface. 15,000×.

Two instances show how these Gamma Coated inserts can result in economies:

- Machinists used tools of C5 and our GC 125 to machine rear axles for tractors. With C5 inserts, one piece was machined per edge. Keeping cutting speed constant and using GC 125, however, enabled them to machine at least ten pieces per edge. (Machine cost: $9.65 per hr.) By increasing cutting speed, they cut the time required for each component from 7.5 to 4.8 min – a reduction of 36%. At this higher speed, cost was reduced from $1.52 to $0.97, also a 36% reduction.
- At another plant, machinists employed a C2 carbide and our GC 315 to turn cast iron flanges.

Mr. Persson is manager, metallurgical research, Coromant Div., Sandvik Steel Works, Sandviken, Sweden.

With C2, they could machine only one component per edge in 14 min. With GC 315, however, two pieces were turned per edge at the same cutting speed. By increasing speed so that only one part per edge could be machined, 10 min were required. In the first instance, total machining cost was reduced 4.3%; in the second, costs were cut 16.5%, and machine time was trimmed 28.6%.

Our studies proved that the coating is pure titanium carbide, chemically inert and with a hardness around Vickers 3,000. Applied uniformly, the coating sticks strongly to base material. This coating also cuts the friction between tool and workpiece, acting to decrease wear and improve the insert's resistance to deformation. In addition, power requirements for the machine tool spindle transmission and feed drive systems are much lower.

AISI M47: An Economical High-Speed Steel With Extra Hardness

By PAUL R. BORNEMAN

DURING THE PAST TEN YEARS, engineers have developed a family of molybdenum high-speed steels which can be heat treated to hardnesses approaching Rc 70. These super-high-speed steels meet requirements of the aerospace industry for machining hard, tough materials. The American Iron & Steel Institute has applied standard designations to several of these grades, including M41, M42, M43, M44, M45, and M46. Laboratory and field tests proved that in cutting applications these steels can improve tool life in machining such materials as superalloys, hardened steels, and titanium.

It soon became apparent, however, that the maximum cutting life was obtained at hardnesses somewhat lower than the maximum obtainable. This fact, along with the very high alloy content and associated high price of the supersteels, suggested that comparable results might be obtained with a less costly alloy.

Mr. Borneman is supervising metallurgist, Tool, Die & Cutting Materials Research, Allegheny Ludlum Steel Corp,. Brackenridge, Pa.

Allegheny Ludlum engineers started a research program, and the result was the introduction of our trademarked Exohard steel – since designated M47 by AISI. Its composition is 1.10 C, 1.60 W, 9.50 Mo, 3.75 Cr, 1.25 V, and 5.00 Co.

Typical Properties

Like other super-high-speed steels, M47 has a sharp demarcation temperature above which rather rapid grain coarsening takes place, as shown in Table I. Impact tests demonstrate the effect of grain coarsening that begins in the critical zone (2,150 to 2,175 F). The table below shows the effect of hardening and tempering temperatures on impact strength as measured with unnotched Izod specimens:

Temperature, F Austenitizing	Temperature, F Tempering	Hardness, Rc	Impact Strength, Ft-Lb
2,150	950	68.0	26.50
2,150	1,000	67.8	29.25
2,175	950	68.4	23.25
2,175	1,000	68.6	17.50

These results indicate that, at these high hardnesses, very small changes in the heat treating temperatures can significantly affect hardness and toughness.

While such toughness differences are certainly of importance in cutting tool performance, they take on added significance in thread rolling difficult-to-deform alloys. This major application area for M47 requires high compressive strength (which is indicated by the steel's high hardness) plus toughness to withstand nonuniform loading.

Triple Tempering Recommended

Like most steels of this type, M47 is usually tempered more than once to develop optimum properties. For our tests, samples were preheated at 1,500 F in salt, austenitized for 3 min at temperatures varying from 2,000 to 2,300 F in salt, quenched in salt at 975 F for 5 min, and air cooled to room temperature. Then, tempering treatments of 2 hr each were applied at temperatures from 800 to 1,200 F for two, four, and six times. Our data showed that austenitizing below 2,200 F, where finer grain sizes are found, developed maximum hardness (Rc 68.0 at 975 F) in four tempering operations. In practice, either three or four tempering treatments are usually applied.

Thread rolling dies of M47 made by Prutton Corp., Cleveland, process hardened high-strength bolts for the automotive industry.

Table I — Grain Size and Hardness of M47 After Quenching

Austenitizing Temperature, F	Shepherd Fracture Rating	As-Quenched Hardness (Air Cooled), Rc	As-Quenched Hardness (Oil Quenched), Rc
2,000	9.5	56.0	65.7
2,050	9.5	62.0	65.5
2,100	9.5	63.1	65.4
2,150	9.0	63.8	64.1
2,175	9.0	63.3	62.6
2,200	7.5	62.4	59.8
2,225	7.5	61.0	59.2
2,250	6.5	60.5	58.9

Budd Co. toolmaker grinds punch of alloy die iron to smooth out marks made by rough machining.

Tooling Alloys for Automotive Dies

By FERDINAND L. EWALD

At the Budd plant, tool steels; cast irons and steels; steels in bars, plates, and forgings; and nonferrous alloys are employed extensively. In this comprehensive article, the author lists do's and don'ts for a great variety of materials in production applications.

OUR IN-PLANT SYSTEM of specifications reduces problems in selecting alloys for body and chassis dies, machine tools, and fixtures. This system, which our laboratory set up, is flexible enough to recognize advances in materials, permitting us to introduce new ones. It also emphasizes low cost materials and their proper application, and considers the capabilities of our plant heat-treating facilities, acting to reduce commerical heat treating costs.

Our system covers four classifications: tool steels; cast materials; bars, plates, and forgings; and nonferrous alloys. In the tool steel category, 18 grades are listed: W1, O1, O2, O6, D2, F2, A2, S1, T1, T2, T5, M1, M2, S7, A6, A10, D7, and a non-AISI wear-resistant die steel. Table I gives the characteristics of each grade. Cast materials include cast iron; die iron; alloy die iron; ductile cast iron (Type 80-60-03); medium-carbon mild steel; high-carbon (0.75%) steel; alloy steel; high-car-

---

Mr. Ewald is assistant manager, Production Laboratory, Budd Co., Philadelphia.

Table I — Characteristics of Tool Steels for Dies

	W1	O1	O2	O6	D2	F2	A2	S1	T1	T2	T5	M1	M2	S7	A6
Cost (base: W1)	1	2.5	2.5	2.5	3.5	3.5	2.5	2.75	6.0	6.25	6.5	6.5	6.75	2.0	3.0
Quenching medium	Brine, water	Oil	Oil	Oil	Air	Water	Air	Oil	Oil, air, salt	Oil, air, salt	Oil, air, salt	Oil, air, salt	Oil, air, salt	Air	Air
Depth of hardening	Shallow	Medium	Medium	Medium	Very deep	Shallow	Very deep	Medium	Deep	Deep	Deep	Deep	Deep	Very deep	Very deep
Nondeforming properties	Poor	Very good	Very good	Very good	Excellent	Poor	Excellent	Very good	Very good	Very good	Very good	Very good	Very good	Very good	Excellent
Wear resistance	Good	Good	Good	Good	Excellent	Excellent	Very good	Good	Excellent	Excellent	Excellent	Excellent	Excellent	Fairly good	Good
Resistance to galling (pickup)	Fair	Fair	Fair	Very good	Excellent	Excellent	Very good	Good	Good	Good	Good	Good	Good	Fair	Good
Toughness	Very good	Fair	Fair	Fair	Poor	Poor	Fair	Very good	Poor	Poor	Poor	Poor	Poor	Excellent	Very good
Weld repair	Excellent	Fair	Fair	Fair	Fair	Fair	Fair	Fair	Fair	Fair	Fair	Fair	Fair	Fair	Fair
Machinability	Excellent	Good	Good	Good	Poor	Fair	Fair	Good	Good	Good	Good	Good	Good	Very good	Good

bon, high-chromium steel, and medium-carbon (0.55 to 0.65%) alloy steel.

The third category contains nine wrought steels. These are low-carbon, hot-worked die steel; low-carbon, cold-worked steel; medium-carbon, hot-rolled steel; medium-carbon steel plates; low-carbon, hardened and tempered alloy steel; low-carbon, annealed alloy steel; medium-carbon, hardened and tempered alloy steel; medium-carbon annealed alloy steel; and medium-carbon, hardened and tempered alloy brake die steel.

The fourth category includes cast manganese-alumium-zinc-copper alloy; cast high-phosphorus tin-lead-copper alloy; and cast aluminum-copper alloy.

The Tool Steels

Being the least costly tool steel, W1 is used whenever possible. It also has the best machinability, and is more easily and safely welded than other tool steels. However, problems with distortion in heat treatment and shallow hardening characteristics limit the use of W1. For example, it is not used for dies with thin or delicate sections requiring high toughness, or where thickness exceeds 3 in. and grinding will be done after hardening. Other grades are also used if the die is to have large holes near edges, sharp inside corners, large changes in section of minimum distortion, or where length-to-thickness ratios exceed 10:1. And forming or coining dies in which high localized pressures occur, or where severe wear or pickup problems exist, are not made of W1.

The steel is used in cutting dies for medium to high production of sheets in thicknesses up to 0.062 in. When production is low (below 50,000 pieces), it is applied in cutting dies for sheets in thicknesses up to 0.250 in. Type W1 is also employed for punches and bushings, forming inserts, flanging dies, and wiper plates.

We generally employ types O1, O2, and O6 for dies that call for greater depth of hardening with more safety and less deformation in hardening. Applications include cutting and forming dies where size and shape requirements are stringent, coining dies where high pressures will be experienced, and dies with holes near edges and great changes in section. But where high toughness, wear resistance, and resistance to galling or pickup are required, these oil-hardening grades are not used.

Type D2 is applied for forming, blanking, and trimming dies where galling or wear is a problem in high production. After being tried out, tools of this steel may be gas or liquid nitrided to increase wear resistance. Where alignment is difficult to maintain, as in slender punches or cutting dies, D2 will not be employed.

Type F2 is used for dies with light sections, simple shapes, generous radii, or holes away from edges. Typical applications include internally quenched draw rings, and heavy dies for forming gages of metal where pickup is a problem. It is never used in coining dies or for dies with sections heavier than 2 in. thick.

Type A2 tool steel is applied for high production cutting dies, and forming dies which are troubled by moderate wear or pickup. Like F2, this grade can be nitrided to improve wear or pickup resistance. Applications lie in forming or coining dies in which high localized pressures develop, but the grade is bypassed when high toughness is required.

Where toughness is the major requirement, S1 is useful. Forming or coining dies that have thin or intricate sections, rivet sets, hammer blocks, and long slender punches are typical applications. However, S1 is not used to make dies for cutting heavy-gage material, or for forming dies where severe wear or pickup can develop. If warranted, tools can be case hardened to increase resistance to wear and pickup.

Type T1 is considered for dies requiring wear resistance. When better wear resistance is required,

A10	D7	Tungsten Die Steel
3.5	7.0	4.5
Air	Air	Oil, air
Deep	Deep	Deep
Excellent	Excellent	Very good
Very good	Excellent	Excellent
Very good	Excellent	Excellent
Fair	Poor	Poor
Fair	Fair	Fair
Good	Poor	Poor

T2 is substituted. Type T5, which has more cobalt, has even greater wear resistance, but is somewhat more brittle than T1 and T2. It is the best steel for roughing cuts or where the temperature is high. The steel also performs well in cutting scaly materials, cast iron, and hard nonferrous materials.

Considered somewhat inferior generally to T1, M1 is used when the other is not readily available. However, M2 is classed as an excellent alternate to T1.

One of the toughest tool steels available, S7 has better wear resistance when case hardened. It is used for rivet sets in hot and cold shock applications; forming or coining dies with intricate, thin sections; long and slender, or highly stressed, punches; hammer blocks; and dies for cutting material with thicknesses over 0.250 in. in low or medium production. It is not employed for forming dies where severe pickup or wear exists, or for dies to cut light-gage carbon steel in medium or high production work.

Type A6 is generally applied where good toughness and wear resistance are needed. Forming dies of the steel are sometimes carburized to improve wear resistance. Heavy-duty cutting dies with intricate or moderately thin sections are made from A6. However, it is not applied in forming or coining dies when severe wear or pickup problems exist or when extreme toughness is required.

Type A10 is used for dies which need good resistance to galling and wear. Applications include coining or forming dies, and parts requiring toughness.

When resistance to wear is the major consideration, D7 is favored. It is used in form or draw dies where severe wear problems exist, but not in dies that need toughness or welding modification.

We apply the nonstandard tungsten die steel when we need the best possible resistance to wear. This steel, which contains 1.5 C, 2.0 Mn, 0.25 Si, 0.90 Cr, 1.0 Mo, and 4.0 W, also resists galling and pickup. Furthermore, its deep hardening characteristics give the steel the same resistance to wear throughout its thickness, allowing for much grinding. However, it is not intended for dies which are to be modified by welding or where toughness is required.

Cast Irons

Conventional cast iron contains 2.6 to 3.3 total C, 0.60 combined C min, 0.50 to 1.0 Mn, and 0.12 S max. We use it for jigs, fixtures, and line dies in which inserts are included to perform such operations as trimming, piercing, and restriking. This cast iron, the lowest in cost, is not used when flanges are formed. Desired hardness is Bhn 150 to 180.

Die iron (2.6 to 3.3 total C, 0.60 combined C min, 0.50 to 1.0 Mn, and 0.12 S max) is applied in sheet metal forming and drawing dies which flange, form, or restrike the part. It is also employed for dies that hold inserts for trimming, forming, or piercing operations accompanied by forming in noninserted areas. Specified structures and hardness between Bhn 195 and 225 are attained by adjusting content of silicon, molybdenum, vanadium, and chromium.

Alloy die iron (2.6 to 3.5 total C, 0.60 combined C min, 0.60 to 0.90 Mn, 0.35 to 0.50 Mo, 0.35 to 0.50 Cr, 1.25 to 2.25 Si, and 0.30 P max) is used for heavy-duty sheet metal forming-drawing dies which form, flange, or restrike on the iron itself. Such dies hold inserts for trimming, forming, or piercing operations. Hardness, usually specified in the Bhn 200 to 250 range, is occasionally raised to the Bhn 225 to 275 range.

Ductile cast iron (Type 80-60-03 containing 3.0 to 3.6 C, 0.20 to 0.60 Mn, 2.0 to 3.0 Si, 0.10 P max, and 1.0 Ni max) is employed when toughness is needed with the antigalling properties of cast iron. It is not used if the casting is to be welded. Composition is adjusted to give the desired hardness and microstructure to the iron. Hardness range is Bhn 200 to 270, and the microstructure essentially consists of a matrix of fine pearlite and ferrite containing randomly scattered sphe-

Table II — Characteristics of Cast Materials for Dies

	Cast Iron	Die Iron	Alloy Die Iron	Ductile Cast Iron	Medium-Carbon Mild Steel	High-Carbon Steel	Alloy Steel	High-Carbon, High-Chromium Steel
Cost (base: cast iron)	1.0	1.05	1.2	—	2.1	2.2	2.5	—
Machinability	Best	Very good	Good	Good	Best	Good	Good	Poor
Wear resistance	Poorest	Very good	Best	Good	Fair	Good	Good	Excellent
Flame hardening properties	Poorest	Good	Best	Good	Poor	Very good	Best	Used as heat treated
Strength	Poorest	Fair	Fair	Best	Good	Good	Best	—
Hardness, Bhn	150-180	195-225	200-250	175-225	120-180	185-250	185-250	269 max
Toughness	Poor	Fair	Fair	Very good	Very good	Fair	Very good	Poor
Weld repair	None	None	None	None	Fair	Poor	Good	Poor

roids of graphite. No massive carbides or flake graphite should be present.

Castings of medium-carbon mild steel (0.20 to 0.40 C, 0.50 to 0.85 Mn, 0.06 S max, and 0.05 P max) are used in inserted dies, fixtures, and jigs that require greater structural strength and toughness than can be obtained from cast iron. It is not employed where forming is done on the steel, but dies can be welded. Hardness range: Bhn 120 to 180.

Castings of high-carbon steel (0.65 to 0.80 C, 0.50 to 0.80 Mn, 0.06 S max, and 0.05 P max) are applied for die inserts, punches, and floaters which must maintain sharp lines and corners. Castings of this type range from Bhn 185 to 250 in hardness, and can be flame hardened in local areas. This steel is not used if galling, seizing, or pickup is anticipated.

Castings of alloy steel (0.40 to 0.60 C, 0.90 to 1.20 Mn, 0.05 S max, 0.045 P max, 0.35 to 0.50 Mo, and 0.90 to 1.25 Cr) are employed in 45° clinching dies, inserts, and punches that require toughness. Dies are often flame hardened in critical areas, and may be fully hardened by heat treatment. They are applied in low and medium production operations for cutting and forming. Hardness is Bhn 185 to 250, and vanadium, nickel, and copper can be added if desired.

High-carbon, high-chromium steel (1.5 to 1.8 C, 0.50 to 0.80 Mn, 0.40 to 0.60 Si, 0.025 S max, 0.025 P max, 0.50 to 1.0 Mo, 0.20 to 1.0 V, and 11.50 to 14.0 Cr) is used for cast inserts in high-production trimming, blanking, drawing, and forming dies. Resisting pickup and wear quite well, this grade is only used in the hardened condition, and never for tools where thickness is not adequate. Maximum hardness is Bhn 269. To give increased wear and pickup resistance, castings are sometimes nitrided.

Steels in Wrought Forms

Low-carbon, hot-worked steel (0.08 to 0.23 C, 0.30 to 1.00 Mn; 45,000 psi tensile strength) in bars, plates, or forgings is used generally for parts of dies which do not undergo severe service. If required, this steel can be case hardened to improve wear resistance. It is weldable without any special precautions.

Low-carbon, cold-worked steel (0.08 to 0.23 C, 0.30 to 1.0 Mn; 70,000 psi tensile strength) has similar applications. It is used when more strength is required, and is not usually case hardened (that operation reduces strength).

Medium-carbon, hot-rolled steel (0.43 to 0.50 C, 0.60 to 0.90 Mn; 90,000 psi tensile strength) is used to make pilot pins, and die and press parts that require extra strength. It, too, can be case hardened to raise wear resistance, and can be employed either heat treated or flame hardened for improved mechanical properties. This steel is readily gas or arc welded with low hydrogen electrodes.

Medium-carbon steel plates (0.40 to 0.50 C, 0.60 to 1.00 Mn) are flame hardened for use in cutting and forming dies, bolster plates, and die parts. This grade can be gas welded to build up local areas. When arc welding a previously flame-hardened die, the processor must preheat and temper the part at 400 to 500 F.

Supplied as bars, plates, or forgings, low-carbon, hardened and tempered alloy steel (0.18 to 0.23 C, 0.50 to 0.70 Mn, 0.20 to 0.35 Si, 3.25 to 3.75 Ni, 0.20 to 0.30 Mo; 135,000 psi tensile strength) is applied to jigs, fixtures, machine tools, die parts, and press parts that require higher strength and toughness than can be obtained with low-carbon steel. This steel is not case hardened, and can be arc welded and cut to shape with oxyacetylene torches.

The low-carbon, annealed alloy steel (0.08 to 0.13 C, 0.45 to 0.65 Mn, 0.20 to 0.35 Si, 3.00 to 3.50 Ni, 1.00 to 1.40 Cr, 0.08 to 0.15 Mo) is generally case hardened after parts are made. It is used for gears, die and jig parts, and press parts which require increased core hardness and wear resistance. It can be gas welded and arc welded; low-hydrogen elecctrodes are employed.

Bars, plates, and forgings of medium-carbon, hardened and tempered alloy steel (0.38 to 0.43 C, 0.75 to 1.00 Mn, 0.20 to 0.35 Si, 0.80 to 1.10 Cr, 0.15 to 0.25 Mo; 125,000 psi tensile strength) are used for jigs, fixtures, machine tools, and other parts that require higher strength and toughness than can be obtained with medium-carbon, hot-rolled steel. This grade is not readily arc weldable, and it will harden excessively when torch-cut. Parts are generally made from as-received material which is Bhn 250 to 321 in hardness.

Medium-carbon annealed alloy steel (0.38 to 0.43 C, 0.60 to 0.80 Mn, 0.20 to 0.35 Si, 1.65 to 2.00 Ni, 0.70 to 0.90 Cr, 0.20 to 0.30 Mo) is never used in the annealed condition. Parts are first made, and then either heat treated to a high hardness or flame hardened. The steel is usually not torch-cut or arc welded. As-received hardness is Bhn 200 max.

Medium-carbon alloy brake die steel, hardened and tempered (0.38 to 0.48 C, 0.70 to 1.00 Mn, 0.20 to 0.40 Si, 1.00 to 1.30 Cr, 0.40 to 0.60 Mo; 150,000 psi tensile strength), is used to make tools and brake dies that require a heat-treated steel with a higher hardness than that of medium-carbon, hardened and tempered alloy steel. Hardness range is Bhn 300 to 325, and hardness must be uniform throughout the part.

Nonferrous Alloys

The cast manganese-aluminum-zinc-copper alloy (Bhn 170 min) is used for high-load, low-speed bearings. (Composition: 60.0 to 68.0 Cu, 0.20 Sn, 0.20 Pb, 2.0 to 4.0 Fe, 3.0 to 7.5 Al, 2.5 to 5.0 Mn, bal Zn.) We employ the cast high-phosphorus tin-lead-copper alloy (Bhn 60 min) for high-speed, low-load bearings for presses and other manufacturing equipment. (Composition: 81.0 to 85.0 Cu, 6.25 to 7.50 Sn, 6.0 to 8.0 Pb, 0.20 Fe max, 2.0 to 4.0 Zn, 0.50 Ni max, 0.15 P.) Finally, the cast aluminum-copper alloy (Bhn 150 min) is applied for high-load, low-speed bearings used in presses and gears. Die inserts constitute another application. (Composition: 83.0 Cu min, 3.0 to 5.0 Fe, 10.0 to 11.5 Al, 0.50 Mn max, 2.5 Ni max.)

How Ford Selects Die Materials

By VICTOR A. KORTESOJA

Metallurgists working with stamping and foundry engineers have devised ways to specify tool steels for flanging and to improve techniques for controlling cast iron die quality.

At the Metal Stamping Div. of Ford Motor Co., engineers strive to cut costs, speed production, and improve the quality of the product (stamped fenders, hoods, and similar parts) by careful attention to shop practices.

Such work often entails much testing to determine which type of tool steel does the best job in a given operation. Also needed are convenient ways to monitor the quality of incoming material and components. Sometimes, appreciable advances can be achieved as we will demonstrate by two case histories.

1. Selecting Steels for Flanging Dies

To make an automobile body stamping, four to six operations are required, including blanking, forming (or drawing), trimming, and flanging. Of these the flanging step, though simple, has usually been most troublesome.

Briefly, the part to be flanged is placed in position on the lower half of the flanging die (called the post). As the upper half moves down, the inner portion (the pad) contacts the blank, and clamps it in place with the edges to be flanged projecting over the sides of the post. Then the outer part of the upper die moves down, and the flanging die contacts the projecting metal, bending and ironing it into the position desired.

Mr. Kortesoja is chief metallurgist, Metal Stamping Div., Ford Motor Co., Dearborn, Mich.

Steel's Job – In an automotive flanging die, flanging steel does the heavy work because it is subject to high pressure and abrasion. Further, as the die slides over the newly formed flange it picks up metal which then scores parts that are subsequently flanged. As a result, the press must be stopped so that the metal pickup can be ground away from the die surface.

When we were faced with this problem, we found that a clear understanding of all factors was needed. A field survey revealed that wiping inserts made of water-hardening tool steels were the most frequent sources of complaints. We also determined that the wiping steel was usually soft because diemakers and tryout men ground the shallow cases completely or almost off. However, dies of water-hardening tool steel did a satisfactory job on easy flanging operations when the surface of the die steel was at the proper hardness, Rc 60 to 64.

In an example where die pickup caused trouble, frame members of SAE 1010 steel 0.106 to 0.136 in. thick were formed and flanged in one die. Because the sheet was thick and excess metal was being squeezed between the die punch and cavity, the die steel (SAE D2) picked up metal.

Testing – We began a test program to determine a more efficient die steel, ordering die sections made from D5, D7, T15 steels, as well as a metal-bonded carbide (Ferro-Tic C) insert. The D7 section tested first was considered inadequate. Metal pickup began within 400 pieces of production. The failure was of interest because it meant that some important characteristic of die steel was missing. Next, sections of T15 steel and the metal-bonded carbide were tested at identical locations on opposite-member dies.

The section of metal-bonded carbide had produced 90,705 pieces when it was removed because of wear (0.012 in.). Examination of the worn areas (Fig. 1) revealed thermal checking and some pickup. Also, tempering had lowered the hardness of the working face from Rc 70 to the Rc 63 to 65 range. Inspection of the tempering curve for this material indicated that 700 to 800 F of heat would produce this hardness.

The section of T15 steel, however, handled 137,000 pieces, while wearing only 0.005 in. Pickup was slight and easily removed. For heavy forming and flanging die inserts, T15 proved to be far superior to any of the other steels tested. However, like the D7 section, the T15 steel section also developed thermal checking in metal-to-metal contact areas.

Problem – From these tests, it became clear that metal pickup on die material was directly related to the resistance to heat softening of the die material. Briefly, the greater this softening resistance, the more that the steel resists pickup.

This finding was confirmed by our results with a small test die which was constructed to evaluate flanging

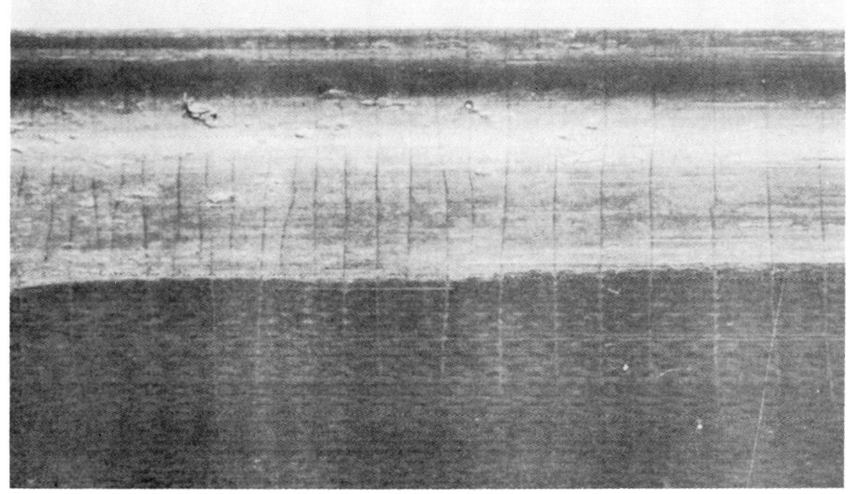

Fig. 1 – When tested as a material for flanging dies, Ferro-Tic C, a metal-bonded carbide, flanged more than 90,000 pieces before being removed from service due to wear. Note thermal cracking and metal pickup; hardness had also dropped from Rc 70 to the Rc 63 to 65 range during use.

die materials. Data are tabulated below:

Tool Steel	Hardness, Rc (Tempered at 1,000 F)	No. of Pieces to Pickup
SAE W110	40	9
SAE O6	40	69
SAE D2	58	75
SAE D7	62	140
AISI T15	66	510 (1st run)
		1,560 (2nd run)
AISI T15 nitrided	66+	2,736

All test sections had hardnesses of Rc 60 to 66 when placed on test. Pickup resistance, it is seen, increases as the resistance to heat softening improves, as shown by the hardnesses of samples tempered at 1,000 F. (It should be noted that hardness of specimens tempered at 1,000 F are listed only to show the relative resistance of die steels to tempering. When tested, all die sections were at or near their maximum hardnesses.)

Figure 2 reveals the degree of metal pickup on faces of die steels W110 and T15 at 10× magnification. It is apparent that both dies have the same relative amount of pickup though the W110 die processed 55 parts while the T15 die handled 510 pieces.

We should note that the chart should not be used as a direct reference because flange contour, metal thickness, height of flange, lubrication, and severity of ironing of the metal all affect pickup. In our test die, conditions are uniform but severe to obtain comparative data.

Findings – Judging by our tests, we can conclude that:

1. Metal pickup on the flanging die face is caused by high frictional heat generated by two metals rubbing together under high pressure, literally welding the flange metal to the die.
2. The length of wipe in the die must be only as long as needed to form the flange properly. Excessive heat is generated if the length of wipe is longer than necessary.
3. Die lubricants should be chosen and applied correctly. When heavy gages are being formed, an E. P. (extreme pressure) die lubricant must be used.

It is important to select the die steel with severity of the flanging operation in mind. Oil-hardening steels (SAE O6 or O1) should be used for making straight, simple flanges up to 1 in. in height on sheet metal thicknesses of 0.035 to 0.075 in. (However, a water-hardening steel can be applied here if care is taken to retain the hard, shallow case.)

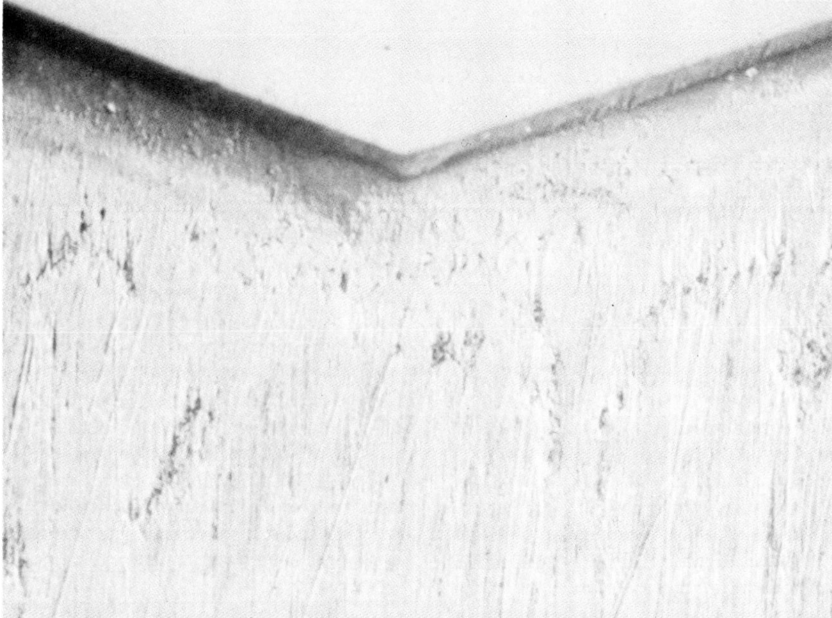

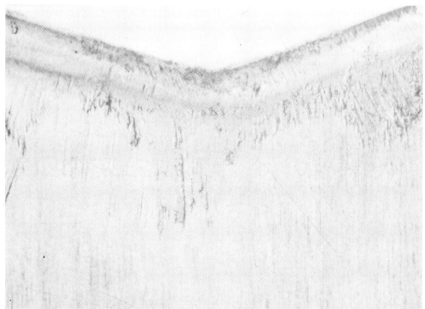

Fig. 2 – Though flanging dies of SAE W110 (top) and AISI T15 display equivalent amounts of metal pickup, the T15 die has processed 510 pieces, nearly ten times the number handled by the other. 10×.

The SAE D7 steel should be used for dies to form sheet 0.035 to 0.075 in. thick going into curved areas such as wheel openings on fenders and quarter panels, and where metal folds are encountered such as at the ends of "character lines," the automotive term used for the accent lines on outsides of body panels. This type of steel is also used to flange 0.035 to 0.075 in. sheet when the flange height is over 1 in.

Finally, AISI T15 steel is the best for flanging sheet over 0.075 in. thick. If pickup still occurs, change to a better lubrication or nitride the die.

2. Assuring Quality in Cast Dies

Dies that draw sheet steel into automobile body parts such as hoods, fenders, and quarter panels are large masses of metal weighing as much as 30,000 lb. Though some dies have tool steel inserts on wear surfaces, most of ours are made from iron castings of the following composition: 3.00 C, 1.00 Mn, 1.75 Si, 0.40 Cr, and 0.40 Mo (Ford designation: M-3A71A). Working surfaces on dies require a hardness of Bhn 201 to 235, and a matrix of fine pearlite with little or no free carbides or ferrite.

We have found when the iron is at

Fig. 3 — To monitor the quality of castings for stamping dies, Ford specifications call for the foundry to provide test coupons "on or adjacent" to the working face. One coupon is tested by the foundry, while the other goes to Ford technicians for confirmatory checks of hardness and composition. The photos at left show typical coupons as cast in place.

the desired hardness, the matrix consists of fine pearlite, the combined carbon is over 0.80%, and flame hardening response is good. (Flame hardening results in a surface hardness of over Rc 60, as measured with calibrated test files.) Also, when the hardness is much above Bhn 255, machining problems increase rapidly; the Bhn 235 aim gives us a margin for error. Should excessive free carbides be present, machining and filing are difficult even at Bhn 201 to 235.

When dies are not flame hardened, we need the maximum hardness level to get the best possible results from the cast dies. However, we prefer to flame harden wear areas.

Experience — These specifications have taken time to develop. For years, incoming castings were not tested. In fact, hardnesses were determined only when die shops complained about machineability. By the time an accurate hardness test could be made, the die was at least 25% completed.

Further, even though samples for compositions were being taken from chips obtained by drilling, we were no longer able to reject the castings when the analysis became available.

Test Program — To correct these two situations, a cast-on coupon was decided upon, the size of which was designed to approximate the average section thicknesses in dies. Specifications called for two 3 by 5 by 6 in. removable coupons to be attached on or adjacent to the working face of the die or punch. (Figure 3 shows coupons on typical castings.) One coupon was to be checked for hardness by the foundry, while the second was to be removed at the Ford die plant or by the vendor die shop for testing at the Ford Motor Co.

This program was a move in the right direction, but it was not enforced or properly implemented. In the fall of 1964, a chief metallurgist was appointed, and his first assignment was to improve the die iron situation. It was decided to continue using the cast-on test coupon, but to correlate and administer the program correctly.

Test Problems — Initial surveys revealed that test coupons were the wrong size, placed incorrectly, or omitted completely, among other things. Such factors are important to assure consistent results.

To explain, the coupon cools faster than the casting because it juts out from it. Correlation between hardness readings of the coupons and the castings required the foundries to make coupons of the right size, locate them at desired areas of castings, and take proper hardness readings from slices removed from coupon centers.

Two years of work were required to bring the coupon system under control so that correlations could be set up between hardnesses of coupons and working faces of finished dies. We have now established that, when all coupon factors are correct, and the die iron has the proper composition, a coupon slice with a Bhn of 217 to 255 will indicate the desired Bhn of 201 to 235 on the working face of the die.

We have also taken care of another problem — delays in receiving reports from foundries. Now, the foundry must clean the casting, knock off the coupon, and check its hardness before the casting is shipped out. A certified chemical analysis is included on the report form, but it is not a primary concern.

With this program, we have developed a quality program with foundries. They now know that the casting meets our specifications before it leaves the foundry. Also, the certification and coupon slice are delivered to the metallurgical department of the Metal Stamping Div. the same day the casting is shipped. If there is any question about the quality of the casting, it is not shipped until we are consulted.

Results — The following results have been achieved from this program:

1. Metallurgical quality has been steadily improving since 1965. On the 1968 die program, the average hardness of all M-3A71A castings was Bhn 190, and 30% of the castings were below Bhn 185. On the 1969 die program, only 1.5% of the castings were below Bhn 185, and the average was Bhn 208. Not only has the average hardness level been raised, but uniformity has been achieved also. (A die program, incidentally, includes about 4,100 castings, averaging about 8.125 lb a piece.)

2. Response to flame hardening has improved in proportion to the improvement in structure.

3. Prior to 1965, tradename irons were being used in a vain attempt to obtain better castings for dies. The use of tradename irons was eliminated when the control program was instituted, cutting die iron costs about 12% per die program. Better die iron is being obtained as well.

Wear Resistance and Impact Strength Key Factors in Selecting P/M Dies

Wear is a problem because powders are abrasive. Other considerations include compressive strength; heat resistance, particularly in hot forming operations; complexity of heat treatment; and length of run.

- Resistance to wear and impact strength are key considerations in the specification of die materials for powder metallurgy.
- D2 is one of the popular cold-work die steels.
- Among hot-work die steels, H13 is generally the choice.
- M43 usage – a high-speed die steel – is on the increase.
- Conventional cemented carbides are getting a lot of attention as a way to solve wear problems.
- One of the newer steel-bonded carbides withstands operating temperatures to 1,000 F.
- Metallic and ceramic coatings applied by detonation gun or a plasma torch bridge the gap between solid carbides and tool steels.
- Surfaces of dies for hot forming may have to withstand repeated temperature cycles ranging from 550 F to about 1,050 F.
- Problems with tool growth after tempering in nitriding can be alleviated by a double high draw at 960 F.

Those were some of the highlights of the information presented at the recent "Powder Metallurgy Tooling Seminar" sponsored by the Powder Metallurgy Equipment Assn. of the Metal Powder Industries Federation at Southfield, Mich.

In the following report, each presentation cited above is developed in greater detail.

Tool Steels – Selection of tool steels is complex if for no other reason than the variety and number available, stated Dennis Huffman of Latrobe Steel Co. There are 1,500 to 2,000 tradenames to contend with. Latrobe alone makes about 200 grades.

Chief considerations in selection are die wear (powder is abrasive) and impact strength (toughness). Other factors include: high compressive strength; heat resistance, particularly in hot forming; price (although metal represents only a small percentage of total die costs); complexity of heat treatment; and length of run (if relatively new parts are made, for example, cheaper die materials may be used).

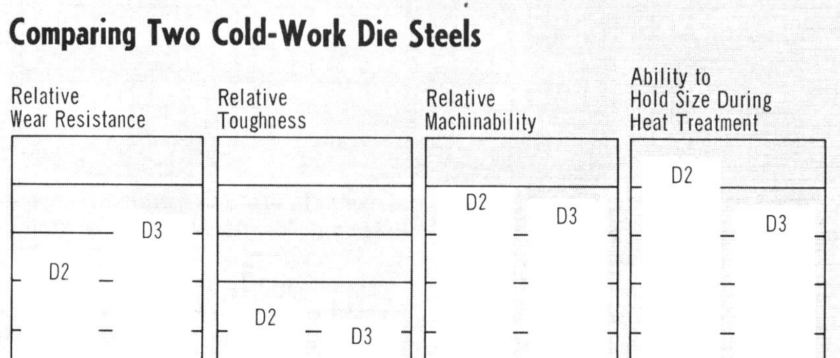

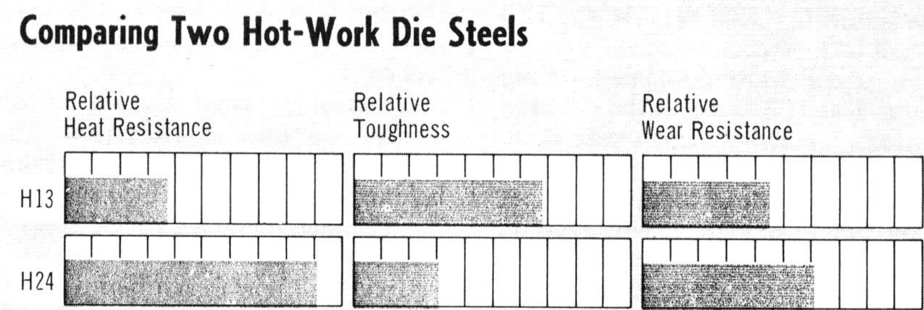

Cold-Work Steels – D2 is the popular grade in this group. It contains 1.50 C, 1.00 V, 12.00 Cr, 0.30 Mn, 0.30 Si, and 0.75 Mo. It is air hardening and nondeforming. Wear resistance is rated excellent in a wide range of applications.

In comparing the properties of D2 with those of D3, it's apparent that there are some tradeoffs (see top chart). D2 has slightly less wear resistance than D3, but it is a little tougher, a little more machinable, and has an edge in holding size during heat treatment.

Such elements as chromium promote hardenability – it forms carbides within the steel matrix. Vanadium is the strongest carbide former.

Cold-work tool steels include air, oil, and water-hardening types. Generally, the higher the alloy content, the higher the heat treatment temperature. The higher total alloy, the higher the wear resistance. The higher the wear resistance, the greater the difficulty in grinding.

Hot Work – H13, the leading grade, provides a good combination of heat resistance, toughness, and wear resistance. It contains 0.40 C, 1.00 Si, 0.40 Mn, 5.25 Cr, 1.20 Mo, and 1.00 V. It is air hardening and can be water-cooled in service. The grade contains few carbides.

Generally, hot-work die steels have a hardness range of Rc 42-52. In die design, notches should be kept to a minimum because they act as stress raisers. Preheating these dies to 600 F will double their toughness. There's only a slight loss in strength.

Hot-work die steels are characterized by long life. Cycling doesn't present any problems as long as upper temperature limits are not exceeded. Above 1,200 F heat checking will occur.

H24 has the highest heat resistance. It's not as tough as H13, but its wear resistance is higher (see bottom chart). H24 contains 0.50 C, 0.25 Si, 0.25

Comparative Properties of Some High-Speed Steels

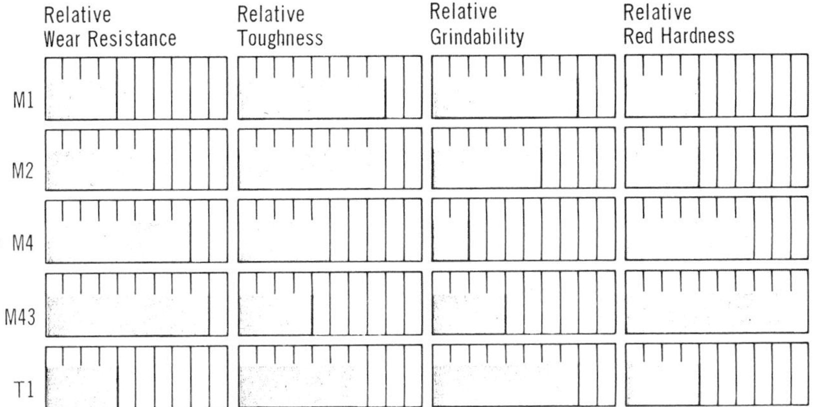

Mn, 15.00 W, 2.80 Cr, and 0.50 V. Tungsten contributes to heat resistance, but it reduces toughness.

High Speed – These are all high alloys. They're heat treated at 2,000 F or higher. They have a hardness range of Rc 60-65. T1 was the original. It has good grindability and relatively good toughness, but other grades surpass it in wear resistance and red hardness (see chart above). Its composition is 0.75 C, 18.00 W, 4.10 Cr, 1.10 V, and 0.70 Mo.

M grades were developed during World War II because of the tungsten shortage. Molybdenum replaced it. One point of molybdenum is the equivalent of 2 points of tungsten.

M1 and T1 are equivalent grades in terms of wear, toughness, grindability, and hot hardness. M1 contains 0.80 C, 1.50 W, 4.00 Cr, 1.15 V, and 8.50 Mo. Carbon and vanadium are the keys to wear resistance.

The T grades, said Mr. Huffman, are dropping out of the picture.

M43, an increasingly popular grade, has high wear resistance and red hardness. It's not as tough or as grindable as M1 or T1. Its composition is 1.20 C, 2.70 W, 3.75 Cr, 1.60 V, 8.00 Mo, and 8.00 Co. The latter helps to maintain red hardness.

Relative toughness of high-speed steel can be varied by heat treatment. Underheating boosts toughness. The same is true in tempering.

Cemented Carbides – These are engineered materials (tungsten carbide plus cobalt) and should be specified in terms of the application, stated Ed Hilty, Carbide Div., Alken Industries.

The more cobalt in a composition, the higher the ductility. Properties of tungsten carbide do not change.

The standard billet, made by P/M methods, is 2 by 10 by 5 in. Billets are cut with diamond wheels. As much preforming and preprocessing as possible are done prior to vacuum sintering. You must allow for shrinkage in sintering.

There are about 20 grades of cemented carbides. Their maximum operating temperature is 1,000 F, which means they won't work in hot forming.

Tips – These materials won't solve breakage problems unless wear is the cause. They're three times stiffer than tool steel; they're stronger in compression but weaker in tension; and their thermal expansion is one-third that of tool steel.

If possible, avoid brazing of cemented carbides to tool steel. The carbide can break. Also, in designing holding devices, the carbide shouldn't be put in tension. Mechanical holding methods are preferred.

Green grit wheels can burn the carbide. This won't show, but die life will be shortened. Diamond wheels should be used.

Care should also be used in EDM. Cobalt can be removed if settings are improper or speeds too high. Also, the material can't be chemically milled.

Misalignment of punches can cause tension problems.

Steel-Bonded Carbides – This is a relatively new family of tooling materials which contain 45 to 55% refractory carbide by volume (such as titanium carbide and tungsten carbide). Like cemented carbides, they are P/M products.

Unlike cemented carbides, pointed out Dr. Kuhmar Mal, Sintercast Div., Chromally American Corp., they can be brazed. Their thermal expansion is closer to that of steel. Also, after annealing, they can be machined with conventional equipment and high-speed or tungsten carbide tooling.

These materials provide the high hardness and wear resistance of cemented carbide. A widely used grade for tools, dies, and wear parts contains 3 Cr, 3 Mo, 0.6 C. However, it has a temperature limitation. Above 400 F, it softens rapidly due to overtempering of the matrix alloy.

A more heat-resistant grade has a high-chromium tool steel matrix. It has a secondary hardening response and is usable to 1,000 F. This grade contains 34.6 WC, 6.6 Cr, 2.0 Mo, 56.0 Fe, and 0.8 C.

In the annealed condition (Rc 42-46), it can be machined with conventional tooling. Finished parts can be hardened and tempered to Rc 68-70. After extensive machining, parts can be relieved by placing them in a protective pack and holding at 1,200 F for 1 hr per inch of cross section.

Processing – In heat treating, it must be kept in mind that the material is susceptible to both carburization and decarburization.

One recommended procedure is to use stainless steel foil for surface protection and heat in a conventional electric or gas furnace without a protective atmosphere. Another alternative is to heat treat in a vacuum with an inert-gas quench.

Proper tempering is required to get a secondary hardening response. This occurs between 900 and 1,000 F due to the transformation of retained austenite to martensite. The 34.6 WC grade (Ferro Tic CM) is always double tempered by holding at 975 for 1 hr, air cooling to room temperature, then retempering to 925 F. The graph below gives hardness curves for

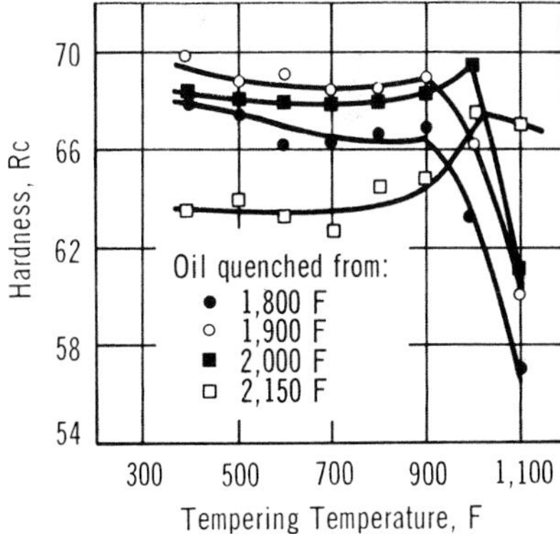

this grade after double tempering (1 hr + 1 hr) at temperatures shown.

Size change in heat treating is predictable, which means that tools can be made close to finish size.

In brazing, a high-temperature brazing alloy is required. It must have the same response to heat treatment as the die material.

In finishing, it should be remembered that the carbide must be exposed to get good wear resistance.

In machining, use heavy feeds, slow speeds, and don't use cutting oils.

In grinding, use fast speeds; soft, open-structured aluminum oxide; and spray mist coolant.

Coatings – Metallic and ceramic coatings are applied by detonation gun process or by plasma torch, reported Jerry Frein, Linde Div., Union Carbide Corp. Over 85 materials, including cermets, are available.

In general, coatings are dense, have bond strengths exceeding 11,000 psi, and are hard (Rc 72-74). Substrates must be about Rc 55. In case of wear, coatings can be stripped and replaced.

Finished coating thicknesses should run 0.002 to 0.003 in. If a high surface finish is wanted, you should provide an additional 0.003 in. thickness for finishing stock. Substrates should have a carbon range of 0.70 to 1. After coating, parts can be hardened in oil or inert gas.

A variety of shapes can be coated; and it is possible to get a number of special effects. For example, when it is necessary to coat a part with two perpendicular surfaces, all corners should be deburred prior to coating.

The most popular coating, which is applied by the detonation gun, is 91 WC, 9 Co. It can replace cemented carbides. Coatings have a diamond pyramid hardness (kg per sq mm 300 g load) of 1,300. As-coated, the surface finish is 150 rms. After it is ground and lapped, it is 1 rms.

Hot Forming – Die life is the key to success in hot forming, said James Hoffmann, Cincinnati Inc. Factors to consider: preform design, lubricant, die temperature, and die material.

Two types of preforms are being used. One is a true blank and considerable forming takes place. In the second approach, the preform is close to the final shape, and forming is minimal. Generally, die wear is lower with the latter process.

Lubricants include graphite (rated the best), mica, glass (rated second best), molybdenum disulfide, molybdenum disulfide combined with lead oxide, phosphates, and synthetics.

Dies tend to soften after production of 300 to 400 parts. By one process, the die temperature is 450 F and the blank is 1,500 F. Die surfaces cycle from 550 F to about 1,050 F.

Little work has been done on die materials. Hot-work and high-speed steels have been used. Some superalloys, such as IN-200, look good. One company is using M4 and H19.

Heat Treatment – D2, A2, and M2 are among the popular tool steels in powder metallurgy, stated Mike Soviak, Commercial Steel Treating.

D2 is hardened at 1,850 F in an atmosphere furnace or salt bath. Ammonia gas is used if parts are small or have fine edges, for example.

Quenching is in hot salt or oil. D2 will provide a higher hardness at 960 F than it does at 550 F. Hardness is Rc 59-61. With the higher draw, you lose only a point or two in hardness.

Another point: many P/M tools are nitrided at 960 to 980 F. Problems with growth can arise following tempering. A double draw at 960 F will solve this, but the heat treater should be told that the part is to be nitrided.

A2 is heat treated in a salt bath or atmosphere furnace; maximum hardness Rc 57-59. Tempering is at 400 F.

M2 is generally salt-bath hardened, then quenched and tempered in salt to avoid decarburization. By slightly underheating, you get additional toughness for powder metallurgy. Hardness runs Rc 61-63.

Trends – Mr. Sovaik stated that Tufftriding is becoming extremely popular. Depths are limited to tenths. The process roughens parts by 5 rms, but they can be polished. Grinding precedes heat treating. Otherwise, the Tufftrided surface can be removed.

Tufftriding takes about 90 min. By comparison, nitriding is slow. A 72 hr cycle, for example, takes 5 days. The first hour doesn't start until you get up to temperature. There's also a cooling period at the end.

Mr. Soviak also noted that surface treatments, such as hard surfacing, are on the increase.

Aluminum Extrusion Tooling

By DENNIS D. HUFFMAN

The 5% Cr hot-work die steels are usually employed for dies and related components. And it's likely they won't be replaced in the foreseeable future. Variations of these grades have improved machinability, wear resistance, or heat resistance. The author also covers maraging steels, superalloys, and the casting of dies.

RATHER EARLY during the use of long, continuous sections of aluminum with uniform cross sections, engineers realized there had to be an easy way to make them. Eventually, they came up with extrusion – the desired shape is produced only once, reversed in a die, and the material is forced through the die opening (as we do with toothpaste), producing the desired section.

Unfortunately, aluminum alloys with the needed strength for structures will resist the extreme distortion required for extrusion. Even when heat is employed to soften a material, high pressures are necessary to force the metal through a relatively small aperture.

Despite such problems, the extrusion of aluminum tubing and shapes had become practical by World War II. Since then, the industry has grown steadily. Aluminum extrusions have proved to be both functional and decorative – a typical application is shown on the next page.

What Die Materials Need

To some degree, heat and pressure act upon all portions of the press which are in contact with, or are exposed to, the materials being extruded. Dies and mandrels are also affected by frictional heat and wear. Extrusion on a production scale thus depends, at least to some degree, on the development of materials which withstand these conditions.

While several types of press equipment are available, all require the same basic types of tooling. Tools which must be replaced with relative frequency include dies, dummy blocks, liner inserts, and mandrels.

Alloys for such parts require resistance to shock, heat, stress, and wear. Because these properties tend to be somewhat divergent – gains in one property are lost elsewhere – compromises are inevitable. The 5% Cr hot-work die steels (H11, H12, and H13) will probably continue to get most of the perishable tooling assignments because they combine good toughness and strength with adequate wear and heat resistance.

Durable tools include backers, bolsters, and containers. Since they are not as directly exposed to extreme temperature or pressure, these tools can usually be formed from relatively low-alloy steels such as 4150 or 4350.

Features of 5% Cr Grades

Several points account for the success of the 5% Cr steels in perishable tooling. Carbon contents are moderate (0.35 to 0.40%) in all three grades, limiting attainable hardnesses and carbide formation, but producing a high level of toughness or resistance to shock. Since the three AISI grades have at least 5% Cr and more than 1% Mo, all will harden in air after being austenitized. Finally, the steels contain tungsten, chromium, molybdenum, and vanadium – all carbide formers which add to wear resistance. Thus, the 5% Cr grades are most suitable for extrusion tooling which must resist high temperatures, thermal shock, abrasive wear, and prolonged high stress.

However, the hot-work die steel of today is not the same as that employed in the past. By improving manufacturing techniques, producers now make a better product even though the chemical composition has not changed greatly. For example, today's die materials are sounder as proved by inspection with sonic, X-ray, and magnetic field tests. Also,

Mr. Huffman is a metallurgical engineer, Latrobe Steel Co., Latrobe, Pa.

Huge rib chords of 7075 extruded at Alcoa's Lafayette, Ind., Works, are for the Boeing 747. Weighing over a ton, they help to attach and secure the wings.

a growing proportion of these steels are vacuum treated, resulting in more uniform properties throughout the finished dies.

Table I lists typical compositions and longitudinal properties of H11, H12, and H13 at Rc 46, a hardness representative of those commonly used. Each grade is heat treatable to a room temperature strength of 200,000 psi or greater, sufficient for most applications. Yield strengths also are closely comparable.

However, extrusion tooling has to operate at 550 to 1,050 F. Figure 1 illustrates the ability of these steels to maintain strength at working temperatures. Note that both tensile and yield strength remain above 100,000 psi until material reaches 1,050 F, the usual limit of the working range for aluminum alloys.

Toughness is also important. Tooling must be able to resist the shock of loading, and other minor irregularities in the operational cycle. Here, the three 5% Cr grades also stand out because, even at the high hardness of Rc 48 to 50, they possess impact strengths of 15 to 21 ft-lb as measured by Izod V-notch tests. Note also in Fig. 2 that resistance to shock increases with rising temperature, particularly to 400 F. Obviously, perishable tooling should be preheated to over 400 F to improve resistance to shock during extrusion.

Precautions Are Advised

Tools must be uniformly preheated because inadequate preheating can produce thermal gradients and differentials in toughness from surface to center. In addition, such a condition can add to stresses already within a die, leading to premature failure. Caution: preheating with a torch is simply not adequate, though it is usually better than no preheating at all.

While preheating times for individual tooling can vary widely, the general rule is 1 hr per inch of minimum thickness in a still air furnace, or 45 min per inch in a circulating air furnace. All times are measured after the loaded furnace reaches the desired temperature.

Design is another factor which influences over-all toughness. Sharp notches, for example, limit the die's ability to resist shock. As the radius at the base of a notch decreases (the notch becomes sharper), the toughness drops, first gradually, then (at 0.02 in.) sharply. Of course, tool, saw, and file marks must be removed because they also can act as notches. This procedure is particularly important when handling large segments because toughness varies inversely with size.

Also of interest is resistance to abrasive wear, a property based at least in part on composition and the level of working hardness. Because working hardness is fixed within a rather narrow range by the operational requirements, higher wear resistance must be achieved primarily by changing the alloy. As Table I reveals, H13 has twice the vanadium of H11. Since vanadium is the strongest carbide former commonly employed in tool steels – and carbides are essential for wear resistance – H13 should be slightly more wear resistant than the other two 5% Cr grades.

Where wear resistance is a major problem, bearing surfaces (those areas which contact moving hot metal) may be nitrided, usually by a two-stage gas process. The resulting case exceeds Rc 64, greatly enhancing the abrasion resistance of the die or mandrel.

It must be remembered, however, that increases in hardness almost invariably result in some loss of toughness. For this reason, it is common practice to mask off all areas of the tool except the wear surfaces to be nitrided. In this manner, the alloy's inherent toughness largely remains while extra wear resistance is added at only critical points. Of course, the processor must make sure that "white" compounds – iron-nitrogen compounds (FeN) which are extremely brittle – do not form a continuous layer or extensive network. Even if nitriding is limited to critical areas of wear, lack of toughness in

Table I — Compositions and Properties of 5% Cr Tool Steels

AISI Designation	Composition, %							Tensile Strength, Psi	Yield Strength, Psi	Elongation, %	Reductio[n] in Area, %
	C	Mn	Si	Cr	Mo	W	V				
H11	0.40	0.30	1.00	5.00	1.30	—	0.50	216,000	201,000	12	43.5
H12	0.35	0.30	1.00	5.00	1.60	1.30	0.30	215,000	200,000	12	42.5
H13	0.40	0.40	1.00	5.25	1.20	—	1.00	218,000	204,000	12	42.5

Note: Room temperature tensile properties, longitudinal.

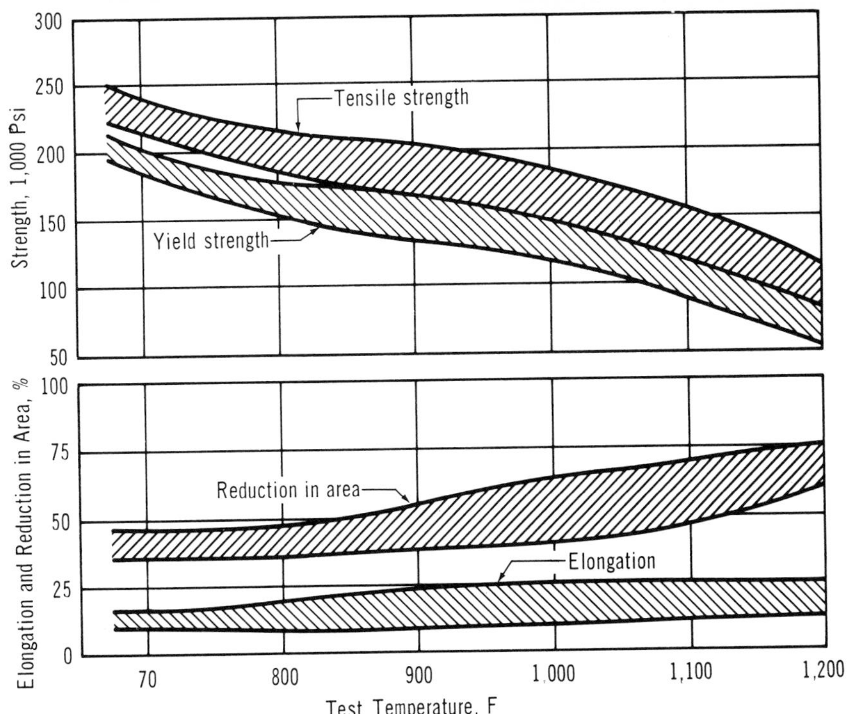

Fig. 1 — Propertie of the 5% Cr hot-work die steels fit them for extrusion of aluminum. Tensile and yield strengths remain high to well over 1,000 F, encompassing the working range for the metal with ease. Ductility rises but slightly within the same range, assuring that dies will retain desired shapes throughout runs.

a white layer can lead to premature cracking and chipping.

Free-Machining Die Steels

In recent years, H13 and other 5% Cr grades have become available in free-machining versions. Manufacturing costs and leadtime are decreased while physical and mechanical properties remain as before. The 5% Cr grades are relatively easy to machine compared with other tool steels, but machining is not simple.

EDM and ECM methols eliminate the need for easily machined grades. However, most extrusion dies will continue to be finished by conventional methods for removing metal for some time. In most instances, therefore, the free-machining grades have a definite advantage. Through applying careful melting and casting techniques, processors can make these grades with short, disconnected sulfides which won't detract from tool performance. Free-machining tool steels contain disconnected sulfides about the size of carbides.

Though free-machining grades do not machine as well as screw stock, they are significantly better than standard tool steels in this property. In many instances, the improvement has made the difference between an impossible machining job and one which has worked out fairly well.

Owing to the relative ease of metal removal, it is possible to form finished tools from blocks prehardened to Rc 45. Using such stock, shops can cut die making time. Though prehardened stock is more difficult to machine, there are advantages. Prehardened sections may be brought to size immediately, eliminating costly, time-consuming heat treatment and finishing. This material can be of particular advantage to smaller shops which must rely on outside heat treaters.

Normally, machinists can handle prehardened stock by reducing cutting tool speed by about 30%, while maintaining the same rate of feed and depth of cut. Since no allowance needs to be made for size changes or surface removal, die dimensions may be cut to the exact values.

Two Wear-Resistant Grades

Other variations of standard grades have also been developed. Tradenames and compositions of two of these grades, derived from H13, are: Alloy P, 0.75 C, 0.40 Mn, 1.00 Si, 5.25 Cr, 1.10 Mo, and 2.50 V; Alloy K, 1.10 C, 0.40 Mn, 1.00 Si, 5.25 Cr, 1.10 Mo, and 4.00 V. Because they have more carbon and vanadium, each grade has greater wear resistance and hardening ability than standard H13, Alloy K being more than twice as wear resistant as Alloy P.

In the extrusion of the more abrasive alloys or for long runs, resistance to wear becomes of paramount importance. Consequently, any die which wears out rather than becoming obsolete is, with certain limitations, a candidate for Alloy P or K. Figure 3 shows that Alloys P and K do not soften as rapidly as H13 above 600 F. Since wear resistance depends on both working hardness and alloy content, the higher percentage of carbides also gives an advantage to the two modified alloys.

Of course, the two modified alloys are somewhat less tough, or less

Fig. 2 — The impact strength of H13 at Rc 45 rises with temperature, an important factor because extrusion dies require good toughness to withstand loading shocks. Test method: Izod V-notch.

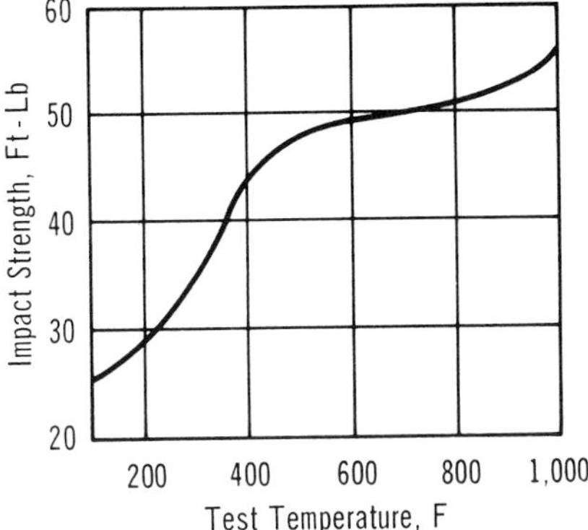

Fig. 3 — Because Alloys P and K retain hot hardness better than standard H13, they will exhibit greater wear resistance. Greater quantities of carbides also aid in this property.

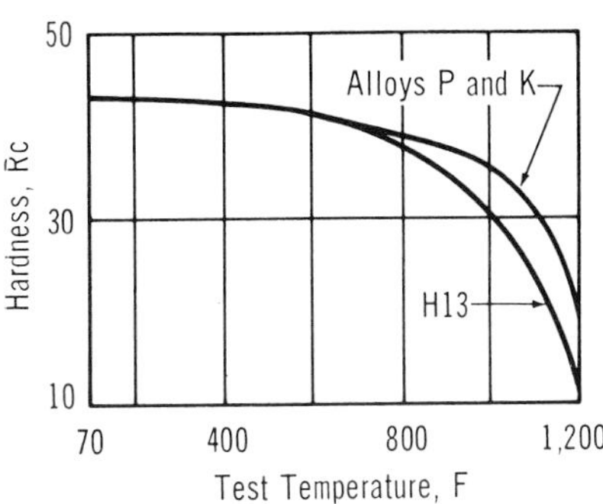

able to absorb shock loading. Impact strengths determined by Izod tests for each of the three materials are shown below:

Hardness, Rc	H13	Alloy P	Alloy K
53.0	13	10	4
49.0	21	11	5
43.0	25	12	8

These materials should not be employed where a history of tool breakage exists. Also, due to their higher alloy content, these grades are more difficult to fabricate.

These alloys might be suitable for button inserts or mandrel caps, particularly with multiple-cavity dies. Used this way, they give the benefits of inherent wear resistance, while the tougher materials employed for holders absorb much of the shock associated with the operation.

Using Maraging Steels

Maraging steels do not appear to be likely candidates for general tooling, but they may be suitable for some high-strength applications, and for dies which require the ultimate in toughness. Furthermore, they are simple to heat treat. But neither the heat nor wear resistance of the maraging grades is particularly outstanding. And measures must be taken to limit the effects of these variables. (Maraging grades tend to soften with long-time exposures at temperatures as low as 850 F). Further, no flexibility in attained hardness is feasible due to the age-hardening mechanism employed with maraging steels.

Maraging grades have potential for particularly difficult applications calling for high strength. Certain processes, for example, require the ultimate in strength for rams or outer liners. For these, nitriding could be applied to counteract the relatively low wear resistance of the maraging grades.

The very high toughnesses of maraging steels may solve the impossible job, one in which a high incidence of tool breakage has precluded any real production. Tooling made of a maraging grade may stay together long enough for some production to be attained. Though the wear resistance will not be equivalent to that of the 5% Cr steels, breakage will be significantly reduced.

Dies of Superalloys

Because some superalloys have proved to be applicable to brass and copper alloy extrusion, a few engineers have felt that these grades would also be beneficial for extruding aluminum. For example, such alloys have gained acceptance in dummy blocks for brass extrusion and, to a lesser extent for, rams and liners.

However, the low wear resistance of superalloys has precluded their selection in almost all mandrel or die applications. Since these grades are also relatively soft at room temperature and lack the strength of the 5% Cr steels, they offer no particular advantage below 1,100 F. Though these grades maintain strength above 1,150 F, such properties are not needed for extruding aluminum billets. And since there are also cost and machining problems associated with high-temperature alloys, it's easy to understand why they have been put to only extremely limited use.

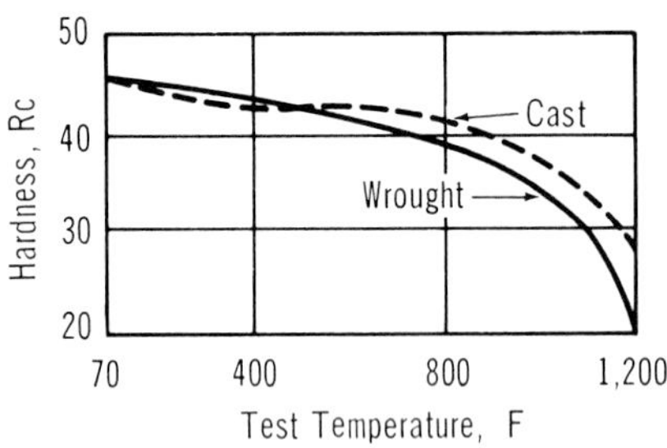

Fig. 4 — Cast H13 retains hot hardness to higher temperatures than wrought H13. Since this advantage increases with temperature, cast dies can last longer at extrusion temperatures.

Casting the Dies

The ever-growing technology of cast-to-shape tooling is becoming significant in the aluminum extrusion field. This process offers significant reductions in waste material and decreased machining and delivery times. In fact, toolmaking processes can be virtually eliminated by casting a finished, or nearly finished, tool.

The initial impetus for trials of precise cast tooling was cost reduction. Cost savings can be realized two ways. First, the amount of machining and finishing required to produce a workable tool is drastically reduced. Second, the weight of metal purchased is significantly lower. In fact, even though the cost per pound for a casting is somewhat higher than that for a forging, this differential can often be largely absorbed simply because less metal is needed.

The use of castings is particularly advantageous when the tooling is repetitive. The cost of patternmaking is spread out over many tools. As an example of economies, a large aluminum extruder reports saving 40% on a popular 8 in. diameter extrusion bridge. Further, company engineers found that cast tooling gives better than a 650% increase in tool life at the same hardness and with the same H13 grade.

For reasons not completely understood, cast tooling seems to have more resistance to operational heat and stress than wrought tooling of the same grade and room temperature hardness.

Castings also appear to be more resistant to heat checking. While minute cracks do occur, they grow at a much slower rate in castings. Furthermore, no problem of directionality exists.

Tensile tests have shown that tensile and yield strengths of cast and wrought H13 are virtually identical all the way from room temperature to 1,100 F. Though cast material was somewhat less tough than wrought material, a definite yield point was noted, illustrating that cast tooling is not inherently brittle. According to hot hardness tests, castings are harder (by 1 to 8 points Rc) above 600 F, when compared with wrought material of the same grade and room temperature hardness. In Fig. 4, note that the advantage of the cast material rises with temperature. While this advantage is not particularly great, it does indicate the stable nature of the cast material.

Cast tooling also has an edge in the leadtime required from order to delivery. Once the pattern is made and stored, a casting can be poured and delivered within a relatively short time, ready for heat treatment and light finishing. Dimensional control also becomes quite consistent once an initial die is made and any necessary corrections are incorporated. For such tooling, reasonable finishing allowances are 0.010 to 0.015 in. on the impression faces, 1/32 to 1/16 in. at the parting line of the mold, and 1/16 to 1/8 in. on the back and outside surfaces.

Sintered Carbides

*By the ASM Committee on Sintered Carbides**

SINTERED CARBIDES are products of powder metallurgy, made of finely divided, hard particles of carbide of refractory metal, sintered with one or more metals of the iron group (Fe, Ni or Co), forming a body of high hardness and compressive strength. The hard particles are tungsten carbide, usually in combination with lesser amounts of other carbides. The additional carbides are those of titanium and tantalum, with some occasional specialized use made of the carbides of columbium, molybdenum, vanadium, chromium, zirconium and hafnium. The auxiliary, or binder, metal is usually cobalt; nickel and iron are used infrequently.

The carbides are present as individual grains, and also as a finely dispersed network resulting from the precipitation, during cooling, of carbide dissolved in the cobalt during sintering.

After solidification the cobalt is present in the interstices as almost pure metal with its original ductility. Solid cobalt dissolves only about 1% of tungsten carbide at low temperature. The limit of solid solubility of tungsten carbide in nickel is 25% and in iron 5%. It is this low solubility of tungsten carbide in cobalt, compared with the solubility in nickel or iron, which accounts for the use of cobalt as the binder metal. The higher solid solubility of carbide in iron or nickel would result in a more brittle auxiliary, or binder, phase. The cobalt also has superior ability to wet the carbide at elevated temperatures, which is important during sintering.

Classifications

The simplest classification of carbides recognizes two broad categories: (*a*) the "straight tungsten carbide" grades, used primarily for machining cast iron, austenitic steel, nonferrous and nonmetallic materials, and (*b*) the grades containing major amounts of titanium and tantalum carbides, used primarily for machining ferritic steel.

The so-called straight tungsten carbide grades usually contain 2 to 5% of titanium and tantalum car-

*ALFRED BORNEMANN, *Chairman*, Professor of Metallurgy, Stevens Institute of Technology; A. B. ALBRECHT, Technical Consultant, Monarch Machine Tool Co.; BENNETT BOVARNICK, Chief, Sintered Metals and Ceramics Branch, Rodman Laboratory, Watertown Arsenal; ERNEST CARLSON, Manufacturing Research Office, Ford Motor Co.; D. F. DICKEY, Director, Research and Development, Firth Sterling, Inc.; R. J. DORR, Manager, Advance Product Engineering, Metallurgical Products Dept., General Electric Co.; EDGAR W. ENGLE, Development Engineer, Kennametal, Inc.

J. W. FOUTCH, Project Engineer, Timken Roller Bearing Co.; W. D. GILDER, Chief Metallurgist, Reed Roller Bit Co.; ROBERT T. HOOK, Chief Metallurgist, Warner & Swasey Co.; L. J. HULL, Chief Metallurgist, Ryan Aeronautical Co.; GEORGE G. LEITCH, Chief Metallurgist, Fellows Gear Shaper Co.; C. P. LITTLE, Metallurgist, Joy Manufacturing Co.; H. M. LUEHMANN, Chrysler Corp.; CARL LUGAR, Ladish Co.; WILLIAM PENTLAND, Research Supervisor, Cincinnati Milling Machine Co.; P. V. SCHNEIDER, Metallurgist, International Business Machines Corp.; K. J. TRIGGER, Professor of Mechanical Engineering, University of Illinois; W. B. WEBER, Heat Treat Engineer, Caterpillar Tractor Co.

bides. The other grades contain about 10 to 40% titanium and tantalum carbides.

A more detailed classification based on composition is shown in Table 1, which also lists typical uses. Groups 1 to 3 are the straight tungsten carbide grades; groups 4 to 9 contain more than 8% TiC plus TaC. Groups 4 to 8 encompass all the steel-cutting grades. Group 9 containing tantalum additions only, is largely confined to heat-resisting and specialized items, such as gages.

Hardness

The exceptional tool performance of sintered carbides results from high hardness and compressive strength. The lowest hardness of sintered carbide is approximately the same as the highest hardness available in tool steel, Rockwell A 85 (Rockwell C 67), and ranges to Rockwell A 93, depending on grade.

Hardness is easily measured and offers some clue to the suitability of a carbide grade for a given application. For instance, groups 1, 2 and 3, having nominal Rockwell A hardness of 92, 89 and 86, are recommended for applications in the following order: (*a*) finishing and light medium cutting of cast iron, (*b*) roughing cuts on cast iron and moderate die applications, and (*c*) high-impact die applications. Each successive application requires higher impact properties, necessitating higher cobalt content, which results in lower hardness. Table 1 shows the magnitude of decrease in hardness with increasing cobalt content. The same need for lower hardness with increased severity of cut exists for the steel-cutting grades of groups 4 to 8.

Hardness measurements alone will not discriminate between the straight WC grades and the TiC or TaC grades. Density, in conjunction with hardness testing, can be used for separating grades, as discussed in the following section. Figure 1 illustrates the range in hardness for individual grades and groups and the effect of increasing cobalt content from 3 to 13% in straight WC grades. Figure 2 illustrates the range of hardness found in a large number of shipments of several commercial grades of one nominal composition (94% WC, 6% Co).

Hardness, among other properties, offers an effective method of evaluating and controlling the quality level of carbide tools. Figure 3 shows the gradual narrowing of the range in hardness of two grades during a period of five years.

Hardness readings should not be taken near cutting edges because the indentation will act as a notch, nor in a straight line for the same reason. In some plants a tool is reground if a large number of hardness readings have been taken in a critical area. Very fine microcracking occurs when carbides harder than Rockwell A 91 are being tested, but this does not appreciably affect the hardness values.

The Rockwell A scale (60-kg load) is used because the 150-kg load of the C-scale would create proportionately deeper indentations and might crack the carbide. No extensive calibration of the A-scale is needed; the only requirement is that the average of five readings on a test block should check within ±0.2 of a hardness number. A detailed statement of recommended practice is available in ASTM B294-54T.

The hardness of sintered carbide is 9 on the Mohs' scale (talc is 1, diamond is 10). Specific hardness comparisons (Knoop indentations) with nonmetallic abrasive materials are shown below:

Material	Knoop hardness number
Tungsten carbide sintered with cobalt binder	1050 to 1500
Topaz	1250
Aluminum oxide	1625 to 1680
Silicon carbide	2130 to 2140
Boron carbide	2250 to 2260
Sapphire	1600 to 2200
Diamond	6000 to 6500

Hardness at elevated temperature is only slightly decreased up to the temperature where rapid oxidation begins (Fig. 4). The difference in hardness among grades at a given temperature is largely dependent on the amount of binder, and to a lesser extent on the grain size, porosity and impurities.

At 1400 F the highest hardness in Fig. 4 is shown by a 6% Co grade of group 1, Rockwell A 85 (Rockwell C 67). Lowest is 13% Co, group 2, Rockwell A 72 (Rockwell C 43).

High-temperature hardness is useful in determining the relative resistance to plastic flow of a carbide at a given temperature, and is important where tool failure is the result of stress combined with temperature, as in hot impact extrusion.

When carbide is exposed to extremely high temperatures for prolonged periods, a grade high in titanium carbide (such as group 4) is recommended because it oxidizes at a slower rate than tungsten carbide.

A specific advantage of using sintered carbide within its temperature range is that repeated cycling between high and low temperature or sustained holding at high temperature has no tempering effect that decreases the hardness. When the metal returns to room temperature, it has its original hardness.

Density

The density is readily determined by weighing the blank in air and then in water, as described in ASTM B311. Nominal density values of the groups are listed in Table 1. Density measurements on 1129 specimens of 65 grades in all nine groups are shown in Fig. 5.

The straight WC grades of groups 1 to 3 have high density, the TiC additions of groups 4 to 6 make these grades lighter, and the TaC additions of groups 7 to 9 result in densities almost equal to the straight WC grades. This is to be expected since the densities of the three pure carbides are as follows: WC 15.6, TiC 4.7, and TaC 14.0 g per cu cm. Cobalt has a density of 8.9 g per cu cm. The effect of TiC is illustrated in Fig. 5, where the lowest density readings were obtained with a grade having 32% TiC, 10% TaC, 7% Co.

A more detailed analysis of seven WC grades is shown in Fig. 6.

Figure 7 shows the range of density values for five steel-cutting grades. Note the increase in density

Table 1. Classification of Sintered Carbides

Carbide group	Composition, % (remainder WC) Co	TaC+TiC	Rockwell A hardness	Density, g per cu cm
Straight Tungsten Carbide				
1	2.5 to 6.5	0 to 3	93 to 91	15.2 to 14.7
2	6.5 to 15	0 to 2	92 to 88	14.8 to 13.9
3	15 to 30	0 to 5	88 to 85	13.9 to 12.5
Added Carbide Predominantly TiC				
4	3 to 7	20 to 42	93.5 to 92.0	11.0 to 9.0
5	7 to 10	10 to 22	92.5 to 90.0	12.0 to 11.0
6	10 to 12	8 to 15	92.0 to 89.0	13.0 to 12.0
Added Carbide Predominantly TaC				
7	4.5 to 8	16 to 25	93.0 to 91.0	12.5 to 12.0
8	8 to 10	12 to 20	92.0 to 90.0	13.0 to 11.5
Added Carbide Exclusively TaC				
9	5.5 to 16	18 to 30	91.5 to 84.0	14.8 to 13.5

Typical Uses

Group 1. Finishing to medium roughing cuts on cast iron, nonferrous metals, superalloys and austenitic alloys; low-impact dies. **Group 2.** Rough cuts on cast iron, especially on planers; moderate-impact dies. **Group 3.** High-impact die applications.

Group 4. Light high-speed finishing cuts on steel. High crater resistance. Low shock resistance. **Group 5.** Medium cuts and speeds on steel. Good crater resistance and moderate shock resistance. Dies with moderate impact involving pickup. **Group 6.** Roughing cuts on steel. Good shock resistance together with wear and crater resistance. Moderate-impact die applications involving pickup.

Group 7. Light cuts on steel where a combination of edge wear and crater resistance is required. **Group 8.** General-purpose and heavy cutting of steel requiring resistance to wear and cratering. Also resists abrasive wear caused by scale.

Group 9. Wear-resistant applications particularly involving heat; gage elements, special machining applications. Special applications involving mechanical shock and heat such as hot trimming of flash.

with decreasing TiC and increasing TaC. Based on nominal composition, the 7 Co – 10 TiC grade should be lighter than the 6 Co – 10 TiC grade, because more cobalt, with a density of 8.9, has replaced the heavier WC, which has a density of 15.6 g per cu cm. The fact that the 7 Co – 10 TiC grade measured heavier is probably due to small amounts of TaC not shown in the nominal composition.

Density determination, along with hardness, is a practical method for identifying tool inserts if the identity of a grade is lost or if grades are inadvertently mixed.

Mercury, with a density of 13.55 g per cu cm, offers an expedient method of identification or separation of the straight tungsten carbide grades from grades having appreciable TiC plus TaC. When placed in mercury, the straight tungsten grades will sink, and the TiC-TaC grades will float. Some of the group 3 grades will float, but such grades are easily separated by a hardness check. Group 9 grades are also an exception. They will sink because the additional carbide is TaC, which is about equivalent to WC in density. However, users would be aware of the presence of these specialty grades. Final identification can usually be verified by hardness.

The most difficult to separate are groups 5 and 8. Both are steel-cutting grades containing TiC with TaC; hardness, density and tool applications are similar.

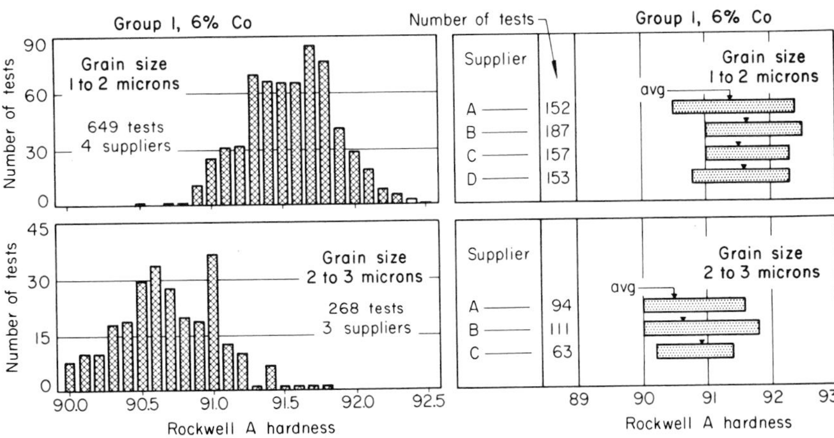

Fig. 2. Distribution of hardness for seven 6% Co grades of group 1 carbide

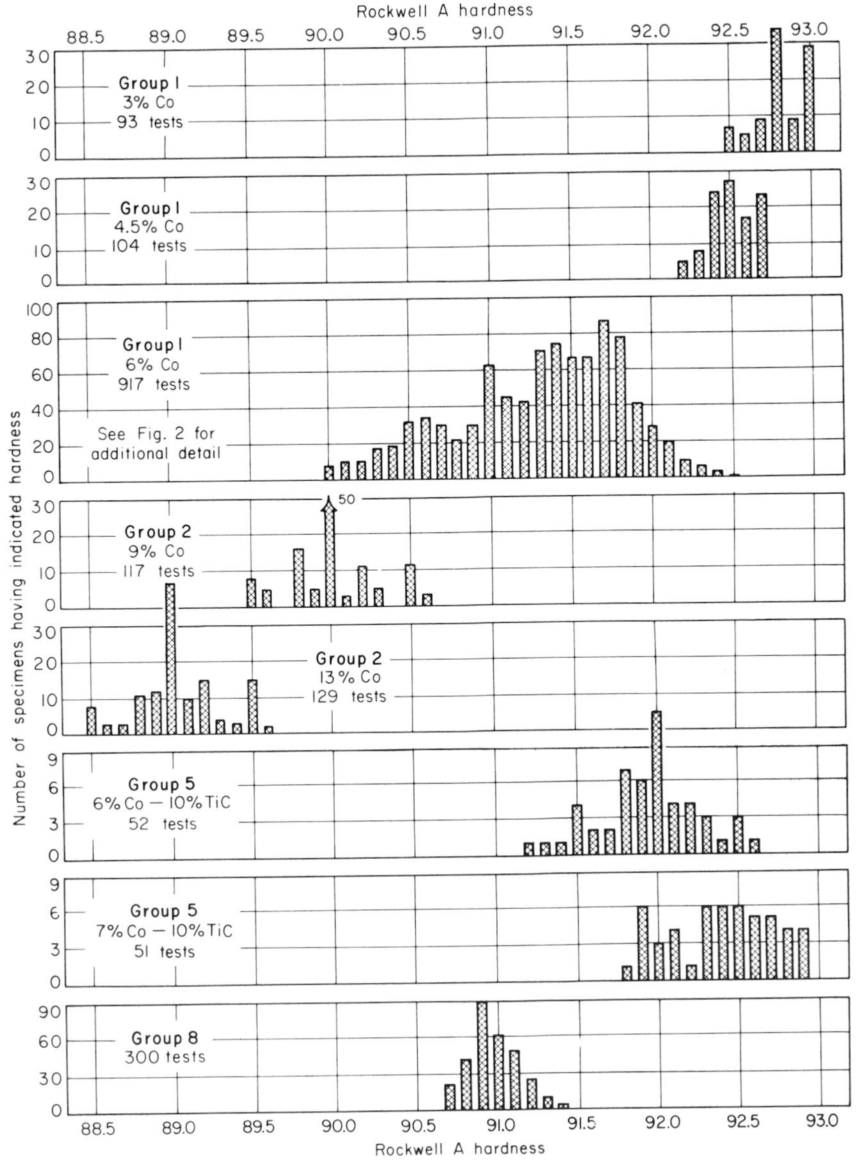

Fig. 1. Hardness of sintered carbides tested in accordance with ASTM B294-54T

Transverse Rupture Strength

Transverse rupture strength of the commercial grades varies from about 115,000 to 485,000 psi, and is useful in determining allowable angles for cutting edges and the amount of support required for tools. Figure 8 gives data on transverse strength by groups.

With increasing cobalt content, strength increases to a maximum at 10 to 15% Co. In greater amounts cobalt acts to separate the carbide grains. The strengthening effect offered by mechanical interlocking of the grains is lessened; more soft cobalt increases the slip paths available for yielding.

Shock resistance also increases when the amount of cobalt in an impact tool is larger. The point of diminished return is at about 30% Co, where hardness and wear resistance become too low for any tool.

Transverse rupture, rather than simple tension, is used to determine the strength of carbides because (*a*) a small sample is less costly, (*b*) the alignment problem is increased with a small tension sample, and (*c*) straight tension testing is less suited for materials of low ductility. In transverse testing, the load is applied to a beam specimen, of either rectangular or round section. Although rupture strength is determined as pounds per square inch by the simple beam formula, the results are not true stress values because of the high magnitude of shear stresses involved. Also, at large deflections, the bending causes deviation from true stress values which is not accounted for by the simple beam formula. This test method provides a useful, reproducible method of comparing strengths among grades.

Various cross sections and center distances are in use. One practice recommended by a large portion of the industry specifies specimen dimensions of 0.200 by 0.250 in. and a distance between centers of 9⁄16 in. (Tentative P-102, as submitted to ASTM for approval, by the Cemented Carbide Producers Assoc).

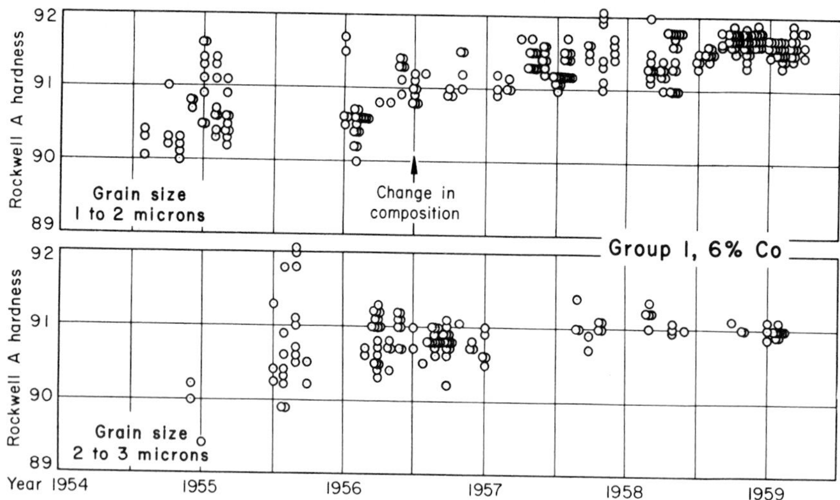

Fig. 3. Change in hardness distribution for two grades of group 1 carbide during a five-year period. The composition change noted in the upper graph was a decrease from 8 to 6% Co. Each of the 385 plotted points represents the average of several hardness readings for a shipment of 100 to 1000 tool inserts; the average shipment was about 750 inserts; hence, about 300,000 inserts are represented by the data plotted above.

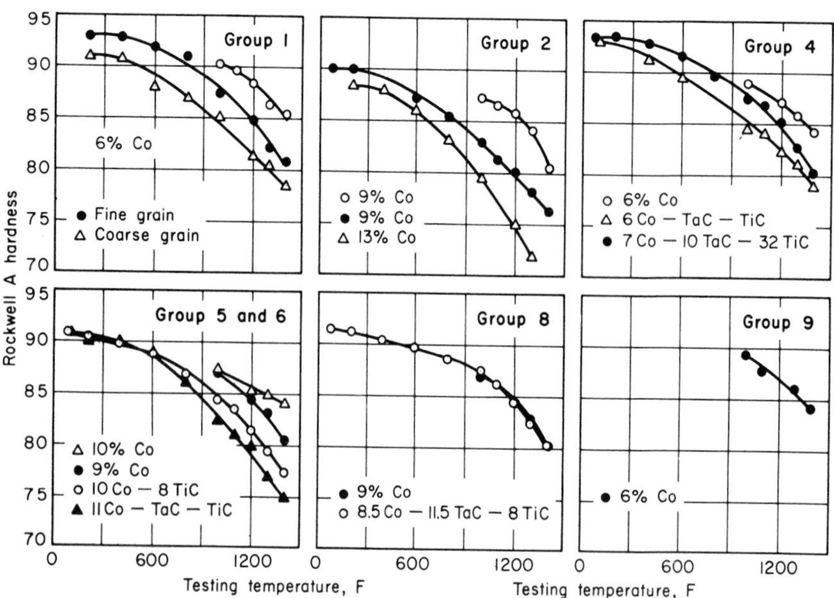

Fig. 4. Effect of temperature on the hardness of sintered carbides. Samples were protected by hydrogen atmosphere during heating, and the hardness readings were taken with an extension penetrator.

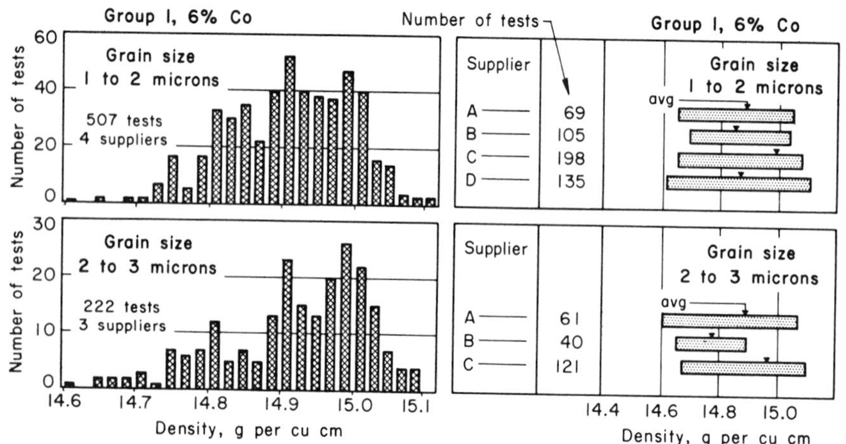

Fig. 6. Distribution of density in seven commercial grades of group 1 carbide. Measurements are the inspection results of one user. Measurements of density are obtained by weighing the blank in air and then in water, as described in ASTM B311.

Compressive strength may be determined according to the practice recommended to ASTM by the Cemented Carbide Producers Assoc, Specification CCPA P-104. The effect of cobalt content on compressive strength for the straight tungsten carbide grades is illustrated in Fig. 9. Compressive strength and elastic limit in compression for most of the carbide groups are listed in Table 2. The decrease in modulus with increasing cobalt content is illustrated in Fig. 9.

Compressive strength values reported in Fig. 9 and Table 2 are, in most instances, the average of three readings taken at one time on three samples ⅜-in. diam and 1 in. long. There is obviously some disagreement relative to the effect of cobalt between Fig. 9 and Table 2. Several variables, including exact composition of the carbide, may contribute to this discrepancy, but the difference is attributed mainly to an insufficient number of tests.

Data recently reported from one plant show the following cobalt-compressive strength relationship

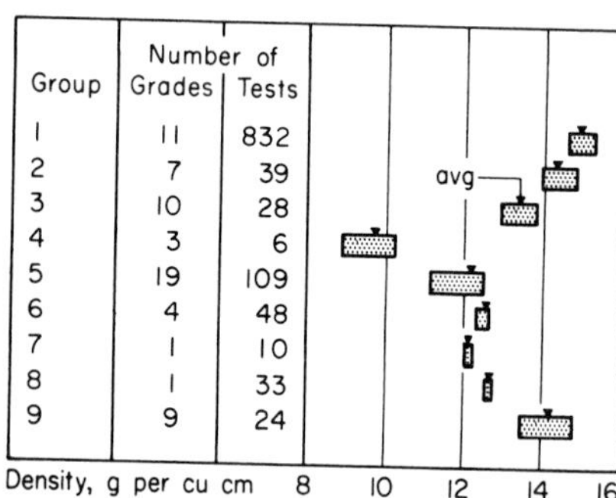

Group
Number of
Grades
Tests
1 11 832
2 7 39
3 10 28
4 3 6
5 19 109
6 4 48
7 1 10
8 1 33
9 9 24
avg
Density, g per cu cm

Fig. 5. Density as determined from incoming inspection measurements by four users of carbides

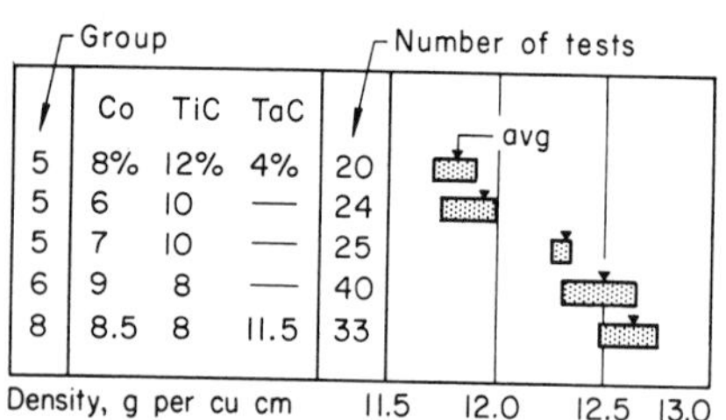

Group
Number of tests
Co TiC TaC
5 8% 12% 4% 20
5 6 10 — 24
5 7 10 — 25
6 9 8 — 40
8 8.5 8 11.5 33
avg
Density, g per cu cm

Fig. 7. Ranges of density observed for five carbide grades used for cutting steel. Incoming inspection measurements by two users.

(average of three tests). These data more nearly agree with those given in Fig. 9, although more tests would undoubtedly show a considerable spread in values.

Co, %	Compressive strength
3	458,000 psi
4.5	530,000
5	450,000
6	500,000
13	414,000
20(a)	391,000
25(a)	354,000

(a) 5% TaC, the remainder WC; others were WC only with Co.

Corrosion Resistance

Commercial grades of sintered carbide have been used under corrosive conditions but usually where a material with high indentation hardness and abrasion resistance was required, so that the more common corrosion-resistant materials are unsatisfactory. For example, carbides are used for bearings and other wear-resisting surfaces in chemical processing and petroleum refining equipment where corrosion resistance is required, but where materials such as stainless steels are too soft. Because sintered carbides are heterogeneous materials the resistance of the carbide and the binder may vary in specific mediums. This variation may result in intergranular attack between the individual particles of carbide.

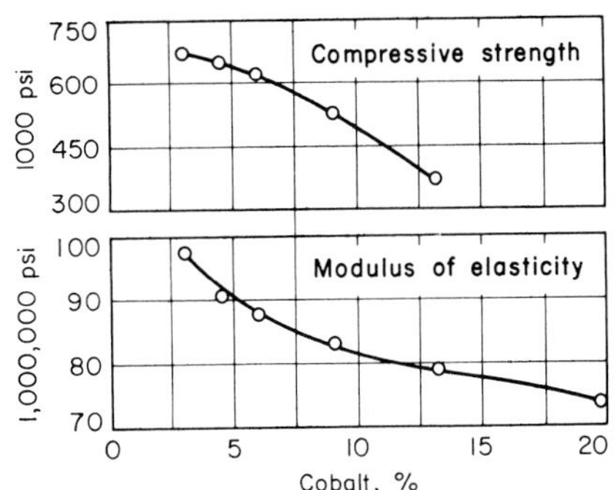

Fig. 9. Effect of cobalt on strength and modulus of sintered tungsten carbide

The straight WC, the WC-TaC, and WC-TiC grades are all resistant to attack by most acids; they are considerably superior to type 302 stainless steel in resistance to attack by reducing acids (such as HCl and H_2SO_4), in both dilute and concentrated solutions. However, carbides have less resistance than type 302 stainless to oxidizing acids such as HNO_3 and to strong alkalis such as 50% solutions of NaOH. Carbides have good, though not complete, resistance to brine solutions, rating about the same as the common 18-8 stainless steels. Stainless steels have shown some superiority to carbides in resistance to attack by sulfur-bearing oils. Sintered carbides show only fair resistance to corrosive attack by the organic acids such as lactic and citric acids.

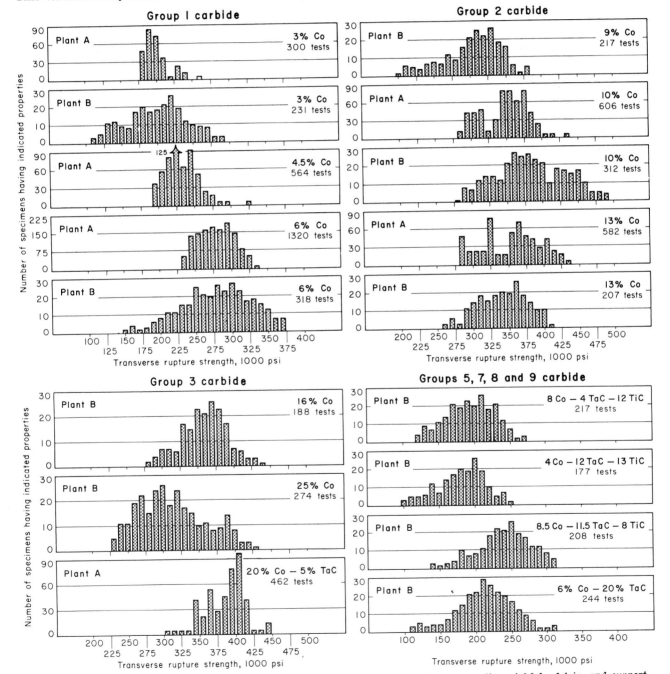

Fig. 8. Transverse rupture strength of carbide groups. Plant A test samples had a cross section of 0.2 by 0.4 in. and support distance between centers of 9/16 in. Plant B test samples, ¾ in. long, had a cross section of 0.2 by 0.375 in. and support distance between centers of 9/16 in.

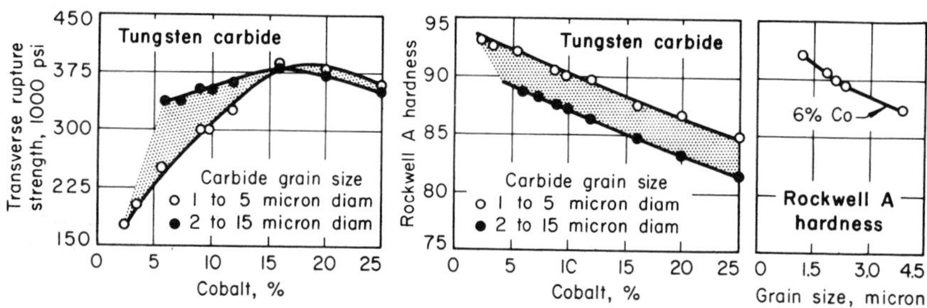

Fig. 10. Effect of cobalt content and carbide grain size on transverse strength and hardness of straight tungsten carbide materials

Sintered carbides are subject to oxidation at elevated temperatures but this is usually limited to a thin film of surface oxide.

Microstructure

Microstructure affects hardness and strength. The size of the carbide particles (grains), their distribution and porosity, and the quality of bond between cobalt and carbide crystals are influential factors. The effect of grain size on hardness was shown in Fig. 2 and 3.

Increasing tungsten carbide grain size lowers the hardness because the cobalt "lakes", which are interspersed between grains, are also larger. The ideal structure for high-strength carbides is one of small carbide grains separated by the thinnest possible layer of auxiliary binder. One experimental study has shown the transverse rupture strength to be a maximum when the average distance between carbide particles is 0.3 to 0.6 microns (0.000039 in.).

Titanium and tungsten form a double carbide, and about 35 to 50% of the double carbide is desirable for tools made from grades having 10 or 15% TiC min. The amount of double carbide is proportional to sintering time and temperature. Hardness and density increase, and porosity decreases with prolonged sintering of the same composition.

A magnification of 1500× is used for microscopic examination and identification of carbide particles. At this magnification WC and TaC crystals are easily identified. The double carbide of TiC with WC is more difficult to distinguish. Polished carbide surfaces are etched in a one-to-one solution of 10% potassium ferrocyanide and 10% potassium hydroxide in water at room temperature.

Heat tinting is another method used to identify the types of carbides. The polished surface is etched with a Na_2CO_3 solution and then heated to 900 F, to give the individual carbides a characteristic yellow or orange color. Columbium carbide is most often used with tantalum carbide, and the two are difficult to separate by analytical or microscopic means.

The effect of grain size on properties is more pronounced in straight WC grades than in the other grades, which normally have coarser grain structures.

Figure 10 shows the effects of cobalt content and carbide grain size on transverse rupture strength and hardness of straight tungsten carbide grades.

Porosity is examined and rated at a magnification of 200×. The method of evaluating and classifying porosity is shown in ASTM B276-54. The *type* of porosity is classified as A, B or C. Type A designates porosity up to 10 microns diam; type B from 10 to 40 microns; and type C includes porosity resulting from uncombined or excess carbon. Increasing *amounts* of porosity of each type are designated by numbers from 1 to 6. Classification is easily accomplished by relating a carbide specimen to the most closely similar microstructure, as illustrated by the series of reference standards (Fig. 11). The results of 912 examinations on seven different grades of group 1 carbide are illustrated in Fig. 12.

Other Properties

Modulus of elasticity and Poisson's ratio are shown in Table 2, as well as some impact values, fatigue limits, thermal expansion, and electrical resistivity values.

A carbide structural element will deflect only 33 to 40% as much as a similar steel one under the same loading conditions, because of the high modulus of elasticity of carbides. Thus they are valuable materials when stiffness is required, as in boring bars for use when hole dimensions impose rigid space limitations. The fact that a carbide section will not deflect as much as a steel one, under the same load, should be considered when composite carbide and steel assemblies are being designed. If this is not taken into account the carbide insert may crack because of lack of stiffness in the steel support. The carbide section should be able to sustain the load placed upon it without exceeding its elastic limit. Under a heavy localized load, the steel support in the composite member has to be designed for enough stiffness to carry about three times the load it is actually called upon to support, to make sure that its deflection will match that of the sintered carbide.

The thermal expansion of carbide is about one third to one half that for steel, an important factor when carbide parts are attached to steel either mechanically or by brazing. If compressive loads are not excessive, shims of copper are inserted between the carbide and steel to supply a low-strength interface that will yield and reduce stresses.

Tungsten carbide and the addition carbides exhibit properties that are largely metallic. The melting point ranges from 5000 to 7000 F. Electrical and heat conductivity and optical reflectivity are in the same range as for metals.

Standardization of Test Methods. Testing methods for hardness, density, transverse rupture, and porosity have been referred to in the earlier sections on these subjects. Other properties for which recommended procedures have been tentatively agreed upon by a majority of the industry, include compressive strength, modulus of elasticity, Poisson's ratio, thermal expansion, electrical resistivity, and metallographic methods.

Methods recommended by the Cemented Carbide Producers Assoc. for determining properties of car-

Table 2. Typical Physical and Compressive Mechanical Properties of Sintered Carbides

Carbide Group	Measured in compression: Compressive strength, psi	Elastic limit, psi	Modulus of elasticity, million psi	Poisson's ratio	Ductility, %	Impact strength, ft-lb (a)	Fatigue limit, 1000 psi (b)	Coefficient of thermal expansion (c)	Electrical resistivity at 20 C, microhm-cm
1 (3% Co)	615,000	500,000	105	0.24	0.60	....	....	....	21.3
1 (6% Co)	614,000	286,000	105	0.28	0.85	0.73	95	5.0	....
2 (10% Co)	600,000	125,000	87	0.20	1.90	1.10(d)	105(d)	5.9(d)	19.6(d)
3 (16% Co)	545,000	95,000	76	0.22	2.70	1.75(e)	....	....	29.2(e)
5	625,000	230,000	78	0.22	1.00	0.60	90	....	....
6	533,000	97,000	80	0.22	2.00	0.40	90	6.8	....
7	635,000	173,000	82	0.21	0.90	....	....	....	....
8	631,000	250,000	81	0.22	1.00	0.60	85	6.00	....
9	705,000	240,000	86	0.22	1.70	0.60	85	6.00	....

(a) Impact values are from unnotched specimens in sections approximately ¼ in. square. (b) Values for fatigue limits are based on 20,000,000 cycles, for specimens of the R. R. Moore rotating-beam type. (c) Average coefficient of expansion for the range 20 to 700 C (68 to 1290 F) in micro-inches per inch per °C. (d) Cobalt content 13%. (e) Cobalt content 20%.

bides are covered by the following CCPA designations:

Density	P-101
Poisson's ratio	P-105
Compressive strength	P-104
Transverse rupture strength	P-102
Microstructure	M-202
Apparent porosity	M-201
Modulus of elasticity	P-106

Carbides Other Than Tungsten Base

Recent developments include the use of metal carbides other than tungsten, of which the most prevalent are the carbides of titanium and chromium, with nickel or a nickel-base alloy as the binder metal. These materials were developed primarily as jet-engine components for operation from 1600 to 2200 F, and are sometimes referred to as "cermets". (The term "cermet" would be as appropriate for the conventional WC-Co system as for TiC-Ni; both are produced by sintering a "ceramic" carbide with a ductile metal.)

The TiC-Ni materials fall into several categories of end products, some of which overlap the common uses of WC-Co grades. In the most common type, the hard phase is predominantly titanium carbide with chromium carbide additions and 30 to 70% of nickel or nickel alloy binder. These grades combine high resistance to oxidation, high hardness, resistance to thermal shock, relatively low density, and good creep rupture properties for the temperatures involved (1800 to 2200 F). The greater solid solubility of the carbide in the nickel alloy results in a binder phase of low ductility having markedly lower impact values than WC-Co material. These range from 1 to 12 ft-lb when tested with a 0.394 by 0.394 by 2.165-in. unnotched impact specimen.

These materials have received some acceptance for high-temperature abrasion-resistant tools such as hot upsetting anvils, hot tube drawing mandrels, tools for trimming hot flash, guide pads or inserts for hot wire and the like.

Hardness and transverse rupture strength of these grades are given in Fig. 13, which shows that decreasing amounts of the nickel binder increase the rupture strength.

A second category of the so-called "cermets" contains the same basic ingredients as the above but the binder content is generally less than 20%. These materials have been applied to cutting both steel and cast iron, particularly in finishing operations at high speed with medium to light chip loads. They are usually produced for mechanical attachment to the shank, but may be joined by brazing.

A third category includes carbides predominantly of chromium, with up to 20% Ni, nickel-chromium or nickel-cobalt binder. These materials have become commercially available because of their improved resistance to oxidation and corrosion at temperatures up to 1800 F.

There are no available service data that compare corrosion resistance of the chromium carbides with other carbides or corrosion-resisting materials. However, laboratory test data indicate that these grades are considerably superior to tungsten carbides in resisting attack by several mediums. It is also indicated that the higher the content of chromium carbide the greater the corrosion resistance. For example, it has been observed that a grade containing 89% CrC and 11% Ni has higher resistance to strongly oxidizing acids (HNO_3 and aqua regia) than grades that contain about 84% CrC and 16% Ni or 70% CrC with 15% TiC and 15% Ni. However, the difference in corrosion resistance among these three grades is far less significant than the difference between the 89% CrC – 11% Ni grade and the various grades of WC. Chromium carbides have more resistance to virtually all of the acids and alkalis than does tungsten carbide. The chromium carbides are completely resistant to brine solutions and sulfur-bearing oils, whereas tungsten carbides are slightly attacked by these mediums.

Chromium carbides can be made to have a coefficient of thermal expansion approximating that of mild steel, and are used for extrusion dies to shape nonferrous metal, pump seal rings, precision gage blocks, valve trim for chemical and food plant use, balls for check valves, and punches for forming plastics. Hardness results for this group are shown in the upper graphs of Fig. 13.

Several other metal carbides, such as molybdenum, vanadium or haf-

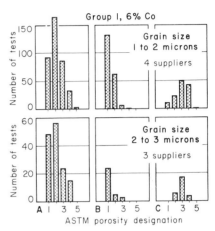

Fig. 12. Porosity ratings of 912 samples of seven carbide grades from group 1, as summarized from the inspection records of one user. For the specific applications involved, the carbides were subject to rejection as follows: C-type porosity of C-3 or greater, B-type porosity of B-3 or greater, and A-type porosity of A-5 (Fig. 11).

Fig. 11. Reference standards for evaluating apparent porosity in sintered carbide (ASTM B276-54). Intermediate amounts of porosity corresponding to A-2, A-4, B-2, B-4, C-2 and C-4 are not shown here. Extreme amounts represented by A-6, B-6 and C-6 are also omitted. The original standards are at a magnification of 200×; those shown here have been reduced to about 135×.

nium, with metal binders of cobalt, nickel, iron or chromium, have achieved limited commercial acceptance for specialized applications, such as high-intensity light sources and rocket nozzles. Other commercially available sintered products with high hardness include metal-bonded chromium boride, boron-bonded zirconium boride, and chromium-bonded aluminum oxide, used for applications such as aluminum vaporizing crucibles and combustion-chamber liner coatings.

Cutting Tools

More carbides are consumed for metal cutting than for any other type of application. Because of their ability to retain a sharp cutting edge, the straight tungsten carbide grades of group 1 are virtually the only tool material used to cut abrasive materials such as Fiberglas and phenolic resins. Carbides of group 1, which have the highest hardness, are also being used for production cutting of white cast iron at Rockwell C 60. Thus, the straight WC grades serve for both extremes of the metal cutting range, as defined by the type of chip produced in the machining operation. The steel cutting grades containing TiC and TaC occupy an intermediate position; the steel chips have ductility and abrasive properties between the two extremes.

The straight tungsten carbides are less suitable for machining steel because the ductility and strength of steel chips cause a severe wiping action along the top surface of the tool. Rapid cratering results in a sharper-angled cutting edge, and the extremely wear-resistant but relatively brittle thin edge fails by minute chipping. Grades of groups 4 to 8 are applied on the basis of their resistance to cratering; grades of group 5 are the most widely used.

The application of carbides in groups 4 to 8 differs in different plants. One specific plant prefers grades of group 4 for high-speed light finishing cuts on steel, group 5 for medium-speed medium-cutting of steel, and group 6 for slow to medium roughing cuts on steel. Grades of group 7 are used for special turning operations such as machining uranium, and group 8 is used more for applications other than cutting tools where a minimum of porosity is required, as for precision gage blocks.

There is also a divergence of opinion on the use of TiC and TaC grades, especially for high-temperature requirements. A choice between the two is frequently resolved on the basis of cost; the TiC grades are cheaper.

Recognition of the opposing nature of cutting tool properties (resistance to edge wear versus resistance to cratering) is the basis for selection of carbide tools; resistance to cratering is gained at the expense of edge wear. Most metal-cutting applications involve a compromise between these two factors, with the hardest grade that will not chip or break being selected.

The reasons for the improved resistance to cratering of TiC and TaC carbides are not well defined. Below a critical temperature range temperature has only a minor effect on wear. Within the sensitive range, temperature is important because it affects the rate of adhesion and diffusion. The critical temperature for a particular grade will depend on both the composition of the work material and the composition of the carbide tool because both contribute material to the diffusion interface.

The hot abrasion resistance of this interface is the determining factor in wear life rather than the hot hardness of the carbide in the as-received condition. For instance, group 1 carbide at 1400 F has a hardness of Rockwell A 85, while group 5 at the same temperature has Rockwell A 81.5. However, the harder group 1 carbide will fail by cratering at speeds where the softer group 5 carbide will perform satisfactorily.

Columbium carbide is utilized in about one fifth of the steel-cutting grades. It is present in group 9 grades in amounts of 0.75 to 2%. In groups 5 and 8, 1.75 to 2.50% CbC is present. Columbium carbides appear to impart essentially the same properties as TaC.

Figure 14 illustrates cutting temperatures in machining 52100 steel with a group 5 carbide.

Full advantage of carbide tooling for metal cutting operations can therefore be gained only by selecting the grade that has the best combination of properties for the specific conditions. For example, carbides may have a high modulus of elasticity, but exhibit little ability to undergo plastic deformation. The material supporting them must have adequate section to endow the assembly with enough stiffness to withstand high localized stresses occurring when heavy cuts are taken. Carbide grades having a hardness of Rockwell A 90 to 91.5 and a transverse rupture strength above 275,000 psi are usually recommended for heavy-duty cutting. Tool failure due to thermal cracks is prevalent in rough cutting with carbides. If cracking prevails it is advisable to use a grade having a lower titanium and tantalum content. Higher cutting speeds require harder and more crater-resistant grades. These grades can be effectively used for rough machining if mechanically held tools and lighter chip loads are

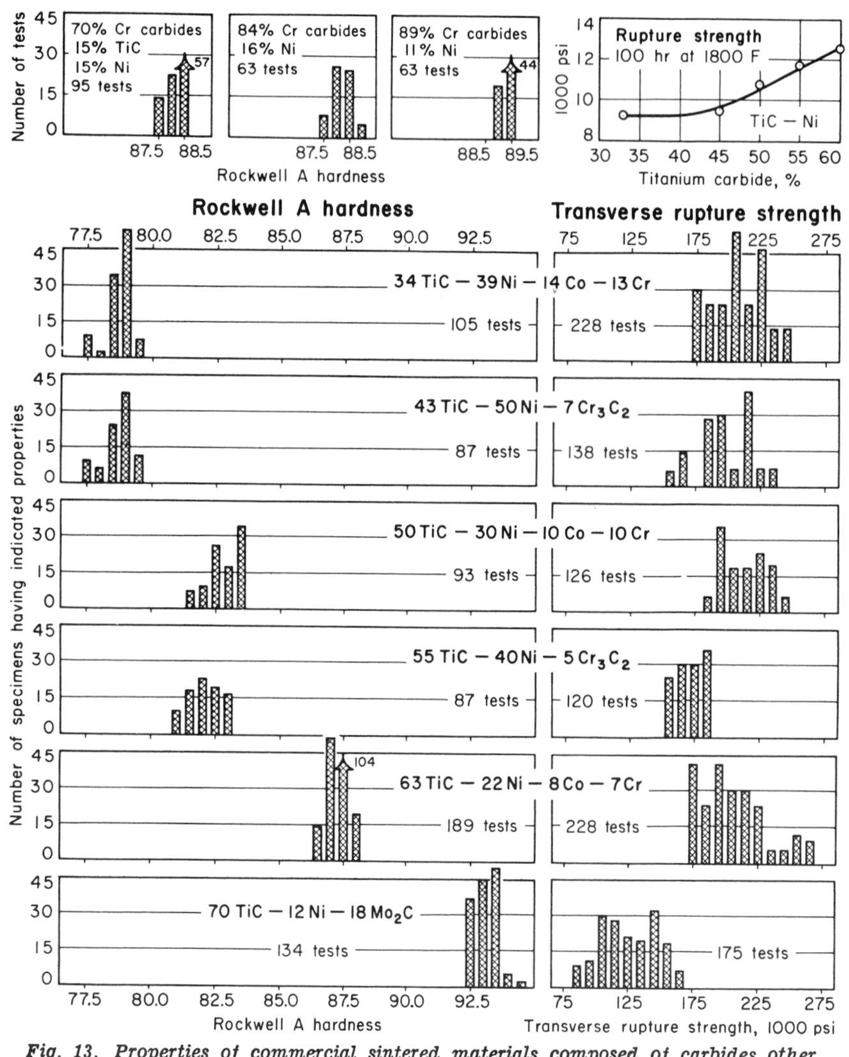

Fig. 13. Properties of commercial sintered materials composed of carbides other than tungsten and a binder phase other than cobalt

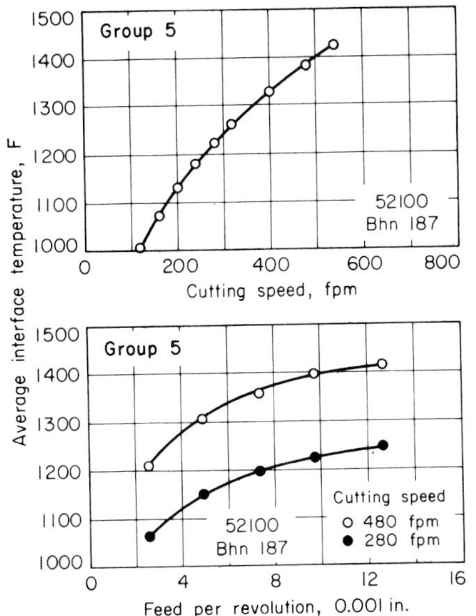

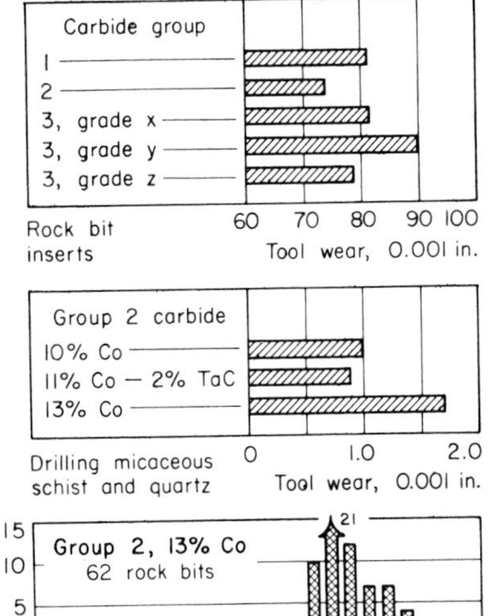

Left. Effect of speed and feed on tool interface temperature for single-point turning. Tool shape is 0- (-2)-7-7-8-0-3/64 in. Effect of speed is for depth of cut of 0.102 in. at a feed of 0.0098 in. per revolution.

Upper Right. Comparison of different carbide groups and grades for rock bit inserts used in drilling pink quartzite

Center Right. Wear of rock bit point per foot of rock drilled in pneumatic percussion drilling with a 3.5-in. drill at 80 psi pressure

Lower Right. Spread in tool life in drilling more than 20,000 ft of granite. Drill inserts were 1⅝-in. diam, used in a 3.5-in. pneumatic drill at 90 psi pressure.

Fig. 14. Machining and impact drilling performance of carbides

employed. Table 3 lists carbide grades that have been found the most satisfactory in one plant for single-point turning under the conditions stated.

Tool life is dealt with in detail in the article on selection of material for cutting tools, in this Handbook.

Earth Drilling

The high hardness and wear resistance of carbides makes them well suited for earth drilling and mining applications, wherein the high-cobalt grades are used exclusively to withstand the impact loads. Typical carbide tool life of drill bit grades is shown in **Fig. 14**.

Various types of specialized rock bits for drilling in extremely hard and abrasive rock formations utilize carbide inserts instead of the conventional hard faced steel teeth. Carbide insert bits are almost mandatory for drilling rocks harder than limestone, and their use has all but made the steel bit for drilling hard rock obsolete, both on a basis of performance and economics. It is estimated that 90% of all pneumatic drilling in hard rock is now done with carbide-insert bits.

The function of the carbide insert is to pulverize the fragmentable formations rather than to pry loose the particles by penetration of the tooth. The cost of this type of bit is usually several times that of the conventional bit. However, the added cost is more than offset by the longer useful life and consequently the reduction in the number of tools needed for this type of drilling. Rock formations containing "chert" are almost impossible to drill except with a carbide-insert bit.

The carbide inserts for rock bits are so small and require such close tolerances that they cannot be cast at reasonable cost. Sintering is the only practical procedure for manufacturing them. Sintered carbide inserts have been used on drag bits, coal mining tools, and in some percussion tools.

The straight tungsten carbide grades most widely used in these applications contain from 10 to 15% Co. Sintered carbides of similar compositions have been used on an experimental basis for manufacturing slush nozzles for jet-type rock bits. The dimensions can be held closely and the surface finish is usually better than that of the presently used cast carbide nozzles. However, these two advantages are of minor importance because the amount of improvement is small and does not contribute to longer life or easier usage.

The cost of the sintered nozzle was at one time approximately twice that of the cast carbide nozzle. However, recent developments utilizing byproduct carbide have considerably decreased this cost differential although the cast nozzles are, at the time of this writing, still less expensive than sintered nozzles.

The composition of cast carbide nozzles varies slightly from different manufacturers — usually it is about 85% W_2C-WC and 15% Co or Ni. In some nozzles, a combination of cobalt and nickel is used because of the lower cost of nickel. The properties of the nozzle are not altered perceptibly by the nickel. Sintered carbide nozzles that are subjected to extreme abrasive wear have proved economical, especially if the shape lends itself more readily to the sintering process than to casting. Certain designs are favorable to sintering because of unavoidable changes in section, tendency toward interdendritic porosity, and retention of certain cast characteristics.

Other developments in earth drilling applications now permit an increase in wear resistance of tool joints and drill collars by coating them with tungsten carbide particles. The tungsten carbide is deposited by a high-temperature flame or by a plasma arc to secure a uniform layer. This layer is then fused or sintered by additional heat from a torch or by heating in a controlled-atmosphere furnace.

Table 4 lists mechanical properties and typical mining applications for five different grades (all from one source) of carbide that are classified as group 2 by cobalt content and density, although the last item shows a hardness value slightly under the normal range for group 2 carbides.

Table 3. Typical Use of Carbides for Machine Turning in One Plant

Carbide group	Cutting conditions: Speed, sfm	Feed, ipr	Depth, in. (max)	Application
6........	125 to 300(a)	0.030 to 0.072	1.125	Heavy turning steel forgings and rolled bar stock
5(b).....	250 to 400	0.018 to 0.033	0.750	General turning steel forgings and rolled bar stock
1(c).....	40 to 80	0.012 to 0.027	0.750	Turning cast steel and chilled rolls
1........	150 to 300	0.018 to 0.040	1.000	Rough turning cast iron

(a) Cast W-Co tools are recommended for cutting speeds below 100 sfm on heavy steel forgings or bar stock. (b) A harder grade should be used if excessive cratering occurs. (c) Bordering on carbide group 2.

Table 4. Some Properties and Typical Mining Applications of Group 2 Carbides (6.5 to 15% Co)

Rockwell A hardness	Other properties: Transverse rupture, psi	Compressive strength, psi	Modulus of elasticity, psi	Density, g per cu cm	Application
90.5	300,000	644,000	82,000,000	14.6	Coal rotary bits
89.2	410,000	643,000	81,000,000	14.6	Coal undercutter bits
89.1	410,000	....	....	14.5	Small rock bits
89.0	410,000	603,000	83,000,000	14.5	Large rock bits, drag bits
86.3	360,000	505,000	80,000,000	14.1	Heavy-impact crushing tools, drag bits

Table 5. Relationship Between Voltage, Current and Stock Removal in Electrolytic Grinding of Two Grades of Tungsten Carbide

Voltage	Current, amp	Rate of metal removal, g per min	Rate of metal removal, cu in. per hr
Carbide with 3% Co			
0....	0	0.04	0.00775
5....	16	0.3	0.0725
10....	24	0.24	0.058
15....	36 to 38	0.36	0.0865
Carbide with 6% Co			
0....	0	0.02	0.0049
5....	16	0.2	0.049
10....	24	0.26	0.036
15....	36 to 38	0.4 to 0.44	0.098 to 0.108

The area cut was ⅜-in. diam rod.

Other Applications

Because of their resistance to abrasive wear, sintered carbides are used for such applications as centerless grinder rests, brick mold liners, facings for hammers in hammer mills, facings for jaw crushers, ball mill linings, balls and seats for check valves, sandblast nozzles, ring and plug gages, gage blocks, and wear pads in machinery.

Because of their high modulus and low coefficient of expansion, carbides are used for structural parts in machines, boring bars and scale pivots. They are also used for thrust where resistance to heat or corrosion, or both, are present, or where lubrication is slight or absent.

Grinding of Carbides

The high hardness of carbides limits the number of materials that can effectively grind the surface; thus the shaping and reconditioning of carbide tools is an expensive procedure. In one abrasion test, using the grinding rate of aluminum oxide as 1.0, silicon carbide abraded the sintered carbide at 5.5 times that rate, boron 12, boron carbide 15, and diamond dust 25 times the abrasion rate of aluminum oxide.

Silicon carbide abrasives are used for roughing and semifinishing, and resinoid-bonded diamond wheels for finishing. The grinding of the steel shanks to which the carbide is attached is done best with an aluminum oxide wheel.

Electrolytic grinding is a process that combines electrolytic and diamond wheel grinding. A metal bonded diamond wheel is used for the cathode. The diamond particles act as insulating spacers separating the work from the wheel, and also serve as scrapers. This process is not readily applicable to cavities or irregular openings such as often occur in dies. Electrolytic grinding has its most practical application where conventional carbide grinding is costly or impossible. The rate of metal removal is a function of the area being cut and the electrical current drawn. Linear removal rates for different grades of sintered carbide have been reported and vary widely. Relationships obtained from electrolytically grinding two grades of ⅜-in. diam straight tungsten carbides are given in Table 5.

The ultrasonic process does not directly use electric current for stock removal. Cutting is accomplished by abrasive action at high frequency and low amplitude and, unlike electrolytic processes, the ultrasonic technique cuts nonconducting materials as well as conducting materials. The cutting or grinding tool is vibrated in the frequency range of 16,000 to 29,000 cycles per second through an amplitude of not more than a few thousandths of an inch. The abrasive (usually boron carbide) is carried in water to the tool and work. Other abrasives such as aluminum oxide and silicon carbide have been used but they do not last as long as boron carbide. Metal removal rates of 0.0035 cu in. per min have been achieved machining tungsten carbide with 240-grit abrasive. Accuracy of ±0.001 with a surface finish of 30 micro-in. can be obtained. Surface finish can be improved to about 8 micro-in. with 800-grit abrasive, but metal removal is much slower.

Table 6. Electrodischarge Removal of Tungsten Carbide at 15 Amp and 25 Volts

Composition of electrodischarge forming tool	Rate of work material removal, 0.001 cu in. per min	Ratio of volume of expended tool metal to volume of carbide removed
Yellow brass	2.0	3.0
20% Ag, 80% W	3.4	0.14
10% Ag, 90% W	3.4	0.09
35% Ag, 65% WC	1.7	0.21

Contour Shaping

Contour shaping of carbide blanks is accomplished by electrodischarge machining, whereby carbide particles are removed by electrical discharges of high current density and very short duration while the discharge tool and carbide are submerged in a dielectric liquid (usually kerosine). Typical rates of removal of tungsten carbide for different discharge tool materials at 15 amp and 25 volts power supply are given in Table 6.

Quality Control

Some users of carbide check all incoming shipments of tools for hardness as a routine procedure and maintain control charts to show variations in production lots of carbides classified as falling within a specific group. Figure 15 shows typical control charts of averages and ranges for 17 consecutive lots of the 13% Co, group 2 carbide of Fig. 1. Figure 15 represents hardness values for successive shipments as received over a period of time. Figure 1 shows the total distribution of hardness readings.

In the upper chart (averages), the UCL_x and LCL_x symbols represent the upper and lower control limits established for this grade. Each plotted point is the average of three hardness readings on samples from a single lot. The average of all the individual averages is 89.02. Plotted directly below on the R chart is the range in hardness for each successive three-sample group. Each R point is plotted directly below the X point for the same lot. The width of the range between upper and lower control limits on the upper chart is 0.8 Rockwell A point. The value of 1.0 at the top of the R chart indicates that the difference between the lowest and highest hardness readings among the three samples should not exceed 1.0 Rockwell A point.

In addition to assisting in the control of individual shipments, the charts show marked deviations in hardness of a particular shipment in relation to previous shipments. The reader may note that for the sixth shipment the average hardness is above the upper control limit. The corresponding R value of zero means that no range in hardness was found — that is, the three readings were identical. The combination of these two points indicates that the shipment is out of control; the hardness is uniformly high. Such a lot would be carefully tested before use or rejection. Normally, new control limits are calculated after 20 consecutive shipments have been tested and plotted.

Costs

The cost of sintered carbide is about \$0.04 per gram or \$17.00 per pound (Feb 1959). Carbides are marketed in two price classifications. The first is largely the straight tungsten carbide groups 1 through 3; the remaining grades are priced about 10% higher. As of Feb 1959, there are 464 individual grades available. Sintered carbides are discussed in several articles on tooling materials. Consult the index for precise references.

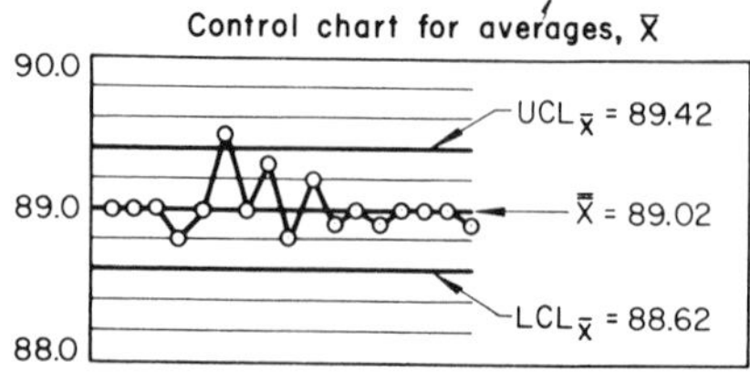

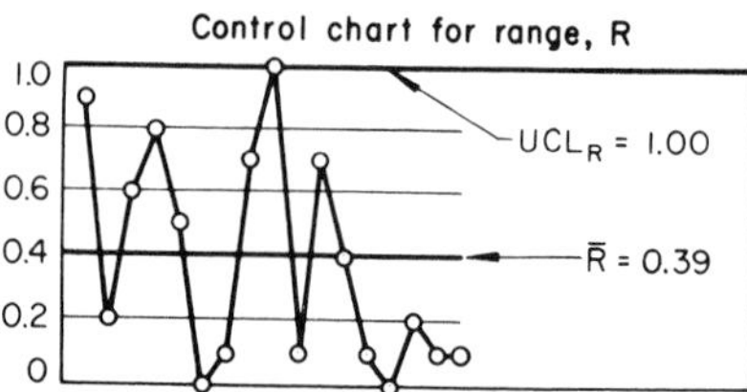

Fig. 15. Quality control charts for hardness of carbide grades. See text for discussion and Fig. 1 for distribution.

Carbide Tools for Cold Extrusion

W. L. KENNICOTT

SELECTION of the most suitable tool material for a cold-forming operation depends upon a number of factors. Among the tool materials commonly used for cold-forming work are the high impact tungsten-base carbides, which can provide unique advantages and performance when appropriately selected and applied. Generally, a carbide tool material is chosen for one of three reasons.

First, carbide tools are useful for long runs where their durability and wear resistance justify their higher cost. Secondly, where the length to diameter ratio of the punch exceeds 4:1, the greater stiffness of carbides (over that of tool steels) provides higher load capabilities in the "slender column" configuration.

Thirdly, carbide tooling is justified where the geometry of the part or the material being worked results in compressive stresses beyond those which can be tolerated by tool steel punches.

Carbide punches, as shown in Fig. 1, have been extensively used in cold extrusion for over 15 years, producing small parts for automotive, farm equipment, and other high production industries. Typical of such parts are bearing cups, valve lifters, wrist pins, and spark plug bodies.

In the last few years, larger parts have been produced by cold extrusion. Primarily, these are cup-like shapes which are used either as-is or subsequently punched out to produce bushings. Another trend encouraging the selection of carbides is the need to handle more difficult alloys in cold extrusion work, or in the warm extrusion of slugs, which may be preheated up to about 1000°F. These materials often require high forming pressures, challenging the capabilities of the tool material.

W. L. KENNICOTT is Vice President, Engineering, Kennametal Inc., Latrobe, Pa. 15650.

ANALYSIS OF REQUIREMENTS

Optimum tool material performance in cold extrusion work requires a combination of critical properties. The first, and perhaps most critical requirement is a high compressive yield strength in the punches. Strain gage measurements on large back-extruded cups have revealed an average cross-sectional compressive load in excess of 330,000 psi. Because this is an average figure across the entire punch area, we can safely assume that localized forces are as much as 50 pct over that due to temporary misalignment, variations in forces during the duration of the working stroke, and the impact occurring when the work material is contacted initially.

Another critical requirement is a high stiffness, or modulus of elasticity. This property is most critical on punches having length-to-diameter ratios of 4:1 or greater. Carbides of the high impact type have $2\frac{1}{2}$ times the stiffness of high alloy tool steels, and thus resist displacement under load to a greater extent.

A third requirement is good wear resistance. This is needed by both punches and dies to handle long runs with a minimum of tool maintenance and minimum of variation in workpiece size.

A fourth requirement, particularly for cold extrusion punches, is a high modulus of rupture, or breaking strength. With the overhang involved in cold extrusion punches and with imbalance in the forces which can occur during a back or forward extrusion operation, there is a bending stress which must be resisted by the punch material. If failure occurs due to such forces, it is on the tensile surface of the punch.

The fifth important factor is shock resistance—a high number of strokes per minute results in high impact at the point of contact.

Finally, good fatigue resistance is required to endure the tens of thousands, or even hundreds of thousands, of strokes expected during the life of such tooling.

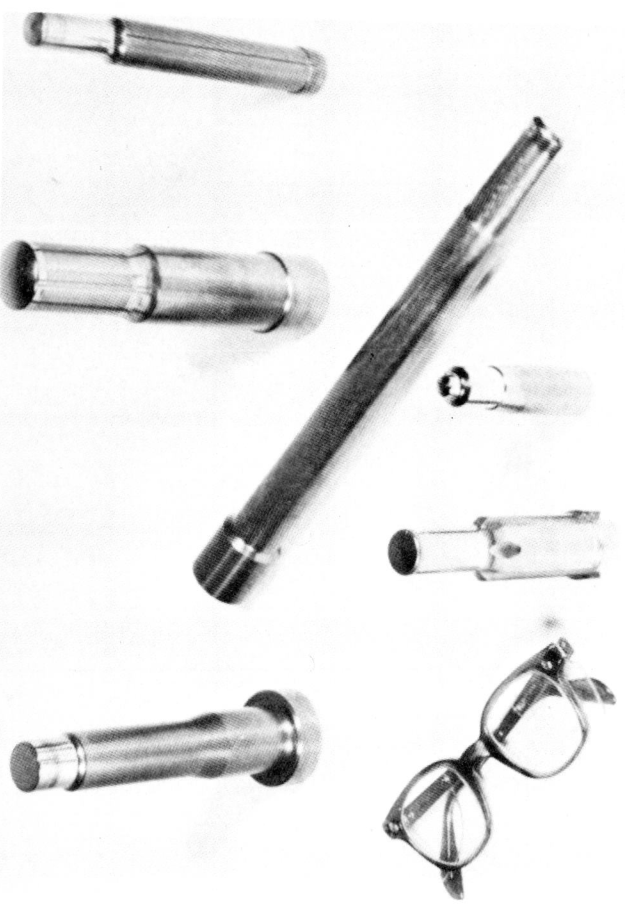

Fig. 1—Carbide punches have been applied for many years.

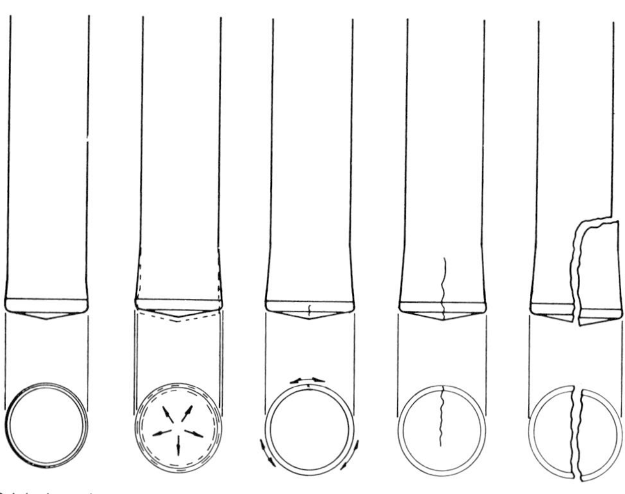

Fig. 2—"Clothespin" cracks occur because of the compression of the punch during use. From left to right, tensile stresses develop, a tensile crack appears, and finally the punch breaks.

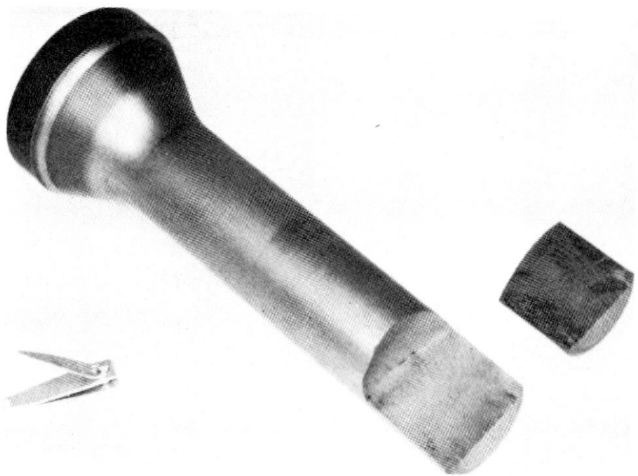

Fig. 3—Typical "clothespin" failure in punch.

Fatigue resistance is closely related to mechanical hysteresis, and the carbides are inherently good in this regard. Given the same deflection, high-impact carbide tool materials will have approximately $\frac{1}{8}$ the mechanical hysteresis of steel. Combining this with their low deflection, due to the high modulus of elasticity, the mechanical hysteresis advantage is approximately 18:1 or 20:1 for a given loading.

LIMITATIONS SHOULD BE CONSIDERED

As with all tool materials, the carbides have certain limitations. To best understand these limitations, it would be well to study the nature of failure in such tooling.

The most common failure in cold extrusion punches is called the "clothespin" crack. As illustrated in Fig. 2, this is a progressive crack. It starts due to compressive yield of the tool material, resulting in the punch becoming somewhat shorter and consequently larger in diameter, particularly at the working end. As the illustration shows, the periphery of the punch increases, placing that surface in tension, and possibly developing one or more tensile cracks at right angles to the tensile stress. With further use, this crack can progress toward the axis of the punch, across that axis, and eventually result in a split punch, Fig. 3.

Such failures obviously come about because the compressive yield strength of the tool material is exceeded. The solution is to either reduce the forces (by changing geometry or properties of the material being worked), or by providing a tool material with higher compressive yield strength.

Another less common type of failure is referred to as "capping". In effect, the working end of the punch is removed as a disc due to tensile stresses that develop as the workpiece is stripped from the punch. This effect usually comes about due to wrong punch geometry or to excessive wear, which has destroyed the original punch geometry so that the punch "hangs up" in the workpiece.

Another type of failure is bending of the punch, with or without complete rupture. This failure would be evidenced by a break at roughly right angles to the axis of the punch and at a section near the holding end.

Fig. 4 shows another type of failure: galling wear. Wear as severe as that illustrated could result in capping, and would certainly produce a rough surface on the workpiece even if the punch continued to function.

NEW DEVELOPMENTS APPEAR

For years, manufacturers of carbides for cold extrusion punches have been fenced in between two critical

Fig. 4—Galling wear also occurs in cold extrusion punches.

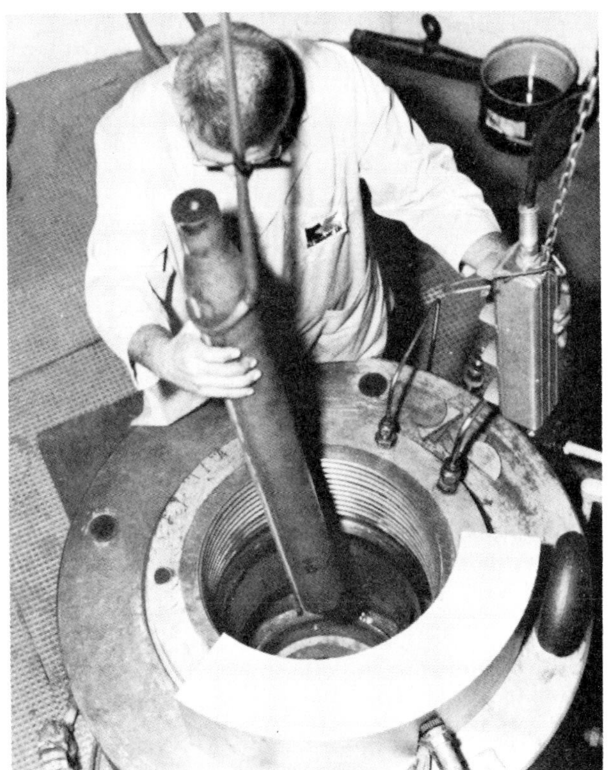

Fig. 5—Hot isostatic pressing produces stronger, tougher carbides for cold extrusion tools.

requirements. If the material was made softer and rougher, it would yield and fail, and if it was made much harder and stiffer, it would shatter due to brittle failure. Given these two limitations, it was sometimes difficult to find the optimum combination for any given set of conditions.

Recently, however, the range has been widened considerably due to new developments in processing methods and equipment. Table 1 shows a tabulation of critical properties for two of the older alloys, two newly developed alloys, and a tool steel, included for comparison. Compared with the more established 3833

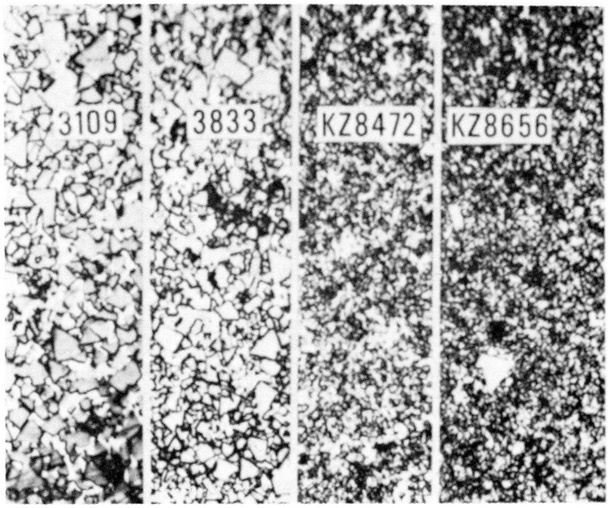

Fig. 6—Newer carbide materials (KZ8472 and KZ8656) have finer grain sizes than established grades (3109 and 3833) due to isostatic pressing at high temperatures.

Table I. Kennicott—MEQ Characteristics of Some Tool Materials

Grade	Cobalt, Pct	Hardness, Rockwell A	Compressive Strength, psi	Modulus of Elasticity, psi	Transverse Rupture Strength, psi	Impact Resistance, in.-lb./in.^{2}
3109	12.2	88.0	560,000	82,200,000	425,000	280
3833, SP143	11.0	89.3	625,000	81,300,000	400,000	280
KZ 8472	15.0	89.6	675,000	75,000,000	500,000	340
KZ 8656	10.0	91.6	750,000	85,000,000	450,000	300
T1, tool steel		84.0, Rc65	560,000	30,000,000		350

composition, for example, KZ8656 has 20 pct more compressive strength, coupled with slightly better transverse rupture strength, impact resistance, and hardness.

These improved properties come about due to a new process—cold isostatic pressing of powders, followed by a vacuum furnace treatment, and finally hot isostatic process in a helium (or argon) atmosphere at 14,000 to 20,000 psi and approximately 2500°F. Fig. 5 shows the equipment required for such an operation. This process reduces or removes micro and macrointernal defects, considerably improving the properties of low yield, high modulus materials. The materials also develop an ultrafine grain size in the finished part. Fig. 6 shows the microstructure of the four carbides of Table I. The greater number of finer grains results in a thinner binding layer between grains, reducing the galling tendency, and minimizing the reduction of yield strength with temperature. This last factor is particularly important on warm extrusion operations.

Thus, new developments in tool materials of tungsten-base carbides for cold and warm extrusion of steel make possible improved performance on existing operations. They extend life of the tool component, and improve its reliability. The improved properties should permit expanded uses in processing higher carbon steel parts, higher alloy steels, and more difficult shapes.

Section VII: Ultrahigh Strength Steels

Many of the critical components of aircraft landing gear are made of ultrahigh strength steels.

The Ultrahigh Strength Steels-I

*By A. M. HALL**

Ultrahigh strength steels — those that can be employed successfully at yield strengths of 180,000 psi or higher — are specified for a variety of critical applications including components of aircraft, missiles and naval vessels, pumps, valves, manufacturing machinery, as well as bolts, pins, fittings, shafts, springs, punches and dies.

One of the fastest growing fields of metallurgy is the technology of ultrahigh strength steels. Many of these materials can develop more strength per pound than can the strong aluminum or titanium alloys. At ultrahigh strength levels, steels can truly be classified as lightweight structural materials.

Heat treated razor blade strip, cold drawn plow steel wire and music wire are yet to be surpassed in tensile strength by any other commercially available material. And, although a few years ago steels in such forms would not have been considered for structural applications, the development of wrapping techniques for the manufacture of high performance components places them squarely in the midst of today's structural materials. Add to this group hard drawn and aged semi-austenitic stainless steel wire, austenitic stainless wire and improved carbon steel wire, and the strength levels achievable by steels in these special forms ranges up to and beyond 600,000 psi.

This article is limited to the various categories of ultrastrong steels which have reached commercial development. Ultrahigh strength steels are defined here as those which can be used successfully in structural applications at levels of at least 180,000 psi yield strength or 200,000 psi tensile strength.

Today's Ultrastrong Steels

The earliest family of materials in the ultrahigh strength structural class are medium carbon, low alloy, quenched and tempered steels. The progenitor of this family was the chromium-molybdenum steel, AISI 4130. This alloy was followed by the chromium-nickel-molybdenum steel, AISI 4340, which came into use where greater strength in thicker sections than could be achieved in AISI 4130 was required.

Numerous modifications of both of these steels have been developed. The goal usually has been to increase toughness at very high strength levels

*Chief, Ferrous and High-Alloy Metallurgy Div., Battelle Memorial Institute, Columbus, Ohio.

and to improve weldability. Modifications have generally taken such forms as increasing the silicon content to avoid 500 F embrittlement on tempering at low temperatures, adding vanadium to promote toughness, reducing the sulfur and phosphorus to improve toughness and transverse ductility, and lowering the carbon content slightly to improve welding characteristics as well as toughness.

In addition to AISI 4130 and 4340 and their variations, nickel-molybdenum and nickel-silicon-molybdenum steels, together with vanadium modifications, have been developed for use at ultrahigh strength levels. At the same time, copper-silicon-molybdenum-vanadium steels also have been added to the list.

Another family of steels which has found application in the ultrahigh strength structural field is the 5% chromium-molybdenum-vanadium steels. These medium alloy, air hardening steels are, in essence, aircraft quality versions of the familiar chromium hot work die steels. Adapted to structural applications, they are available in sheet, strip and other forms quite foreign to tool and die usage. Members of this family are for the most part proprietary and display considerable diversity of composition. However, most of them have the approximate composition of H11 or H13.

High Strength Stainless Steels

Several families of stainless steel, listed below, have members capable of service at ultrahigh strength levels:

- Martensitic
  - Chromium
  - Chromium, low nickel
  - Age hardenable
  - Extra low carbon, age hardenable
- Cold rolled austenitic
  - Chromium-nickel
  - Chromium-nickel-manganese
  - High manganese
- Semi-austenitic

Included are the martensitic chromium steels, AISI 410 and 420 types, martensitic low nickel chromium steels (type 431), as well as age hardenable martensitic stainless steels. The last named are characterized by relatively low carbon contents – which improve their weldability – and the addition of such elements as copper, columbium, titanium, aluminum and molybdenum which render them age hardenable.

One martensitic stainless steel, developed for the Air Force by Crucible Steel Co. and known as AFC 77, contains considerable cobalt along with molybdenum and vanadium. This steel has outstanding strength at temperatures as high as 1200 F.

Table I — Approximate Upper Limit of Strength at Which Selected Steels Are Currently Used

Type of Steel	Yield Strength, Psi	Tensile Strength, Psi	Applications
Low alloy steel	250,000	300,000	Landing gear components
5Cr-Mo-V	240,000	290,000	Airframe structures
Stainless steels			
Martensitic	160,000	180,000	Bolts and rivets
Age hardenable martensitic	170,000	190,000	Clamps and brackets
Extra low carbon, age hardenable martensitic	245,000	257,000	
Cold rolled austenitic	180,000	200,000	Missile tankage for liquid propellant
Semi-austenitic	220,000	235,000	Honeycomb structures
High alloy steels			
18% Ni maraging	295,000	300,000	Fasteners and shafts
9Ni-4Co	240,000	280,000	
Matrix	290,000	360,000	Axles and shafting

Another martensitic stainless steel with extra low carbon has been developed by Carpenter Steel Co. This steel resembles high nickel maraging steels, except that it contains 12% Cr, for corrosion resistance, and less nickel. It can be aged to extremely high strength levels.

The earliest of the cold rolled austenitic stainless steels was the 17Cr-7Ni type which carries the designation AISI 301. Some of the newer types contain as much as 15% Mn together with other elements such as nitrogen, molybdenum and vanadium.

Finally, there are the semi-austenitic stainless steels. The compositions of these materials are so balanced as to render them austenitic as annealed and martensitic through refrigeration or through alteration of the matrix composition by the precipitation of some compound (such as a chromium carbide) on heating at intermediate temperatures. A number of these steels contain such elements as aluminum and molybdenum which promote age hardening.

Special Materials for Ultrahigh Strength Service

Among the various types of high alloy steels in existence, several families have emerged which include members developed expressly for structural applications at ultrahigh strength levels. One family comprises the high nickel maraging steels introduced a few years ago by International Nickel Co., Inc.

Several 18% Ni variants within this family – which also contain substantial additions of cobalt, molybdenum and titanium – have become commercially available in a wide variety of forms. These steels are martensitic as annealed but, because of their extra low carbon and high nickel, the martensite is quite tough and ductile and hence is amenable to forming, welding and machining operations.

In spite of its low carbon content, the martensitic structure of these steels is quite strong, having a yield strength on the order of 100,000 psi. A simple aging treatment at moderate temperatures brings about a tremendous increase in strengh. Depending on composition, yield strengths of 300,000 psi and above can be developed on aging.

Another family of high alloy steels designed for ultrahigh strength service was developed by Republic Steel Corp. Known as 9Ni-4Co steels, they are hardened by quenching and tempering. One grade, containing nominally 0.25% C, can be used at yield strengths up to 200,000 psi. Another grade, with a nominal carbon content of 0.45%, is used at yield strengths up to 250,000 psi.

A third type of ultrastrong high alloy steel has been developed by Vasco Metals Corp. (formerly Vanadium-Alloys Steel Co.). These are essentially high speed steels with the compositions modified in such a way that excess carbides are eliminated from the microstructure. As a result, the strength available can be realized without the brittleness which characterizes high speed steels.

In essence, each of these high alloy grades is the matrix of a high speed steel, and for this reason the group is called Matrix steels by Vasco Metals Corp. Containing about 0.5% C and alloying additions of tungsten, molybdenum, chromium and vanadium, Matrix steels can be heat treated by conventional schedules to tensile strengths of 400,000 psi with sufficient toughness for selected structural applications.

MP FACT SHEET Ultrahigh Strength Steels

Producer and Trade Name	Type of Steel
Allegheny Ludlum Steel Corp.	
12W (type 418)	Martensitic stainless
AM350, 355 and 357	Semi-austenitic age hardenable stainless
A-L 18NiCoMo 250	18% Ni (250) maraging*
A-L 18NiCoMo 300	18% Ni (300) maraging
Armco Steel Corp.	
17-4 PH	Martensitic age hardenable stainless
PH 15-5 Mo	Martensitic age hardenable stainless
17-7 PH	Semi-austenitic age hardenable stainless
PH 15-7 Mo	Semi-austenitic age hardenable stainless
PH 14-8 Mo	Semi-austenitic age hardenable stainless
Bethlehem Steel Corp.	
300 M	Medium carbon, low alloy
Ladish D6A and D6AC	Medium carbon, low alloy
Maraging steels (18% Ni)	18% Ni maraging
Braeburn Alloy Steel Div.	
Ladish D6A and D6AC	Medium carbon, low alloy
Carpenter Steel Co.	
Ladish D6A and D6AC	Medium carbon, low alloy
Pyromet X-12	Martensitic stainless
Custom 455	Extra low carbon martensitic, age hardenable stainless
NiMark I	18% Ni (240) maraging
NiMark II	18% Ni (250) maraging
NiMark III	18% Ni (270) maraging
Crucible Steel Co.	
Hy-Tuf	Medium carbon, low alloy
UHS 260	Medium carbon, low alloy
Crucible 56	5Cr-Mo-V
Crucible 422	Martensitic stainless
AFC 77	Martensitic age hardenable stainless
Firth Sterling, Inc.	
FS Astro 1	
Maraging-250KSI	18% Ni maraging
Maraging-300KSI	18% Ni maraging
Isaacson Iron Works	
Circle I Special	
Jessop Steel Co.	
300 M	Medium carbon, low alloy
Air Steel X200	Medium carbon, low alloy
USS Strux (Mod 98VB40)	Medium carbon, low alloy
HS 220 and HS 260	Medium carbon, low alloy
Ladish Co.	
Ladish D6A, D6AC and D9	Medium carbon, low alloy
Latrobe Steel Co.	
Ladish D6A and D6AC	Medium carbon, low alloy
Dynaflex	5Cr-Mo-V
Lescalloy 5616	Martensitic stainless
Type 422 stainless	Martensitic stainless
Linco (Lapelloy)	Martensitic stainless
Marvac 18LT	18% Ni (240) maraging
Marvac 18	18% Ni (260) maraging
Marvac 18A	18% Ni (280) maraging
Midvale-Heppenstall Co.	
Hy-Tuf	Medium carbon, low alloy
300 M	Medium carbon, low alloy
Ladish D6A, D6AC and D9	Medium carbon, low alloy
HS 220	Medium carbon, low alloy
Midvac 18	18% Ni maraging
Republic Steel Corp.	
HP9-4-25	High alloy steel (200 ksi YS)
HP9-4-45	High alloy steel (260 ksi YS)
Maraging steels (18% Ni)	18% Ni maraging
Standard Steel Div., Baldwin-Lima-Hamilton Corp.	
Ladish D6A and D6AC	Medium carbon, low alloy
Timken Roller Bearing Co.	
Ladish D6A and D6AC	Medium carbon, low alloy
HS 220 and HS 260	Medium carbon, low alloy
United States Steel Corp.	
Air Steel X200	Medium carbon, low alloy
300 M	Medium carbon, low alloy
USS Strux	Medium carbon, low alloy
12MoV	Martensitic stainless
Stainless W	Martensitic age hardenable stainless
USS Tenelon	Cold rolled austenitic stainless
USS 17-5 MnV	Cold rolled austenitic stainless
Maraging steels (18% Ni)	18% Ni maraging
Universal-Cyclops Specialty Steel Div., Cyclops Corp.	
Unimach I	5Cr-Mo-V
Unimach II	5Cr-Mo-V
Unitemp D6AC	Medium carbon, low alloy
Unimar 18 (250)	18% Ni (250) maraging
Unimar 18 (300)	18% Ni (300) maraging
Unimar 20 (250)	20% Ni (250) maraging
Unimar 25 (278)	25% Ni (278) maraging
Vasco Metals Corp.	
Ladish D6AC	Medium carbon, low alloy
Vascojet 1000	5Cr-Mo-V
VascoMax 200	18% Ni (200) maraging
VascoMax 250	18% Ni (250) maraging
VascoMax 300	18% Ni (300) maraging
Vasco M-A	Medium carbon, high alloy
Vasco Matrix II	Medium carbon, high alloy

*Numbers in parentheses indicate strength level in ksi for maraging steel.

Table I highlights the farthest advances, from the standpoint of smooth bar strength, that have been made in applications of ultrahigh strength steels to structural service. In the table, the steels are treated by family, and the data presented are necessarily rather approximate, as the intent is to be indicative rather than definitive.

The forms in which the low alloy and 5Cr-Mo-V steels are generally used are sheet, strip, bar stock and forgings. The martensitic stainless and the age hardenable varieties of this type of steel are made in the form of sheet, strip, small forgings, bar stock and small castings. Cold rolled stainless steels are confined to sheet, strip and foil. The semi-austenitic stainless steels are generally employed in the form of sheet and bar stock, and the high alloy steels find application as forging, bar stock, sheet and plate.

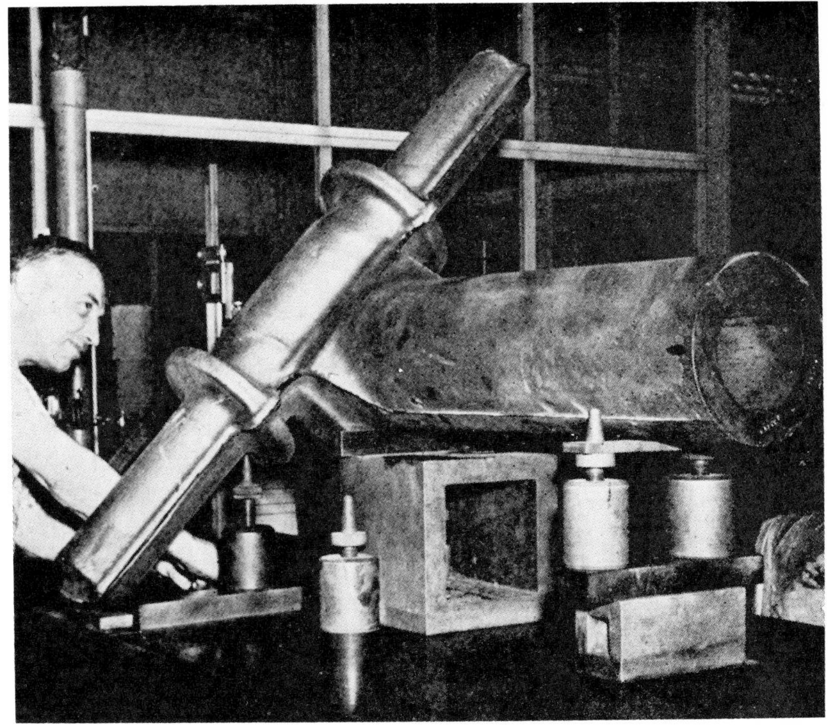

Fig. 1—Technician at Wyman-Gordon Co., Worcester, Mass., checks the dimensions of a forged 4340 landing gear trunnion.

Table II — Applications for Ultrahigh Strength Steels

Characteristics	Applications
Low Alloy Steel	
Well known, moderate cost, high hardenability	Solid propellant rocket motor cases; gun tubes and breech blocks; bolts, pins, fittings and structural components for aircraft; arresting hooks for naval aircraft; axles, gears and shafts; gas storage bottles
5Cr-Mo-V	
Air hardening, which minimizes build-up of residual stress in heat treatment	Actuating cylinders; arresting gear; aircraft engine mounts and landing gear components; shafting, bolts, pins, springs; pump components; fuselage frames, longerons, cargo lug supports and other structural components of aircraft
Martensitic Stainless and Age Hardenable Martensitic Stainlesses	
Moderate corrosion resistance; can be hardened after fabrication by a single heat treatment at moderate temperature	Brackets, clamps, high shear rivets and bolts for aircraft; forgings and sheet metal components; engine mounts; parts for pumps and valves; components of papermaking machinery
Cold Rolled Austenitic Stainlesses	
Moderately resistant to general corrosion and stress corrosion cracking; available only as sheet, strip and foil; limited capability for fabrication and joining	Skin for liquid fueled missiles; tanks for fuel and oxidizers; equipment for hauling food and chemicals
Semi-Austenitic Stainlesses	
Corrosion resistant and readily fabricated	Stiffeners, interior frames, bulkheads and longerons for aircraft; sandwich and honeycomb structures, skins, tanks and ducts for aircraft; boat shafts; nuclear reactor components; skins for spacecraft; compressor disks
18% Ni Maraging Steel	
Can be hardened after fabrication by a single-step heat treatment which causes virtually no distortion or dimensional change; weldable	Solid propellant rocket motor cases; forging and extrusion dies; flexible shafts; aircraft landing gear components; die holders, dummy blocks and extrusion rams
Matrix Steels	
Extraordinary strength available in a quenched and tempered steel; deep hardening	Fasteners, armor, springs, punches and dies

Uses for Ultrahigh Strength Steels

It is axiomatic that applications for ultrahigh strength steels should emphasize extremely high strength to density ratios as a requirement. In many instances, such as airframes and rocket motor cases, vitally important weight savings are brought about by specifying ultrahigh strength steels. In others, ultrahigh strength steels reduce the bulk of the component so that it will fit inside an envelope of fixed size.

Another extremely important reason for selecting steel is cost. Steels have lively competition in mechanical and other properties from numerous materials. However, where the disparity is not vital, steel often wins on the basis of cost. Mill products can be less expensive and the cost of fabrication and construction may be less. As an example, costs are now figuring strongly in the choice of steels for large rocket boosters.

Another advantage in choosing ultrahigh strength steels for "moving structures" is the reduced inertia which accompanies the resulting decrease in weight. This is an important factor in rotating or reciprocating mechanisms such as pumps, engines, actuators and valves. Finally, in components of servomechanisms, the high hardness which accompanies high strength promotes resistance to wear.

Typical applications for the various types of ultrahigh strength steel are listed in Table II. Military and space applications tend to dominate usage patterns. This is natural because such applications provided the impetus for the development of ultrastrong materials. However, in the course of time, it is normal for applications to extend into the civilian economy; such a trend is, in fact, now taking place.

Limitations of the "Ultra" Steels

There are a host of steels capable of attaining ultrahigh strengths which can be considered commercial materials in every sense of the word. Procedures have been worked out to process them in the mill and to fabricate them in the shop and in the field. These steels are in use today at strength levels from 180,000 to 300,000 psi.

On the other hand, it has been demonstrated repeatedly that ultrahigh strength steels are capable of considerably higher strength than the maximum levels at which they are now employed. For example, even by conventional methods of treatment, these steels can develop tensile strengths over 300,000 psi and up to 400,000 psi.

Why the discrepancy between current capability and actual practice? One factor is that the mechanical properties of a material cannot be defined by a series of single values but must be described in terms of scatter bands. The lower portions of these bands are selected in establishing performance specifications. Strength values may therefore be adopted which are considerably lower than the obtainable maximums.

Another limiting factor is notch toughness – ability of a material to sustain a load successfully in the presence of a sharp stress raising notch or defect. The relationship between smooth bar strength and notch strength for iron alloys is such that, while notch strengthening tends to occur at the lower strength levels, at higher strengths the tendency is for notch toughness to decrease with increasing smooth bar strength.

Other factors also limit the strength level at which material can be used with confidence. One of these is the fact that materials and structures as yet fall short of perfection, and hence they tend to contain stress raising defects. The other factor is that current nondestructive inspection techniques have limitations with respect to ability to detect and identify small defects.

Thus it happens that a structure may contain defects too large to be tolerated but too small to be detected so that they can be removed. The only way out of this dilemma is to reduce the strength level at which the steel is used and in this way increase the allowable defect size to the point of detectability.

Resistance to fatigue is another limitation. Endurance limits are seldom more than half the tensile strength at best. Moreover, the ratio of endurance limit to tensile strength decreases markedly as tensile strength rises above 200,000 psi. When the corrosion factor is added to that of cyclic stressing, endurance limits disappear altogether.

Susceptibility to corrosive attack must certainly be considered. High strength steels, including the stainlesses, are subject to attack in numerous commonly encountered media. The stainless steels, of course, are corroded in fewer environments, but they are quite susceptibile to stress corrosion cracking. The nonstainless steels also are subject to stress corrosion cracking in a variety of environments.

Another factor is hydrogen stress cracking. Hydrogen may be introduced into steel during melting, mill processing, electroplating, corrosion or in connection with galvanic protection. Hydrogen stress cracking is a problem to be reckoned with when yield strengths exceed 70,000 psi.

Finally, attention must be called to the fact that joining is a problem with ultrahigh strength steels. Popular joining techniques are brazing and welding. Both invite dimensional changes, distortion, embrittlement and the accumulation of unwanted internal stresses. In addition, there is the ever-present problem of making a joint which realizes the strength and fracture toughness of the parent metal.

This list of limitations might be completely discouraging were it not for three factors. The first is the insistent demand for constructional materials with increased capabilities. Another is that when other materials are employed in high performance applications, numerous problems also arise. Frequently, the problems plaguing other materials are more severe than those encountered with high strength steels. The third factor is that approaches already have been discovered, and are under extensive investigation, for adding another sizable increment to the performance capabilities of steels.

Toward Greater Performance

Substantial increases in smooth bar strength can be attained through thermal - mechanical treatments. These are characterized by the introduction of strain into the structure at some point or points in a cycle which also includes heat treating. A number of treatments of this type, shown in Table III, are being explored.

One of the first thermal-mechanical treatments was simultaneous deformation and transformation from austenite to martensite. The rolling of 17Cr-7Ni stainless steel at room or at very low temperatures (cryogenic) illustrates this type of treatment. Strength levels obtainable in semi-austenitic stainless steels can be increased dramatically by deforming at these very low temperatures followed by aging.

Another type of thermal-mechanical treatment is to deform austenite at comparatively low temperatures where it is metastable, before transformation to martensite or to bainite, and follow this by tempering.

Table III — Thermal-Mechanical Treatments for Ultrahigh Strength Steels

1. Deformation during transformation to martensite at room temperature or at very low temperatures (aging or tempering after deformation)
2. Deformation of martensite followed by tempering
   Deformation at room temperature
   Deformation at tempering temperature
3. Deformation ot tempered martensite followed by retempering
4. Deformation of metastable austenite before transformation to martensite or to bainite
5. Combination treatments

The greatest strength increments can be obtained by combination thermal-mechanical treatments. For example, by deforming a 0.66C-5Ni steel in the temperature range of metastable austenite, quenching to martensite, tempering, straining the tempered martensite and retempering, a yield strength of 520,000 psi and a tensile strength of 526,000 psi can be obtained.

Although development of extremely high smooth bar strength has been the primary reason for the interest in thermal-mechanical processing, there are indications that several of these treatments also can improve the fracture toughness versus yield strength relationship. These treatments can increase toughness at constant strength or maintain toughness while increasing the strength.

Thermal-mechanical processes are as yet being applied commercially to only a limited extent because a number of serious problems can arise. To achieve reproducibility of properties requires close control of the degree of straining and the time and temperature range through which straining is carried out. To achieve the desired amount of deformation throughout a complex shape is a problem, as is anisotropy of mechanical properties and low toughness in the transverse direction.

In addition, power requirements tend to be high because the steel usually exhibits great strength at the deformation temperature. Again, welding such materials is difficult without suffering considerable loss of strength in the area of the joint. At present, only relatively simple shapes and parts which need little final finishing are being fabricated.

There are many other avenues toward improved performance. One of them is steel composition. The principal uses to which alloying elements have been put in the past have been to improve corrosion resistance or increase hardenability. It is now recognized that alloying elements influence fracture toughness. In quenched and tempered steels, there is evidence to suggest that additions of vanadium and nickel can improve toughness. And manganese and silicon seem to be an effective combination promoting toughness.

On the other hand, there are indications that chromium reduces the toughness of quenched and tempered steels. In addition, it is clear that the toughness of this type of steel diminishes as the carbon content is increased.

The development of maraging steels, 9Ni-4Co steels and Matrix steels is a dramatic illustration of the potentialities residing in the realm of alloying. And even now the International Nickel Co. and United States Steel Corp. are working to develop a maraging steel with greater toughness than those already available.

Unwanted Elements in Steel

Another important aspect of steel composition relates to the unintended elements present in the steel. It is clear that this group – sulfur, phosphorus, oxygen, nitrogen and hydrogen – detract from ductility and toughness. Means of making steels containing only small amounts of these elements have come about through the development of advanced melting methods – vacuum induction and consumable electrode, vacuum arc remelting – in conjunction with increased availability of high quality charge materials.

Several avenues which may lead to materials of improved performance combine considerations of composition with other factors. For example, by proper alloying and adjustment of carbon content, tough high strength bainites can be formed on continuous cooling from the austenite temperature region. In the same manner, the M_f temperature may be raised so as to render the steel self-tempering and correspondingly tough on cooling from austenitizing temperature. Developments of these kinds offer the additional opportunity of welding without preheat or postheat. Inland Steel Co. has taken a step in this general direction with the development of low carbon, alloy-free steels which can be "martensitized" in thin gages and are self-tempering.

Steps already taken to reduce the number and size of nonmetallic inclusions as well as the number and size of defects in mill products are encouraging and point to this as another avenue toward improved performance. These steps start with charge materials and go on through melting, ingot practices, breakdown, hot and cold rolling, annealing and indeed, almost all aspects of mill processing.

Improved charge materials and melting methods have already been mentioned. Others include the use of glassy slags to coat and protect ingot surfaces, the development of exothermic hot-top practices, removal of surface defects at intermediate stages in processing, and grinding and polishing the finished product. The importance of efforts along these lines can be stated this way: The smaller the defect, the less fracture toughness required in the material to insure reliable performance at a given strength level, or the higher the reliably usable strength for a given level of toughness.

The Role of Fabrication

Discussion of measures to improve the quality of mill products leads to a consideration of the contribution made by fabricating procedures to the number and severity of defects present in the finished structure. Great strides have been made here also. Discontinuities such as scratches, nicks, gouges, punch marks, undercut welds, mismatched joints and grinding marks may initiate a failure as readily, if not more readily, than a mill defect in the material. Likewise, engineers know the dangers lurking in sharp fillets, slots, small holes and other abrupt changes in section.

It appears that the most serious problem impeding further efforts to reduce the number and severity of stress raising defects in materials and structures resides in the shortcomings of available nondestructive inspection methods. Techniques for locating and identifying internal defects seem to have more limitations than those available to inspect surfaces. Opportunities for further advances appear to be in the direction of increasing the positiveness with which the equipment indicates the presence of a defect, decreasing the minimum size of identifiable defects, improving the capability of finding nonlaminar internal defects and defects not parallel to the surface, and developing means of identifying the type of internal defect located.

The Corrosion Problem

Most of the foregoing discussion has been concerned with means for securing higher strength and toughness. Improvement is also necessary in resistance to general corrosion, stress corrosion cracking, corrosion fatigue and hydrogen stress cracking. The extent of the advances in these areas which can be expected from such approaches as composition modifications, alloy additions or special heat treating or mill processing procedures would seem to be somewhat limited. On the other hand, progress is being made toward an improved understanding of stress corrosion cracking in stainless steels, and the insight gained should lead to the development of stainless steels more resistant to this phenomenon. Likewise, hydrogen stress cracking is becoming better understood.

Substantial opportunities may lie in the field of coatings capable of shielding the metal from a hostile environment. A great deal of progress has been made in the technology of coatings through the years. Investigation of organic coatings has been particularly active in recent years. Such an abundance of compounds has become available that tailoring a coating formulation to a particular application is now considered standard practice.

Advances have also been made in coating methods. Recent processes make use of the fluidized bed, electrostatics, electrophoresis and glow discharge. Much remains to be done and the research effort in this field is considerable.

Requirements to be met by coatings are rigorous: They include freedom from "holidays" and breakthroughs, good ductility, adherence, reliability over wide temperature ranges, inertness, impermeability and ease of application and inspection. However, gains in this field are rewarding. And the very idea of avoiding corrosion and other environmentally influenced deterioration processes by creating an impenetrable barrier between the environment and the metal is a powerful incentive to further efforts and advances because of the inherent simplicity of the concept.

The Ultrahigh-Strength Steels-II

By A. M. HALL

THE TECHNOLOGY OF ultrahigh-strength steels continues to be one of the fast-growing fields of metallurgy, thanks to the steadily increasing demand for these materials and the constant inflow of new knowledge and insights into their behavior. Because of their high strength-to-density ratios, these steels enable users to save much weight in a host of components and structures. Along with these weight reductions come opportunities to increase loads, reduce bulk, decrease inertia in moving parts and structures, and lower the costs for shipping components and erecting large structures. The steels have great wear resistance and extremely high compressive strength, making them useful for tooling and other special applications.

When an ultrahigh-strength steel competes with other materials for a given application in terms of properties, the former is often favored because it is less expensive. In addition, the required mill products may be cheaper and more readily available, and the cost of fabrication and construction can also be lower. Table I lists producers, trade names and types of ultrahigh strength steels produced in the United States today.

Five New Grades

Within recent years, two steels have been added to Republic Steel Corporation's family of 9Ni-4Co steels, bringing the total to four. Nominal compositions of the newcomers, HP9-4-30 and HP9-4-20, are shown in Table II. HP9-4-30 was designed for use at 220,000 to 240,000 psi tensile strength, a range between the 180,000 to 210,000 psi level of HP9-4-25 and the 260,000 to 280,000 psi range of HP9-4-45. The steel also retains strength and toughness after long-time exposure up to 800 F. To develop optimum strength and toughness, normalize at 1,700 F, oil quench from 1,550 F, refrigerate at −100 F, and double temper at 1,000 F.

The second steel, HP9-4-20, was designed for applications requiring comparatively high strength, superior toughness, and good weldability. According to reports, the alloy develops 180,000 psi yield strength with a room-temperature Charpy V-notch value of 50 ft-lb in plate up to 4 in. thick. Actually, the alloy can be used at the same strength level as HP9-4-25, but it has increased weldability and toughness. Recommended heat treatment: normalize at 1,650 F, water quench from 1,500 F, and temper at least 4 hr at 1,025 F.

Compositions of three more maraging steels are given in Table II. One, capable of developing 350,000 psi yield strength, carries the designation 18 Ni (350). Produced in such forms as bar, sheet, and wire, it handles applications requiring high strength and hardness. Examples include fasteners, gears, rotors, flexures, torsion bars, forging dies and inserts, punches, pins, and extrusion tooling. Like other maraging steels, this alloy is annealed at 1,400 F, and aged at 900 F.

The other two, developed by Allegheny Ludlum Steel Corp. and International Nickel Co., are stainless types. Carrying the trade names of Almar 362 (AL) and Marvac 736 (Latrobe Steel Co.) each contains a substantial amount of chromium to give

Mr. Hall is assistant manager, Process & Physical Metallurgy, Battelle Memorial Institute, Columbus Laboratories, Columbus, Ohio.

Table I — Ultrahigh-Strength Steels

Producer and Trade Name*	Type
Allegheny Ludlum Steel Corp.	
D6AC	Medium carbon, low alloy
300M	Medium carbon, low alloy
Potomac A (H11)	5Cr-Mo-V
418 Special	Martensitic stainless
AM350 and AM355	Semiaustenitic age-hardenable stainless
AL-630	17Cr-4Ni age-hardenable stainless
AL-631	17Cr-7Ni age-hardenable stainless
Almar 18 (200)	Maraging
Almar 18 (250)	Maraging
Almar 18 (300)	Maraging
Almar 362	Maraging stainless
Armco Steel Corp.	
17-4PH	Martensitic age-hardenable stainless
15-5PH	Martensitic age-hardenable stainless
PH13-8Mo	Martensitic age-hardenable stainless
PH12-9Mo	Martensitic age-hardenable stainless
17-7PH	Semiaustenitic age-hardenable stainless
PH15-7Mo	Semiaustenitic age-hardenable stainless
PH14-8Mo	Semiaustenitic age-hardenable stainless
Bethlehem Steel Corp.	
300M	Medium carbon, low alloy
D6A and D6AC	Medium carbon, low alloy
Braeburn Alloy Steel Div.	
Braevac H11	5Cr-Mo-V
Cameron Iron Works Inc.	
300M	Medium carbon, low alloy
D6AC	Medium carbon, low alloy
H11	5Cr-Mo-V
18% Marage	Maraging
Carpenter Technology Corp.	
D6A and D6AC	Medium carbon, low alloy
Pyromet X-15	Martensitic stainless
Custom 455	Extra-low-carbon martensitic, age-hardenable stainless
Custom 630	Extra-low-carbon martensitic, age-hardenable stainless
NiMark I	Maraging
NiMark II	Maraging
NiMark 300	Maraging
Consumet 350	Semiaustenitic age-hardenable stainless
Consumet 355	Semiaustenitic age-hardenable stainless
Crucible Steel Corp.	
Hy-Tuf	Medium carbon, low alloy
300M	Medium carbon, low alloy
D6AC	Medium carbon, low alloy
Crucible 218	5Cr-Mo-V
Crucible 422	Martensitic stainless
AFC-77	Martensitic age-hardenable stainless
AFC-260	Semiaustenitic age-hardenable stainless
A. Finkl & Sons Co.	
D6A, D8A, and D9	Medium carbon, low alloy
Green River Steel Corp., Subs. Jessop Steel Co.	
300M	Medium carbon, low alloy
Air Steel X200	Medium carbon, low alloy
USS Strux (98VB40 Mod)	Medium carbon, low alloy
USS Strux High Carbon	Medium carbon, low alloy
D6A	Medium carbon, low alloy
HS 220 and HS 260	Medium carbon, low alloy
Hy-BB	Medium carbon, low alloy
Latrobe Steel Co.	
Lescalloy 300M	Medium carbon, low alloy
Lescalloy D6AC	Medium carbon, low alloy
Lescalloy UT-18	Medium carbon, low alloy
Lescalloy UT-19	Low carbon, low alloy
Jethete M-152	Low carbon, high alloy
Dynaflex	5Cr-Mo-V
Lescalloy HY 180	10Ni, maraging
Lescalloy HP9-4-20	9Ni-4Co, medium carbon
Lescalloy HP9-4-25	9Ni-4Co, medium carbon
Lescalloy HP9-4-30	9Ni-4Co, medium carbon
Lescalloy HP9-4-45	9Ni-4Co, medium carbon
Marvac 12-5-3	Maraging
Marvac 200	Maraging
Marvac 250	Maraging
Marvac 300	Maraging
Marvac 350	Maraging
Marvac 736	Maraging stainless
Midvale-Heppenstall Co.	
Hy-Tuf	Medium carbon, low alloy
300M	Medium carbon, low alloy
D6A, D6AC, and D9	Medium carbon, low alloy
HS220	Medium carbon, low alloy
Republic Steel Corp.	
D6A, D6AC, and D9	Medium carbon, low alloy
Hy-Tuf	Medium carbon, low alloy
300M	Medium carbon, low alloy
98BV40	Medium carbon, low alloy
H11	5Cr-Mo-V
12-5-3	Maraging
18% Ni	Maraging
HP9-4-20	9Ni-4Co, medium carbon
HP9-4-25	9Ni-4Co, medium carbon
HP9-4-30	9Ni-4Co, medium carbon
HP9-4-45	9Ni-4Co, medium carbon
17-4PH	Martensitic age-hardenable stainless
15-5PH	Martensitic age-hardenable stainless
PH13-8Mo	Martensitic age-hardenable stainless
Standard Steel Div., Baldwin-Lima-Hamilton Corp.	
D6A and D6AC	Medium carbon, low alloy
300M	Medium carbon, low alloy
HP9-4-25	9Ni-4Co, medium carbon
HP9-4-45	9Ni-4Co, medium carbon
MA18-200	Maraging
MA18-250	Maraging
MA18-275	Maraging
15-5PH	Martensitic age-hardenable stainless
17-4PH	Martensitic age-hardenable stainless
Timken Roller Bearing Co.	
D6A and D6AC	Medium carbon, low alloy
HS220 and HS260	Medium carbon, low alloy
300M	Medium carbon, low alloy
U. S. Steel Corp.	
Air Steel X200	Medium carbon, low alloy
300M	Medium carbon, low alloy
USS Strux	Medium carbon, low alloy
18Ni (200)	Maraging
18Ni (250)	Maraging
18Ni (300)	Maraging
12MoV	Martensitic stainless
Stainless W	Martensitic age-hardenable stainless
USS Tenelon	Cold-rolled austenitic stainless
Universal-Cyclops Specialty Steel Div., Cyclops Corp.	
Unimach No. 1	5Cr-Mo-V
Unitemp D6AC	Medium carbon, low alloy
Unimar 250	Maraging
Unimar 300	Maraging
Unimar 350	Maraging
Unimar CR-1	Stainless maraging
Unitemp 350 and 355	Semiaustenitic age-hardenable stainless
Vasco, a Teledyne company	
D6AC	Medium carbon, low alloy
VascoJet 1000	5Cr-Mo-V
VascoMax 200 CVM	Maraging
VascoMax 250 CVM	Maraging
VascoMax 300 CVM	Maraging
VascoMax 350 CVM	Maraging
Vasco M-A	Medium carbon, high alloy
Vasco Matrix II	Medium carbon, high alloy
Vulcan Steel Corp.	
Astralloy	Medium carbon, low alloy

* Standard AISI grades such as 4130, 4140, 4340, and 8740 are also classed as ultrahigh-strength steels. These and their AMS counterparts are supplied by many producers.

oxidation resistance and stainless qualities. Allegheny's alloy is annealed at 1,500 F and aged at 900 to 1,000 F to develop a yield strength in the 140,000 to 170,000 psi range. Available in billets, bar stock, special shapes, and wire, the steel is used in a variety of applications, particularly in forged and machined parts. The Inco alloy is annealed at 1,700 F for 1 hr, air cooled, annealed at 1,400 F for 1 hr, and aged at 900 F for 3 hr.

Reliability Is Needed

Factors that determine the reliability with which ultrahigh-strength steels perform continue to be emphasized in research and development programs. Prominent among such factors is fracture toughness. An important approach toward increasing the fracture toughness is through judicious use of alloying elements, and the employment of minimum amounts of carbon commensurate with the strengthening mechanism and the strength level desired. Another approach is through drastic reduction or elimination of detrimental elements (including sulfur, phosphorus, hydrogen, oxygen, and nitrogen) by vacuum-induction remelting, or consumable-electrode vacuum-arc remelting. (Charge materials are, of course, of very high purity.) In fact, it is not uncommon to employ double vacuum-induction melting.

Another factor affecting reliability is the degree of reproducibility of mechanical properties from lot-to-lot and heat-to-heat of a given steel. Here, the goal is to narrow the scatter band for each property by elevating the curve delineating the band's lower limit. To do this, of course, producers need to avoid producing material possessing low values for the property in question. One approach is to melt to very narrow composition ranges. Then, heats with out-of-specification compositions and properties will not be made.

Another approach is to improve the chemical homogeneity of the material. In this way, phenomena that promote low ductility and poor fracture toughness (such as freckles and banding) are less likely to occur. Likewise, property variations from ingot to ingot, as well as from the bottom to the top of individual ingots, are minimized. Chemical homogeneity can be influenced by ingot-making practices, ingot breakdown procedures, homogenizing heat treatments and the steps used to convert ingots to mill products.

Workman assembles hinge system of hatch cover for cargo ship. Because hinge parts require both high strength and corrosion resistance, they are made of Almar 362, a maraging stainless steel.

Finally, steps are being taken to produce mill products with fewer surface and internal stress raisers that detract from load carrying ability. In this category are such items as nonmetallic inclusions, porosity, delaminations, splits, cracks, seams, scabs, and laps.

The composition, size, shape, and number of nonmetallic inclusions are influenced by the quality of the charge materials, the melting and deoxidation practices, the ladle and ingot making practices, and the ingot breakdown procedures. The presence of surface defects is strongly influenced by the procedures used to condition ingots and intermediate mill products such as blooms, slabs, and billets. Equally important are in-process and final inspection practices and equipment.

Table II — Nominal Compositions of New Ultrahigh-Strength Steels

Name	C	Mn	Si	P	S	Ni	Cr	Mo	Co	Other
HP9-4-30	0.30	0.20	0.10*	0.01*	0.01*	7.50	1.00	1.00	4.50	0.10 V
HP9-4-20	0.20	0.30	0.10*	0.01*	0.01*	9.00	0.75	1.00	4.50	0.10 V
18 Ni (350)	0.01*	0.10*	0.10*	0.005*	0.005*	17.50	—	3.70	12.50	1.80 Ti, 0.15 Al
Almar 362	0.03	0.30	0.20	0.015	0.015	6.50	14.50	—	—	0.80 Ti
Marvac 736	0.02*	0.15*	0.15*	0.010*	0.010*	9.50	10.25	2.00	—	0.25 Ti, 0.30 Al

* Maximum.

Typical Compositions, Characteristics, and Uses of Representative Ultrahigh-Strength Steels

Designation *	Composition, %† C	Mn	Si	Cr	Ni	Mo	Other	Characteristics	Uses
Low Alloy									
AISI 4130	0.30	0.50	0.30	0.95	—	0.20	—	Moderate cost; high hardenability; fabricable and weldable; yield strengths to 260,000 psi with adequate toughness.	Solid-propellant rocket-motor cases; gun tubes and breech blocks; bolts, pins, fittings, and structural components for aircraft; arresting hooks for naval aircraft; axles, gears, and shafts; gas storage botles.
AISI 4140	0.40	0.90	0.30	0.95	—	0.20	—		
AISI 4340	0.40	0.70	0.30	0.80	1.80	0.25	—		
AMS 6434	0.35	0.70	0.30	0.80	1.80	0.35	0.20 V		
Ladish D6AC	0.45	0.75	0.25	1.05	0.55	1.00	0.07 V		
300M	0.45	0.80	1.60	0.85	1.80	0.40	0.05 V min		
Medium Alloy									
5CrMoV	0.40	0.30	1.00	5.00	—	1.30	0.50 V	**5Cr types:** Air-hardening, which minimizes buildup of residual stress in heat treatment; strength retained to 1,000 F; high tempering temperature, which provides substantial stress relief; maximum yield strength, 245,000 psi, with about 5% elongation. **HY130/150:** Strong, weldable tough plate steel; up to 150,000 psi yield strength.	**5Cr types:** Actuating cylinders; arresting gear; aircraft engine mounts and landing-gear components; shafting, bolts, pins, springs, pump components; fuselage frames, longerons; cargo lug supports, and other structural components of aircraft. **HY130/150:** Hulls for deep submersibles; nuclear reactor components; large turbine discs; pressure vessels; thick-walled cylinders.
AISI H11	0.35	0.30	1.00	5.20	—	1.50	0.40 V		
AISI H13	0.35	0.30	1.00	5.20	—	1.50	1.00 V		
HY130/150	0.10	0.75	0.25	0.50	5.00	0.50	0.07 V, 0.02 Al		
High Alloy									
HP9-4-20	0.20	0.30	0.10 max	0.75	9.0	0.75	4.50 Co, 0.10 V	Good toughness; good weldability; yield strength of 180,000 psi for 9-4-20, 220,000 psi for 9-4-30.	Hulls for deep submersibles; rocket-motor cases; pressure vessels; armor.
HP9-4-30	0.30	0.20	0.10 max	1.00	7.5	1.00	4.50 Co, 0.10 V		
Maraging									
18Ni (200)‡	0.03 max	0.10 max	0.10 max	—	18.0	3.25	8.5 Co, 0.20 Ti, 0.10 Al	Can be hardened after fabrication by a single-step heat treatment that causes virtually no distortion or dimensional change; weldable; excellent fracture toughness; high strength in thick sections; high compressive strength.	Solid-propellant rocket motor cases; forging and extrusion dies; flexible shafts; aircraft landing-gear components; die holders, dummy blocks and extrusion rams; precision machine components.
18Ni (250)‡	0.03 max	0.10 max	0.10 max	—	18.0	4.90	8.0 Co, 0.40 Ti, 0.10 Al		
18Ni (300)‡	0.03 max	0.10 max	0.10 max	—	18.5	4.90	9.00 Co, 0.65 Ti, 0.10 Al		
18Ni (350)‡	0.01 max	0.10 max	0.10 max	—	17.5	3.75	12.5 Co, 1.80 Ti, 0.15 Al		
Martensitic Stainless									
AISI type 420	0.15 min	1.0 max	1.0 max	13.0	—	—	—	Moderate corrosion resistance; can reach up to 250,000 psi yield strength after fabrication by quenching and tempering at moderate temperature. Some grades, such as 17-4 PH and Custom 455, are further strengthened by precipitation hardening when tempered. Annealed stock is readily cut, formed, machined, and welded.	Brackets, clamps, high shear rivets, and bolts for aircraft; forgings and sheet metal components; engine mounts; parts for pumps and valves; components of papermaking machinery.
AISI type 431	0.20 max	1.0 max	1.0 max	16.0	2.0	—	—		
12MoV	0.25	0.50	0.50	12.0	0.5	1.0	0.30 V		
17-4 PH	0.07 max	1.0 max	1.0 max	16.5	4.0	—	4.0 Cu, 0.30 Cb		
PH 13-8 Mo	0.05 max	0.10 max	0.10 max	12.5	8.0	2.5	1.1 Al		
Pyromet X-15	0.03 max	0.10 max	0.10 max	15.0	—	2.9	20.0 Co		
Custom 455	0.05 max	0.50 max	0.50 max	12.0	8.5	0.50 max	2.0 Cu, 1.1 Ti, 0.3 Cb		
AFC 77	0.15	—	—	14.5	—	5.0	13.5 Co, 4.0 Cu, 0.50 V, 0.05 N		
Semi-Austenitic Stainless									
17-7 PH	0.09 max	1.0 max	1.0 max	17.0	7.0	—	1.0 Al	Corrosion resistant; readily fabricated as annealed; precipitation hardenable to the 260,000-280,000 psi yield strength range with slight ductility (2% elongation).	Stiffeners, interior frames, bulkheads, and longerons for aircraft; sandwich and honeycomb structures, skins, tanks, and ducts for aircraft; boat shafts; nuclear reactor components; spacecraft skins; compressor discs.
PH 15-7 Mo	0.09 max	1.0 max	1.0 max	15.0	7.0	2.5	1.0 Al		
PH 14-8 Mo	0.05 max	1.10 max	0.10 max	15.0	8.5	2.5	1.1 Al		
AM350	0.12 max	0.90	0.50 max	16.5	4.5	3.0	0.10 N		
AM355	0.15 max	0.95	0.50 max	15.5	4.5	3.0	0.09 N		
Austenitic Stainless									
AISI type 201	0.15 max	6.5	1.0 max	17.0	4.5	—	—	Moderately resistant to general-corrosion and stress-corrosion cracking; available only as sheet, strip, and foil; limited capability for fabrication and joining; designed for use as cold rolled; yield strengths to 265,000 psi.	Skin for liquid-fueled missiles; tanks for fuel and oxidizers; equipment for hauling food and chemicals.
AISI type 301	0.10	2.0 max	1.0 max	17.0	7.0	—	—		

*Tradenames are given except for AISI, AMS, and HY steels. †Unless indicated, all percentages are typical. ‡Numbers in parentheses indicate nominal yield strengths in ksi (1,000 psi) of respective grades, as maraged (cooled from austenitizing temperature and reheated to 900 F for 3 hr).

Source: Reprinted by permission of the American Society for Testing and Materials — STP 498, "Introduction to Today's Ultrahigh-Strength Structural Steels," by A. M. Hall, Copyright ASTM.

Properties and Applications of Maraging Steels

R. D. WELTZIN AND C. M. BERGER

THE 18 pct Ni-maraging steels, which are in the family of iron-base alloys, are strengthened by a process of martensitic transformation, followed by age or precipitation hardening. Precipitation hardenable stainless steels are also in this group.

Maraging steels work well in electro-mechanical components where ultra-high strength is required, along with good dimensional stability during heat treatment. Several desirable properties of maraging steels are:

1) Ultra-high strength at room temperature. 2) Simple heat treatment, which results in minimum distortion. 3) Superior fracture toughness compared to quenched and tempered steel of similar strength level (*i.e.*, AISI 4340). 4) Low carbon content, which precludes decarburization problems. 5) Section size is not a factor in the hardening process. 6) Easily fabricated. 7) Good weldability.

These factors indicate that maraging steels could be used in applications such as shafts, and substituted for long, thin, carburized or nitrided parts, and components subject to impact fatigue, such as print hammers or clutches. This study determines the impact-fatigue strength of the 300 and 350 grade maraging steels, and compares their wear resistance with the wear resistance of more commonly used materials.

MATERIALS

The alloys for this study were 18 pct Ni-maraging steels aged to nominal tensile strengths of 300 ksi and 350 ksi. Impact-fatigue loading performance of these steels was compared with impact-fatigue loading performance of AISI 8620 (case hardened), Ti-6 Al-4 V, and Ti-13 V-11 Cr-3 Al. The AISI 8620 is used extensively for fatigue and wear applications, while the titanium alloys have been considered for similar uses. In addition, an aluminum alloy was tested in impact fatigue.

Wear tests were directed at obtaining the comparative wear resistance of materials commonly used for shafts or carburized components.

EXPERIMENTAL PROCEDURES

Comparative wear resistances were determined using the apparatus shown in Fig. 1. In this test, a case-hardened steel disc is abraded against the specimen under test, without lubrication, at a fixed speed of 174 rpm. The load applied resulted in an initial Hertz contact stress of 20,000 psi. Wear rate of the sample was determined by calculating the volume of the wear impression as a function of time.

Impact-fatigue behavior was evaluated using a Wiedemann-Baldwin SF-01-U Universal Testing Machine made by Wiedemann Div., Warner and Swasey Co. A special fixture was designed for this machine to allow a dynamic cyclic impact load to be applied to the test specimen. The specimen geometry used for this testing (Fig. 2) was developed by G. Koves and R. D. Weltzin ("Impact Fatigue Testing of Titanium Alloys," Journal of Materials, ASTM. September 1968).

RESULTS

Impact-Fatigue Test Results

Impact-fatigue data were recorded as applied impact stress versus the number of cycles to failure—that is, in the standard S-N curve form. Fatigue curves for annealed 300 and 350 grades of maraging steel are presented in Fig. 3 and fatigue curves for maraged stock are presented in Fig. 4. Impact-fatigue limits for both grades are 70,000 psi for the maraged material and 60,000 psi for annealed materials.

The case hardened (0.014-in. case depth) AISI 8620 steel used in fatigue loading applications has an impact-fatigue limit of 59,000 psi. The 6 Al-4 V titanium alloy with special surface treatment to improve wear

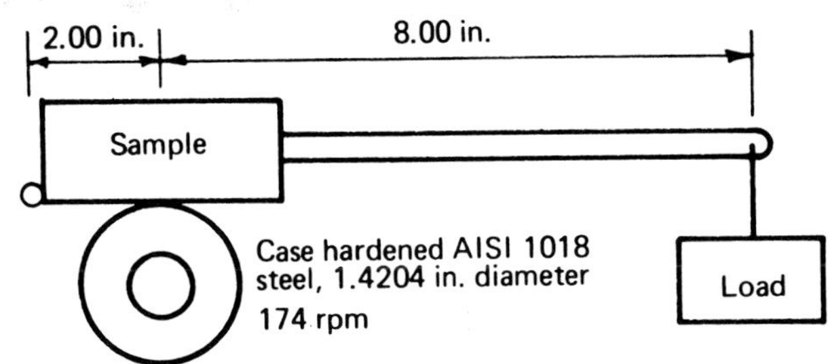

Fig. 1—Schematic of wear apparatus.

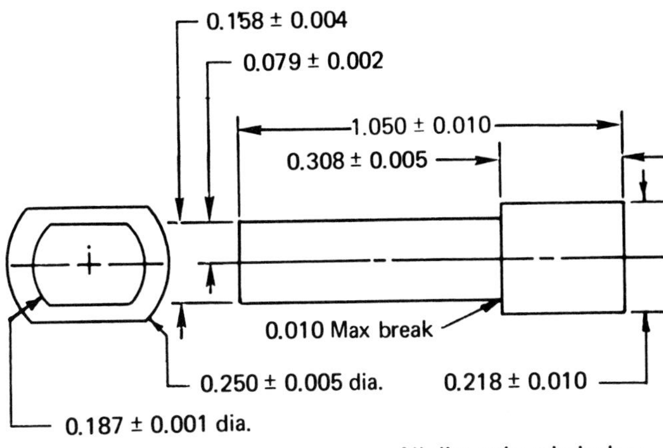

Fig. 2—Impact-fatigue specimen.

R. D. WELTZIN is an Advisory Engineer, Materials and Process Engineering, IBM General Systems Div., Rochester, Minn.; and C. M. BERGER is associated with Taussig Associates, Chicago, Illinois.

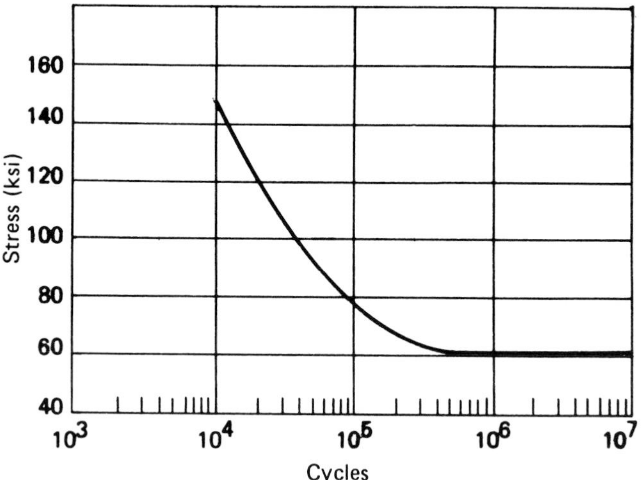

Fig. 3—Impact-fatigue S-N curve for annealed 300 and 350 maraging steel.

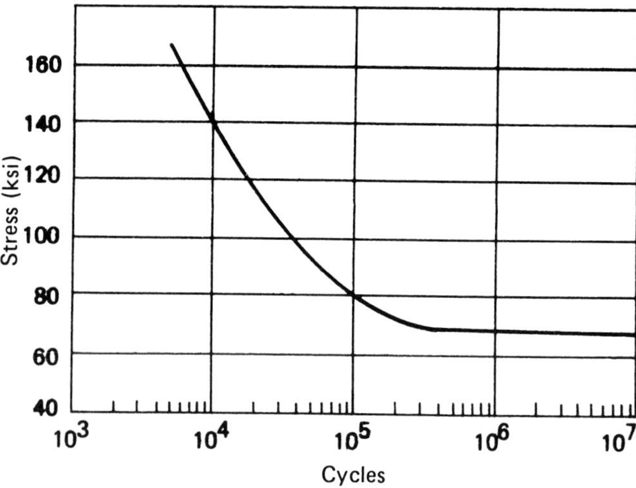

Fig. 4—Impact-fatigue S-N curve for maraged 300 and 350 steel.

resistance had a fatigue limit of 70,000 psi, while the aluminum alloy had a fatigue strength of 30,000 psi at 10^7 cycles.

Table I shows that maraged 18 pct Ni-maraging steel provides an impact-fatigue performance equivalent to any of the other materials tested. This indicates that the maraging steels could be used in applications where high static and dynamic strengths are required, and still hold close dimensional tolerances during fabrication.

Wear Test Results

Fig. 5 presents graphically the abrasion wear test results. This is a wear evaluation of metals often used in shaft applications. Interest in shaft applications arose partially from observation of wear due to sintered bronze bushings on 416 stainless steel shafts, as reported by R. G. Bayer and J. L. Sirico (in "Wear," Vol. 9, 1966, p. 236.). This wear information is also important if maraging steels are considered as replacements for case hardened components.

Analysis of Test Results

The impact-fatigue test results indicate that the maraging steels provide good resistance to fatigue failure caused by repeated impact loading. Table I shows that aged maraging steel possesses superior impact-fatigue strength and superior fracture toughness, when compared to the case-hardened AISI 8620. This mechanical superiority (when combined with dimensional stability of the maraging steels) demonstrates that applications with impact-fatigue loading or other fatigue types of loading, should be pursued.

In several data processing machine applications, the strength considered is the fatigue strength of the material. Certain applications involving high inertial forces (due to speed requirements) suggest that the fatigue strength/density be considered. In these instances, titanium alloys could have considerable application in data processing equipment. In other applications, however, the important factor would be fatigue strength, in which maraging steels stand out as excellent.

Table I. Mechanical Properties of Maraging Steel and Other Alloys

Alloy	Tensile Strength (psi)	Impact Strength (ft-lb)	Impact-Fatigue Strength (psi)
300 CVM annealed	160	14.0	60,000
300 CVM (900°F for 3 h)	280	12.0	70,000
VascoMax* 350 annealed	169	–	60,000
VascoMax 350 (900°F for 10 h)	340	11.0	70,000
8620 (0.014-in. case)	–	8.3	59,000
Ti-6Al-4V	170	8.8	70,000
Ti-13V-11Cr-3Al	190	6.5	52,000
2024 T4 Aluminum	70	4.5	30,000

*Trademark of Vasco, Latrobe, Pennsylvania.

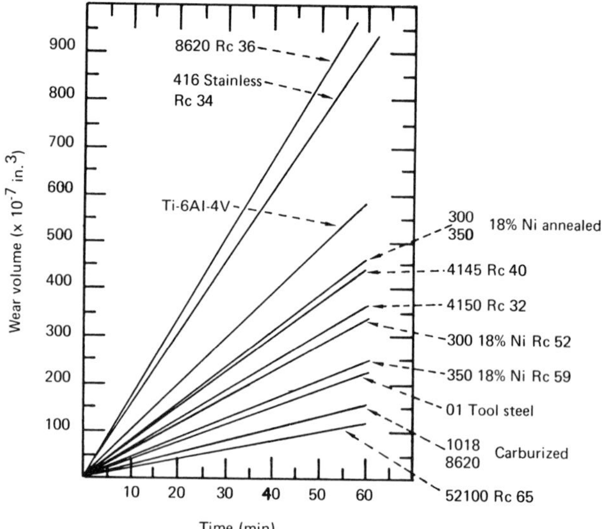

Fig. 5—Abrasion wear test results.

Table II compares several materials for fatigue strength/density and fatigue strength. (This information represents the impact-fatigue tests performed in this study.) Note that, on a fatigue strength/density basis, aluminum compares favorably. However, when considering the cross sectional area of aluminum needed to obtain the required fatigue life through re-

Table II. Impact-Fatigue Comparison of Fatigue Strength/Density and Fatigue Strength

Alloy	Tensile Strength, (psi)	Density (lb/in.^{3})	Fatigue Strength, (psi)	Fatigue Strength/ Density (X 10^4)	Tensile Strength/ Density (X 10^4)
2024 T-4 aluminum	68,000	0.10	30,000	30	68
Ti-13V-11Cr-3Al	190,000	0.16	52,000	32	120
8620 case hardened	150,000	0.28	59,000	21	57
Steel, 300 maraging annealed	160,000	0.28	60,000	21	57
Ti-6Al-4V	170,000	0.16	70,000	43	105
Steel, 300 maraging	280,000	0.28	70,000	25	99
Steel, 350 maraging annealed	170,000	0.28	60,000	21	61
Steel, 350 maraging	355,000	0.28	70,000	25	127

duced stress level, the use of aluminum for most fatigue applications in data processing machines would not be recommended, primarily because of the trend toward miniaturization of components. Because of this trend, the maraging steels and other high strength steels should find extensive applications in the future. Other applications might utilize the ultra-high strength of the maraging steels (up to 350 ksi at room temperature) or their superior fracture toughness.

The study was made of the maraging steels in wear situations because little information was available on their wear resistance compared to other metals. Wear resistance of these high strength alloys is questionable because the maraged structure is a carbonless martensite.

The wear resistance of the maraging steel at RC55 is not as good as that of AISI 52100 steel at the same hardness level. This should be expected because of the difference in microstructure, which results from the difference in carbon content. Some improvements in wear resistance of the maraging steel over that of a through-hardened alloy steel, such as AISI 4145, can be noted. This, along with the wear results on the 416 stainless steel, demonstrates that maraging steels might be beneficial for certain wear applications.

SUMMARY

The properties of maraging steels clearly indicate that these steels have many potential applications in mechanical components of electro-mechanical data processing machines. Use of these steels in shafts that require good dimensional control following heat treatment should be pursued for two reasons. First, maintaining dimensions should be easier because quenching and tempering are not necessary. Second, wear data indicate that equivalent or better wear resistance is obtained from the maraging steel than from the more commonly used shaft materials.

Impact-fatigue strength of 18 pct Ni-maraging steels indicates that these steels could be used in repeated impact loading situations. The good fracture toughness, compared to that of quenched and tempered alloy steels at the same strength level, indicates possible use in high-impact, low-cycle load applications.

Finally, due to the relatively low temperature of aging, the use of the maraging steels for long, thin parts should be considered. Here, their use as a replacement for some case-hardened or nitrided components is indicated, but the potential application should be carefully studied.

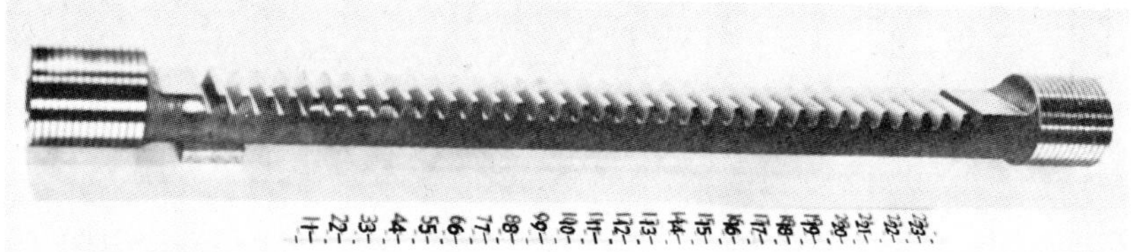

Because little size change occurs in VascoMax 350 when it is hardened, it is suitable for this gear rack. The previously used alloy, carburized AISI 9310, warped too much when heat treated.

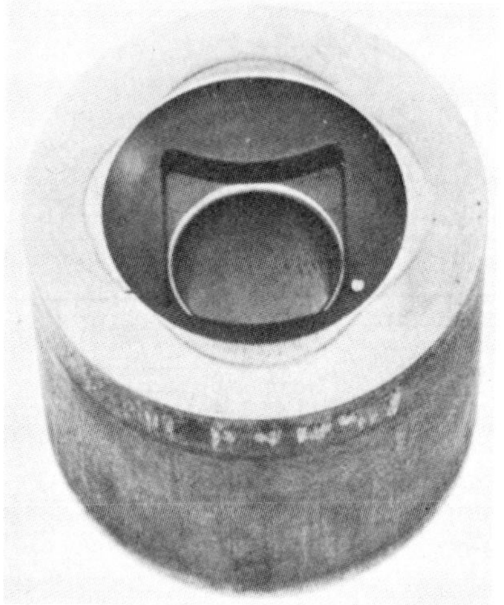

Cold-heading dies of VascoMax 350 endured 8,000 shots; similar dies of AISI D2 failed after 3,000 shots.

Variable wings of the F-111 are positioned by cam followers made of VascoMax 350.

An Extra-Strong Maraging Steel

By RICHARD J. HENRY and ROBERT A. CARY

Devised by balancing the strengthening effects of alloying elements, VascoMax 350 attains a tensile strength of 350,000 psi when solution treated, air cooled, and aged between 900 and 1,000 F. This strength, plus adequate toughness and ductility, makes the alloy useful for such things as dies and extrusion rams.

A MARAGING STEEL with 350,000 psi tensile strength has broad utility. VascoMax 350 CVM, which is based on the 18% Ni system, has a high yield strength-to-weight ratio (Fig. 1) plus good ductility, toughness, and hardness.

The applications in the accompanying illustrations – a cold heading die, cam followers for the swept-wing structure of the F-111, and a gear rack – are typical. Promising hot-work applications include core pins in die-casting dies, forging dies, and extrusion rams.

Welds made by the gas tungsten-arc method develop efficiencies of 90% after aging 3 hr at 950 F without preheat or postheat. Re-solution annealing and aging make it possible to obtain full strength of the base metal in welds.

The Alloy's Characteristics

Metallurgists at our laboratory devised this alloy by optimizing the synergistic strengthening effects of the alloying elements (cobalt, molybdenum, nickel, titanium), while balancing their contents to maintain toughness and ductility. It contains 4.00 to 5.00 Mo, 17.00 to 19.00 Ni,

Mr. Henry is research metallurgist and Mr. Cary is vice president, Technical Services, Vasco, a Teledyne company, Latrobe, Pa.

11.00 to 12.75 Co, 1.2 to 1.45 Ti, 0.05 to 0.15 Al, 0.003 B, 0.02 Zr, 0.05 Ca, and maximums of 0.03 C, 0.10 Si, 0.10 Mn, 0.01 S, and 0.01 P. Judging by initial measurements, fracture toughness is 40,000 to 50,000 psi$\sqrt{\text{in.}}$ at a tensile strength of 356,000 psi.

In tests of one heat, considered representative, 1 hr at 1,475 F was satisfactory for solutioning; this step is followed by air cooling. Subsequent aging 3 hr at 925 F developed 348,600 psi tensile strength, 338,800 psi yield strength, 10.3% elongation, 52.0% reduction in area, and Rc 57.1 hardness.

Figure 2 shows that a maximum tensile strength of 360,000 psi is produced by aging 20 hr at 900 F. Ductility measurements reveal a maximum of 54% reduction in area for specimens aged 1 hr at 950 F, but the underaged structure reaches only 339,000 psi tensile.

This steel ages more slowly than

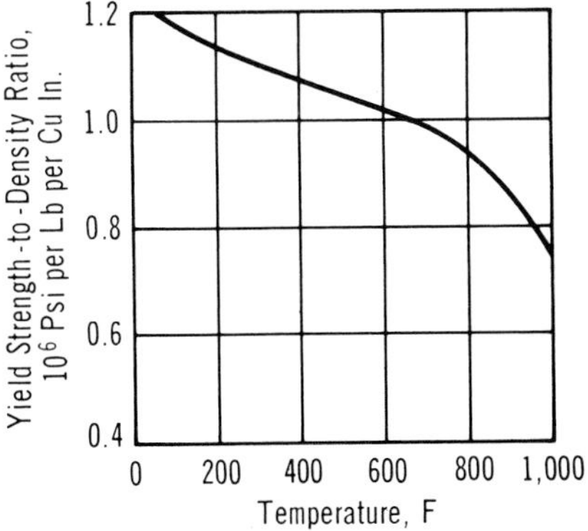

Fig. 1 — Heat treated to the recommended strength, 350,000 psi tensile strength, this alloy has very high strength-to-density ratios to 1,000 F.

18% Ni grades of lower strength (they normally develop peak strengths after aging about 10 hr at 900 F). Though aging occurs much more rapidly at 950 F than at 900 F, aging for at least 6 hr at temperature is desirable for full strengthening.

Billets and Sheets

For the benefit of designers and engineers who prefer data concerning large sections, we present the table below (for material solution treated at 1,500 F, air cooled, and aged 3 hr at 950 F):

	4 by 4 In. Billet	5½ by 5½ In. Billet
Tensile strength, psi	353,600	344,900
Yield strength, psi	337,000	331,200
Elongation, %	5.1	5.9
Reduction in area, %	23.6	25.0

These transverse properties, taken

Fig. 2 — Strengths of VascoMax 350 vary with the aging time and temperature. To assure full strengthening, aging at least 6 hr at temperature is suggested. (These data were obtained by tests of specimens of 5/8 in. bar stock; they were first solution treated for 1 hr at 1,475 F and air cooled.)

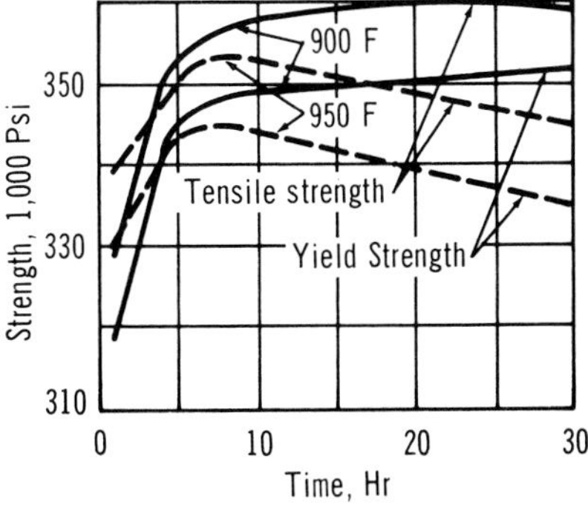

Table I—Mechanical Properties of VascoMax 350 Sheets

Thickness, In.	Tensile Strength, Psi	Elongation, %* 1 In.	Elongation, %* 2 In.	Reduction in Area, %
0.125	362,000	4.7	2.4	15.7
0.066	358,000	4.0	2.0	16.2
0.042	351,000	3.7	2.0	15.3
0.024	349,500	2.5	1.8	14.7

* Apparent reduction in elongation value is due to differences in gage lengths.

at midradius, show that strength and ductility are somewhat reduced. However, these properties display the best combination of strength and ductility available in billets today.

Since properties of sheets are also of interest to designers, we include Table I which gives typical properties of various sizes. Specimens were solution treated at 1,475 F for 1 hr, cooled in air, and aged 6 hr at 900 F. Again, the combination of strength and ductility is excellent.

Notch and Fatigue Properties

Figure 3 reveals the response of the grade to notches of increasing severity. Although it is more sensitive to notches than maraging alloys of lower strength, this alloy has notch properties far superior to those of any conventional martensitic steel at comparable strengths. As the graph shows, the NTS-UTS ratio remains greater than one up to a K_t of 4.

As Fig. 4 indicates, fatigue strength is 110,000 psi for material which was solution treated at 1,475 F, air cooled, and aged 10 hr at

Fig. 3 — Though tensile strength drops as notches become sharper, the ratio remains above 1 until the K_t reaches nearly 5. Samples were solution treated for 1 hr at 1,475 F, air cooled, and aged 3 hr at 950 F.

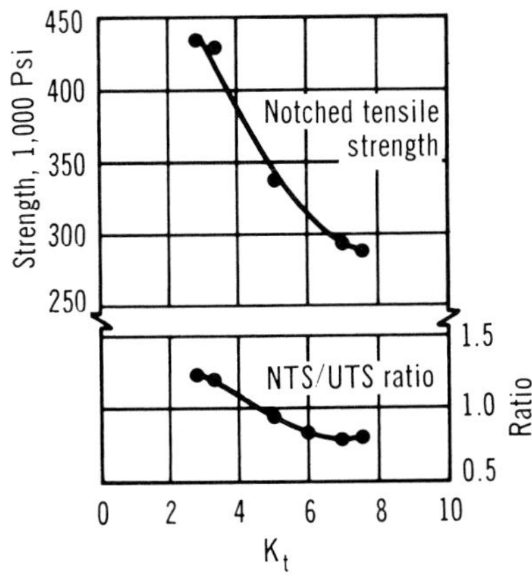

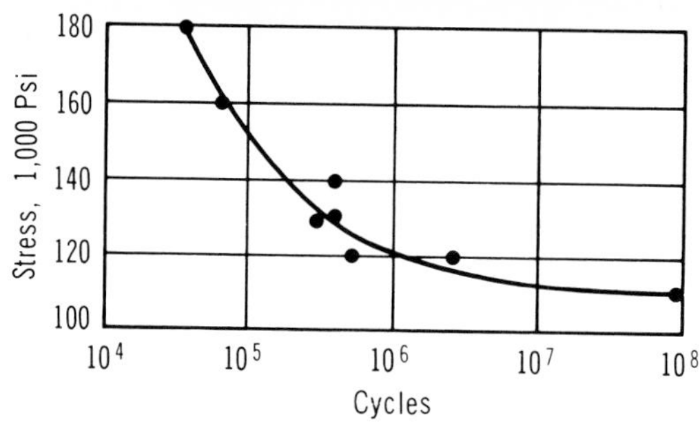

Fig. 4 — At 10^8 cycles, fatigue strength was 110,000 psi for material which was solution treated at 1,475 F, air cooled, and aged at 900 F for 10 hr.

Fig. 5 — VascoMax 350 hds good impact strength (as determined by the Charpy V-notch test) down to room temperature in bar and plate forms. Tensile strength is 350,000 psi, and hardness is Rc 59.

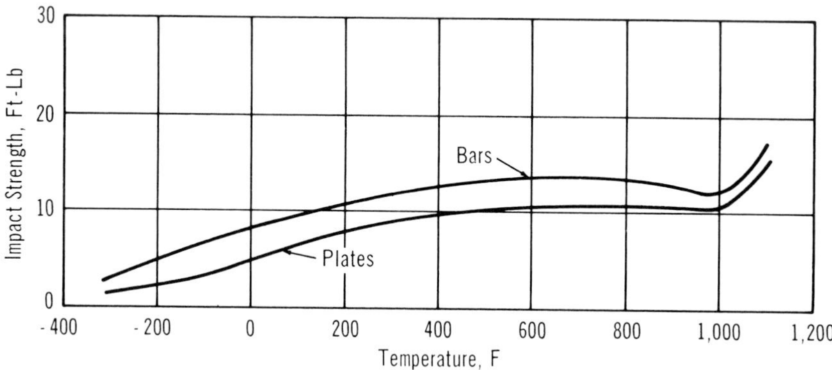

900 F. Work by engineers of International Nickel Co. showed that annealing at 1,400 F, followed by water quenching, resulted in a fatigue strength of 135,000 psi for a laboratory heat.

Figure 5 shows Charpy V-notch impact strength as a function of temperature. Note that no transition temperature is apparent, and that the impact strength at room temperature is 9 ft-lb. Impact values for plates follow the same pattern but at a slightly lower level.

Temperature Changes Properties

Because elevated-temperature properties are important for tooling and aerospace applications, we include Fig. 6, which indicates how mechanical properties vary with temperature. Black lines show data for specimens exposed 200 hr at temperature before being tested at that temperature. A secondary aging reaction probably occurs, raising tensile properties at 700 F. Although exposure for 200 hr at 900 F causes some, although not severe, overaging, long-time aging at 900 F does not embrittle this alloy in any way.

Other tests at high temperature show that yield strength drops gradually to about 300,000 psi at 750 F, then falls more rapidly as the temperature rises. At about 1,000 F, strength is down to 140,000 psi, indicating that the 350 grade offers a little strength advantage over other 18% Ni maraging steels for service at or above that temperature. But all of these maraging steels show better properties than AISI H11 up to 900 F.

Because high tensile strengths indicate high hardnesses, tooling engineers are very interested in stronger materials. In most tooling applications, the three properties of primary interest are toughness, hot hardness, and wear resistance. Toughness properties have already been shown in (Fig. 5). The Charpy V-notch impact strength of 9 ft-lb (at Rc 57) implies toughnesses exceeding 120 ft-lb in the unnotched Izod impact test, the standard toughness test for hard tool steel. Since standard cold work die steels (such as AISI A2 or D2) show unnotched Izod toughness values in the 50 to 80 ft-lb range, the new maraging steel is up to three times as tough as most tool steels tempered to comparable hardnesses.

Hot hardness gradually decreases from Rc 55 at 600 F, through 53 (at 800 F) and 47 (at 1,000 F) to 33 at 1,100 F. In this respect, the steel is better than AISI H13 or AISI M1 high-speed steel at 800 F.

As time goes on, this new maraging grade should continue to replace standard tooling materials in severe applications where its unique combination of strength, hardness, and toughness make it a natural. Also, usage in critical aerospace applications (such as rocket motor cases or even aircraft skin) is within reach. Higher strengths and hardness will give the superstrength maraging steels excellent potential as bearing alloys. Improvements in surface-hardening methods may enable these alloys to replace traditional wear-resistant grades, such as AISI D2 and A2.

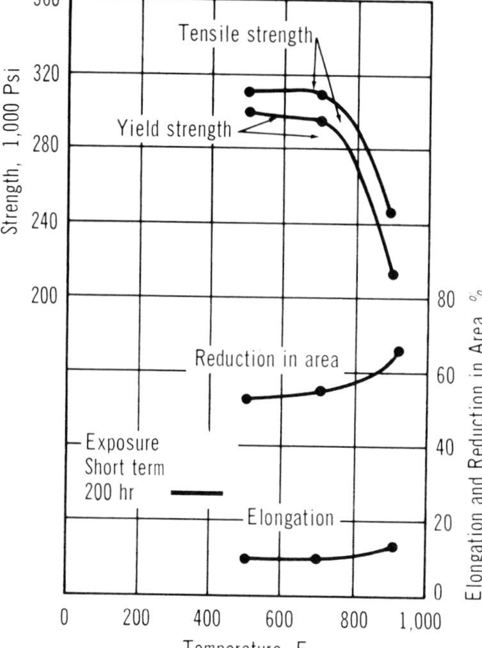

Fig. 6 — The grade retains strength fairly well to the 800 to 900 F level. When aged for 200 hr, a secondary hardening reaction apparently occurs to cause slightly higher strengths to develop at 700 F. Specimens were solution treated for 1 hr at 1,475 F, air cooled, and aged 3 hr at 950 F.

Steels Combine Strength and Fracture Toughness

By L. J. McEOWEN

Ultrahigh strength grades HP 9-4-20, HP 9-4-30, and 300M meet both requirements. They are being applied for aircraft parts such as engine mounts. A higher-strength grade, HP 310, is being developed.

AIRCRAFT AND AEROSPACE engineers are upgrading steel quality by specifying vacuum-arc-remelted (VAR) — also called consumable-electrode-vacuum-melted (CEVM) — steels to replace air-melted steels. Three of these grades, which feature low inclusion content, more isotropic properties (due to the solidification mode), and low gas content, are HP 9-4-20, HP 9-4-30, and 300M. A fourth grade, HP 310, is currently being developed as a potential commercial alloy for applications requiring strengths higher than that of 300M.

Table I lists compositions and Table II gives properties of heat-treated stock. Figure 1 relates tensile strength and fracture toughness for these steels, in comparison with 4330V and 4340. Note that all offer varying combinations of toughness at tensile strengths ranging between 200 000 and 320 000 psi (1380 and 2200 MPa).

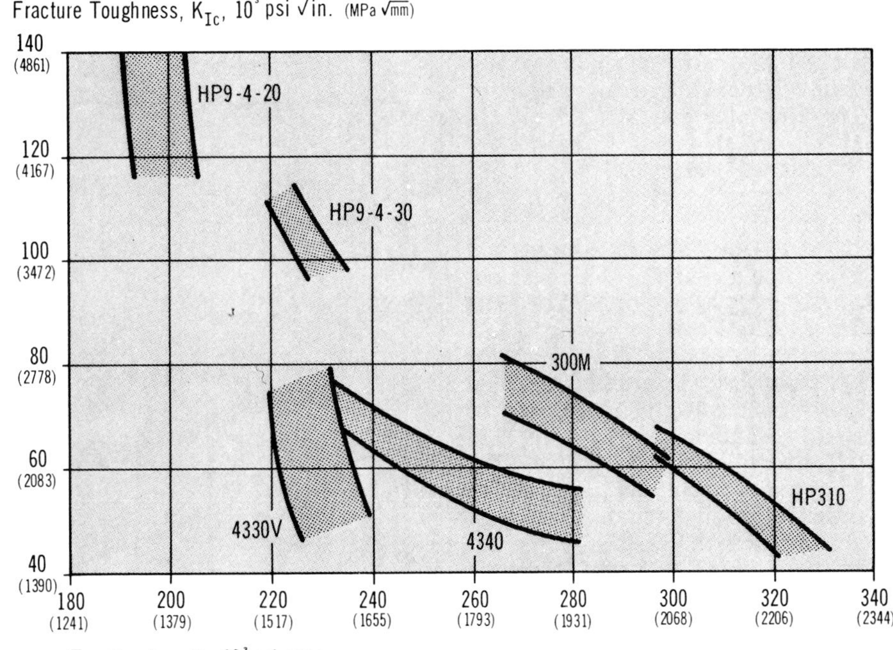

Fig. 1 — Republic's HP grades and 300M combine fracture toughness and strength, both considered essential for critical aircraft and aerospace applications. Properties of 4340 and 4330V are shown for comparison.

HP 9-4-20 — One of two 9Ni-4Co alloy steels developed by Republic Steel Corp., this alloy is for applications requiring optimum combinations of relatively high yield strength and toughness, plus good weldability. The alloy has 180 000 psi (1240 MPa) minimum yield strength in combination with plane-strain fracture toughness of 120 000 psi$\sqrt{in.}$ (4170 MPa$\sqrt{mm}$) minimum. It can be readily welded by the gas tungsten-arc process in the fully heat-treated condition; no preheating or postheating is required.

This alloy has been selected for the horizontal and vertical stabilizer support fittings and for high-pressure actuators of the B-1 bomber. Figure 2 shows the HP 9-4-20 stabilizer fitting being assembled in the first B-1. This alloy is also being used as the rotor-support mast for a new military helicopter. It has great potential for components requiring ability to withstand effects of long cracks plus resistance to fatigue-crack growth. Heat treatment: normalize, 1650 F (900 C), air cool; austenitize, 1525 F (830 C), air cool or oil quench; refrigerate, −100 F (−75 C); temper, 1025-1075 F (550-580 C).

HP 9-4-30 — This is the other commercially available 9Ni-4Co alloy steel developed by Republic. It ranks among the toughest materials in the 220 000-240 000 psi (1520-1650 MPa) tensile-strength level with a guaranteed plane-strain frac-

Fig. 2 — B-1 aft fuselage section has HP 9-4-20 stabilizer support fitting at top.

Mr. McEowen is field metallurgist, Central Alloy District, Republic Steel Corp., Canton, Ohio.

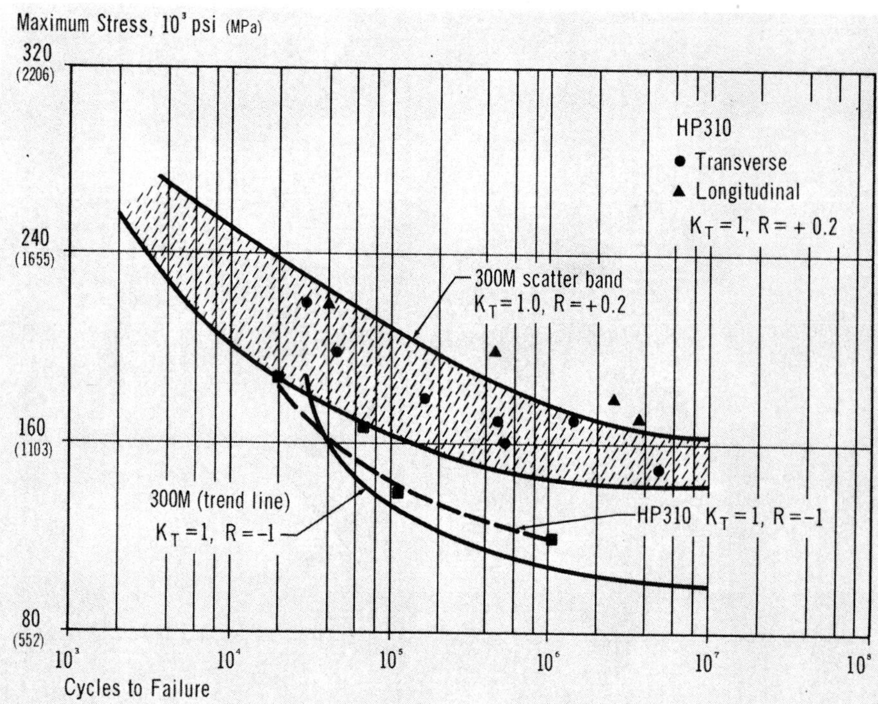

Fig. 3 — Unnotched fatigue strength of HP 310 compares well with that of 300M.

Table I — Nominal Compositions of Aerospace Steels

Alloy	C	Mn	P	S	Si	Ni	Cr	Mo	V	Co
HP 9-4-20	0.20	0.30	0.008	0.004	0.14	9.00	0.75	1.00	0.10	4.50
HP 9-4-30	0.31	0.30	0.008	0.004	0.14	7.50	1.00	1.00	0.10	4.50
300M	0.42	0.75	0.008	0.004	1.70	1.80	0.80	0.40	0.80	—
HP 310	0.41	0.37	0.008	0.004	2.35	1.95	0.97	0.25	0.23	—

ture toughness of 90 000 psi$\sqrt{\text{in.}}$ (3120 MPa $\sqrt{\text{mm}}$). Its high toughness, forgeability, and quench-temper heat treatment have made it popular for aircraft fittings, such as the engine mounts for the Boeing 747. It is replacing less tough steels in many materials lists. Heat treatment: normalize, 1650 F (900 C), air cool; austenitize, 1550 F (840 C), oil quench; refrigerate, −100 F (−75 C); temper, 1000 F (540 C); second temper, 1000-1075 F (540-580 C).

300M — The most widely used VAR alloy steel for aircraft structural applications is 300M, a high-silicon modification of 4340. It has a guaranteed minimum tensile strength of 275 000 psi (1900 MPa), and a guaranteed plane-strain fracture toughness of 50 000 psi$\sqrt{\text{in.}}$ (1740 MPa $\sqrt{\text{mm}}$). This quenched-and-tempered steel has been selected for numerous forged aircraft fittings, such as braces, flaptracks, and slat tracks for many commercial and military aircraft. In addition, it is the steel used for the main landing gear components for the Boeing 747, the Douglas DC-10, and many other commercial and military aircraft, including the B-1 bomber, the C-5A, the F-14, and the F-15. Heat treatment: normalize, 1700 F (930 C), air cool; austenitize, 1600 F (870 C), oil quench; double temper, 575 F (300 C).

HP 310 — An ultrahigh-strength steel developed by Republic Steel under USAF contract, the alloy has a nominal tensile strength of 310 000 psi (2140 MPa) plus the optimum combination of fracture toughness, stress-corrosion cracking resistance, and fatigue strength. The first 90 ton (82 tonne) production heat is currently being evaluated by a major aircraft company. Preliminary results from this heat indicate that the alloy has excellent strength, toughness, and hardenability properties. Furthermore, the fatigue properties of this first heat are very attractive, as shown in Fig. 3. After additional testing and evaluation, HP 310 may become the highest-strength, low-alloy steel commercially available in large section sizes. Heat treatment: normalize, 1700 F (930 C), air cool; austenitize, 1600 F (870 C), oil quench; double temper, 575 F (300 C).

Table II — Properties of Aerospace Steels

	HP 9-4-20	HP 9-4-30	300M	HP 310
Yield strength (tensile), psi (MPa)	190 000 (1310)	195 000 (1340)	240 000 (1650)	271 000 (1870)
Yield strength (compressive), psi (MPa)	200 000 (1380)	205 000 (1410)	255 000 (1760)	—
Tensile strength, psi (MPa)	200 000 (1380)	230 000 (1590)	285 000 (1960)	314 000 (2160)
Reduction in area, %	60	45	30	30
Elongation, % in 1 in. (2.54 mm)	18	13	9	9
K_{Ic}, psi $\sqrt{\text{in.}}$ (MPa $\sqrt{\text{mm}}$)	130 000 (4510)	98 000 (3400)	58 000 (2010)	53 000 (1840)
Hardness, Rc	42	45	54	56
Modulus of elasticity, 10^6 psi (GPa)	28.8 (199)	28.3 (195)	29.0 (200)	29.2 (202)
Shear modulus, 10^6 psi (GPa)	11.1 (76.5)	11.0 (75.8)	11.0 (75.8)	11.3 (77.8)
Poisson's ratio	0.296	0.29	0.320	0.29
Density, lb/in.^{3} (g/cm^3)	0.283 (7.83)	0.283 (7.83)	0.283 (7.83)	0.283 (7.83)

TRIP Steel Is Ready for Use

By EDWARD J. DULIS and VIJAY K. CHANDHOK

Fasteners, surgical needles, armor plate, and high-strength wire and cables are among the applications envisioned. Coils of wire produced by Crucible Steel Corp. are shown at right.

A METALLURGICAL PHENOMENON in high-strength steels, termed transformation induced plasticity (TRIP), was described in the June 1967 issue of ASM's *Transactions Quarterly* by a team of University of California researchers headed by Prof. Earl R. Parker and Prof. Victor F. Zackay. They reported the development of TRIP steels with yield strengths exceeding 200,000 psi, tensile strengths above 240,000 psi, and elongations of more than 35%. Now there is a new chapter in the story. Crucible has completed pilot work on these steels, and it is in a position to supply TRIP wire in commercial quantities.

Background — It is reasonably well known that the formability of type 301 stainless steel can be enhanced through compositional control. As a result, austenite transforms to martensite during deformation. The technique is applied in the manufacture of difficult-to-draw parts such as complex automotive wheel covers. However, in general, the starting condition involves a solution-annealed austenite. In contrast, TRIP steels require a highly deformed austenite as the starting condition, and this condition is developed by severe warm deformation (about 80%) at a temperature above M_d.

Work With Pilot Heats

Crucible initiated its investigation of manufacturing conditions and properties by working with 600 lb heats melted in an air-induction furnace. The nominal composition of the steel studied was: 0.30 C, 2.0 Mn, 2.0 Si, 9.0 Cr, 8.5 Ni, and 4.0 Mo. Ingots were heated to 2,100 to 2,150 F, then forged and/or hot-rolled to intermediate-sized products for TRIP processing. The as-forged billets and sheet bars were conditioned, solution annealed at 2,100 or 2,150 F, water quenched, cleaned, and TRIP processed by rolling, swaging, or drawing to obtain 80% reduction at 800 F. Samples were reheated to 800 F after each pass during processing.

Flat products processed on a pilot rolling mill

Mr. Dulis is director of research and Mr. Chandhok is senior research metallurgist, Research Laboratory, Crucible Steel Corp., Pittsburgh.

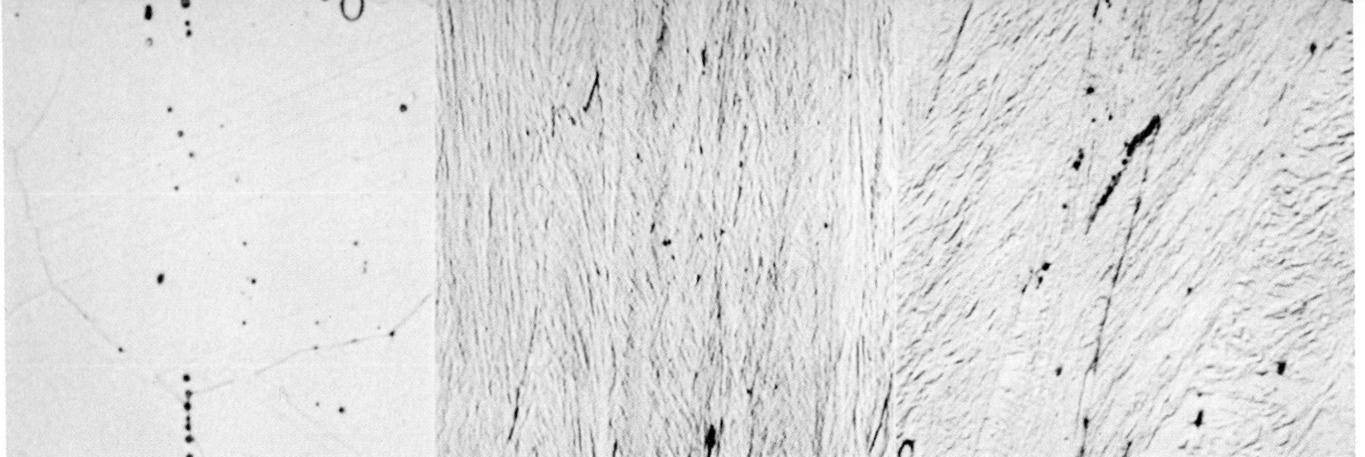

Fig. 1 — The microstructure of three specimens were examined optically: (left) an austenitic structure with some residual carbides after solution annealing at 2,100 F and water quenching; (center) highly deformed austenite after TRIP processing — 80% reduction at 800 F; (right) untempered martensite after TRIP processing and tensile test elongation of 35%; 1,000×.

at Crucible's research laboratory included plates 0.250 and 0.50 in. thick by 4 in. wide and sheets 0.100 in. thick by 6 to 7 in. wide. Bars 0.500 in. in diameter and rods 0.250 in. in diameter were produced on the swaging facility at the Air Force Materials Laboratory at Wright-Patterson AFB. Coils of wire (0.125, 0.100, and 0.053 in. in diameter) were produced by the commercial wire drawing facility at Crucible's Syracuse Div.

Processing Experience

Our pilot studies indicate that no major problems should be encountered in commercial size melting and in hot working. However, TRIP processing (80% reduction at 800 F) of large plate and sheet sections would require a very stiff mill — it would probably have to be specially built. And if the large plate and sheet forms of the current TRIP steel are found to be useful as engineering materials for special applications, consideration could be given to the acquisition of a special rolling mill. The same situation applies to the swaging of bar products.

Wire Is Available

TRIP-processed wire products can be made commercially on existing facilities. Commercial-sized coils can be produced with a special practice that includes effective lubrication at the warm drawing steps. The processing steps include forging the ingots to 3¼ in. square billets, conditioning, heating to 2,100 F, hot rolling to 0.250 in. diameter rod, solution annealing at 2,100 F and water quenching, pickling, coating with a high-temperature lubricant, and warm drawing 80% at 800 F.

Data on tensile properties of the wire at intermediate processing draws (Table I) show that yield strength increases from 174,000 to 221,000 psi and tensile strength from 200,000 to 280,000 psi as the degree of reduction goes from 53 to 84%.

The results of tensile tests on fully TRIP-processed wire at various temperatures (Table II) show that maximum strength and ductility are

Fig. 2 — Three specimens were examined by transmission electron microscopy: (left) the solution annealed structure at 10,000×; (center) TRIP-processed deformed austenite at 22,000×; (right) deformation martensite, TRIP processed and tensile elongated 35% at 22,000×.

Table I — Effects of Warm Drawing on Tensile Properties of TRIP Steel

Wire Diameter, In.*	Warm Drawn, % †	Yield Strength, Psi	Tensile Strength, Psi ‡	Elongation, %	Reduction in Area, %
0.171	53	174,000	200,000	49	53
0.168	55	166,000	198,000	44	52
0.157	61	202,000	221,000	42	59
0.149	65	197,000	210,000	40	55
0.141	68	210,000	224,000	40	58
0.110	81	227,000	245,000	32	55
0.100	84	221,000	280,000	40	54

* All sizes were TRIP processed at 800 to 850 F from 0.250 in. round coils.
† Percentage of reduction in cross-sectional area.
‡ Strain rate of 0.05 in. per in. per min used in testing.

Table II — Tensile Strength of TRIP Steel at Different Temperatures

Test Temperature, F	Tensile Strength, Psi *	Elongation, %	Reduction in Area, %	Magnetic Response †
−320	271,000	8	23	S
−100	305,000	22	52	S
70	248,000	34	55	S
200	237,000	12	46	VW
400	220,000	6	32	VW
600	208,000	7	42	VW
800	196,000	6	39	VW
1,000	188,000	8	38	VW

* TRIP processed by drawing 80% at 800 to 850 F; wire diameter 0.115 in.; strain rate of 0.05 in. per in. per min.
† Tested in gage section. S, strong; VW, very weak.

Table III — Effect of Strain Rate on Properties of TRIP Wire

Test Speed, In./Min	Yield Strength, Psi	Tensile Strength, Psi	Elongation, %	Reduction in Area, %
0.005	210,000	243,000	38	57
0.05	225,000	248,000	34	55
0.10	224,000	249,000	20	63
1.0	223,000	252,000	12	52

Table IV — Tensile Strengths of Sheets

Condition*	Strain Rate, In./Min †	Tensile Strength, Psi	Yield Strength, Psi	Elongation, % in 1.0 in.	Elongation, % in 2.0 in.
		Longitudinal Tests			
TRIP	0.05	220,000	207,000	55	45
TRIP	0.1	220,000	205,000	39	20
TRIP	0.1‡	225,000	206,000	—	32
TRIP + CR	0.05	256,000	234,000	43	30
		Transverse Tests			
TRIP	0.005	242,000	201,000	7	4
TRIP	0.05	242,000	201,000	15	7
TRIP	0.1	248,000	209,000	17	7
TRIP + CR	0.05	289,000	254,000	10	4

* TRIP, 80% reduction at 800 F. CR, cold rolled 16% and aged at 750 F.
† Head speed after yield.
‡ Sample swabbed with alcohol during test to minimize heating.

found in the temperature range of —100 F to room temperature. Temperatures of 200 F and above are greater than M_d so that the austenite to martensite transformation is suppressed, and both strength and ductility are lowered. However, the elevated-temperature strength of the austenite deformed during TRIP processing is high.

The effect of strain rate on room-temperature tensile properties of fully TRIP-processed wire is given in Table III. Tensile strengths between 243,000 and 252,000 psi and elongations between 12 and 38% were found.

Metallographic Examination

Microstructures were examined optically and with electron transmission microscopy after solution treating, after TRIP processing, and after tensile testing.

The optical microstructure of the steel after solution treating at 2,100 F and water quenching (Fig. 1, left) revealed an annealed austenitic structure with some residual carbides. Studies on the effect of solution treating temperatures showed that temperatures higher than 2,100 F were required to take the residual carbides into solution.

After TRIP processing, the structure (Fig. 1, center) consisted of highly deformed austenite shown by the profuse deformation markings. The steel was nonmagnetic in this condition.

After tensile test deformation, the structure (Fig. 1, right) consisted of fine needles of untempered martensite.

Fine structure analysis with transmission electron microscopy showed that the solution annealed alloy was austenitic (Fig. 2, left) and that no fine precipitates existed on grain or twin boundaries. The severe plastic deformation at 800 F in TRIP processing developed a highly deformed austenite containing dense arrays of dislocations on which very fine carbides had apparently precipitated (Fig. 2, center). During tensile testing at room temperature, an additional 35% elongation occurred, which produced the austenite-to-martensite transformation. The structure in Fig. 2, right, exhibited internally twinned (100 to 600 A wide) deformation-induced martensite plates and very fine carbides.

Properties of Sheets

Specimens from 0.100 in. thick sheets were tensile tested in both the TRIP-processed condition (nonmagnetic-austenitic) and TRIP processed plus 16% cold rolling and aging at 750 F (magnetic-partial austenite-to-martensite transformation). The effect of specimen direction, strain rate, and specimen heating during tensile tests were also studied. The

How the Idea for TRIP Steel Evolved

The unusual crack propagation characteristics of steels which exhibit strain-induced transformation suggested the TRIP steel approach, recall Prof. Earl R. Parker (left) and Prof. Victor F. Zackay (right) of the University of California.

The team of researchers they headed up followed two leads in a program sponsored by the U. S. Atomic Energy Commission:

1. There was a change in fracture mode for thin, very tough specimens that was dependent upon testing speed. Crack growth was either in a flat mode or it abruptly changed to a shear mode after an unusually large amount of slow growth in a flat plane.
2. The material was unusually tough, even at —196 C, where plane strain fracture toughness was about 130,000 psi-in.$^{1/2}$ for a steel with a yield strength of 230,000 psi.

The question arose: Can the fracture mode and good toughness be explained in terms of the strain-induced transformation?

The answer turned out to be "Yes."

What's new about TRIP steel is the deliberate use of strain-induced martensitic transformation to prevent necking under tension.

results in Table IV show that both tensile and yield strengths are significantly increased and elongation is moderately decreased by the final 16% cold-rolling step.

Transverse tensile strengths were about 20,000 psi higher than the longitudinal tensile strengths, but the elongations were considerably lower. Varying the strain rate had little influence on the yield and tensile strength, but it did influence the elongation. For the same strain rate of 0.1 in. per min, swabbing with alcohol to minimize heating during testing increased elongation.

Resists Stress Corrosion and Hydrogen Cracking

Stress-corrosion and hydrogen cracking tests were run on 0.100 in. thick sheet specimens both in the TRIP-processed and TRIP plus 16% cold rolled and 750 F aged conditions. In the bent-beam stress-corrosion tests at stresses of 75% of the yield strength in a 5% neutral salt spray, no cracking was found after 500 hr for specimens in both initial conditions. However, substantial general corrosion was found on the cold-rolled and aged specimens where martensite was initially present.

In hydrogen cracking tests, bent-beam specimens were stressed to 75% of yield strength and cathodically charged in a 0.1 normal H_2SO_4 solution, containing 3 mg of arsenic per liter. No cracking was found in TRIP-processed specimens after 62 hr at 0.6 amp per sq in. Current density was then increased to 1.2 amp per sq in., and no cracking was observed after 7 hr. In tests of the TRIP plus cold-rolled-and-aged specimens, cracking occurred between 5 to 12 hr at 0.6 amp per sq in. The specimens showed a laminated type cracking.

The conclusion is that when the steel is in the TRIP-processed condition, it is quite immune to both hydrogen cracking and stress-corrosion cracking. In the TRIP plus cold-rolled-and-aged condition, the steel has a hydrogen cracking tendency somewhat similar to the 12% Cr martensitic stainless steel (type 422). Although higher strength is obtained after subsequent cold working, the strength advantage is offset by poorer hydrogen cracking characteristics.

Outlook for TRIP Steel

The number of requests for additional information and samples reflects considerable interest in these steels. Specific applications have not been defined, but potential uses include fasteners, surgical needles, armor plate, and high-strength wire and cable. Further field trials will eventually establish the significance of this development.

Nickel Steels In LNG Tankers

By GEORGE L. SWALES and ALLEN G. HAYNES

Austenitic stainlesses, 9% and 5-6% Ni steels, and other nickel-containing alloys are playing important roles in a growing number of shipboard pressure vessel and auxiliary equipment applications.

NICKEL-CONTAINING metals have been specified as principal materials for containment tanks for ships having a gross LNG capacity of up to 130 000 m³ (4.6 × 10⁶ ft³ or 815 000 bbl).

The major nickel-containing materials involved are 9% Ni steel, austenitic stainless steels, and 36Ni-Fe low-expansion alloy. Newer alloys, such as 5-6% Ni-Mo steels, are also being evaluated. The austenitic stainlesses, in particular, find additional applications for cargo piping systems, instrumentation, purging gas systems, and other shipboard equipment.

9% Ni Steel

About two-thirds (80% in terms of capacity) of the land-based storage tanks at LNG terminals and peak shaving facilities — in operation, under construction, or on order — have 9% Ni steel inner tanks. The trend to larger terminal storage units is expected to increase this proportion.

For self-supporting shipboard tanks, 9% Ni steel has been used for both cylindrical and spherical designs. It's also a major contender for proposed designs, such as semimembrane and pipe-cell tanks. Other shipboard uses include support structures, such as channel section support bars in the Gaz Transport membrane design.

The material has not been used to date for Conch design rectilinear tanks. However, a study for a 9% Ni steel midships cargo tank for a 130 000 m³ (4.6 × 10⁶ ft³) vessel has been completed and is now undergoing a cost evaluation.

Mr. Swales and Mr. Haynes are senior commercial development officers, European Research & Development Centre, International Nickel Ltd., Birmingham, England. The article is based on a report presented by the authors at the "Shipbuilding and Shipping Record Second LNG Transportation Conference."

The Norman Lady, built at the Moss Rosenberg Shipyard, Stavenger, Norway, is scheduled to carry 540 000 bbl (86 000 m³) of LNG between the Persian Gulf and Japan. The vessel's Kvaerner-Moss spherical tanks are fabricated from 9% Ni steel plate.

The principal advantages of a 9% Ni steel in LNG service include:

- High permissible design stresses coupled with good fracture toughness characteristics.
- Relatively low thermal expansion compared with austenitic stainlesses and aluminum alloys.
- Good weldability by a variety of processes, including shielded metal-arc, gas metal-arc, and submerged-arc.
- High melting point and retention of strength at elevated temperatures (Fig. 1) — important to structural integrity under shipboard fire conditions.

Design Stresses — With the trend to larger land storage and shipboard tanks, the ability to use a high-strength material at a high design stress reduces the weight of material required and avoids many of the fabrication problems likely to be encountered with very thick plates. Because of its relatively high design stress, most current 9% Ni steel applications are met with plate less than 25 mm (1 in.) thick. Thicknesses much greater than 30 to 35 mm (1.2 to 1.4 in.) are unlikely to be needed for most of the proposed larger tanks.

Design stresses for 9% Ni steel, austenitic stainless (type 304), and aluminum alloy 5083 plates are given in Table I. The values are based on properties at 20 C (70 F). If ASME proposals to base cryogenic material design stresses on their properties at low temperatures are adopted, the value for 9% Ni steel would increase from 23 750 psi (164 MPa) to about 30 500 psi (210 MPa) for service at −162 C (−260 F).

In practice, much of the 9% Ni steel used in LNG service has been designed to stresses of 28 500 psi (197 MPa) or less. However, shipping terminal tanks have been built to a design stress of 39 000 psi (269 MPa); and new tanks will use 40 000 psi (276 MPa) — almost twice that of the current ASME code.

It should be noted that for parts of some special designs, such as larger unstiffened spheres, stability against buckling, not design strength, may control plate thickness. In these cases, it may not

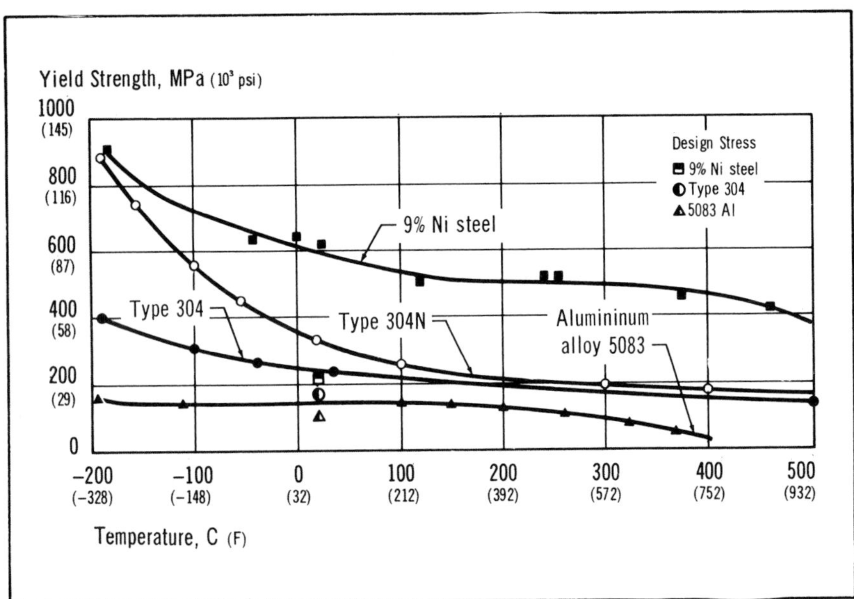

Fig. 1 — Elevated-temperature strength retention by a cryogenic material is critical in the event of a shipboard fire. Performance of 9% Ni steel is outstanding in this respect.

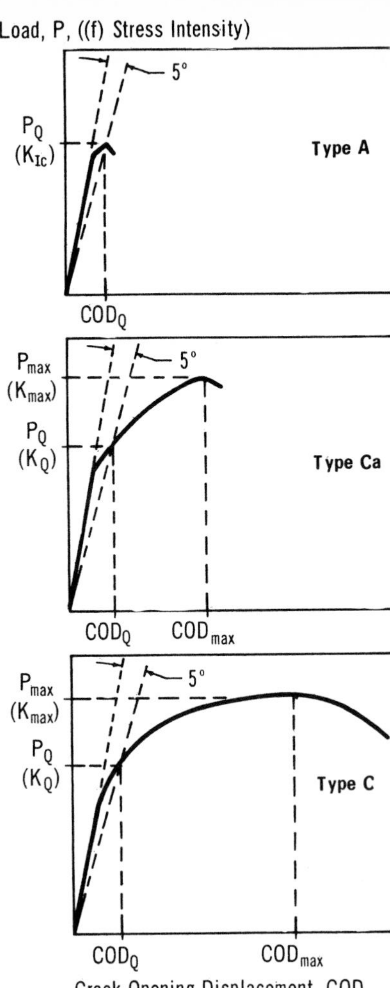

Fig. 2 — No brittle type A failures have been reported for 9% Ni steel. The majority occur by the highly ductile type C mechanism.

be possible to take full advantage of 9% Ni steel's very high design strength.

Fracture Toughness — It's felt that Charpy V-notch test results alone are not reliable enough to predict whether an existing crack of certain size will either grow slowly, causing leakage before failure, or fail catastrophically under the very slow strain rates likely to be encountered in LNG containment vessels.

Two test methods — described in BSI (British Standards Institute) documents DD3 and DD19 — have been developed for assessing the fracture toughness behavior of materials in full plate thickness in the presence of a very sharp crack or notch.

The test procedures are similar: a test piece taken from the full plate thickness and containing a preformed crack is subjected to slow stressing in tension or bending. Simultaneous measurements are made, with increasing load, of the stress intensity (K) at the root of the crack and the crack opening displacement (COD).

In more brittle materials, little or no plastic deformation occurs at the crack root before K reaches the critical K_{Ic} for rapid unstable crack growth. In more ductile materials, yielding will occur and is monitored by changes in the COD value. Typical load-displacement curves that result are given in Fig. 2.

In no instance has a 9% Ni steel failure occurred in the linear elastic region associated with the brittle behavior characterized by the type A curve. In many cases, 9% Ni steel exhibits the highly ductile type C COD curve in which the crack root yields until the stress intensity is equal to the material's full ductile strength. In some cases, failure can occur under ris-

Table I — Design Stresses for Candidate Cryogenic Pressure Vessel Materials

Design Criteria	Design Stress, 10³ psi (MPa) 9% Ni Steel	Type 304	Aluminum 5083
0.25 tensile strength (ASME Section VIII Div. 1)	23.75(a) (164)	18.7 (129)	10.0 (69)
0.33 tensile strength or 0.67 yield strength, whichever is lower (API 620 Q)	31.7 (216)	20.0 (138)	12.0 (83)
0.67 of 0.2% proof strength	37.5(b) or 40.0(c) (258 or 276)	20.0 (138)	13.2(d) (90)
0.67 of 1.0% proof strength	—	23.85(e) (164)	13.3 (92)

(a) Based on weld tensile strength of 95 000 psi (655 MPa). (b) Based on weld metal 0.2% proof stress of 56 250 psi (388 MPa). (c) Based on weld metal 0.2% proof stress of 60 000 psi (414 MPa). (d) Reflects recent API allowance for 0.33 tensile strength. The 0.67 proof stress value is lower: 12 000 psi (83 MPa). (e) British Standard BS 1501-3.

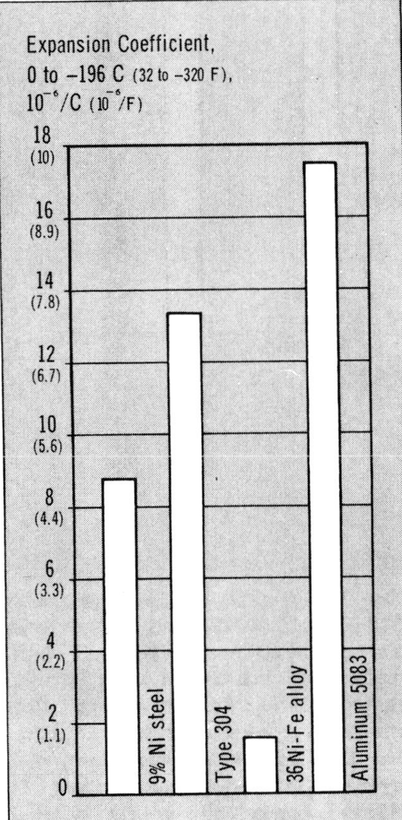

Fig. 3 — Thermal expansion characteristics of 9% Ni steel are better than those of type 304 and aluminum alloy 5083. The low-expansion alloy (36% Ni-Fe) has been tapped for Gaz Transport-design membrane strakes.

ing load conditions (type Ca) at stress intensities slightly below the maximum observed in type C curves.

COD_{max} values are at least 2½ times greater than COD_Q, the value recorded near the onset of yielding. The stress intensity needed to cause ductile fracture, K_{max}, in these steels is 1½ to 2 times larger than that found at the onset of yielding at the crack root, K_Q.

The conservatively estimated maximum length of a through thickness crack that can be tolerated in 9% Ni steel is 160 to 200 mm (6.3 to 7.9 in.) for design stresses of 0.25 UTS (ASME VIII) and 100 to 140 mm (3.9 to 5.5 in.) for designs made to 0.67 YS at room temperature.

In addition, fracture toughness values for 20 to 50 mm (0.8 to 2 in.) thick plate, based on the maximum failure load (K_{max} or K_c) are generally in the range of 160 to 220 MPa(m)½ [145 to 200 ksi(in.)½] at LNG temperatures.

These data indicate an ability for 9% Ni steel to sustain large through-thickness defects, meeting leakage-before-failure requirements.

Thermal Expansion — Dilation characteristics are critical in LNG tank, tank support, and insulation and piping designs in the context of adequately accommodating thermal stresses.

Expansion characteristics of candidate materials for LNG service are compared in Fig. 3. The inherent dimensional movement of 9% Ni steel in cooling to LNG temperature is considerably less than that of type 304 stainless, and less than half that of aluminum alloy 5083. For a 100 ft (30 m) in diameter tank, the change in circumferential length on cooling to −162 C (−260 F) would be about 6 in. (150 mm) for 9% Ni steel and about 12 in. (305 mm) for aluminum.

The relatively low expansion coefficient combined with a high yield strength makes 9% Ni steel a particularly promising candidate for semimembrane-type tank designs. Ship-to-shore pipelines are another potential application. Here, the goal would be to reduce the number of expansion bellows needed, cutting total costs and diminishing the number of potential problem sites.

Fire Conditions — During the critical initial phase of a leak-fed shipboard fire, there must be maximum opportunity to contain and extinguish the fire before it can spread.

Steel equipment would be expected to withstand design stresses and maintain structural integrity for longer periods than nonferrous alloys with lower melting points and with a lower degree of strength retention at even moderately elevated temperatures.

Welding — Until recently, shielded metal-arc welding was used almost exclusively for joining 9% Ni steel. Extensive use is now also made of automatic submerged-arc and pulsed gas metal-arc welding for both field erected land-base storage tanks and for shipboard tanks.

Applicable processes, procedures, and filler metals are detailed in "Joining 9% Ni Steel in Cryogenic Applications," *Metal Progress*, September 1974, p. 113.

Availability — Plates in widths of up to 3 to 3.5 m (9.8 to 11.5 ft) are now generally available. Some mills are offering 5 m (16.4 ft) wide plates. Except for certain spherical tanks, design and handling restrictions will not generally make the wider plates economical.

The availability picture for 9% Ni steel piping and rolled sections has improved substantially, thanks, in the case of piping, to the growing interest in ship-to-shore pipeline applications; and, in the case of rolled sections, to the large amounts specified for support members and stiffeners in the Gaz Transport design.

5 to 6% Ni Steels

Major applications for 5% Ni steel (Table II) are found at service temperatures to −120 C (−185 F). The grade, produced principally in West Germany, has been widely specified for liquefied ethylene cargo tanks on vessels such as the Lincoln Ellsworth and Roald Amundsen.

This steel should not be confused with the lower-nickel-content economical alternatives to 9% Ni steel, such as Armco's Cryonic 5 (5 Ni-Mo) and Nippon Steel's N-Tuf-CR 196 [5.5 Ni-Mn-Mo (Cr)]. These steels (see Table III) require unusual multistage heat treatments. Cryonic 5 is covered by ASTM A645-71 and ASME SA645.

Welding characteristics of the two grades are similar to those of 9% Ni steel. High-nickel alloy welding consumables are preferred. A 37Ni-10Cr-base electrode has been developed specifically for joining N-Tuf-CR 196. Its coefficient of expansion matches that of the base metal.

Potential — The first commercial applications for A645 were for liquid oxygen (−184 C [−300 F]) and ethylene storage. It and the Japanese steel are currently being evaluated for LNG storage tanks, both land-based and shipboard.

First LNG usage is likely to be in conjunction with 9% Ni steel, for thinner sections and for the less critically stressed tank portions. These include the thin upper courses, tank bottoms, and roofs of land-based tanks; some parts of spherical shipboard tanks; stiffeners for rectilinear tanks; stiffeners for tank support skirts; and, possibly, for those portions of semimembrane tanks where plate

Table II — Norwegian Requirements for 5% Ni Steel (DNV-NV20-1-1971)

Composition, %	
Carbon	0.10 max
Silicon	0.15-0.35
Manganese	0.5 max
Nickel	4.75-5.25
Sulfur	0.035 max
Phosphorus	0.035 max
Mechanical Properties at 70 F, min	
Tensile strength, MPa (10^3 psi)	490 (71)
Yield strength, MPa (10^3 psi)	390 (57)
Elongation, %	25
Reduction in area, %	55
Charpy V-notch impact energy at −140 C (−220 F), J (ft-lbf)	34 (25)

The 721 ft (220 m) Descartes holds 310 000 bbl (49 000 m³) of LNG in its six type 304L "waffle pattern," Technigaz membrane-design tanks (left and center). The waffling keeps plate stresses low under thermal contraction. The LNG Tanker Gadinia (right) built by Les Chartier de l'Atlantique, St. Nazaire, France, also utilizes "waffle" membrane tanks. The vessel went into service December 1972.

Table III — Requirements for 5-6% Ni-Mo Type Steels

Composition, %	Cryonic 5 (ASTM A645)	N-Tuf-CR 196
Carbon	0.13 max	0.13 max
Manganese	0.3-0.6	0.9-1.5
Phosphorus	0.025 max	0.03 max
Sulfur	0.025 max	0.03 max
Nickel	4.75-5.25	5-6
Molybdenum	0.20-0.35	0.1-0.3
Aluminum	0.05-0.12	—
Nitrogen	0.02 max	—
Chromium	—	0.6
Typical Three-Stage Heat Treatment*	Austenitize at 855-915 C (1570-1680 F), WQ; temper at 720-760 C (1330-1400 F), WQ; reversion anneal at 620-655 C (1150-1210 F), WQ.	800 C (1470 F), WQ; 670 C (1240 F), WQ; 550-650 C (1020-1200 F), WQ.
Mechanical Properties at 70 F		
Yield strength, MPa (10^3 psi) min	450 (65)	585 (85)
Tensile strength, MPa (10^3 psi)	655-795 (95-115)	690-830 (100-120)
Elongation, % min	20	20‡
Charpy V-notch impact energy, J (ft-lbf) min†		
Longitudinal	34 (25)§	34 (25)‡§
Transverse	27 (20)§	27 (20)‡§

*WQ, water quench. †At −170 C (−274 F) or lower by agreement; 0.38 mm (0.015 in.) min lateral expansion. ‡Nippon Steel reports that N-Tuf-CR 196 meets these requirements for 9% Ni steel (ASTM A553). §Supplementary requirement.

thickness does not exceed 6 to 8 mm (0.2 to 0.3 in.).

Austenitic Stainlesses

Austenitic stainless steels play an important role in LNG ships — for containment tanks, cargo piping systems, instrumentation tubing, purge systems, and a variety of ancillary equipment.

Their popularity is based on satisfactory low temperature properties, availability of required forms, corrosion resistance, fabricability, and elevated temperature strength retention.

Type 304 is the standard grade for cryogenic service; but for deck cargo piping, types 304L, 316, and 316L have also been specified. The more expensive grades are selected for components which may undergo prolonged exposure to seawater.

Type 304L (0.03% C max) has also been specified for its resistance to intergranular corrosion adjacent to welds exposed to warm seawater. A drawback is that type 304L has a lower design strength than type 304. Some LNG ships utilize modified type 304's with

0.05 or 0.06% C max (versus the standard 0.08% C max). These are compromises among corrosion resistance, design strength, and cost.

Types 316 and 316L, even more costly alternatives to type 304, boast improved pitting corrosion resistance in seawater. The steels are not considered necessary for insulated LNG piping systems if reasonable precautions are taken to prevent contact with seawater during construction and service.

A particularly important stainless LNG application is the Technigas membrane design. The standard membrane material is type 304L, selected for its formability, fabricability, and intergranular corrosion resistance. Optimum fatigue characteristics are exhibited by bright annealed sheets given a subsequent skin rolling pass.

Material cost savings of nearly 10% could be realized by switching to type 304. The substitution is deemed possible because of the remote chance of prolonged exposure to warm seawater.

Modifications — The nitrogen-containing, high-proof-strength stainlesses are gaining in importance in cryogenic applications. Europeans are already specifying them in cryogenic processing plant and in liquid oxygen and nitrogen storage and transportation applications. Type 316L+N, for example, is extensively used by some shipyards in chemical tanker construction.

The addition of 0.2% N raises the proof strength by about 40% and moderately increases the tensile strength without sacrificing ductility or fracture toughness. Interestingly, the proof strength of a nitrogen-containing stainless increases more with decreasing temperature than that of its unmodified counterpart. This will be significant if cryogenic pressure vessel codes are altered to base design stresses on properties at operating temperatures rather than at 20 C (70 F).

Fracture toughness (COD) data and impact properties for 35 mm (1.4 in.) thick type 304N plate at −162 and −196 C (−260 and −320 F) are given in Fig. 4. Note the steel's good low-temperature fracture toughness behavior.

Because the nitrogen-strengthened grades carry about a 7% cost premium, they should only be considered for applications that can take advantage of their high yield strengths. Heavy pipe flanges, highly loaded piping system elements, and bolting and shafts in rotating machinery are naturals. Cargo piping systems, however, are not. Here, thin walled piping is used and yield strength is not the main criterion governing thickness.

Semimembrane tank designs are another potential, but doubtful, application. The stainlesses would have to compete with 9% Ni steel, which has a much lower thermal expansion coefficient, a higher yield strength, and a lower basic plate cost.

36% Ni-Iron Alloy

Low-expansion 36% Ni-iron alloy is marketed under various tradenames, including Invar, Nilo 36, and Dilavar. Its traditional cryogenic applications include vacuum-insulated piping systems, flange bolt compensating bushes, and rotating machinery components. Tonnage usage is fairly recent and involves 0.5 to 0.7 mm (0.020 to 0.028 in.) thick, wide-strip primary and secondary barriers in Gaz Transport membrane design LNG tankers. The nickel-iron alloy plate for pressure vessels is now covered in ASTM A658. Gaz Transport suppliers, however, must meet even tighter specifications (see Table IV).

The alloy is also being considered as an alternative to austenitic stainlesses in ship-to-shore LNG pipelines with an eye on reducing the number of expansion bellows or loops necessary. Extensive shipboard pipeline use is not anticipated unless designs require materials with extremely low thermal

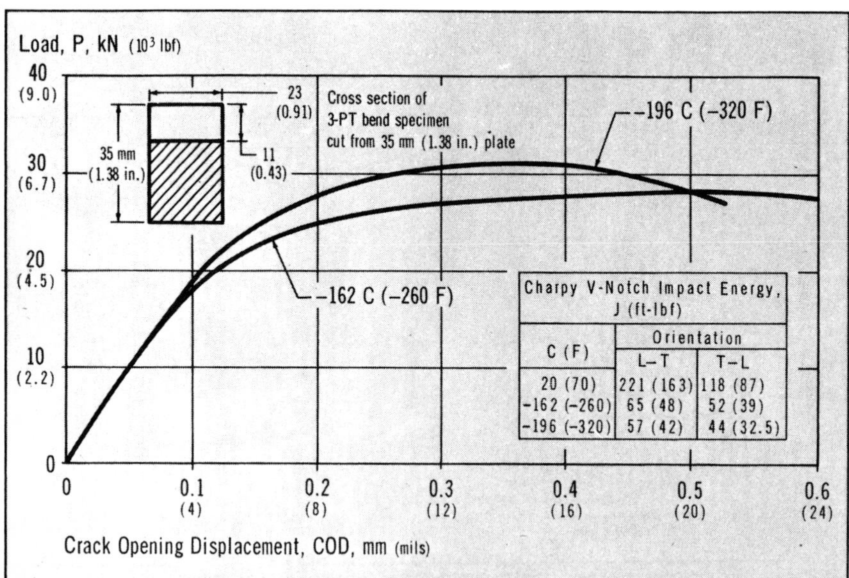

Fig. 4 — Nitrogen-strengthened austenitic stainlesses have excellent low-temperature fracture toughness properties. Heavy and highly stressed components are among the potential applications. These data are for type 304N.

Table IV — Gaz Transport's Specification for Invar M63 36Ni-Fe Alloy*

Composition, %	
Carbon	0.04 max
Silicon	0.25
Manganese	0.2-0.4
Sulfur	0.012 max
Phosphorus	0.012 max
Sulfur + phosphorus	0.02
Nickel	35-36.5
Iron	Bal
Properties	
Thermal expansion coefficient, at −180 and 0 C (−292 F and 32 F), 10^{-6}/K (10^{-6}/F)	1.5±0.5 (0.83±0.2)
Tensile strength at 70 F, MPa (10^3 psi) min†	460 (67)
Yield point or 0.2% proof strength, MPa (10^3 psi) min†	275 (40)
Elongation at 20 and −196 C (70 and −320 F), % min†	30
ISO V-notch impact energy on intermediate rolled slabs, J (ft-lbf)	
20 C (70 F)	156 (115)
−196 C (−320 F)	95 (70)

*Cold rolled and annealed alloy used for tanks on Arctic Tokyo and Polar Alaska.
†Tests run on 0.5 to 1.5 mm (0.02 to 0.06 in.) thick strip.

Table V — Corrosion Resistance of Oil and LNG Tanker Scrubber Alloys

Location	Temperature, C (F)	Alloy	General Rate, mm/yr (mils/yr)	Maximum Pit Depth, mm (mils)
Scrubber bottom	180 (355)	Monel alloy 400	0.05 (2)	0.1 (4)
		Incoloy alloy 825	0.01 (0.4)	Nil
Beneath third stage tray	60 (140)	Monel alloy 400	0.04 (2)	0.05 (2)
		Incoloy alloy 825	0.01 (0.4)	Nil
Water inlet	40 (105)	Monel alloy 400	0.03 (1)	Nil
		Incoloy alloy 825	0.01 (0.4)	Nil

expansion characteristics.

Welding — A feature of the Gaz Transport membrane design tank is the extensive usage of automatic resistance seam welding for joining the flanges of individual 36% Ni-iron membrane strakes. The concept originated in Sweden where it's used to weld stainless steel architectural roofs.

About 90% of total membrane welding is by the resistance seam process; gas tungsten-arc welding is used for the remainder. Initial cracking, porosity, and low ductility fusion welding problems have been overcome thanks to the development, by International Nickel, of a modified composition filler rod containing 3 Mn and 1 Ti. The rod (the Arcos tradename is Invarod) has been used for gas tungsten-arc membrane welds on the Polar Alaska and Arctic Tokyo.

Incoloy Alloy 825

The corrosion problems associated with scrubbers for LNG tanker inert gas scrubbers are basically identical to those encountered by flue gas scrubbers on oil tankers: the combined effects of SO_2 and SO_3 in the seawater used for scrubbing.

Test results (Table V) indicated that Incoloy alloy 825 (45 Ni, 25 Cr, Cu/Mo was a suitable oil tanker scrubber material; behaving somewhat better than Monel alloy 400 (70Ni-30Cu), a material then being specified.

Subsequent oil tanker experience with Incoloy alloy 825 hot gas inlet pipes, baffle plates, and critical lining areas has been extremely good. For example, no problems were encountered after three years' service on the tanker British Sovereign.

The alloy should also be considered for wet scrubber knitted mesh de-misters as an alternative to polypropylene. Compaction and distortion problems plague plastic de-misters. The choice is backed up by a wealth of desalination industry experience.

Cast Air-Meltable High Strength Steels

STEPHEN FLOREEN

This paper describes development of two maraging steels that can be air melted and sand cast. One alloy has been optimized for a 230 ksi yield strength level, and possesses a plane strain fracture toughness (K_{Ic}) value on the order of 95 ksi $\sqrt{in.}$ in sand castings 1 in. thick. A second alloy was developed to give a yield strength of 180 ksi. In a 6 in. thick air melted sand casting, a yield strength of 189 ksi and a K_{Ic} of 118 ksi $\sqrt{in.}$ were obtained. The alloys should be economical, cost-effective materials for many design requirements.

ONE class of materials that has not been used to its full potential is cast high strength steels. The primary reason for this lack of usage is that satisfactory properties often are achieved only in castings that are relatively expensive, largely because vacuum melting and pouring are usually needed to produce such properties. If comparable properties could be achieved in air melted material, cast steels would be much more competitive economically.

The aim of the present work was to develop two new air meltable grades of cast maraging steel. In the early 1960's, a cast version of maraging steel was developed having a nominal composition of 18 pct Ni, 10 pct Co, 4.6 pct Mo, 0.3 pct Ti, 0.1 pct Al, and 0.03 pct max. C.[1] This alloy found some applications, primarily in the aerospace industry. Experience proved, however, that vacuum melting was required to ensure adequate ductility and toughness, and consequently the use of the alloy was restricted. In the initial part of this work, an air meltable composition was sought that had mechanical properties comparable to the original alloy—230 ksi yield strength, 240 ksi tensile strength, tensile elongations of 6 pct or more, and Charpy V-notch impact values of 10 to 15 ft lbs. The work was then extended to develop an air meltable 180 ksi yield strength steel.

STEPHEN FLOREEN is with The International Nickel Company, Inc., Paul D. Merica Research Laboratory, Sterling Forest, Suffern, N. Y.

EXPERIMENTAL PROCEDURE AND RESULTS

Higher Strength Alloy

Vacuum melting of the earlier cast maraging steel probably is necessary because of the presence of reactive elements such as aluminum and titanium, and because of the problems introduced by segregation of titanium and molybdenum.[2] Reducing the level of these elements, therefore, should lessen the need for vacuum melting. These elements, however, are the ones primarily responsible for age hardening in the steel. Thus, reducing their amounts requires some further adjustment in composition to maintain strength. An earlier study[3] has shown that this can be done by increasing the cobalt content. Based on these considerations, a composition range of 17 to 19 pct Ni, 12 to 16 pct Co, 1.8 to 3.5 pct Mo, and 0.1 pct max. Al and Ti each was studied in the initial part of this program aimed at a 230 ksi strength level steel. As discussed later, leaner compositions were then examined for lower strength levels.

Experimental heats were made as 30 lb air induction melts. Charge metals used were electrolytic grades of iron, nickel, and cobalt, and commercially pure molybdenum. The initial melt-down consisted of the iron, nickel, cobalt, and molybdenum. Also added was 0.15C in the form of Fe-C wash metal to produce a carbon boil, experience having proved that a thorough carbon boil was necessary for deoxidation. After the boil, refining additions including silicon, aluminum, titanium, and calcium (in the form of NiCa master alloy, INCOCAL alloy 10) were made. Heats were poured at 2900°F into 1 in. thick keel block molds made of dry sand.

The keel blocks were given the same multiple anneal heat treatment used in the earlier cast maraging steel—2100°F/1 h air cool, 1100°F/1 h air cool, 1500°F/1 h air cool, and a maraging treatment of 900°F/3 h air cool. Several other heat treatments were evaluated, but they did not give as consistently good properties. Castings provided 0.25 in. diam, 1 1/4 in. gage length tensile specimens, Charpy V-notch impact samples, and 1/2 in. thick × 1 in. deep fracture toughness bend samples.

Fig. 1 shows the influence of molybdenum on the yield strength of alloys with 18 pct Ni and either 12.5 or 15 pct Co. With 15 pct Co the strength increased linearly with molybdenum, and the desired 230 ksi yield strength was achieved with approximately 2.8 pct Mo. At the 12.5 pct Co level, strengthening by molybdenum was less effective.

Fig. 2 shows yield strengths *vs* the Charpy V-notch impact values. With increasing yield strength from 200 to 270 ksi, impact values fell off gradually from approximately 20 to 8 ft lbs. At 230 ksi, Charpy values ranged from 10 to 15 ft lbs. In general, at comparable yield strengths, the alloys with 15 pct Co had better impact properties than lower cobalt-higher

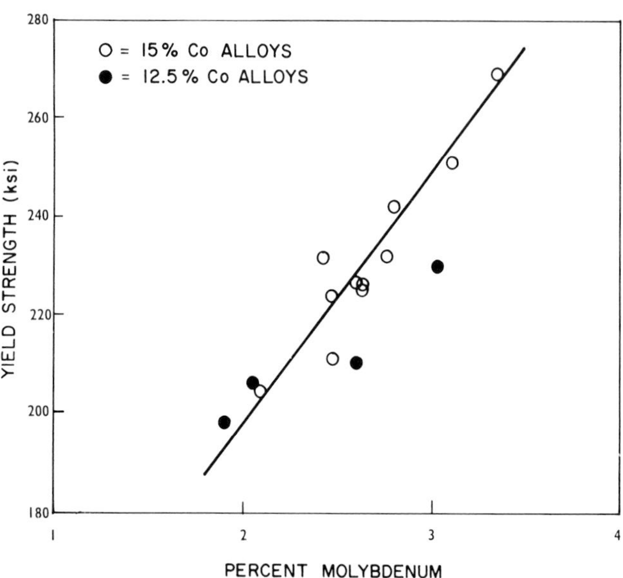

Fig. 1—Yield strength *vs* molybdenum content of 18 pct Ni-0.1 pct Si alloys.

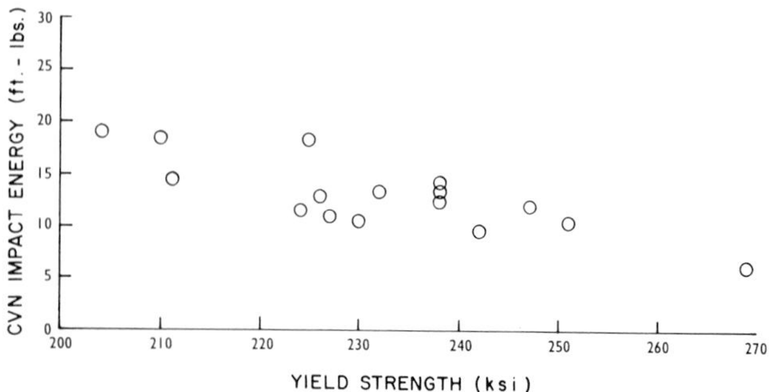

Fig. 2—Charpy V-Notch impact properties *vs* yield strength of higher strength alloys.

molybdenum heats. The nominal alloy composition for the major elements therefore was set at 18.0 pct Ni, 15.0 pct Co, and 2.80 pct Mo.

Achieving the best properties depended very strongly upon proper deoxidation and refining element practice. A good carbon boil plus addition of 0.1 pct Si were essential. Castings of heats made without any silicon, regardless of the deoxidation practice, displayed excessive macroscopic porosity. It seems likely that the porosity resulted from a reaction between the molten steel and the sand molds. Additions of 0.1 pct Si, as metallic silicon, ferrosilicon, or CaSi, eliminated this problem. Excessive silicon contents (above 0.2 pct), however, were detrimental to ductility and toughness. Higher levels of silicon have been found to have similar effects in wrought maraging steels of comparable strength levels.[4] Thus, there is an optimum silicon addition, on the order of 0.1 pct, to control porosity, but higher levels should be avoided.

Comparison of heats with different refining elements showed that small additions of aluminum and titanium were useful for deoxidation purposes, residual levels of approximately 0.05 pct of each element being desirable. An addition of 0.1 pct NiCa also was helpful. Magnesium additions were not helpful. Heats containing up to 0.50 pct Cr had good properties, but comparable amounts of vanadium were harmful. Carbon variations below 0.03 pct had little effect, but higher carbon contents reduced the toughness. Thus, carbon content should be dept at 0.03 pct maximum.

One of the likely production practices for such an alloy is that it would be initially prepared as a vacuum melted master alloy stock, and subsequently remelted in air. Several 100 lb vacuum melted heats therefore were made containing the nominal 18 pct Ni, 15 pct Co, and 2.80 pct Mo along with carbon levels between 0.01 and 0.06 pct. These heats were then remelted in air and sand cast. During remelting, refining additions of 0.1 pct Si, 0.2 pct Al, 0.1 pct Ti, and 0.1 pct NiCa were made. Examination of these heats showed that it was very useful to have a higher carbon content (about 0.05 pct) in the master alloy stock, to protect the alloy from oxygen pick-up during remelting. Final carbon contents in these alloys after remelting in air were about half the initial values. Variations in the other elements were negligible. Mechanical properties of the remelted heats agreed very well with those of heats that were simply air melted.

Fig. 3 shows the plane strain fracture toughness (K_{Ic}) values *vs* yield strength for some of the air melts, and also for the remelted material. The fracture mode in all heats was ductile-fibrous in nature, with fairly small shear lips. Surprisingly good K_{Ic} values were obtained. At 230 ksi yield strength, toughness was on the order of 90 to 100 ksi $\sqrt{\text{in.}}$, which compares favorably with that of wrought maraging steel of this strength level. The toughness is about 20 pct greater than that reported for vacuum melted castings of the earlier cast maraging steel.[5]

The nominal mechanical properties of this alloy are 230 ksi yield strength, 245 ksi tensile strength, 7 pct elongation, 35 pct reduction in area, 13 ft lb Charpy

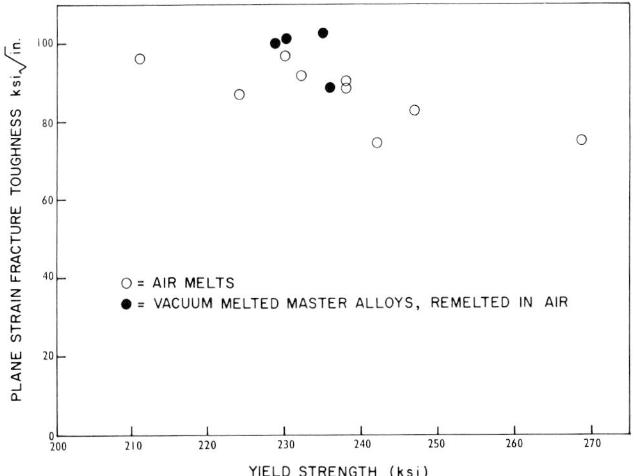

Fig. 3—Fracture toughness *vs* yield strength of higher strength alloys.

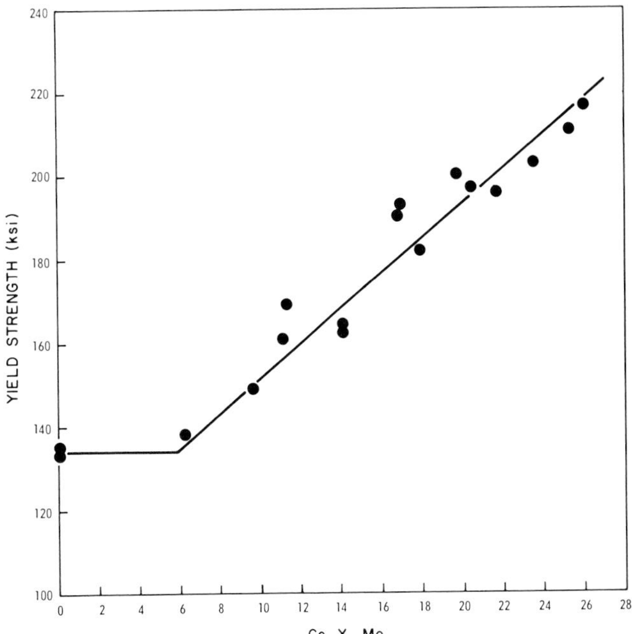

Fig. 4—Yield strength of 18 pct Ni-0.5 pct Si alloys *vs* the products of the cobalt and molybdenum contents.

V-notch impact energy, and 95 ksi $\sqrt{\text{in.}}$ K_{Ic}. Thus, very good mechanical properties can be achieved in air melted, sand cast material. Fracture toughness values in particular are quite high.

Lower Strength Alloy

During the work described above, several lower alloy heats were made that showed good combinations of strength and toughness. One significant difference noted in heats with strengths below approximately 215 ksi was a much different sensitivity to silicon. As noted previously, higher silicon lowered toughness of the high strength heats. In lower strength alloys, however, silicon actually improved the strength-toughness combination.

This is best illustrated by the results from a pair of heats containing 18 pct Ni, 12.5 pct Co, 2.0 pct Mo, and either 0.11 or 0.38 pct Si. The 0.11 pct Si heat had 198 ksi yield strength and a Charpy impact energy of 24 ft lb. Comparable values for the 0.38 pct Si heat were 211 ksi and 24.7 ft lb.

Comparisons of other heats also showed that silicon contents on the order of 0.50 pct raised strength with little change in toughness. In addition to helping the mechanical properties, silicon also simplifies deoxidation practice and improves casting and solidification characteristics.

A second series of heats therefore were examined containing 12 to 18 pct Ni, 0 to 15 pct Co, 0 to 3.6 pct Mo, and 0.5 pct Si. The heats were air melted, sand cast, heat treated, and tested in the same fashion described previously.

The strengthening achieved in these heats is best described by the relationship originally discovered by Decker, Eash, and Goldman[6] that the strength is proportional to the product of the cobalt and molybdenum contents. Fig. 4 shows the yield strengths of 18 pct Ni-0.5 pct Si heats plotted against the pct Co times pct Mo. For Co × Mo contents of six or greater, yield strength increased linearly with the Co × Mo content. As in other maraging steels, these alloys are thought to harden due to the formation of molybdenum containing precipitates, with increasing cobalt lowering solubility of molybdenum.[7] Thus, the Co × Mo product is probably a measure of the amount of precipitation taking place during age hardening.

Reducing the nickel content in these alloys below 18 pct lowered both strength and toughness. Minor variations in aluminum and titanium contents did not appear significant, probably because enough silicon was present to ensure adequate deoxidation. Carbon variations up to 0.05 pct had no strong effect.

At a given yield strength, heats with higher cobalt and lower molybdenum levels had, as before, noticeably better toughness. Thus, to optimize the properties it would be better to start with a higher cobalt content, on the order of 10 pct, and then select the molybdenum content needed to give the desired strength. In the present work, a yield strength of 180 ksi was desired. From Fig. 4, a composition with 10 pct Co and 1.7 Mo therefore was selected.

A recent report from the National Materials Advisory Board[8] has expressed the need for an air meltable cast steel with 180 ksi yield strength and good toughness in section sizes up to 6 in. thick. The report concluded that no alloy was available that fulfilled these requirements. A brief experiment was conducted to see if satisfactory properties could be achieved in this section size with an 18Ni-10Co-1.7Mo-0.5Si composition. A 300 lb air induction melt was prepared and cast into 1, 3, and 6 in. thick keel blocks.

Fig. 5 gives strength and Charpy values of these castings. Also shown for comparison are the properties obtained in some of the earlier 30 lb melts. Values for the 300 lb heat agree reasonably well with those of the smaller heats. Results also indicate an improvement in the strength-toughness combination on going to progressively thicker castings, the reason for which is not clear.

Properties of the 6 in. thick casting are summarized in Table I, fracture toughness being measured in a 1 1/4 in. thick × 2 1/2 in. deep three-point bent beam

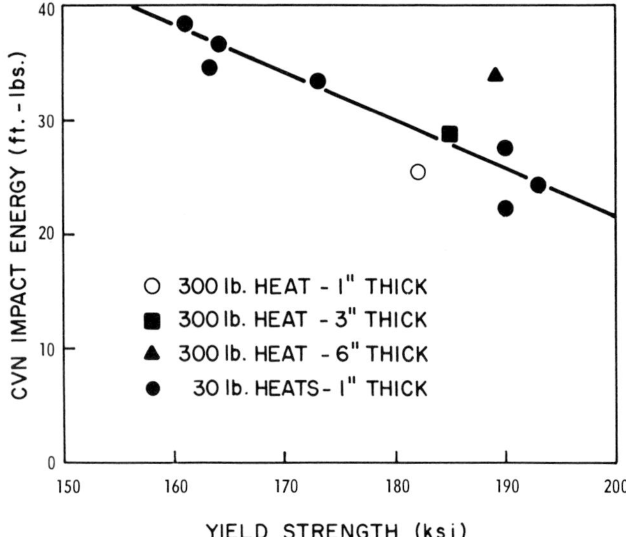

Fig. 5—Yield strength *vs* impact energy of several 30 lb heats and 300 lb heat, lower strength alloys.

Table I. Mechanical Properties from 6 In. Thick Sand Casting

Y.S. (ksi)	U.T.S. (ksi)	pct El.	pct R.A.	Charpy V-Notch Energy (Ft-Lbs) +70 °F	-50 °F	-100 °F	-150 °F	K_{Ic} ksi $\sqrt{in.}$
189	193	10	42	34.0	26.5	24.5	27.5	118

specimen. A yield strength of 189 ksi and a K_{Ic} value of 118 ksi $\sqrt{in.}$ were obtained, which appear to be an excellent combination of strength and toughness for a 6 in. section size. As Fig. 6 shows, the fracture was by a coarse fibrous fracture mode with small shear lips. Charpy results show no sharp ductile-brittle transition temperature down to −150° F.

Table II summarizes the nominal composition and properties of IN-0180, which is the designation given the 180 ksi yield strength alloy. Results show that attractive mechanical properties can be achieved in air melt, sand cast heats in section sizes up to 6 in. thick.

DISCUSSION

One of the most interesting aspects of the present work is that extremely good K_{Ic} values were obtained in cast high strength steels. The K_{Ic} values of the 230 ksi steel are surprisingly high in comparison to the Charpy values. If one uses the Charpy results to predict K_{Ic} values, following the empirical Charpy-K_{Ic} correlations that have been proposed in the literature,[9-11] predicted K_{Ic} numbers are about half the measured ones. The 6 in. section size results with IN-0180 are consistent with the correlations, but other types of cast steels also show relatively high K_{Ic} to Charpy ratios.[12,13] Thus, cast steels may frequently be much tougher than would be thought from Charpy tests.

There are, of course, distinct differences between what is being measured in a Charpy test and a K_{Ic} test. It should be noted, however, that for most applications a K_{Ic} test is probably a more meaningful evaluation of the resistance of a casting to failure in service. If this is so, cast high strength steels such as the present ones should find many useful applications.

SUMMARY

One of the primary obstacles to the usage of cast high strength steels has been the need for vacuum melting and pouring to ensure satisfactory proper-

Table II. Nominal Composition and Mechanical Properties of In-0180-Cast 180 Ksi Steel

Composition (Bal Fe)						
Ni	Co	Mo	Si	Al	Ti	C
18.0	10.0	1.70	0.5	0.05	0.05	0.03

Mechanical Properties in 1 In. Thick Sand Castings Heat Treated 2100°F/1 h AC + 1100°F/1 h AC + 1500°F/1 h AC + 900°F/3 h AC

Y.S. (ksi)	U.T.S. (ksi)	pct El.	pct R.A.	Charpy V-Notch Impact Energy (Ft-Lb)	K_{Ic} (ksi $\sqrt{in.}$)
180	190	12	50	30	118*

*Taken from 6 in. casting–1 in. thick castings not thick enough to give valid K_{Ic} values.

Fig. 6—Fracture surface of K_{Ic} sample from 6 in. thick casting.

ties. In the present work, two new maraging steel compositions have been developed that can be air melted and sand cast.

The higher strength steel is a nominal 18 pct Ni-15 pct Co-2.80 pct Mo-0.1 pct Si-0.05 pct Al-0.05 pct Ti-0.03 pct Max. C composition. This alloy has a yield strength of 230 ksi and a K_{Ic} fracture toughness on the order of 95 ksi $\sqrt{in.}$

The lower strength steel contains 18 pct Ni, 10 pct Co, 1.7 pct Mo, 0.5 pct Si, 0.05 pct Al, 0.05 pct Ti, and 0.03 pct C, and is designed for a yield strength level of 180 ksi. Tests on a 6 in. thick sand casting show an excellent combination of 189 ksi yield strength and K_{Ic} of 118 ksi $\sqrt{in.}$. Results in general demonstrate that very good K_{Ic} values can be obtained in air melted-sand cast high strength maraging steels. Such alloys, therefore, should compete successfully for many industrial applications.

REFERENCES

1. E. P. Sadowski and R. F. Decker: *Mod. Cast.*, 1963, vol. 43, no. 2, p. 26.
2. M. T. Leger and J. P. Aymard: *Rev. Met.*, 1971, vol. 68, p. 783.
3. S. Floreen: *J. Iron Steel Inst.*, 1969, vol. 207, p. 484.
4. C. J. Novak and L. M. Diran: *J. Metals*, 1963, vol. 15, p. 200.
5. W. J. Jackson and H. T. Hall: *Brit. Foundryman*, 1972, vol. 65, p. 286.
6. R. F. Decker, J. T. Eash, and A. J. Goldman: *Trans. ASM*, 1962, vol. 55, p. 58.
7. S. Floreen: *Met. Rev.*, 1968, vol. 13, p. 115.
8. "High Performance Steel and Titanium Castings", NMAB-296, Nat. Acad. of Sciences–Nat. Acad. of Engineering, July, 1973.
9. S. T. Rolfe and S. R. Novak: ASTM STP No. 463, Philadelphia, Pa., 1970, p. 124.
10. W. F. Brown and J. E. Strawley: *ibid*, p. 216.
11. R. H. Sailor and H. T. Corten: ASTM STP No. 514, Philadelphia, Pa., 1972, p. 164.
12. S. Floreen: Int. Nickel Co., Sterling Forest, Suffern, N. Y., unpublished work.
13. M. T Groves and J. F. Wallace: *Steel Foundry Facts*, March, 1973, p. 22.

Section VIII: Selected Data Sheets

SAE Alloy Steel Compositions SAE J404g

SAE No.	Composition,* %								Corresponding AISI No.
	C	Mn	P	S	Si	Ni	Cr	Other	
1330	0.28-0.33	1.60-1.90	0.035	0.040	0.20-0.35	—	—	—	1330
1335	0.33-0.38	1.60-1.90	0.035	0.040	0.20-0.35	—	—	—	1335
1340	0.38-0.43	1.60-1.90	0.035	0.040	0.20-0.35	—	—	—	1340
1345	0.43-0.48	1.60-1.90	0.035	0.040	0.20-0.35	—	—		1345
								Mo	
4012	0.09-0.14	0.75-1.00	0.035	0.040	0.20-0.35	—	—	0.15-0.25	4012
4023	0.20-0.25	0.70-0.90	0.035	0.040	0.20-0.35	—	—	0.20-0.30	4023
4024	0.20-0.25	0.70-0.90	0.035	0.035-0.050	0.20-0.35	—	—	0.20-0.30	4024
4027	0.25-0.30	0.70-0.90	0.035	0.040	0.20-0.35	—	—	0.20-0.30	4027
4028	0.25-0.30	0.70-0.90	0.035	0.035-0.050	0.20-0.35	—	—	0.20-0.30	4028
4032	0.30-0.35	0.70-0.90	0.035	0.040	0.20-0.35	—	—	0.20-0.30	—
4037	0.35-0.40	0.70-0.90	0.035	0.040	0.20-0.35	—	—	0.20-0.30	4037
4042	0.40-0.45	0.70-0.90	0.035	0.040	0.20-0.35	—	—	0.20-0.30	—
4047	0.45-0.50	0.70-0.90	0.035	0.040	0.20-0.35	—	—	0.20-0.30	4047
4118	0.18-0.23	0.70-0.90	0.035	0.040	0.20-0.35	—	0.40-0.60	0.08-0.15	4118
4130	0.28-0.33	0.40-0.60	0.035	0.040	0.20-0.35	—	0.80-1.10	0.15-0.25	4130
4135	0.33-0.38	0.70-0.90	0.035	0.040	0.20-0.35	—	0.80-1.10	0.15-0.25	—
4137	0.35-0.40	0.70-0.90	0.035	0.040	0.20-0.35	—	0.80-1.10	0.15-0.25	4137
4140	0.38-0.43	0.75-1.00	0.035	0.040	0.20-0.35	—	0.80-1.10	0.15-0.25	4140
4142	0.40-0.45	0.75-1.00	0.035	0.040	0.20-0.35	—	0.80-1.10	0.15-0.25	4142
4145	0.43-0.48	0.75-1.00	0.035	0.040	0.20-0.35	—	0.80-1.10	0.15-0.25	4145
4147	0.45-0.50	0.75-1.00	0.035	0.040	0.20-0.35	—	0.80-1.10	0.15-0.25	4147
4150	0.48-0.53	0.75-1.00	0.035	0.040	0.20-0.35	—	0.80-1.10	0.15-0.25	4150
4161	0.56-0.64	0.75-1.00	0.035	0.040	0.20-0.35	—	0.70-0.90	0.25-0.35	4161
4320	0.17-0.22	0.45-0.65	0.035	0.040	0.20-0.35	1.65-2.00	0.40-0.60	0.20-0.30	4320
4340	0.38-0.43	0.60-0.80	0.035	0.040	0.20-0.35	1.65-2.00	0.70-0.90	0.20-0.30	4340
E4340†	0.38-0.43	0.65-0.85	0.025	0.025	0.20-0.35	1.65-2.00	0.70-0.90	0.20-0.30	E4340
4419	0.18-0.23	0.45-0.65	0.035	0.040	0.20-0.35	—	—	0.45-0.60	4419
4422	0.20-0.25	0.70-0.90	0.035	0.040	0.20-0.35	—	—	0.35-0.45	—
4427	0.24-0.29	0.70-0.90	0.035	0.040	0.20-0.35	—	—	0.35-0.45	—
4615	0.13-0.18	0.45-0.65	0.035	0.040	0.20-0.35	1.65-2.00	—	0.20-0.30	4615
4617	0.15-0.20	0.45-0.65	0.035	0.040	0.20-0.35	1.65-2.00	—	0.20-0.30	—
4620	0.17-0.22	0.45-0.65	0.035	0.040	0.20-0.35	1.65-2.00	—	0.20-0.30	4620
4621	0.18-0.23	0.70-0.90	0.035	0.040	0.20-0.35	1.65-2.00	—	0.20-0.30	4621
4626	0.24-0.29	0.45-0.65	0.035	0.04 max	0.20-0.35	0.70-1.00	—	0.15-0.25	4626
4718	0.16-0.21	0.70-0.90	—	—	—	0.90-1.20	0.35-0.55	0.30-0.40	4718
4720	0.17-0.22	0.50-0.70	0.035	0.040	0.20-0.35	0.90-1.20	0.35-0.55	0.15-0.25	4720
4815	0.13-0.18	0.40-0.60	0.035	0.040	0.20-0.35	3.25-3.75	—	0.20-0.30	4815
4817	0.15-0.20	0.40-0.60	0.035	0.040	0.20-0.35	3.25-3.75	—	0.20-0.30	4817
4820	0.18-0.23	0.50-0.70	0.035	0.040	0.20-0.35	3.25-3.75	—	0.20-0.30	4820
5015	0.12-0.17	0.30-0.50	0.035	0.040	0.20-0.35	—	0.30-0.50	—	5015
50B40‡	0.38-0.43	0.75-1.00	0.035	0.040	0.20-0.35	—	0.40-0.60	—	—
50B44‡	0.43-0.48	0.75-1.00	0.035	0.040	0.20-0.35	—	0.40-0.60	—	50B44
5046	0.43-0.48	0.75-1.00	0.035	0.040	0.20-0.35	—	0.20-0.35	—	—
50B46‡	0.44-0.49	0.75-1.00	0.035	0.040	0.20-0.35	—	0.20-0.35	—	50B46
50B50‡	0.48-0.53	0.75-1.00	0.035	0.040	0.20-0.35	—	0.40-0.60	—	50B50
5060	0.56-0.64	0.75-1.00	0.035	0.040	0.20-0.35	—	0.40-0.60	—	—
50B60‡	0.56-0.64	0.75-1.00	0.035	0.040	0.20-0.35	—	0.40-0.60	—	50B60
5115	0.13-0.18	0.70-0.90	0.035	0.040	0.20-0.35	—	0.70-0.90	—	—
5120	0.17-0.22	0.70-0.90	0.035	0.040	0.20-0.35	—	0.70-0.90	—	5120
5130	0.28-0.33	0.70-0.90	0.035	0.040	0.20-0.35	—	0.80-1.10	—	5130
5132	0.30-0.35	0.60-0.80	0.035	0.040	0.20-0.35	—	0.75-1.00	—	5132
5135	0.33-0.38	0.60-0.80	0.035	0.040	0.20-0.35	—	0.80-1.05	—	5135
5140	0.38-0.43	0.70-0.90	0.035	0.040	0.20-0.35	—	0.70-0.90	—	5140
5145	0.43-0.40	0.70-0.90	0.035	0.040	0.20-0.35	—	0.70-0.90	—	5145
5147	0.46-0.51	0.70-0.95	0.035	0.040	0.20-0.35	—	0.85-1.15	—	5147
5150	0.48-0.53	0.70-0.90	0.035	0.040	0.20-0.35	—	0.70-0.90	—	5150
5155	0.51-0.59	0.70-0.90	0.035	0.040	0.20-0.35	—	0.70-0.90	—	5155
5160	0.56-0.64	0.75-1.00	0.035	0.040	0.20-0.35	—	0.70-0.90	—	5160
51B60‡	0.56-0.64	0.75-1.00	0.035	0.040	0.20-0.35	—	0.70-0.90	—	51B60
50100†	0.98-1.10	0.25-0.45	0.025	0.025	0.20-0.35	—	0.40-0.60	—	—
51100†	0.98-1.10	0.25-0.45	0.025	0.025	0.20-0.35	—	0.90-1.15	—	E51100
52100†	0.98-1.10	0.25-0.45	0.025	0.025	0.20-0.35	—	1.30-1.60		E52100
								V	
6118	0.16-0.21	0.50-0.70	0.035	0.040	0.20-0.35	—	0.50-0.70	0.10-0.15	6118
6150	0.48-0.53	0.70-0.90	0.035	0.040	0.20-0.35	—	0.80-1.10	0.15	6150
								Mo	
8115	0.13-0.18	0.70-0.90	0.035	0.040	0.20-0.35	0.20-0.40	0.30-0.50	0.08-0.15	8115
81B45‡	0.43-0.48	0.75-1.00	0.035	0.040	0.20-0.35	0.20-0.40	0.35-0.55	0.08-0.15	81B45
8615	0.13-0.18	0.70-0.90	0.035	0.040	0.20-0.35	0.40-0.70	0.40-0.60	0.15-0.25	8615
8617	0.15-0.20	0.70-0.90	0.035	0.040	0.20-0.35	0.40-0.70	0.40-0.60	0.15-0.25	8617
8620	0.18-0.23	0.70-0.90	0.035	0.040	0.20-0.35	0.40-0.70	0.40-0.60	0.15-0.25	8620
8622	0.20-0.25	0.70-0.90	0.035	0.040	0.20-0.35	0.40-0.70	0.40-0.60	0.15-0.25	8622
8625	0.23-0.28	0.70-0.90	0.035	0.040	0.20-0.35	0.40-0.70	0.40-0.60	0.15-0.25	8625
8627	0.25-0.30	0.70-0.90	0.035	0.040	0.20-0.35	0.40-0.70	0.40-0.60	0.15-0.25	8627
8630	0.28-0.33	0.70-0.90	0.035	0.040	0.20-0.35	0.40-0.70	0.40-0.60	0.15-0.25	8630
8637	0.35-0.40	0.75-1.00	0.035	0.040	0.20-0.35	0.40-0.70	0.40-0.60	0.15-0.25	8637
8640	0.38-0.43	0.75-1.00	0.035	0.040	0.20-0.35	0.40-0.70	0.40-0.60	0.15-0.25	8640
8642	0.40-0.45	0.75-1.00	0.035	0.040	0.20-0.35	0.40-0.70	0.40-0.60	0.15-0.25	8642
8645	0.43-0.48	0.75-1.00	0.035	0.040	0.20-0.35	0.40-0.70	0.40-0.60	0.15-0.25	8645
86B45‡	0.43-0.48	0.75-1.00	0.035	0.040	0.20-0.35	0.40-0.70	0.40-0.60	0.15-0.25	—
8650	0.48-0.53	0.75-1.00	0.035	0.040	0.20-0.35	0.40-0.70	0.40-0.60	0.15-0.25	—
8655	0.51-0.59	0.75-1.00	0.035	0.040	0.20-0.35	0.40-0.70	0.40-0.60	0.15-0.25	8655
8660	0.56-0.64	0.75-1.00	0.035	0.040	0.20-0.35	0.40-0.70	0.40-0.60	0.15-0.25	—
8720	0.18-0.23	0.70-0.90	0.035	0.040	0.20-0.35	0.40-0.70	0.40-0.60	0.20-0.30	8720
8740	0.38-0.43	0.75-1.00	0.035	0.040	0.20-0.35	0.40-0.70	0.40-0.60	0.20-0.30	8740
8822	0.20-0.25	0.75-1.00	0.035	0.040	0.20-0.35	0.40-0.70	0.40-0.60	0.30-0.40	8822
9254	0.51-0.59	0.50-0.80	0.035	0.040	1.20-1.60	—	0.50-0.80	—	—
9255	0.51-0.59	0.70-0.95	0.035	0.040	1.80-2.20	—	—	—	9255
9260	0.56-0.64	0.75-1.00	0.035	0.040	1.80-2.20	—	—	—	9260
9310	0.08-0.13	0.45-0.65	0.025	0.025	0.20-0.35	3.00-3.50	1.00-1.40	0.08-0.15	—
94B15‡	0.13-0.18	0.75-1.00	0.035	0.040	0.20-0.35	0.30-0.60	0.30-0.50	0.08-0.15	—
94B17‡	0.15-0.20	0.75-1.00	0.035	0.040	0.20-0.35	0.30-0.60	0.30-0.50	0.08-0.15	94B17
94B30‡	0.28-0.33	0.75-1.00	0.035	0.040	0.20-0.35	0.30-0.60	0.30-0.50	0.08-0.15	94B30

* Small quantities of certain elements are present which are not specified or required. Considered as incidental, they are acceptable to the following amounts: 0.35 Cu, 0.25 Ni, 0.20 Cr, and 0.06 Mo. †Electric furnace steel. ‡0.0005% B min.

Expected Minimum Mechanical Properties, Conventional Practice, of Cold-Drawn Carbon Steel Rounds, Squares, and Hexagons

The tensile and yield strengths of carbon-steel bars are improved by cold drawing. By comparison, the tensile strength of hot-rolled bars is about 10% less, and their yield strength is some 40% less. For example, a low-carbon steel with a yield-to-tensile-strength ratio of about 0.55 in the form of hot-rolled bars will have a ratio of about 0.85 after cold drawing. While there is some sacrifice in elongation, reduction in area, and impact strength, these changes are relatively insignificant in most structural applications or engineering components.

This improvement is of interest to the design engineer seeking a better strength-to-weight ratio or a reduction in costs by the elimination of alloy contents and heat treatment. The enhanced properties may also be useful in applications involving threads, notches, cut-outs, and in other design requirements that might effect strength adversely.

Turned and polished and turned, ground, and polished bars have the mechanical properties of hot-rolled bars.

AISI No.	As Cold Drawn					Cold Drawn Followed by Low-Temperature Stress Relief					Cold Drawn Followed by High-Temperature Stress Relief				
	Strength					Strength					Strength				
Size, In.	Tensile, 1,000 Psi	Yield, 1,000 Psi	Elongation in 2 In., %	Reduction in Area, %	Hardness, Bhn	Tensile, 1,000 Psi	Yield, 1,000 Psi	Elongation in 2 In., %	Reduction in Area, %	Hardness, Bhn	Tensile, 1,000 Psi	Yield, 1,000 Psi	Elongation in 2 In., %	Reduction in Area, %	Hardness, Bhn
1018, 1025															
⅝ to ⅞ incl.	70	60	18	40	143						65	45	20	45	131
Over ⅞ to 1¼ incl.	65	55	16	40	131						60	45	20	45	121
Over 1¼ to 2 incl.	60	50	15	35	121						55	45	16	40	111
Over 2 to 3 incl.	55	45	15	35	111						50	40	15	40	101
1117, 1118															
⅝ to ⅞ incl.	75	65	15	40	149	80	70	15	40	163	70	50	18	45	143
Over ⅞ to 1¼ incl.	70	60	15	40	143	75	65	15	40	149	65	50	16	45	131
Over 1¼ to 2 incl.	65	55	13	35	131	70	60	13	35	143	60	50	15	40	121
Over 2 to 3 incl.	60	50	12	30	121	65	55	12	35	131	55	45	15	40	111
1035															
⅝ to ⅞ incl.	85	75	13	35	170	90	80	13	35	179	80	60	16	45	163
Over ⅞ to 1¼ incl.	80	70	12	35	163	85	75	12	35	170	75	60	15	45	149
Over 1¼ to 2 incl.	75	65	12	35	149	80	70	12	35	163	70	60	15	40	143
Over 2 to 3 incl.	70	60	10	30	143	75	65	10	30	149	65	55	12	35	131
1040, 1140															
⅝ to ⅞ incl.	90	80	12	35	179	95	85	12	35	187	85	65	15	45	170
Over ⅞ to 1¼ incl.	85	75	12	35	170	90	80	12	35	179	80	65	15	45	163
Over 1¼ to 2 incl.	80	70	10	30	163	85	75	10	30	170	75	60	15	40	149
Over 2 to 3 incl.	75	65	10	30	149	80	70	10	30	163	70	55	12	35	143
1045, 1146, 1145															
⅝ to ⅞ incl.	95	85	12	35	187	100	90	12	35	197	90	70	15	45	179
Over ⅞ to 1¼ incl.	90	80	11	30	179	95	85	11	30	187	85	70	15	45	170
Over 1¼ to 2 incl.	85	75	10	30	170	90	80	10	30	179	80	65	15	40	163
Over 2 to 3 incl.	80	70	10	30	163	85	75	10	25	170	75	60	12	35	149
1050, 1137, 1151															
⅝ to ⅞ incl.	100	90	11	35	197	105	95	11	35	212	95	75	15	45	187
Over ⅞ to 1¼ incl.	95	85	11	30	187	100	90	11	30	197	90	75	15	40	179
Over 1¼ to 2 incl.	90	80	10	30	179	95	85	10	30	187	85	70	15	40	170
Over 2 to 3 incl.	85	75	10	30	170	90	80	10	25	179	80	65	12	35	163
1141															
⅝ to ⅞ incl.	105	95	11	30	212	110	100	11	30	223	100	80	15	40	197
Over ⅞ to 1¼ incl.	100	90	10	30	197	105	95	10	30	212	95	80	15	40	187
Over 1¼ to 2 incl.	95	85	10	30	187	100	90	10	25	197	90	75	15	40	179
Over 2 to 3 incl.	90	80	10	20	179	95	85	10	20	187	85	70	12	30	170
1144															
⅝ to ⅞ incl.	110	100	10	30	223	115	105	10	30	229	105	85	15	40	212
Over ⅞ to 1¼ incl.	105	95	10	30	212	110	100	10	30	223	100	85	15	40	197
Over 1¼ to 2 incl.	100	90	10	25	197	105	95	10	25	212	95	80	15	35	187
Over 2 to 3 incl.	95	85	10	20	187	100	90	10	20	197	90	75	12	30	179

Source: AISI Committee of Hot Rolled and Cold Finished Bar Producers.

Composition-Size Guide for Steel Selection

This Data Sheet is based on tables in ASTM A400 (Recommended Practice for Selection of Steel Bar Compositions According to Section). It is intended to help the designer choose a steel according to the desired section size and mechanical properties for a given part. The three classifications are (1) Steels for severe service requiring 90,000 psi yield strength and over, good ductility, and high notch toughness; (2) Steels for moderate service requiring yield strengths of 75,000 to 185,000 psi, corresponding tensile strengths, and good ductility; (3) Steels for parts requiring yield strengths of 30,000 to 120,000 psi plus fair to good ductility. Steels for severe and moderate service should develop the listed properties when quenched from a suitable austenitizing temperature and tempered at 800 F or higher. Steels for lighter service (low and medium-carbon grades) achieve the indicated properties through hot rolling, cold drawing, or cold drawing with thermal treatment.

Notes: (1) Steels with specified maximum carbon contens of 0.40% or greater, or steels used in conditions requiring hardness of Bhn 293 or greater after heat treatment, are not recommended for applications involving welding; (2) Steels with specified 0.33 C max or greater should not be water quenched in sections less than 1½ in.; smaller sections can quench-crack.

Source: ASTM A 400-69. Additional information was supplied by Climax Molybdenum Co. of Michigan, International Harvester Co., and Republic Steel Corp.

Steels for Severe Service

Desired Minimum Hardness		Equivalent Tensile Strength, Psi	Equivalent Yield Strength, Psi	Minimum As-Quenched Hardness*		Size Class†						
						Diameter of Round (or Distance Between Faces of Square or Hexagonal) Sections, In.						
						To ½ incl	Over ½ to 1 incl	Over 1 to 1½ incl	Over 1½ to 2 incl	Over 2 to 2½ incl	Over 2½ to 3 incl	Over 3 to 3½ incl
						Thickness of Flat Sections, In.						
Bhn	Rc			Bhn	Rc	To 0.3 incl	Over 0.3 to 0.6 incl	Over 0.6 to 1.0 incl	Over 1.0 to 1.3 incl	Over 1.3 to 1.6 incl	Over 1.6 to 2.0 incl	Over 2.0 to 2.3 incl
MODERATELY QUENCHED PARTS (Oil Quenching or Equivalent)												
229 to 293 incl	20 to 33 incl	110,000 to 145,000 incl	90,000 to 125,000 incl	388	42	1330 EX 34 4130 EX 49 5132 8630		94B30				
Over 293 to 341 incl	Over 33 to 38 incl	Over 145,000 to 170,000 incl	Over 125,000 to 150,000 incl	409	44	1335 94B30 4042 5135	4135 EX 36 8640 EX 52 8740	4137		4142	9840	4337 4340
Over 341 to 388 incl	Over 38 to 42 incl	Over 170,000 to 190,000 incl	Over 150,000 to 170,000 incl	455	48	1340 EX 35 4047 4135 5140 8637	4137 EX 37 8642 EX 38 8645 EX 53 8742	4140 94B40		4145 9840	4147 4337 86B45	4340

Over 388 to 429 incl	Over 42 to 45 incl	Over 190,000 to 205,000 incl	Over 170,000 to 185,000 incl	496	51	1345 4140 5145 5150 8640 8642 8645 EX 36 8740 EX 37 9260 EX 38 50B44 EX 52 50B46 EX 53 50B50 EX 56	5147 5155 5160 6150 50B44	4142 EX 40 4145 8655 50B60 51B60 81B45	9840 86B45	4147 4161 4340	4150 4161	E4340
DRASTICALLY QUENCHED PARTS (Water Quenching or Equivalent)												
229 to 293 incl	20 to 33 incl	110,000 to 145,000 incl	90,000 to 125,000 incl	388	42	8625 EX 17 8627 EX 18 EX 26 EX 27	4130 EX 34 5130 5135 8630	94B30 5135	4135 8640 8740		4135	94B30
Over 293 to 341 incl	Over 33 to 38 incl	Over 145,000 to 170,000 incl	Over 125,000 to 150,000 incl	409	44	1330 EX 34 4037 4130 5130 5132 8630	1330 4042 5132 94B30	1335 4135 5135 5140 8637		1340 EX 35 8637 EX 36 8640 8740	4140 5147 8645 50B44	4137 94B40

Steels for Moderate Service

MODERATELY QUENCHED PARTS (Oil Quenching or Equivalent)												
187 to 293 incl	Rb91 to 33 incl	95,000 to 145,000 incl	75,000 to 125,000 incl	388	42	1330 8630 4130 EX 34 5132	8637	3140 8642 8740	4140 94B40		4142	
Over 293 to 341 incl	Over 33 to 38 incl	Over 145,000 to 170,000 incl	Over 125,000 to 150,000 incl	409	44	1335 94B30 4042 4047 5135	4135 EX 36 8640 8740	4137 EX 37 4140 EX 38 8642 8645	81B45	4142	4145	4147 86B45
Over 341 to 388 incl	Over 38 to 42 incl	Over 170,000 to 190,000 incl	Over 150,000 to 170,000 incl	455	48	1340 8637 4047 94B30 4145 5140	1345 50B50 4137 EX 37 4140 EX 38 5150 8642 8645	4142 5147 5155 5160 6150	51B60 8655	4145 8655 EX 40	4147 4337	4150 4340

*Minimum as-quenched hardness for obtaining desired hardness after tempering at 800 F or higher.

†Steels listed as approved for heavier sections or higher strengths may be used in the same conditions for lighter sections and lower strengths. Numbers correspond to SAE, AISI, or ASTM designations.

Composition-Size Guide for Steel Selection (Continued)

Steels for Moderate Service (Continued)

Desired Minimum Hardness		Equivalent Tensile Strength, Psi	Equivalent Yield Strength, Psi	Minimum As-Quenched Hardness*		Size Class†						
						Diameter of Round (or Distance Between Faces of Square or Hexagonal) Sections, In.						
						To ½ incl	Over ½ to 1 incl	Over 1 to 1½ incl	Over 1½ to 2 incl	Over 2 to 2½ incl	Over 2½ to 3 incl	Over 3 to 3½ incl
						Thickness of Flat Sections, In.						
Bhn	Rc			Bhn	Rc	To 0.3 incl	Over 0.3 to 0.6 incl	Over 0.6 to 1.0 incl	Over 1.0 to 1.3 incl	Over 1.3 to 1.6 incl	Over 1.6 to 2.0 incl	Over 2.0 to 2.3 incl
MODERATELY QUENCHED PARTS (Oil Quenching or Equivalent)												
Over 388 to 429 incl	Over 42 to 45 incl	Over 190,000 to 205,000 incl	Over 170,000 to 185,000 incl	496	51	1345 8645 4047 8740 4140 9260 5145 50B44 5150 50B46 8640 50B50 8642 EX 36 EX 37 EX 38	4142 5147 5155 6150	4145 EX 40 5160 8655 50B60 51B60 81B45	86B45 9840	4147 4340 8660	4150	E4340
DRASTICALLY QUENCHED PARTS (Water Quenching or Equivalent)												
187 to 293 incl	Rb91 to 33 incl	95,000 to 145,000 incl	75,000 to 125,000 incl	388	42	1000 series from 1024 to 1040 incl 4130 EX 17 8625 EX 18 8627 EX 26 EX 27	4037 EX 34 4130 5130 5132 8630	5135 94B30	8637	5140 8637 EX 35	4135	8740 8642
Over 293 to 341 incl	Over 33 to 38 incl	Over 145,000 to 170,000 incl	Over 125,000 to 150,000 incl	409	44	1036 to 1045 incl 8630 EX 34	1330 5135 94B30	1335	4135	1340 8637 EX 35	8640 8740 94B30 EX 36	4137 EX 37 4140 EX 38 8642 8645
Over 341 to 388 incl	Over 38 to 42 incl	Over 170,000 to 190,000 incl	Over 150,000 to 170,000 incl	455	48	1335 94B30 4037 4130 5130 5132 5135 8630	4042 4047	1340 5140 8637 EX 35	4137 8642	1345 50B44 4137 50B50 5145 EX 36 5150 EX 37 8640 8642 8740	4140 8645 8742 EX 38	4142 5147 6150 81B45

*Minimum as-quenched hardness for obtaining desired hardness after tempering at 800 F or higher.

†Steels listed as approved for heavier sections or higher strengths may be used in the same conditions for lighter sections and lower strengths. Numbers correspond to SAE, AISI, or ASTM designations.

Steels for Lighter Service

Desired Min. Yield Strength, Psi	Size Class (*) (†)			
	Diameters of Round or Approximately Round Sections, In.			
	To ½ incl	Over ½ to 1 incl	Over 1 to 2 incl	Over 2 to 3 incl
	Thickness of Flat Sections, In.			
	To 0.3 incl	Over 0.3 to 0.6 incl	Over 0.6 to 1.3 incl	Over 1.3 to 2.0 incl
Over 30,000 to 35,000 incl	HR 1016 HR 1020 HR 1018 HR 1019	HR 1016 HR 1020 HR 1018 HR 1022 HR 1019	HR 1018 HR 1019 HR 1021	HR 1018 HR 1022 HR 1030
Over 35,000 to 40,000 incl	HR 1022 HR 1030	HR 1030 HR 1035	HR 1030 HR 1035	HR 1035
Over 40,000 to 45,000 incl	HR 1035 HR 1040	HR 1040 HR 1045	CD 1010 HR 1040 HR 1045	CD 1010 HR 1045 CD 1015 HR 1050 HR 1040
Over 45,000 to 50,000 incl	CD 1010 HR 1040 HR 1045	CD 1010 CD 1015 HR 1045	CD 1015 HR 1045 HR 1137	CD 1020 HR 1050 HR 1137
Over 50,000 to 55,000 incl	CD 1015 HR 1050 HR 1137	CD 1020 HR 1050 HR 1137 HR 1141	CD 1018 CD 1115 CD 1020 HR 1141 CD 1025 HR 1144 HR 1050	CD 1018 CD 1019 CD 1025 HR 1141
Over 55,000 to 60,000 incl	CD 1018 CD 1115 CD 1025 HR 1141 CD 1019 HR 1144 CD 1020 HR 1541	CD 1018 HR 1144 CD 1019 HR 1547 CD 1025	CD 1019 CD 1022 CD 1117 HR 1547	CD 1022 CD 1117 CD 1118
Over 60,000 to 65,000 incl	CD 1022 CD 1117	CD 1022 CD 1117 CD 1118	CD 1030 CD 1118	CD 1030
Over 65,000 to 70,000 incl	CD 1030 CDT 1040 CD 1118	CD 1030 CD 1035 CDT 1045	CD 1035 CDT 1040 CDT 1050 CDT 1137	CD 1035 CD 1040 CDT 1045 CDT 1141
Over 70,000 to 75,000 incl	CD 1035 CDT 1045	CDT 1040 CDT 1050 CDT 1137	CD 1040 CDT 1045 CDT 1141	CDT 1040 CD 1045 CDT 1050 CDT 1137 CDT 1144
Over 75,000 to 80,000 incl	CDT 1040 CDT 1050 CDT 1137	CD 1040 CDT 1045 CDT 1141	CDT 1040 CD 1045 CDT 1050 CDT 1137 CDT 1144	CDT 1040 CDT 1045 CD 1050 CD 1137 CDT 1141
Over 80,000 to 85,000 incl	CD 1040 CDT 1045 CDT 1141	CDT 1040 CD 1045 CDT 1050 CDT 1137 CDT 1144	CDT 1040 CDT 1045 CDT 1050 CD 1137 CDT 1141	CDT 1040 CDT 1045 CDT 1050 CDT 1137 CD 1141 CDT 1144
Over 85,000 to 90,000 incl	CD 1045 CDT 1040 CDT 1050 CDT 1137 CDT 1144	CDT 1040 CDT 1045 CD 1137 CDT 1141	CDT 1040 CDT 1050 CDT 1137 CD 1141 CDT 1444 CDT 1145	CDT 1045 CDT 1050 CDT 1137 CDT 1141 CD 1144
Over 90,000 to 95,000 incl	CDT 1040 CDT 1045 CD 1050 CD 1137 CDT 1141	CDT 1045 CDT 1050 CDT 1137 CD 1141 CDT 1144	CDT 1045 CDT 1050 CDT 1137 CDT 1141 CD 1144	CDT 1050 CDT 1137 CDT 1141 CDT 1144
Over 95,000 to 100,000 incl	CDT 1040 CDT 1045 CDT 1050 CDT 1137 CD 1141 CDT 1144	CDT 1045 CDT 1050 CDT 1137 CDT 1141 CD 1144	CDT 1050 CDT 1137 CDT 1141 CDT 1144	CDT 1141 CDT 1144
Over 100,000 to 105,000 incl	CDT 1045 CDT 1050 GDT 1137 CDT 1141 CD 1144	CDT 1050 CDT 1137 CDT 1141 CDT 1144	CDT 1141 CDT 1144	CDT 1144
Over 105,000 to 110,000 incl	CDT 1050 CDT 1137 CDT 1141 CDT 1144	CDT 1141 CDT 1144	CDT 1144	
Over 110,000 to 115,000 incl	CDT 1141 CDT 1144	CDT 1144		
Over 115,000 to 120,000 incl	CDT 1144			

*HR, Hot rolled; CD, Cold drawn; CDT, Cold drawn and thermally treated.
†Steel listed as approved for heavier sections or higher strengths may be used in the same conditions for lighter sections and lower strengths. Numbers correspond to SAE, AISI, or ASTM designations.

Typical Mechanical Properties of Selected Carbon and Alloy Steels

Using the Data Sheet

This Data Sheet is offered as a guide to show the potential user what to expect from a given grade of steel in the indicated condition. Data were obtained from specimens 0.505 in. in diameter which were machined from 1 in. rounds; gage lengths were 2 in. Average properties of hot-rolled, normalized, and annealed material are listed, while properties of quenched and tempered grades are for single heats. Sources of the data are Bethlehem Steel Corp. and Republic Steel Corp.

Because of the many variables that affect a steel's properties, however, these listed properties should not be considered either as average or typical. Both strengths and ductilities may range up and down from the values given, depending on the compositions of individual heats of the same grade, section sizes, and internal structures. Properties of carbon steels and many alloy steels are also affected by residual elements (particularly nickel, chromium and molybdenum), even though their amounts are limited to maximums by AISI and SAE specifications.

Fine-grained steels normally have better impact strength than coarse-grained types, a factor which should be considered when reviewing the results of Izod tests. Hardness values are not always related to corresponding tensile strengths. In particular, this effect occurs with carbon steels because they are shallow-hardening. Hardness tests were made on surfaces, and these hardnesses will not reflect the tensile strengths obtained with specimens representing bar centers. (Center hardnesses are usually lower than surface hardnesses.)

Hot-rolled properties for alloy steels are not given because these grades are customarily heat treated. Because the samples were small enough to assure full quenching, values indicate strengths and ductilities which may be obtained with hardened, fine-grained steels of a similar section size at room temperatures.

I — Hot Rolled, Normalized, and Annealed

AISI No.*	Treatment	Yield Strength, Psi	Tensile Strength, Psi	Elongation, %	Reduction in Area, %	Hardness, Bhn	Impact Strength (Izod), Ft-Lb
1015	As-rolled	45,500	61,000	39.0	61.0	126	81.5
	Normalized (1,700 F)	47,000	61,500	37.0	69.6	121	85.2
	Annealed (1,600 F)	41,250	56,000	37.0	69.7	111	84.8
1020	As-rolled	48,000	65,000	36.0	59.0	143	64.0
	Normalized (1,600 F)	50,250	64,000	35.8	67.9	131	86.8
	Annealed (1,600 F)	42,750	57,250	36.5	66.0	111	91.0
1022	As-rolled	52,000	73,000	35.0	67.0	149	60.0
	Normalized (1,700 F)	52,000	70,000	34.0	67.5	143	86.5
	Annealed (1,600 F)	46,000	65,250	35.0	63.6	137	89.0
1030	As-rolled	50,000	80,000	32.0	57.0	179	55.0
	Normalized (1,700 F)	50,000	75,500	32.0	60.8	149	69.0
	Annealed (1,550 F)	49,500	67,250	31.2	57.9	126	51.2
1040	As-rolled	60,000	90,000	25.0	50.0	201	36.0
	Normalized (1,650 F)	54,250	85,500	28.0	54.9	170	48.0
	Annealed (1,450 F)	51,250	75,250	30.2	57.2	149	32.7
1050	As-rolled	60,000	105,000	20.0	40.0	229	23.0
	Normalized (1,650 F)	62,000	108,500	20.0	39.4	217	20.0
	Annealed (1,450 F)	53,000	92,250	23.7	39.9	187	12.5
1060	As-rolled	70,000	118,000	17.0	34.0	241	13.0
	Normalized (1,650 F)	61,000	112,500	18.0	37.2	229	9.7
	Annealed (1,450 F)	54,000	90,750	22.5	38.2	179	8.3
1080	As-rolled	85,000	140,000	12.0	17.0	293	5.0
	Normalized (1,650 F)	76,000	146,500	11.0	20.6	293	5.0
	Annealed (1,450 F)	54,500	89,250	24.7	45.0	174	4.5
1095	As-rolled	83,000	140,000	9.0	18.0	293	3.0
	Normalized (1,650 F)	72,500	147,000	9.5	13.5	293	4.0
	Annealed (1,450 F)	55,000	95,250	13.0	20.6	192	2.0
1117	As-rolled	44,300	70,600	33.0	63.0	143	60.0
	Normalized (1,650 F)	44,000	67,750	33.5	63.8	137	62.8
	Annealed (1,575 F)	40,500	62,250	32.8	58.0	121	69.0
1118	As-rolled	45,900	75,600	32.0	70.0	149	80.0
	Normalized (1,700 F)	46,250	69,250	33.5	65.9	143	76.3
	Annealed (1,450 F)	41,250	65,250	34.5	66.8	131	78.5
1137	As-rolled	55,000	91,000	28.0	61.0	192	61.0
	Normalized (1,650 F)	57,500	97,000	22.5	48.5	197	47.0
	Annealed (1,450 F)	50,000	84,750	26.8	53.9	174	36.8
1141	As-rolled	52,000	98,000	22.0	38.0	192	8.2
	Normalized (1,650 F)	58,750	102,500	22.7	55.5	201	38.8
	Annealed (1,500 F)	51,200	86,800	25.5	49.3	163	25.3
1144	As-rolled	61,000	102,000	21.0	41.0	212	39.0
	Normalized (1,650 F)	58,000	96,750	21.0	40.4	197	32.0
	Annealed (1,450 F)	50,250	84,750	24.8	41.3	167	48.0

AISI No.*	Treatment	Yield Strength, Psi	Tensile Strength, Psi	Elongation, %	Reduction in Area, %	Hardness, Bhn	Impact Strength (Izod), Ft-Lb
1340	Normalized (1,600 F)	81,000	121,250	22.0	62.9	248	68.2
	Annealed (1,475 F)	63,250	102,000	25.5	57.3	207	52.0
3140	Normalized (1,600 F)	87,000	129,250	19.7	57.3	262	39.5
	Annealed (1,500 F)	61,250	100,000	24.5	50.8	197	34.2
4130	Normalized (1,600 F)	63,250	97,000	25.5	59.5	197	63.7
	Annealed (1,585 F)	52,250	81,250	28.2	55.6	156	45.5
4140	Normalized (1,600 F)	95,000	148,000	17.7	46.8	302	16.7
	Annealed (1,500 F)	60,500	95,000	25.7	56.9	197	40.2
4150	Normalized (1,600 F)	106,500	167,500	11.7	30.8	321	8.5
	Annealed (1,500 F)	55,000	105,750	20.2	40.2	197	18.2
4320	Normalized 1,640 F)	67,250	115,000	20.8	50.7	235	53.8
	Annealed (1,560 F)	61,625	84,000	29.0	58.4	163	81.0
4340	Normalized (1,600 F)	125,000	185,500	12.2	36.3	363	11.7
	Annealed (1,490 F)	68,500	108,000	22.0	49.9	217	37.7
4620	Normalized (1,650 F)	53,125	83,250	29.0	66.7	174	98.0
	Annealed (1,575 F)	54,000	74,250	31.3	60.3	149	69.0
4820	Normalized (1,580 F)	70,250	109,500	24.0	59.2	229	81.0
	Annealed (1,500 F)	67,250	98,750	22.3	58.8	197	68.5
5140	Normalized (1,600 F)	68,500	115,000	22.7	59.2	229	28.0
	Annealed (1,525 F)	42,500	83,000	28.6	57.3	167	30.0
5150	Normalized (1,600 F)	76,750	126,250	20.7	58.7	255	23.2
	Annealed (1,520 F)	51,750	98,000	22.0	43.7	197	18.5
5160	Normalized (1,575 F)	77,000	138,750	17.5	44.8	269	8.0
	Annealed (1,495 F)	40,000	104,750	17.2	30.6	197	7.4
6150	Normalized (1,600 F)	89,250	136,250	21.8	61.0	269	26.2
	Annealed (1,500 F)	59,750	96,750	23.0	48.4	197	20.2
8620	Normalized (1,675 F)	51,750	91,750	26.3	59.7	183	73.5
	Annealed (1,600 F)	55,875	77,750	31.3	62.1	149	82.8
8630	Normalized (1,600 F)	62,250	94,250	23.5	53.5	187	69.8
	Annealed (1,550 F)	54,000	81,750	29.0	58.9	156	70.2
8650	Normalized (1,600 F)	99,750	148,500	14.0	40.4	302	10.0
	Annealed (1,465 F)	56,000	103,750	22.5	46.4	212	21.7
8740	Normalized (1,600 F)	88,000	134,750	16.0	47.9	269	13.0
	Annealed (1,500 F)	60,250	100,750	22.2	46.4	201	29.5
9255	Normalized (1,650 F)	84,000	135,250	19.7	43.4	269	10.0
	Annealed (1,550 F)	70,500	112,250	21.7	41.1	229	6.5
9310	Normalized (1,630 F)	82,750	131,500	18.8	58.1	269	88.0
	Annealed (1,550 F)	63,750	119,000	17.3	42.1	241	58.0

*All grades are fine-grained except for those in the 1100 series which are coarse-grained. Austenitizing temperatures are given in parentheses. Heat-treated specimens were oil quenched unless otherwise indicated. †Water quenched

Typical Mechanical Properties of Selected Carbon and Alloy Steels

II — Quenched and Tempered

AISI No.*	Tempering Temperature, F	Tensile Strength, Psi	Yield Strength, Psi	Elongation, %	Reduction in Area, %	Hardness, Bhn
1030†	400	123,000	94,000	17	47	495
	600	116,000	90,000	19	53	401
	800	106,000	84,000	23	60	302
	1,000	97,000	75,000	28	65	255
	1,200	85,000	64,000	32	70	207
1040†	400	130,000	96,000	16	45	514
	600	129,000	94,000	18	52	444
	800	122,000	92,000	21	57	352
	1,000	113,000	86,000	23	61	269
	1,200	97,000	72,000	28	68	201
1040	400	113,000	86,000	19	48	262
	600	113,000	86,000	20	53	255
	800	110,000	80,000	21	54	241
	1,000	104,000	71,000	26	57	212
	1,200	92,000	63,000	29	65	192
1050†	400	163,000	117,000	9	27	514
	600	158,000	115,000	13	36	444
	800	145,000	110,000	19	48	375
	1,000	125,000	95,000	23	58	293
	1,200	104,000	78,000	28	65	235
1050	400	—	—	—	—	—
	600	142,000	105,000	14	47	321
	800	136,000	95,000	20	50	277
	1,000	127,000	84,000	23	53	262
	1,200	107,000	68,000	29	60	223
1060	400	160,000	113,000	13	40	321
	600	160,000	113,000	13	40	321
	800	156,000	111,000	14	41	311
	1,000	140,000	97,000	17	45	277
	1,200	116,000	76,000	23	54	229
1080	400	190,000	142,000	12	35	388
	600	189,000	142,000	12	35	388
	800	187,000	138,000	13	36	375
	1,000	164,000	117,000	16	40	321
	1,200	129,000	87,000	21	50	255
1095†	400	216,000	152,000	10	31	601
	600	212,000	150,000	11	33	534
	800	199,000	139,000	13	35	388
	1,000	165,000	110,000	15	40	293
	1,200	122,000	85,000	20	47	235
1095	400	187,000	120,000	10	30	401
	600	183,000	118,000	10	30	375
	800	176,000	112,000	12	32	363
	1,000	158,000	98,000	15	37	321
	1,200	130,000	80,000	21	47	269
1137	400	157,000	136,000	5	22	352
	600	143,000	122,000	10	33	285
	800	127,000	106,000	15	48	262
	1,000	110,000	88,000	24	62	229
	1,200	95,000	70,000	28	69	197
1137†	400	217,000	169,000	5	17	415
	600	199,000	163,000	9	25	375
	800	160,000	143,000	14	40	311
	1,000	120,000	105,000	19	60	262
	1,200	94,000	77,000	25	69	187
1141	400	237,000	176,000	6	17	461
	600	212,000	186,000	9	32	415
	800	169,000	150,000	12	47	331
	1,000	130,000	111,000	18	57	262
	1,200	103,000	86,000	23	62	217
1144	400	127,000	91,000	17	36	277
	600	126,000	90,000	17	40	262
	800	123,000	88,000	18	42	248
	1,000	117,000	83,000	20	46	235
	1,200	105,000	73,000	23	55	217
1330†	400	232,000	211,000	9	39	459
	600	207,000	186,000	9	44	402
	800	168,000	150,000	15	53	335
	1,000	127,000	112,000	18	60	263
	1,200	106,000	83,000	23	63	216
1340	400	262,000	231,000	11	35	505
	600	230,000	206,000	12	43	453
	800	183,000	167,000	14	51	375
	1,000	140,000	120,000	17	58	295
	1,200	116,000	90,000	22	66	252
4037	400	149,000	110,000	6	38	310
	600	138,000	111,000	14	53	295
	800	127,000	106,000	20	60	270
	1,000	115,000	95,000	23	63	247
	1,200	101,000	61,000	29	60	220
4042	400	261,000	241,000	12	37	516
	600	234,000	211,000	13	42	455
	800	187,000	170,000	15	51	380
	1,000	143,000	128,000	20	59	300
	1,200	115,000	100,000	28	66	238
4130†	400	236,000	212,000	10	41	467
	600	217,000	200,000	11	43	435
	800	186,000	173,000	13	49	380
	1,000	150,000	132,000	17	57	315
	1,200	118,000	102,000	22	64	245
4140	400	257,000	238,000	8	38	510
	600	225,000	208,000	9	43	445
	800	181,000	165,000	13	49	370
	1,000	138,000	121,000	18	58	285
	1,200	110,000	95,000	22	63	230
4150	400	280,000	250,000	10	39	530
	600	256,000	231,000	10	40	495
	800	220,000	200,000	12	45	440
	1,000	175,000	160,000	15	52	370
	1,200	139,000	122,000	19	60	290
4340	400	272,000	243,000	10	38	520
	600	250,000	230,000	10	40	486
	800	213,000	198,000	10	44	430
	1,000	170,000	156,000	13	51	360
	1,200	140,000	124,000	19	60	280
5046	400	253,000	204,000	9	25	482
	600	205,000	168,000	10	37	401
	800	165,000	135,000	13	50	336
	1,000	136,000	111,000	18	61	282
	1,200	114,000	95,000	24	66	235
50B46	400	—	—	—	—	560
	600	258,000	235,000	10	37	505
	800	202,000	181,000	13	47	405
	1,000	157,000	142,000	17	51	322
	1,200	128,000	115,000	22	60	273
50B60	400	—	—	—	—	600
	600	273,000	257,000	8	32	525
	800	219,000	201,000	11	34	435
	1,000	163,000	145,000	15	38	350
	1,200	130,000	113,000	19	50	290
5130	400	234,000	220,000	10	40	475
	600	217,000	204,000	10	46	440
	800	185,000	175,000	12	51	379
	1,000	150,000	136,000	15	56	305
	1,200	115,000	100,000	20	63	245
5140	400	260,000	238,000	9	38	490
	600	229,000	210,000	10	43	450
	800	190,000	170,000	13	50	365
	1,000	145,000	125,000	17	58	280
	1,200	110,000	96,000	25	66	235

*All grades are fine-grained except for those in the 1100 series which are coarse-grained. Austenitizing temperatures are given in parentheses. Heat-treated specimens were oil quenched unless otherwise indicated. †Water quenched

AISI No.*	Tempering Tempera-ture, F	Tensile Strength, Psi	Yield Strength, Psi	Elonga-tion, %	Reduc-tion in Area, %	Hard-ness, Bhn
5150	400	282,000	251,000	5	37	525
	600	252,000	230,000	6	40	475
	800	210,000	190,000	9	47	410
	1,000	163,000	150,000	15	54	340
	1,200	117,000	118,000	20	60	270
5160	400	322,000	260,000	4	10	627
	600	290,000	257,000	9	30	555
	800	233,000	212,000	10	37	461
	1,000	169,000	151,000	12	47	341
	1,200	130,000	116,000	20	56	269
51B60	400	—	—	—	—	600
	600	—	—	—	—	540
	800	237,000	216,000	11	36	460
	1,000	175,000	160,000	15	44	355
	1,200	140,000	126,000	20	47	290
6150	400	280,000	245,000	8	38	538
	600	250,000	228,000	8	39	483
	800	208,000	193,000	10	43	420
	1,000	168,000	155,000	13	50	345
	1,200	137,000	122,000	17	58	282
81B45	400	295,000	250,000	10	33	550
	600	256,000	228,000	8	42	475
	800	204,000	190,000	11	48	405
	1,000	160,000	149,000	16	53	338
	1,200	130,000	115,000	20	55	280
8630	400	238,000	218,000	9	38	465
	600	215,000	202,000	10	42	430
	800	185,000	170,000	13	47	375
	1,000	150,000	130,000	17	54	310
	1,200	112,000	100,000	23	63	240
8640	400	270,000	242,000	10	40	505
	600	240,000	220,000	10	41	460
	800	200,000	188,000	12	45	400
	1,000	160,000	150,000	16	54	340
	1,200	130,000	116,000	20	62	280
86B45	400	287,000	238,000	9	31	525
	600	246,000	225,000	9	40	475
	800	200,000	191,000	11	41	395
	1,000	160,000	150,000	15	49	335
	1,200	131,000	127,000	19	58	280
8650	400	281,000	243,000	10	38	525
	600	250,000	225,000	10	40	490
	800	210,000	192,000	12	45	420
	1,000	170,000	153,000	15	51	340
	1,200	140,000	120,000	20	58	280
8660	400	—	—	—	—	580
	600	—	—	—	—	535
	800	237,000	225,000	13	37	460
	1,000	190,000	176,000	17	46	370
	1,200	155,000	138,000	20	53	315
8740	400	290,000	240,000	10	41	578
	600	249,000	225,000	11	46	495
	800	208,000	197,000	13	50	415
	1,000	175,000	165,000	15	55	363
	1,200	143,000	131,000	20	60	302
9255	400	305,000	297,000	1	3	601
	600	281,000	260,000	4	10	578
	800	233,000	216,000	8	22	477
	1,000	182,000	160,000	15	32	352
	1,200	144,000	118,000	20	42	285
9260	400	—	—	—	—	600
	600	—	—	—	—	540
	800	255,000	218,000	8	24	470
	1,000	192,000	164,000	12	30	390
	1,200	142,000	118,000	20	43	295
94B30	400	250,000	225,000	12	46	475
	600	232,000	206,000	12	49	445
	800	195,000	175,000	13	57	382
	1,000	145,000	135,000	16	65	307
	1,200	120,000	105,000	21	69	250

Index

N

O

P

Q

R

T

U

V

W

Z